# Physics and Chemistry of the Solar System

SECOND EDITION

John S. Lewis

Department of Planetary Sciences
University of Arizona
Tucson, Arizona

ELSEVIER
ACADEMIC
PRESS

AMSTERDAM • BOSTON • HEIDELBERG • LONDON
NEW YORK • OXFORD • PARIS • SAN DIEGO
SAN FRANCISCO • SINGAPORE • SYDNEY • TOKYO

Acquisition Editor: Frank Cynar
Project Manager: Sarah Manchester
Editorial Assistant: Jennifer Helé
Marketing Manager: Linda Beattie
Cover Design: Gary Ragaglia
Composition: Integra Software Services Pvt. Ltd
Printer: The Maple-Vail Manufacturing Group

Elsevier Academic Press
200 Wheeler Road, Burlington, MA 01803, USA
525 B Street, Suite 1900, San Diego, California 92101-4495, USA
84 Theobald's Road, London WC1X 8RR, UK

This book is printed on acid-free paper. ♾



**Library of Congress Cataloging-in-Publication Data**
Lewis, John S.
Physics and chemistry of the solar system/John S. Lewis–2nd ed.
p. cm. – (International geophysics series; v. 87)
Includes bibliographical references and index.
ISBN 0-12-446744-X (acid-free paper)
1. Solar system. 2. Planetology. 3. Astrophysics. 4. Cosmochemistry.
I. Title. II. Series.

QB501.L497 2004
523.2–dc22

2003064281

**British Library Cataloguing in Publication Data**
A catalogue record for this book is available from the British Library

ISBN: 0-12-446744-X

For all information on all Academic Press publications
visit our website at www.academicpressbooks.com

PRINTED IN THE UNITED STATES OF AMERICA
04 05 06 07 08 09 9 8 7 6 5 4 3 2 1

# Dedication

This book is dedicated to the founders of Planetary Science:
Rupert Wildt, Gerard P. Kuiper, and Harold C. Urey,
whose thoughts roamed the Solar System before spacecraft did.

# Contents

Foreword xi

I Introduction

Nature and Scope of Planetary Science 1
Guide to the Literature 3
Numbers in Science 4
Dimensions and Units 5
Exercises 6

II Astronomical Perspective

Introduction 7
Distance Scales in the Universe 7
The Big Bang 10
Limitations on Big Bang Nucleosynthesis 14
Galaxy and Star Formation 15
Structure and Classification of Galaxies 16
Classification of Stars 18
Stellar Evolution 25
Star Clusters 27
Stellar Origins 29
Outline of Star Formation 33
Stellar Explosions and Nucleosynthesis 34
Nuclear Cosmochronology 43
Exercises 47

III General Description of the Solar System

Introduction 50
The Sun 50
Orbits of the Planets 52
Changes in Orbital Motion 57
Properties of the Planets 58
Mass and Angular Momentum Distribution 59
Satellites 63
Asteroids 69
Comets 71
Meteors 72
Meteorites 72
Cosmic Dust 73
Cosmic Rays 73
Planetary Science in the Space Age 74

Summary 76
Exercises 76

IV The Sun and the Solar Nebula

Introduction 77
Energy Production in the Sun 77
Energy Transport in the Sun 79
Internal Structure of the Sun 83
Surface of the Sun 84
The Chromosphere 87
The Corona 88
Discovery of the Solar Wind 90
Radio Wave Propagation in Space Plasmas 91
The Solar Wind 92
Chemistry of Solar Material 96
Ionization 97
Dissociation and Molecule Formation 100
Hydrogen and the Rare Gases 101
Oxygen, Carbon, and Nitrogen 102
Magnesium and Silicon 105
Iron 106
Sulfur 107
Aluminum and Calcium 108
Sodium and Potassium 109
Nickel and Cobalt 110
Phosphorus and the Halogens 111
Geochemical Classification of the Elements 111
The Chemistry of Rapid Accretion 116
Kinetic Inhibition 117
Mass and Density of the Solar Nebula 118
Thermal Opacity in the Solar Nebula 121
Dust Opacity 129
Thermal Structure of the Nebula 131
Turbulence and Dust Sedimentation 134
Accretion of Rocks, Planetesimals, and Planets 136
Gas Capture from the Solar Nebula 138
The T Tauri Phase 141
Thermal History of the Early Solar System 143
Exercises 144

V The Major Planets

Introduction 147
Interiors of Jupiter and Saturn: Data 148
Isothermal Interior Models of Jupiter and Saturn 151
Thermal Models of Jupiter and Saturn 154
The Atmospheres of Jupiter and Saturn: Observed Composition 156
Tropospheric Composition and Structure: Theory 159
Cloud Condensation in the $NH_3$–$H_2O$–$H_2S$ System 165
Cloud Physics on the Jovian Planets 174
Galileo Perspectives on Jovian Clouds 179
Ion Production in the Jovian Atmosphere 180
Visible and Infrared Radiative Transfer 183
Horizontal Structure and Atmospheric Circulation 187
Photochemistry and Aeronomy 200
The Jovian Thermosphere 217
Radiophysics and Magnetospheres of Jupiter and Saturn 218
The Interiors of Uranus and Neptune 229
Atmospheres of Uranus and Neptune 238
Perspectives 247
Exercises 247

VI Pluto and the Icy Satellites of the Outer Planets

Introduction 252
Surfaces of Icy Satellites 253
Eclipse Radiometry 256
Surface Temperatures 257
Surface Morphology of the Galilean Satellites 258
Density and Composition of Icy Satellites 265
Internal Thermal Structure of Galilean Satellites 267
Dynamical Interactions of the Galilean Satellites 272
Thermal and Tectonic Evolution of Icy Satellites 275
Minor Satellites of Jupiter 278
Planetary Rings 280
Titan 289
The Intermediate-Sized Saturnian Satellites 293
Minor Satellites of Saturn 296
Satellites of Uranus 299
Satellites of Neptune 303
The Pluto–Charon System 308
The Neptune–Pluto Resonance 311
Spacecraft Exploration 311
Exercises 312

VII Comets and Meteors

Historical Perspectives 317
Nature and Nomenclature of Comets 319

Cometary Orbits 321
Heating by Passing Stars 325
Evaporation and Nongravitational Forces 326
The Nucleus and Coma of P/Halley 328
Chemistry and Photochemistry of Water 328
Further Chemical Processes in the Coma and Tail 332
Behavior of Small Particles 333
Dynamical Behavior of Dust in Space 334
Meteors 336
Cometary Fireballs 343
Cometary Impacts on Jupiter 344
Exercises 347

## VIII Meteorites and Asteroids

Introduction 350
Introduction to Meteorites 350
Meteorite Orbits 353
Phenomena of Fall 355
Physical Properties of Meteorites 358
Meteorite Minerals 362
Taxonomy and Composition of Chondrites 362
Metamorphic Grades of Chondrites 367
Taxonomy and Composition of Achondrites 369
Taxonomy and Composition of Stony-Irons 371
Taxonomy and Composition of Irons 372
Isotopic Composition of Meteorites 375
Genetic Relationships between Meteorite Classes 382
Introduction to Asteroids 384
Asteroid Orbits 386
Stability of Trojan and Plutino Orbits 389
Sizes, Shapes, and Albedos of Asteroids 391
Masses and Densities of Asteroids 393
Photometry and Spectroscopy of Asteroids 394
Thermal Evolution of Asteroids 401
Dynamical Evolution of the Asteroid Belt 406
Centaurs and Trans-Neptunian Objects 409
Relationships among Asteroids, Meteorites, and Comets 412
Radar Observations of Near-Earth Asteroids 415
Asteroid Resources 416
Exercises 419

## IX The Airless Rocky Bodies: Io, Phobos, Deimos, the Moon, and Mercury

Introduction 424
Orbits and Physical Structure of Phobos and Deimos 426
Io: General Properties 430
Io: Surface Processes 430
Io: Internal Energy Sources 432
Io: Geology 433
Io: Atmospheric and Volcanic Gases 435
Io: Escape and the Plasma Torus 437
Io: Genetic Relationships 438
Impact Cratering 438
Motions of the Moon 443
Physical Properties of the Moon 445
Elemental Composition of the Moon's Surface 445
Lunar Rock Types 447
Lunar Minerals 449
Lunar Elemental Abundance Patterns 451
Geology of the Moon 451
Geophysics of the Moon 452
History of the Earth–Moon System 456
Origin and Internal Evolution of the Moon 458
Solar Wind Interaction with the Moon and Mercury 460
The Planet Mercury 461
Motions of Mercury 461
Composition and Structure of Mercury 462
Noncrater Geology of Mercury 463
Geophysics of Mercury 463
Atmospheres of Mercury and the Moon 468
Polar Deposits on Mercury and the Moon 469
Unfinished Business 472
Exercises 474

## X The Terrestrial Planets: Mars, Venus, and Earth

Introduction 477
Mars 478
Motions of Mars 479
Density and Figure of Mars 479
Geophysical Data on Mars 481
Gravity and Tectonics of Mars 483
Geology of Mars 483
Surface Composition 496
Viking Lander Investigations 503
The Shergottite, Nakhlite, and Chassignite Meteorites 505
Atmospheric Structure 508
Atmospheric Circulation 509
Atmospheric Composition 510
Photochemical Stability and Atmospheric Escape 513
Explosive Blowoff 519
Origin and Evolution of the Atmosphere 519

Organic Matter and the Origin of Life 522
Venus 524
Motions and Dynamics of Venus 526
Geophysical Data on Venus 526
Geology of Venus 528
Venus: Atmospheric Structure and Motions 534
Venus: Atmospheric Composition 537
Venus: Atmosphere–Lithosphere Interactions 539
Venus: Photochemistry and Aeronomy 543
Venus: Atmospheric Escape 547
Venus: Planetary Evolution 549
Earth 550
Earth: Motions 551
Earth: Internal Structure 552
Earth: Magnetic Field and Magnetosphere 554
Earth: Surface Geology 554
Earth: Early Geological History 557
Earth: Biological History 559
Earth: Geochemistry and Petrology 563
Weathering in the Rock Cycle 566
Earth: Atmospheric Composition and Cycles 568
Radiocarbon Dating 573
Stable Isotope Climate Records 574
Photochemistry and Aeronomy 575
Escape and Infall 575
Climate History, Polar Ice, and Ice Ages 579
Life: Origins 582
Life: Stability of the Biosphere 587
Exercises 588

## XI Planets and Life around Other Stars

Chemical and Physical Prerequisites of Life 592
The Planetary Environment 595
The Stellar Environment 597
Brown Dwarfs 600
The Search for Planets of Other Stars 603
The Search for Extraterrestrial Intelligence 606
Exercises 608

## XII Future Prospects

Mercury 611
Venus 612
Earth's Moon 612
Mars 613
Asteroids 614
Jupiter 615
Saturn, Uranus, and Neptune 615
Pluto 615
Comets 616
Beyond the Solar System 616

## Appendix I: Equilibrium Thermodynamics 621

Heat and Work 621
Adiabatic Processes and Entropy 622
Useful Work and the Gibbs Free Energy 623
Chemical Equilibrium 623
Exact and Complete Differentials 624
The Maxwell Relations 625

## Appendix II: Absorption and Emission of Radiation by Quantum Oscillators 626

## Appendix III: Exploration of the Solar System 629

## Appendix IV: Basic Physical Constants 634

## Appendix V: Gravity Fields 635

## Suggested Readings

Introduction 637
Chapter I–Introduction 637
Chapter II–Astronomical Perspective 637
Chapter III–General Description of the Solar System 638
Chapter IV–The Sun and the Solar Nebula 638
Chapter V–The Major Planets 638
Chapter VI–Pluto and the Icy Satellites of the Outer Planets 639
Chapter VII–Comets and Meteors 639
Chapter VIII–Meteorites and Asteroids 639
Chapter IX–The Airless Rocky Bodies: Io, Phobos, Deimos, the Moon, and Mercury 640
Chapter X–The Terrestrial Planets: Mars, Venus, and Earth 640
Chapter XI–Planets and Life around Other Stars 641
Chapter XII–Future Prospects 642

## Index 643

# Foreword

At its original conception, this book was based on the structure, scope, and philosophy of a sophomore/junior level course taught at M.I.T. by the author and Prof. Irwin I. Shapiro from 1969 to 1982. Although the content of that course varied greatly over the years in response to the vast new knowledge of the Solar System provided by modern Earth-based and spacecraft-based experimental techniques, the philosophy and level of presentation remained very much the same. The material was brought up to date in 1994 for publication in 1995, and again updated with many corrections and additions for a revised edition in 1997. This second edition was prepared in 2002 to take advantage of the many recent advances in the study of Mars and small Solar System bodies, the discovery and study of more than 100 extra-solar planets, and more mature analysis of the Galileo Orbiter and probe data on Jupiter and its large satellites.

The timing of the various editions of this book has been influenced by the erratic history of planetary exploration. During the 12 years of 1964–1973 there were 87 launches of lunar and planetary spacecraft, of which 54 were involved in the race to the Moon. In the 29 years since the end of 1973, up to the date of this edition in 2002, there have been only 36 additional launches. Both the United States and the Soviet Union experienced prolonged gaps in their lunar and planetary exploration programs: the American gap in lunar exploration extended from Explorer 49 in 1973 to the launch of Clementine in 1994, and the Russian hiatus in lunar missions has stretched from Luna 24 in 1976 to the present. American exploration of Mars was suspended from the time of the Viking missions in 1975 until the launch of Mars Observer in 1992, and Soviet exploration of Mars, suspended after Mars 7 in 1975, did not resume until the launch of the two ill-fated Phobos spacecraft in 1988. Soviet missions to Venus ceased in 1984.

From 1982 to 1986 there was a gap in the acquisition of planetary data by American spacecraft. This drought was interrupted in 1986 by the Voyager 2 Uranus flyby and by five spacecraft encounters with Halley's comet (two Soviet, two Japanese, and one from the European Space Agency), but the drought again resumed until it was broken by the Voyager 2 Neptune encounter and the Soviet Phobos missions in 1989 and the Magellan mission to Venus in 1990. The launch of the Galileo Orbiter and probe to Jupiter, long scheduled for 1986, was severely delayed by the explosion of the space shuttle orbiter Challenger, the resulting 2-year grounding of the entire shuttle fleet, and the subsequent cancellation of the high-energy Centaur G' upper stage intended for launching heavy planetary missions from the shuttle. The European-American Ulysses solar mission, which was not instrumented for intensive planetary studies, flew by Jupiter in February 1992, returning only data on its magnetic and charged-particle environment. The arrival of Galileo at Jupiter, the Galileo Probe entry into

Jupiter's atmosphere in December 1995, the lengthy Galileo Orbiter survey of the Jovian system, and the resumption of small Mars missions (Pathfinder, Mars Global Surveyor, etc.) by the United States have combined with a flood of space-based (Galileo, Near-Earth Asteroid Rendezvous) and Earth-based observations of near-Earth asteroids and Belt asteroids, and intensive Earth-based study of comets, Centaurs, small icy satellites, and trans-Neptunian objects and the highly successful search for dark companions of nearby stars to reinvigorate the planetary sciences. This new resurgence of planetary exploration, with little prospect of Russian participation, has been helped by the active involvement of Japan's NASDA and the European Space Agency in planning and flying unmanned missions to the Moon, Mars, and Venus. The infusion of new data resulting from these several programs creates the necessity of revising this book

In this book, as in that Planetary Physics and Chemistry course in which it was first conceived, I shall assume that the reader has completed 1 year of university-level mathematics, chemistry, and physics. The book is aimed at several distinct audiences: first, the upper-division science major who wants an up-to-date appreciation of the present state of the planetary sciences for "cultural" purposes; second, the first-year graduate student from any of several undergraduate disciplines who intends to take graduate courses in specialized areas of planetary sciences; and third, the practicing Ph.D. scientist with training in physics, chemistry, geology, astronomy, meteorology, biology, etc., who has a highly specialized knowledge of some portion of this material, but has not had the opportunity to study the broad context within which that specialty might be applied to current problems in this field.

This volume does not closely approximate the level and scope of any previous book. The most familiar texts on the planetary sciences are *Exploration of the Solar System*, by William J. Kaufmann, III (Macmillan, New York, 1978 and later), a nonmathematical survey of the history of planetary exploration; *Moons and Planets*, by William K. Hartmann (Wadsworth, Belmont, California, 1972; 1983; 1993), a scientific tour of the Solar System with high-school-level mathematical content; and *Meteorites and the Origin of Planets*, by John A. Wood (McGraw-Hill, New York, 1968), a fine qualitative introduction that is similarly sparing of mathematics and physics. Several other nonmathematical texts are available, including *Introduction to the Solar System*, by Jeffrey K. Wagner (Saunders, Philadelphia, 1991), *Exploring the Planets*, by W. Kenneth Hamblin and Eric H. Christiansen (Macmillan, New York, 1990), *The Space-Age Solar System*, by Joseph F. Baugher (J. Wiley, New York, 1988), and *The Planetary System*, by planetary scientists David Morrison and Tobias Owen (Addison–Wesley, Reading, Massachusetts, 1988).

Another book, comparable in mathematical level to the present text, is *Worlds Apart*, by Guy J. Consolmagno, S. J., and Martha W. Schaefer (Prentice Hall, Englewood Cliffs, New Jersey, 1994). Though much less detailed than the present work, it is well written and appropriate for a one-semester introductory course on planetary science for science majors. The scope of the present text is broader, and the level higher, than any of these books.

As presently structured, this book is a broad survey of the Solar System suitable for reference use or as background reading for any course in Solar System science. The text may for convenience be divided into three parts. The first of these parts contains Chapter I (Introduction), Chapter II (Astronomical Perspective), Chapter III (General Description of the Solar System), and Chapter IV (The Sun and the Solar Nebula). This first part could be called "General Properties and Environment of our Planetary System." It is roughly equivalent to a brief introductory astronomy book emphasizing the concerns of planetary scientists rather than stellar or galactic astronomers. The second part contains Chapter V (The Major Planets), Chapter VI (Pluto and the Icy Satellites of the Outer Planets), Chapter VII (Comets and Meteors), and Chapter VIII (Meteorites and Asteroids), and might fairly be entitled "The Solar System beyond Mars." The third and final part comprises Chapter IX (The Airless Rocky Bodies: Io, Phobos, Deimos, the Moon, and Mercury), Chapter X (The Terrestrial Planets: Mars, Venus, and Earth), Chapter XI (Planets and Life around Other Stars), and Chapter XII (Future Prospects). This part could be called "The Inner Solar System."

Using this volume as a textbook, a planetary sciences course taught in a trimester setting could use one part each term. In a two-semester program, either an inner solar system emphasis course (parts 1 and 3) or an outer solar system course (parts 1 and 2) could be taught. The most ambitious and intensive program, and the most similar to the way the course was structured at M.I.T., would be to teach parts 2 and 3 in two semesters, reserving most of the material in part 1 for use as reference reading rather than as lecture material.

This book is written in appreciation of the approximately 350 students who took the course at M.I.T., and who unanimously and vocally deplored the lack of a textbook for it. These students included both Consolmagno and Schaefer as cited above. I extend my particular thanks to Irwin Shapiro for his many years of cheerful, devoted, always stimulating, and sometimes hilarious collaboration on our course, and for his generous offer to allow me to write "his" half of the text as well as "mine." I am also pleased to acknowledge the helpful comments and suggestions of dozens of my colleagues, but with special thanks reserved for Jeremy Tatum of the University of Victoria, whose detailed comments and physicist's perspective have been invaluable in the preparation of this second edition.

# I. Introduction

## Nature and Scope of the Planetary Sciences

When asked in an interview to give his viewpoint on the frontiers of science, the famous physicist Victor Weisskopf commented that the most exciting prospects fell into two categories, the frontier of size and the frontier of complexity. A host of examples come to mind: cosmology, particle physics, and quantum field theory are clearly examples of the extremes of scale, and clearly among the most exciting frontiers of science. Biology, ecology, and planetary sciences are equally good examples of the frontier of complexity.

When we peruse the essential literature of planetary science, we find that we must, over and over again, come face to face with these same extremes. First, we are concerned with the origin and nuclear and chemical evolution of matter, from its earliest manifestation as elementary particles through the appearance of nuclei, atoms, molecules, minerals, and organic matter. Second, on the cosmic scale, the origin, evolution, and fate of the Universe emerge as themes. Third, we are confronted with the problem of understanding the origin and development of life. In each case, we are brought face to face with the spontaneous rise of extreme complexity out of extreme simplicity, and with the intimate interrelationship of the infinitesimally small and the ultimately large.

Further, our past attempts at addressing these three great problems have shown us that they are remarkably intertwined. The very issue of the origin of life is inextricably tied up with the chemistry of interstellar clouds, the life cycles of stars, the formation of planets, the thermal and outgassing history of planetary bodies, and the involvement of geochemical processes in the origin of organic matter. The connection between life and planetary environments is so fundamental that it has been given institutional recognition: it is not widely known outside the field, but research on the origin of life in the United States is a mandate of the National Aeronautics and Space Administration.

Wherever we begin our scientific pilgrimage throughout the vast range of modern science, we find ourselves forced to adopt ever broader definitions of our field of interest. We must incorporate problems not only on the frontier of complexity, but also from both extreme frontiers of scale. In this way, we are compelled to trespass across many hallowed disciplinary boundaries.

Further, as we seek an evolutionary account of the emergence of complexity from simplicity, we become able to see more clearly the threads that lead from one science to another. It is as if the phenomena of extreme scale in physics existed for the express purpose of providing a rationale for the existence of astronomy.

The other disciplines evolve logically from cosmic events:

The astronomical Universe, through the agency of nuclear reactions inside stars and supernova explosions, populates space with atoms of heavy elements, which are the basis of chemistry.

The course of spontaneous chemical evolution of interstellar matter produces both mineral grains and organic molecules, giving rise to geochemistry and organic chemistry.

Solid particles accrete to form large planetary bodies, and give us geology.

Radioactive elements formed in stellar explosions are incorporated into these planets, giving life to geophysics.

Melting, density-dependent differentiation, and outgassing take place, and atmospheres and oceans appear: petrology, meteorology, and oceanography become possible.

Organic matter is formed, accumulated, concentrated, and processed on planetary surfaces, and biology is born.

Planetary science may then be seen as the bridge between the very simple early Universe and the full complexity of the present Earth. Although it partakes of the excitement of all of these many fields, it belongs to none of them. It is the best example of what an interdisciplinary science should be: it serves as a unifying influence by helping to dissolve artificial disciplinary boundaries, and gives a depth and vibrancy to the treatment of evolutionary issues in nature that transcends the concerns and the competence of any one of the parent sciences. But there is more: planetary science is centrally concerned with the evolutionary process, and hence with people's intuitive notion of "how things work." There is as much here to unlearn as there is to learn.

We, at the turn of the millennium, still live under the shadow of the clockwork, mechanistic world view formulated by Sir Isaac Newton in the 17th century. Even the education of scientists is dedicated first and foremost to the inculcation of attitudes and values that are archaic, dating as they do from Newton's era: viewpoints that must be unlearned after sophomore year. We are first led to expect that the full and precise truth about nature may be extracted by scientific measurements; that the laws of nature are fully knowable from the analysis of experimental results; that it is possible to predict the entire course of future events if, at one moment, we should have sufficiently detailed information about the distribution and motion of matter. Quantum mechanics and relativity are later taught to us as a superstructure on Newtonian physics, not vice versa. We must internally turn our education upside down to accommodate a universe that is fundamentally quantum-mechanical, chaotic, and relativistic, within which our "normal" world is only a special case.

All of these issues come to bear on the central question of the evolution of the cosmos and its constituent parts. Most of us have had a sufficient introduction to equilibrium thermodynamics to know that systems spontaneously relax to highly random, uninteresting states with minimum potential energy and maximum entropy. These are the classical conclusions of J. Willard Gibbs in the 19th century. But very few of us are ever privileged to hear about the development of nonequilibrium thermodynamics in the 20th century, with its treatment of stable dissipative structures, least production of entropy, and systems far removed from thermodynamic equilibrium. Think of it: systems slightly perturbed from equilibrium spontaneously relax to the dullest conceivable state, whereas systems far from equilibrium spontaneously organize themselves into structures optimized for the minimization of disorder and the maximization of information content!

It is no wonder that the whole idea of evolution is so magical and counterintuitive to so many people, and that the critics of science so frequently are able to defend their positions by quoting the science of an earlier century. We often hear expressed the idea that the spontaneous rise of life is as improbable as that a printshop explosion (or an incalculable army of monkeys laboring at typewriters) might accidentally produce an encyclopedia. But have we ever heard that this argument is obsolete nonsense, discredited by the scientific progress of the 20th century? Sadly, there is a gap of a century between the scientific world view taught in our schools and the hard-won insights of researchers on the present forefront of knowledge. The great majority of all people never learn more than the rudiments of Newtonian theory, and hence are left unequipped by their education to deal with popular accounts of modern science, which at every interesting turn is strikingly non-Newtonian. News from the world of science is, quite simply, alien to them. The message of modern science, that the Universe works more like a human being than like a mechanical wind-up toy, is wholly lost to them. Yet it is precisely the fundamental issues of how things work and how we came to be, what we are and what may become of us, that are of greatest human interest. The "modern" artist or writer of the 20th century often asserted modernity by preaching the sterility of the Universe and the alienation of the individual from the world. But this supposed alienation of the individual from the Universe is, to a modern scientist, an obsolete and discredited notion.

The problems of evolutionary change and ultimate origins are not new concerns. Far from being the private domain of modern science, they have long been among the chief philosophical concerns of mankind. Astronomy and astrology were the parents of modern science. The earliest human records attest to mankind's perpetual fascination with origins:

> Who knows for certain and can clearly state
> Where this creation was born, and whence it came?
> The *devas* were born after this creation,
> So who knows from whence it arose?
>
> No one knows where creation comes from
> Or whether it was or was not made:
> Only He who views it from highest heaven knows;
> Surely He knows, for who can know if He does not?
>
> *Rigveda* X 129.6–7
> Circa 3000 BC

Such an attitude, reflective of curiosity, inquiry, and suspended belief, is admirably modern. But today, in light of the exploration of the Solar System, we need no longer regard our origins as complete mysteries. We can now use the observational and theoretical tools of modern science to test rival theories for their faithfulness to the way the Universe really is. Some theories, when tested by the scientific method, are found to give inaccurate or even blatantly wrong descriptions of reality and must be abandoned. Other theories seem to be very reliable guides to how nature works and are retained because of their usefulness. When new data arise, theories may need to be modified or abandoned. Scientific theories are not absolute truth and are not dogma: they are our best approximation of truth at the moment. Unlike dogma, scientific theories cannot survive very long without confronting and accommodating the observed facts. The scientific theories of today are secondary to observations in that they are invented—and modified—by human beings in order to explain observed facts. They are the result of an evolutionary process, in which the "most fit" theories (those that best explain our observations) survive. In planetary science, that process has been driven in recent years in part by the discovery and study of several new classes of bodies both within our Solar System and elsewhere. It is the great strength of science (not, as some allege, its weakness) that it adapts, modifies, and overturns its theories to accommodate these new realities. Our plan of study of the Solar System mirrors this reality.

This book will begin with what little we presently know with confidence about the earliest history of the Universe, and trace the evolution of matter and its constructs up to the time of the takeover of regulatory processes on Earth by the biosphere. We introduce the essential contributions of the various sciences in the order in which they were invoked by nature, and build complexity upon complexity stepwise. Otherwise, we might be so overawed by the complexity of Earth, our first view of nature, that we might despair of ever gaining any understanding at all.

This approach should also dispel the notion that we are about to understand everything. It is quite enough to see that there are untold vistas for exploration, and more than enough of the Real to challenge our most brilliant intellects and most penetrating intuitions.

Let us approach the subject matter covered herein with the attitude that there are a number of fundamental principles of nature, of universal scope, that allow and force the evolutionary process. With our senses at the most alert, willing to entertain the possibility of a host of hypotheses, and determined to subject all theories and observations alike to close scrutiny, we are challenged to grasp the significance of what we see. Let us cultivate the attitude that the ultimate purpose of the planetary sciences is to uncover enough of the blueprints of the processes of evolution so that we will be able to design, build, and operate our own planetary system.

Like it or not, we are assuming responsibility for the continued stability and habitability of at least one planet. The scale of human endeavor has now become so large that our wastes are, quite inadvertently, becoming major factors in global balances and cycles. Soon our scope may be the whole Solar System. The responsible exercise of our newly acquired powers demands an understanding and consciousness superior to that which we have heretofore exhibited. Now is the time for us to learn how planets work.

## Guide to the Literature

It is difficult, as we have seen above, to draw a tidy line around a particular portion of the scientific literature and proclaim all that lies outside that line to be irrelevant. Still, there are certain journals that are more frequently used and cited by practitioners of planetary science. Every student should be aware both of these journals and the powerful abstracting and citation services now available.

Astronomical observations, especially positional measurements, orbit determinations, and the like that are carried out using Earth-based optical, radio, and radar techniques, are often published in the *Astronomical Journal* (AJ). Infrared spectroscopic and radiometric observations and a broad range of theoretical topics often appear in the *Astrophysical Journal* (ApJ). The most important journals devoted to planetary science in the broad sense are *Icarus* and the *Journal of Geophysical Research* (usually called JGR). Two journals are devoted to relatively quick publication of short related papers: *Geophysical Research Letters* (GRL) and *Earth and Planetary Science Letters* (EPSL). Two general-purpose wide-circulation journals also frequently publish planetary science papers, including special issues on selected topics: these are *Science* and *Nature*. The most important western European journal for our purposes is *Astronomy and Astrophysics*.

Russian research papers frequently appear first (or in prompt translation) in English. The most important Soviet journals are *Astronomicheskii Zhurnal* (*Sov. Astron.* to the cognoscenti), *Kosmicheskii Issledovaniya* (*Cos. Res.*), and *Astron. Vestnik* (*Solar System Research*), all of which appear in English translation with a delay of several months.

Other journals containing relevant research articles include *Physics of the Earth and Planetary Interiors* (PEPI), the *Proceedings of the Lunar and Planetary Science Conferences*, the *Journal of the Atmospheric Sciences* (JAS), *Planetary and Space Science*, *Geochimica et Cosmochimica Acta* (GCA), the Russian-language *Geokhimiya*, *Meteoritics*, *Origins of Life*, and perhaps 50 other journals that are usually a bit far from the center of the field, but overlap its periphery.

Many space scientists keep abreast of the politics and technology of space exploration by reading *Aviation Week and Space Technology* (AW&ST), which often prints future news and juicy rumors.

Very valuable service is also rendered by several review publications, such *as Annual Review of Earth and Planetary Science*, *Space Science Reviews*, *Reviews of Geophysics and Space Physics*, and the *Annual Review of Astronomy and Astrophysics*.

Books on the planetary sciences have an unfortunate tendency to become obsolete during the publication process. Nonetheless, many books have useful coverage of parts of the material in the field, and a number of these are cited at the relevant places in the text.

It is often valuable to track down the history of an idea, or to see what recent publications are following a lead established in a landmark paper of several years ago. For these purposes, every scientist should become familiar with the uses of the *Science Citation Index*. Depending upon one's own particular interests, any of a number of other abstracting services and computerized databases may be relevant. The reader is encouraged to become familiar with the resources of the most accessible libraries. Every research library has *Chemical Abstracts*, *Biological Abstracts*, etc.

For the diligent searcher, there will be an occasional gem captured from the publications of the Vatican Observatory, and surely one cannot claim to be a planetary scientist until one has followed a long trail back to an old issue of the *Irish Astronomical Journal*. Be eclectic: have no fear of journals with Serbian or Armenian names. The contents are most likely in English, or if not, then almost certainly in French, German, or Russian, often conveniently equipped with an English abstract.

Many valuable online services have arisen to speed the exchange of scientific data and theories between interested parties, from professional planetary scientists to scientists in other disciplines to the interested public. Never before in history has so much information from all over the world been available in so immediate—and so undigested—a state. These services come, go, and evolve rapidly. Some will be cited at the appropriate places in the text, but the selective use of Web search engines is a more essential part of online research than knowing this month's hottest Web sites. The hazard of this approach to research is that the opinions of professionals, amateurs, ignoramuses, and fanatical ideologues are all weighted equally, and all equally accessible. Never before in history has so much misinformation and disinformation from all over the world been available to mislead the incautious and the gullible. Know your sources!

But planetary science is a genuinely international endeavor. To make the most of the available resources one must be willing to dig deep, think critically, and keep in contact with colleagues abroad. One must be prepared to face the hardship of back-to-back conferences in Hawaii and Nice; of speaking engagements three days apart in Istanbul and Edmonton; of January trips to Moscow balanced against summer workshops in Aspen. I suppose that this is part of our training as thinkers on the planetary scale.

## Numbers in Science

It is assumed that all readers are familiar with scientific notation, which expresses numbers in the format $n.nnnn \times 10^x$. This convention permits the compact representation of both extremely small and extremely large numbers and facilitates keeping track of the decimal place in hand calculations. Thus the number 0.00000000000000000000000000066262, Planck's constant, is written in scientific notation as $6.6262 \times 10^{-27}$, and Avogadro's number, 602,220,000,000,000,000,000,000, is written $6.0222 \times 10^{23}$. Their product is $6.6262 \times 10^{-27} \times 6.0222 \times 10^{23} = 6.0222 \times 6.6262 \times 10^{23} \times 10^{-27} = 39.904 \times 10^{23-27} = 39.904 \times 10^{-4} = 3.9904 \times 10^{-3}$. In some circumstances, where typographic limitations militate against writing actual superscripts and subscripts (as in some scientific programming languages), scientific notation is preserved by writing the number in the form 3.9904E-03.

Numbers are usually written in a form that suggests the accuracy with which they are known. For example, a wedding guest might say "I have traveled 3000 miles to be here today". The literal-minded, after looking up the conversion factor for miles to kilometers, will find that one mile is 1.609344 kilometers, and laboriously calculate that the wedding guest has traveled exactly $3000 \times 1.609344 = 4828.032$ km. One frequently finds such conversions done in newspapers. But this is of course absurd. The guest neither knew nor claimed to know his itinerary to any such precision. He cited his trip as 3000 miles, a number with only one significant figure. The appropriate conversion would then be to round off 4828.032 to the nearest single significant figure, which would be 5000 km.

How then do we represent the results of an accurate survey of a racetrack that finds the length to be 1000 meters with a precision of 0.001 meters? We would then write the length as 1000.000 m. Since measurement uncertainties are seldom so simple, we generally estimate the precision of a measurement by averaging the results of many measurements and reporting the average absolute deviation of the individual measurements from the mean. Thus a series of measurements of the distance between two points made with a meter stick might be 86.3, 85.9, 86.2, 86.6, 86.3, 86.4, 86.0, 86.1, 86.4, and 86.2 cm. The mean of these 10 measurements is 86.24 cm, and the difference of each measurement from that mean are $+0.06$, $-0.34$, $-0.04$, $+0.36$, $+0.06$, $+0.16$, $-0.24$, $-0.14$, $+0.16$, and $-0.04$. The sum of these errors is of course zero; the sum of the absolute deviations (with all the signs positive) is 1.60, and the average deviation is $1.60/10 = 0.16$. Thus we report the result of these measurements as $86.24 \pm 0.16$ cm. The $\pm$ sign is read "plus or minus," and the number following it is called the error limit or the probable error. Note that this is not in fact a limit on the error, but an estimate of the average error of any single measurement. In rare cases a single measurement may deviate from the mean by several times the probable error.

These random measurement errors affect the *precision* (reproducibility) of our measurements. But there is a second important type of error caused by miscalibration or biases in the measurement method. I recall once experiencing a series of strange frustrations in making a bookshelf, caused by the fact that some previous user of the yardstick with which I was measuring had carefully cut the first inch off the scale. Thus two separately measured 9-inch segments, when measured together end to end, totaled exactly 17 inches. Repeated measurement assured me that the total length was $17.00 \pm 0.05$ inches, meaning that the *precision* of the measurement was 0.05 inches. Alas, the *accuracy* (the difference between the measured value and the correct value) was far worse because of the systematic error introduced by the mutilated measurement device.

## Dimensions and Units

Measurements are made in terms of certain fundamental dimensions, such as mass, length, and time. The relationship of certain variables to one another can often be resolved by dimensional analysis, in which the dimensions of the variables are combined algebraically. Supposing one knew that a certain variable, $a$, had dimensions of length/time$^2$, but could not remember the equations linking it to velocity or distance. The correct functional relationship can be deduced by dimensional analysis (except of course for any dimensionless constants) by noting that velocity has dimensions of length/time; therefore (length/time)/time is acceleration, and $v/t = a$. Length is normally denoted $l$, mass is $m$, time is $t$, temperature is $T$, etc., with no measurement units specified. Note that this approach works well for dimensioned constants as well as variables, and can be used for any system of units or for conversions between different systems.

In practice, all measurements are made in convenient or traditional units: length is measured in centimeters in the cgs system, meters in SI, feet in the British system, AU in Solar System astronomy, Ångstrom units in atomic spectroscopy, etc. It is assumed that the reader is generally familiar with "metric" units. These usually fall into one of two categories, Système Internationale (SI) units (meter, kilogram, second) or cgs (centimeter, gram, second). Historically, cgs units were almost universally used in laboratory settings. Physicists have in recent years largely converged on the SI convention. However, planetary science is an eclectic amalgam of physicists, chemists, geologists, astronomers, electronic engineers, meteorologists, spectroscopists, mathematicians, and others. Each of these disciplines brings its own traditions—including traditional units—to the field. Chemists are still intimately familiar with calories, atmospheres, Avogadro's number, Loschmidt's number, amagats, and the cgs system, which was designed for convenience in the laboratory. Some early 20th-century chemistry journals quote measurements without giving units, since "everybody knows" what units are customary (in this case, cgs). Spectroscopists, having recently stopped reporting water abundances in planetary atmospheres in units of micrometers of precipitable water (μm ppt $H_2O$), have moved on in the literature of 2002 to using cm amagats or, even worse, μm atmospheres as the measure of gas column abundances, even though the latter is dimensionally incorrect. Atomic physicists are still replacing Ångstrom units with micrometers and nanometers. The literature on planetary fields and particles is written in a hodgepodge of conventions, perhaps the least of which is SI. The solar wind is usually treated in Gaussian units, and planetary magnetic fields are commonly described in terms of a "magnetic moment" constructed by multiplying the mean surface field times the volume of the planet, often expressed as gauss $cm^3$ or gauss $r_P^3$, despite the fact that these are not the units of magnetic moment.

The scientific study of large explosions has inherited its terminology from engineers and military officers, who traditionally describe explosive power in terms of equivalent mass of TNT (the high explosive trinitrotoluene). The energy released by explosion of one American ton (2000 pounds) of TNT is very close to $10^9$ calories, making it convenient to define the power of explosives in terms of tons of TNT. Nuclear explosives commonly have yields measures in kilotons of TNT, and thermonuclear explosions are measured in megatons of TNT (1 MT TNT $= 10^{15}$ cal $= 4.18 \times 10^{22}$ erg). Geophysicists dealing with explosive volcanic eruptions and planetary physicists studying impact cratering have adopted this strange unit because all the "ground truth" data on large explosions are couched in these terms.

Many astrophysicists routinely use cgs units, or refer mass, luminosity, and radius to the Sun as a standard, and report distances in parsecs. Solar System astronomers routinely use the astronomical unit and Earth's year as standard units, or janskys as a unit of flux. In the same vein, meteorologists diligently strive to describe hydrodynamic processes in terms of dimensionless parameter such as the Rayleigh, Reynolds, Richardson, and Rossby numbers and the Coriolis parameter, although the bar (1 bar $= 10^6$ dyn $cm^{-2}$) is still deeply entrenched as the unit of pressure. The advantage conferred by using dimensionless parameters is largely offset by the necessity of memorizing their names and definitions. Aeronomers deal with rayleighs as a unit of UV flux. Geologists, like astronomers, favor the year (annum) as the unit of time. And all this ignores the persistence of the last dinosaurs of the English system in some backwaters of engineering, where feet, pounds, BTUs, and furlongs per fortnight reign. The task of revising and reconciling all this chaos is beyond the scope of a mere textbook, especially since the purpose of a text is to provide entry to the research literature as it actually exists. Good luck—and watch your units.

## Exercises

### *Guide to the Literature*

I.1 Consult the catalog of your university library or other research library to find out which of the leading planetary sciences journals are immediately available to you. Choose five of these journals and examine their tables of contents, either in hard copy or online, for several recent issues. Write a one-sentence summary of the scope of *Icarus*, *the Journal of Geophysical Research*, the *Astrophysical Journal*, *Geophysical Research Letters*, and *Geochimica et Cosmochimica Acta*. If any of these journals is not available in your library, please substitute another journal from the list.

I.2 Find out which abstracting services in astronomy, space science, physics, chemistry, and geology are available in your library. Which are available online? Familiarize yourself with the use of *Science Citation Index*.

### *Numbers in Science*

I.3 a. Write the following numbers in scientific notation:

0.00054

76,453,000,000,000

4,000,000 × 250,000,000,000

37,194,000/0.000 000 361

b. Write the following numbers in normal notation:

$3.14 \times 10^7$

$6.673 \times 10^{-8}$

$(4.13 \times 10^{-6}) \times (3.77 \times 10^5)$

$4.13 \times 10^{-6}/(3.77 \times 10^5)$

### *Dimensions and Units*

I.4 The ideal gas law relates pressure $P$ (force per unit area = mass × acceleration/area $= ml^2/(t^2l^2) = m/t^2$), temperature ($T$), molar volume $v$ ($l^3$/mol), and the gas constant $R$ [energy/(degree mol) $= ml^2/(t^2T$ mol)]. Use dimensional analysis to write an equation relating these quantities.

I.5 Use dimensional analysis to show how to convert the water flow in a river in units of acre-feet per minute into liters per second. You need not use numerical values for the individual conversion factors (feet/meter, etc.).

# II. Astronomical Perspective

## Introduction

We cannot study the Solar System without some knowledge of the Universe in which it resides, and of events that long predate the Solar System's existence, including the very origin of matter and of the Universe itself. We shall therefore begin by tracing the broad outlines of present understanding of the origin and evolution of the Universe as a whole, including the synthesis of the lighter elements in the primordial fireball, galaxy and star formation, the evolution of stars, explosive synthesis of the heavier elements in supernova explosions, and astronomical evidence bearing directly on the origins of stellar systems and their possible planetary companions. No attempt is made to describe every current theory bearing on these matters. Instead, the discussion cleaves closely to the most widely accepted theories and selects subject matter for its relevance to the understanding of our own planetary system.

## Distance Scales in the Universe

Distances within the Solar System, such as the distance from Earth to the Moon or to the other terrestrial planets, can now be measured by radar or laser rangefinder (lidar) with a precision better than one part in $10^{10}$. The basic yardstick for measuring distances in the Solar System, the mean distance of Earth from the Sun, is called an astronomical unit (AU) and has a length of 149,597,870 km.

To measure the enormously larger distances between the Sun and nearby stars, we must make use of the apparent motion of nearby stars relative to more distant stars produced by Earth's orbital motion about the Sun. Figure II.1 shows how the relative motions of the star and the Sun through space are separated from the effects due to Earth's annual orbital motion. The angular amplitude of the oscillatory apparent motion produced by Earth's orbital motion is called the *parallax* (p), which is inversely proportional to the distance of the star. The parallax of a nearby star is so small that it is conveniently measured in seconds of arc ($''$), and hence the most direct measure of distance is

$$d(\mathrm{pc}) = 1/\mathrm{p}(''), \tag{II.1}$$

where the unit of distance (inverse arc seconds) is called a *parsec* (pc). The distance to the nearest stars is about one parsec. From Fig. II.1 it can be seen that 1 pc is 1 AU/sin ($1''$), or 206,264.8 AU ($3.08568 \times 10^{13}$ km). Since only a handful of nearby stars have parallaxes large enough to be measurable to a precision $< 1\%$, this precision in specifying the size of a parsec is gratuitous: $2 \times 10^5$ AU or $3 \times 10^{13}$ km is entirely adequate for most purposes.

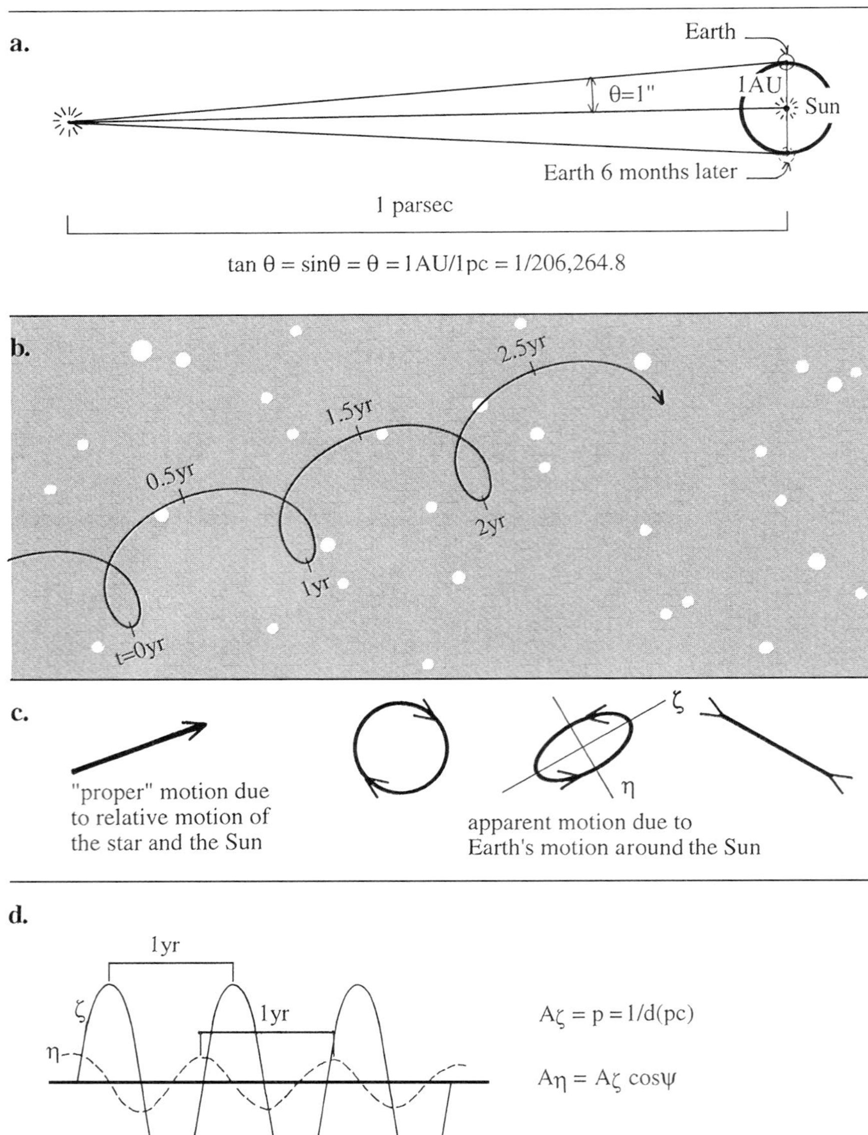


**Figure II.1** Planetary and stellar distance scales. The mean distance of Earth from the Sun, $1.5 \times 10^8$ km, is defined as 1 astronomical unit (AU). The stellar distance unit, the parsec (pc), is the distance from which the radius of Earth's orbit subtends 1 arc sec, as shown in a. The apparent motion of a nearby star against the background of much more distant stars is shown schematically in b. This motion is composed of a "proper" motion due to the relative translational velocity of the Sun and the star, combined with a projected elliptical motion due to the annual orbital excursions of Earth about the Sun (c). A nearby star lying near the plane of Earth's orbit will oscillate back and forth along a straight line in the sky; one close to the pole of Earth's orbit will describe an almost circular path. At intermediate ecliptic latitudes, elliptical paths are seen. When the effect of proper motion is removed, the ratio of the semimajor axis to the semiminor axis of the projected ellipse is easily calculated from the ecliptic latitude of the star, as in d.

We shall see later how such distance determinations permit the calculation of the absolute *luminosities* (erg $s^{-1}$) of stars, and how correlation of spectral properties with luminosity provides a very useful scheme for describing stars in terms of the relationships between their intrinsic properties. For the present it suffices to state that there exists a class of variable stars, called *Cepheid* (SEE-fee-id) variables, whose luminosities have been found to be

a.

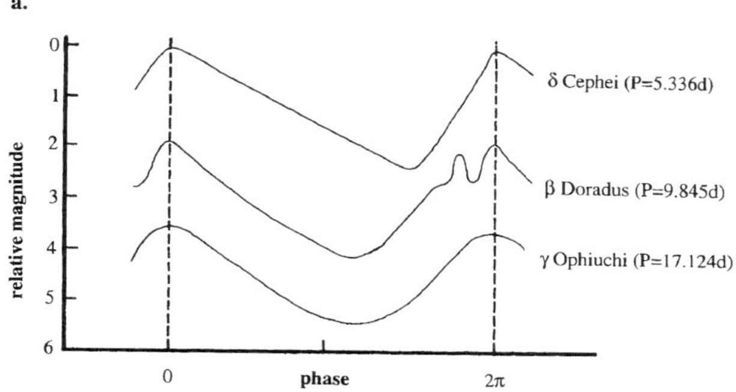


b.

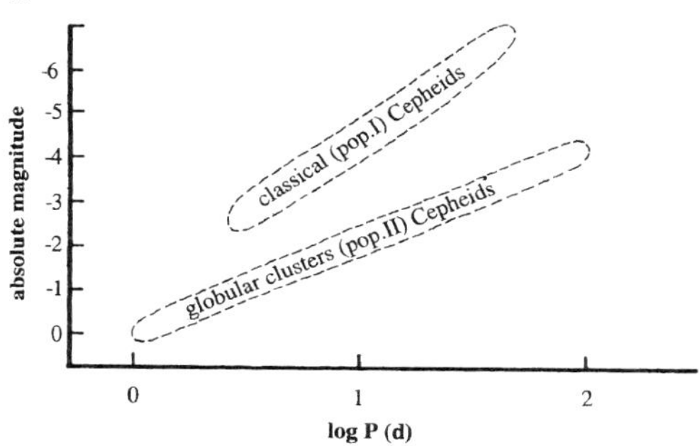


**Figure II.2** Period–luminosity relations for Cepheid variables. The lightcurves, or brightness-vs-time diagrams, for several Cepheids are shown in a. An arbitrary relative magnitude scale is used, and stars with different periods are plotted together on a magnitude-vs-phase diagram (phase = 0 at maximum light) to facilitate intercomparison. The relationships between the lightcurve period and luminosity (as absolute magnitude) are shown for both Pop I spiral arm stars and Pop II globular cluster stars in b.

directly related to their period of light variation (see Fig. II.2). This means that, once we have calibrated this luminosity-period relation for nearby Cepheids, we may then observe a Cepheid that is far too distant for parallax determinations, and use its observed period to calculate its luminosity. Then, from the observed brightness of the star, we can calculate how far it must be from us.

The use of Cepheid variables to determine distances is limited in two ways. First, it is limited in precision by the scarcity of Cepheids, since unfortunately very few are close enough to the Sun for useful distance determinations. Second, this procedure is limited in its range in space, since it can only be applied within that volume of space in which Cepheids can be seen and identified from Earth-based measurements. The former problem limits precision to at best $\pm 20\%$; the latter places a "horizon" for use of Cepheids at a distance of about $2 \times 10^6$ pc = 2 Mpc. Fortunately there are many galaxies, radio sources, and quasistellar objects within this distance, and it becomes possible in principle to apply the same philosophy all over again to extend the distance scale further. For example, we might try to establish the luminosities of one of these classes of objects, or of the very brightest stars in them, by calibrating their distances with Cepheids. We can then use brightness measurements on extremely remote ($\gg$ 2 Mpc) objects to estimate their distances.

In practice this is a very difficult task, fraught with the hazards of making selections between observed objects whose properties are, at best, only poorly understood theoretically.

The most useful type of measurement at present for observing very distant objects is the Doppler shift of their spectra. Let the subscript e denote the point of emission and o the point of observation of light of wavelength. Then the redshift $z$, defined as

$$z = (\lambda_o - \lambda_e)/\lambda_e, \tag{II.2}$$

is related to the relative recession velocity of the source, $v_{rel}$, by

$$\frac{v_{rel}}{c} = \frac{(z+1)^2 - 1}{(z+1)^2 + 1}. \tag{II.3}$$

A redshift of $z = 1$ thus corresponds to $v_{rel}/c = 0.6$, $z = 2$ to $v_{rel}/c = 0.80$, $z = 3$ to 0.88, $z = 4$ to 0.92, etc.

Many measurements of redshifts higher than $z = 3$ have been made for quasistellar objects, and great numbers of galaxies of $z > 1$ have been catalogued. These high redshifts, according to Eq. (II.3), correspond to recession velocities that are a large fraction of the speed of light. Using certain assumptions regarding the luminosities of galaxies at the remote times in the past when they emitted the light now reaching Earth, it is possible to estimate their distances also, and hence to evaluate the dependence of radial velocity on distance. It has been found by this procedure that all distant objects in the Universe are receding from us at velocities which are directly proportional to their distance from us:

$$dR/dt = HR, \tag{II.4}$$

where $R$ is the distance of the object and $H$ is a proportionality constant, called the Hubble constant, which is found to be approximately 75 km $s^{-1}$ $Mpc^{-1}$ with an uncertainty of $\sim 15\%$. Recalling the definition of a megaparsec, 1 Mpc = $10^6$ pc $\times$ 206,000 AU/pc $\times$ $1.5 \times 10^8$ km/AU = $3 \times 10^{19}$ km, and hence $H = 2.5 \times 10^{-18}$ $s^{-1}$.

The reciprocal of the Hubble constant, $1/H$, has dimensions of time and is $4 \times 10^{17}$ s. Since a year contains approximately $3 \times 10^7$ s, the time scale given by the Hubble constant is about $14 \times 10^9$ years = $14 \pm 2$ Ga.

Another way of expressing this result is to say that, some 14 Ga ago, every other galaxy in the Universe was in the same place as our own. At that time, all the matter in the observable Universe must have been hurled outward from some very small volume of space at speeds up to almost the speed of light. Direct evidence of any events that may have occurred before this explosion was presumably eradicated by passage through the extremely dense and energetic "primordial fireball." This ancient and violent explosion, from which all the matter and energy in the Universe originated, is called the "Big Bang."

When we observe objects that have high $z$ and are billions of parsecs away, we are seeing them as they were at the time they emitted the light we now observe, several billion years ago. They are a window on the ancient history of the Universe.

It has long been debated whether the initial explosion was sufficiently energetic to ensure that the galaxies will continue to recede from one another forever (an open universe), or whether their mutual gravitational attraction may eventually slow and stop the cosmic expansion, followed by catastrophic collapse back into a mathematical singularity (a closed universe). The presently known mass of the Universe is insufficient, by about a factor of 10, to stop the expansion, but there are several possible mass contributions that have not been adequately assessed. This missing mass problem also plagues attempts to understand the binding of galactic clusters and the rotation speeds of individual galaxies. Observations by the Hubble Space Telescope (HST) over the past few years suggest that the Universe is open and that the expansion rate is accelerating, a conclusion that hints at a universal force of repulsion beyond the established four forces of gravitation, electromagnetism, and the strong and weak nuclear forces.

However, events in the very earliest history of the Universe are poorly constrained by observation. Production of point-like (black hole) or line-like (superstring) singularities by the Big Bang is avidly discussed by cosmologists, as are the derivation of three-dimensional space from manifolds of higher dimension and "inflation" of space-time. These are exciting topics at the frontiers of research, but their bearing on the solution of observational problems such as the openness of the Universe, the missing mass problem, and the origin of galaxies is as yet very poorly demonstrated. In this book, with its orientation toward explaining the observed properties of the modern Solar System, we may be forgiven for starting a microsecond or two later in our account of the history of the Universe, since by doing so we save several hundred pages of interesting but possibly irrelevant material.

## The Big Bang

The energy density of the Universe during the early stages of the Big Bang was so high that the Universe was dominated by very energetic photons (gamma rays) and neutrinos, plus a varied and rapidly changing population of subatomic particles which were being produced and destroyed with enormous rapidity.

Protons (p), muons ($\mu$), and electrons (e) interacted with the radiation field through both annihilation and creation reactions:

$$2\gamma_p + \nu_p + \overline{\nu}_p = p^+ + p^- \tag{II.5}$$

$$2\gamma_\mu + \nu_\mu + \overline{\nu}_\mu = \mu^+ + \mu^- \tag{II.6}$$

$$2\gamma_e + \nu_e + \overline{\nu}_e = e^+ + e^- \tag{II.7}$$

$$p^+ + e^- = n + \nu_e, \tag{II.8}$$

where $\gamma_p$, $\gamma_\mu$, and $\gamma_e$ are gamma rays carrying the annihilation energies of protons, muons, and electrons, respectively. $\nu_\mu$ and $\nu_e$ are muon and electron neutrinos, and $\overline{\nu}_\mu$ and $\overline{\nu}_e$ are the corresponding antineutrinos, carrying the quanta of spin for the newly produced particles. The positive electron $e^+$ is called a positron, and n is a neutron.

Because of the great mass difference among protons, muons, and electrons, the characteristic gamma ray energies for Reaction (II.5) are much higher than those for Reaction (II.6), which are in turn much higher than those for Reaction (II.7). These energies are equivalent to the masses of the particles formed, in accord with Einstein's principle of mass–energy equivalence. The masses of a number of fundamental particles are given in Table II.1 with their energy equivalents in millions of electron volts (MeV). Those with the greatest rest masses can be formed only during the earliest expansion of the Big Bang fireball, because only then is the temperature high

**Table II.1** Rest Masses of Elementary Particles

| Particle | Rest mass | | Half-life |
|---|---|---|---|
| | MeV | g | s |
| Photon, $\gamma$ | 0 | 0 | Stable |
| Neutrino, $\nu$ | $\sim 0$ | $\sim 0$ | Stable |
| Electron, e | 0.511 | $9.042 \times 10^{-28}$ | Stable |
| Muon, $\mu$ | 105.66 | $1.870 \times 10^{-25}$ | $2.2 \times 10^{-6}$ |
| Pi meson, $\pi$ | 139.58 | $2.470 \times 10^{-25}$ | $1. \times 10^{-8}$ |
| Proton, p | 938.26 | $1.660 \times 10^{-24}$ | Stable |
| Neutron, n | 939.55 | $1.662 \times 10^{-24}$ | 1013 |
| Lambda, $\lambda$ | 1115.6 | | $2.5 \times 10^{-10}$ |
| Sigma, $\Sigma^+$ | 1189.5 | | $8 \times 10^{-11}$ |
| $\Sigma^o$ | 1192.6 | | $< 10^{-14}$ |
| $\Sigma^-$ | 1197.4 | | $1.5 \times 10^{-10}$ |
| Xi $\Xi^o$ | 1314.7 | | $3.0 \times 10^{-10}$ |
| $\Xi^-$ | 1321.2 | | $1.7 \times 10^{-10}$ |
| Omega, $\Omega^-$ | 1674 | | $1.5 \times 10^{-10}$ |
| Heavy baryons | to > 3000 | | $\sim 10^{-22}$ |

enough so that there are significant numbers of photons energetic enough to provide those masses. Production of heavy particles (baryons), such as protons and neutrons, must therefore cease well before meson production ceases, whereas light particles (leptons), such as electrons and positrons, may still be formed at much later times.

The distribution of photon energies in the fireball is described by the Planck function (Fig. II.3):

$$B_\nu = \frac{2h\nu^3}{c^2}\frac{1}{(e^{h\nu/kT} - 1)}, \tag{II.9}$$

where $B_\nu$ is the monochromatic radiance of the radiation field in erg cm$^{-2}$ s$^{-1}$ Hz$^{-1}$, $h$ is Planck's constant, $\nu$ is the frequency, $c$ is the speed of light, and $k$ is the Boltzmann factor. The numerical values of the constants in customary units are

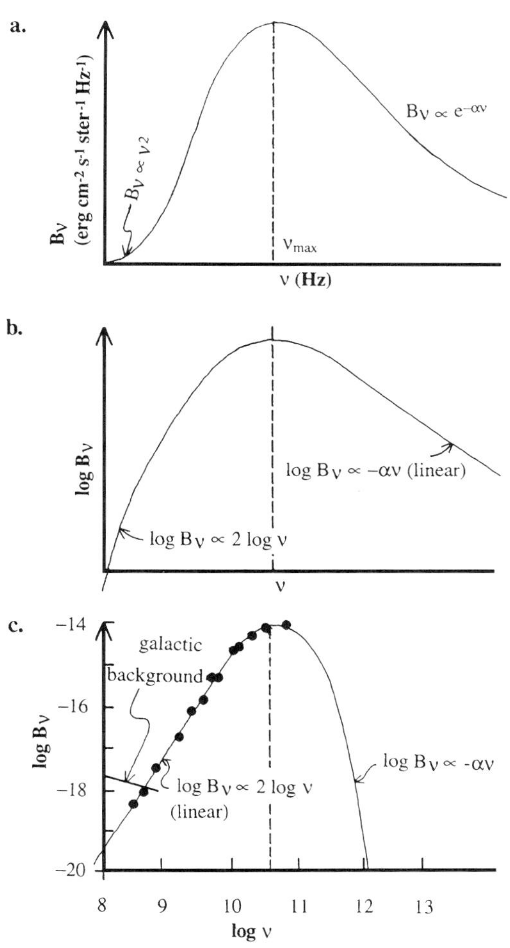


**Figure II.3** The Planck function. The usual linear representation of $B_\nu$ vs $\nu$ is shown in a. Observations at high frequencies well beyond the Planck peak are often graphed as in b, because this plot is linear in that regime. Observations at frequencies below the Planck peak are often graphed on a log–log plot for similar reasons, as we show here in c. The example given shows the observational data from which the 2.7 K background temperature of the Universe is derived.

$$h = 6.625 \times 10^{-27}\ \text{erg s}$$
$$c = 2.997 \times 10^{10}\ \text{cm s}^{-1}$$
$$k = 1.380 \times 10^{-16}\ \text{erg K}^{-1}.$$

It can be shown that a typical photon in this gas has an energy, $h\nu$, which is related to the equilibrium temperature of the radiation field by

$$h\nu \approx 3kT. \tag{II.10}$$

For the typical photon pair to be capable of forming a particle–antiparticle pair, they must carry enough energy to supply the rest masses of the particles,

$$2m_p c^2 = 3kT_p \tag{II.11}$$
$$2m_\mu c^2 = 3kT_\mu \tag{II.12}$$
$$2m_e c^2 = 3kT_e, \tag{II.13}$$

where the subscripts p, $\mu$, and e denote the proton, muon, and electron rest masses and their production temperatures. This conversion of energy into matter is a most practical application of Einstein's principle of equivalence of mass and energy.

Neutrons and protons, with very high masses (Table II.1), are formed together while the temperature is very high, but the products of this synthesis are subject to severe depletion by subsequent reactions. One of these is the mutual annihilation of proton–antiproton pairs [the reverse of Reaction (II.5)], which severely depletes the population of stable baryons. It is not known whether the present Universe contains equal numbers of antiprotons and protons or whether departures from perfect symmetry in the initial conditions led to an unequal production of protons and antiprotons. In addition to this reaction, Table II.1 reveals that the isolated neutron is itself unstable and decays by the reaction [essentially the inverse of Eq. (II.8)].

$$\text{n} \rightarrow \text{p}^+ + \text{e}^- + \nu_e \ (t_{1/2} = 1000\ \text{s}) \tag{II.14}$$

The rate of decay of an ensemble of $N$ radioactive particles (such as neutrons) is

$$dN/dt = -N\lambda, \tag{II.15}$$

where $\lambda$ is the decay constant in units of s$^{-1}$. The half-life is defined as the time required for half the original particles to decay,

$$\int_{N_0}^{N_0/2} \frac{dN}{N} = -\int_0^{t_{1/2}} \lambda dt \tag{II.16}$$

$$\ln\left(\frac{N_0/2}{N_0}\right) = -\lambda t_{1/2} \tag{II.17}$$

$$t_{1/2} = \frac{\ln 2}{\lambda} = 0.69315/\lambda, \tag{II.18}$$

hence

$$\frac{N}{N_0} = e^{-\lambda t} = e^{-0.69315(t/t_{1/2})}. \quad \text{(II.19)}$$

To make the rest mass of the proton requires, according to Eq. (II.11), a temperature of $7 \times 10^{12}$ K, muon formation occurs down to $8 \times 10^{11}$ K, and electrons continue to appear down to about $4 \times 10^9$ K. These temperatures are very much higher than the core temperature of the Sun, which is roughly $10^7$ K.

As the fireball cools through about $8 \times 10^{11}$ K, the rate of meson production very rapidly becomes negligible, and, because both pi and mu mesons are unstable as free particles, they quickly disappear from the system. When electron production is quenched near $4 \times 10^9$ K, mutual annihilation of electron–positron pairs can continue until the populations of these light particles (leptons) are also severely depleted. The same question regarding the possible existence of positron-rich regions of the Universe arises that we earlier encountered with respect to antiprotons; we may combine the two questions and ask whether antimatter regions dominate half the Universe. At present, there is no evidence for such a structure. Antimatter cosmic rays, for example, are unknown.

How much time does it take for the Universe to expand and cool to these several quench temperatures? The time required to cool to $8 \times 10^{11}$ K is only 10 $\mu$s, and $4 \times 10^9$ K is reached in about 10 s for typical models of the Big Bang (Fig. II.4).

During the time in which the temperature is greater than about $4 \times 10^9$ K, the fireball is densely populated by gamma rays, neutrinos, electrons, and positrons, with a significant residual population of baryons as well. Neutrons and protons make up about one part in $10^5$ of the total equivalent energy.

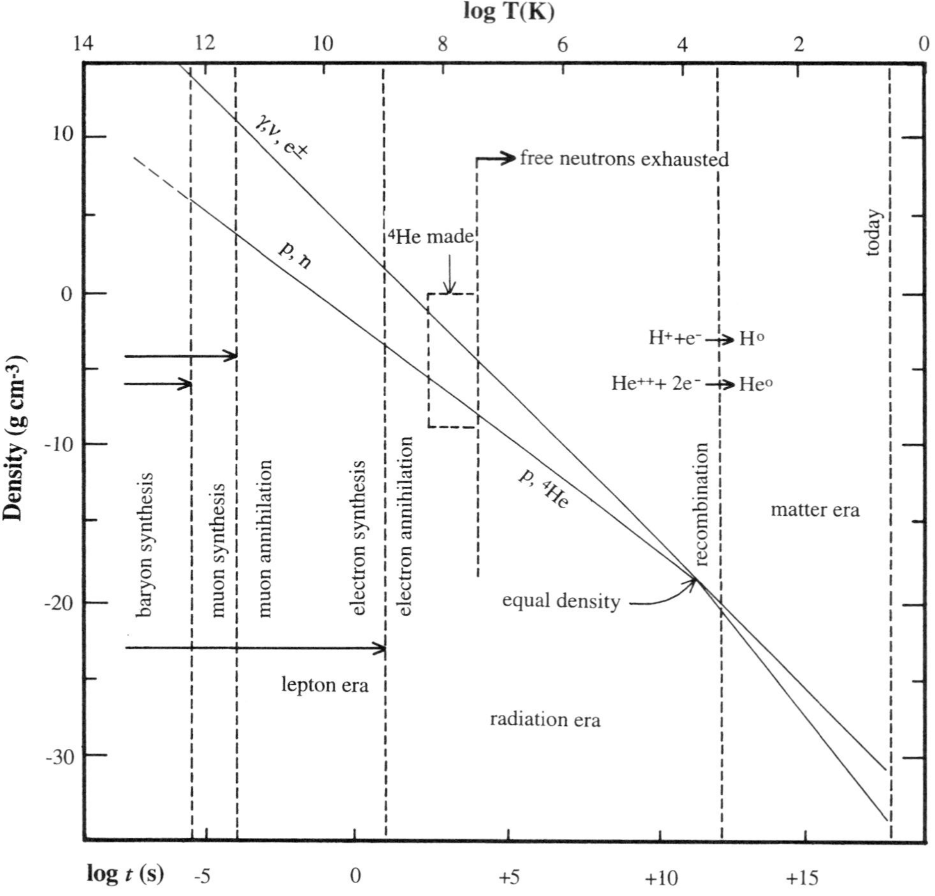


**Figure II.4** Evolution of the Big Bang fireball. The quench points (the times when the temperature first drops low enough to stop production) for the synthesis of baryons, muons, and electrons from the radiation field are indicated, as is the time of electron–ion recombination and the present epoch. The chemical evolution of the system is detailed in Fig. II.5.

There is a rapid interconversion of protons and neutrons by

$$e^+ + n \rightleftharpoons p + \bar{\nu}_e \tag{II.20}$$

$$\nu_e + n \rightleftharpoons p + e^-. \tag{II.21}$$

With both protons and neutrons present, it is possible to synthesize deuterium ($D = {}^2H$), the stable isotope of heavy hydrogen, by

$$p + n \rightarrow D + \gamma; \tag{II.22}$$

however, the inverse reaction, destruction of the deuteron by a gamma ray, is also possible as long as the photon energies are sufficiently large to overcome the nuclear binding energy of the deuteron. Table II.2 gives precisely measured masses for a number of the lighter nuclides, from which we can determine the binding energy of the deuteron. Note that the mass of the deuteron is slightly less than the sum of the masses of its component parts, the proton and the neutron. This "mass defect" is due to the emission of energy by the particles as they join together to form the deuteron. The missing energy, about 0.1 MeV, is the same as that carried by a typical photon at about $10^9$ K. At any higher temperature, therefore, average photons in the environment are energetic enough to reverse the reaction (i.e., destroy deuterium). Thus net formation of deuterium is unimportant until the temperature drops below $10^9$ K some 100 s into the explosion, when destruction of D by gamma rays becomes unimportant. This is much too early for the neutrons to have decayed away (their half-life is over 1000 s), and Reaction (II.22) can thereafter proceed more rapidly than its reverse.

The deuterons that are produced are still extremely reactive at these temperatures, because their nuclear binding energies are not much larger than the thermal energy of the fireball. For example, two deuterons may react,

$$D + D \rightarrow {}^3He + n, \tag{II.23}$$

thus replenishing the waning supply of neutrons as a by-product of the synthesis of helium-3.

Similarly,

$$D + D \rightarrow {}^3H + p \tag{II.24}$$

provides the radioactive isotope tritium (${}^3H$; T), which decays via

$${}^3H \rightarrow {}^3He + e^- + \bar{\nu}_e \tag{II.25}$$

with a half-life of 12.33 years.

Other reactions which are important during this era include

$$p + p \rightarrow D + e^+ + \nu_e \tag{II.26}$$

$$D + n \rightarrow {}^3H + \gamma \tag{II.27}$$

$$D + p \rightarrow {}^3He + \gamma \tag{II.28}$$

$${}^3He + {}^3He \rightarrow {}^4He + 2p + \gamma, \tag{II.29}$$

of which the most interesting is Reaction (II.29). Once produced, ${}^4He$ may be destroyed by collisions with energetic charged particles (spallation reactions) such as

$$p + {}^4He \rightarrow D + {}^3He. \tag{II.30}$$

The cooling of the fireball is so rapid that this is not an important loss process for ${}^4He$, although it does contribute appreciably to the production of deuterium and ${}^3He$.

No elements heavier than helium are produced in significant quantities in Big Bang nucleosynthesis. The abundances of important components of the Big Bang are shown in Fig. II.5 for the critical epoch when the temperature was near $10^9$ K.

**Table II.2** Masses of the Light Nuclides

| Species | $N$ | $Z$ | $A$ | Mass | Binding Energy | | BE per nucleon (MeV/A) |
|---|---|---|---|---|---|---|---|
| | | | | | AMU | MeV | |
| n | 1 | 0 | 1 | 1.00867 | 0 | 0 | 0 |
| ${}^1H$ (p) | 0 | 1 | 1 | 1.00728 | 0 | 0 | 0 |
| ${}^2H$ (D) | 1 | 1 | 2 | 2.01400 | 0.00195 | 1.816 | 0.91 |
| ${}^3He$ | 1 | 2 | 3 | 3.01605 | 0.00719 | 6.690 | 2.23 |
| ${}^4He$ | 2 | 2 | 4 | 4.00260 | 0.02930 | 27.29 | 6.82 |
| ${}^6Li$ | 3 | 3 | 6 | 6.01512 | 0.03270 | 26.15 | 4.19 |
| ${}^7Li$ | 4 | 3 | 7 | 7.01600 | 0.04052 | 37.74 | 5.55 |
| ${}^8Be$ | 4 | 4 | 8 | 8.00530 | 0.05850 | 54.49 | 6.81 |
| ${}^9Be$ | 5 | 4 | 9 | 9.01218 | 0.06029 | 56.16 | 6.24 |
| ${}^{10}B$ | 5 | 5 | 10 | 10.01294 | 0.06810 | 62.23 | 6.22 |
| ${}^{11}B$ | 6 | 5 | 11 | 11.00931 | 0.07911 | 73.69 | 6.70 |
| ${}^{12}C$ | 6 | 6 | 12 | 12.00000 | 0.09570 | 89.14 | 7.43 |

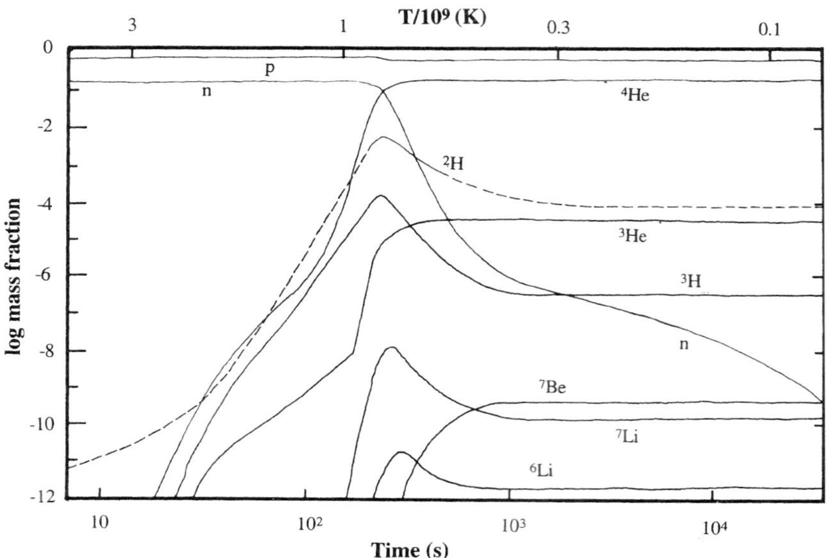


**Figure II.5** Nuclear abundances in the Big Bang fireball. The progress of the reactions that synthesize the lighter nuclides can be followed from "pure" hydrogen to the quenching of synthesis reactions by cooling and the eventual decay of free neutrons. Tritium also decays, but its half-life is much longer than the time covered by this diagram.

## Limitations on Big Bang Nucleosynthesis

> ... what God originally created, that matter which, by dint of His volition, He first made from His Spirit or from nihility, could have been nothing but matter in its utmost conceivable state of—of what?—of simplicity?
>
> Edgar Allen Poe
> *Eureka*

Reactions of elements heavier than hydrogen are strongly inhibited because the reacting nuclei must overcome their mutual electrostatic (Coulombic) repulsion. The rate expressions contain the factor $\exp -(\Delta E^*/kT)$, where $\Delta E^*$, the activation energy barrier, depends on the nuclear charges of the reacting nuclei, $Z_1$ and $Z_2$, as $\Delta E^* = cZ_1Z_2$.

At temperatures below a few million degrees the only nuclear reactions with appreciable rates are the decay reactions (II.14) and (II.24). During this phase of the expansion, photons and neutrinos dominate the Universe, but hydrogen and helium nuclei make up an appreciable fraction of the total energy equivalent:

$$(m_{\mathrm{H}} + m_{\mathrm{He}})c^2 \approx 10^{-2}E_{\mathrm{rad}}. \quad \text{(II.31)}$$

Although conversion of energy into matter by nuclear reactions has ceased, the density of the Universe is still large enough for strong radiation–matter coupling via Compton scattering, the interaction of free charged particles with photons. Thus the energy carried by the radiation field is constantly being fed into the kinetic energy of expansion of the matter in the fireball. The density of the Universe continues to drop, but not as rapidly as the decline in the energy density of the radiation field.

At temperatures of about $10^4$ K the radiation field is cool enough to permit the formation of the first neutral atoms by recombination of free electrons with positive ions of hydrogen and helium. Beyond this point the Universe is, to a good approximation, composed of 28% by mass $^4$He atoms and 72% H atoms. At about $10^3$ K atomic hydrogen can react to make $H_2$ molecules:

$$\mathrm{p} + \mathrm{e}^- \rightarrow \mathrm{H} + h\nu \quad \text{(II.32)}$$

$$\mathrm{He}^{2+} + \mathrm{e}^- \rightarrow \mathrm{He}^+ + h\nu \quad \text{(II.33)}$$

$$\mathrm{He}^+ + \mathrm{e}^- \rightarrow \mathrm{He} + h\nu \quad \text{(II.34)}$$

$$\mathrm{H} + \mathrm{H} \rightarrow \mathrm{H}_2 + h\nu. \quad \text{(II.35)}$$

All this occurs within some $3 \times 10^{14}$ seconds (10 million years) after the beginning of the Big Bang explosion.

The observed helium content of the Universe and the deuterium:hydrogen (D:H) ratio of $1.4 \times 10^{-5}$ are both believed to be direct consequences of the Big Bang. The abundances of these species reflect conditions during the early stages of the explosion. The abundances of D and He in the present Universe suggest a density for the Universe that is not high enough to arrest its expansion and cause it to slow and recollapse.

Another feature of the Big Bang with profound observational consequences is the leftover radiation after the cessation of creation reactions. These photons continue to degrade in energy as the Universe continues to expand. This radiation "cools" from GeV gamma rays to X rays, ultraviolet and visible light, infrared, and finally microwave radiation. One of the crucial experimental confirmations of the predictions of the Big Bang theory has been the detection at microwave (centimeter) wavelengths of an isotropic radiation field with the spectrum of a Planckian emitter [Eq. (II.9)] at a temperature of about 2.7 K (see Fig. II.3c). It seems likely that, given any slight degree of anisotropy in the expansion of the cloud of hydrogen and helium from the fireball, instabilities will develop and propagate. Regions of enhanced density would then eventually give rise to the formation of galaxies and clusters of galaxies. Up to the time of formation of well-defined protogalaxies it is likely that the Universe was devoid of stars and other high-density objects, containing only degraded radiation and cooling hydrogen and helium gas. The sole possible exception might be incredibly dense black holes, composed of tiny portions of the original fireball that never expanded far enough to make what we regard as "normal" matter.

Had the nuclear and chemical evolution of the Universe been arrested at this stage, the entire scope of chemistry would have been limited to the formation of molecular hydrogen. Not only are the chemicals essential to the formation of planets and life absent, but also the very elements essential to their existence have not been formed. How did such a dull and unpromising universe give rise to present complexity?

## Galaxy and Star Formation

A relatively dense gas cloud may collapse if its own gravitational potential energy is greater than its internal thermal energy. This condition, known as Jeans' criterion, after its discoverer, the famous British astronomer Sir James Jeans, is given by

$$\frac{GmM}{r_c} = \frac{3mkT}{2} \tag{II.36}$$

or, substituting

$$M = \frac{4}{3}\pi r_c^3 \rho, \tag{II.37}$$

$$r_c > \left(\frac{9kT}{8\pi G\rho}\right)^{1/2} = c(T/\rho)^{1/2} \tag{II.38}$$

where $G$ is the universal constant of gravitation, $M$ is the mass of the cloud, $r_c$ is the critical unstable radius, and $\rho$ is the density of the cloud. In effect, a molecule in a cloud larger than $r_c$ will have a thermal velocity that is too low for it to escape from the cloud. If the cloud is able to lose energy by radiation, it may then collapse to much higher densities.

As collapse continues, the density of the gas increases and the minimum size of a gravitationally unstable element of the gas also changes. In the most favorable (and reasonably realistic) case, in which the collapsing gas cloud is fairly transparent to infrared radiation, the temperatures of the molecules in it will be governed by exchange of energy with the outside universe (which is changing very much more slowly than the collapsing cloud). The collapse will then be nearly isothermal until the density and opacity of the gas have grown enormously.

The gravitational potential energy of the collapsing cloud accelerates the component helium atoms and hydrogen molecules inward. They collide and partition their increased energy between translational (thermal) motion and internal vibration and rotation of the hydrogen molecule. The energy required to excite vibration of the hydrogen molecule corresponds to a temperature of about 3000 K, whereas pure rotation can be excited by collisions at temperatures near or above 300 K. As the opacity grows, more and more of this energy is stored internally in the cloud.

A molecule is most like a black body (a perfect emitter) at those wavelengths at which it has strong absorption bands (that is, where it is an excellent absorber). Thus once collapse heats the gas to a modest temperature of a few hundred kelvins, the gas will readily become rotationally excited by collisions, and the rotationally excited molecules will emit their excitation energy in the far infrared. As we shall later see, hydrogen gas must be *very* dense before its opacity becomes important. This is why, during the early stages of collapse of a hydrogen gas cloud, the cloud cannot retain much of its collapse energy.

For an isothermal collapsing cloud of constant mass $M$,

$$r_c \leq \frac{2Gm}{3kT} = \frac{2GM}{3kT}\left(\frac{r_c}{R}\right)^3, \tag{II.39}$$

where $R$ is the radius of the parent cloud and $m$ is the mass of the smaller cloudlet formed by fragmentation of the parent cloud of mass $M[m/M = (r/R)^3]$. This leads to the expression

$$r_c \geq \left(\frac{3kT}{2GM}\right)^{1/2} R^{3/2}, \tag{II.40}$$

where the quantity in parentheses is constant. Thus reduction of the radius of the parent cloud by a factor of 4 due to collapse (a density increase by a factor of

$4^3 = 64$) leads to a decrease of $r_c'$ by a factor of 8, which is only half the new radius. The large cloud can therefore fragment into about 4 to 10 smaller cloudlets, each of which will continue to collapse in the same manner until the process is halted by star formation or by the buildup of rotation speed caused by conservation of angular momentum. This phenomenon of hierarchical collapse can produce a large number of levels of structure of many different sizes, ranging in the present case from masses of thousands of times that of our Galaxy down, through that of a small galaxy ($10^{43}$ g), eventually to star clusters and individual stellar systems.

Small gas clouds with quite high densities and low angular momenta will produce first-generation stars with random masses, many of which will be much larger than normal stable stars. We must pursue further the evolution and classification of stars and stellar systems in order to appreciate fully the significance and relevance of these early stages in the evolution of the Universe.

## Structure and Classification of Galaxies

The distribution of matter throughout the known Universe is both sparse and nonuniform. Averaging out all known or suspected galactic matter over the volume of the known Universe (a sphere with a radius of 5 Gpc $= 5 \times 10^9$pc) gives a mean density of $10^{-30}$ g cm$^{-3}$, equivalent to one hydrogen atom per cubic meter. By comparison, the density of matter within our own Galaxy, the Milky Way, is approximately $10^{-24}$ g cm$^{-3}$, some $10^6$ times that of the Universe as a whole.

The characteristic distance scale of the Universe is the Gpc (gigaparsec; $10^9$pc); typical nearest-neighbor intergalactic distances are near 1 Mpc (megaparsec); typical galaxies have dimensions of a few kpc; the distance between neighboring stars in a galaxy is about 1 pc; the diameter of a planetary system is near 1 mpc (milliparsec); distances between neighboring planets are about 1 $\mu$pc (microparsec; $10^{-6}$pc); the size of a planet is about 1 npc (nanoparsec; $10^{-9}$pc). Each step in this scale represents a change by a factor of $10^9$ in the volume, and each step corresponds to an increase in density. The final step brings us to planetary bodies with densities of about 1 g cm$^{-3}$.

On the upper end of the mass scale, even beyond galaxies, there is clustering of galaxies and even clustering of galactic clusters to form superclusters with dimensions up to about 100 Mpc. Many thousands of clusters are known, each typically containing hundreds to thousands of galaxies. One prominent nearby cluster with more than 1000 members is in the constellation of Coma Berenices at a distance of 25 Mpc. Our own Galaxy belongs to the Local Group, a small cluster of which the Magellanic Clouds and the Andromeda Nebula are also members. Millions of galaxies have been photographed, but we know that we can see out to distances so great that only a tiny minority of galaxies are bright enough to be visible at that distance.

Clustering of galaxies extends on down to groups of a mere dozen or so individuals. "Chains" typically containing five or six spiral galaxies connected by streams of stars have been found many times.

Individual galaxies exhibit only a rather limited range of overall morphologies. A "triangular" classification scheme with three main branches suffices to type most galaxies. Figure II.6 displays sketches of spiral, barred spiral, and elliptical galaxies. The two classes of spiral galaxies are each subdivided according to how tightly the spiral arms are wound, whereas elliptical galaxies are classified according to the eccentricity of their projected disks.

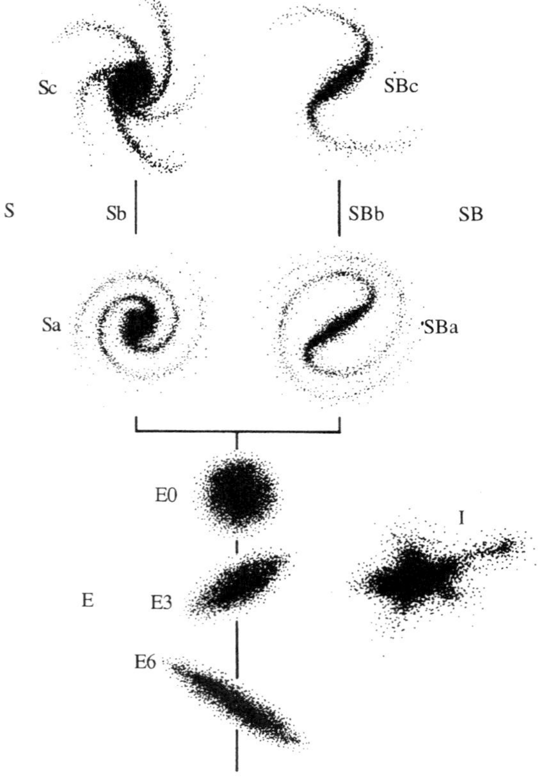


**Figure II.6** Classification of galaxies. Spirals are subdivided according to whether a barlike nucleus is present. Both the barred spiral (SB) and the spiral (S) branches of the diagram are ranked according to how tightly the spiral arms are wound. The highly symmetrical gas- and dust-free elliptical (E) galaxies form the third arm of the diagram. Irregular galaxies, such as the severely distorted Magellanic Clouds, are lumped in yet another category (I).

Note that this classification scheme is of use principally for identification purposes: a spiral galaxy seen edge-on cannot be categorized accurately, and the projected shapes of elliptical galaxies have no simple relationship to their three-dimensional morphologies.

The fundamental distinctions between elliptical and spiral galaxies are, however, unmistakable. Elliptical galaxies are highly symmetrical and almost always completely devoid of gas and dust. Spiral galaxies, on the other hand, have dense gas and dust lanes spiraling outward from their centers. The central region of each large spiral galaxy is usually rather similar to an elliptical galaxy, with a high degree of symmetry, no spiral lane structure, and very little gas and dust. The very centers of the cores of large galaxies often exhibit phenomenal luminosities in the infrared and X-ray regions, frequently accompanied by extremely violent eruptive phenomena and "jets" of extremely hot and fast-moving gas.

Because galaxies frequently form compact clusters or close pairs, it is often possible to measure the radial component of their velocities by means of the Doppler shifts of lines in their spectra and thus to deduce the total mass of the system, and often the masses of the individual galaxies as well. Spiral galaxies are usually found to have masses near $10^{11}$ times that of the Sun ($10^{11} M_{\odot}$), whereas elliptical galaxies are typically a few times less massive on the average. Both classes, however, span factors of about 100 in total mass.

The luminosities of most large spiral and elliptical galaxies are between $10^9$ and $10^{10}$ times that of the Sun ($10^{10} L_{\odot}$). The average luminosity of galaxies in the Coma group is about $0.5 \times 10^9 L_{\odot}$, whereas the average mass is about $4 \times 10^{11} M_{\odot}$. In general, the mass:luminosity ($M$:$L$) ratio for galaxies lies within a factor of 10 of 100:1; that is, average galactic material with a mass of $100\,M_{\odot}$ is required to produce the luminosity of our Sun. Obviously, then, a very large proportion of the mass in a typical galaxy must be either outside of stars or tied up in stars that are enormously less luminous per unit mass than our Sun. The search for this invisible but gravitationally important "missing mass" continues.

The Milky Way, despite the poor perspective we have on its global properties, still provides us with an inside closeup view of many of the processes at work in what appears to be a fairly typical spiral galaxy. The Milky Way, with a mass of $2 \times 10^{11} M_{\odot}$, has an estimated luminosity of $10^{10} L_{\odot}$, for a $M$:$L$ ratio of 20. Our Galaxy is mostly confined to a flattened, disk-shaped volume of space some 30 kpc in diameter and about 8 kpc thick at the center. Away from the center, the disk is only about 4 kpc thick. In addition to the lenticular distribution of stars, gas, and dust which contains the spiral structure, there is the distinct system of stars in the galactic core, and a second spherically symmetrical system of very compact dust- and gas-free globular clusters of stars. Each of these clusters is itself spherically symmetrical and looks like a tiny elliptical galaxy. The "bulges" in the galactic disk near the rotation poles of the Galaxy are due to the central system of non-spiral-arm stars.

The globular clusters associated with our Galaxy typically contain several thousand to a million stars each and occupy a volume of space extending out as far as 40 kpc from the galactic center. The main spiral arm system extends out to about 15 kpc, and the Sun is located roughly 8 kpc from the center. The densities of both the stellar and the globular cluster populations increase rapidly toward the center. Thus the vast majority of the stars in our Galaxy are located within 6 kpc of the center, and fully a third of all the known globular clusters are found within the 2% of the solid angle of the sky closest to the direction of the galactic center, in the constellation Sagittarius. That we are able to see so many clusters in such a small region of the sky is particularly striking in view of the difficulty of observing the central regions of our Galaxy through the intervening lanes of obscuring gas and dust.

Interestingly, the $M$:$L$ ratio of the star populations in globular clusters is higher than that found for spiral-arm star populations, even though interstellar gas and dust are absent. This suggests an important difference between these two major stellar environments.

The Milky Way would collapse under its own gravity in about $10^8$ years if it were not rotating. It is possible to measure the speed of the Sun with respect to selected other bodies by means of the Doppler shift and to deduce from these measurements the approximate orbital speed of the Sun about the galactic center. Measurement of the relative velocities of other nearby stars (which are also in orbit around the galactic center at about the same mean distance) shows us that these stars have random velocities of several kilometers per second. Further, the Sun is found to be moving at a rather high speed relative to the average of the nearby stars: the Solar System is moving roughly toward the star Vega at about 20 km s$^{-1}$.

The speed of the Sun relative to the *average* of the globular clusters is much higher, roughly 200 km s$^{-1}$. Since the distribution and motion of the globular clusters are spherically symmetrical, they do not partake of the orderly rotation of the disk population of stars. As many are moving "forward" as are moving "backward," so their average speed is zero. The Sun's speed relative to them is thus a measure of the orbital speed of the Sun about the galactic core. The direction of this latter motion is in the direction of the star Deneb, which lies in the galactic plane. The stars in the spiral arms of the Galaxy, including all of the Sun's nearest neighbors, orbit in the same direction with roughly circular orbital velocity, about 200 km s$^{-1}$. The average velocity dispersion of these stars is about 10 km $s^{-1}$, corresponding to a typical orbital eccentricity of order $10/200 \approx 0.05$ and a typical orbital inclination of about 10/200 radians $\approx 3°$.

Superimposed on this motion is a random or "thermal" component of a few km $s^{-1}$, which corresponds to modest orbital eccentricities and inclinations. The orbits of the globular clusters are "hot" in that they are as likely to have retrograde as prograde orbits, and the eccentricities of their orbits may approach unity. Not surprisingly, these orbits also extend out quite far from the galactic center. The spiral arm stars and gas clouds thus pursue planet-like orbits, whereas globular clusters have comet-like orbits. The Sun's motion relative to nearby stars, mentioned above, means that its "thermal" velocity is higher than average, about three or four times the average thermal speed found for nearby stars. The Sun takes about 200 million years to complete one revolution about the galactic center; the Solar System has completed fewer than two dozen trips about the galactic core since the origin of the Sun and planets.

Some of the fundamental structural features of the Galaxy can be seen with the unaided eye on any clear, moonless night. The plane of the lenticular star distribution (the Milky Way) is well defined as a band of light girdling the sky, brightest in the direction of the galactic center. In several places the bright background of stars is obscured by dark, dense gas and dust clouds that mark out the plane of the nearby spiral arms. The location of the central plane of the Galaxy is also marked out by the presence of numerous extremely luminous blue–white stars, which are found quite close to this plane. These brilliant blue–white stars are not present in globular clusters or elliptical galaxies. By "brilliant" we of course refer to the intrinsic luminosities of the stars, not simply their apparent brightness as seen from Earth. The luminosity of a star may be given in absolute (erg $s^{-1}$) or relative terms, in units of the luminosity of the Sun, $L_\odot$. The luminosity of the Sun is $4 \times 10^{33}$ erg $s^{-1}$. The blue–white stars, which mark out the galactic plane, have luminosities of 100 to 1000 $L_\odot$ and even higher. These stars make up only an infinitesimal proportion of the population of the Galaxy, but their high luminosity makes them visible over distances of several kpc. They are largely responsible for the low $M{:}L$ ratio of the spiral-arm population of stars. The way in which these luminosities and other intrinsic properties of stars can be determined is most interesting, and deserves further comment.

## Classification of Stars

Historically, stars were first classified solely on the basis of their apparent brightness, ignoring their easily observed color differences. However, we have seen that the spectral distribution of energy versus frequency [Eq. (II.9)] or wavelength

$$B_\lambda = \frac{2hc^2}{\lambda^5}\left(\frac{1}{e^{hc/k\lambda T} - 1}\right) \qquad \text{(II.41)}$$

contains a maximum. Here $B_\lambda$ is the radiation field intensity in erg $cm^{-1}$ $s^{-1}$ per centimeter of wavelength. Differentiating with respect to wavelength, we find that the maximum in $B_\lambda$ occurs where $dB_\lambda/d\lambda = 0$ or

$$5(1 - e^{-hc/k\lambda_m T}) = \frac{hc}{k\lambda_m T}, \qquad \text{(II.42)}$$

where $\lambda_m$ is the wavelength at which $B_\lambda$ is a maximum. The roots of this equation are

$$\frac{hc}{k\lambda_m T} = (0, 4.965114), \qquad \text{(II.43)}$$

of which the first is trivial. For the other, we find

$$\lambda_m T = 2897\ \mu\text{m K}, \qquad \text{(II.44)}$$

which is the Wien Displacement Law.

Note that $B_\lambda$ and $B_\nu$ are not even dimensionally the same. $B_\nu$ is a maximum at

$$T/\nu_m = 1.700 \times 10^{-11}\ \text{K Hz}^{-1} \qquad \text{(II.45)}$$

or

$$\lambda(\nu_m)T = 5098\ \mu\text{m K}. \qquad \text{(II.46)}$$

It is apparent from these considerations that the color of a star (especially the wavelength at which the emitted flux is a maximum) contains valuable information about one important intrinsic property of a star, its surface temperature. On the other hand, the apparent brightness by itself tells us nothing about the intrinsic properties of the star. However, if we had some means of measuring the distances of stars, we could then use the apparent brightness to calculate the absolute brightness (and thus the luminosity) of each star.

The apparent brightness of a star is given by its *apparent* visual magnitude, $m_v$, on a magnitude scale which is logarithmic in flux. The *magnitude* scale, which was first established by naked-eye observations, reflects the logarithmic response of the human eye to radiation intensity. It was customary to describe the brightest stars as "stars of the first magnitude." Slightly fainter stars were then called "second magnitude" stars and so on, down to the limit of detection by the naked eye, sixth magnitude. Thus the magnitude scale decreases to negative numbers for the brightest objects. When a quantitative magnitude scale was established, it was made to conform as closely as possible with the old naked-eye scale. Each step on the magnitude scale is about a factor of 2.5 in flux, and five magnitudes is exactly a factor of 100 in flux. A bright star such as

Vega ($m_v = +1$) therefore delivers to terrestrial observers a light flux 100 times as large as that coming from the faintest naked-eye stars ($m_v = +6$). The brightest star seen in the night sky, Sirius, has an apparent visual magnitude $m_v = -2.6$.

The apparent visual magnitude is approximately given by

$$m_v = -2.5 \log F_v - 10.7, \tag{II.47}$$

where $F_v$ is the total visual (0.4 to 0.8 $\mu$m) flux reaching the observer, in units of $\text{erg cm}^{-2}\,\text{s}^{-1}$. The Sun, which provides $1.37 \times 10^6\,\text{erg cm}^{-2}\,\text{s}^{-1}$ to Earth, has a visual magnitude of $-2.5(\log 1.3 \times 10^6) - 10.7 = -26.1$.

The color, or spectral class, of a star can usually be estimated by photometric comparison of images of the star taken through three or more colored filters that transmit only narrow spectral intervals of light. The most commonly used filters for this purpose are ultraviolet, blue, and the center of the visible region (yellow). This is referred to as the UBV filter system. For more precision, especially with cooler stars, additional filters in the near infrared are added to the set.

On the basis of the UBV photometric classification of stars a number of different color groups can be distinguished. For historical reasons, these color groups form a spectral sequence labeled with an inscrutable sequence of apparently random letters. For the spectral sequence running from violet through the visible region to red, the principal color classes are O, B, A, F, G, K, and M, and the less common classes are R, N, and S. There are endless mnemonics to assist in keeping this sequence intact and in order: my favorite is "Oscar, Bring A Fully Grown Kangaroo: My Recipe Needs Some." (Certain other spectral classes, such as C, are often encountered in the astronomical literature but rarely seen in space.) Thus O and B stars are very strong ultraviolet emitters, blue or violet to the eye, with surface temperatures in excess of 15,000 K. A and F stars, with temperatures near 10,000 and 8000 K, respectively, may be described as white. Our Sun is a representative of the cooler yellow G stars, which have surface temperatures near 6000 K. K stars are orange in color, and M stars, with temperatures below 4000 K, are red.

Given only one further type of data about these stars, their distances from us, it would be possible to construct a two-dimensional (color–luminosity) classification system for stars in which intrinsic properties alone are employed. In fact, as we have already seen, several thousand stars are close enough to the Sun so that the annual motion of the Earth around the Sun causes a measurable displacement in the position of these stars against the background of much more distant stars. Thus, by the simple expedient of comparing photographic images of these stars and their stellar backgrounds on pictures taken 6 months apart, it is possible to calculate their distances. We now can combine the two simplest measurements of the star, its apparent magnitude and its parallax, to determine the *absolute* magnitude of the star, $M_v$. By convention the absolute magnitude is defined as the apparent magnitude the star would have if it were at a distance of 10 pc:

$$\begin{aligned} M_v &= m_v + 5 - 5 \log d(\text{pc}) \\ &= m_v + 5 + 5 \log p('') \end{aligned} \tag{II.48}$$

The absolute magnitude is, like the color, an intrinsic property of the star and is directly related to the star's luminosity by

$$M_v = 6 - 2.5 \log(L/L_\odot); \tag{II.49}$$

therefore, in terms of directly measured properties (parallax and $m_v$), the luminosity of a star is

$$\log(L/L_\odot) = 0.4(1 - m_v) - 2 \log p(''). \tag{II.50}$$

If the several thousand stars with measured parallaxes are used to construct a color vs magnitude diagram, certain very interesting systematics appear at once. Such a plot, known as the Hertzsprung–Russell (H–R) diagram after its inventors (Ejnar Hertzsprung of Denmark and Henry Norris Russell of Princeton University in the United States), is given as Fig. II.7.

The most prominent feature of the H–R diagram is the diagonal strip of stars known as the Main Sequence (MS). A large majority of all the stars near the Sun lie on the Main Sequence, and those stars that are not in the MS lie in very restricted regions of the H–R diagram.

Some stars, usually found enveloped in gas and dust clouds, are too red for their luminosity, compared with the MS. These stars, called T Tauri stars after their prototype, lie parallel to but slightly above the MS. A number of other very luminous stars are much too red for their luminosities relative to the MS (or much too luminous for their temperatures). Such stars achieve their enormous luminosities by having very large surface areas: the total luminosity is proportional to the surface area as well as dependent on the temperature. Such large stars are referred to as giants and supergiants.

We have already seen that the Planck function for a black body emitter, $B_\lambda$, has a shape given by Eq. (II.41) and that the wavelength at which $B_\lambda$ is a maximum is dependent on temperature in accordance with Eq. (II.44). Now we need to know how the total luminosity of a black body depends on its temperature, that is, to evaluate the integral of $B_\lambda$ over $\lambda$. To forestall the necessity of evaluating the integral directly, note

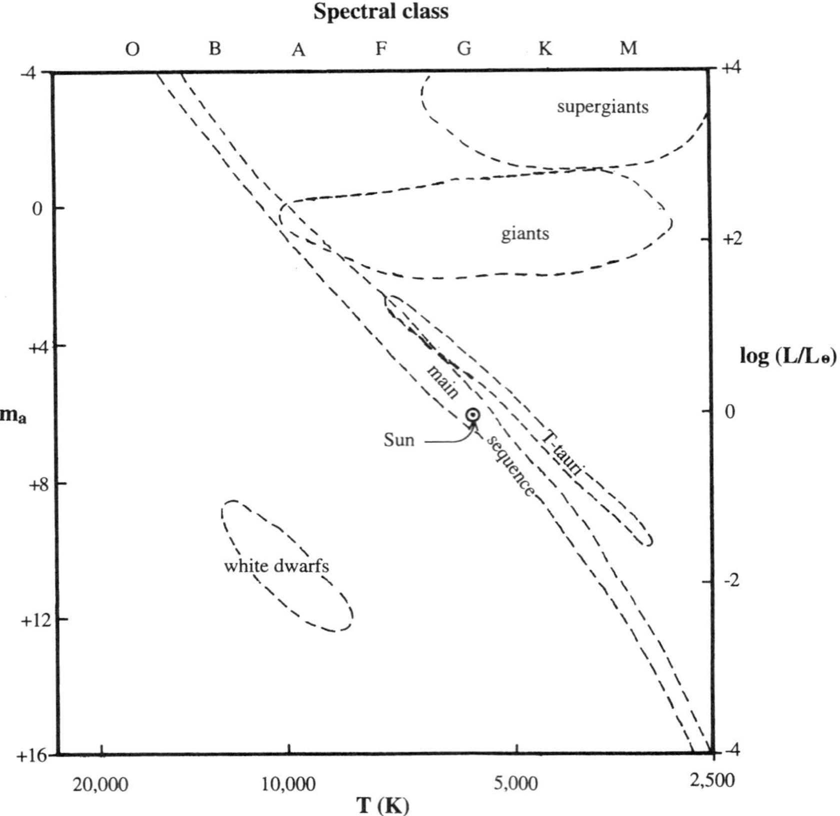


**Figure II.7** The Hertzsprung–Russell diagram. The temperature (spectral class) and absolute magnitude (luminosity) of each of the 10,000 apparent brightest stars seen from Earth are compared. The large majority of the stars lie on the hydrogen-burning Main Sequence (MS).

from Fig. II.8 that the product $B_\lambda \cdot \lambda$ is proportional to the area under $B_\lambda$. Thus

$$\int_0^\infty B_\lambda d\lambda \propto \frac{2hc^2}{\lambda_\mathrm{m}^5}\left(\frac{1}{e^{hc/k\lambda_m T}-1}\right)\lambda_\mathrm{m};\ \lambda_\mathrm{m} = \frac{2897}{T} \quad \text{(II.51)}$$

and

$$\int_0^\infty B_\lambda d\lambda = \sigma T^4, \quad \text{(II.52)}$$

where the proportionality constant, $\sigma = 5.65 \times 10^5\,\mathrm{erg\,cm^{-2}\,s^{-1}\,K^{-4}}$, is called the Stefan–Boltzmann constant.

The luminosity of a star is simply its total radiated power over its entire surface:

$$L = 4\pi R_*^2 \sigma T^4. \quad \text{(II.53)}$$

Thus from Eqs. (II.45) and (II.49), the observed visual magnitude ($m_\mathrm{v}$), the parallax ($p$), and the color allow calculation of the radius ($R_*$). In this way we can calculate that the dimensions of supergiant stars are comparable to those of our entire Solar System. A group of quite hot (blue) but intrinsically faint stars is also found.

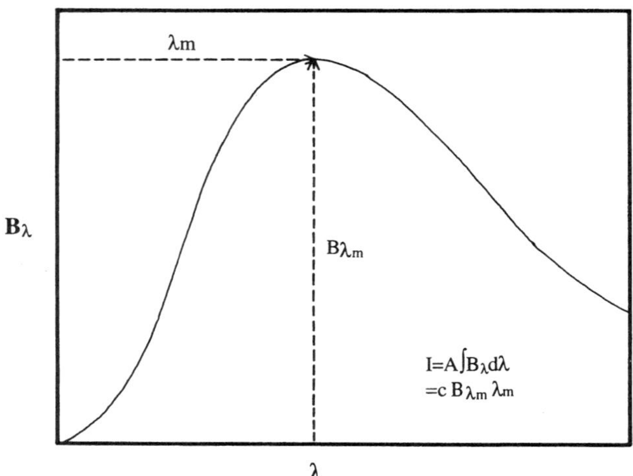


**Figure II.8** The total energy emitted from a Planckian source. Here $\lambda_m$ is the wavelength at which $B_\lambda$ is a maximum.

By the logic applied above to the giants and supergiants, it is clear that these underluminous stars must be extremely small compared with normal (MS) stars; in fact, some of these "white dwarfs" are smaller than the Earth. Astronomers often talk of "red dwarfs" as if they constitute a distinct class of stars; in fact, they are simply the faint red end members of the Main Sequence.

Since it is common to find stars in multiple-star systems, it is often possible to measure the differences between the radial velocities of the components of such a system and thence to determine the orbital velocities of the stars about their common center of mass. Thus, in the same manner that we deduced the masses of galaxies, we may determine the masses of individual stars. In some cases the stellar binary and multiple systems are close enough to us and compact enough so that we may telescopically observe the stars through one or more complete orbits. Through such observations and careful parallax and Doppler measurements it is possible to derive quite accurate masses for the stars. Let us assume for the sake of a brief illustration that we know the parallax (and thus the distance) of a pair of stars that are in circular orbits about their center of mass, the angular amplitude of their orbits (and thus their radii), and their orbital period.

Figure II.9 shows the definitions of the variables: star 1 of mass $M_1$ and orbital velocity $v_1$ orbits at distance $R_1$ from the center of mass. The distance from $M_1$ to $M_2$ is $R = R_1 + R_2$. Equating the gravitational force between $M_1$ and $M_2$ to the centripetal force required to maintain circular motion of each body around the center of mass (CM), we have for body $M_1$:

$$\frac{GM_1M_2}{R^2} = \frac{M_1V_1^2}{R_1}; \quad \frac{GM_2}{R^2} = \frac{V_1^2}{R_1}, \tag{II.54}$$

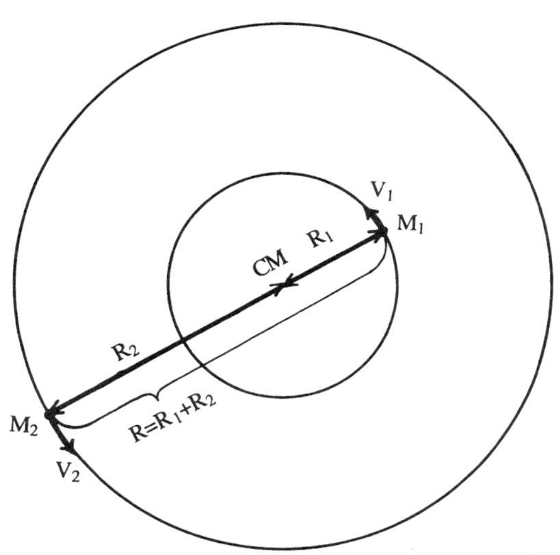


**Figure II.9** Orbits in a binary star system. CM denotes the center of mass of the system. In a two-body system the orbits are of course coplanar; we further assume for simplicity that both orbits are circular. This is in fact the usual circumstance for close binary systems, because tidal interactions between the stars tend to circularize their orbits.

and for $M_2$,

$$\frac{GM_1M_2}{R^2} = \frac{M_2V_2^2}{R_2}; \quad \frac{GM_1}{R_2} = \frac{V_2^2}{R_2}. \tag{II.55}$$

The orbital periods are

$$P_1 = \frac{2\pi R_1}{V_1} \quad P_2 = \frac{2\pi R_2}{V_2} = P_1 = P. \tag{II.56}$$

Substituting for $V_1$ and $V_2$ in Eqs. (II.54) and (II.55) and adding

$$\frac{GM_2}{R^2} + \frac{GM_1}{R^2} = \frac{4\pi^2 R_1^2}{R_1 P_1^2} + \frac{4\pi^2 R_2^2}{R_2 P_2^2} \tag{II.57}$$

or

$$\frac{G}{R^2}(M_1 + M_2) = \frac{4\pi^2}{P^2}(R_1 + R_2), \tag{II.58}$$

By the definition of the center of mass,

$$\frac{M_1}{M_2} = \frac{R_2}{R_1}. \tag{II.59}$$

Thus we have two simultaneous equations in two unknowns, $M_1$ and $M_2$.

An observer on Earth will measure Doppler shifts due to the radial components of the relative velocities $V_1'$ and $V_2'$, which are equal to

$$V_1' = V_1 \sin i; \quad V_2' = V_2 \sin i, \tag{II.60}$$

respectively. The angle $i$ is the inclination of the polar axis of the binary system to the line of sight from Earth. A system viewed nearly pole-on may show no detectable radial velocities. Recall from Eq. (II.56) that $V_1/V_2 = R_1/R_2$ and hence, from Eq. (II.59), $V_1/V_2 = M_2/M_1$. The measured Doppler shift is therefore proportional to $M \sin i$.

If we now determine the masses of a number of nearby Main Sequence stars whose absolute magnitudes and luminosities are known (that is, those stars close enough for accurate parallax measurements), it becomes possible to construct the so-called mass–luminosity ($M$–$L$) diagram given in Fig. II.10. The strong correlation between mass and luminosity is evident. Main Sequence stars above a few tenths of a solar mass follow a law of form

$$L \propto M^{3.5}, \tag{II.61}$$

and a smaller exponent applies for smaller stars, for which we find that

$$L \propto M^2. \tag{II.62}$$

This correlation between luminosity and mass makes it obvious that the Main Sequence is not an evolutionary path followed by stars, but rather a track occupied by stars of different masses. The highly luminous O and B

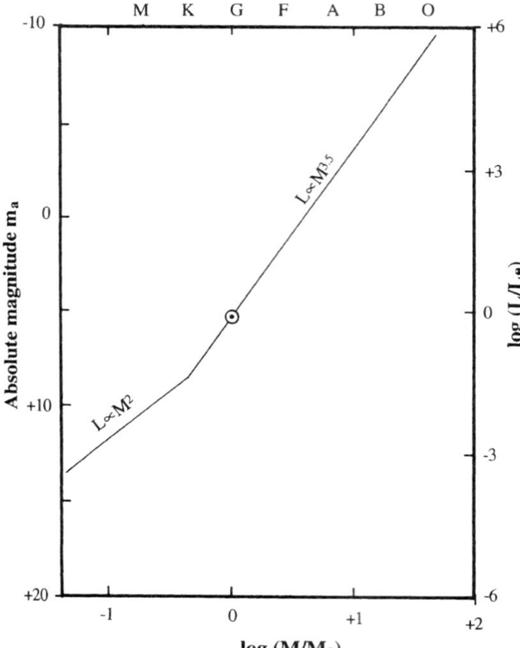


**Figure II.10** The mass–luminosity relationship for Main Sequence stars. The break occurs at spectral class K, where proton–proton chain fusion (lower masses) gives way to the catalytic carbon cycle (higher masses). White dwarfs lie below the MS trend line, whereas giants generally lie quite close to the MS line. The natural dispersion of stars of different ages and compositions about the trend line is roughly 1 mag. Note that, because the mass increases from left to right, the order of spectral classes at the top is the reverse of that in the usual rendering of the H–R diagram.

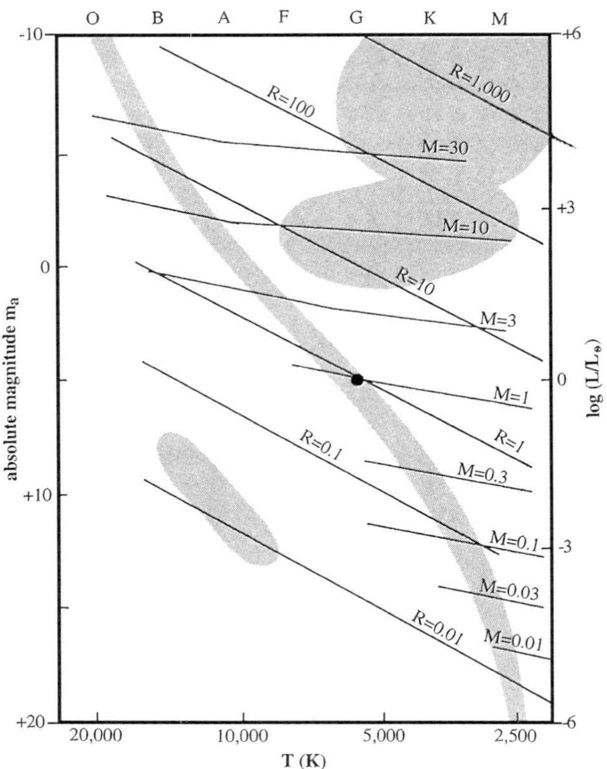


**Figure II.11** The H–R diagram with mass and radius data. Note that white dwarfs are planet-sized bodies, whereas supergiants may have radii of several AU ($1\text{AU} = 213R_\odot$).

stars must be $10^3$ times as massive as M-type red dwarfs. The progression of spectral types along the MS is simply a function of the mass of the star.

Of course this observation has indirect evolutionary implications because it shows that very massive stars may emit energy at $10^{10}$ times the rate of red dwarfs. Thus these very luminous O and B stars must evolve much more rapidly than faint stars and *must on the average be younger* in absolute age (not degree of evolution) than other MS stars. If we combine the H–R and M–L diagrams we may more readily see the correlations among mass, radius, luminosity, and spectral class. Such an overlay is presented in Fig. II.11. The radii plotted in this figure are calculated from Eqs. (II.42) and (II.46) as described above. The densities of stars can now be calculated.

It is interesting to note that the *mean* densities of red supergiants may reach as low as $10^{-9}$ g cm$^{-3}$, whereas white dwarfs may be as dense as $10^6$ g cm$^{-3}$. Nonetheless, conditions near the *centers* of most MS stars are remarkably similar.

It is important to understand exactly what selection criteria have been exercised in the process of constructing the H–R, M–L, and combination diagrams. First of all, in order to attract the attention of terrestrial observers, a star must be either quite close to the Sun or intrinsically extremely luminous. Consider a faint star that is just barely observable at a distance of 10 pc; a star with $10^6$ times its luminosity would appear equally bright even if at a distance of 10 kpc! Thus faint stars must be very close to the Sun to be included in the census, whereas very luminous stars anywhere in the Galaxy may be included.

A much fairer sample of the general population of the Galaxy may be had by examining in detail a representative small volume element of the Galaxy and counting every star in it. By this process we can avoid overestimating the importance of very rare classes of stars that appear in all bright-star counts solely by reason of their enormous luminosity.

The only method available to us for conducting such a representative census is to choose a volume element of space, centered on the Sun, sufficiently small in radius so that even a very faint star of, say, $L = 10^{-4}L_\odot$ would surely be observed and counted if it were anywhere within this volume. It is easy to picture the possibility that an intrinsically faint star only

10 pc away may be so unimposing that no one has ever bothered to measure its parallax. We will attempt to minimize this problem by initially selecting only stars within 4 pc of the Sun.

Figure II.12 is the H–R diagram for stars within 4 pc of the Sun. Several classes of stars prominent in the "bright star" H–R diagram are absent in this more selective sample. Within 4 pc of the Sun, for example, there are no giant or supergiant stars, no O- or B-class stars, and only a few white dwarfs. On the other hand, there is an enormous wealth of faint MS stars, particularly red dwarfs.

Many years ago, Luyten attempted the difficult task of cataloging all stars within 10 pc of the Sun, for the purpose of developing a better statistical idea of the mass and luminosity distributions of stars in the galactic plane. Within this volume of space he was able to find 556 stars, with a total mass of 268 $M_\odot$ and a total luminosity of 235 $L_\odot$. Recent surveys report about 695 stars within 15 pc, but incompleteness at the highest absolute magnitudes is a problem. Gliese has also compiled a catalog of all known stars out to a distance of 25 pc. Luyten's finding of an $M$:$L$ ratio of about 1 conflicts with the far larger value found for many galaxies, generally close to 100:1. The mass density found in visible stars found within 10 pc is 0.064 $M_\odot$ pc$^{-3}$.

The "luminosity function" found for these stars, shown in Fig. II.13, is remarkably strongly peaked, and highly luminous stars are found to be quite rare. Only 4% of the stars surveyed were more luminous than the sun, but these few stars accounted for fully 90% of the light.

Any stars as faint as $10^{-4} L_\odot$ should have been found by the survey, and in fact some stars even fainter than that were found. Such stars are so faint that the census surely missed many of them. It is therefore very probable that the resulting luminosity function is incomplete on the faint end. It should be recalled that such undetected stars would contribute mass but almost no luminosity to the totals, and their hypothetical presence would shift the $M$:$L$ ratio in the direction expected from observations of entire galaxies.

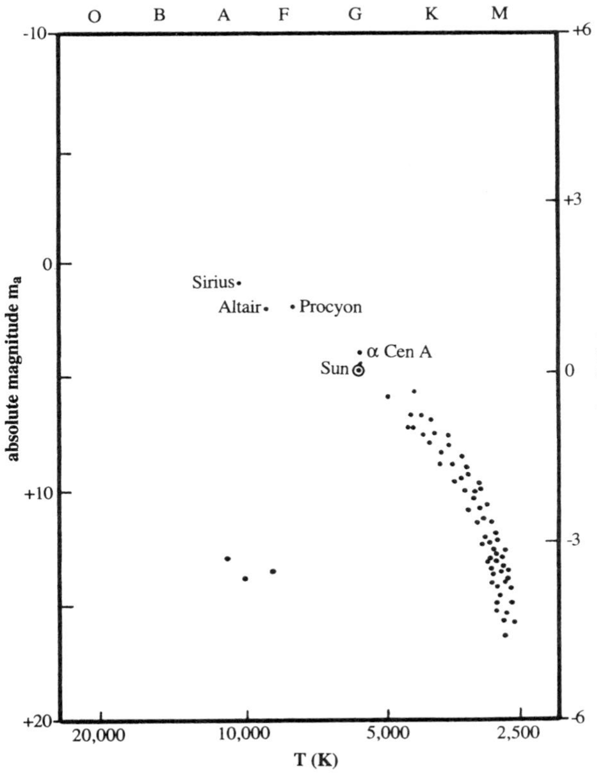

**Figure II.12** The H–R diagram for stars within 4 pc of the Sun. This is intended to be a nearly complete census over a volume of space so small that virtually every star in it will be detected and recognized. Note the absence of giants and highly luminous MS stars and the great abundance of M-class red dwarfs.

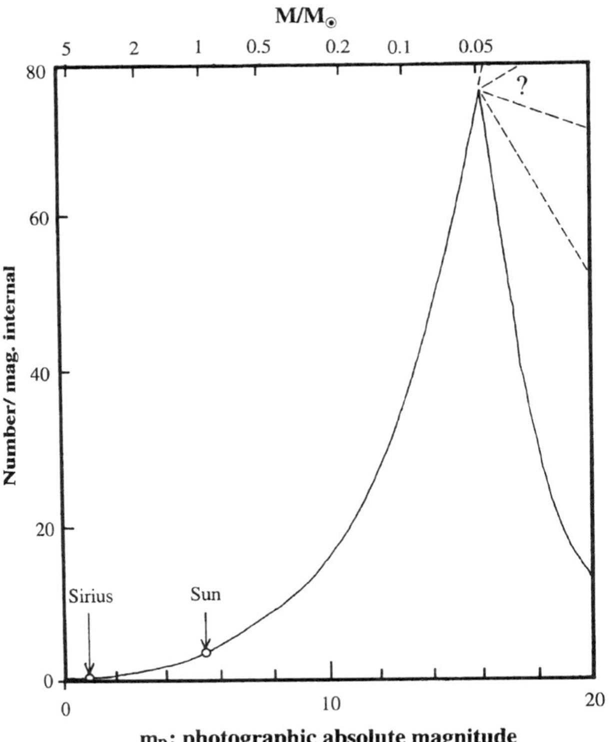


**Figure II.13** Luyten's luminosity function. The number of nearby stars per magnitude interval peaks strongly near a photographic absolute magnitude of +15.7. Stars of higher magnitudes are so faint that the census is incomplete in that range; such stars, if 10 pc distant (the range limit chosen for this survey), would probably not be recognized. More recent mass estimates place the peak of Luyten's function near 0.07 solar masses.

According to Luyten's census, the most probable star type is an M-class dwarf with a (photographic) absolute magnitude of +15.7 and $L = 10^{-4} L_{\odot}$. Such a star would have a mass of only $0.072\, M_{\odot}$, or about 75 times the mass of Jupiter. Less massive bodies, which cannot sustain hydrogen fusion because of the low temperatures and pressures in their cores, are termed *brown dwarfs*. They maintain their feeble luminosity by slow gravitational contraction, not hydrogen fusion. Finding such faint, cool, low-mass bodies is observationally very challenging, requiring searches at infrared wavelengths. Even a nearby brown dwarf may be so faint that it is not seen or not recognized.

It is possible, according to some infrared astronomers, that the brown dwarfs may outnumber the visible stars in our Galaxy. Even so, they cannot contribute an appreciable fraction of the total mass of the Galaxy because of their small individual masses. The detection, properties, and life histories of brown dwarfs are discussed in Chapter XI.

Although accounting in detail for the total mass of the Galaxy is subject to numerous difficulties, accounting for the sources of its luminosity is easy. When visible light is considered, the 4% of the stars brighter than the Sun account for roughly 90% of the total light. At ultraviolet wavelengths the importance of the rare O and B stars is even more overwhelming. Such stars not only are more luminous than the rest of the MS, but also, because of their high surface temperatures, emit most of their enormous energy flux in the ultraviolet. Thus the UV brightness of a galaxy or star cluster is a sensitive index of the abundance of very brilliant O and B stars.

We now know something of the relative frequencies of different star classes in the central plane of the Galaxy, as evidenced by their abundance in the Sun's stellar neighborhood. It would be most interesting to construct H–R diagrams for other samples of stars from elsewhere in the Galaxy for purposes of comparison. Of course, one piece of information we would need for such a study is the distance of the stars in the sample, and it is plainly futile to try to measure the parallaxes of other than very nearby stars. This difficulty may be alleviated if we can find a group of stars that we know are all at the same distance from the Sun. This condition is met by globular clusters, which are small in dimensions compared with their distance from us. All we need is a technique for determining the distance of the entire cluster, obviously some technique other than parallax measurement.

Measurement of the distances of all but the nearest star clusters depends on the existence of Cepheid variables. It will be recalled from our discussion of cosmic distance scales that the pulsation periods of these stars are a smooth, single-valued function of their luminosity. Thus, if a star cluster is found to contain even a single Cepheid of known period, that is sufficient to tell us the absolute magnitude of the variable star. Because we can measure the apparent magnitude of the star, it is then easy to calculate the distance of that star and the cluster in which it resides. Even for clusters that do not contain observed Cepheid variables the *relative* H–R diagram (*apparent* visual magnitude vs color) can still be constructed with confidence.

When stars in globular clusters are studied in this manner, a most startling difference between the H–R diagrams of the cluster stars and galactic-plane stars is found. The H–R diagram for a typical globular cluster is shown in Fig. II.14. The absence of observations of stars with luminosities less than the Sun's is simply attributable to distance: white dwarfs and lower-luminosity MS stars are not bright enough to be observed at such a distance. More striking than this, and far more significant, are the virtual absence of *any* visible MS stars and the abundance of giant and supergiant stars. The entire high-luminosity end of the MS above spectral class F is missing. These are precisely the stars that have the

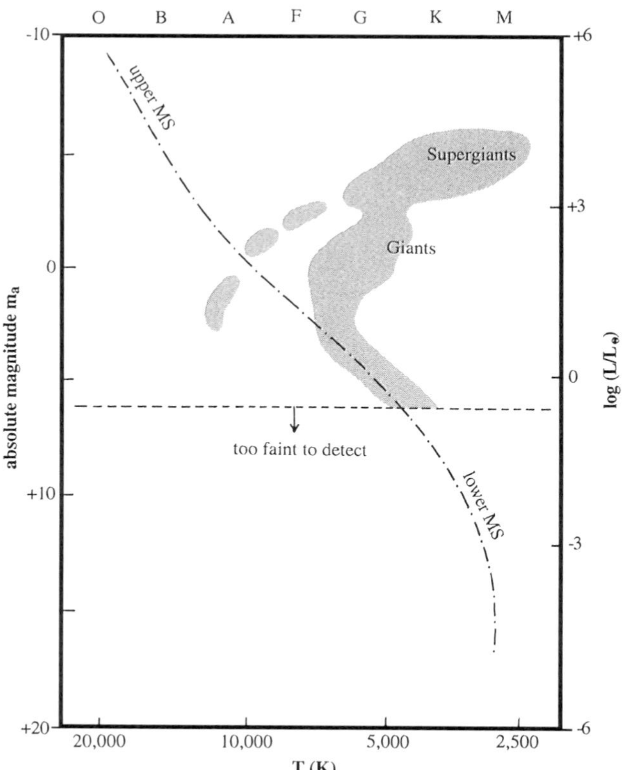


**Figure II.14** The H–R diagram for a distant globular cluster. The diagram is composed so that it may be directly compared with Fig. II.12, the H–R diagram for nearby stars. Note the absence of the upper MS and the unobservability (due to distance) of stars significantly less luminous than the Sun. The giant and supergiant regions are well populated.

shortest lifetimes, and therefore the lowest ages, of all MS stars. It is tempting to conclude that, in the dust- and gas-free globular clusters, no new stars have been formed for some time, whereas in the galactic plane the high density of gas and dust is continuously spawning new stars of all masses.

Because of the profound differences between the local galactic-arm and the globular-cluster populations of stars, it is usual to denote them by distinctive labels. The population of stars characteristic of our Sun's immediate neighborhood is termed Population I, whereas the assortment of stars found in globular clusters (and in the galactic core and elliptical galaxies as well) is termed Population II.

The explanation of the remarkable differences between Population I and Population II does indeed involve consideration of the differences between the star-formation rates in these two different environments, but also requires a considerable amount of theoretical work on the evolutionary pathways of stars. It is impossible to give more than a very brief summary of the theory of stellar evolution here, but some qualitative appreciation of these theories is very helpful in understanding the ancient history of the matter that makes up our Solar System.

## Stellar Evolution

The observational stimulus for much of the theoretical interest in Pop II stars was the discovery that they are remarkably deficient in the heavy elements compared with Pop I stars. For example, a typical Pop II star may contain a hundred times less of the metallic elements than a "normal" Pop I star. Because theoretical studies of heat transport in the Sun showed that the outer layers of stars are not convectively mixed with material from the deep interior, it was considered likely that the observed surface layer was a sample of the primordial matter from which the stars originally formed. In this interpretation, Pop II stars were made of material very close in composition to the debris from the primordial fireball, whereas Pop I stars were formed from material substantially enriched in elements heavier than helium.

These heavy elements are presumably formed by nuclear reactions in stellar interiors; we may thus picture Pop II stars as survivors from nearly the first generation of stars formed in the history of the Universe. As such, they are of unusual interest to theoreticians, who hope to be able to calculate the evolutionary history of stars. From the comparison of these calculated histories with observations of real stars, it should be possible to estimate the age of the oldest stars now found in the Universe.

In the most general terms, a young Pop II star contains hydrogen and helium, in proportions fixed by the Big Bang, mixed completely uniformly throughout. The entire course of future evolution of the star is a function of the mass alone.

A new star is thought to have originated from collapse of an extended gas cloud, late in the hierarchy of fragmentation processes. These processes also gave rise to galaxy clusters, galaxies, and star clusters. During the late stages of collapse of a protostar, heat is produced in vast amounts by liberation of the gravitational potential energy of the cloud. A newly formed star has sources of energy from both nuclear reactions and its own gravitational collapse. Thus the star radiates with a higher luminosity than an MS star of the same mass. After a few million years this excess collapse energy has been lost, and the star has achieved a steady state in which the rate of production of energy by nuclear reactions is exactly balanced by the rate of radiation of energy into space from the emitting surface of the star, the photosphere.

The principal nuclear reactions producing energy in a Pop II star involve the fusion of hydrogen to helium by the proton–proton chain,

$$2(\mathrm{p} + \mathrm{p} \rightarrow {}^2\mathrm{D} + \mathrm{e}^+ + \nu_\mathrm{e}) \qquad \text{(II.63)}$$

$$2({}^2\mathrm{D} + \mathrm{p} \rightarrow {}^3\mathrm{He} + \gamma) \qquad \text{(II.64)}$$

$$\underline{{}^3\mathrm{He} + {}^3\mathrm{He} \rightarrow {}^4\mathrm{He} + 2\mathrm{p} + 2\gamma} \qquad \text{(II.65)}$$

$$\text{Net}: 4\mathrm{p} \rightarrow {}^4\mathrm{He} + 2\mathrm{e}^+ + 2\nu_\mathrm{e}. \qquad \text{(II.66)}$$

The theoretical expression for the rate of energy production by this process is

$$\varepsilon(\mathrm{erg\ g^{-1}\ s^{-1}}) = 0.28\rho X_\mathrm{H}^2 (T_6/13)^{4.1}, \qquad \text{(II.67)}$$

where $\rho$ is the density (g cm$^{-3}$), $X_\mathrm{H}$ is the weight fraction of H in the burning zone (0.72), and $T_6$ is the temperature in millions of kelvins. Nuclear reactions are obviously most rapid in the hottest, most dense point inside the star, its exact center. Central temperatures of about $10^7$ K are required in order to maintain energy liberation rates near 1 erg g$^{-1}$s$^{-1}$, a value typical of stars in the middle of the Main Sequence.

On the MS, hydrogen "burning" produces helium as the main product. As hydrogen near the center of the star is used up to make helium, a core of very pure helium is formed, with hydrogen fusion continuing at the surface of the core. Hydrogen is very rare within the core because the temperature and density at all points within the core must be at least as high as at the surface of the core, and hence any H would be rapidly consumed by fusion reactions. Because the density of helium is much higher than that of the primordial hydrogen-rich mixture, the helium core is stable and does not mix with the outer layers of the star.

The star's structure remains virtually unchanged during the era of hydrogen burning, except that, as the hydrogen fusion zone rises toward the surface of the star, the surface temperature and luminosity increase slightly. The evolution of a Main Sequence star involves about a 10 to 20% luminosity increase over the MS lifetime.

If the star is sufficiently massive, so that the density and temperature at the center of the star can become sufficiently high, then fusion of helium to make heavy elements may occur.

Direct He fusion by the reaction

$$^{4}\mathrm{He} + {}^{4}\mathrm{He} \rightarrow {}^{8}\mathrm{Be} \tag{II.68}$$

is an exceedingly inefficient way to make the heavy elements, because $^{8}$Be is a very unstable nuclide that decays by

$$^{8}\mathrm{Be} \rightarrow {}^{4}\mathrm{He} + {}^{4}\mathrm{He} + \gamma \tag{II.69}$$

with an extremely short half-life of only $10^{-16}$ s. The only feasible route to bridge the instability gap between helium and carbon and to pass over the unstable (missing) isotopes of lithium, beryllium, and boron at masses 5 and 8 involves a reaction that must occur during the transitory existence of the $^{8}$Be nucleus,

$$^{8}\mathrm{Be} + {}^{4}\mathrm{He} \rightarrow {}^{12}\mathrm{C} + \gamma, \tag{II.70}$$

or, essentially, the ternary collision of three alpha particles to make a $^{12}$C nucleus. This can occur only when the density is so high that the collision frequency is very large and when the temperature is so high that the alpha particle has enough thermal kinetic energy to overcome the electrostatic repulsion energy of the transitory Be nucleus. Thus, at some well-defined time, helium burning begins in the center of the helium core. The temperature during helium burning must be near $T_6 = 100$ in order to overcome the Coulombic repulsion between the alpha particles. Only stars with masses of at least 0.8 $M_\odot$ can produce high enough temperatures for the initiation of helium burning and entry into the red giant stage.

Once helium burning begins, it causes a drastic change in the internal structure of the star. For a brief period of time, the outer envelope of the star transports the unaccustomed heat load by convection, heating, and inflating the outer envelope of the star in the so-called "helium flash." Thereafter the star settles down to a short but relatively stable existence as a helium-burning star, lying on a diagonal line on the H–R diagram above and to the right (red) side of the MS. Such stars make up the group that we have so far referred to as giants. Note that we previously rejected the idea that the MS was an evolutionary path pursued by stars and described it instead as the locus of stable hydrogen-burning stars with different masses. Now we slightly expand this interpretation by applying it to the He-burning branch of the H–R diagram as well.

As the helium-burning phase progresses, the star builds up a core of carbon and heavier elements. The helium-burning region stays at the surface of the carbon core, maintaining a temperature of $T_6 = 100$ as it rises closer and closer to the surface. The outer envelope of the star eventually becomes strongly heated, expands enormously, and then, with its nuclear fuel exhausted, collapses to a white dwarf. Stars with masses of 0.8 $M_\odot$ to about 9 $M_\odot$ pursue alpha-particle reactions to an extent that increases with mass, developing extremely tenuous atmospheres large enough to engulf an entire planetary system (as occurs with supergiants near the upper end of this mass range). Mass loss from these atmospheres is facilitated by their very low surface gravity. Such stars then exhaust their nuclear fuel and slide to the left on the H–R diagram (roughly constant luminosity with rapidly shrinking radius and rapidly rising surface temperatures) until they reach and cross the Main Sequence. These O-class highly evolved *Wolf–Rayet* stars have spectra that are very different from those of MS stars of the same color and luminosity. Some, with high abundances of nitrogen and helium and little oxygen, are called WN stars; others, rich in carbon and oxygen but poor in nitrogen, are called WC stars.

We shall return in a more quantitative way to the rates of the alpha-particle reactions in our detailed discussion of nucleosynthesis later in this chapter. For now we concentrate on the structural and observational consequences of evolutionary changes inside stars.

Figure II.15 gives a brief summary of the evolutionary track of a star of one solar mass on the H–R diagram, through the giant, supergiant, and white dwarf stages. The close similarity of Fig. II.15 to Fig. II.14 suggests that a Pop II star cluster is an ensemble with identical age and random masses. Thus we need only postulate that the stars in each globular cluster were all formed at the same time in order to make sense of their present distribution on the H–R diagram. We explain the absence of highly luminous MS stars by stipulating that the cluster is old enough so that these stars have evolved off the MS.

The highest-luminosity MS stars actually observed in globular clusters are precisely those stars whose MS lifetimes are equal to the age of the cluster. The absence of gas and dust to produce new stars and the absence of demonstrably young and overluminous (T Tauri) stars, as well as the absence of O, B, and A MS stars, are in accord with this hypothesis. Those stars observed in the giant and supergiant branches of the diagram are slightly more massive and slightly more highly evolved than the most luminous stars remaining on the MS. We are seeing what looks like an evolutionary sequence because the lifetimes of stars are a function of mass; at any moment

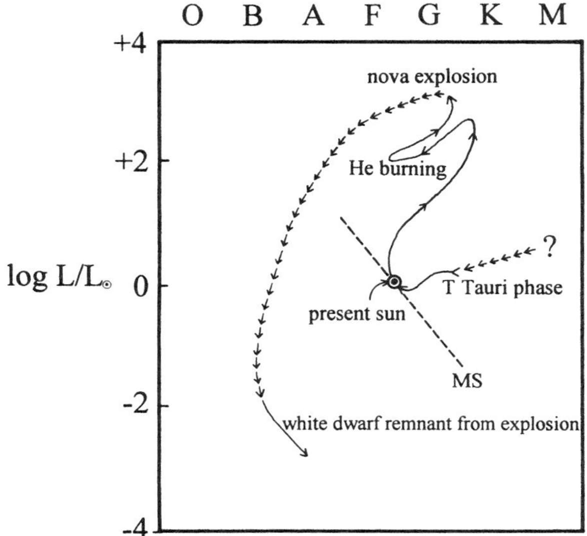


**Figure II.15** Evolutionary track of a star with one solar mass. An early nebular phase with a lifetime of $10^5$ to $10^6$ years collapses to form a T Tauri-phase protostar with a luminosity slightly higher than that of the present Sun. Within about $10^7$ years this young star settles down onto the MS, where it remains for about $10^{10}$ years of quiescent hydrogen burning. It will experience a luminosity increase of about 50% over its MS lifetime. As the hydrogen supply in the deep interior is exhausted, the star expands and becomes more luminous. Ignition of helium burning causes a "helium flash," after which the star settles down for about $10^8$ years on the "helium burning main sequence" as a giant. At an advanced stage of helium burning, the star again flares up as a Wolf–Rayet star, this time violently enough to expel much of the hydrogen-bearing outer envelope in the form of a planetary nebula. The remnant becomes a tiny, very hot white dwarf without a nuclear energy source, which slowly cools for about $10^{10}$ years until it fades into invisibility.

we see stars at all stages of evolution but with the same age.

The complementary conclusion with respect to Pop I stars must be that, because of the presence of young T Tauri stars and short-lived O and B stars, there must be active present-day star formation in the gas- and dust-rich spiral arms of the Galaxy. In round numbers, a star of $5\,L_\odot$ spends $10^9$ years on the MS, whereas a $100\,L_\odot$ star spends only $5 \times 10^7$ years there. Thus, if we observe a star with $100\,L_\odot$, it must be less than $5 \times 10^7$ years old. In fact, MS stars of over $10^4\,L_\odot$ can be observed in our Galaxy. Some of these stars were born since the time of origin of the human species. At the opposite lifetime extreme are the faint red dwarfs, which have such low luminosities that some of them could burn for more than $10^{12}$ years on the MS.

Using the results of the evolutionary studies on globular clusters, it can be shown that the oldest stellar assemblages so far observed are at least 10 Gyr and perhaps 14 Gyr old. This is close to the 14 to 15 Gyr age of the Universe as estimated from the Hubble constant. It is unlikely that any significant star formation has occurred in these clusters since the first few billion years of the history of the Universe.

This simplified discussion of stellar evolution, although sufficient to explain the basic differences between Pop I and Pop II stars, leaves untouched many aspects of the problem, including the crucial question of the pre-MS evolution of stars. Before attempting to define more detailed early evolutionary models, we must first examine in more detail the observational evidence pertaining to young stars, star clusters, and their physical and chemical environments.

## Star Clusters

It has already been remarked that O, B, and T Tauri type stars are found only in gas- and dust-rich regions of the Galaxy, and that these classes of stars are not stable over long periods of time. Because the lifetimes of these stars are very short compared to either the expansion age of the Universe or the evolutionary age of globular cluster stars, we ought to examine closely the available data on such stars for clues to the physical conditions attendant upon star formation.

One observation about O, B, and T Tauri stars which requires comment is that they tend to form rather well-defined clusters in the galactic plane. The so-called "O associations" are rich in O and B stars and contain up to about 100 stars, usually with a "nuclear cluster" of 30 or 40 stars. The stars in these clusters often are found to form chains analogous to the chains of galaxies remarked on earlier. O associations frequently have diameters of 100 to 200 pc. Red supergiants of spectral classes M and S also are found to form clusters centered on O associations.

Clusters of T Tauri stars called "T associations" are also well defined. It is possible that O associations always contain T Tauri stars.

Because of the rotation of the Galaxy, it is difficult to picture how associations with such low masses could avoid being smeared out and dissolved by differential rotation. In fact, measurements of the radial velocities along the line of sight (by Doppler shift of spectral lines) and tangential velocity (by direct observation of changes in the position of the star against the background over many years) confirm that these associations of young stars are in rapid expansion, commonly with velocities of 5 to 20 km $s^{-1}$. When the age of a stellar association is calculated by dividing the radius of the cluster by the expansion velocity, it is found that the youngest clusters are most nearly spherically symmetrical, whereas clusters over a few million years in age have become greatly elongated and diffused by differential galactic rotation. The gravitational potential energy of the star cluster is always found to be far less than the observed kinetic energy of the components; hence the system is not bound, and all "orbits" relative to the center of mass of the cluster are hyperbolic. For these reasons, dissipation of the young cluster and intermingling with the old stellar environment take place on a time scale shorter than a single rotation of the Galaxy.

The best example of an expanding young cluster accessible to us is the nearby Orion Nebula. Among the 578 known stars in this cluster, O, B, and T Tauri stars; red supergiants; and dense gas–dust clouds are prominent. The age of the Orion cluster is about $2.7 \times 10^6$ years, and the expansion is still very nearly spherically symmetrical.

Another nearby group, the Scorpius–Centaurus cluster, is about $70 \times 10^6$ years old and is very dispersed and elongated. The velocity of expansion is unusually low, only $0.7\,\mathrm{km\,s^{-1}}$. A third nearby cluster, in Ursa Major, is about $45 \times 10^6$ years old and also in an advanced state of dispersal.

The young stars seen in our galactic neighborhood are generally associated with rather well-defined expanding clusters. It would be reasonable to presume that stars as a rule originate in dense clusters that rapidly expand and disperse; the origins of all Pop I stars may be intimately bound up in such clusters. Thus the processes occurring in dense gas and dust clouds are of great importance for understanding the origin of all Pop I stars.

In systematically classifying stars by means of readily observable properties, an interesting trend is found when the widths of spectral lines are considered. After correction for Doppler broadening due to the varying surface temperatures of the stars, it is found that the spectral lines of "early type" stars of spectral classes O, B, and A are much broader than those of F stars, whereas G, K, and M stars have consistently narrow spectral lines. An obvious explanation as illustrated in Fig. II.16 is that this broadening is due to a Doppler shift of light originating from widely separated points on the surface of a rapidly rotating star.

For example, in the case of certain "peculiar" stars, notably those of spectral class Ap ("peculiar A stars"), sinusoidal variations appear in the radial velocities of the stars as measured by locating the *centers* of the spectral lines. These shifts are produced by the orbital motion of the Ap star about the center of mass of a multiple-star system, often containing an unseen low-luminosity stellar companion.

The line *widths*, however, seem to show unambiguously that the average rotation velocity of stars is a function of spectral class and that blue young stars have a much higher rotational velocity and angular momentum than older stars of redder spectral class. It is difficult to propose a reason why G, K, and M stars should be formed with lower angular momenta. Attempts to explain these observations have largely fallen into three categories. The first explanation has been that stars possess a mechanism for dissipating their angular momentum and that O, B, and A stars do not live long enough to be fully slowed down. An alternative explanation is that the interiors of all stars rotate rapidly, but the surface regions of cool, long-lived stars are uncoupled from the rotation rate of the deep interior. Only a relatively modest angular momentum loss is then needed to despin the visible outer layers of the star.

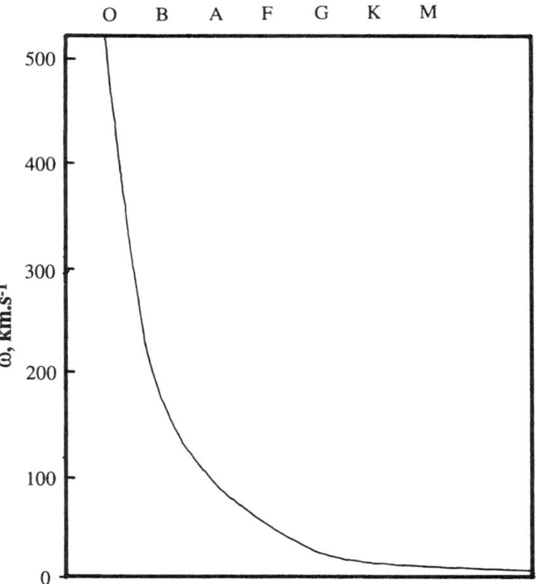


**Figure II.16** Rotational speeds of MS stars. Only G, K, and M stars have succeeded in attaining rotational speeds below about $30\,\mathrm{km\,s^{-1}}$.

Current models for the early histories of stars do in fact predict a mechanism by which angular momentum can be shed by the loss of a very small proportion of the total mass as a sort of very intense solar wind, which remains rigidly tied to the star's rotation (by being embedded in the star's magnetic field) out to large astrocentric distances. The angular momentum that must be added to the solar wind to keep its angular velocity constant is imparted by the magnetic field at the expense of the angular momentum of the star itself. Particles escaping from a magnetosphere of radius 100 $R^*$ and corotating with the star have an angular momentum per unit mass equal to

$$vR = \omega R^2 = 10^4 \omega R_*^2, \tag{II.71}$$

$10^4$ times that of the surface regions of the star. Thus the loss of a fraction of a percent of the star's mass by such a mechanism could virtually despin the star.

Alternatively, we could note that the large majority of the present angular momentum in the Solar System resides in the orbital motions of the planets, not the rotation of the Sun. It has been suggested that the presence of planets provides an angular momentum sink and that slowly rotating stars are simply those stars that have planetary systems. By this interpretation, G, K, and M stars, which make up the overwhelming majority of the stars in the Universe, are as a rule accompanied by planetary systems, whereas the highly luminous "early type" stars are not. Finally, it is possible that collapse of protostellar gas and dust clouds with large angular momenta produces more numerous and less massive stars than do clouds of low angular momentum.

The prediction of the nearly universal distribution of planets implicit in the second explanation of the angular momentum distribution is of course of immense interest and importance. It is appropriate to ask at this point whether there is any observational evidence to support this contention.

Direct optical detection of an extrasolar planet is extremely difficult. We may see the problem by considering a hypothetical planet of Barnard's star. Barnard's star has a parallax of $0.544''$ and thus a distance of 1.84 pc, or $2.06 \times 10^5 \times 1.84 = 3.80 \times 10^5$ AU. The problem of detection of any planetary companion arises from its faintness relative to its star. Let us imagine a Jupiter-sized object with a visual albedo of 0.5 in orbit at $R = 4.7$ AU from Barnard's star. The fraction of the star's light that hits the planet is $r^2/4R^2$, or $(1.3 \times 10^4)^2/4 \times (4.7 \times 1.5 \times 10^8)^2 = 8.5 \times 10^{-11}$, of which only half is reflected. Furthermore, only half of the planet's disk is illuminated when seen from Earth near elongation. Thus the intensity of the planet is $2 \times 10^{-11}$ that of the star at visible wavelengths. This is equivalent to a 26.7 magnitude difference. Since the absolute visual magnitude of an M5 red dwarf is about 13.2, the apparent magnitude of Barnard's star is then (from Eq. II.48) $13.2 - 5 + 5\log(1.84)$ or 9.5 mag, and the planet is 36.2 mag, far beyond the detection limit of any plausible Earth-based optical system.

Some perspective on the prevalence of multiple systems can be gleaned from data on the frequency of pairings of stars of different masses. We have already seen evidence from the luminosity function that objects of mass $0.072\,M_\odot$ or less may be very common. Now we ask whether small objects tend to form pairs with one another, or whether they prefer to form pairs with much more massive objects. In the former case, planet–Sun systems might be rare.

Double and multiple stars are quite common. Visual observations reveal several prominent double stars in the evening sky. The most familiar examples of visual double stars are Mizar–Alcor in Ursa Major (in the handle of the Big Dipper), and the systems of alpha and beta Capricorni. Among the nearby stars roughly 50% are found in stellar multiple systems and 50% as single stars. In "O" associations Blaauw has found 60% single stars, 22% in double systems, and virtually all the remainder in triple systems. Spectroscopic studies reveal that Polaris is a quintuple star and that Castor is a sextuple star. This high degree of pairing and clustering is not a natural result of chance encounters between independent single stars, since calculations show that statistical equilibrium would result in only about 1 double star per $10^9$ single stars. The explanation must be sought in the processes that give rise to young stars.

The mass ratios of stars in double systems have been studied to see whether pairing is independent of mass. If we define $\alpha$ as the ratio of the mass of the smaller star to the mass of the larger (an index running from 0 to 1), the frequency of double star systems as a function of $\alpha$ is found to be roughly constant. Taking an M-class red dwarf as an example of a typical star, it is roughly twice as probable that this star will be a member of a double star system as that it will be accompanied by one or more large planets.

## Stellar Origins

The Orion Nebula, a bright patch visible to the unaided eye, is a nearby dense cloud of gas and dust within which several new stars have appeared in recent history. Brilliant young stars abound in and near the nebula, and the nebula itself emits enormous fluxes of infrared radiation and radio waves. All signs indicate rapid evolution and star formation going on before our eyes.

The stars in the Orion Nebula depart systematically from the Main Sequence. A Hertzsprung–Russell diagram for the Orion Nebula given in Fig. II.17 shows that, although O and B stars in the Nebula lie quite nicely on the MS, fainter stars appear systematically too red for their luminosity (or too luminous for their color). The same effect has been noted in another young cluster, NGC2264. This reddening cannot be attributed to light scattering in surrounding dense interstellar clouds, because such an effect would redden the blue–white MS stars more than the faint red ones.

It is evident that there must be some *intrinsic* property of young low-mass stars that causes them to lie above the MS in the H-R diagram. There are in fact three factors that contribute to the extra luminosity of newly formed stars. First, young stars are still settling down, shrinking in radius as they approach the main sequence. They are still larger than MS stars of the came color and temperature, and therefore have a higher

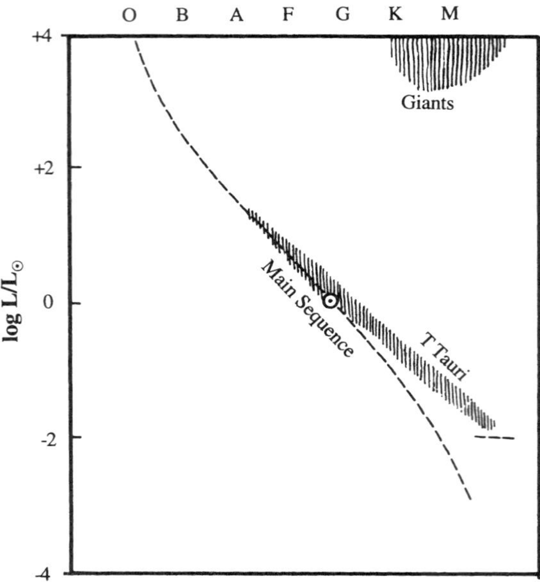


**Figure II.17** H–R diagram for the Orion Nebula and other dark-cloud regions. The samples are so small that only a few stars of high luminosity are seen; however, these stars make up a significant portion of the total population, much more than in the Galaxy at large. These samples always depart systematically from the MS for masses below about one solar mass. The distance of these cloud regions from us puts a luminosity cutoff on the red end of the T Tauri branch.

luminosity due to their larger surface area. Since they are shrinking, there is also a second major source of radiant energy from the conversion of gravitational potential energy into heat. Third, heat is released by the fusion of deuterium in the deep interior of the protostar. The ignition temperature for the reaction

$$^2H + {}^1H \rightarrow {}^3He + \gamma. \qquad (II.72)$$

occurs at temperatures that are too low for proton–proton fusion reactions. However, since the cosmic abundance of deuterium is $<2 \times 10^{-5}$ of the hydrogen abundance, the deuterium fusion energy source can at best be important for a brief period of time. A protostar must have a mass of at least 13 Jupiter masses ($0.0126\,M_\odot$) to initiate deuterium burning. We recall that the minimum mass for normal hydrogen burning is $0.072\,M_\odot$, which is therefore the threshold of the Main Sequence. Bodies in the range 0.0126 to $0.072\,M_\odot$, which experience transient deuterium burning but no hydrogen burning, are too large to be mere planets, but too small to be genuine stars. These, again, are the brown dwarfs.

The energy liberated during collapse of a sphere of uniform density from infinity to radius $R$ is given by

$$E = \int_0^M \int_0^\infty \frac{GM}{R^2}\, dR\, dM = -\frac{GM^2}{R}, \qquad (II.73)$$

where $M$ is the mass and $G$ is the universal gravitational constant. For the Sun this energy release is $4 \times 10^{48}$ erg. Because the luminosity of the Sun once on the MS is $3.90 \times 10^{33}$ erg s$^{-1}$, this collapse energy would suffice to double the luminosity of the Sun for about 30 million years. If released over a shorter period of time, the maximum luminosity would be increased proportionately.

Now consider collapse of a gas cloud to a stable MS star of $100\,M_\odot$. The MS star has a luminosity that, according to the mass–luminosity relationship, is $100^{3.5} = 1 \times 10^7\,L_\odot$. According to the H–R diagram, this star has a surface temperature about four times that of a G0 star, and, because $L \propto R^2T^4$, the radius of the star must be $L^{0.5}/T^2 = 200\,R_\odot$ ($= 1$ AU). The collapse energy of the star is $GM^2/R$, or 50 times that of the Sun, roughly $2 \times 10^{50}$ ergs. Because the luminosity of the $100\,M_\odot$ MS star is $10^7\,L_\odot = 4 \times 10^{40}$ erg s$^{-1}$, we can expect the collapse energy to contribute an appreciable portion of the total luminosity of the star for only $2 \times 10^{50}/4 \times 10^{40} = 5 \times 10^9$ s, which is 150 years. It is thus enormously less probable that we should observe an O class star lying appreciably off the MS than that a G, K, or M star might be seen in such a state.

Because the lifetime of a young cluster is roughly $10^7$ years and the new red dwarfs have enough collapse energy to double their luminosities for at least $10^8$ years, it is not surprising that such superluminous stars are indeed found in expanding clusters. It is also no surprise that T Tauri stars are mostly red.

Dense gas and dust clouds like the Orion Nebula are by no means rare or improbable phenomena. Even naked-eye observation of the sky reveals many dark interstellar clouds silhouetted against the diffuse background light of distant stars in the galactic plane. These dark clouds, which are particularly prominent in the constellation Cygnus, must be viewed as potential sites for star formation.

Even after several billion years of star formation, the central planes of spiral galaxies are rich in raw materials for making new stars. Where did the dust come from? Why has the raw material not been exhausted?

The southern Coalsack in the constellation Crux is illustrative of these small, dense clouds. This cloud, about 170 pc distant from us, has a mass of about $14\,M_\odot$. The density of the dust component in the cloud is about $4 \times 10^{-25}$ g cm$^{-3}$. A radio-frequency search for atomic hydrogen (at the 21-cm *ortho–para* spin transition) fails to show any H emission. Although this observation might be explained by postulating the absence of gas in the cloud or by requiring a very low temperature for the atomic H, the most plausible explanation is that the hydrogen is at low enough temperature and high enough density so that it has mostly combined to form molecular hydrogen.

One of the more interesting nebulae, R Monocerotis, partakes of properties of both nebulae and stars. R Mon is a strong emitter of infrared radiation with a spectral distribution characteristic of a black body at about 700 K. It is embedded in the variable nebula NGC2261. A comparison of the emission spectra of the Sun, R Mon, and T Tauri is given in Fig. II.18. Note the great brightness of T Tauri stars in the UV. This radiation must be emitted by the upper atmosphere of the young star, presumably in a very violently agitated chromosphere and corona. No clear evidence of radiation emitted directly from a stellar photosphere can be found in the spectrum of R Mon. It appears that R Mon could be a very young, very luminous star still mantled by a dense nebular shell. It is possible that R Mon might be a more massive version of the primitive Solar Nebula from which the Solar System formed $4.6 \times 10^9$ years ago.

Perhaps the most striking phenomenon documented by observations of T Tauri stars is the presence of massive, expanding shells of gas about them. Mass loss rates estimated from spectra of T Tauri stars are often above $10^{-7} M_\odot$ per year. Over an estimated pre-MS lifetime of about $10^6$ years, mass losses as high as tens of percent of the total mass of the star may be carried off in the form of a very dense "stellar wind." The extremely high UV fluxes and very intense solar winds associated with T Tauri stars can have profound effects on solid bodies and especially on planets with atmospheres, if bodies

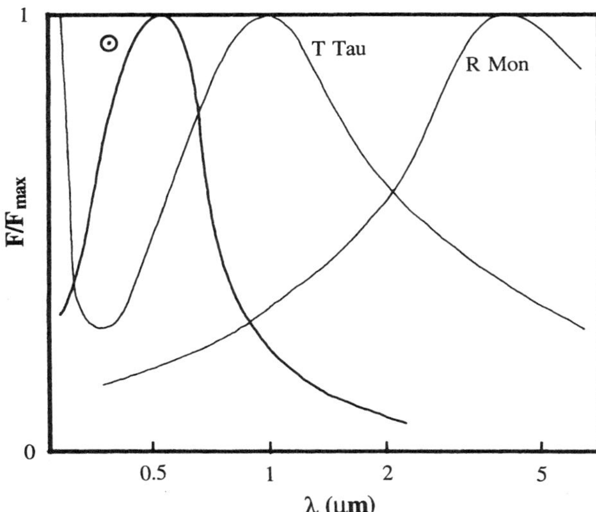


**Figure II.18** IR emission from R Mon, T Tau, and the Sun. The emission spectra are scaled to a peak flux of unity. No clear stellar component is seen in the spectrum of the R Monocerotis variable nebula. The Planck peak for R Mon places its temperature well below 1000 K, clearly so low that dust is the dominant source of opacity. All three bodies display emission spectra that differ significantly from black body emission; the Sun differs least over the wavelength region covered here.

large enough to have atmospheres can accumulate and outgas before the T Tauri phase is over. The magnetic flux carried by the solar wind, when dragged through weakly conducting planetesimals, induces a voltage difference across the body, which in turn drives an electric current. These induced currents, and other related phenomena as well, can heat and melt the interiors of even rather small bodies. This phenomenon, called *solar wind induction heating*, may be very important for asteroid-sized planetesimals. It is explored in Chapter VI in our discussions of asteroid thermal evolution.

In the 1950s an interesting class of bodies was discovered nearly simultaneously by George Herbig in the United States and by the Mexican astronomer G. Haro. These "Herbig–Haro" (HH) objects are exceptionally dense, spheroidal interstellar clouds, always located near T Tauri stars or other young stars. They have spectral features attributable to both reflected light from nearby young stars and shock-heating of gas in their envelopes. Rather than actual forming stellar systems, these seem to be merely fragments of dense molecular clouds that are being strongly heated and irradiated by nearby superluminous young stars. Similarly compact, dense globules are sometimes found in more open space, far from active star-forming regions and without the atomic emission lines that form so prominent a feature of the spectra of HH objects. These isolated clouds are called Bok globules after their discoverer, the Dutch–American astronomer Bart J. Bok.

Dense interstellar gas and dust clouds with temperatures of about 10 to 100 K have been found to contain a stunning variety of gaseous molecules, most of them organic compounds. A list of more than 120 such interstellar molecules, complete through early 2002, is given in Table II.3. Most detections of interstellar molecules are done by microwave spectroscopy of pure rotational transitions. In the ultraviolet, strong evidence for much more massive molecules, including the $C_{60}^+$ ion and a complex assortment of polycyclic aromatic hydrocarbons, has also been found. It is believed that these molecules are produced by gas-phase ion–molecule reactions and possibly also by surface-catalyzed reactions in very dense clouds, where they are available for incorporation into new stellar systems. Whatever their source, these molecules are widely, perhaps universally, distributed in interstellar space wherever the density is high and the temperature low.

The interstellar medium (ISM), the repository of material and energy ejected from giant stars, novae, supernovae, planetary nebulae, and Wolf–Rayet stars, is very complex in structure and dynamics. In modern views of the interstellar medium, most of its volume is dominated by an extremely hot, rarefied plasma derived from stellar explosions, whereas most of the mass is

**Table II.3** Interstellar Molecules

| | |
|---|---|
| Inorganic | $H_2$, $H_3^+$, OH, $H_2O$, SiO, SiS, SiN, $SiH_4$, SO, $SO^+$, $SO_2$, NS, HS, $H_2S$, NH, $NH_2$, $NH_3$, NO, $N_2O$, $N_2$, $N_2H^+$, $H^+$, $H_3O^+$, $H_2D^+$, HNO, $O_3$, HCl, PN, NaCl, AlCl, NaCN, KCl, HF, AlF, MgCN, MgNC |
| Carbon-bearing radicals and ions | CO, $CO^+$, $CO_2$, $C_2O$, $C_3O$, CP, CS, $C_3S$, $C_2$, COS, $C_3$, $C_5$, $C_{60}^+$, $HCNH^+$, $C_2S$, $HCS^+$, SiC, $SiC_2$(*cyclo*), $SiC_3$, $C_4Si$ |
| Hydrocarbons | CH, $CH_2$, $CH_4$, $C_2H$, $C_2H_2$, $C_2H_4$, $CH_3C_2H$, $C_3H$, $C_3H$(*cyclo*), $C_3H_2$, $C_3H_2$(*cyclo*), $C_4H$, $C_4H_2$, $C_5H$, $C_6H$, $C_6H_2$, $C_7H$, $C_8H$, $CH^+$, $CH_3C_4H$, PAHs |
| Organic oxygen compounds: carboxylic acids, aldehydes, alcohols, ketones, esters, and ethers | HCO, $HCO^+$, $HOC^+$, $HOCO^+$, $H_2COH^+$, $H_2CO$, $CH_3CHO$, $NH_2CHO$, HCCCHO, $HC_2CHO$, $(C_2H_5)_2O$, HCOOH, $HCOOCH_3$, $CH_3COOH$, $CH_3OH$, $C_2H_4O$(*cyclo*), $(CH_3)_2O$, $(CH_3)_2CO$(?), $C_2H_5OH$ |
| Organic nitrogen compounds: amines, amides, imines, nitriles, and isocyanides | CN, HCN, HNC, HCNH, $H_2CNH$, $H_2CNH_2$, $HNC_3$, $HC_3NH^+$, $HCONH_2$, $CH_3NH_2$, $H_2NCN$, HCN, HCCN, $C_2CN$, $HC_2CN$, $CH_2CN$, $CH_3CN$, $CH_3NC$, $HC_2NC$, $CH_2CHCN$, $HC_4CN$, $C_4CN$, $CH_3CH_2CN$, $CH_3C_2CN$, $CH_3C_4CN$, $HC_5CN$, $HC_6CN$, $HC_7CN$, $HC_8CN$, $HC_{10}CN$ |
| HCNO compounds | HNCO, $H_2NCH_2COOH$ |
| Thioorganics | HNCS, $H_2CS$, $CH_3SH$, $HCS^+$ |

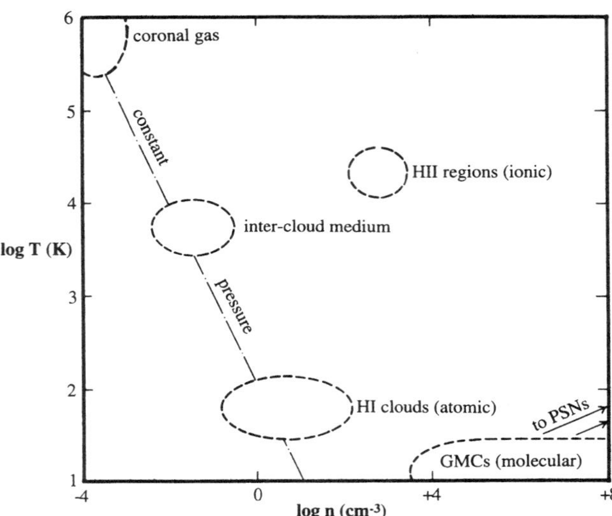


**Figure II.19** Structure of the Interstellar Medium. The temperatures and densities of the five "phases" of the ISM are shown. In order of their total mass, the GMCs and HI regions are comparable and larger than the ICM, which is much more massive than the coronal gas component. In order of their volumes, the coronal bubbles and the ICM are dominant and much larger than the HI regions, which are in turn much larger than the GMCs. HII regions contribute negligibly to the total mass and volume.

concentrated in cool, dense interstellar clouds rich in complex molecules. Figure II.19 shows the five distinctive types of interstellar material. The hottest "phase" in the ISM, at temperatures of millions of degrees, is composed of coronal gas emitted by stars, what we may accurately call *stellar winds* and other hot gases ejected directly by stellar explosions. These very hot, tenuous "bubbles" occupy much of the volume of the Galaxy. Most of the rest of the volume of the Galaxy is occupied by the *intercloud medium* (ICM), with about 100 times the density and 1/100th the temperature of the coronal gas. *HII regions* are small spheroidal regions centered on very hot stars that are potent emitters of UV radiation. HII is astronomical terminology for ionized atomic hydrogen. The spectacular but quantitatively negligible HII regions are kept ionized and heated to temperatures of about $10^4$ K by absorption of UV from the central star. Clouds of much cooler (30 to 100 K) neutral atomic hydrogen, referred to by astronomers as *HI regions*, are common and make up roughly half of the mass of the ISM. Finally, the cold (usually less than 30 K), dense regions in which molecules abound, the *Giant Molecular Cloud* (GMC) complexes, occupy very little volume, but have very large mass. These clouds are the hotbeds of chemical and dynamic activity leading to star formation.

The ISM is wracked by violent dynamical processes. Cosmic rays can penetrate even into the dense GMCs, ionizing the gas and depositing energy. At the very low temperature of GMC interiors, typically about 10 K, these ions are extremely reactive, giving rise to complex molecular species with isotopic compositions that reflect equilibrium at temperatures close to absolute zero. Vast enrichment of deuterium, by a factor of 100 to 1000 or more, occurs in these clouds; DCN is sometimes nearly as abundant as HCN and likewise for other complex organic species. GMCs cool themselves by infrared radiation. Because they are so cool, most of the elements in them are condensed into dust grains, and both dust and polyatomic molecules make very efficient infrared emitters. Shock waves from stellar explosions occasionally traverse the ISM, depositing energy and creating brief excursions to high temperatures. Ultraviolet absorption (mostly from early spectral class MS stars) both drives chemistry and heats diffuse matter. And of course each of the ISM "phases," with its distinctive temperature and density, exerts a pressure on neighboring phases that must be maintained in an approximate steady state that preserves the distinctions among the five phases.

Figure II.20 is a pressure–density plot of the ISM phases. HII regions, being a purely local dynamical

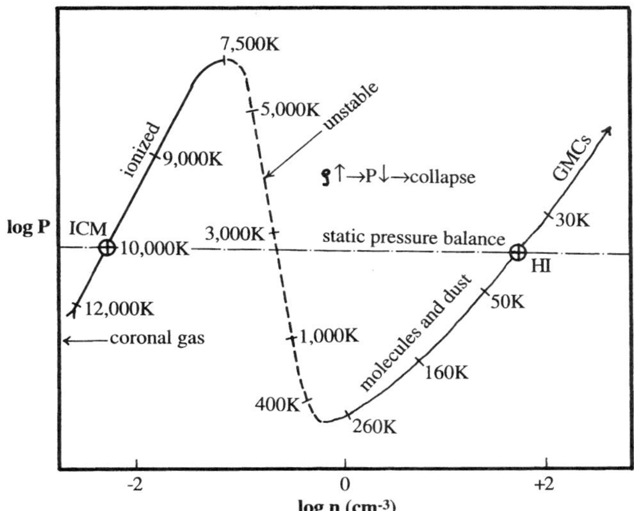


**Figure II.20** Pressure–density–temperature relations for the ISM. The relationships among the coronal, the ICM, the HI, and the GMC components are shown. The line is the locus of solutions for temperature and pressure as a function of density for steady-state dynamic balance of the energy sources and loss mechanisms in the ISM. Coronal and HII regions are source phenomena, not steady-state solutions. The stability of GMCs is due to a new factor that has negligible influence on the ICM/HI pressure balance: gravity. See the text for further explanation.

phenomenon, are omitted. The left-hand branch of the figure is ionized gas, containing coronal gas and the intercloud medium. The central branch (shown with the dashed line) is cool enough to be neutral, and atomic species dominate. In this regime, compression to higher density makes the gas a better radiator, so the pressure *drops* with increasing density. Such behavior is unstable against spontaneous collapse, so this branch of the curve is not populated. The right-hand branch is atomic and molecular, with dust present toward higher densities. Here HI regions and GMCs are found. The ICM, fed by the coronal gas, is in pressure balance with HI regions, with their embedded GMCs. The dynamical balance between the ICM and the HI regions helps maintain the complex multiphase structure. In GMCs, densities are so high that gravity becomes an important factor, leading to further collapse moderated by the best rate at which the cloud can radiate off its collapse energy. Generally, mass flows through the diagram from left to right, shedding energy as it cools. This evolution appears to be a one-way street—but stars, once formed (by collapse of a GMC?), can evolve into mass-shedding giants, which become Wolf–Rayet stars, which produce planetary nebulae. Sufficiently massive stars may even explode. These processes thus provide fresh hot gas, newly processed in stellar interiors, to reenter the diagram from the left!

## Outline of Star Formation

The Infrared Astronomy Satellite (IRAS) has found several thousand dense, warm globules or disks in giant molecular cloud (GMC) complexes where star formation is known to be occurring. There is good reason to believe that these bodies are in fact prestellar disks, detectable in the infrared, but so heavily shrouded in dust as to prevent any internal star, if present, from being seen from outside. IRAS also has found that large numbers of clearly visible stars, such as Vega, are girdled by immense, cold disks of solid debris. As many as 6% of the nearby stars have such massive dust disks.

Combining the various lines of evidence so far presented, we would picture the origin of stars and stellar systems as taking place in this manner. First, dense gas and dust clouds, which are common in spiral galaxies, can collapse under the compressive forces of the hot surrounding plasma, radiation pressure, and self-gravitation. Supernova explosions are most likely to occur in the immediate vicinity of star-forming regions, because the entire lifetime of a massive star can be run through during the lifetime of a typical cloud. The massive star cannot live long enough to get away from the dense cloud region before exploding! Thus supernova shock waves may play a role in stellar formation.

Second, interstellar clouds collapse and fragment while storing much of their potential energy as heat. Third, collapse progresses at a rate limited by the ability of the cloud to dissipate its energy and rotational angular momentum. Fourth, the center of the cloud reaches the ignition point of nuclear reactions.

Fifth, the young, superluminous star lights up and expels the surrounding remnants of the cloud which have not accreted into massive, gravitationally bound bodies. The star passes through a brief period of convective overturn, during which relatively unstable nuclides such as deuterium are burned and depleted throughout the star. The upper atmosphere (chromosphere and corona) of the star is hyperactive during this period and emits an enormous UV flux and an intense T Tauri phase solar wind.

Finally, the star settles down onto the Main Sequence, where it continues stable hydrogen burning for most of the lifetime of the star. Smaller or first-generation stars will run mostly by the proton–proton (pp) chain, whereas more massive and chemically complex stars will, as we shall soon see, be able to use other reaction networks to fuse hydrogen.

The solid material left behind in orbit around the young star is the raw material out of which terrestrial-type planets will be made. Only nebulae with the composition of Pop I stars will contain enough of the heavy elements to make rocky planets. Gravitational collapse processes may be capable of forming Jovian-type gas giant planets about either Pop II or Pop I stars.

One of the immediate consequences of theories of stellar evolution is the conclusion that Pop I stars at formation contain appreciable amounts of the elements heavier than helium, whereas Pop II stars do not. The source of the heavier elements has been sought in theoretical studies of the evolution of stellar interiors; however, it was not at once obvious how the elements heavier than iron were formed or how they were dispersed and mixed into the interstellar medium in order to become available for incorporation into second-generation stars.

We have mentioned without explanation the ejection of processed stellar material by nova explosions. Novae give off only a fraction of a percent of their mass in each explosive episode, but may repeat the explosion many times. Evidence from the study of novae is skimpy, but they contribute only a small minority of the mass ejected by stars. Nova explosions may be limited to highly evolved close double stars with mass exchange between them and hence shed little light on the evolution of single stars.

We have also mentioned another class of astronomical objects that does have obvious relevance to the problem of the disruption of stars and the dissemination of the heavy elements: a number of highly evolved, post-red-giant, post-Wolf–Rayet stars have been observed to be surrounded by extremely hot, tenuous envelopes of gases receding from them at enormous velocities. These shells of high-velocity gas are called *planetary nebulae*, not because of any relationship to planets, but because the blurred, disk-like telescopic appearance of these nebulae has a superficial similarity to a planet. Planetary nebulae form as a natural consequence of cooling and collapse of former Wolf–Rayet stars, which are in turn dying red giants undergoing collapse and mass loss.

About 150 planetary nebulae are known, all characterized by a glowing extended disk and a diagnostic emission spectrum. The nearest planetary nebula, NGC7293 in Aquarius, is 12′ of arc in apparent diameter and has a parallax of 0.050″. The famous Ring Nebula in Lyra is 83″ in diameter, and the Owl Nebula, 200″. As is typical of such nebulae, these objects have at their centers very faint stars that are the only visible remnants of the star whose explosion produced the nebula. Although these central stars are very faint, the nebular envelope glows with considerable brilliance under excitation by ultraviolet radiation from the star. The central stars, although frequently very small, typically have temperatures of 50,000 to 200,000 K, with peak emission in the far ultraviolet and X-ray regions. This emission keeps nebulae with radii of several thousand AU in an ionized (HII) state.

In planetary nebulae, as with the other mass-shedding stars discussed above, we find the very strong suggestion that violent expulsion of matter from stars can occur on a sufficiently large scale to replenish the interstellar medium with stellar material. These nebulae are observed to contain heavy elements in about the same proportions that they are found in Pop I stars. The crucial dilemma then becomes this: if the nuclear processes that make the heavy elements are characteristic of stellar cores and if stellar nova and supernova explosions remove only the outer 0.1 to 30% or so of the stellar mass, how do the heavy elements get out of highly evolved stars?

We know that, in order to form a Pop I star with a planetary system, the collapsing interstellar cloud from which it forms must already contain these heavy elements. Thus we can state as a necessary prerequisite for the existence of Earthlike planets in a stellar system the condition that at least one previous generation of stars has exploded and fertilized space with its debris. It is clear that, in order for us to understand the elemental and isotopic composition of our own planet and solar system, we must look in some detail at the processes of formation of the heavy elements in stars and their dispersal into space.

## Stellar Explosions and Nucleosynthesis

> Aye, for 'twere absurd to think
> That Nature in the Earth bred gold,
> Perfect in the instant.
>
> Ben Jonson
> *The Alchemist*

It has already been remarked that the nuclear "burning" of deuterium to $^3$He can occur at temperatures lower than those needed to fuse hydrogen, by stars with masses as low as about $0.0125M_\odot$. We have also seen that fusion of hydrogen to make $^4$He is the dominant method of energy production in stars. The simplest mechanism for fusing hydrogen, the pp chain, requires temperatures of several million degrees in order to proceed at a significant rate. Any hydrogen-rich body with a mass greater than about $0.072M_\odot$ can become a Main Sequence star (the exact minimum mass depends very slightly on the details of the composition). The rate of energy production by the pp chain, given in Eq. (II.67), is proportional to the temperature raised to the 4.1 power.

The second important mechanism for hydrogen burning, the catalytic carbon cycle (CC), also called the carbon-nitrogen-oxygen (CNO) cycle or the "carbon bicycle," requires the initial presence of small amounts of CNO group elements in the raw material from which the star formed. It competes successfully with the pp chain at temperatures near and above $T_6 = 18$ (18 million K). The reactions involved are

$$^{12}C + H \rightarrow {}^{13}N \qquad + 1.95\,\text{MeV} \qquad (II.74)$$
$$^{13}N \rightarrow {}^{13}C + \beta^+ + \nu_e \qquad + 1.50\,\text{MeV} \qquad (II.75)$$
$$^{13}C + H \rightarrow {}^{14}N \qquad + 7.54\,\text{MeV} \qquad (II.76)$$
$$^{14}N + H \rightarrow {}^{15}O \qquad + 7.35\,\text{MeV} \qquad (II.77)$$
$$^{15}O \rightarrow {}^{15}N + \beta^+ + \nu_e \qquad + 1.73\,\text{MeV} \qquad (II.78)$$
$$^{15}N + H \rightarrow {}^{12}C + {}^{4}He \qquad + 4.96\,\text{MeV} \qquad (II.79)$$
$$^{15}N + H \rightarrow {}^{16}O \qquad (II.80)$$
$$^{16}O + H \rightarrow {}^{17}F \qquad (II.81)$$
$$^{17}F \rightarrow {}^{17}O + \beta^+ + \nu_e \qquad (II.82)$$
$$^{17}O + H \rightarrow {}^{14}N + {}^{4}He. \qquad (II.83)$$

The reaction of $^{15}N$ with $^{1}H$ follows the route of Reaction (II.79) about 99.9% of the time, so that the energetic importance of the second reaction cycle is nil. The half-lives of decay of the radioactive intermediates are 10 min for $^{13}N$, 2 min for $^{15}O$, and 1.1 min for $^{17}F$.

Both cycles result in the same net reaction,

$$4\,{}^{1}H \rightarrow {}^{4}He + 2\beta^+ + 2\nu_e, \qquad (II.84)$$

and both cycles in this "bicycle" restore the participating heavy elements to their original states. For this reason, C, N, and O act as true catalysts in the fusion of hydrogen.

The rate of energy production by the CC process is limited by the rate of the reaction of C with H and is approximately

$$\varepsilon = 1.66\rho X_H X_C (T_6/13)^{20.3}, \qquad (II.85)$$

where $X_C$ is the mass fraction of carbon and the other symbols are as before. Note the dependence of the reaction rate on the 20.3 power of the temperature.

This enormously strong dependence of rate on temperature means that the CC process will be negligible in rate compared with the pp rate at $T_6 = 15$, but strongly dominant at $T_6 = 20$. For Pop I stars, in which the carbon atomic abundance is about $10^{-3}$ of the hydrogen abundance, the two processes produce comparable amounts of energy for stars near $2M_\odot$ ($T_6 = 18$).

One consequence of the strong dependence of the rate of CC on temperature is that even the most brilliant MS stars have core temperatures only 50% higher than sun-like dwarf stars of modest luminosity. One might say that the cores of MS stars are thermostated by the carbon cycle. This phenomenon is sketched out in Fig. II.21.

Now that the crucial role of carbon in hydrogen fusion has been spelled out, it is of great importance to understand how carbon was first made. We have already seen that the crucial step in the synthesis of carbon (and the one which cuts off the course of Big Bang nucleosynthesis) is fusion of the very unstable $^{8}Be$ nucleus with $^{4}He$ [Eq. (II.70)]. Once $^{12}C$ formation begins in the center of the helium core of an evolving star, helium fusion quickly becomes the dominant energy source in the star. The outer envelope goes through the helium flash, and the star departs from the Main Sequence and becomes a giant.

This process of formation of carbon can occur even in Pop II stars once they have evolved to the point where a large helium core has accumulated. Indeed, even a pure hydrogen protostar could follow this evolutionary course if sufficiently massive. The rate of the ternary fusion reaction that produces $^{12}C$ is a very sensitive function of both density and temperature, as we can

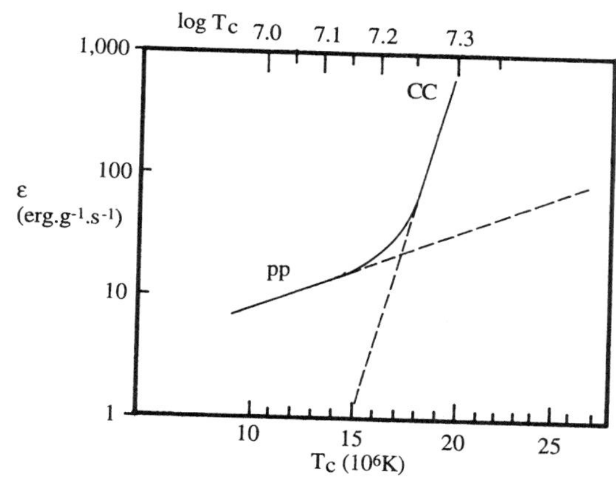


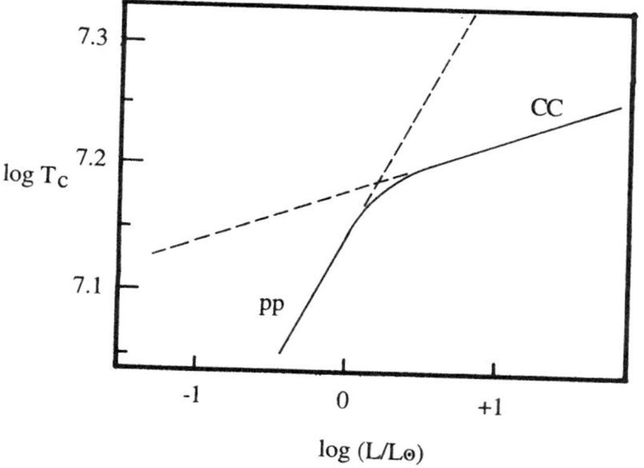


**Figure II.21** Energy production rates and core temperatures of MS stars. The very strong temperature dependence of the rate of the catalytic carbon process (CC) effectively thermostats the core temperature of the star near $(18 \pm 2) \times 10^6$ K.

easily show. Let us reconsider the reactions producing and destroying the intermediate $^8$Be:

$$^4\text{He} + {}^4\text{He} \rightarrow {}^8\text{Be} \quad (\text{II.86})$$

$$^8\text{Be} \rightarrow {}^4\text{He} + {}^4\text{He} + \gamma \; (t_{1/2} = 10^{-16}\,\text{s}) \quad (\text{II.87})$$

$$^8\text{Be} + {}^4\text{He} \rightarrow {}^{12}\text{C} + \gamma. \quad (\text{II.88})$$

The rates of these reactions are

$$dX_{\text{Be}}/dt = k_{86}\rho(X_{\text{He}})^2 \quad (\text{II.89})$$

$$dX_{\text{Be}}/dt = -\lambda X_{\text{Be}} \quad (\text{II.90})$$

$$dX_{\text{Be}}/dt = -k_{88}X_{\text{Be}}X_{\text{He}} = -dX_{\text{C}}/dt. \quad (\text{II.91})$$

Virtually all loss of $^8$Be is by spontaneous decay. We can therefore set these two rates approximately equal to each other:

$$k_{89}\rho(X_{\text{He}})^2 = -\lambda X_{\text{Be}}. \quad (\text{II.92})$$

Because

$$dX_{\text{C}}/dt = k_{91}X_{\text{Be}}X_{\text{He}}, \quad (\text{II.93})$$

we get

$$dX_{\text{C}}/dt \propto X^3_{\text{He}}. \quad (\text{II.94})$$

Any hydrogen within the core would be rapidly consumed by fusion reactions, because the temperature and density in the core are at least as high as they are at the surface of the core. Further, we must take into account the high Coulomb barrier for Reaction (II.91), which causes the reaction to be very slow at temperatures below several tens of millions of degrees. Because of the thermostating effect of the CC process, a hydrogen-containing mixture cannot be this hot. The structure of an evolved star will therefore contain an outer hydrogen–helium envelope, a helium core, and an inner core of carbon.

At slightly higher temperatures, near $T_6 = 100$, further addition of helium nuclei (alpha particles) can occur. The series of products made by successive additions of alpha particles is

$$^{12}\text{C} + {}^4\text{He} \rightarrow {}^{16}\text{O} + \gamma \quad (\text{II.95})$$

$$^{16}\text{O} + {}^4\text{He} \rightarrow {}^{20}\text{Ne} + \gamma \quad (\text{II.96})$$

$$^{20}\text{Ne} + {}^4\text{He} \rightarrow {}^{24}\text{Mg} + \gamma \quad (\text{II.97})$$

and so on. Not only are the nuclides containing multiples of the $^4$He nucleus easy to make by this simple mechanism, but these nuclides are also intrinsically quite stable in the thermodynamic sense.

The most objective measure of stability is to be found in extremely precise measurements of the masses of the nuclides. The mass of any nucleus containing $Z$ protons and $N$ neutrons is always less than the sum of the masses of the constituent particles. The amount by which the mass of the nucleus differs from the sum of the masses of its parts is called the *mass defect*. This mass defect, multiplied by the square of the speed of light, is the *nuclear binding energy*, which is precisely the amount of energy emitted by the formation of this nucleus from its constituent parts:

$$BE = -\Delta mc^2. \quad (\text{II.98})$$

If several nuclei with equal numbers of nucleons [$Z + N = A$ (a constant)] are carefully weighed, these *isobars* will be found to have slightly different masses. The lightest of these is the most stable. It is generally necessary, however, to compare the stability of nuclides that have different atomic weights, that is, nonisobars. It is then most convenient to compare the binding energies per nucleon, $BE/A$, and use this quantity, in some appropriate unit such as grams per nucleon or MeV per nucleon, as our stability yardstick.

Figure II.22 gives the $BE/A$ vs $A$ curve for nuclides from hydrogen to silver. Several prominent features of the graph bear mention. First, the most stable nuclides are the iron-group elements. Second, alpha particle multiples (except $^8$Be with its split personality) are unusually stable, and each successive alpha-multiple nuclide is more stable than the previous one. Third, stability decreases steadily and smoothly beyond the atomic weight of iron.

If nuclear thermodynamic stability were all that mattered, the Universe would consist mostly of iron, with small amounts of nuclides like $^{28}$Si, traces of heavy elements, and no hydrogen. Clearly the Universe is not close to nuclear equilibrium!

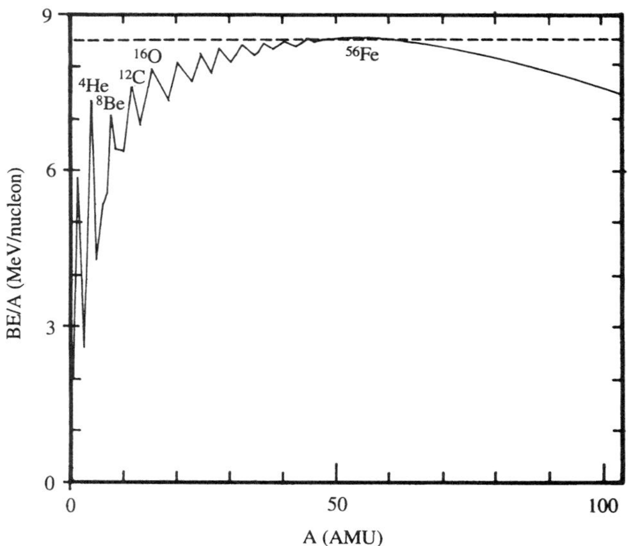


**Figure II.22** Stability of the nuclides. The binding energies per nucleon are graphed as a function of atomic weight. Note the prominence of the alpha-particle multiples and the iron group. The BE/A value for $^8$Be is less than that for $^4$He, so the decomposition of $^8$Be into two alpha particles is spontaneous and very rapid.

We now have sufficient knowledge to examine and interpret the data on the abundances of the elements in Pop I stars, and to see how well and how fully nature meets our expectations. Table II.4 gives the "cosmic" abundances, which are largely derived from studies of the composition of the atmosphere of the Sun and of primitive meteorites, but generally similar to the composition of Pop I stars. Graphs of the abundances of the nuclides with even and odd atomic numbers are given in Fig. II.23. The enormous abundances of the primitive elements H and He are in contrast to the other light elements. The very stable iron-group elements and the alpha-process nuclides from helium burning also stand out. Beyond carbon the sequence of abundant nuclides runs $^{12}$C, $^{16}$O, $^{20}$Ne, $^{24}$Mg, $^{28}$Si, $^{32}$S, $^{36}$Ar, and $^{40}$Ca. Beyond calcium the products of the alpha process are too proton-rich to be stable, and they decay by positron emission or electron capture. Thus the alpha process continues:

$$^{40}\text{Ca} + {}^{4}\text{He} \rightarrow {}^{44}\text{Ti} \quad \text{(II.99)}$$

$$^{44}\text{Ti} \rightarrow {}^{44}\text{Sc e.c.} \quad (t_{1/2} = 47 \text{ years}) \quad \text{(II.100)}$$

$$^{44}\text{Sc} \rightarrow {}^{44}\text{Ca} + e^{+} \quad (t_{1/2} = 4\,\text{h}) \quad \text{(II.101)}$$

$$^{44}\text{Ca} + {}^{4}\text{He} \rightarrow {}^{48}\text{Ti} \quad \text{(II.102)}$$

$$^{48}\text{Ti} + {}^{4}\text{He} \rightarrow {}^{52}\text{Cr} \quad \text{(II.103)}$$

$$^{52}\text{Cr} + {}^{4}\text{He} \rightarrow {}^{56}\text{Fe}. \quad \text{(II.104)}$$

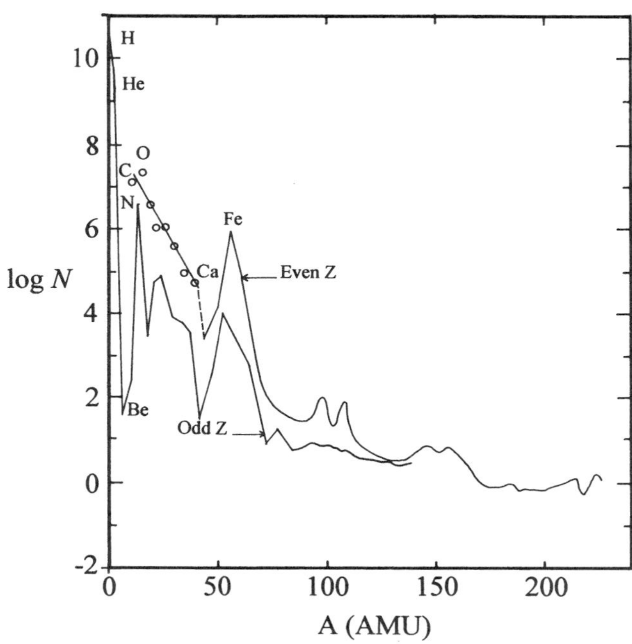


**Figure II.23** Abundances of the elements in the Sun. The abundances are given on a logarithmic scale as the number of atoms of each nuclide per $10^6$ atoms of silicon. The even-$Z$–even-$N$ alpha-process nuclides are marked (○). The lower curve is for odd-$Z$ elements, and the upper curve is for even-$Z$ elements. The pairs of abundance peaks at high A are discussed in the text. The even–odd nuclide $^9$Be clusters with the other unstable isotopes of lithium and boron.

**Table II.4** Abundances of Selected Elements in the Solar System after Grevesse and Anders (1988)

| Element | Abundance | Element | Abundance |
|---|---|---|---|
| H | $27.9 \times 10^9$ | | |
| He | $2.7 \times 10^9$ | F | 843 |
| O | $23.8 \times 10^6$ | Cu | 522 |
| C | $10.1 \times 10^6$ | V | 293 |
| N | $3.1 \times 10^6$ | Ge | 119 |
| Ne | $3.4 \times 10^6$ | Se | 62 |
| Mg | $1.1 \times 10^6$ | Li | 57 |
| Si | $1.0 \times 10^6$ | Kr | 45 |
| Fe | $0.9 \times 10^6$ | Ga | 38 |
| S | $515 \times 10^3$ | Sc | 34 |
| Ar | $101 \times 10^3$ | Sr | 24 |
| Al | $85 \times 10^3$ | B | 21 |
| Ca | $61 \times 10^3$ | Br | 11.8 |
| Na | $57 \times 10^3$ | Zr | 11.0 |
| Ni | $49 \times 10^3$ | Rb | 7.1 |
| Cr | $14 \times 10^3$ | As | 6.6 |
| P | $10 \times 10^3$ | Te | 4.8 |
| Mn | $9.6 \times 10^3$ | Xe | 4.7 |
| Cl | $5.2 \times 10^3$ | Y | 4.6 |
| K | $3.8 \times 10^3$ | Ba | 4.5 |
| Ti | $2.4 \times 10^3$ | Sn | 3.8 |
| Co | $2.2 \times 10^3$ | Pb | 3.2 |
| Zn | $1.3 \times 10^3$ | Mo | 2.6 |

The rates of these reactions are roughly proportional to the temperature to the 30th power! Near the equilibrium stability maximum at $^{56}$Fe the importance of the alpha process nuclides rapidly dwindles away.

Several aspects of the abundance data remain puzzling, however. What is the meaning of the pairs of abundance peaks at large atomic weights, far above the iron-group peak? By what mechanism are the equilibrium (e-process) nuclides formed?

One very useful representation of stability data that helps us answer the first of these questions is to plot the locations of all the stable nuclides on a graph of $Z$ (atomic number) vs $N$ (neutron number). Such a graph is sketched in Fig. II.24. Detailed correlation of the abundances of individual nuclides with the neutron number and atomic number reveal that the abnormally abundant heavy elements beyond iron have certain features in common: the neutron number or the atomic number (or both) is found to be 50, 82, or 126. For the lighter elements, values of $Z$ or $N$ equal to 2, 8, 20, and 28 are found to define the most abundant nuclides. Thus we find large amounts of $^4$He ($Z = N = 2$), $^{16}$O ($Z = N = 8$), $^{40}$Ca ($Z = N = 20$), $^{48}$Ca ($Z = 20, N = 28$), and $^{56}$Ni ($Z = N = 28$). The last is actually a radionuclide,

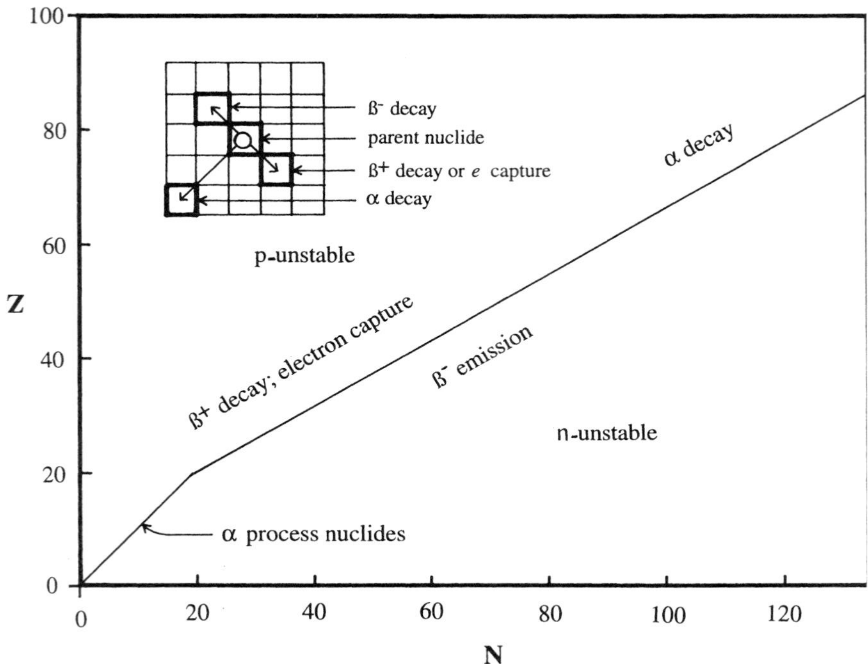


**Figure II.24** Atomic number–neutron number plot of the stable nuclides. The zone within which all stable nuclides occur is outlined. The solid line within this envelope is a smoothed representation of the approximate center of the "energy valley." The modes of decay of nuclides lying outside the stability zone are indicated. The inset shows the direction of each decay mode on the $Z$ vs $N$ plot.

and it rapidly decays to produce the most stable nuclide of all:

$$^{56}\text{Ni} \rightarrow {}^{56}\text{Co} \ (\text{e.c.}; t_{1/2} = 5 \text{ days}) \quad (\text{II.105})$$

$$^{56}\text{Co} \rightarrow {}^{56}\text{Fe} \quad (\text{e.c.}; t_{1/2} = 77 \text{ days}). \quad (\text{II.106})$$

The numbers 2, 8, 20, 28, 50, 82, and 126 are conventionally referred to as "magic numbers" because of their profound and unexpected effect on elemental abundances. The explanation of this phenomenon lies, however, not in magic, but in the fact that the energy levels of nucleons in a nucleus, like those of electrons in atoms and molecules, are discrete and finite in number. Closed shells of neutrons or protons of unusual stability can be formed in the same manner that filled electronic shells of atoms are generated by the Aufbau principle.

When a closed shell of protons is present, all the nuclides with that atomic number are enhanced in stability; we would then expect all the isotopes of this element to be more stable and more abundant than their isobars. As an example, consider atomic number 50, the element tin. Tin has 10 stable isotopes, more than any other element. Similarly, the two neutron numbers that have the largest number of stable nuclides are $N = 50$ and 82.

Closer inspection of the isotopic abundance data reveals that a substantial majority of the stable nuclides contain both an even number of protons and an even number of neutrons. Thus elements with even atomic numbers are generally significantly more abundant than their odd-$Z$ neighbors.

This also means that stable isotopes have a strong tendency to have even atomic weights. Odd-neutron, odd-proton nuclides are rarely stable. Thus not only are odd-$Z$ elements relatively less abundant, but also their stable nuclides virtually all have odd atomic weights. The only naturally occurring odd–odd nuclides (both $N$ and $Z$ odd) are $^{2}$D, $^{6}$Li, $^{10}$B, $^{14}$N, $^{40}$K, $^{50}$V, $^{138}$La, $^{176}$Lu, and $^{180}$Ta. Of these, D, Li, and B are noted for their small binding energies and instability within stars. Several other odd–odd nuclides are unstable against beta decay or electron capture:

$$^{40}\text{K} \rightarrow {}^{40}\text{Ar} + \text{e}^- + \bar{\nu}_\text{e}(t_{1/2} = 1.3 \times 10^9 \text{ years}) \quad (\text{II.107})$$

$$^{40}\text{K} \rightarrow {}^{40}\text{Ar} + \text{e}^+ + \nu_\text{e} \quad (\text{II.108})$$

$$^{40}\text{K} \rightarrow {}^{40}\text{Ar e.c.} \quad (\text{II.109})$$

$$^{50}\text{V} \rightarrow {}^{50}\text{Cr} + \text{e}^- + \bar{\nu}_\text{e}(t_{1/2} = 6 \times 10^{15} \text{ years}) \quad (\text{II.110})$$

$$^{50}\text{V} \rightarrow {}^{50}\text{Ti e.c.} \quad (\text{II.111})$$

$$^{138}\text{La} \rightarrow {}^{138}\text{Ce} + \text{e}^- + \bar{\nu}_\text{e}(t_{1/2} = 1.1 \times 10^{11} \text{ years}) \quad (\text{II.112})$$

$$^{138}\text{La} \rightarrow {}^{138}\text{Ba e.c.} \quad (\text{II.113})$$

$$^{176}\mathrm{Lu} \rightarrow {}^{176}\mathrm{Hf} + \mathrm{e}^- + \bar{\nu}_e(t_{1/2} = 2.2 \times 10^{10} \text{ years}) \quad (\mathrm{II}.114)$$

$$^{180}\mathrm{Ta} \rightarrow {}^{180}\mathrm{W} + \mathrm{e}^- + \bar{\nu}_e(t_{1/2} = 8\,\mathrm{h}) \quad (\mathrm{II}.115)$$

$$^{180}\mathrm{Ta} \rightarrow {}^{180}\mathrm{Hf} \text{ e.c.} \quad (\mathrm{II}.116)$$

One might be justified in regarding $^{14}N$ as a unique nuclide. In fact, nitrogen is the only element whose most abundant naturally occurring isotope is odd–odd.

The existence of these stable closed nuclear shells allows us to understand qualitatively why the elemental abundances should exhibit peaks at masses well beyond the iron group. But why are these peaks located exactly where they are, and why are these peaks double? Phrased differently, why should a particular neutron magic number produce two abundance peaks, neither of which has exactly the magic neutron number?

Our initial approach to this problem is to search for any general mechanism for producing these heavy, neutron-rich nuclides. Let us suppose that there is available a limited neutron flux, produced by certain as yet unspecified nuclear reactions. Because the iron-group and heavier nuclides have much higher neutron capture cross-sections (probabilities) than the CNONe group (often about 10 or more times as large for thermal neutron capture), most neutrons that do not decay will be captured by heavy elements. If the neutron flux is low, a given nuclide may wait $10^3$ to $10^5$ years between successive capture events. Thus unstable isotopes produced by this "neutron activation" process may be present in stars in which a neutron source is present.

These neutron-rich nuclides will tend to decay by beta emission to more stable isobars. Any radioactive product with a half-life less than the mean time between neutron addition events will decay to the isobar with the next higher atomic number, and further neutron addition will continue on that element. Any radioactive product with a half-life longer than the neutron addition time scale will simply add another neutron and will continue doing so until a nuclide with a half-life less than the neutron addition time scale is formed. Each successive n-addition product is farther from the original stable nuclide and less stable, and each has a shorter lifetime than the previous product.

About 50 years ago the element technetium (Tc), with $Z = 43$, was discovered for the first time in nature in the spectra of S-type stars (and, by a strange coincidence, made and discovered almost simultaneously in the laboratory by Charles Coryell). The Canadian astrophysicist A. G. W. Cameron pointed out that the longest-lived isotope of Tc, with a half-life of only $2.6 \times 10^6$ years, must be produced in the outer envelopes of these stars by a slow neutron-addition process.

Figure II.25 shows how the products from slow neutron addition (the s-process) evolve with time on a $Z$ vs $N$ diagram. Assuming a neutron addition time scale of $10^4$ years, isotopes with ever higher atomic weight will be built up from each of the parent nuclides on the stability line until products with half-lives of about $10^4$ years are made. These nuclides lie on a line which is very roughly parallel to and only slightly to the right (high-$N$) side of the stability line. Then beta decay promotes each of these radionuclides to the next higher atomic number, and the process continues.

Whenever a neutron addition event succeeds in filling a neutron shell, the resulting "magic" nuclide will be an abnormally stable and abnormally abundant radionuclide with a half-life of at least $10^5$ years. That abnormally stable nuclide will lie at the intersection of the neutron magic number isotone and the s-process trajectory, somewhat to the right of the stability line. Therefore, if the s-process should terminate, these unusually abundant products would decay back onto the stability line to form an isobar with a slightly higher $Z$ *and a slightly lower* $N$. The resulting stable s-process products display an abundance maximum slightly displaced from the neutron magic number toward lower $N$.

If we now imagine another neutron addition process working on a different time scale, another set of heavy nuclides would be produced with *different offsets of the most abundant nuclides from the neutron magic numbers.* A mixture of the products from these two processes would display double abundance peaks, with the peaks both offset downward in $N$ by differing amounts. The faster the process, the farther the intersection of the evolutionary trajectory and the magic isotone will be from the stability line. The locations of these abundance peaks tell us the time scale of the neutron addition process.

But where do the neutrons needed for the s-process come from? Theory supplies an answer. At the end of hydrogen burning, the most abundant of the CNONe-group nuclides is $^{14}N$, and during the subsequent course of helium burning in the red giant phase, the reactions

$$^{14}\mathrm{N} + {}^{4}\mathrm{He} \rightarrow {}^{18}\mathrm{F} + \gamma \quad (\mathrm{II}.117)$$

$$^{18}\mathrm{F} \rightarrow {}^{18}\mathrm{O} + \mathrm{e}^+ + \gamma \quad (\mathrm{II}.118)$$

lead to a supply of $^{18}O$. Destruction of $^{18}O$ can occur by:

$$^{18}\mathrm{O} + {}^{4}\mathrm{He} \rightarrow {}^{21}\mathrm{Ne} + \mathrm{n} \quad (\mathrm{II}.119)$$

$$^{18}\mathrm{O} + {}^{4}\mathrm{He} \rightarrow {}^{22}\mathrm{Ne} + \gamma \quad (\mathrm{II}.120)$$

$$^{22}\mathrm{Ne} + {}^{4}\mathrm{He} \rightarrow {}^{25}\mathrm{Mg} + \mathrm{n} \quad (\mathrm{II}.121)$$

$$^{22}\mathrm{Ne} + {}^{4}\mathrm{He} \rightarrow {}^{26}\mathrm{Mg} + \gamma \quad (\mathrm{II}.122)$$

$$^{26}\mathrm{Mg} + {}^{4}\mathrm{He} \rightarrow \text{ etc.} \quad (\mathrm{II}.123)$$

At high temperatures $^{25}Mg$ can also "burn" by

$$^{25}\mathrm{Mg} + {}^{4}\mathrm{He} \rightarrow {}^{28}\mathrm{Si} + \mathrm{n} \quad (\mathrm{II}.124)$$

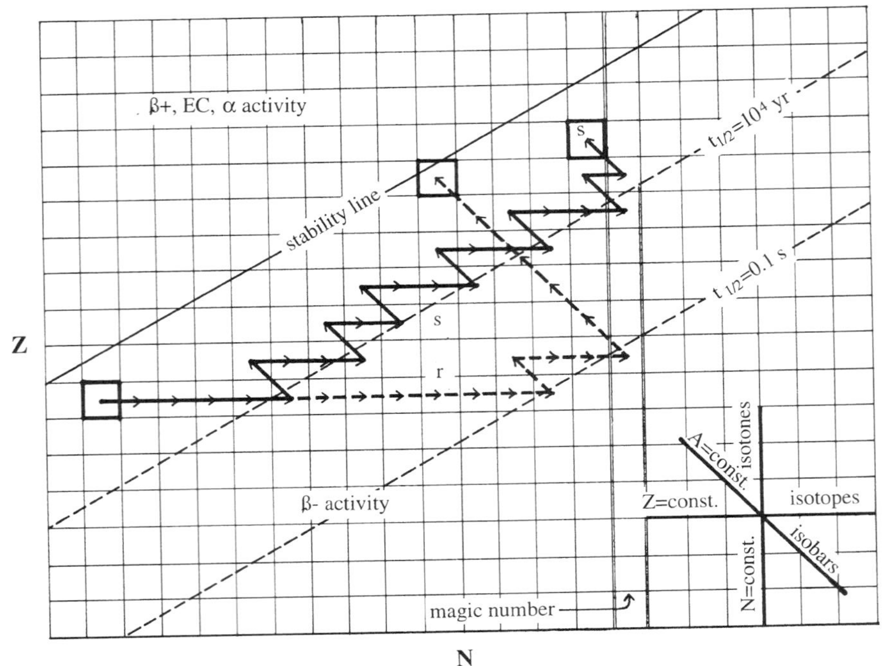


**Figure II.25** Slow and rapid neutron addition processes. The slow (s) neutron addition process is taken to have a time scale of $10^4$ years, whereas the rapid (r) process occurs on a time scale of 0.1 s. The diagonal dashed lines indicate the loci of beta emitters with half-lives of $10^4$ years and 0.1 s, respectively. The evolutionary trajectories for a general stable starting nuclide in both s and r environments are shown up to encounter with a neutron magic number, with subsequent decay back to the stability line. The nuclides with magic neutron numbers are more abundant than the others, and hence their decay products are enhanced in abundance. The existence of a closed neutron shell causes the neutron capture cross-section to be abnormally small.

which provides a large neutron source late in helium burning for all stars massive enough to reach $T_9 = 0.2$ in their cores.

For stars which have exhausted $^4$He in their cores (and which therefore have masses greater than $0.8\,M_\odot$), core temperatures can reach $T_9 = 0.8 - 1.1$. Then another nucleosynthetic process, called carbon burning, becomes possible:

$$^{12}\mathrm{C} + {}^{12}\mathrm{C} \rightarrow {}^{20}\mathrm{Ne} + {}^{4}\mathrm{He} \quad \text{(II.125)}$$

$$^{23}\mathrm{Na} + \mathrm{p} \quad \text{(II.126)}$$

$$^{23}\mathrm{Mg} + \mathrm{n}\ \text{(very slow)} \quad \text{(II.127)}$$

The protons, alpha particles, and neutrons liberated by carbon burning are all very reactive at such high temperatures. Hydrogen fuses at once via reactions such as

$$^{12}\mathrm{C} + p \rightarrow {}^{13}\mathrm{N} + \gamma \quad \text{(II.128)}$$

$$^{13}\mathrm{N} \rightarrow {}^{13}\mathrm{C} + e^+ + \nu \quad \text{(II.129)}$$

$$^{13}\mathrm{C} + {}^{4}\mathrm{He} \rightarrow {}^{16}\mathrm{O} + \mathrm{n}. \quad \text{(II.130)}$$

Such high-temperature proton addition processes are called the p-process. This process alone supplies a large number of neutrons to participate in s-process nucleosynthesis. The only important products of carbon burning are $^{20}$Ne, $^{23}$Na, and $^{24}$Mg, which are produced with the same relative abundances observed in solar system material.

A faster neutron addition process (called the rapid process, or r-process) will, as we have seen, follow a trajectory farther to the $N$-rich side of the stability line. In order to produce the second, higher-$Z$ component of each of the abundance peaks, an r-process contribution with a time scale of 0.1 to 1 s would be required!

The abundance data summarized in Fig. II.23 are compatible with the idea that the heavy elements were formed in two very different types of neutron-rich environment and then mixed in roughly equal proportions at or before the time of origin of the Solar System. There is no evidence for a continuum of different time scales for neutron addition, but rather for one operating on a time scale of $10^3$ or $10^4$ years and the other on a scale of about 1 s. The latter, because of its extremely short time scale, must be characteristic of explosive nucleosynthesis.

Theoretical studies of nuclear reactions during stellar explosions have been most fruitful. It is now possible to sketch out with some confidence the nucleosynthetic mechanisms that gave rise to the present abundances of the elements. The key to understanding explosive nucleosynthesis lies in isolating the classes of nuclear reactions that are involved in triggering violently unstable burning in stellar interiors.

Perhaps the first suggested mechanism which could explain such an instability was the discovery of the violent temperature excursions and structural rearrangements that took place at the beginning of helium burning (the helium flash). In such a core, the density is so high that the electrons are forced to spin-pair. Once the temperature increase caused by the initiation of helium burning has begun, the core warms and dilates, and the degeneracy of the electron gas is lifted. The resultant increase in electron pressure helps to dilate the core further, until the rate of energy generation by helium burning (which is proportional to the cube of the density) drops into a stable steady-state balance with the transport of heat out of the outer envelope of the star. These stable helium-burning stars are the "giant" stars.

A much more serious instability is believed to occur in the interiors of very highly evolved red giant stars that have very high core temperatures. In their cores, the pressures and temperatures are so large that nuclear equilibrium has largely been attained. Reactions have progressed through hydrogen and helium burning, the alpha process, and carbon burning, until the innermost core of the star is largely composed of iron-group elements. At slightly higher pressures, it becomes possible to pack the nucleons more closely by arranging them into the densest nuclide of all, $^4$He, plus assorted free neutrons and neutrinos.

At about the same time, temperatures are so high (near $T_9 = 2$) that photon–photon interactions can synthesize electron–positron pairs with attendant production of neutrinos. Both processes result in the loss of vast floods of energetic neutrinos from the very center of the star. The penetrating power of the neutrinos is so great that the energy is lost from the star with almost perfect efficiency. The core suddenly cools, electron degeneracy greatly increases, the pressure further drops, the core further collapses, and the entire process runs away. The entire remainder of the star then collapses inward in free fall, because the core pressure is no longer adequate to support the overlying layers.

At the surface of each shell of the star, wherever nuclear burning reactions are taking place, the collapse process causes an almost instantaneous heat pulse due to the conversion of the gravitational potential energy into compression heating. The temperature surge in the falling material may be a factor of 2 or more. Because the rates of many of the nuclear reactions are proportional to the 20th to the 30th power of temperature, all the nuclear reactions in the star run away at once. Catastrophic changes in the composition of the star occur on a time scale of about a second, with the liberation of vast amounts of energy. As all the burning shells detonate, the inner layers are crushed into the degenerate core while the outer layers are hurled out into space at a few percent of the speed of light.

Such an explosion, which ejects a significant percentage of the mass of a highly evolved star into the interstellar medium at very high speeds, is called a supernova. Such an exploding star briefly shines with a luminosity that may be greater than that of an entire galaxy. The minimum mass required for a star to evolve into a supernova is about $9\,M_\odot$, so its MS luminosity was about $9^{3.5} = 2200\,L_\odot$. The largest possible MS star is surely less than $100\,M_\odot$, with the limit probably close to $60\,M_\odot$, with a luminosity of over $10^6\,L_\odot$. This, then, is the range of presupernova stars.

Theoretical studies of the explosive burning processes that occur within the various compositional shells of a highly evolved star have successfully reproduced almost every feature of the observed abundances of the elements. A number of astrophysicists have examined the complex reaction networks attendant upon heating a carbon shell to temperatures well above the normal temperature of stable burning, near $T_9 = 2$ at a density of $10^5\ \mathrm{g\,cm^{-3}}$ or more.

The initial reactions include

$$^{12}\mathrm{C} + {}^{12}\mathrm{C} \rightarrow {}^{23}\mathrm{Na} + \mathrm{p} \qquad \text{(II.131)}$$

$$\rightarrow {}^{20}\mathrm{Ne} + {}^{4}\mathrm{He}, \qquad \text{(II.132)}$$

as well as endothermic reactions driven by the absorption of the energy of the heat pulse, such as

$$^{12}\mathrm{C} + {}^{12}\mathrm{C} \rightarrow {}^{23}\mathrm{Mg} + \mathrm{n}. \qquad \text{(II.133)}$$

For slow, stable isothermal carbon burning (at $T_9$ near 1), the neutrons are eventually captured by $^{23}$Mg, so that the net result is the production of only $^{20}$Ne, $^{23}$Na, and $^{24}$Mg in insignificant amounts.

When carbon is burned explosively (at $T_9 = 2$), however, the enormous energy released by the reactions explodes the carbon shell on a time scale of 0.1 s, and the rapid expansion cools and quenches the reaction intermediates long before complete burning can occur. Reactions of the large fluxes of energetic protons, alpha particles, and neutrons with heavier nuclei are also quenched at an intermediate stage. Arnett and Clayton have shown that explosive burning of roughly equal amounts of $^{12}$C and $^{16}$O (the dominant products of He burning) with a trace of $^{18}$O produces a very satisfactory match of the observed $^{20}$Ne, $^{23}$Na, $^{24}$Mg, $^{25}$Mg, $^{26}$Mg, $^{27}$Al, $^{29}$Si, $^{30}$Si, and $^{31}$P abundances. Explosive oxygen burning supplies the isotopes $^{28}$Si, $^{32}$S, $^{33}$S, $^{34}$S, $^{35}$Cl, $^{37}$Cl, $^{36}$Ar, $^{38}$Ar, $^{39}$K, $^{K40}$Ca, $^{41}$K, $^{42}$Ca, and $^{46}$Ti.

The predicted abundances of the neutron-rich elements are sensitive to the assumed abundance of $^{18}O$, which is a product of helium burning in the presence of $^{14}N$, which in turn is a final product of processing of C, N, and O during the final stages of hydrogen burning. Thus the CNONe group abundances in the young star during its hydrogen-burning phase leave a characteristic mark on the products of explosive carbon and oxygen burning at a far later stage of its history. The observed abundances of the n-rich isotopes $^{36}S$, $^{40}Ar$, $^{40}K$, $^{46}Ca$, $^{48}Ca$, $^{50}Ti$, $^{64}Ni$, $^{66}Zn$, $^{70}Zn$, and $^{72}Ge$ suggest that the original CNONe group abundances in the young star were characteristic of Pop I stars. Hence the immediate supernova progenitor of the Sun was itself at least a second-generation star, and the Sun is at least a third-generation star.

As $^{12}C$ and $^{16}O$ are depleted in explosive carbon and oxygen burning the dominant process becomes silicon burning. This process, even more complicated than its predecessors, is almost solely responsible for the synthesis of the ejected iron-group nuclides.

Silicon burning is initiated by the thermal gamma radiation from the initial temperature pulse. The thermal gamma rays photodisintegrate $^{28}Si$ to make alpha particles, which are very reactive in this hot a medium. They react immediately with $^{28}Si$:

$$^{28}\mathrm{Si} + {}^{4}\mathrm{He} \rightarrow {}^{32}\mathrm{S} \tag{II.134}$$

$$^{32}\mathrm{S} + {}^{4}\mathrm{He} \rightarrow {}^{36}\mathrm{Ar} \tag{II.135}$$

$$^{36}\mathrm{Ar} + {}^{4}\mathrm{He} \rightarrow {}^{40}\mathrm{Ca} \tag{II.136}$$

$$^{40}\mathrm{Ca} + {}^{4}\mathrm{He} \rightarrow {}^{44}\mathrm{Ti}. \tag{II.137}$$

Reactions such as

$$^{44}\mathrm{Ti} + {}^{4}\mathrm{He} \rightarrow {}^{47}\mathrm{V} + \mathrm{p} \tag{II.138}$$

$$^{42}\mathrm{Ca} + {}^{4}\mathrm{He} \rightarrow {}^{45}\mathrm{Ti} + \mathrm{n} \tag{II.139}$$

provide protons and neutrons for further equilibration with the alpha-multiple nuclei. The reactions of these small particles with heavy nuclei are enormously more rapid than direct reactions between heavy nuclei because of their much smaller Coulomb barrier. But each of the alpha-multiple nuclei is more easily photodisintegrated than $^{28}Si$ itself. Thus both forward and backward processes are kinetically feasible, and a "quasiequilibrium" can be established.

In effect, then, explosive silicon burning can approach the product distribution of the idealization called the equilibrium (e) process. Several authors have investigated explosive silicon burning in great detail. They have shown that burning at $T_9 = 4.5$ for no more than 1 second produces excellent agreement with observed elemental abundances up to $A = 59$.

There still remains the question of how nuclides much heavier than these are produced in stellar explosions. A clue to the nature of this process is the observation that the heavy nuclides are all neutron-rich and that the rapid neutron addition (r) process is essential in explaining many features of their abundance pattern. All we need do now is discover a sufficiently powerful neutron source that can operate on the time scale of the supernova explosion.

So far our attention has been focused on explosive carbon, oxygen, and silicon burning. Let us now consider the effects of a strong heat pulse, induced by collapse of the stellar core, on the outer hydrogen- and helium-burning shells of a star.

In the hydrogen-burning region of a Pop I star a temperature excursion to $T_6 = 100$ would permit the rapid burning of the ubiquitous traces of $^{12}C$, $^{16}O$, and $^{20}Ne$ by

$$^{12}\mathrm{C} + \mathrm{p} \rightarrow {}^{13}\mathrm{N} \tag{II.140}$$

$$^{16}\mathrm{O} + \mathrm{p} \rightarrow {}^{17}\mathrm{F} \tag{II.141}$$

$$^{20}\mathrm{Ne} + \mathrm{p} \rightarrow {}^{21}\mathrm{Na}, \text{ etc.,} \tag{II.142}$$

with the release of about 2 MeV per proton, or about $10^{17}$ erg g$^{-1}$, for a kinetic energy equivalent to expansion at 4000 km s$^{-1}$. Because the radius of the explosion envelope is initially $10^4$ to $10^5$ km, the characteristic time scale for the explosion must be around 10 s. Thus the radioactive nuclides produced by proton addition will decay within the duration of the explosion only if their decay half-lives are very short.

Among the prominent p-process nuclides,

$$^{13}\mathrm{N} \rightarrow {}^{13}\mathrm{C} + \mathrm{e} + \bar{\nu}\ (t_{1/2} = 10\,\mathrm{min}) \tag{II.143}$$

will not decay appreciably in the time available,

$$^{17}\mathrm{F} \rightarrow {}^{17}\mathrm{O} + \mathrm{e} + \nu\ (t_{1/2} = 1.1\,\mathrm{min}) \tag{II.144}$$

will undergo significant decay, and

$$^{21}\mathrm{Na} \rightarrow {}^{21}\mathrm{Ne} + \mathrm{e} + \nu\ (t_{1/2} = 23\,\mathrm{s}) \tag{II.145}$$

will be substantially reduced by decay. It happens that $^{21}Ne$ is very reactive at these high temperatures and will react within about 1 s after its formation by

$$^{21}\mathrm{Ne} + {}^{4}\mathrm{He} \rightarrow {}^{24}\mathrm{Mg} + \mathrm{n}, \tag{II.146}$$

thus generating about one neutron for every four $^{20}Ne$ nuclei present before the explosion. The neutron capture cross-sections of the iron-group elements are so high that they will absorb neutrons instantaneously until a beta-active isotope with a half-life of about 0.1 s or less is produced. Then beta decay and neutron addition will continue as indicated in Fig. II.24. The production of transuranium isotopes such as $^{254}Cf$ by this r-process path is entirely possible within the time scale of the explosion. Table II.5 summarizes our present understanding of the sources of certain selected nuclides.

**Table II.5** Nucleosynthetic Sources of Selected Nuclides after Cameron (1982)

| Nuclide | Abundance[a] | Sources[a] | Nuclide | Abundance[a] | Sources[a] |
|---|---|---|---|---|---|
| $^{1}H$ | $3.18 \times 10^{10}$ | BB | $^{40}K$ | 0.49 | O, Si |
| $^{2}H(D)$ | $5.2 \times 10^{5}$ | BB | $^{40}Ca$ | $6.99 \times 10^{4}$ | O, Si |
| $^{3}He$ | $3.7 \times 10^{5}$ | BB | $^{54}Fe$ | $4.83 \times 10^{4}$ | Si, e |
| $^{4}He$ | $2.21 \times 10^{9}$ | BB, pp, CC | $^{56}Fe$ | $7.61 \times 10^{5}$ | Si, e |
| $^{7}Li$ | 45.8 | p | $^{58}Fe$ | 2740 | e |
| $^{9}Be$ | 0.8 | spallation | $^{75}As$ | 6.6 | s, r |
| $^{11}B$ | 281.3 | p | $^{80}Se$ | 33.5 | s |
| $^{12}C$ | $1.17 \times 10^{7}$ | $\alpha$(He) | $^{79}Br$ | 6.82 | r, s |
| $^{14}N$ | $3.63 \times 10^{6}$ | pp, CC | $^{81}Br$ | 6.68 | r, s |
| $^{16}O$ | $2.14 \times 10^{7}$ | $\alpha$(He) | $^{82}Kr$ | 5.41 | s |
| $^{20}Ne$ | $3.06 \times 10^{6}$ | $\alpha$(C) | $^{84}Kr$ | 26.6 | s |
| $^{23}Na$ | $6.0 \times 10^{4}$ | C | $^{130}Xe$ | 0.229 | s |
| $^{27}Al$ | $8.5 \times 10^{5}$ | C | $^{132}Xe$ | 1.40 | r, s |
| $^{28}Si$ | $9.22 \times 10^{5}$ | O | $^{197}Au$ | 0.202 | r |
| $^{31}P$ | 9600 | O | $^{206}Pb$ | 0.753 | s |
| $^{32}S$ | $4.75 \times 10^{5}$ | O, Si | $^{207}Pb$ | 0.824 | s |
| $^{35}Cl$ | 4310 | O, Si | $^{208}Pb$ | 2.34 | s |
| $^{36}Ar$ | $9.87 \times 10^{4}$ | O, Si | $^{235}U$ | 0.0063 | r |
| $^{39}K$ | 3910 | O, Si | $^{238}U$ | 0.0199 | r |

[a] Abundances relative to $10^6$ atoms of silicon. BB denotes the Big Bang, pp is the proton–proton chain, CC is the carbon cycle, p is the proton-addition process, $\alpha$ is the alpha process, e is the equilibrium process, and s and r are slow and rapid neutron addition, respectively. C, O and Si denote explosive burning of these elements.

It is fascinating to look back over the modes of origin of the elements in stars with the composition of our own planet in mind. We may picture, as in Fig. II.26, the slow burning of hydrogen in a typical star extending over $10^9$ or $10^{10}$ years. This Main Sequence stage is followed by a briefer interlude about $10^8$ years in duration, during which the principal heat source in the star is helium burning and the ensuing alpha process. Then comes about $10^4$ years of the s-process in parallel with the changeover to a roughly $10^7$-year period of carbon, oxygen, and silicon burning. Then there is a brief ($10^3$ years?) period of stable e-process nucleosynthesis, which builds up an iron-group core.

The composition of the material of the star at this phase of its history has only a faint resemblance to the familiar cosmic abundances, and the various fusion products of these reaction sequences (pp; CC; He burning; alpha process; C, O, and Si burning; e process) are physically separated from each other in the interior of the star. Then some instability inherent in the evolution of massive stars, such as the core collapse scenario mentioned above, causes the star to blow itself apart in a few seconds and to reprocess all the nuclides in its envelope.

The light from the explosion may for a few days outshine the rest of the Galaxy in which the star resides (Fig. II.27). Some substantial portion of the total mass of the star is accelerated to relativistic velocities by the explosion and injected into the interstellar medium. This material then mixes with cold interstellar matter, which ultimately collapses again through a dense GMC stage to form a cluster of next-generation stars. One of the stars made in such a way is our own Sun. The most basic compositional properties of the Earth were shaped in a few seconds of furious thermonuclear explosion, the radioactive debris of which is the clock by which we measure geological time.

## Nuclear Cosmochronology

Certain heavy nuclides (mostly with $A > 56$) produced by explosive nucleosynthesis are radioactive, with very long half-lives. If a naturally occurring radionuclide has a long enough half-life so that its abundance is still measurable, has a short enough half-life so that it can be determined experimentally, and decays to an isotope that is not already very much more abundant than itself, then the decay of that isotope can be used as a clock for measuring geological spans of time.

Consider, for example, the decay of $^{187}$Re by beta emission:

$$^{187}\mathrm{Re} \rightarrow {}^{187}\mathrm{Os} + \mathrm{e} + \bar{\nu} \quad (t_{1/2} = 5 \times 10^{10}\ \text{years}). \qquad \text{(II.147)}$$

Ancient solar-system materials containing rhenium and osmium, such as most classes of meteorites, especially those that have not been subjected to melting or other forms of chemical fractionation since they first condensed as solids, can be used for rhenium–osmium dating. From model calculations of s-process and r-process nucleosynthesis in which the abundances of over 100

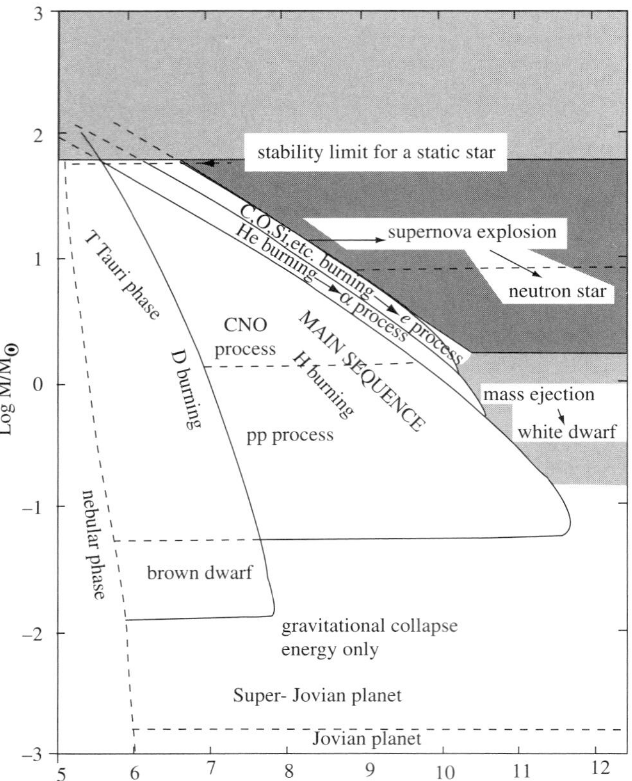


Figure II.26 Evolutionary histories of solar-composition (Population I) bodies. This diagram should serve as a reminder of many of the phenomena discussed in this chapter, but omits much interesting detail. The most massive stable star must be at least 50 solar masses, and surely less than 200 solar masses. Larger stars would have internal radiation pressures so high that the outer envelope would be blown off. More massive stars may exist transiently, but cannot reside on the MS.

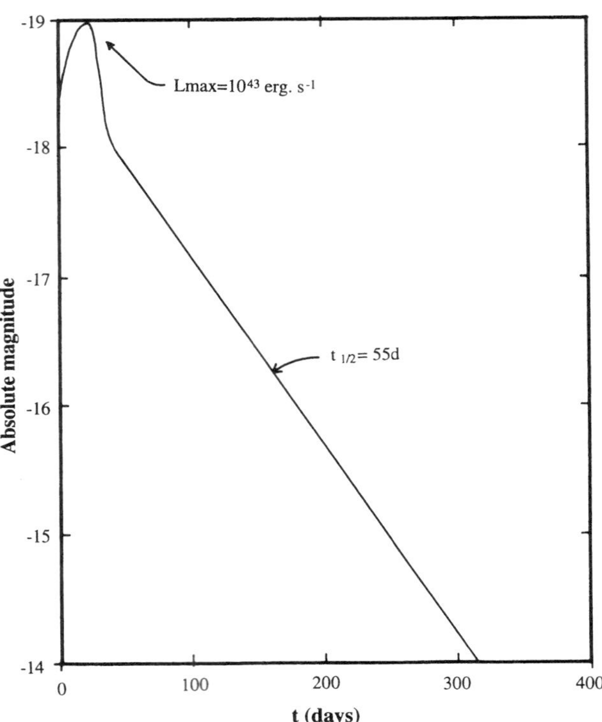


Figure II.27 Lightcurve of a supernova. After about 30 days the light decays exponentially. The luminosity of a fireball would drop more rapidly than this exponential, and a heat source (presumably a radionuclide) in the ejecta would explain the observed lightcurve. The most plausible nuclide (if this is indeed the source of the energy) would be $^{256}$Cf.

other nuclides are also fit, estimates can be made of the original isotopic composition and abundance of both Re and Os. From the study of this system, it can be shown that about $11 \times 10^9$ years has elapsed since the cessation of major nucleosynthesis of present solar-system material by the s-process. Roughly 15% of the mass of the heavy elements in the Solar System has passed through r-process events since that time.

Another, more familiar, example is dating by the beta decay of $^{87}$Rb:

$$^{87}\mathrm{Rb} \rightarrow {}^{87}\mathrm{Sr} + \mathrm{e} + \overline{\nu} \quad (t_{1/2} = 5.2 \times 10^{10}\,\text{years}). \quad \text{(II.148)}$$

Because of the great difference in geochemical behavior between $Rb^+$ and $Sr^{2+}$, the interpretation of rubidium–strontium ages is slightly more complex than for the rhenium–osmium system. Figure II.28 shows how the $^{87}$Sr abundance would evolve with time for four different minerals, A–D, with different initial elemental abundance ratios of Rb and Sr, but with identical initial rubidium and strontium isotopic compositions. Mineral A, with the least Rb content, will therefore show little $^{87}$Sr production, whereas mineral D, rich in rubidium (and hence $^{87}$Rb), will show a large $^{87}$Sr increase at time $t'$. The line $t'$, which links minerals of different compositions but equal ages, is called an isochron. Any isochron extrapolated to Rb/Sr = 0 (the case of no radiogenic contribution of $^{87}$Sr from $^{87}$Rb decay) gives the primordial $^{87}$Sr abundance. The slope of the isochron increases steadily with the elapsed time since the formation of the elements.

The interpretation of such measurements on primitive samples (those which have undergone no chemical fractionation since the minerals came into existence) is therefore rather easy. Complications are introduced if melting and differentiation processes took place at some intermediate time.

Point E′ in Fig. II.28 shows the composition of a melt formed at $t'$ by complete melting of A, B, C, and D, with resulting complete isotopic homogenization. Figure II.28b shows the consequences of recrystallizing this melt and allowing the newly recrystallized minerals

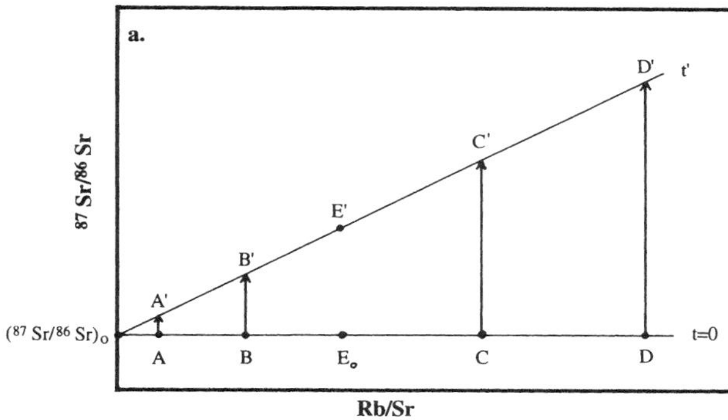


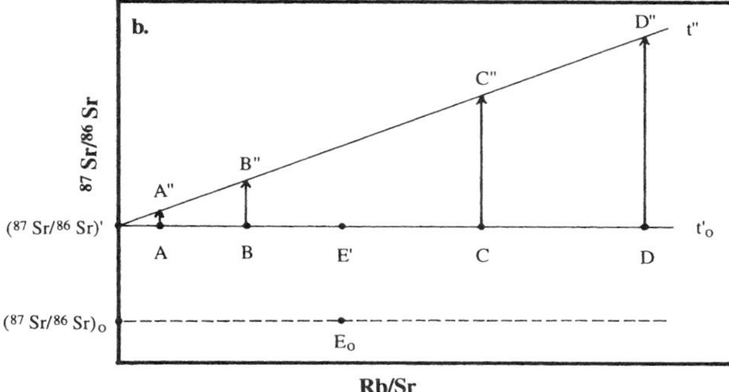


**Figure II.28** Rubidium–strontium dating. An ancient magmatic event $E_0$, with the $^{87}Sr/^{86}Sr$ and Rb/Sr ratios given in a, resulted in the crystallization of four different minerals, A–D, with different elemental compositions but essentially identical isotopic compositions. After time $t'$ the decay of $^{87}Rb$ in each mineral has enhanced the $^{87}Sr$ abundance in proportion to total Rb content. At time $t'$ the isotopic compositions of the minerals will be A′-D′, all lying on a straight line (the isochron) passing through the initial strontium isotope ratio. Remelting of the whole rock in event E′ at time $t'$ gives a homogeneous magma of composition E′. Recrystallization of E′ into minerals A–D and subsequent radioactive decay of $^{87}Rb$ are pictured in b. Note that the new isochron at time $t''$ intersects the Rb/Sr = 0 axis at a new intercept, the location of which fixes E′. E′ in turn gives the *whole rock* age $t'$ *if* the original strontium isotopic composition is known.

to remain without further disturbance until time $t''$. Note that the isotopic homogenization caused by melting event E resets the Rb–Sr clock so that the slope of the isochron thereafter reflects the time elapsed since $t'$, not the original formation time. However, information about the total elapsed time since condensation is preserved, because the new isochron intersects the zero-rubidium line at a new location. If we have an accurate theoretical estimate of the initial $^{87}Sr{:}^{86}Sr$ ratio from nucleosynthesis models, a meteorite that is almost totally devoid of rubidium, or a meteorite that has never been subjected to melting and isotopic homogenization, then we can date the time since the end of nucleosynthesis and the time since initial condensation of the primitive solids.

Study of the Rb/Sr system again shows that the elements making up most of the Solar System are about $11 \times 10^9$ years old, whereas the Rb- and Sr-bearing minerals in most primitive meteorites first condensed about $4.5 \times 10^9$ years ago.

Another very important decay system is

$$^{40}K \rightarrow {}^{40}Ca + e + \nu \qquad (II.149)$$

$$\overset{e.c.}{\rightarrow} {}^{40}Ar + \nu \qquad (II.150)$$

$$\rightarrow {}^{40}Ar + e^+ + \nu. \qquad (II.151)$$

The production of $^{40}Ca$ is useless as a radioactive decay clock because $^{40}Ca$ is a very common alpha-process nuclide whose abundance is not measurably changed by $^{40}K$ decay. The other decay product, $^{40}Ar$, is a gas that can accumulate only in relatively cool mineral grains. Moderate heating will cause complete loss of the radiogenic argon from the grain by diffusion. The $^{40}Ar$ gas-retention ages of the large majority of all meteorites cluster closely around $4.6 \times 10^9$ years, and none are found to be appreciably older.

Several additional dating methods are made possible by the decay of the lanthanide elements thorium and uranium, which can be synthesized only in supernova explosions. Decay can occur by spontaneous fission or by alpha emission. For the present, it will suffice for us to consider the alpha decay mode.

It can be seen from Fig. II.23 that alpha decay of a heavy nuclide causes it to depart from the stability line into the beta-emission instability zone. Thus the emission of a series of alpha particles by a heavy nuclide will be accompanied by interspersed beta emission events as well. Decay of the isotopes of uranium and thorium by such alpha and beta decay chains continues until a stable nuclide is produced, in these cases, isotopes of lead:

$$^{232}Th \rightarrow {}^{208}Pb + 6\,{}^4He + e + \nu \quad (t_{1/2} = 8.8 \times 10^9 \text{ years}) \qquad (II.152)$$

$$^{235}U \rightarrow {}^{207}Pb + 7\,{}^4He + e + \nu \quad (t_{1/2} = 1.8 \times 10^9 \text{ years}) \qquad (II.153)$$

$$^{238}U \rightarrow {}^{206}Pb + 8\,{}^4He + e + \nu \quad (t_{1/2} = 1.0 \times 10^{10} \text{ years}). \qquad (II.154)$$

The U and Th contents can be determined by very sensitive radiochemical techniques, and the trapped $^4He$ content is measured by melting the sample in a vacuum and analyzing the released gases with a mass spectrometer.

It has been found that most meteorites have $^4He$ gas-retention ages that are nicely concordant with the $^{40}Ar$ gas-retention ages at about $4.6 \times 10^9$ years. Some meteorites that show optical evidence of severe mechanical shock damage have concordant but smaller gas-retention ages, and the few meteorites that do not show concordance are always depleted in $^4$He relative to $^{40}$Ar. This is exactly what would be expected to result from diffusive gas loss at moderate temperatures, because helium diffuses much more rapidly than argon. In samples from which gas loss has occurred, the most effective dating technique may be to examine the relative abundances of the lead isotopes by solid-source mass spectrometry.

Several other dating techniques involving very long-lived radioactive parents are discussed in Chapter VIII.

Finally we come to a somewhat more subtle dating technique that is of great interest for other purposes. The radioisotope $^{129}$I decays by beta emission with a half-life of $1.7 \times 10^7$ years:

$$^{129}\mathrm{I} \rightarrow {}^{129}\mathrm{Xe} + \mathrm{e} + \nu. \tag{II.155}$$

Iodine may be retained by certain condensing solids, such as sulfides. However, if the time elapsed between the last r-process production of radioiodine and the condensation of mineral grains was more than a few half-lives long, the iodine would have decayed to xenon. Because xenon does not condense to form minerals, the amount of radiogenic $^{129}$Xe trapped in the grains can be used as a measure of the time gap between the end of nucleosynthesis and the formation of solid presolar grains.

This time interval is found to be 150 to 250 million years for a variety of meteorite types. Interestingly, radiogenic $^{129}$Xe is still being released in tiny traces from Earth's interior.

Using all of the presently available information on the Hubble constant, stellar evolution, and nucleosynthesis, we can sketch out the general history of the evolution of solar system matter shown in Fig. II.29. An essential element of this scheme is the principle of recycling of matter with a resulting increase in the chemical complexity of the Universe. The role of supernova explosions in this recycling process is brought out schematically in Fig. II.30. That figure also hints at the compositional classes of planets which might be formed as companions of first-, second-, and later-generation stars.

We now have defined the broad astronomical context within which the Solar System formed, evolved, and exists today. We have seen that Solar System material is

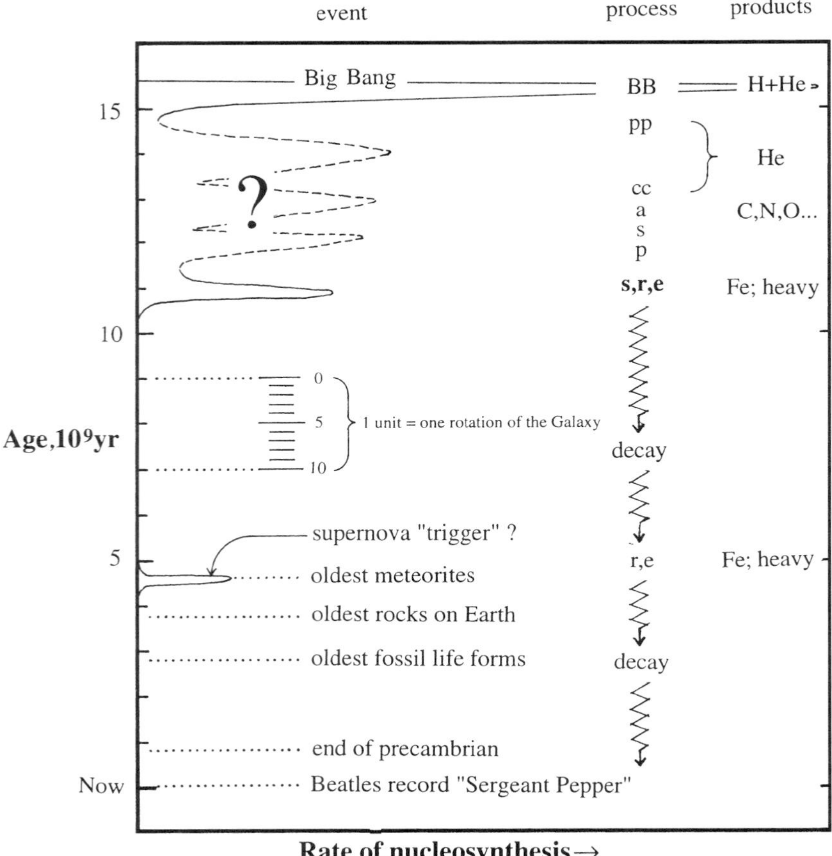


**Figure II.29** Chronology of solar-system material. The general successfully gives birth to the specific.

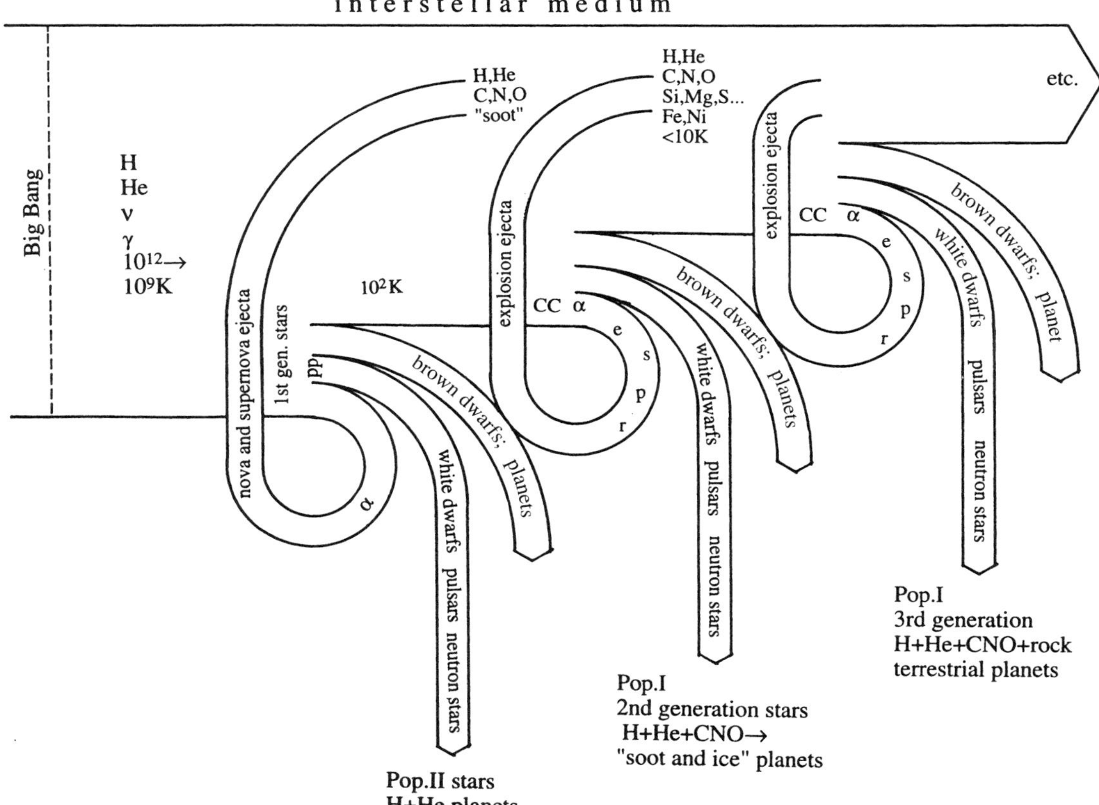


**Figure II.30** Recycling plan.

the culmination of several rounds of cyclic processing involving interstellar gas, collapsing nebulae, hierarchical sequences of nuclear reactions in stellar interiors, cataclysmic disruption of highly evolved massive stars, and fecundation of the interstellar medium by mixing of cold gases with radioactive and chemically complex supernova debris. Although the direct yield of heavy elements is small, these recycling processes act to build up concentrations of both the rock-forming elements and the radioactive heat sources that power the evolution of Earth-like planets.

> Universe has no pollution.
> All the chemistries of the Universe are essential
> To its comprehensive self regulation.
>
> Buckminster Fuller

## Exercises

*Distance Scales in the Universe*

II.1 An amateur astronomer with a small telescope sees a star that is spectrally identical to the Sun, but $10^{12}$ times fainter than the Sun. She has no equipment for measuring parallaxes, but wishes to estimate the parallax of the star. How might she do it?

*The Big Bang*

II.2 High-energy photons are easy to detect from an observatory in space, but infrared photons require more specialized equipment such as cooled detectors. An astronomer wishes to measure the total energy output of a star with a temperature of 10,000 K with an error (due to "missed" photons of long wavelength) of less than 1% of the total incident flux. According to the Planck function, to how long a wavelength must he observe in order to ensure that he covers at least 99% of the flux? *Note*: this problem is to be solved by numerical integration on a computer. Attach your printout.

II.3 A gas in equilibrium with a Planckian radiation field is so hot that photons with the most probable energy are capable of creating electron–positron pairs. What is the temperature of the gas?

II.4 A gas in equilibrium with a Planckian radiation field is hot enough so that only one photon in a million has enough energy to make an electron. Compare the rate of pair production to that in Problem II.3.

II.5 A nuclear explosion in space produces $10^{30}$ free neutrons. After 1 day, how many of the neutrons have not yet decayed?

II.6 Estimate the temperature at which the most-probable photon energy is just sufficient to destroy a D nucleus by photodisintegration.

II.7 How much energy is produced by the nuclear reaction $^{2}H + {}^{10}B \rightarrow {}^{12}C$? Give the answer in MeV per event, grams per gram, and kW hr per tonne. *Note*: the metric tonne is 1000 kg, or 1 Mg.

*Galaxy and Star Formation*

II.8 The derivation of the Jeans criterion assumes dramatically oversimplified initial conditions. Give a qualitative discussion of how the gas behavior might be altered if the gas is dusty and opaque rather than very transparent, as was assumed in the derivation.

II.9 The $M{:}L$ ratio of an assemblage of stars is strongly influenced by the kinds of stars found in it. Compare the $M{:}L$ ratios of a galaxy that consists of B-type Main Sequence stars and a galaxy that contains only M-type Main Sequence stars. Consult Figs. II.7 and II.10 for data.

*Classification of Stars*

II.10 Traditional naked-eye astronomers established a brightness scale for stars and other heavenly bodies, in which the brightest stars were said to be "stars of the first magnitude," those discernibly fainter were "stars of the second magnitude," and so on, down to the limits of visibility to the naked eye, which were "stars of the sixth magnitude." Modern physiologists have found that the human response to many stimuli is logarithmic, not linear, and that the first-magnitude stars were about a factor of 100 times as bright as the sixth-magnitude stars. "Bright" here refers to the flux of visible light from the star in erg $cm^{-2}\,s^{-1}$ as received by the observer. Therefore each magnitude step in the scale represents a ratio in light flux of about a factor of $100^{1/5}$. Astronomical magnitudes are now expressed in terms of a scale in which each magnitude is exactly a factor of $100^{1/5}$, or 2.512...

a. A telescope's field of view contains both Venus, which is at that time at magnitude $-1.0$, and a faint distant field star of magnitude $+24$. What is the intensity of the light flux from Venus relative to that from the star?

b. A red dwarf star orbits about a red giant star that emits 620,000,000 times as much visible light. What is the magnitude difference between these two stars?

II.11 A star with an apparent visual magnitude of 9.6 has a parallax of 0.6 arc seconds (0.6″). What is its visual luminosity?

II.12 Two stars of the same spectral class and mass follow circular orbits about their mutual center of gravity. The diameter of their orbits is 4.3″ and their parallax is 0.2″. How far are they from the center of mass (barycenter) of their binary system?

II.13 A star with an effective temperature of 10,000 K has a luminosity of $10^{30}$ erg $s^{-1}$.

a. What is its radius?

b. What spectral and taxonomic class would you assign it to?

II.14 As early as the 1950s, Luyten showed that within 10 pc of the Sun there is a total stellar mass of 268 Suns and a total luminosity of 235 Suns.

a. Are we justified from these data in concluding that the "typical" star is very much like our Sun? Explain your answer.

b. Why is the concept of an "average star" misleading?

*Stellar Evolution*

II.15 The O-type and M-type MS stars differ enormously in mass, luminosity, color (temperature), and Main Sequence lifetime, but all operate by fusing hydrogen and giving off light. Suppose a typical MS star uses up (fuses) 50% of its hydrogen fuel to helium over its MS lifetime. What fraction of that star's original mass is radiated off as light?

*Star Clusters*

II.16 Rapidly rotating stars with exceptionally strong magnetic fields are especially good at trapping charged particles and transferring angular momentum to escaping particles out to very great distances. Suppose a star with a radius of 800,000 km and a rotation period of 25 hours is able to force ions to corotate with the star out to a distance of $1000R_*$. How fast would the corotating ions be traveling at that distance?

*Stellar Origins*

II.17 An M-class MS star with a 0.1 solar mass, a luminosity of $10^{-4}$ Suns, and a 0.1 solar radius was formed by collapse of an interstellar cloud. How long could it maintain its MS luminosity during its formation phase using only its gravitational potential energy as a source?

*Stellar Explosions and Nucleosynthesis*

II.18 Equation (II.84) shows that the CC process generates positrons. These positrons must annihilate upon collision with electrons, giving an additional source of energy. What fraction of the total energy output from the CC cycle is contributed by these annihilation reactions?

II.19 What fraction of the radioisotope $^{50}$V present at the time of the origin of the Solar System $4.6 \times 10^9$ years ago has decayed since then?

II.20 Natural nuclear explosions (supernovas) can achieve relativistic expansion velocities. Suppose that a rocket engine could be built that draws its power from the fusion of $^3$He and D. Using the data in Table II.2, and making the simplifying (and overly generous) assumption that all the fusion energy release appears as kinetic energy of the products ($^4$He and H), what would the exhaust velocity of the rocket be?

II.21 Suppose a mineral containing uranium, but negligible primordial lead, was crystallized $4.6 \times 10^9$ years ago.

a. Using the data in Table II.5 and Eq. (II.153) and (II.154), calculate what the isotopic composition of uranium was at that time. (The data in Table II.5 refer to "modern" uranium and lead.)

b. Use a simple computer program to model the concentrations of $^{206}$Pb and $^{207}$Pb vs time. Plot the $^{207}$Pb/$^{206}$Pb ratio as a function of time and explain how it, along with the $^{208}$Pb/$^{206}$Pb and $^{208}$Pb/$^{207}$Pb ratios, can be used to provide three independent determinations of the age of the mineral sample.

c. When $^{208}$Pb is used in dating, what further information is required beyond that given above?

*Nuclear Cosmochronology*

II.22 Consider the decay of the radioisotope $^{87}$Rb to $^{87}$Sr in the presence of both primordial $^{87}$Sr and primordial $^{86}$Sr. Show how the isotopic composition of strontium evolves with time by deriving the equation $(^{87}\text{Sr}/^{86}\text{Sr})_t = (^{87}\text{Sr}/^{86}\text{Sr})_0 + (^{87}\text{Rb}/^{86}\text{Sr})(e^{\lambda t} - 1)$.

II.23 Consider two coexisting minerals, A and B, that have very different chemical composition but identical initial isotopic composition, in the same rock. The decay equations are $(^{87}\text{Sr}/^{86}\text{Sr})_{A,t} = (^{87}\text{Sr}/^{86}\text{Sr})_{A,0} + (^{87}\text{Rb}/^{86}\text{Sr})_{A,0}(e^{\lambda t} - 1)$ and $(^{87}\text{Sr}/^{86}\text{Sr})_{B,t} = (^{87}\text{Sr}/^{86}\text{Sr})_{B,0} + (^{87}\text{Rb}/^{86}\text{Sr})_{B,0}(e^{\lambda t} - 1)$. Solve this pair of equations explicitly for the age of the rock, $t$.

# III. General Description of the Solar System

## Introduction

Before addressing the full range of problems presented by our current knowledge of the Solar System it is important that we share some basic body of information on its members. This is particularly important because a great proportion of the "common knowledge" of the Solar System current only a few years ago is now known to be incorrect. We therefore will review in quite condensed form the basic data on the motions and gross physical properties of Solar System bodies, making extensive use of tables and graphs to make comparisons easier. It should be obvious that only a tiny minority of our present factual knowledge of the Solar System can or should be presented in this chapter; this is rather like the map in the front of a road map atlas that shows how the following detail maps fit together.

## The Sun

The Sun, which makes up about 99.9% of the mass of our planetary system, is a typical stable Main Sequence dwarf star of spectral class G2. It pursues an orbit about the center of the Galaxy with a radius of roughly 8 kpc and a period of about 200 million years.

The Sun rotates approximately every 26 days around an axis inclined 7° 10.5′ to the axis of the ecliptic (which is defined by the plane of the Earth's orbit). The rotation period varies with latitude: the shortest period, attained at the equator, is equal to 24.65 days. Spectroscopic velocity determinations indicate a rotation period approaching 34 days near the poles.

Dark features are discernible on the Sun's disk through even the smallest telescopes. These dark spots appear periodically in large numbers, at first at latitudes near ±30° and later toward the equator. These sunspots, which are commonly larger than Earth, are actually only slightly cooler than the surrounding gases. Sunspot outbreaks occur in cycles whose mean period is about 11 years. During eclipses of the disk of the Sun by the Moon, it can be seen that violent fountaining of gas streamers is associated with sunspot activity. These flares commonly span hundreds of thousands of kilometers and often develop velocities high enough to escape from the Sun.

The mass of the Sun is $1.99 \times 10^{33}$ g and its radius is $6.960 \times 10^{10}$ cm, giving a mean density of 1.41 g cm$^{-3}$. The surface gravity of the Sun is 27.9 Earth gravities, 272 m s$^{-2}$. The escape velocity from its surface is 618 km s$^{-1}$.

The luminosity of the Sun is $3.90 \times 10^{33}$ erg s$^{-1}$. From Eq. (II.53), we can calculate the effective temperature of the Sun's luminous surface (the *photosphere*) to be 5800 K.

The flux of solar radiation at the Earth's orbit, called the "solar constant," is usually given in somewhat awkward units as 1.940 cal cm$^{-2}$ min$^{-1}$ or 1370 W m$^{-2}$, equal to $1.37 \times 10^6$ erg cm$^{-2}$ s$^{-1}$. The surface temperatures of Solar System bodies, particularly those devoid of atmospheres, are determined by the amount of solar energy which they absorb, and hence accurate knowledge of the solar constant and its possible changes with time is of value.

Figure III.1 presents a low-resolution approximation of the emission spectrum of the Sun. The spectrum can be approximated as Planckian emission only over limited wavelength intervals. A simple black body interpretation of the solar spectrum is further confounded by the observation that the whole-disk spectrum of the Sun is discernibly different from the spectrum of the center of the disk. These effects occur because there is a vertical temperature gradient in the Sun's atmosphere, and the opacity of the gas is a function of both wavelength and temperature. Thus different wavelengths of observed light originate at different depths in the photosphere.

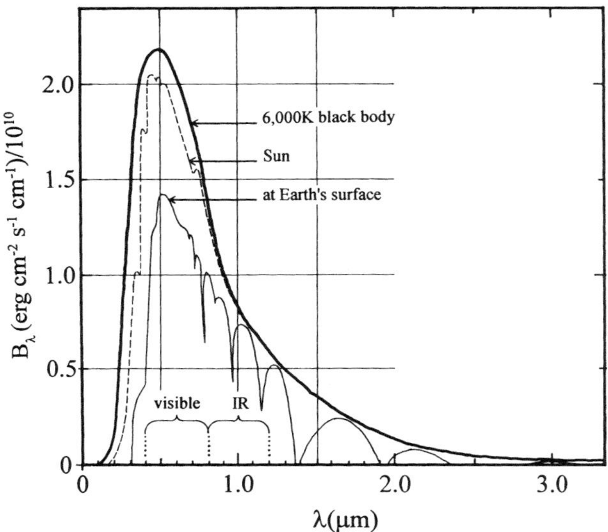


**Figure III.1** Intensity of solar radiation at Earth's orbit and at sea level. The Sun's emission spectrum contains countless sharp atomic absorption features too narrow to appear at this resolution. Absorption by Earth's atmosphere shortward of 300 nm (0.3 $\mu$m) is largely due to ozone. Absorption in the infrared is dominated by water vapor and carbon dioxide. Note the deviation of the solar spectrum from black body emission.

At some wavelengths, such as the middle infrared, the opacity is so great that we can see only a higher, thin layer of quite uniform temperature over the entire disk. At visible wavelengths the radiation originates at greater depth, and attenuation by the intervening atmosphere causes the edge of the disk to look cooler than the center. This effect is called "limb darkening." Curiously, observations in the ultraviolet show limb brightening, which shows that, at the altitudes observed in the UV, the temperature is increasing upward!

In addition to the limb darkening and limb brightening effects observed at UV, visible, and infrared wavelengths, there are easily observable permanent features of the solar spectrum that bear witness to the existence of a vertical temperature gradient: the solar spectrum, when viewed at moderate to high resolution, is found to be interrupted by countless narrow, dark absorption lines, called *Fraunhofer lines* after their discoverer. These lines are due to the presence of a layer of gas, somewhat cooler than the luminous photosphere, interposed in the line of sight. Observations of the edge of the solar disk during eclipses confirm that this layer, the *chromosphere*, is an integral part of the solar atmosphere.

Furthermore, eclipse observations reveal the presence of an extensive, highly rarefied *corona* that extends out at least several solar radii and whose size is directly related to the intensity of sunspot activity. For many years the temperature and composition of the corona were wholly unknown; however, it is now well established that the very peculiar emission lines observed in the corona are due to the familiar elements in extraordinarily highly excited states. The effective temperature of the corona deduced from the observed degree of electronic excitation and ionization must be at least $10^6$ K.

Both Earth-based observations of the streaming of comet tails and direct spacecraft observations in interplanetary space have revealed that a stream of protons, electrons, and heavier ions with velocities of a few hundred kilometers per second is constantly being emitted in all directions by the Sun. This flow of material, called the *solar wind*, emanates from the solar corona and forms an extension of it, extending out far beyond the orbits of the planets.

During intense sunspot activity the Sun emits protons and other charged particles with energies up to several hundred MeV. These particles are termed *solar cosmic rays*. The Sun also emits radio waves at all times, with loud bursts of radio noise often correlated with optically observed sunspot-related flare activity. An introduction to solar physics will be presented in Chapter IV.

## Orbits of the Planets

> People give ear to an upstart astronomer who tries to show that the Earth revolves, not the Sun and the Moon. This fool wishes to reverse the entire science of astronomy....
>
> Martin Luther

The literature on the dynamics of planetary motion is vast, fascinating, and of great historical importance in the evolution of the physical sciences. As early as the third century BC, the Greek philosopher Aristarchus maintained that the Sun was the center of the Solar System. But the Aristotelian synthesis that dominated medieval times, with its acceptance of the geocentric universe of Claudius Ptolemaeus, rejected motion of the Earth about the Sun on the grounds that no apparent motion of other stars could be detected, such as one would expect if we were viewing them from a moving planet.

The Polish astronomer Nikolas Kopernik (1473–1543), usually known by his Latinized name of Copernicus, revived the heliocentric theory of the Solar System. The observational and experimental sciences had long languished under medieval complacency and blind acceptance of Aristotle's dogmas, but Copernicus' revolutionary theory encouraged serious new observational work. Observations by the Danish astronomer Tyghe (Tycho) Brahe (1546–1601) and theory developed by the Johannes Kepler (1571–1630) of Germany provided a strong quantitative basis for the Copernican system. Kepler first formalized and organized the vast body of naked-eye observations of the planets into a precisely stated set of laws of motion, providing a firm descriptive basis for the Copernican heliocentric system. In the following two centuries the problems of dynamical astronomy stimulated much of the life work of Laplace, Bessel, Gauss, Lagrange, and Poincaré, provided the arena for the grandest successes of Newtonian physics, and more than any other single influence contributed to the mechanistic, deterministic world view of the 18th and 19th centuries.

The philosophical ideas of the framers of the American Declaration of Independence and Constitution were very strongly influenced by this view, as was the philosophical school of logical positivism. Marxist dialectic is axiomatically based on a mechanistic world view in which certain forms of social change are regarded as (that is, postulated without proof to be) inevitable manifestations of natural law. These mechanistic ideas, seen in their purest and most elegant form in Newtonian mechanics, still retain their beauty even in the eyes of a generation of physicists instructed in relativity and quantum mechanics and steeped in statistical concepts of causality.

Still, for nearly a century dynamics was regarded as a museum piece. It is only recently that developments in astronautics and radio and radar technology have reopened this field in both theoretical and experimental directions. Every chapter of this book bears testimony to the fruitful application of these ideas, both old and new, to a great range of Solar System phenomena. Because of the artificiality of excising all such discussions from their context and lumping them all together to make a very large (and decidedly heterogeneous) chapter on dynamics, we shall instead present here only a brief and skeletal outline of the classical theory of orbits. All special applications will appear in their appropriate context elsewhere in the book.

The laws of planetary motion as laid out by Johannes Kepler are:

I. Each planet executes a planar elliptical orbit about the Sun, with the Sun at one focus of the ellipse.

II. The area swept out by the radius vector from the Sun to the planet per unit time is a constant.

III. The square of the orbital period divided by the cube of the mean distance from the Sun is the same for all the planets.

The idea of elliptical orbits, especially for the case of Mars, allowed great improvements in simplicity of description compared to earlier systems. The observations of the motions of Mars by Tycho Brahe were good enough to permit description by a simple elliptical orbit, but not good enough to reveal the complicated perturbations of the orbit of Mars by the other planets.

The English physicist Sir Isaac Newton (1643–1727) argued that the gravitational force exerted by the Sun on each planet must be directed toward the center of the Sun, and hence the acceleration produced by this force must be directed along $-r$. He formulated for the first time a universal law of gravitation,

$$F = m\ddot{r} = -\frac{\mu m}{r^2}\hat{r}, \tag{III.1}$$

where $F$ is the gravitational force, $m$ is the mass of the planet, $r$ is the Sun-to-planet radius vector as shown in Fig. III.2, and each superscript dot represents a derivative with respect to time. The magnitude of the vector radius is written as $r$, $\mu$ is a constant for the Solar System

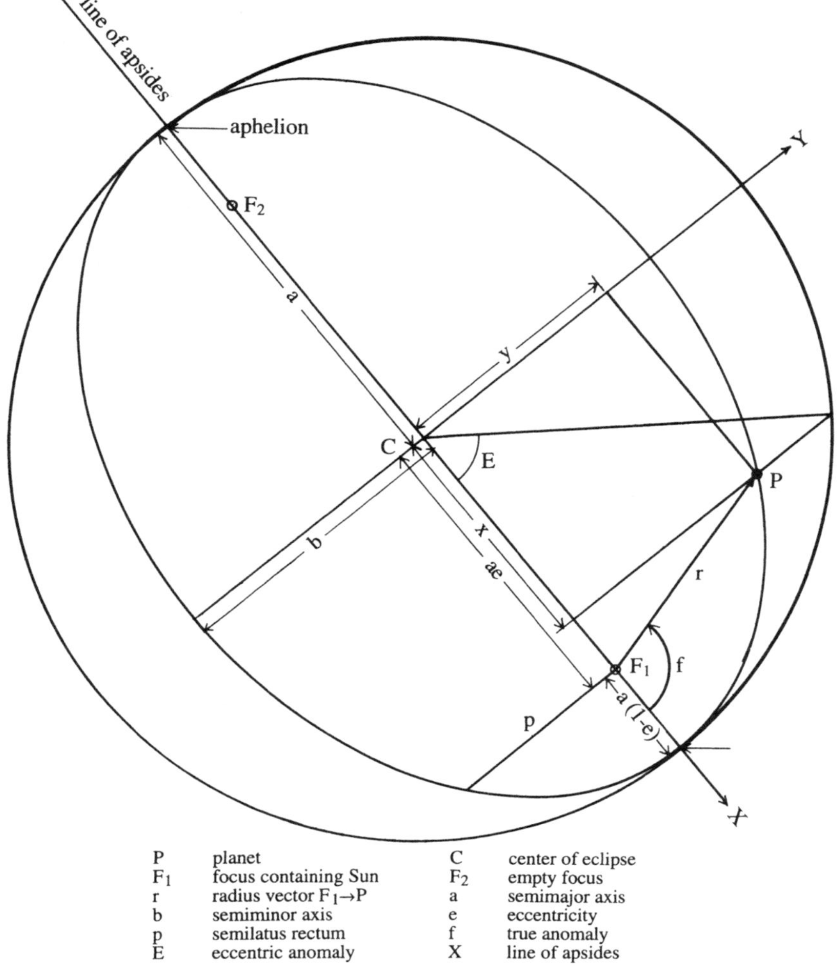


**Figure III.2** Nomenclature for elliptical orbits. The line of apsides passes through the perihelion and aphelion points, as well as both foci and the center of the ellipse. Note the definitions of the true anomaly ($f$) and the eccentric anomaly ($E$) in the diagram.

bviously proportional to the mass of the Sun), and $\hat{r}$ is e unit vector in the direction of $\vec{r}$:

$$\vec{r} = r\hat{r}. \tag{III.2}$$

Strictly speaking, the mass $m$ in the $m\ddot{\vec{r}}$ term of (III.1) is the inertial mass of the planet, whereas $m$ e other term is the gravitational mass. The equiva- of these two masses, which will simply be assumed hereafter, has been tested experimentally since the f Newton (who found agreement to one part in highly sophisticated modern experiments, in greement to one part in $10^{12}$ has been found.

gravitational forces are conservative, being to the gravitational potential ($V$) by

$$F = -\nabla V = -\nabla\left(\frac{-\mu}{r}\right) \tag{III.3}$$

$$\nabla \times F = 0. \tag{III.4}$$

Because the gravitational acceleration and $r$ are aligned together [Eq. (III.1)],

$$r \times \ddot{\vec{r}} = 0. \tag{III.5}$$

But, because

$$\frac{d}{dt}(\vec{r} \times \dot{\vec{r}}) = \vec{r} \times \ddot{\vec{r}} + \dot{\vec{r}} \times \dot{\vec{r}} \tag{III.6}$$

and both terms on the right are zero, we conclude that

$$\vec{r} \times \dot{\vec{r}} = \text{constant} = \vec{h}, \tag{III.7}$$

where $h$ is the angular momentum vector per unit mass, which is clearly conserved.

Now, if we let $\hat{\theta}$ be the unit vector perpendicular to $r$ in the orbital plane, we have

$$\vec{r} \times \dot{\vec{r}} = \vec{r} \times (r\dot{\theta}\hat{\theta} + \dot{r}\hat{r}) \tag{III.8}$$

This cross-product gives us

$$r^2\dot{\theta} = h, \tag{III.9}$$

which, since $r^2 d\theta/2$ is the area element in polar coordinates, is a statement of Kepler's equal area law.

The total energy per unit mass is just the sum of the kinetic and potential energies:

$$E = \frac{1}{2}(\dot{\vec{r}})^2 - \mu/r. \tag{III.10}$$

For closed (elliptical) orbits, $E < 0$.

Now, if we write the cross-product of the gravitational acceleration with the (constant) angular momentum vector,

$$\ddot{\vec{r}} \times \vec{h} = \frac{d}{dt}(\dot{\vec{r}} \times \vec{h}) \tag{III.11}$$

$$= -\frac{\mu}{r^3}\vec{r} \times (\vec{r} \times \dot{\vec{r}}), \tag{III.12}$$

using Eqs. (III.1) and (III.7). Then,

$$\ddot{\vec{r}} \times \vec{h} = -\frac{\mu}{r^3}[(\vec{r} \cdot \dot{\vec{r}})\vec{r} - (\vec{r} \cdot \vec{r})\dot{\vec{r}}] \tag{III.13}$$

$$= -\frac{\mu}{r^3}[\dot{r}\vec{r} - r^2\dot{\vec{r}}] \tag{III.14}$$

$$= \mu\left[\frac{\dot{\vec{r}}}{r} - \frac{\dot{r}}{r^2}\vec{r}\right] \tag{III.15}$$

$$= \mu\frac{d}{dt}\left[\frac{\vec{r}}{r}\right] = \mu\frac{d\hat{r}}{dt}. \tag{III.16}$$

Integrating,

$$\dot{\vec{r}} \times \vec{h} = \mu[\hat{r} + \vec{e}], \tag{III.17}$$

where the constant of integration, $\vec{e}$, is called the eccentricity vector.

If we now form the product

$$\vec{r} \cdot [\dot{\vec{r}} \times \vec{h}] = \mu[r + \vec{r} \cdot \vec{e}], \tag{III.18}$$

then

$$\mu[r + \vec{r} \cdot \vec{e}] = [\vec{r} \times \dot{\vec{r}}] \cdot \vec{h} = h^2. \tag{III.19}$$

Thus, from Eqs. (III.19) and (III.2),

$$r = \frac{h^2/\mu}{(1 + \hat{r} \cdot \vec{e})}. \tag{III.20}$$

From the geometrical properties of ellipses,

$$r = \frac{a(1 - e^2)}{1 + e\cos f} = \frac{p}{1 + e\cos f}, \tag{III.21}$$

where $p$, the semilatus rectum, and $f$, the true anomaly, are defined as in Fig. III.2.

By our present definitions,

$$\hat{r} \cdot \vec{e} = e\cos f; \tag{III.22}$$

hence, from Eqs. (III.20) and (III.21),

$$\mu p = h^2. \tag{III.23}$$

Now let us form the cross-product of $\vec{h}$ with both sides of Eq. (III.17),

$$\vec{h} \times (\dot{\vec{r}} \times \vec{h}) = \mu\vec{h} \times (\hat{r} + \vec{e}) \tag{III.24}$$

$$h^2\dot{\vec{r}} = \mu h[\hat{f} + e\hat{f}_0], \tag{III.25}$$

where $\hat{f}$ is now the unit vector normal to the radius vector in the plane of the orbit and $\hat{f}_0$ is $\hat{f}$ evaluated at a true anomaly of zero (see Fig. III.3a).

It can be seen that Eq. (III.25) resolves $\dot{\vec{r}}$ into two components, one of which (the term in $f_0$) is stationary and the other of which rotates on the $X$–$Y$ plane. This behavior is illustrated in the hodograph in Fig. III.3b.

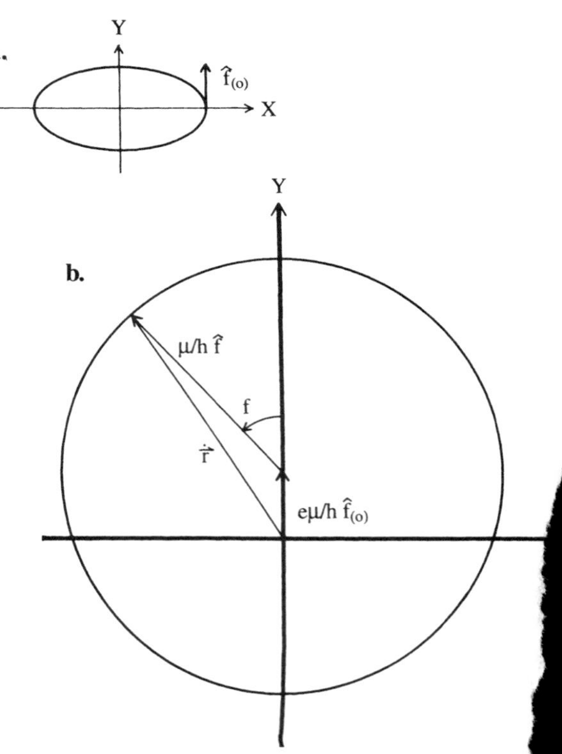


**Figure III.3** Hodograph for elliptical motion. See the explanation.

The total kinetic energy per unit mass can, with the help of Fig. III.3, be resolved into components due to motion in the $X$ and $Y$ directions:

$$v^2 = \dot{\vec{r}} \cdot \dot{\vec{r}} = \left(\frac{\mu}{h}\sin f\right)^2 \tag{III.26}$$

$$+ \left(\frac{\mu}{h}\cos f + \frac{e\mu}{h}\right)^2$$

$$= \frac{\mu^2}{h^2}(1 + 2e\cos f + e^2). \tag{III.27}$$

Here we solve Eq. (III.21) for $e$ cos $f$, substitute, and simplify to get

$$v^2 = \frac{\mu^2}{h^2}(1 - e^2)\left[\frac{2a}{r} - 1\right]. \tag{III.28}$$

From Eqs. (III.21) and (III.23), we can see that

$$(1 - e^2) = h^2/a\mu, \tag{III.29}$$

whence

$$v^2 = \mu\left[\frac{2}{r} - \frac{1}{a}\right]. \tag{III.30}$$

This is called the velocity equation or the *vis viva* equation.

The total energy from Eq. (III.10) is then

$$E = \frac{1}{2}v^2 - \mu/r = -\mu/2a \tag{III.31}$$

for bound elliptical orbits. Thus the total energy per unit mass possessed by any body in orbit around the Sun depends only on the semimajor axis of its orbit.

On the other hand, we have seen from Eq. (III.23) that the orbital angular momentum depends only on the semilatus rectum,

$$h^2 = \mu p = \mu a(1 - e^2). \tag{III.32}$$

Note that, for circular orbits, $r = a$ and $e = 0$; hence Eq. (III.31) collapses to $v_{circ} = (\mu/r)^{1/2}$ and Eq. (III.32) becomes $h^2 = \mu r$.

The full description of the *shape* of a planetary orbit requires only knowledge of two geometrical constants, $a$ and $e$. To describe the *orientation* of the orbit in three-dimensional space we additionally require an angle that specifies the direction of the line of apsides projected on a reference plane (the invariant plane), an angle that gives the inclination of the orbit relative to this plane, and an angle to describe the orientation of the line of intersection of the orbital plane and the reference plane. This line, called the line of nodes, passes through the Sun and through the two points at which the planet crosses the reference plane, the ascending and descending nodes (see Fig. III.4). Finally, we must specify the location of the planet in its orbit at a particular time. These six parameters fully describe the orbit and are equivalent to giving the coordinates of the body in six-dimensional phase space ($x$, $y$, $z$, $\dot{x}$, $\dot{y}$, $\dot{z}$).

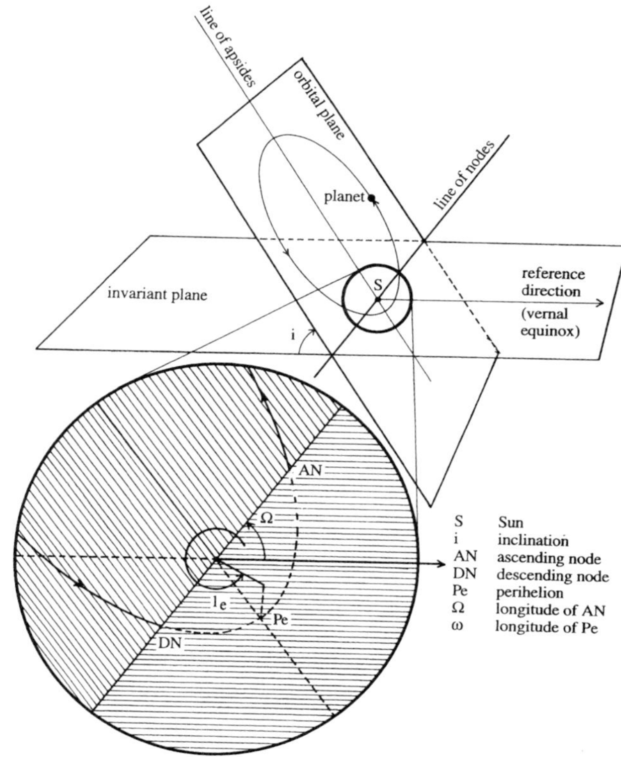


**Figure III.4** Elliptical orbits in three dimensions. The axis of the invariant plane is the direction of the total orbital angular momentum vector of the planetary system. The line of nodes passes through the center of the Sun and through the two points (nodes), where the orbital path crosses the invariant plane. Some reference direction must be chosen to define the orientation of orbits in three dimensions: the direction of the vernal equinox. The longitude of perihelion is measured from that point.

The orbital parameters for the nine known planets of the Solar System are summarized in Table III.1 and depicted graphically in Fig. III.5. There is an apparent regularity in the spacing of the orbits of the planets out to Uranus: the semimajor axes of the orbits (in AU) follow the rule

$$a = 0.4 + 0.3 \times 2^n, \tag{III.33}$$

where $n = -\infty, 0, 1, 2 \cdots 5$. The value of $a = 2.8$ AU found for $n = 3$ corresponds not to the orbit of any planet, but rather to the approximate center of the asteroid belt. Thus, even if some rationale could be invented for the strange appearance of $-\infty$ in the list of allowable values of $n$, there would still remain the fact of the absence of a planet at $n = 3$.

The formula given above, known as the Titius–Bode Law, is far from exact, as can be seen by comparison

Table III.1 Orbits of the Planets

| Planet | $a$ (AU) | $e$ | $i$ | $P$ | $\Omega$ | $\omega$ |
|---|---|---|---|---|---|---|
| Mercury | 0.3871 | 0.20563 | 7.0044 | 87.9686 days | 48.12 | 77.18 |
| Venus | 0.7233 | 0.00678 | 3.3945 | 224.700 days | 76.52 | 131.00 |
| Earth | 1.0000 | 0.01670 | 0.000 | 365.257 days | — | 102.6 |
| Mars | 1.5237 | 0.0934 | 1.8499 | 1.88089 years | 49.42 | 335.7 |
| Jupiter | 5.2042 | 0.0479 | 1.3056 | 11.8620 years | 100.30 | 15.0 |
| Saturn | 9.5751 | 0.052 | 2.4859 | 29.45772 years | 113.50 | 95.4 |
| Uranus | 19.31 | 0.050 | 0.7727 | 84.013 years | 73.99 | 176.0 |
| Neptune | 30.2 | 0.0040 | 1.7725 | 164.79 years | 131.55 | 30 |
| Pluto | 39.91 | 0.2566 | 17.135 | 248.4 years | 110.1 | 223.4 |

with the observational data. Furthermore, despite persistent efforts of generations of scientists to discover a physical basis for this "law," no fully satisfactory explanation has ever been proposed. Nevertheless, so strong has been the fascination of this supposed regularity of nature that it was for many years popular to describe the asteroids as debris left over from the explosion of a hypothetical planet that once orbited where Bode's Law said it should. Our discussion of asteroids and meteorites in Chapter VI will provide ample evidence for the rejection of this conjecture. For now, however,

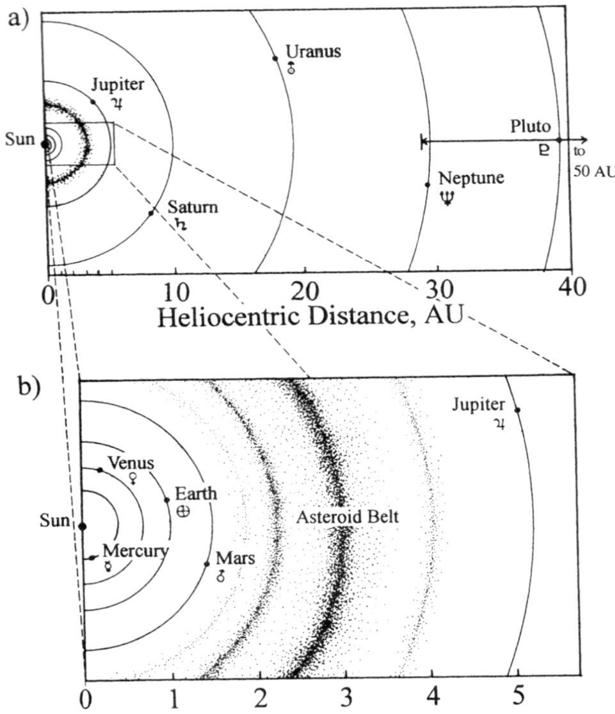


**Figure III.5** Orbits of the planets. The eccentricity of Pluto's orbit, which sometimes brings it inside Neptune's orbit, is indicated in a. b shows the orbits of the inner planets to scale with the asteroid belt and Jupiter's orbit. The detailed distribution of the asteroids in semimajor axis is shown in Fig. III.9.

it is sufficient to remark that the total mass of material in the asteroid belt is orders of magnitude too small to account for the mass of a planet.

Historically, after the discovery of Uranus by the English astronomer Sir William Herschel in 1781, which apparently confirmed Bode's Law, it was generally assumed that another planet might exist beyond the orbit of Uranus at the $n = 7$ position in Bode's formula, at $a = 38.8$ AU. When observations of the motions of Uranus subsequently revealed perturbations by a substantial mass outside Uranus's orbit, J. C. Adams of England and U. J. J. Leverrier of France independently calculated the position and mass of this undiscovered planet, using the assumption that it obeyed Bode's Law. When telescopes were turned to the predicted position the planet known to us as Neptune was quickly discovered. Neptune had, in effect, been discovered by mathematicians who did not know how to use a telescope.

The impact of this *tour de force* of Newtonian physics was considerable even far outside the astronomical fraternity. It was therefore most unsettling when it was found that Neptune was far closer to the Sun than allowable by Bode's Law and that its mass was far less than that derived before its discovery. With the discovery of Pluto by the American astronomer Clyde Tombaugh in 1930, it became clear that Bode's Law was only poorly enforced. Nonetheless, research papers dealing with the possible physical significance of the Law still appear regularly, usually replete with ingenious alibis for the miscreants.

The orbits of the planets are, as can be seen in Table III.1, quite nearly circular. Three planets have orbital eccentricities of significant magnitude: Mercury, Mars, and Pluto. The case of Pluto is particularly noteworthy, because its perihelion distance lies well inside Neptune's orbit. Thus, for a small fraction of each orbit of Pluto (20 out of 248 years), including the years 1979–1999, Neptune is the most distant known planet from the Sun.

The orbital inclination of Pluto is also unique among the planets and is instrumental in preventing collisions between Pluto and Neptune. Possibly the effects which have "regularized" the orbits of the other planets did not act on Pluto. We will pay particular attention to the unique problems presented by Pluto at the end of Chapter VI. The orbital periods of Jupiter, Saturn, Uranus, and Neptune also display interesting near-commensurabilities.

## Changes in Orbital Motion

The basic theory of orbital motion outlined earlier is appropriate for use when the gravitational force is exerted by a mass point or a perfectly homogeneous, spherically symmetrical mass. It also does not, in this form, take into account other forces, such as additional gravitating bodies, rocket propulsion, atmospheric drag, and collisions. Usually such disturbing forces are weak or brief enough that their effects can be treated as small changes (perturbations) on the preexisting orbit. Disturbances that are very brief ($dt << P$), called *impulsive perturbations*, deflect the body into a new orbit essentially at one point, which is clearly the intersection point of the old orbit and the perturbed orbit.

It is mathematically convenient to resolve the perturbing forces into three orthogonal components defined with respect to the orbital plane of the body. The first of these is radial (in the direction of the radius vector $r$), the second is transverse (at right angles to the radius vector in the orbital plane, i.e., in the direction of $\Theta$), and the third is normal (orthogonal to the orbital plane, and hence to both the radial and transverse directions). For a body in circular orbit, the transverse direction is identical to the direction of the velocity vector.

Real physical bodies differ in several respects from the mathematical ideal described. First, bodies large enough to have gravitational fields that are meaningful on the scale of a planetary or satellite system are sufficiently massive to generate high internal pressures. These high pressures lead to self-compression of the interiors of these bodies, even those that are compositionally uniform. Second, planetary and stellar bodies usually are differentiated into compositionally distinct layers: stars build up cores of fusion products composed of elements heavier than hydrogen, and planets separate into compositionally distinct layers according to density. Third, planets too small to have undergone melting and density-dependent differentiation will also depart from a spherical symmetry in complex ways because they are too cold and rigid to relax to spherical or spheroidal shapes. Fourth, any large body that rotates rapidly (i.e., its rotation period is not very much larger than the circular orbital speed at its surface) will have a diminished effective gravitational acceleration near its equator. This causes a distension of the equatorial regions, perceived as a visible oblateness of the planet's disk. This geometrical or optical ellipticity, or oblateness, $f$, is related to the equatorial radius $a$ and the polar radius $c$ by

$$f = (a - c)/a. \qquad \text{(III.34)}$$

The gravitational field of a planet can be represented in a very general way by a spherical harmonic expansion of the potential (the potential energy per unit mass; see Appendix V). The first term in the expansion vanishes if the center of the coordinate system is taken to be the center of mass. The second term, resulting from the oblateness of the planet, is

$$V(r) = -(GM/r)[1 - (a/r)J_2 P_2(\cos\Theta)], \qquad \text{(III.35)}$$

where $P_2$ ($\cos\Theta$) is one of the associated Legendre polynomials, which are tabulated and described in advanced calculus texts. For our purposes, it is sufficient to know that

$$P_2\ (\cos\Theta) = (1/2)[3\cos^2 - 1]. \qquad \text{(III.36)}$$

Here $J_2$ is defined in terms of the moments of inertia of the body about its three orthogonal axes. The principal axis is naturally defined as the rotation axis. Because of the oblateness caused by rotation, the moment of inertia of the planet about this axis, called $C$, is larger than any other. The other two axes of inertia, $B$ and $A$, are orthogonal to the spin axis, and therefore lie in the equatorial plane. If the planet is spherically symmetrical, $C = B = A$. This is usually a good approximation for a nonrotating planet, for which $J_2 = 0$. If the planet is reasonably fluid and rotating, the $A$ and $B$ moments of inertia will be the same ($C > B = A$). This body is then an oblate spheroid. If the planet is distorted by proximity to another body, such as the Moon in Earth's gravity field, then the axis along the direction of orbital motion and that in the radial (Earth to Moon) direction will not necessarily be the same. Such a body is a triaxial ellipsoid with its three moments of inertia ranked $C > B > A$. Students of spherical geometry may wonder about the other possible ellipsoid, the cigar-shaped prolate spheroid ($C = B > A$), but in nature they do not rotate around their long axes. $J_2$ is defined so as to be useful in measuring the departure of oblate spheroids from spherical symmetry; that is, the degree to which $C$ differs from $A$:

$$J_2 = (C - A)/Ma_{eq}^2 \qquad \text{(III.37)}$$

where $a_{eq}$ is the equatorial radius of the body.

The gravitational perturbation exerted on the Moon by Earth's equatorial bulge causes a slow precession of the Moon's orbit. Measurement of this precession rate using very long-term observations of the Moon's orbit (drawing heavily on ancient eclipse observations) gives us an experimentally derived value for $J_2$. High-precision tracking of artificial Earth satellites deep within Earth's gravity well also gives independent determinations of $J_2$.

The effect of these perturbations is to change the orientation of the orbit slowly. Both the argument of perihelion, $\omega$, and the longitude of the ascending node, $\Omega$, drift as the orbit precesses. These torques also affect the mean motion $n = 2\pi/P$:

$$\frac{d\omega}{dt} = -\frac{3}{2}J_2(n/p^2)\left[\frac{5}{2}\sin^2 i - 2\right], \qquad \text{(III.38)}$$

$$\frac{d\Omega}{dt} = -\frac{3}{2}J_2(n/p^2)\cos i, \qquad \text{(III.39)}$$

$$(t - t_0)\frac{dn}{dt} = -\frac{3}{2}J_2(n/p^2) \times (1 - e^2)^{1/2}\left[\frac{3}{2}\sin^2 i - 1\right], \qquad \text{(III.40)}$$

where $p = a(1 - e^2)$ and $t_0$ is the time of perihelion passage.

Note that $J_2$ is a dimensionless measure of the departure of the moment of inertia about the principal axis from that of a sphere, whereas the oblateness is a dimensionless measure of the departure of the geometrical figure from that of a sphere. Both numbers increase with rotation rate, but the two should not be confused with each other. The fractional difference between the maximum and minimum moments of inertia is

$$H = (C - A)/C \qquad \text{(III.41)}$$

The effect of the gravity of Sun and Moon on Earth's equatorial bulge results in a torque that causes the rapidly rotating Earth to precess slowly, like a top. The observed precession rate tells us the magnitude of $H$.

Note that

$$J_2/H = [(C - A)/Ma^2]\times [C/(C - A)] = C/Ma^2. \qquad \text{(III.42)}$$

This ratio, called the *moment of inertia factor*, has a value of 0.4 for a homogeneous sphere. The relationship between oblateness $f$ and $J_2$ is further explored in Appendix V.

## Properties of the Planets

Table III.2 presents the rotation periods, axial orientations, and precession periods of the planets. Particularly noteworthy are the very long rotation period of Mercury, the slow retrograde rotation of Venus, the near equality of the rotation periods of Mars and Earth, the very similar rotation periods of the Jovian planets, and the odd axial inclination of Uranus. Because of this orientation, Uranus presents the meteorologically interesting prospect of a rapidly rotating planet that is sometimes heated at the equator, but also intermittently heated at one or the other pole by some 40 years of continuous sunlight. In 1986, almost coincident with the arrival of the Voyager 2 planetary probe at Uranus, the planet was oriented with one pole pointing almost directly at the Sun.

**Table III.2** Rotation of the Planets

| Planet | Rotation period | Axial inclination | North pole RA | North pole dec | Precession period |
|---|---|---|---|---|---|
| Mercury | 58.646 days | 0.0 | 281.0 | 61.4 | |
| Venus | −243.01 days | 177.3 | 272.8 | 67.2 | |
| Earth | 23h56m4.1s | 23.45 | — | 90.0 | 26,000 years |
| Mars | 24h37m22.6s | 25.19 | 317.6 | 52.82 | 50,000 years |
| Jupiter | 9h50m30s | 3.12 | 268.03 | 64.49 | |
| Saturn | 10h14m | 26.73 | 39.8 | 83.45 | |
| Uranus | 14h42m | 97.86 | 257.18 | −15.08 | |
| Neptune | 18h24m | 29.56 | 295.18 | 40.61 | |
| Pluto | 6.3867 days | 122.5 | 305 | 5 | |

The physical dimensions of the planets are given in Table III.3. The extremely large and rapidly rotating Jovian planets all have accelerations due to rotation that are a significant fraction of their equatorial gravitational accelerations, which act to reduce their effective equatorial gravitational accelerations. It must be noted that the error limits for the radii of the planets are quite varied; Earth's radius is of course best known, and those planets that have been tracked by radar for a number of years (Mercury, Venus, and Mars) also have very well-known radii. Spacecraft occultations have provided great improvements in the measurement of the radii of all the Jovian planets, and stellar occultations had earlier permitted very useful measurements of Neptune and Uranus.

Table III.3 contains the most recent measurements of the masses and densities of the planets. Those planets that have been flown by or orbited by spacecraft (from Mercury out to Neptune) have very well-determined masses. Indeed, for each of these bodies the product $Gm$ is more accurately known than $G$ itself. Thus the limiting factor in determining their masses is the uncertainty in the value of $G$. The mass of Pluto can best be determined from the orbital motions of its satellite. The most recent refinement of the mass of Neptune came from analyzing the data from the 1989 Voyager II spacecraft flyby through its system. Pluto will remain for many years to come the least well-known of the planets.

Table III.3 Dimensior…Aasses, and Densities of the Planets

| Planet | Equat… …adius | Oblateness | Mass ($10^{27}$ g) | Density (g cm$^{-3}$) | Density ($P = 0$) | Geometric albedo |
|---|---|---|---|---|---|---|
| Mercury | …9 | 0 | 0.33022 | 5.43 | 5.30 | 0.160 |
| Venus | 51 | 0 | 4.8690 | 5.24 | 4.00 | 0.65 |
| Earth | …78 | 0.003353 | 5.9742 | 5.515 | 4.05 | 0.367 |
| Mars | 396 | 0.005 | 0.64191 | 3.93 | 3.74 | 0.150 |
| Jupiter | ,850 | 0.061 | 1898.8 | 1.36 | 0.1 | 0.52 |
| Saturn | …,330 | 0.09 | 568.41 | 0.71 | 0.1 | 0.47 |
| Uranus | 5,400 | 0.03 | 86.967 | 1.30 | 0.3 | 0.50 |
| Neptune | …4,300 | 0.03 | 102.85 | 1.8 | 0.3 | 0.41 |
| Pluto | 1,150 | 0 | 0.0129 | 2.03 | 2.0 | 0.6 |

…on of the nine planetary density …l a fundamental distinction of Even our understanding of the Solar values …l inner planets all have densities great …y material, whereas the four large System tiny Pluto all have densities near cha… density is diagnostic of a very low …ght and a very high content of light, …ich as hydrogen and helium.

…us, Earth, and Mars are thus often …ial planets, as distinct from Jupiter, …nd Neptune, which are called the *Jovian* …ies enigmatic Pluto is lumped with the … form a class called the *outer* planets. In …insic properties, however, Pluto has more …ies with the satellites of the Jovian planets …ie Jovian planets themselves.

…ite the clear dichotomization between the …ial and Jovian planets by application of mass density criteria, it would be a serious error to regard …e membership of these classes as at all homogeneous. The observed density variations are unquestionably real. These observed densities are compounded from two effects: the intrinsic zero-pressure density of the bulk planetary material and the mass-dependent effects of compression by the very high pressures prevalent in the deep planetary interiors. Approximate correction for the latter effect has been attempted in the sixth column in Table III.3. These zero-pressure densities should reflect directly the average chemical composition of each planet.

It is also valuable to provide a brief summary of the basic descriptive data on the general appearance of the planets as compiled by many generations of astronomers and one generation of planetary scientists (see Fig. III.6). The major salient features display an almost shocking variety of conditions.

Mercury is a hot, dense, rocky planet devoid of detectable atmosphere.

Venus is red-hot, similar in size to Earth, with a dense and perpetually cloudy $CO_2$-rich atmosphere that presents a nearly featureless face to the Universe.

Earth is distinguished by a water-rich surface whose conditions span the triple point of water. It is shrouded by a moderately dense nitrogen–oxygen atmosphere. Its complex decoration of water clouds is ever changing, and large areas of greenish coloration vary seasonally.

Mars is a small reddish planet that, like Earth, has polar caps and marked seasonal variations. It is lightly coated with a cold, tenuous carbon dioxide atmosphere.

Beyond the asteroid belt we find Jupiter, an enormous hydrogen-rich planet marked with varicolored and unpredictable cloud bands parallel to the equator, bearing in its atmosphere a red blotch the size of Earth.

Saturn, in other respects merely a faded and colder copy of Jupiter, is circled by a thin, brilliant system of rings of small particles, making it perhaps the most stunning of the planets in visual appearance.

Uranus and Neptune, so far from the Sun that visual observations are extremely difficult, are green-tinted gas balls bearing no more than the faintest trace of the cloud bands seen on Jupiter and Saturn. Uranus, like Jupiter, has a very tenuous ring system. Neptune has at least three partial rings.

Little cold Pluto is so unfavorably situated for observation (and for spacecraft visitation) that its 6-day cycle of brightness variation and its low-resolution reflection spectrum have until quite recently been our only clues to the nature of its surface.

## Mass and Angular Momentum Distribution

In surveying the general properties of the Solar System, we have seen that certain commodities, such as total mass, gas abundances, and angular momentum, seem to be distributed very inequitably. We of course note that

**Figure III.6** Visual appearance of the planets. a is a mosaic of Mariner 10 views taken during departure from Mercury, b is a Pioneer Venus UV image of the Venus cloud tops (the planet is featureless at visible wavelengths), and c is an Apollo photograph of Earth. d is a Viking 1 orbiter image of the Olympus Mons on Mars, and e compares the Voyager 1 and 2 approach images of Jupiter, Saturn, Uranus, and Neptune, all to the same scale. There are no high-quality images of Pluto, the only planet not yet visited by spacecraft from Earth.

Figure III.6 *Continued*

This approximation does not consider the distortion of the body from spherical symmetry by rapid rotation.

In Fig. III.8 the Earth–Moon system is plotted as well as the Earth by itself, and the Pluto–Charon system is similarly treated. Note that all the terrestrial planets lie far from the instability line, as does the Sun, whereas the Earth–Moon system and the Jovian planets lie fairly closely on a straight line parallel with and slightly below the instability limit. Pluto, because of its small mass, falls off the figure.

## Satellites

Since Galileo Galilei first turned his telescope on the planets in the early 17th century, it has been clear that a number of other sizeable bodies can be found in the Solar System. In fact, the largest satellites of Earth, Jupiter, Saturn, and Neptune are all of planetary dimensions. Modern observational techniques, culminating with the Voyager missions to the Jovian planets, have provided great amounts of new and tantalizing data on

Figure III.6 *Continued*

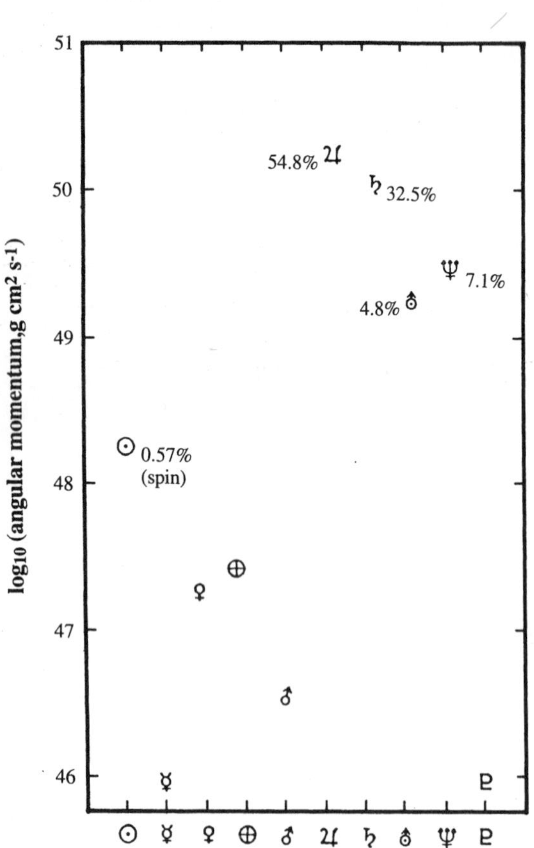


Figure III.7 Orbital angular momenta of the planets. Note the overwhelming importance of the Jovian planets.

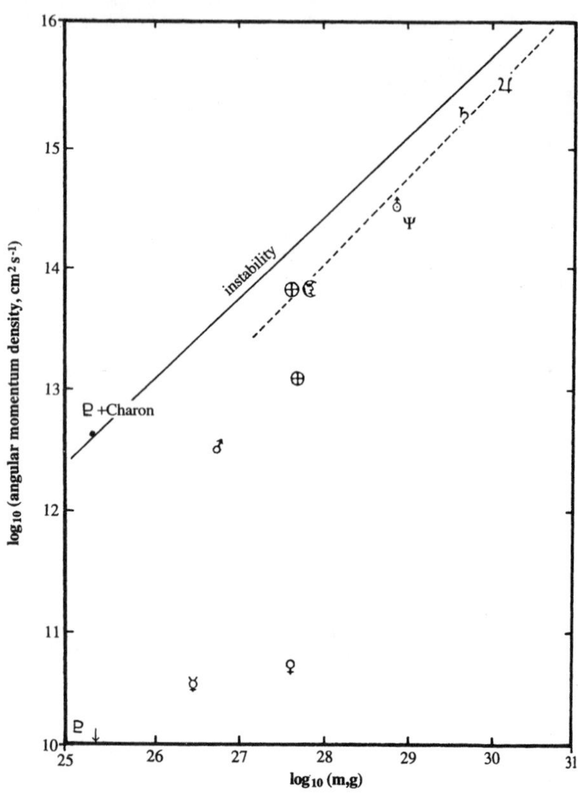


Figure III.8 Angular momentum density of planetary spin. Note the highly despun states of Mercury and Venus. The Jovian planets and the Earth–Moon system parallel the instability boundary at the top. Both Earth and Mars show evidence of effects on their spin, such as tidal transfer of angular momentum to their satellites.

this fascinating class of bodies, showing them to be nearly as diverse in their properties as the planets themselves.

The orbital parameters of most of the known natural satellites are given in Table III.4, and their dimensions, masses, and densities are to be found in Table III.5. We shall briefly discuss the basic natures of these bodies in the order of their heliocentric distance, paying particular attention to trends and oddities in the data.

We note first of all that Mercury and Venus have no known satellites. Mercury's proximity to the glare of the Sun hinders all attempts at observing it. Venus, more favorably situated for observation, almost certainly cannot have a satellite as large as 10 km in diameter. Theoretical studies of the stability of satellite orbits around Mercury and Venus conclude that the solar tidal forces are so strong that such satellites, if ever present,

**Table III.4** Orbits of the Satellites

| Planet | | Satellite | Semimajor axis (10³ km) | Semimajor axis ($R_{pl}$) | Period (days) | Eccentricity[b] | Inclination[c] |
|---|---|---|---|---|---|---|---|
| Earth | | Moon | 384.4 | 60.3 | 27.322 | 0.0549 | 5.15 |
| Mars | MI | Phobos | 9.378 | 2.76 | 0.319 | 0.015 | 1.02 |
| | MII | Deimos | 23.459 | 6.91 | 1.263 | 0.00052 | 1.82 |
| Jupiter | JXVI | Metis | 127.96 | 1.7922 | 0.2948 | <0.004 | 0? |
| | JXV | Adrastea | 128.98 | 1.8065 | 0.2983 | 0? | 0? |
| | JV | Amalthea | 181.3 | 2.539 | 0.4981 | 0.003 | 0.40 |
| | JIV | Thebe | 221.9 | 3.108 | 0.6745 | 0.015 | 0.8 |
| | JI | Io | 421.6 | 5.905 | 1.769 | <.0041 | 0.04 |
| | JII | Europa | 670.9 | 9.397 | 3.551 | <0.01 | 0.470 |
| | JIII | Ganymede | 1,070 | 14.99 | 7.155 | <.0015 | 0.281 |
| | JIV | Callisto | 1,883 | 26.37 | 16.689 | 0.007 | 0.281 |
| | 2000 J1 | | 7,507 | 105.1 | 129.71 | 0.204 | 46[d] |
| | JXIII | Leda | 11,094 | 155.4 | 238.72 | 0.148 | 26[d] |
| | JVI | Himalia | 11,480 | 160.8 | 250.57 | 0.158 | 28[d] |
| | JX | Lysithea | 11,720 | 164.2 | 259.22 | 0.107 | 29[d] |
| | JVII | Elara | 11,737 | 164.4 | 259.65 | 0.207 | 28[d] |
| | 2000 J11 | | 12,557 | 175.9 | 289 | 0.25 | 28.2[d] |
| | 2000 J10 | | 20,174 | 282.5 | 588(R | 0.14 | 166[d] |
| | 2000 J3 | | 20,210 | 283.1 | 584(R) | 0.22 | 150[d] |
| | 2000 J7 | | 21,010 | 294.3 | 621(R) | 0.22 | 149[d] |
| | JXII | Ananke | 21,200 | 296.9 | 631(R) | 0.169 | 147[d] |
| | 2000 J5 | | 21,336 | 298.8 | 632(R) | 0.24 | 149[d] |
| | 2000 J9 | | 22,304 | 312.4 | 683(R) | 0.26 | 165[d] |
| | JXI | Carme | 22,600 | 316.5 | 692(R) | 0.207 | 163[d] |
| | 2000 J4 | | 22,972 | 321.7 | 712(R) | 0.28 | 165[d] |
| | 2000 J6 | | 23,074 | 323.2 | 720(R) | 0.26 | 165[d] |
| | JVIII | Pasiphae | 23,500 | 329.1 | 735(R) | 0.378 | 148[d] |
| | 2000 J8 | | 23,618 | 330.8 | 741(R) | 0.41 | 153[d] |
| | JIX | Sinope | 23,700 | 331.9 | 758(R) | 0.275 | 153[d] |
| | 2000 J2 | | 23,746 | 332.6 | 752(R) | 0.24 | 165[d] |
| | 1999 J1 | | 24,235 | 339.4 | 768(R) | 0.125 | 143[d] |
| Saturn | SXVIII | Pan | 135.6 | 2.220 | 0.576 | 0? | 0? |
| | SXV | Atlas | 137.64 | 2.281 | 0.602 | 0? | 0? |
| | SXVI | Prometheus | 139.35 | 2.310 | 0.613 | 0.0024 | 0? |
| | SXVII | Pandora | 141.70 | 2.349 | 0.629 | 0.00422 | 0? |
| | SXI | Epimetheus | 151.472 | 2.510 | 0.694 | 0.009 | 0.34 |
| | SX | Janus | 151.472 | 2.511 | 0.695 | 0.007 | 0.14 |
| | SI | Mimas | 185.52 | 3.075 | 0.942 | 0.0202 | 1.53 |
| | SII | Enceladus | 238.02 | 3.945 | 1.370 | (0.0045) | 0.02 |
| | SIII | Tethys | 294.66 | 4.884 | 1.888 | 0.0000 | 1.09 |
| | SXIII | Telesto | 294.66 | 4.884 | 1.888 | 0? | 0? |
| | SXIV | Calypso | 294.66 | 4.884 | 1.888 | 0? | 0? |
| | SIV | Dione | 377.40 | 6.256 | 2.737 | (0.0022) | 0.02 |
| | SXII | Helene | 377.40 | 6.256 | 2.737 | 0.005 | 0.2 |
| | SV | Rhea | 527.04 | 8.736 | 4.518 | >0.001 | 0.35 |
| | SVI | Titan | 1,221.8 | 20.25 | 15.945 | 0.0292 | 0.33 |

(*Continued*)

Table III.4 (*Continued*)

| Planet | | Satellite | Semimajor axis (10³ km) | Semimajor axis ($R_{pl}$) | Period (days) | Eccentricity[b] | Inclination[c] |
|---|---|---|---|---|---|---|---|
| | SVII | Hyperion | 1,481.1 | 24.55 | 21.277 | (0.1042) | 0.43 |
| | SVIII | Iapetus | 3,561.3 | 59.03 | 79.331 | 0.0283 | 7.52 |
| | 2000 S5 | | 11,339 | 185.6 | 449 | 0.33 | 46.2[d] |
| | 2000 S6 | | 11,465 | 187.7 | 453 | 0.32 | 46.6[d] |
| | SIX | Phoebe | 12,944 | 214.5 | 550.5(R) | 0.163 | 174.8[d] |
| | 2000 S2 | | 15,172 | 248.4 | 687 | 0.36 | 45.2[d] |
| | 2000 S8 | | 15,676 | 256.6 | 730(R) | 0.27 | 153.0[d] |
| | 2000 S3 | | 17,251 | 282.4 | 825 | 0.27 | 45.5[d] |
| | 2000 S10 | | 17,452 | 285.7 | 858 | 0.47 | 34.7[d] |
| | 2000 S11 | | 17,874 | 292.6 | 888 | 0.38 | 33.1[d] |
| | 2000 S4 | | 18,231 | 298.5 | 924 | 0.54 | 33.5[d] |
| | 2000 S9 | | 18,486 | 302.6 | 939(R) | 0.22 | 167.4[d] |
| | 2000 S12 | | 19,747 | 323.3 | 1037(R) | 0.12 | 175.0[d] |
| | 2000 S7 | | 20,144 | 329.8 | 1067(R) | 0.45 | 175.9[d] |
| | 2000 S1 | | 23,076 | 377.8 | 1311(R) | 0.34 | 173[d] |
| Uranus | UVI | Cordelia | 49.7 | 1.90 | 0.33 | 0? | 0? |
| | UVII | Ophelia | 53.8 | 2.05 | 0.38 | 0? | 0? |
| | UVIII | Bianca | 59.2 | 2.26 | 0.43 | 0? | 0? |
| | UIX | Cressida | 61.8 | 2.36 | 0.46 | 0? | 0? |
| | UX | Desdemona | 62.7 | 2.39 | 0.48 | 0? | 0? |
| | UXI | Juliet | 64.6 | 2.47 | 0.49 | 0? | 0? |
| | UXII | Portia | 66.1 | 2.52 | 0.51 | 0? | 0? |
| | UXIII | Rosalind | 69.9 | 2.67 | 0.56 | 0? | 0? |
| | UXIV | Belinda | 75.3 | 2.87 | 0.62 | 0? | 0? |
| | 1986U10 | | 76.42 | 2.91 | 0.43 | 0.0012 | $0 \pm 0.3$ |
| | UXV | Puck | 86.0 | 3.28 | 0.76 | 0? | 0? |
| | UV | Miranda | 129.783 | 4.95 | 1.413 | 0.0027? | 4.22 |
| | UI | Ariel | 191.239 | 7.30 | 2.520 | 0.0034? | 0.31 |
| | UII | Umbriel | 265.969 | 10.15 | 4.144 | 0.0050? | 0.36 |
| | UIII | Titania | 435.844 | 16.64 | 8.706 | 0.0022? | 0.14 |
| | UIV | Oberon | 582.596 | 22.24 | 13.463 | 0.0008? | 0.10 |
| | UXVI | Caliban | 7,187 | 274.4 | 579(R) | 0.082 | 139.7[d] |
| | UXX | Stephano | 7,960 | 303.9 | 676(R) | 0.146 | 141.5[d] |
| | UXVII | Sycorax | 12,240 | 429.1 | 1289(R) | 0.509 | 152.7[d] |
| | UXVIII | Prospero | 16,150 | 616.5 | 1953(R) | 0.327 | 146.3[d] |
| | UXIX | Setebos | 18,250 | 696.7 | 2345(R) | 0.494 | 148.8[d] |
| Neptune | NIII | Naiad | 48.0 | 1.975 | 0.296 | 0.00 | 0.00 |
| | NIV | Thalassa | 50.0 | 2.058 | 0.312 | 0.00 | 4.5 |
| | NV | Despina | 52.2 | 2.148 | 0.333 | 0.00 | 0.00 |
| | NVI | Galatea | 62.0 | 2.551 | 0.429 | 0.00 | 0.00 |
| | NVII | Larissa | 73.6 | 3.029 | 0.554 | 0.00 | 0.00 |
| | NVIII | Proteus | 117.6 | 4.839 | 1.121 | 0.00 | 0.00 |
| | NI | Triton | 354.3 | 14.0 | 5.877(R) | $<0.0005$ | 159.0 |
| | NII | Nereid | 5513.5 | 219 | 360.16 | 0.75 | 27.6[d] |
| Pluto | PI | Charon | 19.1 | 16.7 | 6.387 | 0? | 94.3 |

[a] R denotes a retrograde orbit.
[b] These bodies have forced eccentricities due to interactions with other satellites.
[c] In degrees, relative to the orbital plane of the planet.
[d] Orbits of distant satellites are strongly perturbed by the Sun.

would not have survived over the age of the Solar System.

Earth's moon has an appearance strikingly similar to Mercury, but a density some 40% lower than Mercury's. It is also enormously more conveniently situated for observation and has been the object of dozens of spacecraft missions in the Soviet Luna, Zond, and Lunokhod programs and the American Pioneer, Ranger, Surveyor, Lunar Orbiter, and manned Apollo programs. About a half ton of lunar surface material

**Table III.5** Physical Properties of Satellites

| Planet | | Satellite | Mass ($10^{20}$ kg) | Radius (km) | Density ($10^3\ kg\,m^{-3}$) | Surface Composition |
|---|---|---|---|---|---|---|
| Earth | | Moon | 734.9 | 1738 | 3.34 | Rocks |
| Mars | MI | Phobos | $1.26 \times 10^{-4}$ | 11[a] | 2.2 | Carbonaceous |
| | MII | Deimos | $1.8 \times 10^{-5}$ | 6.3[a] | 1.7 | Carbonaceous |
| Asteroids[b] | | | | | | |
| Jupiter | JXVI | Metis | – | 20 | – | Rock? |
| | JXV | Adrastea- | | 10[a] | – | Rock? |
| | JV | Amalthea | – | 97[a] | – | Rock with sulfur |
| | JIV | Thebe | – | 50 | – | Rock? |
| | JI | Io | 894 | 1815 | 3.57 | Rock with sulfur |
| | JII | Europa | 480 | 1569 | 2.97 | Ice over rock |
| | JIII | Ganymede | 1482.3 | 2631 | 1.94 | Water ice |
| | JIV | Callisto | 1076.6 | 2400 | 1.86 | Dirty water ice |
| | 2000 J1 | – | – | 8 | – | Carbonaceous? |
| | JXIII | Leda | – | 8 | – | Carbonaceous? |
| | JVI | Himalia | – | 90 | – | Carbonaceous? |
| | JX | Lysithea | – | 20 | – | Carbonaceous? |
| | JVII | Elara | – | 40 | – | Carbonaceous? |
| | 2000 J11 | | – | 2 | – | Carbonaceous? |
| | 2000 J10 | | – | 2 | – | Carbonaceous? |
| | 2000 J3 | | – | 3 | – | Carbonaceous? |
| | 2000 J7 | | – | 4 | – | Carbonaceous? |
| | JXII | Ananke | – | 15 | – | Carbonaceous? |
| | 2000 J5 | | – | 2 | – | Carbonaceous? |
| | 2000 J9 | | – | 3 | – | Carbonaceous? |
| | JXI | Carme | – | 22 | – | Carbonaceous? |
| | 2000 J4 | | – | 2 | – | Carbonaceous? |
| | 2000 J6 | | – | 2 | – | Carbonaceous? |
| | JVIII | Pasiphae | | 35 | – | Carbonaceous? |
| | 2000 J8 | | – | 3 | – | Carbonaceous? |
| | JIX | Sinope | – | 20 | – | Carbonaceous? |
| | 2000 J2 | | – | 4 | – | Carbonaceous? |
| | 1999 J1 | | – | 5 | – | Carbonaceous? |
| Saturn | SXVIII | Pan | – | 15? | – | Water ice? |
| | SXV | Atlas | – | 16[a] | – | Water ice? |
| | SX | Janus | – | 93[a] | – | Water ice? |
| | SI | Mimas | 0.38 | 201 | 1.137 | Water ice? |
| | SII | Enceladus | 0.8 | 251 | 1.2 | Water ice? |
| | SIII | Tethys | 7.6 | 524 | 1.26 | Water ice? |
| | SXIII | Telesto | – | 11[a] | – | Water ice? |
| | SXIV | Calypso | – | 12[a] | – | Water ice? |
| | SIV | Dione | 10.5 | 559 | 1.44 | Water ice? |
| | SXII | Helene | – | 16[a] | – | Water ice? |
| | SV | Rhea | 24.9 | 764 | 1.33 | Water ice? |
| | SVI | Titan | 1345.7 | 2575 | 1.882 | Ices (atmosphere) |
| | SVII | Hyperion | – | 132[a] | – | Dirty water ice |
| | SVIII | Iapetus | 18.8 | 718 | 1.21 | Ice/carbonaceous? |
| | 2000 S5 | | – | 9 | – | Ice/carbonaceous? |
| | 2000 S6 | | – | 7 | – | Ice/carbonaceous? |
| | SIX | Phoebe | – | 110[a] | – | Ice and carbonaceous |
| | 2000 S2 | | – | 12 | – | Ice/carbonaceous? |
| | 2000 S8 | | – | 4 | – | Ice/carbonaceous? |
| | 2000 S3 | | – | 22 | – | Ice/carbonaceous? |
| | 2000 S10 | | – | 5 | – | Ice/carbonaceous? |
| | 2000 S11 | | – | 15 | – | Ice/carbonaceous? |
| | 2000 S4 | | – | 10 | – | Ice/carbonaceous? |
| | 2000 S9 | | – | 4 | – | Ice/carbonaceous? |
| | 2000 S12 | | – | 4 | – | Ice/carbonaceous? |
| | 2000 S7 | | – | 4 | – | Ice/carbonaceous? |
| | 2000 S1 | | – | 10 | – | Ice/carbonaceous? |

(*Continued*)

Table III.5 (*Continued*)

| Planet | | Satellite | Mass ($10^{20}$ kg) | Radius (km) | Density ($10^3$ kg m$^{-3}$) | Surface Composition |
|---|---|---|---|---|---|---|
| Uranus | UVI | Cordelia | – | 20 | – | Water ice? |
| | UVII | Ophelia | – | 25 | – | Water ice? |
| | UVIII | Bianca | – | 25 | – | Water ice? |
| | UIX | Cressida | – | 30 | – | Water ice? |
| | UX | Desdemona | – | 30 | – | Water ice? |
| | UXI | Juliet | – | 40 | – | Water ice? |
| | UXII | Portia | – | 40 | – | Water ice? |
| | UXIII | Rosalind | – | 30 | – | Water ice? |
| | UXIV | Belinda | – | 30 | – | Water ice? |
| | 1986 U10 | | – | 15 | – | Water ice? |
| | UXV | Puck | – | 85 | – | Water ice? |
| | UV | Miranda | 0.7 | 242 | 1.3 | Dirty water ice |
| | UI | Ariel | 13 | 580 | 1.6 | Dirty water ice |
| | UII | Umbriel | 13 | 595 | 1.4 | Dirty water ice |
| | UIII | Titania | 35 | 805 | 1.6 | Dirty water ice |
| | UIV | Oberon | 29 | 775 | 1.5 | Dirty water ice |
| | UXVI | Caliban | – | 10 | – | Dirty ices? |
| | UXX | Stephano | – | 10 | – | Dirty ices? |
| | UXVII | Sycorax | – | 10 | – | Dirty ices? |
| | UXVIII | Prospero | – | 10 | – | Dirty ices? |
| | UXIX | Setebos | – | 10 | – | Dirty ices? |
| Neptune | NIII | Naiad | | 20 | | |
| | NIV | Thalassa | | 20 | | |
| | NV | Despina | | 20 | | |
| | NVI | Galatea | | 20 | | |
| | NVII | Larissa | | 20 | | |
| | NVIII | Proteus | | 20 | | |
| | NI | Triton | 214.2 | 1355 | 2.05 | Ices |
| | NII | Nereid | – | 300 | – | Ices |
| Pluto | PI | Charon | – | 640 | 1.84 | Methane ice |

[a] Mean radii are given for satellites that are markedly non-spherical. Most other radii less than 100 km are inferred from the brightness of an unresolved object.

[b] A number of asteroids have or appear to have satellites. The best documented include the Belt asteroids (243) Ida (satellite named Dactyl), (22) Kalliope, (45) Eugenia, (87) Sylvia, (90) Antiope, (107) Camilla, (762) Pulkova, and (3749) Balam. The Jupiter L5 Trojan asteroid (617) Patroclus has a companion of nearly the same size. The NEA 1998 $ST_{27}$ is a radar binary. All are probably rocky bodies. There have been other recent reports of companion's to 2001 $QT_{297}$, 2001 $SL_9$, and 2002 $BM_{36}$. Also Kuiper belt and plutino objects 2001 $QW_{322}$, 1999 $TC_{36}$, (26308) 1998 $SM_{165}$, 1998 $WW_{31}$, and 2000 $CF_{105}$.

has been returned to Earth; far more is now known about the Moon than about all the other natural satellites combined.

Relative to its primary, the Moon is an unusually massive satellite. It has over 1% of Earth's mass and is surpassed in this respect only by Pluto's satellite Charon. It has a dry, barren, atmosphereless surface that has been heavily cratered by impacts of asteroids and comets and that bears mute evidence of extensive episodic melting during the first billion years of its existence. The heavily cratered highlands are dominantly composed of anorthosite, a rock consisting almost entirely of calcic feldspar (anorthite). The darker, younger plains in the large mare basins are flooded with any of several varieties of basaltic lava.

The lunar orbit is mildly eccentric and inclined about 5.2° relative to the ecliptic (18.2° to 28.6° relative to Earth's equator). The Moon raises tides in Earth's oceans, and frictional dissipation of energy by tidal motions exerts a retarding torque on Earth's rotation and an accelerating torque on the Moon's orbital motion. The time scale for evolution of the Moon's orbit is shorter than the age of the Solar System; the evolutionary history of the lunar orbit is a fascinating topic. The density of the Moon is very much less than that of Earth, suggesting either different modes or different places of origin.

Mars has two known satellites, Phobos and Deimos, both irregular angular lumps of dark, heavily cratered rocky material, with dimensions on the order of 10–20 km. It is interesting that the orbital periods of these satellites bracket the rotation period of Mars, and hence, as seen from the surface of Mars, they cross the sky in opposite directions. These satellites have been closely studied by the Mariner 9 and Viking 1 and 2 Mars orbiters and were the target of the ill-fated Soviet Phobos missions. Their orbital evolution is also of interest and will be discussed in Chapter X.

Jupiter has four enormous satellites (Io, Europa, Ganymede, and Callisto), called the Galilean satellites after their discoverer. Three families of smaller satellites are present. One family lies far below the orbits of the Galilean satellites, close to Jupiter's faint ring. A second family lies far outside the Galilean region, in highly inclined and eccentric orbits. The third family occupies a cluster of very distant, highly inclined *retrograde* orbits (their orbital angular momentum vector lies in their southern celestial hemisphere) that are so susceptible to solar perturbations that the orbits constantly change.

The densities of the Galilean satellites decrease with distance from Jupiter and range so low that Ganymede and Callisto at least must be mostly composed of ices, not rocks. Ice is known to be present on the surfaces of three of these bodies, and the other, Io, displays violent volcanic activity and sulfur magmatism of truly Hadean proportions.

Saturn's largest satellite, Titan, shares with Ganymede the distinction of being larger than the planet Mercury. Titan has a substantial methane-bearing nitrogen atmosphere with a surface pressure higher than that on Mars or Earth. The density of Titan is suggestive of a composition rich in ices, possibly very similar in ice:rock ratio to Ganymede. Iapetus, much smaller than Titan, is divided into two hemispheres that differ in brightness by about a factor of 7. A number of other smaller satellites, some of which participate in an astounding variety of different kinds of orbital resonances and interactions, are also present. Finally, there is the very complex and equally beautiful ring system.

Uranus has five moderate-sized satellites about which little is known, plus a system of small satellites and several very dark and narrow rings which orbit close to the planet. Neptune has two long-known satellites, one of which is so large that it is more easily observable than any of the satellites of Uranus, and six small satellites discovered by Voyager II in 1989. The large satellite, Triton, has the distinction of being the only large satellite of any planet to be found in a retrograde orbit. Pluto has one satellite, appropriately named Charon, which has permitted the first credible determination of the mass and density of Pluto.

In addition to nine planets and about three dozen known satellites with dimensions greater than 10 km, the Solar System contains countless smaller bodies of several fundamentally different classes. Of these, one group, the asteroids, contains virtually all the known bodies in heliocentric orbit.

## Asteroids

In a wide belt between the orbits of Mars and Jupiter there are many thousands of small rocky bodies, most of them in orbits of modest eccentricity and inclination. The orbital elements of these belt asteroids are far from uniformly distributed over the space available. There are many very sharply defined maxima and minima in the distribution of orbital semimajor axes, with many of the minima corresponding closely to resonances of the orbital period of Jupiter. The distribution of asteroid orbits is shown in Fig. III.9.

Careful statistical analysis of the distributions of the orbital parameters of the asteroids reveals that many asteroids are members of "families" with closely similar orbits, each possibly derived from the fragmentation of a single precursor by collision with a high-speed interloper such as a comet or eccentric asteroid.

In addition to these dynamically related families, there are several other classes of orbits of interest. One such class orbits in splendid isolation near 4.1 AU; they are named the Hilda asteroids after their largest member. Two other groups of small, dark asteroids circle the Sun on the orbit of Jupiter, each group close to the apices of the equilateral triangles of which the Sun–Jupiter line is the base. These groups are referred to as the Trojans and the Greeks, although the discerning classicist would detect that some of the individuals are clearly in the wrong camp.

A number of other small asteroids also lie outside the confines of the Belt in the other direction. Some actually cross the orbit of Mars and the other terrestrial planets. The Amor asteroids approach within 1.3 AU of the Sun, but do not cross Earth's orbit because their perihelia lie outside Earth's aphelion distance of $Q = a(1 + e) = 1.016\,\text{AU}$. Fewer yet cross the orbit of Earth (Apollo asteroids), and some of these even cross the orbit of Venus. Only about 1% of the Apollos and Atens reach Mercury's aphelion distance of 0.467 AU. A few (the Atens) have orbital periods less than 1 Earth year. These three groups (Amors, Apollos, and Atens, collectively known as the near-Earth asteroids) cannot persist in such orbits for longer than a few tens of millions of years without colliding with one of the terrestrial planets. They and the comets are responsible for the cratering history of the inner planets.

Since the early 1990s it has become clear that there are large numbers of asteroidal bodies in low-inclination orbits near and beyond Pluto's orbit. Many of these, such as the type example 2060 Chiron and 1992 AD, are in eccentric orbits that cross the orbits of one or more of the giant planets. These bodies, termed Centaurs, are especially vulnerable to strong gravitational perturbations. They may be ejected from the Solar System as a result of a close planetary encounter, or even, like Chiron itself, be diverted from time to time into highly eccentric orbits that pass through the terrestrial planet region.

A related but distinct population of asteroidal bodies forms an extended flattened disk around and

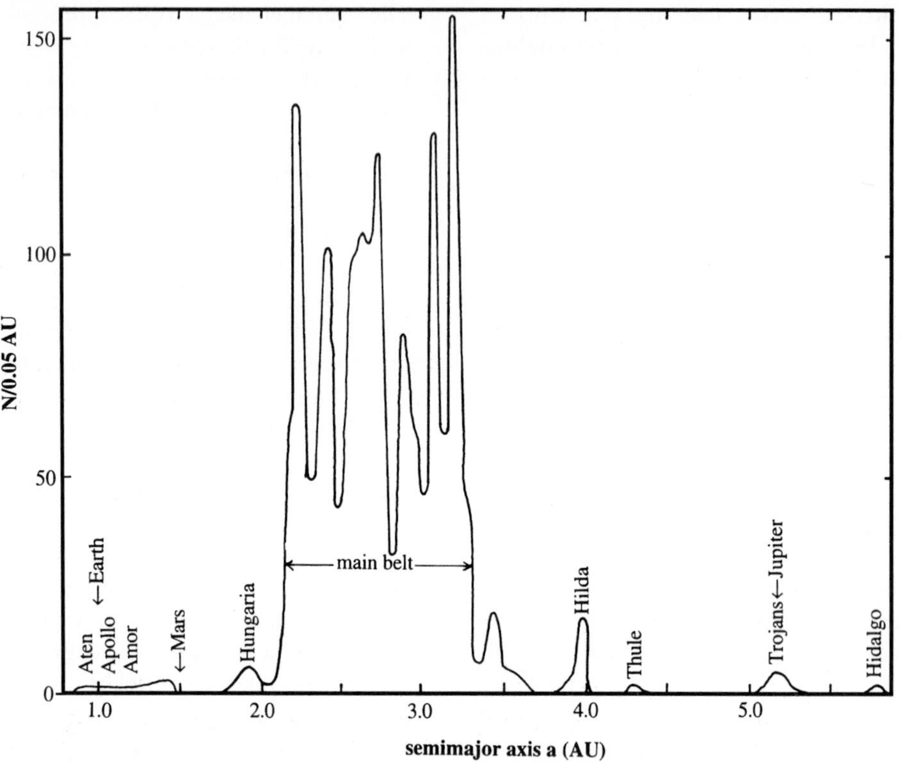


**Figure III.9** Orbital semimajor axes of the asteroids.

beyond the orbit of Pluto, reaching out at least to 50 AU from the Sun. These bodies, which range in size up to several hundred kilometers in radius, constitute the Kuiper belt. Pluto may reasonably be considered as the largest known Kuiper belt object.

The orbital parameters of a few selected large or otherwise interesting asteroids are given in Table III.6. Note that describing these bodies as asteroidal in appearance does not rule out the possibility that they have cometary composition, but lack cometary appearance simply because they are so cold that evaporation of ices is negligible.

Most asteroids are so small and distant that, when seen through a telescope, they look like points of light; this starlike appearance engenders their name (aster-oid). It is therefore difficult to determine the intrinsic properties of these bodies. The masses of only a few very large asteroids which happen to interact gravitationally with smaller neighbors can be determined. Photometric or spectroscopic information is available on more than 600 of the largest and brightest asteroids, and several different spectral classes of bodies, each with its own distinctive distribution of orbital distances, are well known. The intensity of the light from most asteroids varies periodically with a period of several hours, due to the rotation of bodies of jagged and irregular shape. Many small bodies, including Belt asteroids, near-Earth asteroids, and Kuiper belt bodies, have been found to have satellites. Because of the difficulty of observing most of these bodies, and especially the difficulty of detecting small, close companions of distant asteroids, the actual frequency of multiplicity among these populations is poorly constrained; nonethe-

**Table III.6** Orbital Elements of Selected Asteroids

| Asteroid | *a* (AU) | Eccentricity | Inclination (°) | Period (days) |
|---|---|---|---|---|
| 1 Ceres | 2.7675 | 0.0783 | 10.605 | 1681.6 |
| 2 Pallas | 2.7737 | 0.2326 | 34.794 | 1687.2 |
| 3 Juno | 2.6686 | 0.2573 | 13.002 | 1592.3 |
| 4 Vesta | 2.3612 | 0.0892 | 7.143 | 1325.2 |
| 10 Hygiea | 3.1355 | 0.1196 | 3.836 | 2027.9 |
| 31 Euphrosyne | 3.1462 | 0.2274 | 26.343 | 2038.3 |
| 65 Cybele | 3.4363 | 0.1075 | 3.549 | 2326.6 |
| 153 Hilda | 3.9754 | 0.1536 | 7.845 | 2895.1 |
| 279 Thule | 4.2609 | 0.0323 | 2.339 | 3212.5 |
| 324 Bamberga | 2.6815 | 0.3404 | 11.139 | 1603.8 |
| 434 Hungaria | 1.9441 | 0.0738 | 22.510 | 990.1 |
| 511 Davida | 3.1806 | 0.1744 | 15.908 | 2071.8 |
| 617 Patroclus | 5.2338 | 0.1388 | 22.091 | 4373.4 |
| 887 Alinda | 2.5017 | 0.5535 | 9.189 | 1445.2 |
| 944 Hidalgo | 5.8608 | 0.6565 | 42.404 | 5182.3 |
| 1566 Icarus | 1.0778 | 0.8265 | 22.938 | 408.7 |
| 1662 Apollo | 1.4697 | 0.5599 | 6.360 | 650.8 |
| 2060 Chiron | 13.6955 | 0.3786 | 6.923 | 18512.2 |
| 2100 Ra-Shalom | 0.8320 | 0.4365 | 15.755 | 277.2 |

less, it would not be surprising to find that over 10% of the known large asteroids have smaller companions.

Direct measurement of asteroid sizes from Earth is extremely difficult for all but a few of the very largest asteroids. Direct imaging of large bodies such as 1 Ceres by the Hubble Space Telescope is possible; indeed, HST has revealed considerable albedo variation over the surface of Ceres, including large circular dark features suggestive of major impact basins. Indirect techniques based on simultaneous measurement of the visual and infrared brightness of optically unresolved bodies have provided good size estimates for hundreds of asteroids (see Chapter VIII). Polarimetry also provides an independent estimate of the albedos and sizes of many asteroids. A few near-Earth asteroids have been mapped by radar, which provides detailed information on their size and shape. Many NEAs are strikingly nonspherical in shape.

Mass and density data on the asteroids are sparser than size data. Several of the largest asteroids participate in near-resonances with smaller asteroids, permitting determination of the mass of the larger partner from its perturbations on the smaller. The masses of those asteroids found to have orbiting companions can also be calculated from Kepler's laws of motion. Photometric study of mutual eclipses of orbiting asteroid–satellite pairs have also begun to yield size and mass data. Finally, a few asteroids (Mathilde, Ida, Eros) have been visited by the Galileo and Near-Earth Asteroid Rendezvous (NEAR) spacecraft. The combination of spacecraft tracking data and imaging of the asteroids permits calculation of the densities of these asteroids with enough precision to warrant comparison with the densities of meteorites determined in the laboratory.

The size distributions of the NEA and Belt populations of asteroids are well determined, albeit in a complementary way (most known NEAs are so small that they would not be detectable in the Belt). Also, photometric data on the recently discovered Centaur and Kuiper belt populations have begun to reveal the size distribution of their larger and brighter members. It seems clear that the mass of the Belt and NEA populations is dominated by the largest one or two members. The mass of Ceres is larger than the combined masses of all the other Belt asteroids. The total masses of the other populations of asteroids (Trojans, Centaurs, Kuiper belt) are not as well constrained, but it is likely that the Kuiper belt contains at least a few Pluto masses, and possibly much more.

## Comets

One of the most spectacular phenomena in the Solar System is the apparition of a bright comet. Throughout history the sudden appearance and rapid development of bright comets have inspired prophecies of doom and engendered civil disorder during the days or weeks that they are visible in the sky.

Comets fall into two general orbital classifications. First, there are many comets with orbital periods less than 100 years, with orbital eccentricities less than about 0.9° and prograde orbits of inclinations of 10° or 20°. These are called the short-period comets. By a somewhat sloppy habit, they are often called the "periodic" comets. Their names are preceded by the letter P/, followed by the name of the discoverer, as in P/Bradfield. The orbits of some of these bodies differ little from those of known near-Earth asteroids, and indeed the boundary between these classes is determined only by the presence of a coma or tail, not by orbital parameters.

There is a second family of comets with typical orbital periods of millions of years, orbital eccentricities of order 0.9999 and higher, and with orbital inclinations randomly distributed between prograde and retrograde orientations, with their lines of apsides nearly randomly distributed over the celestial sphere. These "long-period" comets, designated by the letter C/, such as C/2001 T3, are of course known from only a single perihelion passage. Their orbital speeds near perihelion are so close to the local escape velocity from the Sun that their orbital paths are often approximated as parabolas, even though (as we shall see in Chapter VII) they almost certainly enter the inner Solar System on bound elliptical orbits.

Comets become extremely brilliant when they approach the Sun to within the radius of the orbit of Mars. The extensive gaseous head and long streaming tails of gas and dust develop by evaporation of volatile material from a small solid nucleus, which is usually no more than a few kilometers in radius. The largest and most active comets may grow tails well over 100 million kilometers long and be bright enough to be visible in the daytime. Data on the orbits of a number of comets are collected in Table III.7, and photographs of several interesting comets appear in Fig. III.10.

The masses and densities of cometary nuclei are unfortunately not known, because no comet has ever been observed to perturb the motion of any other Solar System body. In addition, hardly any comet nuclei are large enough for diameter estimates to be made. Usually only the apparent magnitude of the nucleus can be measured, but the nucleus can only be discerned when it is close to Earth and hence also near perihelion, where the brightness of the head seriously hinders measurement of the nuclear magnitude. The largest and brightest nucleus ever observed is no larger than about 100 km in diameter, very small compared with planets.

**Table III.7** Orbital Elements of Selected Comets

| Comet | *a* (AU) | Eccentricity | Inclination | Period (years) |
|---|---|---|---|---|
| Encke | 2.219 | 0.8463 | 11.93 | 3.31 |
| Grigg–Skjellerup | 2.959 | 0.6657 | 21.14 | 5.09 |
| Tempel 2 | 3.036 | 0.5444 | 12.43 | 5.29 |
| Tempel 1 | 3.116 | 0.5197 | 10.54 | 5.50 |
| Wirtanen | 3.117 | 0.6521 | 11.67 | 5.50 |
| Pons–Winnecke | 3.433 | 0.6347 | 22.31 | 6.36 |
| d'Arrest | 3.441 | 0.6248 | 19.43 | 6.38 |
| Schwassmannn–Wachmann 2 | 3.443 | 0.3984 | 3.76 | 6.39 |
| Kopff | 3.461 | 0.5445 | 4.72 | 6.44 |
| Giacobini–Zinner | 3.516 | 0.7076 | 31.88 | 6.59 |
| Gunn | 3.597 | 0.3164 | 10.38 | 6.82 |
| Arend–Rigaux | 3.604 | 0.5987 | 17.84 | 6.84 |
| Brooks 2 | 3.622 | 0.4907 | 5.55 | 6.89 |
| Holmes | 3.687 | 0.4118 | 19.19 | 7.08 |
| Faye | 3.779 | 0.5783 | 9.09 | 7.34 |
| Harrington–Abell | 3.845 | 0.5421 | 10.25 | 7.54 |
| Arend | 4.005 | 0.5364 | 19.93 | 8.02 |
| Wolf | 4.072 | 0.4068 | 27.51 | 8.21 |
| Whipple | 4.163 | 0.2606 | 9.94 | 8.49 |
| Vaisala 1 | 4.910 | 0.6334 | 11.61 | 10.9 |
| Neujmin 3 | 4.919 | 0.5813 | 3.94 | 10.9 |
| Klemola | 4.931 | 0.6405 | 10.96 | 10.9 |
| Van Biesbroeck | 5.368 | 0.5527 | 6.62 | 12.4 |
| Wild 1 | 5.602 | 0.6471 | 19.90 | 13.3 |
| Tuttle | 5.674 | 0.8241 | 54.69 | 13.5 |
| du Toit | 6.004 | 0.7879 | 18.69 | 14.7 |
| Schwassmann–Wachmann 1 | 6.042 | 0.0447 | 9.37 | 14.9 |
| Neujmin 1 | 6.921 | 0.7756 | 14.17 | 18.2 |
| Oterma | 7.228 | 0.2430 | 1.94 | 19.4 |
| Crommelin | 9.102 | 0.9192 | 29.10 | 27.4 |
| Tempel–Tuttle | 10.337 | 0.9056 | 162.48 | 33.2 |
| Brorsen–Metcalf | 17.075 | 0.9720 | 19.33 | 70.6 |
| Halley | 17.854 | 0.9673 | 162.23 | 76.0 |
| Typical long-period comet | 10,000 | 0.9999 | any | 1,000,000 |

The photochemistry and plasma physics of comet tails are interesting and complex. Abundant spectroscopic evidence on the composition of the vapors evaporated from the nucleus attests to the prevalence of both silicaceous dust and mixed ices in the fresh comet, prior to alteration by the Sun. The only comet to be studied at close range is P/Halley, which was visited by an armada of Soviet, European, and Japanese spacecraft in 1986.

Almost all comets visible to the unaided eye, such as the famous P/Halley, cross the orbits of one or more of the terrestrial planets. The evolution of such orbits, under the disturbing effects of planetary perturbations and nongravitational forces due to gas jetting, and the consequences of collisions between comets and planets are both active areas of research.

## Meteors

The bright, fast-moving streaks of light often seen in the night sky are called *meteors*. They are caused by the entry of small solid particles of cometary or asteroidal debris, called *meteoroids*, into Earth's upper atmosphere at such high speeds that they flash into incandescent vapor in a fraction of a second.. Most of the meteoroids that produce meteor displays on Earth are members of quite well-defined "swarms" that pursue highly eccentric orbits around the Sun and encounter Earth at fixed points on Earth's orbit (fixed dates) at predictable intervals, appearing to radiate from a particular point in the sky called the radiant point. Encounters of Earth with these swarms produce correlated displays of many meteors, collectively called *meteor showers*. Meteor showers are commonly named after the constellation from which they appear to radiate, as in the Leonid, Perseid, and Lyrid showers. In extreme cases, the display may be so intense as to merit the name *meteor storm*

Meteoroids arise from both the evaporation of ices in dust-bearing comets and in collisions on nearby asteroids. Indeed, several prominent meteor swarms are known to be pursuing the orbits once occupied by recently deceased comets or by near-Earth asteroids. Figure III.11 is a time exposure of a brilliant meteor storm in November 1966 in which some observers estimated 100,000 visible meteors in the space of about an hour. A less spectacular reprise of this Leonid shower occurred in November 2001. A number of the best-established and most easily observed meteor showers are listed in Table III.8.

## Meteorites

Meteorites are defined as solid macroscopic bodies that survive entry into Earth's atmosphere. They are not associated in any way with shower meteors and are so rare that they are not often observed to enter the atmosphere. Although an observer can see several meteors per hour on almost any clear, dark night, meteorites of recoverable size fall at an average rate of little more than one per year for each million square kilometers of Earth's surface.

The approximately 10,000 known meteorites are quite varied in their properties, ranging from huge crystals of iron–nickel alloy weighing tens of tons down to 10-mg lumps of clayey material rich in organic matter. Most meteorites are predominantly composed of silicates, sulfides, and metal. Because, aside from the samples returned from the Moon, meteorites are the only samples of extraterrestrial matter available for study in

**Figure III.10** Visual appearance of comets. Comet P/Halley (in the inset on the left) as photographed by the Spacewatch telescope of the University of Arizona in 1986. Comet Ikeya-Seki, in the main picture, photographed by Dennis Milon in Tucson, Arizona, in 1965. The tails of both comets are well developed.

terrestrial laboratories, they play a major role in our ideas about the chemical composition of Solar System material. Also, because most meteorites are much older than the oldest known lunar and terrestrial rocks, they take on great importance in relation to questions of origin.

Meteorites are also a (rather biased) sample of the solid fragmented asteroidal matter that pervades the inner Solar System and that accounts for much of the cratering on the Moon and the other terrestrial planets. That Earth is not immune to such influences is borne out by the photograph of Meteor Crater, Arizona (Fig. III.12), which most certainly was not produced by a mere meteor. Dozens of large terrestrial meteorite craters are now known, several of them accompanied by substantial masses of meteorite fragments.

Some meteorites have the texture of a low-temperature mixture of solids, never severely heated or melted, with ages of about 4.6 billion years (Gyr). These meteorites usually contain small glassy beads of silicates, called *chondrules* after the Greek word for droplet or seed. The meteorites that contain chondrules are called *chondrites*. Other meteorites show textures and compositions diagnostic of melting, density-dependent geochemical differentiation, and subsequent cooling. These range in composition from the stony *achondrites*, through the *stony-irons*, to the dense, metal-rich *irons*. Of these, the oldest and chemically most primitive are the chondrites.

## Cosmic Dust

Every year thousands of tons of dust of extraterrestrial origin fall to the surface of the Earth. The large majority of this dust is produced by aerodynamic ablation of relatively rare large meteors and meteorites, but some of it is dust of meteoric origin that has survived atmospheric entry (micrometeorites). Cosmic dust may be found in deep sea sediments, the polar icecaps, or still suspended in the stratosphere, whence it may be collected by balloon-, airplane-, or rocket-borne filtering apparatus.

## Cosmic Rays

As we mentioned in our brief discussion of the Sun, particles (mostly protons) with energies up to several hundred MeV are emitted by the Sun during periods of intense flare activity. These particles are termed solar

**Figure III.11** Two bright meteors. They were photographed by Donald Pearson during the historic Leonid shower of November 17, 1966, from atop Kitt Peak in Arizona. Photo courtesy of Dennis Milon.

cosmic rays. There is, in addition to this solar source, a steady background of high-energy radiation plainly not of solar origin, which is called galactic cosmic radiation. Galactic cosmic rays with energies up to $10^{21}$ eV have been detected, but energies near 10 GeV are more typical. The low-energy galactic cosmic ray flux can be modulated by fluctuations in the density and energy of the solar wind, but the weak magnetic fields in interplanetary space are unable to deflect the more energetic particles.

Cosmic rays irradiate all atmospheres and exposed surfaces in the Solar System. Their effects, especially production of radioactive nuclides, can be used to measure the time spent by meteorites in interplanetary space. The archaeological tool of radiocarbon dating owes its existence to the production of $^{14}C$ in the Earth's atmosphere by cosmic rays.

## Planetary Science in the Space Age

A very large proportion of human knowledge of the planets, satellites, and other Solar System bodies comes directly from spacecraft exploration of the Solar System.

**Table III.8** Prominent Meteor Showers

| Name | Dates | Source | Radiant | |
|---|---|---|---|---|
| | | | RA | dec (°) |
| Quadrantids | 1–4 Jan | ? | 15h20m | +52 |
| Lyrids | 19–23 Apr | Comet 1861 I | 18h4m | +33 |
| Aquarids | 1–6 May | Halley | 22h16m | −2 |
| Delta Aquarids | 26–31 Jul | ? | 22h36m | −11 |
| Perseids | 10–14 Aug | ? | 3h8m | +58 |
| Giacobinids | 9 Oct | Giacobini–Zinner | 17h40m | +55 |
| Orionids | 18–23 Oct | ? | 6h8m | +15 |
| Taurids | 31 Oct–6 Nov | Encke | 3h40m | +15 |
| Leonids | 14–18 Nov | ? | 10h0m | +22 |
| Geminids | 10–13 Dec | 3200 Phaeton | 7h32m | +32 |

The first rockets capable of launching payloads out of Earth's gravity well were developed from the earliest Soviet and American intercontinental ballistic missiles in the 1950s. Flyby missions have visited every planet except Pluto, plus two asteroids, one comet, and numerous moons of Earth, Mars, Jupiter, Saturn, and Uranus. Atmospheric entry probes have been sent into the atmosphere of Mars, and survivable landers have been placed on the Moon, Venus, and Mars. Manned missions briefly reached as far as the Moon in the late 1960s and early 1970s as part of the Apollo program, during which 12 American astronauts landed on the Moon and returned safely to Earth. Japan and the European Space Agency have begun to launch lunar and planetary probes, and China has the technical capability to do so. Eight lunar missions were attempted in the 1950s, and 79 lunar and planetary missions were launched in the 1960s; 52, in the 1970s; 17, in the 1980s; and 4, in the 1990s. Many individual (successful) missions are mentioned in this book. A table of all known lunar and planetary launch attempts is given in Appendix III.

**Figure III.12** Meteor Crater, Arizona. This mile-wide crater east of Flagstaff, Arizona, was caused by the multimegaton impact of a small (roughly 60 m in diameter) iron asteroid. It most certainly was not caused by a meteor!

## Summary

This brief introductory chapter is intended to familiarize readers who may have had little astronomical background with the basic layout of the Solar System and the nature of the bodies which compose it. The quantitative material summarized in the figures and tables will be of very general use throughout the remainder of the book. The subject areas introduced in this chapter will be treated in detail in the subsequent chapters, in which the general perspective and interrelationships we have just surveyed will be assumed to be familiar.

## Exercises

### *The Sun*

III.1 The escape velocity from the photospheric "surface" of the Sun is 618 km $s^{-1}$. At what temperature would a gas of hydrogen atoms have a mean thermal speed equal to the Sun's surface escape velocity?

### *Orbits of the Planets*

III.2 The *vis viva* equation (III.30) describes the variation of orbital velocity with distance from the central mass. Using the normal symbols for perihelion distance ($q$) and aphelion distance ($Q$), note that Figure III.2 tells us that $q = a(1 - e)$ and $Q = a(1 + e)$. Also note that $\mu$ is proportional to the mass $M$ of the Sun: $\mu = GM$, where the conversion factor $G$ is the universal gravitational constant.

a. Show that $v^2$ (twice the energy density of the orbiting body) at the perihelion point is given by $v_q^2 = (GM/a)[(1 + e)/(1 - e)]$.

b. Derive a similar expression for $v^2$ at aphelion.

c. Consider two circular orbits coplanar with the elliptical orbit under discussion, one osculating (grazing) the elliptical orbit at perihelion, and the other at aphelion. Show that the ratio of $v^2$ for the circular orbit at the perihelion point on the elliptical orbit to $v^2$ for the circular orbit at aphelion is $(1 + e)/(1 - e)$.

III.3 The kinetic energy of a body falling from infinity to the surface of a planet is precisely equal to the gravitational work done on that body during its fall: in other words, the integral of the gravitational force from infinite distance to $r_s$. Reversing the argument, that kinetic energy is precisely equal to the escape energy, $1/2\,mv_{esc}^2$, required for it to just escape from the gravitational well of its primary. Show that, at any distance $r_1$, $v_{esc}^2$ is exactly $2v_{circ}^2$. The orbital velocity at $r$ [Eq. (III.31)] is $v_{circ} = (GM/r_1)^{1/2}$.

### *Properties of the Planets*

III.4 Examine Table III.2. What can you conclude about the polar coordinate system of right ascension (RA) and declination (dec) used therein?

### *Mass and Angular Momentum Distribution*

III.5 Consider Eqs. (III.37 and (III.38). When a fluid planet spins so fast that its equatorial rotation speed is a significant percentage of the circular orbital velocity, serious distortion of the shape of the planet must occur. Compare a fluid planet deformed into an oblate spheroid (A) of equatorial radius $1.414r_0$ and polar radius $0.500r_0$ to a rigid planet (B) of radius $1.000r_0$, both with the same mass, volume, and density.

a. If the two bodies have the same angular momentum, what are their approximate relative rotation periods?

b. What are the equatorial rotation velocities and equatorial escape velocities for these two bodies?

### *Satellites*

III.6 Sketch a histogram of the number of satellites versus log (mass), using the data given in Table III.5. Two or three size bins per decade of mass would be appropriate. Note the gap of about a factor of 20 that occurs between the "midsize" satellites such as Dione and the "lunar-sized" bodies. Many authors have regarded this gap as a very significant clue to early Solar System history.

III.7 Using the data in Table III.7, estimate the velocities of comets Brorsen–Metcalf and Halley as they cross Earth's orbit. Note that their orbits are very similar except for their inclinations.

### *Meteors*

III.8 A long-period cometary meteor strikes Earth's atmosphere at 50 km $s^{-1}$, converting its kinetic energy into heat. The heat capacity of rock is about $10^7$ erg $g^{-1}$ $K^{-1}$, and the melting temperature can be taken as 1500 K. To how high a temperature would the meteor's kinetic energy heat it? What physical processes would occur that would invalidate this temperature estimate?

### *Cosmic Rays*

III.9 A galactic cosmic-ray proton with an energy of $10^{21}$ eV strikes a lead bead with a mass of 1.00 g and deposits most of its energy in that target. Roughly what effect would this energy have on the temperature of the lead bead? Energy conversion factors may be found by looking up "Energy, conversion factors" in the Index.

# IV. The Sun and the Solar Nebula

## Introduction

The physical and chemical properties of solar material play a central role in any discussion of the origin of the Solar System, the properties of the Jovian planets, or the Sun itself. The Sun, in turn, serves as our prototype for the study of all other stars. The raw interstellar material out of which stellar and planetary systems form has a composition essentially identical to that of the atmospheres of all Population I Main Sequence stars. The compositions of planets are directly determined by physical and chemical processes such as the condensation and accretion of solids to make self-gravitating bodies. It is of great importance to understand why some planets form with enormous mass and nearly solar elemental composition, whereas others have vastly smaller masses and rocky or icy compositions that deviate most strikingly from that of the Sun.

Our attempts to understand the chemistry and physics of solar material are assisted by our ability to observe two very different present-day examples. The first, the Sun itself, is typical of Pop I stars of moderate surface temperature. The second example is given by the giant planets Jupiter and Saturn, whose observable envelopes have temperatures of at most several hundred kelvins.

Beyond its importance as our local example of a star, the Sun both directly and indirectly exerts profound effects upon many of the most important properties of the planets. It is the Sun about which they pursue their orbits, which illuminates their surfaces with visible light, irradiates their upper atmospheres with chemically active ultraviolet light, and warms their atmospheres with infrared radiation. The Sun pours out a torrent of electromagnetic radiation over a vast range of wavelengths and a "solar wind" of ions and electrons that dominates the properties of the interplanetary medium out to several tens of AU from the Sun. Our present purpose shall be to explore the chemical and physical properties of solar material and to apply this knowledge to the Sun and to that collapsing fragment of interstellar cloud out of which the Solar System formed, the Solar Nebula.

## Energy Production in the Sun

The major nuclear reactions involved in hydrogen fusion are collected in Table IV.1. These reactions have for the most part been discussed in Chapter II, and a detailed discussion will not be repeated here.

**Table IV.1** Nuclear Reactions in the Sun

The pp chain

$2(p + p \rightarrow {}^2H + e^+ + \nu_e)$
$2({}^2H + p \rightarrow {}^3He + \gamma)$
${}^3He + {}^3He \rightarrow {}^4He + 2p + \gamma$

The catalytic carbon cycle

${}^{12}C + {}^1H \rightarrow {}^{13}N$
${}^{13}N \rightarrow {}^{13}C + e^+ + \nu_e$
${}^{13}C + {}^1H \rightarrow {}^{14}N$
${}^{14}N + {}^1H \rightarrow {}^{15}O$
${}^{15}O \rightarrow {}^{15}N + e^+ + \nu_e$
${}^{15}N + {}^1H \rightarrow {}^{12}C + {}^4He$

${}^{15}N + {}^1H \rightarrow {}^{16}O$
${}^{16}O + {}^1H \rightarrow {}^{17}F$
${}^{17}F \rightarrow {}^{17}O + e^+ + \nu_e$
${}^{17}O + {}^1H \rightarrow {}^{14}N + {}^4He$

Denoting, as before, the rate of energy production per unit volume as $\varepsilon(\text{erg cm}^{-3}\,\text{s}^{-1})$, the radial distribution of energy production within a spherically symmetrical body can be written as $\varepsilon(r)$. The total luminosity of the Sun is found by integration of the energy production rate over the volume of the Sun, which, because of the assumption of spherical symmetry, is most simply represented in terms of thin spherical shells of thickness $dr$. Each of these shells contributes

$$dL(r) = 4\pi r^2 \varepsilon(r) dr \qquad \text{(IV.1)}$$

to the luminosity. As we have seen, $\varepsilon(r)$ is a sensitive function of temperature, density, and composition and is composed of contributions from all the nuclear reactions that can take place within the Sun. Because of the great preponderance of hydrogen fusion over other reactions in the Sun, we will approximate $\varepsilon(r)$ by

$$\varepsilon(r) = \varepsilon_{\text{CC}}(r) + \varepsilon_{\text{pp}}(r) \qquad \text{(IV.2)}$$

$$= 1.66 \rho x_{\text{H}} x_{\text{C}} (T_6/13)^{20.3} + 0.28 \rho x_{\text{H}}^2 (T_6/13)^{4.1}, \qquad \text{(IV.3)}$$

where $\rho(\text{g cm}^{-3})$, $x_{\text{H}}$, $x_{\text{C}}$, and $T_6$ (the temperature in MK) are all functions of $r$.

In a star that has formed from well-mixed interstellar material and that has not evolved into the giant phase and begun carbon production, $x_{\text{C}}$ is essentially constant throughout. From the atomic abundances of the elements $a_i$ given in a previous chapter (Table II.4), we can calculate that $x_{\text{C}}$, the carbon mass fraction, is

$$x_{\text{C}} = A_{\text{C}} a_{\text{C}} / \Sigma_i A_i a_i \approx A_{\text{C}} a_{\text{C}} / (A_{\text{H}} a_{\text{H}} + A_{\text{He}} a_{\text{He}}), \qquad \text{(IV.4)}$$

where $a_i$ is the abundance of element $i$ in atoms per $10^6$ atoms of Si, and $A_i$ is the atomic weight of $i$. The approximation that only hydrogen and helium contribute to the total mass can easily be tested by reference to Table II.4; it is valid to a precision of about 1%. From these data, we find $x_{\text{C}} = 3.84 \times 10^{-3}$. Main Sequence stars burn hydrogen to produce a helium core; hence we expect $x_{\text{H}}$ to remain near 0.78 outside that core but, beneath the surface of the core ($r = r_{\text{c}}$), $x_{\text{H}} = 0$. The zone within which $x_{\text{H}}$ is varying (and energy production is rapid) may be quite limited in radial extent. It will be limited from below by hydrogen exhaustion and from above by the lower temperatures, through the extremely strong temperature dependence of the rates of fusion reactions.

Direct experiments on the energy production mechanisms in stars are almost (but not quite) impossible. Consider the problem: temperatures and densities high enough for nuclear reactions can be produced only in the deep interior of a star. This means that almost the entire mass of a star lies between outside observers and the site of the reactions. In many ways this is a blessing; however, from the point of view of astrophysicists it is either an awkward necessity or a welcome barrier behind which theoreticians may hide, invulnerable to experimental refutation.

The products of hydrogen fusion (from which we must select one or more to observe) are helium, gamma rays, and neutrinos. Short of dismantling the star, a dangerous and costly exercise, the accumulated mass of the helium core is inaccessible to us. Gamma rays have very short mean free paths and cannot reach the surface of the Sun without being degraded to photons of about 1-$\mu$m wavelength. Neutrinos, on the other hand, suffer from the opposite problem: they are fantastically adept at penetrating matter, and even the most cleverly devised neutrino detectors are almost perfectly transparent to neutrinos. Indeed the first deduction of the existence of neutrinos was based on this very property.

Over 50 years ago it became known that beta-decay reactions produced different amounts of detectable energy from event to event and that the total measurable energy was always less than that calculated from the mass difference between the parent and the daughter nuclides. Furthermore, beta decay did not appear to conserve angular momentum! Enrico Fermi postulated that energy (and spin) could be conserved in beta decay only if a hitherto unobserved particle existed, with unusual properties suitable to its task. It must be a moderately energetic (2 MeV), uncharged (noninteractive; hard to detect), small (rest mass, $<< 1$ MeV) particle carrying spin angular momentum as well as energy. This Bilbo Baggins of particles was named the *neutrino*, the "little neutral one."

The cross-section (effective area) for interactions of neutrinos is extremely small, increasing markedly for larger neutrino energies:

$$\sigma_\nu = 10^{-44}\,(E_\nu/0.5\,\text{MeV})^2\,\text{cm}^2. \qquad \text{(IV.5)}$$

Cross-sections for nuclei and nucleons reflect their characteristic dimensions of 1 to 10 fermis (1 fer $= 10^{-13}$ cm) and are commonly near $10^{-24}$ cm$^2$. A convenient unit for nuclear cross-sections is the *barn* (1 barn $= 10^{-24}$ cm$^2$). Photon interaction cross-sections for the electron cloud in atoms and small molecules are often as large as $10^{-16}$ cm$^2$.

The probability of a neutrino collision with a particle over a 1-cm path length interval is

$$P(\text{cm}^{-1}) = \rho\sigma N_0/\mu, \tag{IV.6}$$

where $N_0$ is Avogadro's Number and $\mu$ is the atomic weight. Thus a typical 0.5-MeV neutrino traversing an average sample of solar material will suffer $1.4 \times 6 \times 10^{23} \times 10^{-44}/1 = 10^{-20}$ collisions per centimeter. The mean free path of the neutrino is thus $1 = 1/P = 10^{20}$ cm, which is $10^9$ times the radius of the Sun. One parsec is $1.5 \times 10^{13} \times 2.2 \times 10^5 = 3.3 \times 10^{18}$ cm; hence the neutrino could penetrate a solid shield 30 parsecs thick. Such shielding is certainly not available between us and the Sun!

Neutrinos from the proton–proton chain and the carbon (CNO) cycle are mostly positron-emission [Eq. (II.26)] or electron–positron annihilation [Eq. (II.7)] neutrinos with energies near 0.5 MeV. The proton–proton chain should supply about $6 \times 10^{10}$ neutrinos per square centimeter per second at Earth's orbit if conventional theories of stellar interiors are correct.

Fortunately for our prospects of detecting solar neutrinos, production of traces of the boron isotope $^8$B occurs in the Sun. This isotope decays via

$$^8\text{B} \rightarrow 2\,^4\text{He} + \text{e}^+ + \nu, \tag{IV.7}$$

giving off 14-MeV neutrinos, which have, according to Eq. (IV.5), 780 times the interaction cross section of p-p chain rate. This greatly enhanced detectability largely compensates for their much lower production. Neutrinos are detected by means of the reaction

$$^{37}\text{Cl} + \nu \rightarrow {}^{37}\text{Ar} + \text{e}^-, \tag{IV.8}$$

which takes place in a $10^5$ gal tank of halogenated hydrocarbons (liquid chlorine has only slightly more chlorine atoms per cubic centimeter and is notoriously nasty to work with). The $^{37}$Ar produced is flushed out of the tank and analyzed radiochemically. In order to minimize interference caused by cosmic ray events, the entire apparatus is placed at the bottom of a deep gold mine. At present, the results of this experiment are in disagreement with theory: the rate of detection of neutrinos is about three times less than the rate predicted by the most plausible theories of the solar interior. Until this discrepancy can be resolved it is not certain whether the fault lies in the theory, the experiment, or both. Still, the robustness of the experimental evidence after many years of critical scrutiny, and the great elasticity of theoretical ideas about the deep interior of the Sun, leads us to expect theoretical changes. The most promising approach attributes the discrepancy to "neutrino oscillations", in which neutrinos with very small rest mass freely interconvert between three forms, only one of which is detectable.

In addition to energy production by nuclear reactions, young stars can generate a substantial luminosity from the conversion of their gravitational potential energy into heat during collapse onto the Main Sequence. We have seen [Eq. (II.73)] that this collapse energy is $-GM^2/R$ for a uniform sphere of mass $M$ and radius $R$ formed by collapse from infinity. Thus the change in gravitational potential energy ($\Delta GPE$) is

$$\Delta GPE = -(16/9)\pi^2\rho^2 GR^5, \tag{IV.9}$$

where $\rho$ is the density of the body. The release of this energy can maintain the luminosity of the Sun at its present value for very long periods of time. The rate of mass loss by radiation is, because $L_\odot = \dot{m}_\odot c^2$, just $L_\odot/c^2$, or $4 \times 10^{12}$ g s$^{-1}$, which, prolonged for $5 \times 10^9$ years, gives a fractional mass loss of only 0.03%.

## Energy Transport in the Sun

In order to develop a steady-state model for the structure of the Sun, it is first necessary to examine the mechanisms by which heat may be transported from the site of nuclear reactions in the deep interior out to the radiating surface, the photosphere. Because of the obvious importance of radiation in the Sun, we shall first briefly consider radiative transport of energy.

Following the nomenclature of Eq. (IV.6) and the accompanying definition of the mean free path, we have

$$l = \mu/N_0\sigma\rho, \tag{IV.10}$$

where the dimensions are those appropriate for dealing with individual particles ($\sigma$ is the cross-section of a single particle). It is often more convenient to regard the attenuation of radiation as a property of a continuous fluid characterized by macroscopic variables only. Accordingly, we define the mass opacity ($K$)(cm$^2$ g$^{-1}$) by

$$K = N_0\sigma/\mu. \tag{IV.11}$$

Then,

$$l = 1/K\rho = 1/\alpha \tag{IV.12}$$

is the mean free path. $\alpha$ is the absorption coefficient (cm$^{-1}$). The opacity of a gas varies with wavelength, temperature, and pressure and in general is very different for different gases. Hydrogen and helium are, under normal conditions, commonly thought of as highly transparent, but this is not at all the case in stellar interiors.

Molecules in general may carry translational, electronic, vibrational, and rotational energy. A simple monatomic gas such as helium or atomic hydrogen cannot vibrate or rotate and can carry energy only as translational (kinetic) energy or electronic excitation. In a disordered fluid the translational energy is not quantized, but the electronic energy is. Regardless of the type of excitation, the probability distribution of molecules over states of energy $E$ is given by Boltzmann statistics as

$$N_E = N_0\ \mathrm{e}^{-E/kT}. \tag{IV.13}$$

Simple single-electron systems such as atomic hydrogen and the $He^+$ ion have energy levels given by

$$E = RyZ(1 + 1/n^2) \tag{IV.14}$$

where $Ry$ is the Rydberg constant and $n$ is the principal (electronic) quantum number. Figure IV.1 illustrates the energy levels and some possible transitions between energy levels for H and $He^+$. The zeroes of the energy scales have been shifted by $+cZ^2$ so that the lowest-lying electronic state (the ground state) is assigned zero energy. Note that for $He^+$, for which $Z = 2$, the energies of all electronic transitions are four times those for atomic hydrogen.

The group of lines labeled A represent transitions from the ground state ($n = 1$) to electronically excited states ($n = 2$) of the H atom. This family of transitions is called the Lyman series, and the individual lines are labeled Ly$\alpha$, Ly$\beta$, Ly$\gamma$, etc. The lines labeled B and originating from the $n = 2$ level are called the Balmer series and then Paschen (from $n = 3$), Brackett (from $n = 4$), Pfund (from $n = 5$), etc. Radiation with precisely the energy needed to excite the Lyman series transitions is strongly absorbed by ground state hydrogen atoms. This absorption produces strong absorption lines in a continuous spectrum of light, such as that from a black body, which passes through the gas. The frequencies of these absorption lines are simply given by

$$\nu = \Delta E/h = (dZ^2/h)(1/n_1^2 - 1/n_2^2). \tag{IV.15}$$

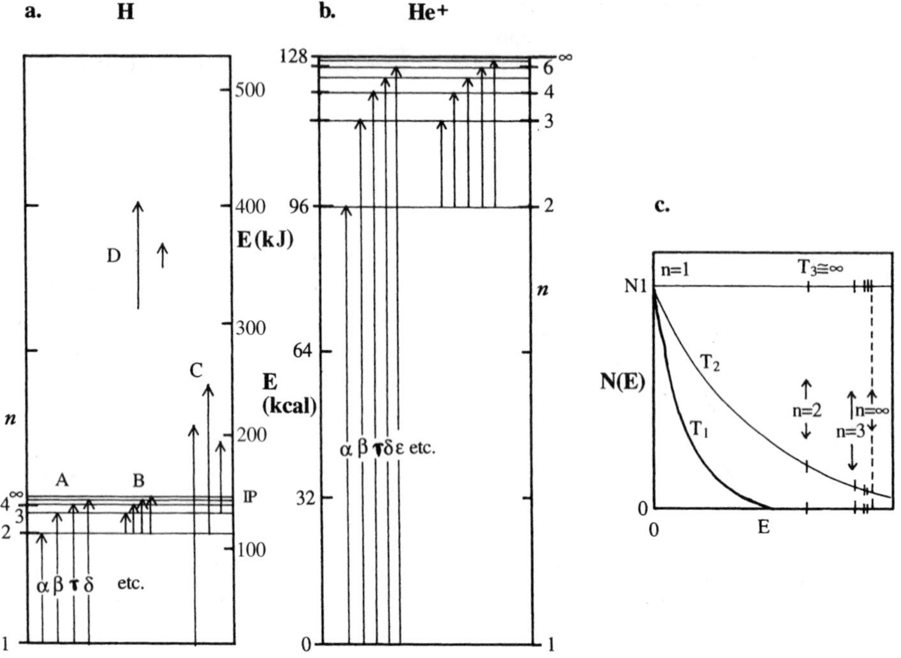


**Figure IV.1** Energy levels and populations of states in H and $He^+$. a shows the energy level diagram of atomic hydrogen. A labels absorptions by the $n = 1$ level to the $n = 2, 3, 4, = \ldots$ levels, the Lyman alpha, beta, gamma, etc., series. B labels transitions from the $n = 2$ level, similarly called the Balmer alpha, beta, gamma, etc., lines. Transitions originating from $n = 3$ (Paschen), $n = 4$ (Brackett), $n = 5$ (Pfund), etc., are also well known. The ionization potential of H is marked IP. Transitions to unbound states over $n = \infty$ are labeled C. These are termed bound–free transitions. Energy absorption that accelerates a free electron (a free–free transition) is illustrated by D. b illustrates similarly the first few transitions of the Lyman and Balmer series of the $He^+$ hydrogenic ion. In c the relative populations of the states of such atoms are sketched for three different temperatures. $T_1$ is so low that $> 99\%$ of the atoms are in the $n = 1$ ground state. $T_2$ is high enough so that the more energetic states are populated and some ionization has occurred. $T_3$ suggests the near-equal populations of states at near-infinite temperatures.

Here the Lyman series is described by the transitions for which $n_l = 1$, etc.

Note how rapidly the spacing of the energy levels collapses for high values of the principal quantum numbers. At $n = \infty$, energy ceases to be quantized. Photons with sufficient energy to raise an electron above this point cause the removal of the electron from the atom into a "free" (unbound) state. This is the phenomenon of photoionization, as illustrated by the lines marked C in Fig. IV.1. Note that transitions such as C do not have quantized energies, because the electron may be left with any amount of translational energy after ionization. Finally, free electrons may be accelerated to higher translational energies by absorption of a photon. These various types of photon–atom interactions are known as bound–bound absorption (type A, type B, and the like), bound–free transitions or photoionization (type C), and free–free transitions (type D).

Free–free transitions may be thought of as the change in energy of a free electron caused by absorption of a photon. There is an analogous process in which a photon is scattered by a free electron, causing a change in both the electron and the photon energies. This phenomenon, when originally described for gamma rays, was called Compton scattering. Scattering that occurs without a change in wavelength is called Thomson scattering [see Eq. (IV.45) and following].

Figure IV.1 also shows the relative populations of the energy levels of a hydrogen atom for three different temperatures. The first temperature shown, $T_1$, with almost all the atoms in the ground state, corresponds to $kT_1 << E_\infty$. At moderate temperatures, where $kT_2 \simeq 0.1E_\infty$, the population of the higher energy states is somewhat larger. At $kT_3 \simeq E_\infty$, many atoms are excited and even ionized. At infinite temperature all states are equally populated. Equation (IV.13) shows that, in any equilibrium system, the most populous state is always the ground state.

Figure IV.2 shows the absorption spectra produced by passing a beam of "white" light through a mixture of H, He, and their ionization products under several different conditions of temperature and pressure. In Fig. IV.2a, where the gas is cooler than the black body temperature of the Planckian radiation source and the pressure is low enough so that few atoms are engaged in collisions at any instant, absorption of radiation by the gas occurs at discrete wavelengths to produce a dark-line absorption spectrum. Absorption due to atomic H is important, and neutral He atoms also absorb at short wavelengths. Absorption by $He^+$ is negligible because the temperature is not high enough to provide a significant population of helium ions.

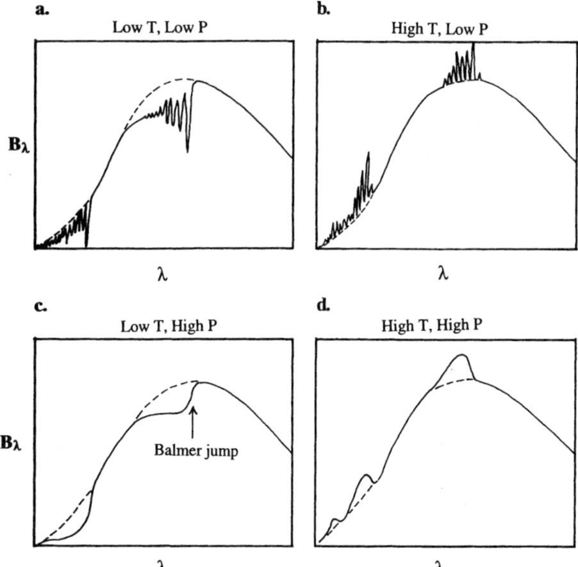


**Figure IV.2** Emission and absorption spectra. We here imagine a thin layer of variable temperature and pressure overlying a hot, dense (Planckian) emitting surface such as the photosphere of a star. A cool, tenuous layer (a) produces a sharp dark-line absorption spectrum, whereas a tenuous gas that is hotter than the background contributes a bright-line emission spectrum (b). With a moderately dense blanketing layer, the lines in the cool absorbing gas (c) and the hot emitting gas (d) are broadened and smeared out by collisional interactions. The dip in c is suggestive of the Balmer jump in stellar spectra.

If the gas has a kinetic temperature larger than the temperature of the radiation field, then the gas will emit energy at wavelengths corresponding to all of the allowed transitions between energy levels. This results in a bright-line spectrum of strong, narrow emission lines superimposed on the transmitted continuous spectrum, as shown in Fig. IV.2b.

At high pressures, a large proportion of the atoms are engaged in collisions at any moment. These collisions occur at energies that are comparable to the spacing between the higher-lying energy levels and result in severe distortion of these high-lying orbitals. The energies of the transitions between these levels then must become very ill defined, and all the absorption lines become smeared out in energy and begin to overlap seriously. In a large ensemble of atoms, a very wide range of collision energies and geometries can all be found at once. Accordingly, the region of the former absorption band of narrow, sharp lines becomes a broad, almost featureless region of absorption (or emission). This collisional broadening, usually called by the less evocative name *pressure broadening*, is illustrated in the last two panels of Fig. IV.2. The spectra of hot stars often exhibit a discontinuity of this sort, as indicated by the label "Balmer jump" in the figure.

In a highly ionized dense gas, continuum (unquantized) states are common, and all vestiges of line and band structure are wiped out. The emission spectrum of the gas approaches that of a perfect black body, that is, a Planckian emitter. This is the case in the interiors of stars. Any external view of a star sees a dense, hot background through the atmosphere of the star, which may have complex and variable thermal structure. Both bright-line and dark-line features, originating at different atmospheric pressure levels, may be seen in the spectrum of a given star.

Let us now imagine a region deep within the Sun in which the opacity is largely due to unquantized free–free transitions and in which the gas emits and absorbs like a black body. There is in this region a radial temperature gradient that is in steady state with the outward flux of energy to the photosphere. The net radiative flux between two parallel horizontal surfaces $dr$ apart and differing in temperature by $dT$ is

$$dF_{\mathrm{r}} = (\sigma/3)[T^4 - (T+dT)^4]$$

$$= (4\sigma/3)T^3 dT, \qquad \text{(IV.16)}$$

where the factor of $\frac{1}{3}$ arises from the fact that only one-third of the radiative flux is in the radial direction. The space between the two surfaces is filled by a good absorber that attenuates this flux by

$$dF_{\mathrm{r}} = -F_{\mathrm{r}}K\rho dr = -F_{\mathrm{r}}dr/1, \qquad \text{(IV.17)}$$

where $K$ is the mass opacity as defined in Eq. (IV.11).

The vertical temperature gradient is related to the radial flux by

$$dT(dT/dr) = -3K\rho F_{\mathrm{r}}/4\sigma T^3 = 3K\rho L_{\mathrm{r}}/16\pi\sigma T^3 r^2. \qquad \text{(IV.18)}$$

Clearly high opacity "insulates" the atmosphere against radiative cooling and leads to a large vertical temperature gradient for a given flux. Thus the higher densities encountered deep within the star would tend, because of their high opacity, to increase the temperature gradient. On the other hand, high temperatures, such as those prevalent throughout the deep interior of the Sun, cause such an enormous growth in the intensity of the radiation field that the "leakage" of even a small proportion of that flux constitutes an important heat loss and serves to decrease the temperature gradient.

The radial temperature gradient is not free to take on any value whatsoever. It is a familiar experience that a gas or liquid heated strongly from below will convect; that is, the density of the hotter, lower layers can be low enough that they become buoyant. The hot material will then rise to the top, carrying its heat with it. Convection generally involves motions over a time scale far shorter than that required for radiative heating or cooling of the gas. We may then regard the pressure–volume work done on the surroundings by the rising, cooling, expanding gas parcel as exactly equal to the decrease in the internal thermal energy of the parcel.

Conservation of energy in this process requires, in the usual statement of the first law of thermodynamics, that

$$C_{\mathrm{v}}dT = -Pdv, \qquad \text{(IV.19)}$$

where $C_{\mathrm{v}}$ is the heat capacity at constant volume, $T$ and $P$ are the temperature and pressure, and $v$ is the molar volume of the gas. The reader who is unfamiliar with the basic definitions and concepts of thermodynamics should refer to Appendix I.

For the simple case of an ideal gas, the equation of state is

$$Pv = RT, \qquad \text{(IV.20)}$$

where $R$ is the universal gas constant, $8.3144 \times 10^9$ erg $\mathrm{K}^{-1}$. In differential form,

$$Pdv + vdP = RdT. \qquad \text{(IV.21)}$$

Combining with Eq. (IV.19),

$$RdT + C_{\mathrm{v}}dT = vdP. \qquad \text{(IV.22)}$$

However, for an ideal gas $R + C_{\mathrm{v}} = C_{\mathrm{p}}$, the heat capacity at constant pressure,

$$vdP = C_{\mathrm{p}}dT. \qquad \text{(IV.23)}$$

If the atmosphere is in hydrostatic equilibrium,

$$dP = -g\rho dr, \qquad \text{(IV.24)}$$

where $g$ is the local gravitational acceleration and $\rho$ is the density of the gas. Again, for an ideal gas, the density $\rho$ is just $\mu/v$, where $\mu$ is the molecular weight of the gas. Thus

$$C_{\mathrm{p}}dT = -\mu g dr. \qquad \text{(IV.25)}$$

This gives us the value of the steepest temperature gradient possible in a static atmosphere, at the onset of spontaneous convective instability:

$$dT/dr = -\mu g/C_{\mathrm{p}}. \qquad \text{(IV.26)}$$

Because this process occurs without gain or loss of heat by the parcel of atmosphere, the process is adiabatic (see Appendix I). Accordingly, this is a process in which the entropy change is zero. We then can write this gradient, called the adiabatic lapse rate, as

$$(\partial T/\partial r)_{\mathrm{s}} = -\mu g/C_{\mathrm{p}}. \qquad \text{(IV.27)}$$

In general, the heat capacity is a weak function of temperature, not a strict constant.

The atmosphere is stable against convection, and the temperature gradient is radiatively controlled, if

$$-\mu g/C_{\mathrm{p}} \geq -3K\rho F_{\mathrm{r}}/4\sigma T^3. \qquad \text{(IV.28)}$$

This condition is satisfied over all of the deep interior of the Sun, but breaks down near the surface of the photosphere.

The third method of transporting heat is conduction. Because collisions are very frequent, electrons and protons achieve local equipartition of energy on very short time scales. The kinetic energy content of the ionized, hydrogen-rich gas is, to first approximation, due to electrons and protons alone,

$$KE = 3NkT/2 \simeq 3(N_e + N_p)kT/2, \quad \text{(IV.29)}$$

where $N$ is the total number of particles, and $N_e$ and $N_p$ are the number of electrons and protons, respectively. In the fully ionized interior of the Sun, the condition of electrical neutrality requires that the number of electrons be equal to the number of protons. Thus the contributions of the electrons and protons to the total kinetic energy are equal. Then

$$\frac{1}{2}m_p v_p^2 = \frac{1}{2}m_e v_e^2, \quad \text{(IV.30)}$$

where $m$ is the mass of the particle and $v$ is its velocity. Because the proton mass is 1836 times the electron mass, the mean electron speed must be $(1836)^{1/2} = 42.85$ times the speed of the proton. Because of this great mobility difference, heat conduction in the Sun must be due almost solely to the electrons. However, electrically charged particles interact strongly and have large collision cross-sections, so that the electron mean free path is only about $10^{-4}$ times the photon mean free path. At local thermodynamic equilibrium, the radiative transport of heat in this ionized gas is much faster than the conductive transport, and the latter can safely be neglected.

## Internal Structure of the Sun

We shall first approach the problem of modeling the interior of the Sun in a very approximate manner. Our purpose will be to show how such order-of-magnitude problems may be approached and how useful estimates of physical parameters may be derived from the simplest of initial assumptions. The first concept needed in this task, called the *Virial theorem*, states that the internal kinetic energy of a system is closely similar to its stored potential energy. From Eq. (II.73), we know that the potential energy is

$$V = -GM^2/R = -4 \times 10^{48}\ \text{erg} \quad \text{(IV.31)}$$

for a homogeneous, spherically symmetric Sun. Equating to the total internal thermal energy, we have

$$4 \times 10^{48} = (3/2)N_\odot kT_\odot, \quad \text{(IV.32)}$$

where $N_\odot$ is the number of particles in the Sun,

$$N_\odot = 2N_\odot M_\odot = 2 \times 10^{57}. \quad \text{(IV.33)}$$

Solving for $T_\odot$, the mean internal temperature of the Sun, we get $8 \times 10^6$ K ($T_6 = 8$). Sophisticated modeling techniques for the physical and thermal structure of the Sun give a central temperature of $T_6 = 14$, and thus our estimate of the mean temperature is quite good.

The approximate pressure profile within the Sun can be quickly estimated if we assume a spherically symmetrical, nonrotating Sun:

$$M(r) = 4\pi \int \rho(r)r^2 dr \quad \text{(IV.34)}$$

or

$$dM(r)/dr = 4\pi r^2 \rho(r). \quad \text{(IV.35)}$$

We also assume hydrostatic equilibrium,

$$P(r) = P(r + dr) + g(r)\rho(r)dr \quad \text{(IV.36)}$$

and

$$dP(r)/dr = -g(r)\rho(r) = -GM(r)\rho(r)/r^2, \quad \text{(IV.37)}$$

whence

$$dP(r)/dr = -(GM(r)/4\pi r^4)[dM(r)/dr], \quad \text{(IV.38)}$$

where we have substituted for $\rho(r)$ from Eq. (IV.35). Multiplying by $dr$ and integrating from the center of the Sun to the surface,

$$\int_0^{r_\odot} dP(r) = -(G/4\pi r^4) \int_0^{r_\odot} M(r)dM(r). \quad \text{(IV.39)}$$

Then

$$P(s) - P(c) = -(G/4\pi r_\odot^4)(M_\odot^2 - 0), \quad \text{(IV.40)}$$

or, because $P(s) = 0$,

$$P(c) = GM_\odot^2/4\pi r_\odot^4 = 4.5 \times 10^{14}\ \text{dyn cm}^{-2}. \quad \text{(IV.41)}$$

The usual cgs pressure units of dyn cm$^{-2}$ and bars (1 bar = $10^6$ dyn cm$^{-2}$) are not very useful, and the SI pressure unit of the pascal (1 Pa = 1 N/m$^2$ = $10^{-5}$ bar) is even less so; indeed, the megabar (Mb) is barely better. This central pressure is 450 Mbar, more than 100 times the pressure in the center of Earth's core.

Of course, the presence of a dense core in the Sun makes this only a lower limit estimate on $P(c)$, and it is not surprising that detailed physical modeling gives a much higher central pressure, near $2 \times 10^5$ Mbar. From Eq. (IV.37) we can see that, in the core, the pressure gradient is proportional to $M(r)\rho(r)/r^2$, or

$$dP/dr = (4/3)\pi G\rho^2(r)r, \quad \text{(IV.42)}$$

from which it is clear that a moderate increase in core density leads to a great increase in central pressure.

Treating the center of the Sun as an ideal gas of molecular weight $\frac{4}{3}$ (fully ionized helium), we can estimate the density to be

$$\rho_c = \mu/v_c = \mu P_c/RT_c$$
$$= 1.33(2 \times 10^{11})/82(1.4 \times 10^7) \qquad \text{(IV.43)}$$

or 230 g $cm^{-3}$. Detailed models give a central density of about 100 g $cm^{-3}$ with perhaps a 15% uncertainty, attesting to the remarkable degree of ideality of so hot a gas. Note that here we have used the universal gas constant $R$ in units of $cm^3$ atm $K^{-1}$ $mol^{-1}$ and have neglected the 1.5% difference between a bar and an atmosphere.

In summary, we have seen both how to make rough estimates of the deep interior conditions of the Sun and how to generate detailed physical models of the internal structure. The detailed models can be constructed by numerically solving a set of equations including Eqs. (IV.1) (energy production), (IV.35) (mass), (IV.37) (pressure), and (IV.18) (temperature), adding the obvious boundary conditions that $P(r_\odot) = 0, L(r_\odot) = L, M(r_\odot) = M, M(0) = 0$, and $L(0) = 0$. A set of typical results from such a detailed numerical model is displayed in Fig. IV.3.

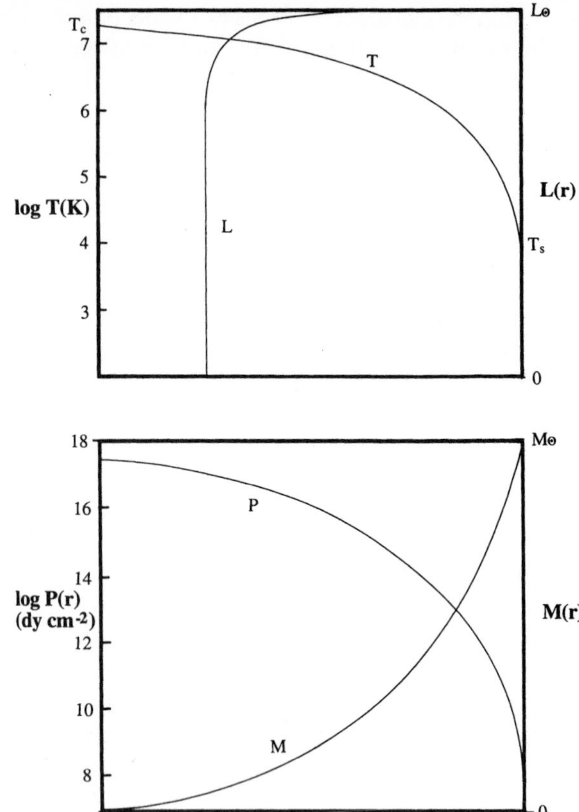


**Figure IV.3** Internal structure of the Sun. The radial distribution of luminosity ($L$), temperature ($T$), and pressure ($P$) are shown for a standard model of the Sun. Note that almost all the luminosity is generated in a thin layer at the surface of the core.

## Surface of the Sun

The radiating surface of the Sun, whose emission spectrum approximates that of a 5800 K black body ($\lambda_{max} \cong 0.5\,\mu m$) is called the *photosphere* (sphere of light) (see Fig. III.1). Because there is very strong anisotropy in the radiation field near the visible surface (the Sun emits quite a bit more energy than it absorbs!), the temperature can drop rapidly with increasing $r$. The temperature gradient is in fact so steep that the outer portion of the photosphere is convective and therefore adiabatic in structure.

The lowest temperature reached is 5000 K at 1.00 $r_\odot$, above which level the solar atmosphere is largely transparent to continuum radiation from below. In this region, the cool, low-density gas causes sharp dark absorption lines in the transmitted light from the photosphere. This region is called the *chromosphere* (colored sphere). The main absorption features formed in this layer, called the Fraunhofer lines, are labeled by letters, such as the D lines. The latter are the familiar pair of lines in yellow light, due to atomic sodium, which give the distinctive yellow color of the sodium flame test and the sodium vapor light.

Within the chromosphere the temperature gradient reverses sign, and temperatures in excess of a million degrees are reached only a few thousand kilometers above the surface of the photosphere. This very rarefied and very hot gas emits both a bright-line spectrum and a smooth continuum, which are best seen when the photosphere is covered by the Moon during a solar eclipse. This hot outer envelope of the Sun is called the *corona*.

The surface of the photosphere is usually marked with numerous dark spots with dimensions of many thousands of kilometers. These spots are commonly found in pairs, in which the leading and following spots have opposite magnetic polarity. Spots are observed to be common for several years at a time and then to be rare for a few years. The overall cycle of spot numbers has a mean period of about 11 years, with about a 4-year rise time to maximum after minimum. Spot pairs in opposite hemispheres have opposite magnetic polarity, with the leading spot having the same polarity as the nearer pole. Successive 11-year cycles have opposite polarity, because the entire magnetic field of the Sun reverses sign after every cycle.

One may thus speak of a 22-year solar cycle. The newest spots at the beginning of each half-cycle appear at relatively high latitudes and, over the lifetime of the spots, drift slowly toward the equator. This pattern is usually fairly symmetrical across the equator. Spots appearing later in the cycle form progressively closer to the equator. The areas of the Sun strongly affected by spot activity between 1900 and 1940 are sketched in the "butterfly diagram" of Fig. IV.4. Sunspots, which are typically several hundred degrees cooler than the surrounding photosphere, often send enormous eruptions of gas out into the corona. These eruptions may escape from the Sun altogether. During the eruption of these *solar flares* the luminosity of the Sun may increase by 0.1 to 1%. Much larger luminosity increases are seen at ultraviolet wavelengths, where the Sun normally exhibits a black body temperature of 3000 to 4000 K. The magnetic fields associated with flares are on the order of 1000 gauss (G) compared to the background surface magnetic field of 1 or 2 G which is typical of the quiet Sun. Sunspots have been recorded for centuries, and it is now generally accepted that, several hundred years ago, the Sun passed through a period of several decades in which there were essentially no sunspots. This prolonged lull in solar activity, the *Maunder Minimum*, is illustrated in Fig. IV.5.

During the years 1650 to 1700 telescopes were in wide use in Europe, yet hardly any sunspots were seen. Associated astronomical phenomena, such as the absence of the intense auroral activity normally associated with high sunspot numbers, and a consistent lack of reports of a complex, structured corona during solar eclipses, are in accord with the apparent scarcity of sunspots.

An interesting side effect of low solar activity is the weakening of the solar wind, which allows higher fluxes of galactic cosmic rays to penetrate into the Solar System. This higher cosmic ray flux is then capable of producing radioactive nuclides at a significantly higher rate. Studies of the radionuclide $^{39}Ar$, which is produced in meteorites by cosmic ray spallation reactions (disruption of nuclei by energetic charged particles), show that the production rate of $^{39}Ar$ over the last two or three half-lives ($t_{1/2} = 269$ years) has been significantly higher than the present production rate, in agreement with the other lines of evidence.

This enhancement of the rate of spallation reactions at times of low solar activity extends also to the production of $^{14}C$ in Earth's upper atmosphere. Reactions such as

$$^{15}N + p \rightarrow {}^{14}C + p + p \qquad \text{(IV.44)}$$

increase the abundance of the 5730-year beta emitter $^{14}C$ at high altitudes. Vertical mixing and horizontal transport by winds homogenize the atmospheric $^{14}C$ distribution in a few years, and dissolution of $^{14}CO_2$ into the surface layer of the oceans occurs on a time scale of about 10 years. The oceans are in turn homogenized by slow convective overturn on a time scale of about 300 years. The dynamics of the atmosphere–ocean system thus tend to average out any changes in the $^{14}C$ production rate over time scales longer than that of the solar cycle. Also, the long half-life of $^{14}C$ helps moderate its concentration fluctuations over time scales up to 10,000 years. There are further complications imposed by the shielding effects of the Earth's magnetosphere: modest variations in the geomagnetic field strength, such as those that have been observed for three centuries,

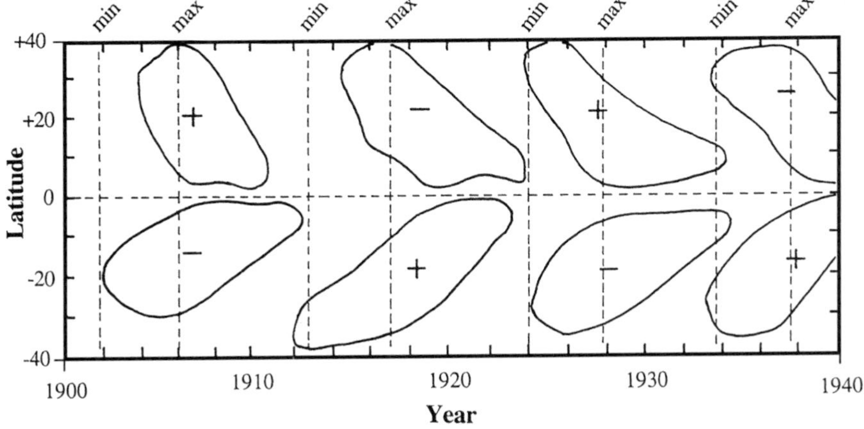


**Figure IV.4** The sunspot "butterfly pattern." In the + regions the leading spot in each bipolar pair is + and the trailing spot is −: the converse holds in the − regions. The leading spot adopts the polarity of the nearer pole. When the tail end of one sunspot cycle overlaps with the beginning of the next cycle at sunspot minimum, as in 1934, the "old" spot pairs near the equator have polarities opposite those in the "new" cycle that begin to appear at high latitudes in the same hemisphere.

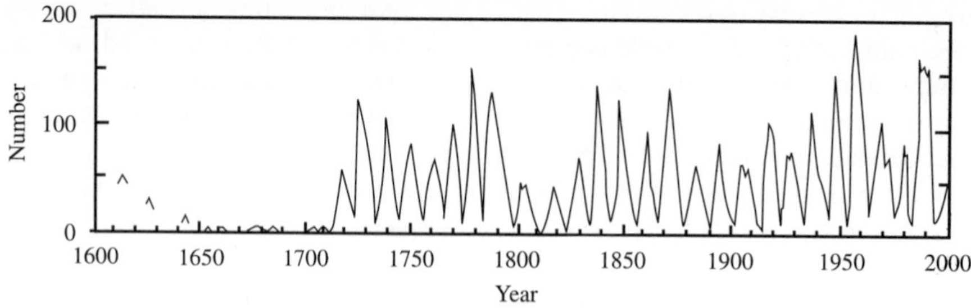


**Figure IV.5** The annual mean sunspot number, 1610–2000. This figure presents the Zurich sunspot numbers (the number of isolated spots plus 10 times the number of sunspot clusters, corrected for the equipment used, site observing conditions, and observer acuity and enthusiasm). The well-documented gap in solar activity in the late 1600s is called the Maunder minimum. A similar gap two centuries earlier is called the Spörer minimum.

impose yet another modulation on the rate of production of $^{14}C$ by cosmic rays. In light of all these complexities, the radiocarbon record of the past few millennia ought to be complex and even ambiguous, but very rich in evidence about a wide variety of interesting and important phenomena.

Figure IV.6 displays the results of $^{14}C$ analyses of some 7000 years of wood samples that have been dated absolutely by tree-ring counting (dendrochronology). We emphasize from the outset that there are two distinct levels of interpretation of the data derivable from studies of the annual tree-ring growth histories. The ring-width "fingerprint" from living 2000-year-old trees such as the bristlecone pine can be matched with tree-ring records from archaeological samples of wood cut in ancient times, but overlapping in age with the oldest living trees. These ancient wood samples in turn extend the record farther back in time. In some arid areas where trees grow slowly and wood does not rot rapidly a complete record exists back to dates as old as 5000 BC. The tree rings therefore give us an absolute measure of the ages of samples of wood which grew within the period of continuous coverage. The second kind of information requires the interpretation of the significance of the variations in the widths of the growth rings in terms of specific climatological variables such as rainfall. For the moment, we shall refer only to the former type of data, the determination of absolute ages by ring counts.

Note that the $^{14}C$ history in Fig. IV.6 contains several discernible types of features. First, we see a very slow, smooth sinusoidal modulation of the initial $^{14}C$ abundance, with a $^{14}C$ maximum (least shielding) near 5000 BC and a minimum (greatest shielding) near 500 AD. Paleomagnetic evidence on the strength of Earth's magnetic field shows a variation in the dipole field strength of the correct amplitude, period, and phase to explain this slow modulation of the $^{14}C$ production rate (dashed line). Second, note the pronounced narrow maxima and minima in the $^{14}C$ abundance. A major maximum coincides beautifully with the Maunder sunspot minimum. A deep minimum occurs near 1200 AD, during a time of abnormally warm climate on Earth. A second $^{14}C$ maximum corresponds to the Spörer sunspot minimum near 1500 AD, a time often referred to as the Little Ice Age. Indeed, both the Maunder and Spörer minima occurred during abnormally cool periods in the northern hemisphere.

We can now see that there are strong and apparently significant correlations between solar activity and climate on Earth. The mechanism and significance of such relationships are under active debate, and we shall return to a review of the related evidence in Chapter X.

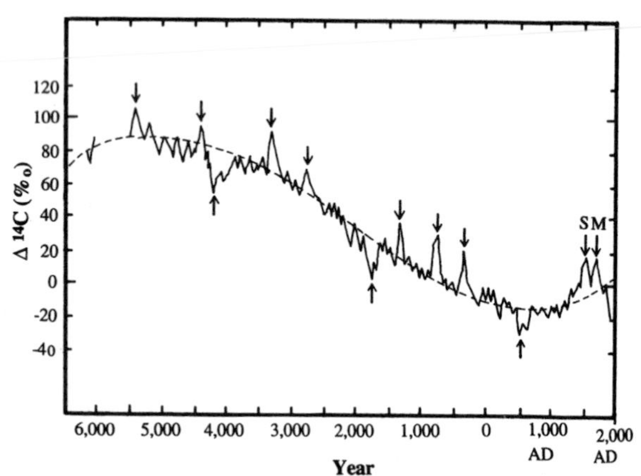


**Figure IV.6** Carbon-14 content of the Earth's atmosphere over the past 7000 years. The $^{14}C$ content of wood samples dated by tree-ring counts has been corrected for decay and for isotope fractionation to find the actual atmospheric isotope composition. Variations of the atmospheric isotopic composition from that in a reference standard are given in parts per thousand (permil; ‰). M and S denote the Maunder and Spörer sunspot minima, respectively.

For now, it must suffice to say that the evidence from dendrochronologically dated $^{14}C$ samples clearly portrays numerous major episodes of abnormally high and low solar activity. Whether these phenomena of the solar surface make themselves felt on Earth through solar luminosity variations, through more subtle spectral changes, or through a solar wind interaction is not clear from the available evidence.

The final feature of Fig. IV.6 is a strong and unprecedented decrease in the $^{14}C$ abundance in the past century. The reason for this phenomenon is well known. Since the beginning of the Industrial Revolution, radiochemically extinct fossil fuels with ages of hundreds of millions of years have been burned in enormous and ever-increasing quantities. This has both raised the global $CO_2$ budget and diluted the atmospheric $^{14}C$ isotopic abundance.

Technical advances in $^{14}C$ chronology should permit refinement and extension of these histories. Previously it was necessary to observe the decay of radiocarbon atoms in order to detect their presence. Now it is possible to separate the carbon isotopes and count atoms of $^{14}C$ without waiting for them to decay. Thus advances in laboratory isotopic chemistry, applied to the study of ancient tree-ring samples, tell us, in a way curiously common in science, far more about the surface activity of the Sun than we could learn from centuries of direct observation!

The tree-ring record has been greatly supplemented and extended by the study of cores drilled through the Greenland icecap and the West Antarctic ice sheet, which retain a record of the concentration of carbon dioxide and other gases over the past 2 million years. Temperatures can be deduced from the isotopic composition of oxygen in snow over this entire span. The oxygen-isotope study of polar ice cores has presented powerful evidence for quasiperiodic climate fluctuations with an approximate 100,000-year period, corresponding to the alternation between the glacial and interglacial periods known from the geological record. The present warm period appears to be one of many that exhibit temperatures similar to or higher than those of the late 20th century, suggesting a powerful and even dominant role of solar variability in governing temperatures on Earth.

An Earth-orbiting spacecraft, the Solar Maximum Mission (SMM), has documented large short-term fluctuations in the intensity of solar radiation at Earth's orbit (the so-called solar constant). Longer-term study of these fluctuations and correlation of them with other solar and terrestrial phenomena should permit us to discover a great deal about the solar surface and the mechanisms by which solar phenomena influence climate.

## The Chromosphere

The thin chromospheric layer is normally detectable because of the absorption lines that it superimposes on the solar spectrum. All told, these lines absorb some 10% or more of the total solar flux. During total eclipses of the Sun, the solar photosphere is covered fully only seconds before the chromosphere is itself eclipsed; however, during this brief interval (and again at the end of totality) a thin, pinkish arc can be seen just above the lunar limb. The dominant color is influenced by the Balmer radiation of atomic hydrogen. The "flash spectrum" seen at this time is extremely rich in emission lines. The temperature of the base of the chromosphere determined by interpretation of the spectrum is about 4300 K, with the temperature rising rapidly toward higher altitudes (Fig. IV.7), reaching about $10^6$ K only 3000 km above the temperature minimum.

Hot jets of gas called spicules arise in the chromosphere, and complex and energetic motions are commonly observed. Because of the very low density of the chromosphere relative to the photosphere, the magnetic fields carried by atmospheric motions below the temperature minimum have a very strong effect on motions in the chromosphere itself.

Turbulent motions in the photosphere often overshoot, bringing vertical momentum into the base of the chromosphere. These motions generate both acoustical waves and gravity waves that propagate upward into a medium of ever-decreasing density. These waves vest a constant momentum flux in ever smaller masses of gas,

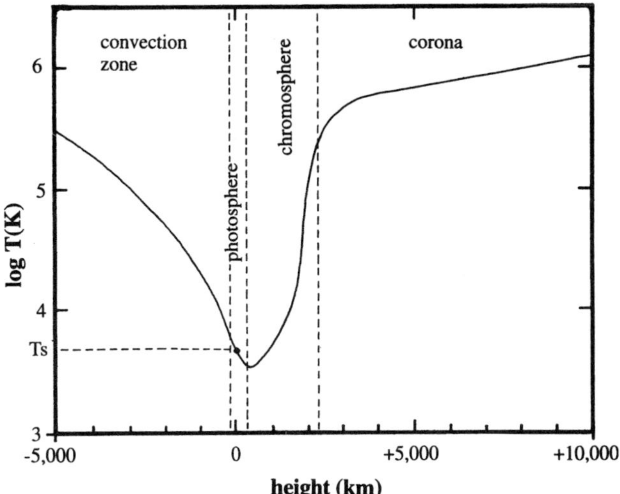


**Figure IV.7** Temperatures near the Solar surface. The effective temperature of the photosphere is 5780 K, whereas the lower chromosphere is as cool as 4300 K. A temperature rise of about a factor of 100 occurs in the chromosphere, mostly within an altitude range of a few hundred kilometers.

and the waves run away to supersonic speeds. This in turn causes shock heating of the overlying gas. This is the cause of the high temperatures in the upper chromosphere and corona.

## The Corona

Over a century ago there were three lines of evidence for a coupling between phenomena on the solar surface and events distant from the Sun. First, geomagnetic variations and auroral displays on Earth often showed the same periodicity as the synodic rotation of the Sun (its rotation as seen from Earth). Second, comet tails were observed to stream out radially from the Sun. Third, the extent of the corona during solar eclipses was surprisingly large and looked very little like a gravitationally bound, spherically symmetrical atmosphere. These observations hinted at some oddities in the upper atmosphere of the Sun, including the extraordinary oddity that that upper atmosphere may extend out to well beyond Earth's orbit and be moving radially outward at high speeds!

Spectroscopic studies of the corona prior to the quantum revolution in the 1920s did little to solve the problem: spectral lines seen in the corona were almost without exception unique, never seen in the laboratory. So severe was the dilemma that astronomers resorted to the postulate that the corona was made of a hitherto unknown element, coronium! The puzzle of the coronal spectrum became tractable, however, when atomic physicists equipped with the newly emerging tools of quantum mechanics began to calculate the wavelengths of the transitions of very highly ionized heavy atoms. It was found that the coronal lines could be explained by emission from a gas so hot that almost all of the electrons were stripped from atoms of atomic number up to 20; species such as the ion $Fe^{20+}$ were responsible for the observed lines. The temperatures calculated for the emitting gas were in excess of $10^6$ K, more than 100 times the maximum temperatures seen on the solar surface. Further, these emissions persisted out to several solar radii with no apparent diminution of temperature. Such a hot gas could not be gravitationally bound to the Sun. Equally clearly, such high temperatures could not be generated and maintained by any known mechanism.

The spectrum of the corona, however, is more complex than even this history relates. Careful spectroscopic study of the lower corona during solar eclipses showed that there were three quite distinct sources of light. The large majority of the light was in fact "white" continuum, not line emission. Closest to the Sun, the scattered light was so white that it did not even show the Fraunhofer line absorption spectrum imposed by the chromosphere! This component was called the $K$ (kontinuierlich) corona. Beyond about 2.5 solar radii the dominant light source was also "white," but was punctuated with the entire family of Fraunhofer lines. This component is photospheric light, filtered by the chromosphere, scattering off of tiny dust particles in orbits about the Sun. The dust-scattered light is called the $F$ (Fraunhofer) corona. Third, and most important close to the Sun, was the extremely hot $E$ (emission-line) corona. The observations of the intensity distributions of these three sources are summarized in Fig. IV.8. Of these three, the E and F corona seem easy to account for in terms of simple physical processes; we may not know where the dust came from or why the corona is so hot, but we can see easy ways for the gas and dust to generate the observed spectra. The case of the K corona is not quite as simple, although the explanation was forthcoming from even the earliest applications of quantum mechanics.

Let us consider an isolated electron, at point e in Fig. IV.9, irradiated by a plane wave of electric field

$$E = E_0 \exp(i\omega t) \qquad \text{(IV.45)}$$

with an energy density of $E^2/8\pi \,\text{erg cm}^{-3}$ and flux

$$F = cE^2/8\pi \text{ erg cm}^{-2}\,\text{s}^{-1}. \qquad \text{(IV.46)}$$

The electron will experience an acceleration of $Ee/m$. The scattered electric field at any nearby point at

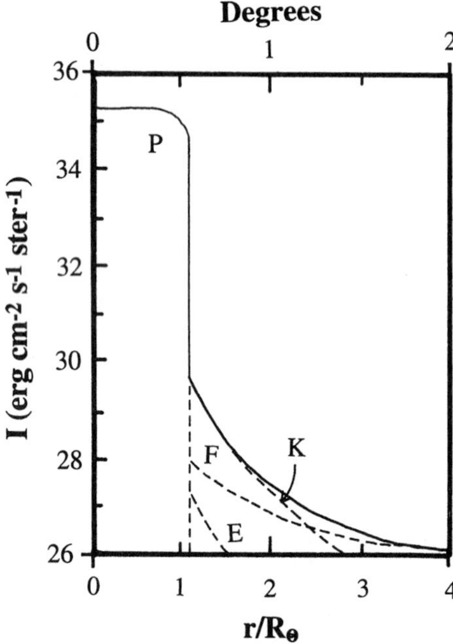


**Figure IV.8** Intensity of light from the Solar photosphere and corona vs distance from the sun's center. $P$ denotes the photospheric emission. $K$ is the continuous (kontinuierlich) Thomson-scattered component, $F$ is the dust-scattered component, and $E$ is the bright-line coronal emission spectrum. Normal daytime scattered light prevents Earth-surface observers from seeing any features beyond about 1.2 $R_\odot$.

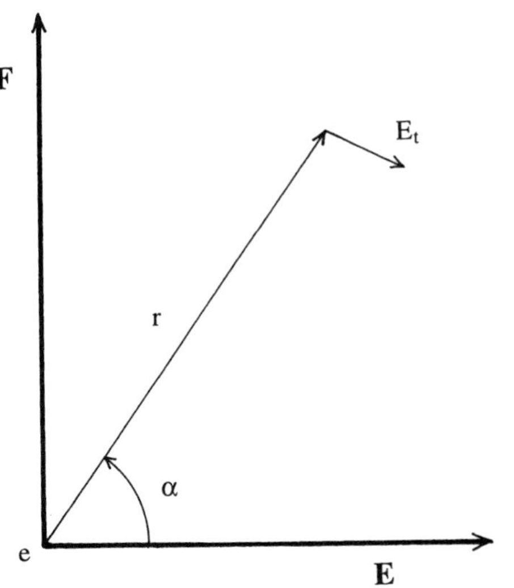


**Figure IV.9** Thomson scattering. The flux of radiation is in direction F, the instantaneous electric field is in the E direction, and the electron is at e. The angle between the incident electric vector and the scattering direction is $\alpha$. The Thomson-scattered electric field at distance $r$ from the electron is $E_t$.

distance $r$ and at a scattering angle of $\alpha$ relative to the incident electric vector is

$$E'(\alpha) = e^2 E \sin\alpha / mc^2 r. \qquad \text{(IV.47)}$$

This sinusoidally varying field also propagates an energy flux of

$$F'(\alpha) = cE_t^2/8\pi \ \text{erg}\,\text{cm}^{-2}\,\text{s}^{-1} \qquad \text{(IV.48)}$$

with a total radiated power of

$$P = \int_0^\pi F'(\alpha) 2\pi r^2 \sin\alpha \, d\alpha \ \text{erg}\,\text{s}^{-1} \qquad \text{(IV.49)}$$

$$= \frac{cE^2}{8\pi} \int_0^\pi \frac{e^4 \sin^2\alpha}{m^2\, c^4 r^2} 2\pi r^2 \sin\alpha \, d\alpha \qquad \text{(IV.50)}$$

$$= \frac{cE^2}{8\pi} 2\pi \left(\frac{e^2}{mc^2}\right)^2 \int_0^\pi (1 - \cos^2\alpha) \, d\cos\alpha \qquad \text{(IV.51)}$$

and thus

$$P = \frac{cE^2}{8\pi} \cdot \frac{8\pi}{3} \left(\frac{e^2}{mc^2}\right)^2 \text{erg}\,\text{s}^{-1}. \qquad \text{(IV.52)}$$

But the scattering of this power from a beam with flux given by Eq. (IV.48) requires that the scatterer (the electron) have an effective cross-section area of $\sigma = F/\text{P}\,\text{cm}^2$; hence

$$\sigma = \frac{8\pi}{3} \left(\frac{e^2}{mc^2}\right)^2. \qquad \text{(IV.53)}$$

This kind of scattering of electromagnetic waves, called *Thomson scattering* after its discoverer, is frequency-independent. The Sun's corona, which contains a moderately dense plasma, is very strongly irradiated with visible and infrared light from the photosphere, so that scattering of continuum radiation can be observed well off the disk of the Sun. This kind of scattering can deposit energy in free electrons and provides a source of heating for a cool plasma. Similarly, hot electrons may return energy to a cool radiation field by such scattering.

The reason for the absence of the Fraunhofer lines in the Thomson-scattered spectrum is simple: the thermal speed of electrons at $10^6$ K is a few percent of the speed of light, so each scattered photon is randomly Doppler shifted in energy by an amount much larger than the widths of the Fraunhofer lines. Dust particles are far slower and hence preserve the Fraunhofer lines. Skylab observations of the solar corona in the mid-1970s revealed a strongly developed "two-phase" structure of the corona. One phase involved relatively dense plasma, magnetically confined between closed field lines, usually located near the solar equator in regions of major sunspot activity. The other phase was very low-density plasma, streaming outward along open field lines. Almost all of these "holes" in the corona first appeared near the equator, grew to substantial size, became connected to the polar region with the same magnetic polarity, and then faded away. Alternating holes, formed about 90° apart in longitude, have opposite polarity. The escaping streams of plasma from these *coronal holes* likewise display alternating magnetic polarity, with a well-defined boundary between neighboring regions. This is the origin of the "magnetic sector structure" of the solar wind. Unusually fast solar wind flows (700 km s$^{-1}$), called monster streams, are observed whenever a large coronal hole exists at low latitudes.

It is likely from the presently available evidence that the polar regions of the Sun normally emit 700–800 km s$^{-1}$ plasma, which streams radially outward. Only when large equatorial coronal holes extend these conditions to low latitudes do we see a monster stream in the ecliptic plane. The nature of the corona and solar wind at high latitudes were to be studied in detail by the ambitious International Solar Polar Mission (ISPM). ISPM, a joint venture of the United States and the Federal Republic of Germany, originally involved launching of two spacecraft into Jupiter flybys, which were to be used to twist the orbital planes of the spacecraft to orbits with about +80° and −80° inclinations, respectively. This would have permitted simultaneous solar imaging at both high southern and high northern solar latitudes, as well as *in situ* study of the high-latitude solar wind. These results could then have been compared with observations by a number of already-launched spacecraft in the ecliptic plane in interplanetary and near-Earth space. Unfortunately, in 1981, a directive of the Office of Management and the Budget (OMB) removed funding for the American spacecraft and forced its cancellation by NASA. The tragedy was compounded in early 1986 when the surviving European spacecraft, renamed Ulysses, was being prepared for launch aboard the American Space Shuttle. When the Challenger shuttle orbiter exploded in flight in January 1986, the shuttle fleet was grounded for 2 years. Shortly thereafter, the Centaur G′ high-energy upper stage needed to launch heavy planetary missions from the shuttle was cancelled for safety reasons. Ulysses was finally launched (alone) in 1990.

Our present phenomenological understanding of the coronal structure and field orientation in the solar wind source region is sketched out in Fig. IV.10. Theoretical development of these ideas is in its infancy, and advances in our understanding may be expected in the near future from both the Ulysses spacecraft and the efforts of theorists.

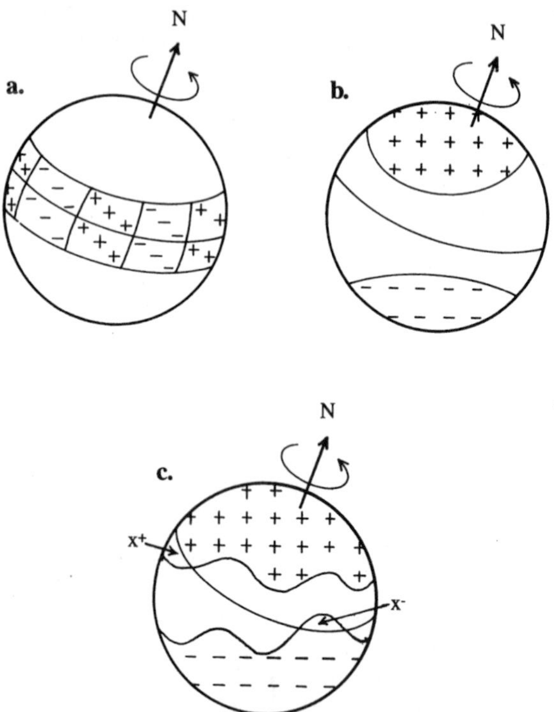


**Figure IV.10** Two-component model of the Sun's magnetic field. a shows the equatorial structure with regions of alternating polarity, and b shows the global dipolar field. The equatorial rotation period is 27 days, but the offset dipole rotates at a rate characteristic of high latitudes, with a period of at least 28 days. The net effect is shown in c. The two regions marked X+ and X− are sources of opposite-polarity fast equatorial streams of solar wind, the so-called monster streams.

We now turn to the lines of evidence mentioned earlier that suggested that the corona could make its presence felt even on Earth and in interplanetary space.

## Discovery of the Solar Wind

The first evidence suggestive of the existence of a radial outflow of very hot ionized gas (plasma) from the Sun was the observation about a century ago that there was a close correlation between solar flares and terrestrial auroral activity, which frequently became prominent a few days after a major flare. The disturbing influence therefore traveled at a typical speed of about 1 AU per 3 days, or $(1.5\times10^{8}\,\mathrm{km})/(3\times10^{5}\,\mathrm{s})=500\,\mathrm{km\,s^{-1}}$. These events also were sometimes seen to be associated with episodes of erratic magnetic activity on Earth (magnetic storms). Also, at about the same time observations of the corona during solar eclipses showed that the intensity of continuum light from regions out to several solar radii was startlingly strong. This was clearly not molecular emission and was attributed to scattering of visible sunlight either by dust particles (Lord Kelvin) or by electrons (Becquerel, Fitzgerald, and Sir Oliver Lodge). The latter case of Thomson scattering required a very substantial electron density and, because of the requirement of electrical neutrality, a large positive ion concentration. Because there was no visible emission from positive ions and because hydrogen is so enormously abundant, the logical candidate was free protons; with no bound electrons, hydrogen ions could contribute both scattering and free–free emission without any distinctive line emission.

The dynamical nature of this dense solar plasma was brought forcibly to the attention of astronomers in the early 1950s by Ludwig Biermann, who showed that the streaming of comet tails radially outward from the Sun was accompanied by a marked acceleration of the cometary plasma up to speeds of several hundred kilometers per second. Biermann correctly argued that corpuscular radiation with about that speed must be emitted by the Sun. Observations of the aberration of the tail direction with variation in the speed of the comet and the acceleration of plasma "knots" in the tail both yielded the same conclusion.

In the 1960s radio and radar techniques were applied to the study of the solar wind plasma. Pointlike extragalactic radio sources were observed to scintillate as a result of the passage of their emissions through the dense, turbulent solar plasma. Careful timing of pulsar signals near the ecliptic plane showed timing shifts when the Sun was nearly in the Earth–pulsar line of sight that could be attributed only to the finite refractive index of the plasma at long radio wavelengths.

But the most extensive body of data on the solar wind was produced by spacecraft launched on lunar and interplanetary missions, which carried them well clear of the Earth's magnetosphere and permitted direct *in situ* observations of the plasma. The speed distribution, direction, temperature, composition, and spatial structure of the solar wind were mapped from a number of spacecraft, most of which did not deviate more than a few degrees from the plane of the ecliptic. They thus mostly sampled the solar wind at low solar latitudes.

The flow of the solar wind was found to average 300 to 400 km $s^{-1}$, directed radially outward from the center of the Sun. At any time, the magnetic field embedded in the solar wind flow was found to exhibit sectorial structure, with "gores" of alternating polarity carrying spiral field lines rooted in the Sun at one end, but apparently open at the other. The plasma clearly originates in some kind of runaway expansion of the corona, but the mechanism and energetics of that expansion were not at all obvious.

The discovery of the Earth's trapped radiation belts by James A. Van Allen in 1958, and the later creation of artificial radiation belts by high-altitude nuclear explosions, caused greatly enhanced interest in the behavior of space plasmas. There was then renewed interest in a model of the behavior of the solar wind that had been proposed by Eugene N. Parker of the University of Chicago several years prior to the first direct studies of the solar wind by lunar and planetary spacecraft. Because of the major role played by radiophysics and space plasma investigations in the early history of the exploration of the Solar System, we shall devote a little more attention to both the basic radiophysics of the solar plasma and the attempts to construct a theory of the dynamical behavior of the solar wind.

## Radio Wave Propagation in Space Plasmas

At present, in addition to *in situ* spacecraft measurements of the local properties of the solar wind, both Earth-based and spacecraft-based radiophysical studies of the large-scale structure are being actively pursued. Interplanetary and lunar spacecraft launched by the United States and the former USSR routinely carry instruments for the study of the interplanetary medium, and some spacecraft, such as the German Helios I and II solar probes, are heavily instrumented for the purpose of comparing observations of the Sun itself with plasma studies. The complement of instruments in such a package usually includes a three-axis magnetometer (on a spinning spacecraft, two axes), energy analyzers, directional detectors to study the motion of the plasma, and mass spectrometers to analyze the positive ion component of the wind both chemically and isotopically. Such observations are extremely sensitive, but they sample each place in space at only one time and at any time can sample only one place; thus either a number of such spacecraft must be in operation in widely separated places simultaneously or some large-scale method of examining the gross properties of the solar wind must be employed.

Because radio and radar signals are influenced by passage through an electron gas, one convenient method of looking at large-scale structure is to study the propagation of these waves. The sources of these radio waves may be transmitters on spacecraft, radar waves transmitted from Earth and reflected off a remote target back to Earth, or natural signals conveniently broadcast by pulsed astronomical radio sources (pulsars).

Let us now consider the propagation of a simple electromagnetic wave through a low-density plasma in the absence of a strong background magnetic field. Electrons, because of their low mass-to-charge ratio, are very sensitive to varying electric fields and are accelerated by the instantaneous electric vector of the radiation. We shall take a sinusoidally varying electric vector:

$$\vec{E} = \vec{E}_0 \exp(i\omega t), \qquad \text{(IV.54)}$$

where $i$ is the square root of $-1$ and $\omega$ is the angular frequency, $2\pi\nu$(rad $s^{-1}$).

Neglecting magnetic forces and collisional interactions, the force balance on the individual electron is, if we define the direction of the $E$ vector as the $x$ direction,

$$m_e a_x = m(\partial^2 x/\partial t^2) = q_e E_x \qquad \text{(IV.55)}$$

or, combining,

$$m_e(\partial^2 x/\partial t^2) = q_e E_{ox} \exp(i\omega t). \qquad \text{(IV.56)}$$

Integrating once with respect to time,

$$(\partial x/\partial t) = q_e E_x/i\omega m_e = V_x, \qquad \text{(IV.57)}$$

which is the velocity of the electron produced by interaction with the radiation field.

Denoting the number density of electrons in the plasma as $n_e$ electrons per cubic centimeter, the velocity given above produces a current density $j_x$(A $cm^{-2}$) in the $x$ direction of

$$j_x = n_e q_e V_x = n_e q_e^2 E_x/i\omega m_e, \qquad \text{(IV.58)}$$

which lags the forcing function $E_x$ by 90°.

From Maxwell's equations,

$$\vec{\nabla} \times \vec{B} = \frac{1}{c^2}\frac{\partial \vec{E}}{\partial t} + \frac{\vec{j}}{c^2 E_0}$$
$$= \frac{1}{c^2}\frac{\partial \vec{E}}{\partial t} - \frac{i n_e q_e{}^2}{c^2 \varepsilon_0 \omega m_e}\vec{E}. \qquad \text{(IV.59)}$$

Taking the curl of both sides, and using the vector identity $\vec{\nabla} \times \vec{\nabla} \times \vec{B} = -\vec{\nabla}^2 \vec{B}$, we get

$$-\vec{\nabla}^2 \vec{B} = \frac{1}{c^2}\vec{\nabla} \times \frac{\partial \vec{E}}{\partial t} - \frac{i n_e q_e^2}{c^2 \varepsilon_0 \omega m_e}\vec{\nabla} \times \vec{E}. \qquad \text{(IV.60)}$$

The term describing the effects of changes of $\vec{E}$ with time can be rewritten using $\vec{\nabla} \times (\partial \vec{E}/\partial t) = \partial(\vec{\nabla} \times \vec{E})/\partial t$. Also, from Maxwell's equations, $\vec{\nabla} \times \vec{E} = -\partial \vec{B}/\partial t$; hence, $\vec{\nabla} \times (\partial \vec{E}/\partial t) = -\partial(\partial \vec{B}/\partial t)/\partial t$, and

$$-\vec{\nabla}^2 \vec{B} = \frac{1}{c^2}\frac{\partial^2 \vec{B}}{\partial t^2} + \frac{n_e q_e{}^2}{c^2 \varepsilon_0 \omega m_e}\frac{\partial \vec{B}}{\partial t}. \qquad \text{(IV.61)}$$

Of course, $\vec{B} = \vec{B}_0 \exp(i\omega t)$, $\partial \vec{B}/\partial t = i\omega \vec{B}$, and $\partial^2 \vec{B}/\partial t^2 = i\omega \partial \vec{B}/\partial t = -\omega^2 \vec{B}$, so that

$$-\vec{\nabla}^2 \vec{B} = \frac{1}{c^2}\left(1 - \frac{n_e q_e{}^2}{\varepsilon_0 \omega^2 m_e}\right)\frac{\partial^2 \vec{B}}{\partial t^2}. \qquad \text{(IV.62)}$$

Thus the phase velocity $v_\phi$ of the oscillation is given by

$$\frac{c^2}{v_\phi{}^2} = 1 - \frac{n_e q_e{}^2}{\varepsilon_0 \omega^2 m_e}, \qquad \text{(IV.63)}$$

but $c/v_\phi$ is just the refractive index $\eta$, and thus, because $\omega^2 = 4\pi^2\nu^2$,

$$\eta^2 = 1 - \alpha n_e/\nu^2, \qquad \text{(IV.64)}$$

where $\alpha$ is a constant factor, $\alpha = qe^2/(4\pi^2 \varepsilon_0 m_e)$.

Thus we can easily use a measurement of the refractive index of a plasma to determine the number density of electrons in it. A signal being propagated over a path whose length is not known *a priori* will be phase shifted to a degree that depends precisely on the frequency we found in the derivation above. Thus we can solve simultaneously for both the electron number density and the distance if we have two or more independent refractive index measurements at two or more different frequencies. Note that a sharply pulsed burst of white noise, as from a pulsar, will be smeared out in time by this plasma effect, with successively lower frequencies arriving successively later in time. This frequency *dispersion* will be most pronounced when the Earth's orbital motion causes the pulsar to pass behind the inner corona.

Note that the refractive index becomes imaginary for $\alpha n_e > \nu^2$, which physically corresponds to zero transmission and efficient reflection of the incident wave. By substituting in the appropriate constants, it is easy to show that the critical frequency for reflection is $9 \times 10^3 n_e^{1/2}$ Hz.

At optical frequencies, $\nu$ is so large that fantastic electron densities are required to get a refractive index discernibly less than unity. Thus there are both upper and lower frequency limits to the use of this technique from a planetary surface. One, a "soft" limit, is due to decreasing sensitivity at high frequencies. The other is a "stone wall" imposed by reflection of low-frequency waves below the critical plasma frequency in the ionosphere of the planet. For typical planetary ionospheres, we shall see that the peak electron number density is between about $10^4$ and $10^6$ cm$^{-3}$, giving a critical frequency of 1 to 10 MHz.

The measurable properties of radio waves propagated through the interplanetary medium (amplitude, phase, and direction) are affected by fluctuations in the electron density along the line of sight, which cause scintillation analogous to the scintillation of starlight caused by density fluctuations in the Earth's atmosphere. Such scintillation measurements provide an additional probe of conditions in the corona and solar wind.

## The Solar Wind

The general picture of the solar wind that has emerged in the course of modern spacecraft measurements is in its essential elements rather simple. The dense, hot solar atmosphere at the base of the corona, deep within the Sun's gravitational potential well, and containing a rather strong embedded magnetic field, expands radially outward, converting its internal energy into kinetic energy of expansion. Heating and acceleration of this coronal gas by waves emanating from the surface of the photosphere, where violent bulk motions are associated with sunspot activity, are also important factors in the energetics of the acceleration region. As observed from Earth's orbit, the solar magnetic field embedded in the outward-streaming gas shows gores of alternating magnetic polarity. This is the so-called magnetic sector structure, as pictured in Fig. IV.11. The usual structure contains six sectors, but as few as four or as many as eight can be seen at times. The reasons for this magnetic structure are not immediately obvious.

The sector boundaries are caused by conditions in the source region of the solar wind, near the surface of the photosphere. Like the streams of water emitted by a rotary garden sprinkler, the paths followed by emitted gas (and by the sector boundaries) are curved even though each particle pursues an almost perfectly radial path. The solar wind, originating in prominences

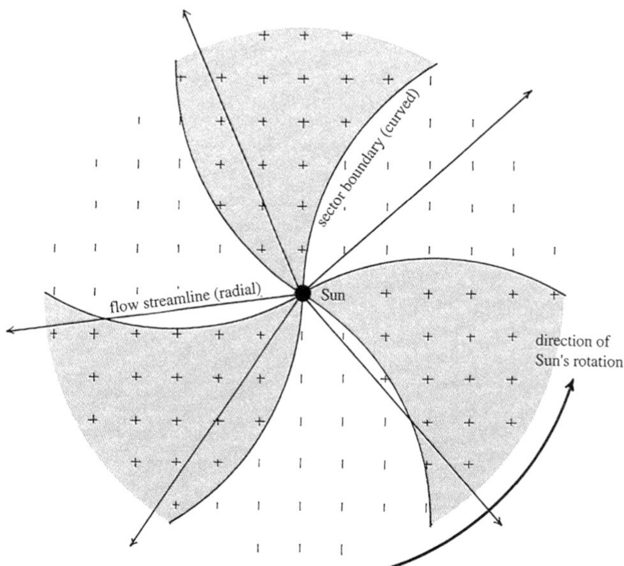


**Figure IV.11** Magnetic sector structure of the solar wind. Several (usually six) gores of alternating magnetic polarity are observed in the plane of the ecliptic. The curved sector boundaries result from almost perfectly radial flow as a consequence of the "garden sprinkler" effect.

rooted in sunspot pairs, always shows a distinct magnetic polarity. As we saw in Fig. IV.4, the polarity is uniform throughout each hemisphere for each sunspot cycle, but the southern and northern hemispheres always have opposite polarity. The overall magnetic field of the Sun somewhat constrains plasma emitted from the equatorial regions, where the field lines in the lower corona are closed arcs. At higher latitudes the field lines are generally open, directed nearly radially outward from the Sun. Charged solar wind particles (mostly protons and electrons, with a substantial admixture of $He^{2+}$ ions and a trace of heavier ions) spiral about these field lines as they progress outward into regions of rapidly diminishing magnetic field strength. Images of the solar corona taken during eclipses show dense regions of trapped arcuate structures at low latitudes, with little evidence of the presence of the corona at high latitudes, where the particles do not accumulate in magnetic traps, but run quickly away to escape. These "coronal holes" are the regions with open field lines.

The hot gases at the base of the corona act to "inflate" the magnetic cavity (the region of closed field line arcs) because the thermal energy density of the gas is comparable to that of the magnetic field ($B^2/8\pi$). The expanding magnetically confined bubble of hot gas eventually bursts open at its weakest point, where the loop is farthest from the Sun, breaking the field lines. The northern hemisphere of the Sun serves as the root of all open field lines with, say, positive polarity, and the southern hemisphere lies at the foot of all field lines of negative polarity. Fields at great distances from the Sun are very nearly radial. At the magnetic equator a peculiar situation occurs: fields of exactly opposite polarities are very close to each other. Therefore, the magnetic field gradient is very strong. In accord with Maxwell's equations, there must be a region of high current density at the interface of these two hemispheres, with current $j = \nabla \times B$. The interface that carries this current (called the *interplanetary current sheet*) is an approximate plane that is magnetically neutral (called the *neutral sheet*). The magnetic equator of the Sun is quite close to the plane of the ecliptic, and the sheet itself is rather floppy. As the Sun rotates, dragging its field around with it, this sheet moves past Earth, flopping from time to time across Earth. Plotting the magnetic polarity of the solar wind as we travel along our orbit (using, for example, spacecraft orbiting outside Earth's own magnetosphere or in orbit about the Moon) we see the magnetic sector structure. If we were observing from significantly above or below the ecliptic plane we would be outside the region swept by the current sheet, and would see no sector structure.

It is useful to have a simple bookkeeping scheme to keep track of the various forms of energy in the solar plasma. The thermal energy, $E_T$, is given by kinetic theory as

$$E_T = 3nkT/2. \qquad \text{(IV.65)}$$

The kinetic energy associated with the bulk expansion velocity ($v$) is

$$E_v = \frac{1}{2}\rho v^2. \qquad \text{(IV.66)}$$

The gravitational potential energy, $E_G$, is

$$E_G = GM_\odot \rho/r, \qquad \text{(IV.67)}$$

and the magnetic energy density in cgs (Gaussian) units is

$$E_M = B^2/8\pi. \qquad \text{(IV.68)}$$

Table IV.2 shows how the total energy is partitioned under quiet-Sun conditions from the base of the corona out to 1 AU. Note that the total internal energy is less than the gravitational binding energy out to about 2.5 $r_\odot$, but that the bulk expansion of the solar wind continues to accelerate until, beyond about 8 $r_\odot$, the wind is no longer bound. Near 1 AU (215 $r_0$), the bulk motion strongly dominates the total energy. The thermal and magnetic energy both drop off smoothly and rapidly with increasing

Table IV.2 Average Quiet Sun Conditions in the Solar Equatorial Plane

| | r/r$_\odot$ | | | | | |
|---|---|---|---|---|---|---|
| | 1.03 | 1.5 | 3 | 5 | 10 | 215 (1 AU) |
| $n_e$(cm$^{-3}$) | $2\times10^8$ | $2\times10^7$ | $4\times10^5$ | $4\times10^4$ | $4\times10^3$ | 7 |
| $T$(K) | $2\times10^6$ | $1\times10^6$ | $7\times10^5$ | $5\times10^5$ | $4\times10^5$ | $4\times10^4$ |
| $B$(Γ) | 1 | 0.4 | 0.1 | 0.04 | 0.01 | $3\times10^{-5}$ |
| $V$(km s$^{-1}$) | 0.6 | 3 | 34 | 130 | 280 | 360 |
| $E_v$(eV cm$^{-3}$) | $4\times10^5$ | $8\times10^5$ | $2\times10^6$ | $3\times10^6$ | $2\times10^6$ | $5\times10^3$ |
| $E_T$(eV cm$^{-3}$) | $9\times10^{10}$ | $4\times10^9$ | $7\times10^7$ | $5\times10^6$ | $4\times10^5$ | 70 |
| $E_M$(eV cm$^{-3}$) | $3\times10^{10}$ | $5\times10^9$ | $3\times10^8$ | $4\times10^7$ | $3\times10^6$ | 20 |
| $E_G$(eV cm$^{-3}$) | $4\times10^{11}$ | $2\times10^{10}$ | $3\times10^8$ | $1\times10^7$ | $8\times10^5$ | 60 |

distance. Note that the acceleration is virtually complete by 10 $r_\odot$. These curious observations demand explanation. In fact, they show that even a detailed theory embracing all these forms of energy would not suffice to provide the observed kinetic energy of expansion.

The first attempts to model the solar wind structure and flow were presented by Eugene N. Parker of the University of Chicago several years before the direct detection of the wind by spacecraft experiments, in response to Biermann's work on the acceleration and aberration of comet tails. This work showed speeds of 300 to 400 km s$^{-1}$ far from the Sun, but cast no light on the acceleration region or the source of energy to drive the expansion.

Parker considered the force balance on each volume element of the solar wind fluid as expressed by Euler's equation; Newton's familiar $ma = F$ is then written

$$\rho(D\vec{V}/Dt) = -\rho\vec{\nabla}U_G - \vec{\nabla}P, \quad \text{(IV.69)}$$

where the two force terms are those due to the gradient of the gravitational potential and the gradient of solar wind gas pressure. In this equation, the time derivative denoted by $DV/Dt$ is constructed for a given element of fluid, not a fixed location in space. Consider any general $F(x, y, z, t)$ that, after time $\Delta t$, becomes $F'(x + \dot{x}\Delta t, y + \dot{y}\Delta t, z + \dot{z}\Delta t, t + \Delta t)$. Then the derivative is

$$DF/Dt = \lim\,(F' - F)/\partial t \quad \text{(IV.70)}$$

$$= (\partial F/\partial x)\dot{x} + (\partial F/\partial y)\dot{y} + (\partial F/\partial z)\dot{z} + (\partial F/\partial t) \quad \text{(IV.71)}$$

$$= \vec{V}\cdot\vec{\nabla}F + (\partial F/\partial t). \quad \text{(IV.72)}$$

Thus

$$\rho(D\vec{V}/Dt) = \rho[\vec{V}\cdot\vec{\nabla}\vec{V} + \partial\vec{V}/\partial t]. \quad \text{(IV.73)}$$

We may profitably begin by constructing a simple static, time-independent model of the corona as a hot atmosphere, preliminary to the construction of dynamic, solar-wind-like models.

In the static case, $\vec{V} = 0$ and the radial velocity gradient $\partial\vec{V}/\partial r$ also vanishes. Euler's equation reduces to

$$\vec{\nabla}P = -\rho\vec{\nabla}U, \quad \text{(IV.74)}$$

which is simply the hydrostatic equation. Because

$$U = -GM_\odot/r, \quad \text{(IV.75)}$$

the static radial pressure gradient is just

$$\vec{\nabla}P = dP/dr = -\rho GM_\odot/r^2. \quad \text{(IV.76)}$$

Protons make up almost all of the density. Because the solar wind must be electrically neutral, the electron number density ($n_e$) must be equal to the proton number density ($n_p$). Thus,

$$\rho = n_e m_p, \quad \text{(IV.77)}$$

where $m_p$ is the mass of the proton, $1.6\times10^{-24}$ g.

The hot plasma is very well described by the ideal gas law:

$$P = (n_e + n_p)kT = 2n_e kT. \quad \text{(IV.78)}$$

The principal remaining problem that must be solved before we can present a quantitative model is the way in which the temperature depends on heliocentric distance.

Because, from Eqs. (IV.76), (IV.77), and (IV.78),

$$dP/dr = -(GM_\odot m_p P/2kT(r)r^2)dr, \quad \text{(IV.79)}$$

we can separate variables to get

$$dP/P = -(GM_\odot m_p/2kT(r)r^2)dr, \quad \text{(IV.80)}$$

or, integrating,

$$P(r) = P_0 \exp\left\{-(GM_\odot m_p/2k)\int_{r_\odot}^{\infty}\frac{dr}{T(r)r^2}\right\}. \quad \text{(IV.81)}$$

Because the pressure at infinite distance must vanish, we must require that the integral evaluated from $r_\odot$ to $r = \infty$ must be infinite. This in turn requires that $T(r)$ must drop off at least as fast as $r^{-1}$.

Any dynamical model for the expansion of the solar wind would depart from this simple static model in that the energy content of the gas at large heliocentric distances would be higher, that is, that $T(r)$ must drop off less rapidly than $r^{-1}$.

Let us then return to the more general approach of the dynamical behavior of the solar wind given by Eq. (IV.73). We shall not force a static ($V = 0$) solution, but will instead impose only the mild constraint that the flow be in a steady state, that is, that the structure of the wind be time independent. In this case, the last term in Eq. (IV.73) vanishes. Next, we can assume spherical symmetry of the flow, with the solar wind everywhere directed precisely radially outward. We then can simplify Eq. (IV.73) to

$$\rho(D\vec{V}/Dt) = \rho V(r)dV(r)/dr. \tag{IV.82}$$

Euler's equation, Eq. (IV.69), then becomes

$$\rho V(r)dV(r)/dr = -\rho dU/dr - dP/dr. \tag{IV.83}$$

This is a differential equation in three unknowns, $P$, $V$, and $\rho$. For a dynamic solar wind with spherical symmetry and radial flow in a steady state, we can add a mass conservation constraint of form

$$4\pi r^2 n(r)V(r) = \text{constant} = 4\pi r_0^2 n_0^2 V_0. \tag{IV.84}$$

The temperature profile for dynamical outflow of the wind must, as we saw above, be less steep than the $r^{-1}$ dependence found for the limiting case of a static "coronal" model. Parker's approximation was to let $T(r)$ be constant. We may now combine Eqs. (IV.77), (IV.78), (IV.83), and (IV.84) for an assumed isothermal gas:

$$n_0\left(\frac{r_0}{r}\right)^2\left(\frac{V_0}{V}\right)m_p V\frac{dV}{dr} = -\frac{GM_\odot m_p n_0}{r^2}\left(\frac{r_0}{r}\right)^2\left(\frac{V_0}{V}\right) - 2kTn_0 r_0^2 V_0\frac{d}{dr}\left(\frac{1}{r^2V}\right). \tag{IV.85}$$

The derivative in the last term is

$$\frac{d}{dr}\left(\frac{1}{r^2V}\right) = \frac{1}{(r^2V)^2}\frac{d}{dr}(r^2V) = \frac{1}{(r^2V)^2}\left(2rV + r^2\frac{dV}{dr}\right). \tag{IV.85a}$$

After about two pages of tedious algebraic manipulation, Eq. (IV.85) reduces to

$$\left(V^2 - \frac{2kT}{m_p}\right)\frac{1}{V}\frac{dV}{dr} = \frac{4kT}{m_p r}\left(1 - \frac{GM_\odot m_p}{kT}\cdot\frac{1}{r}\right). \tag{IV.85b}$$

The ratio $kT/m_p$ is proportional to the square of the mean thermal speed, $v_t$. Likewise, $GM_\odot m_p/kT$ has dimensions of length and can be defined as a critical distance, $r_0$. The most transparent form of Eq. (IV.85) is thus

$$(V^2 - 2V_t^2/3)dV/Vdr = (4V_t^2/3r)[1 - r_0/r]. \tag{IV.85c}$$

Separating variables, we have

$$(V^2 - 2V_t^2/3)(dV/V) = (4V_t^2/3r)(1 - r_0/r)dr. \tag{IV.85d}$$

There are six classes of solutions to this equation, as shown in Fig. IV.12. Some of the solutions have clearly unacceptable forms, such as the double-valued solutions 1 and 2. Solutions 3 and 6 decelerate beyond $r_0$, in contradiction to the observed behavior. Solution 5 passes through a velocity minimum at $r_0$, also contrary to observation. Only the form of solution 4 shows $V = 0$ at the solar surface and high speed at large distances.

Several important questions remain about this model. First, what causes the solar wind to accelerate? The model provides no clue. Indeed, the energy bookkeeping in the model is blatantly incomplete: the magnetic energy, which is comparable to the thermal energy in the source regions of the wind, is entirely omitted. Likewise, wave propagation of energy is ignored. The plasma has been implicitly assumed to have

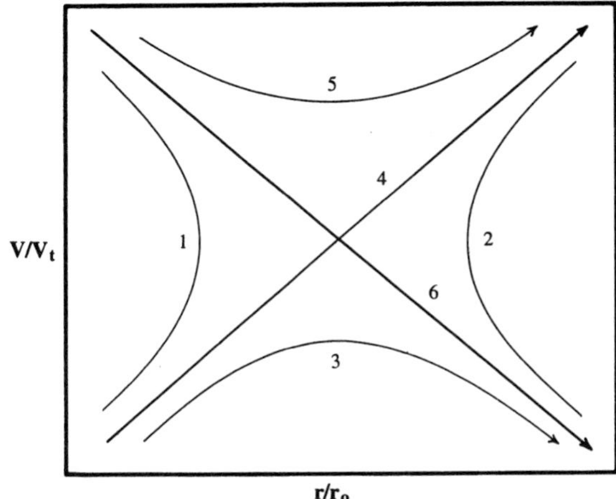


**Figure IV.12** Solutions to Parker's solar wind equations. The only physically acceptable solution that resembles the solar wind (that is, the only one to accelerate out of the Sun and escape at high speed) is solution 4.

an infinite thermal conductivity (the wind was assumed to be isothermal), but this was a mere mathematical convenience, not a physical result. A considerable amount of literature has grown up around attempts to integrate such considerations into models of the solar wind.

One of the most attractive proposals is that turbulence in the solar atmosphere, driven mainly by flare activity with its associated very high magnetic energy densities (fields of kilogauss instead of a few gauss), is a very powerful source of acoustic waves. These waves carry a fixed momentum flux upward into the rarefied upper atmosphere. Since the momentum is vested in ever smaller amounts of gas as the waves travel upward, conservation of momentum requires that the waves accelerate and run away to speeds high enough to heat the corona to over a million degrees. Unfortunately for this idea, several careful studies of the magnitude of this energy source and the temporal and spatial relationships between flare activity and the solar wind structure cast great doubt on the ability of this mechanism to explain the observations.

A second question is whether Parker's solution to the problem is stable: it demands a singular case of Eq. (IV.85d) that has no compelling physical virtue. A more complete and more physical model might avoid the apparent mathematical improbability of this solution.

Recent observations of the solar wind by the NASA/ESA Ulysses spacecraft have provided exciting new insights into the behavior of the wind at high magnetic latitudes. Launched in 1990, Ulysses was injected into an orbit that flew it close to Jupiter. Jupiter's gravity was used to flip the spacecraft up out of the plane of the ecliptic into an eccentric polar orbit about the Sun with an orbital period of 6 years. In 1994, during its first pass over the South Pole of the Sun, Ulysses sampled the solar wind emitted from the southern coronal hole and found it to be significantly faster than the usual 300 to 400 km $s^{-1}$ usually seen near the ecliptic. The density of the gas was low in this region, as was expected from the fact that the foot of this region does indeed look like a hole in the corona, but the total momentum flux (the product of these two factors) turns out to be higher than it is near the ecliptic. These observations were extended to the opposite pole of the Sun in September 1995 with similar results. Further Ulysses polar passages occurred in 2000 over the southern polar regions, and in 2001 over the north pole.

Ulysses also found that the flux of low-energy cosmic rays into the Solar System was lowest close to the plane of the ecliptic, at about 10°S latitude. The integrated solar wind flux throughout the northern hemisphere was significantly larger than in the southern hemisphere.

Although the solar wind departs with a speed in excess of the Sun's escape velocity, the solar wind does not propagate to infinite distance from the Sun: the interstellar medium is pervaded by a hot, tenuous gas with an energy density of about 1 eV $cm^{-3}$, in which the thermal and magnetic energy are comparable. Beyond about 1 AU the solar wind expands at constant speed, so that its energy density drops off as $r^{-2}$. The energy density of the wind at 1 AU is near $5 \times 10^3$ eV $cm^{-3}$, so we can expect the wind to stop somewhere around $(5 \times 10^3)^{1/2}$ AU, or about 70 AU. The situation is rather more complex because of the motion of the Sun through the interstellar medium, which lends a strong asymmetry to this "heliopause" surface. We expect the heliosphere in the interstellar wind to resemble a comet in the solar wind, crowded close to the Sun in the "upstream" direction, and trailing out far "downstream."

Ulysses has also contributed new insight into the structure of the heliosphere. In addition to the asymmetry due to the Sun's motion through the interstellar medium, it was found that the heliosphere has a constricted "waist" in the plane of the ecliptic, which is where the lowest speed solar wind flow is found. The overall shape of the heliosphere could fairly be described as a "windblown peanut."

Even if we had in hand a fully satisfactory model for the behavior of the solar wind, we would still be obliged to test that theory against the truly staggering solar wind energy and density found in T Tauri stars. In the latter case we find a millionfold increase in stellar wind intensity correlated with about a millionfold increase in UV luminosity, emanating directly from the chromosphere of the T Tauri star. Models that invoke violent turbulence, wave transport of energy, and supersonic motions in the chromosphere may lead to explanations of both phenomena.

Even more complex and equally interesting are the many modes of interaction between the solar wind and the atmospheres, surfaces, and magnetospheres of planets, comets, and other bodies in the Solar System. These interactions will be revisited many times in the ensuing chapters.

## Chemistry of Solar Material

We seek to explore the chemistry of material with the same elemental composition as the Sun not only for the insight it can give us into the chemistry of the photosphere and of sunspots (where, because of lower temperatures, molecules are more important). We also hope to learn about the compositions of cool stellar atmospheres, the cooling envelopes of exploding stars, chemical processes in the early nebular gas from which the Sun and the planets formed, and even the chemical behavior of the present atmospheres of the Jovian planets.

The general features of the chemistry of solar material are dominated by the prevalence of hydrogen and helium. Because of the great excess of hydrogen over all the other chemically active elements, the compounds that are chemically stable in a system with solar elemental composition are more reduced (less oxidized) than those encountered in normal experience by residents of planets with oxygen-rich atmospheres.

A typical minor element, which we shall call elementium (El), can be found in any of a great variety of chemical forms and physical states, depending sensitively upon temperature, but also on pressure. Some of these are, in order of decreasing temperature:

| | |
|---|---|
| a. fully ionized plasma | $El^{+z} + ze^-$ |
| b. ionic gas | $El^{2+} + 2e^-; El^+ + e^-$ |
| c. atomic gas | El(g) |
| d. molecular gas | $El_2$(g); ElO(g); etc. |
| e. first condensate | $El_2O_3$(s) |
| f. low-temperature product | $El(OH)_3$ |

The processes a → b, b → c, etc., occur during cooling at well-defined lines in pressure-temperature space, each defined by an equilibrium:

$$\begin{array}{ll} a \to b & El^{+z} + e^- \leftrightarrow El^{+z-1} \\ b \to c & El^{+1} + e^- \leftrightarrow El \\ c \to d & 2El \leftrightarrow El_2 \\ & El + O \leftrightarrow ElO \\ d \to e & El_2 + 3H_2O \leftrightarrow El_2O_3(s) + 3H_2 \\ & 2ElO + H_2O \leftrightarrow El_2O_3(s) + H_2 \\ e \to f & El_2O_3(s) + 3H_2O \leftrightarrow 2El(OH)_3(s) \end{array}$$

For reactions in which gases are converted from one form to another, we define these lines as the P–T loci along which the two El compounds involved in the reaction have equal partial pressures. For a condensation reaction, we define the line as the point of saturation of the condensate, at which its activity becomes equal to unity. For the alteration of a pure mineral into another pure mineral, we choose the point at which the product mineral achieves an activity of 1. We shall denote the activity of a solid or liquid phase by the letter *a* and the elemental abundance of each element (on a scale normalized to a silicon abundance of $10^6$ atoms) by *A*.

Generally, the four different regimes of plasma, neutral atomic gas, molecular gas, and condensate are well separated from each other because the characteristic temperatures of their interconversion processes are very different. These temperatures are dependent on the energies of the various reactions. Ionization requires great energies, typically on the order of 10 eV (light near a wavelength of 100 nm), and hence temperatures that are near

$$T = h\nu/3k = 24,000\,\text{K}. \qquad \text{(II.10)}$$

Thermodynamic tables presented in units of joules (or calories in older tables) show that the dissociation of common diatomic molecules such as $H_2$, NH, CH, and OH requires energies on the order of 250 kJ $mol^{-1}$ (60 kcal $mol^{-1}$)

$$E(\text{cal/mol}^{-1}) = h\nu n_0/j = hcN_0/j\lambda, \qquad \text{(II.43a)}$$

where $N_0$ is Avogadro's number ($6.022 \times 10^{23}$) and $j$ is the conversion factor from ergs to calories ($4.2 \times 10^7$ erg $cal^{-1}$). Thus 60 kcal $mol^{-1}$ is equivalent to a wavelength of about 5000 Å (0.5 $\mu$m) or 2.5 eV, or 6000 K. Heats of vaporization are far smaller, typically a few kilocalories per atom, and hence temperatures of a few hundred to a thousand degrees are expected.

It should be noted that the use of electron volts to measure ionization potentials, Ångstroms to measure the wavelength of ultraviolet light, micrometers for the wavelength of visible light, and kilocalories for the energies of chemical reactions is well entrenched in the literature. The use of micrometers or nanometers for wavelengths is preferred, but some familiarity with each of these systems and their interconversion is unfortunately still essential. A guide to energy interconversions is given for the comfort and convenience of the student in Table IV.3.

In the following discussion we shall generally begin at high (stellar) temperatures and end at low (planetary) temperatures, in keeping with the sequence that we began with our discussion of the Big Bang and stellar nucleosynthesis.

## Ionization

The energy level diagram for the H atom given in Fig. IV.2 showed that the energy required to excite an atom from the ground state to the continuum level (the ionization potential) is exactly $\frac{4}{3}$ times the energy of the Ly $\alpha$ transition from $n = 1$ to $n = 2$. Because the wavelength of Ly $\alpha$ is 1216 Å, the minimum energy photon capable of ionizing ground-state H will have a wavelength of $\frac{3}{4}$ times 1216, or 912 Å, equivalent to 13.6 eV or a metric wavelength of 91.2 nm. From the same figure and from the accompanying text it can be seen that the energy required to ionize $He^+$ (not He!) is exactly

**Table IV.3** Energy Conversion Factors and Definitions

eV, electron volt, the energy acquired by an electron when accelerated through a potential difference of 1 V
erg, the cgs unit of energy, equal to 1 g cm$^2$ s$^{-2}$
cal, calorie, the energy required to raise 1 g of water at 1 atm from 3.5 to 4.5°C
Photon energy in ergs $= h\nu$
$N_0$ particles = 1 mole
$N_0$ photons = 1 einstein
$N_0$ electrons = 1 faraday
$N_0 = 6.0222 \times 10^{23}$ = Avogadro's Number
Energy of 1 einstein of photons $= N_0 h\nu = N_0 hc/\lambda$
Å, wavelength unit $= 10^{-8}$ cm $= 10^{-10}$ m $= 10^{-4}$ $\mu$m = 0.1 nm
Mass equivalence: 1 atomic mass unit (AMU) $= c^2/N_0$ erg = 923 MeV
Conversions
$E$(eV) $= 1.240/\lambda(\mu\text{m})$
1 joule $= 10^7$ erg
1 cal $= 4.2 \times 10^7$ erg
$T(\text{K})\lambda_{\text{max}} = 2897.8\ \mu$m K
$T$(K) = 2340 E (eV)
$c^2 = 8.980 \times 10^{20}$ erg g$^{-1}$ $= 5.55 \times 10^{32}$ eV g$^{-1}$
1 erg $= 6.19 \times 10^{11}$ eV

**Table IV.4** Atomic Ionization Potentials

| Atom | Energy required to remove each electron (eV) 1 | 2 | 3 | 4 |
|---|---|---|---|---|
| H | 13.60 | — | — | — |
| He | 24.58 | 54.4 | — | — |
| Li | 5.39 | 75.29 | 122.4 | — |
| Be | 9.32 | 18.16 | 153.2 | 217.6 |
| B | 8.30 | | | |
| C | 11.26 | | | |
| N | 14.54 | | | |
| O | 13.61 | | | |
| F | 17.42 | | | |
| Ne | 21.56 | | | |
| Na | 5.14 | | | |
| Mg | 7.64 | | | |
| Al | 5.98 | | | |
| Si | 8.15 | | | |
| P | 10.55 | | | |
| S | 10.36 | | | |
| Cl | 13.01 | | | |
| Ar | 15.76 | | | |
| K | 4.34 | | | |
| Ca | 6.11 | | | |
| Fe | 7.90 | | | |

$Z^2$ times larger, corresponding to a wavelength of 22.8 nm (an energy of 54.4 eV). The first ionization potential of helium, which is the energy required to make He$^+$ from He, is 24.58 eV, which, from Table IV.3, can be seen to be equivalent to 50.4-nm (504 Å) light.

From the discussion of the Planck function in Chapter II we can see that typical photon energies are a few times greater than $kT$, and photons several times more energetic than average are present in small numbers. For wavelengths that are so short relative to the Planck peak, the Planck function [Eq. (II.41)] simplifies to

$$B_\lambda = (2hc^2/\lambda^5)\exp(-hc/kT\lambda), \tag{IV.86}$$

so that, at $h\nu = hc/\lambda = 15\,kT$, the exponential factor is $3 \times 10^{-7}$, and $B_\lambda$ is down from the Planck peak by a factor of $5 \times 10^{-5}$, a great reduction indeed. We may say that the photons in the short-wavelength (UV) wing of the black body emission for typical stellar photospheric temperatures will be so energetic that they will be absorbed by neutral atoms, such as those in stellar atmospheres. Thus the flux deficiency (lower black body temperature) of the Sun at UV wavelengths is qualitatively comprehensible.

Table IV.4 lists the first ionization potentials for a number of atoms of cosmochemical interest. Note that hydrogen and helium have rather high ionization energies; indeed, helium is the hardest element to ionize. Elements such as the alkali metals lithium, sodium, and potassium might, because of their ease of ionization, be much more important sources of electrons than H and He at temperatures of a few thousand degrees, despite their low cosmic abundances.

Consider a cool solar-composition gas at 2340 K. The thermal energy yardstick, $kT$, is then equivalent to 1.00 eV. The fraction of H atoms excited to their ionization energy or above is proportional to $\exp(-E^*/kT) = 1.2 \times 10^{-6}$, whereas the fraction of sodium atoms ionized is $\exp(-5.14/1) = 6 \times 10^{-3}$. The relative numbers of ions $n_i$ are

$$\frac{n_{\text{Na}^+}}{n_{\text{H}^+}} \simeq \frac{A_{\text{Na}}\exp -(E_{i(\text{Na})}/kT)}{A_{\text{H}}\exp -(E_{i(\text{H})}/kT)} \simeq 9 \times 10^{-3} \tag{IV.87}$$

where $A_i$ is the atomic cosmic abundance of element $i$ (see Table II.4).

Thus even at this very low temperature, equivalent to $\lambda_{\text{max}} = 1.23\ \mu$m, a very cool red giant or late M-type dwarf, only 1% of the ionization of the envelope is due to sodium. At a temperature of 1640 K (0.7 eV) sodium will account for a quarter of the total ionization and potassium for another 5%. Thus, in stellar photospheres, ionization of hydrogen is normally the main source of free electrons.

Very high photospheric temperatures are required to effect substantial ionization of helium. At 5800 K, the mean temperature of the solar photosphere, the fraction of He ionized to He$^+$ is $\exp(-24.58/2.48) = 5 \times 10^{-5}$, whereas the fraction of He$^+$ that is ionized to He$^{2+}$ is only $3 \times 10^{-10}$. Thus for every $10^{20}$ He atoms there would be $5 \times 10^{15}$ He$^+$ ions and $1.5 \times 10^6$ He$^{2+}$ ions. The $1.5 \times 10^{21}$ H atoms accompanying this amount of He would give $6 \times 10^{18}$ H$^+$ ions and electrons at the same temperature. These electrons are available to react with helium ions and can thereby decrease the degree of ionization of helium.

For more complex atoms with many electrons there will be a long sequence of successive ionizations spread over a very wide range of temperatures. Consider an iron atom heated sufficiently to remove 25 of its 26 electrons. The ionization energy for the last remaining electron is $13.6(26)^2 = 9194\,\mathrm{eV}$! Thermal ionization of iron down to bare nuclei and free electrons would thus require a temperature near $2 \times 10^7$ K, typical of the cores of MS stars. The photon with the minimum energy to lift this electron from the ground state would have a wavelength of 1.35 Å and would thus be a hard X ray.

How can we treat the thermal ionization of a complex mixture in which many different ionization states and many different atomic species with their own ionization potentials are present? We will illustrate the general method with an example. Consider hydrogen and helium, for which the relevant reactions are

$$\mathrm{H} \rightleftharpoons \mathrm{H}^+ + \mathrm{e} \tag{IV.88}$$

$$\mathrm{He} \rightleftharpoons \mathrm{He}^+ + \mathrm{e} \tag{IV.88a}$$

$$\mathrm{He}^+ \rightleftharpoons \mathrm{He}^{2+} + \mathrm{e}. \tag{IV.88b}$$

At equilibrium at temperature $T$, the partial pressures of these species are governed by the relations (see Appendix I)

$$K_{88}(T) = (p_{\mathrm{H}^+} p_{\mathrm{e}})/p_{\mathrm{H}} \tag{IV.88c}$$

$$K_{88a}(T) = (p_{\mathrm{He}^+} p_{\mathrm{e}})/p_{\mathrm{He}} \tag{IV.88d}$$

$$K_{88b}(T) = (p_{\mathrm{He}^{2+}} p_{\mathrm{e}})/p_{\mathrm{He}^+} \tag{IV.88e}$$

In addition to these thermodynamic constraints, we have the mass conservation equations

$$p_{\mathrm{H}} + p_{\mathrm{H}^+} = \text{constant} \equiv \Sigma\mathrm{H} \tag{IV.88f}$$

$$p_{\mathrm{He}} + p_{\mathrm{He}^+} + p_{\mathrm{He}^{++}} = \text{constant} = \Sigma\mathrm{He}. \tag{IV.88g}$$

Further, we know that the sums of H and He are in the same proportion as the elemental abundances of these elements,

$$\Sigma\mathrm{H}/\Sigma\mathrm{He} = a_{\mathrm{H}}/a_{\mathrm{He}}, \tag{IV.89}$$

and we know that the total pressure is

$$P = p_{\mathrm{H}} + p_{\mathrm{H}^+} + p_{\mathrm{He}} + p_{\mathrm{He}^+} + p_{\mathrm{He}^{2+}} + p_{\mathrm{e}}. \tag{IV.89a}$$

Finally, we know that the gas must be electrically neutral:

$$p_{\mathrm{H}^+} + p_{\mathrm{He}^+} + 2p_{\mathrm{He}^{2+}} = p_{\mathrm{e}}. \tag{IV.89b}$$

We now have eight equations in eight unknowns (the six partial pressures and the sums of hydrogen and helium), and we can therefore solve the system for any arbitrary temperature and pressure. If we desire to insert another element, such as oxygen, then

$$\mathrm{O} = \mathrm{O}^+ + \mathrm{e}^- \tag{IV.90}$$

$$K_{90}(T) = (p_{\mathrm{O}^+} p_{\mathrm{e}^-})/p_{\mathrm{O}} \tag{IV.90a}$$

and we need only add a new equation to conserve oxygen,

$$p_{\mathrm{O}} + p_{\mathrm{O}^+} = \Sigma\mathrm{O}, \tag{IV.90b}$$

apply the cosmic abundance data as we did for He, add O and $\mathrm{O}^+$ to the total pressure equation, and add $\mathrm{O}^+$ to the electrical neutrality equation. We then have a system of 11 equations in 11 unknowns, of which the partial pressures of O and $\mathrm{O}^+$ and the oxygen sum are the new variables. Thus the system may be expanded to include as many species as desired, as long as we know the ionization equilibrium constants $K(T)$ and the elemental abundances.

The system rapidly becomes too cumbersome for hand calculations unless intelligent first guesses greatly simplify the task. As an example, we might assume that, in this system, the high ionization potential of helium and the low abundance of oxygen suggest that the total electron pressure could be very well approximated by solving equilibrium (IV.88) alone and then using that calculated electron pressure to solve the other equations singly. In any event, such a system can be solved iteratively by a computer with little difficulty.

One of the most important curiosities of the chemistry of hot hydrogen and helium is the formation of the $\mathrm{H}^-$ ion by electron attachment on a hydrogen atom:

$$\mathrm{H} + \mathrm{e}^- \rightleftharpoons \mathrm{H}^-. \tag{IV.91}$$

This ion is weakly bound, is readily formed in collisions of electrons with H atoms in a dense gas, and is a strong broad-band absorber; the extra electron is so easily removed by a photon that visible light dissociates $\mathrm{H}^-$, and even a small trace of this ion provides considerable opacity across the visible spectrum. Thus, even though the concentration of $\mathrm{H}^-$ is never high, it is a very important source of opacity in the outer envelopes of MS stars. The absorption spectrum of H is sketched in Fig. IV.13.

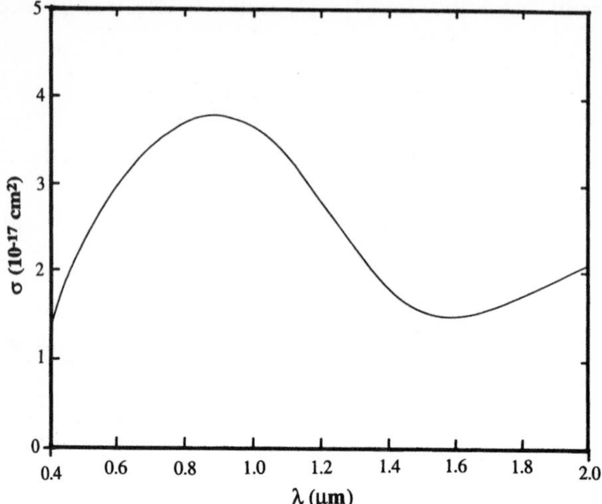


**Figure IV.13** Absorption coefficient of the negative hydrogen ion. Note that $H^-$ provides wide-band opacity across the visible and near infrared spectra. The principal sources of this opacity are bound–free and free–free transitions of $H^-$ and of $H + e^-$, respectively.

## Dissociation and Molecule Formation

At temperatures below about 2000 K, the overwhelming majority of the atoms of every element are neutral, not ionic. In this temperature range $kT$ is comparable to the strengths of some of the strongest chemical bonds, and diatomic and polyatomic molecules may form.

The most universal example of this process is provided by the simplest element:

$$H_2 \rightleftharpoons H + H \qquad \text{(IV.92)}$$

$$K_{92}(T) = (p_H)^2/p_{H_2}. \qquad \text{(IV.93)}$$

Because of the large elemental abundance of hydrogen and helium, the total pressure is well approximated by the sum of the H, $H_2$, and He partial pressures over a wide range of conditions.

If we replace the partial pressures $p_i$ (atm) of each species by the product of its mole fraction ($f$) and the total pressure ($P$), we can write the last equation as

$$P(f_H)^2 = f_{H_2}K_{92}, \qquad \text{(IV.94)}$$

whence it is clear that increasing the total pressure ($P$) causes a decrease in the mole fraction of H and an increase in the mole fraction of molecular hydrogen. A pressure increase opposes the increase in the number of particles caused by dissociation, in accord with *Le Chatelier's principle*:

> Any system, whenever possible, responds to an externally applied stress so as to minimize the internal effects of that stress.

Thus increasing $P$ causes a shift of the equilibrium (the equilibrium constant is indeed constant) in favor of molecular hydrogen, thereby decreasing the internal pressure and thus the internal pressure. Halving the volume of gas with a piston at constant $T$ increases the internal pressure by less than a factor of 2.

The value of the equilibrium constant for Reaction (IV.92) is fairly well given by

$$\log K_{92}(T) = 6.16 - 23{,}500/T \qquad \text{(IV.95)}$$

over the range of temperatures from about 1000 to 4000 K, and thus

$$2\log p_H - \log p_{H_2} = 6.16 - 23{,}500/T. \qquad \text{(IV.96)}$$

The line along which H and $H_2$ are equally abundant thus has the formula

$$\log p_H = 0.5\log p_{H_2} = 6.16 - 23{,}500/T, \qquad \text{(IV.97)}$$

which gives 3815 K when the H and $H_2$ pressures are both 1 atm, 2313 K when they are both $10^{-4}$ atm, and 1660 K when they are $10^{-8}$ atm. The photospheric surface of the Sun, at about 1 atm and 5800 K, is therefore very poor in $H_2$ and very rich in H. Cool MS stars have considerable molecular opacity, so that it is not possible to see as deep as the 1 atm level, and only the coolest stars have photospheric temperatures below 3000 K. The dominance of atomic H is therefore a very general phenomenon.

The most stable diatomic molecules in the stellar environment are those with the greatest bond strengths. It is therefore not surprising that multiple-bonded diatomic molecules such as $N_2$, CO, CN, and $C_2$ are seen in stellar atmospheres up to rather high temperatures.

The most important reactions for making the most stable and abundant of these stellar (and sunspot) molecules are given in Table IV.5. Included in the list are, not surprisingly, the only molecular species observed in sunspots.

A few generalizations can be made about these species. Clearly double- and triple-bonded molecules of abundant elements are dominant. HF, HCl, and OH contain the three strongest covalent single bonds known. Oxides of highly electropositive metals that prefer high valences

**Table IV.5** Formation of Diatomic Molecules

| | | | |
|---|---|---|---|
| $H + H \rightarrow H_2$ | $S + S \rightarrow S_2$ | $P + O \rightarrow PO$ | $Sc + O \rightarrow ScO$ |
| $C + O \rightarrow CO$ | $N + O \rightarrow NO$ | $P + N \rightarrow PN$ | $Zr + O \rightarrow ZrO$ |
| $H + O \rightarrow OH$ | $N + S \rightarrow NS$ | $P + P \rightarrow P_2$ | $Y + O \rightarrow YO$ |
| $C + N \rightarrow CN$ | $H + F \rightarrow HF$ | $C + P \rightarrow CP$ | $Si + H \rightarrow SiH$ |
| $C + S \rightarrow CS$ | $H + Cl \rightarrow HCl$ | $Si + O \rightarrow SiO$ | $Mg + H \rightarrow MgH$ |
| $O + O \rightarrow O_2$ | $H + N \rightarrow NH$ | $Si + S \rightarrow SiS$ | $Ca + H \rightarrow CaH$ |
| $S + O \rightarrow SO$ | $H + C \rightarrow CH$ | $Al + O \rightarrow AlO$ | $Si + F \rightarrow SiF$ |
| $C + C \rightarrow C_2$ | $H + S \rightarrow SH$ | $Ti + O \rightarrow TiO$ | $Mg + F \rightarrow MgF$ |
| $N + N \rightarrow N_2$ | $H + P \rightarrow PH$ | $V + O \rightarrow VO$ | $Sr + F \rightarrow SrF$ |

(3+ or 4+) are also in evidence. It is a curious fact that VO, ScO, TiO, ZrO, and YO are prominent contributors to the dark-band absorption spectra of cool stars.

Most of the molecules in this list are extremely reactive under conditions of lower temperature, at which they will react to form polyatomic gases or condensates. The active metals that form highly refractory (involatile) oxides are generally in the latter category.

Those species from Table IV.5 that react to form polyatomic gases at lower temperatures are listed in Table IV.6. We now have sufficient knowledge of principles to delve into the chemical behavior of solar material in detail, element by element. We shall look for behavior that provides sources of molecular opacity in cool stellar atmospheres and sunspots and for the identity and condensation behavior of the species that form condensates in a solar-composition gas.

## Hydrogen and the Rare Gases

A simple introduction to the equilibrium chemical behavior of solar material is provided by summarizing what we have so far learned about hydrogen. Figure IV.14 shows the major features of the behavior of hydrogen from $10^{-9}$ to 10 bar total pressure over the temperature range from about 10,000 K down to absolute zero. We recall that, above about 10,000 K, hydrogen is highly ionized and is found mainly as a plasma of protons and electrons. From about 10,000 K down to a few thousand degrees, ionization drops rapidly to negligible levels, and the gas becomes almost pure atomic hydrogen. Below 2000 to 3000 K (depending on pressure in the way we discussed in the previous section), molecular hydrogen is dominant. At about 10 K, again depending on the pressure, hydrogen condenses to form solid molecular hydrogen. At pressures of a few tenths of an atmosphere, condensation occurs at a temperature above the melting point of hydrogen, and liquid hydrogen is the stable condensed phase. Finally, at pressures above the critical pressure of hydrogen, it is possible to compress "gaseous" hydrogen to "liquid" hydrogen without ever passing through a phase transition.

**Table IV.6** Formation of Polyatomic Molecules

| | | |
|---|---|---|
| $H + OH \rightarrow H_2O$ | $O + CS \rightarrow COS$ | $H + CH_2 \rightarrow CH_3$ |
| $H + SH \rightarrow H_2S$ | $O + CO \rightarrow CO_2$ | $H + CNO \rightarrow HCNO$ |
| $PO + O \rightarrow PO_2$ | $H + PH_2 \rightarrow PH_3$ | $H + CP \rightarrow HCP$ |
| $H + CO \rightarrow CHO$ | $H + NH \rightarrow NH_2$ | $H + CH_3 \rightarrow CH_4$ |
| $H + CH \rightarrow CH_2$ | $H + CHO \rightarrow CH_2O$ | $CH_2 + CH_2 \rightarrow C_2H_4$ |
| $H + CN \rightarrow HCN$ | $O + CN \rightarrow CNO$ | $CH_3 + CH_3 \rightarrow C_2H_6$ |
| $H + PH \rightarrow PH_2$ | $CH + CH \rightarrow C_2H_2$ | $AlO + H \rightarrow AlOH$ |
| $S + CO \rightarrow COS$ | $H + NH_2 \rightarrow NH_3$ | |

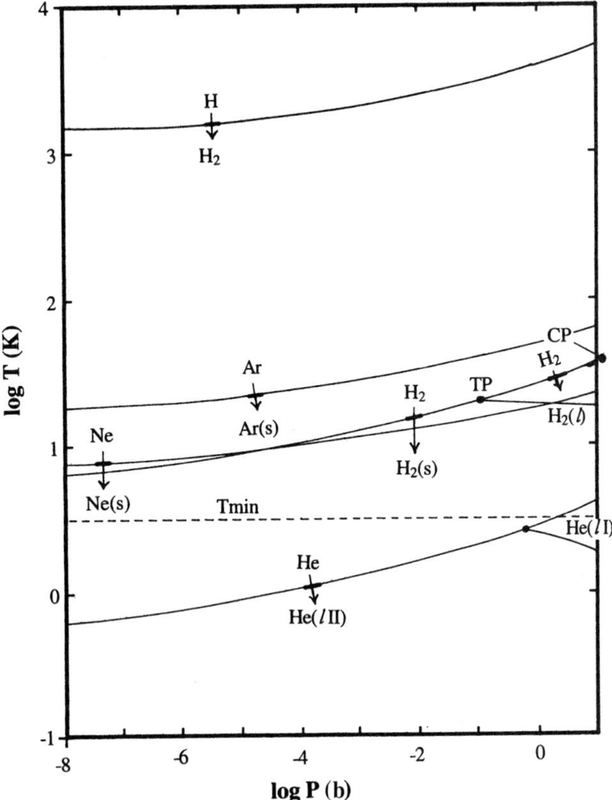


**Figure IV.14** Hydrogen, helium, and neon chemistry. The top line is the locus of equal H and $H_2$ pressures: below it, $H_2$ is the dominant gas. The saturation temperatures for $H_2$, Ne, Ar, and He are illustrated. Note the triple point (TP) and critical point (CP) of $H_2$ and the He gas-liquid I–liquid II pseudo-triple point ("TP"). The horizontal dashed line labeled $T_{min}$ is the microwave background temperature of the Universe. Lower temperatures, although not wholly impossible, require artificial (or natural) refrigeration.

At pressures higher than those included in this graph the freezing line of molecular hydrogen is strongly curved, and at even higher pressures (several megabars) solid molecular hydrogen collapses to form metallic hydrogen, which is an electrical conductor. Because hydrogen makes up over 70% of the mass of the Universe, we would do well to keep its behavior in mind as the backdrop against which all the complex chemistry of the other, rarer elements takes place.

One convenient feature of the chemistry of hydrogen is that it is so much more abundant than any other *chemically active* element (1000 times as abundant as oxygen) that its reactions with the other elements involve a negligible proportion of the total mass of hydrogen present. Also, from about 2000 K down to a few kelvins, hydrogen is present almost exclusively as molecular $H_2$.

Hydrogen and the second most abundant element, helium, together account for about 98% of the mass of solar material. Helium is found exclusively as gaseous He from its ionization temperature (over 10,000 K) down to roughly 1 K, its condensation temperature. Thus, over an immense range of "ordinary" temperatures and pressures (10 to 2000 K; $10^{-9}$ to $10^6$ bar) solar material is well described as a slightly impure mixture of $H_2$ and He gases. We indicate the mole fraction of each gas $i$ in a mixture as

$$f_i = p_i/\Sigma_i p_i = p_i/P. \qquad \text{(IV.98)}$$

This statement may be written quantitatively as $f_H = 0.876$ (76% by mass) and $f_{He} = 0.122$ (22% by mass), totaling to a mole fraction of 0.998 and a mass of 98%. The other 100 elements account for the remaining 0.002 mole fraction and 2% mass.

The chemistry of the other rare gases is nearly as simple as that of helium: they ionize at slightly lower temperatures than He and condense at slightly higher temperatures. Because of their simple behavior, we can usually take their mole fractions to be constant over the "ordinary" range of conditions; they are $f_{Ne} = 2 \times 10^{-4}$; $f_{Ar} = 7 \times 10^{-6}$; $f_{Kr} = 2.8 \times 10^{-9}$ and finally $f_{Xe} = 3.2 \times 10^{-10}$. Note the condensation curves of He, Ne, and Ar in Fig. IV.14.

The heavier rare gases are successively more polarizable and hence more prone to chemical complications. In fact, argon, krypton, and xenon all can form quite stable solid hydrates, as we shall see later. For this reason, the condensation behavior shown in the figure for saturation of pure solid and liquid argon may very well not be achieved.

Unfortunately, with our discussion of hydrogen, helium, and neon we have now exhausted the chemically simple elements. Fortunately, the more difficult elements are more interesting!

We shall now adopt the strategy of treating the chemical behavior of the remaining elements fairly consistently in decreasing order of abundance. In this way, the general features of the chemistry of each element can be grasped without the undue complications introduced by its interaction with less abundant elements. By this we simply mean that, because (for example) silicon is $10^8$ times as abundant as uranium, our discussion of silicon can safely neglect uranium. When we want to understand the chemistry of uranium, for which the chemistry of silicon is of crucial importance, we shall already have that information.

With this general guiding philosophy, we shall now proceed to discuss the fundamental chemical behavior of the other elements in solar-composition material.

## Oxygen, Carbon, and Nitrogen

After hydrogen and helium, the next most abundant element in the cosmos is oxygen. (It is a bit startling to the novice geology student to hear for the first time that the most abundant element in the Earth is oxygen, due to the prevalence of oxide minerals in the Earth.) Immediately after oxygen in cosmic abundance comes carbon, with an abundance $A_C = 0.6A_O$, which reacts strongly with oxygen. Next in order of abundance is nitrogen, with an abundance 0.3 times that of carbon. These three elements are responsible for the large majority of the "icy" materials that can form in the Universe, including water ice, solid methane, ammonia, carbon dioxide, carbon monoxide, and solid gas hydrates. They are closely similar in abundance and highly interactive and coherent in their chemistry, and it is appropriate to discuss them together.

At very high temperatures, near 10,000 K, the atomic ions $O^+, C^+$, and $N^+$ are the dominant forms of these elements. Atomic O, C, and N are important from the ionization region on down to about 3000 K (dependent on pressure). Below 2500 to 3000 K three very stable diatomic molecules, which we have already encountered in the previous sections (CO, carbon monoxide; OH, the hydroxyl radical; and $N_2$, molecular nitrogen) are important. Near 2000 K OH reacts to form the saturated and very stable molecule $H_2O$, water vapor. Below about 600 to 1000 K CO reacts with hydrogen to convert to $CH_4$, methane gas, and finally at a few hundred K molecular nitrogen is similarly reduced to form $NH_3$, ammonia.

Cooling of an H–O–C–N mixture from about 3000 K gives rise to the following sequence of equilibrium reactions:

$$N + N \rightleftharpoons N_2 \qquad \text{(IV.99)}$$

$$H + H \rightleftharpoons H_2 \qquad \text{(IV.100)}$$

$$C + O \rightleftharpoons CO \qquad \text{(IV.101)}$$

$$H + O \rightleftharpoons OH \qquad \text{(IV.102)}$$

$$H + OH \rightleftharpoons H_2O \qquad \text{(IV.103)}$$

$$CO + 3H_2 \rightleftharpoons CH_4 + H_2O \qquad \text{(IV.104)}$$

$$N_2 + 3H_2 \rightleftharpoons 2NH_3 \qquad \text{(IV.105)}$$

$$H_2O \rightleftharpoons H_2O(s) \qquad \text{(IV.106)}$$

$$NH_3 + H_2O(s) \rightleftharpoons NH_3 \cdot H_2O(s) \qquad \text{(IV.107)}$$

$$CH_4 + 7H_2O(s) \rightleftharpoons CH_4 \cdot 7H_2O(s) \qquad \text{(IV.108)}$$

If equilibrium is attained at all temperatures, then formation of the solid ammonia monohydrate uses up only about 10% of the water ice, whereas formation of the methane hydrate uses up all the remaining water ice and leaves substantial excess methane. The methane hydrate is one of a class of compounds called *clathrates*, in which "host" atoms or molecules such as Ar, Kr, $CH_4, N_2$, or CO are physically imprisoned within one of the large natural cavities in the ice lattice. At even lower temperatures, the excess methane that cannot be accommodated in the clathrate hydrate can condense directly as solid methane:

$$CH_4 \rightleftharpoons CH_4(s). \tag{IV.109}$$

These reactions are summarized in Fig. IV.15.

Because the abundance of C is only slightly more than half that of O, the abundances of CO and OH after Reaction (IV.102) are roughly equal, as are those of CO and $H_2O$ after Reaction (IV.103). After the reduction of CO in Reaction (IV.104) the water abundance is roughly doubled and hence rises to about twice the methane abundance.

The boundary between the CO- and the $CH_4$-rich regions is very important for several reasons. The activity of graphite attains a maximum on this line,

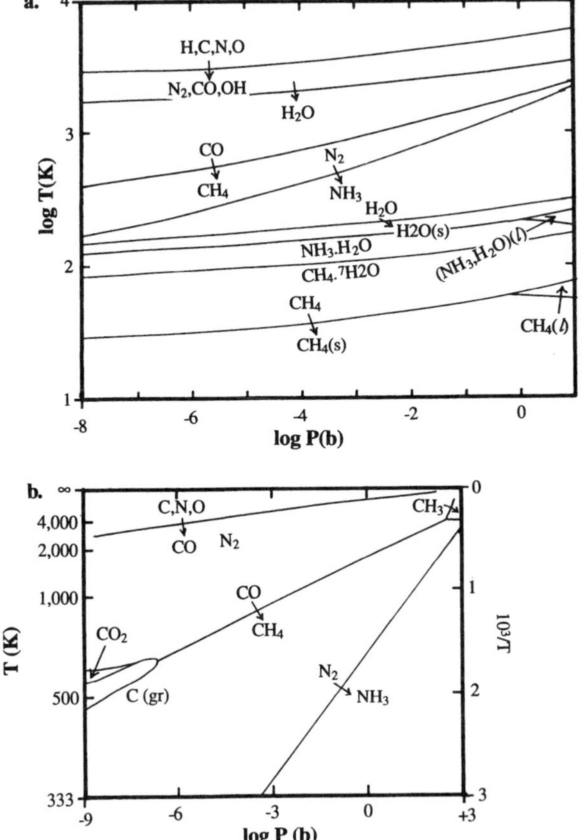


**Figure IV.15** Carbon, nitrogen, and oxygen chemistry. Atomic H, C, N, and O combine to form OH, CO, and $N_2$ and thence stepwise to form $H_2O$, $CH_4$, and $NH_3$ as the temperature falls. The condensation processes are, in order, condensation of water ice I, partial conversion of ice I to ammonia monohydrate, conversion of all remaining water ice to the methane clathrate hydrate, and condensation of the leftover methane. b illustrates the equilibrium regions of dominance of the various compounds of C, N, and O. Note the region of thermodynamic stability of graphite at low pressures ($< 10^{-7}$ bar).

and the abundance of hydrocarbon gases is a maximum here. The abundance of the common gas $CO_2$, which is familiar to residents of highly oxidized planets, is generally rather small in a solar composition gas, except within a narrow range of conditions shown in Fig. IV.15.

The melting behaviors of methane and of the ammonia–water system are indicated in the figure. The latter is of great importance in the atmospheres of the giant planets and in the interiors of icy satellites and will be discussed in detail in Chapters V and VI. For now, it suffices to note that, at pressures near 10 bar, gaseous ammonia may react with water ice to produce an aqueous solution of ammonia. Because dissolved ammonia may depress the freezing point of water to as low as −100°C, this reaction introduces some interesting complications in the phase behavior.

We emphasize that we are assuming strict thermodynamic equilibrium. There are several ways in which this assumption might be faulty. First, the reactions that reduce carbon monoxide and nitrogen to methane and ammonia may be too slow to take place in a reasonable amount of time. Second, these reactions may make unstable or metastable products in the presence of catalytically active mineral grains. Third, entry of ammonia and methane into cold ice to form their solid hydrates may be severely diffusion limited and may not reach equilibrium in the available time. With these caveats, let us look more closely at the reactions by which carbon monoxide and nitrogen are reduced to methane and ammonia [Reactions (IV.104) and (IV.105)].

We shall show how to find the boundaries between the CO and the $CH_4$ regions and between the $N_2$ and the $NH_3$ regions, that is, the lines along which the partial pressures of these respective pairs of gases are equal. The pressure and temperature dependence of these two reactions can be written down in simple form:

$$K_{105}(T) = \frac{(p_{NH_3})^2}{p_{N_2}(p_{H_2})^3} = \frac{(f_{NH_3})^2}{f_{N_2}(f_{H_2})^3}\frac{1}{P^2} \tag{IV.110}$$

$$\log K_{105}(T) = -12.23 + 5720/T \tag{IV.111}$$

$$K_{104}(T) = \frac{p_{CH_4}p_{H_2O}}{p_{CO}(p_{H_2})^3} = \frac{f_{CH_4}f_{H_2O}}{f_{CO}(f_{H_2})^3} \tag{IV.112}$$

$$\log K_{104}(T) = -13.21 + 11740/T. \tag{IV.113}$$

Approaching the $N_2{-}NH_3$ boundary from the low-$P$ side along an isotherm ($T$ is a constant; therefore, $K(T)$ is a constant), the mole fraction of molecular nitrogen is essentially at the level $f_{N_2} = f_{H_2}(A_N/A_H) = 1.18 \times 10^{-4}$ until the boundary is closely approached. In this temperature–pressure region the mole fraction of molecular hydrogen is constant. Then

$$f_{NH_3} = P(K_{105} f_{N_2} (f_{H_2}^3)^{1/2} \propto P \qquad \text{(IV.114)}$$

On the high-pressure side of the boundary, the ammonia mole fraction is constant at twice this level, and the $N_2$ mole fraction is a strong function of pressure:

$$f_{N_2} = \frac{(f_{NH_3})^2}{(f_{H_2})^3 K_{105}} \frac{1}{P^2} \propto P^{-2}. \qquad \text{(IV.115)}$$

Similarly, for the boundary between the CO-rich region and the $CH_4$-rich region, we can calculate the values of the CO and $CH_4$ mole fractions as a function of $P$. Equation (IV.112) shows that this may be done readily if the water vapor mole fraction is constant. Actually, however, the water vapor abundance nearly doubles close to the boundary as CO is reduced to $CH_4$ and $H_2O$ by increasing pressure. On either side of the boundary, however, the water mole fraction is quite closely constant. In the CO-rich region almost all the carbon is CO, and hence 40% of the total oxygen is found as $H_2O$. Thus, in the CO region, $f_{CO} = 0.8 \times A_O/A_H = 4.8 \times 10^{-4}$. In the methane region, $f_{H_2O} = 2A_C/A_H = 1.19 \times 10^{-3}$. Thus, in these two regions,

$$f_{CH_4} = \frac{K_{104} f_{CO} (f_{H_2})^3}{f_{H_2O}} P^2 \propto P^2 \qquad \text{(IV.116)}$$

and

$$f_{CO} = \frac{f_{CH_4} f_{H_2O}}{(f_{H_2})^3 K_{104}} \frac{1}{P^2} \propto P^{-2}. \qquad \text{(IV.117)}$$

For both the carbon and the nitrogen systems we can sketch out very easily the values of the mole fractions of the gases found along an isotherm. These results are given in Fig. IV.16 for $T = 1000$ K. The slopes on the $\log f$ vs $\log P$ plot of $-2$ for CO and $N_2$, $+2$ for $CH_4$, and $+1$ for ammonia are all clearly seen. Also, the slope for the mole fraction of atomic H is $\frac{1}{2}$, as we expect from Equation (IV.97).

Figure IV.16 was in fact constructed by detailed chemical equilibrium calculations on a system containing hundreds of compounds of hydrogen, oxygen, carbon, and nitrogen. It is fascinating to see the stark simplicity of the results. Species such as HCN, $C_2H_2$, $C_2H_4$, $C_2H_6$, $CH_2O$, and HCNO are not abundant enough to appear on the diagram. The chief among these is ethane, $C_2H_6$, which reaches a maximum mole fraction of $7 \times 10^{-11}$ at 1000 K. Even the very stable $CO_2$ has an abundance everywhere smaller than 1 ppm because of the enormous excess of $H_2$.

Our general conclusion regarding organic compounds must be that, at temperatures below the $CH_4$–CO boundary, the chemical stability of $CH_4$ is so great that other carbon compounds are suppressed. In the CO region from about 1000 to 3000 K there is enormous complexity, in that tiny traces of large numbers of radicals and small polyatomic molecules are present, especially those listed in Tables IV.5 and IV.6. There are, however, no large abundances of interesting products. Above about 3000 K the chemistry again becomes simple as even species such as CO, CN, and OH decompose into their constituent atoms.

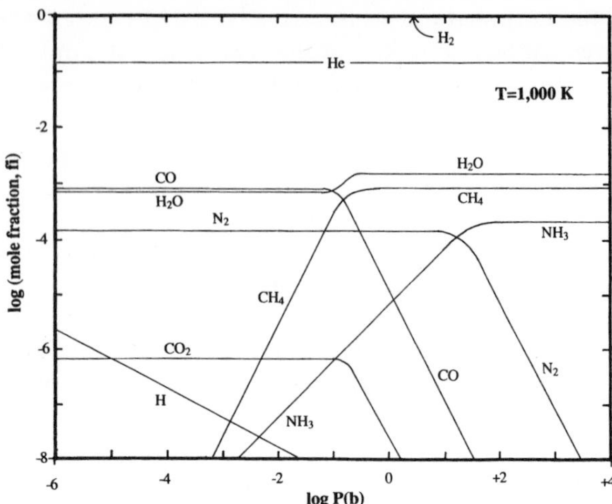


**Figure IV.16** Mole fractions of H, C, N, and O gases along an isotherm. The crossover points for ammonia and methane at 1000 K are emphasized. Note also the $CO_2$ and H abundances.

So far, we have considered the system H, He, O, C, N, Ne, Ar, Kr, and Xe. We have encountered only a few equilibrium condensates that might be important above about 20 K. These are water ice, solid ammonia monohydrate, solid methane, and the solid clathrate hydrates of methane, argon, krypton, and xenon. At temperatures above about 200 K, only graphite is a possible condensate. However, graphite is thermodynamically stable only at temperatures so low that reactions that might form it would have negligible rates. Graphite precipitation from a solar composition gas does not seem to be a reasonable process in stellar atmospheres or in circumstellar gas clouds.

Of the ices found to be thermodynamically stable in the Universe at large, water ice is the only one found as an abundant terrestrial mineral. Solid clathrate hydrates of light hydrocarbons have been found in permafrost beds overlying natural gas deposits in high-latitude settings such as Siberia, northern Canada, and Alaska. Gas hydrate deposits are also found in many locations on the ocean floor, where they are stabilized by the low temperatures (close to 4°C) and high pressures. Related hydrates, including the propane clathrate hydrate $C_3H_8 \cdot 7H_2O$, are also formed in natural gas pipelines when moist gas is subjected to high pressures.

The terrestrial planets are generally too warm for the other ices and are generally very deficient in methane, ammonia, and heavy rare gases. The outer Solar System presents a suitable environment for the formation and survival of such ice minerals.

In discussing the chemistry of hydrogen and helium, we covered the entire range of chemistry accessible to the early Universe and to the systems of Pop II stars, reflecting only the nucleosynthetic contributions of the Big Bang and perhaps the pp chain and the carbon cycle. In extending our discussion to carbon, nitrogen, oxygen, and neon, we treated the chemistry of the elements at the heart of the catalytic carbon (CNO) process, the products of helium burning, and the early steps of the alpha process. We may conceive of second-generation planetary systems in which this chemistry occurs and is the whole story; heavier elements may be negligible in abundance in them or even totally absent. Such systems could contain gas-giant planets or icy planets, but terrestrial (rocky) planets and satellites would be impossible. To build such objects requires the abundant presence of rock-forming elements, such as the next two alpha-process nuclides beyond $^{20}Ne$, $^{24}Mg$ and $^{28}Si$. We shall now turn our attention to these elements.

## Magnesium and Silicon

Traces of SiO and SiS gases can be found even at temperatures of 3000 to 4000 K. Near 2000 K the dominant gaseous species are Mg, SiO, SiS, and Si. At very high pressures $SiH_4$ (silane) gas is also important.

Condensation of these elements begins with the formation of solid forsterite, $Mg_2SiO_4$, near 1400 K, followed at most pressures by the formation of enstatite, $MgSiO_3$, at only slightly lower temperatures. The major reactions are

$$SiO + 2Mg + 3H_2O \rightleftharpoons \underset{\text{forsterite}}{Mg_2SiO_4(s)} + 3H_2 \qquad (IV.118)$$

$$SiS + H_2O \rightleftharpoons SiO + H_2S \qquad (IV.119)$$

$$Si + H_2O \rightleftharpoons SiO + H_2 \qquad (IV.120)$$

$$SiO + \underset{\text{forsterite}}{Mg_2SiO_4(s)} + H_2O \rightleftharpoons \underset{\text{enstatite}}{2MgSiO_3(s)} + H_2. \qquad (IV.121)$$

Here, as elsewhere, solids (s) and liquids (l) are marked as such, or mineral names are given, whereas gases are left unmarked.

Because the abundance of magnesium is almost exactly equal to the abundance of silicon, cooling to the point of complete condensation (about 1100 K) produces almost pure enstatite.

At much lower temperatures, below about 400 K, it is possible for enstatite (or forsterite, if present) to react with water vapor to produce hydroxyl silicates such as talc:

$$\underset{\text{enstatite}}{4MgSiO_3} + 2H_2O \rightleftharpoons \underset{\text{talc}}{Mg_4Si_4O_{10}(OH)_4}. \qquad (IV.122)$$

Through such reactions water may be retained in a chemically bound state in minerals such as talc, serpentine, or chlorite at temperatures far too high for direct condensation of water or ice.

The reactions of silicon and magnesium are summarized along with those of iron in Fig. IV.17. Note that increases in pressure favor forsterite condensation, its alteration to enstatite, and the eventual formation of talc in accordance with

$$K_{118}(T) = \frac{a_{fo}(p_{H_2})^3}{p_{SiO}(p_{Mg})^2(p_{H_2O})^3}. \qquad (IV.123)$$

We can then solve for the activity of forsterite and express the gases in terms of mole fractions rather than partial pressures:

$$a_{fo} = \frac{K_{118} f_{SiO}(f_{Mg})^2(f_{H_2O})^3}{(f_{H_2})^3} P^3. \qquad (IV.124)$$

Reactions (IV.121) and (IV.122) can be treated in the same manner. In the absence of condensation, $f_{SiO} \approx f_{Mg}$ = constant, and the activity of forsterite increases as $P^3$. When the activity reaches 1.00, condensation occurs. The activity of the condensate is set equal

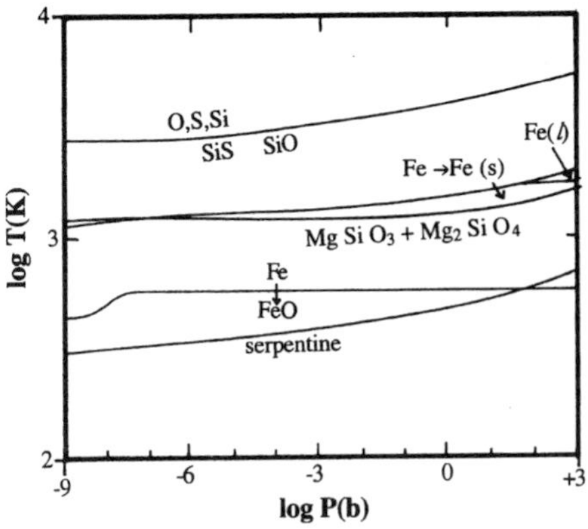


**Figure IV.17** Silicon, magnesium, and iron reactions. The major high-temperature event is the reaction and condensation of SiO, Mg, Fe, $H_2O$, and other gases to produce metallic iron and the magnesium silicates. The endpoint in the oxidation of metallic iron to form FeO-bearing silicates and the point of conversion of ferromagnesian silicates to serpentine are also given.

to 1.00, and the gas pressures in equilibrium with the condensate are recalculated.

The course of nucleosynthesis beyond $^{28}Si$ passes first through the alpha process products $^{32}S$ and $^{36}Ar$ and then through a tortuous series of reactions to the iron-group (the e-process) abundance peak. The next two elements after Si and Mg in abundance are Fe and S, which interact strongly with each other.

## Iron

The only gaseous iron species of any consequence is atomic iron vapor. Upon cooling to about 1500 K this vapor condenses directly to metallic iron. At total pressures of several bars iron condenses above its melting temperature, and liquid iron is the stable phase. The relevant reaction is

$$\mathrm{Fe} \rightarrow \mathrm{Fe(s, l)} \tag{IV.125}$$

$$a_{\mathrm{Fe}} = K_{125} p_{\mathrm{Fe}} = K_{125} f_{\mathrm{Fe}} P, \tag{IV.126}$$

from which it is obvious that increasing total pressure favors direct condensation of minor constituents.

The vapor pressure of metallic iron is given rather well by [see Appendix I, Eqn. (AI.46)]

$$\begin{aligned} \log p_{\mathrm{Fe}}(\mathrm{atm}) &= -\Delta G^{\circ}_{\mathrm{vap}}/2.303RT \\ &= (T\Delta S_{\mathrm{vap}} - \Delta H^{\circ}_{\mathrm{vap}})/4.576T \\ &= \Delta S_{\mathrm{vap}}/4.576 - \Delta H^{\circ}_{\mathrm{vap}}/4.576T, \end{aligned} \tag{IV.127}$$

where $\Delta G^{\circ}$ (cal mol$^{-1}$) is the standard Gibbs free energy of vaporization, $\Delta H^{\circ}$ (cal mol$^{-1}$) is the standard enthalpy of vaporization, $R$ is the gas constant (cal mol$^{-1}$ K$^{-1}$), and $\Delta S$ (cal mol$^{-1}$ K$^{-1}$) is the standard entropy of vaporization (see Appendix I). It can be seen that log $pFe$ increases linearly with $-1/T$. At the point of saturation of iron vapor the activity of solid iron reaches 1.00, and $\log pFe = -\log K_{125}$. Because $\Delta S_{\mathrm{vap}}$ and $\Delta H^{\circ}_{\mathrm{vap}}$ are very slowly varying functions of temperature,

$$\log K_{125} = a/T - b, \tag{IV.128}$$

where $a$ and $b$ are constants. This very general expression is applicable to a wide range of chemical equilibria, as we saw in our earlier discussion of thermal dissociation of molecular hydrogen. Condensation of metallic iron occurs slightly above the condensation temperature of magnesium silicates at most pressures, but the two condensation curves virtually coincide below $10^{-6}$ bar total pressure.

At lower temperatures the chemistry of iron becomes rather more complex. The details of its behavior are of great importance for understanding the chemistry of meteorites, asteroids, and the terrestrial planets.

Metallic iron can be oxidized by water vapor at low temperatures according to the reaction

$$\mathrm{Fe(s)} + \mathrm{H_2O} = \mathrm{FeO(s)} + \mathrm{H_2}. \tag{IV.129}$$

Pure solid FeO (wüstite) of unit activity cannot be formed in a solar composition system above a temperature of about 400 K. If we solve the equilibrium expression for Reaction (IV.129) for the activity of FeO, we find

$$a_{\mathrm{FeO}} = K_{129} a_{\mathrm{Fe}} p_{\mathrm{H_2O}}/p_{\mathrm{H_2}} \approx K_{129} f_{\mathrm{H_2O}}/f_{\mathrm{H_2}}, \tag{IV.130}$$

which is a function of temperature alone. There is no explicit dependence of FeO activity on pressure. We can recall, however, from our discussion of Reaction (IV.104), that the mole fraction of water changes rather abruptly by about a factor of 2 when the $CO-CH_4$ dominance boundary is crossed. Thus the FeO activity is pressure-independent except for a jog at the $CO-CH_4$ boundary.

By an interesting coincidence of nature the $Fe^{2+}$ ion is an almost perfect match for $Mg^{2+}$ in both charge and ionic radius. Thus FeO may freely replace MgO in a wide range of minerals, including both enstatite and forsterite. In our discussion of silicon and magnesium, we argued that the near equality of their cosmic abundances caused enstatite to be the dominant magnesium silicate after completion of condensation. But we now must add FeO to this system at somewhat lower temperatures. This must cause a large increase in the ratio of divalent cations to silicon. In fact, because the cosmic abundance of iron is also almost equal to those of Mg and Si, complete oxidation of metallic iron to FeO (below about 400 K) would provide $SiO_2$, MgO, and FeO in nearly equimolar amounts, precisely what is needed to make $MgFeSiO_4$, an iron-oxide-bearing analogue of $Mg_2SiO_4$. Minerals of this stoichiometry and structure, regardless of the Fe:Mg ratio, are called olivine. Pure $Fe_2SiO_4$ is fayalite, and all naturally occurring olivines are solid solutions of the two end members, forsterite and fayalite. The general formula for olivine which reflects the allowed range of compositions is $(Fe, Mg)_2SiO_4$.

Substitution of FeO for MgO also occurs in enstatite. Solid solutions ranging in composition from $FeSiO_3$ (ferrosilite) to $MgSiO_3$ (enstatite) are members of the pyroxene family, which we shall soon revisit when we discuss another divalent cation, calcium. *Pure* $FeSiO_3$ is unstable with respect to decomposition into a mixture of FeO and $SiO_2$; pure ferrosilite does not exist.

Because FeO can be present in minerals as solid solutions of $FeSiO_3$ and $Fe_2SiO_4$ even when pure FeO cannot be made, FeO is able to enter silicates in appreciable quantities even well above 400 K. Indeed, it is now possible for the silicates to accommodate *all* the iron as

FeO compounds in solid solutions before the temperature drops below 490 K. Thus metallic iron must disappear by this temperature, and pure wüstite never forms. From about 1400 K down to 490 K, Fe metal coexists with FeO-bearing silicates.

The oxidation of Fe metal involves incorporation of FeO in silicates by

$$\underset{\text{kamacite}}{\mathrm{Fe(s)}} + \mathrm{H_2O} + \underset{\text{enstatite}}{\mathrm{MgSiO_3(s)}} \rightleftharpoons \underset{\text{olivine}}{\mathrm{FeMgSiO_4(s)}} + \mathrm{H_2} \quad \text{(IV.131)}$$

$$\mathrm{FeMgSiO_4(s)} + \mathrm{MgSiO_3(s)} \rightleftharpoons \mathrm{(Mg,Fe)_2SiO_4(s)} + \mathrm{(Mg,Fe)SiO_3(s)}. \quad \text{(IV.132)}$$

If we denote the mole fraction of $Fe^{2+}$ silicate in the olivine as $x(\mathrm{FeO}) = n(\mathrm{FeO})/(n(\mathrm{FeO}) + n(\mathrm{MgO}))$, then the concentration of FeO depends on temperature approximately as

$$\log x_{\mathrm{FeO}} = -5.85 + 2775/T, \quad \text{(IV.133)}$$

where $x_{\mathrm{FeO}}$ is close to but not identical to $a_{\mathrm{FeO}}$ in Eq. (IV.130).

Below 490 K the ferromagnesian silicates olivine (now dominant) and pyroxene (now minor) can be altered by reaction with water vapor in much the same way that enstatite alteration produces talc. The product in the present case is a layer-lattice ferromagnesian silicate, such as serpentine or chlorite, with essential hydroxyl groups. Formation of serpentine can be written

$$\underset{\text{pyroxene}}{\mathrm{(Fe,Mg)SiO_3(s)}} + 2\mathrm{H_2O} \rightleftharpoons \underset{\text{quartz}}{\mathrm{SiO_2(s)}} + \underset{\text{serpentine}}{\mathrm{(Fe,Mg)_3Si_2O_5(OH)_4}}. \quad \text{(IV.134)}$$

The main reactions relevant to the system Fe–Mg–Si–O–H is solar material are summarized in Fig. IV.17. The dependence of the FeO content of the silicates on temperature is illustrated in Fig. IV.18. The extremely low FeO content of freshly condensed $Mg_2SiO_4$ (0.01 mol% FeO) shows that our discussion of Mg silicate condensation neglecting iron is still reliable. We still expect that FeO-poor enstatite will be the dominant silicate mineral from the end of condensation to fairly low temperatures (600 K), at which conversion to FeO-bearing olivine takes over.

We have seen in Fig. IV.15b that the activity of graphite reaches $a(\mathrm{gr}) = 1.00$ only at temperatures below about 450 K, so cold that attainment of equilibrium would take an impossibly long time. We have also seen that, at temperatures of 600 K or higher, where reaction rates become appreciable, the activity of graphite is less than 0.1, and hence graphite cannot form. However, our experience with FeO should teach us to ask whether

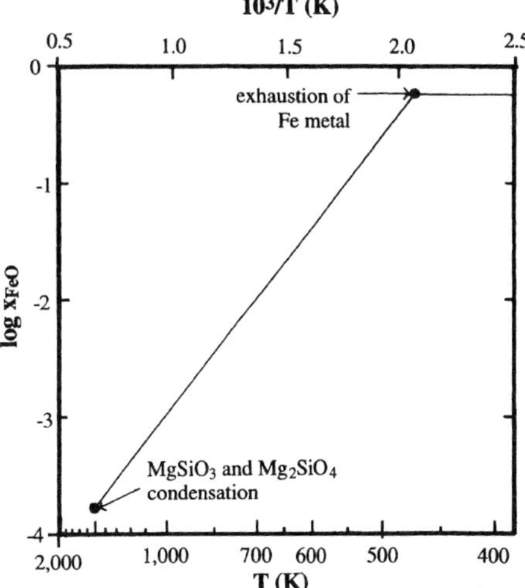


**Figure IV.18** Iron oxidation and FeO entry into silicates. Freshly condensed magnesium silicates (mostly enstatite) are virtually devoid of FeO. Most oxidation of metal occurs below 600 K, and metal is exhausted near 490 K.

there might not be some host phase present at these higher temperatures within which carbon might be accommodated in solid solution. In fact, one of the best solvents known for carbon is liquid iron, and the solubility of carbon in solid iron is significant at higher temperatures. We shall return to this matter when we discuss the chemistry of nickel.

With this discussion of silicon, magnesium, and iron we have now covered the four most abundant elements in the terrestrial planets. After O, Si, Mg, and Fe, the next most abundant element is sulfur.

## Sulfur

Near 2000 K, sulfur is found largely as SiS and the SH (sulfhydryl) radical, although traces of SO, COS, and CS are also present. Solid sulfides of silicon are quite unstable, and hence SiS does not condense. Instead, SiS (and SH) are converted to $H_2S$ during cooling.

The first and most important sulfide to form is FeS, which is produced by corrosion of metallic Fe by $H_2S$ at and below 680 K. The reaction by which FeS, the mineral troilite, forms is

$$\mathrm{Fe(s)} + \mathrm{H_2S} \rightleftharpoons \mathrm{H_2} + \mathrm{FeS(s)}, \quad \text{(IV.135)}$$

where, analogous to the case of FeO formation,

$$a_{\mathrm{FeS}} = K_{135} a_{\mathrm{Fe}} p_{\mathrm{H_2S}}/p_{\mathrm{H_2}} \approx K_{135} f_{\mathrm{H_2S}}/f_{\mathrm{H_2}}. \quad \text{(IV.136)}$$

As with FeO, the activity of FeS is therefore a function of temperature, but independent of pressure. However, unlike FeO, there is no substance present in which FeS can dissolve when it has an activity less than unity. Therefore troilite appears abruptly at 680 K, at which, during cooling, $a_{FeS}$ first reaches 1.0. FeS formation is essentially complete a mere 80 K lower. Because the sulfur abundance is less than the iron abundance, $H_2S$ is fully removed from the gas by equilibration with metal below about 600 K. Troilite, stoichiometric FeS, should not be confused with the terrestrial mineral pyrrhotite, $Fe_{1-x}S$, which has a deficiency of iron. Troilite is formed in the presence of metallic iron and can have no such deficiency.

We may now reflect a bit more upon the chemistry of iron, in light of the fact that Fe metal, FeO-bearing silicates, and FeS are all important products in a solar-composition system. We may say that iron exhibits siderophile (metal-loving), lithophile (rock-loving), and chalcophile (sulfur-loving) geochemical traits simultaneously, an unusual and nearly unique degree of versatility. This has profound significance for the terrestrial planets, because iron is *by mass* the second most important element in them. It is also interesting that these chemical complexities are results of the introduction of yet another nucleosynthetic class of products, the e-process elements.

Figure IV.19 summarizes the chemistry of the iron-bearing minerals. The temperature scale is linear, not a $1/T$ scale as in Fig. IV.18, and emphasizes the enormous importance of FeO at low temperatures as well as the sharp onset of FeS formation.

## Aluminum and Calcium

Calcium and aluminum oxides are very refractory substances with high melting points and low vapor pressures that are commonly used as the major constituents of firebrick. They are also, not surprisingly, among the earliest condensates to form during the cooling of solar material. Their dominant gaseous species near 2000 K, AlO and Ca, react near 1800 K to produce a complex series of refractory oxide condensates that are very poor in Si and Fe. Corundum $Al_2O_3$, spinel $MgAl_2O_4$, perovskite $CaTiO_3$, and several Ca aluminosilicates such as gehlenite ($Ca_2Al_2SiO_7$) and anorthite ($CaAl_2Si_2O_8$) condense over a temperature range of only 200 K.

Below the condensation temperature of the magnesium silicates it is possible for CaO to enter extensively into the pyroxenes as the end-member $CaSiO_3$ (wollastonite). Because of their nomenclatural complexity, we have collected the naming conventions for the pyroxenes and olivine into Fig. IV.20. The calcic pyroxenes crystallize in the monoclinic system and are called clinopyroxenes. Members of the enstatite–ferrosilite solid solution series usually crystallize in the orthorhombic system and are called orthopyroxenes. Occasionally, however,

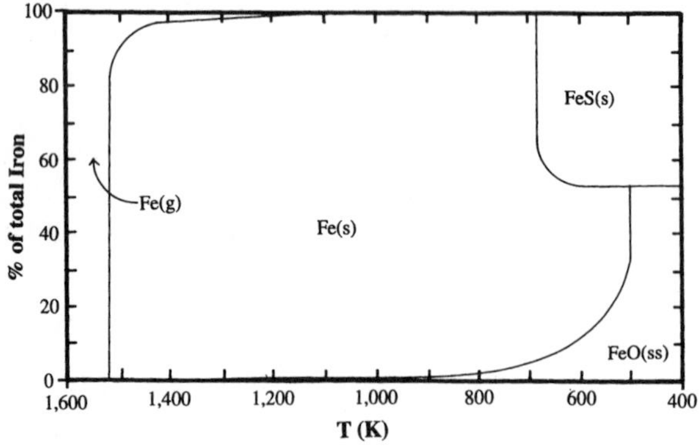


Figure IV.19 Distribution of iron between its minerals. The fraction of total iron in the monatomic iron vapor, in solid metallic iron, in solid FeS, and in FeO solid solutions in silicates are shown from above the condensation point of metal down to 400 K.

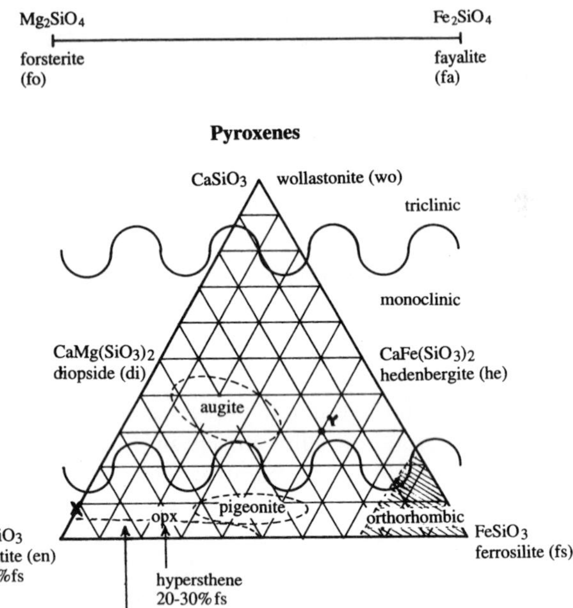


Figure IV.20 Olivine and pyroxene nomenclature. Pure ferrosilite is unstable (shaded region). High-temperature condensates lie deep in the diopside and enstatite corners of the di–he–fs–en pyroxene quadrilateral. At lower temperatures, oxidation of metallic iron by water vapor produces bronzite and hypersthene, and alters much of the pyroxene to olivine of intermediate composition. Minerals close to wollastonite composition, which crystallize in the triclinic system, are found only in highly alkaline rocks and metamorphosed limestones. Pyroxenes found in meteorites generally lie in the three regions labeled augite, pigeonite, and orthopyroxene. It is common for all three of these compositions to coexist. See text for further explanation. (See Fig. IX.22.)

clinoenstatite or clinobronzite is encountered. The point marked X near the $MgSiO_3$ corner gives the solar Ca:Mg ratio. Point Y indicates a pyroxene composition of 20% enstatite, 30% wollastonite, and 50% ferrosilite, a very improbable composition, but a useful clue to how to read the figure. The mole fractions of the Ca, Mg, and Fe silicates must of course add up to 1.00; there are therefore only two independent compositional variables, and we can represent an isothermal, isobaric phase diagram in only two dimensions. The shading of the $FeSiO_3$ corner of the diagram is intended to suggest the instability of pure ferrosilite.

At much lower temperatures, calcium-bearing minerals may react with water vapor to produce hydroxyl silicates, especially the very stable amphibole $Ca_2Mg_5Si_8O_{22}(OH)_2$, tremolite. It appears that tremolite is the first phase capable of retaining water in a cooling system of solar composition. Ultimately tremolite formation, which is limited by the small abundance of Ca, is a far less important water trap than serpentinization of the ferromagnesian minerals. It may be, however, the most accessible water-bearing mineral in the vicinity in which the terrestrial planets accreted.

The chemistry of calcium and aluminum is summarized along with that of sodium and potassium in Fig. IV.21.

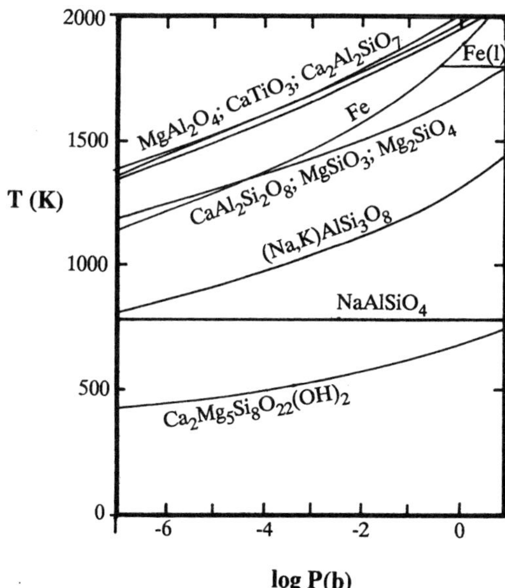


**Figure IV.21** Aluminum, calcium, sodium, and potassium chemistry. The refractory oxide condensates, including $MgAl_2O_4$ (spinel), $CaTiO_3$ (perovskite), and $Ca_2Al_2SiO_7$ (gehlenite), are followed by conversion of most of the calcium minerals to $CaAl_2Si_2O_8$ (anorthite) near the condensation temperature of the magnesium silicates. Small traces of alkali feldspars dissolve in the anorthite at high temperatures; however, 50% condensation of K and Na is achieved only at the alkali feldspar line shown. Some $NaAlSiO_4$ (nepheline) forms near 775 K. The first hydroxyl silicate to become stable is tremolite.

## Sodium and Potassium

Atomic sodium and potassium vapor, Na and K, remain in the gas phase until completion of the condensation of enstatite. At slightly lower temperatures it becomes possible for the alkali metal vapors to react with aluminum-bearing minerals to produce alkali aluminosilicates. A simple conceptual example would be

$$\underset{\text{corundum}}{Al_2O_3(s)} + \underset{\text{quartz}}{6SiO_2(s)} + 2Na + H_2O \rightleftharpoons \underset{\text{albite}}{2NaAlSi_3O_8(s)} + H_2. \quad \text{(IV.137)}$$

However, pure solid $Al_2O_3$ and $SiO_2$ cannot coexist, and quartz is never a stable phase at equilibrium in a solar-composition system. A better description of the process would be

$$\underset{\text{spinel}}{MgAl_2O_4(s)} + \underset{\text{enstatite}}{13MgSiO_3(s)} + 2Na + H_2O \rightleftharpoons \underset{\text{forsterite}}{7Mg_2SiO_4(s)} + \underset{\text{albite}}{2NaAlSi_3O_8(s)} + H_2. \quad \text{(IV.138)}$$

Thus consumption of silicon by alkali aluminosilicate formation will generate 3.5 moles of forsterite for every mole of albite. Also, a system with an overall Mg:Si ratio of 1.00 *must* contain olivine once the alkali metals condense, even neglecting the effects of oxidation of Fe metal to FeO. A similar effect occurs due to entry of CaO into aluminosilicates (gehlenite and anorthite).

Anorthite ($CaAl_2Si_2O_8$) can form solid solutions with both albite ($NaAlSi_3O_8$) and orthoclase ($KAlSi_3O_8$). Such aluminosilicate solid solutions are collectively called feldspars. Because of the large ionic size of potassium, the low-temperature solubility of orthoclase in the soda-lime feldspar is not large, although feldspars are fully mutually soluble at elevated subsolidus temperatures. Note that the great mutual solubility of anorthite and albite occurs despite their superficially very different formulae. This is because replacement of $Ca^{2+}$ by $Na^+$ *combined with* the replacement of $Al^{3+}$ by $Si^{4+}$ leads to no net change in electrical charge, and the ionic size differences are not as large as that between $Na^+$ and $K^+$. It is interesting that, by these same arguments, pyroxenes, (Ca, Mg, Fe)$SiO_3$, should be to some degree mutually soluble with corundum, $Al_2O_3$. Indeed, at high temperatures pyroxene does assume a significant content of alumina. Pyroxenes close to the diopside–hedenbergite line in Fig. IV.20 that also contain some alumina are called augite.

The nomenclature and stability relations of the feldspars are shown in Fig. IV.22. Note especially the complex terminology for the albite–anorthite solid

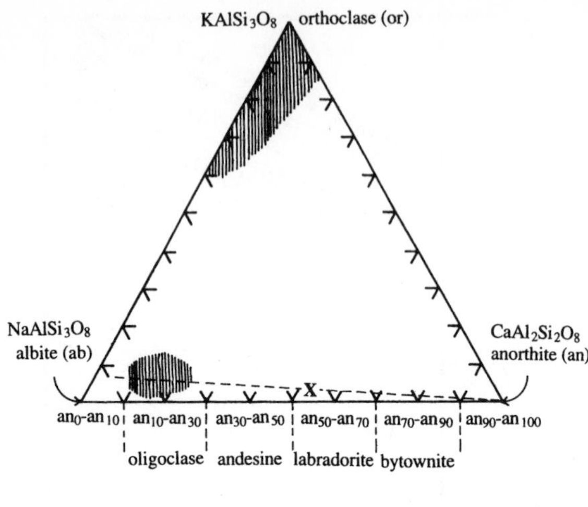


**Figure IV.22** The feldspars. Natural feldspars form a single solid solution at elevated temperatures; however, assemblages equilibrated at low temperatures often contain a potassium-rich feldspar of approximate orthoclase composition coexisting with a plagioclase (albite–anorthite solid solution) phase. The large ionic size of K discourages extensive substitution of orthoclase into plagioclase. However, solar proportions of K, Na, and Ca (marked X in the diagram) contain so little K that the low-temperature solubility limit for K-spar in plagioclase is not exceeded. Because Ca is also an important component of pyroxenes, the ratio an/(ab + or) in the feldspars is shifted along the dashed line away from the anorthite corner. Meteoritic feldspars (see Chapter VIII) are generally found in the shaded area of oligoclase composition. Where differentiation and fractional crystallization processes have greatly enhanced the or:ab ratio, a second feldspar of nearly pure orthoclase composition (lightly shaded area near the or corner) can coexist with plagioclase.

solution series, usually called the plagioclase series. Because of the small but adequate low-temperature solubility of orthoclase in plagioclase and the low cosmic abundance of potassium relative to calcium and sodium, the dominant feldspar compositions are expected to be close to the plagioclase line, and only a single feldspar is normally present. It should also be mentioned that pure potassium feldspar can occur in two other crystal structures distinct from that of orthoclase, sanidine and microcline. These three materials (collectively termed K-spar) can be abundant in settings in which magmatic processes have led to a large enrichment of alkalis relative to calcium, such as in Earth's continental crust.

Retention of potassium is of great importance to the terrestrial planets, because $^{40}$K decay by electron or positron emission is a very large source of heat, sufficient by itself to heat solid planetary material up to the melting point. The half-life for $^{40}$K decay is $1.3 \times 10^9$ years, and thus both the abundance and the rate of decay of $^{40}$K were greater by a factor of factor of $2^{(4.5/1.3)} = 11$ at the time of origin of the Solar System. The $^{40}$K nuclide, a product of explosive silicon burning in supernova explosions, thus captures a small portion of the energy of an exploding star for later use; it introduces the possibility of planetary thermal evolution extending over periods of billions of years.

As we have seen, $^{40}$K decay also provides us with a useful nuclear clock for dating geological events. Accumulation of $^{40}$Ca produced by beta decay of $^{40}$K is not a useful clock because $^{40}$Ca is already a very abundant product of explosive O burning and because potassium normally occurs in solid solution in plagioclase feldspar with a very large calcium content. However, $^{40}$Ar is a rare s-process nuclide, very inefficiently retained during condensation of minerals because of its high volatility. Therefore accumulated $^{40}$Ar is a useful clock for any mineral or rock that contains potassium.

The chemistry of calcium, aluminum, sodium, and potassium is summarized in Fig. IV.21.

## Nickel and Cobalt

The chemical properties of Ni and Co are closely similar to those of iron. Both are slightly enriched in the first metal to condense, but then, in the temperature regime from about 1400 K down to 680 K, their ratios to iron in the metal phase are very close to the solar Ni:Fe abundance ratio.

Several different phases can be formed in the Fe–Ni system. Of particular importance are the cubic $\alpha$ (alpha-iron) phase, which bears the mineralogical name kamacite, and the $\gamma$ (gamma-iron) phase, which is called taenite. The phase relations in the binary Fe–Ni system are shown in Fig. IV.23. The dotted line traces the locus of the composition of the metal phase during cooling from first condensation to the point of disappearance of the last metal at 490 K. When the dotted line lies within the $\alpha + \gamma$ two-phase region these two phases form an intricately intergrown structure. Note that below the appearance temperature of the $\alpha$ phase there is a narrow temperature region near 700 K in which the pure $\alpha$ phase may possibly be stable, whereas just prior to the disappearance of the last metal the pure $\gamma$ phase might be stable. Of course, it may take a very long time for equilibrium to be attained within the metal at such low temperatures.

When the concentration (and activity) of nickel has been raised to very high levels by the oxidation of metallic iron and its removal from the metal, it becomes possible for nickel to enter into sulfides:

$$8\text{FeS(s)} + \text{Ni(s)} = \underset{\text{pentlandite}}{(\text{Fe, Ni})_9\text{S}_8}. \qquad \text{(IV.139)}$$

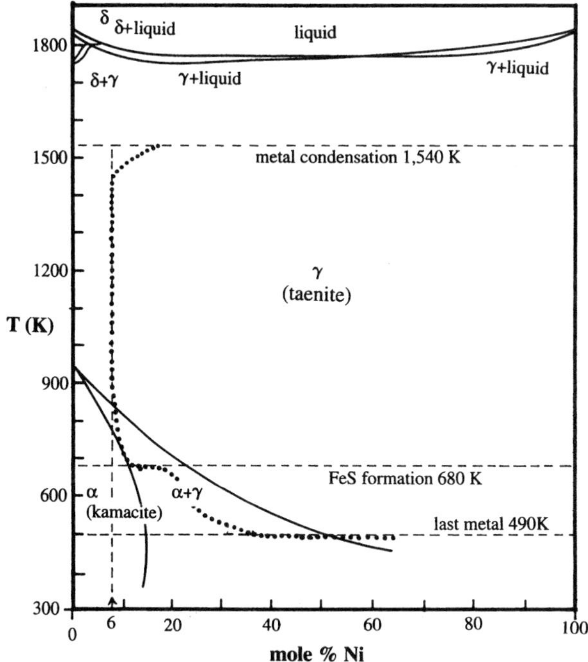


**Figure IV.23** The iron–nickel system. The stability fields of liquid metal, delta, gamma, and alpha iron are shown for 0.01 atm. Freshly condensed metal (1540 K) is slightly Ni-rich. The dotted line traces the metal composition during cooling of a solar-composition gas. Note that the metal enters the taenite-plus-kamacite two-phase region near 870 K and probably stays inside it through FeS condensation and substantial oxidation of iron. The last remaining metal phase is very Ni-rich, with 60% Ni (kamacite) composition. This metal phase disappears due to entry of nickel into sulfides such as pentlandite, $(Fe, Ni)_9S_8$, near 490 K.

With the understanding of the chemistry of iron, nickel, and cobalt that we have gained, it is now possible to use the data on the thermodynamics of solid solutions of carbon in iron–nickel alloys to calculate how much carbon should be dissolved in the metal phase over the pressure and temperature range of interest to us. We will recall that the carbon activity on any isotherm or isobar reaches a maximum at the $CO-CH_4$ equal-abundance line and that solid graphite is actually a stable phase at low pressures and temperatures on that boundary. The calculated concentration of dissolved carbon is displayed in Fig. IV.24.

## Phosphorus and the Halogens

There are several gaseous phosphorus compounds that are important at elevated temperatures, notably PN, PO, PS, and P. Phosphorus, like carbon, enters the metal phase in extremely small quantities (0.01% of total P) at elevated temperatures. Solid schreibersite, $Fe_3P$, is the only important carrier of phosphorus down to temperatures near 750 K, at which conditions first favor

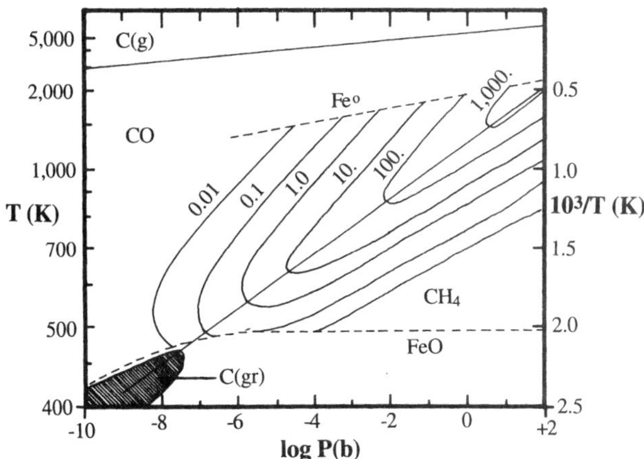


**Figure IV.24** Dissolved carbon in Fe–Ni alloys. The equilibrium concentrations of carbon in ppm by weight are given over the entire pressure–temperature range of stability of the free metal phases. Note the maximum in carbon content near the $CO/CH_4$ equal-abundance line.

oxidation of phosphide to phosphate. There whitlockite, $Ca_3(PO_4)_2$, and fluorapatite, $Ca_5(PO_4)_3F$, form nearly at the same temperature.

Chlorine, which exists as a mixture of HCl, NaCl, and KCl gases above 900 K, reacts to form the complex halide–silicate sodalite, $Na_4Al_3Si_3O_{12}Cl$, at that point. Sodalite remains stable down to low temperatures. Fluorine, which is found mainly as HF gas at high temperatures, condenses first as fluorapatite.

Thermodynamic data on the heavy halogens Br and I are of such low quality and so incomplete that no predictions of their behavior can be made with confidence. The condensation behavior of phosphorus, chlorine, and fluorine is shown in Fig. IV.25.

## Geochemical Classification of the Elements

We have now discussed the main features of the equilibrium chemistry of 20 of the 23 most abundant elements in the Sun, plus the heavy rare gases krypton and xenon. If the principle of equal time were applied to the remainder of the periodic table, we would become obliged to convert this chapter into a book in its own right. Instead, we shall recognize that, because there *is* a periodic chart, there are certain important familial resemblances in chemical and geochemical behavior.

To date we have recognized explicitly three major types of geochemical affinity, largely based on how these elements behave on Earth:

*lithophile* = rock-loving: oxides and silicates
*chalcophile* = sulfur-loving: sulfides
*siderophile* = metal-loving: free metals.

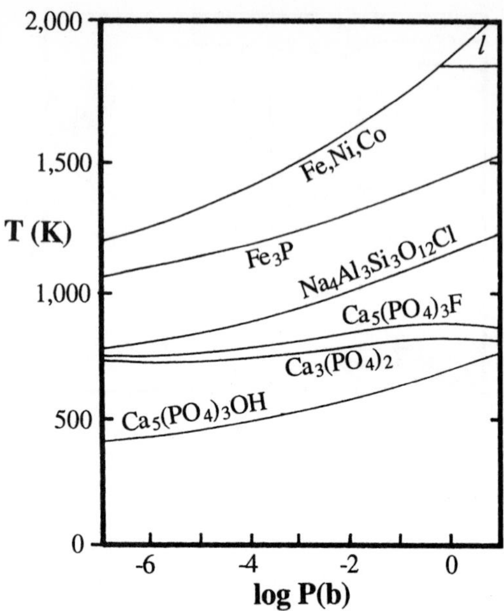


**Figure IV.25** Phosphorus, chlorine, and fluorine chemistry. The first important mineral condensate in this group is the iron phosphide schreibersite. The most stable chlorine mineral is sodalite, and fluorine enters preferentially into the phosphate fluorapatite. A minor water-bearing phase, hydroxyapatite, forms at about the same temperature as tremolite.

To these we add formally the volatile elements, such as the rare gases and nitrogen, which do not belong to any of these groups:

*atmophile* = air-loving: volatiles.

The shortcomings of this classical approach are already becoming partly evident. For example, what is iron in this scheme? Apparently it can be a siderophile, lithophile, and chalcophile all at the same time. Further, changes in temperature, pressure, or oxidation state may cause iron to change categories. Nickel, which usually behaves as a siderophile in solar-composition systems, becomes a chalcophile at low temperatures and turns lithophilic under mild oxidation.

For the special and limited case of a solar-composition system we may choose a classification scheme suited to the task of describing the behavior of the elements during condensation. To describe the geochemical differentiation of a planet, we may choose yet another scheme; indeed, we shall. The classification scheme that we shall use is as follows.

1. Refractory siderophiles. These are metals that condense at temperatures well above the condensation point of iron–nickel alloys. The only elements in this category are of extremely low cosmic abundance, notably W, Os, Ir, and Re. These are the first condensates to form during cooling of stellar material.

2. Refractory oxides. These include many of the aluminum and calcium compounds that we have already encountered, such as gehlenite, corundum, spinel, and perovskite. Those highly electropositive elements which have valences of 2+ or higher almost all fall into this category. These include the oxide moieties $Al_2O_3$, CaO, $Ti_2O_3$, $TiO_2$, $V_2O_3$, $VO_2$, $Sc_2O_3$, $ZrO_2$, SrO, $Y_2O_3$, and BaO and the family of rare earth element oxides (REEOs), including the major radioactive elements U and Th. Uranium and thorium are two of the three (with K) most important heat-producing elements. The mass of refractory oxides is only some 5% of the total mass of silicates, so the concentration of U and Th in early condensates is 20 times that in a planet or meteorite.

3. Iron–nickel metal. Among the minor elements that are present in this round of condensation are Co, Cu, Au, Pt, Ag, and a number of other metals, as well as those nonmetals that have an appreciable solubility in hot iron. These include P, N, and C, with small traces of sulfur, germanium, and perhaps chlorine.

4. Magnesium silicates. In this group we also find small amounts of certain other predominantly lithophilic elements, including Mn, B, F, and some Cr and Li.

5. Alkali metals. Na and K are accompanied at slightly lower temperatures by rubidium and cesium and perhaps by Cl.

6. Moderately volatile chalcophiles. Accompanying and following FeS formation we find Zn, Pb, Ga, Ge, Se, Te, and As and possibly also Br and I.

7. Loosely bound oxygen and –OH in minerals. This category includes the chemically labile oxygen in FeO and its compounds, as well as the hydroxyl silicates such as amphiboles, serpentine, and chlorite.

8. Ice minerals. These are water ice itself; the solid hydrates of ammonia, methane, and rare gases; and solid methane.

9. Permanent gases. These are the three gases that are, under natural conditions, virtually uncondensable: $H_2$, He, and Ne.

For convenient reference, a flow chart for the major elements is presented in Fig. IV.26. The "staircase" formed by the solid line indicates the condensation temperatures for the elements listed across the top of the figure. By following the arrows, the chemical history of each element may be traced from the gas phase at 2000 K to the solid state at 1 K. The bulk composition of the solids after each step is given in Table IV.7.

We have also collected together the main chemical processes of the major elements in Figs. IV.27 and IV.28,

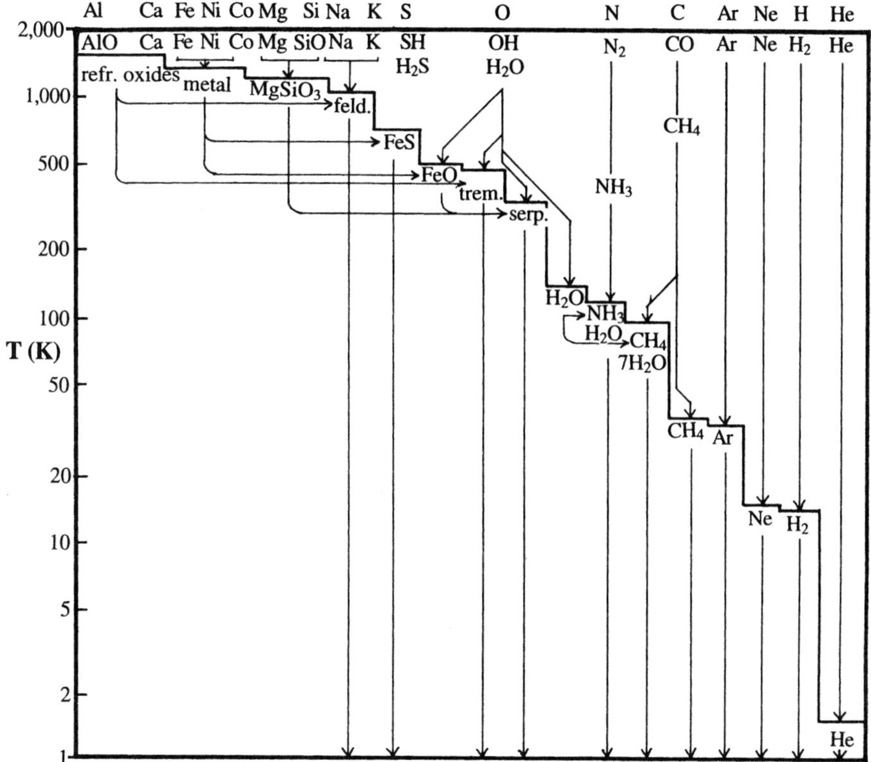

Figure IV.26 Major element flow chart for equilibrium condensation. The sequence of reactions of gases (above the solid line) and condensates (below the solid line) in solar material cooling at equilibrium. A pressure of $10^{-2}$ bar is assumed.

in which the condensation curves for many important reactions are displayed in a linear temperature format. The reactions emphasized here are those that have an appreciable effect on the overall composition, density, and volatile-element content of the condensed solids.

It will be our eventual goal to interpret observed compositional features of the Solar System in terms of such a chemical model. In doing so, we will necessarily make frequent reference to the content of Figs. IV.26 through IV.28 and to Table IV.7. At present, armed only with the information on the densities of the planets and satellites given in Tables III.3 and III.5, we can see that a general perspective on the densities of Solar System bodies is emerging from this kind of chemical modeling. The three broad classes of Solar System bodies proposed by Harrison Brown in 1952, *rocky*, *icy*, and *solar*, are now not only reproduced in general terms, but also described in considerable detail.

It is important to realize that the chemical model we have explored so far, the equilibrium condensation model, contains a wealth of explicit and quantitative predictions about the bulk composition, density, major-element chemistry, oxidation state, volatile-element content, and radioactivity as a function of formation conditions. There are only two relevant variables, temperature and pressure, that may be usefully constrained even by rather simple physical models of the conditions in the vicinity of the Sun at the time of the formation of the planets. By their

Table IV.7 Mineral Assemblages for Equilibrium Condensation of Major Elements

| Step | Materials present |
|---|---|
| 1 | $Ca_2Al_2SiO_7$, $MgAl_2O_4$ |
| 2 | List 1 + (Fe, Ni) |
| 3 | List 2 + $MgSiO_3$ |
| 4 | (Fe, Ni) + $MgSiO_3$ + $CaMgSi_2O_6$ + plagioclase |
| 5 | List 4 + FeS |
| 6 | (Fe, Ni) + $MgSiO_3$ + plagioclase + FeS + tremolite |
| 7 | $(Fe, Mg)_2SiO_4$ + plagioclase + $(Fe, Ni)_9S_8$ + amphibole |
| 8 | Serpentine + plagioclase + sulfide + amphibole = rock |
| 9 | List 8 (rock) + $H_2O$ |
| 10 | Rock + $H_2O$ + $NH_3 \cdot H_2O$ |
| 11 | Rock + $NH_3 \cdot H_2O$ + $CH_4 \cdot 7H_2O$ |
| 12 | Rock + $NH_3 \cdot H_2O$ + $CH_4$ + $7H_2O$ + $CH_4$ |
| 13 | Rock + $NH_3 \cdot H_2O$ + $CH_4 \cdot 7H_2O$ + $CH_4$ + Ar = rock + ice |
| 14 | List 13 (rock plus ice) + Ne |
| 15 | Rock + ice + Ne + $H_2$ |
| 16 | Rock + ice + Ne + $H_2$ + He = everything (solar) |

*Note.* Elements considered for this list: H, He, O, C, N, Ne, Si, Mg, Fe, S, Ar, Ca, Al, Na, Ni, and K. The total abundance of all other elements is less than the abundance of nickel.

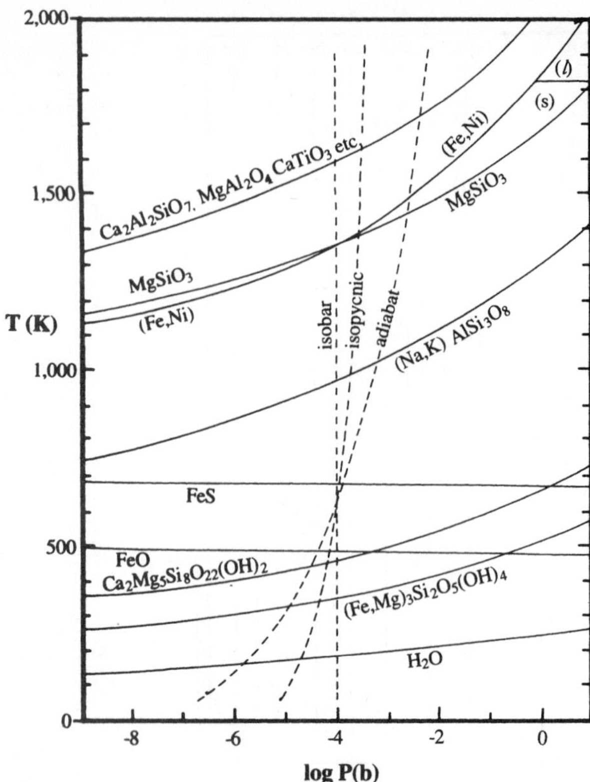


Figure IV.27 High-temperature equilibrium condensation. The minerals formed by the major elements are sketched. The three different temperature–pressure profiles correspond to isobaric (constant pressure), isopycnic (constant density), and adiabatic (constant entropy) structures. Note the extreme similarity of the condensation sequences for these very different sets of conditions. Thus condensate composition can be used to place strong constraints on temperature, but some other source of information is needed to constrain the pressures in the nebula.

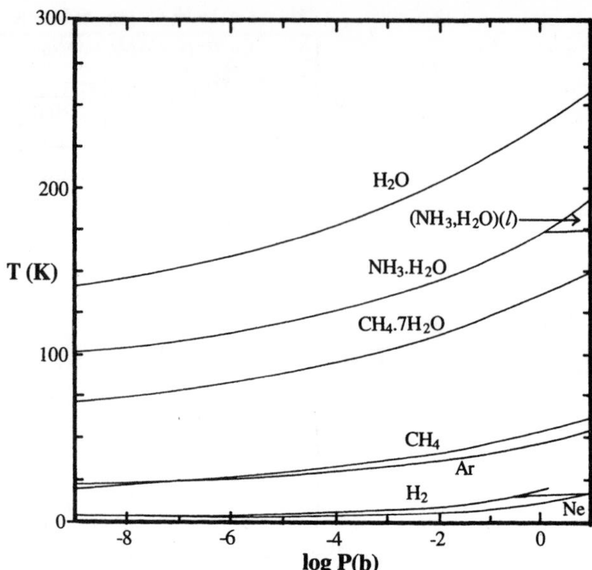


Figure IV.28 Low-temperature equilibrium condensation. The ice condensates formed at equilibrium in a system of solar composition are shown. Note the eutectic melting of water ice I plus ammonia monohydrate at 173 K. Pure water ice melts at 273 K, just off the right-hand edge of the diagram.

very nature, equilibrium calculations give us the molecular and mineralogical composition of a system as a *state function*; that is, the results are a function of the temperature, pressure, and elemental abundances alone and do not depend in any way on the past history of the system. We therefore say that equilibrium calculations give us results that are *path-independent*. This is both a strength and a weakness of the equilibrium approach; if a natural system is indeed well equilibrated, then its behavior may be calculated in great detail from theory, and the conditions under which equilibration occurred can be deduced with great confidence. But if the system was not well equilibrated it will not be well matched by equilibrium calculations; *path-dependent* compositional features will be observed reflecting the ($P$, $T$) path by which the final equilibration conditions were approached. This presents us with a much harder task with much greater rewards. We must understand the thermodynamic and *kinetic* (reaction rate) behavior of the system in order to understand the details of its composition quantitatively. (I refer here to incomplete alteration of previously formed grains, diffusional limitations on the entry of water, oxygen, sulfur, and other materials into metal and silicate grains, etc.) If we succeed in this difficult task, we gain insight into the history and evolution of the system, not just a snapshot of its state at one point in time.

If we now look again at Figs. IV.27 and IV.28, we can see a most useful feature of the equilibrium results: if we should happen to know the pressure at which condensation occurred within a factor of 10 or even a factor of 100, then the *sequence* of chemical reactions taking place during cooling is virtually immutable. Further, a factor of 20 error in estimating the pressure would lead to only about a 6% error in estimating the condensation temperature! Thus the *relative* condensation temperatures are excellently known, and even the *absolute* condensation temperatures should be reliable to a precision of a few percent. All that is required to achieve this level of precision is to develop a means of calculating the pressure of the medium out of which planetary material condensed or in which it was equilibrated. This need only be done within a factor of 10 or 20.

It can also be seen from inspection of Fig. IV.27 that very different assumed functional relationships between pressure and temperature have very little effect on the sequence of reactions. Isobaric (constant pressure), isopycnic (constant density), and adiabatic (constant entropy) profiles are shown.

In order to facilitate more detailed comparison between the observations of densities of Solar System bodies and the predictions of the equilibrium condensation model, it is desirable to use the compositional information derived from the model to make explicit predictions of the dependence of bulk condensate density on condensation temperature. To do this, we shall run through the reactions listed in Table IV.7 to determine their effects on the bulk (cumulative) density of the condensed material. Although, as usual, we shall run through these reactions from high temperatures to low, and although this is usually called the "condensation sequence," it is essential to keep in mind that these results are derived from the assumption of thermochemical equilibrium and contain no dependence on the history of the material.

To assist in visualizing the details of the density variations, we shall display them in Fig. IV.29. A nominal pressure of $10^{-4}$ bar is used to calculate the densities in this figure. Even substantially different pressures, ranging from $10^{-2}$ to $10^{-6}$ bar, cause virtually no change in the shape of the density-vs-temperature curve.

Along the sequence from step 1 to step 9 in Table IV.7 there are several general trends of note. Both the oxidation state of the rock (as measured by the $Fe^{2+}$: $Fe^0$ ratio) and the volatile content ($S + H_2O$) increase as the temperature drops. Thus, after metal condensation (step 2), the general trend is for a decrease of density with each step. There is one important exception to this trend: step 5,

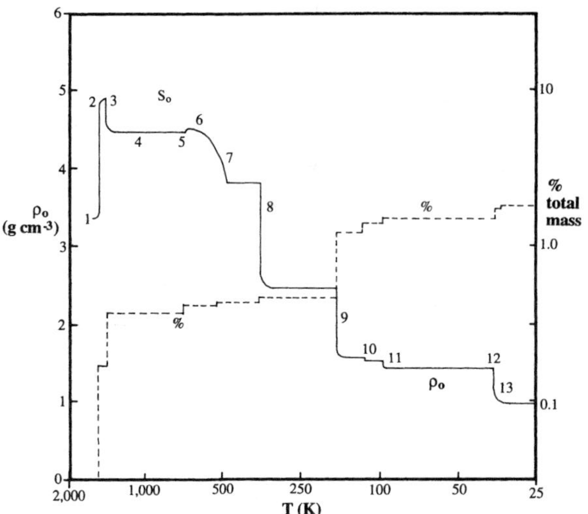


**Figure IV.29** Bulk density of equilibrium condensate and percentage of total mass condensed. The uncompressed density of the total condensate along the equilibrium condensation sequence as listed in Table IV.7 is shown by the solid line, and the total condensed mass as a percentage of total solar-composition mass is shown as the dashed line. The temperatures given are for a particular adiabatic model of the nebular structure (600 K at $10^{-4}$ bar), but the results look very similar for a wide range of assumptions regarding the pressure distribution.

the formation of troilite, actually adds to the rock an element, sulfur, whose atomic weight is greater than the average atomic weight of the rock. Thus retention of the volatile element sulfur actually leads to an *increase* in density.

Steps 9 through 13, the condensation of ices, have a profound effect on the bulk density of condensate. Although the rock density varies from a maximum of about 5 g cm$^{-3}$ (after step 2) down to about 2.5 (after step 8), condensation of $H_2O$ ice alone suffices to drop the density to about 1.7. Step 12, condensation of low-density methane ice, causes a further density drop to about 1.0, which is virtually unchanged by Ar condensation (step 13).

Hydrogen condensation, step 15, combined with the virtually simultaneous condensation of neon in step 14, would lower the bulk density of condensate to about 0.2 g cm$^{-3}$. Because no solid Solar System bodies with densities less than 1.0 are known and because it is questionable whether temperatures low enough for hydrogen condensation are plausible during the collapse phase of the protosun, we shall not consider further the consequences of steps 14 through 16, and we do not include them in Fig. IV.29.

Up to now, aside from a few caveats, we have treated the chemistry of solar material by assuming that the condensates and gas are strictly in equilibrium with each other at all temperatures and pressures. Simultaneously, we have mentioned the sequence of reactions that takes place as a result of equilibration during a cooling process. We have emphasized the path independence of these equilibrium calculations and have clearly shown that the results require that intimate contact between gas and condensed solids must be preserved in order that many of these reactions might be able to occur. Many of these reactions involve reactions between the gas phase and previously condensed (that is, high-temperature) minerals. This model implicitly assumes that, if the solar-composition gas was cooling down, the rates of chemical reactions were fast compared with both the rate of cooling and the rate of accretion of the solid condensates into large bodies of negligible surface-to-volume ratios, which essentially ceased to interact with the gas. Alternatively, temperatures may have dropped off systematically with distance from the Sun, but did not change with time. The equilibrium approach is thus a model for the chemical behavior of highly dispersed small grains in a solar-composition nebula, with no presumption that the materials outside the orbit of Mercury were ever fully evaporated. There is no need to postulate that all these species actually "condensed" in the solar nebula, only that they were formed by local equilibration.

## The Chemistry of Rapid Accretion

Now that we have explored the conceptually (and computationally) simplest model for the chemical behavior of solar material, let us examine the consequences of a completely different assumption regarding the relative rates of accretion and chemical equilibration. Heretofore we have assumed that accretion of solids to form large bodies is very slow compared with the rate of chemical reactions. This assumption results in a model in which the products of equilibration are homogeneous in composition and in distribution within the planetesimals which eventually accrete. The "condensation temperature" of the solids in any large body simply reflects conditions in the nebula at the last time that gas-grain reactions were possible. This might be, for example, the time of dispersal of the nebular gases. However, the dispersal of the nebula involves such massive flow of gas that small dust grains would surely be embedded in the gas and dispersed along with it. Bodies with sizes of at least a few meters must have been present in order that solids might remain behind to form planets after the gas dissipation. This consideration obliges us to consider how the accumulation of such bodies might alter the chemistry of condensation.

We may address this problem by phrasing a rather simple question: what would be the composition of a solid body that grows by accretion in the nebula at a very rapid rate, so that any grain, once condensed, is instantly accreted onto the surface of a large body? Rapid burial of freshly condensed grains prevents any significant reequilibration of condensates with the cooling gas.

This model leads at once to the image of a cooling solar-composition gas in which refractories nucleate, condense, and rapidly accrete to form tiny cores. Iron–nickel alloy next nucleates in the gas, condenses to form particles, and accretes to form a massive metallic layer atop the refractory core. Next follows a layer of magnesium silicates and so on. This results in the formation of onion-layered planetesimals whose structures are very grossly analogous to the present core–mantle structure of the Earth. Because the planets are accreted in a layered fashion (rather than out of a homogeneous, equilibrated mixture of solids), this model is usually called the *nonhomogeneous accretion* model.

Two features of the assumptions of this model should be noted from the outset. First, the various layers are condensed at very different temperatures and may be far removed from chemical equilibrium with each other. Second, this model tacitly assumes that one planetesimal formed at the present location of each planet and grew by dust accretion to its present size. If, as seems far more likely, thousands or millions of such onion-layered planetesimals formed, then the accretion of these bodies to form terrestrial planets would have been so violent that all vestiges of layering would have been erased by their collisions, and homogeneous planets would have resulted. A third (semantic) criticism of this model is that it should properly be named the *heterogeneous accretion* model.

We shall now briefly summarize the results of such a nonhomogeneous accretion process operating during the cooling of a solar-composition gas. Many of the detailed features of the chemistry of solar material that we discussed in connection with the equilibrium condensation process will be omitted from this discussion for the sake of brevity. One important point must be made at the outset: because this is no longer an exact equilibrium process, the results are *path-dependent*. Although isobaric, isopycnic, and adiabatic cooling processes fortunately give closely similar results, more complex trajectories in $P$–$T$ space will give results that are not solely dependent on the final values of $T$ and $P$. The composition and structure of the condensed bodies is no longer a thermodynamic state function.

Adapting the geochemical classification scheme that we used to treat equilibrium condensation, we can similarly follow the progress of the nonhomogeneous accretion condensation sequence.

1. Refractory siderophiles. Essentially identical to the equilibrium condensation behavior.

2. Refractory oxides. Somewhat simpler than the equilibrium condensation case, but the differences are minor and need not concern us here.

3. Iron–nickel. The first metal nucleated is rather nickel-rich and is buried after accretion by nickel-poor iron. The metal layer will therefore have a gradient of nickel content. Note that this metal layer is accreted on top of the much less dense and highly radioactive refractory oxide layer.

4. Magnesium silicates. The first condensate in this group will be forsterite, $Mg_2SiO_4$, followed quickly by enstatite, $MgSiO_3$. The gas phase then contains an excess of Si over Mg after forsterite begins to condense; hence, completion of enstatite condensation would leave an excess of silicon in the gas. This condenses as $SiO_2$ (quartz, tridymite, and cristobalite) with the last of the enstatite. Thus a strong gradient in MgO content will be evident, dropping from about 57% at the base of the layer to a few percent at the top.

5 and 6. Alkali metals and chalcophiles. At temperatures low enough to form alkali aluminosilicates, all Al (and calcium) compounds will already have been deeply buried. Lithium will probably follow the Mg silicates as before, but sodium will condense as the sulfide $Na_2S$ and

as the salt NaF at lower temperatures. Because Na alone is about seven times as abundant as all the halogens combined, halide condensation will be minor relative to precipitation of $Na_2S$. Potassium, rubidium, cesium, and perhaps some magnesium will condense mainly as $K_2S$, $Rb_2S$, $Cs_2S$, and MgS, respectively, along with the moderately volatile chalcophiles. The notable exception is FeS. Fe metal is deeply buried and inaccessible to attack by gaseous $H_2S$; hence, FeS never forms. The other sulfide-forming elements are in sum much less abundant than sulfur, and the great majority of the $H_2S$ therefore remains in the gas phase to well below 680 K.

7. Mineral-bound O and OH. Because of the burial of metallic iron, FeO does not form. Pyroxene, olivine, and calcium-bearing minerals are likewise not in contact with water vapor, and no phyllosilicates form.

8. Various ammonium salts, especially ammonium dihydrogen phosphate ($NH_4H_2PO_4$) and $NH_4Cl$, condense next.

9. Ice minerals. Direct condensation of $H_2O$ ice occurs as before, but in somewhat larger quantities; the water that was removed by formation of FeO and hydroxyl silicates in the equilibrium model is still in the gas phase in the nonhomogeneous accretion model. Once water ice condenses and accretes it presents a negligible surface area to the gas. Thus $NH_3 \cdot H_2O$ forms in tiny amounts, and the methane clathrate hydrate does not form. The $H_2S$ gas left over from sulfide condensation, a weak acid, reacts with ammonia, a weak base, to precipitate the salt ammonium hydrosulfide, $NH_4SH$, on top of the ice layer. The cosmic abundance of N is about seven times that of S, so the excess $NH_3$ must condense at yet lower temperatures as solid ammonia. Then all methane and argon condense as their solids, and the heavy rare gases follow.

10. Permanent gases. The behavior of hydrogen, helium, and neon remains unchanged.

A flow chart for the major elements in the nonhomogeneous accretion model is given in Fig. IV.30. Note that all condensates are either pure one-component condensates of preexisting gases (water, methane, iron, ammonia, etc.) or products of gas–gas reactions (perovskite, enstatite, ammonium hydrosulfide, etc.).

In keeping with the rapid-accretion model, reactions between the gas phase and already-formed solids are prohibited. Because a solid, once condensed, is invulnerable to further reactions, the bulk mineralogy at any point is simply the list of all minerals so far produced. This is reflected by the absence of complicated arrows below the condensation "staircase" in Fig. IV.30 and makes a table of mineral assemblages like that in Table IV.7 unnecessary.

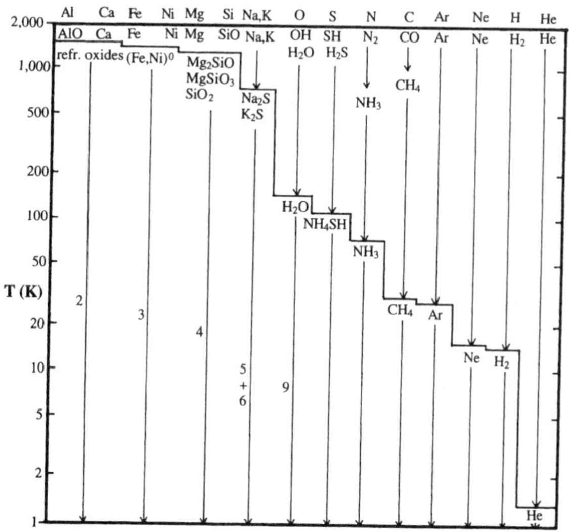


**Figure IV.30** Major element flow chart for nonhomogeneous accretion. The mineral assemblages for this model can be simply determined: all the minerals so far condensed are present as layers, accreted in the order of condensation.

The condensation thresholds are graphically represented in Fig. IV.31 for comparison with the equilibrium results in Figs. IV.27 and IV.28. Methane, argon, etc., are omitted from Fig. IV.31 because they are the same as those in Fig. IV.28.

The density of bulk condensate as a function of temperature for the nonhomogeneous accretion model is given in Fig. IV.32 for comparison with the equilibrium condensation results in Fig. IV.29. The numbers are the labels of the condensation reactions given in Fig. IV.31. The most striking feature is the very wide range of temperatures over which the bulk density remains constant, from about 1400 K down to about 170 K.

Now that we have developed two rather detailed descriptions of the chemistry of solar material, with particular emphasis on the 15 most abundant elements, we should briefly explore the most likely ways in which the condensation process may be kinetically hindered from achieving these results. It will then be appropriate to turn our attention to the construction of physical models of the solar nebula and to apply our hard-won chemical insight to such models.

## Kinetic Inhibition

If the solids in the inner Solar System in fact result from unidirectional cooling from an extremely hot gas (rather than, say, equilibration at some maximum

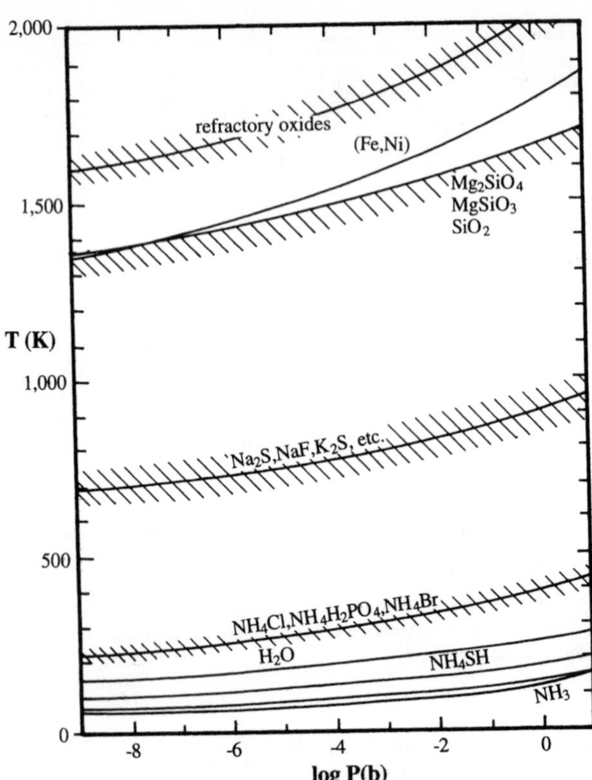


Figure IV.31 Condensation in the nonhomogeneous accretion model. This figure should be compared with the equilibrium condensation diagram in Figs. IV.27 and IV.28.

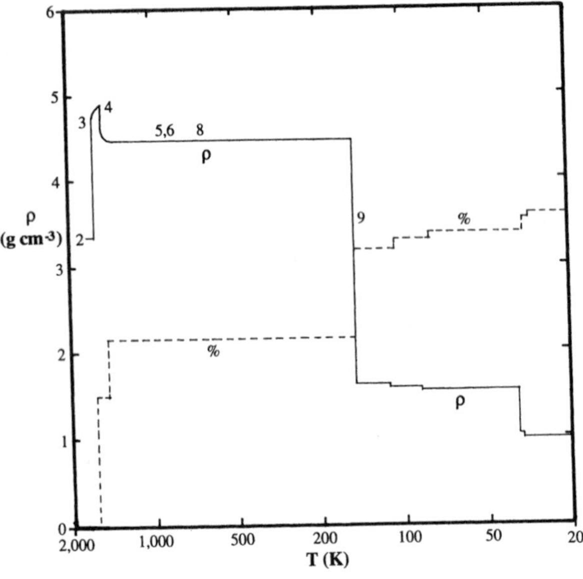


Figure IV.32 Bulk density of condensate and percentage of total mass condensed in the nonhomogeneous accretion model. This figure should be compared with the equilibrium condensation results in Fig. IV.29. Note the featureless density history from about 1400 to about 170 K.

temperature that varied smoothly with heliocentric distance), then one type of departure from equilibrium would be the coating of high-temperature grains with lower-temperature condensates, a kind of microscopically nonhomogeneous accretion process. Such coating is actually sometimes observed in high-temperature grains that make up a very small mass fraction of two rare classes of carbonaceous meteorites, but it is by no means a common feature in meteorites and seems not to have been important.

Similarly, the formation of FeS, phyllosilicates, alkali feldspars, FeO-bearing silicates, and gas hydrates during a cooling process involves alteration of previously condensed grains by gases. It should then be common to see metal particles with alteration rims of FeS or the iron oxides magnetite ($Fe_3O_4$) or hematite ($Fe_2O_3$), olivine and pyroxene particles with surface coatings of serpentine, etc. In fact, such phenomena are almost never seen in meteorites.

On the other hand, many chondritic meteorites are clearly not at internal chemical equilibrium. Mutually incompatible mineral grains and metastable glass particles (chondrules) abound, and the chrondrites appear to be poorly equilibrated low-temperature mechanical mixtures of materials that have experienced diverse thermal histories. The most striking departures from equilibrium seen in the ordinary chondrite types involve the presence of glass and the coexistence of minerals such as olivine with wide grain-to-grain variations in FeO content.

Another important kinetic problem presented by the chemistry of a presolar nebula concerns the interconversion of CO, $CH_4$, $CO_2$, $N_2$, and $NH_3$. If the hot interior of a turbulent nebula contains $N_2$ and CO and if $CH_4$ and $NH_3$ are the stable forms of carbon and nitrogen at low temperatures, then the rate of outward mixing of high-temperature gases may be greater than the rate at which sluggish low-temperature equilibration reactions can make methane and ammonia. In order to assess this possibility, we will need to know a great deal more about the thermal structure and dynamics of the primitive solar nebula.

## Mass and Density of the Solar Nebula

We have seen from our discussion of the chemistry of solar material that the fraction of the total mass of a solar-composition gas that is "rocky" is very small, only about 0.4% of the total mass. About 98.2% of the mass is accounted for by the "permanent" gases, and 1.4% by "ices." This means that the minimum mass of raw solar material required to make the terrestrial planets is about 100/0.4 times their present total mass, for a grand total of 500 Earth masses. This is more than the total mass of all the Jovian planets combined! At the very least, the solar nebula must have contained 1000 Earth masses of solar-composition material in the region from the orbit of Mercury out to about 40 AU. This comes to at least 0.3% of the mass of the Sun.

In order to make a more precise estimate of this lower limit on the mass of the nebula and to determine how this mass was distributed as a function of distance from the center of the nebula, we shall ascribe approximate compositions to each of the planets based on their observed densities, calculate the mass of primitive solar material that was required to make that planet, and spread that material out over an annulus bracketing the present orbit of the planet. We shall somewhat arbitrarily choose the boundaries of these annuli to lie halfway between the orbits of the planets, and we shall count the asteroid belt as one planet. It is unlikely that this method will introduce errors as large as a factor of 2 for any planet.

For Mercury, because of its high uncompressed density, we are led to suspect complete condensation of iron, but incomplete condensation of silicates. From Fig. IV.29 we estimate a solar nebula mass about 350 times the planetary mass. For Venus, Earth, and Mars, factors of about 230 are appropriate, and a somewhat smaller factor, perhaps about 200, should be used for the asteroids. Jupiter and Saturn, as we have seen, are only slightly enriched in heavy elements over solar proportions, whereas Uranus and Neptune are 60 to 90% ice-plus-rock material. For the Jovian planets we take enhancement factors of 5, 8, 15, and 20, respectively. Table IV.8 gives the planetary and nebular masses, annulus radii and areas, surface densities (grams of nebular mass per square centimeter of projected area in the ecliptic plane), and the minimum total nebular mass. The surface densities are also displayed in Fig. IV.33 to help determine the dependence of surface density on heliocentric distance. The total nebular mass for this model is about 1200 Earth masses ($3.5 \times 10^{-3} M_{\odot}$).

Dashed lines mark the effects of several rather gross rearrangements of mass: shrinking Mercury's annulus from a radial range of 0.33–0.83 to $0.58-0.83 \times 10^{13}$ cm; lumping all the mass of Mars, the asteroids, and Jupiter together and spreading it over the total area of their annuli; and lumping the masses and annular areas of Neptune and Pluto. It seems that there is little that can be done to change the general characteristics of the mass distribution. Generally, the surface density of the nebula, $\sigma$, drops off as $r^{-1.8}$. There is a serious local deficit of mass in the asteroid belt, and Mars is also somewhat deficient in mass, albeit by a smaller factor. The details of the diagram at and inside Mercury's orbit are presumably overwhelmingly dominated by the material that became the Sun. The most plausible means of changing the slope of the $\sigma(r)$ vs $r$ curve would be to put more heavy elements inside Uranus, Neptune, and especially Saturn. However, this is unlikely to decrease the slope by more than about 0.3 units. Similarly, slopes steeper than $-2.1$ seem unlikely.

A disk of surface density $\sigma(r)$ with a hole of radius $r_h$ in the center has a total mass of

$$M = \int \sigma(r) dA = \int_0^{2\pi} \int_{r_h}^{r_s} \sigma(r) r dr d\theta$$
$$= 2\pi \int_{r_h}^{r_s} \sigma(r) r dr. \qquad \text{(IV.140)}$$

If $\sigma(r) = 3300 r^{-1.8}$, then

$$M = 6600\pi \int_{r_h}^{r_s} r^{-0.8} dr, \qquad \text{(IV.141)}$$

which diverges as $r$ approaches infinity. The integral still diverges logarithmically for $\sigma = cr^{-2.0}$. If the surface density varied as $r^{-2.1}$, then the integral would converge:

$$M = 6600\pi \int_{r_h}^{r_s} r^{-1.1} dr$$
$$= 2.1 \times 10^5 (r_s^{-0.1} - r_h^{-0.1}). \qquad \text{(IV.142)}$$

**Table IV.8** Minimum Mass of the Primitive Solar Nebula

| Planet | Mass ($10^{26}$) | F[a] | $M_{solar}(10^{26}$g) | $r_{ann}(10^{13}$cm) | $A_{ann}(10^{26}$cm$^2$) | $\sigma = M/A$(g cm$^{-2}$) |
|---|---|---|---|---|---|---|
| Mercury | 3.3 | 350 | 1,160 | 0.33–0.83 | 1.82 | 637 |
| Venus | 48.7 | 270 | 13,150 | 0.83–1.29 | 3.06 | 4300 |
| Earth | 59.8 | 235 | 14,950 | 1.29–1.89 | 6.00 | 2500 |
| Mars | 6.4 | 235 | 1,504 | 1.89–3.20 | 20.95 | 72 |
| Asteroids | 0.1 | 200 | 20 | 3.2–6.0 | 80.9 | 0.25 |
| Jupiter | 19,040 | 5 | 95,200 | 6.0–11.0 | 267 | 355 |
| Saturn | 5,695 | 8 | 55,560 | 11.0–21.5 | 1072 | 42.4 |
| Uranus | 870 | 15 | 13,050 | 21.5–36.8 | 2802 | 4.7 |
| Neptune | 1,032 | 20 | 20,640 | 36.8–52.0 | 4240 | 4.9 |
| Pluto | 0.1 | 70 | 7 | 52–70 | 6900 | 0.001 |

[a] *F* is the factor by which the planetary mass must be multiplied to adjust the observed material to solar composition.

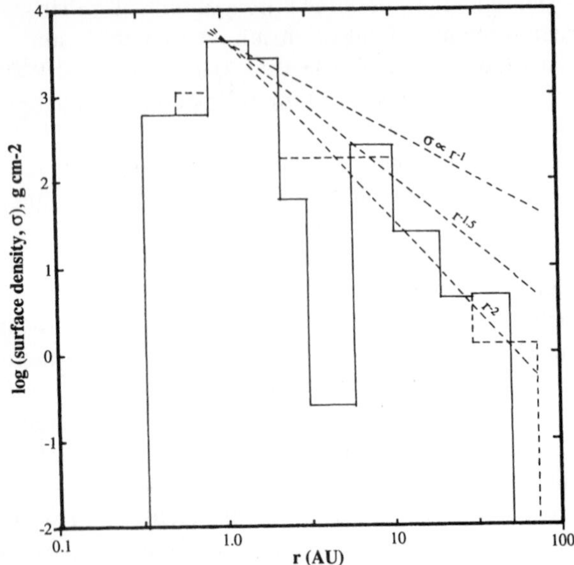


**Figure IV.33** Mass distribution in the solar nebula. A mean slope of $r^{-1.5}$ to $r^{-2.0}$ is suggested. The inner and outer edges appear sharply truncated. The inner edge is certainly due to the infall of matter from that region into the forming Sun. The outer edge may be due to a finite scale size of the original nebular condensation at the time of its last Jeans instability.

Logarithmic divergence (or worse) is not necessarily a mortal sin in this case, because the solar nebula was presumably derived from fragmentation of a massive interstellar cloud by Jeans instability [Eqs. (II.36)–(II.40)], and hence the nebula may have formed with a rather well-defined characteristic dimension, $r_J$, saving us from the awkward necessity of integrating out to infinity.

We may picture the nebula as a highly flattened disk of thickness $<0.1r$. Each volume element of the nebula is subject to three major forces, as shown in Fig. IV.34:

a. Gravitational, directed toward the center of the nebula,

b. Centrifugal, directed radially outward from the rotation axis of the nebula, parallel to the central symmetry plane of the disk, and

c. Pressure gradient, directed generally "upward" (in the $z$ direction, perpendicular to the disk plane) and "outward" (in the $r$ direction). For the purposes of an approximate steady-state model, we shall assume that the pressure gradient force exactly balances the $z$ component of the gravitational force.

Then, to order of magnitude, hydrostatic equilibrium implies

$$g_z = g \sin\phi \approx [(GM/(r^2+z^2)](z/r) \approx GMz/r^2 \quad \text{(IV.143)}$$

$$dP = -\rho g_z dz = -(P\mu/RT)(GMz/r^3)dz. \quad \text{(IV.144)}$$

where $\phi$ is the heliocentric ecliptic latitude of the volume element. Rearranging and integrating, assuming for now that $T$ is constant,

$$\int_{P_c}^{P_z} P^{-1}dP = -(\mu GM/RTr^3)\int_0^z z dz, \quad \text{(IV.145)}$$

or

$$P_z = P_c \exp - [\mu GM/2RTr^3]z^2. \quad \text{(IV.146)}$$

This isothermal pressure distribution is extremely flat for small values of $z$, but drops off very rapidly at large $z$.

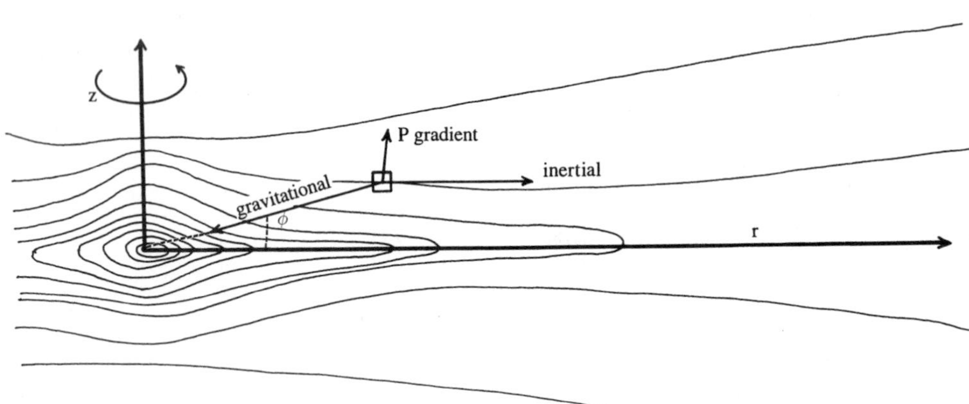


**Figure IV.34** Forces on a volume element in the solar nebula. The three main steady forces on a gas parcel are the gravitational, pressure-gradient, and centrifugal forces. There is also a turbulent acceleration that time-averages to zero. Solid bodies in the nebula are so dense that they scarcely feel the pressure-gradient force and thus tend to sediment down onto the symmetry plane of the nebula and to spiral inward under the influence of gas drag. Note that the pressure gradient force is *not* directed in the $+z$ direction, but somewhat outward from it.

The pressure $P_z$ drops to one half of its value at the central plane, $P_c$, when

$$\ln 0.5 = -0.7 = -[\mu GMz^2/2RTr^3]. \quad \text{(IV.147)}$$

Inserting $1.1\,M_\odot$ for $M$, a molecular weight ($\mu$) of 2.3 (solar material), $r = 3\,\text{AU}$, and $T = 300\,\text{K}$ (an average point in the nebula, in the asteroid belt), then $z' = 3 \times 10^{12}\,\text{cm} = 0.2\,\text{AU}$. At $r = 1\,\text{AU}$ and $T = 600\,\text{K}$, $z' = 0.05\,\text{AU}$. Thus $z/r$ is indeed small for the large bulk of the gas. The dependence of gas pressure on $z$ is shown in Fig. IV.35; note that, at a value of $z$ only three times that of the $0.5P_c$ level, $P$ is down to $10^{-3}P_c$!

Continuing in the spirit of order-of-magnitude calculations, let us approximate the nebula by a uniform slab of pressure $0.6P_c$ and half-thickness $z' = 0.5 \times 10^{13}\,\text{cm}$, as indicated by the dashed lines in Fig. IV.35. From Fig. IV.33 we can see that the surface density of the nebula was near $300\,\text{g}\,\text{cm}^{-2}$ at 3 AU; hence, we can equate the total mass of gas to this figure and derive an approximate value of $P_c$,

$$\sigma/2z' = [0.6P_c\mu/RT] = \rho\ (\text{g}\,\text{cm}^{-3}), \quad \text{(IV.148)}$$

or $P_c = 0.5\,\text{dyn}\,\text{cm}^{-2} = 0.5\mu\text{bar}$. More precisely,

$$\sigma = \int_{-\infty}^{\infty} [P(z)\mu/RT]dz \quad \text{(IV.149)}$$

$$= (\mu P_c/RT)\int_{-\infty}^{\infty} \exp$$

$$= \frac{\mu P_c}{RT}\int_{-\infty}^{\infty} \exp - \left(\frac{\mu GM}{2RTr^3}\right) z^2 dz, \quad \text{(IV.150)}$$

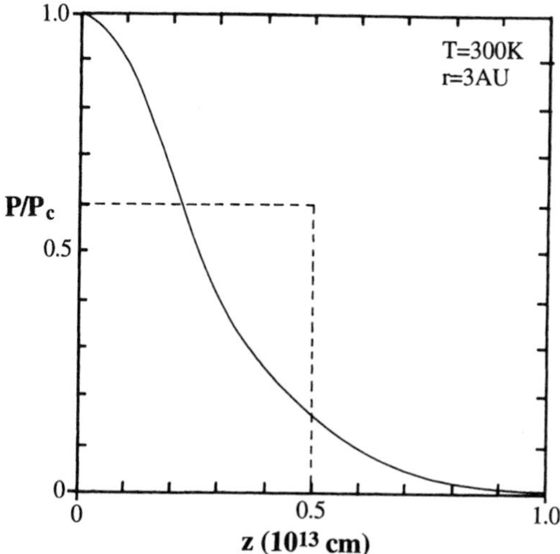


**Figure IV.35** Pressure dependence on height above the nebular plane in an isothermal nebula. The effective thickness of the nebular disk as seen from the Sun is only a few degrees, comparable to the range of orbital inclinations of planets and asteroids. The dashed line treats the nebula as a uniform slab for the purpose of making approximate calculations easier.

which is a standard definite integral,

$$\sigma = (\mu P_c/RT)(2\pi RTr^3/\mu GM)^{1/2}, \quad \text{(IV.151)}$$

or

$$\sigma/P_c = (2\pi\mu r^3/RTGM)^{1/2} \quad \text{(IV.152)}$$

$$= 2160[r(AU)^{3/2}/T^{1/2}, \quad \text{(IV.153)}$$

whence

$$P_c = \sigma T^{1/2}/[2160r(AU)^{3/2}] = 3300T^{1/2}r^{-1.5}/2160r^{1.5}$$
$$= 1.5T^{1/2}r^{-3}, \quad \text{(IV.154)}$$

using the dependence of $\sigma(r)$ on $r$ found in Fig. IV.33.

It is good to recall at this point that we are still discussing a minimum-mass nebula, just able to provide the present masses of the planets. If any solid matter has been lost from the Solar System since the formation of the solar nebula, then this model will underestimate the pressures and densities.

However desirable it may be to assign an upper limit to the mass of the nebula, there are very few scraps of relevant information on which such a limit might be based. The best available method seems to be to use observations of T Tauri stars to place limits on the total mass that a young star can eject. Of course most of the ejected mass would normally come from inside 0.1 AU, but at least we can get a serviceable, if generous, upper limit on the mass of the nebula. It has been estimated that a Sun-like star emits 1 to $3 \times 10^{-8} M_\odot$/year over a T Tauri stage lifetime of about $3 \times 10^6$ years, for a total mass loss of $0.06M_\odot$. This number is surely uncertain by a factor of 2 or more. We will consider the effect of a mass loss of up to $0.1M_\odot$, some 30 times the lower limit of $0.0035M_\odot$ that we found necessary to make the planets. We shall then settle on a mean value of the nebular mass of $0.02M_\odot$ for the purpose of constructing models. The uncertainty on this mass estimate we shall take as a factor of 6; however, that represents the full range of plausible mass estimates. We should bear in mind that most of the mass of the T Tauri phase solar wind probably originates in the corona of the central star, not in the nebula. If we now had some idea of the dependence of temperature on either the midplane pressure or the heliocentric distance, we would then be able to sketch out a crude but serviceable three-dimensional model of the pressure, temperature, and density distributions in the nebula.

## Thermal Opacity in the Solar Nebula

We have seen that collapse of an interstellar cloud to form a star releases enormous amounts of energy. This energy is liberated throughout the interior of the nebula and is radiated off into space from its surface. Instant collapse is prevented by the high internal temperatures and pressures caused by the heat of contraction, as well as the difficulty of transporting heat from the deep interior of a massive cloud out to its photospheric surface. The rapid radiative loss of this heat is prevented solely by the high opacity of the gas to solar radiation. This opacity is due in part to infrared absorption by molecules, and partly to dust. We have already seen the nature of the electronic absorption spectrum of atomic hydrogen. We have also examined the chemistry of hydrogen over a very wide range of temperatures and pressures and have crudely assessed both the temperature and the pressure in the regions of the nebula where rocky (but not icy) minerals are stable. We can conclude that conditions in the nebula near where the material of the terrestrial planets condensed were such that hydrogen would have been almost solely found as the $H_2$ molecule. Electronic transitions of $H_2$ lie in the ultraviolet part of the spectrum. The peak of the Planck function for a typical range of nebular temperatures, 1000 to 100 K, lies between about 3 and 30 $\mu$m. It is clearly necessary for us to discover what species are important sources of opacity in this wavelength region, in a nebula with solar elemental composition. As a rule, electronic transitions of molecular gases usually lie at ultraviolet wavelengths. Such transitions are therefore not promising sources of opacity in the infrared. Molecules, however, can also absorb energy to excite internal vibration. Vibrational transitions must have energies no greater than the electronic excitations lest the molecule dissociate rather than vibrate.

In order for photons to change the vibrational level of a molecule, the oscillating electric field of the photons must couple with the molecule in such a way as to stretch and compress or bend a chemical bond. Figure IV.36a illustrates a symmetrical diatomic molecule exposed to radiation of wavelength $\lambda$. The scale of the figure is grossly distorted so as to make the molecule visible; typical dimensions of diatomic molecules are about $2 \times 10^{-8}$ cm, whereas a typical near-infrared photon has a wavelength of $3\,\mu\text{m} = 3 \times 10^{-4}$ cm. We can see that the effect of the changing E field is the same on both atoms. Because the two atoms are the same (a homonuclear diatomic molecule) and the molecule is electrically neutral, the effect is zero. Thus a symmetrical diatomic molecule does not interact with the radiation field even if the field is oscillating precisely at the natural vibration frequency of the molecule.

Figure IV.36b shows an asymmetrical diatomic molecule such as CN, CO, OH, or SiO similarly subjected to an oscillating electric field. These molecules have small to moderate dipole moments, so that the ends of the molecule have small opposite charges. In this case, the instantaneous E field will exert oppositely directed forces on the two atoms whenever E is nonzero. If the molecule is driven at its resonant frequency, it will readily absorb energy from the radiation field and undergo vibrational excitation, absorbing one or more quanta of radiation in the process.

In quantum mechanics, it can be shown that the allowed characteristic energies of a diatomic oscillator are

$$\varepsilon_v = (h/2\pi)(K/m)^{1/2}(v + 1/2), \qquad \text{(IV.155)}$$

where $K$ is the spring constant of the molecule ($K = F/x$) and $m$ is the reduced mass of the molecule, $m = m_a m_b/(m_a + m_b)$. The term $v + \frac{1}{2}$ describes the quantization of the energy levels; $v$ is called the vibrational quantum number and can range over $v = 0, 1, 2 \cdot \infty$. The initial increment of $\frac{1}{2}$ is of interesting origin. Consider a molecule that had absolutely no motion; the uncertainty in its momentum would be zero. However, from the Uncertainty Principle, the product of the uncertainties in the momentum and the position of the molecule is constant,

$$\Delta p \Delta x = h/2\pi = \hbar \qquad \text{(IV.156)}$$

and therefore $\Delta x$ would be infinite, and the molecule would be lost forever! The molecule therefore cannot be stationary and still be somewhere knowable. There thus arises a vibrational ground state for $v = 0$ of energy

$$\varepsilon_o = (\hbar/2)(K/m)^{1/2} \qquad \text{(IV.157)}$$

which is the irreducible minimum energy of the molecule at absolute zero, called the zero-point energy.

A classical oscillator, as illustrated in Fig. IV.36e, experiences a restoring force of $-Kx$ when the displacement of the atoms from their mean position increases their separation by a distance $x$. This force produces a restorative acceleration

$$F = Kx = -m\ddot{x}. \qquad \text{(IV.158)}$$

This simple differential equation is satisfied by

$$x = \sin(K/m)^{1/2} t. \qquad \text{(IV.159)}$$

The frequency of this sinusoidal oscillation is

$$\nu_1 = (1/2\pi)(K/m)^{1/2}, \qquad \text{(IV.160)}$$

and the quantum-mechanical energy associated with an oscillator that has this frequency is $h\nu_1$. Thus we can see the significance of both the classical and the quantized parts of Eq. (IV.155).

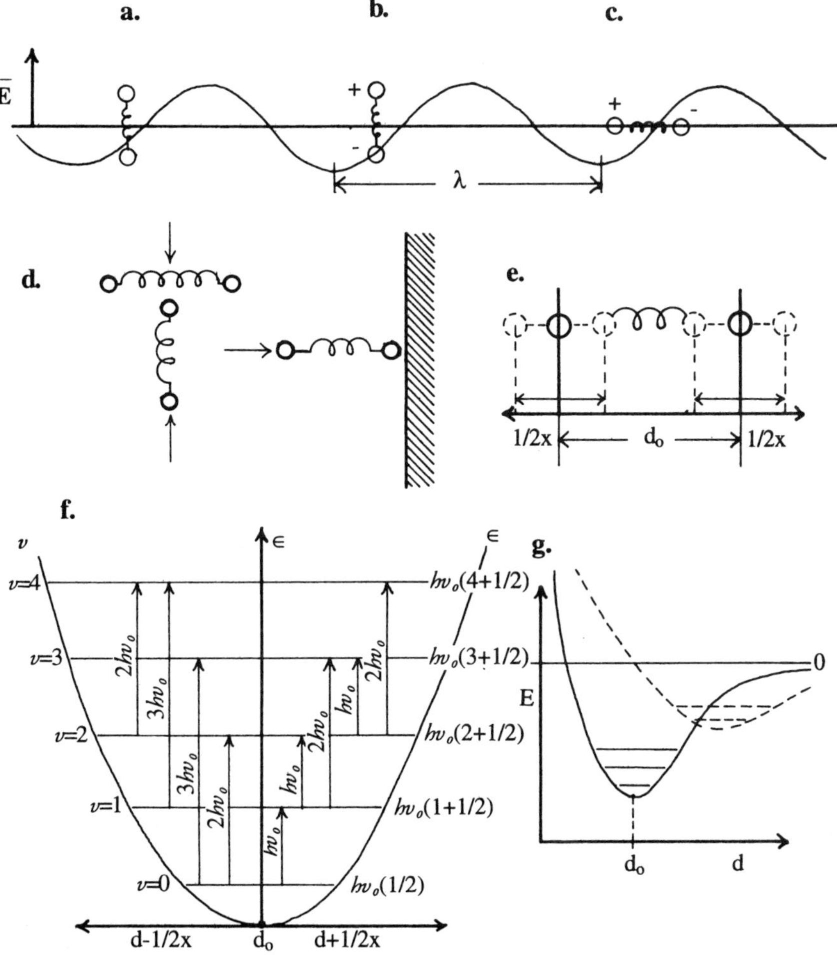


**Figure IV.36** Vibrational excitation of molecules. a illustrates a symmetrical molecule irradiated by an oscillating dipole electric field. There is essentially no interaction. In b, a molecule with a small electric dipole moment will be alternately stretched and compressed as the external field oscillates; if the field oscillates at the vibration frequency of the molecule, the molecule will absorb some of the radiation and become vibrationally excited. In c, a molecule with a dipole moment aligned with its axis along the direction (Poynting) vector of the radiation is not excited by the radiation. d illustrates schematically the collisional excitation of molecular vibration by intermolecular and wall collisions. e shows a molecule of mean bond length $d_0$ oscillating with atomic displacements of $\pm x$. f shows the parabolic potential well of a classical Hooke's Law oscillator, with the quantum-mechanical allowed vibrational energy states superimposed. The vibrational quantum number ($v$) and the energy of each state are given, as are a number of possible transitions resulting from absorption of radiation. Note that all the transition energies are integral multiples of the fundamental frequency $\nu_0$. g shows the potential surfaces for the ground electronic state (solid line) and the first excited electronic state (dashed line) of a typical real molecule. Note that only the lower portion of the potential well of each state is close to parabolic. The departure of the well from parabolic shape is said to cause "anharmonicity" of the vibrational levels.

In Fig. IV.36f we have indicated a number of possible energy-absorbing transitions for a quantum oscillator with a parabolic potential well. (Why? Show that the potential energy of a classical oscillator varies as $x^2$.) As always in equilibrium systems at finite temperatures, the most populous vibrational state is the ground state, with the populations decreasing toward higher energies as $\exp(-\varepsilon/kT)$. Note that the spacings of all energy levels are the same, $\Delta\varepsilon = h\nu_1$, where $\nu_1$ is called the fundamental frequency of the oscillator. In a real molecular oscillator the potential well departs from a parabola at moderate values of $\nu$, due to both very strong short-range Coulombic repulsion when the atoms are close and weakening of the force constant at high values of $x$, at which the overlap of the atomic orbitals of the atoms becomes small. This is indicated schematically in Fig. IV.36g for one particular electronic state of the molecule, such as the ground state. The dashed lines in the same drawing suggest what the potential energy surface and vibrational energy levels might look like in the first electronically excited state of the molecule. Note that, for certain vibrational levels in these two electronic states, there is no possibility of a transition from one electronic state to another because these configurations cannot both exist at the same interatomic spacing. The rule prohibiting such transitions is called the *Franck–Condon principle*.

Typical molecules have force constants and reduced masses that yield vibrational fundamental frequencies near $10^{14}$ Hz or a wavelength of several micrometers. Very massive atoms will form molecules with very closely spaced vibrational levels; in the limit, two macroscopic masses connected by a spring must behave classically.

It is of great importance to our efforts to realize that many polar molecules are strong absorbers in the several-micrometer wavelength region, the so-called thermal infrared. Now it remains to discover which molecules are important sources of thermal opacity in solar material.

In Table IV.7 we found that several diatomic and polyatomic molecules may have appreciable abundances at temperatures in the present range of interest, about 1500 K down to 50 K. These include $H_2O$, CO, $H_2S$, $CO_2$, $PH_3$, $NH_3$, $CH_4$, and PN. Of these, the strongest absorber, the one with the largest dipole moment, and the most abundant molecule are one and the same, $H_2O$. Water vapor has several strong fundamental and combination bands in the near infrared. Because the water molecule is nonlinear and triatomic, it has more than one possible vibrational mode. In fact, it has three, each with a characteristic frequency.

Any diatomic molecule can both vibrate and rotate. Table IV.9 enumerates the different independent (orthogonal) modes available to a wide variety of molecules.

The energies allowed to a quantum mechanical rigid rotator are given by

$$\varepsilon_r = J(J+1)\hbar^2/(2I) = e_r^0 J(J+1) \qquad \text{(IV.161)}$$

where $J$ is the rotational quantum number and $I$ is the moment of inertia of the molecule about the rotation axis.

For linear molecules, we can see from simple geometrical considerations that there are two rotational modes, about two axes orthogonal to each other and to the axis of the molecule, which will have equal values of $I$ and hence precisely equal energies. In spectroscopy, two levels that have exactly the same energy are said to be *degenerate*. Note that rotation about the molecular symmetry axis corresponds to a case of near-zero moment of inertia and hence to inaccessibly high excitation energy. This mode can therefore not be excited. Conversely, for very massive molecules the moment of inertia becomes very large, and the spacing between the rotational levels vanishes. Thus quantum mechanics imposes no practical constraints on the rotation of baseballs; it is not necessary for the catcher to specify the rotational quantum number of a pitch, merely that he wants spin. (Besides, imagine how long it would take him to signal the desired value of $J$! Consider a 260-g ball 4 cm in radius rotating at 100 Hz. What is $J$?) Note that, unlike vibrational levels, the rotational levels are not uniformly spaced. Note also that Eq. (IV.161) was derived for a rigid rotator. But rapidly rotating molecules are subject to centrifugal distortion, which increases $I$ about the rotation axis and therefore decreases the energy spacing between high-lying rotational levels. Symmetrical diatomic molecules do not couple well with electromagnetic radiation, and hence molecules like $H_2$ have forbidden pure rotational and pure vibrational transitions.

Typically the rotational levels are very closely spaced compared to vibrational levels, corresponding to resonant frequencies on the order of $10^{12}$ Hz or wavelengths of several hundred micrometers. Pure rotation transitions would thus supply useful thermal opacity only at temperatures near $T = 2900/\lambda = 10$ K. At this temperature the only uncondensed gases are He, Ne, and (depending on the pressure) $H_2$. Helium and neon of course have no rotational or vibrational absorption, and $H_2$, although it rotates (rotational fundamental at $1.7 \times 10^{13}$ Hz, or about 17 $\mu$m), cannot couple directly with the radiation field. Also, because of the very small

**Table IV.9** Internal Degrees of Freedom of Molecules

| No. of atoms | Structure | Example | No. of degrees of freedom: Total | Trans | Vib | Rot |
|---|---|---|---|---|---|---|
| 1 | Monatomic | He | 3 | 3 | — | — |
| 2 | Diatomic | CO | 6 | 3 | 1 | 2 |
| 3 | Linear | $CO_2$ | 9 | 3 | 4 | 2 |
| | Bent | $H_2O$ | 9 | 3 | 3 | 3 |
| 4 | Linear | HCNO | 12 | 3 | 7 | 2 |
| | Planar | $CH_2O$ | 12 | 3 | 6 | 3 |
| | Pyramidal | $NH_3$ | 12 | 3 | 6 | 3 |
| 5 | Linear | $HC_2CN$ | 15 | 3 | 10 | 2 |
| | Planar | $H_2CCO$ | 15 | 3 | 9 | 3 |
| | Tetrahedral | $CH_4$ | 15 | 3 | 9 | 3 |
| 62 | Chain | $C_{20}H_{42}$ | 186 | 3 | 180 | 3 |
| | Linear | $C_{62}$ | 186 | 3 | 181 | 2 |

value of $I$ for this minimal molecule, it would provide no opacity in the 200- to 1000-$\mu$m region even if the rotational transition were not forbidden by symmetry.

The energy levels of a typical diatomic molecule, taking into account electronic, vibrational, and rotational excitation, are portrayed in Fig. IV.37. Three different electronic states are represented: $E_1$ is the ground electronic state, $E_2$ is the first electronically excited state, and $E_3$ is an unbound (dissociative and repulsive) state with energy greater than zero. In $E_3$ there cannot be either vibration or rotation, and therefore no quantized sublevels are seen. A number of vibrational levels of $E_1$ and $E_2$ are schematically indicated.

Transitions may occur between different vibrational and rotational levels according to certain selection rules,

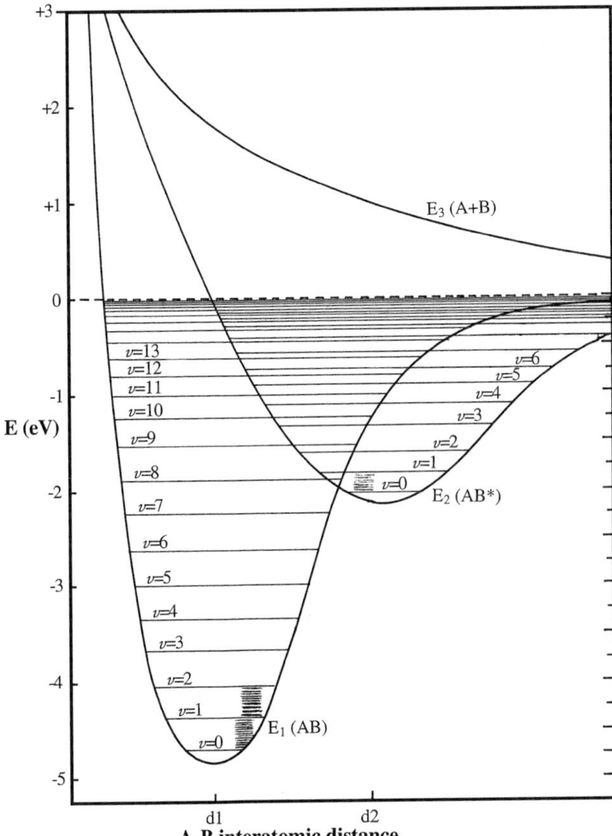


**Figure IV.37** Electronic, vibrational, and rotational energy levels. The potential energy curves are given for three electronic states: $E_1$ is the electronic ground state, $E_2$ is an electronically excited state with a weaker bond and larger mean interatomic distance, and $E_3$ is an unbound (predissociative) state in which the Coulombic repulsion force exceeds the electronic binding force at all separations. Absorption of radiation is essentially instantaneous: the nuclei cannot move on the time scale of the transition, and only electrons may be perturbed. Thus radiation could not excite this molecule from the $v = 0$ level of the electronic ground state to the $v = 1$ level of the first excited state because these two states cannot both occur at the same internuclear distance. Note that some levels coincide almost perfectly in energy.

which specify what changes in the vibrational and rotational quantum numbers are allowed. Simultaneous changes in electronic, vibrational, and rotational energy may be brought about by absorption of a single photon.

Rotational fine structure is sketched in for only a few low-lying vibrational levels of $E_1$, although it is present for all bound states. The spacing of the rotational levels has been substantially exaggerated so that they can be seen; recall that their actual spacing is of order $10^{12}$ Hz, or 0.005 eV. The characteristic energies required for ionization, dissociation, electronic excitation, vibrational excitation, and rotational excitation, along with the characteristic temperatures $[T(K) = 2897/\lambda(\mu\mathrm{m}) = 2897\nu/c]$ of these modes, are shown in Fig. IV.38. Also given for comparison is the energy distribution of the solar spectrum.

Several points become clear from this diagram. First, ionization and dissociation can absorb only a very small fraction of the solar flux. Second, the peak of the Planck function for the Sun lies in a region in which gases are quite poor absorbers. Third, direct heating of gases by the Sun is quite effective in the infrared. Fourth, at gas temperatures from a few kelvins up to about 200 K the kinetic energy carried by gas molecules (and thus the distortion energy stored as potential energy of bending or twisting during collisions) is adequate to excite rotation of all molecules except $H_2$. Fifth, at temperatures of several hundred kelvins, the $C_V$ of hydrogen is $5R/2$ and the $C_P$ is $7R/2$. The translational energy of the average gas molecule can directly excite molecular vibration.

The total internal energy of a gas depends on how many of its modes are excited. At low temperatures, the thermal energy of collision of molecules may be insufficient for rotational, vibrational, or electronic excitation of the molecule. Then, according to kinetic theory, the total energy of each molecule is just $mv^2/2$, and this mean kinetic energy is just $3kT/2$. We should add, from our knowledge of the quantum mechanical treatment of molecular vibration, that there is also a constant, small zero-point vibration energy present. The rate of increase of internal energy when the temperature is raised, the heat capacity, is

$$C_V = d(3kT/2)/dT = 3k/2 \quad \text{(IV.162)}$$

per molecule, or $3R/2$ per mole. In Appendix I, we prove that the heat capacity at constant pressure, $C_P$, is $C_V + R = 5R/2$.

The rotational fundamental of hydrogen gas is at 17 $\mu$m, which, from Fig. IV.38, requires a collision with a molecule at about 200 K to excite. Above about 200 K, therefore, heating the gas through $dT$ requires supplying energy for both translation and rotation, and the heat capacity must be larger. We have seen that the molar

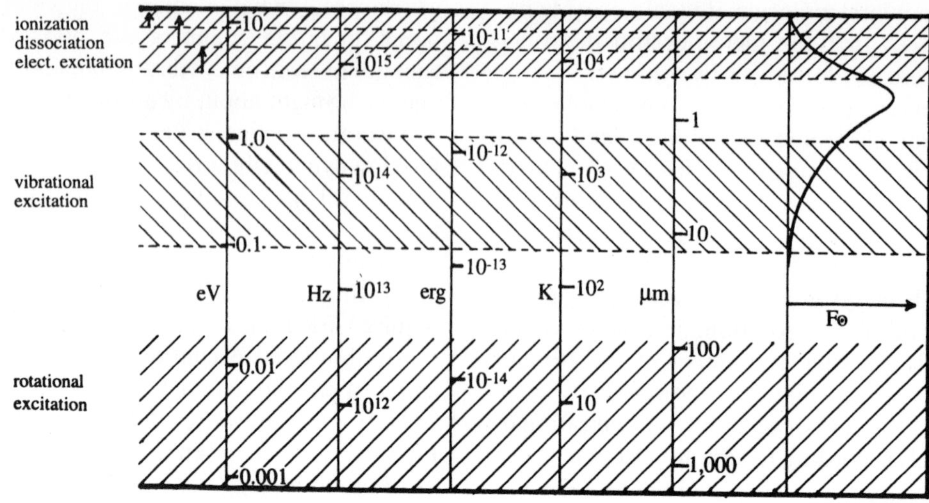


**Figure IV.38** Molecular energy levels, solar radiation, and thermal energy. The energy scales of particles, radiation, thermal motion, and the solar emission spectrum are compared with the absorption and excitation energies of a typical molecule.

heat capacity for translation is $R/2$ for each ($x$, $y$, $z$) degree of freedom of the molecule, and we can recall that $H_2$ has two orthogonal but energetically degenerate rotation modes. Each of these modes also contributes $R/2$ to the heat capacity, so that, at several hundred kelvins, the $C_P$ of hydrogen is $5R/2$ and the $C_V$ is $7R/2$. The vibrational fundamental of $H_2$ at 1.7 $\mu$m becomes excited only near 2000 K. Each vibrational mode has a kinetic part and a potential part, and $C_P$ and $C_V$ increase by $R/2$ per mode. Because hydrogen begins to dissociate near this temperature, $C_P$ never attains its full possible value of $9R/2$. Figure IV.39 shows the behavior of the heat capacities of $H_2$, He, and a solar-composition mixture. Hydrogen behaves in a very different way from most other molecules: because of the low masses of the H atoms, it is the most quantum mechanical (least classical) of molecules and has by far the largest vibrational excitation temperature of any molecule.

Let us now imagine a cloud of $H_2$ gas containing a trace of water vapor, exposed to solar illumination. The near-IR tail of the solar emission can be readily absorbed by the vibrational bands of water vapor. The vibrational energy can be converted into rotational and translational energy by collisions. A rapidly vibrating water molecule can collisionally excite rotation of $H_2$ molecules, even though the direct excitation of $H_2$ rotation by the Sun is forbidden. If the trace of water vapor is condensed, the gases in the cloud essentially cease to interact with the Sun.

Now let us consider the significance of the fact that typical collision energies may be far larger than the spacings of the rotational levels of molecules. We shall consider the fundamental vibrational transition for a general gas in which the individual molecules are distributed over a substantial number of different rotational states both before and after the vibrational transition. To first approximation, the transition energies of all of these molecules are just $h\nu_1$, where $\nu_1$ is the fundamental frequency. To be more precise, we must allow for the energy differences caused by changes in the rotational quantum number $J$ simultaneous with the change in $v$. Individual molecules that are excited from, say, the $v = 1$ and $J = 8$ level may end up in any of a large number of $v = 2$ states with different (lower or higher) values of $J$. Thus the absorption spectrum for this band (the $v$ band) will contain a large number of lines whose spacings are determined by the energy differences between the rotational levels. From Eq. (IV.161) we can see that the individual levels lie at energies $\varepsilon_r^0$ (0, 2, 6, 12, 20, 30, 42, 56...) above the rotational ground state. It is clear that all possible transitions (except of course for anharmonicities caused by the centrifugal distortion of high-lying rotational states) have energies that are multiples of $\pm 2\varepsilon_r^0$, and all multiples are possible. Thus we expect a manifold of rather evenly spaced lines, the spacing of which provides us the means to deduce the rotational moment of inertia of the molecule. These lines lie both above and below $h\nu_1$ in energy. Such a rotation–vibration absorption band is pictured in Fig. IV.40. Each of the absorption lines is of finite width, because the uncertainty principle allows $\Delta E = 0$ only for states with infinite lifetime and because the motions of the molecules impart a Doppler broadening to the spectral lines of large ensembles of molecules. Still, the line width is small compared with the spacings between lines.

The envelope of the rotational lines contains valuable information on the temperature of the absorbing gas, because the relative intensities of the lines depend on the relative populations of the rotational levels. In a hot gas

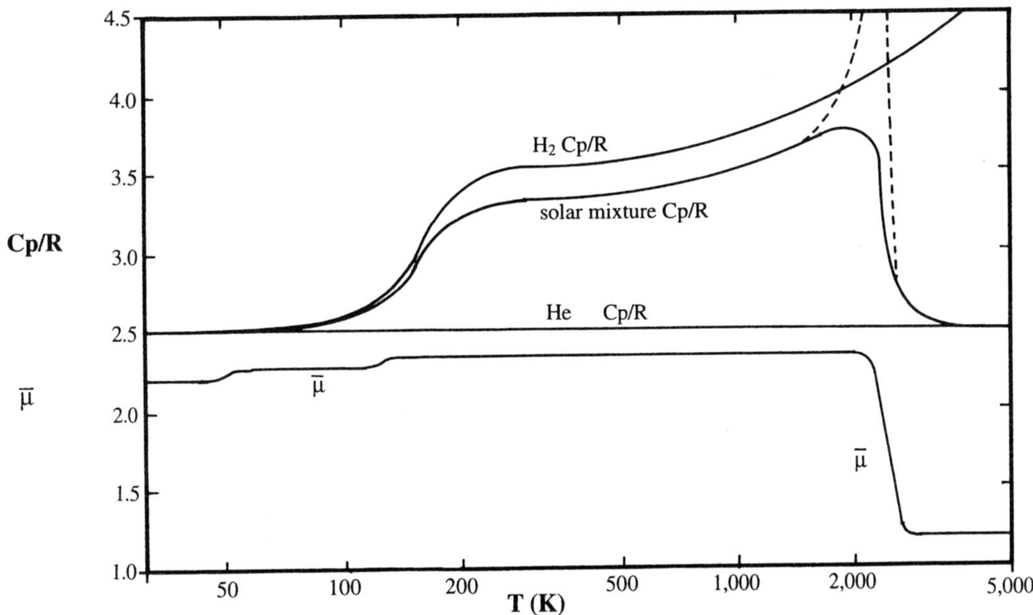


**Figure IV.39** Heat capacity of hydrogen, helium, and solar material. The low-temperature heat capacity is that of a monatomic gas, because temperatures are too low for intermolecular collisions to excite even the rotation of molecular hydrogen. Above 1000 K the vibrational excitation of hydrogen begins to appear, but the energy of the vibrational fundamental is so close to the dissociation energy of the molecule that the molecule falls apart before the full vibrational contribution to the heat capacity can be realized. The dashed line indicates the contribution to the apparent heat capacity due to the absorption of the dissociation energy of hydrogen; the solid line indicates the heat capacity, ignoring dissociation. Clearly the latter is an excellent approximation under almost all conditions, but in the vicinity of the dissociation temperature it is a disastrously bad approximation. A handy measure of the degree of dissociation is the mean molecular weight.

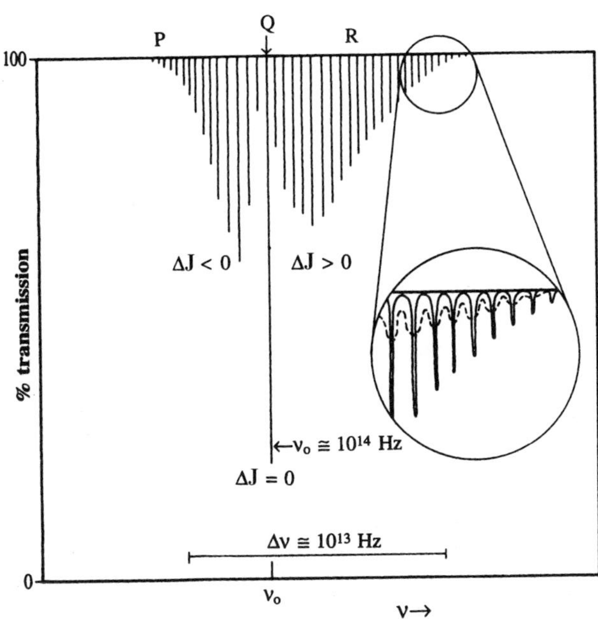


**Figure IV.40** A rotation–vibration absorption band. The rotational structure of the band is complex and far from symmetrical. The inset illustrates the effect of an increase in pressure on the spectrum (dashed curve).

exp $(-\varepsilon/kT)$ will be nearly unity, and the molecules in the gas will be distributed widely and almost uniformly over dozens of rotational levels. This will produce a vibration–rotation band with a very flat envelope. With regard to Fig. IV.40, it should be mentioned that the $\Delta J = 0$ spike (the $Q$ branch) will be absent for certain gases.

The spacings of the individual rotation lines are about $10^{12}$ Hz, the envelope of the band spans about $10^{13}$ Hz, and the frequency of the $Q$ branch is about $10^{14}$ Hz. The thermal energy of intermolecular collisions at 300 K is equivalent to about $3 \times 10^{13}$ Hz. Thus any molecule that has recently been involved in a collision will have suffered distortion of its energy levels by an amount that can easily smear out the rotational structure in the spectrum. This phenomenon of broadening (and, in a dense gas, eventual obliteration) of rotational lines is called *collision broadening* or, less precisely, pressure broadening. At high pressures, ever larger proportions of the molecules are suffering collisional perturbations at any moment. The wings of the individual lines expand and overlap, closing down the transmission windows between the lines. The insert in Fig. IV.40 shows the effect of increased pressure on the line profiles. Thus a high-resolution spectrum in which individual line profiles can be measured provides information on the pressure of the gas. Another point of direct relevance to the present discussion is that a low-pressure gas can "leak" radiation even in the heart of a strong absorption band, if the wings of the lines are weak. This principle may also be expressed by saying that the *continuum opacity* of the gas may be very low at low pressures, even when the *line opacity* is fairly high.

Line widths that are not Doppler limited are determined by the lifetimes of the undisturbed molecules, essentially the times between collisions. A state that persists undisturbed for $\Delta t$ seconds will have an energy uncertainty of

$$\Delta E \Delta t = \hbar/2 \tag{IV.163}$$

The lifetime of the undisturbed state is the mean free path of the molecule ($l$) divided by the mean thermal velocity ($V$). For hydrogen at normal temperatures and pressures, $l \approx 10^{-5}$ cm and $V \approx 10^5$ cm s$^{-1}$, so $\Delta t = 10^{-10}$ s and $\nu = 10^{10}$ Hz. (For Doppler broadening under the same conditions, $\Delta\nu = \nu V/c = 10^{14}(10^5/3 \times 10^{10}) =$ only $3 \times 10^8$ Hz.) Because

$$l = kT/\pi r^2 P \tag{IV.164}$$

and

$$V = (kT/m)^{1/2}, \tag{IV.165}$$

we get

$$\begin{aligned} \Delta E &= h/2\Delta t = hv/l = hV/l \\ &= \pi h a^2 P/(mkT)^{1/2}, \end{aligned} \tag{IV.166}$$

where $a$ is the radius of the molecule.

One consequence of molecular collisions is that the collision shifts and distorts the electron shells relative to the nuclei. Thus an $H_2$ molecule, while distorted by collision with another molecule, will have a temporary collision-induced dipole moment. It will then be able to absorb resonant photons that would, for the isolated molecule, be symmetry forbidden. The duration of a collision is approximately the ratio of the radius of the H molecule to its thermal speed, about $10^{-8}/10^5 = 10^{-13}$ s, or about $10^{-3}$ of the mean undisturbed lifetime between collisions. Thus all energy levels are smeared out to a width of about $10^{13}$ Hz during a collision. This width is equivalent to a wavelength of about 30 $\mu$m, comparable to the wavelength of the $H_2$ rotational fundamental itself (17 $\mu$m). Therefore dense hydrogen gas will absorb over a very broad spectral range of $\Delta\nu/\nu \approx 1$ centered on the 17-$\mu$m wavelength. This absorption increases in strength in proportion to the fraction of molecules involved in collisions and thus is not proportional to $P$, as in ordinary transitions, but to $P^2$. This broad feature will contribute important thermal opacity at pressures $> 1$ bar if the temperature is such that the Planck peak is anywhere near 17 $\mu$m,

$$T = 2900/\lambda_{max} \approx 170 \text{ K} \tag{IV.167}$$

within about a factor of 2.

For the case of the solar nebula, pressures are too low for this to be an important source of opacity in the region of the nebula where the Planck peak is near 17 $\mu$m, and regions of the nebula dense enough to provide this opacity are so hot that thermal decomposition has destroyed the $H_2$ molecule. This has not been a futile exercise, however, because this phenomenon is of central importance in the atmospheres of the Jovian planets.

Now that we have seen the ways in which pressure may affect the process of light absorption, let us return to Fig. IV.40 to see how the path length traversed through a gas will affect the spectrum. We have already seen that the individual absorption lines can be narrow and extremely strong in their centers while the average band transmittance is still high. This is because the wings of the line drop off in strength very rapidly away from the line centers. The absorption coefficient is given by

$$\alpha(\lambda) = f(e^2/4\varepsilon_0 mc^2)\lambda_0^2\Delta\lambda/[4\pi^2(\lambda-\lambda_0)^2 + (\Delta\lambda/2)^2], \tag{IV.168}$$

where $\lambda_0$ is the line center and $\Delta\lambda$ is the natural line width:

$$\Delta\lambda = (\lambda_0^2/hc)\Delta E = (\lambda_0^2/c)(1/\Delta t_1 + 1/\Delta t_2). \tag{IV.169}$$

Here $\Delta t_1$ and $\Delta t_2$ are the lifetimes of the two states between which the transition occurs. Weak lines with unsaturated cores often have a Lorentz line profile, so that their integrated strength is equal to the product of their depth times their width. Rather than deal with the width and the depth of a line, it is often convenient to use an overall measure of the strength of an absorption feature that is independent of the spectral resolution with which we observe the line. We find the equivalent width of a line, $w$, by integrating the transmitted intensity over frequency. For a homogeneous slab of gas,

$$w = \int_0^\infty (1 - \exp \alpha x)d\nu \tag{IV.170a}$$

or, for the case of an optically thin gas ($\alpha x << 1$),

$$w = x\int_0^\infty \alpha d\nu = xS, \tag{IV.170b}$$

where $S$ is a measure of the intrinsic strength of the line. The equivalent width of a line is defined as the width of a rectangular absorption line profile with zero transmittance and the same area as the observed line, as illustrated in Fig. IV.41.

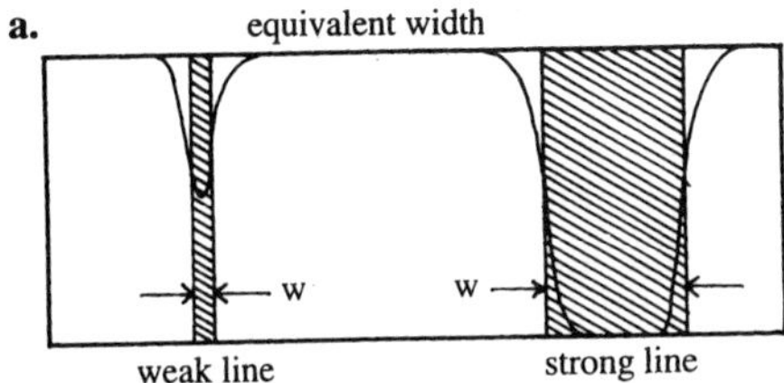


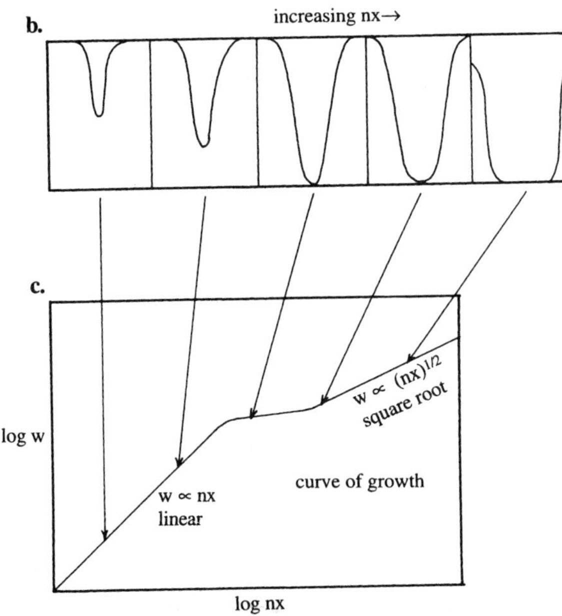


**Figure IV.41** Equivalent width and the curve of growth. a shows the equivalent width of a weak line and a strong (saturated) line. b shows schematically the changes in line shape that occur as the line strength is increased by growth in either the gas number density or the path length through the gas. The resulting equivalent widths for this sequence of gas abundances are shown in c. Note the linear, nearly flat, and square root portions of this curve of growth.

For weak lines, doubling the amount of gas or the path length through the gas will double the equivalent width. At each point on the absorption line, the absorbed intensity is

$$dI = -I\sigma n dx, \qquad \text{(IV.171)}$$

where $\sigma$ is the molecular cross section ($cm^2$/molecule), $n$ is the number density of the absorbing gas ($cm^{-3}$), and $dx$ is the increment of path length through the gas. Thus

$$\int_{I_0}^{I_x} I^{-1} dI = \ln(I_x/I_0)$$
$$= -\sigma n \int_0^x dx = -\sigma n x. \qquad \text{(IV.172)}$$

For small amounts of absorption, $I_x = I_0(1 - nx)$. Also, note that the product $\sigma n$ is the same as $\varepsilon N$, where $\varepsilon$ is the molar extinction coefficient and $N$ is the number of moles of absorber per cubic centimeter.

Because we are holding the pressure constant and using a Lorentz line profile, $w$ is proportional to $I_0 - I_x$, or $w \propto nx$. Thus $w$ increases linearly with the column abundance of the absorber along the line of sight. The relation expressing how $w$ grows with the column gas abundance is called the *curve of growth*, and we can say that, for weak Lorentz lines, the curve of growth is linear.

From Eq. (IV.172) we can see that, for large values of $\sigma n x$, the intensity at the absorption line center will drop only as

$$I_x = I_0 \exp -(\sigma n x) \qquad \text{(IV.173)}$$
$$I_0 - I_x = I_0[1 - \exp -(\sigma n x)], \qquad \text{(IV.174)}$$

whereas the wings of the line (small $\sigma$) are growing as $x$. When $I(x)$ becomes very small at the line center, doubling $\sigma n x$ will have negligible effect on the transmitted intensity. We then say that the center of the line is *saturated*. When the core of a line is saturated, adding more absorber to the line of sight contributes to the opacity almost solely by increasing the absorbance in the wings of the line. The equivalent width, which increases as the product of width times depth, will now grow due only to increasing line width. For a Lorentz line profile, this means that $w$ will now grow as $(\sigma n x)^{1/2}$ in this domain. This is called the *square root* portion of the curve of growth.

The changes of line shape and equivalent width with $\sigma n x$ are summarized in Fig. IV.41. Note that, at very high gas abundances, the wings of the individual lines merge and begin to lower the continuum. This effect is wholly independent of the collision broadening phenomenon discussed earlier and is simply due to the presence in the line of sight of such immense column abundances of gas $nx$ that even very small molecular absorption cross-sections will provide significant $\sigma n x$ products.

If we were to compress a fixed column abundance of gas so that exactly the same mass of absorber is observed at higher pressures, $Nx$ will remain constant and the absorption in the spaces between the lines would remain constant until pressure broadening becomes detectable. Thereafter, the absorption between the lines would increase as $n^2$.

## Dust Opacity

Although condensates make up a very small fraction of the mass of the nebula, they can be important sources of opacity if they are good absorbers or scatterers. As with gases, we can analyze the absorption of electromagnetic energy in terms of electronic, vibrational, and rotational transitions; however, in solids it is very unusual for free rotation of a molecule or portion of a molecule to take place. Vibrational excitation of solids occurs at particular wavelengths (lattice bands) that may be located, especially in the case of weakly bonded solids, very close to the wavelengths at which the vapor absorbs. Silicates generally exhibit a band near 10 $\mu$m, which is an important source of opacity near 300 K.

A very large proportion of the continuum opacity in the nebula is due to the presence of metal grains. Electrical conductors contain, by definition, mobile *conduction band* electrons that are free to move about. In a process analogous to free–free absorption of energy by electrons in a plasma, these electrons may couple with an oscillating electromagnetic field and be accelerated. They in turn collide with and are scattered by lattice atoms and dissipate their energy as heat. Metallic iron–nickel alloy is thus a moderately strong and very broad-band absorber.

Solids that do not contain conduction band electrons, but that have an unoccupied delocalized (conduction) level not too high in energy above the highest-lying occupied levels, may become conductors if sufficiently energetic radiation is present. The ultraviolet (or visible) radiation promotes electron transfer across the band gap into the conduction band, and the material will become a broad-band absorber. Such substances are called *semiconductors*. Typical naturally occurring semiconducting minerals include a number of metal oxides, such as FeO. It is interesting that such semiconducting surfaces are also extremely effective catalysts for a wide variety of chemical reactions, including the production of complex hydrocarbons from CO and hydrogen. This reaction, called the *Fischer–Tropsch process*, was the basis of the production of synthetic fuels by Germany during World War II.

In the nebula, conducting particles, especially iron–nickel alloy, can interact most effectively with the radiation field if their circumferences are at least as large as the wavelength of the incident light:

$$2\pi r > \lambda. \qquad \text{(IV.175)}$$

Because metallic particles are stable in the nebula from about 1500 K down to below 500 K, they will be imbedded in a medium for which the Planck peak will lie between about 2 and 6 $\mu$m. Thus metal particles with radii of about 1 $\mu$m or larger are most effective in providing opacity. Clearly, however, there is nothing to be gained by making the particles larger than this; the total cross-section area of a given mass of spherical metal particles is inversely proportional to the radius of the particles:

$$A(\text{cm}^2)/M(\text{g}) \leq (3\pi a^2)/(4\pi a^3 \rho_{\text{Fe}}) = 3/(4a\rho_{\text{Fe}}) = K\ (\text{cm}^2\text{g}^{-1}). \qquad \text{(IV.176)}$$

The absorption coefficient of the nebula is just

$$\alpha = K\rho_{\text{n}}(\text{cm}^{-1}), \qquad \text{(IV.177)}$$

where $K$ is the mass opacity and $\rho_n$ is the density of the nebula.

The mass fraction of metal particles in the nebula, which depends on the molecular weights ($\mu_i$) and atomic abundances ($A_i$) of iron, nickel, and the major elements, can be approximated as $\mu_{\text{Fe}}A_{\text{Fe}}/(\mu_{\text{H}}A_{\text{H}} + \mu_{\text{He}}A_{\text{He}})$, or $1.14 \times 10^{-3}$, and $K$ in units of square centimeters per gram of nebula is $0.034/(4\rho_{\text{Fe}}a) = 10^{-4}/a$. Thus $\alpha$ is $10^{-4}(\rho_n/a)$. Assuming, as usual, ideal gas behavior,

$$\alpha = 10^{-4}P\mu/(aRT^{1/2}) \qquad \text{(IV.178)}$$

We have seen that [Eq. (IV.154)], at 1 AU from the center of the nebula,

$$P_{\text{c}} = 1.5T^{1/2}r^{-3} = 1.5T^{1/2}; \qquad \text{(IV.179)}$$

hence

$$\alpha = 1.5 \times 10^{-4}\,\mu/(aRT^{1/2}) \qquad \text{(IV.180)}$$

Taking $T = 600$ K, $R = 8.2 \times 10^7$, and $\mu = 2.3$, we find that $\alpha$ is $1.7 \times 10^{-13}/a$. The mean free path of a thermal photon is then

$$l = 1/\alpha = 5.8 \times 10^{12}a. \qquad \text{(IV.181)}$$

For $a = 1\ \mu\text{m} = 10^{-4}$ cm, $l = 5.8 \times 10^8$ cm, or 5800 km. As we saw, the half-thickness of the nebula is about 0.05 AU [Eq. (IV.147) and others], or $7.5 \times 10^{11}$ cm, more than 1000 photon mean free paths. The nebula is thus optically thick in the thermal IR near Earth's orbit.

The thermal opacity of iron would essentially vanish if the mean free path of thermal photons approached 0.1 times the half-thickness of the nebula. For that to happen, either 99% must be accreted into large planetesimals or all the iron particles must be near 150 $\mu$m in radius. A "smoke" of iron particles with sizes much less than 1 $\mu$m could also decrease the opacity.

Infrared radiation with mean free path $l$ will penetrate through the nebula in the $z$ direction with an intensity

$$I(z) = I_0 \exp(-z/l) = I_0 \exp(-\tau). \qquad \text{(IV.182)}$$

The quantity $\tau$ is called the *optical depth*. This and many other definitions related to the attenuation of light are collected in Table IV.10 as a guide to the necessary terminology.

We have seen that most of the mass of the nebula lies within $z = 0.05r$ of the nebular symmetry plane and, thus, at the level at which $P = 0.6P_{\text{c}}$, $z/l = \tau = 1500$. The fraction of transmitted radiation is then

$$I(z)/I_0 = \exp\,(-1500) = 10^{-650}. \qquad \text{(IV.183)}$$

A 100-times smaller optical depth would still be a very effective opacity blanket.

Although we do not at present know how to assign a size to particles in the Solar Nebula, we have seen that it

**Table IV.10** Light Absorption and Transmission Terminology

$\sigma$ = particle cross-section area (square centimeter per particle)
$K$ = mass opacity (square centimeter of area per gram of absorber)
$\alpha$ = absorption coefficient = $K_\rho$ (square centimeter of area per cubic centimeter of absorber) ($cm^{-1}$)
$\varepsilon$ = extinction coefficient = $K/\mu$ (square centimeter of area per mole of absorber) (used especially by chemists)
$A$ = absorbance = optical depth[$I/I_O = \exp(-\tau)$] (used by chemists)
$T$ = transmittance = $1 - \exp(-A) = 1 - \exp(-\tau)$

is altogether plausible that the portion of the nebula that contains metallic iron could be highly opaque to thermal infrared. For nebular temperatures between about 150 and 400 K, the dominant sources of opacity would be water vapor and solid oxides and sulfides of Fe.

Ferrous oxide (FeO) in solid solution in silicates exhibits moderately strong but narrow ($\Delta\lambda/\lambda \approx 0.1$) absorption bands at 0.95 $\mu$m and near 1.90 $\mu$m, with the exact wavelength dependent on the structure of the mineral in which FeO resides. Because of their short wavelengths, these bands contribute little thermal opacity at relevant nebular temperatures: FeO becomes abundant only below about 700 K, at which the Planck peak is already beyond 4 $\mu$m. At temperatures of a few hundred kelvins, the most effective contributors to thermal opacity are bands near 10-$\mu$m wavelength. Silicates have a strong feature in this region.

Minerals that contain two different oxidation states of the same element usually exhibit very strong *charge transfer* bands due to exchange of electrons between the high-lying energy levels of the two different charge states. In the solar nebula the only important elements that can assume more than one oxidation state in minerals are Fe, Ti, and Ni. Iron is of course the most important of these. Titanium can be found in the 3+ and 4+ oxidation states, and, in silicate glasses, intense coloration can be produced by a few percent of titanium. Iron can be oxidized at low temperatures to form a mineral of composition $Fe_2O_3 \cdot FeO$, usually written $Fe_3O_4$, called magnetite. Exposure of metallic iron to a solar gas below 400 K can produce magnetite, although we should recall that metal is unstable with respect to formation of FeO-bearing silicates at this temperature. Magnetite is very intensely black, but is a weaker absorber than metal in the thermal IR.

## Thermal Structure of the Nebula

We have seen, in our discussion of solar physics, that the temperature gradient in a gas is related to its opacity and the heat flux through it by

$$(dT/dr) = -3K\rho F(r)/4\sigma T^3 = 3\alpha F(r)/4\sigma T^3. \quad \text{(IV.18)}$$

Taking $\alpha = 10^{-7}, \sigma = 5.67 \times 10^{-5}$, and $T = 1000$ K, we get

$$(dT/dr) = 1.3 \times 10^{-12} F(r). \quad \text{(IV.184)}$$

The gravitational collapse energy of an infinite gas cloud of mass 1 $M_\odot$ to a radius of 0.1 AU is

$$E = -GM_0^2/r = 1.80 \times 10^{47}\ \text{erg}. \quad \text{(II.73)}$$

The time scale for collapse is not well known, but is estimated to be about $10^5$ years. Thus the luminosity of the solar nebula due to release of the gravitational potential energy of collapse is about $2 \times 10^{47}/(10^5 \times 3 \times 10^7) = 7 \times 10^{34}$ erg $s^{-1}$. This luminosity is spread unevenly over the surface of a disk about 40 AU in radius, of total surface area $2\pi r^2 = 2 \times 10^{30}$ $cm^2$. The mean heat flux through the surface of the nebula is $3 \times 10^4$ erg $cm^{-2}$ $s^{-1}$. If the opacity were very low, then the heat would be radiated more quickly; both the heat flux and effective temperature would be higher.

The effective temperature of the radiating layer in this example is $(F/\sigma)^{1/4} = 160$ K, or about the temperature of water ice condensation. An error of a factor of 10 in the time scale for collapse would change this temperature by a factor of about 1.7 to allow a range of 100 to 300 K.

We can now derive a mean temperature gradient in the $z$ direction of $1.3 \times 10^{-12} \times 3 \times 10^{-4} = 4 \times 10^{-7}$ K/$cm^{-1}$, or 0.4 K/km. Most of the heat is liberated within the half-thickness $z'$ [Eq. (IV.147)], and near $z'$ the gas pressure drops by a factor of 2 over a $\Delta z$ range of $1.5 \times 10^{12}$cm. In this region the opacity is high enough to provide a temperature difference of $4 \times 10^{-7}$ $dz = 6 \times 10^5$K!

We must suppose that such an enormous temperature gradient, which is what would be required to deliver the necessary heat flux through such a large opacity, could not possibly build up without driving rapid convective overturn. The adiabatic gradient is

$$(\partial T/\partial z)_S = -\mu g/C_p, \quad \text{(IV.27)}$$

which, for solar-composition gas, is usually near $-2.3g/(7R/2)$, or about $-10^{-8}g$. Substituting the expression for $g$ in the nebula from Eq. (IV.143),

$$(\partial T/\partial z)_S = -10^{-8}GMz/r^3 = -1.5 \times 10^8 z/r^3. \quad \text{(IV.185)}$$

For $r = 1$ AU and $z = 0.05$ AU, the adiabatic gradient is then about $3 \times 10^{-10}$ K $cm^1$ or $3 \times 10^{-5}$ K $km^{-1}$. Thus the radiative temperature gradient would be over 1000 times as large as the adiabatic gradient, and it is unavoidable that rapid, efficient convective transport of heat will dominate the cooling of the interior of the nebula. It is very difficult to model the details of convective heat transport, but we can mention for the moment that convective speeds are unlikely to reach a Mach number greater than about 0.3, or about $300\,\mathrm{m\,s^{-1}}$ in warm hydrogen. (Mach 1 is the speed of sound in the medium.) This turbulence causes rapid transport of gases and easily blows about small dust grains, helping to keep fine dust particles uniformly distributed in the gas. Note also that with this adiabatic gradient, $\Delta T$ becomes about 600 K, which is certainly of the right magnitude for a region where the midplane temperature is less than 1000 K and the photospheric temperature is about 200 K.

We may now collect a complete set of equations to describe the vertical thermal structure of the convective portion of the nebula. These include Eq. (IV.27) above, plus

$$(dP/dz) = -P\mu GMz/RTr^3 \qquad \text{(IV.142)}$$

$$g(z) = GMz/r^3. \qquad \text{(IV.143)}$$

The most important differences between this model and the isothermal first approximation shown in Fig. IV.35 are that the central regions of the disk (low $z$) are hot and isothermal because $gr = 0 = 0$, whereas regions far from the central plane are isothermal at low temperatures because of the low opacity far from the central plane. The entire region out to $P \approx 10^{-3}P_c$ is convective. Because temperature decreases with $z$, pressure will decline much more rapidly with $z$ at high $z$ than pictured in Fig. IV.35. A sketch of such a numerically integrated ($P$, $T$) profile across the disk is given in Fig. IV.42 along with an isothermal comparison model with the same midplane temperature and pressure.

There are many possible approaches to modeling the radial structure of the nebula. These approaches generally give results that are both divergent and difficult to test observationally. We shall be content with the semiempirical method already employed, whereby observational and theoretical limits on $\sigma(r)$ are used to constrain the pressure and density at midplane. Given that the surface density varies as $r^{-1.5}$ (Fig. IV.33),

$$P_c = \sigma[RTGM/2\pi\mu r^3]^{1/2} = 1.5T^{1/2}r^{-3} \qquad \text{(IV.152)}$$

for the minimum-mass nebula with isothermal vertical structure. For the mean model, $P_c$ should be about 5 times larger, and for the massive model, $P_c$ would be about 30 times larger. The effect of an adiabatic vertical structure is to decrease $P_c$ by 10 to 25% for any given $\sigma$, which is a very minor correction. From Appendix I, the relationship between temperatures in an adiabatic process is

$$P/P_0 = (T/T_0)^{C_p/R}. \qquad \text{(AI.26)}$$

The value of $C_p/R$ for the solar nebula between about 150 and 2000 K is close to $\frac{7}{2}$, whence $T$ varies as $P^{0.288}$. $P_c$ will then vary as $r^{-4.1}$, and $T_c$ will vary as $r^{-1.16}$. The radial and out-of-plane pressure structure of such a model is shown in Fig. IV.43. An explanation of the deduced dependence of $\sigma$ upon $r$ must be sought in detailed physical models of the solar nebula. One promising approach, in which steady mass flow into the nebula is assumed, has been developed by A. G. W. Cameron, T. D. Lin, G. Morfill, and others. This accretion disk model was inspired by recent advances in the study of astronomical X-ray sources.

We have now used our limits on $\sigma(r)$ deduced from the mass distribution in the Solar System in combination with our deduction of adiabatic structure for the nebula to predict the radial variations of the midplane temperature and pressure, $P_c(r)$ and $T_c(r)$. We will now turn to the available data on the densities and bulk compositions of the planets as we discussed them in Chapter III and the theoretical calculations of the temperature and pressure dependence of the bulk condensate density in order to test our model for the nebula, especially the functional form of $T(r)$.

First, we must compare the uncompressed (zero pressure) densities of planets and satellites, as deduced from their observed masses and radii, with the theoretically calculated densities. Both the strict equilibrium condensation and the strict rapid (nonhomogeneous) accretion scenarios are summarized in Fig. IV.44. The observed densities of solar system bodies are located in the diagram by the simple assumption that the high zero-pressure density of Mercury requires formation at a temperature just below the condensation temperature of metal, before magnesium silicates are fully condensed.

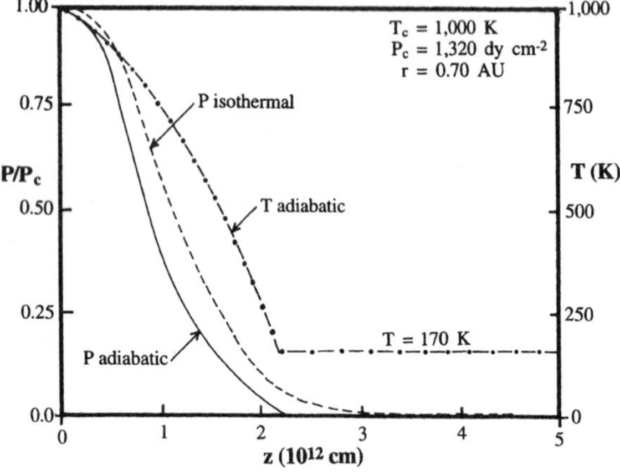


Figure IV.42 Vertical temperature and pressure profiles in an adiabatic nebula. The adiabatic model with any fixed central-plane temperature has a lower mean temperature and smaller scale height than the corresponding isothermal model. This model corresponds to a section through the nebula near the orbit of Venus.

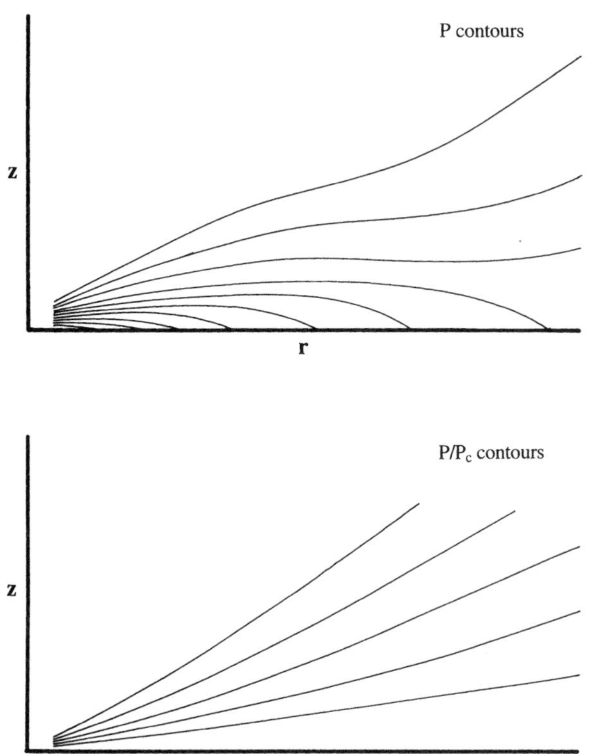


Figure IV.43 Isobaric contours in an adiabatic nebula. This figure provides an overview of the pressure structure of an adiabatic nebula constructed according to the discussion in the text.

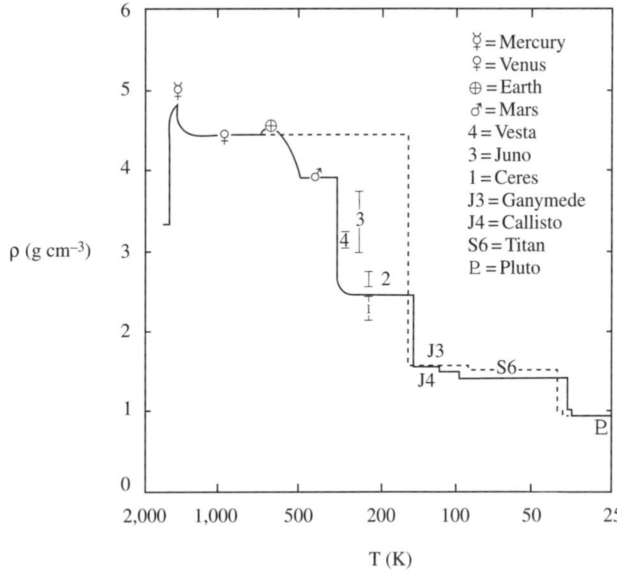


Figure IV.44 Uncompressed densities of solid bodies vs heliocentric distance. The densities of a number of solar system bodies are compared with the predictions of both the pure equilibrium and pure nonhomogeneous accretion schemes. The density of Mercury can be only approximately fit at best, and any allowance for accretion of Mercury out of materials formed over a range of heliocentric distances would make the problem worse.

The theoretical estimates of the density of Mercury both fall somewhat short of the observed density, but the behavior of the other planets and asteroids, as well as the distant satellites of the Jovian planets, is remarkable. The nebular model fits the zero-pressure densities of the planets with remarkable precision, providing that the preplanetary solids have been thoroughly equilibrated with the local nebula over narrow and well- defined temperature intervals. The equilibrium assumption yields predicted densities that differ dramatically from those of the nonhomogeneous accretion model for Mars and the asteroids; we must, if we accept the principle of equilibration of solids with gas in a nebula with a steep radial temperature gradient, conclude that almost all of the asteroid belt except its innermost edge is dominated by highly oxidized, serpentinized material. This is a prediction that was made by the equilibrium condensation model in 1970, prior to the first compositional characterization of any asteroid by infrared and visual photometry, spectroscopy, and polarimetry. Observations described in Chapter VIII provide a powerful test of this idea.

We must conclude that, based on the limited set of data so far considered, our semiempirical nebular model with adiabatic structure, with a temperature dropping off as $r^{-1.1}$, does a very good job of explaining the densities of solid Solar System bodies. When we take into account the similarity of the satellite systems of Jupiter, Saturn, and Uranus to little Solar Systems, we see that radial temperature gradients should have been present within these systems during the nebular stage. Thus a study of the dependence of the density and bulk composition of these satellite systems on distance from their primaries ought to be very rewarding.

We should remark that the corrections of the observed planetary densities to zero pressure are very small for Mercury and Mars, and hence the densities given must be quite reliable. The corrections for the self-compression of Venus and Earth are much larger, in excess of 1 g $cm^{-3}$, but Venus and Earth are so similar in mass and composition that the *relative* zero-pressure densities can be quite reliably compared. It is likely, but not certain, that the uncompressed density of Venus is smaller than that of Earth by about 1%. It is interesting that the equilibrium model predicts that condensates at the orbit of Earth ought to be about 1.5% denser than those at the orbit of Venus, due to FeS retention at Earth. It would be fascinating to know whether this is a case of art emulating nature or of random variations emulating data.

It is in some ways a bit startling that the differences between the terrestrial planets are so sharply defined. It is entirely plausible that the accretion of a planet should involve sampling of solids originating over a rather wide range of heliocentric distances. This would inevitably lead to an averaging of compositions and a blurring of distinctions between planets. The *observed* densities of the planets attest to the preservation of primordial differences. It is thus valuable to explore the dynamics of accretion of vast swarms of solid bodies to see whether planets can in fact be assembled without erasing evidence of these differences.

Our discussion has emphasized that equilibrium condensation with accretion of local condensates offers a far more satisfactory prediction of the densities of Solar System bodies than the sequential-accretion model. Nonetheless, all studies of the planetary accretion process agree that each planet must accrete a significant fraction of its mass from distances of at least a few tenths of an AU away, both outward from and inward toward the Sun. A planet cannot be composed solely of locally condensed material. If, say, 15% of the mass of Earth were accreted from Mars' region and another 15% from Venus' region, the bulk density of Earth would be very little affected. Accretion by Earth of 1% of volatile-rich material from the asteroid belt likewise would have no detectable effect on Earth's density, but could be an extremely important source of volatiles, possibly dominating Earth's inventory of water, carbon, nitrogen, etc. Comparison of Earth's volatile-element inventory with those of Mars, Venus, and meteorites may therefore provide a sensitive test of models of planetary accretion.

## Turbulence and Dust Sedimentation

As long as the opacity of the nebula remains high, it will maintain a steep temperature gradient and be convective. If the dust grains, which are the principal source of continuum infrared opacity, settle quickly into the central plane of the nebula, then the thermal opacity will vanish and turbulence will virtually cease. On the other hand, the distribution of small grains of dust in the nebula is governed by the interplay of several important forces, including gravity, coagulation, and turbulence. The principal homogenizing agent for dust is turbulence, and the continued existence of turbulence requires a nearly homogeneous distribution of dust! Not all the forces affecting dust have been modeled simultaneously, but calculations on the sedimentation and coagulation of dust particles in a *nonturbulent* nebula have been reported. We shall quickly summarize the major physical phenomena associated with the behavior of dust in the nebula. We begin with what little can be said regarding turbulent speeds in the nebular gas.

At various stages in the evolution of the nebula, up to four different driving forces for turbulence may be active. The first of these is the chaotic deposition of mass and angular momentum in the solar nebula by infalling matter. This is especially important during the growth of the nebular disk. It is customary to model disk formation as occurring symmetrically and smoothly; no one believes this is the case, but this simplification permits us to evade an area of near-total ignorance with minimal loss of face.

The second major source of turbulence is the thermal energy liberated in the interior of the nebula by its gravitational collapse. Although we are not sure exactly how large a heat flux was provided by this source, we can make a reasonable guess. We shall simply estimate the amount of gravitational potential energy released by the collapse of an interstellar gas cloud from infinite distance to, say, a size of a few astronomical units. This amount of energy is just $-GM^2/r$, or about $10^{42}$ erg. This energy is radiated from the surface of a disk with a total surface area of $2\pi r^2 \approx 10^{28}\,\text{cm}^2$, and the lifetime of the nebula is limited by various modes of instability to about $10^{13}$ s. The mean thermal flux is thus about $10\,\text{erg}\,\text{cm}^{-2}\,\text{s}^{-1}$. By comparison, the total potential energy available from the collapse of the Sun to a radius of $10^6$ km is about $10^{49}$ erg. Some portion of this is surely transmitted to and through the nebula by turbulent mixing. We shall then take the flux through the nebula as $10^4\,\text{erg cm}^{-2}\,\text{s}^{-1}$, with an uncertainty of a factor of $10^3$. This means that the lower limit on the photospheric temperature of the disk, corresponding to the lowest possible heat flux, is about 20 K, whereas the highest possible temperature (for $10^7\,\text{erg cm}^{-2}\,\text{s}^{-1}$) would be 600 K. Since, based on chemical evidence alone, we would expect a mean disk surface temperature on the order of 100 to 200 K, we can accept these limits as reasonable and probably overly generous.

We shall assume that the heat transported from the vicinity of the central plane of the disk to the nearest region of the disk surface is carried by free convection. From mixing length theory, the heat flux ($F$) carried by convection is related to the mean vertical turbulent speed ($v_z$) and to the density of the gas by

$$v_z = (F/\rho)^{1/3}. \qquad \text{(IV.186)}$$

Near 1 AU the density not far from the midplane is about $10^{-8}\,\text{g cm}^{-3}$, and hence a flux of $10^4\,\text{erg cm}^{-2}\,\text{s}^{-1}$ would require a mean vertical wind speed of $10^4\,\text{cm s}^{-1}$. Note that, if our flux estimate is in error by a factor of $10^3$, the estimated turbulent speed is changed by only a factor of 10. We now know the characteristic size and characteristic turbulent speed, and we can calculate a mixing time of about $10^8$ s.

Sound speed in the nebula is about $10^5$ cm s$^{-1}$, so we are here considering turbulent wind speeds of Mach 0.1. Wind speeds in planetary atmospheres rather often reach this Mach number, and even the turbulence in the stem of the mushroom cloud from a large thermonuclear explosion does not exceed Mach 0.3.

Now let us consider a dust particle of mean radius $a$ (keep in mind that real dust particles are almost never spherical!) falling through a gas of density $\rho$ with thermal speed $V_t$ in a gravitational field of acceleration $g$. The instantaneous speed of fall of the particle relative to the gas is $V_z$. Because of the motion of the particle through the gas, collisions with gas molecules are more frequent and more severe on the front of the grain. The pressure exerted by the local gas on a stationary surface is

$$P = nRT/V = \rho V_t^2/3, \qquad \text{(IV.187)}$$

and the pressure difference between the front and the rear surfaces of a moving grain is

$$\Delta P = (\rho/3)[(V_t + V_z)^2 - (V_t - V_z)^2]. \qquad \text{(IV.188)}$$

The acceleration produced by this pressure difference is just the force ($\Delta P \times$ area) per unit mass:

$$\begin{aligned} g_z &= (\rho/3)(4V_tV_z)(3\pi a^2/4\pi\rho_s a^3) \\ &= (\rho/\rho_s)(V_tV_z/a). \end{aligned} \qquad \text{(IV.189)}$$

The acceleration vanishes at the terminal velocity of the grain:

$$V_z = (\rho_s a/\rho V_t)(GM_z/r^3). \qquad \text{(IV.190)}$$

Thus a particle of $a = 1\ \mu$m($10^{-11}$ g) falls at $10^{-2}$ cm s$^{-1}$, and a terminal velocity of $10^4$ cm s$^{-1}$ is reached by bodies of radius 100 cm ($10^7$ g). Thus a 1-$\mu$m particle has a fallout time of $10^{14}$ s, some 10 times longer than the expected lifetime of the nebula. In the presence of turbulence with a mean vertical speed of $10^4$ cm s$^{-1}$, particles up to $\approx$ 100 cm in size will be blown along with the gas and effectively homogenized over a region extending some $10^{12}$ cm above and below the central plane.

Bodies with dimensions of about 10 cm to 1 m will experience a force balance quite different from that which acts on a parcel of atmosphere. The meter-sized rock is so small and dense that the pressure-gradient force is negligible. In the absence of turbulence, the centrifugal force would be balanced exactly by the gravitational force. Any displacement of the body from the central plane would cause the body to execute an inclined Keplerian orbit, which crosses the dense central plane region twice per orbit. The out-of-plane ($z$) component of the momentum of the body would decay rapidly because of frictional drag. This drag can be calculated from Stokes' Law, because the size of the body (>10 cm) is now larger than the gas mean free path (0.1 cm). To an adequate approximation, however, the body will halve its momentum relative to the gas after it sweeps out a volume of gas with a mass equal to its own. If the body moves over a range of $10^{11}$ cm in $z$ in half an orbital period ($P = 1$ year), the mean vertical speed of the body will be about $10^4$ cm s$^{-1}$. With a cross-section area of $3 \times 10^4$ cm$^2$ and a speed of $10^4$ cm s$^{-1}$ in a gas with density $10^{-8}$ g cm$^{-3}$, the body will sweep out mass at the rate

$$m = \rho A v_z, \qquad \text{(IV.191)}$$

or about 3 g s$^{-1}$. With its mass of about $10^7$ g, the body will damp out most of its out-of-plane momentum on a time scale of only $3 \times 10^6$ s. Note that, if the mean vertical turbulent wind speed is $10^4$ cm s$^{-1}$, it will accelerate the body sufficiently to maintain vertical excursions of this magnitude on the same time scale.

However, because the meter-sized body is essentially in a Keplerian orbit and because it is moving through a rarefied gas for which the inward gravitational force is slightly offset by the outward-directed component of the pressure-gradient force, the body must be moving about the Sun more rapidly than the gas. The force on a body due to solar gravity, the centripetal force necessary to maintain circular motion, the pressure gradient force, and the gas turbulence force are, respectively,

$$F_g = GMm/r^2 \qquad \text{(IV.192)}$$

$$F_c = mV_K^2/r \qquad \text{(IV.193)}$$

$$F_P = (m/\rho_{body})(dP/dr) \qquad \text{(IV.194)}$$

$$F_t = -c_D\rho a^2\rho(V - V_t)^2. \qquad \text{(IV.195)}$$

The forces on a 100-cm body are $6 \times 10^6$, $6 \times 10^6$, $10^{-4}$, and $10^4$ dyn, respectively, whereas those on a cubic centimeter of gas at the same heliocentric distance are $10^{-8}$, $10^{-8}$, $10^{-11}$, and $10^{-11}$ dyn, respectively. Thus the body must move through the gas with a relative speed of about $10^{-11}/10^{-8}$ times the Keplerian speed, or $10^4$ cm s$^{-1}$, and a frictional drag term must be added to the force balance on the body. The body will accordingly constantly lose momentum to the gas and constantly fall sunward to lower-energy orbits. Eventually, such motion could cause a solid body to spiral into the Sun. Note, however, that the randomly directed turbulent drag is as large as this noncorotation drag.

At sizes of $10^4$ to $10^5$ cm the drag acceleration on rocky bodies becomes so small that their orbital evolution in response to drag becomes negligible. It is often convenient to describe turbulent mixing in terms of a single number, analogous to the molecular diffusion coefficient, $D(\text{cm}^2\,\text{s}^{-1})$. Random turbulence in the $z$ direction may be thought of as a one-dimensional random walk, for which the mean displacement of the gas parcel from its initial position increases as the square root of time. That is, its linear displacement ($\Delta z$) squared is equal to a constant times the elapsed time, $\Delta t$. The constant, $K$, is called the *eddy diffusion coefficient*:

$$K = (\Delta z)^2/\Delta t. \tag{IV.196}$$

If the gas has a characteristic dimension such as the *scale height* ($H$) (the altitude interval over which the pressure changes by a factor of $e$), then $K$ can be written as $H^2/t$, where $t$ is called the mixing time: it is the mean time required for a molecule to be transported a distance ($H$) by random turbulent mixing. Alternatively, we can equate $K$ to $Hv_z$, where $v_z$ is the mean vertical speed of the gas due to turbulence. Except at very low gas densities, at which the molecular mean free path is very large, eddy mixing is almost always much faster than molecular diffusion. We may see this easily by writing the relationship between the molecular diffusion coefficient and the characteristic length and speed scales of molecular diffusion: $D = lv_t/3$, where $l$ is the mean free path and $v_t$ is the thermal speed of the molecules. In the nebula at 1 AU, $K$ is about $10^{17}$ and $D$ is about $10^4\,\text{cm}^2\,\text{s}^{-1}$.

There are several important chemical consequences of such large eddy mixing rates. First, we have seen that CO and $N_2$ are stable gases at high temperatures, but that these gases are thermodynamically unstable relative to $CH_4$ and $NH_3$ at low temperatures. We have already seen that the conversion of molecular nitrogen to ammonia in the solar nebula becomes thermodynamically possible only below about 330 K, and it is surely difficult to carry out this reaction at such low temperatures unless infinite time is available. But if turbulent mixing is rapid, gases that have recently been equilibrated in the hot, dense interior of the nebula, where reactions are very fast, may be blown outward into regions where their chemical lifetimes are very much longer than the convective mixing time. Thus the high-temperature gases will be replaced by mixing far more rapidly than they will be destroyed by reactions. Further, any methane or ammonia made by the slow low-temperature reduction process will not be allowed to accumulate, but will instead be quickly mixed inward into hot regions where it is unstable and is rapidly destroyed. As a result, the entire solar nebula was probably rich in CO and $N_2$, with only a few percent of the total carbon and nitrogen abundance in the form of methane, ammonia, or carbon dioxide. The process by which high-temperature gases are preserved under nonequilibrium conditions by rapid cooling is called *quenching*.

Our discussion of turbulence in the nebula also permits us to make a crude estimate of the lifetime of the nebular gas disk. The eddy diffusion coefficient for radial motions in the nebula can be approximated as the product of the mean turbulent speed $v$ and the radial scale height $H$: $K = 10^4 \times 10^{13} = 10^{17}\,\text{cm}^2\,\text{s}^{-1}$. The time scale for radial transport of angular momentum is set by the time it takes for turbulent mixing to "diffuse" from the heart of the nebula to its periphery, a distance of about 32 AU ($5 \times 10^{14}$ cm). That characteristic radial diffusion time is just the square of the distance scale ($l$) divided by the eddy diffusion coefficient. Thus the time scale for extensive structural evolution of the gas disk is about $(5 \times 10^{14})^2/10^{17} = 2.5 \times 10^{12}$ s, or nearly $10^5$ years.

## Accretion of Rocks, Planetesimals, and Planets

We have seen that gas motions can move meter-sized rocks. This implies that the dissipation of the nebula will remove meter-sized chunks of rock from the Solar System. Clearly the present bodies in the Solar System must have accreted to at least that size while the nebula was still present. But how rapid could accretion be, and how large could the objects become? What were the mechanisms by which individual grains grew to rocks, mountains, asteroids, and finally to planets? Could full-sized planets have been present within the solar nebula?

The first step in the accretion process is for individual grains to encounter each other and collide. This occurs inside a convective, turbulent medium, with wind speeds on the order of $100\,\text{m s}^{-1}$. The density of the nebula is only about $10^{-8}\,\text{g cm}^{-3}$ near 1 AU, and the density of rocky material is only 1% of the total. If the average particle is about 1 $\mu$m in radius (a mass of about $2 \times 10^{-11}$ g), then there were a few grains per cubic centimeter. The gas mean free path was about 0.1 cm, and thus the mean particle size was about 100 times smaller than the mean free path. Under these conditions, dust particles behave like large molecules, not aerodynamic bodies. The drag acceleration for these small particles is that given in Eq. (IV.189). Dust particles of somewhat different sizes entrained in the same eddy will have a relative motion due to the different terminal velocities of the grains; the larger grain will move faster and will therefore tend to sweep out smaller particles.

As an example, consider particles of radii 0.1 and 1 $\mu$m, with terminal velocities of about $10^{-3}$ and $10^{-2}\,\text{cm s}^{-1}$, respectively. The larger particle is thus moving at about $10^{-2}\,\text{cm s}^{-1}$ through a diffuse cloud of finer dust. The particle sweeps out a volume of $\pi a^2 v_z$ cm$^{-3}$ s$^{-1}$, or about $3 \times 10^{-10}$ cm$^{-3}$s$^{-1}$. There is about $10^{-10}$ g of dust per cubic centimeter of gas, equivalent to $10^4$ 0.1-$\mu$m grains. The mean time between collision events is about $3 \times 10^5$ s, and the larger particle will sweep out about enough mass to double its own mass in about $3 \times 10^8$ s. If all collisions result in sticking, then centimeter-sized particles with masses of about 10 g will be common after about 4000 years. Further growth over the range $a = 1$ to 100 cm would take another 2000 years if all collisions continued to result in accretion. Bodies in this size range have achieved speeds of $10^3$ to $10^4$ cm s$^{-1}$, and collisions of small particles with them will be rather violent. Even a collision of a 10-cm object with a 100-cm object will occur with a collision speed of about $10^4$ cm s$^{-1}$. It is not obvious that such collisions result in net accretion, because these speeds are sufficient to crush a mass of material much larger than the mass of the smaller colliding body. Complete shattering of hard rock requires an energy of about $3 \times 10^7$ erg cm$^{-3}$. Thus a 10-g chunk of radius $a = 1$ cm colliding with a piece of hard rock at $10^4$ cm s$^{-1}$ will bring in a kinetic energy of $5 \times 10^8$ erg, enough to shatter 16 cm$^3$ (50 g) of rock. It seems likely that such collisions must be highly erosive. This conclusion is greatly strengthened by the consideration that sub-meter-sized bodies assembled in slow collisions must have very low crushing strengths, probably comparable to those of clods of dirt or carbonaceous meteorites ($10^6$ erg cm$^{-3}$). Similar impacts upon such weak targets could crush and expel a mass a thousand times that of the impacting grain.

It is therefore by no means certain that slow collisions of small grains will result in efficient accretion. Hard, dry single grains of silicates would surely tend to rebound without accretion. There are, however, several other types of intergrain forces that favor accretion. One of the more obvious of these is magnetic attraction. Grains below their Curie temperatures with permanent dipole moments will tend to orient themselves so as to minimize their potential energy. This results in strong intergrain attraction at small distances. The meteoritic materials that are capable of such magnetic attraction are magnetite ($Fe_3O_4$) and the metallic phases kamacite and taenite.

Another important intergrain force is electrostatic interaction. Typically, most grains will be positively charged or neutral. Because silicate grains are dielectrics, a positively charged grain will induce an image dipole in a neutral collision partner by polarizing the electron distribution in it. This results in a small net attractive force. This force is even sufficient to overcome the overall electrostatic repulsion of weakly charged dielectric grains with like charge. A third type of sticking force is essentially a strong van der Waals attraction, usually associated with surface coatings of highly polar materials or long-chain organic molecules.

A fourth type of sticking force is due to partial fusion of a grain at the point of impact with another. For cold grains, shattering is energetically much more probable than fusion. This fusion mechanism probably works only for grains that are already at a temperature very close to the melting point prior to collision. However, very few materials ever experience temperatures close to their melting temperatures in a solar-composition system at reasonable pressures. The most promising materials are metallic iron–nickel, water ice, solid ammonia hydrate, and solid methane. All of these phases require total pressures on the order of 1 bar or higher for their liquid to be stable and hence are of limited utility in the solar nebula.

We may conclude that accretion of bodies to at least meter size must have occurred in the nebula, but that there were certain constraints on the growth process. Over a period of about $10^4$ years, some particles accumulated to sizes of several centimeters. Collisions of these bodies with much smaller (gas-following) grains became so disruptive that the larger bodies were able to accrete only other bodies with sizes similar to their own, that is, bodies with low relative velocities. During this phase, collisions of small grains with large rocks would serve to greatly increase the number of small grains. Because the thermal opacity is due almost entirely to small grains, the opacity presumably remained large, and the gas remained convective.

Next, we must consider the effects of meter-sized bodies embedded in the nebula, but traveling $10^4$ cm s$^{-1}$ relative to the gas. The drag force exerted on the gas by such a body is

$$F_D = \tfrac{1}{2} C_D \pi a^2 \rho V^2, \qquad \text{(IV.197)}$$

where $C_D$ is the drag coefficient (about 0.44 in this case). The Earth formation zone contains some $6 \times 10^{27}$ g of solids and about $10^{30}$ g of gas. The number of $10^7$ g ($a = 100$ cm) bodies was on the order of $6 \times 10^{20}$. Each body experiences a drag force of $6 \times 10^3$ dyn and dissipates a power of $FV = 6 \times 10^7$ erg s$^{-1}$. The entire planetesimal swarm then dissipates $3.6 \times 10^{28}$ erg s$^{-1}$. The surface area of the disk in the Earth formation zone is about $6 \times 10^{26}$ cm$^2$, so the steady-state energy flux out of the zone must be about 60 erg cm$^{-2}$s$^{-1}$. This power is dissipated deep within the disk, near the central plane, by bodies that cross the central plane at small angles. The mean vertical turbulent wind speed, as estimated from mixing length theory, is then just $V_z = (F/\rho)^{1/3}$, or about 1600 cm s$^{-1}$. Thus it is an unavoidable consequence of accretion that the nebula be strongly stirred. This conclusion is not sensitive to the sizes assumed for the largest bodies; if, instead of $10^7$ g, we assume that the mass is apportioned between bodies with masses of $10^{17}$ g, the calculated turbulent wind speed becomes 200 cm $s^{-1}$. This is fast enough for the turbulent gas to carry along bodies up to centimeter size. It should be noted that these wind speeds are markedly subsonic.

We do not presently know to what size bodies could grow while the nebula was still present, but it seems likely that, over a time scale of about $10^4$ years, much of the mass accreted into bodies with sizes of meters to kilometers. If bodies with dimensions of thousands of kilometers were able to form in certain regions of the nebula, due perhaps to the local availability of an extraordinarily effective sticking agent or to the rapid settling out of the dust grains which provide opacity and turbulence, then these bodies could interact gravitationally with the nebular gas in interesting ways. We shall return to this possibility below.

Once the nebula has been dissipated, after perhaps $10^5$ to $10^6$ years, the accretionary evolution of the Solar System proceeds without any complications imposed by aerodynamic forces. We may picture a vast swarm of kilometer-sized bodies in slightly eccentric and slightly inclined orbits, colliding at low speeds and accreting. As larger bodies with appreciable gravitational fields build up, the sampling of many small accreting bodies causes the orbital eccentricity and inclination of the large body to average out to ever smaller values. At the same time, the ever-increasing gravity of the body causes it to become ever more efficient at perturbing the smaller bodies and pumping them up to ever higher inclinations and eccentricities. This in turn causes the orbit of each body to cross the orbits of more other bodies. Collisions thus become both more numerous and more energetic. Collisions between roughly equal-sized bodies may become disruptive, whereas collisions between bodies of disparate sizes may almost always result in net accretion. Orbital inclinations of 10 or 20° and eccentricities of about 0.15 can be built up during this era, as the size of the largest body grows from about 500 to 6000 km in radius (asteroidal to planetary size).

The acceleration of smaller bodies by gravitational interactions with the large planetesimals means that the material accreted by each growing planet, although dominated by solids from near by, also contains appreciable contributions from well outside and well inside the orbit of the planet. Further, we must not forget that the orbital decay caused by aerodynamic drag in the nebula has already moved a large amount of material inward to various degrees. Thus each planet must sample materials with a range of formation locations and conditions. Indeed, computer simulations of the terminal stages of accretion suggest that 10 to 25% of the mass of each planet may come from regions closer to the next inner and next outer planet, and a fraction of a percent of the mass of a planet like Earth could come from as far away as the asteroid belt. Such sampling will have little effect on the major-element chemistry and bulk density, but may be extremely important in determining the volatile-element abundances of a planet. As little as 0.1% of water-rich low-temperature material from the asteroid belt could provide all the water in Earth's oceans, crust, and mantle!

But such accretion of volatile-rich minerals pales to insignificance compared with the amount of volatiles that might be captured by a planet if the planet were present with its full present mass in the Solar Nebula. We therefore must briefly explore the mechanisms by which a massive body can capture nebular gases.

## Gas Capture from the Solar Nebula

Imagine a gravitating body that is accreting while it is embedded in a continuous nebula. Over dimensions of, say, 100 radii of the body, the density, temperature, and pressure are essentially constant except for the effects of the gas distribution caused by the gravitational influence of the body. From the Boltzmann equation for the distribution of molecules over energy levels $E$, the population of particles at the surface of the body relative to that at large distance is

$$n_s/n_\infty = \exp E_{esc}/kT, \qquad \text{(IV.198)}$$

where $n_s$ and $n_\infty$ are the gas density in $cm^{-3}$, and $E_{esc}$ is the gravitational energy difference between the surface and "infinity," which we take to mean "greater than 100 body radii." The energy required for the escape of a molecule of mass $m$ is $E_{esc} = mv_{esc}^2/2 = GMm/r_s$. Thus, for an isothermal gas,

$$n_s/n_\infty = \exp\,[2GMm/r_skT_\infty]. \qquad \text{(IV.199)}$$

Substituting $M = 4\pi\rho r_c^3/3$ and $m/k = \mu/R$, where $\mu$ is the gram molecular weight, we have

$$n_s/n_\infty = \exp\,[8\pi\rho G\mu r_s^2/3RT_\infty]. \qquad \text{(IV.200)}$$

We may then simply specify the molecular weight of solar material as $\mu = 2.3$ and then use the radius and density of any solar system body in conjunction with its estimated nebular temperature ($T$) to calculate the resulting gas pressure distribution. We base the estimated nebular temperatures on Fig. IV.44, which gives us the highest temperature at which the solid components of each body were in contact with the gas. Note, however, that this probably grossly overestimates the ability of a body to capture gas, because the equilibration temperatures refer to a dispersed state of solids in the nebula, not to contact between a few large bodies and the nebula.

One convenient way of parameterizing this problem is to regard the escape speed of the body and either the thermal speed or the temperature as the independent variables. Thus,

$$n_s/n_\infty \propto p_s/p_\infty \propto \rho_s/\rho_\infty = \exp\left[\mu V_{esc}^2/2RT_\infty\right]. \qquad \text{(IV.201)}$$

Figure IV.45 displays the results of these calculations for a wide range of escape speeds and temperatures relevant to the interaction of massive bodies with the Solar Nebula. It can be seen clearly that lunar-sized bodies are ineffectual at capturing a solar composition atmosphere at temperatures much above 100 K and that Mercury, even if fully accreted in the nebula, would have been far too hot to have captured any significant amount of gas. Mars, with an enhancement factor of $10^3$ to $10^4$, embedded isothermally in a medium with a pressure of about $10^{-5}$ bar, would have a surface pressure near $10^{-1}$ bar of solar composition gas. The most extreme examples, Venus and Earth, could have captured masses of solar material comparable to the mass of the planet. This would of course require that we take into account the contribution of the captured gas to the gravitational potential, which would in turn have increased the mass of captured gas. It would seem that isothermal gas capture from the solar nebula would have turned Earth and possibly Venus into Jovian planets.

The planets, however, follow near-Keplerian trajectories through a nebula in which the radial pressure gradient force partly offsets gravity. Accordingly, the planets move about the Sun faster than the gas, with a relative velocity of about 0.1 km s$^{-1}$, close to the speed of sound. Thus the gas flows through the planetary potential well rather rapidly, and it is no longer clear that it can radiate out the energy it picks up as it falls into the gravitational potential well so as to remain isothermal. Consider gas not far from the surface of the Earth at a level at which the pressure is 1 bar. The local density is about $10^{-4}$ g cm$^{-3}$, and the scale height of the gas distribution (the altitude interval over which the pressure falls by a factor of $e$) is $H = RT/\mu g$ or about $2 \times 10^7$ cm (200 km). We now need to know the opacity of solar material (Fig. IV.46). The opacity is about $10^{-3}$ cm$^2$ g$^{-1}$ under these conditions, so the vertical gas column is above the 1-bar level, which contains $\rho H = 2 \times 10^3$ g cm$^{-2}$ of nebular material. The opacity is then about 2 cm$^2$ per square centimeter of surface area, that is, an optical depth $\tau$ of 2. In addition to the damping effect of this opacity, there is another constraint imposed by the relative time scales of infall of the gas and emission of radiation. A speed of $10^5$ cm s$^{-1}$ at high altitudes and a gas density of $10^{-8}$ g cm$^{-3}$ at these altitudes mean that the mass influx rate will be about $10^{-3}$ g cm$^{-2}$ s$^{-1}$. The heat content of this gas after capture is near $v_{esc}^2/2$ per gram, or just under $10^{12}$ erg g$^{-1}$. This means the flow will bring in a heat load of about $10^9$ erg cm$^{-2}$ s$^{-1}$. In order for a perfect black body to radiate out this large a heat load, the temperature must be $(F/\sigma)^{1/4}$ or nearly 2000 K. Thus the presence of a massive body embedded in the solar nebula requires either that the body heat up continuously or that the envelope must, even at high

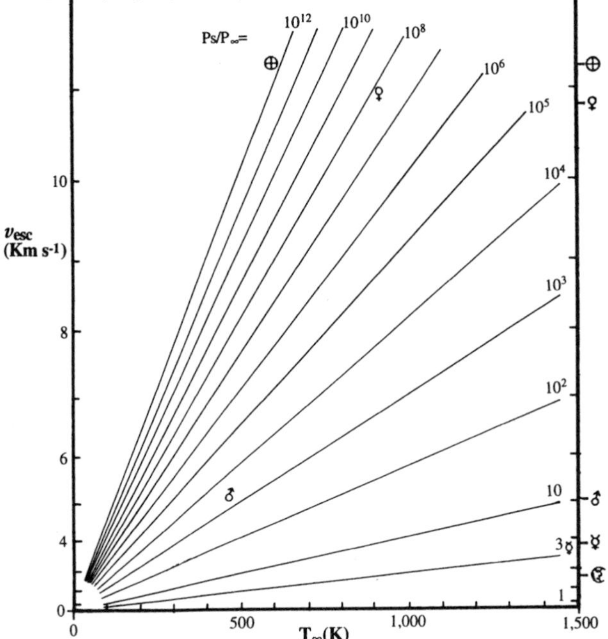


**Figure IV.45** Isothermal gas capture from the solar nebula. Contours of surface gas pressures relative to the nearby unperturbed central plane nebular pressure are given.

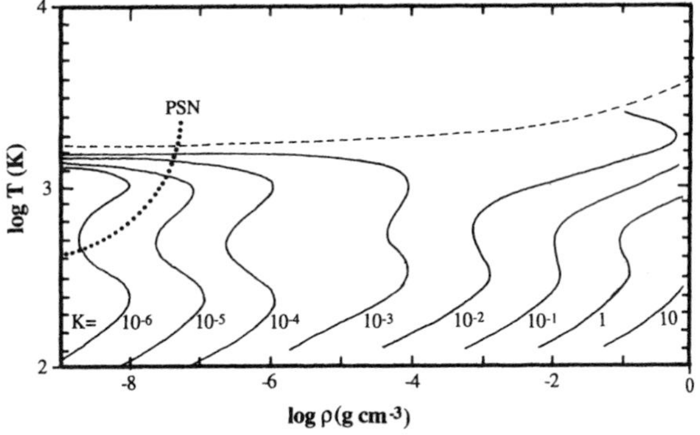


**Figure IV.46** Opacity of solar material. The total opacity of solar material, $K$ (cm$^2$ g$^{-1}$), is given for the temperature and pressure range of interest. At high temperatures, all solids evaporate and all polyatomic molecules decompose, so the mean opacity becomes dominated by the negative hydrogen ion, as in stellar atmospheres.

altitude, be so hot that solids are fully evaporated. Evidently we must look more carefully at the energetics of gas capture and leave out the clearly inappropriate assumption of isothermal capture.

In this case we shall assume hydrostatic equilibrium,

$$dP = -g\rho dr = -(P\mu GM/RTr^2)dr, \qquad \text{(IV.202)}$$

adiabatic structure, and ideal gas behavior:

$$C_P dT = v dP = (RT/P)dP. \qquad \text{(IV.203)}$$

Solving the latter equation for $dP$ and substituting into the former,

$$C_P dT = -(\mu GM/r^2)dr. \qquad \text{(IV.204)}$$

Integrating from ($r_S$, $T_s$) to infinity, we find

$$C_P(T_s - T_\infty) = \mu GM/r_s = \tfrac{1}{2}\mu v_{esc}^2. \qquad \text{(IV.205)}$$

The surface temperatures found are given in Fig. IV.47 along with the corresponding pressures at the surfaces of the massive bodies. It should be recalled that at 2000 K on the Solar Nebula adiabat even refractory oxides vaporize. Iron and magnesium silicates vaporize below 1500 K. Thus we see that the surfaces of both Mars and Mercury would have been hot enough to vaporize silicates, had these planets been present in the nebula. Earth and Venus are another matter altogether: adiabatic gas capture by either of them would have produced surface temperatures in excess of the photospheric temperature of the Sun.

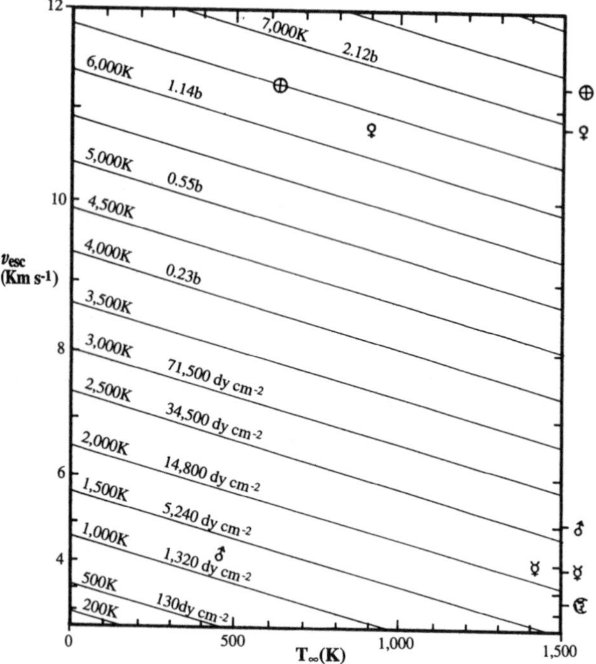


**Figure IV.47** Adiabatic gas capture from the solar nebula. The surface temperatures and pressures for strictly adiabatic gas capture from the standard solar nebula model are shown.

The conclusion seems to be that very rapid planetary accretion (which is of course essential in order to have planets present while the nebula is still intact), because it delivers such huge heat loads to the surfaces of the accreting planets, will keep planetary atmospheres convective and adiabatic up to altitudes at which the optical depth becomes small. This conclusion, however, must be tempered by the realization that most of the outer envelope of such a body would be tenuous and transparent, so that the outer atmosphere would be able to radiate off its capture energy and remain roughly isothermal. The atmospheric structure would be radiatively controlled above that altitude and would quickly become adiabatic below that level. Thus nature would conspire to establish a structure intermediate between the isothermal and the adiabatic cases, which are the two easiest to calculate. Such a *radiative–convective* thermal structure will again be of interest when we turn to the atmospheric structure of Jupiter and Saturn.

With respect to quantitative modeling of gas capture by embedded planets, we point at two further difficulties. First, because of the ram pressure due to the motion of the planet through the nebula, spherically symmetrical hydrostatic equilibrium solutions cannot be correct. Further, the relative speeds are close to sound speed; a bow shock may be present. An acceptable solution to this problem may require use of a supersonic hydrodynamic model, a task of formidable proportions in light of our very modest knowledge of the conditions present in the nebula and the physical properties of the bodies moving within it. It makes little sense to invest a great deal of time and energy in this problem unless there exists some evidence that such a primitive captured solar atmosphere was in fact once present on one or more of the inner planets. In our treatment of the atmospheres of the terrestrial planets (in Chapter X) we shall search diligently for any such evidence and (alas) fail to find it.

Despite these difficulties, it is clear that some of the planets *must* have formed in the presence of the nebula, because vast quantities of solar gas were captured by them. These of course are the Jovian planets Jupiter, Saturn, Uranus, and Neptune. We now have in hand a large body of information on the chemical and physical behavior of solar-composition material, atmospheric structure, and heat transport, gained mostly in very hot, dense stellar interiors or very tenuous gaseous nebulae. We are now in an excellent position to survey the current state of our knowledge of these planets.

There is a further bonus that we may collect upon study of the Jovian planets and their satellites: the systems of Jupiter, Saturn, and Uranus each contain a number of satellites in prograde, coplanar orbits of low inclination and eccentricity. In short, they resemble miniature Solar Systems. Thus study of the physical and chemical systematics of these satellite families may provide us with independently derived insights into the workings of the Solar System at large, including the processes that gave it birth.

## The T Tauri Phase

Like all good things, even the solar nebula came to an end. The events surrounding the ignition of the Sun may have been violent, because a slight overshoot in the collapse process may have caused a "hydrogen flash" analogous to the helium flash we encountered in our survey of stellar evolution. Whether or not such a heat pulse occurred, the Sun entered the T Tauri phase with two or three times its present luminosity and subsided over about $10^7$ years into a stable early MS star. The effective temperature of the T Tauri Sun was probably close to the present value.

The stellar winds of T Tauri stars have been observed to have radial velocities of 200–300 km s$^{-1}$ at two or three stellar radii. This is about half the escape velocity of the star at that level, but is presumably still well within the acceleration region. Mass loss rates of about $10^{-8} M_0$ year$^{-1}$ extending over about $10^7$ years seem to be typical and will be assumed for the current discussion.

The kinetic energy carried by this dense solar wind at distances greater than about 20 stellar radii is

$$KE/m = \tfrac{1}{2}v^2, \qquad \text{(IV.206)}$$

about $10^{15}$ erg g$^{-1}$, and the mass loss rate is then $7 \times 10^{17}$ g s$^{-1}$. The mass flux at distance $r$ (AU) is then $2.5 \times 10^{-10}/r^2$ g cm$^{-2}$ s$^{-1}$. This is a proton flux of $1.5 \times 10^{14}$ cm$^{-2}$ s$^{-1}$ at 1 AU.

The solar wind power is

$$p = \tfrac{1}{2}v^2\dot{m}, \qquad \text{(IV.207)}$$

or $2.5 \times 10^5 r^2$ erg cm$^{-2}$ s$^{-1}$. The momentum flux is 0.01 dyn cm$^{-2}$ at 1 AU. This dynamic solar wind pressure is some $10^6$ times the present solar wind pressure. This force is directed radially; however, the solar wind impact on a particle suffers aberration due to the orbital speed of the particle, which is typically a few percent of the solar wind speed. The radial component of the incident momentum simply serves to offset a small part of the solar gravity. The tangential component tends to decelerate the particle and cause it to spiral into the Sun.

A small solid particle of radius $a$ exposed to this wind for time $t$ will gain a tangential momentum of

$$p = m\Delta V_r = \pi a^2 tP/r(AU), \qquad \text{(IV.208)}$$

which, over the duration of the T Tauri phase, is $1{-}10^{12}(a/r)^2$. The mass of the particle is $4\pi\rho a^3/3$, and hence the radial velocity accumulated by a particle near 1 AU is approximately $v_r = 5 \times 10^{10}/a$ cm s$^{-1}$ for a typical solid grain of density 4 g cm$^{-3}$. The orbital velocity of the grain is about $3 \times 10^6$ cm s$^{-1}$. The energy that must be imparted to the grain in order for it to fall into the Sun is just half the energy required to lift the grain to escape velocity from the Sun:

$$\int_0^\infty F dr = -GMm\int_0^\infty r^{-2}dr = GMm/r_0. \qquad \text{(IV.209)}$$

That work is of course exactly equal to the kinetic energy the particle would assume by falling from infinity to $r$, which in turn is exactly the kinetic energy required for the object to escape from $r$ to infinity:

$$E_{\text{esc}} = \tfrac{1}{2}mV_{\text{esc}} = GMm/r, \qquad \text{(IV.210)}$$

or

$$V_{\text{esc}} = (2GM/r)^{1/2} = (2\mu/r)^{1/2}. \qquad \text{(IV.211)}$$

We recall from Eq. (III.31) that a body in circular orbit has an orbital speed

$$V_{\text{circ}} = (\mu/r)^{1/2} = V_{\text{esc}}/2^{1/2} \qquad \text{(IV.212)}$$

The escape speed is then $4.22 \times 10^6 r^{-1/2}$ cm s$^{-1}$. The escape energy per unit mass is $8.9 \times 10^{12} r^{-1/2}$ erg g$^{-1}$, from which we see that each gram of the T Tauri solar wind carries enough energy for the dissipation of about 100 g of nebular material from 1 AU to infinity. Only about 5% of this energy is directed at the inner edge of the nebular disk, but, if this energy were used with perfect efficiency to remove gas, it could sweep out 0.2 solar masses of gas and entrained solids. Nebular masses much less than this would be easy to remove.

Close to the Sun, out to perhaps 20 solar radii, the magnetic field strength is high enough to effectively tie the solar wind to the solar surface and force the entire plasma out to that distance to corotate with the Sun. The radial velocity at the surface of the corotation region is about 400 km s$^{-1}$ and the corotation speed is $2r_{\text{co}}/t_0$, where $r_{\text{co}}$ is the distance from the Sun and $t_0$ is the solar rotation period. For $r_{\text{co}} = 20R_\odot$ we have $v = 40\pi R_\odot/t_0$. One of the main effects of such a dense solar wind is the radial transport of angular momentum and hence the slowing of the Sun's rotation. The corotation speed at $20R_\odot$ is of course 20 times the speed of rotation of the surface of the Sun, and the angular momentum per unit mass of material exiting from the corotation region is $v_{\text{co}}r_{\text{co}} = 20v_\odot \times 20R_\odot$, or 400 times that of the equatorial photosphere of the Sun itself. Loss of less than $0.01M_\odot$

of this material would be sufficient to essentially despin the Sun. If the Sun formed with an equatorial rotation rate typical of early spectral class (O, B, and A) MS stars, we can see from Fig. II.16 that the equatorial surface rotation speed might have been 50 to 500 km s$^{-1}$. The escape speed from the surface of the present Sun is

$$V_{esc} = (2GM_{\odot}/R_{\odot})^{1/2} = 618\ \mathrm{km\,s^{-1}}. \quad \text{(IV.213)}$$

For an early T Tauri phase Sun of 1.2 solar masses, $3L_{\odot}$, and the same temperature as the present Sun, we can estimate $R = 1.7R_{\odot}$. The escape velocity would then be 520 km s$^{-1}$. A rotation speed greater than 300 km s$^{-1}$ would involve great geometrical distortion and be on the verge of rotational fission. A maximum rotation speed on the order of 100 km s$^{-1}$ is more reasonable.

Another important consequence of the removal of the nebula is the changeover of the thermal environment of solid bodies from an adiabatic gas ($T = cr^{-1.1}$) to radiative control by the Sun. A rotating solid body of radius $a$ will intercept solar energy with a flux of $L_{\odot}/4\pi r^2$ with a cross-section area of $\pi a^2$. A fraction ($A$) of the incident light will be reflected back into space, and the remainder will be absorbed and eventually reemitted as thermal black body radiation from the entire surface of the body, $4\pi a^2$. At steady state,

$$(1 - A)(\pi a^2 L_{\odot}/4\pi r^2) = 4\pi\varepsilon a^2\sigma T^4, \quad \text{(IV.214)}$$

where $\varepsilon$ is the thermal emissivity of the body (usually close to 1.0) and $A$ is the Bond albedo. It can then be easily seen that

$$T = [(1 - A)L_{\odot}/16\pi\varepsilon\sigma]^{1/4} r^{-1/2} \propto r^{-1/2} \quad \text{(IV.215)}$$

which is a much weaker dependence of $T$ on $r$ than that found for the convective nebula.

Also, we saw that, in the presence of the nebula, the temperature was near 600 K at 1 AU. From Eq. (IV.215) we see that, in the presence of a T Tauri Sun of luminosity $3L_{\odot}$, a body of albedo 0.2 at 1 AU from the Sun would have a temperature of 347 K.

Figure IV.48 shows typical nebular and T Tauri phase temperatures vs distance from the Sun. Because the albedo depends on the composition of the solids, we have sketched in a schematic temperature dependence for radiative steady state between the Sun and likely local solids, using high albedos for refractory oxides and ices, but low albedos where metallic iron or iron oxides are important.

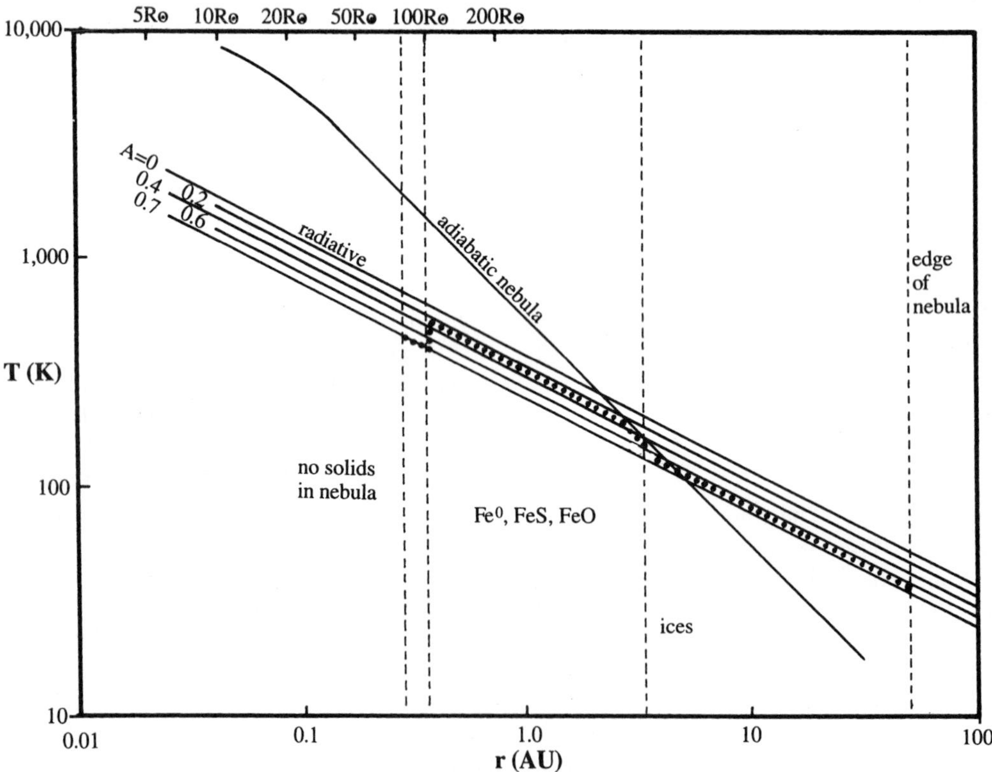


**Figure IV.48** Radial temperature profiles in the nebula and during the T Tauri phase of the Sun. A solar luminosity of 3 $L_{\odot}$ is assumed for the T Tauri phase. Note that, in the inner Solar System, removal of the nebula would cause cooling, whereas far from the Sun, the opposite would occur. A high albedo is used both for icy condensates and for refractory oxide minerals, whereas metal, sulfide, and iron-bearing mineral grains are assumed to be dark.

It is interesting that the entire inner Solar System would be dramatically cooled by dispersing the nebula and turning on the superluminous T Tauri phase! This is of course directly attributable to the dissipation of the thick insulating blanket of gases and dust that hindered radiative cooling while the nebula was in place. From about 4 to 10 AU the thermal effects of removing the nebula would not have been great, and beyond about 10 AU solar radiation would cause a significant increase in the temperature, perhaps to the serious detriment of any solid methane or solid argon that may have originally condensed. Their vulnerability to evaporation depends on the size of the solid bodies: particles can evaporate very readily, while planets with deep potential wells would persist.

## Thermal History of the Early Solar System

When one considers the complexity of the thermal environment in the solar nebula, with gravitational energy being turned into heat by collapse; with infall of interstellar gas and dust; with condensation, sedimentation, and accretion of solids; with temperature structure and energy fluxes dependent on opacity, which varies with position and time; and with the eventual ignition of the Sun, the T Tauri phase, deuterium burning, etc., it is astonishing that so simple an idea as equilibrium could work at all. We have wholly neglected variations of temperature with time, the luminosity history of the nebula, radial and vertical transport of particles, and many other effects as well. Was the nebula really so simple?

The answer can be only that it certainly was *not* simple, but it may be that, for certain crude uses such as estimating bulk composition and density, it is good enough. Also, for the purposes of an introductory text, it is much more important to teach the principles by which these processes can be modeled and understood than to attempt a final solution to all these difficult problems! A more stringent test of the model would involve the study of very ancient meteorites, many of which carry virtually unaltered evidence regarding physical and chemical processes in the nebula. Another test would be to examine closely the planetary inventories of highly volatile elements, which are sensitive indicators of radial or vertical mixing processes; can pure equilibrium condensation explain these abundances, or must long-range sampling be invoked? Both of these areas are complex and interesting and will be dealt with in Chapters VIII and X, respectively.

For the moment, there seem to be only a few weak generalizations about the thermal history of the nebula that we can assert with reasonable certainty. The temperature at any heliocentric distance first rose during the period of most rapid collapse and then cooled preparatory to or in concert with the dissipation of the nebula, over a time of about $10^5$ years. The interplanetary medium was then swept with a powerful T Tauri phase solar wind lasting about $10^7$ years, which gradually subsided into an early MS phase with luminosity about 20% lower than that at present. During the T Tauri phase the Sun was highly luminous in the ultraviolet. The local thermal history at particular points in the Solar System may have been considerably more complex than this, and the genealogy of the grains that end up in a particular planet, after a gas-free accretionary era of about $10^8$ years, may be even more complex again. What, then, is the significance of the "formation temperatures" deduced from the chemical compositions of solid bodies?

We have emphasized that the equilibrium condensation model is path-independent and ascribes a single temperature to each distance from the Sun. This makes the model tractable, and it turns out to work surprisingly well. The nonhomogeneous accretion model we explored is just one, perhaps the simplest, of the path-dependent models. It is surprising that this alternative model works so poorly. We must suppose that the temperatures in the nebula passed through a well-defined maximum in the presence of the full solar complement of gas and prior to any substantial accretion of large ($>10$ m) planetesimals. The chemistry at each distance from the Sun then reflects most closely the conditions at that time when *the rates of chemical reactions were at a maximum for that location*. Either subsequent alteration at lower temperatures was very brief or it failed to leave a substantial imprint due to the extreme sluggishness of reactions at lower temperatures. This argument would suggest that sulfidization and hydration of grains in the nebula, if unimportant in the cooling stages of the dense inner nebula, might have been even less important in the much less dense gas farther out. But if formation of sulfides, hydroxyl silicates, FeO, etc., is not important farther from the Sun, how do we explain the observed densities of Mars and the asteroids? There are two possible answers. The presolar grains were highly disordered, glassy particles with substantial stored potential energy, which were highly reactive in the nebula. The second possibility is that the oxidized and volatile-rich minerals did not need to be formed in the nebula because they themselves are presolar; the nebular phase simply cooked and reequilibrated grains in the innermost, warmer regions inside the inner edge of the asteroid belt. In either case, there is no evidence that the nebula was ever hot enough for wholesale evaporation of the major rock-forming minerals anywhere outside the orbit of Mercury. We shall present isotopic evidence to this effect in our discussion of meteorites in Chapter VIII.

At this point in our survey, we may look back over the events that transpired from the time of the Big Bang through the era of formation of stars and galaxies, culminating in the condensation of solid preplanetary material in a stellar nebula and its accretion into planets. Figure IV.49, although it mentions only hydrogen, should evoke the similar journeys traveled by the other elements as well. From a universe populated only by photons, neutrinos, and energetic particles, we have seen the progressive emergence of leptons, baryons, hydrogen–helium plasma, neutral atomic gas, molecular hydrogen, and finally solids. Our pursuit of cosmology and nuclear physics gave us entry into the realm of atomic physics, then chemistry, and finally geochemistry. So far we have dealt almost exclusively with events and processes that affected at least the whole Solar System, even the whole Universe. We now can turn our attention to the component bodies of our planetary system. We shall begin with those that we are currently best equipped to face: the Jovian planets, with their nearly Sun-like composition. In so doing, we may reencounter at a deeper level of understanding many ideas already introduced qualitatively or semiquantitatively. Perhaps some day we may know enough to confront our own planet. . . .

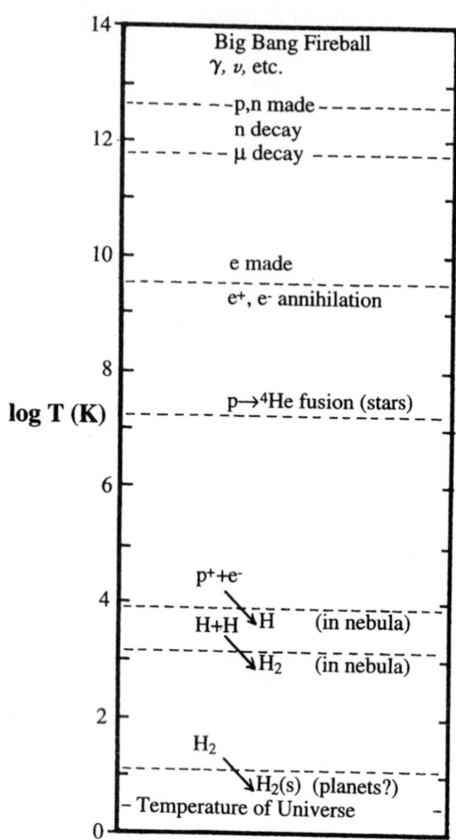


**Figure IV.49** Hydrogen: from the Big Bang to stars and planets. The history of the dominant element in the Universe is sketched out to serve both as a parable for the others and as a reminder of the threads linking the previous chapters of this book.

## Exercises

### *Energy Production in the Sun*

IV.1 If a star's composition differed from that of the Sun only in that it contained absolutely no carbon, the star is not precluded from fusing hydrogen via both the pp chain and the catalytic carbon cycle. Explain why.

IV.2 Suppose a detector containing a million gallons (call it 3.6 million liters) of carbon tetrachloride detects one neutrino per month derived from the decay of $^8$B in the Sun. How many neutrinos pass through the tank each second without reacting?

### *Energy Transport in the Sun*

IV.3 Calculate the ionization potential of the $C^{5+}$ ion. This ion is present in the Sun's corona. Tables IV.4 and V.2 contain some useful data on a number of other species.

IV.4 Is Thomson scattering quantized or continuous?

IV.5 A chemically homogeneous, nonrotating, self-gravitating fluid sphere happens to have a density profile that can be rather accurately described by the empirical equation $\rho = \rho_c(1 - r/r_s)$. The density outside the surface radius of the body ($r > r_s$) is everywhere zero. What are the pressures at the surface of the body and at its center?

IV.6 In the derivation of Eq. (IV.53), the effect of protons has been neglected. How large a change in the Thomson scattering cross-section would result from including protons along with electrons in the derivation?

### *Radio Wave Propagation in Space Plasmas*

IV.7 Using Equation (IV.64), calculate the lowest frequency that could be used by a radio transmitter on a Venus entry probe to send data back to Earth. Estimate the peak electron density in the upper atmosphere of Venus by reference to Fig. X.35.

### *The Solar Wind*

IV.8 Show by *dimensional* arguments that the energy density of a gas (erg cm$^{-3}$) is equivalent to the pressure exerted by the gas (dyn cm$^{-2}$).

IV.9 The pressure of sunlight, like that of the solar wind, depends on the energy density:

$$P_{\text{light}} = E/V = F_{\odot}/c.$$

Compare the pressure of sunlight to the solar wind dynamic pressure at Earth's orbit.

*Ionization*

IV.10 Show how to include the $H^-$ ion in Eqs. (IV.88f), (IV.89a), and (IV.89b).

*Hydrogen and the Rare Gases*

IV.11 What is the average molecular weight of a mixture of helium and molecular hydrogen in a cosmic (solar) composition gas?

IV.12 Figure IV.15 shows that the maximum chemical complexity of carbon compounds occurs at a rather well-defined temperature and pressure. List the four carbon species found in comparable abundance at that point, and give its pressure and temperature.

IV.13 The line describing the partial pressure of atomic H in Fig. IV.16 has a constant slope. Derive the equation for that slope.

*Magnesium, Silicon, and Iron*

IV.14 Several authors have proposed that processes in the early solar nebula may have caused selective transport of water. Qualitatively indicate what changes in the condensation sequence described in reactions (IV.118) to (IV.134) might occur as a consequence of

a. a tenfold enhancement of the water abundance above solar,
b. a tenfold depletion of the water abundance.

*Iron and Sulfur*

IV.15 As a test of your understanding of these sections, sketch out the course of condensation behavior of iron and sulfur in a simplified chemical system that contains hydrogen, helium, carbon, nitrogen, oxygen, neon, iron, and sulfur in their solar proportions, but no silicon or magnesium. How would Fig. IV.19 be changed?

IV.16 Assuming a solar composition gas at temperatures low enough for tremolite to form, what is the maximum number of moles of tremolite that could be formed per mole of silicon?

*Geochemical Classification of the Elements*

IV.17 a. What are the liquid phases that can be condensed out of a solar-composition gas at pressures of less than 100 bars?
b. Under what circumstances might a liquid phase condense in a gas of solar elemental abundances?
c. Could this happen in the Solar Nebula?

*The Chemistry of Rapid Accretion*

IV.18 To follow the equilibrium chemical behavior of the elements in a system of solar composition, we have, for didactic purposes, presented the results in order of decreasing temperature, starting at high temperatures at which no solids are present, and ending at low temperatures with a complex mineral assemblage. But equilibrium calculations are rigorously independent of path: the calculated mineral assemblage at any given temperature–pressure point is a state function, unaffected by whether the system was previously warmer, previously colder, or previously at a different pressure. If one mistakenly assumes that the entire mineral content of the early Solar System was in fact made by cooling and sequential condensation from a hot gas, then one would also mistakenly predict that the oxygen isotopes of all solids would be thoroughly homogenized. If, however, one assumes that preexisting low-temperature solids were partially equilibrated during a warm phase in the solar nebula, with the maximum temperatures and pressures experienced only close to the Sun, then equilibrium would be closely approached near the Sun, reflecting temperatures and pressures dropping off with heliocentric distance, but also that equilibration beyond the orbit of Mars would occur at temperatures too low for equilibrium to be closely approached.

Using the low-temperature rocky condensate composition given in step 8 of Table IV.7 as the raw material present throughout the inner Solar System, describe how the products of partial equilibration might depend on distance from the Sun, and assess whether such a process might help satisfy the constraints mentioned in the section on Kinetic Inhibition.

*Mass and Density of the Solar Nebula*

IV.19 Note that slope of the nebular surface density plot in Fig. IV.33 could be affected by the presence of a large undiscovered mass at or beyond Pluto's orbit.

a. We would probably have already detected a Uranus-like planet as far out as 200 AU. What would be the effect of the discovery of such a body on the slope?
b. How many 100-km ($10^{21}$ g) icy bodies in roughly circular orbits just beyond Pluto would be necessary to cause us to change the slope of the diagram? *Note*: This region is referred to as the "Kuiper Belt."

*Thermal Opacity in the Solar Nebula*

IV.20 Consider a uniform sphere with a radius of 4 cm and a density of $1.0\,\mathrm{g\,cm^{-3}}$ rotating at 100 Hz. What is its rotational quantum number $J$?

IV.21 The heat of dissociation of molecular hydrogen is about 100 kcal per mole. If most of the dissociation occurs, as in Fig. IV.39, over a 500 K range of temperature, how large is the inflection of the apparent heat capacity (the dashed line)?

*Dust Opacity*

IV.22 Explain clearly in words why several-micrometer iron particles provide far more thermal opacity than the same mass of iron distributed as either 0.1-$\mu$m or 10-cm particles.

*Thermal Structure of the Nebula*

IV.23 The adiabatic temperature profile presented in Fig. IV.42 completely neglects radiative transport of heat, which is most important at low pressures (i.e., low opacities). Sketch qualitatively how the temperature profile might look for a fully *radiative-convective* model that takes proper account of both forms of heat transport.

*Turbulence and Dust Sedimentation*

IV.24 An auditorium 10 m in height, width, and depth is sealed and maintained in a perfectly isothermal state. A plastic bag of air spiked with a trace of radiocarbon (in the $CO_2$ component of the air) is gently opened by remote control at the front of the auditorium. The diffusion coefficient can be taken as about $10^3\,\mathrm{cm^2\,s^{-1}}$.

a. What is the time scale for molecular diffusion of the radioactive gas to the back of the room?

b. If the air in the room were being circulated by fans, what would be the required mean turbulent wind speed for eddy diffusion to beat molecular diffusion in dispersing the radioactive gas?

IV.25 Volcanic dust particles with radii of 0.05 $\mu$m are injected into Earth's stratosphere at an altitude where the atmospheric pressure is 0.1 bar. What is the mean sedimentation rate of these particles?

IV.26 Solid bodies must accrete to significant size in order to survive the dissipation of the nebula. Reasoning from the data in Table IV.8, estimate how large solid bodies near Earth's orbit must have been in order to avoid being blown away during the dissipation phase of the nebula.

*Gas Capture from the Solar Nebula*

IV.27 Adiabatic capture from a continuum background gas is a very marginal process, since removal of the background gas (in the present context, dissipation of the solar nebula) impacts the structure of the entire atmosphere. Discuss this issue in light of Eqs. (IV.202) through (IV.205), and indicate why the time scale for loss of the nebula is an important factor.

*The T Tauri Phase*

IV.28 According to Fig. IV.48, removal of the solar nebula by the T Tauri phase of the Sun will expose the inner planetary region to the high-luminosity early Sun, resulting in a *decrease* in the surface temperatures of solid bodies in that region. Either planet-sized ($10^{26}$ g) bodies have accreted by that time in the inner planet region, or they have not. Suppose that, for example, only 10-m bodies were present in the Mercury region (0.4 AU from the Sun). How long would it take for them, radiating at their nebular temperatures, to cool to their T Tauri steady-state temperature?

# V. The Major Planets

## Introduction

The outer Solar System presents an astoundingly diverse panorama. The four giant planets subdivide naturally into two classes, Jovian and Uranian. The Jovian planets, Jupiter and Saturn, which are not very far from the composition of the Sun or of other Population I stars, together have more than 100 times the combined masses of the terrestrial planets. The Uranian planets, Uranus and Neptune, are far denser than solar material and present strong but ambiguous evidence regarding large-scale fractionation processes in the outer regions of the solar nebula. Comets, which spend nearly all their lifetimes in interstellar space, may be the least altered and most ancient material in our Solar System. The Asteroid Belt, which defines the boundary between the inner and outer regions of the Solar System, contains surviving rubble from a planet or planets that failed to grow, but instead was ground into dust and boulders. Cometary meteors and asteroidal meteorites provide samples for study on Earth, permitting the application of sophisticated analytical techniques to samples from bodies which have not yet been visited by spacecraft. Comets and asteroids, as small bodies, may preserve evidence of conditions and processes during the time of the formation of the Solar System, unaltered by thermal activity since that time. The Centaurs and trans-Neptunian objects provide important clues to chemical and dynamical processes in the early Solar System, including the interrelationships among these "asteroidal" bodies, short-period comets, and the ice-bearing satellites of the outer planets.

Each of the four giant planets is accompanied by a satellite system of some complexity. Those of Jupiter, Saturn, and Uranus are reminiscent of small Solar Systems, while that of Neptune bears witness to profound and catastrophic evolutionary changes. All four of these giant planets also have at least rudimentary ring systems. Pluto, about which little was known until recently, and the only planet in the Solar System not yet visited by a spacecraft, is odd in a number of ways and seems easier to relate to the satellites of the outer planets than to the planets themselves.

The satellites of the outer planets are extremely complex and interesting bodies in their own right and are of great intrinsic interest. Some of them are so large that, if they orbited directly around the Sun, we would not hesitate to call them planets. At least fifteen of these satellites are larger than the largest asteroid, 1 Ceres. Many of these satellites have such low densities that ices must be a major constituent of their interiors. The comparative study of the satellites is potentially as difficult (and as rewarding) as the study of the planets themselves. There are 10 times as many known satellites as planets in our Solar System, representing a wide variety of types. We must in addition consider the ring systems of Jupiter, Saturn, and Uranus, which present almost all the structural properties of the asteroid belt, plus a number of other traits peculiar to themselves.

As with our initial approach to the study of the Solar System, we shall begin with an overview of the origin and global properties of the Jupiter and Saturn systems. The detailed treatment of the giant planets will occupy all of Chapter V. We shall devote Chapter VI to Pluto and the satellites of the outer planets. The next two chapters are dedicated to the study of small, primitive solid bodies: Chapter VII is devoted to comets and meteors, and Chapter VIII to asteroids and meteorites. The Centaurs and trans-Neptunian objects, including the Kuiper Belt, have such strong affinities with Pluto, large icy satellites, comets, and the most distant asteroids that they might sensibly be relegated to Chapter VI, VII, or VIII. So that we may have the broadest perspectives on these bodies, we shall delay discussing them until Chapter VIII, where we will have the benefit of the relevant satellite, comet, and asteroid lore from Chapters VI–VIII.

The main themes of Chapter V are the internal composition and structure of the giant planets, the structure, composition, and motions of their atmospheres and clouds, the effects of ultraviolet sunlight and lightning, and the nature of their magnetospheres. This discussion draws heavily from the results of the *Pioneer 10*, *Pioneer 11*, *Voyager 1*, *Voyager 2*, *Ulysses* and *Galileo* missions. Where our present understanding is poor, we shall emphasize principles and survey the available data.

## Interiors of Jupiter and Saturn: Data

There are several important sources of information regarding the internal composition, structure, and dynamics of the Jovian planets. Among these are the masses, radii, and densities of the planets; the oblateness of their disks; their rotation periods; their internal mass distributions (departure from spherical symmetry) as deduced from studies of the orbits of their natural satellites and planet-orbiting spacecraft (the *Galileo Orbit*er), and the trajectories of flyby spacecraft; their external magnetic fields as studied by Earth-based radio astronomy and by *in situ* spacecraft; and the elemental, isotopic, and molecular composition of their atmospheres deduced from both remote sensing and the Galileo atmospheric entry probe.

Historically, the abundances of several of the major components of the atmospheres of the giant planets gave us crucial insight into the possible nature of their interiors. Before the era of spacecraft exploration of the Solar System, compositional information on these planets was most easily obtained via spectroscopic studies of the molecular constituents of the outer fringes of their atmospheres.

In the case of both Jupiter and Saturn, the penetration of visible and near infrared radiation into their atmospheres is blocked at pressures of about 1 bar by one or more dense layers of clouds. Rupert Wildt, working at Göttingen in the 1930s, was the first to show that the visible reflection spectrum of Jupiter could be explained as being dominated by "combination bands" of methane and ammonia. The vibrational–rotational fundamental frequencies of these molecules lie in the infrared, with energies far too small to be seen in the visible part of the spectrum, but multiple simultaneous transitions in which several such transitions are excited by a single energetic photon can occur. Such "combinations" are quite rare in a tenuous gas and generally can be observed readily only when there is a vast amount of absorbing gas in the line of sight. Wildt thus concluded that there must be immense amounts of methane and ammonia in the atmosphere of Jupiter and that methane must be enormously abundant on all the giant planets.

Wildt, however, did not stop there. He pointed out that both methane and ammonia are readily destroyed by ultraviolet radiation, yet both are still present in the outer Solar System. Because the only feasible way to restore the methane and ammonia seemed to be equilibration with large amounts of hydrogen at high pressures (deep within the planetary atmospheres), Wildt postulated that there must be deep, convective, hydrogen-rich envelopes on all the Jovian planets. The Nobel-prize-winning Canadian spectroscopist Gerhard Herzberg later confirmed the presence of very large amounts of hydrogen on all of the Jovian planets. By 1952, Gerard Kuiper of the University of Chicago proposed that the main cloud layer on Jupiter was made of tiny crystals of solid ammonia and hence that ammonia had been partially depleted by condensation in the portion of the atmosphere observable from above. Although our present knowledge of the atmospheres of Jupiter, Saturn, Uranus, and Neptune is vastly superior to that available at that time, all of these fundamental conclusions have survived the test of time.

That atmospheric observations had profound implications for our understanding of the interiors of these planets was first pointed out by Wildt in 1934. He showed that bulk planetary compositions very rich in hydrogen were apparently required to explain the low densities of Jupiter and Saturn and proposed that both the atmosphere and the interior might have the same composition. That composition might very well be the same as that of the Sun. This possible equality of atmospheric and bulk composition meant that careful quantitative interpretation of the spectra of their atmospheres might provide input data for theoretical models of their interiors.

As high-resolution infrared spectroscopy advanced during the 1960s and 1970s the abundances of the major elements H, C, and N were determined in the Jovian atmosphere. The best determinations of the solar H:C ratio are near 2700 (Table II.4), whereas the latest estimates of the $H_2$:$CH_4$ ratio on Jupiter and Saturn lie in the range from 450 to 1400 (H:C = 900 to 2800). Thus the carbon abundance could be solar or could range up to as much as three times solar. Independent interpretations of the Jovian IR spectrum give methane to hydrogen ratios that range over a factor of three because of uncertainties in the optical models of the Jovian atmosphere that are used to interpret the spectra, not because of major differences in the observations themselves.

The ammonia abundance above the Jovian cloudtops, as determined by infrared spectroscopy, gives an N:H ratio that is several times lower than solar. This is of course expected if the clouds contain solid ammonia, with the gaseous ammonia abundance above the clouds limited by its vapor pressure at these low temperatures (below 120 K). As we would also expect, Saturn, which is farther from the Sun and colder than Jupiter, shows much less ammonia vapor above its cloudtops. There is no direct infrared detection of ammonia vapor on the even colder Uranus and Neptune.

If the elemental abundances in Jupiter and Saturn are in fact close to those in the Sun, then the second most abundant element in these planets, both by number and by mass, must be helium. The helium atom of course has no vibrational or rotational features in its spectrum. Direct detection of electronic features due to helium is possible only in the far ultraviolet, a wavelength region in which Jupiter's atmosphere is a very strong absorber and Earth's atmosphere is completely opaque.

With the advent of spacecraft flybys of Jupiter (Pioneer 10 and 11, Voyager 1 and 2, and Ulysses) it became possible to determine the helium abundance in the upper atmosphere of Jupiter. The measured H:He ratio, deduced from the effects of helium–hydrogen collisions on the absorption of long-wavelength infrared radiation, is $17.2 \pm 2$ for Jupiter and $32 \pm 8$ for Saturn, compared with the solar value of 14.4. Jupiter's atmosphere is thus probably slightly depleted in helium, whereas Saturn's atmosphere is apparently depleted, relative to the Sun.

Two other elements are represented by tiny traces of gases in Jupiter's atmosphere: phosphorus, a component of the gas phosphine, $PH_3$, and germanium, in the gas germane, $GeH_4$. The observed abundances of both elements are close to their solar abundances. These observations, like the others detailed above, probe to only rather shallow depths in the Jovian atmosphere, to levels where the pressures do not exceed a few bars. These and other minor and trace constituents of the atmospheres of Jupiter and Saturn have chemical and dynamical interest and will be discussed later in this chapter.

Is there any other kind of compositional information that we might be able to derive from remote sensing observations? Because we are able to observe only the outermost, coldest, and most rarefied fringes of the atmosphere, we are limited to observing only highly volatile, relatively abundant species.

As we can recall from our discussion of the equilibrium chemistry of solar material in Chapter IV, the only species that remain uncondensed at the saturation temperature of ammonia are hydrogen, methane, and the noble gases. From our discussion of light absorption by gases, we have seen that the noble gases, being monatomic, can absorb light only by electronic transitions. These transitions require so much energy that they lie far in the ultraviolet region of the spectrum, at wavelengths far too short to penetrate the Earth's atmosphere. Hydrogen, as a symmetrical diatomic molecule, can rotate and vibrate quite freely, but interacts very poorly with radiation because the rotational and vibrational excitations are formally forbidden because of the molecular symmetry. A temporary dipole moment may be induced in a hydrogen molecule during a collision with another molecule, and thus $H_2$ becomes a good absorber only at high gas densities, at which collisions are frequent. Thus, at the level of our present knowledge, the composition expected for a solar composition atmosphere at the condensation temperature of ammonia is fully in accord with our meager observational data on Jupiter. Saturn departs most strikingly from Jupiter in its enhanced abundances of helium (about twofold) and phosphine (about sixfold).

Aside from spectral data, there are several other lines of evidence that bear on the physical state of these atmospheres. We have, for example, already seen how the relative intensities of the rotational lines in a near infrared vibration–rotation band permit us to determine the temperature of the absorbing gas and how the shapes of the individual absorption lines contain information about the gas pressure.

In addition, we may make use of the long-wave thermal emission from the planet to determine its effective temperature. For a black body emitter, the effective temperature will be independent of wavelength: $B_\lambda$ is Planckian. However, real objects (and especially gases) deviate from black body behavior in complex and informative ways. Thus a spectrum of $B_\lambda$ vs $\lambda$ or $T$ vs $\lambda$ will convey useful data on thermal structure and on the identity and abundance of specific absorbing and emitting species.

The "bolometric" (integrated over $\lambda$) emission from both Jupiter and Saturn is very interesting. In principle, if a planet is in thermal steady state with the Sun's radiation, we should be able to calculate the planet's temperature knowing only the fraction of incident light absorbed and the distance of the planet from the Sun.

The flux of energy from the Sun at 1 AU, 1.98 cal cm$^{-2}$ min$^{-1}$, is called the solar constant, denoted $F_{\odot,\oplus}$. In more useful units, the flux at distance $r$ from the Sun is

$$F_{\odot,r} = 1.375 \times 10^6/r(\mathrm{AU})^2 \ \mathrm{erg\,cm^{-2}\,s^{-1}}. \qquad \text{(V.1)}$$

Denoting the fraction of the total incident energy (integrated over all wavelengths) that is reflected back into space as the albedo ($A$), the amount of energy received by a planet of radius $a$ is $(1 - A)\pi a^2$. If the planet attains a steady state with the Sun, then the total amount of energy absorbed per second will be equal to the total amount of thermal emission per second over the entire surface of the planet,

$$F_{\odot,r} = (1 - A)\pi a^2 = 4\pi a^2 \ \sigma\varepsilon T^4, \qquad \text{(V.2)}$$

where $T$ is the mean temperature of the planet and $\varepsilon$ is the emissivity. If the emissivity is not known, we can let $T_e^4 = \varepsilon T^4$, where $T_e$ is the effective temperature of the planet, the temperature a perfect black body would have to have in order to radiate the observed total thermal flux. We then find that

$$T_e = [1.38 \times 10^6(1 - A)/4\sigma r^2]^{1/4}$$
$$= 280[(1 - A)/r^2]^{1/4}. \qquad \text{(V.3)}$$

The visible and near-infrared reflection spectrum of Jupiter is well known, and the fraction of all incident sunlight that is reflected is found to be $A = 0.44$, which gives $T_e = 106$ K for Jupiter. For Saturn, $A = 0.62$, $r = 9.539$ AU, and thus $T_e = 71$ K.

However, the observations of Jupiter and Saturn in the thermal infrared have a very different story to tell, as Fig. V.1 shows. The effective temperature is observed to be 126 K over a very wide range of wavelengths, and the planet is not at any wavelength as cold as 106 K! The turn-up in brightness temperature below 5 $\mu$m is simply due to reflected sunlight, but the large emission spike, with effective temperatures of 220 to 310 K at 5 $\mu$m, is definitely intrinsic to the planet. Hydrogen, methane, and ammonia are all very weak absorbers at this wavelength, and this narrow spectral region may represent a "transmission window" within which we can see unusually deeply into the atmosphere. At long wavelengths,

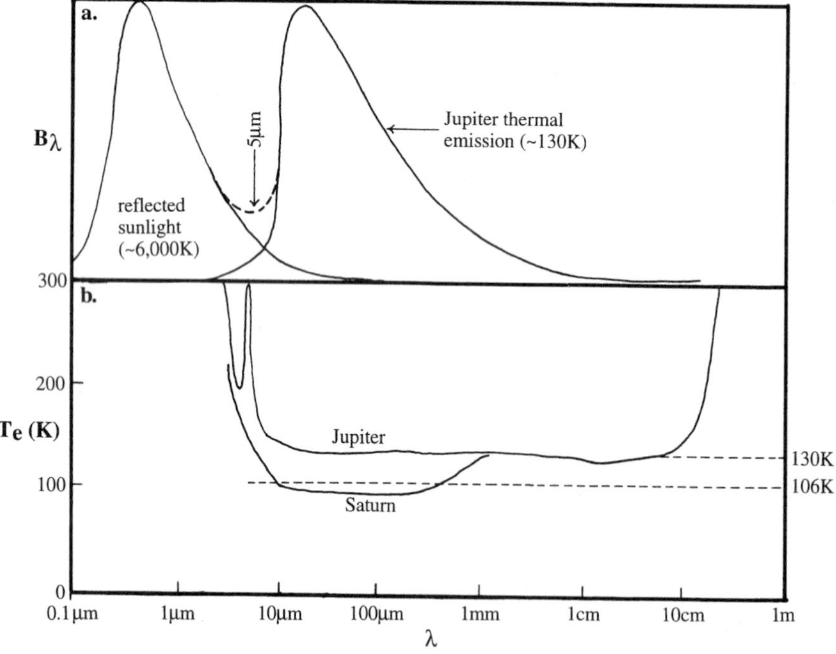


**Figure V.1** Reflection and emission spectrum of Jupiter. a shows the smoothed reflection spectrum and thermal emission spectrum of Jupiter on a plot of log $B_\lambda$ vs log $\lambda$. The peak of the reflected intensity lies near the Planck peak for the Sun, whereas the thermal emission peak lies near the Planck peak for a body with the temperature of Jupiter's cloud tops (i.e., at a wavelength of about 20 $\mu$m). b illustrates the effective temperatures of Jupiter and Saturn as functions of wavelength. Note the 5-$\mu$m "window" on Jupiter and the rapid turnup of brightness temperatures beyond about 10 cm.

ammonia is a good absorber near 1.3 cm, but all the gases expected to be present in a solar-composition mixture quickly become transparent at $\lambda > 3$ cm. The region from 5 $\mu$m to 1 cm is full of highly collision-broadened rotational and vibrational bands and is quite opaque.

The emitted thermal flux from Jupiter is $\sigma T_e^4 = 1.6 \times 10^4$ erg cm$^{-2}$ s$^{-1}$, and the thermal component due to the reemission of absorbed solar energy is $\sigma(106)^4 = 0.7 \times 10^4$ erg cm$^{-2}$ s$^{-1}$. Thus the total energy output of Jupiter is 2.3 times as large as the amount of energy received from the Sun. This observation shows that Jupiter has a thermal net luminosity of $5.6 \times 10^{24}$ erg s$^{-1}$. Coupled with the evidence in the 5-$\mu$m window and the centimeter wavelength region for substantially higher temperatures not far beneath the cold (130 K) cloudtops, we can see that we have several strong constraints on the thermal structure of the Jovian atmosphere.

Our observations of the thermal emission from Saturn tell a closely similar story. Over the wavelength range nearest the Planck peak, Saturn has an effective temperature of 95 K. The total emitted flux is therefore $(95/71)^4 \approx 3.2$ times the reemitted solar flux. The total flux is $\sigma T_e^4 = 4600$ erg cm$^{-2}$ s$^{-1}$, of which 1400 erg cm$^{-2}$ s$^{-1}$ is due to the Sun and 3200 erg cm$^{-2}$ s$^{-1}$ to an internal heat source.

One powerful constraint is that the atmospheric structure must be capable of transporting the large observed flux upward through the atmosphere. Another is that the high 5-$\mu$m and centimeter-wavelength brightness temperatures must be accounted for by a reasonable model for the altitude and wavelength dependence of opacity. Thus these observations give us some constraints on the thermal structure of the atmosphere and on dynamical heat-transport processes in it. Because of the importance of dynamics, we should also be cognizant of the voluminous literature on the motions of cloud systems on Jupiter; there are many strong indicators of rapid overturn of cloud features, even those the size of North America, over a time scale of $10^6$ to $10^7$ s. Indeed, even the structures and colors of the major bands may change radically in a few weeks to months, and severe local changes may occur in a day or two. There is thus strong evidence for large-scale overturn of the Jovian atmosphere. The source of energy that drives these motions must be found, and it is logical to ask whether the large internal heat source of Jupiter might drive convection in the atmosphere at reasonable rates. Further, if convective mixing is indeed strong and rapid, is the planet nearly homogeneous in composition? If it is, then the spectroscopic data on the composition of the atmosphere are directly relevant to the deep interior as well. With these observational constraints in mind, we shall now turn to models of the interiors of Jupiter and Saturn.

We have already seen how to model the internal structures of adiabatic or isothermal ideal gas bodies of solar composition in hydrostatic equilibrium. Modeling objects as dense as planets, however, takes us far into the realm of nonideal behavior of gases. This disadvantage has associated with it a corresponding advantage: the densities become so high that the molecules are squeezed together against a very steep part of their potential energy curve and the molar volume accordingly becomes much less sensitive to temperature. We must accordingly connect the asymptotic low-pressure case of ideal gas behavior to the region of "simple" high-pressure behavior through an intermediate pressure regime in which the behavior of the fluid is strongly nonideal. Finally, we must be concerned with the effects of extremely high pressures in the cores of the Jovian planets, because pressures there are comparable to the maximum that can be withstood by the molecular orbitals of the bonding electrons in hydrogen. We must therefore develop some understanding of the equations of state of important planetary materials over a wide range of pressures in order to model the Jovian planets successfully.

## Isothermal Interior Models of Jupiter and Saturn

We shall begin the task of modeling Jupiter and Saturn by considering isothermal planetary structures at 0 K. This is of course a disastrously bad approximation for the portion of a planet that we can observe directly, but is useful in the deep interior, where the density is comparable to that of a solid.

The equation of state of molecular hydrogen at low pressures is that of the ideal gas law, $Pv = RT$, where $v$ is the molar volume of the gas, $V/n$. Hence,

$$P = RT/v = R\rho T/\mu. \qquad \text{(V.4)}$$

Real molecules deviate from the mathematically convenient infinitesimal hard spheres used in the derivation of the ideal gas law in that they have intermolecular potentials that vary in a complex way with distance. Figure V.2 illustrates the potentials for a mass point, a finite hard sphere, and a real molecule. Note that the potential for the hard sphere is discontinuous. The real molecule gets "stiffer" the closer one gets to the center because of the quantum-mechanical prohibition against particles with the same quantum numbers occupying the same space. The principal deficiencies of the ideal gas law are that it neglects both the volume occupied by molecules (which becomes important at densities close

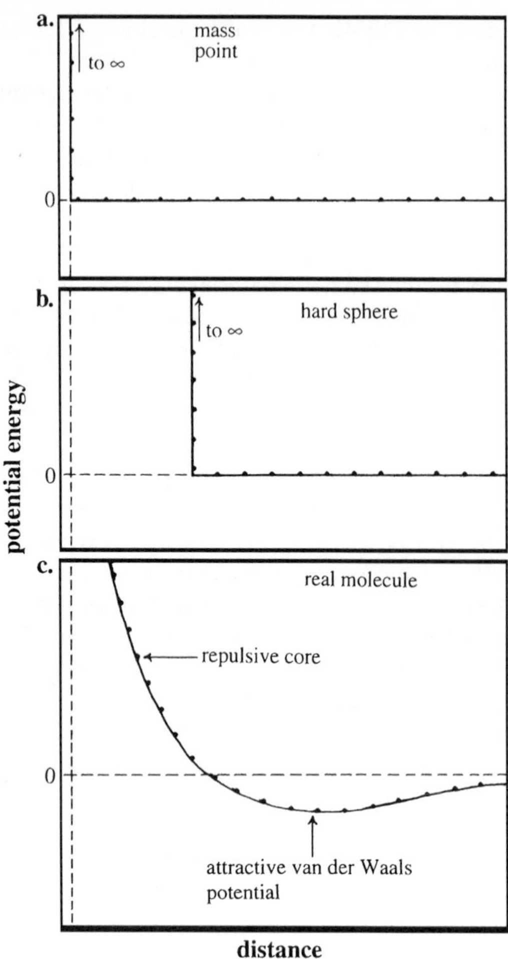


**Figure V.2** Intermolecular potentials. a shows the potential vs intermolecular distance $D$ for a mathematically idealized mass point with an infinite repulsive potential of zero extent, an approximation appropriate only for collisionless gases. b shows the potential for a hard sphere, with an infinite repulsive potential of finite extent, appropriate for modeling the behavior of ideal gases. c is a representative potential for real molecular collisions. The potential is weakly attractive at distances of a few Ångstroms (because of the polarizability of the molecules) and passes through a minimum near 1 or 2 Å, thus permitting the formation of stable solids and liquids at low temperatures. The potential becomes strongly repulsive because of electrostatic (Coulomb) repulsion of the nuclei at very small distances and reaches up to MeV energies for interatomic distances that place the nuclei in close proximity to each other.

to that of a liquid) and the "stickiness" of molecules at small separations from each other. The latter force, which causes the potential to be mildly attractive for spacings near 1 or 2 Å, is due to the finite polarizability of all real molecules: electric fields in one molecule induce oppositely polarized charges in its neighbors. These *van der Waals forces* permit the formation of liquids and solids.

A number of equations of state have been proposed to treat gases that have been compressed to high enough densities or cooled to low enough temperatures that these factors have a significant influence on the $P$–$v$–$T$ relationships. Perhaps the most famous is the van der Waals equation of state,

$$P = RT/(v - b) - a/v^2, \qquad \text{(V.5)}$$

where $a$ and $b$ are constants for a given gas.

This equation of state is useful for densities up to those of a liquid, but its accuracy deteriorates very rapidly toward the high-density limit of its range. Dense, hot molecular hydrogen at pressures greater than the critical pressure and temperatures far above the critical temperature (usually termed *supercritical hydrogen*), and especially mixtures of dense molecular hydrogen with other gases, is not very well understood.

In the immediate vicinity of the critical point the van der Waals equation becomes very unreliable. In that region the Dieterici equation of state,

$$P = RT(e^{-a/vRT})/(v - b), \qquad \text{(V.6)}$$

is much more accurate, but this equation is not widely used because of its more complex algebraic form. A simpler equation, and one that offers significant improvement over the van der Waals formulation, is the Berthelot equation,

$$P = RT/(v - b) - a/Tv^2. \qquad \text{(V.7)}$$

Because of the importance of the critical point conditions $T_c$, $P_c$, and $v_c$ in these equations, it is often desirable to express the equation of state in terms of the *reduced variables* $\pi = P/P_c, \tau = T/T_c$, and $\phi = v/v_c$. The modified Berthelot equation,

$$P = (RT/v)[1 + (9/128\tau - 27/64\tau^3)\pi]$$
$$\{\pi = [128\tau/9(4\phi - 1)] - 16/3\tau\phi^2\}, \qquad \text{(V.8)}$$

permits somewhat better description of the gas behavior at the cost of greater complexity. Even more complex, less physical, and more accurate are the virial equation of state,

$$Pv = RT(1 + b/v + c/v^2 + d/v^3 + \cdots), \qquad \text{(V.9a)}$$

and the Beattie–Bridgeman equation,

$$Pv = RT + B/v + C/v^2 + D/v^3 + \cdots, \qquad \text{(V.9b)}$$

where $B$, $C$, and $D$ are themselves functions of $T$.

There are relatively straightforward and reliable techniques for calculating the equation of state of a very cold solid of "normal" density, and at some point in Jupiter's interior it becomes preferable to model the equation of state as that of a zero-temperature solid perturbed by elevation of the pressure and temperature rather than as a very dense nonideal gas very strongly perturbed by intermolecular forces. Because the large preponderance of the masses of Jupiter and Saturn is hydrogen and helium, helium and molecular hydrogen are clearly very important species.

Also, a massive body rich in hydrogen will undergo a phase transition at a few megabars ($1\,\text{Mbar} = 10^{12}\,\text{dyn}\,\text{cm}^{-2}$), in which molecular hydrogen is compressed to the point at which the electrons can enter into a metallic-type delocalized energy level, called a conduction band. The protons then form a regular lattice, similar to the lattice of any other metal, and the electrons form a completely delocalized electron gas that, as we have seen, is an excellent conductor of both heat and electricity. The metallic hydrogen layer begins to form at the center of a cold hydrogen body at a particular well-defined critical size. Unfortunately, our experimental knowledge of the location of the phase transition is poor, due to the inability of static high-pressure systems to maintain pressures of megabars. The transition to metallic hydrogen lies above 2.5 Mbar, so high that only shock compression by violent explosions can generate the necessary pressures and then only for times on the order of a microsecond. The best available estimates of the transition pressure under Jovian temperatures are near 3 megabars (300 GPa).

The natural energy scale for dealing with hydrogen and helium atoms is $e^2/a_0$, where $e$ is, as usual, the electron charge ($4.80 \times 10^{-10}$ esu) and $a_0$ is the Bohr radius. The atomic unit of pressure (force per unit area or energy per unit volume) is then $e^2/a_0^4$, which is $2.94 \times 10^{14}\,\text{dyn/cm}^2$, or 294 Mbar. Thus pressures of about this magnitude would be sufficient to crush the electron orbitals in atoms and molecules and delocalize the electrons, making hydrogen an excellent conductor of electricity. Ultimately, sufficiently high pressures can crush the proton–electron fluid to nearly the density of nuclear matter. Such extreme conditions are of fundamental importance in Big Bang cosmology and in the advanced stages of stellar evolution (neutron stars), but are not readily attainable in bodies the size of Jupiter.

An approximation of the central pressure corresponding to a given mass and radius can be simply had by the method we used for the Sun in Chapter IV:

$$P_c = GM^2/8\pi r^4. \qquad \text{(IV.41)}$$

Thus $P_c$ is approximately $1.4 \times 10^{13}\,\text{dyn}\,\text{cm}^{-2}$ (14 Mbar) for an ideal gas Jovian planet. With inclusion of a core made of material intrinsically denser than hydrogen and helium, a higher core density and higher central pressure would be expected. Nonetheless, it is clear that the central pressures of both Jupiter and Saturn are high enough so that metallization of hydrogen should occur. The case is rather different for Uranus and Neptune. Equation (IV.41) tells us that $P_c$ is proportional to $M^2/r^4$, so the first-order (lower limit) estimate of the central pressures of the Uranian planets is about 3 Mbar. The effect of changing our assumption of a compositionally uniform interior to a density-stratified structure with a core is to increase the central pressure several megabars more, but to exclude hydrogen from the vicinity of the center.

Thus by these crude arguments alone, we may expect the critical size for metallization of hydrogen to lie in the size gap between the Jovian and the Uranian planets. In order to be more precise, we would need to develop more detailed and complex models.

As the next step in this direction, let us now consider the results from calculations using equations of state for pure hydrogen at low temperatures. The radius of a cold pure-hydrogen planet as a function of its mass is shown in Fig. V.3. It can be seen that the largest cold hydrogen body that can be built has only a few times the mass of Jupiter. Beyond that mass, the central pressure becomes so large that all atomic orbitals are collapsed, and fermions (protons and electrons) pair their spins. Because each microscopic energy state of this proton–electron fluid is doubly occupied, the state is said to be "degenerate." Beyond about $5\,M_J$, the addition of $M$ grams of mass to the outer molecular hydrogen layer causes more than $M$ grams of hydrogen to become degenerate. The density of the degenerate core is very high, and the core occupies a negligible volume, so the entire planet shrinks. Note that these cold hydrogen models are not intended to be realistic per se; they merely help us grasp the principles that we shall later apply to more complex and more realistic models. For example, bodies above $70\,M_J$ will become stars, and it is clearly impossible to accrete a body of $10^{30}$ g while still keeping it at 0 K. The gravitational potential energy that must be released in collapse of a $10^{30}$ g body from infinite radius to radius $a$ is $GM^2/a = 10^{43}$ erg. Taking a mean molar heat capacity of $C_p = 5R/2$ and a molecular weight of 0.5 for hot hydrogen, this is enough energy to heat the planet to 25,000 K and involves compression to a mean density of only $0.7\,\text{g}\,\text{cm}^{-3}$, orders of magnitude less than that in a degenerate core.

Thus much higher temperatures are a logically necessary consequence of accretion. One important effect of higher temperatures is to excite fermions into higher-lying energy states and lift the degeneracy of the core material.

In Fig. V.3 we can see that the giant planets are all smaller than pure cold hydrogen planets of equal mass. They must therefore be composed of material that is intrinsically more dense than hydrogen, especially because we know that they are not at 0 K! Uranus and Neptune, though less massive than Jupiter and Saturn,

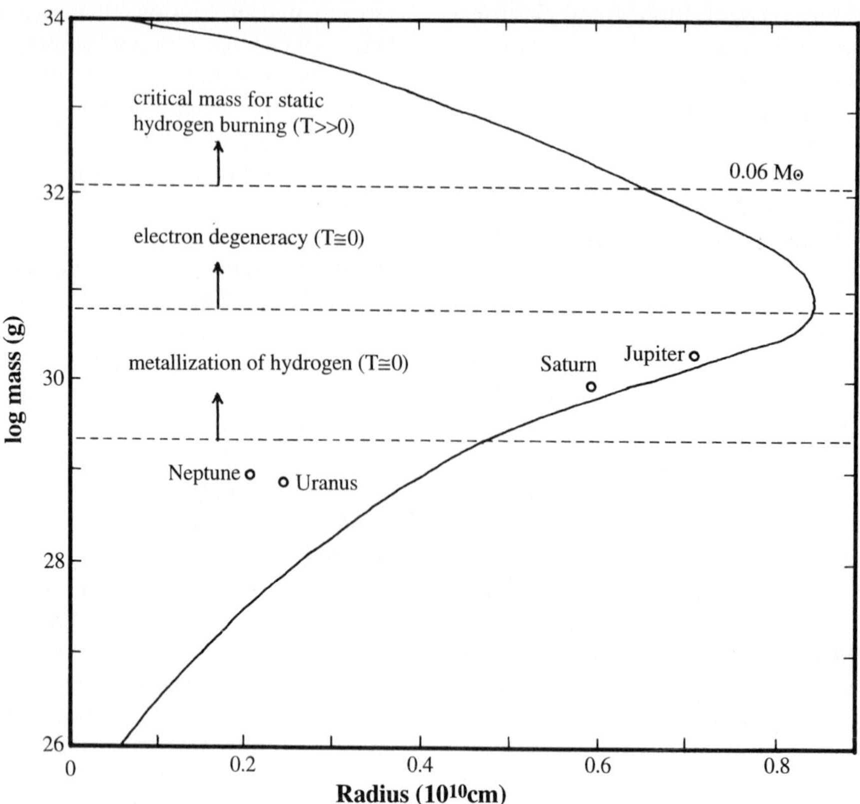


**Figure V.3** Mass–radius diagram for cold hydrogen. Spherical, nonrotating bodies of pure hydrogen show a monotonic increase of radius with mass up to approximately $10^{31}$ g and radii of about $9 \times 10^9$ cm. Beyond that size the central pressures are so high that they overcome the crushing strength of the hydrogen atom, and a very dense core of collapsed hydrogen with no distinct atoms or molecules begins to form. Bodies at nonzero temperatures would be larger (thermal energy helps resist compression) and hence lie to the right of the solid curve for pure hydrogen. The fact that the giant planets all lie to the left of the curve tells us that all of them must have mean atomic weights significantly heavier than pure hydrogen. Note that Jupiter and Saturn are much closer to the curve (and to pure hydrogen) than are Uranus and Neptune.

are denser. They therefore must be made of material with a higher density than that composing Jupiter and Saturn. The separation of the giant planets into two pairs, Jovian and Uranian, can be seen from Fig. V.3 to be valid for composition as well as mass and radius.

The next reasonable step to improve our planetary models is to use a solar-composition mixture rather than pure hydrogen. When this is done, it is found that the model planets become too dense and also exhibit a higher degree of central condensation of mass (lower rotational moment of inertia) than do Jupiter and Saturn. Bearing in mind these difficulties and the existence of strong internal heat sources in both planets, it seems clear that we must now include the planetary thermal structure and its heat transport mechanisms in the model.

There are at present several rather different approaches to modeling Jupiter and Saturn that emphasize different aspects of their origin, composition, or structure. We will qualitatively combine all these models so as to reflect as faithfully as possible our current level of knowledge.

## Thermal Models of Jupiter and Saturn

A necessary feature of any successful model of the Jovian planets is that it explain their observed thermal emission and account for the transport of heat from their deep interiors up to observable levels of their atmospheres. There are three possible mechanisms for transport of heat: radiation, conduction, and convection. We have already remarked on the enormous thermal opacity of dense hydrogen, and indeed it is clear from our thermal IR observations that these atmospheres become opaque at 10- to 25-$\mu$m wavelength at pressures of only 1 bar. Also, we have seen that this collision-induced opacity increases as the square of the total pressure.

Deep in the molecular hydrogen envelope of Jupiter the density is that of solid $H_2$ or higher. The heat flux conducted through any material is proportional to $K$, its thermal conductivity, and to the temperature gradient,

$$F_r = K(\partial T/\partial r), \quad \text{(V.10)}$$

where $K$ has dimensions of erg cm$^{-2}$ s$^{-1}$ K$^{-1}$. The thermal conductivity of solid hydrogen, which is a very good conductor, is $10^8$ erg cm$^{-2}$ s$^{-1}$ K, and the temperature gradient required to transport the observed surface heat flux of $0.9 \times 10^4$ erg cm$^{-2}$ s$^{-1}$ through the shell of $r = 0.5r$ is $F/K = 3.6 \times 10^{-4}$ K cm$^{-1}$, or 36 K km$^{-1}$. The temperature difference between the surface and the center of the planet would then be of order $\Delta T = rF/K$, or about $2 \times 10^6$ K! Thus conduction of heat through even solid hydrogen would yield starlike central temperatures, at which no solid could survive. It seems certain from this consideration alone that a very hot, convective fluid would be present instead of a nonconvective solid. Fortunately, we can calculate the size of the temperature gradient that would have to be present in the atmosphere of Jupiter in order for convection to take place. We have seen in Chapter IV that, assuming conservation of energy in a moving parcel of gas, hydrostatic equilibrium, and ideal gas behavior, the temperature gradient at which convection begins is the adiabatic lapse rate,

$$\partial T/\partial r = -\mu g/C_p. \quad \text{(IV.27)}$$

We already know the molecular weight $\mu$ and heat capacity $C_p$ of a solar-composition gas (Fig. IV.39). At levels in Jupiter at which the heat capacity is very large, $\partial T/\partial r$ will be a minimum. Because of the large pressures within a planet, the thermal dissociation of hydrogen is inhibited, and $C_p$ becomes 4.3 $R$ for the solar-composition mixture at 5000 K. The molecular weight, for the same reason, stays near 2.28, not 1.2. Then the adiabatic gradient will be approximately $-2.28g(r)/4.3\ R$, and, using half the equatorial surface gravity for $g(r)$, the vertical temperature gradient becomes $\partial T/\partial r = -0.7$ K km$^{-1}$. Thus convection will occur even for quite small values of the temperature gradient.

The theoretical equation of state of hot ($\approx$6000 K) metallic hydrogen is well approximated by

$$P\ (\text{Mbar}) = 9.95\rho^{5/3}(1 - 0.909\rho^{-1/3} + 0.614\rho^{-2/3} - 0.021\rho^{-1}). \quad \text{(V.11)}$$

Experiments with shock-wave production of metallic hydrogen suggest that this theory is probably good to 1 or 2% precision. By this approach, with detailed consideration of the thermodynamic properties of dense molecular hydrogen and metallic hydrogen, it is possible to conclude that the entire interiors of both Jupiter and Saturn must be convective. It also becomes possible to show that the central temperatures of both planets are roughly 100 times lower than that required in the conductive case (in which convection was arbitrarily omitted), approximately $10^4$ K rather than $10^6$ K. We therefore are now in a position to appreciate the value of fully adiabatic, solar-composition models of the interiors of Jupiter and Saturn.

To some degree our choice of models for investigation depends upon our conception of the mechanism of formation of the planet. If we prefer hydrodynamic gas capture onto a massive solid core, then we shall favor models of Jupiter with massive cores; if we favor gravitational instability of the nebula, then we shall prefer an almost completely gaseous or fluid planet with exactly solar composition. It now appears that the best available equations of state of molecular $H_2$, He, and metallic hydrogen are good enough so that uncertainties in the solar H:He ratio become important. Also, there is every reason to expect that the molecular $H_2$-to-metallic-H transition discriminates against incorporation of helium in the metallic phase; thus the question of the partitioning of helium between "atmosphere" (molecular $H_2$) and "mantle" (metallic H) arises. Finally, of course, there is the question of the disposition of the heavy elements, even in a solar-composition planet. A dense heavy-element core, once present, may be impossible to destroy because the large density difference between it and the overlying H/He mantle may make convective mixing across the interface impossible. Elements of moderate volatility, such as the ice-forming elements, may possibly be confined by certain accretion scenarios to the planetary core. But in other plausible formation scenarios these elements may be spread nearly uniformly throughout the planet.

From the best available equations of state and dynamical data on the density and rotational moment of inertia of Jupiter, it seems most likely that Jupiter has a dense core making up 3 to 4.5% of the planetary mass and that the metallic H "mantle" is completely melted and depleted in helium relative to solar material. The helium excluded from the mantle should enhance the He:$H_2$ ratio in the molecular envelope above the solar value. The values of the gravitational field harmonic coefficients $J_2$ and $J_4$, determined by tracking of the Pioneer 10/11 and Voyager spacecraft, seem to require a density appreciably higher than that of the usual recipe for solar material, in keeping with such a helium enrichment. However, this effect can be removed if the surface of the metallic hydrogen is hot enough, $>$9000 K. Then the solubility of He would become high enough so that it would not be excluded from the mantle and enriched in the envelope. One point of agreement among all models now available is the central temperature of Jupiter, which is close to 20,000 K. Because present estimates of the melting curve of metallic hydrogen lie so low, it seems certain that the entire metallic portion of Jupiter is melted. Thus Jupiter is in its entirety a fluid, convective planet with adiabatic structure.

A core of mass $>0.5\%$ of the planetary mass can be provided by the rocky material present in an exactly solar-composition planet. All the ice-forming and rock-forming elements combined would total 3% of the planetary mass. Yet our observations of a solar (or larger) abundance of methane and nearly solar abundance of ammonia in the atmosphere make it hard to accept that all the icy material could be buried in a core. Some enrichment of the heavy elements in Jupiter is very likely.

The structure of Saturn is somewhat less interesting because of the lower mass and internal pressure and because of the relative paucity of useful observational data. It now appears that metallic hydrogen is marginal (but probably not absent) in Saturn and that there is a larger enrichment of heavy elements than in Jupiter.

For the purposes of modeling the outer envelopes of Jupiter and Saturn, it is of great interest to know what the composition of the "excess" heavy material was. One plausible suggestion is that this is no more than the general solid material stable at that heliocentric distance in the Solar Nebula. In that case, the elements enriched in Jupiter would include all the rock-forming elements, including S, plus $H_2O$ ice, and possibly ammonia. Saturn would have enhanced abundances of all of these, plus possibly methane. If this simple explanation were correct, then the only elements that might be enriched in accessible parts of Jupiter would be sulfur and oxygen (as $H_2S$ and $H_2O$) and possibly N as ammonia. Of course, these species and possibly methane as well could be enriched in the outer envelope of Saturn. The other heavy elements are probably of little consequence for modeling observable portions of either planet.

The presence of an extensive liquid, electrically conducting mantle on Jupiter has obvious similarities to the liquid, electrically conducting outer core of the Earth, which is the source of Earth's magnetic field. In addition, ionization of hydrogen by the high temperatures and pressures in the molecular mantle may provide quite significant electrical conductivities even there.

Based on plausible internal structures and the convective mechanism for heat transport, it is reasonable to attribute the source of heat to the continuing tail-end of the collapse phases of both Jupiter and Saturn. It is possible that the heat is derived from the growth of the metallic hydrogen mantle, which is associated with slow shrinkage of the entire planet at a rate of 1 mm per year; indeed, these two mechanisms are in no sense mutually exclusive or even competitive, either being a logical consequence of the other.

A typical internal model of Jupiter is given in Fig. V.4a. The density discontinuity at the molecular $H_2$ to metallic H transition at 0.76 $r_J$ is clearly visible, and the core of heavy elements is obvious. Only about 0.01% of the volume of the planet experiences a gravitation acceleration less than the "surface" (0.7-bar level) value.

Figure V.4b is a comparable internal model of Saturn. Note the much lower pressures deep in the gaseous envelope. Even complete absence of a metallic H mantle would not prevent the electrical conductivity from being large, because, as in the envelope of Jupiter, high pressures decrease the width of the gap between the filled molecular orbital of $H_2$ and the lowest-lying (unfilled) conduction band. Further, species much more easily ionized than $H_2$ are present in trace amounts deep in the lower atmosphere. When high temperatures are combined with high pressures, thermal excitation of electrons across the narrowed band gap becomes possible. Thus even the absence of a metallic H region on Saturn would not preclude the generation of magnetic fields by an internal convectively driven dynamo.

## The Atmospheres of Jupiter and Saturn: Observed Composition

We have known since the pioneering work of Rupert Wildt in the 1930s that hydrogen, methane, and ammonia are present in Jupiter's atmosphere. Since 1966, however, there have been enormous advances in spectroscopic instrumentation. These advances have greatly multiplied the number of known species and correspondingly enriched our understanding of the processes affecting atmospheric composition.

Conventional spectroscopy disperses light according to its wavelength, by either refraction through a prism or diffraction off a grating. The dispersed light is then swept across a small detector, which converts the photon beam to an electric current. This current is recorded on magnetic media, or amplified and used to drive a chart recorder. The width of the wavelength interval intercepted by the detector at any time, $\Delta\lambda$, is called the *spectral resolution* of the instrument. A typical broad-coverage spectrum, such as the entire visible or entire near infrared region, must therefore consist of approximately $\lambda/\Delta\lambda$ separate measurements. Thus a spectrum of resolution $\lambda/\Delta\lambda = 10^4$ must contain $10^4$ samples of the spectrum, and the detector must therefore waste 99.99% of the incident light during sampling of any single wavelength interval. To some degree, this problem may be offset by using several detectors simultaneously (called *multiplexing*), but it is still

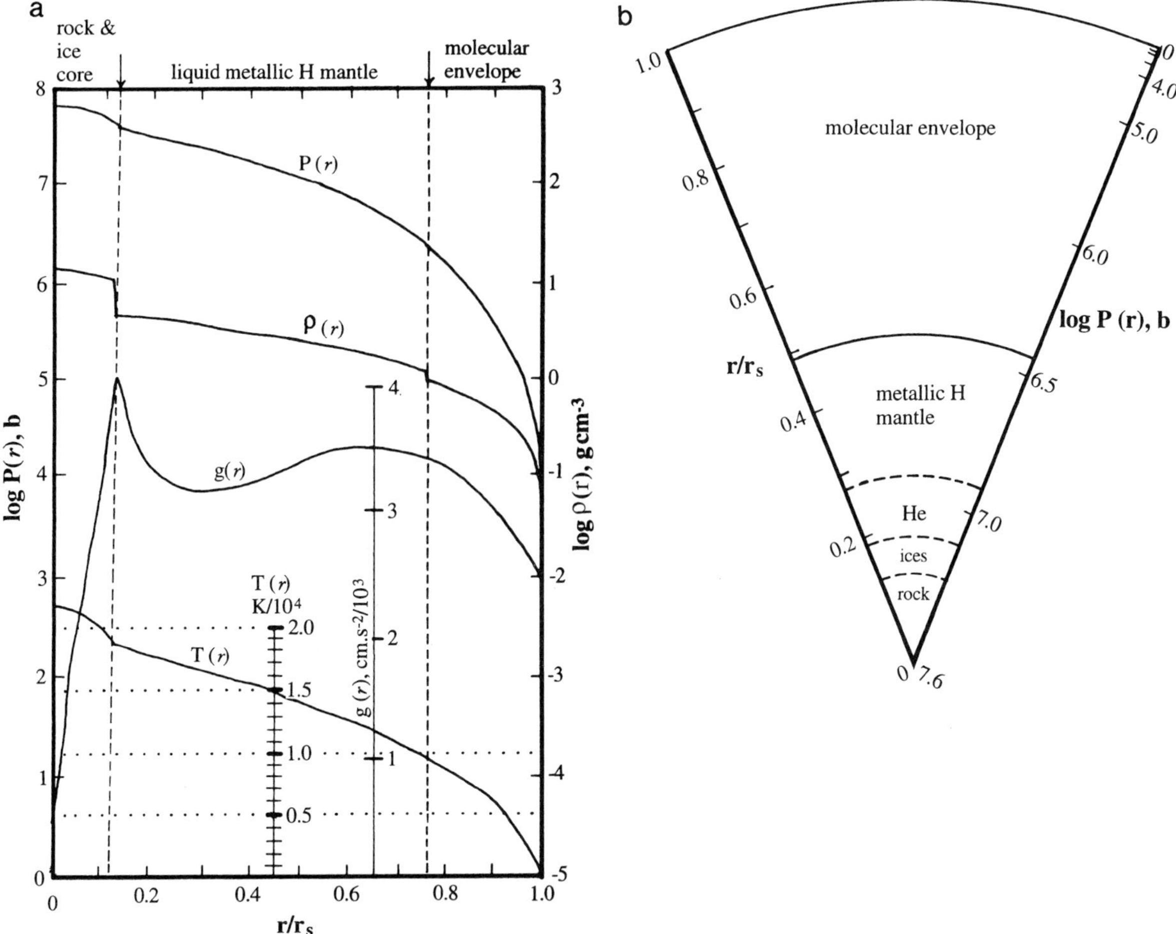


**Figure V.4** The interiors of Jupiter and Saturn. The pressure, density, gravitational acceleration, and temperature are given as a function of radius for typical internal models. a illustrates the great importance of metallization of hydrogen on Jupiter. b is a "pie" diagram of a likely internal structure for Saturn. Note the smaller metallic hydrogen region and the helium-rich layer formed because of the limited solubility of helium in fluid metallic hydrogen. It is not clear whether helium would mix with the icy component or how well the ice-forming elements will mix with (dissolve) the rocky core materials. The boundaries between these layers are thus merely schematic, and no distinct boundaries may exist.

obvious that taking a high-resolution spectrum of a faint source will require long observing times or very large telescopes with great light-gathering power.

There are, of course, strong reasons for desiring high spectral resolution. A low-resolution spectrum that resolves only the envelope of rotation–vibration bands without resolving the individual lines is inadequate for estimating rotational temperatures. A medium-resolution spectrum that separates the lines but does not resolve individual line profiles can provide useful temperature and abundance data, but will not allow the collision-broadening of the spectral lines to be measured. Only a high-resolution spectrum with $\Delta\lambda$ several times less than the line widths can permit full extraction of the information inherent in the spectrum. Depending on the species observed and on the wavelength region, resolutions of $10^4$ to $10^6$ may be needed.

The spectroscopic technique that has revolutionized planetary astronomy is interferometry, sometimes called *Fourier transform spectroscopy*. In this technique, a wide spectral region is admitted undispersed but well collimated into the spectrometer (Fig. V.5). There the beam is passed through a diagonal half-silvered (uniformly but incompletely reflective) mirror called a *beam splitter*. One half of the beam is reflected off a fixed mirror and traverses a constant path, whereas the other half of the beam reflects off a moveable mirror. These two reflected beams are then recombined to interfere with each other. The resultant intensity of the combined beams, measured as a function of the path-length difference between the two beam paths, is called an *interferogram*. When the paths followed by the two beams differ by an integral number of wavelengths of light of a particular frequency, then that light will interfere constructively with

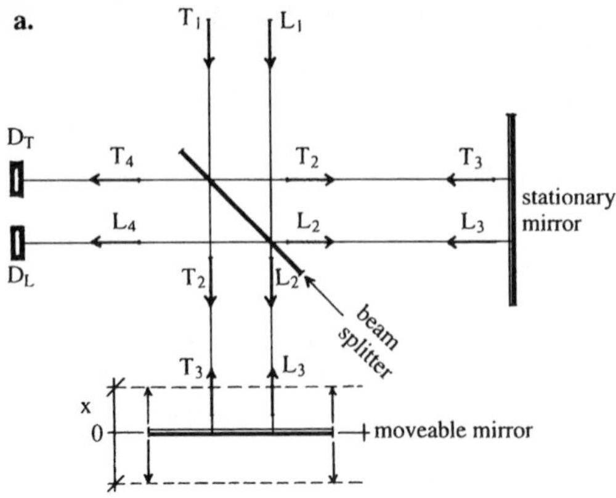


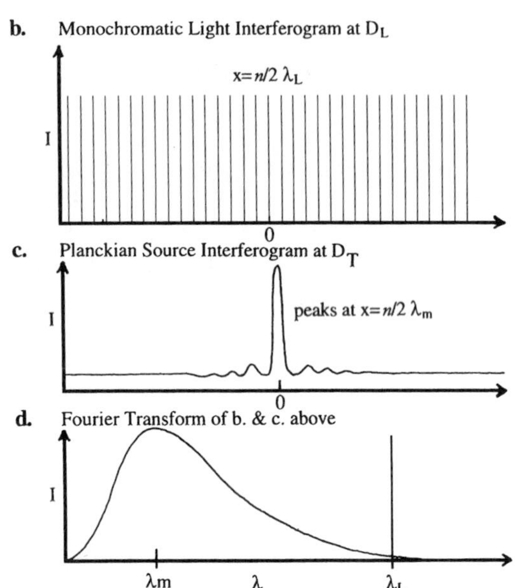


**Figure V.5** Interferometric spectroscopy. a depicts an interferometer accepting two parallel beams from a telescope (T) aimed at a Planckian source and an internal calibration laser (L). The beams strike a lightly silvered beam splitter, and half of the intensity of each beam is reflected off a stationary mirror and thence to the two detectors $D_T$ and $D_L$, separately. The other half of the beam intensity passes through the beam splitter and is reflected off a traveling mirror. The laser-source and Planckian-source interferograms are shown in b and c. The laser source is used to monitor the precise position of the moveable mirror. In practice it is sufficient to use a single detector for both beams together. The original spectrum, reconstructed by taking the Fourier transform of the sum of the two interferograms, is shown in d.

itself and be fully represented in the observed interferogram. Other wavelengths which do not meet this criterion will interfere destructively.

To determine the interferogram precisely, it is essential that the mirror position be monitored equally precisely. This is done by inserting a very pure single-wavelength (*monochromatic*) beam of light into the instrument with a small laser.

In Fig. V.5 we describe the mirror position by the variable $x$, which is zero when the light paths in the two arms of the interferometer are equal. When $x = 0$, *all* light of any frequency whatsoever interferes constructively with itself at the detector. This is called a "white light spike" in the interferogram. A mirror displacement of $x$ in either direction will increase the total light path by $2x$, and a monochromatic light source of wavelength $\lambda$ will interfere constructively with itself and give a large signal at detector D whenever $2x$ is some exact integer multiple of $\lambda$: $x = n\lambda/2$ (see Fig. V.5b).

An interferogram for a Planck function is given in Fig. V.5c. Note that low multiples of $\lambda_m$ show up as peaks in the interferogram, but each repetition becomes more smeared out, with the intensity eventually reaching a nearly constant level.

The original spectrum can be reconstructed from the interferogram by taking the Fourier transform of the interferogram. Because of the symmetry of the interferogram about $x = 0$, the Fourier series may be expressed using only cosine terms as

$$F(i) = \Sigma_i \; a_i \; \cos \; ix, \qquad \text{(V.12)}$$

where the index $i$ runs from zero to some maximum $I$. The larger the value of $I$ (the number of samples in the interferogram), the higher the resolution of the spectrum. The value of $x$ is precisely monitored by counting the laser interference fringes.

The advantage of an interferometer is that every photon in the entire spectral region under study is used at all times in constructing the interferogram. Thus no light is ever wasted, and even high-resolution spectrograms can be made using reasonable-sized telescopes with observing times of about 1 second per interferogram. A dispersive spectrometer on the same telescope, with a resolution of $10^6$, would need $10^6$ detectors in order to operate as efficiently! This strong point of the interferometer is called the *multiplex advantage*. When even greater spectral resolution or sensitivity is desired, the interferogram may be scanned more slowly, and large numbers of interferograms may be added coherently (that is, they are aligned at $x = 0$ before adding).

Such high-resolution interferometric spectra of Jupiter and Saturn have been available for a number of years, and the information in them is immense. After methane, ammonia, and hydrogen, which had already been identified as constituents of both Jupiter and Saturn prior to the introduction of Fourier-transform spectroscopy, a number of other species have been discovered. These include ethane ($C_2H_6$), acetylene ($C_2H_2$), ethylene ($C_2H_4$), monodeuteromethane ($CH_3D$), carbon-13 methane ($^{13}CH_4$), HD, phosphine ($PH_3$), water vapor ($H_2O$), germane ($GeH_4$), hydrogen cyanide (HCN), and carbon monoxide (CO). The $^{13}C$:$^{12}C$ ratio has been found to be indistinguishable from the terrestrial or meteoritic value, whereas the D:H ratio is much lower than in Earth's oceans, but very similar to that believed characteristic of primitive solar nebula material. The $^{15}N$:$^{14}N$ ratio in ammonia has been found to decrease upward in the stratosphere by about a factor of 2 relative to the ammonia isotopic composition seen in the 5-$\mu$m "hot spots." Photolysis would destroy isotopically light ammonia slightly more readily than heavy ammonia, contrary to observation, but fractional distillation of ammonia would leave a vapor that is slightly enhanced in the lighter isotope. The other chemical species require chemical explanation of their sources, stability, and observed abundances. In addition, large numbers of other species have been sought but not found in high-resolution spectra. Their absence is a great help in that it places constraints on chemical models of the Jovian planets: a model not only must pass the test of explaining the abundances and altitude distribution of the observed species, but also must not predict abundances of other species in excess of their observational detection limits. The abundances of known species and upper limits on some other species of interest are given in Table V.1.

**Table V.1** Abundances in the Atmospheres of Jupiter and Saturn

| Species | Formula | Notes | Abundance (Mole fraction) Jupiter | Saturn |
|---|---|---|---|---|
| Hydrogen | $H_2$ | | 0.85 | 0.84 |
| Helium | He | | 0.15 | 0.16 |
| Methane | $CH_4$ | | $1 \times 10^{-3}$ | $2 \times 10^{-3}$ |
| Ammonia | $NH_3$ | a | $8 \times 10^{-4}$ | $6 \times 10^{-4}$ |
| Water vapor | $H_2O$ | b | $1 \times 10^{-3}$ | $2 \times 10^{-3}$? |
| Hydrogen sulfide | $H_2S$ | b | $8 \times 10^{-5}$ | ? |
| Hydrogen deuteride | HD | | $6 \times 10^{-5}$ | $6 \times 10^{-5}$ |
| Neon | Ne | c | $2.2 \times 10^{-5}$ | ? |
| Argon | $^{36}Ar + ^{38}Ar$ | d | $1.6 \times 10^{-5}$ | ? |
| Phosphine | $PH_3$ | | $1 \times 10^{-6}$ | $6 \times 10^{-6}$ |
| Monodeuteromethane | $CH_3D$ | | $2 \times 10^{-8}$ | $2 \times 10^{-8}$ |
| Krypton | Kr | d | $1.5 \times 10^{-9}$ | ? |
| Hydrogen chloride | HCl | | $< 10^{-9}$ | $10^{-9}$ |
| Germane | Ge | | $6 \times 10^{-10}$ | $< 10^{-9}$ |
| Xenon | Xe | d | $2 \times 10^{-10}$ | ? |
| Carbon monoxide | CO | e | $1 \times 10^{-10}$ | $< 10^{-10}$ |
| Ethane | $C_2H_6$ | f | $4 \times 10^{-6}$ | $5 \times 10^{-6}$ |
| Acetylene | $C_2H_2$ | f | $3 \times 10^{-8}$ | $1 \times 10^{-7}$ |
| Ethylene | $C_2H_4$ | f | $1 \times 10^{-9}$ | $< 10^{-9}$ |
| Hydrogen cyanide | HCN | f | $1 \times 10^{-9}$ | $< 10^{-9}$ |
| Methyl amine | $CH_3NH_2$ | f | $< 10^{-9}$ | $< 10^{-6}$ |
| Hydrazine | $N_2H_4$ | f | $< 10^{-9}$ | $< 10^{-9}$ |

*Notes*: a: lower tropospheric abundance from microwave opacity; b: solar abundance assumed for Jupiter below the water and $NH_4SH$ cloud bases; c: Ne:H on Jupiter was measured by the Galileo Probe as 10 times the solar Ne:H ratio; d: the heavy noble gases Ar, Kr and Xe were measured by the Galileo Probe at about 2.5 × solar; e: CO is probably transported up to the tropopause by turbulent mixing; f: photochemical products detected in the stratosphere but of negligible abundance in the troposphere.

## Tropospheric Composition and Structure: Theory

Our approach shall be to describe the chemical behavior of solar material along a Jovian pressure–temperature profile. We have already seen that the structure of the lower atmosphere is adiabatic, and we need only fix some ($P$, $T$) point in order to calculate the entire adiabat from theory. This convectively mixed, adiabatic portion of the atmosphere is called the *troposphere*. We shall investigate the gas-phase and condensation chemistry of the troposphere as a function of altitude, first for a solar-composition model and then for a composition in which certain of the heavy elements are enriched above their solar abundances.

The starting point for our adiabatic model shall be the temperature and pressure determined from near-IR spectroscopy and thermal-IR radiometry of the cloud-tops. We will take the cloud-top temperature on Jupiter as 130 K and the pressure as 0.5 bar at that level. Using the heat capacity data in Fig. IV.39, we can easily calculate the adiabat down to levels where the density approaches 0.1 g cm$^{-3}$. The adiabatic atmospheric models we shall henceforth use are shown in Figs. V.6 and V.7. The Jupiter model in Fig. V.6 is carried down to the 700-K,

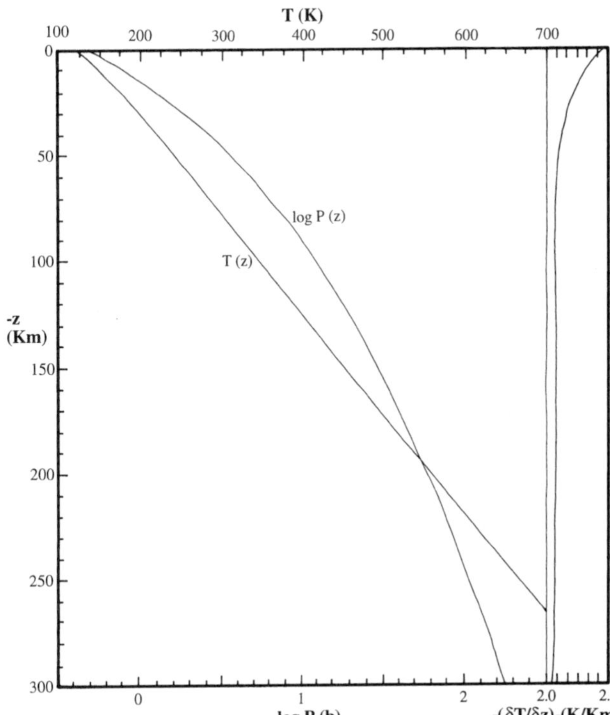


**Figure V.6** Adiabat for the atmosphere of Jupiter. The vertical profiles of pressure, temperature, and temperature gradient are calculated for an adiabatic model of the Jovian atmosphere consistent with Pioneer and Voyager data on the topmost 50 km. The altitude scale is depth beneath the ammonia clouds.

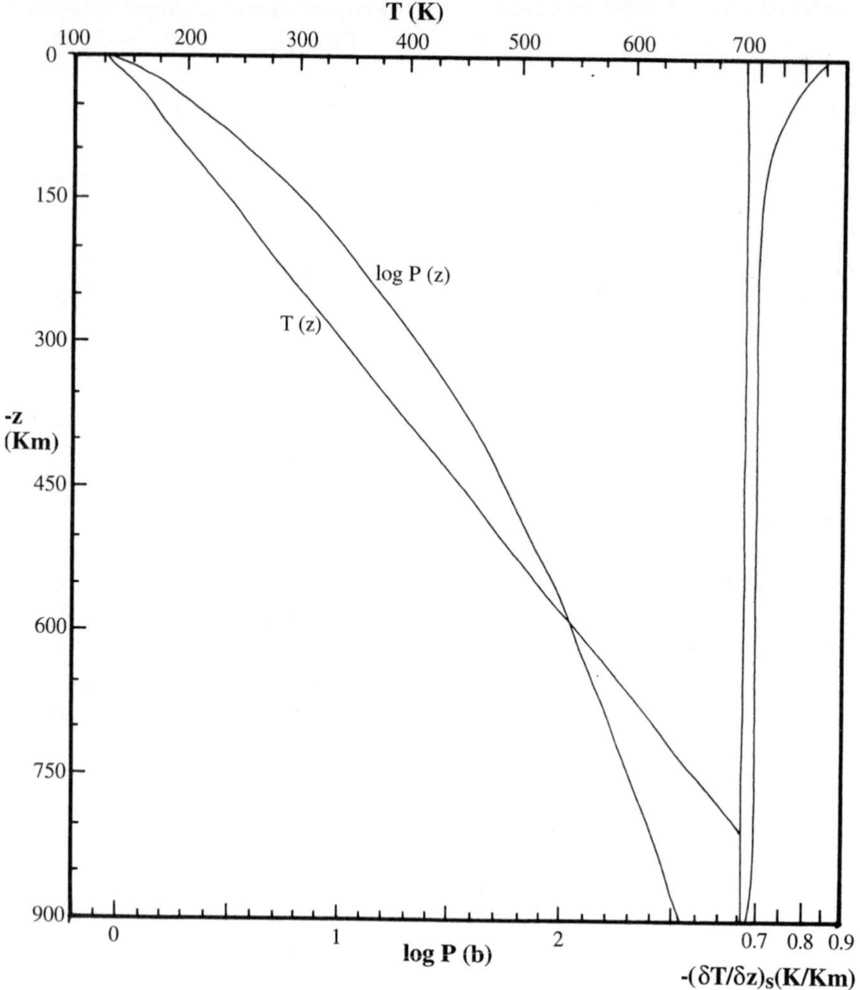


**Figure V.7** Adiabat for the atmosphere of Saturn. The vertical profiles of pressure, temperature, and temperature gradient are calculated for an adiabatic model of the Saturnian atmosphere consistent with Pioneer and Voyager data on the topmost 50 km. The altitude scale is depth beneath the 1-bar pressure level.

174-bar level. Note the strong variation of the adiabatic lapse rate with altitude at temperatures below 300 K, due to the contributions of the rotational modes of molecular hydrogen to the specific heat, and the further decrease in the gradient beginning near 700 K, due to excitation of the vibrational modes of $H_2$.

From Eq. (IV.11) we see that hydrostatic equilibrium, which is an exact balance between the pressure-gradient force $P$ and the gravitational force $-\rho(r)g(r)$, is analytically integrable only if the temperature is constant or a simple function of $r$. For the normal range of altitudes ($Z$) in a planetary atmosphere it is usually quite accurate to take $g =$ constant. Thus, for an isothermal atmosphere,

$$dP/P(Z) = d \ln P = -(\mu g/RT)dZ, \qquad \text{(V.13)}$$

which integrates to

$$\begin{aligned} P &= P(Z_0) \exp\left[-(Z - Z_0)/(RT/\mu g)\right] \\ &= P(Z_0) \exp -(Z - Z_0)/H, \end{aligned} \qquad \text{(V.14)}$$

where $H = RT/\mu g$. $H$, the *scale height*, is the altitude increment over which the pressure drops by a factor of $e$. We can see from Fig. V.6 that, over a very wide range of altitude, the Jovian troposphere has a constant temperature *gradient* of $-2.1\ \mathrm{K\,km^{-1}}$; thus, $T = T_0 - 2.1(Z - Z_0)$, and

$$dP/P(Z) = -\{\mu g/[RT_0 - 2.1R(Z - Z_0)]\}dZ, \qquad \text{(V.15)}$$

whence

$$\begin{aligned} P(Z) &= P(Z_0)[1 - 2.1(Z - Z_0)/T_0]^{\mu g/2.1R} \\ &= (T/T_0)^{C_p/R}. \end{aligned} \qquad \text{(V.16)}$$

This explicit relationship between $P$ and $Z$ when the lapse rate is constant should be applied only with care, to ensure that $g$ and $C_p$ *are* constant. Figure V.8 provides a bridge between atmospheric models per se and interior models. Only Jupiter is shown here, because the only transformations needed to convert to Saturn are to multiply the Jovian altitude scale by $g_J/g_S$ (about 3) and to double the pressure at any given temperature. Figure V.8 extends down to 0.9 $R_J$ from the center of the planet, a depth of 7000 km.

Note that we have in no case extended the atmospheric models upward above the 130-K level. We shall reserve description of this region, which has a physical and chemical structure radically different from that of the lower atmosphere, for later detailed description. Also, it should be recalled that we have in each case calculated and graphed atmospheric structures for the equatorial regions, where the distance from the center of the planet is the greatest and the "centrifugal force" due to rotation is greatest. This is done because Earth-based observations select strongly in favor of the equatorial regions, and it is there that we have the best data to develop and to test models. Because $g$ is everywhere else greater than at the equator, the temperature and pressure gradients will be greater, and constant-pressure surfaces will become ever more nearly spherical at greater depths.

It is upon these temperature and pressure gradients that the chemistry of the atmosphere is acted out. Although our treatment of the chemistry of solar material in Chapter IV provides a very useful introduction to processes in Jupiter and Saturn, there are also many significant differences.

First, we saw in the case of the Solar Nebula that, at the high-temperature, high-pressure end of an adiabat, all solids are completely evaporated. In a planetary atmosphere with a solar composition and an adiabatic structure, pressures are a million or more times higher than those in the nebula, and the condensation curves are driven up to much higher temperatures.

Several major species in the solar nebula calculations, including refractory oxide minerals, metallic iron, and magnesium silicates, condense at temperatures in excess of 1500 K, and in the interior of Jupiter they

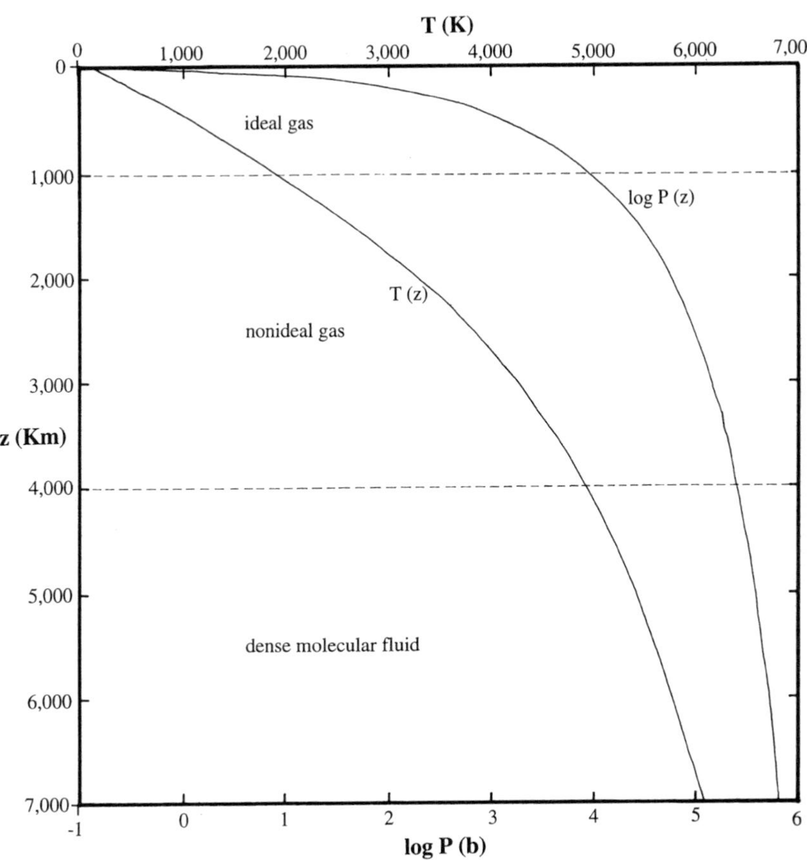


**Figure V.8** Structure of the deep atmosphere of Jupiter. The ideal-gas upper troposphere is here connected to the dense molecular lower troposphere through a highly nonideal region with a poorly constrained equation of state.

would have condensation temperatures above 2000 to 3000 K. From Fig. V.8 we can see that the pressure at the 2000-K level is already 10 kbar, which gives a gas density that, in the ideal gas case, is $\mu P/RT$, or $0.14\,g\,cm^{-3}$. This is close to the density of liquid hydrogen, and we are clearly out of the ideal gas realm altogether.

The chemistry of nonideal gases is generally rather poorly known, although the thermodynamic properties of hydrogen and helium are fairly well understood. In chemical thermodynamics, the reactivities of solids and liquids are given according to the convention that the reactivity (or thermodynamic *activity*) of a pure solid or pure liquid at 1 atm of pressure is unity. At high pressures (100 atm or thereabouts, depending upon the material), compression of the solids and liquids causes the atoms to be displaced inward against the steep mutually repulsive (Coulomb) part of the potential curve, and it takes less energy to remove an atom or molecule from the condensed particle than it normally does at lower pressures. We may say that the *activity* of the solid has been increased above the defined low-pressure reference value of 1.00. This activity increase is usefully thought of as an increase in vapor pressure. A solid compressed to the point at which its vapor pressure is $p'_v$ instead of the standard 1-atm value of $p_v$ at the same temperature has an activity

$$a = p'_v/p_v. \tag{V.17}$$

(All pressures in chemical thermodynamics are conventionally in units of atmospheres, a convenience on the surface of the Earth, but an awkward inconvenience elsewhere. By one of those curious coincidences with which nature abounds, 1 atm is very close to the metric unit of pressure, the bar: 1 atm = 1.015 bar. We shall almost invariably lose no precision if we substitute the two units freely for one another.) A solid or liquid compressed over the pressure interval from 1 to $P$ atm, and undergoing a continuous volume change (no phase changes), has an activity defined by

$$RT \ln a = \int_1^P v(P)dP, \tag{V.18}$$

where $v(P)$ is the molar volume of the condensed substance in $cm^3\,mol^{-1}$. When the pressure is not too high, $v(P)$ is not much less than the 1-atm standard value, $v(1)$, and we have

$$a = e^{-v(1)(P-1)/RT} = e^{\mu(P-1)/\rho RT}. \tag{V.19}$$

Thus we can see that materials with very low densities show nonideal behavior even at very modest pressures of a few bars. The best example of such a low-density material is hydrogen—but we will not have solid $H_2$ present inside Jupiter!

We can see from the approximate relationship in Eq. (V.19) that iron metal ($\mu = 56, \rho = 7.5$) at 2000 K has an activity of 1.004 at 10 atm, 1.04 at 100 atm, 1.5 at 1000 atm, and 60 at $10^4$ atm. Water ($\mu = 18, \rho = 1$), which is far more compressible and far less dense, will at 300 K have an activity of 1.08 at 10 atm, 2.08 at 100 atm, and 1500 at 1000 atm! Figure V.9 illustrates the dependence of the activity of water on pressure at a temperature of 300 K. In general, ices have low densities and great sensitivity to pressure, but the ices are precisely those substances which condense at lowest pressures along any adiabat.

We shall begin our discussion of the chemistry of the atmosphere of Jupiter as we began our description of the nebula: with the most abundant elements at high temperatures. The two most abundant elements, hydrogen and helium, have very simple chemistry over the range of temperatures and pressures we are now considering. The only important reaction of hydrogen we need consider at pressures less than several megabars is the dissociation of molecular hydrogen to atomic H gas:

$$H_2 \rightleftharpoons H + H \tag{IV.92}$$

$$\log K_{92}(T) = 6.16 - 23{,}500/T \tag{IV.95}$$
$$= 2\log p_H - \log p_{H_2}$$
$$= 2\log f_H - \log f_{H_2} + \log P. \tag{IV.96}$$

The equation for the Jovian adiabat in the region of interest (2000 to 5000 K; $10^4$ to $5 \times 10^5$ bar) is approximately

$$P = P_0(T/T_0)^{C_p/R} \tag{V.20}$$

or

$$\log P = 4 + 4.2\log T - 4.2\log 2000$$
$$= 4.2\log T - 9.86. \tag{V.21}$$

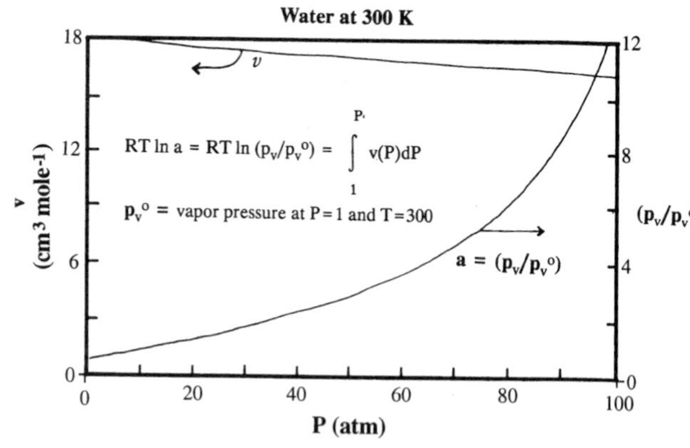


**Figure V.9** Activity of a liquid or solid at high pressure. The thermodynamic activity ($a$) (a measure of the escaping tendency or vapor pressure of the condensed phase) of liquid water at 300 K is shown here for pressures ranging from 1 to 100 atm. Note the *accelerating* elevation of the vapor pressure with increasing total pressure.

Combining Eqs. (IV.96) above with (V.21), we have

$$6.16 - 23{,}500/T + \log f_{H_2} - 2\log f_H = 4.2\log T - 9.86. \qquad (V.22)$$

When $H_2$ is largely undissociated, $f_{H_2} = 0.876$, and Eq. (V.22) reduces to

$$\log f_H = 7.98 - 11{,}750/T - 2.10\log T, \qquad (V.23)$$

from which $f_H = 1.5 \times 10^{-5}$ at 2000 K, $5.8 \times 10^{-4}$ at 3000 K, $3.0 \times 10^{-3}$ at 4000 K, and $7.2 \times 10^{-3}$ at 5000 K. Thus thermal dissociation of $H_2$ may provide enough H to make it the third most abundant gas above 4000 K, but the dominant role of $H_2$ is never threatened.

The most stable compounds of oxygen, carbon, and nitrogen are, not surprisingly, $H_2O$, $CH_4$, and $NH_3$, respectively, although appreciable quantities of CO and $N_2$ may also be present. The ammonia equilibrium is given by

$$\log [f_{NH_3}^2/(f_{N_2} f_{H_2}^3 P^2)] = 5720/T - 12.23, \qquad (IV.114–IV.115)$$

and the Jupiter adiabat between 500 and 2000 K can be approximated by

$$P = 825(T/1000)^{3.6}, \qquad (V.24)$$

whence, using $f_{H_2} = 0.876$ as above, we obtain

$$\log(f_{NH_3}^2/f_{N_2}) = -27.82 + 5720/T + 7.20\log T. \qquad (V.25)$$

The total abundance of nitrogen compounds is $f(\Sigma N) = f(NH_3) + 2f(N_2) = 2.06 \times 10^{-4}$. When $NH_3$ is dominant, $\log f(NH_3) = -3.686$, and we calculate that $f(N_2)$ on the Jupiter adiabat is $3.8 \times 10^{-11}$ at 500 K, $1.35 \times 10^{-7}$ at 1000 K, and $6.6 \times 10^{-7}$ at 2000 K. Thus even at 2000 K, $f(N_2)$ is only 0.3% of the total nitrogen abundance. The mole fraction of $N_2$ reaches a maximum value of $6.8 \times 10^{-7}$ at 1830 K.

For the equilibrium between $CH_4$ and CO, we have

$$\log [f_{CH_4} f_{H_2O}/(f_{CO} f_{H_2}^3 P^2)] = 11{,}740/T - 13.21. \qquad (IV.116–IV.117)$$

We can then see how the ratio $f_{CH_4}/f_{CO}$ varies along either segment of the adiabat approximated by Eqs. (V.20) and (V.24). As above, we find

$$\log (f_{CH_4}/f_{CO}) = -25.88 + 11{,}740/T + 7.20\log T. \qquad (V.26)$$

Thus $f(CH_4)/f(CO)$ is $1.1 \times 10^{17}$ at 500 K, $2.9 \times 10^7$ at 1000 K, and $5.7 \times 10^3$ at 2000 K. The ratio reaches a minimum at 3220 K, at which $f(CH_4)/f(CO) = 1870$, as can be seen by setting the derivative of Eq. (IV.26) equal to zero and solving for $T$. Thus CO is never more abundant than roughly 1 ppm of the total gas pressure, but is close to this abundance over a very wide range of temperatures, from 2000 K up to at least 5000 K. Likewise, $f(CO)$ is less than $10^{-9}$ below 1000 K.

The domination of the gas phase by $H_2$, He, $H_2O$, $CH_4$, and $NH_3$, already evident at low temperatures in the solar nebula, is extended to high temperatures by the very high pressures deep within Jupiter.

After H, He, O, C, N, and Ne, the next most abundant elements are Fe, Si, and Mg, the three major rock-forming elements.

At the 3000-K level on the Jupiter adiabat the pressure is already so high (50 kbar) that it is probable that no condensed phases could be present. The vapor pressures of refractory oxide and metal phases, although very poorly known at these pressures, almost certainly become large enough for complete evaporation before temperatures of 3000 K are encountered on the adiabat. Perhaps it is easiest to picture the portion of the atmosphere that has pressures ranging from 50 kbar up to about 3 Mbar as a dense, luminous liquid, with the temperature of the photosphere of a star, in which all solids are dissolved. Figure V.10 shows how rapidly the vapor pressure of liquid iron increases due to the high pressures. It can be seen that iron grains would evaporate near 2700 K rather than the 3000 K expected from simple extrapolation of the low-pressure vapor pressure equations of iron.

These considerations do not disprove the existence of a metal or silicate core in Jupiter by any means, because it is the unknown solubility of these materials in liquid metallic hydrogen that is relevant in the deep interior. Furthermore, an accretion history that begins with a massive rocky core as a nucleus for gravitational gas capture may well allow persistence of the remains of that core until the present, even if it is chemically unstable, simply because of the enormous density contrast between "rocks" and hydrogen, which may suffice to prevent mixing of the dense core materials into the metallic H mantle even if solubility were no problem.

After metallic iron, the next major condensates would be the magnesium silicates. At high pressures, the gaseous silicon species will no longer be simply SiO and SiS:

$$SiO + 3\,H_2 = H_2O + SiH_4 \qquad (V.27)$$

$$SiS + 3\,H_2 = H_2S + SiH_4. \qquad (V.28)$$

The silane ($SiH_4$) partial pressure is clearly enhanced by large total pressures:

$$K_{27} = p_{H_2O} p_{SiH_4}/(p_{H_2}^3 p_{SiO}) \qquad (V.29)$$

$$p_{SiH_4}/p_{SiO} = f_{H_2}^3 K_{27} P^2/f_{H_2O} \qquad (V.30)$$

and similarly for SiS reduction [Eq. (V.28)].

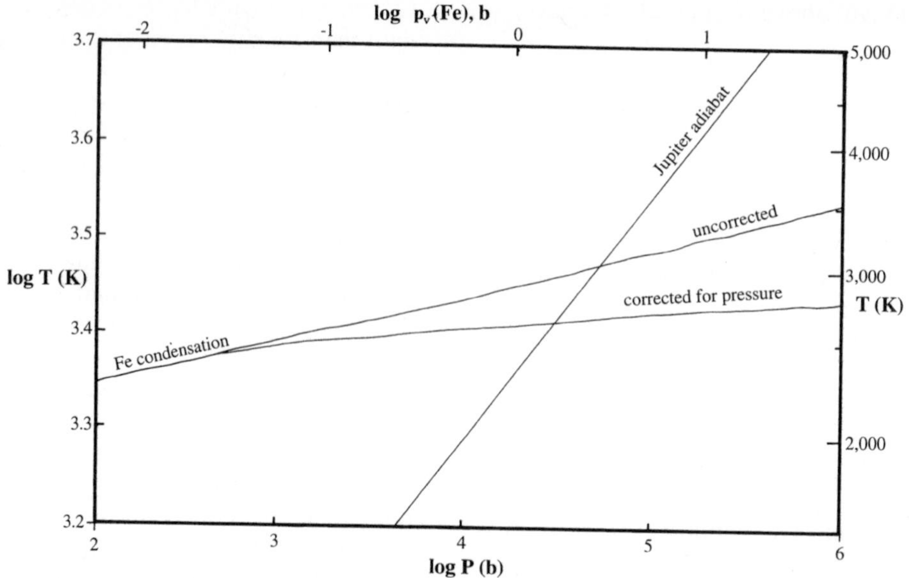


**Figure V.10** Effect of high pressures on the condensation of iron. Iron, which is denser and less compressible than water, shows little change of its activity and vapor pressure until pressures pass 1 kbar. Beyond that point, iron is substantially more volatile than it is at low pressures. Similar phenomena affect the condensation temperatures of silicates, but their gaseous species and condensation reactions are so complex that it is difficult to quantify their behavior. Fortunately, the most important silicates condense at lower temperatures than iron and hence form at much lower pressures.

The reaction of SiO (at low total pressure) to make solid magnesian olivine ($Mg_2SiO_4$; forsterite; fo) is, as we have seen,

$$2\ Mg + SiO + 3\ H_2O = \underset{\text{forsterite}}{Mg_2SiO_4(s)} + 3\ H_2 \qquad (V.31)$$

whereas, at high pressures, we will have

$$2\ Mg + SiH_4 + 4\ H_2O = Mg_2SiO_4(s) + 6\ H_2. \qquad (V.32)$$

At low pressures, we see that $a_{fo} \propto P^3$, whereas at high pressures $a_{fo} \propto P$. Thus increasing pressures (up to the point of takeover of nonideal behavior) tends to favor $Mg_2SiO_4$ condensation, but in a far less marked manner than at nebular pressures. At very high pressures, Eq. (V.19) shows that the activity required to effect condensation quickly becomes enormous.

Condensation of enstatite at low pressures proceeds via

$$Mg + SiO + 2\ H_2O = \underset{\text{enstatite}}{MgSiO_3(s)} + 2\ H_2 \qquad (V.33)$$

so that $a_{en} \propto P^2$, whereas at higher pressures (1 to 10 bar) $a_{en}$ ceases to depend on pressure at all. At very high pressures ( $> 1$ kbar), nonideality again takes over.

The direct condensation of MgO,

$$Mg + H_2O = \underset{\text{periclase}}{MgO(s)} + H_2, \qquad (V.34)$$

is unimportant at low pressures because of the ease of formation of magnesium silicates, but it can be seen that $a_{MgO} \propto P$ at all pressures. At 10 kbar the activity of pure solid periclase already is 300 (compared with $> 10^{11}$ for forsterite), and we conclude that, just below 2000 K and $10^4$ bar, MgO is condensed, whereas $Mg_2SiO_4$ is not. At slightly lower temperatures $Mg_2SiO_4$ also condenses, leaving a marked excess of silicon over magnesium and other divalent metals in the gas.

Direct condensation of this silicon as $SiO_2$ is hindered by the fact that the reaction

$$SiH_4 + 2\ H_2O = SiO_2(s, 1) + 4\ H_2 \qquad (V.35)$$

gives a silica activity, $a_{SiO_2}$, which varies as $P^{-1}$, the opposite of the low-pressure behavior, where the starting gas is SiO or SiS. Accordingly, $SiO_2$ condensation is largely uncoupled from MgO condensation and occurs at lower temperatures and pressures. Of course, if the periclase condensate is present as a fine dust at the level where quartz is becoming stable, magnesium silicates will form instead. The particle sizes for such a condensate are very difficult to estimate, but given some knowledge of them we can calculate the fallout velocity directly. Any particle will reach a terminal (steady-state) velocity of fall when the frictional retarding force exactly equals the weight of the particle:

$$6\pi\eta a v_z = mg = 4\pi\rho a^3 g/3, \qquad (V.36)$$

where $\eta$ is the coefficient of viscosity and $a$ is the radius of the particle. From this,

$$v_z = 2\rho a^2 g/9\eta. \tag{V.37}$$

This equation also permits us to calculate the size of the largest particle that can be blown upward by an updraft of velocity $v_z$, which will be of use to us when we have some information regarding rates of vertical motions on Jupiter. For the moment, we can see that a 10-$\mu$m-radius MgO particle ($a = 10^{-4}$ cm, $\rho = 3.2\,\mathrm{g\,cm^{-3}}$) deep in the atmosphere of Jupiter ($g = 2630\,\mathrm{cm\,s^{-2}}$; $\eta = 3\times 10^{-4}\,\mathrm{g\,cm^{-1}\,s^{-1}}$) will have a fallout velocity of $10\,\mathrm{cm\,s^{-1}}$.

Condensation of new droplets and particles is influenced by the fact that new (and therefore small) particles are mostly surface. Therefore the surface tension is an unusually large part of the total energy content of the particle, and the vapor pressure is affected. The work done to create a surface of area dA (the surface energy) is clearly just $w = -\gamma dA$, where $\gamma$ is the surface tension in erg cm$^{-2}$. Applying Eq. (V.18) to the vapor pressure of a droplet,

$$RT \ln a = RT \ln p_v/p_v^0 = v(P)\Delta P. \tag{V.18}$$

Imagine a droplet of radius $a$. Its surface energy is clearly $E = 4\pi a^2\gamma$, and a change of radius of $da$ would cause a change in its surface energy of $dE = 8\pi a\gamma da$. Thus shrinking causes the total surface energy to decrease. At equilibrium, a pressure differential ($\Delta P$) must exist across the surface such that $\Delta P 4\pi a^2 da = 8\pi a\gamma da$, whence $\Delta P = 2\gamma/a$. Combining with Eq. (V.18), we get

$$RT \ln p_v/p_v^0 = v\Delta P = 2\gamma v/a. \tag{V.38}$$

Thus when we approach the state of saturation from a "clean" gas free of preexisting condensation nuclei (*homogeneous nucleation*), the first particles to form, because they are very small, have vapor pressures that are significantly elevated over those of large droplets. Thus a partial pressure of vapor significantly larger than the normal vapor pressure must be present before small droplets can nucleate. It is thermodynamically easier to nucleate large particles, but kinetically easier to nucleate small ones!

When these factors are balanced, we find that, for Jovian conditions of present interest, typical new particles are 10 to 30 $\mu$m in radius and fall velocities approach $1\,\mathrm{m\,s^{-1}}$, which is very rapid indeed. Accordingly, we shall ignore vertical transport of cloud particles for the purposes of the current discussion and return to this matter in more detail later. With this assumption, then, we may see that the remainder of the condensation sequence will be no other than the nonhomogeneous accretion sequence, because reactions between the gas phase and already present condensates are impossible. This condensation sequence takes on a clear meaning in a planetary atmosphere; a rising parcel of gas, cooling adiabatically, undergoes sequential condensation of Fe, MgO, $Mg_2SiO_4$, $MgSiO_3$, and $SiO_2$, leaving each fresh condensate in turn as a distinct cloud layer. The cloud particles do not travel along with the rising gas because they are too large; if they fall below the cloud base, they evaporate and their vapors are swept back upward to recondense. Thus discrete cloud layers are maintained.

After condensation of $SiO_2$, sodium oxide and sulfides of potassium, rubidium, and cesium condense. Next, other moderately volatile chalcophiles such as lead, zinc, thallium, and cadmium precipitate, followed by several nonmetals, including selenium, germanium, and tellurium.

At a temperature near 400 K, ammonium halides, especially $NH_4Cl$, begin to precipitate. Because of the small cosmic abundances of the halogens compared with the abundance of nitrogen, all the halogens are removed without significant depletion of ammonia.

Phosphine, $PH_3$, is oxidized by water vapor near the 800-K level to make $P_4O_6$ gas, which in turn reacts with ammonia near the 400-K level to make solid ammonium salts. At equilibrium, $PH_3$ should be unobservably rare in the vicinity of the cloudtops. The main reactions are

$$4\ PH_3 + 6\ H_2O = P_4O_6 + 12\ H_2 \tag{V.39}$$

$$P_4O_6 + 10\ H_2O = 4\ H_3PO_4 + 4\ H_2 \tag{V.40}$$

$$H_3PO_4 + 2\ NH_3 = (NH_4)_2HPO_4(s). \tag{V.41}$$

The only species remaining in the gas at the 300-K level are $H_2$, the rare gases, $CH_4$, $NH_3$, $H_2O$, and $H_2S$. As we have seen, the sequence of condensates will now be $H_2O$, $NH_4SH$, and $NH_3$; the temperature required for $CH_4$ condensation is too low to be reached on Jupiter.

The structure of the cloud layers which are expected to form at temperatures above the saturation point of $H_2O$ is shown schematically in Fig. V.11. The structure of the corresponding layers on Saturn is well enough approximated by multiplying the altitude scale by 3 and the pressures by 2, leaving the temperature scale unchanged.

Condensation of clouds formed from a mixture of $H_2O$, $NH_3$, and $H_2S$ is complex enough to require more detailed treatment.

## Cloud Condensation in the $NH_3$–$H_2O$–$H_2S$ System

Saturation of $H_2O$ occurs when the partial pressure of water vapor in the atmosphere is equal to the local vapor pressure of water or of $H_2O$ ice, whichever is lower. For the reaction

$$H_2O(s, 1) = H_2O(g) \tag{V.42}$$

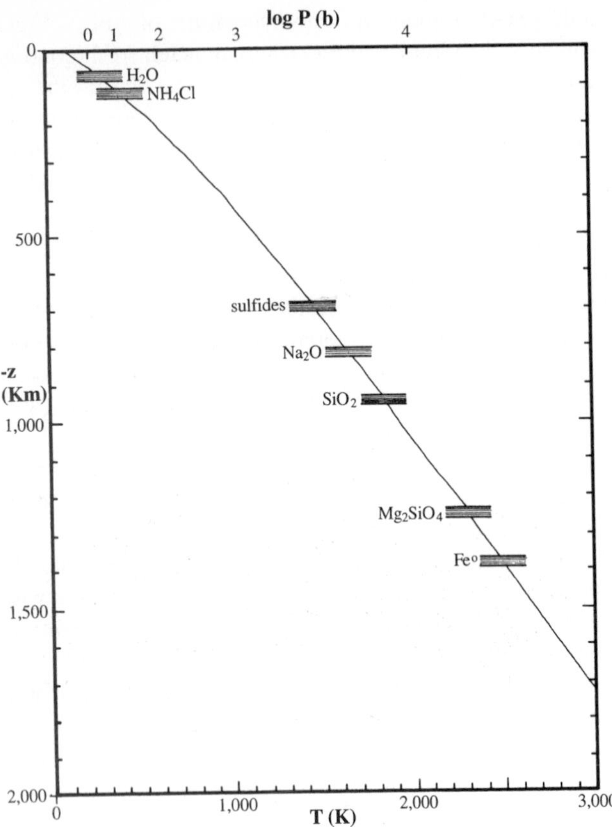


**Figure V.11** Cloud structure deep in Jupiter's atmosphere. The approximate condensation locations of a number of species in a solar-composition model of Jupiter's upper troposphere, from the water clouds (base at about 6 bar) on down. The atmosphere is by no means transparent in the intercloud regions because of the opacity contributed by Rayleigh scattering and collision-induced absorption.

the equilibrium constant is

$$\ln K = \ln(p_{H_2O}/a_{H_2O}) = \Delta G^\circ_{vap} \quad (V.43)$$

or

$$\ln(p_{H_2O}) = \ln(a_{H_2O}) + \Delta H^\circ_{vap}/RT - \Delta S_{vap}/R. \quad (V.44)$$

Because of the low density of ice, we should not in general neglect the effects of pressure on activity if the total pressure is greater than about 10 bar. Thus, for generality, we do not set $\ln a(H_2O) = 0$, but instead use

$$\ln a(H_2O) = (1/RT)\int_1^P v(P)dP. \quad (V.18)$$

At pressures of 10 bar, $v(P)$ is very close to the standard molar volume of ice at $P = 1$ atm, and we can approximate Eqs. (V.44) and (V.34) by

$$\ln(p_{H_2O}) = v\,(P-1)/RT + \Delta H^\circ_{vap}/RT - \Delta S_{vap}/R. \quad (V.45)$$

Below the level at which saturation occurs,

$$P_{H_2O} = f_{H_2O}P = A_O P/(0.5A_H + A_{He}) = 1.18 \times 10^{-3}P, \quad (V.46)$$

and along a Jovian adiabat near 300 K, where $C_p/R = 3.37$, $P = 7.4(T/300)^{3.37}$. Thus saturation of $H_2O$ on the Jupiter adiabat occurs, where

$$3.37\ln T - 23.96 = v(3.32 \times 10^{-8}\ T^{3.37})/RT + \Delta H^\circ_{vap}/RT - \Delta S_{vap}/R. \quad (V.47)$$

Figure V.12 shows the vapor pressure equations of $H_2O$ ice and liquid water and adiabats for Jupiter and Saturn. In both cases, the intersection of the adiabat with the vapor pressure curve calculated from Eq. (V.46) lies above the melting temperature of ice, and the cloud base is therefore made of droplets of water.

At pressures high enough to affect the thermodynamic properties of ice, the melting temperature as well as the vapor pressure are affected. We wish to be able to describe the dependence of the melting temperature, $T_m$, on pressure.

Consider melting of a general solid substance $A$(s) to form a liquid $A$(l). At any point on the equilibrium melting curve $T_m(P)$ the molar Gibbs free energies of the solid and liquid are equal (Appendix I). Therefore moving between any two points (1 and 2) on the melting curve, the change in Gibbs free energy of the liquid, $G(1) = G(1,2) - G(1,1)$, is exactly equal to the change in Gibbs free energy of the solid, $G(1) = G(s) = G(s,2) - G(s,1)$.

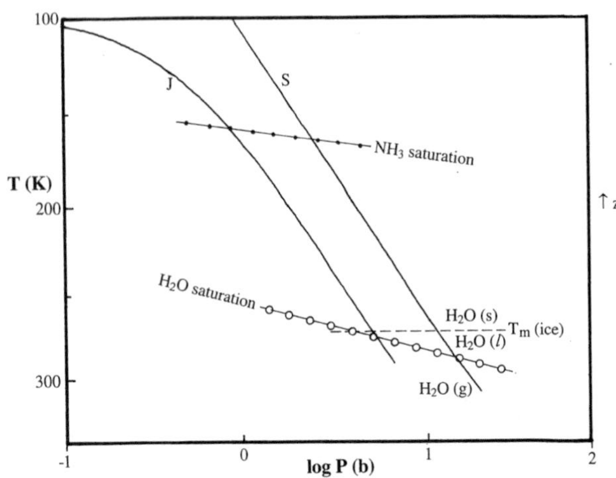


**Figure V.12** Water and ice condensation. The vapor pressure of water and ice, assuming a solar proportion of oxygen, is plotted along with the adiabats for the upper tropospheres of Jupiter and Saturn. Saturation first occurs on both planets at cloud-base temperatures above the melting point of ice.

The differential expression for a change in Gibbs free energy due to pressure and volume changes is

$$dG = VdP - SdT, \quad \text{(AI.38)}$$

and, on the phase boundary, $dG(\text{s}) = dG(\text{l})$:

$$v_s dP - S_s dT = v_l dP - S_l dT. \quad \text{(V.48)}$$

Thus the slope of the melting curve, $dT_m/dP$, is

$$dT_m/dP = [v_l - v_s]/[S_l - S_s] = \Delta v_m/\Delta S_m \quad \text{(V.49)}$$

On the melting curve, because $G(\text{l}) = G(\text{s})$ at each temperature (because the solid and liquid are in equilibrium at each point on the melting curve), then

$$\Delta G_m = G_l - G_s = 0 = \Delta H_m^\circ - T\Delta S_m. \quad \text{(AI.40)}$$

Thus, combining with Eq. (V.49),

$$dT_m/dP = T\Delta v_m/\Delta H_m^\circ. \quad \text{(V.50)}$$

This is a form of the Clausius–Clapeyron Equation. Over pressure ranges in which $\Delta v_m$ does not change very much, this equation may be integrated to give

$$\ln(T_m/T_m^\circ) = \Delta v_m(P - P^0)/\Delta H_m^\circ, \quad \text{(V.51)}$$

where of course $P^0$ is 1 and $T_m^\circ$ is the triple point temperature of water, 273.16 K. Normal water ice (ice I) is very unusual in that $\Delta v_m$ is negative; that is, ice floats. This is also true of bismuth, but every other common material has a melting temperature that increases with pressure. Addition of solutes such as ammonia leads, as might be expected, to substantial complications.

A system that contains $\Phi$ different phases with $C$ components can be fully described if we know the composition of every phase. Because the sum of all the mole fractions in any phase is 1, we can describe the composition fully with only $C - 1$ mole fractions for each of the $\Phi$ phases, making a total of $\Phi(C - 1)$ composition variables. In addition, we must know both the temperature and the pressure of the system. Thus the total number of variables needed is $\Phi(C - 1) + 2$. At equilibrium each component must be in equilibrium with each phase, and each equilibrium is described by an equation in which the equilibrium constant is given as a function of $T$ and $P$. For each component, the number of such equilibria is one less than the total number of phases present, for a total of $C(\Phi - 1)$ constraints over all $C$ components. Thus, when all the constraints are considered, the number of remaining independent variables $f$ is $\Phi(C - 1) + 2 - C(\Phi - 1)$:

$$f = C - \Phi + 2. \quad \text{(V.52)}$$

This equation is called the *Gibbs Phase Rule*.

Thus, when the only component is water ($C = 1$), the region in which the only phase present is $H_2O$ vapor ($\Phi = 1$) is characterized by $C - \Phi + 2 = 2$ degrees of freedom. This means that there is a one-phase region in $P - T$ space in which either pressure or temperature may be varied independently. When a phase condenses, then we may have liquid water and water vapor coexisting. Then $C = 1, \Phi = 2$, and $f = 1$. This means that, when these two phases coexist, pressure and temperature cannot be varied independently; there must be a functional relationship $P(T)$ between them so that only one degree of freedom is available. Such an equation $[P(T)]$ is called a *vapor pressure equation*. When water, $H_2O$ ice, and vapor coexist (the triple point), then $\Phi = 3$ and $f = 0$. This means that only a single set of values of temperatures and pressure can characterize the triple point, with no freedom for variation.

If we add an inert gas that does not react chemically with water, but that contributes to the total pressure, then we need a new variable to describe the composition of the system. Such a variable might be $x(H_2O)$, the mole fraction of water vapor, or the total pressure ($P$). Under these circumstances, the "triple point" at which water, ice, and vapor coexist is described by $C = 2, \Phi = 3$, and $f = 1$. This means that the triple point actually follows a line $[P, T, x(H_2O)]$ on which the value of any one of these parameters precisely fixes the values of the other two. For instance, at 10 bar total pressure, we can calculate the melting temperature of ice from the Clausius–Clapeyron Equation and the vapor pressure of either ice or water from Eq. (V.47).

This behavior can be seen in Fig. V.13. Water vapor is everywhere present because the vapor pressure is never zero; hence, the freezing curve of ice given here is actually the triple line of water in the $H_2O +$ other gas system.

Adding ammonia means adding one more component and one more composition variable, $x(NH_3)$. We now have $H_2O + NH_3 +$ inert gas, so that $f = 5 - \Phi$.

Because it is easier to understand two-component systems (not because it is sufficient to understand them) we will discuss the binary (two-component) $NH_3 - H_2O$ system first. Here, as we have seen, $f = 4 - \Phi$. But, unlike the $H_2O$-inert gas system, $NH_3$ and $H_2O$ interact strongly. They form solutions, and solid ammonia hydrates are quite stable. An example of the complexity of this system is given in Fig. V.14, in which the freezing behavior of aqueous $NH_3$ solutions at $P = 1$ atm is displayed. Once we have specified $P$ we have thrown away one degree of freedom of the system, and $f = 3 - \Phi$. The Gibbs Phase Rule then assures us that when only one phase is present, such as the solution, there are two degrees of freedom, $T$ and the mole fraction of $NH_3$ (or water), in the solution,

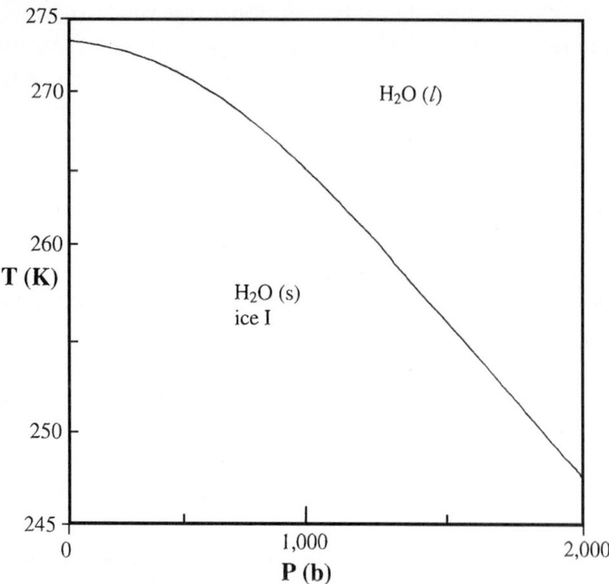


**Figure V.13** Water ice/water/vapor equilibria. The melting curve of pure water ice is calculated from the Clausius–Clapeyron equation. Beyond about 2 kbar the behavior is made much more complex by the formation of high-pressure (dense) crystal structures of ice. We shall treat the high-pressure behavior in Chapter VI.

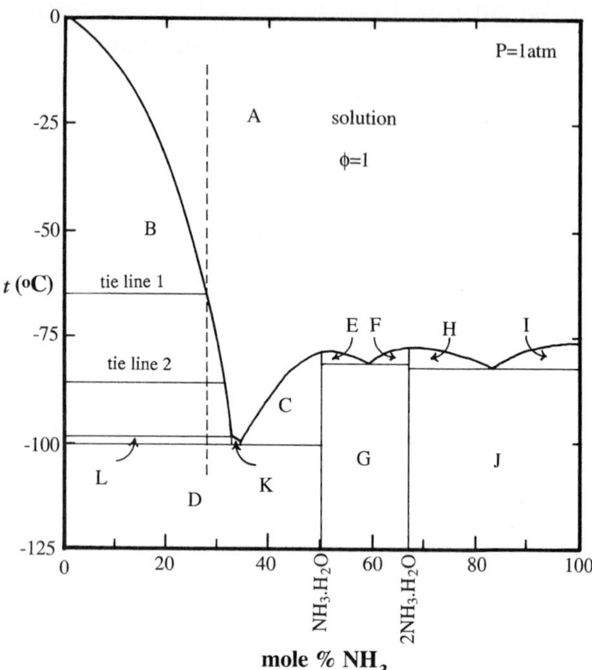


**Figure V.14** Freezing behavior in the $NH_3$—$H_2O$ system. Dissolving ammonia in water dilutes and lowers the activity of liquid water and hence depresses its vapor pressure and freezing temperature. In region A a stable single-liquid solution of ammonia and water is present. B is a two-phase region containing water ice plus a solution of composition given by the illustrative tie lines (isotherms) 1 and 2. Cooling from 1 to 2 with constant total composition (vertical dashed line) causes ice to form and the concentration of ammonia in the residual liquid to increase. Regions C, E, F, H, I, and K are also two-phase regions ($\phi = 2$) in which an ice phase coexists with a solution. Regions D, G, J, and L are two-phase regions occupied by two coexisting ices (such as ammonia hydrate and water ice in region D). Note the deep temperature minima (*eutectic* points) in the melting curve at which three phases, a solution and two ices, coexist ($\phi = 3$). Above the scalloped line only liquid is present; this curve is properly called the *liquidus*. Below the lines atop regions D, G, and J only solids are present; this is termed the *solidus*.

which we shall denote $X(NH_3)$ (or $1 - X(H_2O)$. Wherever two phases, such as solution and $H_2O$ ice, coexist, there is only one degree of freedom, and each temperature corresponds to a particular value of the concentration of the solution in equilibrium with the ice. The line along which solution is in equilibrium with solid phases is called the freezing curve. When three phases coexist, then $f = 0$, a triple point. Clearly multicomponent systems may contain more than one triple point. Now let us reflect upon Fig. V.14. We specified $P = 1$ atm for this diagram, but at these low temperatures the vapor pressures of $H_2O$ and $NH_3$ in equilibrium with the condensed phases will generally be less than 1 atm. This means that, under a piston exerting a pressure of 1 atm, there will be no gas phase whatsoever until the combined equilibrium pressures of $NH_3$ and $H_2O$ become larger than 1 atm. Thus no gas phase coexists with the triple points we have described, and they are not *quadruple points* unless we relax the constraint that $P = 1$ atm and allow total pressures low enough for a gas phase to appear. If we were to redraw Fig. V.14 subject to the condition that the total pressure ($P$) was simply the sum of the $NH_3$ and $H_2O$ equilibrium pressures, then there would be no striking changes in the diagram (why?), but the names of some of the features would change; for example, the triple points now become quadruple points.

Now let us add an inert gas to the mixture of ammonia and water. Now $f = C - \Phi + 2 = 5 - \Phi$. When no condensate is present, $\Phi = 1$, and the system has four degrees of freedom, $P$, $T$, $x(H_2O)$, and $x(NH_3)$, for example. When a solid phase is present in equilibrium with the gas ($\Phi = 2$), then there are only three degrees of freedom. There must therefore be a functional relationship between some of the state variables. It is easy to see that this is the vapor pressure equation of the solid phase. When two solids coexist with the gas we have two vapor pressure equations, only two remaining independent variables, and so on.

When a liquid solution phase, characterized by a composition variable $X(NH_3) = 1 - X(H_2O)$, is present, then the new apparent degree of freedom $X(NH_3)$ is accompanied by two new vapor pressure equations, one for $NH_3$ in equilibrium with the solution and one for $H_2O$ vapor. Thus when $\Phi = 2$, we still have $f = 3$

($P$, $T$, and whichever variable of $X(NH_3)$, $X(H_2O)$, $x(NH_3)$, $x(H_2O)$, or $P(NH_3)$ we choose). Thus we can see how the composition of the condensed phases can be related to the composition of the gas as a general function of $P$ and $T$.

In the $NH_3 - H_2O$ system we shall neglect the solubility of inert gas in the condensed phases, which is a very good approximation to reality. The condensates that are known to be possible at equilibrium in this system at pressures below 1 kbar include aqueous $NH_3$ solution, solid $H_2O$ ice, solid $NH_3$ ice, and the solid hydrates $NH_3 \cdot H_2O$ and $2NH_3 \cdot H_2O$.

In Fig. V.14 at $P = 1$ atm ($f = 3 - \Phi$) it is clear that, within the solution stability field (region A), there are two degrees of freedom, $T$ and $X(NH_3)$. The scalloped line is the freezing curve, at which a cooling solution of concentration $X(NH_3)$ will begin to crystallize. The vertical dashed line corresponding to $X(NH_3) = 0.27$ shows that freezing will begin when the solution is cooled to $-65°$ C (208 K). Any bulk composition, $X(NH_3)$, and temperature, $t(°\,C)$, that lies within region B will, at equilibrium, have two condensed phases present, pure $H_2O$ ice and solution. The two phases can be at equilibrium with one another only if their temperatures are the same, so that the compositions of the two coexisting condensates can be linked together by a horizontal *tie line* (an isotherm). Tie line 2 in Fig. V.14 shows that, at $-85°$ C, a sample that contains a total of 27 mol% $NH_3$ will, at equilibrium, consist of a mixture of pure $H_2O$ ice and an aqueous $NH_3$ solution of concentration $X(NH_3) = 0.32$. By conservation of mass, we can see that the amounts of the two condensates are fully determined; the conservation equations for $NH_3$ and $H_2O$ are

$$X(NH_3)_{total} = X(NH_3)_{H_2O\ ice} M_{H_2O\ ice} + X(NH_3)_{sol'n} M_{sol'n} \qquad (V.52)$$

$$X(H_2O)_{total} = X(H_2O)_{H_2O\ ice} M_{H_2O\ ice} + X(H_2O)_{sol'n} M_{sol'n}, \qquad (V.53)$$

where $M_i$ is the fraction of the total number of moles of material which is present in phase $i$. For the present case,

$$0.27 = 0 \times M_{H_2O\ ice} + 0.32\, M_{sol'n} \qquad (V.54)$$

$$0.73 = 1 \times M_{H_2O\ ice} + 0.68\, M_{sol'n}, \qquad (V.55)$$

whence $M_{sol'n} = 0.844$ and $M_{H_2O\ ice} = 0.156$.

The amounts of the two phases present obey a simple "lever rule," with the fulcrum at the intersection of the tie line and the bulk composition ($X(NH_3) = 0.27$) lines. Each region labeled B, C, E, F, H, or I is a two-phase region containing one solid ice phase and liquid. Each of these regions can be decorated with tie lines just as we have done for region B.

The minima in the freezing point curve correspond to the lowest temperatures at which any liquid can be present. They are called eutectic points, and their coordinates are the eutectic temperature and the eutectic composition. Note that a solution with composition exactly equal to the eutectic composition will freeze all at once at the eutectic temperature to form two coexisting ices. Any solution of $NH_3$ and $H_2O$ with noneutectic composition (say, $X(NH_3) = 0.27$) will freeze progressively as it is cooled from the freezing curve (where an ice first appears) to the relevant eutectic temperature (where the last liquid freezes).

It can be seen that each pair of ices ($H_2O + NH_3 \cdot H_2O$; $NH_3 \cdot H_2O + 2\,NH_3 \cdot H_2O + 2\,NH_3$ $H_2O$; and $NH_3$) exhibits eutectic behavior, and we may speak of the $H_2O$ ice $+ NH_3 \cdot H_2O$ eutectic, etc. When one speaks of the "eutectic in the ammonia–water system," this imprecise statement must be taken to refer to the $H_2O + NH_3 \cdot H_2O$ eutectic, which is the lowest temperature at which a liquid can exist in the $NH_3 - H_2O$ system.

In many cases it is desirable to specify whether a system is fully solid, partly solid, or fully liquid. For this reason, the temperature above which no solids appear is called the *liquidus* curve, and the temperature below which no liquids appear is called the *solidus*. The liquidus in Fig. V.14 is the scalloped line, which is usually loosely called the freezing curve, and the solidus consists of the three lines across the tops of regions D, G, and J. Clearly only solid $H_2O$ and $NH_3 \cdot H_2O$ exist in region D, only the solid hydrates exist in region G, and only solid $2\,NH_3 \cdot H_2O$ and solid ammonia exist in region J. Each of these regions is a two-phase region within which tie lines may be drawn and the lever rule applied.

Note that, at each of the eutectic points, two solid phases and the solution can coexist. There $\Phi = 3$ and $f = 0$. If we replace the constraint that $P = 1$ atm with one that there is a vapor of $NH_3$ and $H_2O$ in equilibrium with the eutectic, then $f = C - \Phi + 2 = 2 - 4 + 2 = 0$, and we can call each eutectic point a quadruple point. There is then a particular gas-phase composition corresponding to the eutectic point. There is no reason why $x(NH_3)$ in this gas should be equal to $X(NH_3)$ in the liquid: the eutectic composition of the liquid is as a rule different from the composition of the gas phase with which it is in equilibrium.

Now let us consider how to relate the composition of the gas phase with the composition of the condensed phases. After the trivial case in which no condensate is present ($\Phi = 1$), the simplest cases are those in which either solid $H_2O$ ice or solid $NH_3$ ice is present. In such a case, the partial pressure of the condensing substance in the gas is constrained by the condition of vapor pressure equilibrium with the solid, $P(NH_3)(T)$ or $P(H_2O)(T)$. Then $f = C - \Phi + 2 = 2 - 2 + 2 = 2$, and $T$ and the partial pressure of the *noncondensing* gas are the independent variables.

Next, consider the circumstance in which pure solid $NH_3 \cdot H_2O$ is in equilibrium with the gas (this occurs only on the vertical line where $X(NH_3) = 0.50$ below the solidus at $-100°C$ in Fig. V.14). The equilibrium between solid and gas can be written

$$NH_3 \cdot H_2O = NH_3(g) + H_2O(g), \qquad (V.56)$$

for which the equilibrium constant is

$$K_{56} = p(NH_3)p(H_2O)/a(NH_3 \cdot H_2O) \qquad (V.57)$$

or, at pressures low enough so that the activity of the solid hydrate is 1,

$$\begin{aligned}\log p(NH_3) &+ \log p(H_2O) \\ &= \log K_{56} \\ &= S_{vap}/2.303R - H_{vap}{}^{\circ}/2.303RT. \qquad (V.58)\end{aligned}$$

Here $T$, $P(H_2O)$, and $P(NH_3)$ are variables related by Eq. (V.58), so that any one of them can be eliminated. Thus $f = 2$.

When solid 2 $NH_3 \cdot H_2O$ is present, we have

$$2\,NH_3 \cdot H_2O(s) = 2\,NH_3 + H_2O \qquad (V.59)$$

$$\begin{aligned}K_{59} &= p(NH_3)^2 p(H_2O) \\ &\div a(2\,NH_3 \cdot H_2O) \\ &= p(NH_3)^2 p(H_2O), \qquad (V.60)\end{aligned}$$

thus

$$2\log p(NH_3) + \log p(H_2O) = \log K_{59}, \qquad (V.61)$$

and the arguments regarding the application of the Gibbs Phase Rule are the same as above.

In a slightly more complex case, consider a solution of concentration $X(NH_3)$ coexisting with the gas:

$$H_2O(sol'n) = H_2O \qquad (V.62)$$

$$NH_3(sol'n) = NH_3 \qquad (V.63)$$

$$K_{62} = p(H_2O)/a(H_2O)(sol'n) \qquad (V.64)$$

$$K_{63} = p(NH_3)/a(NH_3)(sol'n). \qquad (V.65)$$

For each gas there will be a vapor-pressure equation $p_i(X_i, T)$, not necessarily of simple form, because the enthalpy and entropy of vaporization and the activity of each solute will be functions of $X_i$ and $T$ as well. Using these two vapor pressure equations, there are two remaining degrees of freedom, $T$ and $X_i$.

When two compatible ices coexist, then we can apply two vapor pressure constraints simultaneously, such as Eqs. (V.56) and (V.59), in the region in which the two hydrates coexist. Under these circumstances, the equilibrium expressions [Eqs. (V.57) and (V.60)] yield

$$K_{59}/K_{56} = p(NH_3), \qquad (V.66)$$

which is a function of temperature alone. Then, because we know $P(NH_3)(T)$, we can use either Eq. (V.57) or (V.60) to calculate $P(H_2O)(T)$. This situation, in which two condensed coexisting phases in a two-component system give unique partial pressures of both gases for each temperature, is analogous to a vapor pressure equation for a single condensed phase in a one-component system. The general case of $n$ components and $n$ condensed phases ($n + 1$ total phases, including the gas) gives $f = C - \Phi + 2 = n - (n + 1) + 2 = 1$, and the sole free variable is temperature. This situation, in which knowledge of the temperature alone (and the identity of the solid phases) fully characterizes the gas partial pressures, is called *buffering*. The *buffer reaction* corresponding to Eq. (V.66) is

$$2NH_3 \cdot H_2O(s) = NH_3 \cdot H_2O(s) + NH_3 \qquad (V.67)$$

$$K_{67} = p(NH_3) = K_{59}/K_{56}. \qquad (V.68)$$

This brings out the formal similarity to simple vapor pressure equilibrium, the simplest ($n = 1$) kind of buffer.

The next three cases are (a) pure solid $NH_3$ or $H_2O$ coexisting with liquid, (b) pure solid hydrate with liquid, and (c) two solid phases with liquid ($\Phi = 4$ counting the gas). These are only slightly more complex than the examples we have already considered and may be worked through for practice by those who have had no previous experience with phase equilibria.

We have so far avoided explicit statements of the vapor pressure equations $[p_i(X_i, T)]$ for the solution phase because of their complexity; however, it is easy to present these relationships graphically. We shall, in Fig. V.15, represent the phase diagram of the $NH_3 + H_2O$ system on the $P(NH_3) - P(H_2O)$ plane. High total vapor pressures (top right) of course correspond to high temperatures.

Wherever pure $H_2O$ ice exists, $P(H_2O)$ is a function of $T$ alone, and the isotherms lie along lines of constant $P(H_2O)$. Where pure $NH_3$ ice exists, $P(NH_3)$ is a function of $T$ alone, and isotherms lie along lines of constant $P(NH_3)$.

When the only solid phase is $NH_3 \cdot H_2O$, Eq. (V.58) assures us that an isotherm is a line of slope $-1$ on the $\log P(NH_3) - \log P(H_2O)$ plane. Similarly, when only 2 $NH_3 \cdot H_2O$ is present, Eq. (V.67) shows that the isotherms must have a slope of $-\frac{1}{2}$ in the $\log P(NH_3)$ vs $\log P(H_2O)$ diagram.

In the liquid solution region the isotherms are curves that become horizontal in the extreme of $X(NH_3) \approx 1$ (liquid ammonia) and vertical when $X(H_2O) \approx 1$ (pure

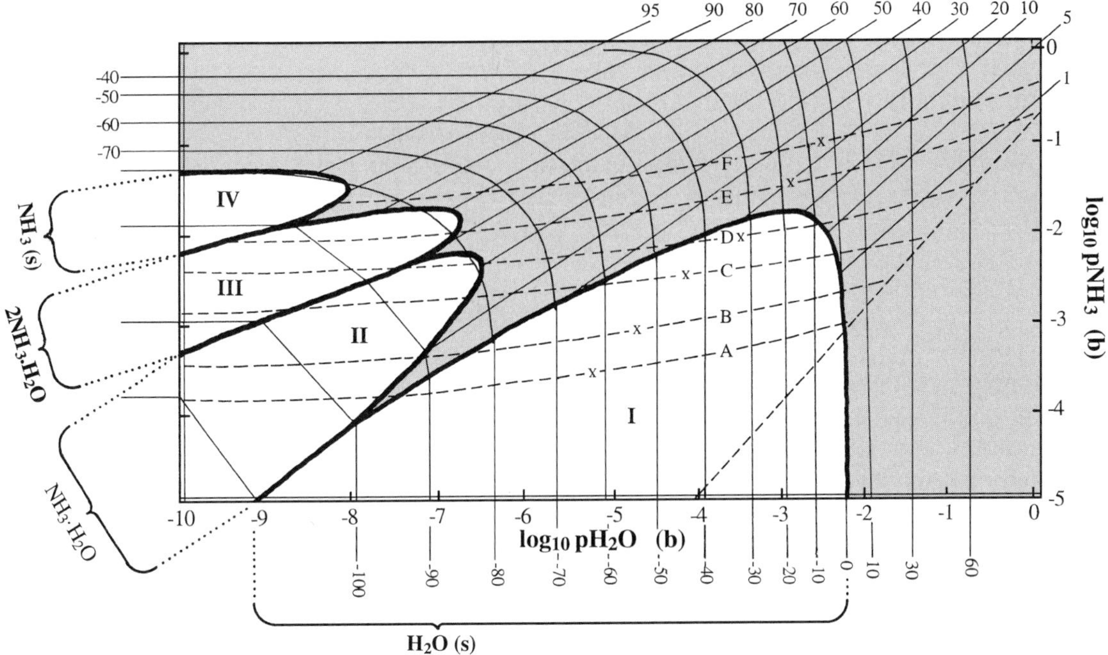


**Figure V.15** Vapor pressures in the $NH_3$-$H_2O$ system. This diagram, graphed in the log $p(NH_3)$ vs log $p(H_2O)$ plane, relates the compositions of condensed water- and ammonia-bearing phases to the composition of the coexisting vapor. Region I contains water ice, II contains ammonia monohydrate, III contains diammonia hydrate, and IV contains solid ammonia. Contours of constant temperature and solution composition (in mol % $NH_3$) are superimposed. Aqueous ammonia solution exists in the stippled region.

liquid water). These extremes are approached when $P(NH_3) >> P(H_2O)$ and $P(NH_3) << P(H_2O)$, respectively, in the top left and bottom right of the diagram. It is instructive to follow the $-80°$ C isotherm and explain each of the seven branches of the curve in this way. The scalloped curve is again the freezing curve, and the four regions under it contain, from right (water rich) to left ($NH_3$ rich), $H_2O$(s), $NH_3 \cdot H_2O$(s), 2 $NH_3 \cdot H_2O$(s), and $NH_3$(s), respectively.

The solution concentration $X(NH_3)$ above the freezing curve will generally increase with increasing $P(NH_3)$: $P(H_2O)$ ratios. Both isotherms and, within the solution region, *isopleths* of constant $X(NH_3)$ are shown in Fig. V.15.

Now we need only have some knowledge of the way in which $P(NH_3)$, $P(H_2O)$, and $T$ vary with altitude in an atmosphere in order to calculate what condensed phases, if any, may be in equilibrium with the gas. However, for Jupiter and Saturn we know the cosmic abundances of the elements and hence know $P(NH_3)$ and $P(H_2O)$ as functions of $P$ before condensation occurs. We know a $(P, T)$ point on the adiabat, and hence we know $P(T)$. Because there is no observable "ground level" our altitude scale is arbitrary. We thus define $Z(T)$ by setting $Z = 0$ at some particular temperature level, such as 130 K or 300 K.

Finally, in order to describe the entire inert gas–$NH_3$-$H_2O$-$H_2S$ system, we must allow for condensation of $H_2S$ compounds. Hydrogen sulfide, as a weak acid, readily reacts with the weak base ammonia to form the salts $NH_4SH$ (ammonium hydrosulfide) and $(NH_4)_2S$ (ammonium sulfide). $H_2S$ is soluble in basic ammoniacal solutions,

$$NH_3(\text{sol'n}) + H_2S = NH_4^+(\text{sol'n}) + HS^-(\text{sol'n}), \quad (V.69)$$

and solid $H_2S$ must also be included.

The vapor pressure relations for the solution and solid phases now must be graphed in three dimensions, with a log $P(H_2S)$ axis added to Fig. V.15. Solution isotherms and isopleths now become two-dimensional intersecting surfaces. The vapor pressure equations for the three-component solution now become even more complex, because, for example, $P(NH_3) = P(T, X(NH_3), X(H_2S))$. Also, Fig. V.15 was constructed for the case in which $P = P(NH_3) + P(H_2O)$. In the real Jovian case $P$ is very much larger than $P(NH_3) + P(H_2O) + P(H_2S)$ because of the great abundances of $H_2$ and He, and we must expand the graph to four dimensions. Fortunately we are spared the necessity of attempting this by the fact that the solubilities of hydrogen and helium are very low. Mapping out the entire ($P$, $T$, $P(NH_3)$, $P(H_2O)$) volume would be a formidable task, but, if solar composition is assumed, this reduces to mapping out the ($P$, $T$) plane, which is readily tractable.

Below the level of $H_2O$ saturation, $f(NH_3)$ and $f(H_2O)$ are constant, so that $P(NH_3)/P(H_2O)$ is constant. The dashed line of slope +1 on the $\log P(NH_3) - \log P(H_2O)$ plane in Fig. V.15 represents the compositions allowed for the gas phase in the *absence* of condensation. The nearly horizontal dashed lines show the ($P(NH_3)$, $P(H_2O)$) trajectories followed above the altitude of water saturation, where $P(H_2O) = P_{vap}(H_2O)$. It can easily be seen that the ammonia concentration in the solution is small, less than $X(NH_3) = 0.2$ (approximately 1 mole per liter), and that the first condensate on both Jupiter and Saturn is dilute aqueous $NH_3$ solution. Condensation alone removes 0.02 mole of $NH_3$ from the gas for every 0.8 mole of water, and adiabatic expansion alone reduces both $P(NH_3)$ and $P(H_2O)$ by the same factor. Because the vapor pressure depends on temperature more strongly than the total pressure along an adiabat depends on temperature, the effect of condensation is to reduce the $H_2O$ partial pressure far more rapidly than the $NH_3$ pressure. On both planets, water saturation occurs at temperatures not very far above the freezing curve (or freezing line, because both $P$ and $X(NH_3)$ are small enough to have little effect on $T_m$).

Thus only a few kilometers above the water-droplet cloud base, the temperatures are low enough to freeze the droplets. Figure V.14 shows that the solid phase produced by freezing such dilute $NH_3$ solutions must be pure $H_2O$ ice. (The concentration of dissolved $H_2S$ is several times less than the $NH_3$ concentration and can be safely neglected.)

Let us imagine a region in the atmosphere in which both aqueous $NH_3$ solution droplets and $H_2O$ ice crystals coexist with the gas. In the $H_2O$–$NH_3$–inert gas system we have three components and three phases, leaving two degrees of freedom, a line in ($P$, $T$) space. However, the adiabat for each atmosphere is a line in ($P$, $T$) space, and these two lines can only intersect at a point. Thus each two-condensed-phase region has zero vertical extent, being confined to a single ($P$, $T$) point on the adiabat. Thus the aqueous solution cloud tops will indeed give way abruptly to a cloud layer made of crystals of $H_2O$ ice at the 270-K level.

Condensation of ammonium hydrosulfide occurs via the equilibrium

$$NH_4SH(s) = NH_3 + H_2S \tag{V.70}$$

$$K_{73} = p(NH_3)p(H_2S)/a(NH_4SH), \tag{V.71}$$

with saturation occurring at 215 K. At this temperature, water vapor has already been depleted by about a factor of 300 by condensation, so that water-bearing clouds of any sort, whether composed of ice or droplets, have very little mass.

Above the $NH_4SH$ condensation level the $H_2S$ and $H_2O$ partial pressures drop off so rapidly that the cloud masses are negligible down to the point of saturation of ammonia, below 160 K. There a third dense cloud layer, composed of crystals of solid ammonia, will condense. The structure of the major cloud layers formed below 500 K is shown schematically in Fig. V.16.

We have earlier mentioned the possibility that sulfur, oxygen, or both might be present in Jupiter with abundances markedly in excess of their expected solar value. It is now possible for us to describe the abundances of a wide variety of gases, both condensible and "permanent," in Jupiter's atmosphere for both solar and nonsolar compositions and to investigate the sensitivity of observable properties in the cloudtop region to the detailed assumptions regarding composition of the lower atmosphere.

The principles for calculating these profiles are simple, although the practice may sometimes become tedious. Each element keeps its solar mole fraction as it follows an adiabat down to its condensation temperature.

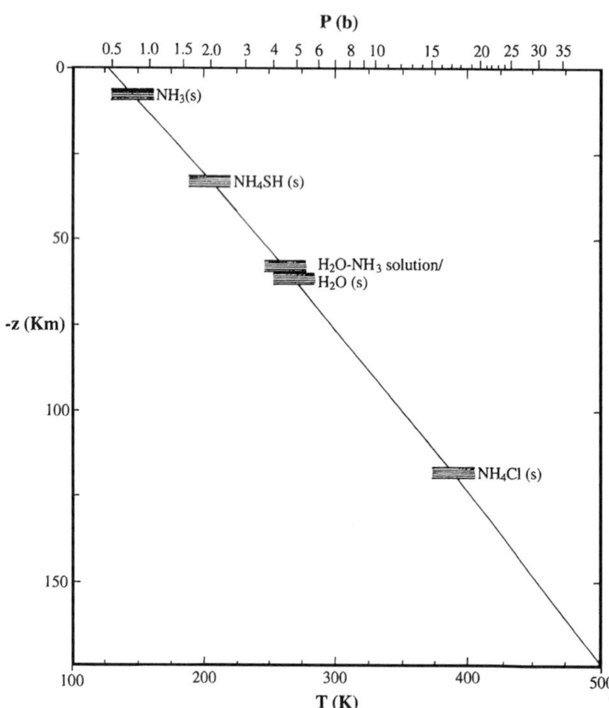


**Figure V.16** Cloud structures in Jupiter's upper troposphere. The saturation levels of ammonia, ammonium hydrosulfide, aqueous ammonia solution and water ice, and ammonium chloride are shown as calculated for a solar composition gas.

Then the partial pressure ceases to be given by $p_i = cP$ and is given instead by a vapor pressure equation, $p_i = p(T)$. Figure V.17 shows how the partial pressures of a number of gases vary along a Jupiter adiabat. The results for Saturn are very similar and do not merit a separate figure.

From Fig. V.16 we may deduce the general effect of increasing the abundance of oxygen: saturation will occur at rather high temperatures, and a thick, massive cloud layer of very dilute aqueous $NH_3$ solution clouds will be present. A 30-times-larger $H_2O$ abundance would lead to saturation at the 13-bar, 335-K level instead of the 5.5-bar, 280-K level for the solar-composition Jupiter model. As before, only a small fraction of the total amount of ammonia will dissolve. The $NH_3$ partial pressure at the base of the droplet clouds is 4.0 mm of mercury (5.3 mbar), and the partial pressure of $NH_3$ at all higher (colder) levels in the atmosphere is of course less. Thus the trajectory followed by $P(NH_3)$ and $P(H_2O)$ in Fig. V.15 goes through a wide expanse of the $H_2O$ ice region, entering it at 271 K and leaving it near 200 K. By this point, of course, the water has been depleted to about one millionth of its original partial pressure (or $10^{-5}$ of its original mole fraction). Ammonium hydrosulfide and solid ammonia will condense very much as before, at only slightly lower temperatures. Of course, condensation of so much water has other effects that we will need to consider in detail, especially the thermal effects of heat of condensation released by the water vapor as it condenses into droplets.

Varying the sulfur abundance has much more significant impact on observable properties of the planet. Because the cosmic nitrogen abundance is only about seven times the sulfur abundance, $NH_4SH$ precipitation will completely remove $NH_3$ and leave leftover $H_2S$ gas above the $NH_4SH$ cloudtops if $H_2S$ is enhanced by a factor of more than 7. Thus the entire $NH_3(s)$ cloud layer would vanish, the "visible" clouds would be a dense $NH_4SH$ layer with its base near the 225-K level, $NH_3$ would not be observable in the gas above the clouds, and the mole fraction of $H_2S$ above the clouds would be equal to the value in the lower atmosphere minus the $NH_3$ mole fraction in the same region. It is easy to sketch out quickly a whole suite of models that have enhanced sulfur abundances.

Simultaneous enhancement of water and $H_2S$ by any factor over 7 will of course produce observable effects very little different from those produced by simply enhancing $H_2S$ alone.

We have seen that observations of the thermal flux emitted by Jupiter in the 5-$\mu$m window show a whole-disk temperature near 200 K. It is very interesting that this emission is very much stronger, with effective temperatures up to $\approx$300 K seen very locally in certain dark

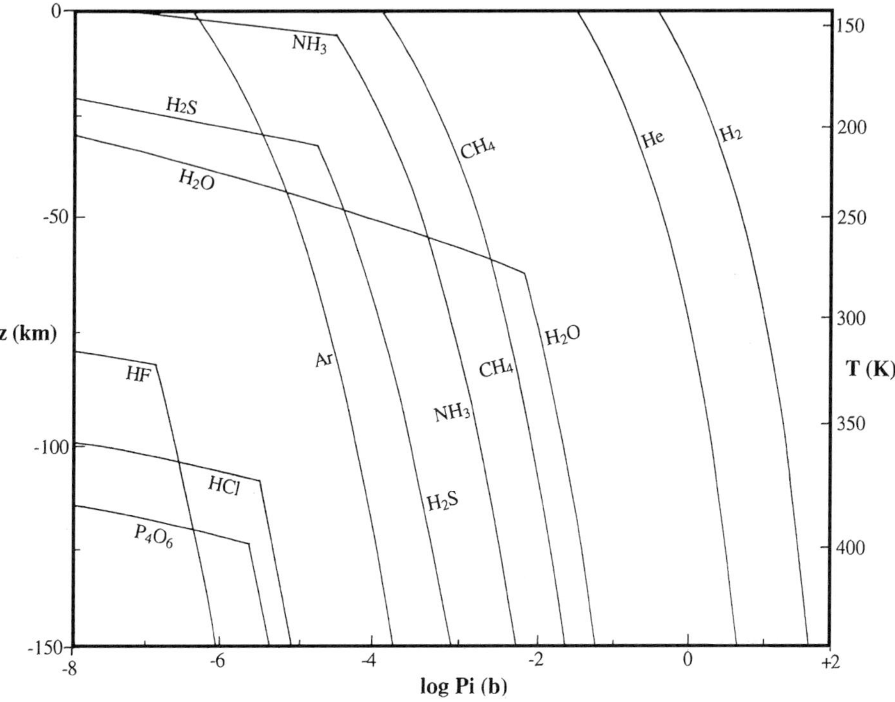


**Figure V.17** Gas abundances in Jupiter's upper troposphere. The partial pressures of a number of major constituents are shown as a function of temperature and depth beneath the ammonia cloud layer. The calculations assume solar elemental composition.

bluish spots, in the North Equatorial Belt (NEB). The hottest spots are generally the darkest spots on the planet visually. Most of these are found in the intensely colored (brown to orange) NEB.

If we wish to postulate a clearing of the $NH_3$ and $NH_4SH$ cloud layers sufficient to permit the eye to see to pressures greater than 2 bar and to permit unattenuated 5-$\mu$m emission from the level of the water clouds or deeper, then we must construct an optical model that takes into account relatively subtle sources of opacity. It is by no means obvious that these requirements of low opacity are easily met.

For example, even a perfectly clear, cloud-free, nonabsorbing gas attenuates light by Rayleigh scattering. The scattering efficiency of any molecular scatterer is proportional to the wavelength of the light to the $-4$ power,

$$\sigma_R = c\lambda^{-4}, \qquad \text{(V.72)}$$

where $\sigma$ is the cross-section for Rayleigh scattering in square centimeters per molecule. Thus visible light of $\lambda = 0.5\mu$m will be scattered $10^4$ times as efficiently as the 5-$\mu$m thermal emission. The constant $c$ for hydrogen is large enough so that pressures of a few bars will suffice to scatter blue sunlight many times over while allowing almost all the red end of the solar spectrum to penetrate or be absorbed. Thus a very deep, perfectly clear atmosphere over a dark surface would look blue due to scattered sunlight. Similarly strong scattering of 5-$\mu$m light would require enormously higher densities. We conclude that such a cloud clearing is an acceptable explanation of the dark blue 5-$\mu$m hot spots, but that a careful assessment of other sources of opacity at 5 $\mu$m is urgently needed.

We shall now consider in some detail the thermodynamics and physics of cloud condensation and stability.

## Cloud Physics on the Jovian Planets

Any chemical reaction, condensation process, or phase change in a planetary atmosphere involves the release or absorption of heat. If the mole fraction of the substances undergoing these transformations is large enough, then the heat change due to the process may have an important effect on atmospheric structure.

Let us first imagine a container which holds 1 mole of liquid water at $P = 1$ atm. Heat is being withdrawn from the container at a constant rate of $-dq/dt$ J s$^{-1}$. The water, with heat capacity $C_p$(l), cools at the rate of

$$dT/dt = (dq/dt)/(dq/dT) = (dq/dt)/C_p(1). \qquad \text{(V.73)}$$

Cooling will continue at this rate until the freezing point of water is reached. We have now a one-component system containing two coexisting phases, with the additional constraint that $P = 1$ atm. Thus $f = C - \Phi + 1 = 0$. In other words, as long as liquid and solid coexist at $P = 1$ atm, the temperature cannot change.

The heat of fusion of ice at 1 atm pressure, $\Delta H^\circ_m$, is just $-\Delta H^\circ_f$, the heat of freezing. In order to withdraw this amount of heat from the water and freeze all the water, we must cool the water for a time, $t_f$, at a constant rate of $-dq/dt$:

$$\Delta H^\circ_f = -(dq/dt)t_f \text{J mol}^{-1}. \qquad \text{(V.74)}$$

After all the water has been frozen, the further withdrawal of heat simply has the effect of cooling the ice at a rate $(dq/dt)/C_p$(ice). This sequence is illustrated in Fig. V.18. Somewhat more complex, but also more relevant to processes in planetary atmospheres, is the condensation of a vapor to form a liquid (or solid) phase. Consider a container held at $P = 1$ atm, containing 1 mole of water vapor and some inert gas at a temperature above the normal boiling point of water. Withdrawal of heat from the water vapor at the constant rate $-dq/dt$ will cool the vapor at the rate $dT/dt = (dq/dt)/C_p$(gas), where $C_p$ is the bulk heat capacity of the gas at constant pressure. $C_p = f_{\text{inert}} C_{p_{\text{inert}}} + x(H_2O)C_p(H_2O)$. At 100° C, condensation of liquid water begins. Inert gas must now be added to maintain a total pressure of 1 atm. The system contains two components and two phases, and the pressure is constrained to be 1 atm, so $f = 2 - 2 + 1 = 1$. Thus the temperature and the partial pressure of the inert gas are related by the equation $P = 1 = p_v(H_2O)(T) + P_{\text{inert}}$. Either one may be varied freely, but not both at the same time. We may choose to say that temperature is the free variable. In other words, the two phases, water vapor and liquid water, may coexist at any of a wide variety of different temperatures, and condensation of water does *not* occur all at one temperature. (The case of a *one*-component system contained by a piston exerting 1 atm pressure is very different and should be thought through and compared with the case we are considering here.)

Thus continued withdrawal of heat from water vapor and liquid water below the saturation temperature has the dual effect of cooling the system and of lowering the partial pressure of water vapor by condensation. Letting $\Delta H^\circ_v$ denote the heat of vaporization of water at $P = 1$ atm, and $dn_v/dT$ the number of moles of vapor condensation during cooling through $\Delta t$,

$$dq/dT = C_p(g)n_v + C_p(1)n_1 - \Delta H^\circ_v dn_v/dT + C_p(\text{inert})n_{\text{inert}}. \qquad \text{(V.75)}$$

The third term, the heat released by condensation, clearly depends on how much vapor is available to condense. When saturation first occurs the vapor pressure

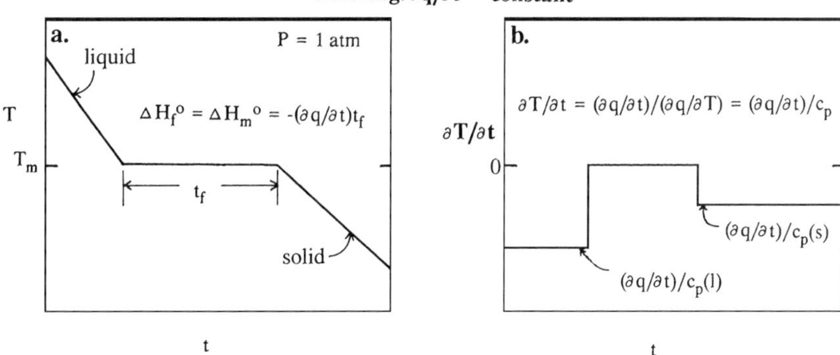


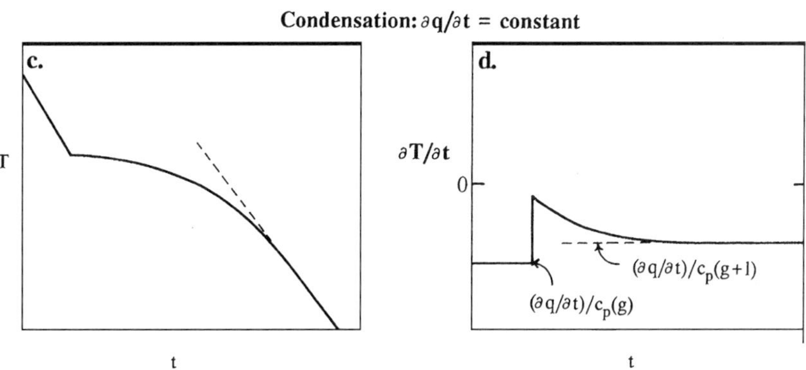


**Figure V.18** Changes of state: freezing and condensation. a displays temperature vs time for cooling of a liquid through the freezing temperature, assuming a constant rate of withdrawal of heat, $dq/dt$. The duration of the halt in the temperature curve is proportional to the heat of melting. b shows the rate of change of temperature, which is zero while both liquid and solid coexist and varies inversely with the heat capacity for pure phases. c is a $T$ vs $t$ plot for condensation of a minor gas constituent at a total fixed pressure of 1 atm. There is no halt in the cooling rate, only a decrease proportional to how much vapor is condensing in each temperature interval. d shows the asymptotic limit approached at low temperatures, at which the vapor pressure of the condensible gas is negligible. Note the dramatic decrease of $dT/dt$ at the point of saturation of the condensate.

is very large, and cooling through a few degrees condenses a great amount of liquid. At 200 K the vapor pressure is so low that even cooling to absolute zero condenses a negligible number of moles of additional material. Thus the magnitude of $dq/dt$ may be very small when condensation begins. This behavior is sketched out in Fig. V.18. Note that at very low temperatures Eq. (V.75) simplifies to

$$dq/dT = C_p(1) + C_p(\text{inert})n_{\text{inert}} \tag{V.76}$$

because $n_v + n_l = 1$ mole, and $n_v$ vanishes as $T \rightarrow 0$. A rising parcel of atmosphere containing hydrogen, helium, and a trace of water vapor will release heat in a similar manner, with the greatest heat release occurring immediately upon saturation at the cloud base. The rate of cooling of the parcel of atmosphere by adiabatic expansion will reach a local minimum at this level, and the vertical temperature gradient may be markedly decreased.

We shall now rederive the equations for the ($P$, $T$, $Z$) structure of an adiabatic planetary atmosphere in hydrostatic equilibrium, but including the effects of condensation of minor constituents. The statement of conservation of energy, the first law of thermodynamics, is now, in general for 1 mole of gas,

$$Pdv + C_v dT + \Sigma_i \lambda_i dn_i = 0, \tag{V.77}$$

where $v$ is the volume occupied by Avogadro's number ($N_0$) of gas molecules, $dn_i$ is the change in the number of moles of substance $i$ in the gas phase due to condensation, and $\lambda_i = -\Delta H_v^\circ(i)$ is the molar heat of condensation of substance $i$. For simplicity we will consider a single condensable substance, so $dn_i = dx$:

$$Pdv + C_v dT + \lambda dx = 0. \tag{V.78}$$

From the ideal gas law,

$$Pdv + vdP = RdT, \qquad (V.79)$$

and therefore

$$vdP - \lambda dx = (C_v + R)dT = C_p dT. \qquad (V.80)$$

The vapor pressure equation for the condensable substance is

$$\ln p = \Delta S/R - \lambda/RT, \qquad (V.81)$$

where

$$d\ln p/dT = \lambda/RT^2 = dp/pdT, \qquad (V.82)$$

which is another form of the Clausius–Clapeyron Equation.

We shall assume that the atmosphere is in hydrostatic equilibrium:

$$dp = -P(\mu g/RT)dz. \qquad (V.83)$$

We still need an expression for $dn$ [in Eq. (V.78)] in terms of the pressure variables $P$ and $p$, but $x = p/P$ because we have a total of 1 mole of gas, and thus

$$dx = dp/P - (p/P^2)dP, \qquad (V.84)$$

where the first term reflects the change in the number of moles of condensable gas due to changes in the vapor pressure and in the molar volume, whereas the second term corrects for the change in molar volume caused by changes in the total pressure.

Combining Eqs. (V.82) and (V.84) to eliminate $dp$,

$$dx = p\lambda dT/PRT^2 - (p/P^2)dP; \qquad (V.85)$$

substituting for $dx$ in Eq. (V.80),

$$vdP - p\lambda^2 dT/PRT^2 + (P\lambda/P^2)dP = C_p dT, \qquad (V.86)$$

and for $dP$ in Eq. (V.83),

$$[RT/P + p\lambda/P^2][-P\mu g/RT]dz = [C_p + p\lambda^2/PRT^2]dT; \qquad (V.87)$$

and rearranging,

$$(dT/dz)_S = -\mu g[1 + p\lambda/PRT]/C_p[1 + p^2\lambda^2/C_p PRT^2]. \qquad (V.88)$$

It is convenient to use the dimensionless energy variable $Y = f\lambda/RT$ as a measure of the strength of the heat contribution from condensation:

$$(dT/dz)_S = -\mu g(1 + Y)/C_p[1 + RY^2/xC_p], \qquad (V.89)$$

for $H_2O$ condensation near 280 K, $Y \approx 10^{-3} \times 10^4/(2 \times 280) = 0.018$. At this temperature $C_p \approx 3.4R$, and the last term in the denominator becomes $(0.018)^2/(3.4 \times 10^{-3}) = 0.094$. Then the moist adiabatic lapse rate differs from the dry adiabatic lapse by a factor of $1.018/1.094 = 0.93$. Thus water condensation decreases the temperature gradient by 7% at the water cloud base. For ammonia, $Y \approx 0.04$ and $C_p \approx 3.0R$; hence the factor is $1.014/1.027 = 0.987$. Thus condensation of the less abundant $NH_3$ with its smaller $\lambda$ produces a larger effect on the temperature gradient than water condensation does because $NH_3$ condensation occurs over a much narrower altitude range. (Note also that for substances of $x < 10^{-3}$, it is a reasonably good approximation to let the factor $(1 + d)/(1 + e)$ be $1 + d - e$ or even (when $e >> d$) $1 - e$.)

There are very few species that are abundant enough in a solar-composition gas to have an appreciable thermal effect upon condensation; water and ammonia do, as we have seen. Methane, which saturates at 60 K, will have a large effect, $Y \approx 10^{\neq 2}$, $C_p = 2.5R$, and $x = 5 \times 10^{-4}$, and thus the factor is $1.01/1.08 = 0.935$. Hydrogen sulfide ($x = 3 \times 10^{-5}$), will have $Y =$ only 0.0013 and a factor of $1.0013/1.02 = 0.982$. Silicon, magnesium, and iron, with large heats of condensation, have abundances below $10^{-4}$ and condense over wide altitude ranges; for iron, $\lambda = 1.2 \times 10^5$, $Y = 0.0013$, and the lapse rate factor is $1.0013/1.007 = 0.994$, so close to unity that it is not necessary to make the correction.

Another interesting effect that also, like latent heat, profoundly influences the effective heat capacity and structure of the atmosphere is contributed by hydrogen. The protons in the nuclei of the two H atoms in molecular hydrogen may be aligned parallel (*ortho*) or antiparallel (*para*). At high temperatures the statistical weight of the *ortho* state is three times that of the *para* state. Thus, above about 200 K, the equilibrium *o:p* ratio is 3. At lower temperatures the equilibrium *o:p* ratio drops rapidly, approaching zero below 25 K. But equilibrium is not easy to achieve! A paramagnetic gas or surface can catalyze the conversion, but neither is common on the Jovian planets. "Normal" hydrogen, with an *o:p* ratio of 3, can readily be cooled to 25 K in the laboratory without any detectable conversion of *ortho* to *para* hydrogen. The energy of spin pairing is quite significant, and "normal" and equilibrium hydrogen must therefore have markedly different heat capacities. The thermal structure of a cold hydrogen atmosphere is sensitive to the rate of conversion of *ortho* to *para* in rising parcels of atmosphere. Curiously, observations of the cloudtop atmospheres of the Jovian planets suggest that equilibrium is closely approached on all three planets. On Jupiter, for example, the *o:p* ratio reflects equilibration at 130 K, typical of the heart of the ammonia cloud layer. This seems to require efficient catalytic conversion on aerosol grain surfaces, probably on high cloud particles whose surfaces have been damaged by solar UV light.

Now that we know how to calculate vertical temperature profiles in the presence of condensation processes or chemical reactions, we may address specifically the question of the vertical distribution of cloud mass.

To a first approximation, cloud particles will be most abundant where they condense, and the maximum density will occur at the cloud base, just above the saturation level. We may establish a simple *relative* cloud mass profile by raising a column of atmospheric gas along a Jupiter adiabat in a number of small height increments, $dz$. After each increment the temperature and pressure changes for each volume element of the column are computed. The difference between the value of the mole fraction of each condensable gas before that step and that after it is the amount of material that must condense over the altitude interval $dz$. That material is accumulated at that altitude and not moved along with the gas when the column is raised through its next height increment. The total cloud masses derived in this manner may be excessively large, because no fallout mechanism is provided. The *relative* masses of the different cloud layers are, however, likely to be reasonable reflections of reality. The procedure, then, is to select a volume element of gas, such as $N_0$ molecules. This amount of gas presents a vertical column of $1\,cm^2$ cross-section and 35 m length at the water cloud level or 100 m at the ammonia cloud level. The partial pressure of condensable gas in each of these 1-mole parcels stacked in the $1\text{-}cm^2$ column is calculated from Eq. (V.82). Then each gas sample is moved upward into the next box, the partial pressure ($p$) is recalculated, and the decrease in $p$ due to adiabatic expansion is calculated via $dp/p = dP/P$, where $P$ comes from the adiabatic condition and the hydrostatic equilibrium condition. The decrease in $p$ in excess of that caused by expansion gives the amount by which condensation has changed the gas pressure. This pressure change divided by $P$ is $x$, and $x\mu$ is the mass of condensate formed. The density of condensate is then simply $\mu\Delta x/v$. Each volume element of gas can be expanded in this manner for enough $dz$ steps of 35–100 m to reach the $NH_3$ cloudtops (1000 steps). Each volume element individually gives a cloud-density profile that faithfully reflects where condensation occurs, but not the absolute mass of condensate or its altitude profile in the presence of realistic atmospheric motions.

The relative cloud masses calculated in this manner for the nominal Jupiter adiabat are given in Fig. V.19, along with the corresponding moist adiabatic lapse rate. If the deviations from the dry adiabatic lapse are neglected, the deduced atmospheric structure will still be only very slightly in error. If a moist adiabat and a dry adiabat are both run up to the 0.5-bar level, starting from identical conditions at 300 K, the moist adiabat will be only about 2 K warmer above the $NH_3$ clouds.

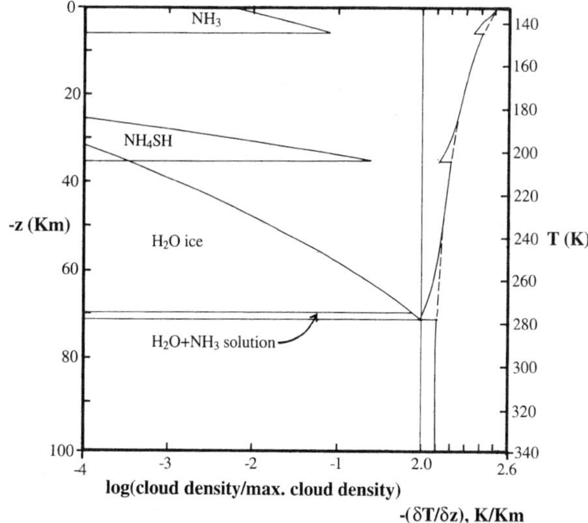


**Figure V.19** Relative cloud masses and moist adiabatic lapse rate in Jupiter's atmosphere. Note the decreases in the lapse rate caused locally by the condensation of cloud materials. An adiabatic region with the same base temperature and pressure but no condensibles would have a higher average lapse rate and hence be cooler at high altitudes by about 2 K. This temperature and density difference can drive atmospheric motions.

That temperature difference, however, is enough to produce a 1.5% larger density in the top of the dry adiabatic column. This condition will propagate, with the dry column sinking and the moist column rising, driven by the latent heat released by condensation. Such a mechanism has been suggested for maintaining the clear regions in the North Equatorial Belt and the bright white, apparently fresh cloud particles in the neighboring zones.

A realistic treatment of cloud density profiles would involve mechanistic consideration of droplet size distributions, dependence of vapor pressure (and droplet composition) on particle size, the speed of advection currents (updrafts), droplet sedimentation rates, sweep-up of small droplets by faster-falling large ones, the freezing behavior of two- and three-component solution droplets, the viscosity and surface tension of droplets of different sizes, aerodynamic and collisional breakup of the largest droplets, and a number of other factors. The very low surface tensions and high vapor pressures characteristic of ammonia–water solutions will cause clouds made of them to behave very differently than terrestrial water clouds. In addition, the horizontal variation of cloud density caused by rising and falling columns of atmosphere and by the general planetary circulation remind us that the clouds are a three-dimensional, not a one-dimensional, phenomenon.

Cloud models for the Jovian atmosphere were not seriously affected by the observations of the Galileo

Probe, which unfortunately fell into a rare cloudless region. A dry, descending gas column with severely depleted water content was traversed by the Galileo Probe. Because of the lack of confirming data, only the principal features of these cloud models for Jupiter (temperature of the cloud base; general decrease of average cloud density with altitude) should be taken seriously.

Saturn may have similar features, but there are serious observational problems that make this question difficult to investigate from Earth. First, Saturn receives 3.7 times less sunlight per unit area than does Jupiter; second, the maximum solid angle subtended by Saturn (observed at opposition) is 6.6 times smaller than that of Jupiter at opposition. Saturn's slightly higher albedo still leaves it some 20 times harder to observe than Jupiter.

With respect to cloud structure, Saturn's slightly colder ("lower") adiabat means that all cloud layers will have their bases at somewhat higher temperatures than the corresponding layers on Jupiter. Although aqueous $NH_3$ solution clouds are marginal on Jupiter, they are expected to be massive on Saturn. Condensation of droplet clouds should begin at 20 bar and 295 K, and the freezing point should be 271 K; solid $NH_4SH$ should saturate at 8 bar and 220 K, and solid $NH_3$ should saturate at 3 bar and 160 K. The *relative* cloud masses are very insensitive to the adiabat chosen as long as the gas composition below the clouds is strictly solar.

If water is enriched severalfold relative to solar proportions, then the solution cloud base extends downward to warmer saturation levels and the mass of the $H_2O$-bearing condensate increases by a similar factor. Enhancement of $H_2S$ causes a similar increase in the saturation temperature and total mass of $NH_4SH$ clouds. Figure V.20 shows Jupiter's cloud structure for 10-fold enrichment of both $H_2O$ and $H_2S$ (S:N > 1). Here, the large latent heat released by condensation (and lowering of the mean molecular weight by removal of the heavy condensates) combine to suppress vertical motions. We remarked earlier on the absence of the $NH_3$ clouds when the $H_2S$ abundance exceeds that of $NH_3$. In this case, an excess of $H_2S$ survives above the $NH_4SH$ cloudtops.

The actual vertical structure of the Jovian clouds is only poorly known from observations and presents a variety of complex theoretical problems to the would-be modelist. The first of these is the mechanism and rate of nucleation of condensates. Second, the vapor–condensate interactions during particle growth must be understood. Third, coalescence and fragmentation of cloud particles should be considered. Fourth, sedimentation rates and fallout of large particles must be included in the model. Fifth, the calculation of sedimentation behavior demands a knowledge of mean

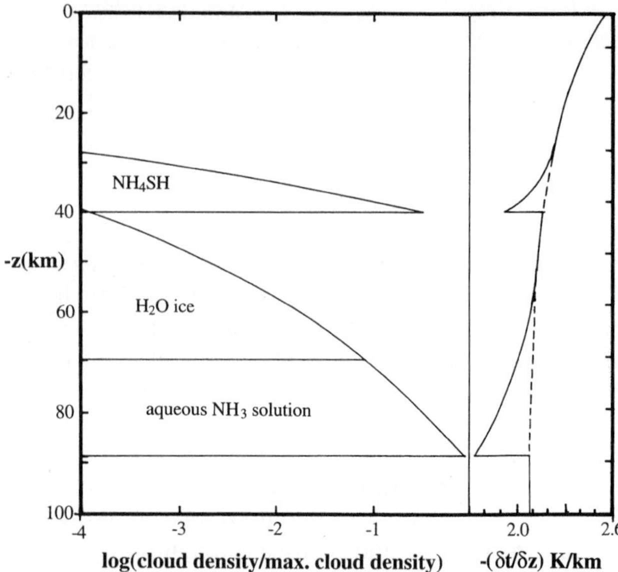


**Figure V.20** Effects of 10-fold $H_2O$ and $H_2S$ enrichment on cloud structure. Addition of ice-forming elements to a solar-composition atmosphere leads to a great increase in the importance of aqueous ammonia solution and further lowers the lapse rate in the region of cloud condensation. These consequences are especially important for Uranus and Neptune.

vertical wind speeds, for which we have only indirect quantitative estimates. Finally, even when we have mastered all these difficulties for one local set of conditions, we still have horizontal variations to consider. It is all very well to understand processes in updrafts, but a planetary atmosphere that has all updrafts and no downdrafts is meteorologically unappealing (except, perhaps, for glider enthusiasts).

Nucleation of cloud particles, the first of the problems listed above, is a problem only because, as we mentioned earlier, very small particles have vapor pressures much higher than large particles of the same composition. Thus the first two problems above are intimately associated.

Let us imagine 1 mole of liquid (or solid) arranged as droplets (or particles) with a narrow size spectrum, of typical radius $a$. The total volume of liquid is simply

$$V = \mu/\rho_l = 4\pi a^3 N_0/3, \qquad \text{(V.90)}$$

where $N_0$ is the number of particles per mole. Here we have for convenience assumed a spherical geometry for the particles.

The Gibbs free energies of 2 moles of droplets having droplet sizes differing by $da$ will differ because of the work that must be done to make the extra surface area of the smaller droplets. This work, called the *surface work* or *surface energy*, is $\gamma dA$, where $\gamma$ is the work done to make a unit area of surface and $dA$ is the change in surface area of the mole of liquid as the mean droplet size changes by $da$.

$$dG = -\gamma dA = -8\pi\gamma N_{\rm p} a da = (3\mu\gamma/\rho a^2)da, \quad \text{(V.91)}$$

where we have eliminated $N_0$ using Eq. (V.90).

Integrating from drop size $a = a_1$ to $a = a_2$,

$$G_2 - G_1 = \Delta G = dG = (3\mu\gamma/\rho)a^{-2}da \quad \text{(V.92)}$$

$$\therefore \Delta G = -(3\mu\gamma/\rho)(1/a_2 - 1/a_1). \quad \text{(V.93)}$$

Thus the molar Gibbs free energies of the two ensembles of droplets differ by an amount that varies as $1/a$. In the limit of large droplet size, the contribution of the surface work to the free energy clearly vanishes. In this limit, we can equate the standard vapor pressure of the droplet to that when $a_1 = \infty$. Then the molar Gibbs free energy of an assemblage of small droplets must be elevated and they must have a vapor pressure that is larger than the standard vapor pressure:

$$\Delta G_{\rm vap}(r) = \Delta G^{\circ}_{\rm vap} - 3\mu\gamma/\rho r \quad \text{(V.94)}$$

$$\ln p_a = -\Delta G^{\circ}_{\rm vap}/RT + 3\mu\gamma/\rho aRT \quad \text{(V.95)}$$

$$= \ln p^0 + 3\mu\gamma/\rho aRT \quad \text{(V.96)}$$

or

$$p = p^0 \exp(+3\mu\gamma/\rho aRT). \quad \text{(V.97)}$$

For typical liquids, $a$ must be $0.1\mu$m or less for the surface work term to become important. In the size range $r < 0.1\ \mu$m a gas may be markedly supersaturated with respect to the equilibrium vapor pressure of the condensate without the pressure being adequate to permit the spontaneous condensation of small droplets (homogeneous nucleation). It is easy to see that, in an ensemble of particles of very different sizes, the ones smaller than $0.1\ \mu$m will readily evaporate. The extra vapor will quickly condense on the larger particles.

If suitable condensation nuclei are present, *heterogeneous nucleation* of a new condensate on the preexisting material may occur. Water-soluble substances, especially hygroscopic materials, are ideal as condensation nuclei. In the atmospheres of Jupiter and Saturn, solid ammonium, sodium, and potassium salts are available, although of these, ammonium phosphates and $NH_4Cl$ are likely to be most important at the level of water saturation. The availability of these nuclei of course depends on where they condense, how large their particles are, and whether the vertical component of the wind is strong enough to lift them to the level where their services are needed.

Small particles of condensate (0.1 to 1 $\mu$m) are most easily lifted by turbulent eddies. In the Jovian stratosphere, at pressures near $10^{-3}$ bar, the gas mean free path ($l$) is $10^{-2}$ cm (100 $\mu$m), and small particles are thus far smaller than the mean free path ($a < 1$). This is precisely the case we treated in our discussion of dust transport in the solar nebula [Eq. (IV.189)]. Larger particles deeper in the atmosphere ($a >> 1$) will experience viscous (Stokes) drag, as described in Eq. (IV.197). For stratospheric fine particles,

$$(\rho_{\rm g}/\rho_{\rm s})v_{\rm t}v_z/a = g \approx 2700\ {\rm cm\ s^{-2}}, \quad \text{(V.98)}$$

which shows that the sedimentation speed of the smallest stable condensate particles is essentially equal to the mean vertical wind speed. Thus only such particles may be abundant in the stratosphere.

## Galileo Perspectives on Jovian Clouds

The Galileo Probe spacecraft was designed and instrumented to solve the fundamental compositional problems presented by the atmosphere and cloud layers. The Galileo mission fought extraordinary obstacles from the very start. Its launch was repeatedly delayed before 1986 by a host of problems. The spacecraft was finally cleared to be launched by a Centaur G′ stage carried into Earth orbit by the Space Shuttle. But, just weeks before the scheduled launch, on January 26, 1986, the Challenger orbiter exploded, killing its crew and bringing the NASA manned spaceflight program to a 32-month halt. As part of the safety reviews surrounding the Challenger disaster, NASA decided to prohibit carrying the hydrogen–oxygen Centaur stage aboard future Space Shuttle missions. Thus, even when the remaining shuttle orbiters were returned to flight status in 1987, Galileo was left without a booster. On October 17, 1989, Galileo was finally launched aboard an upper-stage stack based on the IUS stage and carried into orbit by the Atlantis shuttle orbiter. The lower performance launch meant that Galileo required additional energy to reach Jupiter. This energy could only be gained by a series of flybys of Earth and Venus, using gravity assist from both planets to pump up the eccentricity of its orbit about the Sun. After observing Earth, Venus, and the Moon on these flybys, Galileo passed through the asteroid belt on its was to Jupiter, observing first the main-belt asteroid Gaspra, then Ida and its previously undiscovered satellite, Dactyl. The Galileo Probe and orbiter separated long before encounter with Jupiter, the orbiter taking up a highly eccentric initial orbit about Jupiter that gives it repeated close flybys of the Galilean satellites Europa, Ganymede, and Callisto. From its vantage point high above the clouds, the orbiter would track the entry probe and store the brief blizzard of probe data on its tape recorder. Then, over a period of weeks, the remaining (lower gain) antenna on the orbiter would trickle this information back to Earth. But just weeks before the probe entry into Jupiter, the tape recorder aboard the orbiter began behaving erratically, forcing changes in the mode of operation of the recorder and raising fears of the possible loss of crucial probe data. Amidst all these concerns and difficulties, the Galileo Probe entered the atmosphere of Jupiter on December 7, 1995.

The probe instruments were designed to measure the atmospheric structure down to levels well below the base of the water clouds. Because a dense multiple cloud layer covers some 98% of the surface area of Jupiter, an entry probe that traverses a wide range of altitudes provides an ideal platform for the study of cloud structures. The centerpiece instrument aboard the probe, the mass spectrometer, was to determine the gas and cloud composition over a range of altitudes beginning well above the top (ammonia) cloud layer. Physical data on the thermal structure and cloud density profiles in the atmosphere could then be correlated with the mass spectrometer data to afford a very complete picture of the atmosphere–cloud system.

By an almost incredible coincidence, the probe happened to enter a rare "warm spot," a region where the 5-$\mu$m flux seen from above is far in excess of that associated with a 140 K black body. Such regions often have effective temperatures of 200 to even 300 K. The high infrared flux is caused by the near absence of cloud opacity in the "warm spots," which is in turn caused by the general absence of rising motions. It is rising gas that both elevates condensable gases to altitudes colder than the saturation temperatures of ammonia, ammonium hydrosulfide, and water and also levitates small condensate particles to maintain dense cloud layers. Thus the probe fell into a region with only a very thin haze, without the well-developed cloud layers that cover 98% of the planet. The subsiding gases in this region must, by reason of continuity, have risen in the upwelling, cloud-rich regions and deposited their content of condensable gases as condensate particles, which are left behind by the upwelling gas. Thus the descending gas must be depleted in condensables relative to the rest of the Jovian atmosphere.

The probe, during its 57 minutes of active life, found a water abundance of about 0.2 times that expected for solar proportions of the elements, compared to two to five times the solar O:H ratio expected based on previous observations. Several other chemically active gases, including methane, ammonia, hydrogen sulfide, and traces of hydrocarbons, were also reported. Precise estimates of their abundances are given in Table V.1. In addition, helium was found to be present at about half the solar abundance level, almost certainly due to partial phase separation of helium from hydrogen deep in Jupiter's atmosphere, where hydrogen is metallized but helium is not. The denser helium-rich phase would "rain out" of the metallic hydrogen fluid. The heavier inert gases, neon, argon, krypton, and xenon, were all detected. The latter two, the gases most easily condensed (and those most likely to enter into solid clathrate hydrates), are distinctly enriched compared to the others. Thus the idea that Jupiter contains a substantial enhancement of all the "cometary" or "iceball" elements, combined with the idea that the most condensable gases are locally depleted from the region probed by Galileo, is apparently compatible with the available evidence. However, the preliminary estimate of the methane abundance is also only half the expected value, close to the amount expected for a solar C:H ratio. The reason why the methane abundance reported by Galileo lies so far below that given by both Earth-based and Voyager spectroscopic data is not understood.

The probe also measured a large vertical flux of infrared radiation that affirms the unusually low abundance of water vapor in the atmosphere below the entry site. Presumably that region is also subsiding and therefore unusually dry.

Lightning was sought using two different experiments, an optical flash detector and a radio-frequency "click" detector. In the region accessible to the optical sensor, the lightning rate was, not surprisingly, about a factor of 10 below terrestrial lightning rates. The radio device, with its much larger range, detected some 50,000 discharges. The Solid State Imager (SSI) instrument on the Galileo Orbiter tracked optical flashes from 26 lightning storms, all of them in areas of cyclonic shear and embedded within west-bound currents. The brightest single flash delivered $1.6 \times 10^{10}$ J of optical energy. The optical signature of the flashes indicate an origin close to the base of the water clouds, where theory predicts the highest cloud density and the greatest rate of latent heat release. The average optical power from Jovian lightning was about $3 \times 10^{-7}$ W m$^{-2}$, nearly the same as on Earth.

There are at present no plans for future probe or orbiter missions to Jupiter. Although the Galileo Orbiter will continue to watch the Jovian clouds from above, most of its future work will center on the Galilean satellites.

## Ion Production in the Jovian Atmosphere

Even in the total absence of solid condensation nuclei, supersaturation can be avoided if ions are reasonably abundant. For example, water, with its large dipole moment, will readily form several layers of molecules about a single p or e charge. The "cluster ion" series $H^+$, $H_3O^+$, $H_5O_2{}^+$, $H_7O_3{}^+$, $H_9O_4{}^+$, etc., can even develop at water partial pressures far short of saturation.

Positive ions and free electrons may be produced either internally (by radioactive decay of K, U, or Th) or externally (by cosmic rays and dumped energetic particles from radiation belts) in the absence of solid condensation nuclei. The radiation belts of Jupiter are, as we shall see, a veritable hotbed of energetic particles. Protons with energies of 1–10 MeV and higher are available and are constantly being precipitated into the atmosphere at unknown rates, probably at moderately high latitudes. These high-energy protons are fairly effective at ionizing anything they strike, and a single 10-MeV proton can produce very large numbers of ions before coming to rest. The energy lost in producing a (+, −) ion pair in hydrogen is roughly 2.5 times the first ionization potential of $H_2$, 15.6 eV. The ionization efficiency in rare gases is somewhat higher (60%) because the rare gases have no capacity for absorbing energy to excite vibration, rotation, or dissociation; only ionization and translational excitation are possible. In a solar-composition gas mixture, $2.6 \times 10^4$ ion pairs can be produced by a single 1-MeV proton.

A high-energy ion loses most of its energy to ionization and electronic excitation of the target gas. At lower energies, when the velocity of the ion is comparable to the Bohr "orbital velocity" of the inner K-shell and L-shell electrons of the target, the bombarding ion can pick up electrons from the target atoms. The resulting energy loss is essentially purely electronic excitation. At even lower bombarding energies, elastic collisions are dominant. In the special case of protons bombarding hydrogen, the "K-shell" energies are indistinguishable from the valence electron energies, and hence the charge-exchange and elastic collisions mechanisms closely coincide.

The efficiency of ion-pair production clearly must depend on the proton energy if for no other reason than that the interaction mechanisms are different in different energy regimes. Also, a fast ion of velocity $v$ will exert a Coulomb force on a valence electron that is independent of $v$, but the time of interaction will vary as $1/v$. The change in momentum caused by the interaction is the integral of the Coulomb force over time and hence is far less for high-velocity ions. The amount of energy lost to the target by the fast ion is proportional to the square of the momentum change ($\Delta E = \Delta p^2/2m_p$) and thus proportional to $1/v^2$. Thus $\Delta E \propto 1/E$, and a highly energetic ion will penetrate deeply and lose energy slowly. When it has lost 90% of its original energy to numerous weak interactions, it will be able to lose 10 times as much energy per interaction, and it will then very quickly be stopped, depositing a large fraction of its total energy over the last few percent of its path.

From Fig. V.21 we can see that the Coulomb force between the energetic proton and an electron in the target material can be resolved into two components, $F_\parallel$ and $F_\perp$, and that the effect of integrating $F_\parallel$ over the entire interaction ($-\infty < x < \infty$) is zero. The nonvanishing interaction is due to $F_\perp$:

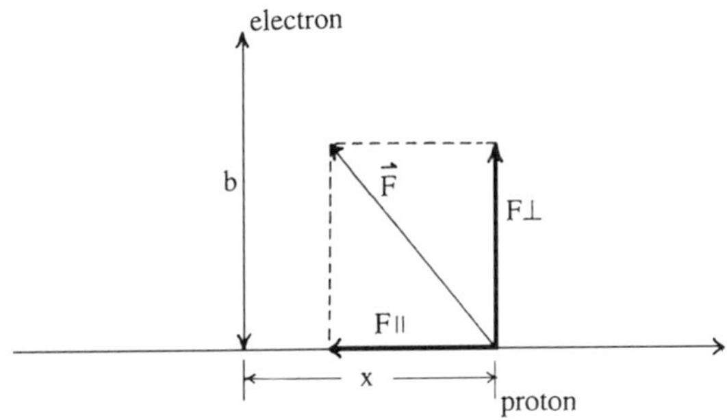


**Figure V.21** Proton–electron collision nomenclature. A fast-moving proton encounters a "stationary" electron. Impact parameter $b$ is the minimum encounter distance. Coulombic interaction force $F$ is decomposed into two components, one parallel to the motion of the proton (which averages out over the course of the encounter) and one orthogonal to its motion.

$$F_\perp(x) = [e^2/(x^2+b^2)][b/(x^2+b^2)^{1/2}]. \tag{V.99}$$

The momentum transferred to the electron over the entire encounter is

$$\Delta p = \int_0^\infty F_\perp(x)dt = (1/v)\int_{-\infty}^{\infty} F_\perp(x)dx \tag{V.100}$$

$$= (e^2 b/v)\int_{-\infty}^{\infty}(x^2+b^2)^{-3/2}dx$$

$$= 2e^2/bv, \tag{V.101}$$

where $v$ is the proton velocity and where $b$ and $x$ are defined in the figure.

The energy transferred to the electron is $(\Delta p)^2/2m_e$:

$$E = 2e^4/m_e b^2 v^2. \tag{V.102}$$

The number of electrons approached to an impact parameter between $b$ and $b+db$ over a differential proton path length, $dx$, is just $2\pi bndbdx$, where the number of electrons per unit volume in the target is $n$.

The total energy loss rate is now the integral of the local loss rate over all values of $b$:

$$-dE/dx = 4\pi ne^4/m_e b^2 v^2 \int_{b_{min}}^{b_{max}} bdb. \tag{V.103}$$

The limits of integration are imposed by two simple criteria. First, virtually all the interaction occurs during the period of time that $x < b$; that is, $t \approx 2b/v$. Because the derivation assumes that the electron is essentially stationary during the encounter, this time must be less than the "orbital period" ($P$) for the Bohr model, which is $1/\omega$, the Bohr angular frequency. Thus the upper limit on $b$ is $b_{max} = v/2\omega$. The other limit follows from the maximum velocity that an electron can gain from a collision with a heavy ion: from conservation of momentum, this is simply $2v$, which, from Eq. (V.102), corresponds to impact parameter $b_{min} = e^2/m_p b^2$. Then evaluation of the integral of $d \ln b$ in Eq. (V.103) and substitution of the upper and lower limits on the impact parameter give

$$-dE/dx = (4\pi e^4 n/m_p v^2) \ln(m_p v^3/2e^2\omega). \qquad \text{(V.104)}$$

This result was first derived by Niels Bohr in 1913. The much more complex (nonrelativistic) quantum-mechanical treatment of the problem gives the same initial factor, but with the logarithmic term replaced by

$$\ln(2m_p v^2/I), \qquad \text{(V.105)}$$

where $I$ is the ionization potential of the target atoms.

The range of an ion is easily calculated from

$$X_{stop} = \int_{E_0}^{0} (dx/dE)dE, \qquad \text{(V.106)}$$

where the distance $X_{stop}$ is called the stopping distance. As we have seen, much more ionization will occur per distance interval at low energies, immediately prior to stopping.

In a planetary atmosphere, the useful distance parameter is the column mass of gas traversed, especially for vertical or diagonal rather than horizontal propagation. In the horizontal (constant $n$) case, the column mass of gas is related to $n$ by

$$\sigma = \mu Px/RT(\text{g cm}^{-2}) = n\mu/2N_0, \qquad \text{(V.107)}$$

where $N_0$ is Avogadro's number, $6.023 \times 10^{23}$, and 2 is the number of electrons per molecule in both hydrogen and helium.

For an isothermal plane-parallel atmosphere, a proton propagating at angle $\phi$ relative to the zenith will pass through a column of unit cross-section area containing a mass per unit area of

$$\sigma(z) = -\int_{\infty}^{z} \sec\phi\, \rho(z)dz \qquad \text{(V.108)}$$

$$= -\mu \sec\phi/RT \int_{\infty}^{z} \exp(-z/H)dz, \qquad \text{(V.109)}$$

where $H$ is the isothermal scale height $RT/\mu g$. Integrating and substituting for $H$,

$$(z) = \sec\phi/g \exp(-\mu gz/RT), \qquad \text{(V.110)}$$

where $Z$ is now the altitude relative to the 1-bar pressure level. Figure V.22 shows the stopping depths of vertically incident protons of energies 0.1 to 100 MeV in a hydrogen–helium isothermal atmosphere at 110 K. It can be seen that even 100-MeV protons stop at pressures near 30 mbar. Only relativistic ($E > 0.1m_p c^2$) protons can penetrate to the troposphere, whereas typical 1- to 10-MeV radiation belt protons will penetrate only to pressures less than 1 mbar, 100 km above the 1-bar level. The protons, once stopped, and the $H_2^+$ ions made by them will undergo charge exchange with atmospheric gases to minimize the potential energy:

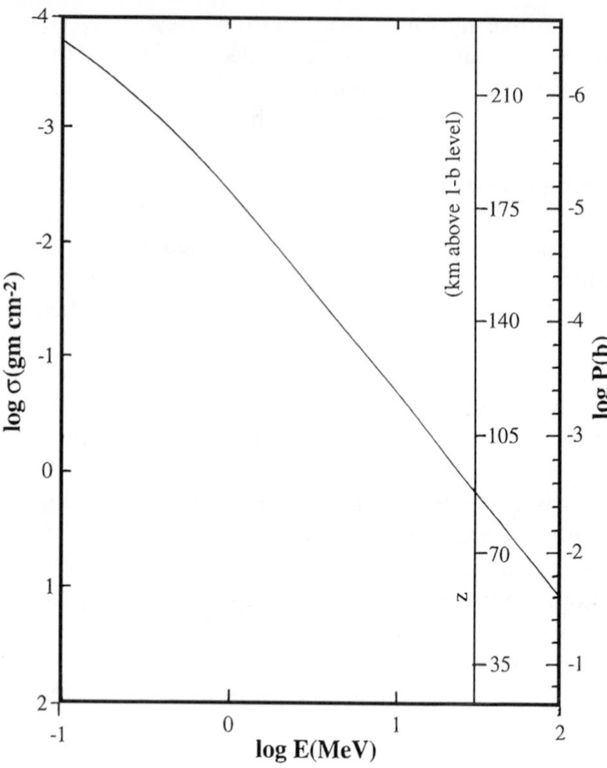


**Figure V.22** Penetration of energetic protons into an isothermal hydrogen–helium atmosphere. Radiation-belt or cosmic-ray protons in the 0.1- to 100-MeV energy range cannot penetrate into the Jovian troposphere ($P > 300$ mbar).

$$H_2^+ + N_2 \rightarrow N_2^+ + H_2 + 0.1\,\text{eV} \qquad \text{(V.111a)}$$

$$H_2^+ + CH_4 \rightarrow CH_4^+ + H_2 + 1.1\,\text{eV} \qquad \text{(V.111b)}$$

$$H_2^+ + NH_3 \rightarrow NH_3^+ + H_2 + 4.8\,\text{eV} \qquad \text{(V.111c)}$$

$$H_2^+ + Kr \rightarrow Kr^+ + H_2 + 1.7\,\text{eV} \qquad \text{(V.111d)}$$

$$H_2^+ + H \rightarrow H^+ + H_2 + 2.0\,\text{eV}. \qquad \text{(V.111e)}$$

The ionization energies of these and other relevant species are tabulated in Table V.2. Note that H+ has a relatively low ionization potential and is thus able to ionize only $NH_3$ among the chemically active gases on Jupiter:

$$H^+ + NH_3 \rightarrow NH_3^+ + H + 2.8\,\text{eV}. \qquad \text{(V.112)}$$

Table V.2 First Ionization Potentials of Molecules

| Species | First ionization potential (eV) | Species | First ionization potential (eV) |
|---|---|---|---|
| NO | 9.5 | Kr | 13.9 |
| $H_2S$ | 10.4 | CO | 14.1 |
| CS | 10.6 | $CO_2$ | 14.4 |
| $NH_3$ | 11.2 | $CH_4$[a] | 14.5 |
| $CH_2O$ | 11.3 | HCN | 14.8 |
| $C_2H_2$ | 11.6 | $N_2$ | 15.5 |
| Xe | 12.1 | $H_2$[a] | 15.6 |
| $O_2$ | 12.5 | Ar | 15.7 |
| $H_2O$ | 12.6 | HF | 17.7 |
| H | 13.6 | Ne[a] | 21.5 |
| HCl | 13.8 | He[a] | 24.5 |

[a]Actually present in the ionospheres of the Jovian planets.

Ammonia, however, is essentially absent at very high altitudes due to condensation and photolysis. It is likely that the most important chemical consequences of photoionization lie in the reaction $H_2^+ + CH_4 \rightarrow H_2 + CH_4^+$ and the subsequent hydrocarbon chemistry.

It is likely that these ions play a major role in the chemistry of the atmosphere and the nucleation of condensates only at the altitudes at which very strong absorption of solar UV actually occurs. The principal agents of nucleation within the $H_2O$, $NH_4SH$, and $NH_3$ cloud layers are likely to be the material of the next lower layer.

We have now treated the clouds' composition and gross vertical structure as if they were simply consequences of an invariant and given thermal structure. So far, except for general arguments regarding the necessity of convection and a detailed treatment of heat release by condensation, we have not considered the role of cloud condensates in modifying the radiative environment. We shall therefore next turn our attention to vertical radiative (solar and planetary) and mass fluxes and their dependence upon cloud structure.

## Visible and Infrared Radiative Transfer

The transmission, absorption, emission, and scattering behavior of light constitute the area of study called *radiative transfer* theory. Our purpose here shall be to draw together in a qualitative way the main ideas and phenomena subsumed under this name as they relate to planetary atmospheres.

First, in the simplest case, we consider a gas interacting with light of a frequency far removed from the resonant frequencies of the gas, such as visible or infrared radiation in helium. The interaction cross-section in its general form is

$$\sigma_s = (8\pi a^2/3)[\nu^4/(\nu^2 - \nu_0^2)^2], \qquad \text{(V.113)}$$

where $a$ is the effective radius of the molecule, $\nu$ is the frequency of the electromagnetic radiation, and $\nu_0$ is the nearest resonant frequency of the gas. If, as above, we take $\nu_0 >> \nu$ then $\sigma_s = \sigma_R = c\nu^4 = C\lambda^{-4}$, which is our equation for Rayleigh scattering, Eq. (V.72). At very high altitudes in an atmosphere, where highly energetic solar X rays and extreme ultraviolet light (EUV) are absorbed, free electrons are produced in large numbers. At these levels where photoionization is important, the free electrons can interact with the radiation field directly. But the free electrons do not have quantized energies and thus do not have resonant absorption frequencies. Therefore $\nu_0 = 0$, and the cross-section reduces to

$$\sigma_s = \sigma_T = 8\pi a^2/3. \qquad \text{(V.114)}$$

The cross-section $\sigma_T$ defined here for free–free scattering is the *Thomson scattering* cross-section, which, as we have seen, is independent of frequency. Its magnitude is only $10^{-26}$ cm$^2$ per particle, or $10^{-2}$ cm$^2$ per mole. The Thomson scattering equation was derived from basic principles in Eqs. (IV.45) to (IV.53).

If some cooperation between the scattering effects of spherical particles can be achieved, the scattering cross-section per unit mass can be greatly enhanced. Consider an alternating $E$ field of peak strength $E_{max}$ being scattered by $n$ different particles. The amplitude of the scattered wave is proportional to the square of the scattered $E$ component; thus, two particles stuck together will, if their sizes are very much less than the wavelength of the incident light, scatter twice as large a part of the $E$ field, giving four times the scattered amplitude. There are now $n/2$ independent particles, each scattering with four times the amplitude of the single particle.

Two attached particle or molecules need not scatter high-frequency light in a mutually reinforcing way if they are so far apart that there is a large phase difference between the two scattered $E$ components, in which case destructive interference is as likely as constructive interference. Particles will therefore be strong scatterers if $a << \lambda$. Beyond $\lambda = 2\pi a$ there are very complex relationships among scattered intensity, scattering angle, wavelength of incident light, particle size, and particle shape. The theory of scattering by such particles, called *Mie scattering*, cannot in general be worked out for realistic cases. [Note that, for Mie scattering of visible light ($\lambda \approx 0.6\,\mu$m) a particle radius of $a = \lambda/2\pi \approx 0.1\,\mu$m is required. Since a typical atom has a radius of about $2 \times 10^{-4}\,\mu$m, a Mie-scattering particle must be about

$2 \times 10^3$ atoms in diameter. For spherical particles, that is equivalent to about $4 \times 10^9$ atoms per particle.] Mie theory is tractable, however, given certain simplifying assumptions, such as a "monochromatic" (single-radius) size distribution of purely spherical particles. Often the directional dependence of scattered intensity $I(\phi)$ is simply assumed to have a particular shape, such as isotropic ($I(\phi) = \text{constant}$) or smoothly forward scattering ($I(\phi) = (4A_I/\pi)\cos\phi$, where $A_I$ is the *isotropic single-scattering albedo*).

A general feature of all kinds of scattering can be appreciated from Fig. V.23. Let us imagine a photon propagating in the $+Z$ direction with arbitrary polarization of the electric vector in the $x$ and $y$ directions. A photon scattered in the $+X$ direction cannot be polarized in the $+Z$ direction because there can be no polarization of the original photon in the direction of its motion. Similarly, once scattered in the $+X$ direction the photon cannot have any polarization in the $+X$ direction either. It follows that the scattered photon must be *linearly polarized* in the $y$ direction, with the electric vector describing oscillations only parallel to the $y$ axis.

Clearly in most natural circumstances in which scattering is important, multiple scattering will occur. Because most particles are rather strong forward scatterers, photons that end up heading back toward the light may very well have participated in a large number of scattering events and may have followed a complex trajectory within the scattering cloud layer. Because the abundances of the condensible gases responsible for forming clouds increase approximately exponentially with depth in the cloud (actually, $\log p \propto \exp(A - B/T) = \exp(A + C/Z)$), this means that multiply scattered photons may travel through greatly enhanced quantities of absorbing gas compared to single scattered ("reflected") photons. An optical model of an atmosphere in which a homogeneous gas overlies an isotropically scattering (*Lambert scattering*) surface is a convenient simplification, but can resemble reality only if the cloudtop is dense and extremely well defined in altitude. Thus *reflecting layer* models may be used effectively only when most of the emergent photons are singly, not multiply, scattered.

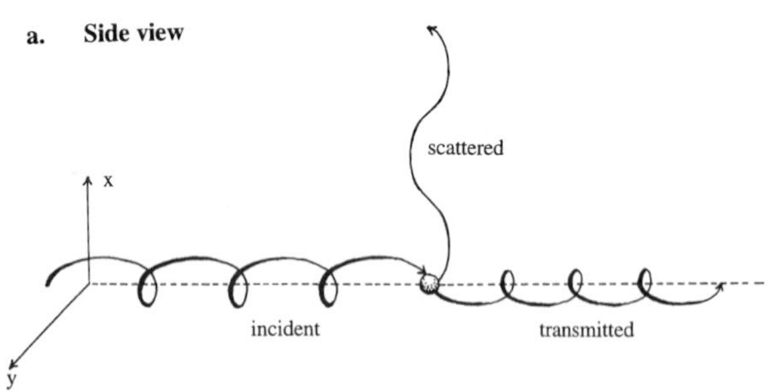


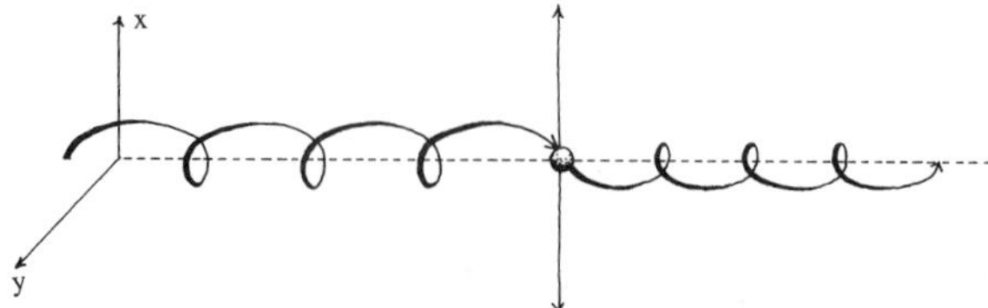


**Figure V.23** Polarization of scattered light. a shows a side view of the scattering process. Note the reduced amplitude of the circularly polarized transmitted wave emerging from the event. Looking back on the event from along the path of the scattered photon (b) we see that the plane of polarization of the scattered light is orthogonal to the direction of propagation.

Up to this point we have not explicitly considered the absorption of photons by particles. Often clouds are formed of fresh icy condensates with isotropic single-scattering albedos of 0.99 or higher. Nonetheless, highly forward-scattering particles with this albedo may funnel photons in the forward direction so effectively that those photons that end up heading back toward the Sun (on Jupiter or Saturn, this means back toward Earth!) may have undergone dozens of scattering events. The probability of surviving, say, 16 such events is $(0.99)^{16} = 0.85$, and we would therefore judge such a multiple-scattering cloud to have an albedo of 0.85. Hence even very low concentrations of absorbing material ($A < 0.9$) can lend very strong color to such a cloud.

Submicrometer-sized particles will contain a number of molecules proportional to $a^3$, and the scattering power of these particles will go as $a^6$ until $a$ approaches the wavelength of the incident light. There is then a transition region centered near $\lambda = 2\pi a$, and then, for even larger particles, the scattering cross-section goes simply as the geometrical cross-section, $\pi a^2$. The scattering efficiency *per unit mass* clearly must grow as $a^3$ until $\lambda = 2\pi a$ and then decline as $a^{-1}$ for larger sizes. This is shown in Fig. V.24. Thus the particles that are best for scattering visible light have $a \approx 0.1$ $\mu$m and circumferences close to one wavelength. Conversely, observations made in visible light are very sensitive to 0.1-$\mu$m particles and slightly less sensitive to 1-$\mu$m particles, but very insensitive to 0.01-$\mu$m particles if they are present along with the others. Interestingly, surface-work effects cause significant enhancement of the vapor pressure in this same size regime ($a < 0.01$ $\mu$m), and hence such particles are likely to have very short lifetimes against evaporation as well as being hard to detect. We are not in general free to observe at arbitrarily small $\lambda$ in order to see these small particles, because atmospheres are generally very good absorbers of UV light. Also, we recall that the

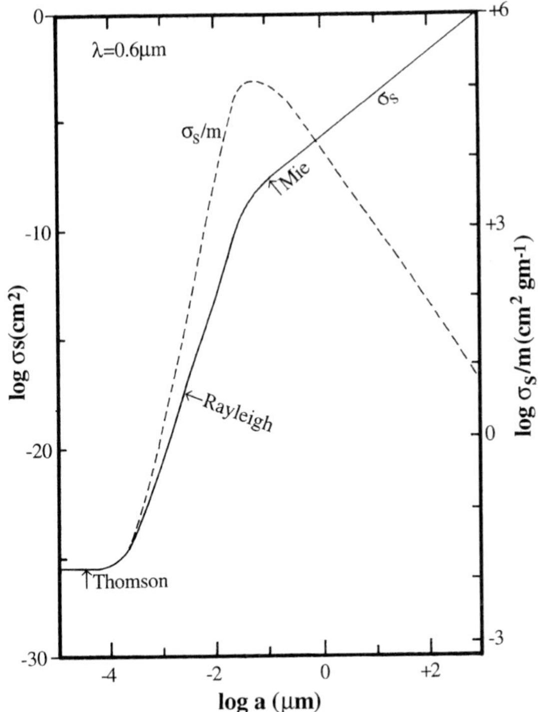


**Figure V.24** Scattering versus sizes of particles. The scattering cross-section of a single particle of radius $a$, ranging from the size of an electron up to 1000 $\mu$m (1 mm), is shown as the solid line. The dashed line gives the scattering cross-section per unit mass. Note the peak in scattering efficiency per gram near a radius of 0.1 $\mu$m. The Thomson, Rayleigh, and Mie regimes are indicated. All calculations are for sunlight near the Planck peak, at a wavelength of 0.6 $\mu$m.

Rayleigh-scattering cross-section grows as $\lambda^{-4}$ in the UV region, and even a weakly absorbing gas will attenuate light very strongly at 0.1 $\mu$m. Of course this behavior blows up at $\nu = \nu_0$, where strong resonant absorption (and resonant scattering) occur. Thus, although the Sun is moderately bright in the hydrogen Lyman-$\alpha$ line, high-altitude atomic H on Jupiter and Saturn will interact extremely strongly with this radiation and prevent its deep penetration into the atmosphere.

Absorption and emission processes for molecules are treated in Appendix II, where the interrelations between spontaneous emission, stimulated emission, and absorption are discussed briefly. Deep in a planetary atmosphere collision rates are so high that collisional deexcitation is dominant, causing severe broadening of the energy levels and associated decreases in lifetime of the excited states. A different way of expressing this phenomenon is that if the mean life between collisions, $\tau_{\text{coll}}$, is much longer than the half-life for spontaneous emission, $\tau_{\text{A}}$ (see Appendix II), then the energy state of the molecule may be dominated by the externally imposed radiation field, in this case, the Sun. Deep in the atmosphere, where $\tau_{\text{coll}} << \tau_{\text{A}}$, thermal equilibrium will be obtained quickly, and the internal level of excitation of the molecule will reflect the same temperature as the translational motion. Furthermore, because of serious broadening of infrared (vibrational) and microwave (rotational) levels by collisions, the continuum opacity in these regions will be so large that the thermal radiation field will locally be in equilibrium with all the other modes. The solar contribution to the radiation field will be absorbed strongly near the 1-bar level and will not be important at greater depths. Because collision broadening at a given temperature and pressure is more effective at infrared wavelengths than in the visible region, ($h\nu_{\text{IR}} \approx kT$, whereas $h\nu_{\text{vis}} >> kT$), a rising parcel of gas that is initially in thermal equilibrium will first begin to admit visible light from the Sun and then begin to radiate out into space its own internal thermal energy. At low pressures, the collision lifetimes become very long, and radiative equilibrium with the Sun takes over. Note that these arguments are causally related to the role of convection in atmospheric heat transport, because the buoyancy that drives convection will be quickly lost when the opacity of the gas gets low enough to allow radiative cooling into space.

The effects of scattering opacity on this scenario are interesting. First of all, cloud particles capable of scattering Jovian thermal radiation strongly should have radii near $\lambda_{\text{m}}/2\pi$, where $\lambda_{\text{m}}$ is the peak in the thermal Planck function, $B_\lambda$. Thus

$$a = 2897/2\pi T = 461/T\ \mu\text{m}, \tag{V.115}$$

which is 1 to 3 $\mu$m for typical Jovian temperatures. These are entirely reasonable particle sizes, being neither so small as to be unstable against evaporation nor so large as to be badly depleted by rapid fallout. Thus clouds act to help "insulate" the atmosphere against radiative cooling and allow the adiabatic structure to persist up to higher altitudes than would otherwise be possible. At the same time, penetration of sunlight near the peak of the solar Planck function is strongly inhibited by 0.1-$\mu$m particle scattering. The presence of even small amounts of absorbing matter in the topmost cloud layer will cause strong local absorption of solar energy.

For the observational astronomer interested in selecting an appropriate optical model for the atmosphere to help interpret spectroscopic observations, the situation is unpleasantly complex. The conventional assumption of a reflecting layer model to interpret abundance data is perhaps usable over limited wavelength ranges well outside of regions where the continuum is severely depressed by gaseous absorption, but even in this case the interpretation of observed line strengths as abundances can be badly upset by even a thin, high-altitude haze layer. The *ratios* of abundances of two species can be better estimated from a reflecting-layer model, but only if these species are well mixed over the entire region of the atmosphere visible from above. Consider the effects of even minor height or albedo variations in the $NH_3$ clouds upon the deduced $CH_4$:$NH_3$ ratio: the $CH_4$ mole fraction is a constant over a very wide range of altitudes, and near the $NH_3$ cloudtops $p(CH_4)$ drops vertically with the same scale height as the total atmosphere, $H = RT/\mu g \approx 17\,\text{km}$. On the other hand, the $NH_3$ "scale height" is given in this region by the saturation vapor pressure curve:

$$(d \ln p(\text{NH}_3)/dT)(dT/dz) = \Delta H^\circ_{\text{vap}} \mu g / C_\text{p} R T^2 \qquad \text{(V.116)}$$

$$H(\text{NH}_3) = RT^2 C_\text{p} / \lambda \mu g = 2\,\text{km}. \qquad \text{(V.117)}$$

Thus 2-km vertical excursions in the height of the "reflecting layer" would change the estimated $NH_3$:$CH_4$ ratio by a factor of $e^{2/2}/e^{2/17} = 2.42$!

Above the region of condensation of $NH_3$ the atmosphere rapidly becomes transparent to visible light, because the cloud particle scale height must be much smaller than the gas scale height, and the visible opacity of $H_2$ goes as $P_2$. [Vertical mixing below the tropopause is rapid, with an eddy diffusion coefficient of $K = 16^6$ to $10^9\,\text{cm}^2\,\text{s}^{-1}$, whereas the stratosphere is very quiescent ($K \approx 10^4\,\text{cm}^2\,\text{s}^{-1}$). The mean vertical wind speed in the stratosphere is roughly $w = K/H \approx 10^{-2}\,\text{cm}^2\,\text{s}^{-1}$, which is, as we saw earlier, barely sufficient to support the smallest condensate particles.] The main way in which the atmosphere can interact with solar and planetary radiation in this region is through absorption of sunlight in the $\lambda < 2\,\mu\text{m}$ combination bands of methane and ammonia, such as the $3v_3$ band of $CH_4$ at $1.1\,\mu\text{m}$. These bands are weak, have limited spectral coverage, and absorb only 1% of the incident solar flux. If there were no other means of heating the atmosphere above the cloudtops, then this would correspond to an energy input of only $10^2\,\text{erg}\,\text{cm}^{-2}\,\text{s}^{-1}$ averaged over the entire surface of the planet. A perfect black body emitter at only 36 K would be in steady state with this flux; however, the gas is obviously not even remotely like a black body emitter. If the emissivity of the gas is $\varepsilon << 1$, then the gas would have to be at a temperature far above 36 K in order to emit an equal flux and achieve a steady state.

Emission from a $H_2$–$CH_4$–$NH_3$ mixture below 130 K will occur largely in the $CH_4$ and $NH_3$ vibrational fundamental bands, which are relatively "black" and relatively close to the Planck peak ($3.5\,\mu\text{m}$ for $CH_4$ vs $25\,\mu\text{m}$ for $\lambda_\text{m}$). Of course, even a tiny trace of a species that is a good emitter (and absorber!) near $20\,\mu\text{m}$ will have a large effect on the cooling rate. Theoretical estimates of the temperature of this cold region (the *stratosphere*) are very dependent on the details of the assumptions made, and can run anywhere from 80 to 130 K. Observations of Jupiter now agree reasonably well that a minimum temperature of 100 K is found at the 100 mbar pressure level. At higher altitudes the temperature again rises ($T/Z > 0$).

A structure of this type is called a *thermal inversion*, and the warm region at and above the temperature maximum is called the *mesosphere*. Figure V.25 shows the most recent estimates of the thermal structure in the adiabatic (tropospheric), cold positive-gradient (stratospheric), and hot negative-gradient (mesospheric) layers.

The small thermal fluxes involved in the stratosphere–mesosphere have another consequence: the day–night cycles will have little effect on the temperature. The *maximum* amount of heat that can be absorbed during the day is 1% of the solar constant at Jupiter. This heat is spread over a wide range of altitudes. The mass of gas affected can be approximated by the mean density of gas in the stratosphere multiplied by the scale height (why?), so the rate of temperature rise is

$$\Delta T/t = \Delta q/m C_\text{p} = 0.01\, L_\odot / 4\pi r_\text{J}^2 C_\text{p} \rho H, \qquad \text{(V.118)}$$

where $\rho H$ is $(P\mu/RT)/(RT/\mu g) = P/g$. Thus $\Delta T/t \approx 100\,g/C_\text{p}P$. Taking a stratospheric temperature of 110 K, $C_\text{p} = 2.65\,R$, $P = 0.1\,\text{bar} = 10^5\,\text{dyn}\,\text{cm}^{-2}$, and $\Delta T/t = 10^{-8}\,\text{K}\,\text{s}^{-1}$. The length of the Jovian day is $1.8 \times 10^4\,\text{s}$, and thus $\Delta T$ is only $1.8 \times 10^{-4}\,\text{K}$. In a steady state, the heat lost at night is exactly equal to the heat gained during the day. Given these rates of absorption and emission of heat, the time required for the

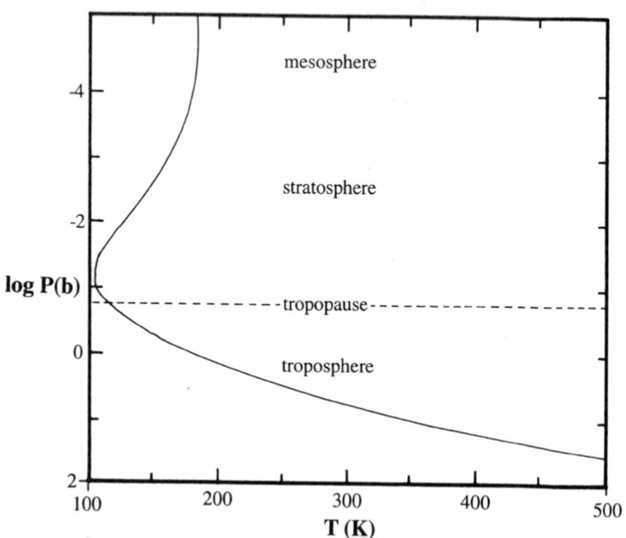


**Figure V.25** Thermal structure of the upper troposphere and stratosphere–mesosphere of Jupiter. The temperature minimum, 110 K, is near the 100 mb level. Galileo found 155 K at the 10 mb level, 159 K from 10 mb to 10 $\mu$b, and 185 K at 1 $\mu$b (1 dyn cm$^{-2}$).

temperature to change by a factor of e would be $\Delta \ln T = 1 = 10^{-8}\Delta t$, or $\Delta t = 10^8$ s. Thus we say that the *radiative time constant* in this region, $\tau_{rad}$, is $10^8$ s (3 years). In order for the temperature to be controlled radiatively, the time constant for turbulent mixing and convective overturn, $\tau_{dyn}$, the *dynamic time constant*, must be longer than $\tau_{rad}$. In order to explain the large amount of heating required by these observations, we shall find it necessary to look closely into the production and destruction of minor constituents of the upper atmosphere.

## Horizontal Structure and Atmospheric Circulation

The visual appearance of Jupiter has for three centuries provided a fascinating and puzzling stream of observations. Rapid and extensive changes often take place, with areas larger than Earth sometimes changing markedly in color and morphology in a few days. Day-to-day changes are often visible, even though the practical limit of resolution (1″) corresponds to 3000 km on Jupiter at opposition and to 450 km at superior conjunction. Seasonal variations are not seen or expected, because of the very small axial tilt, 1.3°.

Despite the rapid changes, certain general patterns persist for long periods of time or recur. First in importance and in persistence is the dichotomy of the face of Jupiter into two areas: the quiescent polar regions (at latitudes above 145°) and the active, banded low latitudes. The two polar regions rarely display any activity that is obvious to a terrestrial observer, although occasional bright and dark spots appear. Occasionally the polar regions seem to vary subtly in color over long periods of time, but they generally are a neutral gray.

The banded portion of the planet is divided by convention into *belts*, which are typically dark gray, brown, orange, or muddy reddish brown, but sometimes very dark blue, and *zones*, which are quite bright and pale yellow or cream colored. The number, width, color, latitude, and activity of the belts and zones evolve continuously, so that the nomenclature is constantly being revised. In general, however, there is usually a bright *Equatorial Zone* (*EZ*), bracketing the equator, often asymmetrically. The neighboring dark belt to the north is called the North Equatorial Belt (NEB) and then the North Tropical Zone (NGZ), the North Tropical Belt (NTrB), the North Temperate Zone (NTeZ), the North Temperate Belt (NTeB), and the North Polar Region (NPR). All of this terminology is illustrated on the line drawing of Jupiter in Fig. V.26. Note that the nomenclature can accommodate features that are not always present; the figure portrays Jupiter with essentially all of the recurrent structures ever seen as if they were present simultaneously. Usually, Jupiter looks appreciably different, usually simpler, with only approximate north–south symmetry.

In 1973 and 1974, at the time of the Pioneer spacecraft flybys, the STe and SSTe belts were fairly distinct and dark, whereas the South Branch of the SEB was very indistinct and scarcely discernible as different from the bright STrZ. The northern edge of the North Branch of the SEB was very dark orange–brown. The Equatorial Zone was very active with extensive systems of white and dark streaks, and the NEB was moderately dark over its entire width. The NTrZ and NTeB were chaotic, but the NTeZ was bright and wide. The most striking and easily recognized feature of Jupiter is the Great Red Spot (GRS), a huge region with about the same surface area as the Earth. The GRS has been under essentially continuous study for a century and a half, and credible accounts of a feature similar in appearance and location to the GRS go back about three centuries. As with most of the studies of Jupiter's markings, these observations have largely been carried out by extensive and systematic coordination of observations by amateur astronomers, not professionals. Amateurs have also greatly contributed to the body of knowledge in several other areas, such as meteor mapping, comet discovery, and observations of the lightcurves of variable stars.

The GRS is not without its repertoire of tricks. Sometimes the color is an extremely intense "brick red" and sometimes it is so pale and gray that it is hard to identify. The dimensions of the GRS vary also, or at least the apparent dimensions vary in the sense that the GRS appears smaller when the color contrast between it and the STrZ is weakest. Finally and most startling of all, in view of its very long life, is the tendency of the Spot to wander in longitude. If we choose as the origin the mean position of the GRS in a coordinate system that has the same *average* rotation period as the GRS, we can then represent the motion of the Spot so as to *minimize* its longitude excursions. In this coordinate system the motions of the GRS are less than in any other system of constant rotation rate. This motion is displayed in Fig. V.27. Note that, within a period of a little over a century, the GRS has wandered *at least* three full circumferences of the planet! Long stretches of steady drift in longitude, corresponding to constant periods of rotation, are evident, as are three briefer episodes in which the longitude in this odd coordinate system is nearly constant.

Other curious behavior of the rotation period of the planet is well established. For example, the equatorial zone rotates in 9h50m30.003s (called System I rotation), whereas the rest of the planet generally rotates in

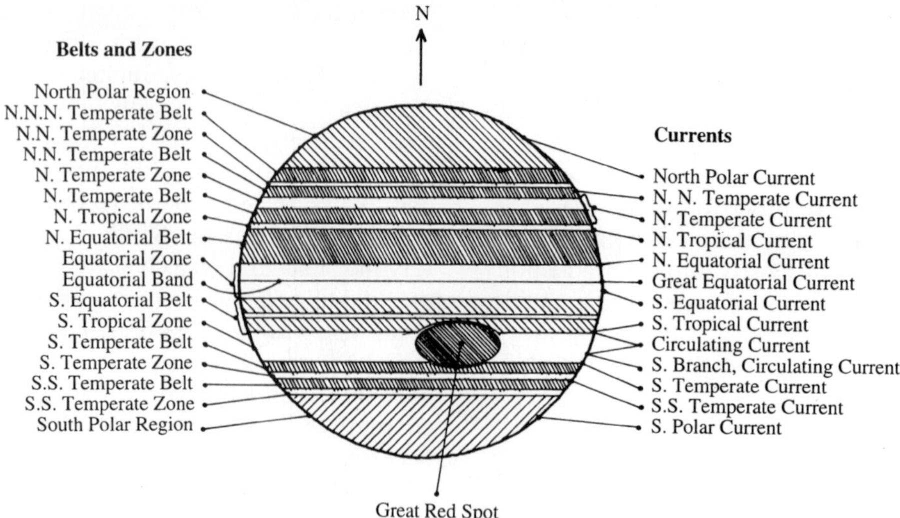


**Figure V.26** Nomenclature for Jupiter's cloud bands. Some of the most frequently seen and most useful bands seen in Jupiter's atmosphere are shown here. The bright bands (zones) and the dark bands (belts) alternate in a clearly asymmetrical manner. The most prominent feature in the southern hemisphere is the Great Red Spot (GRS) in the South Tropical Zone (STrZ). The most striking feature of the northern hemisphere is the dark, active North Equatorial Belt (NEB), in which dark reddish "barges" of high temperature can often be seen.

9h55m40.63s (System II). The rotation rate deduced from periodicities observed in the radio wavelength emission is 9h55m29.73s, obviously very close to System II, except that the error limits on either determination are 0.02 s or less. This figure applies both to the microwave (decimetric) emission and to the longer-wave sporadic radio-wavelength bursts in the 10-m (decametric)-wavelength region, which are both clearly of nonthermal origin and are associated with the planetary magnetosphere. We must conclude that the System II period is rather close to the rotation period of the interior derived from the radio data, whereas the equatorial zone (EZ) is clearly traveling much faster. The excess velocity of the EZ relative to System II or to the radio period (System III) is 106 m s$^{-1}$, a truly impressive wind speed.

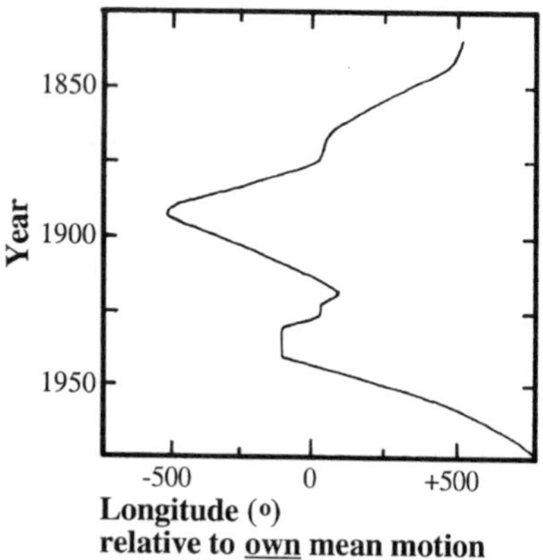


**Figure V.27** Wandering of the Great Red Spot. The longitude of the GRS *relative to its own mean motion* is shown for a 140-year period. Note that the amplitude of its drift even in the coordinate system that minimizes its motion is over 1200° of longitude. During that time, no clear motion in latitude was observed.

Although the wind speeds near the equator are the largest seen on the planet, observations of the motions of transient bright and dark spots have revealed complex wind patterns with many well-defined and reproducible "currents" confined to limited ranges of latitude. The main currents are shown in Fig. V.28.

Many of the wind speeds measured in this way are in excess of 10 m s$^{-1}$, and the regions of different wind speeds seem to be separated from one another by only the thinnest of boundaries. It is likely that the wind shear along the meridian (line of constant longitude) ($dv_x/dl$) is locally very large, and the local meteorological conditions may be very unstable. The Pioneer and Voyager photographs show complex structure in regions of high meridional wind shear, with chains of vortices aligning along the interfaces between belts and zones, often large enough to give a strongly scalloped appearance to the boundary.

The natural unit for measuring speeds of motions in a continuous fluid is the speed of sound. Dissipation of energy becomes very severe near the speed of sound, and natural circulation in an atmosphere cannot usually approach this speed very closely. The speed in units of the speed of sound, $v/c_s$, is called $M$, the Mach number. Because, by terrestrial standards at least, winds of over

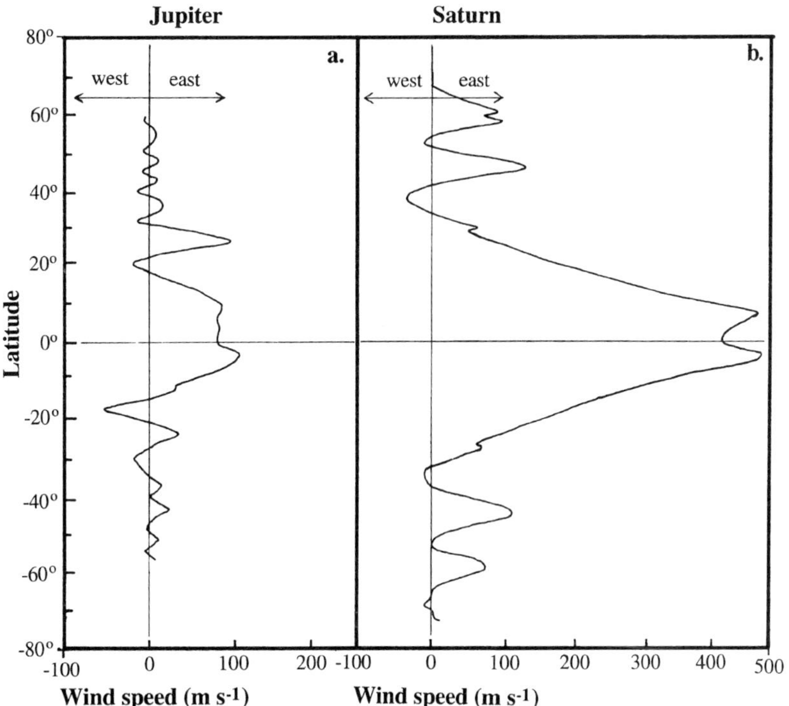


**Figure V.28** Speeds of currents in the atmospheres of Jupiter and Saturn. The speeds of currents deduced from the motions of observed bright spots on Jupiter (the east–west or *zonal* wind speeds) are shown in a. b uses the far smaller data set on spot motions on Saturn to construct a similar figure.

$100\,\mathrm{m\,s^{-1}}$ sound very fast indeed, it is worthwhile comparing these observed speeds with the speed of sound on Jupiter. How does the speed of sound depend on the physical and thermodynamic properties of the medium?

Consider a sound wave passing through a one-dimensional gas. The sound wave has associated with it an increase in the pressure and density of the gas above the background level by small increments of $P$ and $\Delta\rho$:

$$P = P_0 + \Delta P; \quad \rho = \rho_0 + \Delta\rho. \tag{V.119}$$

We also will have an equation of state for the medium, relating pressure and density: $P_0 = P(\rho_0)$. Then we can write

$$\begin{aligned} P_0 + \Delta P &= P(\rho_0 + \Delta\rho) \\ &= P(\rho_0) + \Delta\rho(dP/d\rho), \end{aligned} \tag{V.120}$$

where we have expanded $\rho$ about $\rho_0$. Then

$$K = \Delta P/\Delta\rho = (dP/d\rho)\rho_0, \tag{V.121}$$

where the constant $K$ is the proportionality constant between $\Delta P$ and $\Delta\rho$.

The position of a volume element of the medium before encountering the sound wave is $x$, and the displacement caused by the sound wave arriving at time $t$ is $X(x,t)$, so that the new position is $x + X(x,t)$. A neighboring volume element, originally at $x + \Delta x$, will be displaced by the sound wave to a new position, $x + \Delta x + X(x + \Delta x, t)$, at the same instant. The column density of air in the resting gas ($\mathrm{g\,cm^{-2}}$) between $x$ and $x + \Delta x$ is $\rho_0 dx$. But, when the sound wave is passing through, this same mass of air is displaced and lies between $x + X(x,t)$ and $x + \Delta x + X(x + \Delta x, t)$. The new density, $\rho$, is then calculable from

$$\begin{aligned} \rho_0\Delta x &= \rho[x + \Delta x + X(x + \Delta x, t) - x - X(x,t)] \\ &= \rho[(dX/dx)\Delta x + \Delta x], \end{aligned} \tag{V.122}$$

which rearranges to

$$\rho_0 = (\rho_0 + \Delta\rho)(dX/dx) + \rho_0 + \Delta\rho \tag{V.123}$$

$$= -\rho_0(dX/dx) - \Delta\rho(dX/dx). \tag{V.124}$$

Typically, for normal sound waves, $\Delta\rho$ is $\approx 10^{-6}\rho_0$, and we can neglect the last term.

Now we need to know the relationship between the displacement of the gas and the pressure-gradient force (proportional to $dP/dx$) that causes it. This is easy: consider the slice of gas between $X$ and $x + \Delta x$, of density $\rho_0$. The mass of this slice is $\rho_0\Delta x$, its acceleration is $d^2X/dt^2$, and the force on it is just the product of the mass times the acceleration. This force is, however, the pressure-gradient force, which is the force *difference* due to the pressure difference between $x$ and $x + \Delta x$:

$$P(x, t) - P(x + \Delta x, t) = -(\partial P/\partial x)\Delta x. \quad \text{(V.125)}$$

However, the pressure ($P$) is just a constant, $P_0 + \Delta P$, and any change in $P$ is due to changes in $\Delta P$, so

$$\begin{aligned} -(\partial P/\partial x)\Delta x &= -(\partial \Delta P/\partial x)\Delta x \\ &= \rho_0(\partial^2 X/\partial t^2)\Delta x. \quad \text{(V.126)} \end{aligned}$$

Canceling $\Delta x$, letting $\Delta P = K\Delta\rho$, and then replacing $\Delta\rho$ with $-\rho_0(dX/dx)$, we have the *wave equation*,

$$\partial^2 X/\partial t^2 = K(\partial^2 X/\partial x^2). \quad \text{(V.127)}$$

Dimensionally, K is $cm^2\,s^{-2}$, and we shall replace it with $K = c_s^2$, where $c_s$ is called the *speed of sound*. Thus we have

$$c_s^2 = (\partial P/\partial \rho)\rho_0. \quad \text{(V.128)}$$

We know that gases at 1 bar are usually very poor emitters of infrared energy, and the $\Delta T$ associated with a sound wave is only on the order of $10^{-6}$, so the compression energy stored in the high-pressure part of the sound wave is infinitely more likely to be passed on as bulk motion than to be radiated as heat. Thus we expect the passage of the sound wave to be adiabatic, not isothermal.

For an adiabatic process occurring in an ideal gas, $\rho \propto P/T$ and $T \propto P^{(\gamma-1)/\gamma}$. Thus $\rho \propto P^{1/\gamma}$ and $dP/d\rho = \gamma P/\rho$. The speed of sound is then

$$c_s = (\gamma RT/\mu)^{1/2} = [C_p RT/\mu(C_p - R)]^{1/2} \quad \text{(V.129)}$$

for an ideal gas.

For a gas of solar composition, the sound speed depends on only $C_p$ and $T$. We know that $C_p$ is a function of $T$ alone, and we know how to evaluate that function. Thus the speed of sound in a solar-composition gas is a function of temperature alone. This function is shown in Fig. V.29. The $T^{1/2}$ dependence is much stronger than the temperature dependence of $C_p$, and thus the graph shows only minor offsets at the temperatures at which excitation of the rotational and vibrational modes of $H_2$ causes changes in $C_p$. Note also that $C_pR/(C_p - R)$ approaches 1.3 $R$ at high temperatures. We can see from Fig. V.29 that the speed of sound in the stratosphere–mesosphere is $800\,m\,s^{-1}$, compared with $890\,m\,s^{-1}$ in the ammonia cloud layer at $T = 140\,K$.

If we imagine an 800-Hz ($\lambda = 1\,m$) sound wave in the stratosphere, the time taken for a pressure pulse to travel past a given point is $1/800\,s \simeq 10^{-3}\,s$. But we have already seen that the radiative relaxation time at this level is $10^8\,s$. This means that ordinary sound changes the local temperature and returns it to the original value some $10^{11}$ times faster than radiative cooling can work to remove the temperature excess. Therefore our assumption of adiabatic transmission of sound is secure. Sound waves of $10^{10}\,m$ could dissipate energy in this region by radiation, but this extraordinary kind of sound would have a wavelength 20 times the circumference of Jupiter! Long-wave disturbances associated with weather systems can indeed dissipate energy into the upper atmosphere, but by a wholly different mechanism.

This brief excursion into the physics of sound transmission has assured us that observed flow speeds on Jupiter

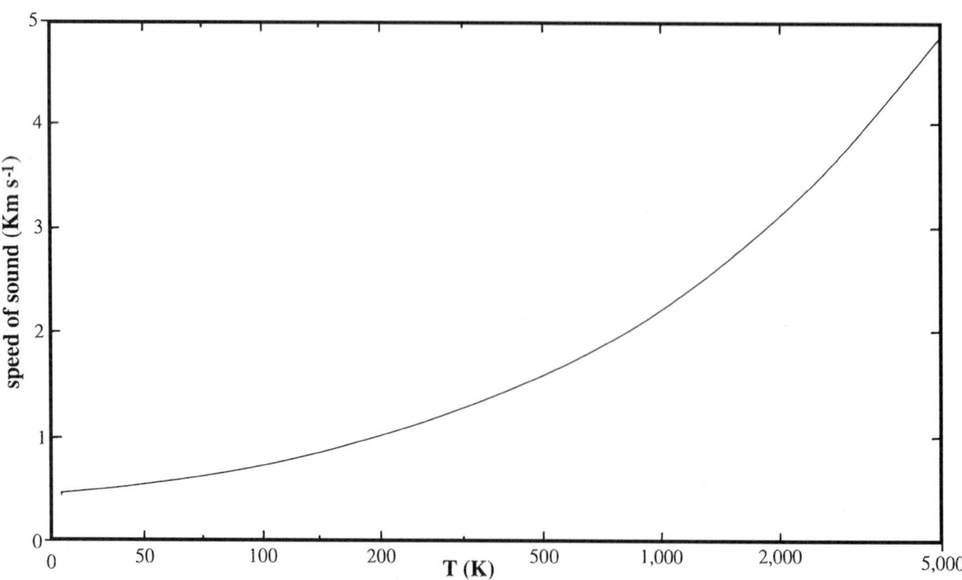


**Figure V.29** Speed of sound in a solar-composition gas. The speed of sound for a solar-composition atmosphere is calculated from Eq. (V.129). The fastest zonal wind speeds seen on Jupiter and Saturn are well in excess of 0.1 $c_s$ (Mach 0.1).

are no larger than Mach 0.2. On the other hand, it has given us no insight into the reason for such large speeds.

Also, we have not exhausted the phenomena observed in connection with motions of the GRS and other spots. For example, observers at the University of New Mexico saw small white spots appear far "behind" the GRS, engage in a lengthy stern chase, and then, as they are about to reach the trailing end of the GRS, accelerate abruptly. They then have been observed to dash counterclockwise around the circumference of the GRS vortex like runners around a track, sometimes persisting for more than 1.5 revolutions before fading to invisibility. While visible, the spots traveled at a speed of several tens of meters per second.

Similarly situated bright and dark spots were also observed beginning in 1902 on the interface between the STrZ and the SEB, appearing 90° ahead of the GRS with a 21-s shorter period. This complex of spots, called the South Tropical Disturbance (STD), caught up with the GRS in June of 1902, accelerated abruptly, and passed it by in 5 days, rather than the 42 days that would have been required before the spots accelerated. The STD survived this encounter, with a new period closer to that of the GRS, drifted all the way around the planet again, and repeated its performance at its next meeting with the GRS. This phenomenon was repeated *nine* times until, in 1940, the periods of the GRS and the STD became equal. But the story does not end here.

In 1940 the STrZ also produced several bright white spots that followed very peculiar paths. These spots oscillated in longitude with smoothly decreasing amplitude while also drifting relative to System II. They followed, so far as can be determined, a path describable as a damped sine wave! These oscillating spots did not persist, however. The longitude oscillations have been regarded by some as due to *vertical* damped oscillations of a buoyant "bubble" of gas that rose to the tropopause, overshot, and finally dissipated its energy and came to rest. (If the bubble of gas conserves angular momentum, how large must its vertical oscillations be to allow its period to change by 10 s from the mean?)

Such anecdotes regarding peculiar happenings on Jupiter are merely some of the more extravagant representatives of a vast and fascinating literature. Large numbers of such observations are collected in a book entitled *The Planet Jupiter*, by the English astronomer B. M. Peek. Any conceivable meteorological theory could draw support from the observations catalogued therein, and indeed many have done so. But the great preponderance of present data on motions in the Jovian atmosphere comes from the Voyager 1 and 2 flybys. These observations are rich in detail and complexity, but the best of the photographs span only about a week of time. The lower-resolution coverage, spanning several months before and after the Jupiter encounter, demonstrates that the odd phenomena seen by Earth-based astronomers are real and do not do full justice to Jupiter's diversity.

There is a much smaller literature on Earth-based observations of motions in Saturn's atmosphere, partly due to the observational difficulties discussed earlier, and partly to a less chaotic and spectacular meteorology. A few bright spots have been seen, and some useful data have been collected from the Voyager flybys regarding cloud motions. Perhaps the most striking difference in motions between Jupiter and Saturn is the high symmetry of east–west (zonal) zones across the equator. Jupiter exhibits far less symmetry (and vastly more visible structure) despite its very small axial tilt and Saturn's pronounced obliquity. Jupiter's atmospheric motions also depart greatly from axial symmetry. Saturn's most unique feature is a jet stream with a speed of approximately Mach 0.6, a speed record for the Solar System.

The general moral to be drawn from the cloud band morphology studies is that Jupiter has a very active meteorology. We already know that there is a large heat flux from the deep interior to drive convection throughout Jupiter's envelope and that a similar but smaller heat source is present in Saturn. We would hope eventually to understand the mechanisms for vertical heat transport, horizontal motions, and dissipation in terms of a unified global model of atmospheric dynamics. Although a number of the pieces of that puzzle are now available, the unified model still eludes our grasp.

The internal heat source, which maintains the temperature gradient infinitesimally above the adiabatic lapse rate, is responsible for strong vertical mass fluxes, rising in some regions, but of course balanced by subsiding fluxes elsewhere. The typical vertical velocities in the presence of a vertical heat flux can be shown in Prandtl mixing length theory to be

$$v_z = [R\phi/C_p\rho]^{1/3} \tag{V.130}$$

with uncertainties of fully a factor of 10. We lose no precision by taking a mean value of $C_p/R = 3.93$ for the entire troposphere of Jupiter. Then, assuming ideal gas behavior and adiabatic structure, with T(1 bar) = 165 K,

$$v_z = [\phi T/0.1\,P]^{1/3} = 625\phi^{1/3}T^{-0.78}, \tag{V.131}$$

where $\phi$ is in erg cm$^{-2}$ s$^{-1}$ and $v_z$ is in cm s$^{-1}$. We are justified in taking $\phi = 1.2 \times 104$ erg cm$^{-2}$ s$^{-1}$ as a planetary mean value, because this is intermediate between the internal source strength of $0.8 \times 10^4$ and the total emitted flux of $1.5 \times 10^4$ at the top of the troposphere, with the reemitted solar heat included. Thus, along a Jupiter adiabat, $v_z = 1.5 \times 10^4\,T^{-0.78}$ cm s$^{-1}$, whereas on Saturn, because of the lower flux and lower adiabat, $v_z$ is lower by ≈50%.

For cloudtop temperatures of 130 K, $v_z = 340$ cm s$^{-1}$, whereas at the 2000-K level, $v_z = 40$ cm s$^{-1}$. Thus we are entirely justified in taking the mean vertical velocities in upwelling regions as being on the order of 1 m s$^{-1}$. Of course, by continuity (conservation of mass), the velocities in subsiding regions are comparable in size if the rising and subsiding regions have similar areas. If the subsiding regions are limited to the NEB and other dark Belts, then the velocities of subsidence $v_z$ would have to be >10 m s$^{-1}$. At the tropopause, a typical sample of cloudtop material may travel $10^4$ km before encountering a downdraft, which will require a time on the order of $10^4/V_{xy}$, where $V_{xy}$ is the mean horizontal wind velocity. Taking $V_{xy}$ as close to 10 m s$^{-1}$, the lifetime of a parcel of cloudtop material is $10^6$ s before it is recycled into the interior. We previously estimated a radiative time constant of $10^8$ s for the 100-mbar level in the stratosphere. If we recall the $P^2$ dependence of molecular hydrogen thermal opacity (and emissivity!), we would then expect a radiative time constant ($t_{\mathrm{rad}}$) of $10^6$ s at the 1-bar level, at which we observe a dynamical time constant ($t_{\mathrm{dyn}}$) of $10^6$ s. Below the 1-bar level the atmosphere between that level and empty space is opaque ($\tau > 1$), and we can no longer calculate radiation loss rates so simply. Essentially all the heat flux at these depths will be carried by bulk motions.

It is desirable to have a phenomenological measure of rates of vertical mixing analogous to the diffusion coefficient to describe the *average* rate of mixing over the planet. This number, called the *eddy diffusion coefficient*, $K$ (cm$^2$ s$^{-1}$), may be used pragmatically and evaluated observationally even when the actual mechanism of mixing is obscure. Into it we will lump the mixing effects of all processes involving bulk motions, from small-scale eddies up to motions on the scale of the planetary circulation. This constant implies an acceptance of a random-walk mixing process with step sizes that are small compared with the characteristic dimensions of the atmosphere and will clearly be inapplicable if all mixing is due to very large-scale overturn. Over an altitude interval of one scale height, $H$, in the stratosphere, where the lifetime against vertical mixing is $T$,

$$K = H^2/t_{\mathrm{dyn}} = v_z H. \qquad \text{(V.132)}$$

For the Jovian stratosphere, where $H = RT/\mu g \approx 16$ km and $t \approx 10^8$ s, $K \approx 3 \times 10^4$. It should be kept in mind that this is a very approximate estimate, and we should be open to any opportunity to constrain the mixing rate observationally. For the moment, it suffices to point out that the very short dynamical time constants (large values of $v_z$) caused by convective heat transport in the troposphere require eddy diffusion coefficients over $3 \times 10^8$.

The vertical radiative and dynamical fluxes are summarized as functions of altitude in Fig. V.30. All fluxes given are averages over the entire planet.

We must now confront the problem of describing and explaining the motions of the Jovian atmosphere in terms of quantitative physics. In order to give a fair appreciation of the magnitude of this task, it is necessary to begin at a very elementary level and introduce realistic considerations ("complications") one at a time.

The logical place to begin is with a brief return visit to hydrostatic equilibrium to indicate the ways in which this very useful approximation breaks down. Imagine, as we have several times before [Eq. (III.169), for example], a stationary gas in a gravitational field, with pressure increasing with depth:

$$dP = -\rho g dz. \qquad \text{(III.169)}$$

Generalizing to three dimensions (not constraining the pressure gradient to lie in the $-Z$ direction), we have

$$dP/dx + dP/dy + dP/dz = -\rho g(x, y, z). \qquad \text{(V.133)}$$

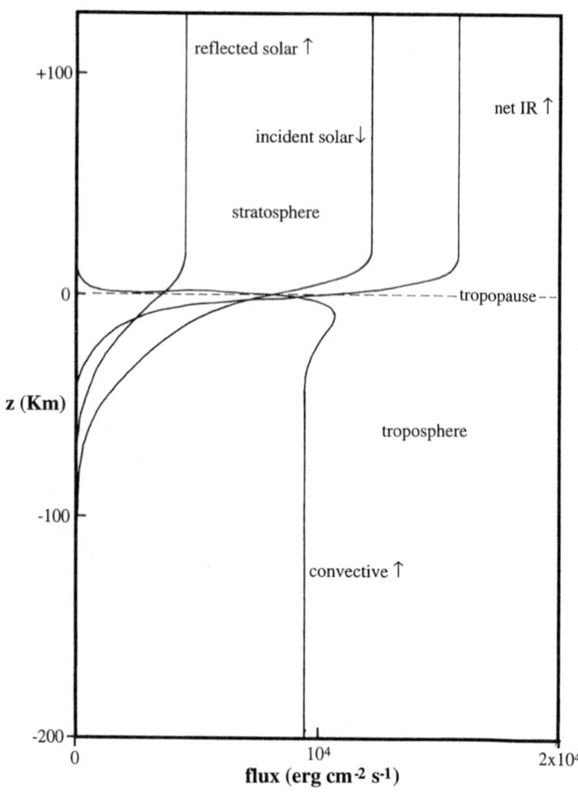


**Figure V.30** Radiative and convective fluxes in Jupiter's atmosphere. The magnitudes and directions (arrows) of the major energy fluxes in the Jovian atmosphere are given. These are intended to represent planet-wide averages, not conditions at any particular point.

The acceleration due to gravity is now more handily expressed as the local derivative of the gravitational potential ($U$):

$$dP/dx + dP/dy + dP/dz = -[dU/dx + dU/dy + dU/dz]. \quad \text{(V.134)}$$

We can save much time in writing such expressions by using the gradient operator $\nabla$. This operator is resolved into $x$, $y$, and $z$ components, and hence even when this operator operates upon a scalar the result is a *vector*. Then, defining the vector as

$$\vec{\nabla} = (d/dx + d/dy + d/dz), \quad \text{(V.135)}$$

we have from Eq. (V.134)

$$\vec{\nabla}P = -\rho\vec{\nabla}U. \quad \text{(V.136)}$$

In other words, we can say that the *pressure gradient force* ($P$) is exactly balanced by the gravitational force, which we now see as a *gravitational potential gradient force*. Any given spatial distribution of the scalars $\rho$, $P$, and $U$ may be selected, but these combinations will not in general satisfy Eq. (IV.136). We would say that almost all possible configurations are *unstable against convection*.

Our progress would be aided by having functional relations between the parameters describing the state of the fluid. Usually we would think first in terms of an equation of state $\rho = \rho(P, T)$, but very often in the treatment of fluid flow the assumption is made that the fluid is *incompressible* ($\rho =$ constant). Next, in a moving fluid, we need some way of keeping track of the fluid mass: an expression of conservation of mass. If the velocity of fluid flow normal to the surface of a unit volume of fluid is $v$, then the mass flux out of that volume is $\rho v$, and the rate of decrease of density is

$$d\rho/dt = -[(\partial/\partial x)\rho v_x + (\partial/\partial y)\rho v_y + (\partial/\partial z)\rho v_z]. \quad \text{(V.137)}$$

This operator operating upon $\vec{v}, \vec{\nabla}$ is called the *divergence* operator. The result of the operation is a *scalar*. In the case of incompressible flow (called the *Boussinesq* approximation by meteorologists), $\rho =$ constant and $d\rho/dt = 0$; hence

$$\vec{\nabla} \cdot \vec{v} = 0, \quad \text{(V.138)}$$

and the velocity has zero divergence.

Next, we must account for the effects of forces on the motions of each volume element of fluid. In general, a fluid will contain both pressure gradients and potential gradients and in addition will experience shear forces due to viscous transport of momentum. The overall force balance is

$$\rho a = -\vec{\nabla}P - \rho\vec{\nabla}U + F_{\text{visc}}, \quad \text{(V.139)}$$

where $a$ is the net acceleration and $F_{\text{visc}}$ is the force due to viscosity. A common simplification is to assume *inviscid flow*; i.e., $F_{\text{visc}} = 0$. Real (physical) fluids generally require inclusion of the effects of viscosity, but mathematical fluids can be strikingly simple. Now we need to substitute into Eq. (V.139) a simple expression for the acceleration ($a$).

A given volume element of a fluid that has velocity $\vec{v}$ and position $(x, y, z)$ will, after time $\Delta t$, have velocity $\vec{v} + \Delta\vec{v}$ and position $(x + v_x\Delta t, y + v_y\Delta t, z + v_z\Delta t)$. The new velocity, $\vec{v} + \Delta\vec{v}$, is then

$$\begin{aligned} v(x + v_x\Delta t, y + v_y\Delta t, z + v_z\Delta t) &= v(x, y, z, t) + (\partial\vec{v}/\partial x)v_x\Delta t \\ &\quad + (\partial\vec{v}/\partial y)v_y\Delta t + (\partial\vec{v}/\partial z)v_z\Delta t \\ &\quad + (\partial\vec{v}/\partial t)\Delta t \end{aligned} \quad \text{(V.140)}$$

by the definition of partial derivatives.

The acceleration is then

$$\begin{aligned} a &= \Delta\vec{v}/\Delta t \\ &= v_x(\partial v/\partial x) + v_y(\partial v/\partial y) \\ &\quad + v_z(\partial v/\partial z) + \partial v/\partial t, \end{aligned} \quad \text{(V.141)}$$

which we can write in operator notation as

$$a = (\vec{v} \cdot \vec{\nabla})\vec{v} + \partial\vec{v}/\partial t \quad \text{(V.142)}$$

Now, combining Eqs. (IV.139) and (IV.142) for inviscid flow,

$$\partial\vec{v}/\partial t + (\vec{v} \cdot \vec{\nabla})\vec{v} = -\vec{\nabla}p/\rho - \vec{\nabla}U. \quad \text{(V.143)}$$

It can be shown in vector analysis that

$$(\vec{v} \cdot \vec{\nabla})\vec{v} = (\vec{\nabla} \times \vec{v}) \times \vec{v} + \vec{\nabla}(\vec{v} \cdot \vec{v})/2, \quad \text{(V.144)}$$

where the vector cross-product $\vec{\nabla} \times \vec{v}$ is defined as having the components

$$(\vec{\nabla} \times \vec{v})_z = \nabla_x v_y - \nabla_y v_x = \partial v_y/\partial x - \partial v_x/\partial y \quad \text{(V.145a)}$$

$$(\vec{\nabla} \times \vec{v})_x = \nabla_y v_z - \nabla_z v_y = \partial v_z/\partial y - \partial v_y/\partial z \quad \text{(V.145b)}$$

$$(\vec{\nabla} \times \vec{v})_y = \nabla_z v_x - \nabla_x v_z = \partial v_x/\partial z - \partial v_z/\partial x. \quad \text{(V.145c)}$$

The vector $\vec{\nabla} \times \vec{v}$, called the *curl* of $\vec{v}$, is clearly orthogonal to both $\vec{\nabla}$ and $\vec{v}$.

In the study of the motion of fluids it is often convenient to consider the vector $\vec{\nabla}\times\vec{v}$, and we therefore introduce a quantity $\vec{\Omega}=\vec{\nabla}\times\vec{v}$, called the *vorticity*. For some purposes, it is adequate to neglect the vorticity ($\vec{\Omega}=0$). Such flow is called *irrotational*.

If we consider a closed loop drawn in a fluid, a useful measure of the rate of circulation of any vector field $V$ around the loop can be had by forming a line integral of $V$ times the length element of the loop, $ds$, and integrating all the way around the loop. It is the component of $V$ in the $ds$ direction *only* that contributes to the circulation. The cyclic line integral is then

$$\oint V_s ds = \oint \vec{V}_s \cdot \vec{ds}. \tag{V.146}$$

This integral is called the *circulation* of the vector field $V$. Let us now consider a square region of space with corners at $(x, y)$, $(x+\Delta x, y)$, $(x+\Delta x, y+\Delta y)$, and $(x, y+\Delta y)$ in the vector field $\vec{V}$. Starting in the $+x$ direction from $(x, y)$, we will calculate the circulation of $V$ about the square. The contribution of the first leg of the square to the circulation is the product of the $x$ component of $\vec{V}$ times $\Delta x$ or $V_x\Delta x$, which is almost (but not quite) balanced by the contribution from the third leg. Actually, $V_x(3) = V_x(1) + (\partial V_x/\partial y)\Delta y$ in general, allowing for spatial variations in $V_x$.

Similarly, for side 2 we have $V_y\Delta y$ and, for side 4, $V_y(4)\Delta y = [V_y(2) + (\partial V_y/\partial x)\Delta x]\Delta y$. The net circulation around the entire square is

$$V_x(1)\Delta x - [V_x(1) + (\partial V_x/\partial y)]\Delta y + V_y(2)\Delta y - [V_y(2) + (\partial V_y/\partial x)\Delta x]\Delta y \tag{V.147}$$

or simply

$$[(\partial V_y/\partial x) - (\partial V_x/\partial y)]\Delta x\Delta y. \tag{V.148}$$

The product $\Delta x\Delta y$ is just the area of the square, $\Delta A$, and the bracketed term is a vector in the $Z$ direction, normal to the square; in fact, it is just the $Z$ component of the curl of $V$! In general,

$$\oint_{\text{loop}} \vec{V}\cdot\vec{ds} = \oint_{\text{surface}} (\vec{\nabla}\times\vec{V})_n dA, \tag{V.149}$$

where $n$ denotes the direction normal to the surface. This relationship of curl $\vec{V}$ to the circulation of the vector field ($\vec{V}$) is called *Stokes' Theorem*.

We can now see that the circulation per unit area is equal to the vorticity ($\vec{\Omega}$) as defined above: the vorticity is the circulation in the plane normal to $\vec{\Omega}$.

From Eqs. (V.143) and (V.144) we have

$$\partial\vec{V}/\partial t + \vec{\Omega}\times\vec{V} + \vec{\nabla}v^2/2 = -\vec{\nabla}P/\rho - \vec{\nabla}U. \tag{V.150}$$

Taking the curl of both sides for the case of incompressible flow ($\rho$ = constant and $\vec{\nabla}\cdot\vec{v}=0$) and setting the curl of every gradient to zero (why?), we get

$$\partial\vec{\Omega}/\partial t + \vec{\nabla}\times(\vec{\Omega}\times\vec{v}) = 0, \tag{V.151}$$

which, with the definition $\vec{\Omega}=\vec{\nabla}\times\vec{v}$ and the conservation equation $\vec{\nabla}\cdot\vec{v}=0$, serves to describe the velocity field for inviscid flow of an incompressible fluid completely.

We can now speak of another simple concept that can often prove very useful, the idea of *steady flow*. By "steady" we imply only that the local velocity field is unchanging: $\vec{v}(x, y, z) \neq v(t)$. If we follow the flow lines (draw a path always parallel to the local velocity vector), we draw what is termed a *streamline*. Thus we can say that steady flow is the case in which the streamlines are stationary. We may then take the steady-flow condition to be $\partial\vec{v}/\partial t = 0$ and drop the first term from Eq. (V.150). In the case of unsteady flow, the streamlines are free to move about, and no such simplification may be made.

In the usual circumstance in which rotation can occur, the measure of the strength of the circulation is the vorticity. Let us indicate the direction of the vorticity ($\vec{\Omega}$) by lines running parallel to $\vec{\Omega}$, and (as with magnetic fields) let us indicate the magnitude of the vorticity by the density of these lines in space. Such lines are called vortex lines and are as useful in rotational flow as streamlines. Because $\vec{\Omega}=\vec{\nabla}\times\vec{v}$, $\vec{\nabla}\cdot\vec{\Omega}$ is always zero, meaning that vortex lines are conserved in the sense of being unending and closed, like lines of magnetic force. Equation (V.151) can be taken to mean that vortex lines move along with the fluid: particles do not cross vortex lines or streamlines. This follows from conservation of angular momentum.

In the general case of viscous flow, Eqs. (V.139) and (V.142) give us

$$\partial\vec{v}/\partial t + (\vec{v}\cdot\vec{\nabla})\vec{v} = -\vec{\nabla}P/\rho - \vec{\nabla}U + F_{\text{visc}}/\rho. \tag{V.152}$$

In the absence of flow there are of course no shear forces and no viscous dissipation; however, if flow occurs in the neighborhood of a boundary there will be a velocity gradient normal to the surface of the boundary. The flow velocity will in fact vanish at the boundary. Two parallel boundaries $d$ apart and differing in velocity by $v$, with steady flow between them (parallel, straight streamlines), will exert a shear force per unit area of

$$F = \eta v/d, \tag{V.153}$$

where $\eta$ is the coefficient of viscosity. In the general case, in which the rate of change of the local shear strain is composed of components in both the $x$ and the $y$ directions, the shear stress $S_{xy}$ is

$$S_{xy} = \eta(\partial v_y/\partial x + \partial v_x/\partial y). \tag{V.154}$$

For a compressible fluid, another term must be included to take into account higher derivatives of the velocity

$$S_{ij} = \eta(\partial v_i/\partial x_j + \partial v_j/\partial x_i) + \eta'\delta_{ij}(\vec{\nabla} \cdot \vec{v}), \qquad (V.155)$$

where $\eta'$ is called the second coefficient of viscosity and $\delta_{ij}$ is the Kronecker delta ($\delta_{ij} = 1$ if $i = j$; $\delta_{ij} = 0$ if $i \neq j$). The general expression for the viscous force is found to be

$$F_{\text{visc}} = \eta\nabla^2\vec{v} + (\eta + \eta')\vec{\nabla}(\vec{\nabla} \cdot \vec{v}), \qquad (V.156)$$

where $\nabla^2\vec{v} = \vec{\nabla} \cdot \vec{\nabla}\vec{v}$ is called the Laplacian operator. The general equation for motion of the fluid then becomes

$$\rho\{\partial\vec{v}/\partial t + (\vec{v} \cdot \vec{\nabla})\vec{v}\} = -\vec{\nabla}P - \rho\vec{\nabla}U + \eta\nabla^2\vec{v} + (\eta + \eta')\vec{\nabla}(\vec{\nabla} \cdot \vec{v}). \qquad (V.157)$$

Letting $\vec{\Omega} = \vec{\nabla} \times \vec{v}$ as before,

$$\rho\{\partial\vec{v}/\partial t + \vec{\Omega} \times \vec{v} + \tfrac{1}{2}\vec{\nabla}v^2\} = -\vec{\nabla}P - \rho\vec{\nabla}U + \eta\nabla^2\vec{v} + (\eta + \eta')\vec{\nabla}(\vec{\nabla} \cdot \vec{v}). \qquad (V.158)$$

Taking the curl of both sides,

$$\partial\vec{\Omega}/\partial t + \vec{\nabla} \times (\vec{\Omega} \times \vec{v}) = (\eta/\rho)\nabla^2\vec{\Omega}. \qquad (V.159)$$

The import of this equation is that, unlike the inviscid flow described in Eq. (V.151), we have a much more complex case in which vorticity no longer comoves with the liquid, but rather can "diffuse" through it because of viscous coupling. Another way of phrasing this is to say that a vortex tends to exchange mass with its surroundings and thus spreads a given amount of vorticity out over an ever-larger volume. This is analogous to the decay with time of a magnetic field; vortex lines of course remain continuous and closed, but get ever farther apart.

Is a planetary atmosphere incompressible, inviscid, and irrotational? Typically it is none of these; however, a glance at the horrendous complexity of Eq. (V.158) makes it very clear why grossly simplified assumptions are made in the application of fluid dynamics theory to meteorological problems.

One approach to making the mathematics of fluid dynamics tractable is to reduce the infinite number of different conceivable combinations of variables to manageable size by astute combinations of variables into a smaller number of nondimensional parameters. A good example of this approach is the reformulation of Eq. (V.159) for incompressible viscous flow,

$$\partial\vec{\Omega}/\partial t + \vec{\nabla} \times (\vec{\Omega} \times \vec{v}) = (\eta/\rho VD)\nabla^2\vec{\Omega}, \qquad (V.160)$$

where $\eta$ is the coefficient of viscosity and $\rho$ is the density as before, but $\vec{V}$ is a characteristic reference velocity and $D$ is a characteristic dimension. The new time scale is set by the ratio of $D/\vec{V}$, and every variable has been nondimensionalized and coalesced into the factor in front of the Laplacian. The value of that parameter is alone required in order to characterize the nature of the flow! The factor is called the Reynolds number, defined as $\text{Re} = \rho VD/\eta$.

Of course it must be borne in mind that this approach is valid only for incompressible flow and certainly cannot be applied to high velocities. At high speeds, similarity in both the Reynolds number and the Mach number is required for the flow to remain similar.

The drag force exerted by fluid flowing past a body is proportional to the kinetic energy density of the fluid and to the cross-section area of the body. The proportionality factor, CD, is called the drag coefficient and is dependent in a complex way on both the geometry of the body and the Reynolds number:

$$F_{\text{drag}} = C_{\text{D}}\rho V^2 A/2. \qquad (V.161)$$

As a rule, for most convex shapes (spheres, hexagonal prisms, Apollo Command Modules) the drag coefficient is on the order of 1, but may vary by about a factor of 3 from this typical value. As a rule, $C_{\text{D}}$ is largest ($>2$) at low Reynolds numbers ($\text{Re} < 10$) and smallest at very high values ($\text{Re} \approx 10^6$). At very low Reynolds numbers ($\text{Re} < 1$) the drag coefficient varies as $1/\text{Re}$.

Motions in the atmosphere of Jupiter are unfortunately more complex to describe than even the worst case we have yet encountered, that of viscous incompressible flow. First of all, we have seen that flow velocities can be very high (up to Mach 0.1 at least), and we are well aware that there are large vertical density gradients in any atmosphere that lies in a gravitational field. Observationally, we can see that the flow is not steady, but strongly turbulent (high Reynolds number), and that enormous features such as the GRS behave as vortices, or at least generate them. We know from the observed heat emission that the vertical temperature gradient is superadiabatic by a small amount (perhaps by $10^{-5}\,\text{K km}^{-1}$) and that strong convection is occurring. Because of the unevenness of heating by the Sun, there must be meridional (poleward) transport of heat from the equatorial regions by atmospheric motions. Our observations of the NEB strongly imply that belts are clear, subsiding regions, whereas the zones are cloudy, upwelling regions; thus the temperature gradients in these adjacent bands are different due to the heat release by condensation in the zones. Thus the densities at equal altitudes in a belt and a zone can at best be the same at one level. Finally, and not the least of the complexities, the planet is a very rapid rotator. Any gas parcel moving along a surface of constant altitude will change its distance from the rotation axis whenever there is a northward or southward component to its velocity. Conservation of angular momentum requires that a torque be exerted by the moving parcel. Even at the equator, where this *Coriolis force* vanishes for horizontal motions, any radial motion (convection!) will involve this effect. If the angular momentum of a gas parcel, $L = mvr = m\omega r^2$, remains constant as the parcel changes its distance from the rotation axis by $\Delta r$, then the new angular velocity $\omega'$ is related to the old $\omega$ by

$$\omega'(r + \Delta r)^2 = \omega r^2 = (\omega + \Delta\omega)(r + \Delta r)^2, \qquad \text{(V.162)}$$

from which $\Delta\omega/\Delta r = 2\omega/r$. If there are strong viscous forces that keep $\omega$ constant, then the torque that must be applied is

$$\tau = dL/dt = 2m\omega r(dr/dt). \qquad \text{(V.163)}$$

Thus a rising parcel of air (anywhere except at the poles) will extract angular momentum from the rest of the planet. Of course, we must have continuity of motions; mass must be conserved. If gas is rising somewhere, then somewhere else a return flow must take mass back to the place of origin of the parcel. A good approximation for most purposes is the steady-state assumption, in which the net mass flux through any $r$ is zero, and thus the net torque vanishes. However, every rising flow will be deflected westward, and every subsiding flow will accelerate eastward. The "cloud plumes" seen in some of the Voyager pictures of Jupiter may be striking examples of this phenomenon. As a result, cyclic flow will not be simply steady flow along closed streamlines. There will be diffusion of vorticity out of each such cell and constant input of energy as hot, moist gas rising from below.

A convenient way to describe the effect of the Coriolis force, which vanishes at the equator, is with the *Coriolis parameter*, $f$, defined by $f = 2\omega \sin l$, where $l$ is the latitude. Where the Coriolis parameter is smallest, the bands are widest. The banded structure breaks down altogether at latitudes near $\pm 45°$. The usual meteorological criterion for judging the importance of Coriolis forces is the Rossby number,

$$\text{Ro} = v_x/fL, \qquad \text{(V.164)}$$

where $v_x$ is the eastward component of the wind and $L$ is the characteristic distance scale for the horizontal motions. The effect of Coriolis forces in causing rotational motion is important if Ro is very small, $\ll 1$. For large horizontal distance scales rotation becomes ever more important, up to the largest allowable scale sizes, $L \approx R_J$. The only obvious exception is the immediate vicinity of the equator. As we have seen, large-scale horizontal flows are a common feature on Jupiter.

In a steady-state flow situation (not necessarily steady flow), the horizontal pressure gradient forces and the Coriolis forces will be in balance. This condition is called *geostrophic* flow.

Previously, in discussing the nature of atmospheric motions near belt–zone boundaries, we mentioned that a dry adiabatic column of atmosphere in hydrostatic equilibrium alongside a moist adiabatic column in hydrostatic equilibrium can at most have the same density at a single altitude. The lower part of the moist adiabatic column produces heat from the latent heat of condensation, rises buoyantly to the tropopause, radiates off heat, and falls on a dry adiabat to approximately close the loop. Now we can regard the rather more complex circumstances that occur away from the equator of a rapidly rotating planet. The rising gas in the zones experiences a Coriolis force that accelerates the gas westward on the side of the zone closer to the equator and eastward on the side closer to the nearer pole. Viewed from above, we would say that the rising gas has clockwise vorticity in the northern hemisphere and counterclockwise vorticity in the southern hemisphere. This kind of rotation, associated with warm, rising, high-pressure areas, is called *anticyclonic*. In the belts, the subsiding cooler gas is twisted in the opposite sense and will take on clockwise vorticity in the southern hemisphere, and counterclockwise vorticity in the northern. This is called, not surprisingly, *cyclonic* motion.

Thus we can see that the east–west component of the wind, $v_x$, will increase with altitude, with the vertical shear, $\partial v_x/\partial z$, driven by the *meridional* (north–south) temperature gradient, $\partial T/\partial y$. This effect is described by the *thermal wind equation*:

$$f\,\partial v_x/\partial z = -(g/T)(\partial T/\partial y). \qquad \text{(V.165)}$$

Because these thermal winds blow at right angles to the temperature gradient they are not to first order effective in transporting heat horizontally. It is only because of viscous effects and turbulent instability that such occurs.

Now let us consider how we would distinguish which of two experimentally sampled parcels of air is the more buoyant and thus which will rise and which will sink. We have a conceptual definition already in terms of "higher" vs "lower" adiabats. We can make the definition much more compact by referring to the temperature that the gas parcel at $(P, T, Z)$ would have if *compressed adiabatically* to a particular standard pressure, such as the surface pressure of Earth or the 10-bar level of Jupiter. This temperature is called the *potential temperature*, $\Theta$, and is simply defined by

$$\Theta = T(P_0/P)^{(\gamma-1)/\gamma} = T - (\partial T/\partial z)_{\text{ad}}z. \qquad \text{(V.166)}$$

Clearly a column of air is marginally stable if $\Theta$ is constant, which corresponds to the precise condition of adiabatic and hydrostatic equilibrium. If $\Theta$ increases with altitude or, more precisely, if $\partial\Theta/\partial z$ is positive, then the gas is stable against overturn, and $\partial\Theta/\partial z < 0$ corresponds to free convection with a superadiabatic lapse rate. The vertical gradient of potential temperature is called the *static stability, S*,

$$S = \partial\Theta/\partial z = \partial T/\partial z - (\partial T/\partial z)_{\text{ad}}, \qquad \text{(V.167)}$$

which is just the difference between the actual lapse rate and the adiabatic lapse rate. (Recall that $(\partial T/\partial z)_S$ is *negative*, and $\partial T/\partial z$ usually is also.)

In meteorology a dimensionless quantity called the *Richardson number*, Ri, is used as the measure of static stability:

$$\text{Ri} = (gS/T)/(\partial v_x/\partial z)^2. \qquad \text{(V.168)}$$

If the vertical motions have a characteristic size of one scale height, then $\partial v_x/\partial z \approx v_x/H$, and

$$\text{Ri} = gSH^2/Tv_x^2. \qquad \text{(V.169)}$$

Very low static stabilities (negative $S$ and negative Ri) correspond to rapid small-scale turbulent overturn, with cells about one scale height in all dimensions; the large-scale circulation pattern is overpowered and irrelevant. This of course contributes to large values of the eddy diffusion coefficient, $K$. This rapid small-scale overturn is called *free convection*.

Static stabilities only slightly different from zero correspond to states in which the large-scale circulation dominates the local effects. We may refer to this case as *forced* convection. Somewhat more stable structures, with Richardson numbers between 0.25 and 0.95, correspond to situations in which the thermal wind supplies energy to drive mixing via small-scale turbulent eddies. This mixing both smoothes out the vertical wind shear and allows horizontal transport of heat in the north–south direction, which of course makes the atmosphere more stable. This is called *inertial instability*. For values of $\text{Ri} > 0.95$ the only instability mechanism, called *baroclinic instability*, depends upon the sensitivity of the (stable) structure to *horizontal* perturbations, in which a parcel of stable air from one band is inserted into a neighboring band (also stable) at a level at which the inserted parcel either sinks or rises. The associated horizontal motions, which transport heat down the temperature gradient, may be many times larger than a scale height.

The last case, of very large static stabilities, corresponds to the simple case of pure thermal winds, with only accidental horizontal heat transport to relieve the gradient. Differential solar heating leads to net poleward transport of heat carried by second-order viscously driven motions. There will be a general rising in the equatorial regions, poleward winds at high altitudes, sinking at high latitudes, and return flow to the equator at low altitudes. This kind of circulation, called the *Hadley regime*, obviously has a horizontal scale comparable to the radius of the planet. It would be expected only in very stable atmospheres on slowly rotating planets and is clearly irrelevant to Jupiter.

Because of the presence of a large internal heat source, we would expect a very low static stability for Jupiter's atmosphere; however, differential solar heating has the exact opposite effect!

Two other types of instability can be important, neither related in a useful way to the Richardson number. The first of these, called *barotropic instability*, is defined by the condition

$$\partial^2 v_x/\partial y^2 = \partial f/\partial y \equiv \beta. \qquad \text{(V.170)}$$

Thus, this instability appears when the horizontal shear gradient becomes very large, and the instability draws kinetic energy from the mean flow to drive horizontal eddy mixing and lower the shear gradient. This particular instability can be investigated quite directly, because $\partial f/\partial y$ is trivially calculable and the gradient of the shear can be estimated from studies of the speeds of the "currents" in Jupiter's atmosphere. It is found that this instability should be present on the edges of both the Equatorial Current and the North Temperate Current, and quite conceivably elsewhere as well. The vortices seen at the interfaces between belts and zones in the Pioneer and Voyager pictures are strong evidence for the widespread presence of barotropic instability.

The other instability mechanism, *radiative instability*, is based on the idea that the condensate clouds may locally enhance the greenhouse effect due to their high infrared opacity, which causes deposition of solar energy under the clouds, which drives more rapid rising motions, etc. This idea meets some difficulty in that the radiative heating time constant would be very long compared with the typical dynamical time constant, so that we would not expect this effect to have time to develop itself.

A variety of particular features of the circulation, such as the Equatorial Current and the Great Red Spot, require special comment. The transport of momentum toward the equator has been suggested to be due to inertial instability, baroclinic instability, and convection in a rotating frame. Inertial instability is unable to account for acceleration relative to the low-level motions, whereas baroclinic instability can if $\text{Ri} \approx 1$. The convection mechanism produces cells whose motions are deflected to the right (in the northern hemisphere) and therefore contribute to an increase in westward wind speed toward the equator. This mechanism would be effective if the convection were present down to great depths, as is entirely probable.

The Great Red Spot was long believed to be the top of a stagnant column of air (a *Taylor column*) above a surface obstacle. There are two rather fundamental objections to this. One is that the GRS does, after all, drift widely in longitude compared to any uniformly rotating interior (it drifts westward relative to System III), and the drift speed is variable. Second, no modern model for Jupiter has a surface, let alone mountain ranges or other obstacles the size of the Earth. It has been argued by A. Ingersoll of the California Institute of Technology that the GRS, like the zones, is a rising, densely cloud-covered region of anticyclonic vorticity. Infrared images of Jupiter obtained by flyby spacecraft show that the zones are colder than the belts (because the cloudtops are high in the zones, and clearing in the belts allows heat to escape from much deeper) and that the visually dark-brown belts are the warmest features on the planet, whereas the GRS is as cold as any feature on the planet, even a little colder than the zones. This strongly suggests that the GRS is an active precipitating storm system of immense size and with an unusual degree of vertical turbulence. When we next examine the photochemistry of the Jovian atmosphere, these distinctions will be important. Figure V.31 compares two images of the Great Red Spot taken

**Figure V.31** Voyager 1 and 2 views of the Great Red Spot. a, taken by Voyager 1 on 1 March 1979, shows the GRS, its turbulent wake, and a large, bright white oval in the dark South Temperate Belt (STeB). Note the highly elongated smaller spot being swirled around the GRS (below). b was taken by Voyager 2 on 7 July 1979, showing the same region (and a different white oval) just 4 months later. Aside from the GRS, every visible feature has changed strikingly. The circulation of the GRS can be seen, but at this time it was not entrapping passing smaller spots.

**Figure V.31** *Continued*

4 months apart by the Voyager 1 and Voyager 2 flybys. Figure V.31a, taken by Voyager 1 at a distance of about 5 million km, shows the GRS engulfing a smaller red spot as a large white oval passes by on the lower right. Figure V.31b, taken by Voyager 2 at a range of about 6 million km, shows another white spot passing the GRS. The features in the turbulent wake of the GRS can change dramatically in a week. The global photomosaics shown in Fig. V.32 in cylindrical projection compare the planetary scenes as viewed by Voyager 1 (top; February 1979) and Voyager 2 (bottom; June 1979). The color differences are somewhat exaggerated; nonetheless, the differences between the features seen in these two views are striking. Note, for example, the very different appearance of the string of "dark barges" in the North Equatorial Belt and the white ovals south of the GRS.

We should add that many "little red spots" have been reported in the past; their existence was confirmed by the Pioneer imaging experiment and voluminously documented by the Voyager flybys. The GRS is not unique, merely the largest and probably the longest-lived member of its family. These spots are generally located in zones, where the vorticity is known to be anticyclonic. Even Saturn bears a few such spots (see Fig. V.33).

Figure V.34 shows one of the few bright markings seen on Saturn, which are sought avidly because tracking

**Figure V.32** Voyager 1 and 2 photomosaic maps of the Jovian clouds. The two images are aligned so that the planetocentric longitude (in the radio system) is the same. Note the westward drift of the GRS over the 4-month interval. The white ovals have moved eastward and changed substantially in appearance in the same time. Note also the red "barges" in the NEB that have also drifted eastward and changed dramatically (See color plate 1).

their motions is the only way to determine wind speeds in Saturn's atmosphere. This photograph and the two other Voyager 1 views in Fig. V.35 show wavelike features in bright zones. Compared with the atmosphere of Jupiter, that of Saturn shows very subdued color contrast and structure. The polar regions on Jupiter are subdued in appearance and irregularly mottled, whereas the poles of Saturn show even fewer features (Fig. V.35a).

With respect to the polar regions, the spacecraft infrared measurements clearly show that the *average* temperature of Jupiter is essentially independent of latitude. Because the reemitted solar flux arises from an input that is distributed like cos $l$, the curious conclusion is that the internal meridional transport of heat must occur in just such a way as to release most of the internal heat in the polar regions or that meridional transport of heat deposited by the Sun is somehow perfectly efficient. The observed temperature gradient (less than 3 K between equator and pole) is not sufficient to drive baroclinic instabilities. Ingersoll suggests that minor cooling of the clouds in the polar regions steepens the lapse rate enough to decrease the static stability and accelerate the upward convective transport of internal heat. Interestingly, although no banded structures appear near the poles in the excellent Pioneer 11 high-latitude images, there are abundant faint "bubbles" or "doughnuts" that suggest active convection.

Further progress in understanding the atmospheric circulation on Jupiter and Saturn must await much more detailed photographic coverage, with high enough spatial resolution to see convection cells about one scale height in size, with the longest possible time base and widest possible spatial coverage. High-resolution thermal imaging for comparison with the visual images might prove very valuable. Also, of course, *in situ* studies of the structure and motions of the atmosphere by survivable atmospheric entry probes would help enormously. The first opportunity for long-term study of the circulation regime by a Jupiter orbiter was opened with the arrival of the *Galileo* Orbiter in December 1995. Global observations of Jupiter over a period of several Jovian years permit a correlation of mean local properties with global behavior, placing the *Galileo* probe data in the context of the entire planet.

## Photochemistry and Aeronomy

We have up to now paid attention to the propagation of both visible and infrared radiation in solar-composition atmospheres, but ignored the effects of ultraviolet radiation. This is not because the effects of UV light are unimportant, but rather because molecules are such strong absorbers of UV that very little of it penetrates

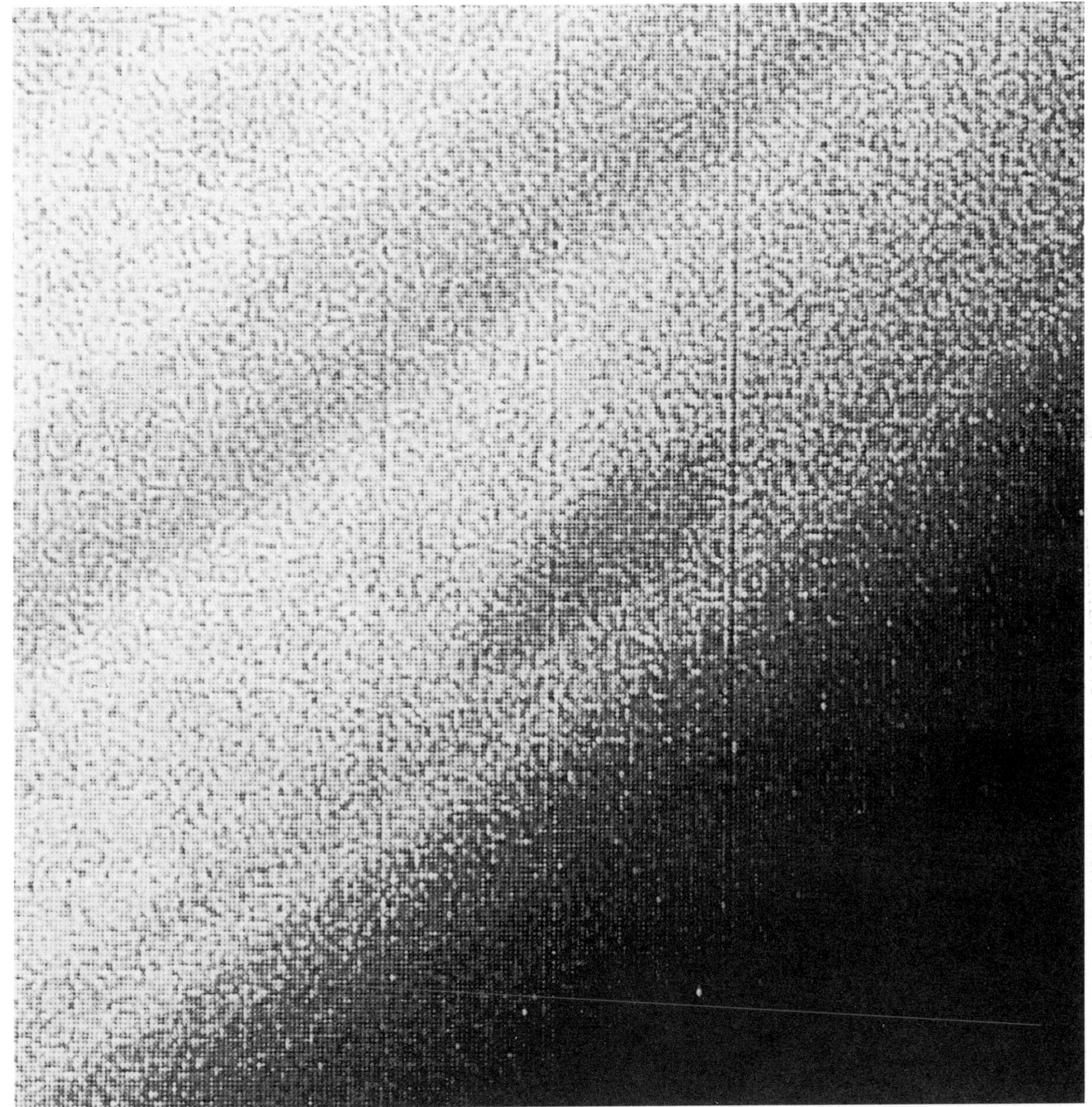

**Figure V.33** Saturn's largest red spot. This Voyager 1 image, with enhanced color contrast, shows the Saturnian red spot amid a very subdued pattern of belts and zones. None of the dramatic structure seen around Jupiter's GRS is evident.

into the troposphere. As we turn our attention to higher altitudes, the stratosphere, mesosphere, and even higher, the physical and chemical effects of UV light become extremely important. Ultraviolet radiation may effect any of three different types of processes: electronic excitation, dissociation (symmetrical bond-breaking), and ionization. In order to address the compositional effects of UV irradiation, we must know the solar UV emission spectrum, the absorption spectra of the gaseous species present ($H_2$, He, $CH_4$, $NH_3$, Ne, $PH_3$, and $H_2S$), the vertical distribution of the absorbing species, the products of photolysis, and the rates of reactions between the photolysis products. In order to estimate the steady-state abundances of complex photolysis products such as hydrocarbons, we also must identify and quantify the major processes responsible for their loss and destruction. This procedure becomes most complex at high altitudes, at which the photochemical lifetime of species such as $CH_4$ is short and the vertical distribution of the absorbing species may be profoundly altered by photolysis.

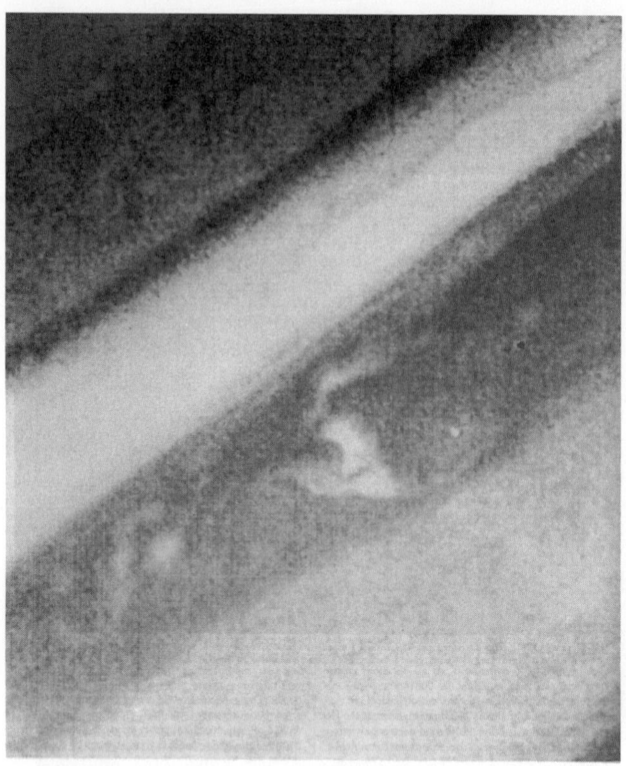

**Figure V.34** A rare bright cloud feature on Saturn. This Voyager 1 image, strongly computer enhanced in color contrast and brightness, shows one of Saturn's most prominent bright-cloud features. Note also the wave disturbance in the adjacent zone.

The solar UV flux at Jupiter is given in Fig. V.36. The spectrum is given as a running average over 5-nm (50-Å) intervals. (Note that the aeronomy literature commonly uses Ångstroms as the unit of wavelength.) The $He^+$ 304-Å Lyman $\alpha$ line and the H 1216-Å Ly $\alpha$ line dominate the extreme ultraviolet (EUV) shortward of 1500 Å. The spectrum is more like that of a black body at longer wavelengths, but the effective temperature is nearer to 4000 K than the 5800 K deduced from fitting a Planck function to the visible emission spectrum. This is, as we have seen, due to absorption in the cooler chromospheric layer of the Sun, with only a few high-altitude emission features of hot H, He, C, N, O, etc., superimposed.

The absorption coefficients of $H_2$, He, and the chemically interesting gases $CH_4$, $NH_3$, $PH_3$, and $H_2S$ are given in Fig. V.37. The most important features are the rather sharply defined cutoff wavelengths for UV absorption. Hydrogen and helium are somewhat weaker absorbers than the other gases in the extreme UV, but are of course enormously more abundant. Helium ceases to absorb at 500 Å, and hydrogen, at 1000 Å. Methane, which must be the ultimate parent molecule for the production of organic molecules, absorbs out to 1600 Å, and thus can be photolyzed by the intense solar H Ly $\alpha$ emission line at 1216 Å. From Fig. V.36, we can see that hydrogen and helium will absorb mostly solar $He^+$ Ly $\alpha$ at 304 Å and solar He emission at 584 Å, which are 100 times lower in peak intensity than the H Ly $\alpha$ peak. Methane, however, cuts off at too short a wavelength to take advantage of the exponentially increasing solar flux beyond 1500 Å. Note that absorption by $H_2$ and He to a large degree shields methane from photolysis shortward of 1000 Å because of their enormous abundance and good absorption ability. Ammonia and phosphine, with their similar structures and absorption spectra, cut off near 2350 Å, where the available solar flux is 100 times higher than the H Ly $\alpha$ peak and thus may be photolyzed at rates greatly surpassing that of $CH_4$ photolysis. Also, this means that shielding of ammonia and phosphine by $H_2$, He, and $CH_4$ combined will scarcely reduce their rate of photolysis. Because ammonia is a condensible gas, we expect that the $NH_3$ abundance will increase steeply as the solid-$NH_3$ clouds are approached from above and that $PH_3$ may be the dominant absorber only at altitudes at which $NH_3$ has been severely depleted by condensation. This we expect to be the case only a few kilometers above the $NH_3$ cloud base (Fig. V.17). Finally, $H_2S$, which is a reasonably good absorber even beyond 2700 Å, is limited severely in vertical extent by condensation of $NH_4SH$. Accordingly, $H_2S$ photolysis can be significant *only* if UV photons of 2350 Å $< \lambda <$ 2900 Å penetrate through the $NH_3$ cloud layer. When and where such clearing occurs, $H_2S$ photolysis may easily surpass all other photolysis rates combined because of the very large available solar flux. The main obstacle to $H_2S$ photolysis would then be the vapor pressure of ammonium hydrosulfide; hydrogen sulfide gas would be very severely depleted by condensation at $T < 200$ K and pressures less than about 2 bar. This limits the presence of gaseous $H_2S$ to levels at which the intensity of solar UV light has been dramatically diminished by Rayleigh scattering. Nonetheless, solid particles of $NH_4SH$ lifted by atmospheric turbulence may themselves be photolyzed and darkened by solar UV exposure.

This argument clearly shows that, before we can construct a detailed photochemical model of Jupiter or Saturn, we must have a good idea of the vertical distribution of the absorbing gases and aerosols. Further, we require that Fig. V.17 be extended to much higher altitudes, certainly at least through the stratosphere and mesosphere. For example, consider absorption by $H_2$ and He with an absorption coefficient of $10^2$ cm$^{-1}$ shortward of 1000 Å. The optical depth is proportional to the product of the absorption coefficient $\alpha$, the distance ($d$) traveled through the gas, and the local gas number density ($n$). At standard conditions, $\tau = 1$ corresponds

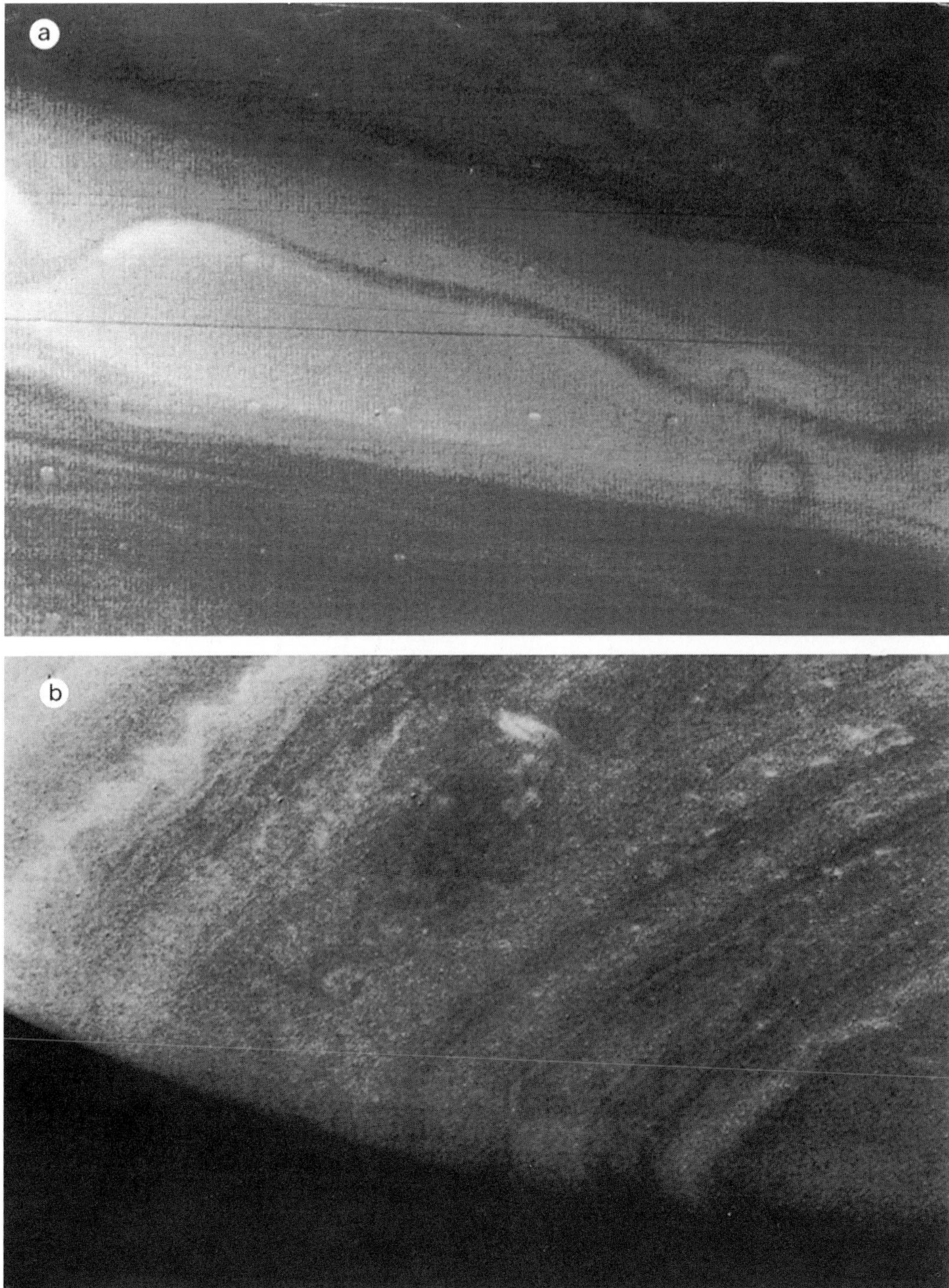


**Figure V.35** Wave disturbances in Saturn's atmosphere. Note also the fading away of the belt-zone structure toward the pole in the lower image.

to $\alpha N_0 d = 10^2 N_0 d = 1$, where $N_0$ is 1 amagat. Thus the incident light would be attenuated by a factor of $e^{-1}$ after traversing only $10^{-2}$ cm! Because the distance scale appropriate to Jupiter is one scale height, or very roughly $2 \times 10^6$ cm, the number density at which incident photons reach $\tau = 1$ will lie at a density of $10^{-8}$ amagat. This is near $10^{-8}$ bar for temperatures of a few hundred kelvins. At these extremely low pressures, at which the absorption rates of hydrogen and helium are highest, the mean free paths of atoms and molecules are

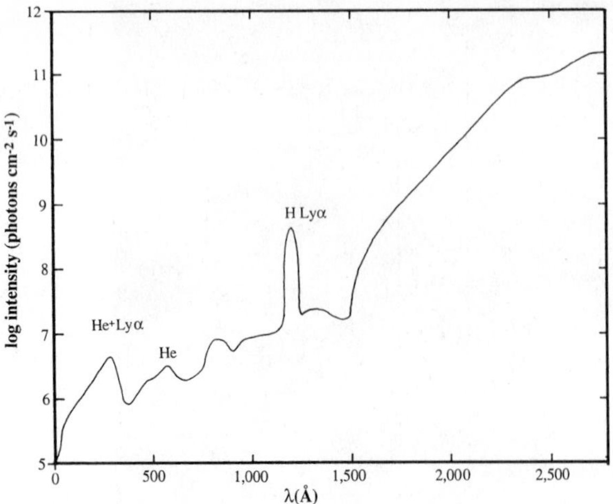


**Figure V.36** Solar ultraviolet flux at Jupiter. The smoothed flux over 50-Å wavelength intervals shows the great importance of the hydrogen Lyman $\alpha$ line and the rapid increase of flux toward the Planck peak for the solar photosphere. Emissions shortward of 1500 Å are mostly line features originating in the solar chromosphere and lower corona and reflect a range of temperatures. Note that 1,000 Å is equal to 100 nm.

immense. The mean free path is just $1/n\sigma$, where $\sigma$ is the collision cross-section (cm$^2$) and $n$ is the local density (cm$^{-3}$). At the level at which the density is $10^{-8}$ amagat ($10^{-8}$ times Loschmidt's number $= 10^{-8} \times 2.689 \times 10^{19}$ cm$^{-3}$), a typical small molecule with dimensions of a few Ångstrom units ($\sigma = 5 \times 10^{-16}$ cm$^2$) will have a mean free path of $1/(3 \times 10^{11} \times 5 \times 10^{-16}) = 6 \times 10^3$ cm, nearly 0.01 scale heights! A random walk with a net vertical displacement of one scale height in units of 0.01 scale heights typically takes $10^4$ steps, and, at a molecular speed near 1 km s$^{-1}$, this will take only 10 min. Thus diffusion is extremely rapid at these high altitudes, and bulk mixing on a time scale of a few minutes over a whole scale height would be required to defeat the effects of diffusion. The required eddy diffusion coefficient is $H^2/10^3$ s $\approx 10^{10}$ cm$^2$ s$^{-1}$, which is very large.

Now imagine a mixture of hydrogen, helium, and methane at an altitude ($Z_0$) at which the mean free path is comparable to a scale height ($l > 10^{-4} H$). From the Boltzmann equation, we know that there will be equipartition of energies between these molecules: the mean speeds of $H_2$, He, and $CH_4$ will be in the ratio 2.8:2:1, and the distribution of the vertical velocities of these molecules will be

$$N(v_z) = \exp(-\mu v_z^2/2kT). \qquad \text{(V.171)}$$

Each vertical velocity ($v_z$) corresponds to a particular kinetic energy, $\mu v_z^2/2$, which in turn corresponds to a particular maximum *altitude* at which all the kinetic energy has been converted into potential energy:

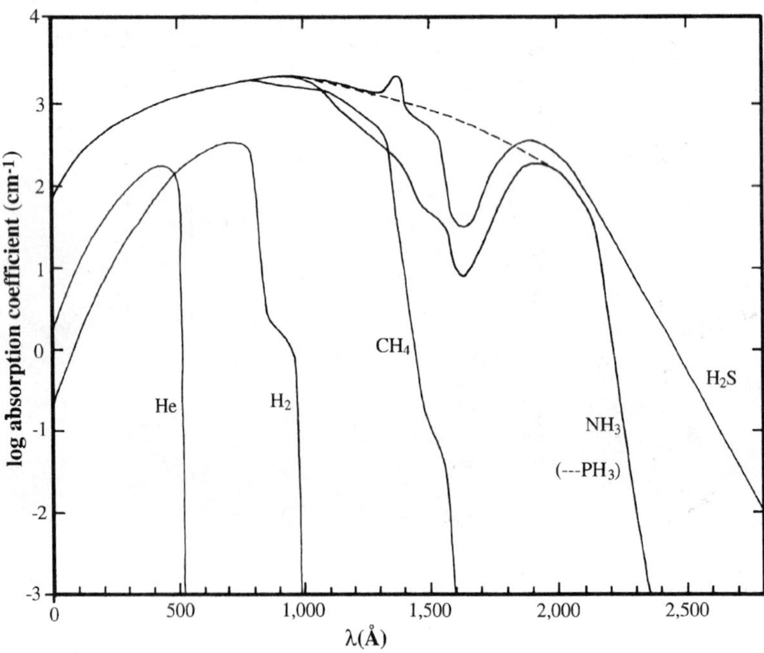


**Figure V.37** UV absorption coefficients of selected gases. These data are smoothed over 50-Å wavelength intervals. The absorption spectrum of phosphine is closely similar to that of ammonia and is omitted here for clarity. Comparison of this figure with Fig. V.36 reveals that helium and hydrogen absorb very little solar UV flux. Methane photolysis is primarily attributable to H Ly $\alpha$, and ammonia, phosphine, and hydrogen sulfide are decomposed by the enormously larger solar flux available longward of 1600 Å.

$\mu g(z_{max} - z_0) = \mu v_z^2/2$, or $z_{max} = z_0 + v_z^2/2g$. But by equipartition of energy, $v_z^2$ is proportional to $kT/\mu$, and hence $z_m = z_0 + kT/\mu g$. In short, the vertical distribution of *each species separately* depends on its own molecular weight, and we expect the atmosphere to approach a condition of hydrostatic equilibrium in which each species has its own scale height, $kT/\mu g$. The methane scale height will be eight times smaller than the $H_2$ scale height; while $H_2$ is dropping by a factor of $e$, the $CH_4$ pressure is dropping by a factor of $e^8$, and the $CH_4$ mole fraction ($f_{CH_4}$) will drop by $e^7 = 1100$ times! Such separation is easily prevented by even gentle vertical turbulence where the atmosphere is dense and the mean free path is small, but diffusive separation becomes extremely important at high altitudes. The methane scale height of 3 km at the level at which diffusive separation takes over is comparable to that deduced for the $NH_3$ scale height in the region of its condensation!

The rate of thermal diffusion in a gas, $J$ (molecules per square centimeter per second), is dependent upon the speeds of the molecule and the gradient in the concentration of the diffusing species, $dn/dz$; $J_z = -lv(dn/dz)/3$, or in the more general three-dimensional vector form,

$$\vec{J} = -lv\vec{\nabla}n/3 = -D\vec{\nabla}n. \qquad \text{(V.172)}$$

The factor $D = lv/3$ is called the *diffusion coefficient*, which has the same dimensions as the *eddy diffusion coefficient, K*.

In a steady-state situation, in which the local upward and downward fluxes of each species are in perfect balance, the diffusive separation effect is balanced by turbulent mixing. Obviously the vertical distribution of a trace heavy species $i$ will be very easily described if mixing is rapid and hydrostatic equilibrium is achieved for the gas as a whole; then $x_i$ = constant. If diffusion is dominant, however, then

$$\begin{aligned} x_{i,d} &= p_{i,d}/P \\ &= p_i(0) \exp \\ &\quad - [g(\mu_i - \mu_{H_2})(z - z_0)/RT], \end{aligned} \qquad \text{(V.173)}$$

where the level $z = z_0, P = P_0, p_i = p_i(0)$ is the base of the diffusively segregated region. This altitude level is called the *turbopause* or *homopause*. Above this level, diffusion is so fast that molecules can move freely, and deviations from this profile are quickly erased. If we describe the $x_i$ profile relative to the diffusive mole fraction profile $xi,d$, then we can define a new variable, $x_i^* = x_i/x_{i,d}$. The diffusion flux is proportional to the gradient of this quantity,

$$J_z = -D_i n_i(\partial x_I^*/\partial x), \qquad \text{(V.174)}$$

whereas the flux carried by eddy diffusion is just

$$J_z = Kn(\partial x_i/\partial z). \qquad \text{(V.175)}$$

As we have seen, $K$ is generally a function of $Z$, and $D$ of course depends on the temperature and pressure ($v$ and $l$). Thus it becomes necessary to insert the known or expected $D$ and $K$ profiles into Eqs. (V.174) and (V.175) and solve the two equations simultaneously, taking as a boundary condition the composition at some level in the atmosphere, usually the turbopause. At high altitudes, $K$ usually varies roughly as $n^{-1/2}$.

The simplest photochemical models still require one other type of information, the chemistry of the photolysis products. Certainly the simplest chemistry would be that of the rare gases, of which we shall consider only helium. The photoionization of helium,

$$He + h\nu \rightarrow He^+ + e^- - 24.5\,eV, \qquad \text{(V.176)}$$

can be followed by recombination,

$$He^+ + e^- + M \rightarrow He + M + 24.5\,eV. \qquad \text{(V.177)}$$

The function of the third body, $M$ (any gaseous species), is to help carry off the recombination energy of $He^+ + e^-$. If $M$ is He, then the two helium atoms produced in Reaction (V.177) carry off translational and electronic energy equal to that originally required for photoionization. If $M$ is $H_2$ or some other polyatomic gas, then it will be able to absorb vibrational and rotational excitation energy as well and will be more effective in causing *three-body recombination*. Molecular hydrogen can be ionized by UV light with wavelength less than 804 Å (80.4 nm):

$$H_2 + h\nu \rightarrow H_2^+ + e^- - 15.6\,eV \qquad \text{(V.178a)}$$

$$\rightarrow H^+ + H + e^-. \qquad \text{(V.178b)}$$

From Table V.2 we can see that the first ionization potential of helium is larger than that of $H_2$, and hence an electron can be lost by $H_2$ to $He^+$ during a collision. This is called a *charge exchange* reaction,

$$H_2 + He^+ \rightarrow H_2^+ + He + 8.9\,eV \qquad \text{(V.179a)}$$

$$H^+ + H + He. \qquad \text{(V.179b)}$$

This is the most important loss process for $He^+$ ions.

Photons longward of 804 Å (80.4 nm) but shortward of 845 Å can directly photodissociate $H_2$ via

$$H_2 + h\nu \rightarrow H + H, \qquad \text{(V.180)}$$

but an unstable (*predissociative*) state of $H_2$ may be made by any photon of $l < 106.9$ nm,

$$H_2 + h\nu \rightarrow H_2^* \rightarrow H + H, \qquad \text{(V.181)}$$

where the asterisk denotes an electronically excited state.

The $H^+$ and $H_2^+$ ions can react with the abundant molecular hydrogen by

$$H^+ + H_2 + M \rightarrow H_3^+ + M \quad (V.182a)$$

$$H_2^+ + H_2 \rightarrow H_3^+ + H, \quad (V.182b)$$

where M is any nonreactive collision partner capable of absorbing some of the energy of an exothermic reaction; in this case, preferably an $H_2$ molecule. The $H_3^+$ ion, in which two electrons are shared by three protons, is destroyed (dissociated) by electron addition (charge recombination). This is usually expressed by saying that the $H_3^+$ ion can undergo *dissociative recombination* with free electrons,

$$H_3^+ + e^- \rightarrow H_2 + H, \quad (V.183)$$

which releases more atomic hydrogen.

The main loss process for H is three-body recombination:

$$H + H + M \rightarrow H_2 + M. \quad (V.184)$$

These are the main sources and sinks involved in the photochemistry of He and $H_2$ on the outer planets. The solar flux available for $H_2$ and He photolysis at Jupiter is $10^9$ photons $cm^{-2}\,s^{-1}$. Direct photoionization of *either* $H_2$ or He can be seen to result in formation of two H atoms, so the overall rate of destruction (and reconstitution) of $H_2$ is equal to the available solar flux shortward of 1000–1100 Å.

Generally we can write the rate expression for a chemical reaction involving a three-body collision as the rate-limiting step as

$$A + B + C \rightarrow AB + C \quad (V.185a)$$

$$d(n_{AB})/dt = k_{185} n_A n_B n_C, \quad (V.185b)$$

where $n_i$ is the number density of species $i$ and $k$ is the rate constant of the reaction in $cm^6\,s^{-1}$. For Reaction (V.177), letting $x_H$ denote the mole fraction of H,

$$dn_{H_2}/dt = k_{184} x_H{}^2 n^3, \quad (V.186)$$

where $n$ is the total number density. Obviously three-body Reaction (V.184) is enormously accelerated by increasing density, because the rate of $H_2$ formation is proportional to the cube of the total number density. Thus we can expect that photolytic H production at high altitudes will be balanced by H recombination lower in the atmosphere. The steady-state column abundance of H will depend sensitively on the rate of vertical mixing, which is essentially given by molecular diffusion above the turbopause and by eddy diffusion below it. If we had a means of measuring the H column abundance, we could locate the turbopause and, from that deduce the rate of eddy mixing directly.

But, fortunately for us, the Sun is a strong emitter of H Lyman $\alpha$ radiation, and the H 1216-Å Ly $\alpha$ emission from Jupiter in response to the known solar excitation can be observed from rockets and spacecraft out of Earth's atmosphere. The Pioneer UV photometer observations of Jupiter show a surprisingly low Ly $\alpha$ intensity, requiring very rapid vertical mixing, and thus a high turbopause altitude. The present estimates of the eddy diffusion parameter ($K$) near the turbopause range from $10^6$ to $10^9\,cm^2\,s^{-1}$, with a preferred value near $10^8$.

Using estimates of the temperature profile at high altitudes, direct calculation of diffusion rates, and this deduced value for the eddy diffusion rate, we can calculate an atmospheric profile for $H_2$, He, and H abundances, neglecting for the moment the photolysis of other species. Using the reaction network we have so far considered, which we summarize in Fig. V.38, we calculate an atmospheric structure such as that given in Fig. V.39. The plots of log $n_i$ vs $Z$ would be straight if the atmosphere were isothermal. The rapid dropoff of $H_2O$, $H_2S$, and $NH_3$ due to condensation and that of Ne and $CH_4$ due to diffusive separation are evident. The profiles of photochemically active gases such as $CH_4$, $NH_3$, and $PH_3$ shown here neglect their destruction by UV photolysis at high altitudes. These gases are in reality strongly depleted by photolysis.

Methane, which is present in great quantities high above the cloudtops, is not a very effective UV absorber because of its UV cutoff at $\lambda \approx 1600$ Å, where the solar flux is still low. As we have seen, the main energy source for methane photolysis is solar H Ly $\alpha$. We have already commented on the absorption and scattering of solar H Lyman $\alpha$ radiation by atomic hydrogen, and it is valuable to know whether this radiation is depleted to a serious degree by the atomic H on Jupiter above the level at which incoming photons first encounter methane (i.e., the turbopause). The peak number density of H is only $10^9\,cm^{-3}$, and the column density of H is thus only $10^{11}\,cm^{-2}$ (recall that H must have a scale height twice that of $H_2$ above the turbopause). An atomic hydrogen density of $10^4\,cm^{-3}$ is $10^4/3 \times 10^{19} = 3 \times 10^{-16}$ amagat, and the optical thickness of the atomic H layer is $\tau$ and $and \approx 3 \times 10^{-7}$. Thus the Jovian exosphere is almost perfectly transparent to H Ly $\alpha$, and 1216-Å radiation will survive to drive photolysis of methane. It is worth mentioning also that the solar chromosphere and the base of the corona are at temperatures of $10^4$ to $10^6$ K, and thus the *width* of the solar H Ly $\alpha$ line will be broadened by the Doppler effect due to the large thermal speeds of the emitting atoms. The H absorption feature on Jupiter will have a Doppler width reflecting the local exospheric temperature of a few hundred kelvins and thus is far too

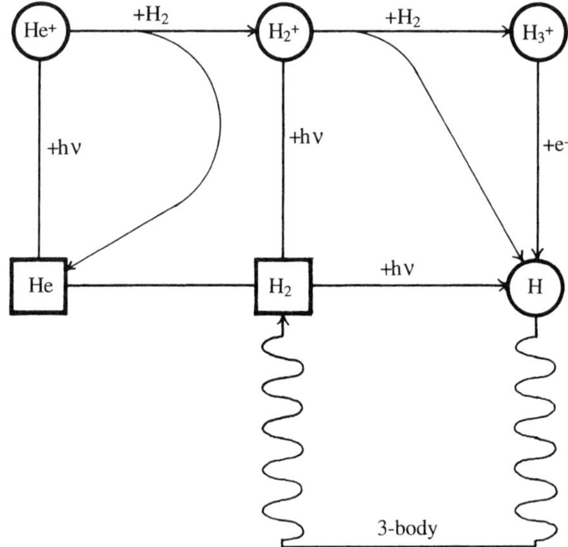


**Figure V.38** Hydrogen and helium photochemistry. The main reactions involved in the photolysis and recombination of helium and hydrogen in the Jovian atmosphere are related by a reaction network. Other species, such as the $H^-$ ion, are present in only negligible quantities.

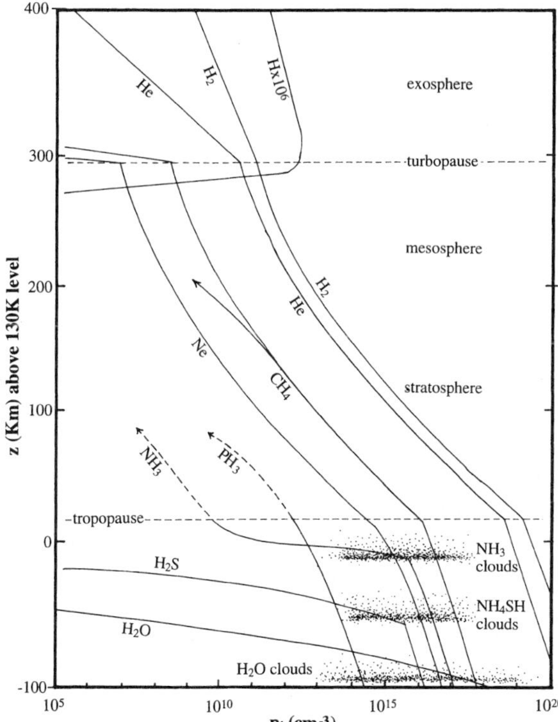


**Figure V.39** Upper atmosphere structure of Jupiter. Species are well mixed in the troposphere, having constant mole fractions except where certain species are depleted by condensation. No condensation occurs in the stratosphere–mesosphere, and changes in mole fractions are due only to photochemically driven chemical reactions.

narrow to eclipse the solar H line even if the Jovian column density of H were much larger. Molecular $H_2$ is not a significant source of opacity beyond a wavelength of about 950 Å and does not affect the transmission of H Ly $\alpha$ light except through Rayleigh scattering.

Because of the shielding of methane from solar UV of $\lambda < 84.5\,\text{nm}$ by enormous amounts of strongly absorbing hydrogen and helium, photons shortward of the methane photoionization threshold only rarely encounter methane. Irradiation by the solar 121.6-nm Ly $\alpha$ line produces a small amount of $CH_3^+$ ion along with the main product, the $CH_2$ radical with all electron spins paired. For spin quantum number $S$ the *multiplicity* of the state is $2S+1$, which is 1 when the spin is zero. This is said to be a *singlet* state, indicated by the superscript 1: $^1CH_2$. The photolysis reactions producing $CH_3{}^+$ (methyl ion) and $^1CH_2$ (methylene) are

$$CH_4 + h\nu(\lambda < 121.6\,\text{nm}) \rightarrow {}^1CH_2 + H + H \quad (V.187a)$$

$$^1CH_2 + H_2 \quad (V.187b)$$

$$CH_3^+H + e^-, \quad (V.187c)$$

where 92% of the total yield from H Ly $\alpha$ photolysis is singlet methylene. The remaining 8% yield is in the production of the methyl radical, $CH_3$, which has a single unpaired electron of $S = 1/2$ and a multiplicity of $2S+1 = 2$; the product is *doublet methyl*:

$$CH_4 + h\nu \rightarrow {}^2CH_3 + H. \quad (V.188)$$

The methylene radical can be deexcited to a state in which the geometry is like that of a methane molecule with two hydrogen atoms removed by symmetrical bond breaking, with consequently two unpaired electrons, for multiplicity $2(1)+1 = 3$. This molecular radical is called triplet methylene. $^1CH_2$ is often written simply $CH_2^*$ to denote that it is electronically excited; hence triplet methylene may be written $CH_2$.

Singlet methylene reacts readily with $H_2$,

$$^1CH_2 + H_2 \rightarrow CH_3 + H, \quad (V.189)$$

and it can abstract a hydrogen atom from methane:

$$^1CH_2 + CH_4 \rightarrow CH_3 + CH_3. \quad (V.189a)$$

Collisional deexcitation of $^1CH_2$ produces the radical $^3CH_2$, which is fairly resistant to reaction except with other radicals, such as with $CH_3$ to make ethylene:

$$^3CH_2 + CH_3 \rightarrow C_2H_4 + H. \quad (V.189b)$$

The methyl radical usually returns to methane by the reaction

$$CH_3 + H + M \rightarrow CH_4 + M, \quad (V.190)$$

and only about 10–15% of the $CH_3$ is lost in the formation of higher hydrocarbons, largely by formation of ethane:

$$CH_3 + CH_3 + M \rightarrow C_2H_6 + M. \qquad \text{(V.190a)}$$

Photolysis of both ethane and ethylene gives rise to production of acetylene, $C_2H_2$, which can in principle polymerize to make more complex hydrocarbons. However, Darrell Strobel of Johns Hopkins has examined the chemistry of these simple hydrocarbons in some detail and has been unable to find a credible mechanism for making polyacetylenes under Jovian conditions.

Note that the small amount of $CH_3^+$ ion made at high altitudes by Lyman $\alpha$ irradiation can react with methane to make heavier hydrocarbons:

$$CH_3^+ + CH_4 \rightarrow C_2H_5^+ + H_2 \qquad \text{(V.190b)}$$

which, upon recombination, make neutral gases that condense readily under cold stratospheric conditions. These hydrocarbon ions can also serve as condensation nuclei for neutral species such as ammonia, hydrazine, and hydrocarbons.

All carbon compounds are, however, thermodynamically unstable relative to methane under Jovian conditions of temperature and pressure. Mixing of photochemically produced hydrocarbons down to the 1000-K level will eventually be followed by hydrogenation reactions, which reconstitute methane.

The reaction network worked out for hydrocarbons, largely by Strobel, is shown in Fig. V.40. Strobel estimates that very roughly 10% of the $C_2$ hydrocarbons may go on to form $C_3$ hydrocarbons.

If we take the total rate of production of organic matter on Jupiter as $10^9$ molecules $cm^{-2}\,s^{-1}$, the rate of production of mass will be $10^{-14}\,g\,cm^{-2}\,s^{-1}$. This is equal to the steady-state downward flux of photochemically produced organic matter at the tropopause. If we take an eddy diffusion constant $K = H^2/t$, based on observations of the lifetime of cloud features, or estimate $v_z$ from Prandtl mixing length theory applied to the observed thermal emission and let $K = Hv_z$, we in either case derive an eddy diffusion coefficient near $10^8\,cm^2\,s^{-1}$. The total convective mass flux of $\rho(z)v_z$ is $10^{-2}\,g\,cm^{-2}\,s^{-1}$ in the upper troposphere, and the mass mixing ratio $M_{org}/M_{tot}$ is thus $10^{-11}$, for a mole fraction of $(CH_2)_x$ of $10^{-12}$. Thus in rapidly mixed regions the concentration of organic matter will be completely negligible. This conclusion does not necessarily follow for the slowly mixed stratosphere and lower mesosphere, which serve as bottlenecks for loss of hydrocarbons into the lower atmosphere. It would be useful to place bounds upon the value of the eddy diffusion coefficient in this region in order to assess the probable concentration of heavier hydrocarbons. We have already argued that the dynamic time scale in the stratosphere must be $\tau_{dyn} > 10^8$ s in order that the temperature may be controlled by radiation rather than dynamics; this required that the eddy diffusion coefficient be no larger than $3 \times 10^4$ in this region ($P \approx 100$ mbar).

The Faint Object Spectrometer on the *Hubble Space Telescope* (HST) has used the vertical distribution of ammonia on Jupiter to calculate the eddy diffusion coefficient in the Jovian stratosphere from the competition between upward turbulent transport and top-side photochemical depletion of that gas. The HST results show an eddy diffusion coefficient of about $K = 2 \times 10^4\,cm^2\,s^{-1}$ in the ammonia cloudtop region (500 mb total pressure), with $K$ dropping off rapidly to only $5 \times 10^2$ near the 200 mb level, followed by a slow rise to about $K = 10^3$ at 40 mb. Local concentrations of biologically interesting (organic) compounds in the stratosphere near the 250 mb pressure level then reflect a vertical turbulent-mixing speed of about $v_z = K/H = 10^3/3 \times 10^5 = 0.003\,cm\,s^{-1}$. The mean steady-state downward flux of "heavy" hydrocarbons, estimated from the photolysis rate of methane, is, as before, about $5 \times 10^{-14}\,g\,cm^{-2}\,s^{-1}$, which predicts a lower stratospheric mass fraction of organic matter, principally $C_2$ hydrocarbons, of about $3 \times 10^{-7}$. The total mole fraction of all $C_2$ hydrocarbons at that level then must be near 0.03 ppm. Higher in the stratosphere the abundance could reach a few ppm; indeed, Galileo reported 1 to 4 ppm of acetylene near the 0.01 mb level and 3 to 6 ppm of ethane in the 0.4 to 10 mb region.

In the source region of organic molecules the concentration may be even higher. Strobel estimated theoretically that the most abundant product, ethane, may reach a mole fraction of $x_{C_2H_6} \approx 10^{-5}$ for an assumed smooth variation of $K$ from $2 \times 10^4\,cm^2\,s^{-1}$ in the stratosphere to $10^8\,cm^2\,s^{-1}$ at the turbopause. Eddy mixing in this altitude interval is driven by waves propagating upward from disturbances (weather) in the troposphere. Under these circumstances, the upward momentum flux is constantly being passed on to smaller masses of atmosphere. This agitates the atmosphere and causes the eddy diffusion coefficient to increase upward with $K \propto n^{-1/2}$.

Ammonia, as an abundant condensible gas with a large absorption cross-section out to 2350 Å, will photolyze in a manner that is very sensitive to the assumed temperature near the "cold-trap" level at 100 mbar. A low stratospheric temperature, say, 90 K, would freeze ammonia out very efficiently, so that solar UV radiation of a wavelength longer than the methane cutoff (1600 Å) would have to nearly reach the tropopause before encountering an absorber, as we have seen. However, a more likely average value of the temperature at the 100-mbar level is 100–110 K. There is an observational test relevant to the vertical distribution of ammonia; the UV albedo of Jupiter

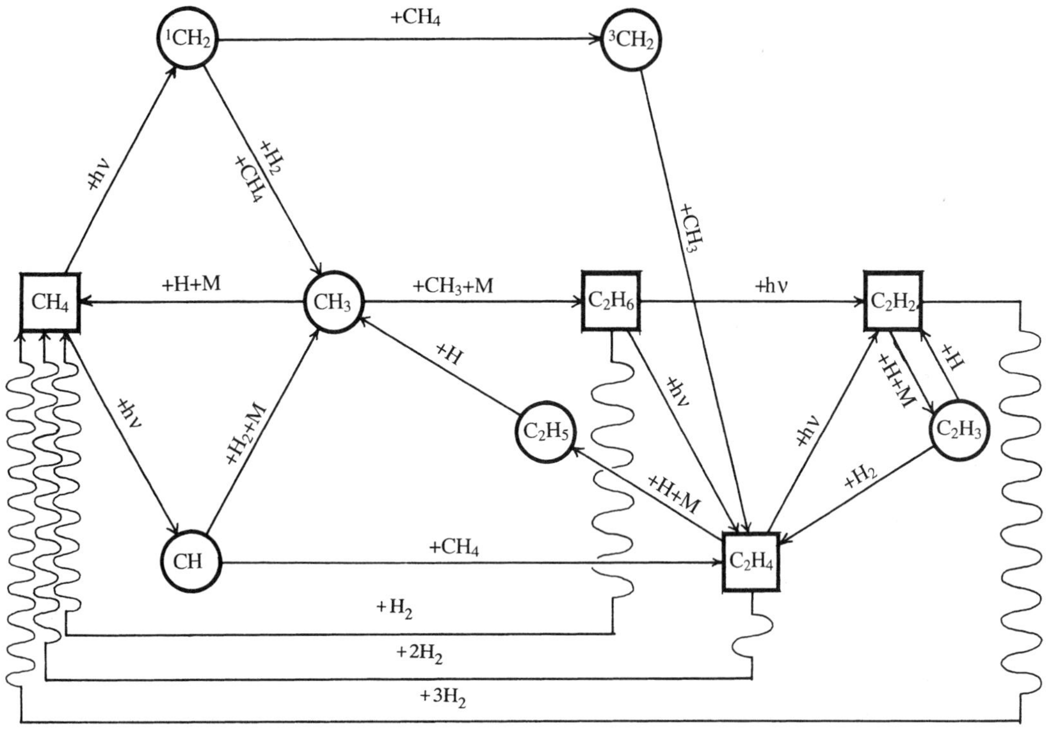


**Figure V.40** Methane photochemistry on Jupiter. This summary, by R. G. Prinn, shows the main features of hydrocarbon chemistry on Jupiter as developed in work by him and D. Strobel. Production of heavier hydrocarbons is a minor process. The lower part of the diagram suggests reduction of heavier hydrocarbons by hot, high-pressure hydrogen in the deep atmosphere.

from 1600 to 2350 Å does not show signs of very strong absorption by gaseous $NH_3$, and $NH_3$ must be far more severely depleted at high altitudes than $NH_3$ condensation through a 110-K cold trap could provide. Interestingly, the UV albedo remains low out to well beyond 3000 Å. This cannot be explained by any gas we have so far considered.

The sole important photolysis reaction for ammonia is

$$NH_3 + h\nu \rightarrow NH_2 + H. \qquad (V.191)$$

The amino radical can undergo the three-body reaction

$$NH_2 + H + M \rightarrow NH_3 + M \qquad (V.191a)$$

to reconstitute ammonia or may react with itself via

$$NH_2 + NH_2 + M \rightarrow N_2H_4 + M \qquad (V.191b)$$

to form $N_2H_4$, hydrazine.

Gaseous hydrazine is vulnerable to UV photolysis to about the same degree as ammonia, but it also has a vapor pressure comparable to that of water; at 110 K, hydrazine is extremely easy to condense. The supply of condensation nuclei at the 100-mbar level due to ionizing radiation from above is very small. From Fig. V.22 we may see that even a 100-MeV proton will be stopped above the 100-mbar level. It is thus likely that the gas will supersaturate and nucleate hydrazine homogeneously to form a photochemical haze.

The main photochemical loss process for hydrazine is conversion to nitrogen. The reaction network developed by Strobel and by Ronald G. Prinn of M.I.T. is given in Fig. V.41.

Ultimately, solid hydrazine particles and gaseous nitrogen will be cycled through high temperatures and high pressures deep in the atmosphere and revert to their more chemically stable parent, $NH_3$.

The hydrazine particle haze, which will be rapidly produced by UV photolysis, can quickly build up to the point of having important optical effects (absorption and scattering) at wavelengths extending out to 4000 Å. Simultaneously, $NH_3$ gas is severely depleted at higher altitudes. The resulting UV albedo can readily be fitted to the observed value when a stratospheric temperature minimum of 110 K and an eddy diffusion coefficient of $2 \times 10^4$ are used. For lower values of K, ammonia gas will be seriously depleted at higher altitudes and Rayleigh scattering would cause the planet to look brighter than it is. Except for the UV region, ammonia photochemistry provides no strongly absorbing species and hence cannot explain the colors on Jupiter.

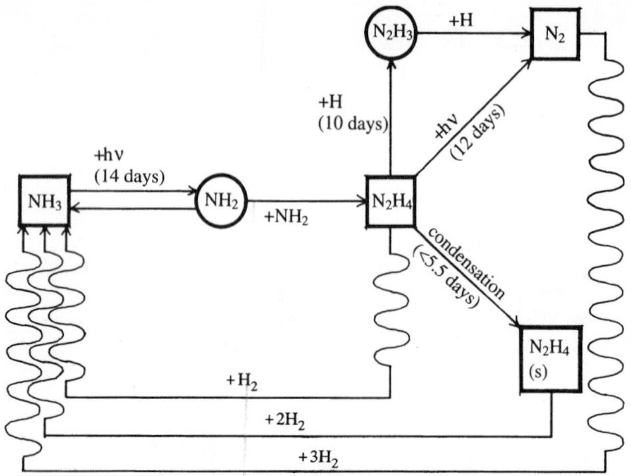


**Figure V.41** Ammonia photochemistry on Jupiter. This summary, by R. G. Prinn, shows the reaction paths to the important products $N_2$ and $N_2H_4$ (hydrazine). Note the condensation of hydrazine.

We have so far ignored the mutual reaction of $CH_4$ and $NH_3$ photolysis products, even though there is some overlap of the altitude ranges of $CH_4$ and $NH_3$ photolysis. It is likely that very small amounts of species such as $CH_3NH_2$ (methylamine) are produced as side products. Hydrogen cyanide, HCN, has often been suggested qualitatively as a likely photochemical product, and a simple mechanism for making it on Jupiter has been suggested. In any event, HCN cannot accumulate in the gas phase, where $NH_3$ gas is present, because of the low vapor pressure of the salt $NH_4CN$, ammonium cyanide, which is roughly as volatile as $NH_4SH$. Methylamine has a relatively high vapor pressure and may be more abundant. It has never been observed, but small traces of HCN have been found to be present near the cloudtops.

The total supply of solar photons capable of effecting methane photolysis is about 300 times smaller than the flux that can dissociate ammonia. Were it not for competition from $PH_3$, $NH_3$ photolysis would totally dominate all the photochemistry above the $NH_3$ clouds.

Phosphine has been reported to be present on Jupiter with an abundance very close to that expected from solar elemental abundance considerations. Our expectation from the assumption of thermochemical equilibrium was that phosphine would be oxidized at the 800-K level and precipitated as oxysalts such as ammonium phosphates near the level of the water clouds. However, if the oxidation of phosphine is sufficiently sluggish, rapid vertical mixing up to cooler levels (perhaps $3H$ higher, at 500 K) will completely quench the reactions destroying $PH_3$. If $K \approx 10^9$ $cm^2\,s^{-1}$ at these levels, the lifetime for mixing from the 800-K to the 500-K level will be $\approx 9H^2/K \approx 10^6$ s, and the reaction would have to have a time scale at least this long for $PH_3$ to survive. It is not known whether the reactions are indeed this slow.

However, given that $PH_3$ really is observed to be present in the upper troposphere, we can then calculate the photochemical consequences of its presence quite readily. Because $PH_3$ is less abundant that $NH_3$, but not condensible, it will compete most effectively with $NH_3$ for UV photons at altitudes well above the $NH_3$ cloud tops. To some degree, $PH_3$ might compete with $CH_4$ for photons of $\lambda < 1600$ Å at high altitudes, but the absorption coefficients are similar and methane is $\approx 10^3$ times as abundant, so $PH_3$ cannot shield methane appreciably. As with $NH_3$, the large flux of available photons at $\lambda > 1600$ Å means rapid photolysis, hundreds of times faster than $CH_4$ photolysis.

Because of similar molecular structure and stability, the photochemistry of $PH_3$ is very similar to that of $NH_3$,

$$PH_3 + h\nu \rightarrow PH_2 + H, \qquad (V.192)$$

and three-body recombination slowly replenishes $PH_3$ by

$$PH_2 + H + M \rightarrow PH_3 + M. \qquad (V.192a)$$

Phosphine is reconstituted by disproportionation of $PH_2$ via

$$PH_2 + PH_2 \rightarrow PH_3 + PH, \qquad (V.192b)$$

which is very slow, and the three-body formation of diphosphine,

$$PH_2 + PH_2 + M \rightarrow P_2H_4 + M. \qquad (V.193)$$

Diphosphine will photolyze about as well as $NH_3$, $PH_3$, or $N_2H_4$ and will produce $P_2H_2$ and $H_2$, $P_2$ and $2H_2$, or two $PH_2$ radicals.

Atomic H produced by either $NH_3$ or $PH_3$ photolysis (and $NH_2$ from ammonia photolysis) can abstract a hydrogen atom from phosphine to generate more $PH_2$:

$$PH_3 + H \rightarrow PH_2 + H_2. \qquad (V.194)$$

PH undergoes a series of reactions terminating in the formation of elemental phosphorus, and $P_2H_2$ from diphosphine photolysis can readily yield free phosphorus. Solid elemental phosphorus exists in a variety of forms (usually termed white, yellow, red, and black), but these forms themselves can have diverse colorations depending upon the temperature, mode of formation, and particle size. The products predicted by theory and observed in some laboratory photolysis experiments on $PH_3 + H_2$ mixtures are solid triclinic $P_4$, generally called red phosphorus, and diphosphine. The photolysis of diphosphine is likely to be the main source of elemental phosphorus. It has been suggested that phosphorus formation can be inhibited by free-radical scavenging reactions of acetylene, but acetylene production rates on Jupiter (and Saturn) are so low that this is at best a minor effect. The overall reaction network for $PH_3$ is given in Fig. V.42.

It appears that almost all of the $PH_2$ produced by direct photolysis of $PH_3$ is made into diphosphine *in situ* in the stratosphere–mesophere, and much of this ends up as solid particles of phosphorus. As with $C_2H_6$, $C_2H_2$, $N_2$, and $N_2H_4$, solid phosphorus must be transported down to high-temperature and high-pressure regions to be reworked into $PH_3$. Calculations suggest that, for eddy diffusion profiles of the sort that we have used to date, the optical thickness of the phosphorus particle haze over the planet should typically be small ($\tau < 0.1$), but in regions of enhanced dynamical activity (such as the GRS?), where $K = 10^6$ well up into the stratosphere, the optical depth may reach $\tau = 1$, which is quite red indeed. The model of the GRS as an intense storm system would seem to be nicely compatible with the idea of red phosphorus as the local chromophore.

An interesting complication arises from the low (but very poorly known) vapor pressure of diphosphine. Condensation of diphosphine is probably important

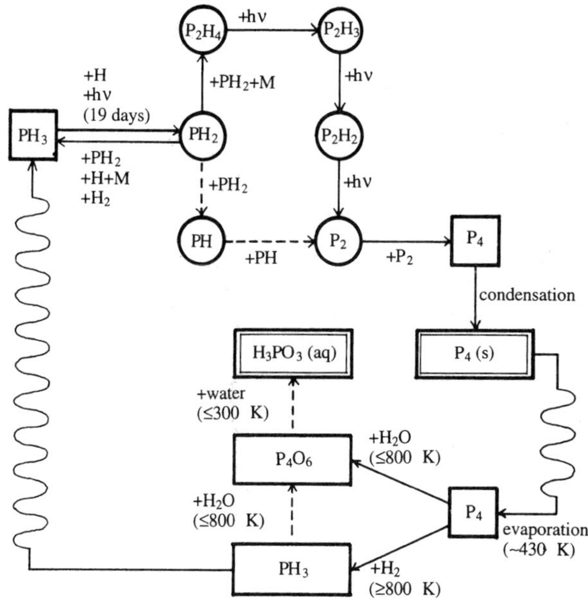


**Figure V.42** Phosphine photochemistry on Jupiter. This summary, incorporating the work of R. G. Prinn, D. Strobel, and the author, shows the principal reactions in the production and destruction of phosphine on Jupiter. The condensation of $P_2H_4$ (diphosphine) should be very important at stratospheric temperatures on Saturn, and the rate of formation on solid phosphorus should be greatly reduced. Slow processes are denoted with dashed arrows.

near the tropopause on Saturn, but not on Jupiter. If diphosphine is indeed an important intermediate in the formation of red chromophores, then the very subdued colors on Saturn are readily comprehensible.

The other intensely colored areas on Jupiter, especially the NEB, may have no dynamical or chromatic kinship to the GRS. The NEB is hot at 5$\mu$m and partly cleared of major particulate opacity down to at least the 200-K level (for $\approx$ 1-$\mu$m particles) and appears to be a subsiding region. The GRS is or has associated with it a huge Earth-sized vortex; it shows no sign of high temperatures at 5$\mu$m or any other IR wavelength; and it is generally described as reddish, brick red, or pink, whereas the NEB is often termed orange–brown with very dark cloud-free "blue" spots (Rayleigh scattering) with 5-$\mu$m effective temperatures from 290 to over 300 K.

Penetration of solar UV light to levels below the $NH_3$ cloud base is limited to wavelengths longward of the $PH_3/NH_3$ cutoff at 2350 Å. It is just below the $NH_3$ condensation level where $H_2S$ gas, with its UV cutoff at $\lambda > 2700$ Å, may first be found. If the $NH_3$ clouds are locally absent, then the large solar UV flux from 2350 to 2700 Å will survive, diffused but not seriously attenuated by Rayleigh scattering, down to the level at which $H_2S$ is present. The flux available for $H_2S$ photolysis is roughly 10 times that for $NH_3$ and $PH_3$ photolysis. Not only gaseous $H_2S$, but also solid $NH_4SH$, will be vulnerable to destruction by the long-wave UV flux.

The principal photolysis reaction of gaseous $H_2S$ is

$$H_2S + h\nu \rightarrow HS + H. \qquad (V.195)$$

In addition, thermal H atoms can abstract a hydrogen atom from $H_2S$ via

$$H_2S + H \rightarrow HS + H_2 \qquad (V.196)$$

Thus each absorbed photon can make two hydrosulfuryl radicals. These radicals can recombine via

$$HS + HS \rightarrow H_2S + S \qquad (V.197)$$

$$HS + H \rightarrow H_2 + S \qquad (V.198)$$

$$HS + HS \rightarrow H_2 + S_2 \qquad (V.199)$$

$$HS + H + M \rightarrow H_2 + M. \qquad (V.200)$$

The sulfur radicals in turn can join together to produce lengthy chains of sulfur atoms. These may be in the form of the higher homologues of $H_2S$, $H_2S_x$ (hydrogen polysulfides), polymeric sulfur, $S_x$, or $NH_4^+$ salts of polysulfide anions such as $HS_x^-$ and $S_x^{2-}$. These solid sulfur compounds range from pale yellow ($H_2S_2$) to gold, orange, and brown for the longest chains. (Amorphous sulfur made by quenching liquid sulfur is red.) In addition, UV photolysis of solid $NH_4SH$ particles imparts a yellow–orange color to them by reactions apparently similar to those that take place in the gas phase. The ultimate fate of the solid sulfur and polysulfides is to fall to levels where hydrogenation to $H_2S$ is rapid.

The reaction network for gaseous $H_2S$ is given in Fig. V.43. The downward flux of solids is balanced on the average by the upward gaseous $H_2S$ flux. The total $H_2S$ photolysis rate, $10^{13}\,\mathrm{cm^{-2}s^{-1}}$, corresponds to about 0.02 g of sulfur per square centimeter per year. A milligram of sulfur per square centimeter (3 weeks' accumulation) should give an easily visible color.

Cross-products among $NH_3$, $PH_3$, and $H_2S$ have not been considered in detail. Clearly, because of the great overlap of the $NH_3$ and $PH_3$ photolysis regions, products such as $PH_2NH_2$ may be produced. Also, the atomic hydrogen atoms produced by photolysis of any hydride by photons with a large excess of energy are fired off with substantial kinetic energy. These "hot atoms" can abstract H atoms from saturated molecules if they collide before the hot H atom has engaged in too many collisions and taken on the local thermal energy. These *thermalized* H atoms can, as we have seen, abstract H from $H_2S$ or $PH_3$, but can do little else except recombine with other radicals, including of course other H atoms.

Consider a "hot" atom with a speed several times its local equilibrium thermal speed. We may regard the background gas, almost pure $H_2$ and He, as standing still while the hot H atoms dash through it. The physics

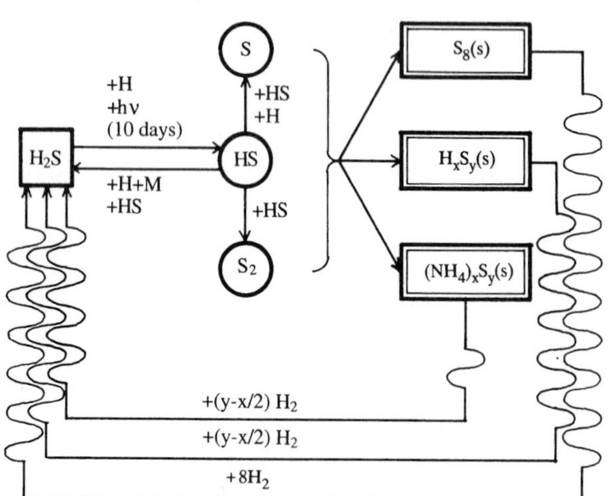


**Figure V.43** Hydrogen sulfide photochemistry on Jupiter. After a summary by R. G. Prinn. Solid ammonium hydrosulfide ($NH_4SH$) undergoes similar photolysis processes to make yellow, orange, red, and brown chromophores.

of the collision of a hot atom with a "stationary" molecule is easiest to see in the center-of-mass (CM) coordinate system, as shown in Fig. V.44. Denoting the hot atoms with a subscript 1, the target atoms with a subscript 2, and conditions *after* the collision with a prime, we have, from conservation of momentum,

$$m_1v_1 + m_2v_2 = m_1v_1' + m_2v_2' = p_{\mathrm{tot}}, \tag{V.201}$$

where $p_{\mathrm{tot}}$ is the (conserved) total momentum. Because we have chosen the CM system,

$$m_1v_1 = -m_2v_2 \quad \text{and} \quad m_1v_1' = -m_2v_2', \tag{V.202}$$

and $P_{\mathrm{tot}} = 0$. By conservation of energy, neglecting any energy stored internally in atom 1 or molecule 2,

$$\tfrac{1}{2}mv_1^2 + \tfrac{1}{2}m_2v_2^2 = E_{\mathrm{tot}}. \tag{V.203}$$

The speeds $v_1$ and $v_2$ are just the magnitudes of the vectors $\vec{v}_1$ and $\vec{v}_2$, $v = (v_x^2 + v_y^2)^{1/2}$. Combining,

$$E_{\mathrm{tot}} = \tfrac{1}{2}m_1v_1^2(1 + m_1/m_2). \tag{V.204}$$

Figure V.44 shows the evolution of the encounter with time; the plane containing the particle motions is translated normal to itself to give the time dimension. The path of the origin of a *stationary* frame (originally located on particle 2 before the collision) is shown by a dashed line. In the most extreme encounter geometries, $v_1$ may be unchanged in the CM system (grazing impact; large impact parameter), or, in the opposite extreme, $v_1$ may become $-v_1$ (head-on collision; zero impact parameter). Usually, $m_2 = 2$ and $m_1 = 1$ (using 1 AMU as the mass unit), so $v_1 = -2v_2$ and $v_1' = -2v_2'$. The total kinetic energy is $3v_1^2/4$ in the CM system and is clearly conserved. The locus of the allowed $v_1$ vectors is given by Fig. V.44b for the CM system for the case of an H–$H_2$ collision and an H–He collision. These velocities are transformed into the stationary ("laboratory") frame in Fig. V.44c. This figure does not present the scattering *amplitude* as a function of scattering angle $\phi$, just the particle *velocity* as a function of $\phi$. If we adopt the $+x$ axis as the direction of $v_1$, then $\sin\phi = v_{1_y}/v_{1_x}$.

It can readily be seen from Fig. V.44c, in which the velocities have been transformed into the laboratory system ($v_{1L}$ is the incoming H atom velocity relative to the stationary frame), that H atoms suffer far greater momentum loss per encounter upon collision with $H_2$ than with He. The *average* velocity of atom 1 after encounter in the laboratory system is $2v_{1L}/3$ for $H_2$ collisions and $4v_{1L}/5$ for He collisions, so that the energy carried by H (or its temperature) drops by a factor of $(3/2)^2 = 2.25$ for every $H_2$ collision and $(5/4)^2 = 1.56$ for every He collision. There are about 10 times as many $H_2$ collisions, so the mean loss of energy per collision is a factor of 2.2. Three collisions decreases the energy by an order of magnitude, from several thousand degrees

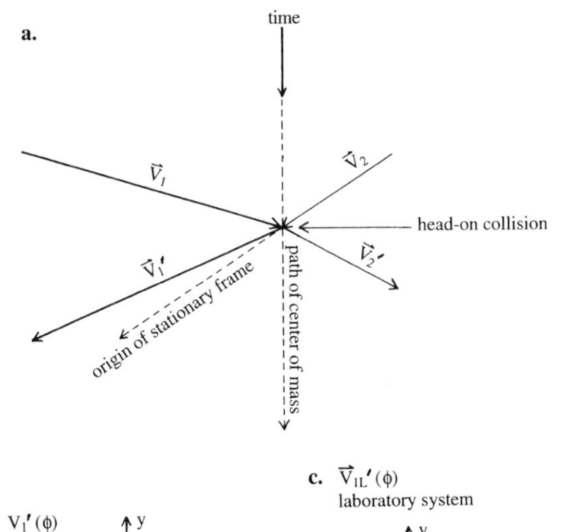


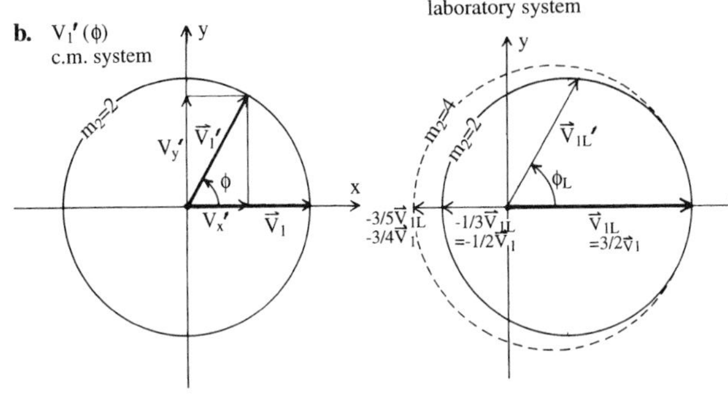


**Figure V.44** Atomic collisions in center-of-mass and laboratory frames. a shows the nomenclature of a collision on a time–distance plot. Time increases downward. It is assumed that a light, fast atom (typically atomic H) is fired into a "stationary" gas, typically $H_2$, which is moving at thermal speeds that are negligible compared with those of the incident hot atom. b shows the velocity of the scattered hot atom (circle), as seen in the center-of-mass system, as a function of the scattering angle $\phi$. c shows the velocity distributions in the stationary ("laboratory") system for collisions with molecular hydrogen and helium.

(capable of abstracting H atoms from $CH_4$, $NH_3$, etc.) to a few hundred degrees, at which temperature only recombination reactions and $H_2S$ destruction are significant. With an abundance of 1/3000 that of $H_2$, methane will be encountered within its first three collisions by only $10^{-3}$ of all the hot H atoms produced. Even then the probability of reaction is not 1. We therefore expect that production of $CH_3$ and other organic radicals by

$$H^* + CH_4 \rightarrow CH_3 + H_2, \qquad (V.205)$$

where $H^*$ is *translationally* excited atomic hydrogen, will have a rate below $10^{-3}$ of the $H_2S$ photolysis rate. This is very small, but still may be a stronger source of organic matter than direct $CH_4$ photolysis! Laboratory experiments have been done by Bishun Khare on near-UV irradiation of *nearly equimolar* mixtures of $H_2S$, $NH_3$, $CH_4$, and $C_2H_6$ in order to assess the chemical effects of hot H-atom reactions. The number of collisions required to lower the H-atom velocity by a factor of 10 in a gas of mean molecular weight 20 will be about eight, and the probability of encountering a reactive molecule per collision is 1, so we expect that virtually every H atom in this laboratory context will cause an abstraction reaction before being thermalized, unless the probability that a collision will react is much less than 1. Because the hot H atom source was irradiation of this mixture with the 2537-Å emission line of atomic mercury, only $H_2S$ will photolyze directly, and only Reaction (V.195) will occur, just as on Jupiter. We would expect 1 HS radical to cause between $\frac{1}{2}$ and 1 sulfur atom to precipitate out as a solid, and the 1 H atom should produce 1 $CH_3$, $NH_2$, $C_2H_5$, or SH radical or $\frac{1}{2}$ an organic radical per photolysis event. [If all H atoms were thermalized before reacting, only sulfur would be produced, but with a yield of 1.5 to 2 sulfur atoms per photon absorbed because of Reaction (V.196).] In fact, Khare found that a reddish brown polymer was produced in copious amounts. Analysis of the product shows it to be 96% sulfur and only 4% of the other elements (H, C, and N) combined. We conclude that nonreactive scattering of H atoms by these reactive gases is more probable than reactive collisions and that H is effectively thermalized even in a system optimized for rapid reactions. On Jupiter, we may expect that no more than $10^{-4}$ organic molecules are produced by $H_2S$ photolysis per $H_2S$ molecule destroyed.

Let us now summarize the mechanisms by which photolysis may produce colored species ("chromophores") on Jupiter and Saturn. The dominant chromophore in regions of clearing of the $NH_3$ cloud layer is sulfur and its compounds. Sulfur chains, hydrogen polysulfides, ammonium polysulfide salts, and covalent polysulfides such as $HS_xCH_3$ are all possible. On Saturn, no 5-$\mu$m window has been observed, and no strongly colored cloud bands visually similar to Jupiter's Belts have been reported. The general coloration of Saturn is a subdued yellow and may be due to photolysis of tiny $NH_4SH$ or $P_2H_4$ particles transported up to the $NH_3$ cloud level by vertical mixing or to some of the other colored materials that can be produced by $NH_3$, $PH_3$, and $CH_4$ photolysis.

Photolysis of $PH_3$ and $NH_3$ should occur on both planets in a comparable fashion, except of course that all time scales are longer at Saturn. Hydrazine should contribute to the long-wave UV opacity on Saturn as on Jupiter, and phosphorus should have little effect on the visible color unless a violently turbulent region (analogous to our favored explanation of the GRS) were present.

Production of condensible and colored organic matter may occur in may occur in the regions of the atmosphere in which direct $CH_4$ photolysis is taking place, or in which photolysis of $H_2S$ (and, to a lesser degree, $PH_3$ and $NH_3$) is producing hot H atoms. These two sources probably have comparable strengths (in molecules $cm^{-2}s^{-1}$) for reasons already discussed. The carbon in Khare and Sagan's "red polymer" is a very minor constituent which probably consists mainly of $CH_3$ groups, which contribute nothing to the color. The main photolysis products at high altitudes, $C_2H_6$ and $C_2H_2$, are also not colored, but it is conceivable that $C_2H_2$ could polymerize to form the red polymer cuprene $(CH)_x$, as was suggested decades ago by Harold Urey. The problem here is one of rates, because even the $C_2H_6$ production rate is 2000 times slower than sulfur production. The rates of production of all these species are summarized in Fig. V.45.

The locally produced chromophores seem to be causally associated with dynamically idiosyncratic features such as the NEB and GRS. Chromophores that are produced on a truly global scale (such as the postulated photochemical source of red organic matter) leave the observed association of colors with dynamical features unexplained.

The chemistry discussed so far omits mention of oxygen compounds, especially water. The UV cutoff of water vapor is 1950 Å, shortward of the $NH_3$ and $PH_3$ cutoffs, and therefore water vapor will photolyze only if it is present above the $NH_3$ cloudtops, where the temperatures are 110 to 130 K. There is such strong drying of the atmosphere by condensation at these temperatures that an $H_2O$ mole fraction less than $10^{-14}$ must be expected. Oxygen-containing complex molecules must, if present, be produced by other mechanisms and in other locations.

Figure V.46 summarizes the rates of photolysis for $H_2$, He, $CH_4$, $PH_3$, $NH_3$, and $H_2S$ for the eddy diffusion profile presently preferred by aeronomers. The depth of penetration and chemical effects of solar UV photons as functions of wavelength are given for comparison in Fig. V.47. The major photochemical products are indicated for each distinct altitude and wavelength regime. The "shielded" region is that which sees less than 1% of the monochromatic intensity $B_\lambda$ incident at the top of the atmosphere.

This discussion of photochemistry on Jupiter and Saturn has omitted oxygen and its compounds because water is fully condensed out of both atmospheres at levels far below the photochemically active region in the stratosphere and upper troposphere. Nonetheless, external sources such as infall of meteoroids and cometary debris may inject significant traces of water into the upper atmospheres of both planets, especially in the form of icy ring debris in the equatorial regions of Saturn.

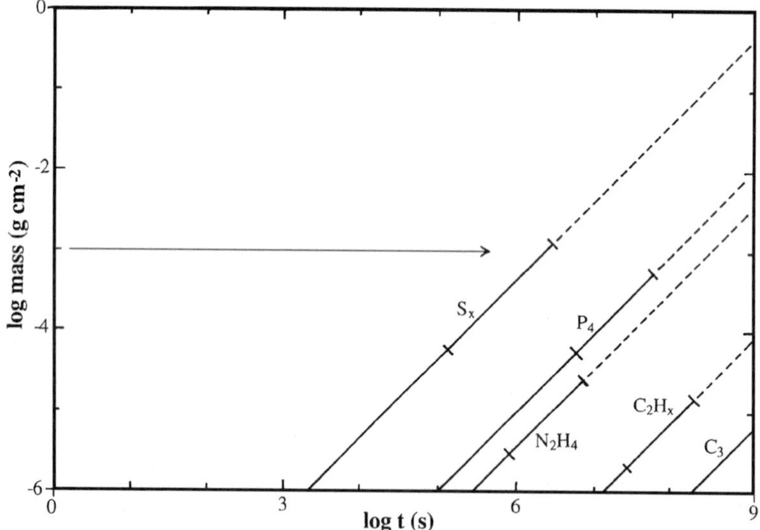


**Figure V.45** Accumulation of photolysis products on Jupiter. The mass of each product per unit area of Jupiter is shown as a diagonal line with a slope equal to the production rate. The time scale available for accumulation of each product is not infinite, put approximately equal to the overturn time of the atmosphere in the region in which that product forms. The overturn time is given fairly well by $\tau_{dyn} = H^2/K$, where $H$ and $K$ are the local scale height and eddy diffusion coefficient, respectively. The two marks on each accumulation line give the likely lower and upper limits of the accumulation time of that species. The most abundant products, sulfur and phosphorus, are brightly colored and about 100 times as abundant as the (colorless) $C_2H_x$ products of methane photolysis. On Saturn hydrazine and diphosphine condense readily, and red phosphorus is harder to form. Sulfur could be made only by photolysis of solid particles of $NH_4SH$ lifted to the cloudtops by winds.

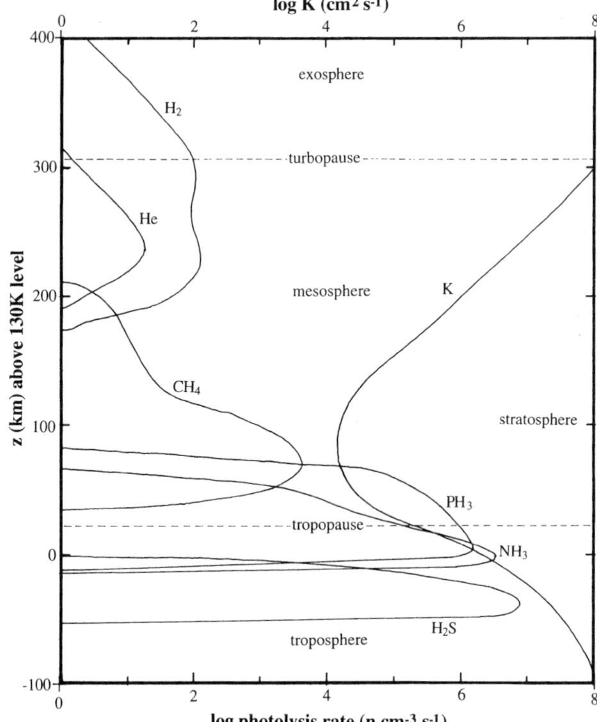


**Figure V.46** Photolysis rates of major gas species on Jupiter. The photolysis rates (number of molecules per cubic centimeter per second) are sketched out from the vicinity of the water cloud base, where no UV photons can penetrate, to a level in the exposure at which the number density is so small that photolysis rates virtually vanish. The eddy diffusion coefficient drops from about $10^8$ $cm^{-2}$ $s^{-1}$ near the water cloud base to about $3 \times 10^5$ at the tropopause and reaches a minimum of about $10^4$ in the stratosphere. This figure considers absorption and scattering by the gas but neglects shielding of hydrogen sulfide by overlying cloud particles. Most of the UV light reaching the $H_2S$ photolysis region has been Rayleigh scattered in the forward direction two or three times before being absorbed in a photolysis event.

Thermochemical and photochemical processing of the ablation vapors will convert most of the oxygen into CO, although small traces of water vapor, $H_2CO$, and $CO_2$, limited by their vapor pressures, may also survive.

Vertical mixing of the atmosphere can also lift minor constituents of the lower troposphere up to levels where they, though out of chemical equilibrium with the local atmosphere, are observable from above. The best example is CO, which is present at equilibrium at high temperatures and pressures, and which is destroyed by hydrogen reduction reactions as the gas rises to observable levels. The rates of the reactions destroying CO drop off exponentially with $-1/T$, and there is some altitude above which the time scale for vertical turbulent mixing is shorter than the chemical reaction time scale. At that point, the surviving CO ceases to react and the reduction reaction becomes "frozen in." CO survives out of equilibrium in the cooler gas at higher altitudes simply because its destruction is *quenched* by the low temperatures. The result of such mixing is that it becomes very difficult to distinguish between internal and external sources of CO observed in the cloudtop regions. This dilemma is especially severe on Saturn, where traces of stratospheric CO may derive from both sources. There have been occasional suggestions of other disequilibrium energy sources besides photolysis for producing organic matter and visible-wavelength chromophores. The most widely discussed of these mechanisms is electrical discharge activity, especially lightning.

The initial problem is to identify the ultimate source of the electrical charges and fields in the atmosphere. The changes may in principle be due to deposition of energetic cosmic rays or precipitation of ions from radiation belts or to radioactive decay within the lower atmosphere. We shall briefly estimate upper limits to the rate of ion production by each of these mechanisms and thus show their unimportance.

For a solar-composition Jupiter, the major radioactive energy source, $^{40}$K, is also the most volatile of the radioactive elements. The rate of energy generation by $^{40}$K decay is $3 \times 10^{-10}$ erg per second per gram of total mass, or $6 \times 10^{20}$ erg $s^{-1}$ for the entire planet. This provides a mean energy flux of 1 erg $cm^{-2}$ $s^{-1}$. Assuming 16 eV is needed to produce an ion pair, and inelastic dissipation of 1 MeV per decay event by $\beta$ particles, then each decay event provides $10^5$ ions, and the production flux is $2 \times 10^{11}$ ions $cm^{-2}$ $s^{-1}$. Most of these ions will be lost rather quickly by three-body recombination reactions, but some may be transported by turbulent mixing from the level at which they are formed ($T \approx 1500$ K) to the upper troposphere ($T < 300$ K).

From our discussion of the solar wind in Chapter III, we can estimate the solar wind particle density at 5.2 AU as 0.5 $cm^{-3}$ (half protons and half electrons) and the solar wind velocity as 500 km $s^{-1}$. The rate of energy input into the surface of Jupiter's magnetosphere is $nm_p v^3 A/3$, where $A$ is the cross-section area of the magnetosphere. As we shall see, the Pioneer magnetometer and particle experiments found the radius of Jupiter's magnetosphere to be 100 times Jupiter's radius (100 $R_J$). The power input is $4\pi(0.5)(0.4 \times 10^{-24})(5 \times 10^7)^3(7 \times 10^{11})^2 = 1.5 \times 10^{23}$ erg $s^{-1}$, or 0.025 erg $cm^{-2} s^{-1}$. Only a tiny fraction of this energy makes its way into Jupiter's radiation belts, and only a fraction of the belt particles are dumped into the upper atmosphere. If both of these processes were *perfectly efficient*, then the power being dumped into Jupiter's upper atmosphere as MeV protons would be $0.025(100R_J/R_J)^2 = 250$ erg $cm^{-2}$ $s^{-1}$ averaged over the entire planet. This would be about the same as the solar UV flux shortward of 3000 Å, but is deposited typically at altitudes far above the tropopause (see Fig. V.25),

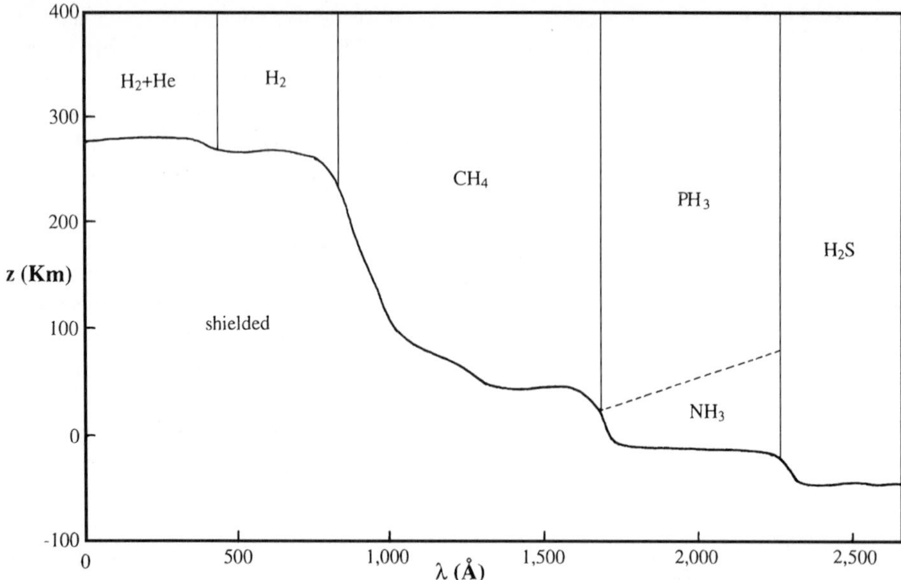


**Figure V.47** Depth of penetration of UV light on Jupiter. The main absorbing species are indicated as a function of wavelength and depth in the Jovian atmosphere.

mostly above the $10^{-5}$-bar level. This gives an upper limit on the rate of ion formation of $5 \times 10^{13}\,\mathrm{cm^{-2}\,s^{-1}}$. This flux is of course many orders of magnitude larger than the flux due to cosmic rays.

These are not satisfactory mechanisms for generating charge carriers in the vicinity of the visible cloud layers.

Lightning, however, is dependent on the existence of mechanisms for separation of charges on a large scale and on the consequent development of extremely large potential differences. The charge separation mechanism is apparently due to preferential charging of falling cloud particles. Given such a mechanism, the development of large vertical electrical potentials is inevitable. Thus the production of lightning depends on the presence of particles, which (if they are dust) must be raised by vertical winds, or (if condensates) must be formed in rising columns of air. This seems to imply that lightning is most common in Jupiter's zones, which are, as we have seen, essentially colorless.

On Earth, the currents carried to ground by lightning in a thunderstorm are far larger than the supply of ions originally available. It appears that a "bootstrapping" mechanism for multiplying ions is essential. Perhaps a small degree of charge separation raises the field strength to the point at which electrons are pulled off of particles' surfaces and accelerated through the atmosphere, generating many ion pairs before being stopped. In any event, large numbers of "new" ions must be made, and the ultimate source of energy for their production must be atmospheric motions. Indeed, it is common experience that violent rainstorms are electrically active, whereas lightning is rare in snowstorms (where, because of the much lower temperatures, $p_v$ is low and the heat released by condensation is small).

On Earth, the efficiency of generation of lightning by atmospheric motions is only $\approx 10^{-5}$ ergs of lightning per erg of convectively transported energy, despite the widespread occurrence of dense rain clouds and the enormous amount of latent heat released by condensation of water vapor. On Jupiter, with its lower fractional abundance of volatiles and lower average latent heat of condensation offset by a higher degree of cloud cover and high total abundance of condensable gases, a grossly similar efficiency would not be surprising. If we allow a conversion efficiency of about $10^{-5}$ and use the convective flux values for the upper troposphere from Fig. V.30, then we would expect a lightning energy dissipation rate of about $0.1\,\mathrm{erg\,cm^{-2}\,s^{-1}}$. Analyses of the conversion efficiency from Voyager data suggest an efficiency of about $3 \times 10^{-5}$. Galileo SSI data cited earlier suggest an optical emission of about $3 \times 10^{-7}\,\mathrm{W\,m^{-2}}(3 \times 10^{-4}\,\mathrm{erg\,cm^{-2}\,s^{-1}})$, which would be compatible with a luminous efficiency of about $3 \times 10^{-3}$ for Jovian lightning discharges.

Laboratory experiments on the chemical effects of electrical discharges on mixtures of methane, ammonia, and water vapor have shown that a wide variety of moderately complex molecules of biological interest are made. The best yields are about $10^{-14}$ g of hydrocarbons per erg of lightning discharge energy, found in experiments on *pure methane* using 60-kV spark potentials. If we again are generous and assume that electrical discharges in a gas containing 0.03 mol% $CH_4$ produce organic matter with an efficiency *equal* to that found in laboratory experiments on *pure methane*, then the total production rate of organic matter by lightning is $10^{-14}\ g\,cm^{-2}\,s^{-1}$, several times smaller than the photochemical production rate of ethane and acetylene. More reasonable assumptions would give far lower yields.

The possibility of the existence of life on Jupiter and the probability that terrestrial organisms, if inadvertently introduced into the Jovian atmosphere, might thrive there are popular and exciting themes, but of very limited scientific credibility. With respect to growth of terrestrial life forms, we find that the most favorable conditions would be found in the vicinity of the $H_2O-NH_3$ solution clouds; however, the combination of high pH and high $NH_3$ concentration would be lethal to any known terrestrial microorganism. The question of a food source is an interesting one; $10^{-11}$ g/g of total organics, almost all of which is ethane, of course cannot support any known organism. The eddy diffusion constant must be roughly $10^8\ cm^2\,s^{-1}$ near the water cloud base, so that mixing times to the 400 K level will be very short, and the organisms will be killed by heat. For an adiabatic lapse rate of $2\ K\,km^{-1}$, sterilizing temperatures will be reached 50 km below the water clouds, and $\tau_{dyn} = (5 \times 10^8)^2/10^8 = 3 \times 10^5\ s = 3$ days will be expected.

In addition, there are serious hazards attendant upon upward mixing, such as killing by UV irradiation and freezing. We may further compound the difficulties by requiring that all necessary trace elements be present in adequate amounts (but not enough to be toxic) and by pointing out that typical microorganisms require vitamins, sugars, or other special forms of room service, such as the company of other particular microorganisms, for their survival.

The moral of this story is very simple: organisms and environments have evolved to fit one another like key and lock, and expecting a random terrestrial population of microorganisms to make it on their own on Jupiter is like assuming that your front door key from home will start a randomly selected Moscovitch. There may be some amusement derivable from the assumption, but, once you really are in Moscow, you should be emotionally prepared to spend a lot of time riding the Metro.

Another crucial aspect of environments favorable to life is the simultaneous presence of cyclic processes of energy transport with stable, long-lived local environments, where bugs are not broiled, frozen, UV irradiated, pH cycled, squeezed, freeze dried, and chemically insulted daily. The element of stability is absent on all of the giant planets. What would life be like on Earth without a surface?

The fascinating question of the likelihood of *native* Jovian lifeforms, ideally adapted to conditions at some level in the atmosphere, cannot be resolved either by theory or by existing observations. Furthermore, it is not clear that detailed theoretical treatment of a question of this sort, wherein nearly every qualitative question is unresolved, is of value. Nature is assuredly both cleverer and more subtle than our wildest imaginings. After all, against all odds *we* exist, but would our nature be forseeable by Jovian theoreticians?

Life as we know it—that is, carbon-based life using liquid water as a medium—would seem to encounter its worst obstacle at the first step: it is very difficult to discover the source of the organic compounds out of which it may evolve. The best sources, weak as they are, seem to be UV photolysis of methane and $H_2S$. It is also conceivable that ion reactions caused by precipitation of radiation-belt protons and electrons into the atmosphere could contribute. Perhaps the best way to assess the importance of such a mechanism is to look closely at the structure and energetics of the very uppermost portion of the atmosphere, above the turbopause. In this region, heating by solar EUV and X radiation and by dumped protons and electrons will be taking place. Theoretical estimates of the steady-state temperature of this region, the *thermosphere*, based on solar heating alone yield temperatures of 750 K.

## The Jovian Thermosphere

The Pioneer 10 and 11 and Voyager 1 and 2 spacecraft carried stable transmitters, whose signals were monitored as the spacecraft passed behind Jupiter. This *radio occultation* experiment provides two types of information about the upper atmosphere structure: the *intensity* of the signal and its *phase shift* can be measured. The reason for the phase shift is simply that the refractive index of any material is different from the refractive index of empty space. Thus a signal that passes through ever longer path lengths of ever denser atmosphere as the spacecraft is occulted (hidden) by Jupiter will suffer ever greater phase shifts until finally attenuation of the signal by absorption and scattering becomes so great that the signal can no longer be detected. In the upper fringe of the thermosphere, the rate of photoionization is large and the rate of recombination reactions is small, and ions and free electrons are therefore common. This region is for obvious reasons called the *ionosphere*. Radio waves traversing the ionosphere in a grazing path from spacecraft to Earth will, as we have seen, be very sensitive to the presence of electrons. The local refractive index will be given by

$$\eta^2 = 1 - \alpha n_e/\nu^2, \quad \text{(III.60)}$$

and, for suitably high frequencies, $\eta = 1 - \alpha n_e/2\nu^2$. For a given spacecraft-to-Earth distance, $D$, the difference in transit time between a signal *in vacuo* and a signal passing through the medium of refractive index $\eta$ is just

$$\Delta t = D/c - D/\nu = D/c - D/\eta c \quad \text{(V.206)}$$
$$= (D/c)(1 - 1/\eta)$$

or, because $\eta$ is close to unity,

$$\Delta t = -(D/c)(\alpha n_e/2\nu^2), \quad \text{(V.206a)}$$

and a given cycle arrives earlier after passing through the electron gas by an amount proportional to the product of the electron density and the path length; in other words, the column density of electrons. If, over a time interval, $t$, we move a planetary ionosphere into the path that was once in a vacuum, then the *frequency* of the signal received at Earth must change. The number of cycles by which the signal is shifted by the change in the product $n_e D$ is $\Delta t\nu$, and the new frequency is $\nu + \Delta t\nu/t = \nu(1 + \Delta t/t)$, where $\nu\Delta t/t$ is the frequency change $\Delta\nu$. Thus measurement of the frequency of the received signal as a function of time gives a precise measure of the column density of electrons. From this, the gradient of the refractivity is calculated, the refraction of the radio path can be calculated, and finally the column density of electrons along the path can be associated with a particular altitude of tangency to the planet. The vertical profiles of $n_e(z)$ needed to fit the observed occultation profile can then be calculated. Even when the absolute value of $n_e$ is not known, the *rate* of frequency shift with altitude gives the scale height of the electron distribution and thus the electron temperature.

From the Pioneer, Voyager, and Galileo radio occultation results, a peak number density of $3 \times 10^5\,\mathrm{e\,cm^{-3}}$ is found at the base of the ionosphere, and an ionospheric temperature of 750 K is deduced. This is several times larger than the value expected for solar heating alone and may be due to heating by dumped energetic particles from the magnetosphere. We shall return to the fascinating and complex Jovian magnetosphere again in some detail.

One final aspect of the upper atmosphere is worth noting: the thermal energy of ionospheric H atoms is $2kT = 2 \times 10^{-13}$ erg per atom compared with the gravitational potential energy of $2 \times 10^{-11}$ erg/atom. The thermal velocity is $3.5\,\mathrm{km\,s^{-1}}$, whereas the escape velocity is $50\,\mathrm{km\,s^{-1}}$. For a Maxwellian distribution, where the most probable speed is

$$\alpha = (2kT/\mu)^{1/2} \quad \text{(V.207)}$$

the number of molecules with speed $v$ is related to the number with the most probable speed by

$$n(v)/n(\alpha) = (v/\alpha)^2 \exp[(\alpha^2 - v^2)/\alpha^2], \quad \text{(V.207a)}$$

and hence the number with $v > v_{esc}$, or with $v/\alpha > 14.4$, is only $2 \times 10^{-88}$. Exospheric particles, with mean free paths of about one scale height and scale heights of 250 km ($T = 750$ K; $\mu = 1$), have a mean time between collisions of $H/\alpha = 70$ s. The time such an H atom must wait before randomly acquiring escape velocity is then $70/(2 \times 10^{-88}) = 3.5 \times 10^{89}$ s, $10^{72}$ times the age of the Universe! Thermal escape of even atomic hydrogen is therefore impressively impossible. Thus our study of the composition, structure, and dynamics of the atmospheres of Jupiter and Saturn terminates with the conclusion that these atmospheres are indefinitely stable. No evolution of Jovian planets by loss of light gases is possible, and hence they cannot evolve in the direction of becoming more similar to the inner planets by shedding their massive gaseous envelopes.

The Jovian planets will remain intact until, about 5 billion years from now during the red giant phase of the evolution of the Sun, the solar luminosity will rise to about $10^4$ times its present value while the photospheric temperature drops to about 3000 K. The surface area must accordingly rise by a factor of $1.6 \times 10^5$. The red giant Sun will then have a radius 400 times as large as at present, or about 2 AU, completely engulfing the terrestrial planets. The overall density of the Sun will drop to about $2 \times 10^{-4}\,\mathrm{g\,cm^{-3}}$. From Jupiter, the distended Sun will subtend an angle of 45°. The gray-body temperature of Jupiter will increase to over 1000 K, causing its entire atmosphere to expand enormously. The upper atmosphere will begin to blow off under the influence of the luminosity and intense solar wind activity of the Sun. No careful study of this phase has been attempted, but even massive Jupiter and Saturn may be torn apart by the stresses of this era.

## Radiophysics and Magnetospheres of Jupiter and Saturn

In 1955, in the course of radio observations of the supernova remnant called the Crab Nebula, Burke and Franklin of M.I.T. were surprised to observe short, intense bursts of 22-MHz radio noise. Because of the limited spatial (angular) resolution of their receiver antenna, about 2.5°, short runs of data could not decisively identify the source. But over a period of 9 months the source of these noise bursts drifted continuously with respect to the Crab Nebula, showing a motion that matched that of Jupiter. The source's flux density, $S_\nu$, defined by the integral of the source intensity over solid angle,

$$S_\nu = \int I_\nu d\Omega, \qquad (V.208)$$

was found to be $10^{-19}$ W per square meter of receiver per unit frequency ($W\,m^{-2}\,Hz^{-1}$). The intensity, $I_\nu(W\,m^{-2}\,Hz^{-1}\,sr^{-1})$ of the source, which is the flux density per unit solid angle subtended by the source as seen by the observer, could not be estimated directly without some assumption about the angular size of the emitting region, because the emission seemed to come from an area too small to be resolved by the radio telescope. In the delicate nomenclature of radio astronomy, in which 1 W per square centimeter is like the tread of battalions of marching elephants, the usual unit of the flux density measured at the receiver is the *flux unit* (FU), defined as $10^{-26}\,W\,m^{-2}\,Hz^{-1}$. On this scale most bright astronomical sources have flux densities of 1 FU, whereas Jupiter weighs in at $10^7$ FU.

One simple explanation of this observed flux density might be thermal emission from Jupiter. As we have seen, the Planck function for the spectral distribution of energy with frequency for a black body emitter is

$$B_\nu = (2h\nu^3/c^2)/[\exp(h\nu/kT) - 1]. \qquad (I.3)$$

For radio wavelength observations of bodies at "normal" temperatures, $h\nu$ is very small compared with $kT$, and, because we may replace $e^x - 1$ with $x$ for small $x$, we have the simple long-wavelength Rayleigh–Jeans form of the Planck function:

$$B_\nu = 2kT\nu^2/c^2 = 2kT/\lambda^2. \qquad (V.209)$$

For a very cold body like Pluto, $T \approx 40$ K, and $h\nu/kT$ is 0.05 for $\nu < 40$ GHz. (Because $e^{0.05} - 1 = 0.0513$, this amounts to only a 2.5% error in $B_\nu$.) Typically, observations are made at $\nu < 10$ GHz, where the error is at most a fraction of 1%.

From Eq. (V.209) it is possible to convert an observed emission intensity into a *brightness temperature*, $T_b$, which is the temperature at which a black body would emit the observed intensity at the given wavelength:

$$T_b = c^2 B_\nu / 2k\nu^2. \qquad (V.209a)$$

Note that this method of determining the temperature by observation of a single wavelength interval requires knowledge of the solid angle subtended by the source. If we now calculate the *expected* thermal flux from a black body Jupiter at 130 K, we find that, at 22 MHz, Jupiter should be providing only $10^{-9}$ FU! Interpreted as thermal emission from the disk of Jupiter, the observed value of $10^7$ FU would require the temperature of the planet to be $10^{18}$ K! However, it should not be supposed that this radio flux is an important part of the total emission from Jupiter. After all, at wavelengths so very far removed from the Planck peak, even minuscule energy sources show up as enormous brightness temperatures. Taking $10^{-19}\,W\,m^{-2}\,Hz^{-1}$ as the observed flux density and assuming (incorrectly, as we shall see) that it comes out of Jupiter's interior, then the emitted flux at Jupiter is $10^{-19}(D/R_J)^2\,W\,Hz^{-1}$, where $D$ is the distance of Jupiter from Earth and $R_J$ is the radius of Jupiter. The bandwidth of the receiver was 2 MHz, so the total emitted flux in the 22-MHz region was $2.3 \times 10^{-5}\,W\,m^{-2} = 2.3 \times 10^{-3}\,erg\,cm^{-3}\,s^{-1}$ at Jupiter. This is only $1.6 \times 10^{-7}$ of the total thermal emission from the planet. On the other hand, over the entire planet this comes to an impressive radiated power of $4\pi R_J^2 \times 2.3 \times 10^{-5} = 1.4 \times 10^{12}$ W!

We have seen in Fig. V.1 how the thermal–infrared and microwave emissions of Jupiter behave out to wavelengths of 30 cm. At even longer wavelengths, the linear temperature scale of Fig. V.1 ceases to be convenient. In Fig. V.48 we give the currently available data on a logarithmic flux density scale, which, as we have seen, is equivalent to a logarithmic $T_B$ scale. At high frequencies the emission fits nicely to a black-body curve ($E_\nu \propto S_\nu \propto \lambda^{-2}$) out to nearly 10-cm wavelength (3 GHz), but shifts from a constant $T_B$ to a constant $S_\nu$ curve beyond 30 cm (1 GHz), in what is called the decimetric region. At wavelengths between 1 and 7 m the galactic background, which is rising toward long wavelengths, obscures the Jovian emission. From 10 to 70 m the decametric burst emission is clearly visible. We have plotted two curves, one for the typical burst emission and the other for the peak emission rates observed. The turndown beyond 10 MHz was first securely measured by the Radio Astronomy Satellite (RAS). This wavelength region is almost precisely on and beyond the critical frequency for reflection of radio waves by the Earth's ionosphere, which we have already calculated approximately via Equation (III.60). Despite the enormous fluctuations in the decameter radiation, neither the decimetric nor the thermal component has been found to be irregularly variable.

Detailed studies of the decimetric radiation have been made, both of its spatial distribution around the center of the visible disk of Jupiter and of its polarization. It has been found that the decimetric radiation is linearly polarized and that both the *angle* of the direction of polarization relative to the rotation axis of Jupiter and the *intensity* of the signal vary in a very regular and nearly sinusoidal manner, with a period that has remained precisely constant at 9 h 55 m 29.73 s ± 0.02 s since discovery (see Fig. V.49). This is the so-called radio

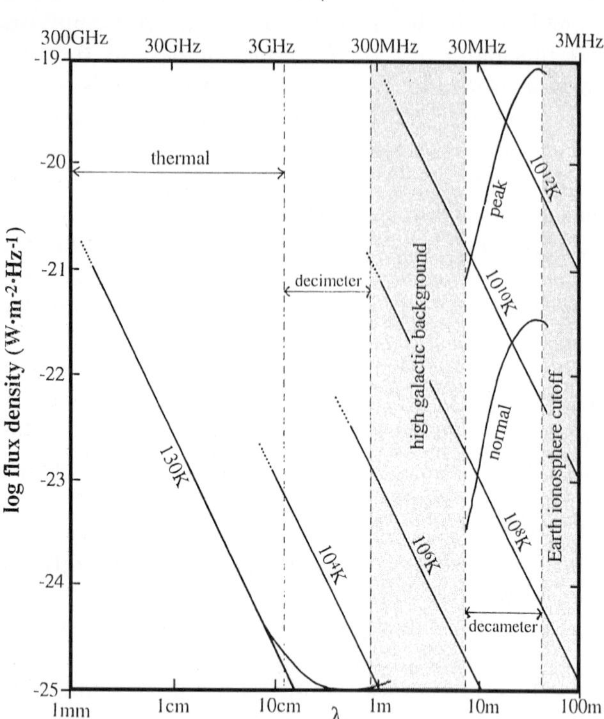


**Figure V.48** Radio-wavelength emission from Jupiter. At wavelengths less than a few centimeters the emission from Jupiter corresponds rather well to a black body spectrum with a temperature of about 130 K in the infrared, rising to about 300 K near 10 cm. This is a thermal (unpolarized) emission from the atmosphere. The 1- to 10-m region is obscured by intense galactic emissions. Beyond about 10 m the emission is highly nonthermal (polarized and non-Planckian) with a black body equivalent temperature of $10^7$ to $10^{13}$ K, originating from an area much larger than the visible disk of the planet.

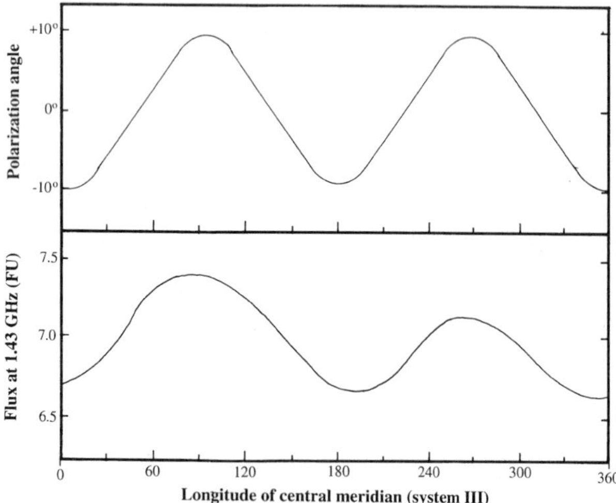


**Figure V.49** Polarization and intensity of the Jovian decimetric radiation. The periodicity of this nonthermal radiation tells us the internal (System III) rotation period of Jupiter.

rotation period of Jupiter, or *System III* rotation. There is every reason to believe that this represents the "true" rotation period of Jupiter, unperturbed by wind currents. But what is the source of the decimetric emission, and why should it be such a good indicator of the rotation period of the interior of the planet?

It has long been known that accelerating charged particles emit electromagnetic radiation. In space about Jupiter, as about the Earth, there is a large volume dominated by the magnetic field of the planet in which there are great populations of trapped protons and electrons. In a *uniform* magnetic field, moving ions and electrons of charge $q$ experience a deflection force (called the Lorentz force) orthogonal to both the $B$ and the $v$ directions:

$$\vec{F} = q\vec{v} \times \vec{B}. \tag{V.210}$$

Because of the constant bending of the $v$ direction by the deflection force, the charged particle will circulate about the magnetic field lines of force. The centripetal acceleration due to the magnetic force equals the acceleration required to maintain circular motion:

$$q\vec{v} \times \vec{B} = mv^2/r_c \tag{V.210a}$$

or

$$r_c = mv/qB. \tag{V.211}$$

Here $r_c$ is the *cyclotron resonance radius*, and the frequency of gyration $\nu$ is $v/2\pi r_c$,

$$\nu = qB/2\pi m, \tag{V.212}$$

the *cyclotron resonance frequency*. Electromagnetic energy of this frequency can be absorbed by a nonrelativistic gas of ions embedded in a magnetic field, and the circulating ions of course produce an alternating electric field of the same frequency and hence function as transmitters of long-wave radio emission. For example, a 1-MeV particle in a magnetic field of 1 gauss ($B = 1$ G) will produce an oscillating field with a frequency of 16 kHz ($\lambda = 19$ km). Such emission, which might be produced by nonrelativistic protons trapped in radiation belts in Earth's (or Jupiter's!) magnetic field, is called *cyclotron emission*. Viewed from the pole of motion of the ion (along the magnetic field lines) the $E$ vector circulates and the radiation is circularly polarized. In the plane of motion the $E$ field oscillates sinusoidally in one dimension, and the radiation is linearly polarized. Viewed from any other angle, the radiation is elliptically polarized.

An interesting and important change occurs at moderately relativistic velocities ($KE \approx m_0c^2$); because the speed of the particle is now close to that of light, the $E$ field produced by the particle in the direction of instantaneous motion will be "piled up" and will become peaked to form a pulse appreciably shorter than the period of the entire cycle. A highly relativistic particle ($KE >> m_0c^2$) with a total energy of

$$E = (p^2c^2 + m_0^2c^4)^{1/2} \qquad \text{(V.213)}$$

(which means an energy $E \approx pc$ when $pc >> m_0c^2$) will, with respect to a frame moving at the speed of light, have very large accelerations when viewed head-on. This is illustrated in Fig. V.50. From electrodynamics, the electric field associated with an accelerating charge at distance $R$ is

$$E_x = -(q/4\pi\varepsilon_0 c^2 R)(\partial^2 x'/\partial t^2). \qquad \text{(V.214)}$$

As can easily be seen in Fig. V.50b, the apparent motion $x'$, as mapped onto a frame moving at the speed of light, develops cusps whose second derivatives become very rapidly varying with time and very large in magnitude. In effect, the particle radiates only cyclotron radiation in the direction of the pole of its motion, but emits powerful high-frequency bursts in all directions lying in the plane of motion. This high-frequency radiation is called *synchrotron emission*. In the example given in Fig. V.50, a compression of a factor of 2.5 in the pulse provides an electric field oscillation at $(2.5)^2$ times the cyclotron frequency. Electrons of 5-MeV energy (10 times the rest-mass conversion energy, $m_0c^2$) have energy

$$\begin{aligned} E &= m_0c^2(1 - v^2/c^2)^{-1/2} \\ &= 5\,\text{MeV} \approx 10^{-5}\,\text{erg}, \end{aligned} \qquad \text{(V.215)}$$

equivalent to momentum $p = E/c = 3 \times 10^{-16}\,\text{g cm s}^{-1}$. From Eq. (V.215), we can calculate that $E = 10m_0c^2$ at $v = 99.5\%$ of the speed of light, so that a compression factor of 200 would be expected for the apparent motion time scale. The second *time* derivative of course feels this

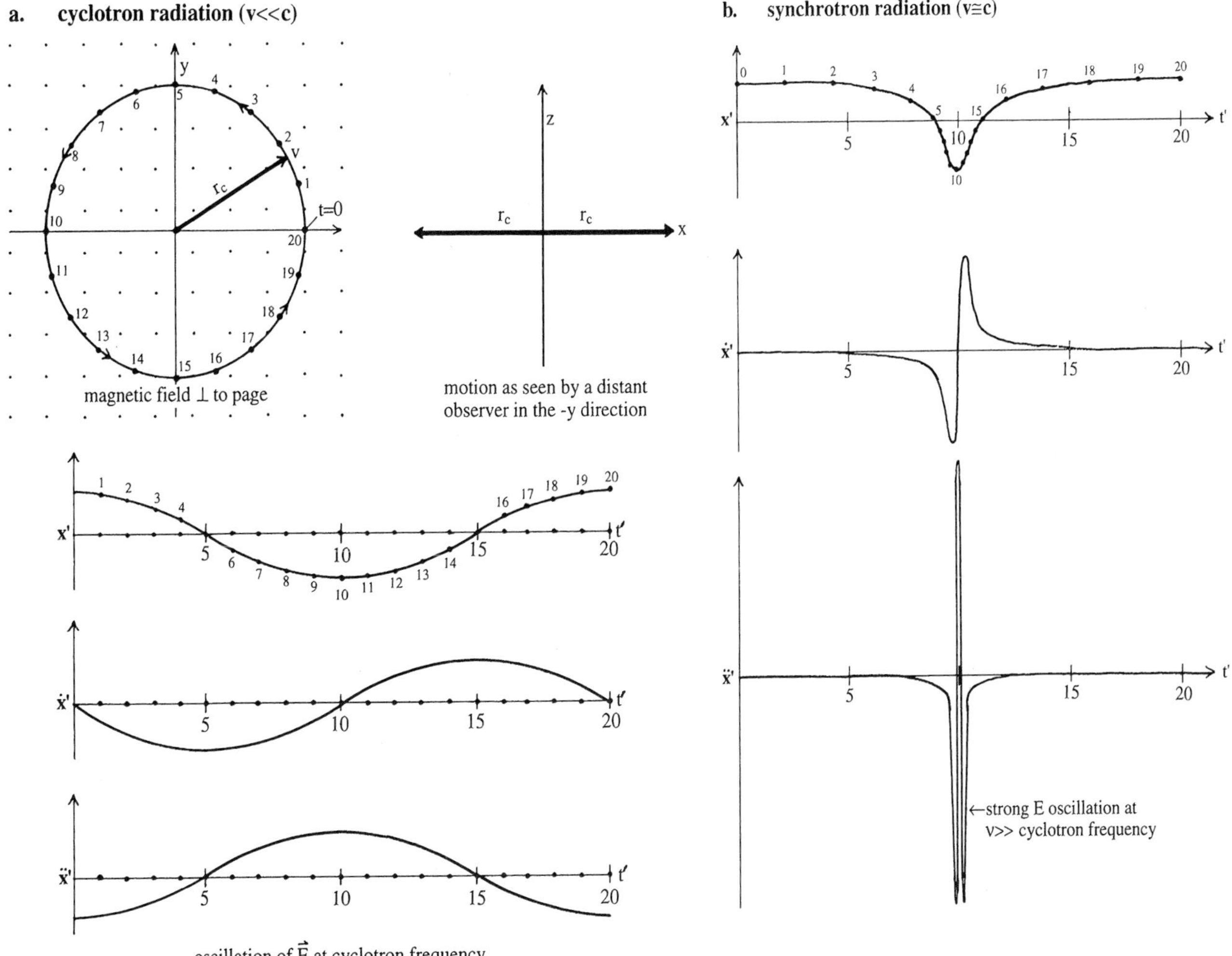


**Figure V.50** Cyclotron and synchrotron radiation. a shows the cyclotron motion of a charged nonrelativistic particle ($v << c$) trapped in a uniform magnetic field. The motion as seen from along the magnetic field lines and from the side, as well as the position, velocity, and acceleration of the particle, is given. b shows the apparent position, velocity, and acceleration of a relativistic ($v$ near to $c$) ion as seen from the side. Note that the synchrotron radiation has a very much higher frequency than the cyclotron radiation.

compression factor twice, so that $\nu_{synch} \approx (200)^2 \nu_{cycl}$, or about 600 MHz. Multi-MeV electrons would thus emit at frequencies in the decimetric region.

If all ions had the same mass, charge, and energy and were trapped in a constant magnetic field of fixed strength, then this would suffice to fully define the expected synchrotron emission. We could then simply find the time when the intensity of received decimetric radiation is a maximum and identify this with the time when Earth passed through the magnetic equator of Jupiter (of course allowing for the Jupiter-to-Earth trip time!). We could also identify the polarization angle with the tilt of the Jovian magnetic dipole relative to the rotation axis and thus estimate the orientation of the magnetic field relative to the center of the planet without ever leaving Earth. In practice, the magnetic field of a planet has a strength that depends on both the distance from the planet and the magnetic latitude, and the trapped particles are also not monoenergetic. A schematic planetary magnetosphere is shown in Fig. V.51. It can be appreciated that the spatial distributions of the intensity and polarization of the synchrotron emission will be valuable clues to the structure of the magnetic field, but that, because of this large information content, spatially resolved (interferometric) studies of the radiation would provide a rather demanding assortment of data to the interpreters. Even if the field is a pure dipole, the location of the center of the dipole (three coordinates), its orientation in space (two coordinates), and polarity (one coordinate) must be determined. Because of the symmetry of cyclotron and synchrotron radiation about the field lines, the polarity cannot be determined in this manner.

The principle of the radio interferometer may be appreciated by considering that the angular resolution of a collector of waves is at best

$$\Delta\Theta = \lambda/d, \qquad \text{(V.216)}$$

where $d$ is the diameter of the receiver. A huge (100 m) radio telescope operating at 30-cm wavelength has a resolution of $3 \times 10^{-3}$ rad, or 10 min of arc. The diameter of the Sun or Moon as seen from Earth is about 30min($'$) of arc, whereas Jupiter is only $0.6'$. However, two electronically linked radio telescopes 3000 km apart on the Earth's surface would have a resolution of 30 cm/3000 km, or $10^{-7}$ rad, which is 0.02 s($''$) of arc. This compares favorably with the 200-in. (508 cm) Palomar telescope operating at 0.5 $\mu$m($5 \times 10^{-5}$ cm), the theoretical limit of resolution, which is also close to $5 \times 10^{-5}/500 = 10^{-7} = 0.02''$. The optical telescope, however, can never perform at its theoretical limit because of the effects of Earth's inhomogeneous and turbulent atmosphere.

It was through the interpretation of radio-interferometric observations of the synchrotron (decimetric) radiation from Jupiter that the existence, strength, position, and orientation of the Jovian magnetic field were determined. As we shall see, the *in situ* magnetometer studies of Jupiter's inner magnetosphere by Pioneers 10 and 11 fully confirmed these interpretations, and they and the Voyager 1 and 2 flybys added a vast wealth of detail.

But now we realize that our simple description of the circular motion of charged particles in a magnetic field is only a first approximation to reality. We need now to consider two other major factors: the effect of gradients in the strength of the $B$ field in the "radial" direction (i.e., in the plane of motion of cycling particles) and the effect of injection of particles whose motions are not confined to the plane of the magnetic equator.

We shall define the component of the particle velocity that is parallel to the $B$ field as $v_{\|}$ and the component orthogonal to $B$ as $v_{\|}$. The angle between the $B$ field and the total velocity vector $\vec{v}$ will be called the *pitch angle*, $\alpha$, defined by

$$\tan\alpha = v_{\perp}/v_{\|} = v\sin(\alpha/v)\cos\alpha. \qquad \text{(V.217)}$$

Because the motion about the field lines conserves energy,

$$|\vec{v}|^2 = v_{\perp}^2 + v_{\|}^2 = \text{constant}, \qquad \text{(V.218)}$$

and the cyclotron radius follows as

$$r_c = mv_{\perp}/q\mathrm{B} = (mv\sin\alpha)/qB. \qquad \text{(V.219)}$$

In the absence of an electric field and with no variations of the magnetic field with time, each particle cycles along a path that encloses a constant magnetic flux, $\Phi$,

$$\Phi = \oint \vec{B}\cdot d\vec{S} = \text{constant}, \qquad \text{(V.220)}$$

where the integration is around one loop executed by the particle. If the $B$ field varies only very slightly over the scale of the cyclotron radius, the flux is just $\Phi = B\pi r_c^2$, where $B$ is the average magnetic field strength within the path. The constant $\Phi$ is called the *first adiabatic invariant*.

Next, combining with Eq. (V.219),

$$\pi B r_c^2 = \text{constant} = (\pi/B)[(mv\sin\alpha)/q]^2, \qquad \text{(V.221)}$$

from which it follows that the pitch angle is related to the local magnetic field strength by

$$(\sin\alpha)^2/B = \text{constant}. \qquad \text{(V.222)}$$

Thus for any given particle of fixed energy, there exists a critical value of the magnetic field strength at which $\sin\alpha = 1$ (the pitch angle vanishes) and the particle is "reflected."

We shall assume that the overall geometry of the magnetic field is that of a dipole. From Maxwell's equations

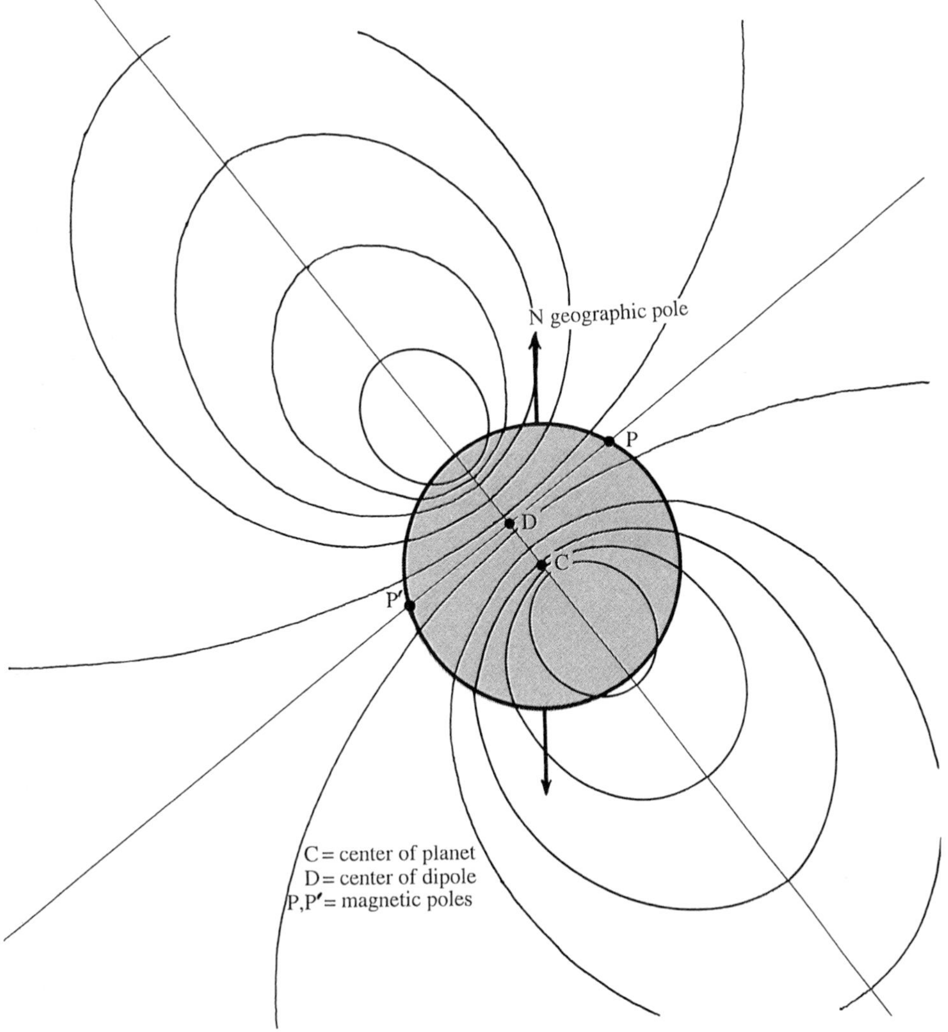


**Figure V.51** A planetary dipole magnetic field. The dipole may generally be tilted relative to the geographic (spin) pole and offset from the center of the planet.

$$\vec{\nabla} \times \vec{B} = \vec{J}/\varepsilon_0 c^2 + (1/c^2)\partial E/\partial t, \qquad \text{(V.223)}$$

but we are neglecting $E$ fields, and currents tend to be very small in the magnetosphere; hence $\vec{\nabla} \times \vec{B} = 0$. This is true if $B$ is the gradient of some *magnetic potential V*: $\vec{B} = \vec{\nabla}\vec{V}$. This potential can generally be expressed as a spherical harmonic expansion of form

$$V = \sum_{n=1}^{\infty}\sum_{m=1}^{\infty}\left(\frac{r_0}{r}\right)^{n+1} P_{nm}(\cos\Theta) \{g_{nm}\cos \mathrm{m}\phi + h_{mm}\sin m\phi\}, \qquad \text{(V.224)}$$

where $\Theta = 90 - l_{\mathrm{mag}}$ is the colatitude (angle from the magnetic pole) and $\Phi$ is the magnetic longitude. The $P_{nm}$ $(\cos\Theta)$ are the associated Legendre polynomials.

For the simple and often useful case of a pure dipole field, $V$ becomes independent of longitude and is simply

$$V = (M\cos\Theta)/r^2, \qquad \text{(V.225)}$$

where $M$ is the planetary dipole moment in G cm$^3$.

The radial and latitudinal components of the field are then

$$B_r = -\partial V/\partial r = (2M\cos\Theta)/r^3 \qquad \text{(V.226)}$$

$$B_\Theta = -(1/r)\partial V/\partial\Theta = (M\sin\Theta)/r^3 \qquad \text{(V.227)}$$

The magnitude of the total field is just

$$|B| = (B_r^2 + B_\Theta^2)^{1/2} = (M/r^3)(1 + 3\cos^2\Theta)^{1/2}. \qquad \text{(V.228)}$$

The local direction of the magnetic field lines is specified by the values of $B_r$ and $B_\Theta$:

$$B_r/B_\Theta = 2\cos\Theta/\sin\Theta = 2\,\mathrm{ctn}\,\Theta. \qquad (V.229)$$

From Fig. V.52 we can see that the local tangent to the field lines defines $B_r/B_\Theta = dr/rd\Theta$; hence

$$B_r/B_\Theta = 2\cos\Theta/\sin\Theta = dr/rd\Theta \qquad (V.230)$$

or

$$dr/r = 2(d\sin\Theta/\sin\Theta), \qquad (V.231)$$

whence

$$d\ln r = 2d\ln\sin\Theta, \qquad (V.232)$$

and thus

$$r = r_0\sin^2\Theta = r_0\cos^2 l, \qquad (V.233)$$

where $l$ is the *magnetic* latitude. The constant of integration, $r_0$, is the distance of that field line from the center of the planet, where it crosses the magnetic equator ($\sin^2\Theta = 1$). Often, $r_0$ is expressed in units of the planetary radius, $r_s$, and $r_0/r_s$ is called the *magnetic shell number*, *L*.

Now, combining Eqs. (V.228) and (V.233) and letting the integrated field strength of the planet $(M) = B_0 r_0^3$, we get

$$B(l) = (r_s/r_0)^3 B_0[(1 + 3\sin^2 l)/\cos^6 l]. \qquad (V.234)$$

At the maximum latitude reached by the spiraling particle, where $\sin\alpha = 1$ and the particle is "reflected" back toward the equator into regions of lower magnetic field strength, we have $\sin^2\alpha/B = 1/B(l_{max})$. Denoting the pitch angle of the particle when it crosses the magnetic equator as $\alpha_0$ and the equatorial field as $B_0$,

$$(1/B)\sin^2\alpha = (1/B_0)\sin^2\alpha_0 = 1/B(l_{max}). \qquad (V.235)$$

Then, from Eqs. (V.234) and (V.235),

$$B(l_{max})/B_0 = 1/(\sin^2\alpha_0) = [1 + 3\sin^2 l_{max}]^{1/2}/(\cos^6 l_{max}). \qquad (V.236)$$

Thus a particle injected at the magnetic equator with a pitch angle of 0 will simply follow the field line into the pole ($l_{max} = 90°$) and be dumped into the planet's atmosphere; if the particle has no $v_\perp$ at the start, it never will have any. An initial (equatorial) pitch angle of 90° would of course give the particle no poleward component of velocity, and thus the maximum magnetic latitude reached will be 0°. Particles with pitch angles will work their way poleward slowly and will be mirrored at low latitudes, whereas particles with small pitch angles will travel far more rapidly in latitude and have a longer trip between mirror points. It is amusing to evaluate the times taken for particles to travel from one mirror point to the other as a function of their initial (equatorial) pitch angle, $\alpha_0$. If we denote the arc length traveled by a particle along a field line as $s$, then the half-trip time (from bounce 1 to bounce 2) is

$$\tau_{1/2} = \int_1^2 ds/v_\parallel = 2\int_0^1 ds/v_\parallel. \qquad (V.237)$$

where the integral is now from the equator to the mirror point. The element of arc length is, from Fig. V.52, given by $ds^2 = dr^2 + r^2 dl^2$, whence

$$ds = [(dr/dl)^2 + r^2]^{1/2} dl. \qquad (V.238)$$

From Eq. (V.218), $v_\parallel^2 = v^2\cos^2\alpha = v^2(1 - \sin^2\alpha)$, and, with Eq. (V.235) and the geometrical relation $dr = -2r_0\sin l\cos l\,dl$, we have

$$\tau = 2\tau_{1/2} \qquad (V.239)$$

$$= \frac{4r_0}{v}\int_0^{l_{max}} \frac{\cos l[1 + 3\sin^2 l]^{1/2}}{\left\{1 - \frac{\sin^2\alpha_0}{\cos^2 l}[1 + 3\sin^2 l]\right\}} dl. \qquad (V.240)$$

This truly depressing integral (call it $I(\alpha_0)$ to save time) can be evaluated numerically and is found to be quite slowly varying, ranging from 1.4 at $\alpha_0 = 0°$ to 0.75 near $\alpha_0 = 90°$. The integral can be reasonably well approximated by $I(\alpha_0) = 1.3 \times 0.56\sin\alpha_0$. To a good first approximation, in fact, we may say that the bounce time depends strongly on $r_0$ and $v$ and hardly at all upon $\alpha_0$! Further, the bounce time does not depend on mass: a proton and an electron in the same magnetic shell with the same velocity will bounce with the same frequency.

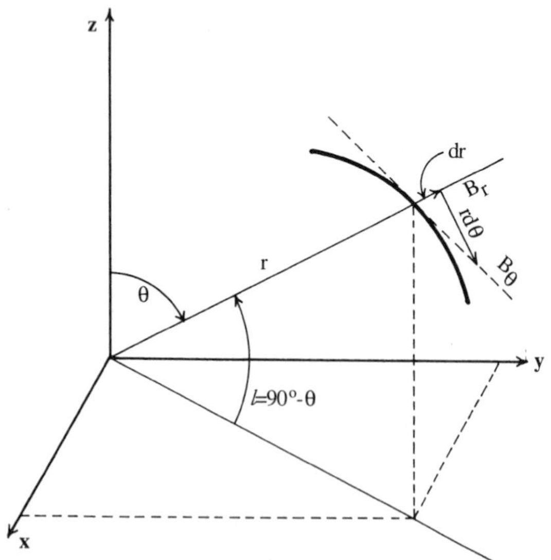


**Figure V.52** Direction of magnetic lines of force. $l$ is the latitude; the colatitude $\theta$ is $90° - l$.

Finally, for a planet of magnetic moment $M$, $r_0$ is just $(M/B_0)^{1/3}$ and thus rather weakly dependent on the magnetic field strength. For Jupiter, typical bounce times for relativistic particles ($v \approx c$) in the $r_0/r_s = 2$ shell are $8r_s/c$, which are near 2 s, compared with the cyclotron resonance period of $10^{-4}$ s.

The third major kind of motion possible for a trapped particle is due to forces that we have not considered so far, such as gradients in the magnetic field strength. Figure V.53 compares the particle gyration trajectories for three cases: a constant $B$ field; two regions of constant but different $B$ fields; and a region with a smooth gradient of $B$. Clearly, if the field is not homogeneous the radius of curvature of the electron path will be smaller in regions of high $B$-field strength, and the path will be an epicycloid that *drifts* relative to the field.

It is convenient in treating the drift of a particle in an inhomogeneous magnetic field to recognize explicitly the magnetic moment ($\mu$) associated with the cyclotron gyrations. The moment is the product of the electric current (carried by a single particle!) times the area ($A$) of the near-circle about which it cycles,

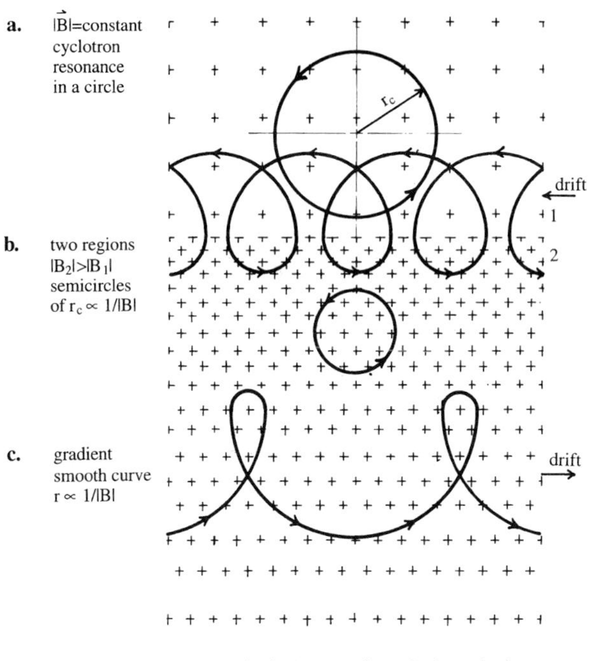


**Figure V.53** Charged particle motion in homogeneous and inhomogeneous magnetic fields. a shows cyclotron motion in a homogeneous field. b shows motion across a discontinuity in magnetic field strength. The cyclotron radius is much smaller in the higher field, leading to drift of the particle at right angles to the field gradient (following the boundary between the weaker and the stronger fields). In c the field strength has a smooth gradient. The particle motion then follows a smooth curve with a constantly changing radius of curvature.

$$\mu = JA = (v_\perp q/2\pi r_c)\pi r_c^2$$
$$= v_\perp q r_c/2 = (v_\perp q/2)(mv_\perp/qB) = \mathrm{KE}_\perp/B, \qquad \text{(V.241)}$$

where we have substituted in for the current ($J$), the area ($A$), and the cyclotron radius ($r_c$) from Eq. (V.219). The symbol $KE_\perp$ denotes the kinetic energy associated with cyclotron motion. This is relativistically correct if $m$ is taken to be the actual particle mass, not the rest mass:

$$m = m_0/(1 - v^2/c^2)^{1/2}. \qquad \text{(V.242)}$$

The potential energy of the magnetic moment ($\mu$) in a magnetic field is

$$U = -\vec{\mu} \cdot \vec{B}, \qquad \text{(V.243)}$$

and the force on the particle is just $-\nabla U$,

$$F_\perp = \mu \nabla_\perp B, \qquad \text{(V.244)}$$

where we denote with subscripts the fact that the force is perpendicular to the direction of $B$. Rewriting Eq. (V.219) in vector notation,

$$\vec{r_c} = (m/qB^2)(\vec{v} \times \vec{B}) \qquad \text{(V.245)}$$

or, in terms of the momentum $p = mv$,

$$\vec{r}_c = (1/qB^2)(\vec{p} \times \vec{B}). \qquad \text{(V.246)}$$

Now, if there is a change, $\Delta p$, in the momentum (i.e., if there is a *force* acting), then the corresponding change in $r_c$ is

$$\Delta\vec{r}_c = (1/qB^2)(\Delta\vec{p} \times \vec{B})$$
$$= (1/qB^2)(\vec{F}_\perp \times \vec{B})\Delta t, \qquad \text{(V.247)}$$

equivalent to stating that $\Delta p/\Delta t = F$. Because we already know the force ($F$) exerted on the dipole moment by the magnetic field from Eqs. (V.245) and (V.242), we can combine to get an explicit equation for the rate of displacement of $r_c$:

$$\Delta\vec{r}_c/\Delta t = -(\mu/qB^2)(\nabla_\perp\vec{B} \times \vec{B}) \qquad \text{(V.248)}$$

$$= -(mv_\perp{}^2/2qB^3)(\nabla_\perp\vec{B} \times \vec{B}). \qquad \text{(V.249)}$$

The rate of motion of $r_c$ is called the *drift velocity*, $V_D$, which we have now shown to be orthogonal to both the $B$ field and its gradient. Thus the drift direction is such that it follows contours of constant magnetic field strength. Because of the dependence on the gradient of $B$, the drift rate is quite slow, taking times on the order of an hour for drift of a 1-MeV particle about Jupiter.

Thus the typical trapped particle in the Jovian magnetosphere makes $10^4$ cycles about the field lines per bounce and about $10^3$ bounces per drift cycle around the planet.

Despite the apparent complexity of the description so far, we still have assumed a strictly dipolar planetary field and have neglected all other forces such as electric fields and the effects of rotation of the planet. It is instructive to look at some *in situ* data on the Jovian magnetosphere to assess the magnitude of the task remaining.

Pioneer 10 and 11, Voyagers 1 and 2, and the Ulysses spacecraft carried a variety of experiments designed principally to define the magnetic and particle environments of Jupiter and (except for Pioneer 10) Saturn. These spacecraft encountered Jupiter's bow shock (where the solar wind becomes disrupted by the planetary magnetosphere) at over 100 Jupiter radii from the planet, and the spacecraft passed in and out through the bow shock and the magnetopause several times. As Jupiter rotated, the magnetosphere, which as we have seen must be tipped about 10° relative to the rotation pole, wobbled north and south relative to the spacecraft, and the location of the magnetopause moved in response (Fig. V.54). Once within the magnetosphere, the magnetic field oscillated in intensity with Jupiter's rotation period. Deep within the inner magnetosphere the populations of trapped radiation-belt particles were found to have distinct minima at the orbital radii of Io and Europa, probably due to sweeping out of the ions in these magnetic shells by collision with the satellites. The magnetic field close to Jupiter was found to be tilted 9.6° in the direction of System III longitude 231°, with a surface field of about 14 G at the North Pole and 10.7 G at the South Pole. The center of the effective dipole was found to be offset 8000 km from the center of Jupiter, almost in the equatorial plane but a little to the north. The magnetic field strength integrated over the volume of the planet was calculated to be $4.25\,\mathrm{G}\, r_J{}^3 = 1.5 \times 10^{30}\,\mathrm{G\ cm}^3$, about $10^4$ times that of Earth. Interestingly, the inner magnetosphere is only poorly described by a dipole field: the data from the two Pioneer flybys suggested an integrated strength of the quadrupole component of the field that is about 22%, and the octupole component about 18%, of that of the dipole field. Thus the structure of the inner magnetosphere will require far more than a few flybys for adequate exploration, and we can expect the physics and energetics to differ markedly from our experience in Earth's magnetosphere.

The inner magnetosphere in the vicinity of the orbit of Io was found to be populated by enormous fluxes of energetic ions and electrons. The local energy flux

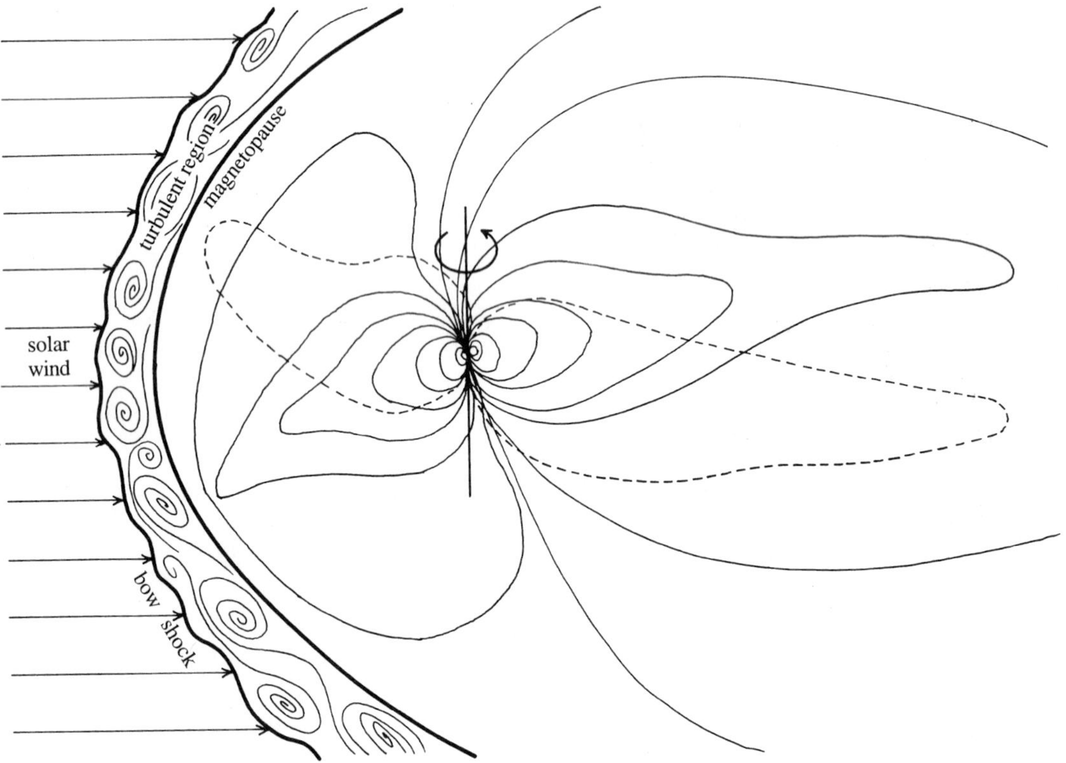


**Figure V.54** The magnetosphere of Jupiter. The inner field, close to a dipole, is distorted by Jupiter's rotation and by the dynamic pressure exerted by the solar wind (Sun to left). Because of the 10° tilt of the magnetic field of Jupiter, the outer magnetosphere "flops up and down like a hat brim" with every rotation of the planet, as suggested by the dashed lines. This causes conditions at the magnetopause to vary markedly every Jovian day and forces the bow shock to oscillate in and out.

due to energetic particles reaches a maximum of $10\,\mathrm{erg\,cm^{-2}\,s^{-1}}$, about $10^{-3}$ of the solar flux at Jupiter. The abundance of energetic particles diminishes dramatically with increasing energy.

The peculiar shapes of the field lines in the outer magnetosphere are due to a combination of circumstances also not found on Earth. Because of the 27-fold dilution of the solar wind at Jupiter relative to Earth and because of the enormous magnetic moment of Jupiter, the magnetic field in the outer magnetosphere can be so low that the thermal energy density of the particles can exceed the energy density carried by the magnetic field. Since the magnetic field strength drops off with $r^{-3}$, whereas the force necessary to maintain circular motion drops off only as $r^{-1}$, the outer regions of the magnetosphere are severely distended by the rapid rotation of the planet.

From Eq. (V.228), we can see that the magnetic field far above the magnetic equator has a magnitude

$$|B| = (B_r{}^2 + B_\Theta{}^2)^{1/2} = (M/r^3)(1+3\cos^2\Theta)^{1/2} = M/r^3. \quad \text{(V.250)}$$

For a mean surface magnetic field strength of 4.25 G, the external field strength in the nearly dipolar portion of the field is just $4.25(r_J/r)^3$. (The integrated field strength, misleadingly called the "planetary dipole moment" by some space scientists, is often quoted as $4.25\,\mathrm{G}r_J{}^3$.) At large distances from the planet the higher-order fields are completely negligible, because, as we can see from Eq. (V.225), the potential falls off with distance more rapidly for such fields.

The solar wind pressure near 5 AU is, as we have seen, approximately $mnv^2/2$, and we would expect a purely dipolar Jovian magnetosphere to terminate on the sunward side at the point, where the solar wind dynamic pressure just equals the magnetic pressure from Jupiter's field. In Gaussian units,

$$B^2/8\pi = 0.72(r_J/r)^6 = n_J m v^2/2 = n_E m v_\infty^2/2R_J{}^2/R_E)^2, \quad \text{(V.251)}$$

where $n_E$ is the solar wind number density at 1 AU ($10\,\mathrm{cm^{-3}}$), $R_J/R_E$ is Jupiter's heliocentric distance in units of Earth's distance (i.e., in AU), and $v_\infty$ is the "thermal" velocity of the solar wind beyond 1 AU, about $4\times10^7\,\mathrm{cm\,s^{-1}}$. The solar wind pressure is then $5\times10^{-10}\,\mathrm{dyn\,cm^{-2}}$, and this will be balanced by the Jovian dipole field when $r = 33.5r_J$. The distension of the Jovian magnetosphere out to over 100 $r_J$ is thus due to a planetary magnetic field that drops off much more *slowly* than a dipole at large distances from the planet.

Because the corotation speed due to approximate solid-body rotation of the magnetosphere increases with distance from Jupiter and the orbital velocity of satellites decreases with $r^{-1/2}$, there must be a distance from Jupiter at which the field would corotate with an orbiting satellite. This occurs when $v_{orb} = v_{corot}$, or, from Eq. (II.32),

$$(GM/r)^{1/2} = \omega r. \quad \text{(V.252)}$$

The angular rotation rate of Jupiter in the relevant frame (System III) is $1.7585\times10^{-4}\,\mathrm{rad\,s^{-1}}$, and we find the equality satisfied at $r = 1.60\times10^{10}\,\mathrm{cm} = 2.29\,r_J$. This is almost exactly on the orbit of J5, Amalthea. Therefore all the more distant satellites of Jupiter are behind swept from behind by a "magnetospheric wind." Both JXV Adrastea and JXVI Metis, which orbit much closer to Jupiter than JV Amalthea, orbit at a higher angular speed than the magnetosphere, and thus experience a magnetospheric "headwind." Even Amalthea, of course, does not simply bore a hole in the particle population near it, because the oppositely directed drift of electrons and protons will refill the region of the field lines passing through Amalthea in a matter of minutes. What then happens is that the particle population in the *entire magnetic shell* on which Amalthea lies will be depleted appreciably, just as with the Galilean satellites.

It is most interesting that the existence of strong interactions between the satellites and the trapped particles has been known since 1964, from Earth-based observations. In that year the distribution of Jovian decameter bursts in time was found to be strongly modulated by the position of JI, Io, in its orbit. The relationship is shown in Fig. V.55. Figure V.55a shows the convention for defining the orbital phase angle ($\phi$) of the satellite. The position of Jupiter is described by the longitude of the central ("sub-Earth") meridian at the time of the observation. Figure V.55b shows that the distribution of decameter bursts over all possible orientations of Jupiter and Io is very nonuniform. When Io is at elongation from Jupiter ($\phi = 90$ and 270°) the probability of burst emission is greatly enhanced. Bursts can, on the other hand, occur as almost any Jovicentric longitude crosses the sub-Earth meridian.

A number of theoretical explanations of this effect have been offered. Perhaps the most attractive is that due to Peter Goldreich and Donald Lynden-Bell, who pointed out that, as the magnetosphere flows past Io, the huge $(\vec{v}\times\vec{B})$ forces developed between the sub-Jupiter and anti-Jupiter points on Io may short out if Io is a good enough electrical conductor. Then currents carrying as much as $10^8$ W may flow through Io, down the Jovian field lines, and on to complete the circuit. This model therefore predicts very large electric currents in the "flux tube" defined by those Jovian field lines which pass through Io at any moment. Further, the surface of Io

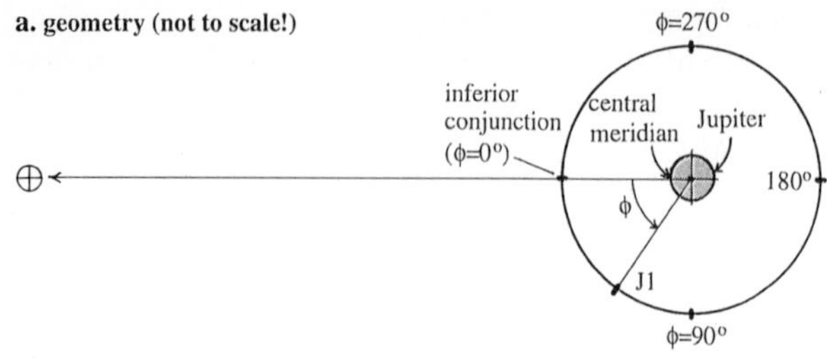


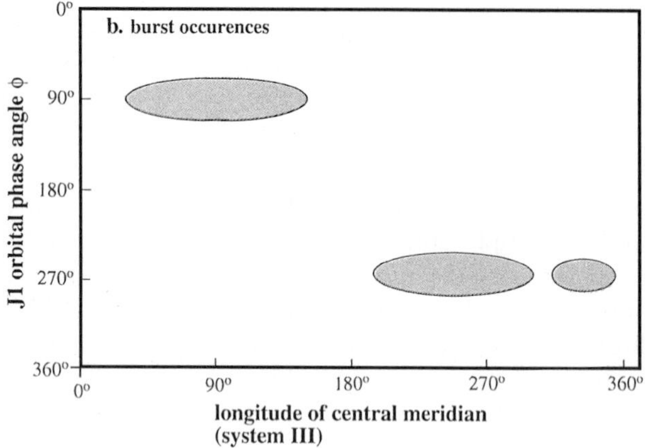


**Figure V.55** Modulation of Jupiter's radio emission by Io. a defines the geometry, with the orbital phase angle of Io being measured from the central meridian of Jupiter as seen from Earth. b shows the regions around Io orbital phase angles of 90° and 270°, at which Jupiter's radio emission is stimulated.

and the upper atmosphere of Io at the feet of the Io flux tubes will be subjected to violent particle bombardment. The necessary conductivity could be provided by either an electrically conducting interior or an ionosphere on Io.

The precise source for the decametric radiation may well be associated with dumping of magnetospheric particles into Jupiter's upper atmosphere and onto Io, but it is unlikely that anything will be known for certain until decametric measurements are made with high spatial resolution or, better yet, an appropriately instrumented spacecraft is flown through the Io flux tube.

Theoretical attempts to explain the presence of the Jovian magnetic field are generally centered about the presence of a massive region of fluid metallic hydrogen in Jupiter, and it seems that any mechanism dependent on the presence of metallic hydrogen would make it likely that the field on Saturn would be weaker. However, Roman Smoluchowski has pointed out that, at high pressures, the energy gap between the filled molecular orbital of $H_2$ and the (empty) conduction band is greatly decreased and that even $H_2$ may conduct appreciably. Ionic impurities, such as alkali metal ions, would further enhance the conductivity. Qualitatively, thermal stresses in an electrically conducting fluid produce bulk motions that can amplify and sustain the effects of small initial fields (or currents).

Prior to the arrival of the first spacecraft flybys, there were at least three tentative reports of detection of radio emission from Saturn, but the state of the magnetosphere of Saturn was almost wholly unknown. The 1979, 1980, and 1981 arrivals of Pioneer 11, Voyager 1, and Voyager 2, respectively, led to an immense increase in our knowledge.

Saturn's magnetic field was found to have strong dipole character, with a much smaller contribution of higher-order terms than the Jovian field. As with Jupiter (but unlike Earth) the Saturnian field is oriented with the north magnetic pole aligned toward the north geographic pole. Indeed, the Saturnian field is unique in that the alignment of the magnetic and geographic poles is extremely close. The field, with an integrated field strength of 0.2 G $R_S^2$, is tilted less than 1° relative to the geographic pole (Fig. V.56). The center of the dipole is also extremely close to the center of the planet, being offset by no more than 0.01 $R_S$ in the equatorial plane.

At large distances from the planet, over 10 $R_S$, the magnetic field strength suddenly begins to depart markedly from that of a perfect dipole, in that the field is too strong. Multipole components in the planetary field would give an excess strength very near the planet, but cancel themselves beyond a few planetary radii, and thus cannot be responsible for this phenomenon. The excess field is provided by a ring current of trapped ions circulating about the planet in the equatorial plane, in the regions marked in Fig. V.57.

As with Jupiter, the energy spectrum of the trapped particles is extremely steep. The Voyager 1 proton energy measurements at Saturn are summarized in Fig. V.58. Protons with energies of 10 MeV are about $10^{10}$ times less abundant than those with energies of $10^{-2}$ MeV. Note the distinct feature for proton energies of 16 to 200 MeV: these are produced by the decay of free neutrons that are produced in nuclear reactions caused by cosmic rays striking atoms in the Saturn system, such as the rings. In the outer Saturnian magnetosphere there is an important contribution of heavy ions (oxygen or nitrogen) that has evidently been ejected from the Saturnian satellites.

As with Jupiter, Saturn's radiation belts are strongly affected by the presence of satellites and rings. The Pioneer 11 electron and proton (80 MeV) flux energies are shown in Fig. V.59. All charged particles are removed in the vicinity of the A ring, and there is a dip in the electron population at the radial distance of the F ring. Because the bounce frequency of such close electrons is high, they are compelled to make frequent passes through the equatorial plane, where they are at hazard of collision with ring particles. Note that there are deep gaps in the proton populations at the orbits of the minor satellite 1979-S2

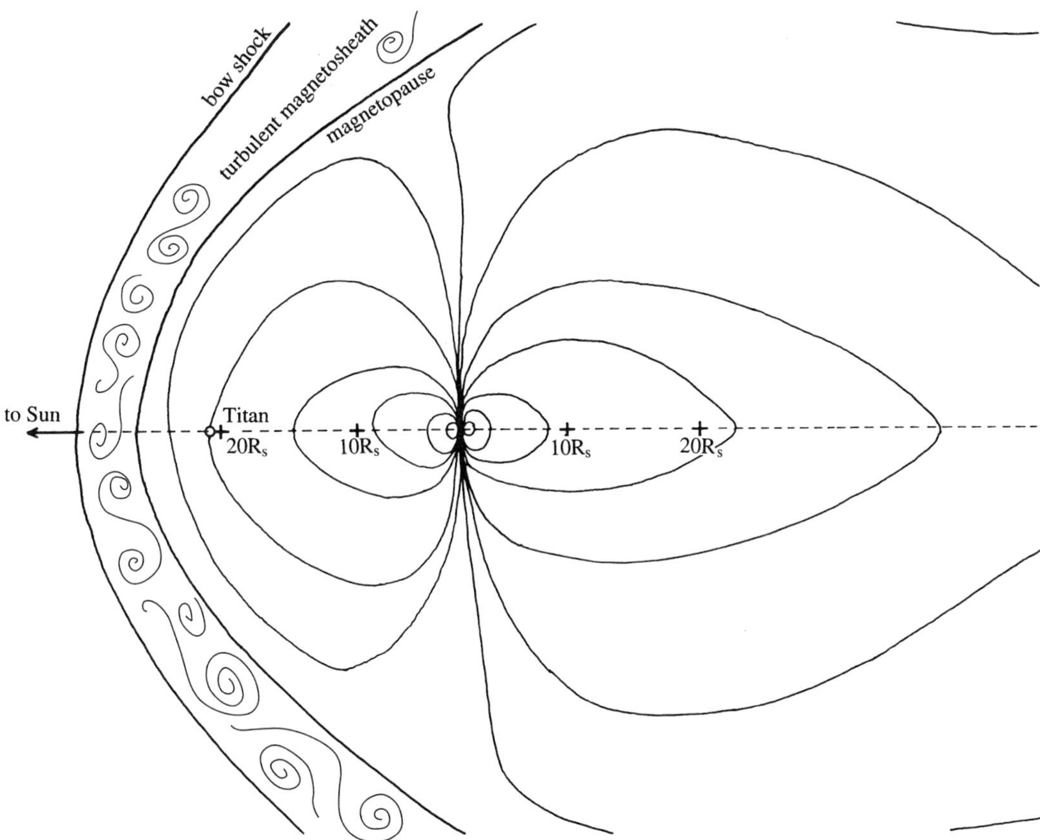


**Figure V.56** The magnetosphere of Saturn. The inner field, close to a dipole, is distorted by centrifugal force and by the dynamic pressure exerted by the solar wind (Sun to left). Because of the small tilt of the dipole field there is little disturbance of the field structure near the magnetopause as the planet rotates. However, because of the large tilt of the spin axis of the planet, the solar wind stagnation point (where it strikes the bow shock head-on) oscillates between a latitude of 30° N and a latitude of 30° S over the Saturnian year.

and Mimas, whereas electrons seem not to notice these satellites. These energetic protons, which are made by decay of neutrons from the cosmic-ray bombardment of the rings mentioned above, have very long diffusion times and cannot refill the gaps made by the satellites.

The inner magnetospheres of Jupiter and Saturn are generated and maintained by processes in the deep interiors of the planets, and the detailed mapping (over time) of these magnetospheres at very low altitudes could provide extremely useful information about the location and nature of the processes that generate these fields. The main obstacles to carrying out such experiments are, first, the immense energetic cost of placing satellites in low orbits about such massive planets and, second, the severe radiation environment maintained by the presence of the fields that we wish to measure. In Saturn's inner magnetosphere, where the rings absorb the charged particles, the rings constitute a severe hazard! It is amusing, and typical of the planetary sciences, that study of the most distant and rarefied aspect of Jupiter's structure, the magnetosphere, leads us directly back to processes in the planet's core. It is this complexity of interaction, in which all levels of structure interact, that makes the study of the planets so challenging and rewarding.

## The Interiors of Uranus and Neptune

Because of the small angular diameters of Uranus and Neptune (3.7″ and 2.3″, respectively, at opposition), it is extremely difficult to make useful observations of their properties from Earth. With respect to the first-order variables defining the properties of their interiors, mass, radius, rotation period, oblateness, luminosity, and composition, the pre-Voyager data left much to be desired. The Voyager 2 flyby of 24 January 1986 has of course given us a vastly improved knowledge of the Uranus system, and the Voyager 2 Neptune flyby on 25 August 1989 has had an even more profound impact on our knowledge of Neptune and its rings and satellites.

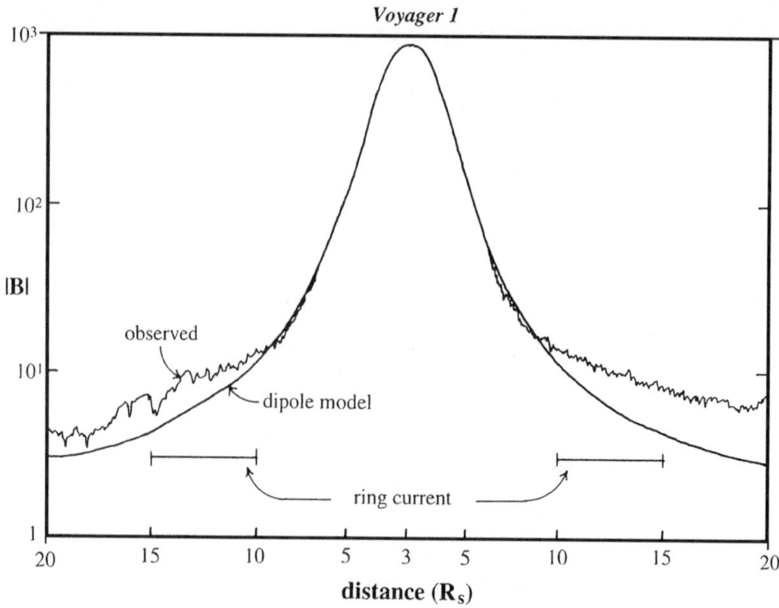


**Figure V.57** The Saturnian magnetic field. The Voyager 1 flyby of Saturn observed an inner magnetosphere with a clean dipole character, but with a marked excess of field strength beyond about 10 $R_S$. This additional field is produced by a ring current of charged particles trapped in the magnetosphere near the equatorial plane.

All mass determinations depend on a knowledge of the universal gravitational constant, $G$, but unfortunately this is the least precisely known of the fundamental physical constants. The product $GM$ (and the ratio of masses of two bodies) can be determined with far greater precision than $M$ itself. The quoted errors on the masses of bodies therefore often do not reflect the (much smaller) errors in measuring $GM$, but rather the error in determining $G$. Relative masses ($GM_1/GM_2$) are therefore much more accurately known than individual masses ($M_1$, $M_2$). From the tracking of Voyager 2, $GM_U$ was found to be $5793939 \pm 60\,\text{km}^3\,\text{s}^{-2}$ ($5794547 \pm 60$ for the entire Uranian system) and $GM_{Nsys}$ was found to be $6836534 \pm 20\,\text{km}^3\,\text{s}^{-2}$. The mass of Uranus, determined from the motions of its satellites, is $14.51 \pm 0.10$ Earth masses, or $87 \times 10^{27}$ g, whereas that of Neptune is $17.26 M_E (103 \times 10^{27}\,\text{g})$ with about the same precision. The corresponding Voyager results are $M_U = 86.827 \times 10^{27}$ g and $M_N = 102.43 \times 10^{27}$ g. The equatorial radius of Neptune (at the 1-bar reference level in the atmosphere) was determined from observations of the occultation of a faint star in 1970 to be 25,220 km. The radius of Uranus was determined by data from the stellar occultation events in which the Uranian rings were discovered and further studied, and more recently the radii of both Uranus and Neptune have been improved by data from the Voyager 2 encounters. The equatorial radius found in this way for Uranus is 25,660 km at the 1-bar pressure level. The oblateness has been determined from analysis of photographs taken by the Stratoscope balloon-borne telescope to be $0.022 \pm 0.001$. The polar radius can be calculated from the oblateness,

$$e = (a - b)/a = 0.022, \tag{V.253}$$

giving $b = 25050 \pm 60$ km. The mean radius is $(2a + b)/3 = 25{,}460$. The oblateness ($e$) of the visible disk of Neptune was estimated from astronomical observations and simple theory to be 0.021 to 0.026. The polar radius of Neptune was found by the Voyager 2 radio occultation experiment to be $24{,}340 \pm 30$ km and the equatorial radius, $24{,}764 \pm 20$ km, for an oblateness of $e = 0.0171$.

The density of Uranus is calculated to be $1.267\,\text{g cm}^{-3}$, and that of Neptune, $1.640\,\text{g/cm}^{-3}$. Both planets are much denser than Saturn (0.686), even though they are far smaller and hence less compressed by their own gravitation. The "surface" acceleration of gravity of Uranus neglecting centrifugal force and oblateness is then $900\,\text{cm s}^{-2}$, with $g = 1125\,\text{cm s}^{-2}$ for Neptune. The escape velocities of the planets are, to the same approximation, $21.4\,\text{km s}^{-1}$ for Uranus and $23.6\,\text{km s}^{-1}$ for Neptune.

Uranus is noteworthy for its peculiar spin: the rotation axis lies almost in the plane of its orbit, tilted 98° with respect to the normal to the ecliptic. The orientation of the spin axis can be deduced without recourse to direct observations of the planet, because the orbits of the five satellites of Uranus are very precisely coplanar, and tidal forces would quickly disrupt the satellites' orbits if the equatorial plane did not coincide closely with their orbital plane.

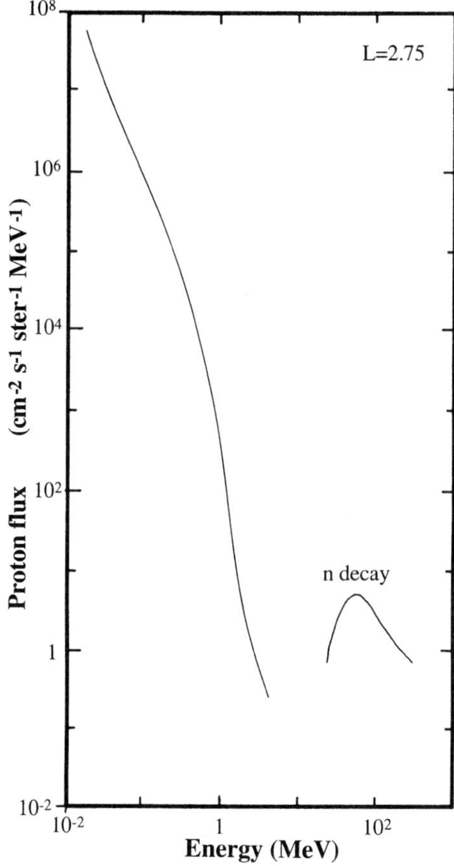


**Figure V.58** Proton energy spectrum in Saturn's inner magnetosphere. These Voyager measurements near magnetic shell number $L = 2.75$ show trapped protons, perhaps derived from the solar wind, and protons from cosmic ray albedo neutron decay (CRAND), from free neutrons produced by cosmic ray bombardment of the nearby rings.

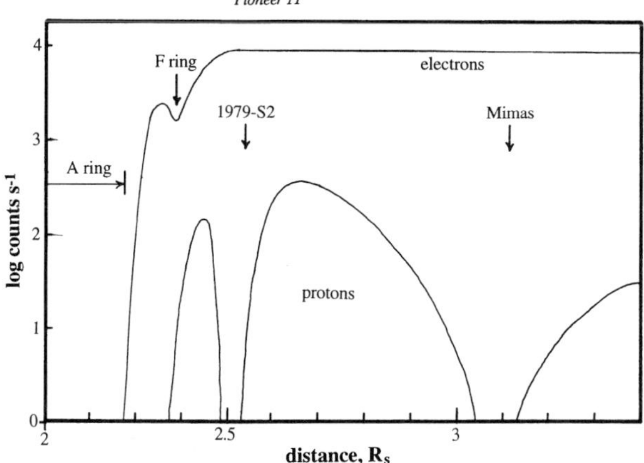


**Figure V.59** Absorption of magnetospheric particles by rings and satellites. These Pioneer 11 data from the Saturn flyby of 1979 show deep minima in the particle abundances at the radial distances of two satellites and the rings.

Direct observations of Uranus for the purpose of determining the rotation period and axial orientation are very difficult. There are several mentions of faint parallel bands on the disk in the older literature. The Stratoscope pictures showed no signs of bands or any other features, despite their 5× better resolution, and the incomparably superior Voyager 2 pictures disclosed only a very faint polar darkening and a barely discernible circumpolar band (Fig. V.60). A few small, low-contrast bright features were seen at low latitudes, showing velocities slightly lower than that of the atmosphere at higher latitudes.

Historically, in the absence of visible features, the Doppler shift of reflected solar spectral lines was first used to determine the rotation speed. In this technique, a spectrometer slit is laid across the diameter of the planet, and the spectrum is dispersed at right angles to the slit (Fig. V.61). If the slit lies along a meridional diameter, all points imaged by the slit will have the same radial velocity, and the solar Fraunhofer lines in the spectrum will form images precisely parallel to the slit. If the slit is placed along the equator, the opposite ends of the slit image will be Doppler shifted in opposite directions by the planetary rotation: the approaching limb is blue shifted, and the receding limb is red shifted. The images of the solar Fraunhofer lines will then be *tilted* relative to the slit, with the angle of tilt proportional to the rotation rate. By trying a number of slit orientations, the orientation of a previously unknown spin axis may be determined, as well as the sense of spin. In the unique case of Uranus, this technique does not always work: when the spin axis points at Earth *all* apparent diameters of the disk are meridians, and no Doppler shift can be observed. But of even greater concern was the fact that rotation periods determined in this manner for Uranus varied widely, from about 11 to over 24 h.

Photometric attempts to determine the rotation period of Uranus gave about 24 h, but this periodicity is suspiciously similar to that imposed by the viewing conditions on Earth. Arguments based on Uranus' gravitational moments and oblateness of its disk give a rotational period of about 16 h. Neptune shows strong diurnal brightness changes due to cloud structure, permitting a reliable photometric determination of the rotation period of about 18.3 h. Near-infrared images of Neptune taken within methane bands (to suppress scattered sunlight and emphasize regions with high, bright clouds) have been taken by Bradford Smith of the University of Arizona. These images clearly show large, bright cloud features that show the planet's rotation directly. But these observations refer to specific latitudes and altitudes in a dynamically complex atmosphere. The "true" rotation period of the planet is not that observed for any small and moving sample of the outer millionth of the planetary mass, but that of the planetary interior.

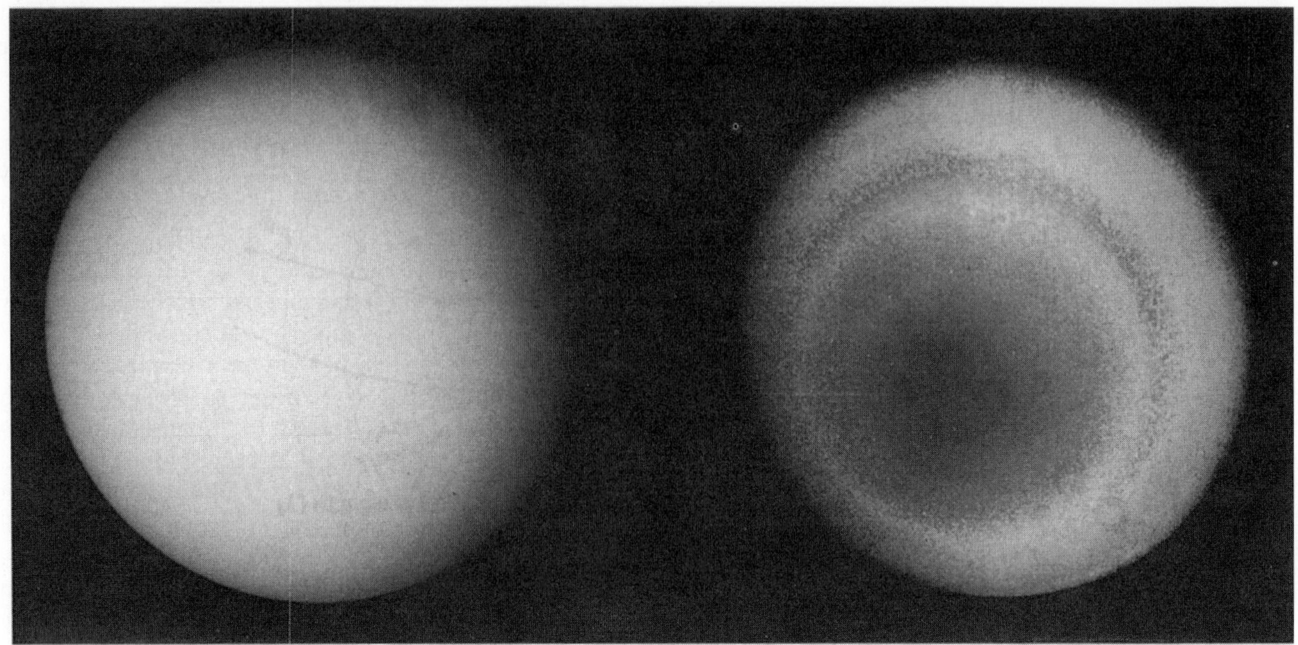

**Figure V.60** Voyager 2 views of the polar region of Uranus. The image at the left is a true-color picture. The image at the right has been computer enhanced to exaggerate color and intensity differences. A slightly darkened polar region and a faint circumpolar band can be seen in the enhanced version. Note that the illuminated hemisphere is almost centered on the pole.

Uranus was last nearly pole-on to the Sun (and Earth) in 1985. (Unfortunately, Uranus was very close to the central plane of the Galaxy at that time, which makes sensitive photometric studies very difficult.) With its 87-year orbital period, it was last in ideal geometry for study of atmospheric motions (subsolar point on the Equator) in 1964 and will next be so situated in 2006. The geometry of Uranus' spin axis and the Voyager 2 encounter is shown in Fig. V.62.

Voyager 1 was able to determine the rotation speed of the interior of Uranus, but not its atmosphere, by the periodicity of the magnetic field observed during Voyager's traverse of the Uranian magnetosphere. The measured interior rotation period (analogous to the Jovian System III rotation rate) was found to be $17.24 \pm 0.01$ h. All of the bright spots seen by Voyager are moving in the direction of the planetary rotation at speeds 50 to 200 m s$^{-1}$ faster than the speed given by the "System III" rotation period. The rotation period found by Voyager 2 for the interior of Neptune (from the periodicity in magnetospheric and radio emission phenomena) is $16.11 \pm 0.01$ h.

Thermal infrared observations of Uranus and Neptune in the 20-$\mu$m window curiously seem to show that Neptune is at least as warm as Uranus despite Neptune's much greater distance from the Sun: however, this conclusion is based on only a single narrow infrared wavelength interval. The measured 20-$\mu$m temperature for Uranus is $55 \pm 3$ K, whereas the rotating gray-body temperature for Uranus with an albedo of 0.35 is $57 \pm 3$ K. The infrared interferometric spectrometer (IRIS) experiment on Voyager, again with incomplete coverage of the thermal region, put an upper limit of 59.4 K on the effective temperature of Uranus and $59.3 \pm 0.8$ K on Neptune. The Planck peak for a 57-K black body is at 50 $\mu$m, but this is a very difficult spectral region in which to observe. In any case, the $57.2 \pm 1.6$ K 24-$\mu$m temperature observed for Neptune is different enough from the radiative steady-state temperature of $46.6 \pm 1.1$ K to be taken as strong evidence for an internal heat source. The emitted IR flux observed by Voyager 2 exceeds the absorbed solar flux by a ratio of 3.61($\pm$ 0.28):1 (total flux, 3.6 units per unit of reemitted solar energy; net internal heat flux, 2.6).

The generally close similarity in physical properties between Uranus and Neptune seems to suggest that, if either planet has an internal heat source, they both ought to. It is easy to accept the apparent internal heat source on Neptune of about 2.7 times the solar input as representing the same phenomenon we have observed on Jupiter and Saturn, presumably release of heat by slow collapse of the planet. In that case, it is Uranus that is anomalous. A possible explanation lies in the unique axial tilt of Uranus: the hemisphere presently being turned toward Earth has cold soaked for 42 years, whereas the hemisphere that has been heated strongly for the last few decades is now turning away from the Sun. We *expected* the thermal emission of Uranus might be anisotropic, especially if, as seems likely, heat transport across the equator is very inefficient.

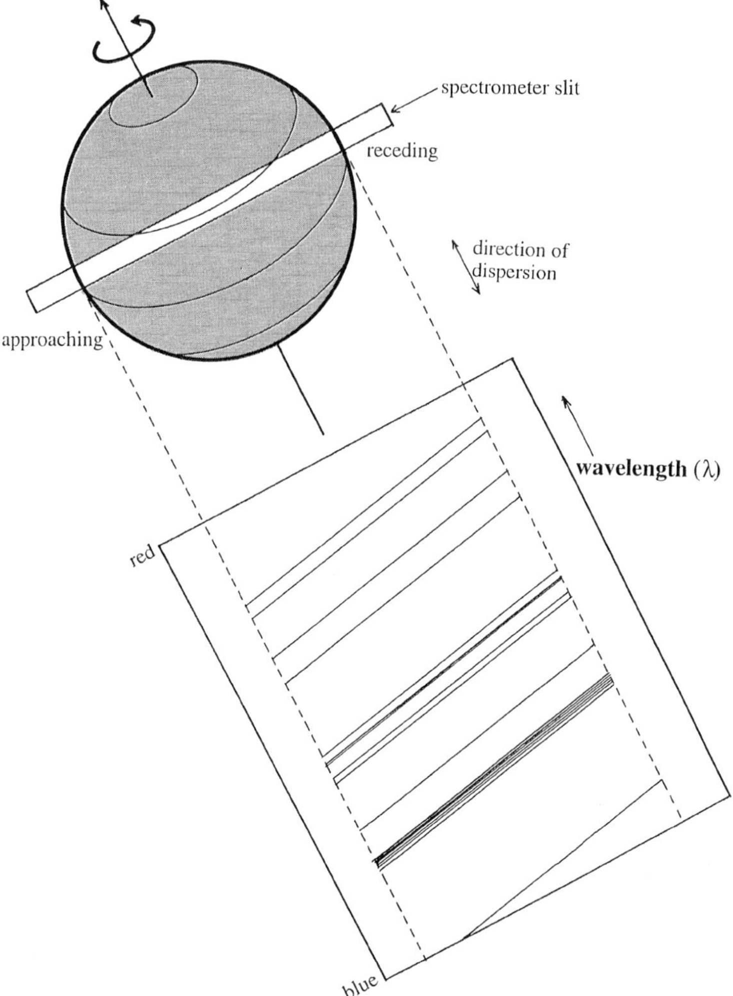


**Figure V.61** Spectroscopic determination of rotation rates. The spectrometer entrance slit is laid across the planet, and the light is dispersed by a prism or grating in the direction orthogonal to the slit length. The tilt of the spectral lines reveals the spin velocity of the planet. Note that this technique can be applied even to a featureless, unbanded planet by simply taking a series of spectra with different slit orientations.

We would have to observe Uranus' thermal emission from pole to pole in order to determine the true thermal luminosity of the planet—and we must do this all at one time. Simply looking at all geometries as Uranus proceeds around its orbit is obviously not enough. A flyby spacecraft (or orbiter) can provide the needed geometrical coverage to establish whether or not an internal heat source is present. Unfortunately, the IRIS experiment on Voyager was designed for use near the Planck peaks of the much warmer Jupiter and Saturn, and the spectral coverage was very incomplete. Nonetheless, the spectral coverage available was sufficient to show that the emission almost certainly cannot have the spectral shape of a Planckian emitter warmer than 59 K and cannot have large variation with latitude ($\Delta T < 2$ K). This casts serious doubt on the ability of the axial tilt to influence the overall thermal state of the planet in a major way.

Certain aspects of the atmospheric composition are important as guides to internal composition. However, attempts to determine the composition of the atmosphere of Uranus by infrared spectroscopy have been limited by several factors. First, Uranus is in visible light a sixth magnitude object, but even visually it appears blue–green because of encroachment of IR absorption features of $H_2$ and $CH_4$ deep into the red end of the spectrum. A low-resolution spectrum of Uranus is shown in Fig. V.63: the usually informative near-infrared region is nearly black due to absorption in the short-wavelength wings of the hydrogen rotational fundamental and in methane combination bands.

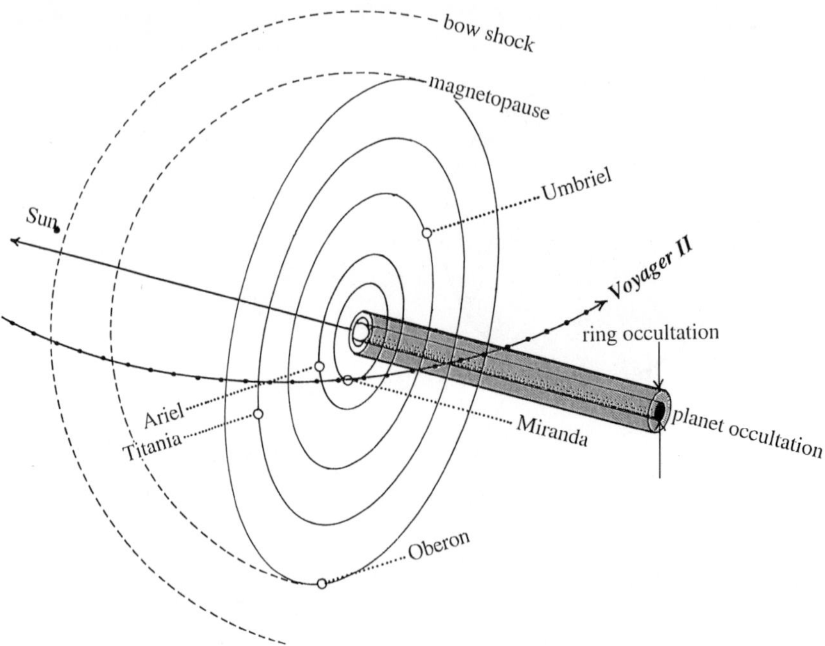


**Figure V.62** Geometry of the Voyager 2 Uranus flyby of January 1986. The pole of the satellite system and the rotation axis of the planet point almost at the Sun, and Voyager approaches a "bull's eye" of satellite orbits. With such geometry, only one satellite may be flown by at close range (in this case, Miranda). The magnetosphere extends Sunward about 18 Uranus radii. After its closest approach to Uranus and Miranda, Voyager flew through the shadow of the planet and was also occulted as seen from Earth.

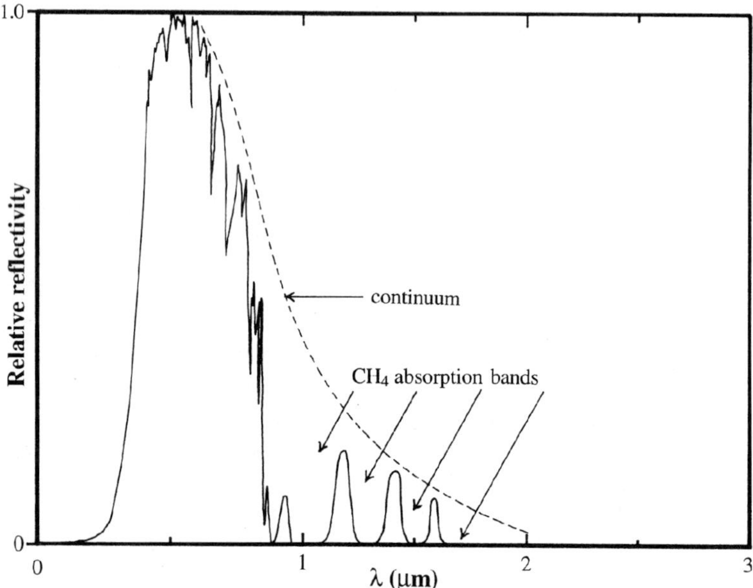


**Figure V.63** Reflection spectrum of Uranus. All of the strong absorption bands seen in the visible and near infrared spectra are produced by methane, although the wings of the hydrogen collision-induced dipole absorption also contribute to the broad absorption in the infrared. Photographs of Uranus in the methane bands show only a black planet, but similar images of Neptune show bright regions, which can be due only to high-altitude clouds or haze.

Fortunately the infrared spectra of the Jovian planets acquired have given us useful constraints on the helium abundance through the indirect effect of helium on producing collision-induced absorption by hydrogen. The helium mole fraction determined in this manner for the atmosphere of Jupiter is $0.180 \pm 0.040$. The Galileo Probe mass spectrometer and dedicated helium analysis experiment have revised this number to $0.156 \pm 0.010$. The best present estimate for Saturn $0.16 \pm 0.05$, compared to $0.26 \pm 0.05$ for Uranus, $0.32 \pm 0.05$ for Neptune, and 0.27 for the Sun. A conservative interpretation of these data would attribute the atmospheric helium contents of Uranus and Neptune to direct capture of nebular gases of solar composition. Jupiter and Saturn, on the other hand, seem to require some separation of hydrogen and helium. As we have seen, theory suggests that exsolution of helium from metallic hydrogen at very high pressures and low temperatures should be moderately effective in Jupiter and Saturn, resulting in the precipitation of a dense helium-rich fluid layer. Uranus and Neptune lack pressures high enough to metallize hydrogen in their gaseous envelopes. Neon, which the Galileo Probe found to be depleted by about a factor of 10 relative to solar abundances, would readily dissolve in the dense helium layer. Interestingly, the more polarizable heavy noble gases, argon, krypton, and xenon, are actually *enhanced* in the Jovian troposphere by a factor of 2.5 or 3 above their solar proportions. This is about the same enrichment factor as found for carbon and sulfur. Phosphorus (in phosphine) and germanium (in germane) may be slightly above solar proportions, but probably not by as large a factor as C or S. There remains the question of whether water vapor is similarly enriched. "Cometary" ices far from the Sun, bearing clathrate hydrates of the heavy noble gases, are the likely source, but require tens of Earth masses of very cold ice in Jupiter, probably too cold to have ever been stable at Jupiter's present distance from the Sun. Scenarios involving inward migration of the Jovian planets during the accretionary era are compatible with these observations, as is the possibility that growing proto-Jupiter swept up and swallowed one or two Uranian planets in eccentric orbits. An *in situ* analysis of the noble gases in the atmosphere of Uranus or Neptune, which would permit a direct test of this idea, is not presently included in any nation's future plans for Solar System exploration.

As in the case of both Jupiter and Saturn, estimates of the methane and hydrogen abundances have been made for the purpose of determining whether the H:C ratio is nearly solar. Earth-based IR spectroscopy of Uranus estimates the $CH_4$ abundance at 30 to 100 times larger than the mole fraction of methane found on Jupiter. The hydrogen abundance deduced from weak quadrupole features in the visible region is also immense. The atomic H:C ratio is 40 with a probable error of no more than a factor of 2, implying a $H_2$:$CH_4$ number ratio of 70:1 (i. e., 1.4% methane by volume), compared to the solar H:C ratio of about 2800:1. Thus methane appears to be enhanced by 20× relative to its solar abundance. It must be emphasized, however, that this is a very tricky observational and interpretive problem, and this factor could be significantly in error. However, Voyager IRIS and radio occultation data confirm a Uranus methane abundance of about 2% with a probable error of less than a factor of 2. Methane cloud condensation complicates the interpretation of the data, but a similar methane enhancement on Neptune is very likely.

In our discussion of the interiors of Jupiter and Saturn, we compared spectroscopic determinations of the H:C and H:N ratios in the upper troposphere with theoretical equation-of-state studies of the structure and density of solar-composition planets and were able to show that the two approaches led to a mutually satisfactory model. In part, this procedure depends on the lack of ambiguity with which "solar composition" can be defined, for it is the overwhelming prevalence of hydrogen and helium in Jupiter and Saturn that makes such an approach possible. If, as in the case of Uranus or Neptune, the density of the planet is definitely far too high for solar material, then we must ask which elements (compounds) are enriched and in what proportions. If we assume that only the rock-forming elements are enriched, then a relatively small fraction of the volume of Uranus or Neptune in the form of a rocky core could account for a large proportion of the overall mass. Then almost all of the *volume* of the planet could be solar material like Jupiter and Saturn. In the opposite extreme, *every* other element might be enriched relative to hydrogen. In this case, the closest one could come to matching the observed properties of Uranus would be a planet that was roughly 70% He, 20% ice, and 10% rock by mass, with hydrogen present only in minor quantities.

If we consider condensation–accretion scenarios that might have effected the necessary elemental fractionations, then we find that not all compositions are equally probable. The simplest scenario for formation of a gas-giant planet would postulate some form of instability in a gas of solar composition, leading to the one-step formation of a gravitationally bound gaseous protoplanet. After several billion years of radiative cooling and shrinkage, a solar-composition planet is found. But, because of the disturbing effect of the Sun's gravity, this scheme may not work at all, and it certainly cannot explain the radically non-solar compositions of Uranus and Neptune. In the second case, we can imagine two massive solid protoplanetary cores, composed of those substances that were condensed in the Solar Nebula at 20 AU, growing in the presence of the nebula to the point at which hydrodynamic capture of nebular gases begins. The core mass would be about the same for each Jovian planet, about 10 Earth masses. The capture does not go to completion before dissipation of the nebula by the ignition of the Sun. The finished planet is found to be enriched in all the condensed species appropriate to the local conditions in the nebula. A third, and more appealing, scenario would call for slow accretion of vast numbers of solid bodies so that only the largest two beyond Saturn were large enough to capture nebular gas. The minimum mass needed for such bodies to capture gas is about 0.1 Earth masses. These two bodies then assumed solar composition, but, after the dissipation of the nebula, large masses of solid material were brought in by capture of the hosts of smaller bodies that were large enough to avoid being blown away along with the nebula ($a > 1$ to 10 m), but far too small to capture $H_2$ and He themselves ($r < 1000$ km).

From Fig. IV.28 we saw that temperatures near 20 AU in the nebula should have been so low that complete condensation of $H_2O$ and $NH_3$ must have taken place. In the equilibrium condensation model, $CH_4 \cdot 7H_2O$ has also condensed everywhere beyond 7 AU. Solid methane begins to condense near 16 AU in both models. We have seen in Fig. IV.48 that, beyond 10 AU, dissipation of a continuous, nearly adiabatic nebula would lead to an *increase* in local temperatures, so that small bodies containing solid methane that condensed at 40 K might find themselves in a vacuum with a gray-body steady-state temperature of $\approx 50$ K, with a $CH_4$ vapor pressure of $10^{-5}$ bar. Evaporation of the solid methane might then occur at a rate limited only by the rate at which the latent heat of vaporization can be supplied by absorption of sunlight. We will treat this situation in detail in our discussion of comets. For now, it suffices to say that such small methane bodies will evaporate at the rate of centimeters per year at 20 AU.

Chemical histories for the solar nebula that take into account not only thermodynamic factors, but also the kinetic barriers standing in the way of equilibration (Chapter IV), suggest that, even far from the Sun, methane and ammonia ought to be minor components compared with CO and nitrogen. Thus retention and preservation of methane from the solar nebula ought to be of rather minor importance. Methane synthesis within a dense planetary subnebula, however, might be very important.

Because of the presence of a substantial methane atmosphere on Saturn's satellite Titan and in consideration of the high degree of methane enrichment in the atmosphere of Uranus, we will especially consider two types of compositional models for Uranus and Neptune. Both will assume that two distinct chemical components, one of them of solar composition and the other an equilibrium condensate, have been mixed to make these planets. We shall assume that the condensate was, in the first case, the equilibrium condensate after $CH_4 \cdot 7H_2O$ formation and, in the second, the equilibrium condensate after condensation of solid $CH_4$. We can then construct models in which the condensate:solar mass ratio is varied to match the observed density of the planet. The rotational moment of inertia of the planet, as measured by the $J_2$ coefficient in the spherical harmonic expansion of the planetary gravitational field, can then be matched by varying the radial distribution of condensate in the planet.

From the point of view of these radial distributions, there are several types of assumptions that might be made and tested. Of these, the one giving the largest principal moment of inertia will be a chemically homogeneous, single-layer planet, in which the composition is independent of position. Second, we could separate the gas (largely $H_2$ and He, but probably $CH_4$ as well) from the icy and rocky material, dividing the planet into two layers, a gaseous envelope and a homogeneously mixed ice-plus-rock core. Alternatively, we could imagine that, because of high internal temperatures, there is a rocky core and a fluid envelope of homogeneously mixed $H_2$, He, $CH_4$, $NH_3$, and $H_2O$, which undergoes fractionation (by condensation) only in a very thin superficial layer of the atmosphere, very near the visible cloud layer. Finally, we may have a family of low-moment-of-inertia models in which the permanent gases, ices, and rocks form three distinct layers.

With respect to the assumed thermal structure, the compositional and structural variations of these models are so immense that we need not be so careful about our assumptions. Perhaps the safest and most reasonable first guess, even for a Uranus with no large internal heat source, is an adiabatic thermal structure.

Two such models are illustrated in Fig. V.64 for both Uranus and Neptune. Such models have the heavy elements enriched $\approx 200\times$ relative to the Sun. In both cases, compositionally homogeneous (one-layer) models have much too large a value of $J_2$, whereas three-layer models have too small a $J_2$. We may imagine either a high mutual solubility of rocky and icy materials with a distinct atmosphere or, more probably, a rocky core with a deep ocean of supercritical water that merges gradually into the hydrogen- and helium-rich envelope, with polar gases such as ammonia and water fractionated into the denser fluid and nonpolar species such as the rare gases and methane fractionated into the atmosphere.

It is good to remember that there are polar compounds of several major elements in the interiors of these planets that have comparable abundances and significant mutual solubility at high pressures. As in our discussion of the elevation of the vapor pressure and solubility of phases subject to high pressures inside Jupiter, we must consider the real possibility that there are no sharply defined compositional discontinuities within Uranus and Neptune.

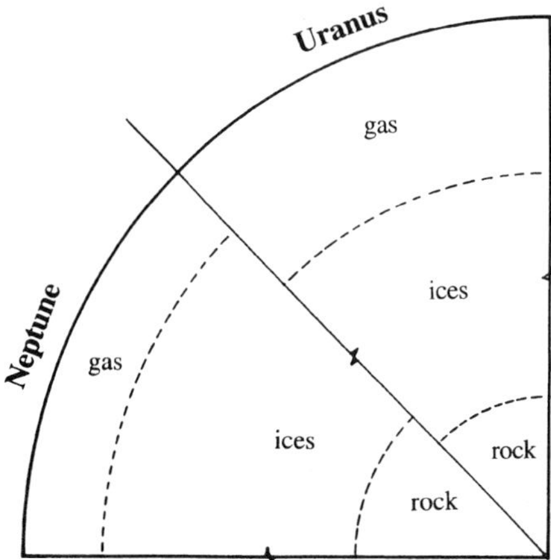


**Figure V.64** Schematic internal structures of Uranus and Neptune. Fully separated three-layer models have too small a rotational moment of inertia, whereas homogeneous models have too small a $J_2$. There must be a strong radial composition gradient, although there may not be well-defined interfaces between the layers.

Instead, the low-molecular-weight outer envelope may grade into a supercritical "ocean" of complex composition, which in turn grades into a dense ocean very rich in dissolved mineral species. The main limiting factor for the dissolution of minerals in supercritical water is the density: at planetary interior pressures, the density is actually greater than that of normal liquid water, and its solvent powers are proportionally enhanced. There is an important question not addressed by layered physical models: namely, is such a planet chemically stable? If we can picture the "ice layer" as a very hot, dense supercritical fluid rich in water and ammonia, then it is very likely that silicates will be rather soluble. Silica is well known to be soluble in high-pH solutions at high temperatures and pressures. For example, enstatite may undergo

$$H_2O + 2\,NH_3 + MgSiO_3 = SiO_4^{4-} + 2\,NH_4^+ + Mg^{2+}, \qquad (V.254)$$

and ions such as $Fe^{2+}$ and $Mg^{2+}$ may be dissolved as aquoammoniates such as $(NH_3)_2Fe(H_2O)_2^{2+}$. Aluminum, calcium, and alkali metal ions are all very soluble in basic ammoniacal solutions, as are nickel, cobalt, and other transition metals. Troilite may dissolve to form sulfide, sulfite, or even sulfate ions with possible release of hydrogen gas if oxyanions are formed. At the least, FeS will be reduced by hydrogen-bearing gases to make $H_2S$, which should be fairly soluble in ammoniacal water because of the formation of $NH_4^+HS^-$. There is of course great uncertainty whether the solubility of the mineral phases would suffice to dissolve them completely. Theory is unable to offer much help, because the chemical system contains at least six major components ($SiO_2$, $H_2O$, $NH_3$, MgO, FeO, and FeS) at temperatures and pressures for which relevant thermochemical data are not available. Even if the data existed, it is very difficult even to imagine (let alone draw!) an eight-dimensional phase diagram.

Whatever the detailed chemistry, there is one important aspect associated with the presence of rock-forming elements that should not be neglected: heating due to radioactive decay of $^{40}$K, $^{235}$U, $^{238}$U, and $^{232}$Th. The rock component generates about $S = 5 \times 10^{-8}$ erg g$^{-1}$ s$^{-1}$ of heat, and roughly 4 $M_E$ of rock should be present. The resulting heat flux from radioactive decay alone is thus

$$\Phi = SM_{\text{rock}}/4\pi r_U{}^2, \qquad (V.255)$$

or about 25 erg cm$^{-2}$ s$^{-1}$. Using the thermal conductivity of ice, about $10^6$ erg cm$^{-1}$ s$^{-1}$ K$^{-1}$, the temperature gradient needed inside Uranus to transport this heat flux outward at steady state is given by

$$\Phi = -K\nabla T = -10^6(\partial T/\partial z) = 25, \qquad (V.256)$$

for a temperature gradient of $2.5 \times 10^{-5}$ K cm$^{-1}$, or 2.5 K km$^{-1}$. Thus, over a depth interval of 10,000 km, the temperature would increase by 25,000 K! But the small abundance of radioactive nuclides would provide energy sufficient to raise the temperature from 0 K to only about 500 K, even if all the decay energy were stored internally. Also, quite irrespective of the nature of the internal heat source, the temperature gradient required to transport internal heat by conduction would require complete melting of the interior. All heat transport would then be by convection. This is true even though we have wholly neglected the possibility of a contraction heat source analogous to those of Jupiter and Saturn, which are $\approx$400 times larger.

If we imagine the emitted thermal flux from Uranus to be composed only of a reemitted solar component and a component from radioactive decay, a 57-K effective temperature due to radiative steady-state with the Sun (a flux of 600 erg cm$^{-2}$ s$^{-1}$) would be perturbed by only +0.6 K by the radioactive decay heat. On a planet with such incomplete spectral coverage of the thermal emission, radiometry is unable to detect so small a perturbation. However, a flux sufficient to keep the atmosphere convective is surely present, and has important consequences for atmospheric models.

Finally, studies of the magnetosphere and charged-particle environments of Uranus and Neptune have led to a very interesting insight into their internal processes.

Uranus was found to have a magnetosphere extending Sunward to a distance of 18.0 $R_U$. The bow shock created by the magnetosphere was first encountered at 23.7 $R_U$. The field is only roughly represented by a dipole, with higher-order terms contributing importantly, but approximates a dipole with a tilt of 60.1° to the rotation axis. The effective center of the dipole is offset 0.3 $R_U$ from the center of mass. The surface field strength varies from about 0.1 to 1.1 G, and the field strength integrated over volume is 0.23 G $r_U^3$. Thus the effective seat of the planetary field is in that region that we would characterize as a deep, chemically complex aqueous mantle, a probable good conductor of electricity. The plasma trapped in the magnetosphere has an energy density low compared with that of the magnetic field itself ($B^2/8\pi$), and hence distortion and "inflation" of the field by the plasma pressure, such as were so important with Jupiter, are not a factor at Uranus. A "cool" ($\approx 10$ eV) plasma component is dominant inside about $L = 6$, and a "hot" (1000 eV) plasma dominates farther from the planet. $L$ here, called the "magnetic shell number" by planetary physicists, is simply the distance from the center of the planet in units of the planetary radius. The plasma is dominated by $H^+$ and electrons, with $H_2^+$ also present to the extent of 0.1%. Heavier ions were not detected.

Neptune's bow shock was encountered by Voyager 2 at a distance of 34.9 $R_N$, and a "fuzzy" magnetopause was crossed near 26 $R_N$. The field was again, as for Uranus, found to be very complex and rich in quadrupole and octupole components. The surface fields, including the higher-order components, are modeled to range from about 0.1 to 1.0 G, and the integrated field strength comes to 0.133 G $r_N^3$. The tilt of the equivalent dipole was found to be 46.8°, almost as large as that of Uranus, with the center of the equivalent dipole offset by an astonishing 0.55 $R_N$ from the center of the planet, placing it well within the expected aqueous mantle. At the surface the higher-order terms in the field are larger than the dipole component; however, because the dipole strength drops off as $r^{-3}$ and the higher-order terms drop off at least as fast as $r^{-5}$, the dipole field dominates over almost the entire volume of the magnetosphere. The plasma in the magnetosphere contains about 1300 protons per $H_2^+$ ion (plus about 0.1 $He^+$). Some heavy ion, apparently $N^+$ from Triton, is seen higher in the magnetosphere. Radio observations from Earth had already established the existence of a trapped radiation belt about Neptune near $L = 7$. Below that altitude absorption of charged particles by the rings and satellites greatly reduces the particle populations.

Typical internal models for Neptune give an aqueous mantle mass of about 10 Earth masses. Clearly the partitioning of other species, such as ammonia, methane, and hydrogen sulfide, between the atmosphere and the ocean is of great importance in determining the physical properties and composition of both layers.

## Atmospheres of Uranus and Neptune

We have already mentioned that the only tropospheric species so far identified in the spectra of Uranus and Neptune are methane and $H_2$. The present estimates of the $CH_4$ abundance on Uranus are near 225 km amagat, with more apparently present on Neptune. The unit of gas density, the amagat (agt), is the density of a gas with just Loschmidt's number of molecules per cubic centimeter at normal temperature and pressure, $N/V = P/kT = 10^6/(273.16 \times 1.38 \times 10^{\neq 16}) = 2.65{\times}10^{19}\ \mathrm{cm}^{-3}$. The unit of gas column abundance, the cm agt, is then $2.65 \times 10^{19}\ \mathrm{cm}^{\neq 2}$. The methane abundance on Uranus is 3 km agt, and that on Neptune, $\approx 6$ km agt. Both planets thus appear to have H:C ratios greater than 140, compared with the solar ratio of 2800. Given the extremely low infrared temperatures of both planets, the absence of ammonia near the tropopause is not at all surprising. Somewhat more surprising is the absence of any clear absorption by ammonia in the microwave region, which probes the Neptunian atmosphere down to levels with temperatures near 300 K. The atmosphere of Uranus shows more opacity than Neptune at wavelengths of about 1 cm, again suggesting the presence of abundant ammonia on Uranus. Paradoxically, the Voyager radio occultation experiment suggests more ammonia opacity on Neptune, approximating saturation.

The reflection spectrum of Uranus, Fig. V.63, shows such strong effects of collision-induced dipole absorption by hydrogen that it is obvious from simple visual inspection that sunlight is penetrating without serious cloud scattering down to levels with pressures of several bars. Indeed, it is interesting to ask what a *pure* $H_2$ atmosphere, totally devoid of clouds and other absorbing gases, would look like from above. The only important effects to consider are Rayleigh scattering and collision-induced absorption. It is not hard to see that, if it were not for Rayleigh scattering, every incident photon would penetrate to a level at which it would be absorbed. It turns out, in fact, that even *with* Rayleigh scattering a pure $H_2$ planet would look much darker than Uranus in the red and infrared. Thus some additional scattering source is required to reflect more light back to space. It is reasonable to try to explain this in terms of cloud particles, because particles are such efficient scatterers and because Rayleigh scattering by other gases has the same wavelength dependence as that for $H_2$ and will not help to brighten the red end of the spectrum relative to the rest.

Small and poorly determined traces of acetylene in the upper atmosphere of Uranus have been found by ultraviolet observations, and there has been an infrared detection of ethane on Neptune. Both are surely, as on Jupiter and Saturn, products from the photochemical processing of methane. The Voyager ultraviolet spectrometer (UVS) experiment reports a high-altitude acetylene abundance of about $10^{-7}$ on both planets. The ethane abundance is about 25 (+22, − 7) times larger. Photochemical models for production and loss of $C_2$ hydrocarbons on Neptune predict an ethane:acetylene ratio less than 7:1, and the reason for the discrepancy is unclear.

The upper atmosphere of Uranus exhibits a UV airglow, stronger than that seen on Jupiter and Saturn, produced by electron bombardment of hydrogen, which produces about $10^{29}$ H atoms per second. Escape of about half of these fresh H atoms produces an extended hydrogen corona. It is speculated that the energy to maintain this excitation comes ultimately from upward-propagating momentum from tropospheric motions, similar to the theory for generation of the solar wind discussed earlier. The temperature of the upper atmosphere on both planets is maintained at about 750 K.

Besides $H_2$, helium, and methane, the only other constituent of the lower atmosphere so far detected on either planet is the HD molecule. The D:H ratio found is at least $2 \times 10^{-5}$ but not more than $4 \times 10^{-4}$, only slightly larger than the solar D:H ratio. If isotopic equilibrium were achieved during the condensation of ices at temperatures near 100 K, a much larger enrichment of deuterium would be expected in the icy component of both Uranus and Neptune. However, there are strong chemical kinetic arguments that the rates of the isotopic-exchange reactions at such temperatures are far too low to allow equilibrium to be approached.

We shall now make a simple first step toward modeling the atmospheric and cloud structures of Uranus and Neptune by considering how a solar-composition atmosphere would look at such low temperatures. Then we may add complications such as the enhancement of abundances of heavy elements, and we may use the available observational data to suggest improvements and changes in the model. The atmospheres are assumed to be adiabatic, and the 1-bar pressure level is set at 80 K for Uranus and 65 K for Neptune. The pressures and temperatures for condensation of $H_2O$, $NH_4SH$, $NH_3$, $CH_4$, and Ar, calculated for solar-composition atmospheres with 10-fold enrichment of condensibles, are shown in Fig. V.65. The cloud structures for Jupiter and Saturn are just those we encountered previously. At 20 AU, the tropopause temperature is low enough for methane condensation, whereas at Neptune's orbit the temperature would be low enough for condensation of solid argon if it were not for Neptune's internal heat source.

Because the spectroscopic observations of both Uranus and Neptune show large enhancements of methane, it is difficult to avoid the conclusion that the other less-volatile ice-forming materials (water, ammonia, and carbon oxides) are at least as strongly enriched in Uranus. We have calculated cloud-mass profiles for reasonable

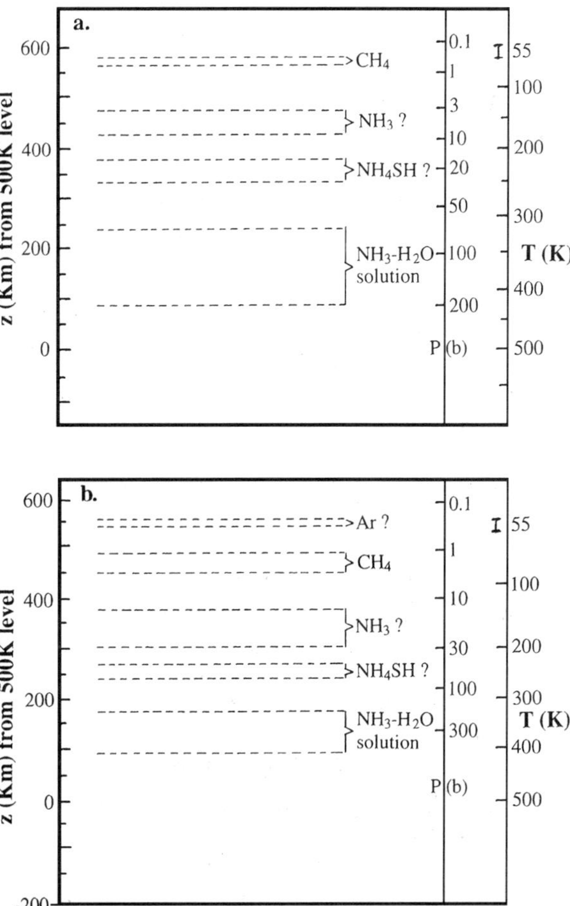


**Figure V.65** Cloud condensation on Uranus and Neptune. a (Uranus) shows a high methane cloud. The condensation regions for ammonia and ammonium hydrosulfide are marked as questionable for reasons discussed in the text. Deep, massive water clouds containing dissolved ammonia and hydrogen sulfide are suggested; they may lie atop a deep, hot ocean that makes up most of the mass of the planet. The Neptune cloud model (b) may differ in that a slightly denser methane cloud or a high-altitude argon haze may be present. It is such high clouds that are responsible for the detection of "weather" on Neptune.

Uranus and Neptune temperatures with a 10-fold enrichment of $H_2O$, $NH_3$, $H_2S$, $CH_4$, and Ar relative to solar abundances. The Uranus model, Fig. V.65a, shows a very deep massive cloud layer of aqueous $NH_3$ solution droplets, extending down to levels at which $P = 150$ bar and the temperature is $\approx 390$ K. Less dense, but still substantial, cloud layers of solid $NH_4SH$ and solid $NH_3$ crystals are found, and finally, 120 km above the $NH_3$ cloud base, a slightly less massive layer of solid $CH_4$ clouds is predicted. The scale height of the atmosphere at the $CH_4$ cloud level is $H = RT/\mu g = 30$ km, and the mass of hydrogen above this level is $M = P/g = 725\,\mathrm{g\,cm^{-2}}$. The mass of hydrogen above the methane clouds, $725\,\mathrm{g\,cm^{-2}}$, is equivalent to $725N_0/\mu(H_2) = 2.18 \times 10^{26}$ molecules per square centimeter, or 82.4 km agt. Similarly, the column density of hydrogen above the $NH_3$ cloud base is 1240 km agt. Methane, with 10 times its solar abundance (now at H:C = 280), would have a column abundance of 9 km agt above the ammonia cloud base and less than 0.6 km agt in and above the solid-methane clouds. The infrared temperature clearly does not refer to a layer of ammonia clouds, but could quite reasonably be attributed to the solid ammonia clouds. However, if the methane clouds were optically thick at both visible and IR wavelengths, then the observed amounts of $H_2$ and $CH_4$ (225 and 3 km agt) may be explainable as due to multiple scattering in the $CH_4$ clouds. Alternatively, the estimates may be somewhat low and may refer to light penetrating through a thin methane haze to a well-defined "reflecting layer" cloud top of ammonia crystals.

The problem becomes further diffused when we turn to the radio-wavelength thermal emission of Uranus for clues to the vertical distribution of ammonia. These data, shown in Fig. V.66, show temperatures at wavelengths of 1 to 10 cm that, curiously, are *higher* than the temperatures of Jupiter and Saturn at the same wavelength! The main source of opacity at $\lambda \approx 3$ cm is obviously not Rayleigh scattering: 3000-Å light penetrates to 2 bar, so 3-cm radio waves ($3 \times 10^8$ Å!) should penetrate to $(10^5)^4 = 10^{20}$ times as high a column density! Particle scattering is effective only if $\lambda \approx 2\pi a$, so that particle circumferences of $2\pi a = 1$ cm are required. This is possible only deep in the heart of the water clouds because of the fragility and rapid fallout speed of such large drops. The opacity above that level is in fact largely due to the inversion transition of the ammonia molecule. The $NH_3$ molecule has three H atoms in a triangle, with the N atom above the center of the triangle to one side of the plane. The energy required to shove the N atom through to the other side of the plane ("invert the umbrella") corresponds to $\lambda = hc/E = 1.2$ cm. We saw in Fig. IV.38 that the thermal motion of a gas at 100 K is sufficient to perturb seriously transitions in the thermal infrared during collisions. The very-low-energy $NH_3$ inversion transition is therefore *extremely* vulnerable to collision broadening. At pressures of only a few hundred millibars the absorption feature at 1.2 cm becomes badly broadened. This supplies continuum opacity throughout the microwave region. It is therefore somewhat startling that theoretically calculated emission spectrum of Uranus in this region can be made to resemble the observed spectrum only if ammonia is essentially absent above the 210-K level! This is precisely the behavior we would expect if $H_2S$ were more abundant than ammonia in the lower atmosphere, and $NH_4SH$ cloud precipitation exhausted the supply of $NH_3$ so that some $H_2S$ survived above the $NH_4SH$ clouds instead of ammonia, as on Jupiter. The hydrogen sulfide would then either condense (at 100 K) or be depleted badly by photolysis, depending on the relative rates of upward turbulent mixing by eddy diffusion and photolytic destruction of $H_2S$. In this case, Uranus would have a sulfur haze or $H_2S + H_2S_x$ cloud layer rather than an ammonia cloud layer, and the observed $H_2$ and $CH_4$ abundances may refer to the gas column above that layer.

An excess of $H_2S$ over $NH_3$, if real, would have important cosmogonic repercussions. The solar sulfur abundance (Table II.4) is roughly one seventh of the N abundance. Addition of a condensed component to a solar-composition component cannot increase the S:N ratio in the planet unless the condensed material is free of ammonia. This requires a nebular condensation temperature of $T_{cond} \geq 130$ K, which is far too high for retention of methane. Thus a two-component compositional model cannot simultaneously both provide more $H_2S$ than $NH_3$ and provide a large enhancement of methane.

An alternative method of depleting ammonia at high altitudes is to consider the dissolution of almost all of it in deep, warm clouds of water droplets. This places no narrow or unusual constraints on the abundances of any condensible species.

Other ways of depleting ammonia in the atmosphere are not easily found, and the weight of the microwave evidence is substantial. This effect is also seen clearly on Neptune, which exhibits a brightness temperature of 225 K at $\lambda = 6$ cm. A theoretical cloud-density profile for Neptune with 10-fold enhancements of $H_2O$, $H_2S$, $NH_3$, $CH_4$, and Ar above their solar abundances is given in Fig. V.65b, showing a cloud structure similar to that of Uranus except for the possible addition of a thin haze layer of solid argon at the tropopause. The arguments we have presented above for Uranus apply to Neptune as well, as the rather fragmentary IR and microwave brightness temperature data in Fig. V.66b make clear. Figure V.67 compares the thermal structure of the atmosphere of Uranus, as deduced from Earth-based and Voyager 2 observations, with the best available thermal structure models for the Neptunian atmosphere derived from Earth-based infrared studies. The similarity of these structures is due to the presence of an internal heat source in Neptune, which would otherwise be markedly cooler as a consequence of its greater distance from the Sun.

We still have too little compositional data on the atmosphere of Neptune to be useful in constraining the condensation–accretion history of the outer planets. Until the Voyager Neptune encounter, our best source of information on the stability of condensates in the outer parts of the solar nebula was the satellites of the outer planets.

The physical and meteorological behavior of the atmospheres of Uranus and Neptune is also the subject of much interesting, but inconclusive, speculation. If, as we see, several bars of hydrogen can be reached by

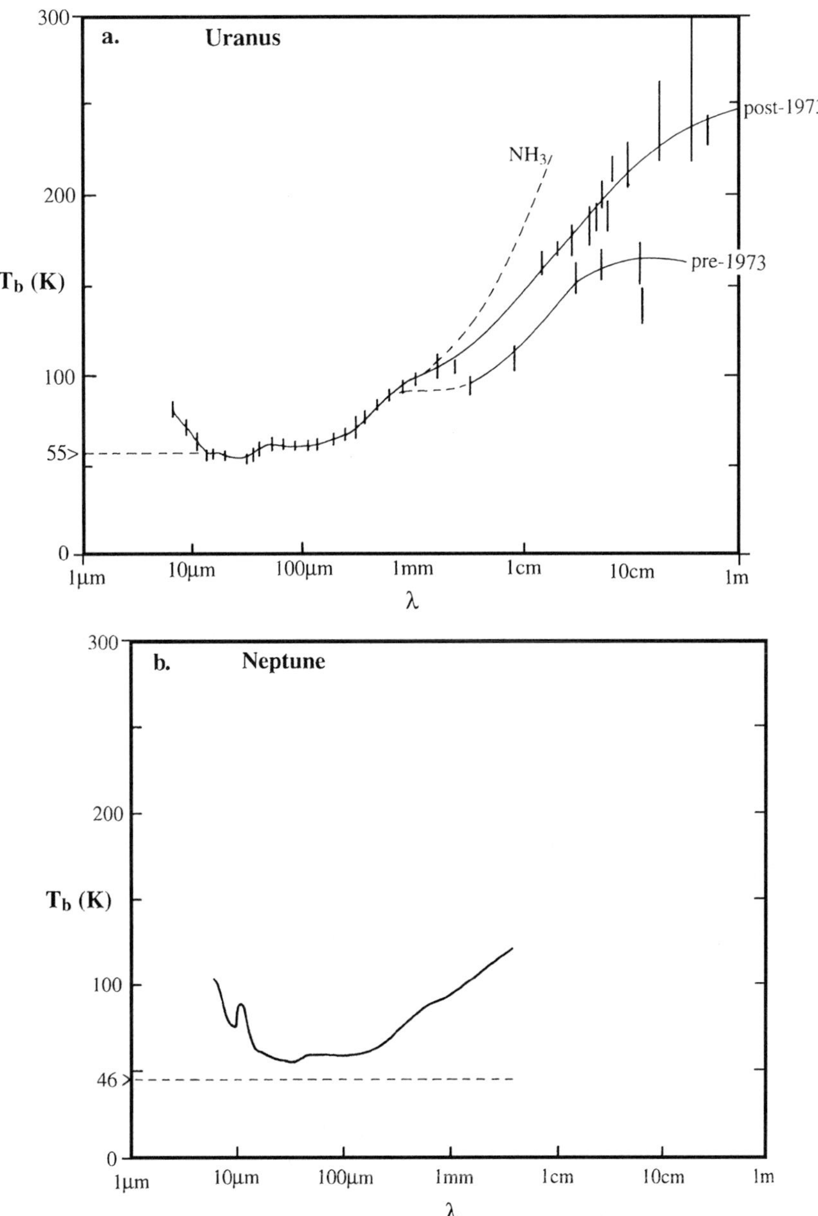


**Figure V.66** Infrared and radio brightness temperatures of Uranus and Neptune. a illustrates the thermal infrared temperature of about 55 K and the rapid rise of $T_b$ toward longer wavelengths. The effective temperature of the planet is dominated by the infrared because the Planck peak for a 60-K body is near 50 $\mu$m. Note the increase in the microwave brightness temperatures measured since 1973. The opacity in the microwave region is dominated by ammonia gas, and the spectral changes shown suggest a reduction in the ammonia abundance to about 1 ppm. b illustrates the greatly inferior data set for Neptune. The dashed line at 46 K illustrates what the brightness temperature of Neptune would be if the planet had no internal heat source.

visible sunlight, then roughly 60% of the incident solar flux will be absorbed by the gas. The solar constant at 19.18 AU is 3750 erg cm$^{-2}$ s$^{-1}$. The illumination averaged over the entire planet is exactly one quarter as much, and the absorbed energy is then $0.6 \times 3720/4 =$ 560 erg cm$^{-2}$ s$^{-1}$(0.56 W m$^{-2}$), corresponding to an effective temperature of 57 K. The amount of heat required to heat or cool a 1 cm$^2$ column of the visible atmosphere by temperature increment $\Delta T$ is just

$$\Delta H = C_p P \Delta T / \mu g, \qquad \text{(V.257)}$$

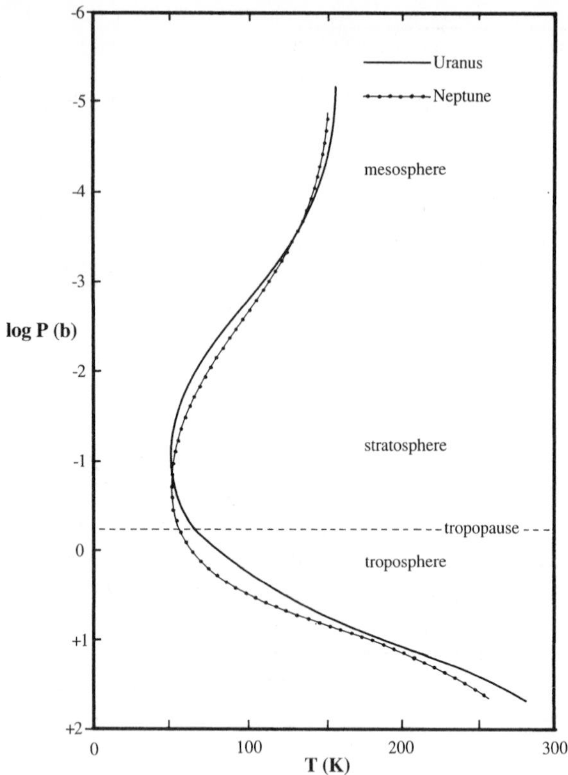


**Figure V.67** Temperature–pressure profiles for the atmospheres of Uranus and Neptune. The data for pressures less than about 1.5 bar are derived from Voyager observations, and the $T$ vs $z$ plot for higher pressures is calculated from an adiabatic model. Note that the temperature minima on the two planets are closely similar.

which, for a 10° change at the 1-bar level, is $10^{12}$ erg. The radiative time constant, $\tau_{\rm rad}$, is then

$$\tau_{\rm rad} = \Delta H/\phi_{\odot,\rm U} = C_{\rm p} P \Delta T/\mu g \sigma T^4, \qquad \text{(V.258)}$$

or $2 \times 10^9$ s, which is 70 years. Here $\phi_{\odot,\rm U}$ is the solar radiant flux at Uranus' distance from the Sun. Thus, over the 42-year-long polar night of Uranus, the temperature of the atmosphere down to the 8-bar level *at the most* changes by $42 \times 10/70 = 6$ K. We see that thermal emission at 57 K is so feeble that it cannot cool a large mass rapidly. If any large diurnal or seasonal temperature excursions occur, they must be restricted to the vicinity of the tropopause. The close correspondence between the observed thermal IR flux and that expected from a solid-$CH_4$ cloud layer suggests that even the dayside of the planet is shrouded with an optically thick (at thermal IR wavelengths) insulating blanket. Further thickening of this layer at night by radiative cooling of the tropopause region may not have much effect on the thermal IR emission. This in turn suggests that the effective temperature of Uranus may not vary much with phase angle and hence that the observed heat flux is typical, and Uranus has no large internal heat source due to contraction. If the thermal IR flux were found from Earth-based measurements to vary from one side of the visible disk to the other, with the newly "risen" side colder, then it would be plausible to think of sunlight "burning off" most or all of the methane haze to allow increased thermal IR emission. If such a phase effect is not found, then the case for a large (>100 erg cm$^{-2}$ s$^{-1}$) internal heat source is severely weakened. Indeed, no such effect is found in the infrared, and Voyager confirms that the effective temperature of the planet is remarkably uniform, to $\pm 2$ K. The microwave data on Uranus, presented earlier in Fig. V.66a, show an increase in the centimeter-wavelength brightness temperature over earlier years, reflecting an increase in the opacity of the atmosphere near the 200-K level. This opacity is located at too low a temperature to be associated with absorption by water vapor or scattering by large aqueous cloud droplets and is most reasonably attributable to a modest decrease in the mole fraction of ammonia vapor, from about $10^{-4}$ to $10^{-6}$.

An interesting complication was introduced by 1999 laboratory studies of the microwave absorption spectrum of phosphine. The opacity was larger than expected, suggesting the existence of an inversion mode in phosphine similar to the "umbrella mode" of ammonia. Phosphine is a nearly planar molecule, for which there was no previous evidence of an inversion mode. If this laboratory work stands up to scrutiny, phosphine may be an important contributor to the microwave opacity on all the Jovian planets, requiring a reinterpretation of the microwave data. The effect is likely to be largest on the planets with the greatest enrichment of heavy elements, Uranus and Neptune.

If we accept 59 K for Neptune as a valid estimate of its effective temperature, we require a total emitted flux of 700 erg cm$^{-2}$ s$^{-1}$, of which 270 erg cm$^{-2}$ s$^{-1}$ is re-emitted solar energy. We then could estimate the internal heat fluxes ($\phi_{\rm int}$) of Jupiter, Saturn, and Neptune as 5440, 2010, and 433 erg cm$^{-2}$ s$^{-1}$, respectively. There are at least two simple and purely empirical trends that could conceivably fit these numbers: one is that $\phi_{\rm int}$ drops off with the incident solar flux, $f_\odot/R^2$, and the other is that it is proportional to the planetary mass, $\phi_{\rm int} \propto M_{\rm P}$. Note that any correlation between the rate of liberation of internal energy and the rate of heating by the Sun would have to be attributed to unknown physics.

The proportionality of $\phi_{\rm int}$ to the planetary mass makes some qualitative sense; because the total gravitational collapse energy is proportional to $M^2$, the resulting heat per unit mass (which must be radiated away) is proportional to $M$. Still, the mechanism for determining the *rate* of heat release is not evident. Table V.3 illustrates how these two purely empirical rules work for Jupiter, Saturn,

**Table V.3** Internal Heat Sources in the Giant Planets

| $\phi$(erg cm$^{-2}$s$^{-1}$) | J | S | U | N |
|---|---|---|---|---|
| $\phi_\odot/4 = \phi_0$ | 12,700 | 3,800 | 940 | 380 |
| Absorbed: $(1 - A)\phi_0$ | 9,220 | 2,880 | 738 | 297 |
| Observed: $\phi_{IR}$ | 14,660 | 4,890 | 696 | 730 |
| $\phi_{int}$ | $5,440 \pm 430$ | $2,010 \pm 140$ | $42 \pm 47$ | $433 \pm 46$ |
| Model | | | | |
| $\phi_{int} \propto 1/R^2$ | (5,440) | 1,604 | 398 | 160 |
| $\phi_{int} \propto M_P$ | (5,440) | 1,600 | 250 | 293 |
| $\phi_{nuclear}$ | 5 | 6 | 25 | 25 |
| $T_{bb}$(calc) | $109.5 \pm 1.4$ | $82.4 \pm 0.9$ | $58.2 \pm 1.0$ | $46.6 \pm 1.1$ |
| $T_{eff}$(obs) | $124.4 \pm 0.3$ | $95.0 \pm 0.4$ | $59.1 \pm 0.3$ | $59.3 \pm 0.8$ |
| Geometric albedo | 0.274 | 0.242 | 0.215 | 0.215 |

and Neptune and shows the predictions of both rules for the magnitude of the internal heat source in Uranus.

The observed IR flux and the deduced internal heat source of Neptune are parenthesized to remind us that the observations do not span the peak of the Planck function. The models $\phi_{int} \propto 1/R^2$ and $\phi_{int} \propto M_P$ are normalized for the planet whose heat fluxes are best known, Jupiter. It can be seen that both empirical models give reasonable agreement with existing observations of Saturn and Neptune, and both predict an internal heat source for Uranus of $360-570\,\text{erg}\,\text{cm}^{-2}\,\text{s}^{-1}$, which we shall just call 500. Thus the average emitted flux of Uranus over all phase angles is predicted to be $1100\,\text{erg}\,\text{cm}^{-2}\,\text{s}^{-1}$, for an effective temperature of $66 \pm 1\,\text{K}$. If half the planet were thermally blanketed by a $55 \pm 1\,\text{K}$ emitter ($CH_4$ clouds), then the other half of the planet would have to emit $1700\,\text{erg}\,\text{cm}^{-2}\,\text{s}^{-1}$ to maintain a steady state, which requires an effective temperature of $74 \pm 3\,\text{K}$. Such temperatures and fluxes are, in light of Voyager measurements, quite impossible.

Assuming chondritic proportions for the heavy-element (ice plus rock) component of each of the giant planets, we can calculate a radioactive decay energy flux ($\phi_{nuclear}$) of $25\,\text{erg}\,\text{cm}^{-2}\,\text{s}^{-1}$ for Uranus, which is a lower limit for the heat flux through the lower atmosphere. An upper limit on the radioactive decay flux would be to assign the heavy element cores to rocky matter exclusively, which would yield a flux of $100\,\text{erg}\,\text{cm}^{-2}\,\text{s}^{-1}$ (and cause serious problems with the rotational moment of inertia and figure of the planet). We then apply Prandtl mixing length theory, as we did on Jupiter, to estimate the rate of vertical motions produced by this heat flux:

$$v_z = [R\phi/C_p\rho]^{1/3}. \qquad \text{(V.130)}$$

For an ideal, adiabatic, solar-composition gas near 100 K, with $T(P = 2\text{ bar}) = 100\,\text{K}$,

$$v_z = 100\phi^{1/3}T^{-1/2}. \qquad \text{(V.259)}$$

Near the 200-K level, the predicted vertical velocity for the above range of heat fluxes is from 20 to $35\,\text{cm}\,\text{s}^{-1}$. Because the velocities deduced from Prandtl mixing length theory are subject to errors of a factor of 10 or even 30, we can interpret this to mean that vertical velocities from $1\,\text{cm}\,\text{s}^{-1}$ to $10\,\text{m}\,\text{s}^{-1}$ are reasonable. With this caution in mind, we shall use $30\,\text{cm}\,\text{s}^{-1}$ for illustrative purposes.

The scale height at the 200-K level on Uranus, $H = RT/\mu g$, is about 90 km, decreasing to 30 km near the cloud tops. Writing the eddy diffusion coefficient as

$$K = H^2/\tau_{dyn} = v_z H, \qquad \text{(V.132)}$$

we calculate a typical value of $K = 3 \times 10^8\,\text{cm}^2\,\text{s}^{-1}$. The dynamic time scale for vertical mixing over one scale height is then $\tau_{dyn} = H/v_z = \text{H}^2/K$, or $3 \times 10^5$ s. Reinserting error limits of a factor of 30, this yields dynamic overturn times of a few hours to 100 days, centered near 3 days.

An interesting dynamical puzzle is posed by the observation that the *ortho:para* ratio for molecular hydrogen near the clouds of both Uranus and Neptune is close to local equilibrium. It seems clear that the cloud particles are acting as catalysts for the conversion, probably through the agency of photochemically produced unpaired electrons on the grain surfaces. The energy released by equilibration is an important factor in the heat budget of the atmosphere, and how and where it is released have obvious consequences for the dynamics of the upper troposphere.

Meteorological activity ("weather") on Neptune has been known for a number of years, but the Voyager imaging results provide infinitely more detail on the visual appearance of both Uranus and Neptune. Briefly, because the contrast over the disk of Uranus is only about 5%, it has proved difficult to detect meteorological phenomena even from Voyager (see the inset in Fig. V.68). A faint banded structure axisymmetric about the rotation pole can be discerned in contrast-stretched images. The topmost cloud layer (methane) has a base at about 1.3 bar and 81 K. The few spots and streaks observed seem to move zonally (east–west). The zonal velocity seems to decrease with altitude. No lightning or auroral activity was detected

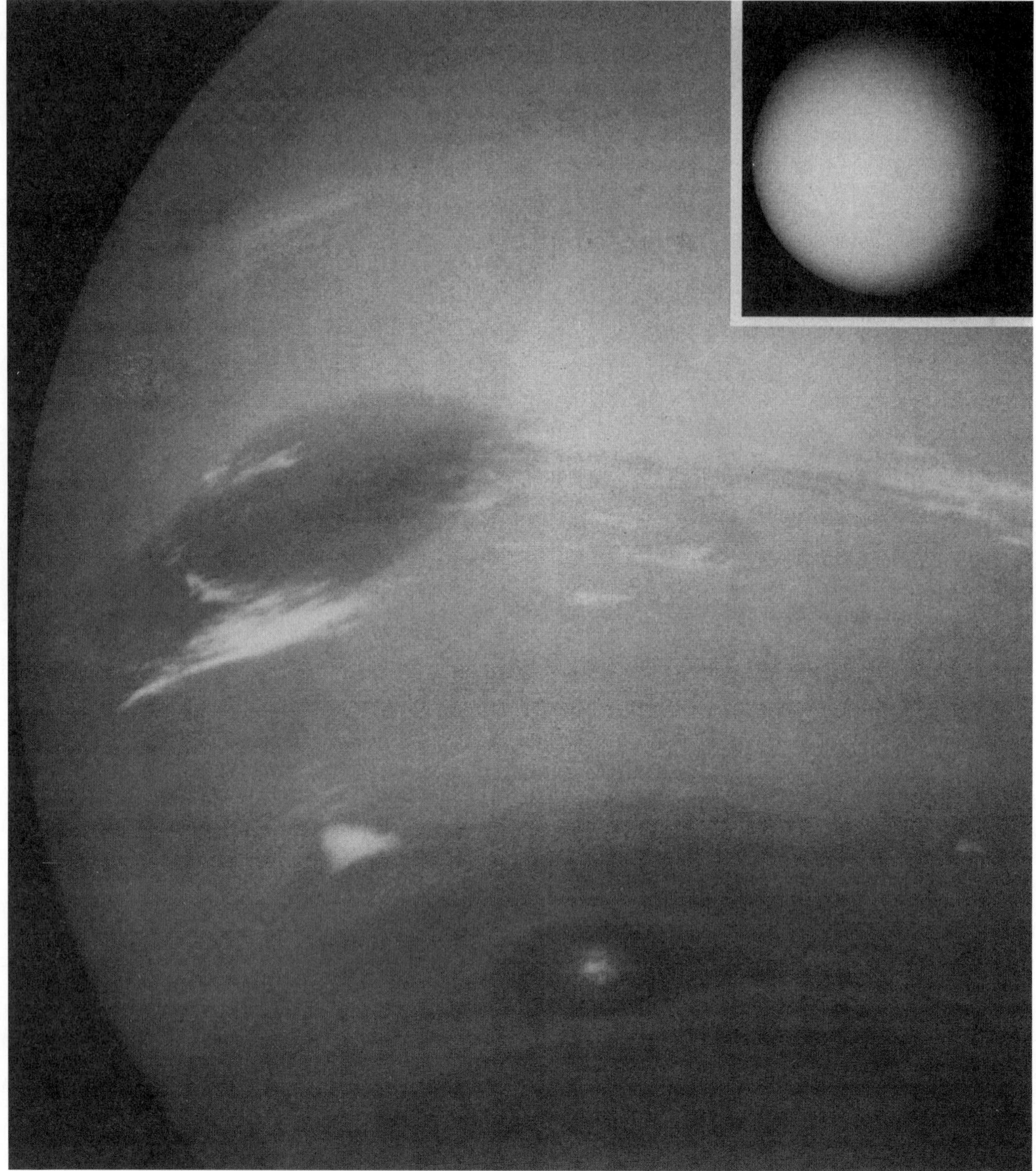

**Figure V.68** Voyager images of Uranus and Neptune. The day side of Uranus is shown in the inset. Contrast levels are so low that the faint banded structure is not discernible. The main figure is the day side of Neptune as seen by the Voyager 2 spacecraft, with the Great Dark Spot (GDS) at center left, a little dark spot (with a bright core) at the bottom, and the wedge-shaped Scooter between them. Note the bright streaks of high-altitude clouds.

by the imaging experiment, although the radio science experiment detected whistlers, which are audiofrequency electromagnetic emissions caused by lightning. The name arises from the fact that the frequency-dependent refractive index of the upper atmosphere and magnetosphere disperses the originally impulsive signal into a sound like a whistle of steadily decreasing pitch.

Neptune proved far more interesting, as Fig. V.68 attests. The planet has the Great Dark Spot (GDS) at latitude 20° S, reminiscent in scale and shape of the

Jovian Great Red Spot. The GDS rotates in the counterclockwise sense and has associated streamers and high-altitude bright haze. A thin methane cirrus cloud (at $P = 1.5$ bar) overlays most of the planet at an altitude of 50 to 100 km above the main cloud layer ($P = 3$ bar), and even more tenuous photochemical haze layers are seen at higher altitudes. A small dark spot is seen at 55° S, and an irregular and variably shaped cloud formation called "the Scooter" is seen at 42° S. Winds appear to be zonal with speeds from 20 m/s (retrograde) to 325 m/s (prograde) near 70° S. Most of the planet is covered by a retrograde jet extending from about 50° S to 45° N. Meridional mixing would move gas with a smaller angular momentum content into the equatorial regions, leading to weak retrograde circulation. Weak nightside auroral emission was seen, and no images of lightning have been reported. Faint whistler activity was observed by the Voyager radio science experiment.

The photochemical time scale for production of organic molecules on Uranus is longer by a factor $(R_U/R_J)^2$ than that at Jupiter: a downward mass flux of $\phi_m = 4 \times 10^{-15}\,\mathrm{g\,cm^{-2}\,s^{-1}}$ of ethane and other hydrocarbons is expected at steady state. The expected concentration of photochemically produced organic matter in the troposphere is then $\phi_m/v_z = 10^{-10}\,\mathrm{g\,cm^{-3}}$, or a mole fraction $f(C_2H_x) = 10^{-17}$. All heavier organic species would be less abundant.

The photochemistry becomes substantially more interesting if $H_2S$ is more abundant than $NH_3$ in the lower troposphere. Then the entire solar flux from 1650 to 2700 Å can go into $H_2S$ photolysis, and inorganic chromophores may be produced. The probable greater concentration of methane makes hot-atom production of organics from $H_2S$ photolysis more likely by about a factor of 10, but the colder background gas will thermalize H a little more rapidly and partially offset this advantage. At about the 3-bar level, 2700-Å light is effectively blocked by Rayleigh scattering alone, so that such photolysis must take place near or below 100 K, close to the $H_2S$ saturation temperature. Hydrogen polysulfides, $H_2S_x$, are likely to be major photolysis products. There is, however, a sensitive upper limit on the amount of hydrogen sulfide present above the cloud tops on Uranus, at a level less than the actual cosmic abundance of sulfur. Thus, if ammonia is enriched by a factor of 20 on Uranus, and hydrogen sulfide is present in sufficient amounts to remove all the ammonia as ammonium hydrosulfide, an enrichment of hydrogen sulfide by a factor of 100 is required. If the factor were 99, then enough ammonia would survive ammonium hydrosulfide condensation to be readily observable; if the factor were 101, enough hydrogen sulfide would survive to be readily observable. The random achievement of such a perfect equality between ammonia and hydrogen sulfide is so improbable as to be unreasonable. Near-total extraction of both gases into a hot, dense "ocean" seems a much more reasonable explanation. Nonetheless, sufficient traces of hydrogen sulfide may be present near the tropopause to make its photolysis possible.

The minimum temperature in Uranus' stratosphere is near 55 K, a temperature so low that almost any conceivable gas will be condensed. This has serious consequences for the accumulation and photochemical reworking of even so simple a molecule as ethane. There may very well be a significant photochemical haze layer composed of solid ethane particles.

The presence of a radioactive heat source and the possible prevalence of an electrically conducting supercritical $H_2O$-rich fluid in the interior suggest the possibility of planetary magnetic field. When Uranus is pole-on to the Sun the solar wind impinges directly on the polar regions, where open field lines extend far out from the planet. Merging of the magnetospheric field with the solar wind field *must* occur over a substantial region, and solar wind ions will be efficiently funneled into the magnetosphere.

Even for weak planetary dipole moments, the bow shock can lie very far from the planet. The dynamic pressure of the solar wind at the orbit of Uranus is only $3 \times 10^{-11}\,\mathrm{dyn\,cm^{-2}}$, and, because the magnetosphere can stand off the solar wind when

$$B^2/8\pi = P_{\mathrm{dyn}} = (M_U^2/8)(R_U/R)^6, \qquad \text{(V.260)}$$

a volume-integrated magnetic field strength of $M_U = 0.03\,\mathrm{G}\,R_U^3$ will stand off the solar wind out to 10 planetary radii! A planetary field of $3\gamma (= 3 \times 10^{-5}\,\mathrm{G})$ $R_U^3$ will already produce a tiny magnetospheric cavity outside the atmosphere. But a large Uranian magnetosphere with low field strengths and with the plane of the magnetic equator very far from Earth is almost impossible to detect from Earth.

During the approach of Voyager 2 to Uranus, a bow shock was encountered at 23.5 $R_U$, and its magnetosphere was entered at 18 $R_U$. The magnetosphere departed very substantially from earlier expectations. After observing a Jovian field tilted by about 10° to the spin axis and a Saturnian field aligned within 1° of the spin axis, it came as a shock to find that the Uranian field is tilted by 60° to the spin axis, with the center of the effective dipole offset by 0.3 $R_U$ from the center of the planet. This unusual geometry suggests that the field is produced in an electrically conducting (ionized) aqueous solution rather than in a metallic core. As discussed earlier, the surface field varies widely, from about 0.1 to 1.1 G, as a direct consequence of the dipole offset.

**Figure V.69** The night side of Neptune. A thin crescent of hazy atmosphere is illuminated by the Sun, as seen by the Voyager 2 spacecraft as it departed the Solar System in August 1989 on its way into interstellar space. Note the large satellite Triton at right.

## Perspectives

So far in this chapter we have treated the Jovian planets as independent bodies. There are two important extensions of this discussion that are needed to place them in their context within the Solar System.

The first point is that these planets form a sequence of compositions with increasing heliocentric distance: the mass of the condensed (rock plus ice) component varies little, whereas the mass of uncondensible gases (largely hydrogen and helium) drops off dramatically from nearly 300 Earth masses at Jupiter to about 75 Earth masses at Saturn and a few Earth masses at Uranus and Neptune. There are few hints from the spectroscopic study of planetary atmospheres what compositional differences may be associated with the condensed component. It is reasonable to expect the elements that are intermediate in volatility between water and hydrogen to have abundances in the condensed material that increase radically as one goes outward from Jupiter. We see as evidence for such a trend the large atmospheric enrichment of carbon relative to hydrogen in Uranus and Neptune. Future Galileo entry-probe studies of Jupiter may extend our understanding of the fractionation processes that formed Jupiter from solar material, but the greatest breakthrough would be the determination of the atmospheric abundances of several elements on Uranus and Neptune. It will require a substantial improvement in our database to resolve the current debate about such fundamental issues as the relative bulk abundances of H, C, N, and O in Uranus.

The second area for fruitful exploitation is the satellite systems of the Jovian planets. By simultaneous study of these planets and their extensive and fascinating satellite and ring systems we may come to an understanding of the physical and chemical processes by which these systems formed. This exercise is closely analogous to the study of the origin of the Solar System; the analogy between the planetary system and the regular satellite systems of Jupiter, Saturn, and Uranus is provocative, and we would be remiss if we did not seek out the parallels among the mass distribution, compositional gradients, and evolutionary histories of all four of these systems. Accordingly, we shall now turn our attention to the satellites of outer planets. But as a last reminder of Voyager 2's profound contributions to the understanding of the inner three Jovian planets and in remembrance of its ultimate study of Neptune, we conclude this chapter with a picture taken by Voyager as it departed the Neptune system, heading outbound on the last leg of its epic voyage into interstellar space (Fig. V.69).

## Exercises

*Interiors of Jupiter and Saturn: Data*

V.1 A perfectly black spacecraft with an albedo of 0 and an emissivity of 1.00, orbiting a planet with no internal heat source and a temperature of 200 K at low altitudes, is passing over the subsolar point on the planet.

a. What is the total incident light flux on the spacecraft, including direct sunlight, reflected sunlight, and thermal emission from the planet?

b. What surface temperature would the spacecraft have in steady-state balance with this radiation field?

V.2 How does the observed thermal luminosity of Jupiter's internal heat source compare with that predicted by a literal application of the mass-luminosity relation for lower-MS stars [Eq. (II.62)]?

V.3 a. Calculate the central pressures of cold pure-hydrogen bodies with masses of $10^{26}$, $10^{28}$, $10^{30}$, $10^{32}$, and $10^{34}$ g, using the mass-radius diagram in Fig. V.3 as a first approximation.

b. Explain why this may be a significant underestimate of the real central pressures of cold solar-composition bodies with these masses.

*Thermal Models of Jupiter and Saturn*

V.4 Present in your own words the argument by which the observed heat flux from Jupiter is used to deduce the presence of a deep convective atmosphere. The heat flux from Jupiter's interior could in principle be carried by conduction, radiation, or convection.

*Tropospheric Composition and Structure: Theory*

V.5 The newly discovered planet Gwena has a noxious, torpid atmosphere composed (according to a popular theory) of industrial by-products. All that is known about the planet is that the atmosphere is rich in hydrocarbons, that the temperature at the 200-mb level in its atmosphere is 273.16 K, and the temperature on Gwena's surface is very close to 546.32 K. (The latter figures are known only because they are reported on their 11 p.m. news broadcasts.) The four most popular theories among terrestrial planetologists (who agree that the atmosphere is probably adiabatic) are that Gwena's atmosphere is composed of (i) 50% by volume Xe and 50% methane, (ii) pure propane ($C_3H_8$), (iii) pure tetraethyl lead $Pb(C_2H_5)_4$, and (iv) 50% by volume He and 50% methane.

a. What is the surface atmospheric pressure on Gwena for each model?

(i) 50% Xe and 50% methane
(ii) propane
(iii) tetraethyl lead
(iv) 50% He and 50% methane

b. In what important respect will atmospheres (i) and (iv) differ?

V.6 Consider the dissociation equilibrium of hydrogen, $H + H \leftrightarrow H_2$.

a. On a plot of log $P$ versus $-1/T$, covering the pressure range from $10^{-8}$ to $10^4$ bar, find the line of equal abundance (partial pressure) of H and $H_2$. (Solve for at least three temperature points.)

b. To the left of this line (lower pressures), write the expression for log $f(H_2)$ as a function of log $P$ at constant $T$.

c. To the right of this line (higher pressures), write the expression for log $f(H)$ as a function of log $P$ at constant $T$.

d. Sketch a graph of the mole fraction of H and the mole fraction of $H_2$ as a function of log $P$ over the stated pressure range, assuming constant $T$.

V.7 Suppose that iron condenses as liquid droplets at the level on Jupiter where the temperature is 2700 K and the pressure is $3 \times 10^4$ bars.

a. What is the scale height of the atmosphere in this region?

b. Given that the eddy diffusion coefficient is about $10^9\ cm^2\ s^{-1}$ at this level, what are the mean vertical wind speeds there?

c. Liquid iron droplets may coagulate readily on contact. How large would these droplets have to be in order for them to fall even opposed by the mean vertical wind speeds calculated above?

### *Cloud Condensation in the $NH_3$–$H_2O$–$H_2S$ System*

V.8 a. Why is it that, when Fig. V.14 is redrawn for the case $P(\text{total}) = p(H_2O) + p(NH_3)$ instead of $P = 1$ atm, the appearance of the figure is essentially unchanged?

b. One new line, the boiling curve ($P_v = 1$ atm) is added to the figure when it is redrawn. Using the data in Fig. V.15, sketch on Fig. V.14 the approximate location of the boiling curve.

V.9 The boundary between solid ammonia and diammonia hydrate in Fig. V.15 defines a buffer relation. Show how one of the gas pressures is directly buffered by the coexistence of this pair of phases, and show how the pressure of the other gas is also fully determined.

V.10 Prove that, in a laboratory vessel in which solid ammonium hydrosulfide and solid ammonia monohydrate coexist, the ratio of the water vapor partial pressure to the hydrogen sulfide partial pressure is a function of $T$ alone.

V.11 At the surface of the Earth-sized planet Awnge about 10% of the violet (0.45 $\mu$m) radiation from its Sun is scattered during transit through its atmosphere.

a. If the atmosphere has a pressure of 2 atm at the surface and the mean molecular weight is 35, estimate the Rayleigh scattering cross-section for the atmospheric gas mixture.

b. Estimate the fraction of incident red (0.90 $\mu$m) light that is scattered by passing through the same atmosphere when the Sun is 30° above the horizon. (*Hint*: Treat the atmosphere as plane-parallel, and neglect the curvature of the planet.)

V.12 An isothermal layer of pure "snow" (a solid atmospheric condensate of unspecified composition) initially consists mostly of 0.1-$\mu$m particles for which $p/p_0 = 1.20$. After a time, this layer will recrystallize into larger particles with slightly lower vapor pressures.

a. Recalling the approximation that, for small values of $x$, $\ln(1 + x) = x$, calculate the change of vapor pressures $\delta p/p_0$ of particles with sizes from 0.1 $\mu$m up to 1 cm.

b. Empirically, we find that very old glacial ice has extensively recrystallized to make centimeter-sized crystals. Why might the recrystallization process become ineffectual at sizes of 1 cm or greater?

### *Ion Production in the Jovian Atmosphere*

V.13 A massive water cluster ion, $H_{11}O_5^+$, encounters and captures a free electron in Earth's lower stratosphere, a region that is close to saturation with water. Explain what should happen as a result of neutralization of the cluster-ion charge.

V.14 The H Lyman-alpha (Ly-$\alpha$) emission line in the UV spectrum of the Sun (1216 Å) illuminates the upper atmosphere of Jupiter.

a. What species present above the ammonia clouds (Fig. V.17) could be ionized by this radiation?

b. Which of the species present in this region might be ionized by $He^+$ Ly-$\alpha$radiation from the Sun (304 Å)?

c. Even if the $He^+$ Ly-$\alpha$ line were *not* absorbed by chemical reactions at high altitudes, it will not penetrate to the ammonia cloud level. Why?

### *Visible and Infrared Radiation Transfer*

V.15 Interstellar space is pervaded by a very tenuous, hot gas. Suppose that a particular region of space has an electron density of 1 particle per cubic centimeter. The Thomson scattering regime in

Fig. V.24 applies to this gas. In principle, light traversing a great enough distance through the interstellar gas will be seriously scattered.

a. If we observe a globular cluster on the opposite side of our Galaxy (but high enough above the galactic plane so that we do not need to worry about intervening dust clouds) at a range of 10 kpc, what fraction of the 0.5 $\mu$m light from that cluster would be scattered before reaching the Solar System?

b. How much reddening would result from this scattering? That is, compare the attenuation at 0.5 $\mu$m in the visible to that at 1.0 $\mu$m in the near IR.

V.16 Suppose that, as a result of a low-latitude impact event, a huge mass of fine black dust is injected into the ammonia cloud layer on Jupiter, sufficient to lower the albedo of the clouds over the affected region to 0.5. This black dust can absorb equally well at all wavelengths. Estimate the radiative time constant for the affected region and compare it to that for an unaffected region of the same cloud layer.

V.17 A hydrogen plasma consisting of fully ionized atomic hydrogen at a temperature of $10^6$ K occupies the lower corona of the Sun. Estimate the mean atomic weight, heat capacity, ratio of specific heats, and speed of sound in this region.

V.18 Recent observations of the cloud layer temperatures and colors on Jupiter have suggested that no more than a few percent of the area of the planet is covered by descending columns of gas. Taking a typical estimate of 97% upwelling (cloudy) and 3% subsiding (clear) areas, estimate the mean vertical wind speed in the subsiding regions.

V.19 Why is the curl of every gradient equal to zero?

V.20 Two adjacent cloud bands on Jupiter differ in temperature by 5 K at the same altitude. The temperature change occurs over a 1000-km-wide band near 20° N. How strong a wind shear would be developed by this temperature difference?

*Photochemistry and Aeronomy*

V.21 A "bubble" (called a "parcel" by meteorologists) of warm gas near the equatorial tropopause of Jupiter rises to—and overshoots—the tropopause, falls back, and oscillates vertically until its motion is damped out by friction.

a. Why does the rapidly rising parcel stop rising in the stratosphere?

b. Assuming conservation of angular momentum, how large an east-west velocity difference would be caused by a 5-km overshoot into the stratosphere?

V.22 Using dimensional analysis, derive the functional form of the Prandtl mixing length equation, expressing $v_z\,(\mathrm{m\,s^{-1}})$ as a function of $\phi\,(\mathrm{J\,m^{-2}\,s^{-1}})$ and $\rho\,(\mathrm{kg\,m^{-3}})$.

V.23 a. By reference to Fig. V.17, V.36, and V.37, offer some general conclusions about how effective the gases in V.37 are in shielding one another from photolysis. (*Example*: Hydrogen overlies ammonia in enormous quantity, but is incapable of shielding ammonia from photolysis to any significant extent because hydrogen does not absorb over the wide wavelength interval from 1000 to 2350 Å in which the Sun is very bright and ammonia absorbs strongly.)

b. Water has an absorption spectrum intermediate between those of methane and ammonia in Fig. V.37. What gases, if any, are capable of shielding water vapor from solar UV photolysis on Jupiter?

V.24 Because of diffusive separation, the methane abundance will drop off dramatically at high altitudes, above the turbopause, much in the way that the abundance of ammonia drops off at its level of condensation (Fig. V.17). Assume that diffusion is faster than turbulent mixing at gas densities of less than $10^{12}\,\mathrm{cm^{-3}}$, and that methane, hydrogen, and helium are present in their solar proportions at that level. Sketch their density profiles for the region from $n = 10^{13}$ to $10^{10}\,\mathrm{cm^{-3}}$.

V.25 Sketch or describe in words the locations of the eight electrons in both the singlet and triplet methylene radicals.

V.26 Thermalized H atoms from the region of $H_2S$ photolysis on Jupiter are capable of removing H atoms from $H_2S$ [Reaction (V.196)], but the "fresh" unthermalized H atoms from $H_2S$ photolysis, which are capable of attacking methane, phosphine, and ammonia, are ineffectual because they are so quickly thermalized. But the same trace gases in an argon atmosphere would be much more conducive to hot-atom reactions. By means of arguments similar to those given just before Eq. (V.205), compare the cooling of a 3000 K H atom in a pure-hydrogen gas to cooling in $^{40}$Ar. Give the effective kinetic "temperature" ($T \propto v^2$) as a function of the collision number for both cases, ending with the thermalization of H to an effective temperature of 300 K.

V.27 The estimated abundance of photochemically produced organic material in the water-cloud layer on Jupiter, $10^{-11}$ g/g, contains a small fraction of condensable, water-soluble products such as $CH_3NH_2$. Very generously assuming that 10% of the organic matter is soluble in water, estimate the concentration of dissolved organics in the water-cloud layer and the mass of these organic "nutrients" in a 10-$\mu$m raindrop.

*The Jovian Thermosphere*

V.28 a. Suppose that an M-class MS star with a luminosity of $10^{-4}$ Suns has a photospheric temperature of 3000 K and an even cooler chromosphere. How would you expect the electron density in the ionosphere of a Jupiter-like planet 5.2 AU from that star to differ from that in Jupiter's ionosphere?

b. During a flare the luminosity of the M star doubles, and 20% of the total luminosity emerges as ultraviolet light. Redo the calculation in part a for flare conditions.

*Radiophysics and Magnetospheres of Jupiter and Saturn*

V.29 A radio telescope on Earth observes Saturn in two perpendicular polarizations. The following flux density measurements are obtained:

| | **Flux Density B (FU)** | |
|---|---|---|
| **Frequency (MHz)** | **Polarization 1** | **Polarization 2** |
| 300 | $4.4 \times 10^{-2}$ | $4.0 \times 10^{-2}$ |
| 1,000 | $9.2 \times 10^{-2}$ | $8.4 \times 10^{-2}$ |
| 3,000 | 0.24 | 0.22 |
| 10,000 | 2.1 | 2.1 |

a. For which frequency do we expect the measured brightness temperature to be most representative of the thermal emission from the disk of the planet? Why?

b. All of the above decimetric observations were made at a distance from Saturn of $\Delta = 8.5$ AU. Assuming that all the radiation is emitted uniformly from the disk of the planet, calculate the brightness temperature at each of the observed frequencies.

c. The observed infrared brightness temperature of Saturn (at 10 $\mu$m) is $T_b = 95$ K. On a graph of log $B_\lambda$ versus log $\lambda$ (from 1 to 100 cm), plot the observational data given above and draw in the expected behavior of $B$ for a black body emitter at 95 K. Explain the discrepancy between these two curves.

V.30 Explain clearly why electrons orbiting at 16 kHz in a magnetic field can produce strong emission at frequencies of several hundred MHz.

V.31 Distinguish clearly between the magnetic shell number $L$ and the distance from the planet in units of $R_s$. Refer to Fig. V.51 for convenience.

V.32 Write a computer program to evaluate the integral in Eq. (V.240). Verify that the approximations mentioned in the text are valid.

V.33 Figure V.57 shows an enhancement of the magnetic field of Saturn beyond $10R_S$, due to enormous quantities of trapped charged particles in the radiation belts. An otherwise astute chemist, upon hearing of this phenomenon, responds that the excess field strength might be caused by higher-order components of the planetary magnetic field. Use physical arguments to show the error in this line of reasoning.

*The Interiors of Uranus and Neptune*

V.34 The spectroscopic method of determining rotation rates (Fig. V.61) must sometimes be applied to bodies that have completely unknown axial orientations or severe limb darkening. Describe how you would deal with these two difficulties.

V.35 The extensive He fractionation on Saturn implies a very massive metallic hydrogen layer in the deep interior, with a proportionately smaller, shallower layer on Jupiter. Reconcile this with the data on the location and orientation of the effective planetary dipoles given in the previous section.

V.36 How could Earth-based observations have established the existence of a trapped radiation belt around Neptune near $L = 7$ before any spacecraft had visited the system?

*Atmospheres of Uranus and Neptune*

V.37 A roughly spherical hot-air balloon is designed to float at the 1-bar, 60 K level in the atmosphere of Uranus (or Neptune). The empty weight of the balloon and payload is 30 kg. The vehicle carries a nuclear power source to heat the air in the balloon.

a. What balloon volume and radius is required for "hot-air" temperatures of 100, 200, and 300 K?

b. If most heat loss from the balloon is by radiation from its opaque surface, what heater power levels are required to maintain the floater at each of the three given temperatures? Remember that the atmosphere is radiating back at about 60 K.

V.38 If the mole fraction of hydrogen sulfide below the 210 K level on Uranus does exceed the ammonia mole fraction, then $H_2S$ photolysis will occur with high efficiency near the tropopause.

a. Scaling the photolysis rate of Jupiter to the location of Uranus, and assuming $K = 10^4$ to $10^8$ cm$^2$ s$^{-1}$ at the tropopause, estimate the local steady-state concentration of condensed sulfur-bearing photolysis products.

b. Can photolysis limit the column abundance of hydrogen sulfide at the 1-bar level so severely that IR spectroscopy of the region might not detect it? (Detection requires an $H_2S$ mole fraction of at least $10^{-6}$.) (*Hint*: Equate the upward eddy flux of $H_2S$ to the photolysis rate.)

c. If we also take into account the absence of ammonia and phosphine above the $H_2S$ photolysis region, will the rate of destruction of $H_2S$ be substantially increased? Explain.

d. In the case outlined in part (c), would the rate of hot-atom reactions resulting from photolysis be materially affected? Explain.

V.39 Two independent studies of the reaction of two molecular species both show a plot of the logarithm of the reaction rate vs $-1/T$ that consists of two straight-line segments, apparently due to two competing reaction mechanisms with different temperature dependence. One study shows the slope of log(rate) vs $-1/T$ to be higher above $T = 300\,\mathrm{K}$; whereas the other study shows the slope to be lower at the higher temperatures. Show that one of these findings is physically impossible and explain the behavior of the other.

# VI. Pluto and the Icy Satellites of the Outer Planets

## Introduction

Our close study of the Jovian planets has presented us with clear evidence that their compositions range from nearly solar (Jupiter) to nearly the composition of a solar-proportion mixture of the condensible ice-forming and rock-forming elements (Neptune and Uranus). We earlier saw, in Chapter III, that the satellites of the outer planets and Pluto have densities characteristic of condensed mixtures of ices and rock, ranging from essentially rocky (Io) to very ice-rich (several of the minor satellites of Saturn). The trends exhibited in Chapter III bore out a clear and systematic decrease of the densities of Jupiter's Galilean satellites with increasing distance from Jupiter, and we would not be rash to attribute this trend to radial temperature gradients in a primitive circumjovian nebula. These trends may be closely analogous to those seen in the densities of the planets themselves, reflecting the radial temperature gradient in the primitive solar nebula. In a sense the study of the satellites of the outer planets merely extends the range of compositions seen in the Jovian planets themselves to ever less volatile-rich bodies, culminating in rocky Io, a body with great affinities with the terrestrial planets.

We therefore undertake the study of ice-plus-rock bodies with the hope that we may be able to deduce more about the physical and chemical processes that formed them, so that we may then apply this understanding to the planetary system as well. Included in this chapter are the large icy satellites, including the three outer Galilean satellites of Jupiter as well as Titan and Triton, the Pluto–Charon system, and the smaller icy satellites and ring systems of the four giant planets. The smaller satellites have the best chance of having escaped the vigorous reprocessing caused by internal thermal activity and hence may bear evidence of very ancient processes that occurred during the time of the formation of the Solar System.

Further, we study these ice-bearing bodies because of their enormous and largely unexpected richness and diversity, as fascinating bodies in their own right. We see them now, since the Voyager flybys, as individuals with rare or unique properties, not simply as anonymous and nondescript members of a single vaguely defined class. The best-studied of these satellites have been found to be at least as diverse as the entire set of terrestrial planets and their satellites. As our exploration of the Solar System has reached out to Jupiter and then Saturn,

Uranus, and Neptune, we have found at each step that new phenomena await us.

## Surfaces of Icy Satellites

All of our knowledge of the composition, structure, and history of the surfaces of icy satellites is based on remote observations of the reflection and emission of electromagnetic radiation. A certain body of terminology must be mastered before these data can be interpreted usefully.

Consider a surface irradiated by monochromatic radiation from the Sun. The surface will not in general be smooth or perfectly reflective. Instead, some of the incident radiation will be absorbed, some will be transmitted into the interiors of the more transparent surface materials, and some will be reflected one or more times off of the surface before being lost back to space. This single or multiple scattering, occurring as it does on a rough, irregular surface, returns radiation to the entire hemisphere of sky, but not in a uniform manner. The total amount of radiation returned to space at each wavelength can be determined by integration over the entire sky, and the fraction of the monochromatic incident radiation thus reflected can be calculated. This fraction, called the *monochromatic Bond albedo*, $A(\lambda)$, must be less than or equal to unity. The integral of the monochromatic emitted flux over wavelength divided by the integral of the incident flux over the same wavelengths is called the *Bond albedo, A*, which also must be less than unity. It is this parameter that describes the response of the surface to solar heating; the absorbed flux is just the product of the incident flux $F_{in}$ times $(1 - A)$. It is this absorbed flux that heats the surface. In a radiative steady state the emitted thermal flux in the infrared will be exactly equal to this absorbed solar flux.

Although this seems very simple in principle, it is not usually a simple matter to deduce the Bond albedo from remote brightness measurements. This is because the angular distribution of the scattered radiation is not simple and is not easy to determine. Consider a satellite of Uranus observed from Earth; the entire range of observable scattering angles is limited by the angular excursions of the Earth as seen in the sky of Uranus. This is a range of only $\pm 3°$. The only way to extend the observations to larger angles is to send a spacecraft to fly by Uranus. Thus such observations could only be carried out by the Voyager 2 spacecraft.

Let us define the phase angle $\phi$ as the Sun–satellite–observer angle. The observed flux at phase angle $\phi$ is $F(\phi)$. By definition, the body is said to have a *phase integral* of

$$q = 2\int_0^{\pi} I(\phi)\sin\phi\, d\phi. \qquad \text{(VI.1)}$$

For colored materials the phase integral is a strong function of wavelength.

The simplest reflecting case for textured surfaces is a *Lambert surface*, which is a diffuse perfect reflector at all wavelengths. The *geometric albedo* may be thought of as the amount of radiation scattered from a body (of spherical or ellipsoidal figure) relative to that from a flat Lambert disk with the same cross-section area as the observed body. We may easily see how to compare the local reflectivity ($r$) at the subsolar point (the point at which the Sun is exactly at the zenith) with the reflectivity of a Lambert surface, but the significance of treating a spheroidal surface in this manner may seem obscure. Our only excuse is that, by defining the geometric albedo ($p$) in this way, we are able to write

$$A = pq. \qquad \text{(VI.2)}$$

Rough surfaces composed of very reflective materials are good sources of multiply scattered light. The shadows that may lie behind grains when they are observed from directions other than the direction of illumination are partly obliterated by multiply scattered light. Rough dark surfaces, on the other hand, have well-defined shadows, and their apparent brightness drops off rapidly as the viewing angle departs from the illumination angle. The associated apparent "surge of brightness" when the Sun, body, and Earth are aligned is called the *opposition effect*. The rate of variation of brightness with phase angle is called the *phase coefficient*. Large phase coefficients are associated with very rough and dark surfaces.

Until the Pioneer and Voyager flybys of Jupiter, Saturn, and Uranus, several serious limitations prevented us from determining accurate geometric albedos, phase integrals, and Bond albedos. These included the very limited phase-angle coverage available from Earth, the very poor knowledge of the radii of most of the satellites of these planets, and the lack of nonimaging methods to estimate these radii. However, recent decades have seen not only an explosive growth in spacecraft technology, but also great improvements in Earth-based astronomical observational techniques. One of the most useful of these techniques for studying small bodies is simultaneous visible and infrared radiometry. The visual brightness is proportional to the albedo ($A$) and to the cross-section area, whereas the infrared brightness (which is due to the reemission of absorbed solar radiation) is proportional to the cross-section area and to $(1 - A)$. Thus, when simultaneous measurements of visual and infrared flux are available, both $A$ and the radius can be calculated. For rapidly rotating small bodies the thermal emission is spread over the entire surface, whereas the absorption of solar energy occurs over the cross-section area of the body. Neglecting latitude-dependent effects for the moment, we may approximate the steady-state balance between absorbed and emitted radiation by

$$(1 - A)F_{\odot}/R^2 = 4\varepsilon\sigma T_s^4 = 4\sigma T_e^4, \qquad \text{(VI.3)}$$

where the term on the left side is the absorbed solar energy and the term on the right is the thermal emission. $T_s$ is the surface temperature of the body averaged over its entire surface, $\varepsilon$ is the thermal emissivity (usually very close to 1.0), $T_e$ is the effective temperature, and $\sigma$ is the Stefan–Boltzmann constant, $5.67 \times 10^{-5}$ergcm$^{-2}$s$^{-1}$K$^{-4}$. It can be seen that radiatively controlled surface temperatures for objects with similar albedos should drop off inversely with the square root of the heliocentric distance.

As a rule, most plausible and cosmically abundant surface materials exhibit absorption bands in the visible or near infrared parts of the spectrum. These bands complicate the interpretation of the albedo by introducing wavelength dependence of the geometric albedo, but of course they provide us with very valuable compositional information.

Until recently, few bodies in the outer Solar System could be examined with high spectral resolution because of the very low intensity of their reflected radiation at Earth. It was conventional to characterize the spectra of these bodies using the same system of broad-band color filters commonly used in stellar astronomy. These consisted of the U (ultraviolet) filter centered near 360 nm, the B (blue) filter centered near 420 nm, and the V (visible) filter centered near 540 nm. Each of these filters has a bandpass of at least 80 nm. Absolute brightnesses in each of these spectral bands are expressed in magnitudes, and colors are expressed in terms of magnitude differences between these bands. Higher magnitudes of course denote lower brightness. The U–B colors of the Galilean satellites, for example, show a striking variation from Io (large positive U–B; very "red") to Callisto (U–B very close to that for the Sun; "white").

Over the past two decades it has become possible to examine the reflectivities of many of the satellites of the outer planets over the entire range of wavelengths from the ultraviolet to the near infrared (where reflected solar radiation gives way to reemitted thermal radiation) with photometers having far higher spectral resolution than the conventional UBV filter system. Not only crude colors, but detailed spectra of surface materials can be obtained. In the case of a few of the brighter satellites, high-resolution infrared spectra can be obtained by terrestrial observers. In addition, high-resolution interferometric IR spectrometers were flown through the Jupiter, Saturn, and Uranus systems on the Voyager missions. Most of these investigations span the spectral region in which many rock-forming and icy minerals have distinctive and diagnostic absorption bands. The absolute reflection spectra of these satellites, represented as a plot of the geometric albedo vs wavelength, show a number of very interesting features that we shall cover in detail in our specific discussions of the individual satellites. For now, it must suffice to sketch out a few of the more interesting features of these spectra.

First, Io, Titan, Amalthea, and Europa are strikingly red. Io is extremely bright in the red and infrared. Europa is nearly as bright as Io out to about 1.2 $\mu$m, but is quite dark farther in the infrared, with strong absorption bands of water ice near 1.6 and 2.0 $\mu$m. Ganymede is much darker in the visible and near IR, also with strong water ice absorption features. At visible wavelengths, Callisto is the darkest of the Galilean satellites. It also exhibits water ice bands in the infrared. Titan and Triton display no features due to water ice, but have strong bands due to methane in the near IR and the red end of the visible spectrum. Saturn's rings and satellites, as well as the satellites of Uranus, appear to be dominated by water ice. The evidence is quite compelling for Saturn's rings, less so for Rhea, Dione, and the smaller Saturnian satellites, and sparse but suggestive for the Uranian system.

A number of the satellites of the outer planets exhibit marked variation of their optical properties with orbital phase. For example, the UBV magnitudes (and hence both the geometric albedos and the color) of the Galilean satellites vary systematically and reproducibly with orbital phase. It was concluded long ago from this evidence that these satellites must be rotationally locked on to Jupiter, each keeping one face perpetually pointed at the planet. Titan apparently reveals to us only a hazy, dense atmosphere, without a direct view of its surface and without discernible periodic brightness patterns. The most startling example of the phenomenon is Iapetus. At the V filter Iapetus exhibits a brightness variation of 2 magnitudes! Curiously, the U–B, B–V, and U–V colors are virtually independent of orbital phase, with peak-to-peak variations of no more than 0.1 magnitude. Thus the spectrum of the surface is everywhere very similar, whereas the geometric albedo varies with longitude by a factor of 7!

Another valuable source of information on the surfaces of the satellites is the polarization state of the light reflected from their surfaces. The polarization measurements can be made with considerable sensitivity and precision, and laboratory studies of a number of materials with different albedos, grain sizes, scattering angles, and packing densities are available for comparison. The limiting factor in the interpretation of the astronomical data is the absence of any satisfactory theoretical understanding of the detailed relationships between the properties of the scattering surface and the polarization properties of the observed light. This is in part due to the fact that polarization measurements can be made with considerable sensitivity, often to a precision of 0.1%. The complexity of the physical and optical model of a surface necessary to make predictions good to this degree of precision is simply prohibitive. It is therefore necessary to compare astronomical polarization measurements to laboratory data in order to interpret the observations.

Figure VI.1 is a schematic portrayal of a typical observed dependence of the degree of polarization, $P(\%)$, on the phase angle. The polarization usually displays a rather simple phase angle dependence. Light scattered at zero phase is essentially specularly reflected or symmetrically scattered, so that no net polarization is observed. At small phase angles the observed light is slightly polarized in the scattering plane (small negative polarization). At a scattering angle of about 2° to 12°, depending on the surface, the polarization again vanishes. At higher scattering angles, the light is observed to be polarized in the direction at right angles to the scattering plane (positive polarization), with the polarization increasing almost linearly with increasing phase angle. The polarization curve can usually be rather satisfactorily described by only four parameters: the degree of polarization at the point of greatest negative polarization, $P_{min}$; the phase angle at which this polarization occurs, $\phi_{min}$; the phase angle at which the polarization crosses zero again, $\phi_{cross}$; and the slope, $h$,

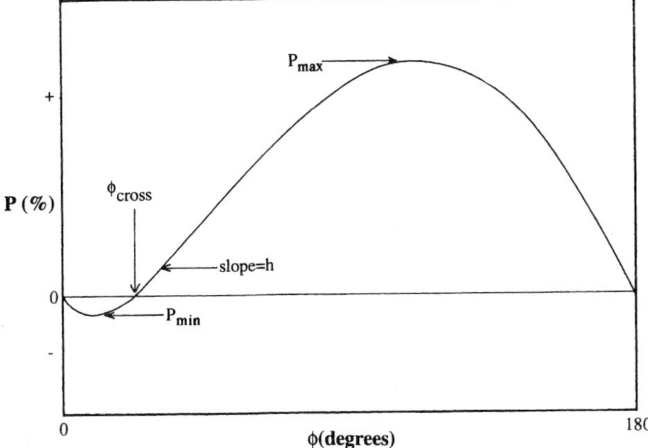


**Figure VI.1** Dependence of polarization of scattered light on phase angle. This figure displays the general features of the polarization behavior of light scattered from a rough surface. The phase angle is defined as the source–surface–observer angle. See text for explanations.

of the polarization curve at the point at which it crosses zero.

There is a well-established correlation between the reflectivity of the surface and $P_{min}$: the value of $P_{min}$ becomes smaller (more negative) as the reflectivity decreases. There exists a similar but less diagnostic relationship between slope and reflectivity, with the slope generally being greater for the least reflective surfaces; the main limitation to our application of this "slope–albedo relationship" to satellites arises simply from the limitations on the phase angle coverage available from Earth-based observations. This relationship is of great importance in the study of asteroids, which are close enough to the Sun so that they can be observed out to large phase angles from Earth. It also follows that the parameters describing the "negative branch" of the polarization curve are of much greater utility in the study of the satellites of the outer planets.

A further useful generalization is that dark surfaces (i.e., those for which multiple scattering is unimportant) exhibit polarization properties that depend only on the scattering angle and not on the separate values of the angles of incidence and emergence. Transparent grains exhibit more complex behavior, and it is in general no longer sufficient to merely add the incident and emergent angles and report the polarization as a function of their sum (the scattering angle). Instead, the polarization must be measured as a function of both of these angles separately. This places much more severe demands on laboratory measurements, because it adds another parameter that must be explored. Even worse, the resultant polarization curves of all bright, transparent materials are likely to be very similar. This ambiguity can be resolved only by extending the polarization measurements to wavelengths at which the "transparent" solids have absorption bands (at, we hope, different wavelengths!).

The polarization of light reflected from the Galilean satellites has been studied by a number of astronomers. All four of these satellites are rotationally locked on to Jupiter, and all of them display at least modest variations of color and albedo with orbital phase. Io and Callisto, in particular, exhibit large polarization variations with phase, closely correlated with the color and albedo variations. The literature is rife with interpretations of polarimetric data, but it is fair to say that we would not be able to deduce anything useful about these satellites by considering the polarimetric data in isolation. By combining the results of multicolor photometry and spectroscopy with those of polarimetric studies, it can be concluded that the materials diagnosed by polarimetry as "low-opacity" grains on Io and Europa are very different, the former apparently dominated by sulfur and absorbed sulfur dioxide and the latter by water ice. Ganymede, like Europa, displays the reflection spectrum of water ice. Polarimetry implies a "stony" or "moonlike" surface for Callisto, but this line of evidence alone is not diagnostic of composition, but merely suggestive of albedo. Other darkening agents besides the transition metals in basalts are easily imagined, and spectroscopy of Callisto fails to prove the presence of any constituent other than water ice and the unknown darkening agent.

The satellites of Saturn are the present frontier of Earth-based polarimetric study of satellites. Only a few of the larger Saturnian satellites are bright enough for study, and the range of phase angles observable from Earth (only $\pm 6^\circ$) is insufficient to permit observation of the crossover (zero polarization) angle or the slope of the polarization curve near the crossover. Iapetus, which varies greatly in albedo from one side to the other, displays a very deep negative polarization branch ($P_{min} = -1.3\%$). The bright satellites Rhea and Dione (and the bright hemisphere of Iapetus) show, as we might expect, much shallower negative branches ($P_{min} = -0.2$ to $-0.4\%$). These qualitative expectations, although generally correct, leave much unexplained. For example, the phase coefficient of the bright side of Iapetus is actually larger than Rhea's, suggesting that the bright hemisphere of Iapetus should have a deeper negative polarization branch than Rhea does (assuming similar geometric albedos, which seems reasonable). In fact, Rhea's negative branch is about twice as deep. In general, the colors of the Saturnian satellites are quite similar, with dark surfaces in the UV and IR compatible with the presence of water-ice surfaces with small and variable amounts of darkening agents. By contrast, the Galilean satellites are very dark in the UV, a consequence of the presence of sulfur on their surfaces.

Titan exhibits a polarization behavior quite unlike that of the other satellites: its polarization over the entire accessible range of phase angles is positive. This effect cannot be attributed to scattering either from a solid surface or from a deep molecular atmosphere. The only sensible interpretation is that Titan has an atmosphere containing a dispersed haze of tiny particles. Thus polarization and spectral observations do not refer to a surface. Titan presents such a peculiar and interesting case that we shall treat it separately later in this chapter.

No polarimetric data were available for the satellites of Uranus and Neptune until the Voyager 2 flyby. The five major satellites of Uranus (Miranda, Ariel, Umbriel, Titania, and Oberon) were found to have colors similar to those of the Saturnian satellites but with lower albedos, attesting to the presence of a darkening agent in water ice. Nereid is similar to the satellites of Uranus in both color and albedo.

## Eclipse Radiometry

As we have seen, spectroscopic and polarimetric data combine to give us reliable information about the presence of a very few constituents of satellite surfaces. Some information about the physical state of these surfaces on the scale of 1 $\mu$m can be deduced as well, because the depths of absorption bands in transparent solids depend upon particle size.

Another important source of information on the physical properties of surfaces is the response of the thermal emission to changes in the amount of sunlight received. This may be observed by watching the infrared flux from a fixed surface region drop off during local sunset or by measuring the thermal emission during the eclipse of a body by its primary. The former method cannot be applied to the satellites of the outer planets by Earth-based observers because the phase angle variations as seen from Earth are too small to give us a view of their dark sides. Spacecraft supply a platform from which high-phase-angle observations can be made quite readily, permitting the measurement of surface temperatures over the entire range of latitudes and longitudes. However, we should recall that most of these satellites are known to be rotationally locked onto their primaries. Spacecraft flybys are of very short duration compared with the orbital period of the satellite, and it is difficult to follow any small, discrete region across the terminator into night. For example, Callisto rotates at only about 1° per hour.

Fortunately, eclipses of the Galilean satellites are common. Moreover, these satellites commonly enter the shadow of Jupiter at a large enough angular distance from Jupiter so that measurements may be made without difficulty. The time required for a Galilean satellite to enter the shadow is determined by its diameter and orbital speed and where it enters the shadow of Jupiter and is less than a half hour even for Callisto. Thus all insolation is cut off in a very short time, and even the rapid response of the surface temperature to the loss of sunlight can be followed quantitatively. In fact, eclipse cooling is very rapid indeed on all of the Galilean satellites. The results of a typical eclipse observation are shown in Fig. VI.2.

The eclipse measurements carried out on the Galilean satellites show that their surfaces have very low thermal inertias, indicative of a very poorly packed frost layer covering the surface. Any extensive exposures of solid ice or hard rock would show up strongly in the

**a.**

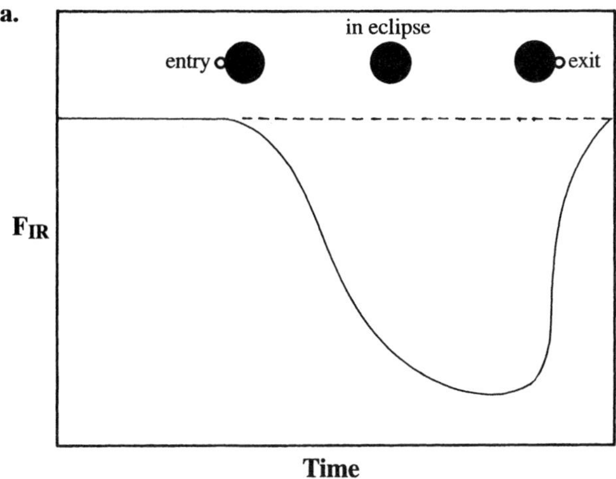


**b.**

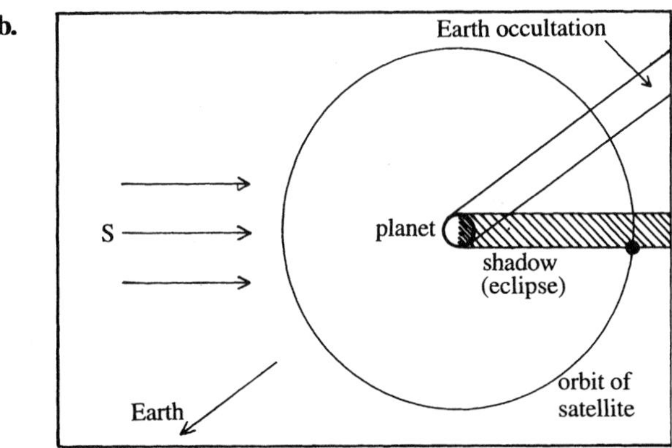


**Figure VI.2** Infrared observations of eclipse cooling. (a) The infrared flux is given as a function of time for the passage of a satellite through the shadow of a planet. b illustrates the geometry of a typical eclipse event for a satellite of, say, Jupiter. Note that the occultation of the satellite by Jupiter as seen from Earth does not in general correspond with the eclipse of the satellite. The emitted thermal flux from the satellite decays continuously during the eclipse as the satellite cools. Warming occurs over a short period of time immediately after the satellite exits from the planet's shadow.

thermal emission during the eclipse, because they would have very large thermal inertias and hence cool slowly. Such an effect has not been seen on any of the Galilean satellites.

Because of the large inclination of the orbits of Saturn's satellites to the plane of the ecliptic, eclipses of the larger satellites are impossible most of the time. Even when these eclipses do occur, the angular separation of the eclipses from the planet is small, and the rings interfere seriously with attempted observations. For these reasons, plus the coldness of the Saturnian satellites and their small angular sizes, no measurements of their eclipse cooling have been carried out.

## Surface Temperatures

In general, the temperature at the surface of an airless body within the Solar System is dominated by the absorption and reemission of sunlight. Imagine a unit element of surface area on the surface of a sphere, illuminated by the Sun. We define $\zeta$ as the zenith angle of the Sun, measured as the Sun–surface–zenith angle. The incident flux of solar illumination on the surface area element at heliocentric distance $R$ is

$$F_{\text{in}} = (F_{\odot} \cos \zeta)/R^2, \tag{VI.4}$$

and the local surface temperature is regulated by a radiative steady-state balance between the absorbed component of the solar radiation and the reemitted thermal flux,

$$F_{\text{abs}} = (1 - A)F_{\text{in}} = F_{\text{out}} = \sigma\varepsilon T_s^4, \tag{VI.5}$$

whence

$$T_s = [(1 - A)F_{\odot} \cos \zeta/\sigma\varepsilon R^2]^{1/4}. \tag{VI.6}$$

Figure VI.3 shows the geometry of illumination and the resulting distribution of temperature over the surface. Note that $T_{\max}$ occurs at the subsolar point, at which the zenith angle is zero ($\cos \zeta = 1$).

This approximation is very useful for the dayside hemisphere of a slowly rotating, poorly conducting, airless body. In general, rotation introduces a sinusoidal variation of illumination when the zenith angle is less than 90° and zero illumination when the zenith angle is greater than 90° (i.e., when the Sun is below the horizon). The surface material will conduct some heat downward when the illumination level is high and conduct the same heat outward again when the illumination level is low or zero. This introduces a time lag in the thermal response of the surface to the incident energy and drives a temperature wave into the surface.

The wave equation for a three-dimensional thermal wave is

$$\begin{aligned}\partial T/\partial t &= -K\nabla^2 T \\ &= -K[(\partial^2 T/\partial x^2) + (\partial^2 T/\partial y^2) + (\partial^2 T/\partial z^2)],\end{aligned} \tag{VI.7}$$

which reduces to

$$\partial T/\partial t = -K(\partial^2 T/\partial z^2) \tag{VI.8}$$

in one dimension. We know that the forcing function (solar heating) is periodic and that the temperature will vary with depth according to some function, $f(z)$. Let us then substitute $T(z, t) = f(z) \exp(i\omega t)$ into the previous equation:

$$K(d^2 f(z)/dz^2) = i\omega f(z). \tag{VI.9}$$

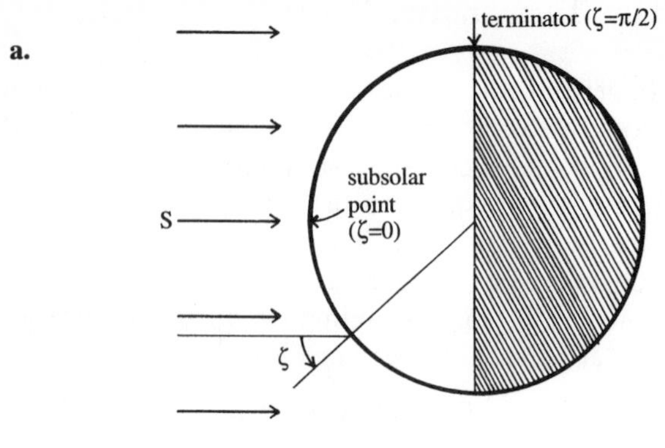


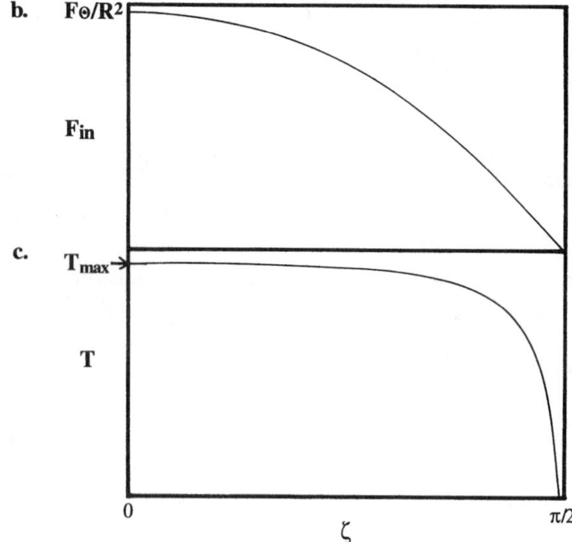


**Figure VI.3** Illumination and temperature distribution over a spherical surface. (a) Geometry of illumination. (b) The incident flux on the surface as a function of the solar zenith angle (cosine law). (c) Steady-state surface temperature vs solar zenith angle for a nonrotating spherical body with uniform albedo ($\cos^{1/4}$ law).

The general solution to this differential equation is composed of a periodic and an exponential term,

$$f(z) = Ae^{az} + Be^{-az}, \qquad \text{(VI.10)}$$

but the exponential part is not physically meaningful in the present case (the temperature excursions clearly do not get larger the deeper we are below the surface!), so $A = 0$. The temperature will then vary as

$$T(z, t) = Be^{-az}e^{i\omega t}. \qquad \text{(VI.11)}$$

The solution to this equation is

$$T = T_o e^{-cz} \cos(\omega t - cz), \qquad \text{(VI.12)}$$

which describes an attenuated wave in the $z$ direction, and $c$ is just $(\omega/2K)^{1/2}$, where $\omega$ is the angular frequency (rad $s^{-1}$). The wave is an exponentially damped cosine wave with a speed of $\omega/c = (2\omega K)^{1/2}$. The period of the wave is $2\pi/\omega$ and the wavelength is $2\pi/c$. Note that the speed of the wave is proportional to the square root of the frequency of the incident wave. This means that annual illumination variations (caused by axial tilt or orbital eccentricity) drive a wave that travels much more slowly than the diurnal wave and, because of the exponential factor in the last equation, penetrates much more deeply. The dependence of temperature on depth is illustrated for diurnal variations (Fig. VI.4a) and for a combination of diurnal and annual waves (Fig. VI.4b). For normal materials, even the annual wave is unimportant below a depth of about 10 m. Exceptionally insulating materials, such as a fine dust layer in a vacuum, may damp out the annual thermal wave at even shallower depths.

Features of the structure of the topmost few meters of the surface may be influenced by the diurnal and annual thermal waves; even melting and evaporation may occur close to the surface. But the large-scale structure of the surface bears witness to "geological" processes rooted deep in the interior of the body, related to its internal thermal state and structure, not to its surface environment.

Daytime surface temperatures on the Galilean satellites are close to 100 K, and the Saturnian satellites are near 70 K. The Uranian satellites were visited by Voyager 2 at a time of very unusual geometry, when the direction of the poles of their orbits (and hence their rotational poles also) was almost directly toward the Sun. As a consequence, absorption of sunlight over the Sunward hemisphere of each satellite (a cross-section area of $\pi r_s^2$) is reradiated over the entire illuminated hemisphere (with area $2\pi r^2$), not over the whole surface of the satellite ($4\pi r^2$). This elevates the mean temperature of the illuminated hemisphere by a factor of $\sqrt{2}$ over that for a body with rapid spin and its spin axis normal to the direction of the Sun and causes the night side (caught in a decades-long night) to drop to extremely low values. The dayside temperature of Miranda, the best-observed Uranian satellite, reaches $86 \pm 1$ K. The thermal luminosity of the nightside is so low, and its Planck peak is so far removed from the wavelength interval covered by the IRIS experiment, that it cannot be detected. The surface temperature of Triton, the coldest body so far seen in the Solar System, was found to be $38 \pm 4$ K. The smaller Neptunian satellites could not be measured directly, but the imaging experiment suggests somewhat lower albedos (and hence higher temperatures) for them.

## Surface Morphology of the Galilean Satellites

As the Voyager 1 far-encounter photographs in Fig. VI.5 attest, there is a great diversity of surface landforms on the Galilean satellites. At one extreme, Callisto and

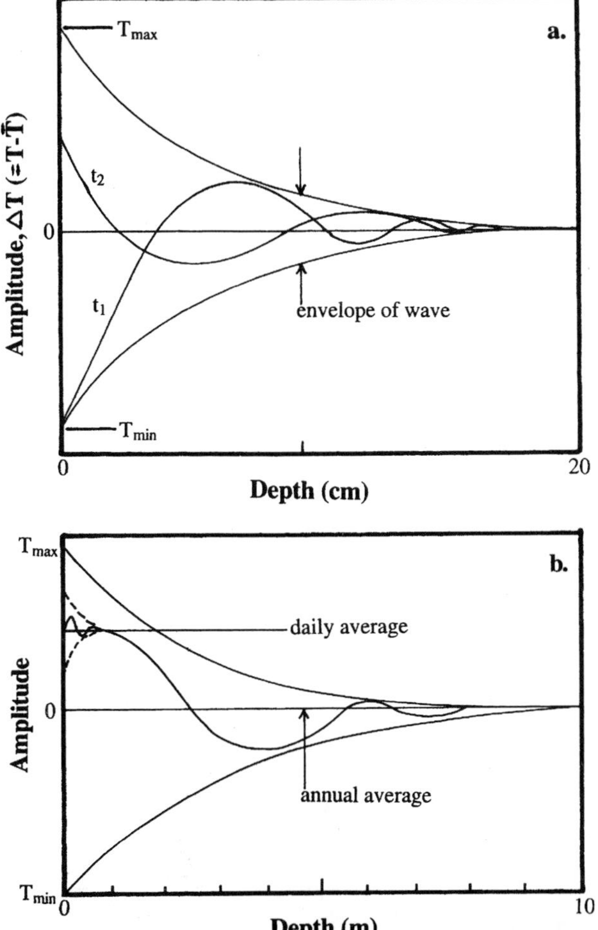


**Figure VI.4** (a) Penetration of a diurnal thermal wave. A sinusoidally varying surface temperature drives an exponentially damped short-wavelength cosine wave into the surface. The dawn ($t_1$) and midmorning ($t_2$) temperature profiles are illustrative. The local temperature oscillates between upper and lower temperature envelopes. For normal geological materials and normal length of day, the diurnal wave damps out at a depth of a few tens of centimeters. (b) Combined diurnal and annual thermal waves. On the surface of a rotating body with either axial tilt or large orbital eccentricity there is a slow seasonal variation of insolation (solar illumination) in addition to the daily illumination cycle. The temperature at a depth of about 1 m shows no detectable diurnal variation, but does vary substantially with the season. The solid curve illustrates the temperature profile shortly after midnight on a day in late spring. The diurnal wave is clearly visible in the top 0.5 m. The seasonal wave, with an angular frequency hundreds of times smaller than the diurnal wave's, has less than 1/10th of the propagation speed and more than 10 times the depth of penetration. At a depth of about 10 m the annual wave damps down to the local mean annual temperature. Deeper in the crust there are no periodic temperature variations, only a slow general increase in temperature with depth.

Ganymede are rather heavily cratered. On Callisto, in particular, cratering is by far the dominant process in shaping landforms. At the opposite extreme, Io is virtually devoid of craters. The few craters seen on Io are almost all unambiguously attributable to volcanism, not to impact. The smaller Jovian satellites are relatively poorly studied. The surfaces of the satellites of Saturn, Uranus, and Neptune were studied by the Voyager mission, and the satellite of Pluto has yet to be imaged by a spacecraft flyby. The satellites of Saturn, studied by the Voyager spacecraft, are generally quite heavily cratered, although there is also evidence of other processes at work. Titan, whose surface is of extraordinary interest because of the presence of a dense atmosphere, is endowed with an atmospheric cloud and haze of such density that we are completely unable to see its surface in visible and infrared light. Thus our comparisons of the surface structures on the satellites of the outer planets are limited to only about 10 of the over 100 known bodies of that class.

Although the crater densities on the small and medium-sized Saturnian satellites and on the icy Galilean satellites are high, there is good evidence that they have not simply undergone bombardment by asteroids and comets exactly in the same way that Mercury, the Moon, and Mars did. The great distance of the Jovian and Saturnian systems from the Sun ensures that the proportions of these classes of bombarding objects and their impact velocities must be very different from those at 0.4 to 1.6 AU. The populations of extinct short-period comets and Earth-crossing asteroids that strike our Moon are very unimportant at Jupiter or Saturn, whereas comets in Jupiter's family (which closely approach Jupiter and frequently have aphelia very close to Jupiter's orbit), Jupiter- and Saturn-crossing asteroids, and long-period comets are the important cratering projectiles in the Jovian system. On Callisto, where evidence of competing tectonic processes is quite meager, the craters seen probably extend very far back in Solar System history, to a time when the bombardment flux was much greater than it is today.

One important factor in the distribution of craters on the Galilean satellites is the anisotropy of the infalling flux of bodies caused by the large orbital velocities of the satellites. The satellites for which this effect is most profound (Amalthea and Io) provide little relevant evidence, because Amalthea has not been observed with high resolution and Io displays a very young surface, unmarked by impacts. Because all these satellites are rotationally locked onto their primaries, their trailing sides are struck at least 10 times less often than their leading sides. This effect causes the crater population on the leading side to have a shorter lifetime, because of the obliteration of older impact scars by more recent impacts. The longitude distribution of impact craters on the large icy satellites, however, has not been easy to explain quantitatively in terms of impacts by Sun-orbiting bodies.

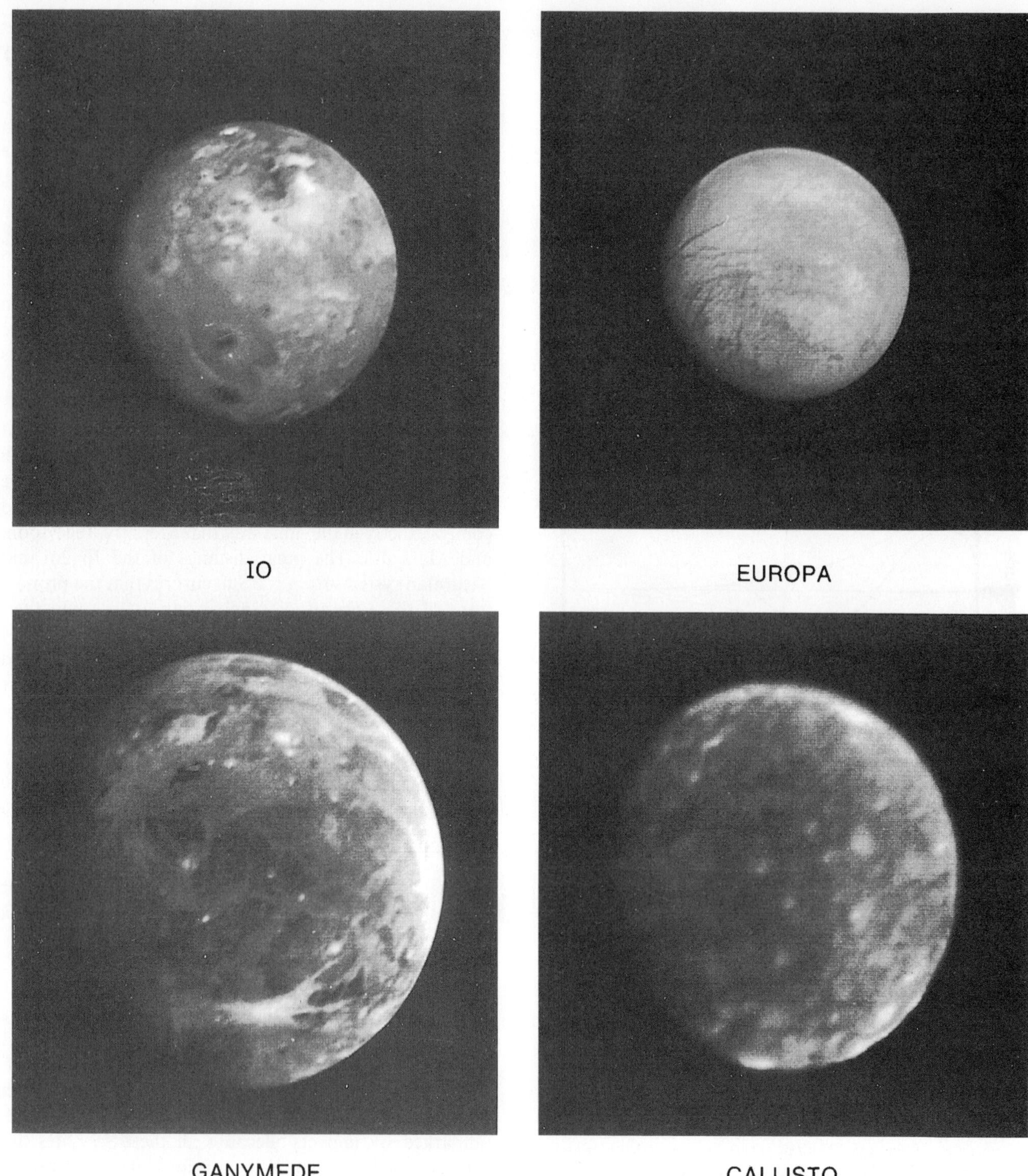


**Figure VI.5** General comparison of the Galilean satellites (to scale). The intense orange coloration of Io (here exaggerated) is faintly mirrored by Europa. Callisto is most heavily cratered and "moonlike" in appearance (See color plate 2).

The Voyager and Galileo missions have provided us with detailed coverage of the surfaces of three ice-rich Galilean satellites, Europa, Ganymede, and Callisto. Curiously, the surface structures seen on these bodies differ strikingly from one another. Callisto's densely cratered surface, with its paucity of internally produced tectonic features remarked upon above, stands in sharp contrast to the slightly larger Ganymede. The latter displays a remarkable global system of rifts filled with parallel bands of ice, attesting to severe tensional fracturing of the surface (Fig. VI.6). The degree of expansion needed to open the observed rifts cannot be provided by simple thermal expansion of the interior or by phase changes in the icy interior; rather, there must have been

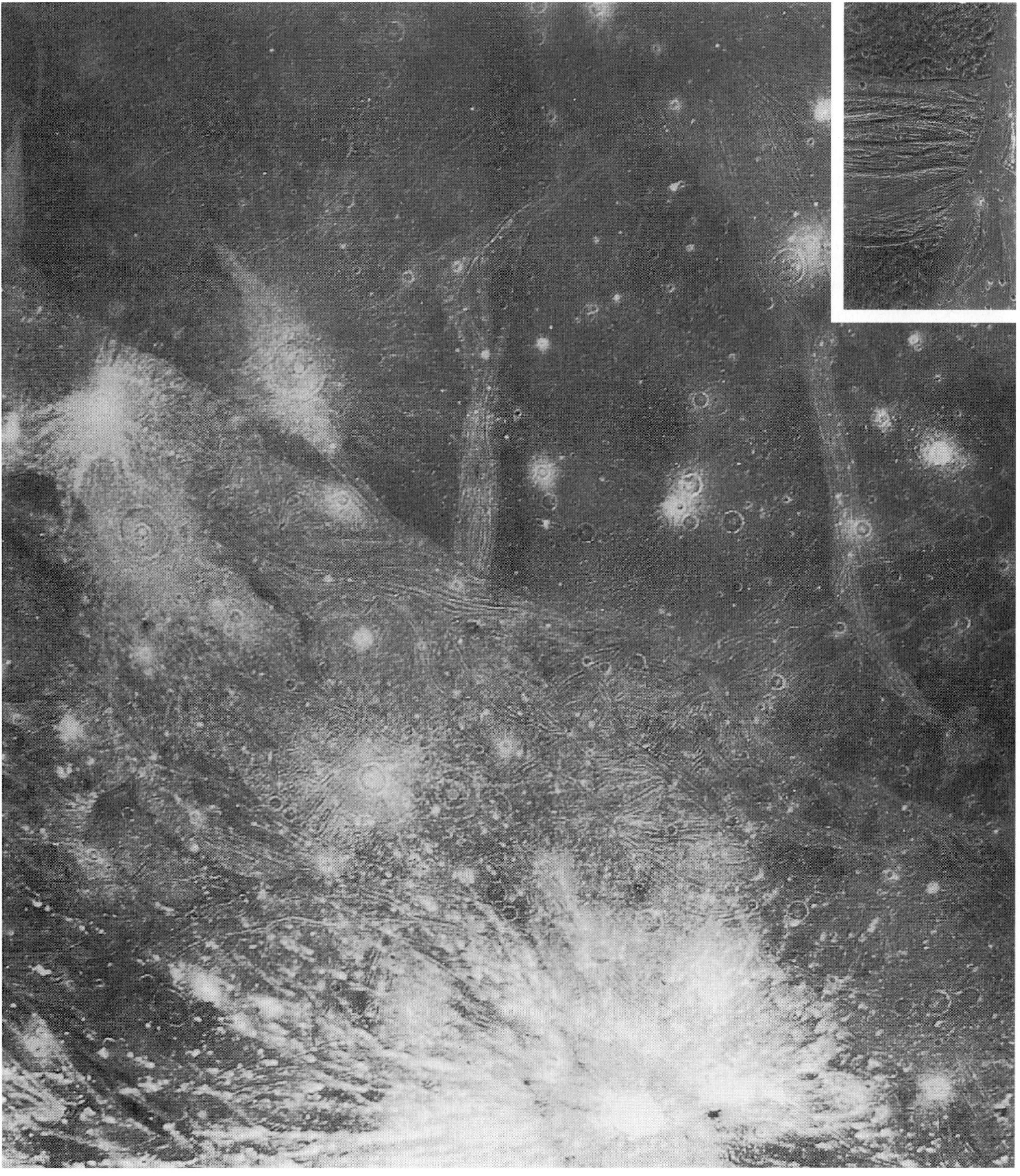

**Figure VI.6** Voyager 2 image of Ganymede. Note the many cracks and fissures. The brilliant spots and streaks are "snow" from violent impact cratering events. The surface of Ganymede has very subdued relief. Inset: A closeup of the grooved terrain region called Erech Sulcus, terminated on the right by the smoother Sippar Sulcus. This Galileo Orbiter image shows the intersection of these two belts of crustal extension, grossly analogous to the Great Rift Valley in Africa.

tectonic activity capable of stressing the crust and opening the large rifts. The surface of Ganymede divides readily into two very different major terrain types. Nearly half of the surface is dark in color and very heavily cratered, dating back to a very early period in Solar System history when the flux of cratering objects was much higher than it is at present. The largest of these craters have essentially no remaining topographic relief.

The remainder of the surface area of Ganymede is dominated by younger, much less heavily cratered terrain. It is usually strongly marked with parallel bands or grooves, often with an appearance reminiscent of multilaned highways. Loss of about half of the ancient crust and production of the younger banded terrain were evidently accompanied by tensional deformation of the crust, because the edges of the banded regions frequently conform almost perfectly to each other in shape despite the wide intervening expanse of banded terrain. The bands have evidently been produced by upwelling at spreading centers in a manner similar to the production of new oceanic crust on Earth at midocean spreading centers. It is reasonable to suppose that convective motions in the underlying mantle have been responsible for transmitting the driving forces for these motions to the base of the crust. The vertical relief in the banded terrain is a few hundred meters, with the bands usually spaced 3–10 km apart. It follows that, at the time of formation of the grooved terrain, the ice crust of Ganymede must have been no more than about 10 km thick.

The darkness of the more ancient terrains suggests that they contain some significant admixture of primordial solids, such as the dark material that makes up carbonaceous chondrites. The darkening agent might be tarry organic heteropolymers, magnetite, or other low-albedo minerals inherited from the nebula out of which the planets and satellites accreted. Alternatively, the dark materials might be organic or mineral matter altered or formed within the interior of Ganymede during its earliest evolution. In Chapter VIII, as part of our study of meteorites, we shall consider the evidence that many of the distinctive dark materials in low-temperature meteorites may have originated not in the solar nebula, but in the interior of a large asteroid. Similar processes might have occurred on Ganymede.

This interpretation of the albedo differences between ancient, heavily cratered terrain and younger, grooved terrain is sensible if Ganymede accreted as a more or less homogeneous mixture of rocky and icy materials, but then warmed and differentiated from within. The heat sources available to drive this heating, with eventual melting of ice and sedimentation of a rocky core, include radionuclide decay, heating by an early high-luminosity phase of the Sun or of Jupiter, and dynamical interactions with the other members of the Jovian system. In any event, the surface temperature of Ganymede during the time of formation of the features we now observe could not have reached or even closely approached the melting point of water ice. Melting of the interior, beneath a thin, permanently frozen exterior, would leave a brittle crust containing comparable amounts of rock and ice particles. The heated region will rapidly drop its burden of dense silicates, oxides, and sulfides to form a core, leaving the thin primitive crust "floating" on a mantle of nearly pure liquid water. The water of course has a much lower density than the crust, and foundering of the crust should occur very readily. Later in Ganymede's history, after thorough extraction of dense materials from the crust into the core, water reaching the surface at spreading centers would be very clean and bright.

With these considerations in mind, it is informative to look at Callisto again (Fig. VI.7). Callisto is smaller, darker (a visual albedo about half that of Ganymede), and farther from Jupiter and the other large bodies in the Jovian system. Every heat source that we have mentioned would be of equal or lesser importance on Callisto than on Ganymede. In view of the very ancient, heavily cratered appearance of Callisto and the apparent absence of younger, grooved terrain of tectonic origin, we find it easy to believe that the darkening agent on both bodies is an ancient, dense material from the accretionary era. Callisto, less strongly heated than Ganymede, has preserved intact (except for damage inflicted by the impacts themselves!) a thick, rigid, and very ancient crust with a large proportion of rocky solids and primordial organic polymers. Remote spectroscopic observations sample the surface only to depths of a few micrometers and hence see materials that have been thoroughly exposed to solar heating, solar UV irradiation, cosmic rays, magnetospheric particles, and micrometeoroid bombardment. It is therefore not possible to accept the observed surface as a representative sample of the bulk composition of Callisto's crust. Still, the geological evidence that it is very ancient and very little altered by tectonic processes is convincing and has obvious compositional implications. Infrared spectra of the surface of Callisto by the Galileo spacecraft show features attributable to carbon dioxide, sulfur, and CN groups in organic compounds. These materials attest to the presence of cometary or carbonaceous chondrite materials, mixed in with the surface ice, and not segregated by melting and differentiation.

The Galileo magnetometer experiment finds a distinctive signature for Callisto, which affects the surrounding medium as if it were a fairly conductive sphere. The simplest interpretation is that there is a conducting ocean beneath the ice. Thus we are left with

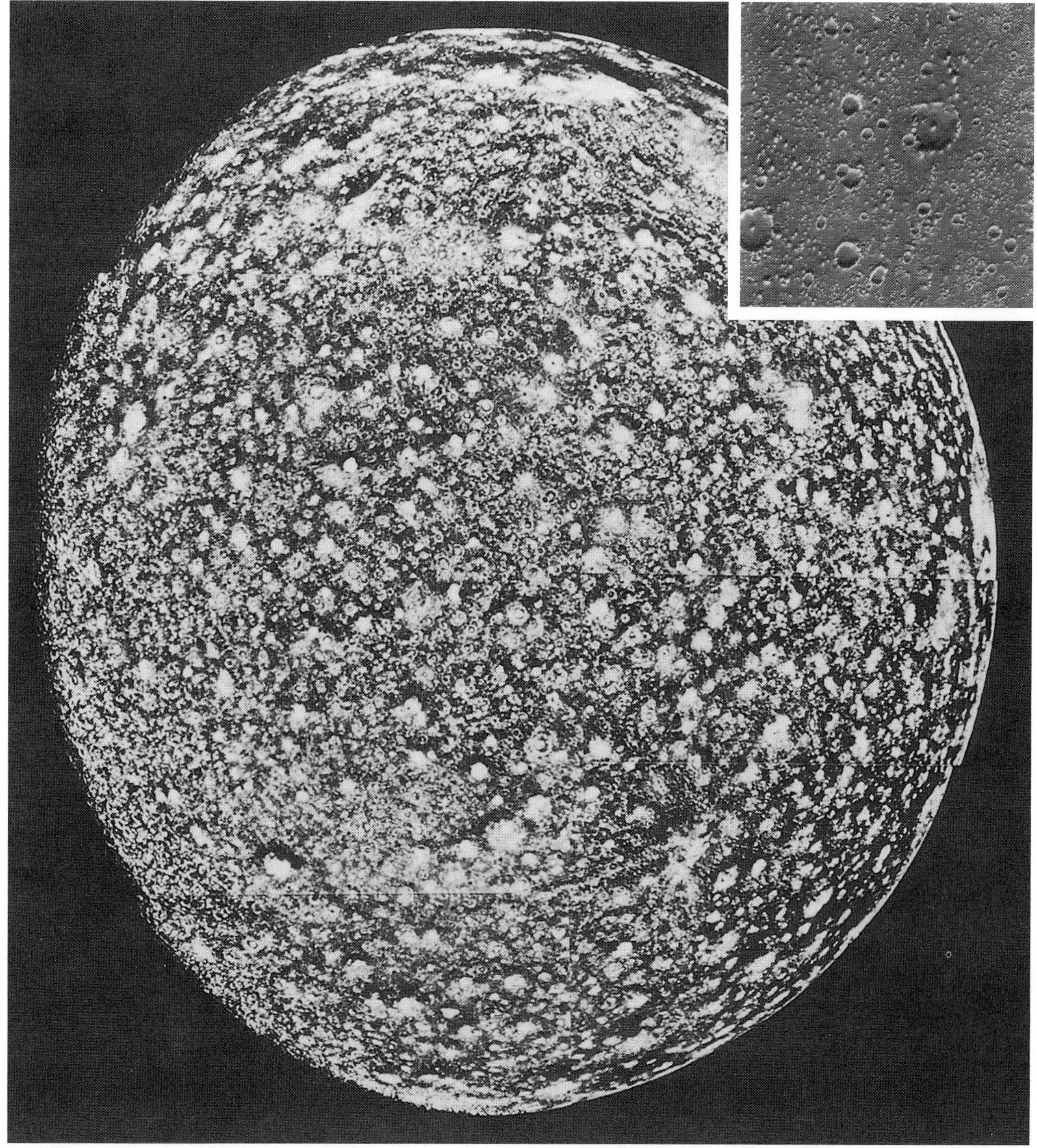

**Figure VI.7** Callisto as seen by Voyager 2. The contrast in this black-and-white mosaic has been computer-enhanced to bring out surface details. Note the universal heavy cratering. The faint straight lines visible in the image are the joints between the individual frames used in assembling the mosaic. Inset: This Galileo Orbiter closeup of typical cratered terrain on Callisto reveals the absence of small craters and the widespread mantling of the surface by a dark blanket of fine "dust." The two phenomena may be related.

an implausible scenario in which the deep interior of Callisto melts, but a surface shell of ancient dirty ice survives.

Bright, fresh craters with associated high topographic relief are found widely distributed on Ganymede. The fresh, high-relief craters that exhibit bright ray systems are, curiously, far more common on the trailing side than on the leading side. This is almost certainly due to the fact that the rays are thin sprays of material and are erased by the much more numerous impacts of bodies that are far smaller than the original projectile that formed the rayed crater. The flux of these small bodies is of course far higher in the leading hemisphere.

Europa, located yet closer to Jupiter, has even less surface relief than Ganymede and Callisto. The surface is marked by vast numbers of narrow, sinuous cracks, ranging in length from about one radius of the satellite down to the limit of resolution. The surface is pale yellow (Fig. VI.8). The general impression is that of a thin, brittle ice layer afloat on a dynamic liquid interior. The Galileo Orbiter images show many examples of scalloped cracks in the ice surface. These cycloids are best attributed to tidal stresses near the subsolar point on a thin ice shell that slowly drifts relative to the interior. Also evident in many places on Europa are areas of chaotic terrain that, according to Richard Greenberg and coworkers, represent places where oceanic convection currents or plumes have thinned or even penetrated the crust, leaving an expanse of floating ice fragments that have refrozen and become immobilized.

There has also been much interesting speculation regarding the possible content of dissolved salts in Europa's ocean, encouraged by the detection of a trace atmosphere of potassium atoms. Potassium provides a source of both radiogenic argon and radiolysis products such as hydrogen, oxygen, and hydrogen peroxide.

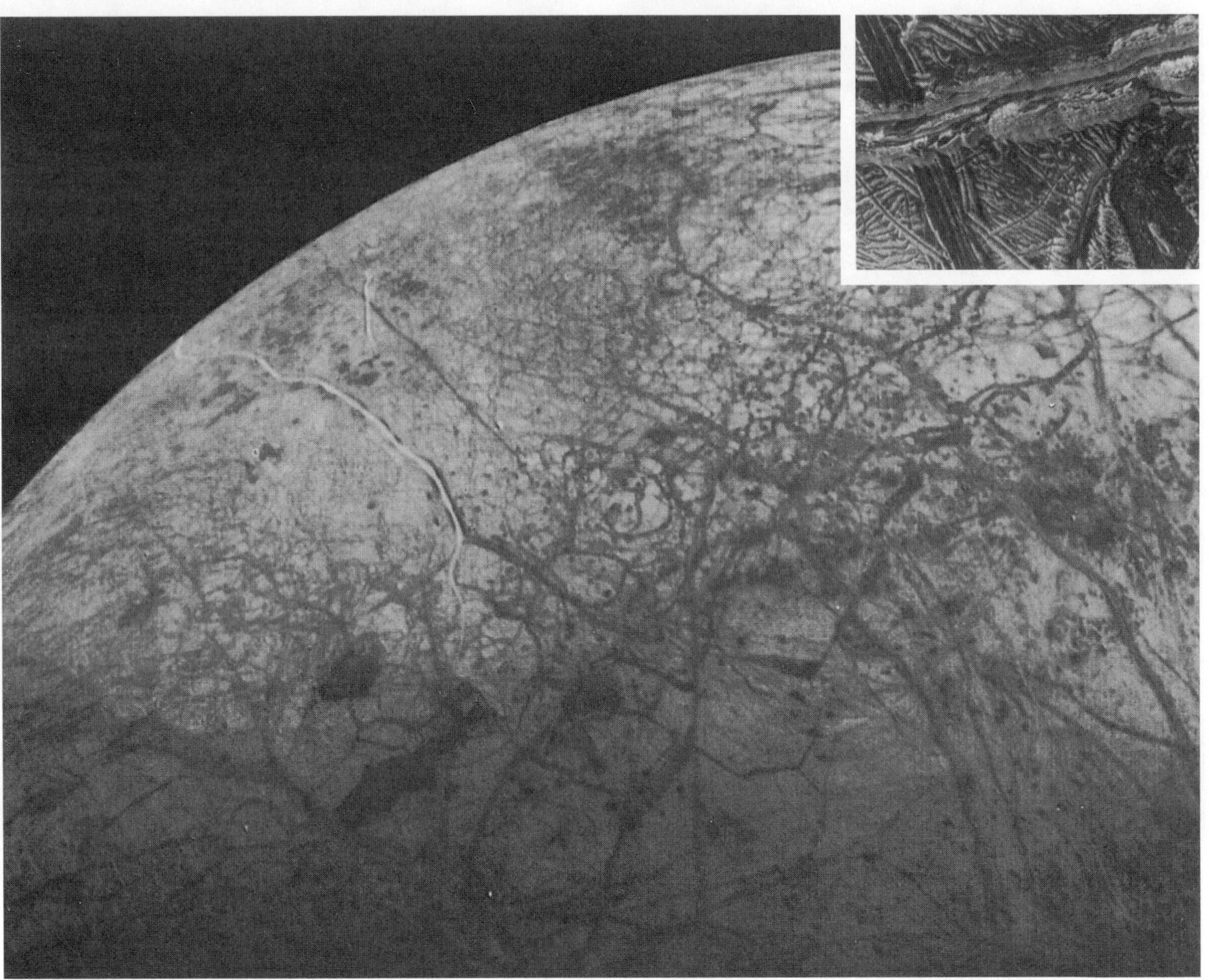

**Figure VI.8** A Voyager 2 image of the surface of Europa. The network of fine, curving fissures covers the surface much more densely than the broad "freeways" seen on Ganymede. The pale yellow color, here slightly exaggerated, is due to the ejection of sulfur from Io. Inset: This Galileo high-resolution image shows a system of ridges and valleys cutting older features, including what may once have been a closely similar structure (far left). The dark, relatively smooth area at lower right may be a product of flooding by water or warmer (less viscous) ice.

## Density and Composition of Icy Satellites

The diameters of many of the icy satellites have been determined by spacecraft occultation measurements or by high-resolution photographic mapping by the Voyager 1 and 2 flyby missions to Jupiter and Saturn and the Voyager 2 flybys of Uranus and Neptune. These diameters were summarized in Table III.5.

The masses of many of the larger icy satellites were first determined, before the era of spacecraft exploration, from their long-term gravitational interactions. Very precise tracking of the Pioneer 10 and 11 and Voyager 1 and 2 spacecraft as they flew through these satellite systems has yielded a great improvement in many of the mass estimates and hence has made it possible to calculate the densities of these satellites to useful precision for the first time. The mass determinations and the calculated bulk densities are also given in Table III.5.

One of the most striking features of the data in Table III.5 is the existence of a class of satellites that are so large that we could reasonably view them as planetary bodies. This class includes the icy satellites Europa, Ganymede, and Callisto about Jupiter, Saturn's largest satellite Titan, and Neptune's large satellite Triton. There are also two rocky (ice-free) moons in this size class, Earth's Moon and Jupiter's Io.

Second, we note the striking decrease of the densities of the large satellites outward from Jupiter (Table VI.1). Such density differences betray profound compositional differences, ranging from rock (Io, the nearest) to roughly solar-proportion mixtures of ice and rock (Ganymede and Callisto, the farthest). This density trend seems to require a strong radial temperature gradient outward from Jupiter at the time of the formation of the satellites, possibly analogous to the solar nebula out of which the planets themselves formed. Third, we see that the satellites of Saturn and Uranus generally have low densities, without any clear evidence of a radial composition trend about either planet. Fourth, the satellite systems of Jupiter, Saturn, and Uranus have structures similar to that of the Solar System itself, whereas the Neptune system seems, by comparison, to be severely disturbed.

**Table VI.1** Radii, Masses, and Densities of the Galilean Satellites

| Satellite | Radius (km) | Mass ($10^{24}$ g) | Density (g cm$^{-3}$) |
|---|---|---|---|
| Io | 1820 | 89.1 | 3.53 |
| Europa | 1552 | 48.7 | 3.03 |
| Ganymede | 2635 | 149.0 | 1.93 |
| Callisto | 2410 | 106.5 | 1.83 |

The materials making up the icy component of these satellites may be made by any of the various chemical schemes we discussed in Chapter IV. First, if the ices in these satellites were produced in the solar nebula under conditions permitting a close approach to chemical equilibrium, the dominant and first-condensing ice (near 160 K) would be water ice. At somewhat lower temperatures (farther from the Sun) ammonia hydrate, $NH_3 \cdot H_2O$, would form, exhausting all nitrogen from the gas phase. Next, at even lower temperatures, methane clathrate hydrate, $CH_4 \cdot 7\ H_2O$, would form by the complete conversion of the remaining water ice by reaction with gaseous methane. Finally, near 40 K, the leftover methane gas would condense as solid methane.

As we discussed in Chapter IV, however, there is strong reason to doubt that ammonia and methane are in fact the dominant forms of carbon and nitrogen in the solar nebula, equilibrium behavior notwithstanding. At temperatures low enough for methane and ammonia to be thermodynamically stable ($<680$ and $<360$ K, respectively), the chemical reactions that synthesize these gases from CO and molecular nitrogen are so slow that they cannot approach equilibrium. At temperatures high enough so that the chemical reactions can take place during the lifetime (or turbulent mixing time) of the nebula, about 1000 K, no more than about 1% of the carbon and nitrogen would be found in the form of methane and ammonia. Thus CO, $N_2$, and $CO_2$ will be at least as important gases as the reduced forms expected from equilibrium theory. The low-temperature condensates then begin with water ice and small amounts of ammonium carbonate and carbamate salts, followed by mixed clathrates of nitrogen, carbon monoxide, and methane. Carbon and nitrogen are so abundant that not all of these species can be accommodated in the clathrate, so they condense separately at lower temperatures.

In planetary subnebulae, however, the gas density may be many orders of magnitude higher than in the surrounding solar nebula. This high density enormously increases the molecular collision rate, and thus accelerates the attainment of equilibrium.

Yet another variant of the gas and condensation chemistry can occur if solid ice grains, once condensed from a cooling gas, are immediately accreted into larger bodies and therefore cease to interact effectively with the nebular gas at lower temperatures. In such a scenario, for example, Fe metal is wholly accreted and deeply buried long before temperatures fall low enough to permit FeS formation. Thus most of the $H_2S$, deprived of the ability to form sulfide condensates, remains in the gas phase down to low temperatures. It then condenses as $NH_4SH$ (if ammonia is abundant) or as $H_2S$ at low temperatures. In a dense, cold, but equilibrated gas, the sequence is water ice, ammonium hydrosulfide, ammonia, and then methane and argon. This is the sequence seen in the atmospheres of the Jovian planets, in which gravity plays the part of removing (precipitating) the condensed solids from the rising, cooling gas stream.

Given our uncertainties about the thermal and dynamic history of the materials in the icy satellites, we cannot leap to any certain conclusions about precisely which ices are present (and in what proportion) within the icy satellites. We do, of course, expect that high-pressure reequilibration in planetary subnebulae will permit a much better approach to equilibrium than in the solar nebula; therefore even the observational evidence for the importance of carbon oxides and the near absence of ammonia and methane in cometary ices (presumably formed in the low-density Solar Nebula) does not constitute proof that methane and ammonia were absent in the regions in which the icy satellites formed.

A second major factor that complicates the interpretation of observed densities in terms of bulk composition is the self-compression of icy materials at the pressures prevalent in the interiors of large satellites. Now that we have accurate density data for many of the icy satellites, there is no excuse for neglecting the effects of self-compression on the density. As an example, consider the behavior of spheres of cold ice as sketched out in Fig. VI.9. Note that the interior pressures are too small to matter for 100-km bodies, but that self-compression rapidly escalates in importance toward larger sizes. We showed beginning with Eq. (IV.34) that the central pressure of a uniform spherical body is given by

$$P_c = GM^2/4\pi r^4, \qquad \text{(IV.41)}$$

which, in terms of $\rho$ and $r$, is just

$$P_c = 4\pi G r^2 \rho^2/9. \qquad \text{(VI.13)}$$

For a uniform sphere of cold water ice, with a density of 0.9, the central pressure is $7.55 \times 10^{-4} R^2$, where the pressure is in bars and $R$ is in kilometers. An ice ball with a radius of 1 km has a central pressure of only 755 dyn cm$^{-2}$, a 10-km radius sphere reaches 0.0755 bar at its center, and a 100-km sphere reaches only 7.55 bar. But when the radius is 1000 km the central pressure reaches 755 bar, sufficient for significant compression, and a planet-sized ice ball ($R = 10,000$ km) will attain a central pressure of 75.5 kbar.

However, ice is a substance noted for its extraordinarily low density. It is one of the most compressible solids known and one of the very few known substances whose density is less in the solid than in the liquid state. Further, its melting behavior and its structure are sensitive functions of pressure. Compression of ice I

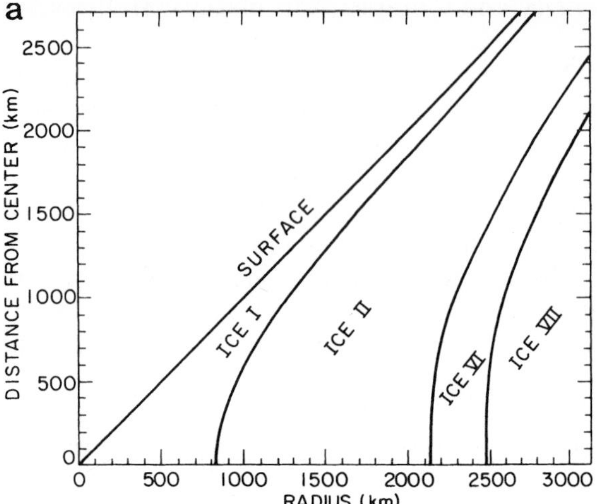


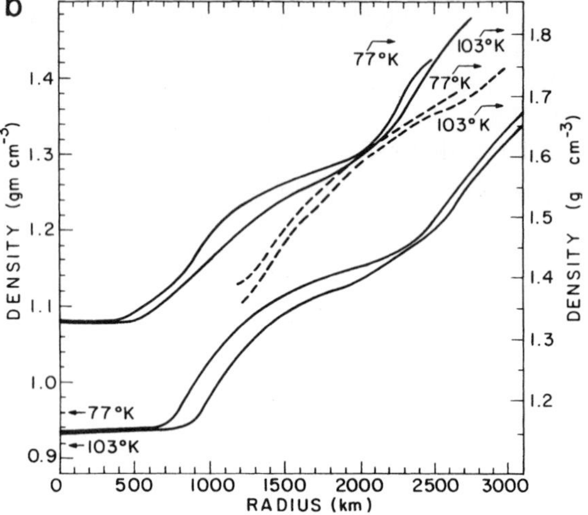


**Figure VI.9** Self-compression of ice spheres. (a) The phase stability of ice is shown as a function of depth for isothermal pure ice spheres at 103 K. Each vertical section through the diagram gives the vertical section through a satellite of that radius. High-density forms of ice are absent for bodies with radii less than 820 km, but high-density forms of ice dominate in bodies over 1000 km. (b) The density of cold, nonrotating spheres of water ice is shown for a wide range of radii at 77 and 103 K. Note the rapidly escalating importance of self-compression at radii above 800 km. The upper curves in b are for solar-proportion mixtures of ice and rock.

causes it to approach the density of the liquid state; indeed, compression of ice that is well below its normal freezing point by concentrating a large force on a small area causes the phenomenon called *regelation*. The most familiar example of regelation is ice skating, in which the skater actually rides on a thin pressure-melted film of water at temperatures below 0°C.

If the temperature of an isothermal ice ball is close to the normal (1 bar) melting point of ice, 273 K, then pressures in its interior may be sufficient to cause melting. Consider melting of any general solid ($A$) by the reaction $A(s) = A(l)$. The phase boundary between the solid and liquid follows some curve in $(P, T)$ space. Imagine two nearby points on this melting (freezing) curve, called 1 and 2, as in Fig. VI.10. Because the two phases are in equilibrium with each other on the phase boundary, the Gibbs free energy of formation of the solid at each point must exactly equal that of the coexisting liquid at the same point. The Gibbs free energy change in any phase in moving along the freezing curve from 1 to 2 can be calculated from the elementary relation

$$dG = V\,dP - S\,dT. \qquad \text{(AI.38)}$$

Because the two phases are in equilibrium with each other at both points (equal values of $G$) the change in $G$ in the solid phase (s) must be exactly equal to the change in $G$ in the liquid phase (l),

$$V_s\,dP - S_s\,dT = V_l\,dP - S_l\,dT \qquad \text{(VI.14)}$$

or, along the phase boundary,

$$\partial P/\partial T_m = (S_l - S_s)/(V_l - V_s) = \Delta S/\Delta V, \qquad \text{(VI.15)}$$

whence

$$\partial T_m/\partial P = \Delta V/\Delta S. \qquad \text{(VI.16)}$$

The entropy change upon reversible melting, $\Delta S$, is just the latent heat of fusion divided by the temperature, $\Delta H^\circ_m/T$ (see Appendix I). The slope of the melting curve is then

$$\partial T_m/\partial P = T\Delta V/\Delta H^\circ_m. \qquad \text{(VI.17)}$$

This is one form of the Clausius–Clapeyron Equation.

For water ice, the molar volume change upon melting is negative and the entropy change is positive. The slope of the melting curve is therefore negative. As compression proceeds, the changing values of $\Delta V$ and $\Delta S$ cause the phase equilibrium line to curve slightly. The melting temperature decreases steadily until it reaches −22.0°C at a pressure of 2115 bar. Further compression causes the formation of a new crystal form of ice, ice III, which has a positive $\Delta V$ of melting.

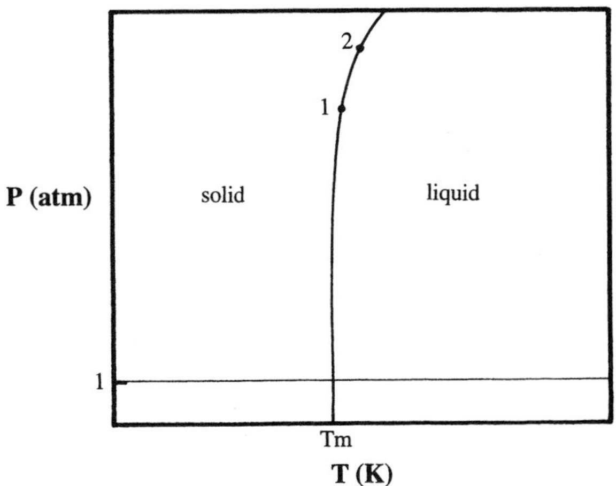


**Figure VI.10** Solid–liquid phase equilibrium. The curved line is a phase boundary between two condensed phases. The denser phase is favored at higher pressures.

Accordingly, the slope of the ice III melting curve is positive, and the melting temperature rises toward higher pressures. Figure VI.11 shows the equilibrium stability field of a number of phases in the pure-$H_2O$ system. Note that, according to the Gibbs Phase Rule,

$$f = C - \phi + 2, \qquad \text{(V.52)}$$

which, for a one-component system, gives $f = 3 - \phi$. Wherever in the phase diagram one finds three phases coexisting the number of degrees of freedom ($f$) is zero. This is called a *triple point*. Wherever two phases coexist, their equilibrium has a single degree of freedom, which is a line in $(P, T)$ space. Single phases may exist stably over finite ranges of $T$ and $P$.

From the ice phase diagram in Fig. VI.10 we can see that an isothermal ice ball at any temperature below −22.0°C (251.16 K) would not be melted at any depth, but all warmer bodies can enter the liquid field if the central pressure is high enough to lie to the right of the ice I–liquid water melting curve. Note that pure water bodies with temperatures over 273 K can be frozen into high-density forms of ice at great depths. The ice V + ice VI+ water triple point lies at +0.16°C and 6380 bar, so such a structure would require a pure-water body with a radius of about 3000 km. Note that these temperatures are much higher than the surface temperatures of bodies at and beyond Jupiter's distance from the Sun.

The presence of a dense rocky core of course permits much higher gravitational accelerations and allows these high pressures and exotic, dense forms of ice to form at shallower depths. The radioactive elements present in the rocky component are important sources of heat that are capable of melting ices, permitting the separation of the rocky and icy components according to their densities (a process called *differentiation*) and maintaining a substantial temperature difference between the deep interior of a large satellite and its surface.

It is now amply clear that the interior structure of an icy satellite must depend in a sensitive way on the details of its internal thermal state and composition. We therefore must develop more complex structural and thermal models in order to understand their present structures.

## Internal Thermal Structure of Galilean Satellites

A mixture of water ice and rocky solids with the same elemental proportions as those in solar material will have the composition given in Table VI.2, roughly

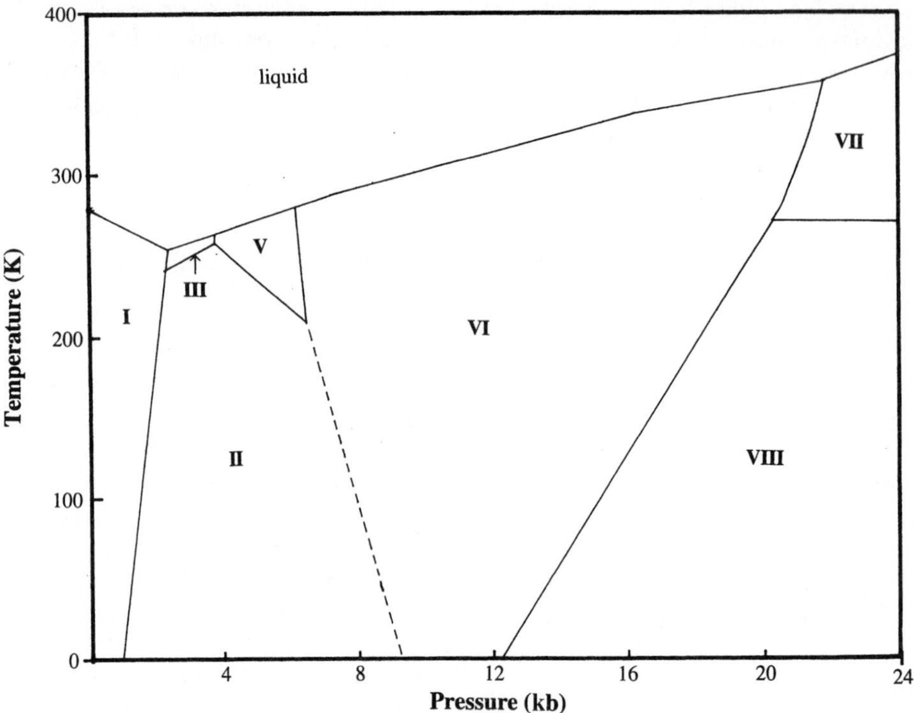


**Figure VI.11** Phase diagram of water. The stability fields of liquid water, ordinary ice (I), and a number of high-density forms of ice are shown.

60% ice and 40% rock. Ices produced by equilibrium condensation in planetary subnebulae at even lower temperatures will contain ammonia and methane, which condense in the form of ammonia hydrate, methane clathrate hydrate, and solid methane. The mass percentages for these various composition classes are also given in Table VI.2.

The principal long-lived radionuclides present in the rocky component of the condensate are $^{40}$K, $^{232}$Th, $^{235}$U, $^{238}$U, and $^{87}$Rb. The half-lives for all of these nuclides are comparable to the age of the Solar System, and their abundances are sufficient to permit them to play a major role in the production of heat inside planets and satellites. Each of these isotopes decays at a rate proportional to the abundance of the parent radionuclide, according to the decay equation

$$dN_i = -\lambda_i N(t)\, dt. \qquad \text{(VI.18)}$$

The constant $\lambda_i$ is called the *decay constant* for the isotope $i$. Rearranging this equation and integrating,

$$\ln(N_i/N_0) = -\lambda_i(t - t_0), \qquad \text{(VI.19)}$$

where the subscript 0 indicates the beginning of decay. The time interval over which half the parent radionuclide decays is given by

$$\ln(1/2) = -\lambda_i(t_{1/2}) = -0.693, \qquad \text{(VI.20)}$$

whence the half-life ($t_{1/2}$) is readily seen to be a constant. The present-day abundances, half-lives, and heat production rates for these isotopes in rocky material are given in Table VI.3. Note that the most important modern heat source, $^{40}$K, also has a rather short half-life of about $1.4 \times 10^9$ years. This means that, in the distant past, $^{40}$K was an even stronger source of heat than it is today. The isotope $^{235}$U, which contributes only 1.4% of the modern radioactive heat production, has a half-life of only 0.72 Ga (billion years) and hence was more important by a factor of $2^{(4.6/0.72)} = 84$ at the time of planetary formation, 4.6 Ga ago. Figure VI.12 shows how the size of the major internal radiogenic heat sources has declined from the time of the origin of the planets to the present.

In an initially homogeneous planet composed of a mixture of rock and water ice the decay of these nuclides will at first provide heat at the rate of about $S = 4 \times 10^{-7}\,\mathrm{erg\,s^{-1}}$ per gram of rock, or $1.6 \times 10^{-7}\,\mathrm{erg\,s^{-1}\,g^{-1}}$ for the mixture. This rate of heat production will decline by about a factor of 9 over the history of the Solar System. The heat capacity of the ice–rock mixture is readily calculated: a weight fraction of 0.6 for ice, with a heat capacity of $0.5\,\mathrm{cal\,g^{-1}\,K^{-1}}$, and a weight fraction of 0.4 for rock, with a heat capacity of $0.25\,\mathrm{cal\,g^{-1}\,K^{-1}}$, give an overall heat capacity of $0.4\,\mathrm{cal\,g^{-1}\,K^{-1}}$. Recalling that there are $4 \times 10^7$ ergs per calorie, this is $1.6 \times 10^7\,\mathrm{erg\,g^{-1}\,K^{-1}}$. To heat the mixture

**Table VI.2** Compositions of Condensates in Low-Temperature Solar Material

| Element | Abundance (Si = $10^3$) | Component | Component mass | Mass % Dry rock | Rock + S | Wet rock | Rock + ice |
|---|---|---|---|---|---|---|---|
| Mg | 1,060 | MgO | 42,400 | 21.658 | 20.331 | 17.309 | 8.210 |
| Fe | 1,050 | FeO | 75,600 | 38.616 | 36.247 | 30.861 | 14.638 |
| Si | 1,000 | $SiO_2$ | 60,000 | 30.649 | 28.768 | 24.495 | 11.618 |
| Al | 85 | $Al_2O_3$ | 4,335 | 2.214 | 2.078 | 1.769 | 0.839 |
| Ca | 72 | CaO | 4,032 | 2.060 | 1.933 | 1.646 | 0.780 |
| Na | 60 | $Na_2O$ | 3,720 | 1.900 | 1.783 | 1.518 | 0.720 |
| Ni | 48 | Ni | 2,832 | 1.447 | 1.358 | 1.156 | 0.549 |
| All others[a] | 51 | Oxides | 2,850 | 1.456 | 1.366 | 1.163 | 0.551 |
| S | 400 | S | 12,800 | 0 | 6.136 | 5.224 | 2.479 |
| O | 2,000 | Bound $H_2O$ | 36,400 | 0 | 0 | 14.859 | 7.049 |
| O | 15,100 | Water ice | 272,000 | 0 | 0 | 0 | 52.567 |
| N | 3,700 | Ammonia | 63,000 | 0 | 0 | 0 | 0 |
| C | 11,800 | Methane | 189,000 | 0 | 0 | 0 | 0 |
| Total | | | | 100.000 | 100.000 | 100.000 | 100.000 |

| Component | Mass % Wet Rock | Rock + Ice | $+NH_3$ | $+CH_4$ clathrate | $+CH_4$ ice |
|---|---|---|---|---|---|
| Rock | 85.14 | 41.00 | 36.46 | 34.78 | 26.55 |
| Water | 14.86 | 59.00 | 53.79 | 51.30 | 39.17 |
| Ammonia | 0 | 0 | 9.75 | 9.26 | 7.07 |
| Methane | 0 | 0 | 0 | 4.66 | 27.21 |
| Total | 100.00 | 100.00 | 100.00 | 100.00 | 100.00 |

[a] Elements less abundant than nickel are lumped together in solar proportions.

from an original temperature of, say, 100 K to the lowest (*eutectic*) melting point of water ice, at −22°C (251 K), requires about $C_P\Delta T = 2.4 \times 10^9$ ergs per gram, which, at the early heat production rate, would take $2.4 \times 10^9/1.6 \times 10^{-7}$ s, which is $5 \times 10^8$ years. The average intensity of the slowly decaying heat source over the first 0.5 Ga is about 85% of the initial rate used in our calculation, which means we should revise our estimate to about $5.8 \times 10^8$ a. At that time, the entire interior of the body has produced enough heat to raise itself to the ice I + ice III + liquid triple point and begin melting. This will occur, however, only in bodies so large that the heat produced by internal radioactive decay cannot reach the surface of the body and be radiated off into space within those 580 million years or in smaller bodies that accrete in planetary subnebulae at temperatures far above 100 K. These bodies will have had an opportunity

**Table VI.3** Long-Lived Radioactive Heat Sources[a]

| Element | Isotope | Abundance (ppm wt) | Half-life ($10^9$ years) | Heat production rate (erg $g^{-1}$ $s^{-1}$) |
|---|---|---|---|---|
| K | | 815 | | |
| | $^{40}K$ | 0.097 | 1.47 (to $^{40}Ca$) 11.8 (to $^{40}Ar$) | $2 \times 10^{-8}$ |
| Th | $^{232}Th$ | 0.04 | 13.9 | $1 \times 10^{-8}$ |
| U | | 0.012 | | |
| | $^{235}U$ | 0.00009 | 0.72 | $7 \times 10^{-10}$ |
| | $^{238}U$ | 0.012 | 4.51 | $1.2 \times 10^{-8}$ |
| Rb | | 2.5 | | |
| | $^{87}Rb$ | 0.7 | 50. | $2 \times 10^{-9}$ |

[a] Present-day heat production rates per gram of rocky material.

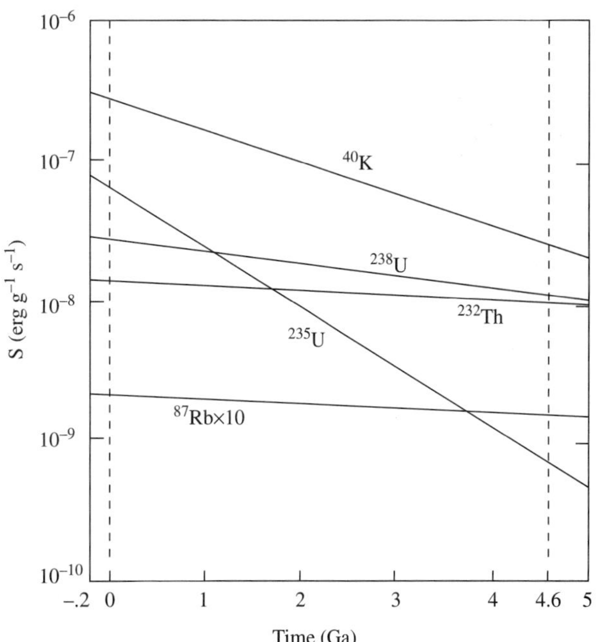


**Figure VI.12** Time dependence of long-lived radiogenic heat sources in rocky material. Note the early importance of the relatively short-lived isotopes $^{40}K$ and $^{235}U$. After Lewis and Prinn (1984).

to melt and undergo density-dependent differentiation into a rocky (actually, muddy!) core and an outer shell of water and ice.

Consider a present-day satellite with a thick, cold ice crust overlying a rocky core. We shall initially assume, for convenience, that the body is in a thermal steady state, in which the rates of energy production and loss are exactly equal. The thermal luminosity of the body is, like that in a Main Sequence star, localized strongly in its core. At all levels in the aqueous mantle outside the core, the total luminosity (erg $s^{-1}$) is constant at $L = SM = S\rho V$. The flux through each spherical shell of radius $r > r_c$ (the core radius) is $\Phi = L/4\pi r^2$. The approximate near-surface heat flux for Ganymede is then 2.5 erg $cm^{-2}$ $s^{-1}$.

The near-surface region of a satellite of one of the giant planets will have a temperature controlled by solar radiation. Depending on the Bond albedo and the heliocentric distance, this temperature will usually lie in the range from 40 to 110 K. With such low surface temperatures, the crust must be frozen to a substantial depth.

In general, for a spherical body in radiative steady state, with only solar and radiogenic heating at work,

$$dE/dT = 0 = S\rho V - \varepsilon\sigma T_s^4(4\pi r_s^2) + (1 - A)F_\odot(\pi r_s^2/R^2). \qquad \text{(VI.21)}$$

The first two terms on the right are the radioactive heating and radiative cooling terms, and the third term is the absorbed (not reflected) portion of the solar flux. Substituting in for the volume ($V$) and simplifying,

$$4r_sS\rho/3 + (1 - A)F_\odot/R^2 = 4\varepsilon\sigma T_s^4. \qquad \text{(VI.22)}$$

Normally the solar flux greatly exceeds the internal heat flux. Even for a Ganymede-sized body with a radiogenic surface heat flux of 2.5 erg $cm^{-2}$ $s^{-1}$, the solar flux will be 5000 times larger at 10 AU and 50 times larger at 100 AU. Thus within the planetary system the surface temperature of a solid body is controlled by solar illumination. At distances beyond about 500 AU the internal heat source controls the surface temperature. A black body surface radiating the radiogenic flux will have a surface temperature given by

$$r_sS\rho = 3\varepsilon\sigma T_s^4, \qquad \text{(VI.23)}$$

which, for a Ganymede, would be 16 K. For an ice-plus-rock sphere 4 km in radius, the internal heat source could maintain a surface temperature of only 3.2 K. Bodies of this size or smaller, such as comet nuclei, would be so cold at great heliocentric distances that heating by the Big Bang black body radiation would prevent their temperatures from falling below about 3 K.

Let us suppose that the crust of a large icy satellite is, as suggested by our observations of Ganymede, made of nearly pure water ice. The thermal luminosity of the satellite, $S\rho V$, is lost uniformly through the entire surface area of the body. The mean surface temperature, regulated by the absorption of sunlight, is about 110 K. These temperatures are so low that radiation cannot carry any significant part of the internal heat flux, and ice at 110 K is so rigid that convective transport of heat can easily be ruled out. Thus the internal heat flux is carried to the surface entirely by conduction. Equating the rate of heat production to the rate of heat loss by conduction (called the *thermal steady-state approximation*),

$$S\rho V = -\kappa(4\pi r_s^2)(\partial T/\partial z), \qquad \text{(VI.24)}$$

and substituting for $V$ as before, we can solve for the vertical temperature gradient in the crust:

$$\partial T/\partial z = -S\rho r_s/3\kappa. \qquad \text{(VI.25)}$$

Using the value of $S$ typical of bodies containing solar proportions of rock and water ice ($S = 1.7 \times 10^{-8}$ at present) with a density similar to that of Ganymede (about 2 g $cm^{-3}$) and using $6 \times 10^5$ erg $cm^{-1}$ $s^{-1}$ $K^{-1}$ for the heat conductivity of ice at 110 K, we find that the near-surface temperature gradient is $-1.9 \times 10^{-14} r_s$ K $cm^{-1}$. If $r_s$ is expressed in kilometers and the temperature gradient in K $km^{-1}$, this can be written $-1.9 \times 10^{-4} r_s$ K $km^{-1}$. Ganymede, with a radius of 2600 km, would have a temperature increase of 0.5 K per kilometer of depth.

However, the thermal conductivity of ice decreases as the temperature increases: at 170 K the conductivity has fallen to $4 \times 10^5$ erg $cm^{-1}$ $s^{-1}$ $K^{-1}$, and at the normal melting point of ice (at 1 bar pressure) the conductivity is only $2.25 \times 10^5$. Therefore the temperature gradient becomes steeper with depth. If conduction were the only means of heat transport in the ice crust, then the average temperature gradient in the crust would be about 1 K $km^{-1}$. The normal melting point of ice would be reached at a depth of about 160 km.

We have seen, however, that the melting temperature of ice decreases with increasing pressure. We can calculate the dependence of pressure on depth by the hydrostatic equation

$$dP = -\rho g(r)\, dr, \qquad \text{(VI.26)}$$

and the near-surface acceleration of gravity, *g*(*rs*), is just

$$g(r_s) = GM/r_s^2, \qquad \text{(VI.27)}$$

whence $dP\ dr = -28.5$ bar $km^{-1}$. At a depth of 80 km the pressure is already 2280 bar, above the pressure at which the melting point of pure water ice is depressed to its lowest value, 251 K. By Eq. (VI.25), we can calculate that the temperature at that depth is about 170 K. Thus, although the surface temperature (110 K) is 160 K colder than the melting temperature of ice at the surface (273 K), the temperature at 80 km depth is only about 80 K below the local melting point of ice. This has a profound effect on the physical behavior of the ice: at temperatures far from the melting point, ice is extremely strong, but it becomes enormously weaker and more subject to viscous flow as the melting temperature is approached. The viscosity depends roughly exponentially on the temperature:

$$\eta = \eta_0 \exp\{c(T_m - T)/T\}. \qquad \text{(VI.28)}$$

For water ice at low pressures, $\eta_0$ = about $10^{14}$ P and $c$ is about 26. These constants will need to be redetermined for highly compressed ice I near 250 K; nonetheless, we can expect that convective heat transport will rapidly become important at great depths in an ice crust.

The physical case presented by convective transport of the internal heat flux differs radically from that of conductive transport. Because the thermal conductivity of ice is so low, rather high temperature gradients are required to transport the steady-state heat flux. But, in a regime in which convection is dominant, the convective motions must take place on a time scale shorter than the conductive transport time. This means that each parcel of convecting ice must behave nearly like a closed system. It will interconvert its thermal energy and its compression energy readily, but it will gain or lose heat to its surroundings only very slowly. To a good approximation, the convecting ice assumes an adiabatic structure, just as we saw for the atmospheres of stars and planets. But there is a great difference: ice is only very slightly compressible, and its thermodynamic behavior must be very different from that of an ideal gas!

Let us look at the thermodynamics of an adiabatic process in a very general way, appropriate to gases, solids, and liquids alike. First, suppose that the material under consideration has an equation of state, $v(T, P)$, that is known. In general, then, small changes in the temperature and pressure affect the molar volume ($v$) of the material:

$$dv = (\partial v/\partial T)_P\, dT + (\partial v/\partial P)_T\, dP. \qquad \text{(VI.29)}$$

Defining $\alpha$ as the coefficient of thermal expansion and $\beta$ as the isothermal compressibility,

$$\alpha = (1/v)(\partial v/\partial T)_P \qquad \text{(VI.30)}$$

$$\beta = -(1/v)(\partial v/\partial P)_T, \qquad \text{(VI.31)}$$

where $\beta = 1/B_T$, the isothermal bulk modulus, we have

$$dv = v\alpha\, dT - v\beta\, dP. \qquad \text{(VI.32)}$$

Recall that, in general, the heat capacity at constant pressure, $C_p$, is $(dq/dT)_P$ (see Appendix I), and $dq_{rev} = TdS$, whence $dq/dT = T(dS/dT)_P$. Then

$$(\partial S/\partial T)_P = C_P/T. \qquad \text{(VI.33)}$$

Also, from the Maxwell relations (Appendix I),

$$(\partial S/\partial P)_T = -(\partial v/\partial T)_P. \qquad \text{(AI.57)}$$

In an adiabatic process the entropy is a crucial variable. In general, the entropy will depend on the state functions $T$ and $P$:

$$dS = (\partial S/\partial T)_P\, dT + (\partial S/\partial P)_T\, dP. \qquad \text{(VI.34)}$$

Substituting in for the two partial derivatives and applying the equation to an adiabatic process ($dS = 0$), we then have

$$dS = (C_P/T)\, dT - (\partial v/\partial T)_P\, dP = 0, \qquad \text{(VI.35)}$$

whence

$$(\partial T/\partial P)_S = v\alpha T/C_P. \qquad \text{(VI.36)}$$

For solids and liquids, the hydrostatic equation gives the dependence of pressure on depth [Eq. (VI.26)], and we can determine the adiabatic temperature gradient by the chain rule:

$$(\partial T/\partial P)_S(\partial P/\partial z) = (\partial T/\partial z)_S = -\mu g T\alpha/C_P. \qquad \text{(VI.37)}$$

This is the adiabatic lapse rate for solids, liquids, or gases. The temperature gradient calculated for an adiabatic solid or liquid with a density or about 1 on Ganymede is only a few thousandths of a degree per kilometer. Thus *convection virtually wipes out the vertical temperature gradient* in the convecting region.

This means that, in a differentiated large satellite with a rocky or muddy core and an ice mantle, the temperature at the surface of the core is at most a few degrees warmer than the temperature at the point in the crust at which the viscosity first drops low enough to permit convection. This temperature is well below the normal melting point of ice. Further, pressures deep within the body are many kilobars, so high that the freezing point is elevated above 0°C. Thus convection at any depth within the body tends to thermostat the convecting region at a temperature well below the local melting point. The thin crustal layer that has a conductive temperature profile will be too cold and brittle to convect along with the warmer, much less viscous ice below it. The cold, brittle surface layer of a planet or satellite is generally termed the *lithosphere*. The lithosphere is subject to substantial stresses imposed on it by the convecting material below and will, if it is not too thick, display very large-scale faulting and relative motion of different blocks of crust. Large coherently moving regions of crust that are separated from other regions by such faults are called *crustal plates*. If the cold crust is too thick, it will resist these stresses and not show signs of plate motions. If, as sometimes happens, the crust is quite thin, then plates may crumple or slide over and under each other when thrust together by the motions of the convecting mantle.

Figure VI.13 is a somewhat schematic thermal structure model of a large Galilean satellite in a thermal steady state (the rate of heat production is equal to the rate of heat loss from the surface) with present-day levels of radioactivity. There are several important circumstances that would make the body warmer throughout and possibly make liquid water an important constituent of the interior. First, the total heat production rate by long-lived radionuclides was nearly a factor of 10 larger early in the history of the Solar System. This increased the heat flux through the crust by the same factor. The mean temperature of the crust was also raised, which in turn lowered the heat conductivity of the crustal ices. Thus the near-surface temperature gradient was then higher by more than a factor of 10, or about 12 K km$^{-1}$. The temperature would then have risen from the mean surface temperature of 110 K to the minimum melting temperature of 251 K at a depth of only about 12 km, where the pressure was about 340 bar. Thus even if the temperature was constant at all greater depths, melting would have unavoidably occurred in the 2-kbar region.

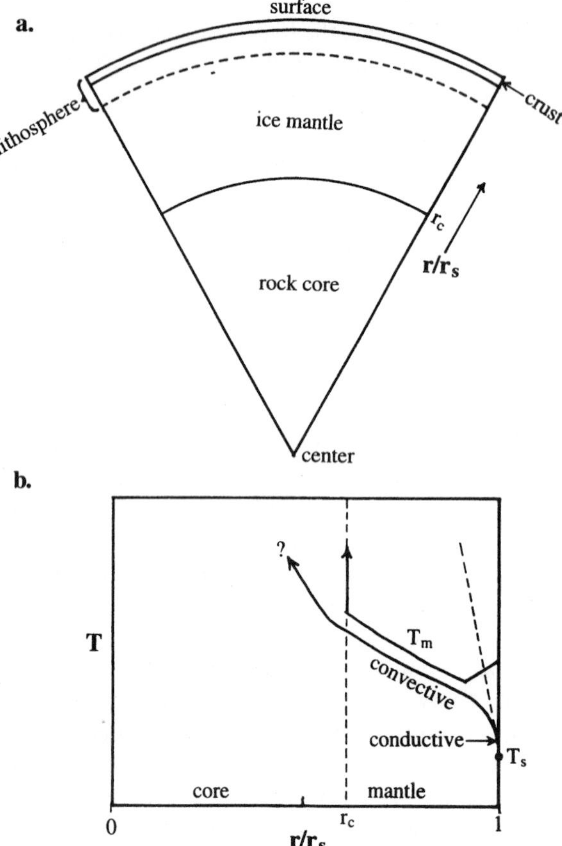


**Figure VI.13** Thermal structure of a large icy satellite. The structure is given in the pie diagram in a. The upper solid line in b is the melting curve, and the lower is the temperature profile. This model represents a present-day thermal model in steady state, with due allowance made for convective transport of heat by solid ices. In the absence of convective transport (dashed line) the model would follow a conductive steady-state temperature profile, leading to melting at shallow depths.

Second, early in the history of a satellite there are several possible additional heat sources that may either provide additional internal heat or raise the surface temperature of the satellite. These include radiative heating from an enveloping protoplanetary nebula, gravitational potential energy and kinetic energy liberated by the accretion of smaller bodies, and decay of short-lived radionuclides, especially $^{26}Al$. For satellites at great distance from their planets or for large ice-plus-rock bodies in heliocentric orbit, both inductive heating by the T Tauri phase solar wind and radiative heating by the T Tauri phase of the Sun may be important.

Third, the addition of small quantities of soluble substances such as ammonia or salts can lower the freezing temperature of water dramatically, to as low as −100°C in the case of aqueous ammonia solutions.

Fourth, as in the case of Europa, the body may accrete at the edge of the ice stability field and contain a much higher ratio of rock to ice than the solar proportion found in Ganymede and Callisto. This leads to much less dilution of the radioactive heat sources by radiochemically inert matter and hence a higher heat production rate per unit of mass.

Fifth, when heat sources are active on time scales shorter than the conductive time scale of the crust, the rate of heat loss from the surface cannot keep pace with the rate of heat production, and the steady-state assumption breaks down. Global warming or cooling will then occur, and careful time-dependent modeling will then be needed to describe the thermal structure of the body.

Finally, gravitational (tidal) interactions between the large satellites in a compact satellite system may provide a brief but very intense source of heat at an early stage in its evolution. In order to assess the importance of such interactions, we need to consider the dynamical interactions of the satellites of the outer planets.

## Dynamical Interactions of the Galilean Satellites

The first natural satellite studied, Earth's Moon, always keeps the same face toward Earth. The Moon rotates at a constant rate, but, because of its orbital eccentricity of 0.0549, it does not revolve around the Earth at a constant rate: as seen from Earth, the Moon "rocks" back and forth on its axis once every month, allowing terrestrial observers to see a total of about 60% of the lunar surface over the course of a month. The equality of the rotational and orbital periods of the Moon can scarcely be a matter of coincidence, any more than the resonant rotation of Mercury (exactly three rotations for every two orbital periods) could be the result of pure chance. Tidal interactions are responsible for these spin states.

Consider a secondary body (a satellite) with a deformable surface layer (such as a crust, ocean, or atmosphere) in circular orbit about a much more massive primary body (a planet). The center of mass of the moon follows a circular orbit under the influence of the gravitational force on it, $(-GMm/r^2)$. The fictitious centrifugal force (which is simply the size of the *centripetal* force needed to keep the moon in its circular orbit) increases with distance. The point on the surface of the moon at which the planet is directly overhead moves around the planet with a speed fixed by the motion of its center. But this point is closer to the planet than the center of the moon is; the gravitational force exerted by the planet is *stronger* there and acts to cancel part of the moon's gravity. Thus things weigh less there, and the surface layer will bulge upward in that region. A *tidal bulge* will decorate the planetward side of the moon (Fig. VI.14).

The region on the opposite side of the moon is traveling faster than the orbital speed of a body at the same distance from the planet. The gravitational force that would be needed to hold it in place at that distance is thus *greater* than that at the center of the moon—but the gravitational force of the planet is actually *smaller* there than at the center of the moon. This deficiency of gravitational force allows the surface layer on the far side to bulge outward to form a second tidal bulge. These two bulges elongate the moon along a radial line from the planet.

Because the moon and planet are actually orbiting about their common center of mass, the same phenomenon occurs on the planet. However, the mass of the planet is commonly 100 to 10,000 times that of the moon, and the force exerted by the moon on the planet is therefore very small relative to the planet's surface gravity. The bulges raised on the planet are therefore much smaller.

Now let us suppose that the planet is rotating many times per orbital period of the moon (many days per month) and that both motions are in the same direction in the same plane. Then the rotation of the planet will tend to "drag" the planet's tidal bulge ahead of the position of the moon. In this configuration, the moon's gravitational pull on the tidal bulge no longer lies precisely along the moon–planet line. Instead, there is a significant component of the tidal force that operates *against the spin of the planet*. This force is called a *tidal torque*. By Newton's laws, an equal and opposite force must be exerted on the moon by the gravitational pull of the planet's tidal bulges. This force leads to a *secular acceleration* of the moon, which of course causes it to retreat from the planet into a higher, longer-period orbit. Over very long periods of time, energy will be extracted from the planetary spin and added to the moon's orbital energy until either the new length of the day becomes equal to the new length of the month or the moon escapes altogether from the planet, leaving the planet still spinning slowly. Pluto and its satellite Charon have achieved a perfect lock in which the planetary day and month are equal. In this state the tidal bulges are perfectly aligned along the planet–moon line, and the tidal torque vanishes.

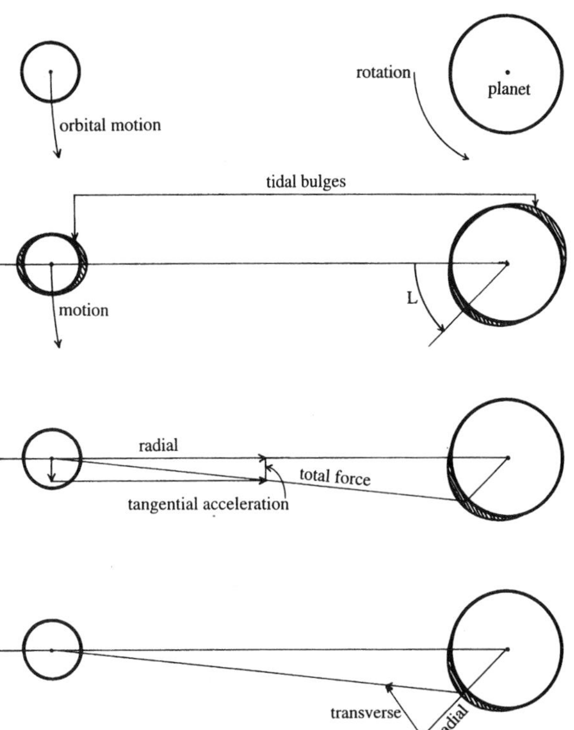


**Figure VI.14** Tidal torques. A massive satellite orbits a planet, with each body raising tides on the other. The tidal bulge on the planet is dragged ahead of the position of the satellite by its rotation. The moon–planet–bulge angle [the lunitidal interval ($L$)], typically about a degree, is exaggerated here for clarity. The gravitational attraction of the bulge for the satellite exerts a force that can be resolved into two components, radial (planetocentric) and tangential. The latter force exerts an accelerating torque on the moon's orbital motion, causing it to move out to a higher, longer-period orbit. The necessary energy is sapped from the rotational energy of the planet, causing its spin rate to slow.

We have so far neglected the spin of the satellite. With a little thought we can see that, because of the very large gravitational force exerted by a planet, the tidal bulges raised by the planet on the moon will exert very important torques on the moon opposing its rotation.

Consider for a moment the outer deformable layer of the planet. If it is infinitely flexible and has zero viscosity, then it will not be dragged off the planet–moon line by rotation. If it is a real substance, however, it will interact viscously with the interior of the planet and hence be dragged along and also dissipate energy. Not all of the energy extracted from the rotation of planets and satellites by tidal forces goes into angular velocity changes. Some significant fraction is dissipated within the bodies as heat.

A satellite formed with an initial spin period of a few hours will not only be despun by tidal interactions in a geologically short time, but also heat up strongly while doing so. A large, rapidly rotating body may store enough rotational energy to heat the entire body by over 100°C.

While such a temperature increase is not important for rocky bodies, it may profoundly affect the thermal evolution of ice-rich bodies. The large satellites of the outer planets are without exception rotationally locked on their primaries, and almost all of them clearly contain vast amounts of easily melted ices. Tidal interactions between these satellites and their planets are unquestionably important.

The large Galilean satellites of Jupiter not only are rotationally locked onto Jupiter (a *spin–orbit resonance*), but they also are involved in a complicated mutual orbit–orbit resonance. This means that any force exerted on one of these bodies would be transmitted by means of gravitational interactions to the other participants in the resonance. The easiest way to state the nature of the resonance is in terms of the *mean motion*, *n* (degrees day$^{-1}$), of the satellites in their orbits about Jupiter; the proper motion is just $360/P$, where $P$ is the orbital period in days. The resonance is then described by $n_1 - 3n_2 + 2n_3 = 0$, where the subscripts 1, 2, and 3 denote Io (J1), Europa (J2), and Ganymede (J3), respectively. This unusual triple resonance, first discovered by the great French mathematician Pierre-Simon Laplace in 1805, is still known by his name.

The first important question regarding orbital resonances is why they occur. Almost all of the satellites in the Solar System have orbital periods that are longer than the length of their planet's day. Thus they will continuously be accelerated by the tidal torques exerted by their primaries and will continuously retreat from the planet to higher, slower orbits. Close satellites will raise very large tidal bulges on their primaries, and the gravitational effect of that tidal bulge on the nearer satellites will of course be large. Further, massive satellites will raise much larger tidal bulges than small satellites. We must expect all satellite orbits to evolve and to do so at different rates. For this reason, adjacent satellite pairs may wander into resonant relationships. But how can we expect pairs of satellites to *stay* in resonance as they drift outward at different rates?

Laplace's answer to this question was to consider a massive satellite (S) in a circular orbit about its primary. Farther out is a second, less massive satellite (s) in an eccentric, coplanar orbit. We must consider what would happen if these satellites were to accidentally evolve into a resonant relationship (meaning that the positions of the planet, S, and s repeat at regular intervals that are integral multiples of their orbital periods). When s and the planet are in opposition as seen from S (180° apart in the sky) the satellites may in general be at any point on their orbits. This exact configuration will repeat regularly. No particular orientation of the two orbits is preferred at the time that the resonance is first established (see Fig. VI.15).

But every time S passes under s it is accelerated slightly along its orbit up to the point of conjunction, and decelerated thereafter. If both orbits were perfectly circular, these interactions would be symmetrical and lead to no net torque on either satellite. But, when one of the orbits is eccentric, the only interactions between S and s that are symmetrical are those that take place at the periapsis and apoapsis of the orbit of s. Any other orientation causes a small unbalanced net torque that perturbs the satellites toward that stable alignment—and

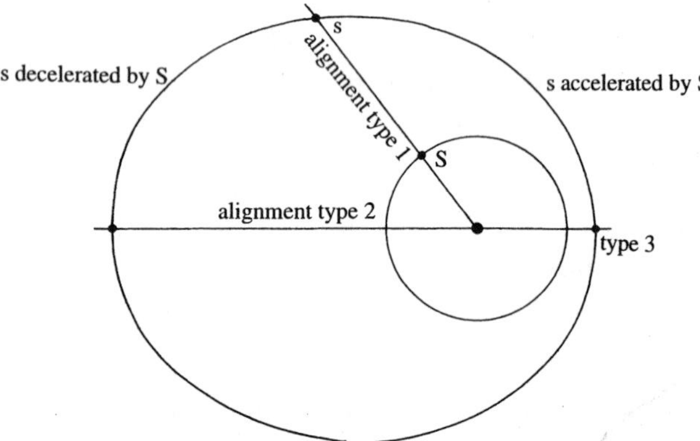


**Figure VI.15** Orbit–orbit resonance. Satellites S and s are in resonant orbits, which means that the exact configuration of the system repeats at regular intervals; the orbital periods of the two satellites are in a ratio of small whole numbers, such as 2:1 or 3:5. The more massive satellite, S, is taken to be in a circular orbit, and the orbit of s is assumed to be eccentric. The s–S–planet alignment (syzygy) may occur at any point on the orbit. Three examples are shown: type 1 (random), type 2 (at s apogee), and type 3 (at s perigee). Types 2 and 3 are stable, but any other alignment will result in repetitive unbalanced torques being applied to s by S.

overshoots it! It is entirely possible for the planet–S–s line at conjunction to oscillate slowly back and forth past the periapsis–planet–apoapsis line (the line of apsides) of the orbit of s. This slow oscillation, like that of an ancient set of scales, is called *libration* after the Latin word for scales, *libra*. A very slight departure of s and S from resonance would cause the line of conjunctions to rotate slowly about the planet, to *circulate* rather than librate about the stable configuration. This way of looking at resonant or near-resonant motions in terms of the positions of the bodies at conjunction only can be regarded as a *stroboscopic* perspective.

Libration about the stable resonance would persist forever if there were not some means of slowly damping out the motions of the bodies. But we have already seen that, in any real material, the constantly shifting tidal forces cause the flexure of imperfectly elastic materials, which dissipates an important amount of heat. Thus, like the slowly damped oscillation of a pendulum under the dissipative influence of air friction, the libration of a resonant satellite pair is slowly damped by viscous dissipation of energy in the interiors of the satellites.

Thus the gravitational interaction of a satellite pair can lead to formation of a stable lock between them. A small force (such as tidal torque) acting to disturb the satellites unequally will *force their orbits to evolve together*. This pair may then eventually drift into a configuration in which S and s have a resonant relationship with a third moon, and the entire process can repeat itself to establish a three-satellite stable resonance. Such resonances are not *unconditionally* stable, because a strong disturbing force (such as a major impact event) may break the lock and permit the satellites to evolve independently, at least as far as the next orbital resonance.

In the Io–Europa–Ganymede resonance, the most surprising feature has long been the almost unobservably small libration—and this in a very complex three-body resonance in which there is energy flow among all three satellites and Jupiter. It was just weeks prior to the encounter of Voyager 1 with Jupiter in 1979 that Stanton J. Peale of the University of California at Santa Barbara and his colleagues realized the implications of this observation: there must be strong energy dissipation in the interiors of the Galilean satellites, especially Io, in order to keep the libration so small. As we shall see in Chapter X, when we discuss the rocky planetary bodies, this huge energy source has had a profound effect on the evolution of Io.

The same mechanism provides a much smaller but still important amount of heat inside Europa. It is conceivable that the heat flux is large enough to maintain the lower portion of Europa's thin $H_2O$ surface layer in the liquid state.

## Thermal and Tectonic Evolution of Icy Satellites

To trace the thermal history of a satellite from the time of its formation to its present state and to determine whether a particular ice-plus-rock body has differentiated into a dense rocky core and an ice envelope, we need to have a substantial amount of information about the conditions under which the satellite originated and what heat sources have acted upon it. In general, we do not have sufficient information about any satellite to make this possible. Instead, we have had to be content with running suites of thermal models for a wide range of possible assumptions and checking the predictions of these models against our observations of the satellites. Some comments on our progress are appropriate.

First, it is relatively easy to calculate conductive models for satellites heated only by long-lived radionuclides. Each volume element of the satellite produces heat at a rate that is dependent on the concentration of rock-forming elements and upon time in the manner we discussed above. If convection is ignored, then these bodies will heat up to the melting point and begin to differentiate if they have radii greater than a few hundred kilometers. In general it would seem that larger bodies (with better insulated interiors) should be easier to melt than small bodies. It is certainly true that the smallest bodies will have thermal diffusion time scales that are short compared with the half-lives of the major radioactive heat sources, and hence heat will not be accumulated readily. The thermal diffusivity of ice at about 100 K is $K = \kappa/\rho C_P = 7 \times 10^5/(0.94 \times 2 \times 10^7) = 0.034\,\mathrm{cm^2\,s^{-1}}$. This is about 10 times larger than the thermal diffusivity of loose terrestrial soils, and 3 times that of packed soils. The thermal diffusion time scale over a distance, $d = 10\,\mathrm{km}$ (a 10-km-radius body), is $d^2/K = 3 \times 10^{14}\,\mathrm{s}$, or about 1 million years. The diffusion time is equal to the decay half-life of $^{40}K$ when the body is about 360 km in radius. Larger bodies might then accumulate heat despite conductive cooling.

However, all is not that simple. For one thing, bodies large enough to have central pressures over 2 kbar will have melting points that increase with pressure. Also, once temperatures rise to 70–80% of the melting temperature at any depth, convective heat transport will become more rapid than conduction.

The usual criterion of convective instability in a fluid layer is that the Rayleigh number,

$$\mathrm{Ra} = \alpha g d^3 \Delta T / K\nu, \qquad \text{(VI.38)}$$

be larger than about 1000. Here $\alpha$ is the coefficient of thermal expansion [see Eq. (VI.30)], $g$ is the local acceleration of gravity, $d$ is the layer thickness, and $\Delta T$ is the temperature difference across the layer in excess of the adiabatic temperature difference (which will usually be extremely small—remember, convection becomes important where $\Delta T$ is roughly equal to the steady-state *conductive* gradient, which is 100 or more times larger than the adiabatic gradient). $K$ is the thermal diffusivity (obviously, large diffusivities make it harder to get convection going), and $v$ is the kinematic viscosity, $\eta/\rho$. The surface–center temperature difference for a uniform conductive body is roughly $S\rho d^2/\kappa$, and $\Delta T$ is just this amount less the adiabatic temperature difference, $\Delta T_{\mathrm{ad}}$.

This convective cooling mechanism arrests the warming of bodies that are larger than a few hundred kilometers in radius and makes internal melting much harder to achieve by means of long-lived radionuclide decay. The best way to melt such a body is to accrete it warm; the second-best way is to provide a strong heat source (such as tidal dissipation in Io and Europa) that provides more heat than potassium, uranium, and thorium can. Unfortunately, we know very little about accretion-related heat sources and ambient temperatures at the time of formation. Thus our models are hard to constrain.

The melting of a massive icy satellite, with a central pressure far above 2 kbar, is hindered by the high melting temperatures of dense forms of ice. First melting is likely to occur at a pressure near the minimum in the melting temperature at 2 kbar. Much of the outer regions will then undergo density-dependent separation of rocky and icy materials, as in Fig. VI.16. The surface and a thin layer beneath it will be too cold to melt and too stiff to convect downward and will remain as a primitive ice–rock mixture. At slightly greater depths there will be a low-density pure ice layer ($\rho = 0.94$) floating on liquid water with a density of about 1. At the base of the water layer will be a thick layer of rocks or mud, ranging in density from about 2.4 to as high as 3.5. Beneath the mud layer is the undifferentiated core, made of dense forms of ice still mixed with rock, with a net density of about 2.0. Precipitation of the mud involves the fall of dense materials in a gravitational field; it releases some gravitational potential energy as heat, and the heat produces further melting.

The structure that is the immediate product of early melting and differentiation is unfortunately unstable, because the mud layer is denser than the undifferentiated core on which it rests, and the cold, primitive crust is denser than the ice I on which it rests. At some early date during melting, accidental fracturing of the crust should permit truly stupendous geysering of liquid water through the crust, followed by foundering, breakup, and melting of the primitive crust. A pure ice I layer will quickly form on the surface. The unstable interior must eventually come to terms with gravity, presumably by an intense episode of single-cell convection, in which the primitive core and its outer mud shell switch places. Both the core inversion and the crustal foundering are energetic, exothermic processes. Either one could trigger the other. Note that the largest bodies have the thinnest primitive crust and thus would undergo these processes most easily.

One of the central dilemmas presented by the geology of the Galilean satellites is the great difference in albedo and surface morphology between Ganymede and Callisto. Ganymede's surface is heavily reworked and shows strong evidence of global-scale tectonic processes. The surface has the spectral signature of almost pure water ice. Callisto, however, has a densely cratered and very ancient surface, virtually devoid of any suggestion of internal thermal activity, and has a high concentration of dark nonice materials in its crust. Since Callisto is somewhat smaller than Ganymede, it is tempting to attribute these differences to the persistence of a thick, primitive crust on Callisto. Ganymede, with its higher heat flux, must have suffered an impact sufficient to penetrate its thinner crust. The dark, primitive material then foundered and was replaced by water.

This is a plausible explanation, but is not the only tenable one. Perhaps the most reasonable alternative is that conditions at the time of formation of the Galilean satellites were dominated by a strong radial temperature gradient outward from Jupiter. Ganymede accreted out of warmer material in a warmer medium and (because of its greater mass) liberated more potential energy during accretion. This was sufficient to push Ganymede over the edge into a history of extensive melting and tectonic activity, whereas the smaller, colder Callisto at best underwent limited layer melting at depths much deeper than the thickness of Ganymede's crust.

There is another interesting but poorly explored phenomenon associated with the heating of icy satellites by radioactive decay. That is the chemical destruction of water by beta and gamma radiation, a process called *radiolysis*. This will liberate a constant supply of hydrogen and oxygen gas that, upon melting of the ice, would be available to form bubbles and drive volcanic eruptions of cold water. In differentiated bodies with warm, dry rocky cores, almost no water is present in the region where the rocks are. Radiolysis would then be important only if some of the radioactive material, such as potassium, had been leached out of the rocks and resided in aqueous solution in the mantle. Dissolution of such materials would also have the effect of lowering the freezing point of water.

Ices exposed at the surface of any icy satellite are also subject to radiolysis driven by cosmic rays and by protons from the solar wind and from radiation belts. Laboratory experiments on plausible ice mixtures have

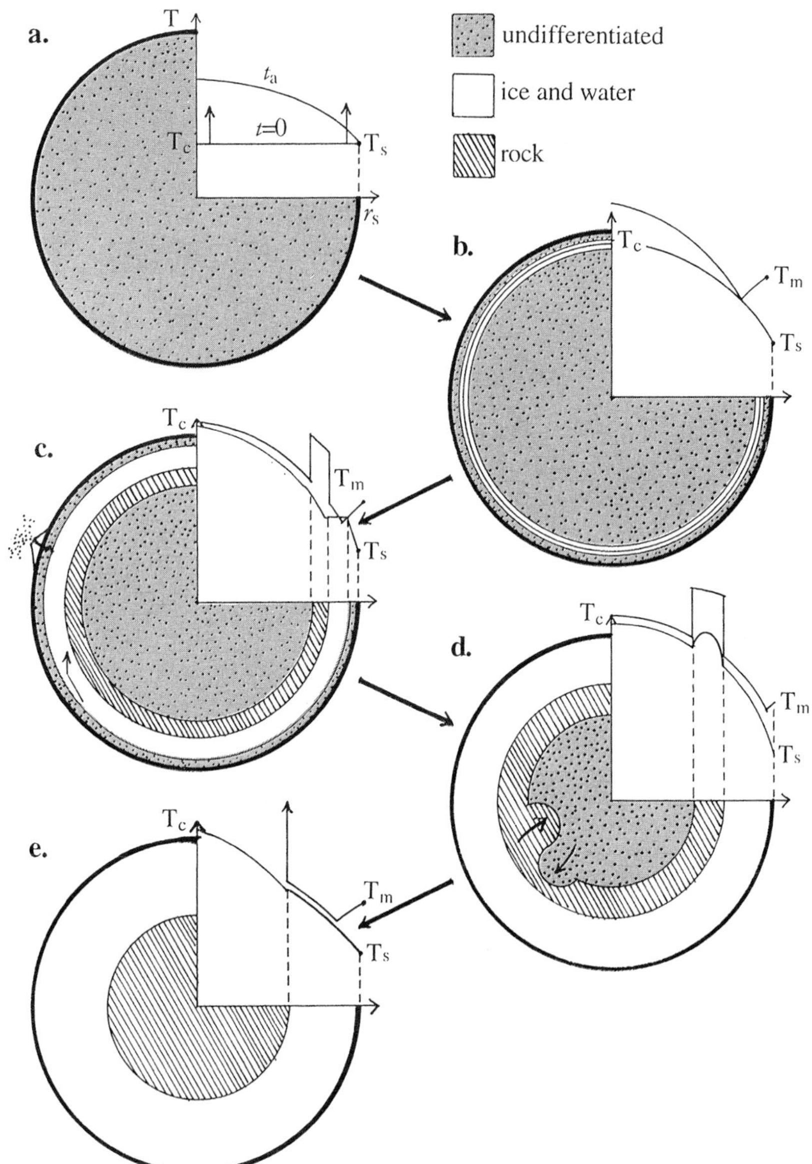


**Figure VI.16** Differentiation of large ice-plus-rock bodies. (a) An initially homogeneous isothermal mixture of rock and water ice (here conservatively assumed to be at the surface temperature of the body) begins to heat up. (b) The internal temperature profile first reaches the melting curve near the 2-kbar pressure level, at which ice is easiest to melt. A thin layer segregates by density into dirt and water. (c) After more extensive melting, only an inner core and a thin crust remain undifferentiated. The crust, being denser than the pure water on which it floats, may founder as a result of faulting, impact puncturing, or volcanism. (d) The outer shell has now differentiated completely, leaving a thick outer layer of convecting pure water ice (with or without liquid water) atop a dense layer of rocky "mud," which in turn rests on a less-dense undifferentiated core. Convective instability inverts the interior structure, liberating energy that may complete the differentiation process (e). In general, the larger the body, the farther it can progress along this evolutionary path.

demonstrated a fascinating ability to produce diverse and biochemically significant organic products. Some experiments start with the ices expected to be prevalent in the outer solar nebula and planetary subnebulae (water, CO, nitrogen, $CO_2$, methane, ammonia), some also include reactive molecules made efficiently by a wide range of high-energy processes (formaldehyde and HCN), and some add molecules known to be abundant in cometary ices or in the interstellar medium (such as methanol).

Laboratory irradiation of methanol–water ice mixtures produce formic acid and ethylene glycol; irradiation of HCN, ammonia, and water ice makes amino acids, H–C–N polymers, and the organic base adenine, $(HCN)_5$; formaldehyde, ammonia, and water ice yield $NH_2$–$[CH_2O$–$]_x$–H polymers, and so on. Irradiation of pure water ice makes hydrogen peroxide and liberates hydrogen. Indeed, the VIMS experiment on the Galileo Orbiter has detected approximately 0.13% $H_2O_2$ in the surface of Europa. This chemistry is also relevant to all ice-bearing small bodies, including comet nuclei, Centaurs, trans-Neptunian objects, and small ice-rich satellites.

The solid materials of these satellites, especially at accessible depths, will be dominated by igneous minerals, of which by far the most important is water ice I. The interiors will be dominated by ice III, V, and VI. The cores may preserve the mineralogy of carbonaceous meteorites (clays, sulfides, sulfates, carbonates, and magnetite), or they may have warmed sufficiently to release volatiles and solutes into aqueous solution (Europa). We are far from having any reliable knowledge of the composition, temperature, or density of any satellite core.

The best evidence for cores in the Galilean satellites comes from experiments on the Galileo Orbiter. Intense radio noise was found in the immediate vicinity of Ganymede, marking the point of entry into a magnetosphere centered on Ganymede's deep interior. The magnetometer data were used to infer the existence of a dense rocky sphere of radius about 1500 km, topped by a largely frozen ocean of ices about 700 km deep. Most interpretations of these data have spoken of the field as generated in a metallic core within the rocky interior, although it is important to keep in mind that a deep electrically conducting (saline) ocean may confuse the interpretation of the magnetometer data.

Callisto, slightly smaller than Ganymede and farther from Jupiter, shows no clear evidence of magnetic field generation, complicated by puzzling hints of the presence of an ocean. The moment of inertia ratio $C/MR^2$ for Callisto is $0.355 \pm 0.004$, too low for a self-compressed ice-rich body of uniform composition, but too high to suggest complete differentiation. It is possible that a thick outer shell of dirty ice on Callisto has never melted and differentiated. Here again, as with Ganymede, models are usually presented in terms of layers of ice, silicates, and metal. Paradoxes abound: presumably Ganymede and Callisto, with their closely similar densities, are made of virtually identical materials. Should we then picture Callisto as a mixture of grains of ice, iron metal, and silicates? This makes no geochemical sense: iron and ice belong to widely separate parts of the Solar System and are a chemically unstable mixture. A core made by melting outer solar system material would be made of FeS and other sulfides, not metallic iron. Perhaps one could rationalize the production of iron metal by autoreduction of Fe oxides by carbon in a water-free carbonaceous source rock, via the reaction $Fe_3O_4 + 2C \rightarrow 3\ Fe^o + 2CO_2$. In this case, a 700-km metallic core in Ganymede, with a mass of $10^{25}$ g, would be formed by release of $5 \times 10^{24}$ g of carbon dioxide, enough to coat Ganymede's surface with a layer of dry ice 200 km thick. Where is it? Clearly, iron does not appear to be a tenable core material on Ganymede.

The second dilemma is why such a slowly rotating body, locked as it is in a 1:1 spin–orbit resonance, would generate a dipolar field. This is to my mind not a serious objection, since Coriolis forces are still significant in a body rotating in an inertial frame every 7 days. Third, there is the even more interesting issue of how a liquid conducting core could be maintained for billions of years in a small body equipped with a vast supply of cooling fluid. The best answer to this dilemma seems to be that, by virtue of its 1:2:4 orbit–orbit–orbit resonance with Io and Europa, enough tidal power is dissipated inside Ganymede to keep the interior warm. It then remains only to explain why this energy is dissipated deep inside Ganymede's interior, in the putative sulfide core.

Finally, we must remark that the four Galilean satellites seem to have evolved in four different ways. Ice-rich Ganymede and Callisto may have followed divergent paths since their formation. Ice-poor Europa cannot have more than about a 100-km skin of water, and some of this may be melted even today. Io, however, is even more distinctive in its properties. It is an ice-free rocky body, racked by intense geological (or should we say "iological"?) activity driven by the orbital resonant interactions of the three inner Galilean satellites. It is such a different case that we will treat it (in Chapter IX) as a variety of terrestrial planet, not as an enormously aberrant icy satellite. The compositional sequence from Io to Callisto, like the overall compositional trends seen in the Solar System, contains valuable information about the origins of regular satellite systems. It will be most valuable to survey the other two regular satellite systems, those of Saturn and Uranus, as further clues to the conditions under which the Solar System and its component planetary satellite systems were formed.

## Minor Satellites of Jupiter

The minor Jovian satellites fall naturally into three families. One family, orbiting well inside the realm of the Galilean satellites, consists of Amalthea (JV, read

"Jupiter five") and its smaller sisters Metis (JXVI), Thebe (JXIV), and Adrastea (JXV). Of these, only Amalthea is widely enough known so that it is usually called by its name instead of its number. (Amalthea also has the double distinction of having been named by the great French popularizer of astronomy, Camille Flammarion, and having served as the site of a classic science fiction story by Arthur C. Clarke, entitled "Jupiter Five.") These satellites have very small orbital eccentricities ($e < 0.02$) and inclinations ($i < 1°$). Two of them (Metis and Adrastea) orbit so close to Jupiter that they have orbital periods of less than a Jovian day, so that, as seen from Jupiter, they rise in the west and set in the east, opposite to the apparent motions of the Galilean satellites. Metis and Adrastea have almost identical orbital periods and orbit just outside Jupiter's tenuous, narrow ring.

These inner small satellites may be regarded as members of Jupiter's regular satellite system, and therefore as genetically related to the immensely larger Galilean satellites. The most massive of these small inner satellites, Amalthea, probably has about 0.01% of the mass of Io (Fig. VI.17). The visible spectrum of Amalthea is only very poorly known, but seems to resemble that of Io.

There is a wide gap just outside the regular inner satellite system, commencing at the orbit of Callisto ($a = 1.883 \times 10^6$ km) and extending to a distance of over $11 \times 10^6$ km, within which there is only a single known satellite, S/2000 J1. This small satellite has an orbital inclination of 46° (all the regular inner satellites have $i < 1°$) and an eccentricity of $e = 0.204$ (compared to $e < 0.02$ for all the inner satellites). Its diameter is only about 16 km, making it comparable in size to the Martian satellites Phobos and Deimos.

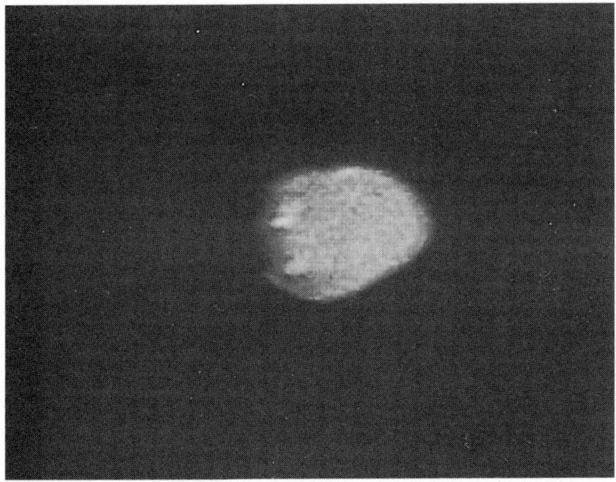

**Figure VI.17** Jupiter's satellite Amalthea (J5). This Voyager 1 image shows the dark, reddish (color exaggerated) surface of this irregular satellite. Impact craters are marginally observable at this resolution.

The two outer families of Jovian satellites are anything but regular. Both families orbit at great distances from the planet, one at about 11.5 million kilometers (six times as far as Callisto) and the other at about 23 million kilometers.

The 11-million km family contains four named members, Leda (JXIII), Himalia (JVI), Lysithea (JX), and Elara (JVII), and one recent discovery, S/2000 J11. Only the most devout aficionados bother to learn their names, and only a select and classically educated subset of them learn how to spell or pronounce them correctly. Normally they are identified by their numbers. The largest of these, JVI, is only about the size of Amalthea, with a diameter of roughly 180 km. JVII is less than half as large, and the others are smaller still. Almost nothing is known of their properties. The visual magnitude is used to estimate size, assuming low and equal albedos compatible with highly carbonaceous D and P asteroids in Jupiter's Trojan (L4 and L5) families. They are too faint for useful spectrophotometry.

These satellites orbit so far from Jupiter ($a$ ranging from 11.084 million km and 12.557 million km) that they have orbital periods of $265 \pm 25$ days. Indeed, their orbits are strikingly similar. Their orbital eccentricities are between 0.1 and 0.25, and their orbits are inclined $28 \pm 1°$ to the equatorial plane of Jupiter. The spectrum of JVI Himalia is similar to the asteroid spectral type F (see Chapter VIII), a subtype of the carbonaceous asteroids.

The last family of Jovian satellites, like the others, also consists of four named members, Ananke (JXII), Carme (JXI), Pasiphae (JVIII), and Sinope (JIX). New and as yet unnamed satellites found in 1999 and 2000 bring the list of known bodies in this family up to a total of 14. The four named members have diameters of roughly 30 to 70 km, compared to 4 to 10 km for the ten newly discovered bodies. It would not be surprising to find many more small satellites if the sensitivity limit of the search program can be extended to the 1–2 km size range.

These outermost satellites are also closely clustered in their orbital properties. All fourteen pursue *retrograde* orbits with orbital periods of about 2 Earth years (588 to 768 Earth days) and inclinations of $154 \pm 11°$. Their orbital eccentricities range from about 0.13 to 0.41. Because of their small sizes they are all very hard to study from Earth. Even a spacecraft in close proximity to Jupiter is at a disadvantage because of the immense size of the Jovian system: the diameter of the orbit of S/1999 J1 is 48.5 million kilometers, compared to the minimum distance of Earth from Venus of 41.5 million km! Further, the surface illumination of these bodies by the Sun is 27 times less than that on Earth's Moon, and they are covered with some of the darkest materials in the

Solar System. Because of their small size, strange orbital inclinations, and immense distances from Jupiter these and their neighboring family all evaded study by the Pioneer 10 and 11 and Voyager 1 and 2 flybys and the Galileo Orbiter.

There are only sparse data available on the physical and chemical nature of the small Jovian satellites. Of the 24 small satellites, only the largest, Amalthea, has been photographed with sufficient resolution to provide any information about surface features. Although the other three inner small satellites were first detected by the Voyager imaging experiment, they are seen only as tiny, slightly irregular specks of light. All four of these satellites appear to be irregular and distinctly nonspherical in outline. Amalthea has a rugged, heavily cratered surface that has experienced a bombardment history comparable to that of the lunar highlands. The 20 outer satellites, since they have been observed only from Earth, are seen only as unresolved points of light. Their estimated diameters neglect the high probability that they, like all the other small ($r < 200$ km) bodies elsewhere in the Solar System, deviate markedly from spherical shape. The composition, cratering history, and physical state of their surfaces remain conjectural. Unfortunately, testing these conjectures will not be easy.

Mass, density, and compositional data are lacking for the small Jovian satellites. While it is easy to believe that the inner satellite family formed at somewhat higher temperatures than Io, and hence are rocky in composition, we cannot point to direct evidence that this is true. Our situation with regard to the outer satellite groups is even worse. We cannot be sure that these satellites, in their very high, loosely bound orbits, are even native to the Jovian system. Although there is no consensus on their origin, capture from heliocentric orbit is a real possibility, and dynamical exchange with the Trojan clouds is a reasonable probability. The wide, unpopulated gap between the (regular) orbit of Callisto and the nearest of the (irregular) outer satellites suggests a lack of continuity in their origins. We would even be justified in suspecting that the outer satellites may be fragments of debris from the asteroid belt or from the vast spaces between the orbits of Jupiter and Saturn.

The only compositional data we have on the small Jovian satellites is the "reddening" of Amalthea, which lends it a coloration very similar to that of Io. Metis, Amalthea, and Thebe all have a hemispheric asymmetry in brightness, with the leading hemisphere brighter on all thee bodies, possibly due to the effects of impact of heliocentric cosmic dust. Recall that Io's outward nearest neighbor, Europa, also is visibly stained with Io's color. Here is an interesting puzzle: is it really possible for Io to "paint" its neighbors? If so, how?

## Planetary Rings

***Jupiter.*** The last members of the Jovian system awaiting our description are the myriad of small objects that make up the modest, rather simple ring that lies between about 1.72 and 1.80 $R_J$ from the center of the planet (Fig. VI.18). The ring was unknown before the arrival of the first planetary spacecraft at Jupiter. Indeed, it was Pioneer 10's discovery of a striking, unexpected drop in the fluxes of magnetospheric particles inside the orbit of Amalthea that first suggested the possible presence of a ring. Both of the Voyager spacecraft were successful in imaging the ring. A grand total of 24 images, all but one from Voyager 2, provide us with all the detailed information we have on its structure.

Most of the matter in the Jovian ring is found in a very thin disk of rocks and dust with a rather sharp outer edge just inside 1.81 $R_J$ and a fuzzy inner edge at about 1.72 $R_J$. The visible-wavelength optical depth of the ring is very low, roughly $3 \times 10^{-5}$. The inner edge gives way to an even more diffuse, fat (extended in the $z$ direction, out of the plane of the thin disk) "halo" of fine dust that extends inward to roughly 1.3 $R_J$. There is some marginal evidence for an even more diffuse outward extension of the thin disk, perhaps as far as to the orbit of Amalthea.

The thin ring is much brighter in forward-scattered light than in backscatter, an observation that is impossible to reconcile with the light-scattering behavior of rocks or moonlets. Clearly most of the light from the rings has been scattered by particles that are close in size (grain circumference) to the wavelength of light. Several analyses of the Voyager measurements of ring brightness at different phase angles agree that the particles are typically about $a = 2\,\mu$m in radius. Such particles not only scatter efficiently in the forward direction, but also have very large cross-section areas per gram. (See Fig. V.24.) Attributing most of the area to small particles should not be interpreted as meaning that large particles are absent; most of the mass could well be present in the form of rocks or larger bodies with very small cross-section areas per gram.

The 10-km-radius satellite JXV, Adrastea (at 1.8064 $R_J$), with no detectable eccentricity or inclination, and the 20-km satellite JXVI, Metis (1.7922 $R_J$), lie close to the outer edge of the main ring or perhaps just within it.

***Saturn.*** One of the most interesting conclusions from the Voyager data on the Jovian ring is how radically it differs from the famous ring system of Saturn. Instead of a single faint, diffuse band, Saturn's ring system is so brilliant and extensive that it was detected, but not understood, by the first telescopic observers in the 17th century (beginning with Galileo in 1610). The Pioneer 10 flyby paved the way for close approaches to Saturn by

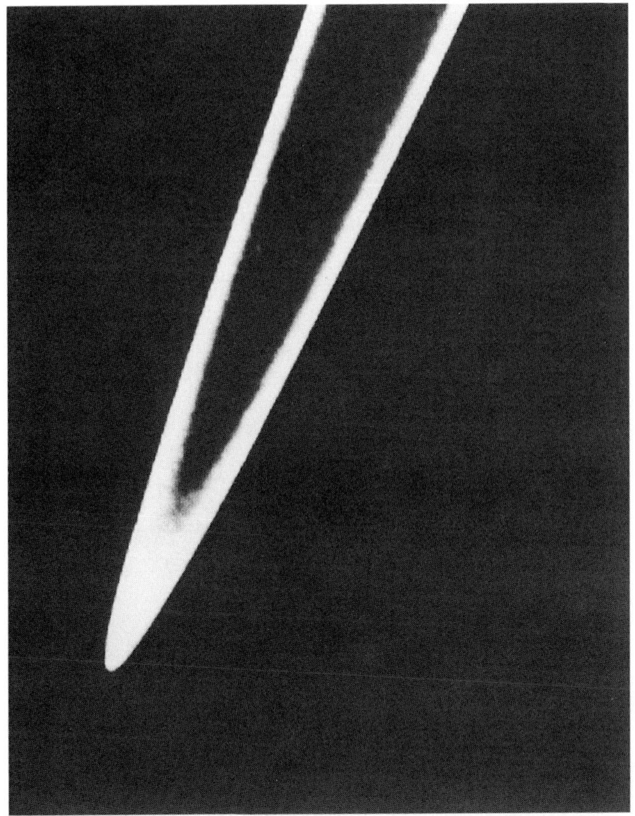

**Figure VI.18** The Jovian ring. This Voyager 2 view shows the rather sharp outer edge and the diffuse inner edge as seen from slightly out of the ring plane at a range of 1.5 million km.

the Voyager spacecraft with their superior camera systems, and it fell to the Voyagers to discover an incredible wealth of features and phenomena in the rings, sufficient to stimulate an explosive growth of theoretical modeling of ring systems.

The brightest, and earliest known, rings of Saturn reach from 1.525 to 2.267 $R_S$. The dark gap at 1.949 to 2.025 $R_S$ between these two rings, the Cassini Division (discovered in 1675), serves to separate the A ring, outside the Division, from the B ring on the inner side (Fig. VI.19). Much later (in 1850) a diffuse band extending from the inner edge of the B ring in to 1.235 $R_S$ was discovered and named the C ring in 1852. Until that time debate raged between those who believed that each ring was a solid, continuous disk and those who claimed that each ring was made of myriads of small, separate solid bodies orbiting independently about Saturn. By 1853 it was known that the edge of Saturn's disk could be seen through the C ring, and the single-body hypothesis was generally abandoned. In 1859 no less an authority than James Clerk Maxwell showed that a single-body ring would be dynamically unstable against even infinitesimal disturbances and hence that the rings must comprise swarms of particles. By 1911 there was photographic proof that even the A and B rings were partially transparent.

The structure of the rings as we now know it is summarized in Table VI.4. Note the locations of the orbits of the several small satellites that have been found interleaved between the rings. The main new features are the very faint D ring which lies internal to A, B, and C; the satellite SXVIII Pan embedded in the A ring and SXV Atlas just outside the A ring; and a strange narrow and multiple-stranded, braided-looking F ring with its bracketing satellites, SVI Prometheus (formerly S/1980 S27) and SXVII Pandora (née S/1980 S26) (Fig. VI.20). Next come the coorbital satellites Epimetheus and Janus, which we shall describe separately, and then the extremely tenuous G ring, Mimas, the E ring, and finally Enceladus, which lies very close to the core of the E ring, perhaps even within it. The smallest of the known satellites within the rings are little more than 10 km in radius. There may be a host of unrecognized 1-km bodies embedded in the denser rings, too small to be detected by Voyager or from Earth.

Earth-based observations of the rings early established that they disappeared completely when seen edge-on and that they therefore must be extremely thin. Voyager confirmed that the rings are as thin as 50 to 100 m.

As thin as the rings are, they are also subject to the disturbing gravitational forces of Titan (which is massive and has a nonzero inclination and eccentricity) and the Sun. They combine to warp the plane of the rings by up to several hundred meters.

The finite thickness of the rings, combined with the substantial optical depth, attests to a complex and interesting dynamics for the ring particles. If we consider a particle in a nearly circular orbit in the midst of the rings with sufficient orbital inclination to take it as high as 100 m from the ring plane (a much larger than average excursion), the out-of plane component of the particle's speed at the time of ring-plane passage, $V_z$, is, relative to the circular orbital speed $V_c$,

$$V_z/V_c = 2\sin(i/2). \tag{VI.39}$$

For inclinations below a few degrees this is very close to sin $i$, and for angles much less than 1° the sine is very well approximated by sin $i = i$ (with the angle given in radians). Our particle rises as much as 100 m above the ring plane at a typical distance from Saturn of about $10^5$ km, for an inclination of $10^{-6}$ rad (1 $\mu$rad). The circular orbital velocity of course varies with position in the rings, but is everywhere on the order of 10 km s$^{-1}$. Thus $V_z$ is about $10^{-6}$ times $10^6$ cm s$^{-1}$, or 1 cm s$^{-1}$. Most particles of course move more slowly.

Thus each particle will cross the ring plane twice on each orbit (at the ascending and descending nodes) at a speed of about 1 cm per second. If the normal optical depth has a typical value of 1 (0.6 is average in the A ring and 1.5,

**Figure VI.19** The Voyager spacecraft returned images with fairly conventional phase angles, but of quality superior to any Earth-based pictures, even while it was many millions of kilometers from Saturn. This view shows several of the satellites (note the reddish Titan at the top), the classical outer A ring, the Cassini Division, and the inner B ring. The shadows of the rings on the planet and of the planet on the rings can both be clearly seen.

in the B ring), then a small particle has a probability of about $1/e$ of traversing the ring in the normal direction without a collision. Because low-velocity collisions of ice particles will always involve some dissipation of acoustic energy, some crushing, and some chance of capture, it is astonishing that the observed thickness of the rings could persist for very long. Some source of energy must be available to "stir up" the rings and keep them from total collapse

**Table VI.4** Saturn's Rings and Ring Satellites

| Feature | Distance $R_S$ | Distance km |
|---|---|---|
| | 1.11 | 66970 |
| D | \| | |
| | 1.235 | 74510 |
| C | \| | |
| | 1.525 | 92000 |
| B | \| | |
| (corotation) | 1.861 | 112300 |
| B | \| | |
| | 1.949 | 117580 |
| Cassini Division | \| | |
| | 2.025 | 122170 |
| A | \| | |
| | 2.210 | 133300 |
| Enke Division | \| | |
| Pan (SXVI) | 2.220 | 135600 |
| | 2.220 | 135600 |
| A | \| | |
| | 2.267 | 136780 |
| Atlas (1980 S28) | 2.282 | 137670 |
| Prometheus (SXVI) | 2.310 | 139350 |
| F | 2.324 | 140180 |
| Pandora (SXVII) | 2.349 | 141700 |
| Epimetheus (SXI) | 2.510 | 151420 |
| Janus (SX) | 2.511 | 151470 |
| G | 2.82 | 170100 |
| Mimas (SI) | 3.075 | 185540 |
| | 3.08 | 186000 |
| E | \| | |
| Enceladus (SII) | 3.946 | 238040 |
| E | \| | |
| | 8.0 | 483000 |

to a featureless monolayer. The persistence of sharp ring boundaries and of distinct fine structure within the rings also is hard to understand without such a source of energy.

For all except the very smallest particles, the only force of consequence is gravity. The gravity of Saturn accounts for the orbital motion, and the Sun's gravity contributes to the warping of the ring plane, but it is the gravity of the inner satellites of Saturn that provides the solution to our problem. As we emphasized above, there are a number of small satellites that orbit near or amid the rings. These satellites interact in complex ways with the ring particles, producing tightly wrapped spiral density waves, bending waves, and tightly confined narrow ringlets. The quantitative understanding of these interactions is beyond the scope of this book. Generally, most of the mass of the ring particles is found in the size range from about 1 cm to 10 m. The particle size-frequency ($n_r$ vs $r$) relation is well approximated by a simple power law,

$$n_r = n_0 r^{-3}, \qquad \text{(VI.40)}$$

so that the number of 10-m particles is about $10^{-9}$ times the number of 1-cm particles. Because a 10-m particle has $10^9$ times the mass of a 1-cm particle, this means that there is about the same total mass in each radius interval. Clearly this situation cannot extend to all conceivable sizes (the total mass would then become infinite!). Indeed, the rings lie so close to Saturn that large bodies (held together by gravity) are not stable, but are torn apart by Saturn's gravity. Also, particles smaller than about 1 cm are relatively depleted because they can be swept up by and stuck to the surfaces of larger particles.

The stability of gravitationally bound small bodies with zero tensile strength near a large gravitating mass was first studied by the French mathematician R. A. Roche in 1847. Suppose two small spherical satellites, each with mass $m$, radius $a$, and density $\rho_s$ are located touching each other on a radial line from a larger planet of mass $M$, radius $A$, and density $\rho_P$. The gravitational attraction between the two satellites is then $Gm^2/(2a)^2$. The planetary gravitational field gradient at the location of the satellites is the distance derivative of $-GM/r^2$, which is $2GM/r^3$, and the force acting to separate the two bodies is the distance between their centers ($2a$) multiplied times the field gradient, or $4GMma/r^3$.

The bodies will experience a net force separating them if their mutual gravitation is less than the opposing force due to the field gradient,

$$Gm^2/4a^2 < 4GMma/r^3 \qquad \text{(VI.41a)}$$

or

$$r^3 < 16Ma^2/m. \qquad \text{(VI.41b)}$$

However, the ratio of masses $M/m$ is $\rho_P R^3/\rho_s a^3$, and the criterion for breakup becomes

$$r/R < (16\rho_P/\rho_s)^{1/3} = 2.520\ (\rho_P/\rho_s)^{1/3}. \qquad \text{(VI.42)}$$

For a synchronously rotating fluid satellite, subjected to distortion by both inertial and tidal forces, Roche found that a satellite would suffer disruption if it were closer to the planet than 2.456 $(\rho_P/\rho_S)^{1/3}$ planetary radii. This critical distance for disruption is called the *Roche limit*. Thus, in the absence of "stickiness," the ring region would be conducive to the fragmentation of satellites into small particles.

There is indeed a substantial population of very small particles, with diameters comparable to a wavelength of visible light, in the B ring. These are especially obvious in forward-scattered light (that is, when the spacecraft is looking through the rings back toward the Sun). Immense sprays of fine particles, called spokes, are often seen in the B ring, crossing the ringlet pattern at

**Figure VI.20** Two Voyager views of the F ring of Saturn. The broader view (right) shows striking brightness variations along the ring. The high-resolution view at left shows the bizarre braiding of filaments within the ring. Two small moons shepherd the F ring.

sharp angles and propagating radially at high speeds (Fig. VI.21). The longest duration of a spoke is about half of an orbital period, strongly suggesting that the spokes are sprays of fine particles knocked into slightly inclined orbits by energetic events within the ring plane. These small ice particles interact readily with energetic protons and electrons to take on small electric charges, and the $q\mathbf{V} \times \mathbf{B}$ forces on these small particles are an appreciable fraction of the gravitational force.

Magnetospheric charged particles are rather efficiently sopped up by dense rings, but a weak and all-pervasive source of electrons and protons is available even in the most sheltered part of the ring system; cosmic rays have high enough energy to penetrate the magnetic field of Saturn and strike ice particles of all sizes. These interactions produce neutrons that will then slowly decay to protons and electrons, with a half-life of about 15 min. Thus these "splash albedo neutrons" provide charged elementary particles that can in turn produce charges on small ice particles. The ice particles can also become charged by photoejection of electrons by solar ultraviolet light. These charges cannot build up indefinitely. Most small ice particles have very irregular shapes, and field emission can bleed off high charges from sharp points. Alternatively, the charge density may become high enough so that the mutual electrostatic repulsion of the different parts of a grain become sufficient to fracture it.

**Figure VI.21** Spokes in Saturn's B ring. These six Voyager images show rapid changes in the dust distribution over less than an hour.

Small charged ice particles will feel $q\mathbf{V} \times \mathbf{B}$ (Lorentz) forces in proportion to the speed that they have relative to the magnetic field. But the synchronous rotation altitude (at which the ring particle Keplerian orbital frequency equals the rotation frequency of the planet and its attached magnetic field) occurs within the B ring. The spokes usually show a distinctive wedge shape with the apex of the wedge located at or near synchronous orbit, where the Lorentz forces vanish.

The mechanism for the production of spokes is presently under debate, with no resolution yet in sight. The Voyager flybys have proved a powerful stimulant to the study of planetary rings, and many important questions are under investigation today. Some of the outstanding unresolved problems are the mass spectrum of ring particles (including ringmoons), the radial compositional variation of ring particles, and the stability of ring structures. The structure of the ring is, however, so complex (see Fig. VI.22) that we cannot expect all of its mysteries to be solved soon. Most tantalizing is the prospect that dynamical studies of the rings may provide new insights into the behavior of solids in the early Solar System at the time of formation of the planets. The rings bear the infrared signature of water ice, although the albedo attests to an admixture of dark dust. The mid-IR spectrum of the rings is featureless and Planckian, giving a temperature of $90.3 \pm 0.9\,\mathrm{K}$ for the A ring and $90.5 \pm 0.6\,\mathrm{K}$ for the B ring.

***Uranus.*** The rings of Uranus were discovered during an Earth-based study of the occultation of a star by the Uranian system in 1977, only 1 year after the indirect detection of Jupiter's ring by Pioneer 10 and

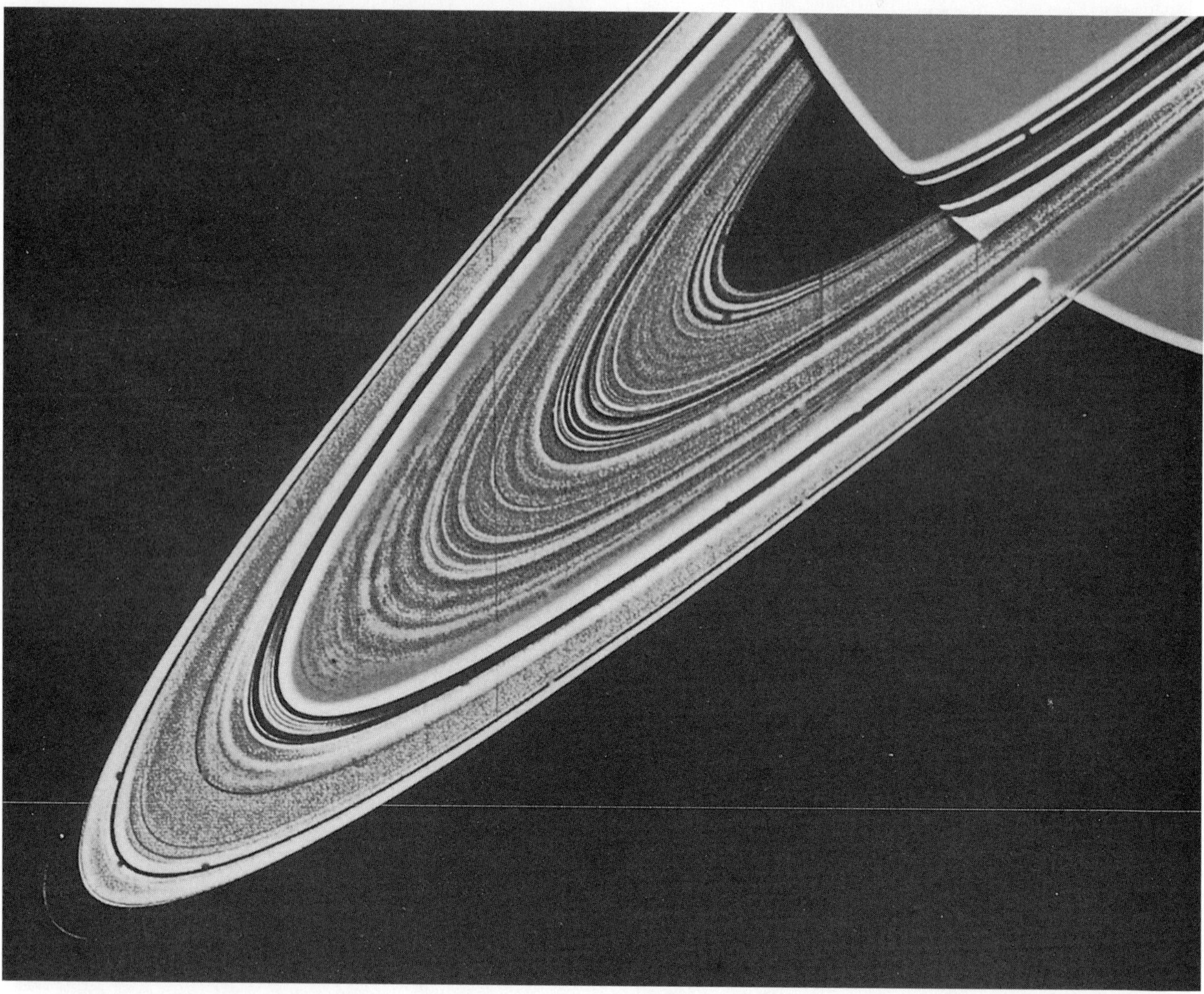

**Figure VI.22** Detailed structure of Saturn's rings. Even in this distant Voyager view, computer contrast enhancement reveals approximately 100 identifiable ringlets. Note the complex structure even within Cassini's Division (where the four bright, evenly spaced ringlets are seen). The F ring is barely discernible at elongation (lower left).

2 years before the first imaging of Jupiter's ring by Voyager 1. More recently, the Uranian rings were studied by Voyager 2 during its flyby in 1986. A schematic comparison of the three well-established ring systems is shown in Fig. VI.23.

The rings of Uranus form a striking contrast with those of Saturn. Instead of broad bright bands punctuated by narrow dark gaps, the Uranian system comprises a set of nine very narrow rings separated by wide gaps, almost like a photographic negative of Saturn's rings (Fig. VI.24). Eight of these rings have widths of about 1 to 10 km. The distance of these rings from Uranus spans a range of about 10,000 km. Table VI.5 presents the orbital data on the system; note that the rings actually cover only about 1% of this region.

The optical depths of the rings are on the order of 1. The rings are even more difficult to photograph than the small surface area would suggest. It is evident that the albedos of the ring particles are very low, again in sharp contrast with the high-albedo ices seen in Saturn's rings.

The Uranian rings have a number of unusual and interesting properties. We have noted that they are narrow and sharply defined, not diffuse like the Jovian ring. More surprising, they pursue noticeably eccentric orbits that are not exactly coplanar.

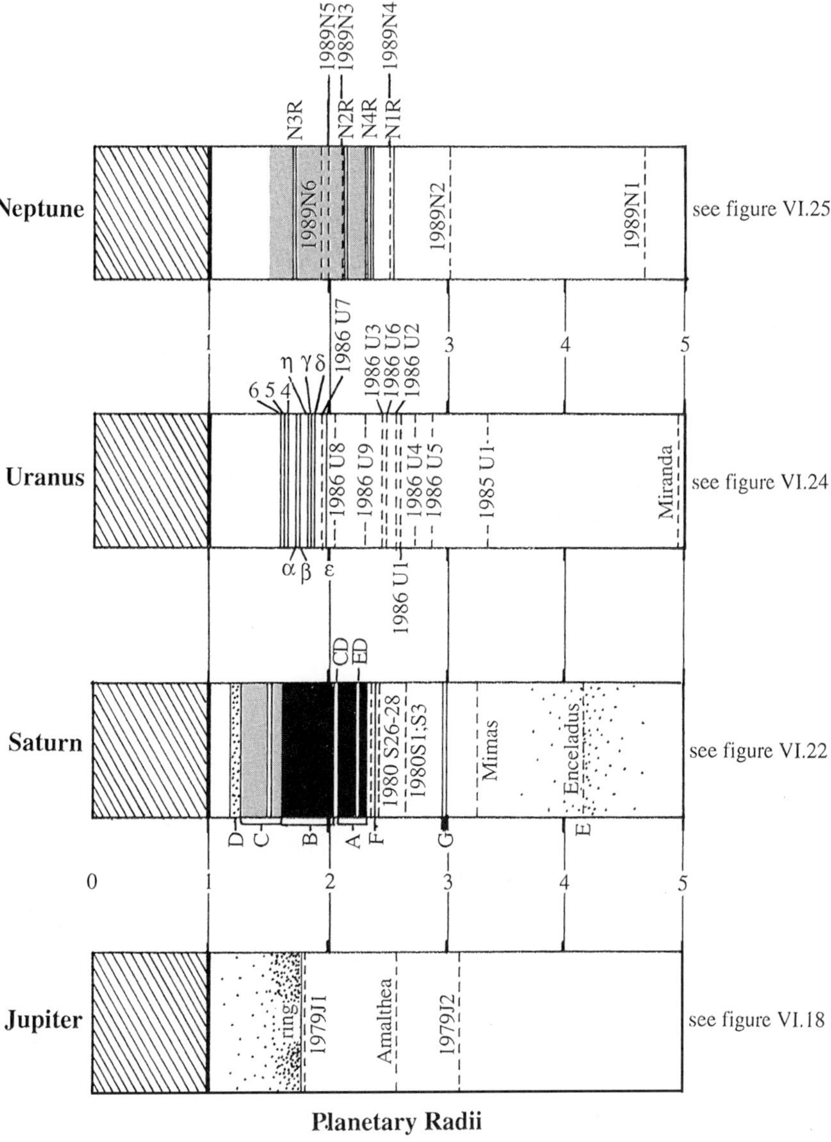


**Figure VI.23** Comparison of the ring systems of Jupiter, Saturn, Uranus, and Neptune. Satellite orbits are marked with dashed lines; dense rings, with solid black; and thin dust rings, with light shading.

a.

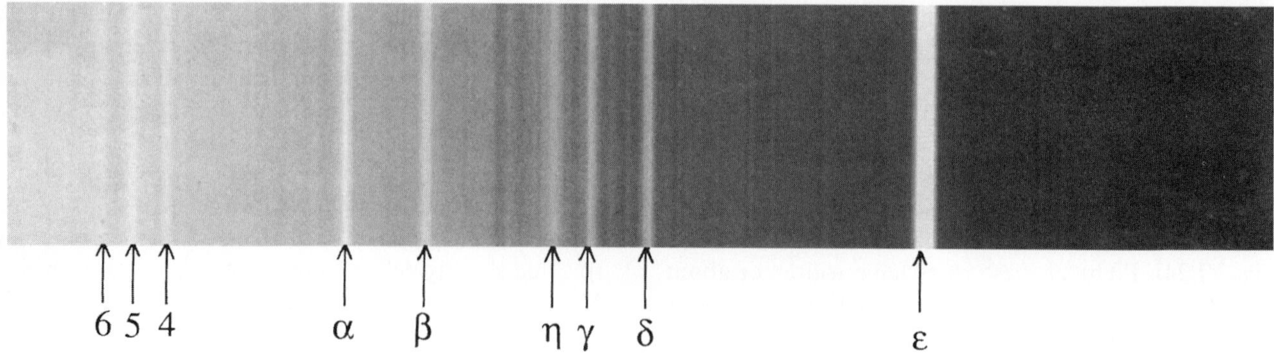


b.

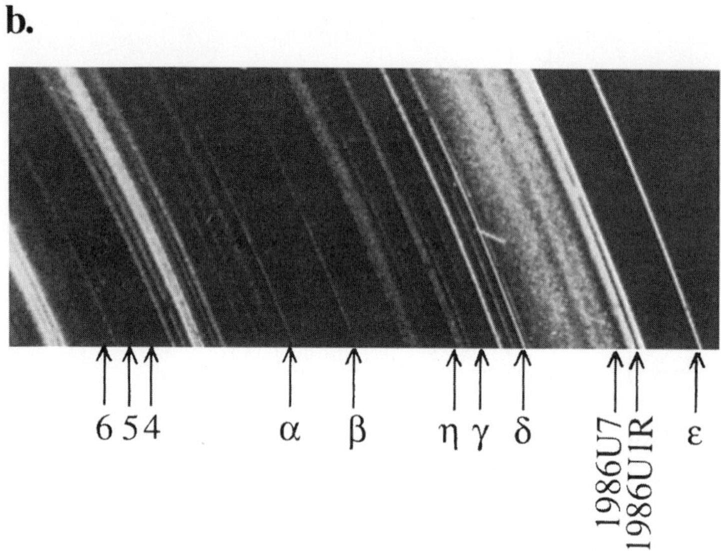


**Figure VI.24** Cross-section of the rings of Uranus. (a) Several Voyager images taken through color filters have been combined to make this representative "slice." The nine major rings are marked. Most of the background brightness is due to scattered light from Uranus or to side effects of the computer processing. (b) The rings seen at a phase angle close to 180°, which reveals the distribution of micrometer-sized dust particles.

**Table VI.5** Orbits of the Uranian Rings

| Ring | Semimajor axis ($a$)(km) | Eccentricity ($e$) | Inclination ($i$)(°) | Width (km) |
|---|---|---|---|---|
| 6 | 41,877.3 | 0.00101 | 0.066 | $<2$ |
| 5 | 42,275.2 | 0.00185 | 0.050 | $<2$ |
| 4 | 42,609.6 | 0.00115 | 0.022 | $<2$ |
| $\alpha$ | 44,758.3 | 0.00078 | 0.017 | 5–10 |
| $\beta$ | 45,701.0 | 0.00043 | 0.006 | 5–11 |
| $\eta$ | 47,214.9 | 0 | 0 | $<2$ |
| $\gamma$ | 47,666.3 | 0 | 0.006 | 3 |
| $\delta$ | 48,338.7 | 0.00006 | 0.012 | 2.5 |
| $\varepsilon$ | 51,188.1 | 0.00794 | 0 | 20–96 |

The three inner rings (6, 5, and 4) are all very narrow, and they all remain unresolved (that is, narrower than about 2 km) in the Earth-based occultation observations. They could, however, not be narrower than a few hundred meters and still intercept as much light as they are observed to do. The $\alpha$ (alpha) ring clearly varies in width azimuthally about Uranus. The widest portion of the $\alpha$ ring sometimes is clearly bimodal, with two well-resolved density maxima. The $\beta$ (beta) ring varies similarly in width, but no bimodal character has been reported.

The next ring in order of distance from Uranus, $\eta$ (eta), has a broad, diffuse background component plus a dense, narrow component at its outer edge.

$\eta$ looks like a mirror image of Saturn's narrow F ring, which has a dense region at its inner edge. The next feature, the $\gamma$ (gamma) ring, is very dense and narrow, and the next, the $\delta$ (delta) ring, looks similar to the $\eta$ ring except that it may be accompanied by a minor ringlet about 12 km below it. These three rings have negligible eccentricity.

Finally, the outermost or $\varepsilon$ (epsilon) ring has a large and variable width, the largest eccentricity of any of the Uranian rings, and a complex internal structure. Over most of its longitudinal extent the ring has distinct density maxima near both its inner and its outer edges. The total light absorption by the $\varepsilon$ ring dwarfs that of all the other Uranian rings combined. If particle sizes are similar in all the rings, then the $\varepsilon$ ring must account for more than 80% of their total mass. Moreover, the orbit of the $\varepsilon$ ring precesses about the planet in a coherent manner, preserving the width variation with planetocentric distance. Peter Goldreich and Scott Tremaine have shown that the self-gravitation of the ring can hold it together in this manner if the $\varepsilon$ ring contains about 99% of the total ring system mass. This mass, combined with the observed optical depth of the $\varepsilon$ ring, suggests that the mean particle size is about 20 cm.

The spectral properties of the Uranian rings contain valuable information on their particle sizes and compositions. Earth-based observations are almost impossible at wavelengths at which Uranus is bright, and it is therefore usual to observe the rings at wavelengths at which methane is a strong absorber (and Uranus is quite black). The rings, observed in these favorable wavelength intervals (2.0 to 2.4, 3.0, 3.5, and 3.9 $\mu$m), have albedos of only 0.02 to 0.03, with no clear evidence of absorption features due to ring materials. The extremely low albedo is consistent with the dominance of black sooty or tarry carbonaceous matter in the Uranian satellite system.

***Neptune.*** There is Earth-based evidence for a ring system about Neptune, but it is rather difficult to interpret. Occultation measurements have indeed found evidence of a ringlike feature, but other observations of similar quality at other longitudes show no sign of such a feature. Indeed, of some 100 observed stellar traverses of the vicinity of the Neptunian ring region, covering about 50 distinct occultation events, fully 90% found no evidence for rings. Longitudinally continuous dense rings are incompatible with this evidence, but tenuous or very clumpy ringlike features, perhaps better described as arcs than rings, are indicated. The Voyager 2 Neptune encounter in 1989 shed considerable light on this issue: an outer, extremely clumpy ring (imaginatively named 1989N1R) was found at 62,900 km from the planetary center, and a fainter, more uniform ring (named 1989N2R) was found at 53,200 km. N1R shows three major clumps, and high-resolution images show a streaked or braided appearance in these clumps. A very faint and diffuse ring (1989N3R) was also found at 41,900 km, and an extremely tenuous ring was found reaching from the edge of 1989N2R out to about 59,000 km. Some of the highest-resolution images show a narrow, clumpy ring feature slightly interior to 1989N1R. There is also evidence of a broad, diffuse dust band close to the planet, well inside the orbits of these rings. Figure VI.25a shows the main ring features, and Fig. VI.25b clearly displays the extreme clumpiness of 1989N2R. The direction of orbital motion in Fig. VI.25b is clockwise, and the longest clump is therefore the one in the trailing orbital position.

The stability of these clumpy features presents an interesting dynamical problem. Why do the ring particles not "diffuse" along the ring and smooth out the longitudinal mass distribution?

## Titan

The Solar System contains seven very large satellites, with masses on the order of $10^{26}$ g. These are Earth's Moon, the four Galilean satellites of Jupiter (J1, Io; J2, Europa: J3, Ganymede; and J4, Callisto), Saturn's largest moon (S6, Titan), and Neptune's large and strange moon (N1, Triton). They span a range of composition from mostly ice (Ganymede, Callisto, Titan, and presumably Triton) to dry rock (Io and the Moon). Titan is of special interest for several reasons: it is only slightly smaller than Ganymede, the largest known moon; it is by far the largest satellite of Saturn; and it has a massive atmosphere. It has been studied in detail by two flyby spacecraft and is therefore relatively well known. Titan's massive atmosphere is very cloudy, preventing any view of its surface at visible and infrared wavelengths. The infrared spectrum of the atmosphere in and above the clouds was examined in 1944 by Gerard Kuiper, who found strong absorption features due to methane gas. The relative strengths of the methane bands, however, seemed to suggest that most of the atmospheric pressure was due to some other, as yet unobserved, gas. There simply was not enough methane seen to provide the observed broadening of the methane absorption features.

Spectroscopic searches for other constituents were made with increasing sensitivity in the last few years prior to the Voyager flybys of the Saturn system in 1980 and 1981. These spectroscopic searches generated a tentative detection of hydrogen, a failure to

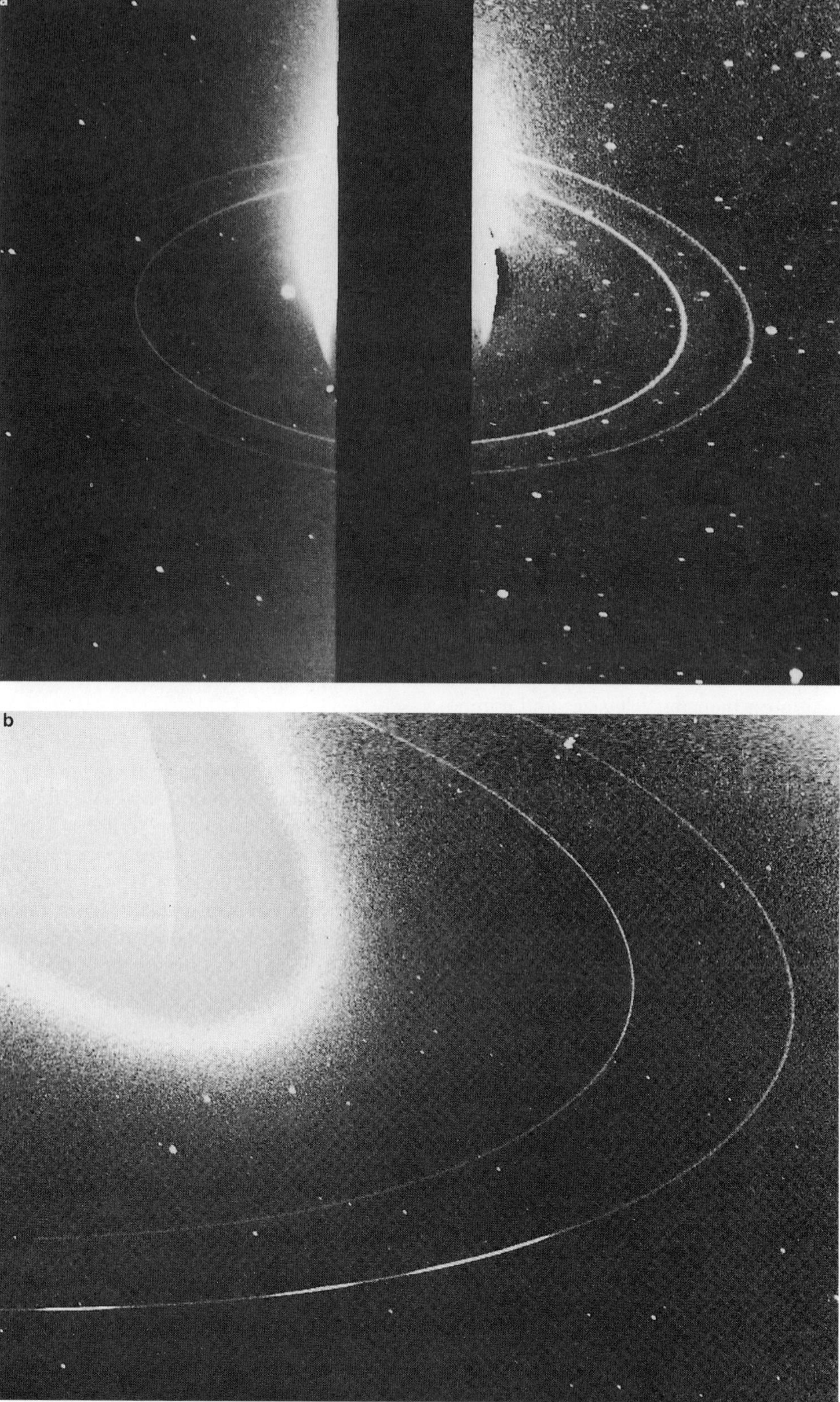


**Figure VI.25** The rings of Neptune. (a) Voyager 2 viewed the entire ring system during approach to Neptune. This view has the disc of the planet blocked out to avoid washing out the faint images of the rings. The innermost ring (1989N3R) is clearly visible. The most intense ring is 1989N2R, and the outer distinct ring is 1989N1R. The other rings are fainter (see Fig. VI.23). (b) The very lumpy structure of the outermost ring, 1989N1R, is clearly visible.

detect ammonia gas (not surprising in light of the very low temperature of Titan's atmosphere), and the tentative identification of several strong emission bands in the thermal infrared. Initially, the thermal IR observations had such low spectral resolution that it was not possible to determine whether the high flux was due to discrete band emission in the warm upper atmosphere or to heat emitted from the surface, leaking out through the atmosphere. We now know that the emission is due to warm ethane ($C_2H_6$), ethylene ($C_2H_4$), and acetylene ($C_2H_2$).

Opinion was divided on the meaning of these measurements. John Caldwell adopted a WYSIWYG (what you see is what you get) model, in which the observations were assumed to penetrate to the surface. The surface pressure was then only about 20 mbar, and the surface temperature was close to 86 K. The atmosphere was assumed to be pure methane except at high altitude, at which photochemically produced $C_2$ hydrocarbons were present in a layer heated by absorbed solar radiation. Donald M. Hunten of the University of Arizona emphasized the potential importance of nitrogen gas, a major constituent of the atmosphere. In his model, the infrared opacity of the methane–nitrogen mixture provided a "greenhouse effect" that elevated the surface temperature to perhaps as high as 200 K. The pressure of nitrogen gas at the surface could then be as high as 20 bar.

Curiously, Titan seems to have assimilated some of the main features of both models. The actual surface pressure and temperature of Titan were determined by monitoring the phase shift and attenuation of radio signals from Voyager 1 as it was occulted (hidden) by Titan. This radio occultation experiment provided two vertical profiles through the atmosphere, reaching all the way to a surface at 94 K and 1.5 bar. The density of the atmosphere was so high that nitrogen had to be the principal constituent.

Voyager imaging of Titan showed a nearly featureless, smoggy atmosphere with only very subtle color and intensity variations. The entire atmosphere is very red at all times. At the time of the Voyager 1 encounter, the northern (spring) hemisphere was slightly darker and browner than the southern (fall) hemisphere. A dark haze hood was found over the north pole, and photographs of the limb showed a diffuse high-altitude haze well above the level of the main cloud deck. The atmosphere is partially transparent at wavelengths between the strong methane bands. The Hubble Space telescope has been used to produce a crude image of Titan's surface as seen through one of these spectral "windows" (Fig. VI.26), but interpretation of the image in terms of surface features and processes is as yet very preliminary.

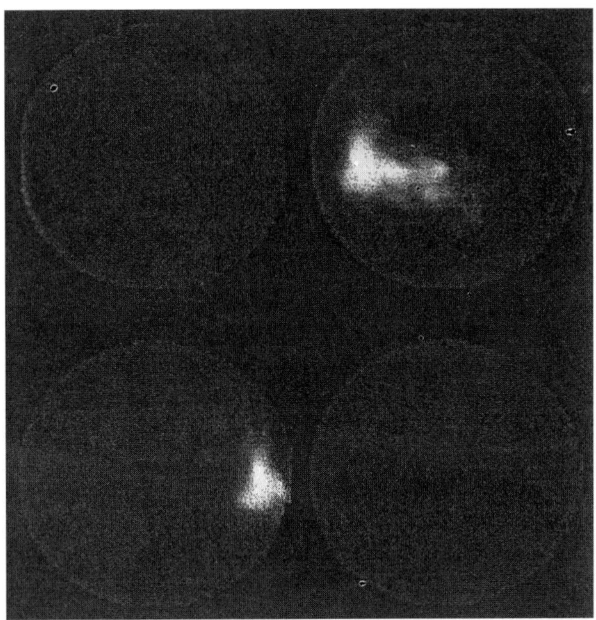

**Figure VI.26** Hubble Space Telescope image of the surface of Titan as seen through a near-infrared transmission window. The meaning of the features seen is unclear. Courtesy of the Space Telescope Science Institute.

The infrared interferometer spectrometer (IRIS) experiment on the Voyager spacecraft provided a rich new insight into the composition of Titan's atmosphere. It is now clear that methane makes up several percent of the total atmospheric pressure, with nitrogen accounting for almost all of the remainder. The hydrogen abundance is about 0.2%. There is about 100 ppm of carbon monoxide and about a part per billion of carbon dioxide. In addition to the hydrocarbons already identified from Earth-based spectroscopic studies, propane ($C_3H_8$), methyl acetylene ($H_3C$–CCH), and diacetylene (HCC–CCH) were found by IRIS. Three organic nitrogen compounds, hydrogen cyanide (HCN), cyanogen (NC–CN), and cyanoacetylene (HCC–CN), were found. The composition of Titan's atmosphere is summarized in Table VI.6. The nitrogen-dominant composition with about 6% methane as the principal organic species suggests a photochemically active regime in which the production of ethane, ethylene, acetylene, and nitrile (–CN) species occurs with good efficiency. The Sun provides about $1.5 \times 10^{10}$ photons per second per square centimeter of Titan's surface area at wavelengths capable of dissociating methane ($\lambda < 1650$ Å). The main products are ethane (80%) and acetylene (20%), with a total yield of about $4 \times 10^{-13}$ g cm$^{-2}$ s$^{-1}$ of hydrocarbons. Thus photolysis of methane over the age of the Solar System ($1.4 \times 10^{17}$ s) will provide about $4 \times 10^{-13} \times 1.4 \times 10^{17} = 5.6 \times 10^4$ g cm$^{-2}$ of accumulated

**Table VI.6** Composition of the Atmosphere of Titan

| Species | Mole fraction | Species | Mole fraction |
|---|---|---|---|
| $N_2$ | $\sim 0.94 \pm 0.04$ | $C_2H_4$ | $4 \times 10^{-7}$ |
| Ar | $< 0.25$ | HCN | $2 \times 10^{-7}$ |
| $CH_4$ | $0.06 \pm 0.04$ | $CH_3CCH$ | $3 \times 10^{-8}$ |
| $H_2$ | $2 \times 10^{-3}$ | $HC_4H$ | $\sim 3 \times 10^{-8}$ |
| CO | $5 \times 10^{-5}$ | $HC_3N$ | $\sim 3 \times 10^{-8}$ |
| $C_2H_6$ | $2 \times 10^{-5}$ | NCCN | $\sim 3 \times 10^{-8}$ |
| $C_3H_8$ | $4 \times 10^{-6}$ | $CO_2$ | $\sim 2 \times 10^{-9}$ |
| $C_2H_2$ | $2 \times 10^{-6}$ | | |

$C_2$ hydrocarbons. This corresponds to about 600 m of liquid ethane.

The hydrocarbon and organic nitrogen content of Titan's atmosphere can be sensibly explained by energetic processes, such as solar UV irradiation and electron bombardment of a $N_2$–$CH_4$ mixture. Any source of oxygen, such as infall of "meteoritic" ice from collisional debris orbiting within the Saturnian satellite system or outgassing of primordial carbon oxides from solid hydrates, would provide the observed proportions of CO and $CO_2$ through photochemical reactions in the upper atmosphere.

The fate of condensible photochemical products is surely complex. Aerosols in the atmosphere are probably rich in ethane and other organic products. According to work done by Jonathan Lunine of the University of Arizona, these species will eventually rain out onto the surface of Titan, where they may participate in the formation of "lakes" or even a shallow "ocean" of liquid methane–ethane–nitrogen solution. Allowing for the mutual solubility of methane, ethane, and nitrogen, he estimated that total ocean depths of 0.7 km to perhaps 9 km are possible.

Interestingly, the present mass of methane in the atmosphere is only about $0.08\,\text{bar} \times 10^6\,\text{dyn cm}^{-2}\,\text{bar}^{-1}/g_s$, or about $500\,\text{g cm}^{-2}$. Here $g_s$ is, as usual, the surface gravitational acceleration. At the present photolysis rate of $4 \times 10^{-13}\,\text{g cm}^{-2}\,\text{s}^{-1}$, this methane would last only about $1.2 \times 10^{15}$ s, or $4 \times 10^7$ years, a mere 1% of the age of the Solar System. Thus either there is a massive surface reservoir that continues to release methane into the atmosphere or we are privileged observers, lucky enough to see the last 1% of Titan's original methane reservoir just before it disappears. Post-Voyager, Earth-based observations lead to complications for this plausible story. The thermal emission of Titan has been measured at centimeter wavelengths and found to be $80.4 \pm 0.6$ K. Given that the surface temperature is constrained by Voyager measurements to be $97 \pm 5$ K, this implies a surprisingly high microwave emissivity for the surface ($0.85 \pm 0.05$), higher than that of the icy Galilean satellites except for the brightest part of Callisto. Radar studies by Duane Muhleman of California Institute of Technology show that Titan also exhibits a surprisingly high radar reflectivity, with a bright spot that (if it is real) reveals a nonsynchronous rotation of Titan with a period about 1 h less than the orbital period. Finally, high-resolution infrared observations of Titan in spectral gaps at about 1.3, 1.6, and 2 $\mu$m between strong methane bands provide some information on the reflectivity and color of the surface. A rapid increase of albedo toward shorter wavelengths is seen, suggesting water ice but not hydrocarbons as the dominant surface material. The issue of the areal extent and depth of the expected "ocean" is thus in doubt and cannot be resolved without new data. The Cassini joint mission of the European Space Agency and NASA, expected to drop a survivable entry probe into the atmosphere of Titan, will provide an excellent opportunity to test our theories of Titan's surface and evolutionary history.

Argon and neon have both been suggested as major atmospheric constituents, but atomic spectral lines of these elements are absent from the ultraviolet emission spectrum of the upper atmosphere, and the mean molecular weight of the atmosphere seems, based on the radio occultation experiment results, to be close to 28. Neon suffers a serous credibility problem, because no one has ever suggested a sensible reason why it ought to be present. Two sources of argon are conceivable: primordial (nebular) argon retained in solid clathrate hydrates and radiogenic argon from decay of $^{40}$K in the interior of Titan after its formation. The former is a mixture of $^{36}$Ar with some $^{38}$Ar, whereas the latter is pure $^{40}$Ar. The amount of radiogenic argon on Titan is unknown; however, assuming a composition class consisting of rocks + water ice + ammonia (Table VI.2), chondritic composition of the rock component would give a K content of 815 ppm (Table VI.3) for a present-day $^{40}$K content of 0.097 ppm by mass. Given an age of 4.5 Ga, a $^{40}$Ca half-life of 1.31 Ga, and a branching ratio of 12:1 for $^{40}$Ar production vs $^{40}$Ca production, we estimate an original $^{40}$K content of $2^{4.5/1.3} = 10$ times the present amount, or 1 ppm. Since 90% of the original 1 ppm has decayed, and since 1/12th of the decay product is $^{40}$Ar, we have 0.08 ppm of $^{40}$Ar produced in the rock component. Our compositional model of Titan is 36.46% rock, or 0.029 ppm $^{40}$Ar in bulk Titanian material. Since the mass of Titan is $1.3457 \times 10^{26}$ g and its surface area is $2.277 \times 10^{17}\,\text{cm}^2$, the maximum possible present-day loading of $^{40}$Ar on the surface of Titan is $m/A = 2.9 \times 10^{-8} \times 1.3457 \times 10^{26}/2.277 \times 10^{17} = 17$ $\text{g cm}^{-2}$. The acceleration of gravity on Titan's surface is $g_T = GM_T/r_s^2 = 135\,\text{cm s}^{-2}$, and the surface partial pressure due to $^{40}$Ar is $P = F/A = mg_T/A = 135 \times 17 =$

2295 dyn cm$^{-2}$ = 2.3 mbar. If the total pressure is 1500 mbar, then the mole fraction of radiogenic argon that could be released by perfect outgassing is only $1.5 \times 10^{-3}$.

It is also worth recalling that this same source of energy can drive radiolysis of water and its solutes by beta and gamma radiation from potassium decay, generating significant traces of strong oxidizing agents such as oxygen and hydrogen peroxide.

One of the additional benefits of the Voyager radio occultation data on Titan is a firm measurement of its radius. Since the mass is well known (both from its perturbations on the other Saturnian satellites and its effect on the trajectory of Voyager) we can calculate the density of the "solid body" of Titan directly. The radius is 2575 km and the density is 1.88 g cm$^{-3}$, compared with 2631 km and 1.94 g cm$^{-3}$ for Ganymede. Corrected to zero pressure, these two densities might easily be compatible with identical compositions. The small observed difference in compressed densities can be fully accounted for by the slightly larger mass and slightly higher internal pressures in Ganymede. There is thus no observational basis for the (otherwise reasonable) belief that the ices that participated in the formation of Titan were significantly different from those present in the outer part of Jupiter's Galilean system. We remain without any direct evidence for the presence of large amounts of ammonia, methane, carbon monoxide, carbon dioxide, and nitrogen in the interior of Titan. The atmospheric mass is of course a negligible part of the total ice mass.

Surface temperatures on Titan are low and stable, far below the melting point of ice. Even depression of the freezing point of water by dissolved ammonia (eutectic temperature of 173 K) could not possibly make an aqueous solution stable on the 94-K surface of Titan. It is hard to rule out the possibility of extrusive or eruptive volcanism of ammonia–water solution, but nothing in our knowledge of Titan demands it. If it has ever occurred, however, it would supply atmospheric ammonia, which would photolyze to make nitrogen. It is possible that ammonia photolysis, not release of primordial nitrogen, is responsible for the present atmospheric nitrogen inventory. Such photolysis is surely negligible at present because the surface temperature is so low that the vapor pressure of ammonia would be discouragingly small.

During the early history of Titan, however, accretion heating and impact shock events may have not only elevated surface temperatures and made the atmospheric composition more complex, but also provided enormous masses of organic matter and nitrogen from the shock-wave heating of the early atmosphere. Work by Thomas D. Jones at the University of Arizona suggests that all the observed nitrogen, plus thick deposits of organic polymers of acetylene, HCN, and the like, could have been readily made during the terminal accretion phase of the Saturnian satellites.

The Cassini mission mentioned earlier, presently en route to Saturn, will drop a heavily instrumented probe, Huygens, into Titan's atmosphere to measure the composition and physical properties of its atmosphere, and possibly confirm the presence of an "ocean" on its surface. Arrival at Saturn will be on July 1, 2004.

## The Intermediate-Sized Saturnian Satellites

Our study of Jupiter's system was somewhat hindered by the absence of satellites with masses below $48 \times 10^{24}$ g but above $0.012 \times 10^{24}$ g. None of the Jovian satellites are close to the critical size for melting, and none are in the transition region between the "small satellites" with radii less than about 200 km, which preserve heavily cratered surfaces and very irregular figures, and the large, smooth bodies in which internal processes have been important. However, the Saturnian system provides us with a variety of satellites in this size range.

The second-largest Saturnian satellite, Rhea, has a mass of $2.5 \times 10^{24}$ g, and (in descending order of mass) Iapetus, Dione, Tethys, Enceladus, and Mimas follow. Mimas weighs in at $0.038 \times 10^{24}$ g. The "mass gap" of the Jovian system is thus populated by six different Saturnian satellites—and different they are!

Rhea has a closely spherical figure, but also has an ancient, heavily cratered surface, at first sight reminiscent of Callisto's (see Fig. VI.27). There is a systematic variation of Rhea's brightness with its position in its orbit: Rhea is rotationally locked onto Saturn, and its leading hemisphere is slightly and nearly uniformly brighter than its trailing hemisphere. The darker hemisphere is decorated with numerous curved, wispy markings that have no close counterpart on the Galilean satellites. It is not known whether these features are produced by internal or external processes.

Rhea images with resolutions of about 1 km returned by Voyager 1 show Moon-like sharp craters with more surviving vertical relief than we see on the Galilean satellites, presumably because the lower temperature gradient and lower surface temperature of Rhea prevent viscous relaxation of topography with relief of a few kilometers. Many of these sharp craters have bright white patches of unknown origin on their walls.

Iapetus, only slightly smaller than Rhea, has one of the oddest appearances of any satellite in the Solar System. Like Rhea, it varies systematically in brightness with orbital phase angle. However, the leading hemisphere

**Figure VI.27** Voyager photomosaic of Rhea. This view of the bright hemisphere shows a very heavily cratered surface with little evidence of even weak internal thermal activity.

is *much darker* than the trailing hemisphere. Much of the leading hemisphere has an albedo of only 0.02 to 0.04, whereas the trailing hemisphere has a mean albedo of 0.5! The former is typical of a very dark or carbonaceous rock; the latter is typical of clean, fresh ices. Voyager 1 photographed Iapetus from great distance with low resolution, and Voyager 2 observed the trailing hemisphere under somewhat better conditions, approaching as close as 0.9 million km. These images confirmed that the dark region is symmetrically arrayed around the apex of its orbital motion and that the albedo distribution, rather than grading smoothly from high to low as the apex is approached, is strikingly bimodal, with only a narrow transition zone between the two. The bright regions are heavily cratered, but the dark region is so black that there was insufficient light received to make a good image for use in crater counting.

The spectrum of Iapetus (which is of course dominated by its bright icy side) shows the presence of water ice. The spectrum of the dark side, corrected for some contribution by bright icy material, is as dark and red as the carbonaceous polymeric material in meteorites. Indeed, the carbonaceous meteorites provide a good match to the spectrum of the dark side.

The radius of Iapetus was refined by the Voyager images to $718 \pm 18$ km. The density works out to $1.2 \pm 0.1$ g cm$^{-3}$. Pre-Voyager calculations of the self-compression of cold ice spheres by Mark Lupo and me showed that the mass of Iapetus must be strongly dominated by ice; indeed, a pure water-ice composition cannot be ruled out.

Explanations for the hemispheric asymmetry of Iapetus center around either the selective accretion of dust or the preferential removal of ices (and the concentration of dark, involatile native dust) by impacts in the same region. The albedo variation evidently has little or nothing to do with intrinsic geological and tectonic processes, but rather is intimately related to collisional or accretionary processes. It is possible that dust eroded from the more distant satellite Phoebe by impact events has been preferentially swept up by the leading hemisphere of Iapetus, that impact vaporization of ice preferentially enriched a dark, dusty component on the leading side, or even that dark polymeric organic matter is somehow synthesized in impacts. It seems that impact experiments on dust/ice mixtures might be able to resolve this issue.

The third of these intermediate-size satellites, Dione (Fig. VI.28), looks very much like Rhea. It is a spherical, heavily cratered object with wispy bright markings. Its density is rather tightly constrained as $1.43 \pm 0.04$ gcm$^{-3}$, high for Saturn's satellites, but still significantly less than the uncompressed density of Titan, Ganymede, or Callisto. The albedo is high (0.5), typical of clean ices, and, like Rhea (and unlike Iapetus), it is about 0.5 mag brighter on its leading side. Like the larger Saturnian satellites, Dione is rotationally locked on Saturn. The white wisps show up most strongly on the trailing, dark side, where they have an albedo about three times larger than that of the background dark materials. Narrow valleys and ridges, similar in appearance to terrestrial graben/horst structures, are seen on the bright side, some stretching for hundreds of kilometers.

Large craters ($d > 30$ km) are rare on Dione, and the crater density varies regionally. Both phenomena suggest a more extensive or more recent geological reworking of Dione than anything we can see on Rhea.

Tethys, about the same size as Dione, is distinguished by its extremely high albedo (0.8). The mass is poorly determined, and no useful constraints can be placed on the density.

Tethys has two impressive surface features of very large size. The first, the giant crater Odysseus, has a diameter of 400 km (the radius of Tethys is only 530 km). Tethys could not have survived a significantly larger impact. The second, a great valley or rift complex evocatively named Ithaca Chasma (Fig. VI.28) after a slightly warmer fissure that dissects the Cornell University campus, stretches fully $\frac{3}{4}$ of the way around Tethys, reaching a width of about 100 km. Crater counts on the Ithaca Chasma show that it is only a little younger than the surrounding terrain.

Enceladus (Fig. VI.28) presents its own set of astonishing properties. First, the satellite is extraordinarily reflective and strongly backscattering. The geometric albedo in the visible spectrum is about 1.05, and the Bond albedo is 0.9, making it the most reflective of Saturn's satellites, brighter than fresh snow. Because the surface is bright and rough, multiple scattering of light is important. But this means that, in order to have so high a Bond albedo, the single-scattering albedo for the average grain must be an astonishing 0.99. This is possible only for very clean ices or snows. One consequence of the high reflectivity is a very low surface temperature.

The second striking feature of Enceladus is its complex and diverse surface morphology. Some of the surface units are essentially free of craters. Huge systems of curved, kilometer-high ridges spread across the surface. The different terrain types have distinct cratering histories and ages, but the optical properties of the surface are very uniform. To confound matters further, Enceladus lies very close to a density peak in the E ring. There has been discussion of the possibility that the fresh surface "snow" of Enceladus and the E-ring material were produced as recently as a few thousand years ago by a violent event of unknown nature. Coupled with the

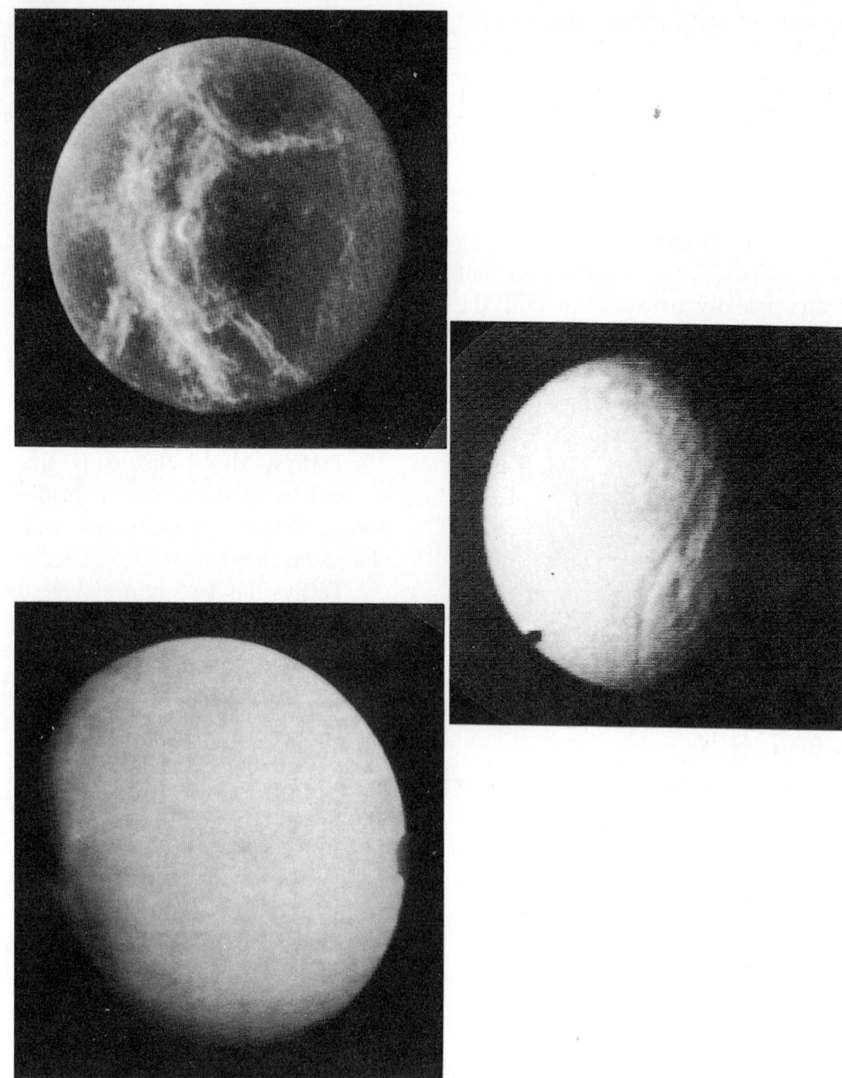

**Figure VI.28** Dione, Tethys, and Enceladus. These intermediate-sized Saturnian satellites display an interesting variety of surface features. Dione (top) has wispy markings reminiscent of the darker side of Rhea. Large craters are rare. The image of Tethys (right) shows the huge Ithaca Chasma structure. Enceladus (bottom) is extremely bright and rough with very few craters, suggesting extensive, and probably violent, reworking of its surface.

extraordinary tectonic activity of Enceladus, the simplest explanation seems to be to associate these phenomena with an "eruption" on the satellite. Unfortunately, it is very hard to see what energy source might account for all this activity. The silicate (and radioactive element) abundances are not unusually high, as we can see from the bulk density of about 1.2. There are also no known resonances or tidal interactions that could be energetically important.

Mimas, slightly smaller than Enceladus, orbits closer to Saturn than any of the other "classical" (pre-1960) satellites. The heavily but nonuniformly cratered surface is dominated by the very large crater Herschel, about 130 km in diameter (Fig. VI.29). The clear proof of powerful internal activity seen on Enceladus is absent on Mimas.

## Minor Satellites of Saturn

We shall for convenience define the small satellites as those with radii less than that of Mimas (about 200 km). We shall treat these in several groups in the order of their distance from Saturn. All of these satellites are irregular in shape, too small for internal thermal activity to permit them to relax into spheres. It will be

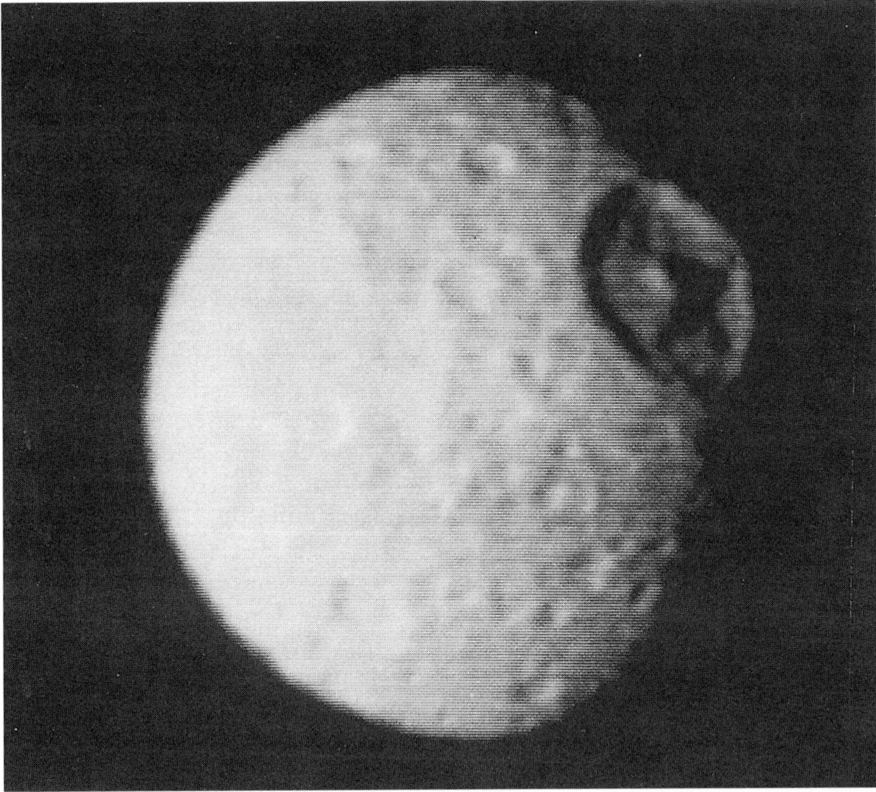

**Figure VI.29** Mimas. The giant crater Herschel stares out into space. A slightly larger impact event would have disrupted Mimas into a cloud of fragments.

useful to refer to Tables III.4, III.5 and VI.4 as background for the following discussion.

The first of these small satellites, SXVIII Pan, orbits in the Cassini division in the A ring. The next two lie just outside the edge of the bright A ring (2.267 $R_S$). They are SXV Atlas, SXVI Prometheus, and SXVII Pandora, whose orbits have eccentricities of 0.002 to 0.009 and inclinations less than 0.1°.

Second come three varieties of coorbital satellites: that is, satellites that share an orbit with one or more other satellites. These include the SXI Epimetheus–SX Janus pair and the SXIII Telesto–SXIV Calypso pair, which circle Saturn *on the orbit of SIII Tethys*. These both have near-zero eccentricity and inclination. Finally, there is SXII Helene, which follows the orbit of SIV Dione.

Figure VI.30 shows a system consisting of a large central body, a much smaller secondary body in circular orbit, and a third tiny body, as seen in a coordinate system that is rotating so as to keep the apparent positions of the primary and secondary bodies fixed. There are five singular points in this system at which particular conditions are met. The first of these is the point of balance of the gravitational forces of the primary and the secondary between them. The second is the point outside the orbit of the secondary where, under the combined gravitational pull of the two larger bodies, a small body would have an orbital period equal to that of the secondary. The third is close to the orbit of the secondary, but on the opposite side from the secondary itself. All of these balance points are unstable, since even an infinitesimal displacement of the small body would cause it to accelerate away from the balance point.

There are two other points, 60° ahead of and behind the primary on its orbit, respectively, that are actually stable against small disturbances. Bodies placed near these fourth and fifth points will execute tadpole-shaped "orbits" about them. These five locations are the Lagrange points L1 (inner), L2 (outer), L3 (opposite), L4 (leading) and L5 (trailing) after their discoverer, the great 18th-century French mathematician Joseph-Louis Lagrange. We shall return to the Lagrange points in Fig. VIII.21 and the associated text.

These points, which were originally calculated for motions in the Sun–Jupiter system, have direct relevance to motions of tiny satellites within the Saturn system. For example, Telesto and Calypso follow tadpole orbits about the L4 and L5 points on the orbit of Tethys. Helene orbits near Dione's L4 point.

Another class of orbits is possible, as illustrated by the horseshoe-shaped orbit in Fig. VI.30. This path

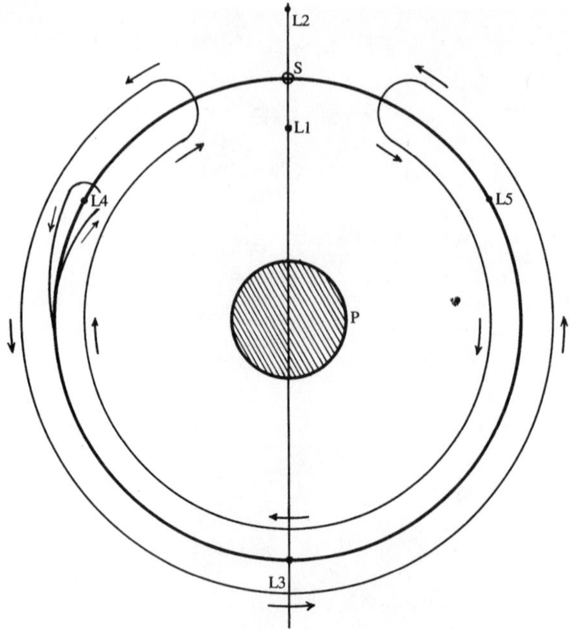


**Figure VI.30** Lagrangian satellite orbits. The dark circle is the orbit of the secondary body, S, about the primary body, P. The system is viewed in a rotating reference frame in which P and S are stationary. The Lagrange points are labeled L1 through L5. A tadpole orbit about the leading Lagrange point (L4) is illustrated. A horseshoe orbit spanning 320° of longitude is also included.

would be followed by a tiny coorbital satellite that is moving fast enough relative to the L4 or L5 point so that it can slip past the L3 point at the "tail of the tadpole." The resulting horseshoe orbit permits the tiny satellite to follow a slightly faster orbit on an inside track until it approaches the secondary, which accelerates it to a speed too high for it to stay on the inside track. It then switches to an outside track, where its orbital period is longer than that of the secondary. It therefore drifts back around the planet until it again feels the gravity of the secondary "behind" it in its orbit. This of course results in deceleration of the tiny satellite, dropping it into the inside track again.

A slightly more complicated behavior occurs when the secondary and tertiary bodies have comparable masses. Then the gravitational forces of the tertiary body on the secondary are no longer negligible, and the secondary executes a small horseshoe orbit of its own, "dancing" with the tertiary. Such is the case with Janus and Epimetheus.

The next small satellite is SVII Hyperion, with an orbit outside Titan's but well inside the orbit of Iapetus. Hyperion's orbital period is in a 4:3 resonance with Titan. Also, because of the very irregular shape of Hyperion, torques exerted on it by Titan during close approaches will tend to impart unpredictable, chaotic rotation to it. The view of the Solar System as seen from the surface of a chaotically rotating satellite would be truly bizarre. Hyperion has a reflection spectrum between 0.4 and 1.1 $\mu$m that looks exactly like that of the dark side of Iapetus. However, we find that the albedo of Hyperion is about three times as high as that of the dark side of Iapetus. It is possible that the materials that darken both Hyperion and Iapetus are the same, and that they are mixed with different proportions of a bright icy phase on these two bodies.

Beyond Hyperion ($a = 1.481$million km) there is a wide gap before Iapetus ($a = 3.561$ million km), then an even wider gap to the next pair of small satellites, S/2000 S5 and S/2000 S6, both near 11.4 million km. These two small bodies, probably of dirty ice (cometary) composition, both have eccentricities of about 0.33 and inclinations of $46.4 \pm 0.2°$. They have estimated radii of 9 and 7 km, respectively. We would expect the gravitational influence of these bodies on each other to be completely negligible. The most plausible explanation of the similarity of their orbits is that they are derived from fragmentation of a single body or dissociation of a binary body. The orbits of the liberated components then drift apart because of differential precession of their lines of apsides.

SIX Phoebe was for many years the only known Saturnian satellite that emulated the outermost family of Jovian satellites: it has an orbital eccentricity of 0.163 and a retrograde orbit inclined about 5.2° to the equatorial plane of Saturn. It takes 550.5 Earth days to complete a single orbit. It is also much darker than the other Saturnian satellites, with an albedo (0.06) similar to that of the leading hemisphere of Iapetus. Interestingly, Iapetus also has a spectral feature apparently due to water ice at wavelengths near 2 $\mu$m, implying that the surface is an ice darkened with substantial carbonaceous matter. In some scenarios, Phoebe is the source of dark material that falls onto Iapetus and darkens its front side. Since 2000, however, five more retrograde Saturnian minor satellites have been discovered, all outside Phoebe's orbit.

The next small satellite beyond Phoebe, S/2000 S2, 15.17 million km from Saturn, has an orbital inclination and eccentricity that are closely similar to those of S/2000 S5 and S6. Like these two bodies, S/2000 S2 follows a prograde orbit.

Next in order of distance, at 15.68 million km, is S/2000 S8, in a retrograde orbit that takes 2 Earth years to complete. At a paltry 4 km radius, it is very close to the limiting magnitude limit for detection.

Next comes a loose grouping of four small prograde satellites, S/2000 S3, S10, S11, and S4, with eccentricities of 0.27 to 0.54, orbital inclinations of 33.1 to 45.5°, and distances ranging from 17.25 to 18.23 million km.

Immediately beyond these prograde bodies, at distances of 18.49 to 23.08 million km, lie four more retrograde satellites, S/2000 S9, S12, S7, and S1, all with radii of 4 to 10 km. Again, the radii are inferred from the visual magnitude and an assumed albedo, with no spectral data available. Affinities to the short-period comet (Chapter VII) and Centaur (Chapter VIII) populations is plausible, but untested. The orbital period of the most distant of these bodies, S/2000 S1, is 3.59 Earth years.

Dynamically, the outermost satellite family of Saturn is apparently volatile, with capture and loss events occurring from time to time. The two outer families of satellites, prograde and retrograde, almost overlap in semimajor axis and overlap almost completely in instantaneous planetocentric distance. Frontal collisions at roughly 2 km $s^{-1}$ are possible, but the mean time between collisions for any individual body is of order $10^{11}$ a. Even more intriguing is the region from $11 \times 10^6$ km to $16 \times 10^6$ km, with its sequence of two prograde (nearly coorbital), one retrograde, one prograde, one retrograde, and then, beyond $17 \times 10^6$ km, four prograde satellites. With so many small satellites available, the mean time between collisions anywhere in the swarm becomes roughly equal to the age of the Solar System. Note that the periapsis distance of the most distant known satellite, S/2000 S1, is $q = 23.076 \times (1 - e) = 15.23$ million km, and its nearest neighbor, S/2000 S7, reaches in to 11.08 million km, inside the mean distance of the S/2000 S5 and S6 pair. The entire outer satellite swarm beyond Iapetus occupies spatially overlapping orbits that have no overlap with Iapetus or any of the closer satellites: it is precisely those inner satellites that have near-circular, near-coplanar orbits, attesting to formation from a common circumplanetary nebula. As search programs are extended to fainter bodies of even smaller sizes, the number of such bodies is likely to increase dramatically, possibly making collisions a statistically significant factor in the evolution of the outer satellite cloud.

## Satellites of Uranus

The orbits of the five largest satellites of Uranus were rather well characterized before the Voyager 2 flyby in January 1986, but almost nothing was known of the intrinsic properties of the satellites. Because of the difficulties inherent in observing the faint and distant Uranian system from Earth, Voyager's contribution to our knowledge of Uranus and its satellites has been even more profound than its earlier exploration of the Jovian and Saturnian systems.

The five large (pre-Voyager) satellites were named Miranda, Ariel, Umbriel, Titania, and Oberon. All five are distinctly smaller than the "big seven" satellites (the four Galilean satellites plus the Moon, Titan, and Triton). Miranda, the innermost of the five, is also an order of magnitude less massive than the other Uranian satellites, with a radius of 242 km. Ariel and Umbriel are next in order of both distance and size (580 and 595 km in radius, respectively), and the outer two, Titania and Oberon, are larger again (805 and 775 km). The masses of the smaller satellites are less accurately known from the Voyager tracking data, but the densities of all five large satellites are probably in the range 1.2 to 1.7. (Fig. VI.31). A density of 1.55 would be well within the error bars for all five satellites.

Because of the large axial tilt of Uranus, the poles of the planet can point to within a few degrees of the Sun. Indeed, at the time of the Voyager 1 Uranus encounter, the approaching spacecraft saw the orbits of the satellites arrayed like a bull's-eye. The geometry of approach committed the spacecraft to diving through the plane of the satellite orbits almost at right angles to the plane, like an arrow aimed at a bull's-eye. Thus it was possible to carry out a close flyby of only one satellite.

A second important constraint on the Voyager spacecraft was that it had to use the gravitational kick of Uranus to redirect the spacecraft to Neptune. This requirement placed very narrow constraints on the distance of closest approach to Uranus, which in turn compelled the selection of Miranda as the satellite to be examined at close range. Of course almost nothing was known of the surface states of these satellites before the encounter, except that the four largest satellites had very large opposition brightness surges (characteristic of fluffy surfaces) and albedos distinctly lower than that of pure ice. They and Miranda all showed absorption bands due to water ice in their spectra. Because Miranda, the smallest of the five satellites then known, seemed to be different in its properties, there was some ground for concern that the flyby would visit the least evolved, dullest satellite in the system.

Fortunately, and astonishingly, this is just the opposite of what was found. Miranda turned out to have an extremely diverse, complex, and interesting surface with overwhelming evidence of internal thermal activity (Fig. VI.32). Interpretation of the geology of Miranda is still at an early state, but it seems that the easiest way to expedite internal melting in so small a body is for there to be a substantial amount of ammonia in its interior. Melting would then commence at the ammonia–water eutectic temperature, about 174 K. Methane, which, because of its low melting temperature and viscosity, is even more mobile than ammonia–water melts, can be ruled out as a major constituent of the other four

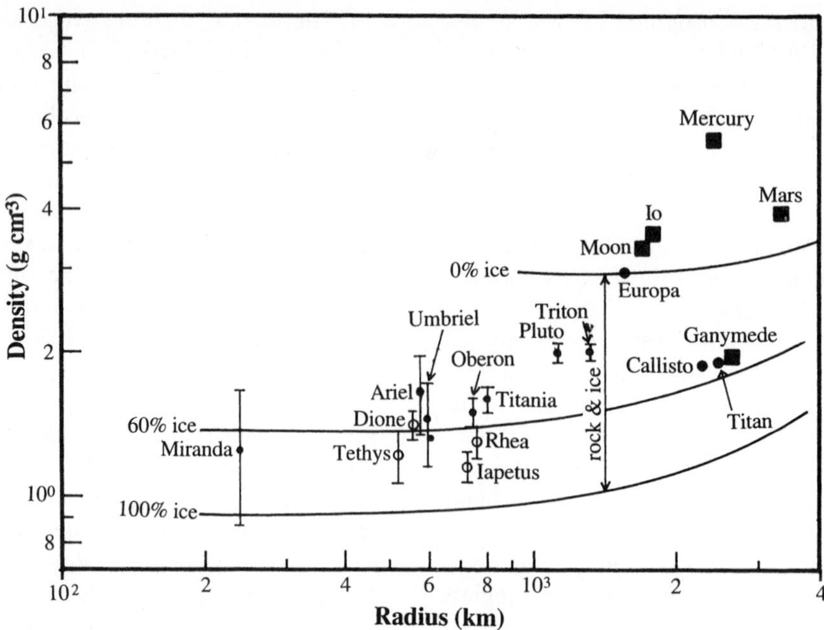


**Figure VI.31** Densities of the satellites of the outer planets. A solar-proportion mixture of water ice and rock (60%), with an uncompressed density of 1.3 to 1.5, could be reconciled with many of these data. The Saturnian satellites show a wide spread of densities without any clear radial trend.

Uranian satellites on density grounds, but Miranda's mass and density are so poorly known that we cannot prove the same for it. Nonetheless, it seems most improbable that the satellite closest to the planet would be the one with the most volatile composition. More likely, Miranda has melted and differentiated more thoroughly than the larger, more distant satellites. Of course, the reason for more extensive melting of the smallest of the five satellites cannot be internal radiogenic heating.

The photographs in Fig. VI.32 show that old, heavily cratered terrain is interleaved in a seemingly chaotic manner with heavily altered terrain. Deep faults crisscross the surface, sometimes exposing deep layers with a variety of albedos. Dark albedo features of unknown origin also mark the surface in broad, nearly parallel streaks.

Ariel, observed under circumstances not quite as favorable as Miranda, has a lightly cratered surface cut by numerous immense fissures (Fig. VI.33). These valleys are hundreds of kilometers long and several kilometers deep and often have the appearance of a dropped block of crust between two roughly parallel faults (a structure that, on Earth, is called a *graben*, the German word for grave). Fresh, very bright craters spot the surface nonuniformly. The grabens have been partially filled by later deposits, some of which seem to have been cut and reshaped by sinuous streamlike flows.

Umbriel, a near twin to Ariel in size, has clearly had a much less active interior. The surface is densely and almost uniformly cratered, and the vast fissure systems that are seen on Ariel (and that dominate the smaller Miranda) are absent (Fig. VI.34). Even the impact craters are generally dark, suggesting the ubiquitous presence of abundant dark, carbonaceous material in the crust. Curiously, there is one very large (140 km diameter) crater that has a brilliant rim and one somewhat smaller (110 km diameter) crater with a bright central peak but a very dark rim. These features are not understood.

Titania, much larger and much farther from Uranus (and from Voyager) than Umbriel, looks like a hybrid of Ariel and Umbriel (Fig. VI.34). There are several very large rift systems, but the crater density on the surface is very high and nearly uniform. A brief, intense, and early thermal episode, possibly associated with incipient melting, seems to be indicated. Oberon, about as large as Titania but even farther from the planet, seems to preserve a primitive, heavily cratered surface with no clear evidence of internal thermal activity (Fig. VI.34).

In both the Ariel–Umbriel and the Titania–Oberon pairs of similar-sized satellites, it is the member that lies closer to Uranus that has undergone the more extensive thermal activity. Of these five satellites, the smallest (and closest) has seen the most extensive activity. A strong connection either to tidal heating or to a radial gradient in their formation temperatures is suggested.

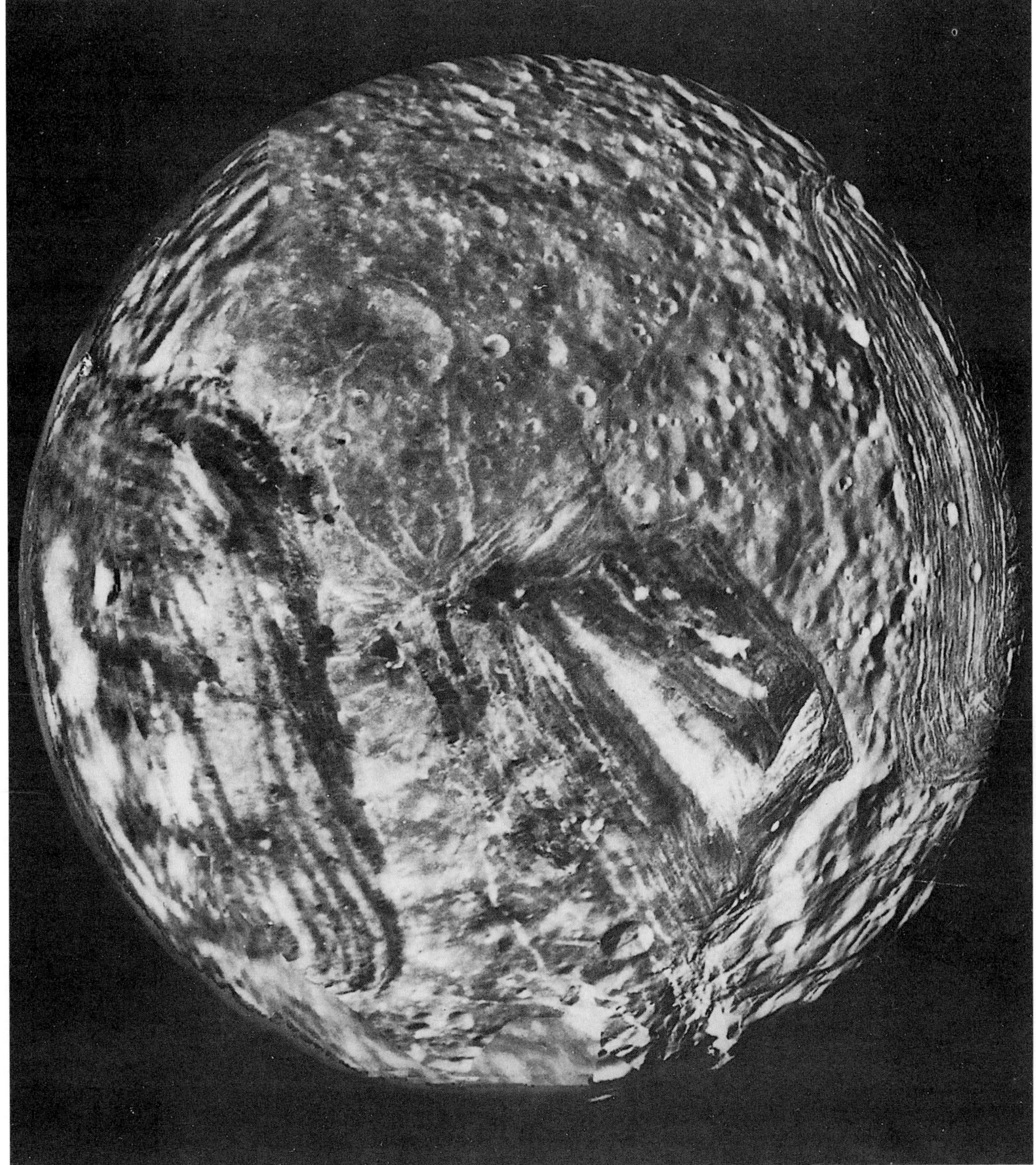

**Figure VI.32** Miranda. This Uranian satellite serves as a kind of sampler of geological processes in ice-rich bodies. Several strikingly different types of terrain are in evidence, ranging from very ancient, heavily cratered units reminiscent of the lunar highlands to the strange chevron with its very low crater count and strong albedo contrasts.

In addition to these five large satellites, there are, interior to Miranda's orbit, 10 small satellites that were discovered by Voyager 2, and one small satellite, S/1986 U10, that was discovered in 1986 and then lost until 2001. These satellites have inferred sizes ranging from a radius of about 20 km to 85 km for UXV Puck and are suspected to be ice-rich. All 16 inner satellites out through Oberon are in nearly coplanar, low-eccentricity orbits in the equatorial plane of Uranus. From Oberon ($a = 582{,}596$ km) to the next satellite, UXVI Caliban

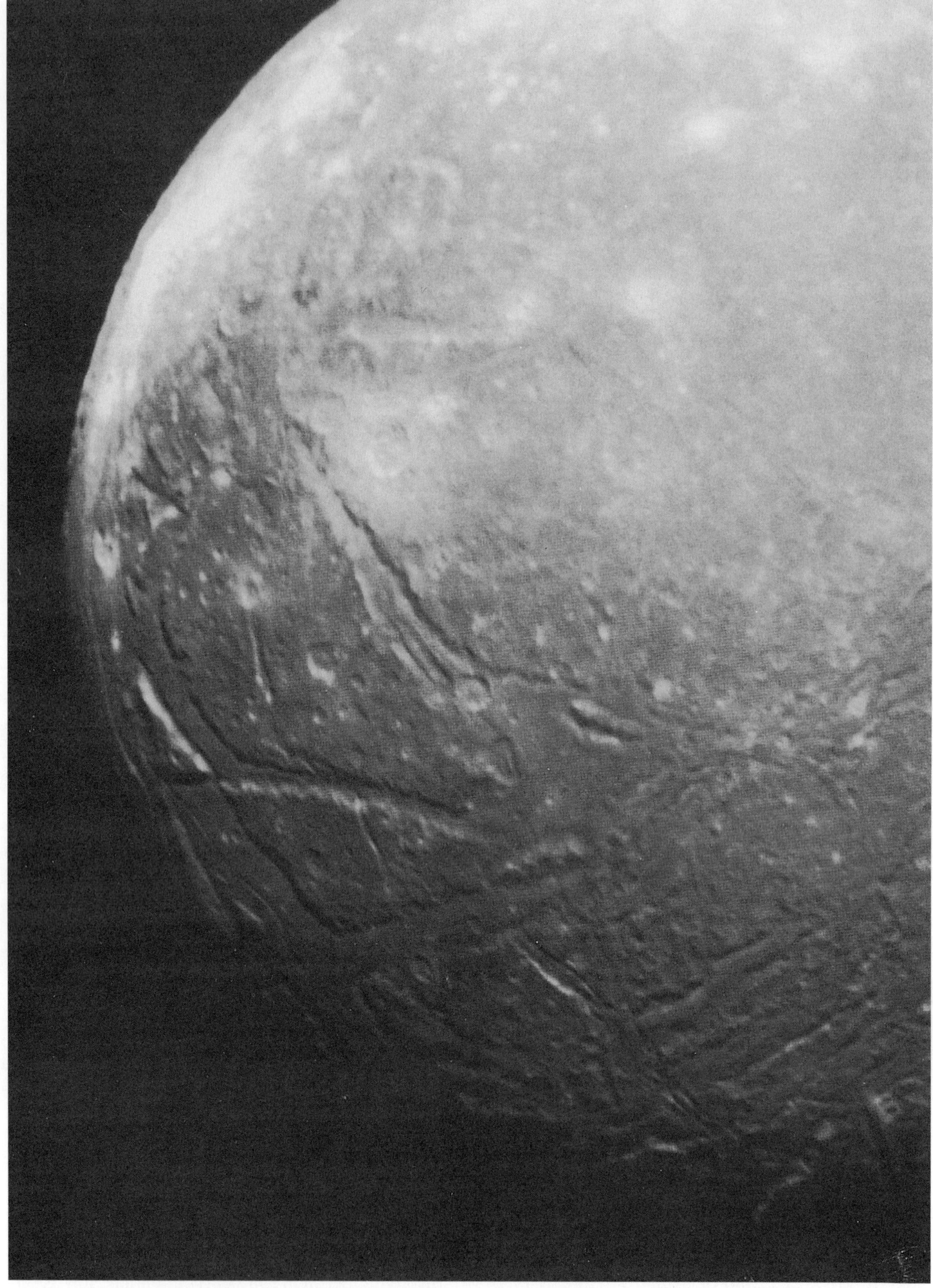

**Figure VI.33** Ariel. Visually suggestive of a body intermediate in evolution between Callisto and Ganymede, Ariel displays an extensive system of wide, deep fissures. Note the very low crater density on the valley fill deposits near the terminator (right).

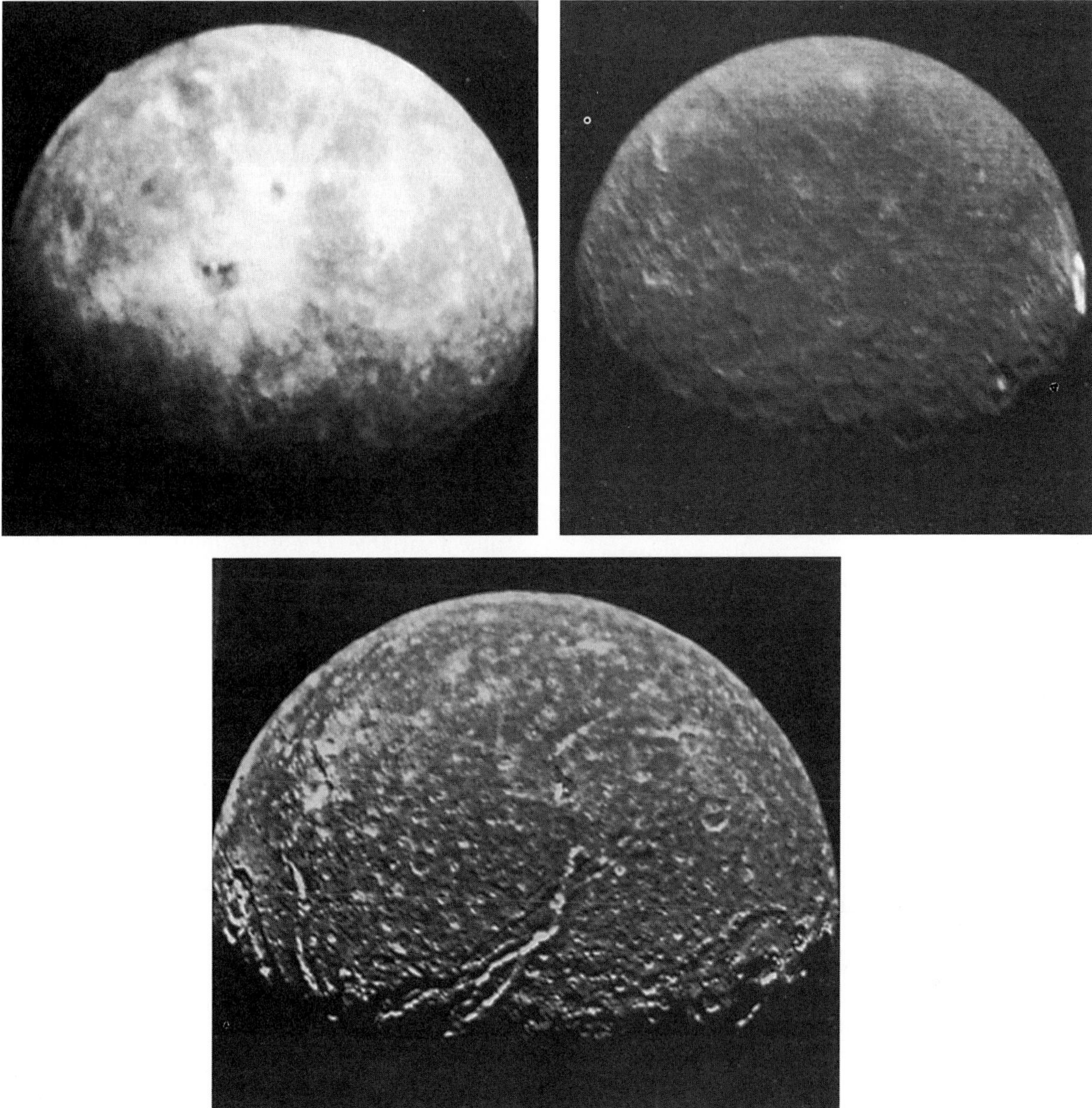

**Figure VI.34** Umbriel, Titania, and Oberon. Umbriel (top) is heavily cratered and shows little evidence of internal tectonic activity. Note the bright feature (crater rim?) at the top. Titania (right), the largest satellite of Uranus, has a system of giant fissures similar to those on Ariel (Fig. VI.31). Oberon (bottom) is heavily cratered. Several large craters are prominent in this view.

($a = 7.187$ million km), is a very wide, unpopulated gap. The five outer satellites, all of which have been discovered since 1999, are in eccentric retrograde orbits. The most distant known Uranian satellite, UXIX Setebos, takes 2345 Earth days (6.42 Earth years) to orbit Uranus. All are presumed to have cometary composition, and all have radii on the order of 10 km. None of the small satellites have measured masses or densities. Again, as with the outer satellite swarms of Jupiter and Saturn, there is a clear demarcation and clean separation between the "primordial" inner satellites derived from planetary subnebulae and the transient, distant population of asteroidal or cometary small bodies. See Figure VI.35.

## Satellites of Neptune

Our pre-Voyager knowledge of the orbits of Nereid and Triton, the two "classical" satellites of Neptune, was fairly good, but the intrinsic properties of the satellites remained very poorly known. A renaissance in our

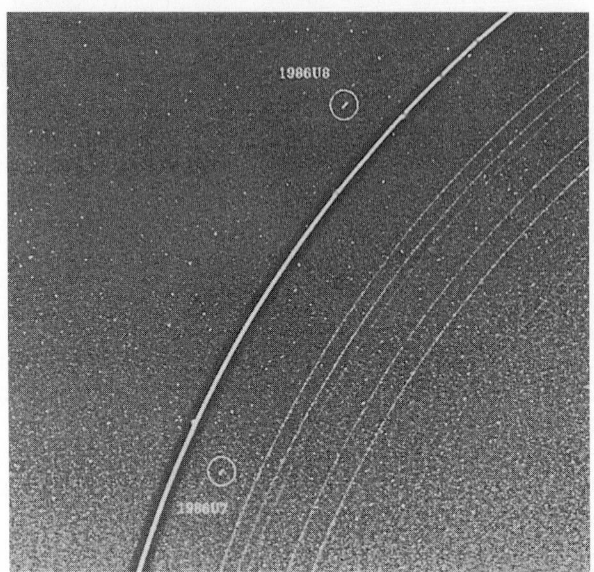


**Figure VI.35** Shepherds watch the rings by day. The two small satellites UVI Cordelia (formerly S/1986 U6) and UVII Ophelia (formerly S/1986 U7), both discovered by Voyager 2, are seen here tending the epsilon ring of Uranus. It is likely that the Uranian rings are kept confined to such narrow limits by the gravitational effects of several small satellites.

knowledge of Neptune and its satellites occurred with the August 1989 encounter of Voyager 2 with Neptune and its system.

Although we knew little about Neptune's system, what we *did* know was very peculiar indeed. The two satellites, Nereid and Triton, differ dramatically in their size and orbital properties.

Triton, one of the "big seven" satellites in the Solar System, has an almost perfectly circular orbit at a distance of 354,500 km from Neptune—but the orbit is *retrograde*, with an inclination of 159°. The orbital period is 5.877 Earth days, and the rotation of Triton is rotationally locked onto Neptune.

The best Earth-based estimate of the radius of Triton, based on a marginal detection of thermal emission and the assumption of a thermal steady state for the Sunward hemisphere of Triton, was 1750 km with an uncertainty of $\pm 15\%$. The mass, determined from its perturbations of the motion of Nereid, was uncertain by $\pm 20\%$. The nominal mass, combined with the mean radius estimate of 1750 km, gave a density of 6.0, a number that seems altogether ridiculous. According to the condensation scenarios we discussed earlier, a typical icy body the size of Triton, if condensed in the poorly equilibrated Solar Nebula (where pressures are too low for extensive reduction of CO and $N_2$ to methane, water vapor, and ammonia), ought to have an uncompressed density near 2.0. If condensed in a well-equilibrated dense planetary subnebula (in which reduction of CO and $N_2$ occurs much more easily), the uncompressed density should be near 1.6.

Triton's retrograde orbit is clearly unstable. Tidal bulges raised on Neptune by Triton would have to move around Neptune against the planetary spin. Any frictional coupling between the tidal bulge and the interior of Neptune would cause the bulge to be displaced by a degree or so in the prograde direction from the sub-Triton point on Neptune. This is *behind* Triton in its orbital motion about Neptune (Fig. VI.36) and will inevitably lead to a small retarding torque on Triton. This torque lowers the energy of the orbit so that it shrinks toward Neptune's surface. The satellite, if it has a low density, will eventually be stripped apart by Neptune's tidal forces at Neptune's Roche limit. If it is too dense or strong to be torn apart in this manner, the satellite will fall onto Neptune intact in a catastrophic collision.

The fractional rate of energy dissipation is used to define a variable, $Q$, essentially the time scale for energy dissipation,

$$Q = 2\pi E/\Delta E, \qquad \text{(VI.43)}$$

where $E$ is the maximum amount of energy stored in the body at any point during a tidal cycle and $\Delta E$ is the amount of energy dissipated per cycle. The planeto-

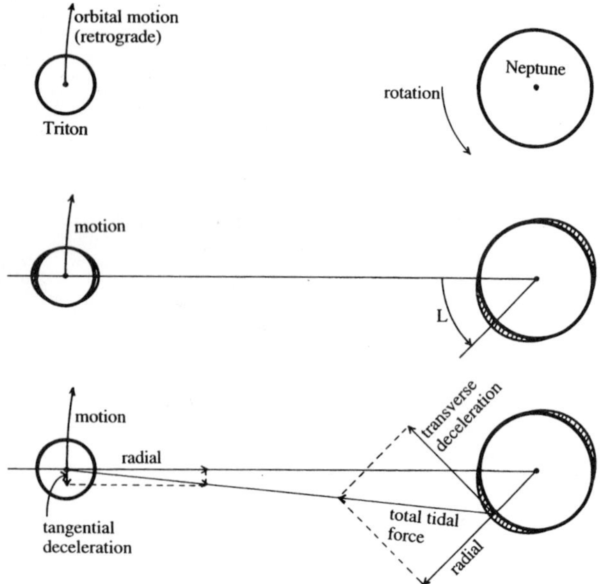


**Figure VI.36** Tidal torques on Triton and Neptune. Because of the retrograde orbit of Triton and its large size, tidal torques exerted on it by the tidal bulge it raises on Neptune act strongly to retard the orbital motion of Triton. The lunitidal interval, $L$, is greatly exaggerated in this drawing. This torque will partially despin Neptune as Triton's orbit continually drops closer to Neptune. The process will lead to the tidal disintegration of Triton at its Roche limit. Much of its mass will soon fall onto Neptune. The rest will form a massive ring system that will be unstable in the long run (cf. Fig. VI.14).

centric lag angle, $L$, between the sub-Triton point and the tidal bulge is

$$\sin L = Q^{-1}. \tag{VI.44}$$

The angle $L$ is also called the *lunitidal interval*. Large values of $Q$, according to these equations, mean very low dissipation rates, ineffective coupling, long lifetimes against dissipation, and small lag angles. From studies of the satellites of Uranus, $Q$ may be as large as $3 \times 10^4$. If the same value of $Q$ also applies to Neptune, the time scale for the decay of Triton's orbit would be many billion years, perhaps even longer than the Main Sequence lifetime of the Sun.

Under the influence of Solar gravity, the plane of Triton's orbit precesses with a period of about 640 years. The latitude of the subsolar point on Triton can range from about 50°S to about 50°N. It is presently near 40°N and increasing, leaving the south pole in perpetual darkness for about 100 years. There must be a global temperature minimum near Triton's south pole, with a sizeable area cold enough to serve as a cold trap to capture planetary volatiles.

Triton is bright enough to permit spectral observations over the visible and near infrared wavelengths. Figure VI.37 shows a composite spectrum for the wavelengths from 0.8 to 2.5 $\mu$m, as presented by Dale Cruikshank of the University of Hawaii. Laboratory spectra of solid methane and water ice are included for comparison. Several features deserve comment. First, in the visible region, Triton is distinctly red. There are several infrared bands that are clearly due to methane, but it may be that solid methane suffices to produce them. There is also a strong, sharp absorption feature at 2.16 $\mu$m that is clearly not due to methane. Interestingly, dense gaseous nitrogen and liquid nitrogen absorb at precisely that wavelength. It is perhaps hazardous to conclude too much from a single spectral feature, but we cannot help remembering the nitrogen–methane atmosphere of Titan. Simply cooling Titan to Triton's temperature would condense the atmosphere almost completely and provide a spectrum strikingly similar to that actually seen on Triton. It is reasonable to suppose that the surface of Triton contains abundant solid methane and solid nitrogen. Water ice has also been detected in its spectrum. Recently, over a period of several months in 1998, Triton seems to have experienced a brief warming event of unknown cause. There was an abrupt, distinct reddening associated with a 50% increase in the gaseous nitrogen abundance. It is of course reasonable that volatilization of a large mass of nitrogen would better expose a less volatile surface pavement of red, photochemically produced organic matter, and that reddening and darkening of the surface would raise temperatures and hence raise the gaseous nitrogen abundance. The presence of mobile, volatile frosts is also compatible with Triton's brightness surge near opposition, when the shadows of surface grains become invisible from Earth.

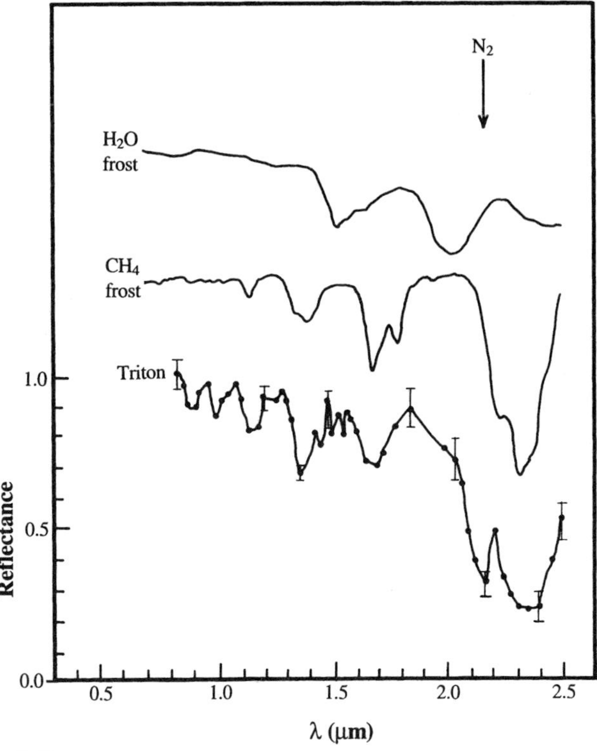


**Figure VI.37** Visible and near-infrared spectrum of Triton. Features due to solid methane, water ice, and nitrogen are indicated. The 2.16-$\mu$m feature strongly suggests the presence of condensed nitrogen.

Gaseous methane and ammonia must be present in vapor-pressure equilibrium with the surface. If the atmospheric pressure is small, then local temperatures will vary substantially, reaching a maximum close to the subsolar point and a minimum at the winter (south) pole. But the vapor pressure is an extremely strong function of the temperature; indeed, at these temperatures ammonia would be virtually absent from the atmosphere. We would expect heat absorption near the subsolar point to evaporate methane and nitrogen, whereas condensation in the dark south polar regions will release latent heat. Horizontal transport of vapor will effect an important transport of latent heat from the dayside to the nightside. Thus the assumptions under which the radius was estimated from the measured dayside thermal flux may be seriously in error. Triton may be emitting much of its heat on the unobserved side, and hence the dayside may be cooler than was assumed. To provide the observed 25-$\mu$m thermal flux, the satellite would have to be substantially larger than 1750 km. This in turn removes the density paradox, permitting Triton to have a sensible density of only 1.5 to 2, not 6. So matters stood until the Voyager 2 flyby.

As on Titan, solar ultraviolet photolysis of methane gas must surely occur. But, at the low temperature of Triton, heavier ($C_2$ to $C_4$) hydrocarbons would readily condense and hence not be present in observable concentrations in the gas phase. As on Titan, reddish organic H–C–N compounds may dominate the visible spectral shape. At Triton's distance of 30 AU from the Sun (vs 9.5 AU for Titan) the rate of methane photolysis is limited to $4 \times 10^{-13} \times (9.5/30.0)^2 = 4 \times 10^{-14}\,\mathrm{g\,cm^{-2}\,s^{-1}}$ if the atmosphere is optically thick (or less if most photons are reflected from or absorbed by the surface).

Voyager 2 contributed an incomparably more detailed perspective on Triton. Voyager 2 provided an estimate of $1355 \pm 7$ km for the radius of Triton and measured its mass as $GM_{\mathrm{NI}} = 1428.5 \pm 4.5\,\mathrm{km^3\,s^{-2}}$, suggesting a density of $2.05\,\mathrm{g\,cm^{-3}}$. A well-observed stellar occultation of a star by Triton in 1997 showed that the occultation radius (in the atmosphere well above the surface) is $1422 \pm 2$ km. This measurement provides a very firm lower limit on the density of Triton of $1.74\,\mathrm{g\,cm^{-3}}$. Using the surface radius of 1355 km, the surface gravity $g_{\mathrm{NI}} = GM_{\mathrm{NI}}/r_s^2 = 77.8 \pm 0.5\,\mathrm{cm\,s^{-2}}$ and the escape velocity is $1.41\,\mathrm{km\,s^{-1}}$. The surface pressure of Triton's atmosphere is a meager $16 \pm 3\,\mu$bar at a temperature of 38 K, and the atmospheric mass is $P/g_{\mathrm{Tr}} = 0.21\,\mathrm{g\,cm^{-2}}$. The atmosphere is dominantly nitrogen, with a methane mole fraction of about $10^{-4}$, suggesting only $2 \times 10^{-5}$ g methane per square centimeter ($6.02 \times 10^{23} \times 2 \times 10^{-5}/16 = 0.75 \times 10^{18}$ molecules per square centimeter). The photochemical lifetime of the methane in Triton's atmosphere is $2 \times 10^{-5}/4 \times 10^{-14} = 5 \times 10^8$ s, which is only 16 Earth years. Because the atmospheric methane abundance matches the equilibrium vapor pressure of solid methane at 38 K, replenishment of gaseous methane by evaporation from a vast surface reservoir of solid methane is required.

The Voyager imaging of Triton showed a spectacular diversity of surface features (Fig. VI.38). The surface is devoid of heavy cratering and therefore young. Seasonal $N_2$ snow cover brightens the surface in the winter hemisphere, but with a slight reddish tinge characteristic of organic polymers. The western (trailing) hemisphere has an extensive "cantaloupe terrain" with a coarsely textured, almost crater-free surface. The eastern (leading) hemisphere has several large caldera-like basins filled with ice. The low crater density suggests complete resurfacing of Triton some time after the period of heavy bombardment in early Solar System history. The unique orbit of Triton about Neptune implies severe disturbance of its motion, possibly capture from heliocentric orbit. Supposing capture from an orbit with a semimajor axis of 30 AU and an eccentricity of 0.2, the orbital velocity of Triton relative to Neptune was about $1\,\mathrm{km\,s^{-1}}$. Thus a kinetic energy density of at least $0.5 \times 10^{10}\,\mathrm{erg\,g^{-1}}$ must have been released in a very short time in order to effect capture by Neptune. Further energy dissipation must have occurred as Triton evolved into its present orbit. The heat capacity of cold water ice is about $10^7\,\mathrm{erg\,g^{-1}}$, and hence the capture heat pulse must have dissipated enough energy through the tidal interaction to heat Triton by several hundred degrees almost instantly. This is enough to melt all the ices in Triton and release virtually all of its internal methane and nitrogen inventory to the surface. Complete differentiation of these volatiles, ices, and rock must have occurred on the time scale of a single passage through the Neptune system, about $10^5$ km at a speed of about $1\,\mathrm{km\,s^{-1}}$, or about 1 Earth day.

The Voyager photopolarimeter experiment found that particles on Triton's surface have single-scattering albedos of about 0.96 at a wavelength of 0.75 $\mu$m, typical of ices, on an unusually compacted surface, similar to that of Europa. This suggests rather effective compaction by a strong heating event late in the cratering history. Note from Fig. VI.38 that neither of the two radically different terrain types seen by Voyager is heavily cratered.

A number of black wind streaks are seen on the icy surface, with probable ages of less than 1000 years. In addition, two geyser-like plumes were observed in the process of eruption. These eruptions rise roughly 8 km and stream out in the prevailing wind, showing a dense streak of dark dust reaching as far as 100 km from the eruption site. These plumes are apparently caused by fountaining of nitrogen trapped in the surface.

Neptune's only other classical satellite, Nereid, follows a highly eccentric ($e = 0.75$) and inclined ($i = 27.6°$) orbit at great distance from Neptune ($a = 5.515 \times 10^6$). The orbit of Nereid is sufficiently unusual to suggest severe perturbation by Triton during and after its capture. Its orbital period is 360.16 Earth days. Its radius was poorly known from Earth-based observations, but its brightness (19th mag at mean opposition), combined with a guessed albedo, suggested a radius of about 200 km. Voyager 2 found Nereid to be distinctly nonspherical, with a radius of $170 \pm 25$ km. The geometric albedo of Nereid is about 0.2 and the phase integral about 0.5 (a Bond albedo of about 0.1), so it is a fairly bright small satellite, probably an ice–dust mixture. In its color and photometric properties, Nereid is similar to the smaller Uranian satellites. Water ice has been identified on Nereid's surface by means of water ice absorption features at 1.54 and 2.05 $\mu$m. Photometry of Nereid in the U, B, R, and I filters finds variability of up to 1.83 mag in brightness, with a flat spectrum reminiscent of carbonaceous

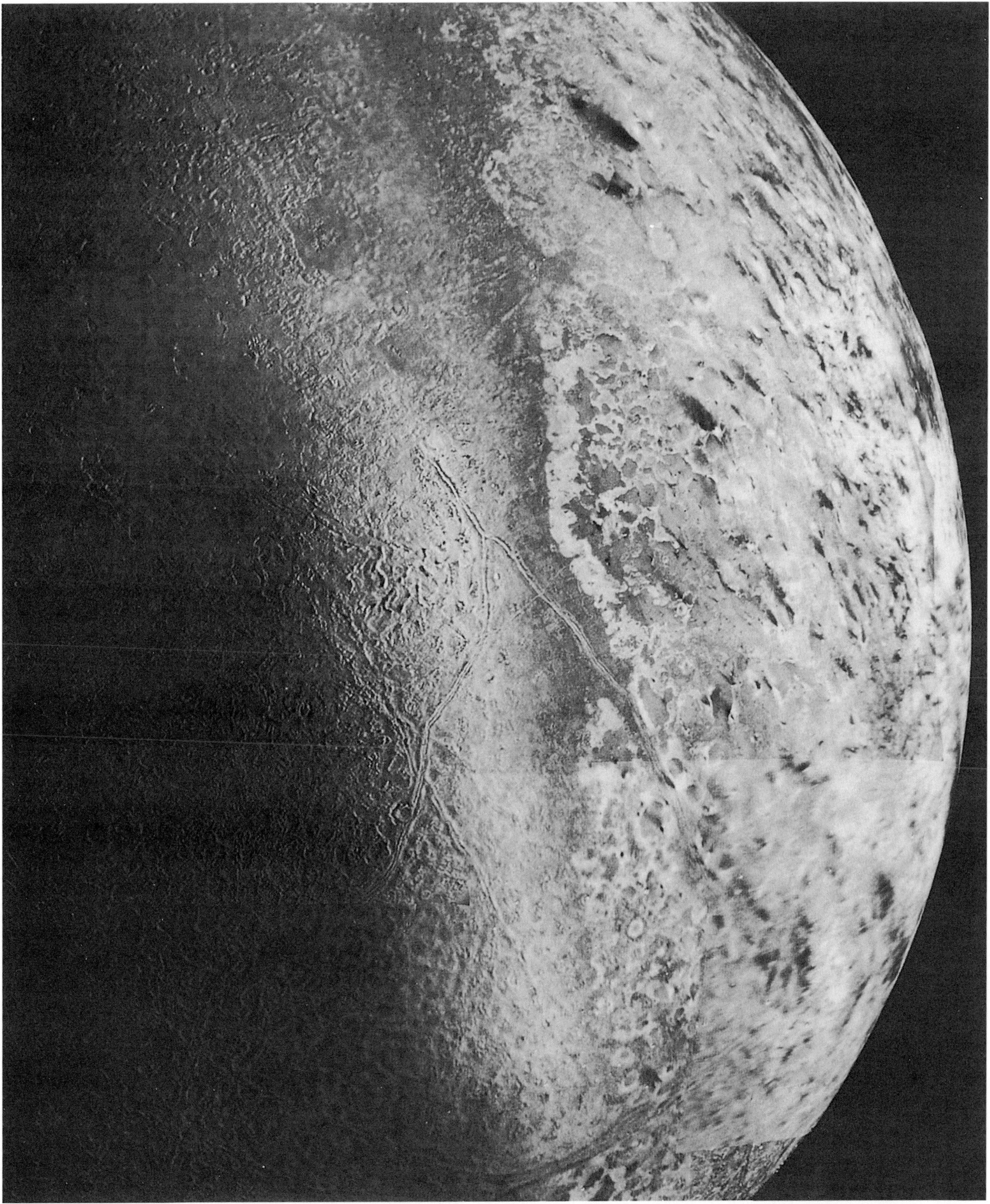

**Figure VI.38** Triton. This Voyager 2 image shows the contact between Triton's two very different types of terrain. Some of the frames in this mosaic are of lower resolution, causing the apparent blurring in the lower right quadrant.

asteroids, distant satellites of Jovian planets, and Centaurs. It has been suggested that Nereid may be a captured Centaur. Nothing is known of its rotational period or axis orientation, as Earth-based and Voyager data on its light curve are apparently incompatible.

Voyager not only observed Triton and Nereid, but also discovered six previously unknown small satellites orbiting in an orderly system close to the planet, far below Triton's orbit. Because Voyager flew by close to Neptune, it had a far closer view of these new satellites than it did of Nereid. The largest of these, NVIII Proteus (formerly S/1989 N1), shown in Fig. VI.39, is $208 \pm 8$ km in diameter, making it Neptune's second-largest satellite. The others, Larissa, Galatea, Despina, Thalassa, and Naiad, orbit even closer to Neptune than Proteus. Their albedos are low, and their photometric and spectral properties are very similar to those for the inner small satellites of Uranus. The low albedo (and "flat" spectra) of these six satellites suggest carbonaceous material and are compatible with evolution by impact processes, not thermal differentiation. The spin states of these small bodies are not well known. Proteus and (probably) Larissa seem to spin in 1:1 resonance with their orbital periods, and it would not be surprising if all six did so.

## The Pluto–Charon System

Pluto was discovered by Clyde Tombaugh at Lowell Observatory in Flagstaff, Arizona, in 1930, and its satellite Charon was discovered in 1978. Pluto and Charon orbit closely about their common center of gravity in circular orbits. Charon is, relative to its primary, the largest satellite in the Solar System. The two constitute

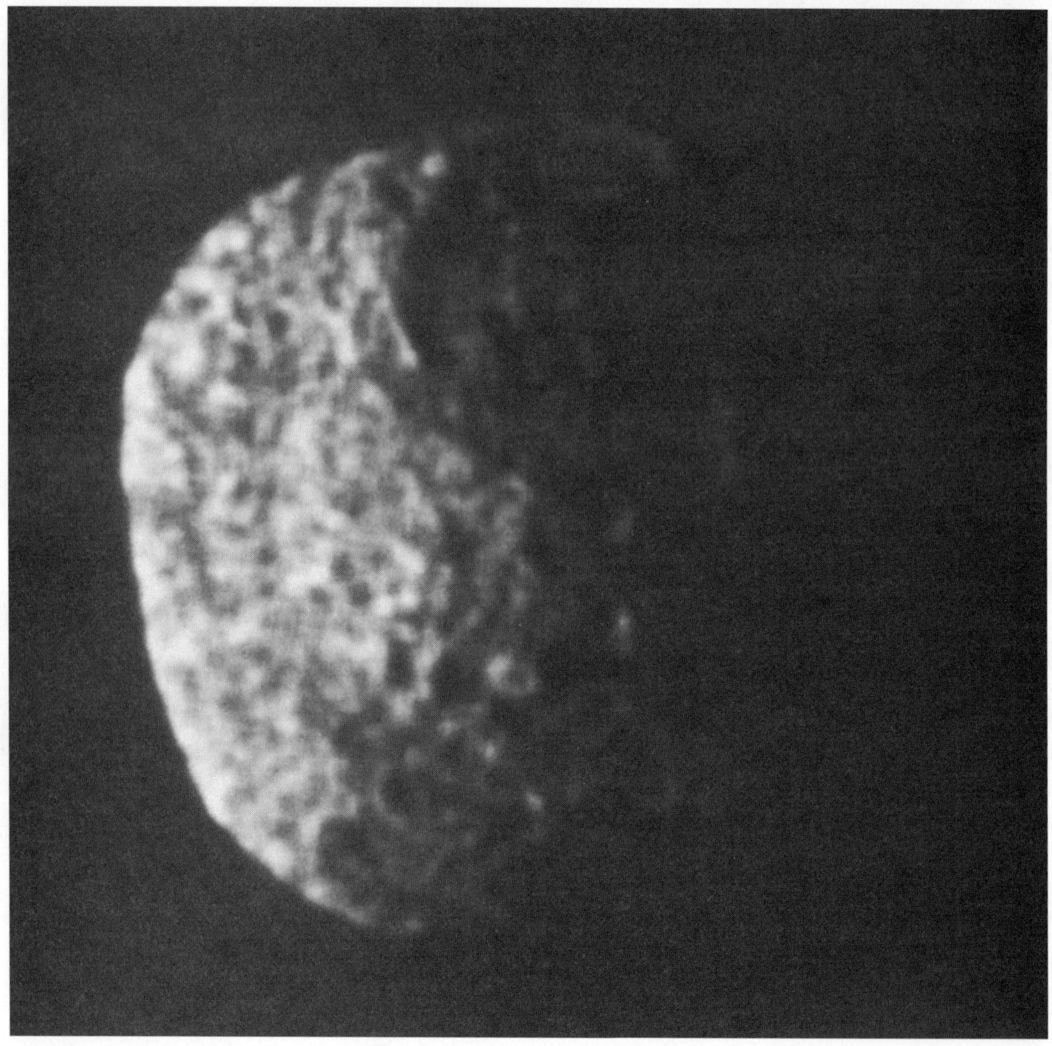

**Figure VI.39** Neptune's small satellite, Proteus. This is one of six small Neptunian satellites discovered by Voyager.

the closest known approximation to a double planet. The rotation of both planet and satellite is locked on to their orbital period of 6.387 Earth days. Thus the lunitidal interval is zero, and no tidal evolution of the orbit can occur. Any eccentricity induced by an impact or any other cause would be erased quickly.

The mean distance of Charon from the barycenter of the system is about 19,640 km. The inclination of the plane of motion of the satellite to Pluto's orbit is about 99°, and Pluto's orbit is inclined 17° with respect to the ecliptic. The pole of the Pluto–Charon system precesses under the influence of solar torques with a period of 3.7 Ma about the normal to its orbit, so a pole-on orientation to the Sun at perihelion is possible, but is not a permanent feature of the system.

Pluto is about $1151 \pm 6$ km and Charon is $593 \pm 13$ km in radius, suggesting that the volume of Charon is close to 12% of Pluto's. The total mass of the system is $1.47 \pm 0.08 \times 10^{25}$ g. Note that both bodies are well within the range of sizes encountered in the satellite systems of the Jovian planets, far smaller than the planets. Indeed, Pluto is smaller than the "big seven" satellites (the Moon, Io, Europa, Ganymede, Callisto, Titan, and Triton). For this reason and because of the expected ice-rich composition of bodies so far from the Sun, we choose to treat Pluto as a deviant icy satellite rather than as a deviant planet. Visual comparison of Pluto to other ice-rich bodies is hampered by the limited resolution available in Pluto images. Since no spacecraft has yet visited Pluto, we are limited by the resolution of the Hubble Space Telescope. Fig. VI.40 shows the surface of Pluto as imaged by the HST. Association of albedo features with topographic relief or slopes is not yet possible.

Because of the high axial tilt of the system (and of both of the bodies in it) and because of the high eccentricity of Pluto's orbit about the Sun, temperatures on the surfaces of both bodies can vary greatly. The Infrared Astronomical Satellite (IRAS), observing the Pluto–Charon system around the time of their outward crossing of Neptune's orbit, found subsolar temperatures on Pluto ranging from 55 to 61 K, depending on the local albedo. Charon, which is slightly darker, was found to have a subsolar temperature of about 64 K. Near perihelion, Pluto and Charon cross the orbit of Neptune and spend about 20 years inside Neptune's orbit before again venturing out into the even more frigid wastes beyond. Pluto passed Neptune's orbit inbound in 1979 and passed it again outbound in 1999. Perihelion passage was in November 1989. Another consequence of the high axial tilt is that, twice in each orbit of Pluto about the Sun, the plane of Charon's orbit about Pluto crosses the Sun. Observers on Earth can then observe repeated mutual eclipses of these two bodies over a period of several years, as indeed happened in 1985–1990. Careful photometry of the system during such an "eclipse season" permits determination of the radii, masses, spectra, albedos, and temperatures of both members. The average density of the system was found to be $2.03 \pm 0.2\,\mathrm{g\,cm^{-3}}$, suggestive of solar nebula ice–rock mixtures with minor methane and ammonia and major amounts of nitrogen and carbon monoxide.

The visible and near-infrared spectra of Pluto (actually including Charon) are shown in Fig. VI.41. In the visible region, Pluto is quite bright and red, but a comparison with the Triton spectrum shown above reveals that the shape of the visible spectrum is quite different from that seen on Triton. Features in the

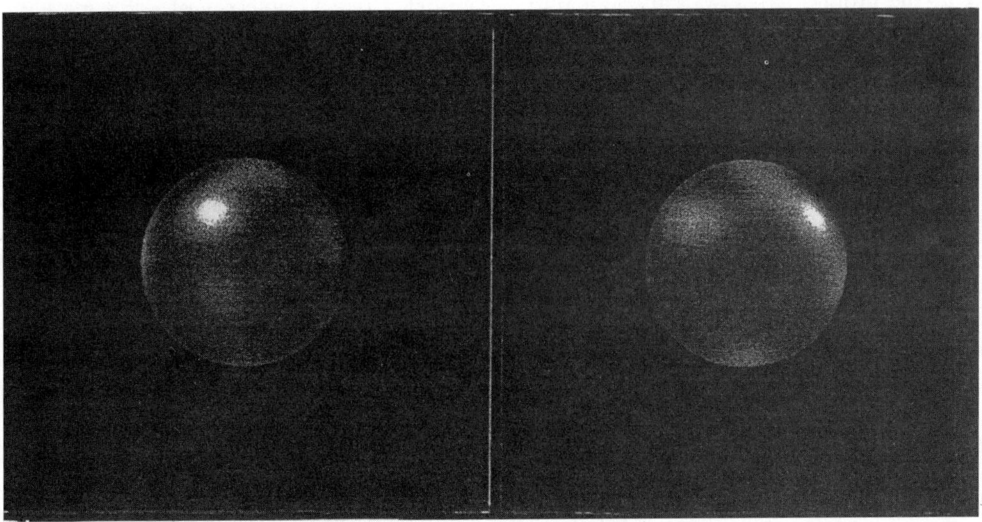

**Figure VI.40** The two hemispheres of Pluto as imaged by the Hubble Space Telescope. Courtesy of the Space Telescope Science Institute.

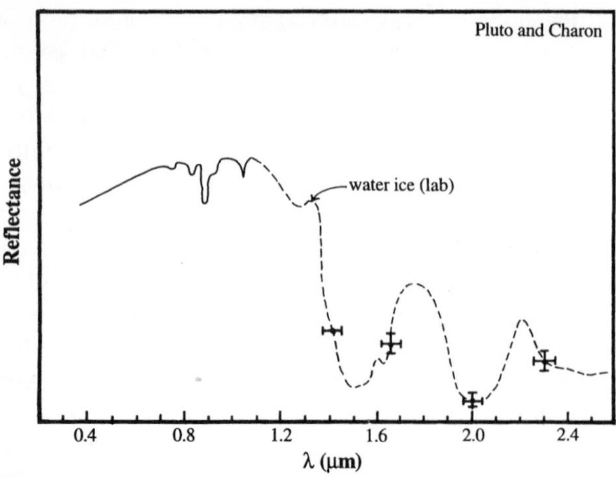


**Figure VI.41** Visible and near-infrared spectrum of Pluto and Charon. Features due to solid methane are evident (cf. Fig. VI.37).

infrared again clearly show solid methane. Tobias Owen and co-workers have detected solid CO and solid $N_2$ in Pluto's spectrum, in keeping with the idea of a Solar Nebula origin. The Bond albedo is poorly constrained because we have such limited phase-angle coverage of Pluto from Earth, but the geometric albedo is much higher than Triton's, suggesting a fresh snowy surface. Indeed, methane frost has also recently been detected on Pluto, apparently with a spotty, regional coverage. Increased methane snow cover can be expected as Pluto retreats from perihelion, overlying the red material and water, nitrogen and CO ices; however, the lowest global temperatures would be expected somewhat after perihelion passage because of the thermal lag of the surface. It is likely that during each orbital "winter," when Pluto is beyond Neptune's orbit, the atmosphere largely condenses and falls to the ground as snow. Pluto not only is brighter at these times, but also has less variation of brightness with rotational phase. During approach to perihelion the hemisphere containing the Sunward pole loses its snow cover through sublimation, which causes the observed darkening trend of the past 40 years.

The atmospheric inventory of methane on Pluto is presumably regulated by vapor pressure equilibrium with the surface, as on Triton. Because the two bodies are presently nearly the same distance from the Sun, the geometric albedo of Pluto suggests a low temperature; however, the temperature is poorly constrained observationally to range from about 35 to 60 K. Photochemical processing of atmospheric methane must destroy $CH_4$ at about the same rate as that on Triton, which is sufficient to remove Pluto's present atmospheric mass some $10^5$ times over the age of the Solar System. Thus the argument we developed for a massive methane reservoir on Triton must also apply to Pluto.

In 1980 a stellar occultation by Charon was observed, and Pluto occulted another star in 1988. Photometric observations of these occultations help us constrain the sizes of both bodies, but also reveal the presence of a trace of atmosphere and a thin haze layer on Pluto. James Elliot and co-workers at M.I.T. found a very tenuous, distended atmosphere that extended up to an altitude of about 1000 km above the surface. The column abundance of methane is clearly less than that found in a 1-m path length of methane gas at normal temperature and pressure (1 m agt), for a surface pressure little over 1 $\mu$ bar. Roger Yelle and Jonathan Lunine of the University of Arizona have examined the radiative balance in Pluto's tenuous atmosphere and attribute the attenuation observed in the stellar occulation to a temperature gradient in the atmosphere. They derive a temperature of about 104 K at roughly 50 km above the surface and use the extinction scale height deduced from the stellar occultation to calculate a mean molecular weight of $25 \pm 3$ for the atmosphere in this region. Of the candidate heavy gases, CO and nitrogen (which have been independently verified as present on the surface) are the most plausible. Argon is also a possibility. Elliot's latest interpretations suggest 0.2 to 0.4 m agt of methane, with a surface pressure of 0.8 to 2.5 $\mu$ bar. Escape of methane from a distended atmosphere about a low-mass planet is unavoidable, as was first argued by Larry Trafton and S. Alan Stern at the University of Texas, who concluded that hydrodynamic escape (but continuous fluid flow, not molecular diffusion) would rapidly remove all of Pluto's atmosphere. Donald Hunten and A. J. Watson looked into the energetics of this escape process and concluded that atmospheric escape must be energy limited, so that at most a few kilometers of solid methane could be lost over the entire history of the Solar System. Current models show Pluto's atmosphere to be so distended that it is almost cometary in nature. A search for an extended gaseous envelope is marginally possible with existing Earth-based instrumentation.

It has so far not been possible to resolve Charon both spatially and spectrally at the same time, and hence we know little about its composition and surface color. Subtracting a Pluto-only spectrum (taken when Charon is in full eclipse) from a Pluto-plus-Charon spectrum (taken at a different orbital phase and hence a somewhat different part of Pluto) reveals no evidence of a methane absorption in the spectrum of Charon. The mutual eclipse studies imply that the albedo of Charon is about $\frac{2}{3}$ that of Pluto and is spectrally neutral. Generally, the limited data suggest that Charon is more similar in its spectral and photometric properties to the Uranian satellites than it is to Pluto.

## The Neptune–Pluto Resonance

The image of two planets with crossing orbits raises the prospect of a planetary collision. Is it possible for Pluto to collide with Neptune? If so, how soon might it happen? If not, how can they avoid collision?

The semimajor axis of Neptune's orbit is 30.058 AU, and that of Pluto is 39.440 AU. The ratio of their orbital periods is then, by Kepler's second law, $(39.440/30.058)^{3/2}$, or 1.503, which we shall write as 3.006:2. This is so close to a resonant 3:2 relationship that we are obliged to carry out long-term computer calculations of their orbits, including planetary perturbations, in order to see how their motions may "drift" so as to allow close encounters or even collisions. Such calculations have been done, with most interesting results. It is found that, if one simply looks at the positions of the two planets every three Neptune periods and every two Pluto periods ("stroboscopically"), when Neptune is at inferior conjunction as seen from Pluto, the relative positions of the planets rock back and forth rather than evolving continuously in the same direction; Neptune can pass Pluto only when Pluto is near aphelion. Thus the Pluto–Neptune resonance is *librating*, with the period oscillating slowly back and forth from just over 3:2 to just under 3:2. It is impossible for Pluto and Neptune, in their present orbits, to get within many AU of each other.

The unusual orbital geometry of Neptune and Pluto has encouraged speculation that there may be a genetic relationship between them. A popular suggestion is that Pluto and Charon may have escaped from orbit about Neptune. This origin for Pluto, however, does not account for the present librating resonance. Some major mechanism for dissipation of energy would be needed to permit Pluto and Neptune to fall into such a resonance, but the only mechanism capable of providing strong enough forces (tidal interaction) also requires very close repetitive encounters, a condition incompatible with the present orbits. The origin of the Pluto–Neptune resonance remains an interesting puzzle.

By the same arguments, attempts to account for the orbital oddities of the irregular Neptunian satellite system by flying Pluto through (or out of) the system seem most unlikely to be correct: Pluto and Charon together have much too little mass to flip Triton into a retrograde orbit, and ejecting two satellites as a "double planet" together seems hopelessly improbable. The recent discovery of great numbers of small "plutinos," many with satellites of their own, in 3:2 resonance with Neptune removes the Pluto–Charon system from the "special creation" category and motivates an evolutionary discussion of Pluto as a member of a large dynamic class. One recently discovered Kuiper Belt body, 2002 $LM_{60}$, named Quaoar, is half the diameter of Pluto but follows a low-eccentricity, low-inclination (7.9°) orbit with a period of 288 years. Its albedo is about 0.1, compared to about 0.6 for Pluto, which approaches the Sun closely enough to mobilize frost-forming methane and nitrogen. It is, like Pluto itself, a transition object between "asteroids" and "planets," but it is not a plutino. The plutinos are discussed in Chapter VIII.

Alan Stern, of the Southwest Research Institute, has proposed that the improbabilities associated with the capture of Triton into a retrograde orbit about Neptune, the formation of the Pluto–Charon pair, their fortuitous wandering into a stable resonance with Neptune, and even the extreme axial tilt of Uranus can all be mitigated by postulating a "cleanup" phase late in planetary accretion, when hundreds to thousands of small icy planets orbited the Sun in the giant planet region.

## Spacecraft Exploration

Five spacecraft, including Pioneer 10 and 11, Voyager 1 and 2, and Ulysses, have flown through the outer Solar System. Voyager 2 completed the planetary phase of its mission in 1989. The Galileo Orbiter and Probe missions have, despite daunting technical obstacles and equipment failures, successfully completed their in-depth exploration of the Jovian system. Missions to Pluto, which have several times reached the advanced planning stages, are presently on hold awaiting the approval of NASA and Congress. The joint NASA/ESA Cassini mission is, at the time of this writing, still en route to Saturn, due to arrive and drop the Huygens probe into the atmosphere of Titan in 2004. The Cassini Orbiter has the potential to observe the Pluto-Charon system at phase angles far too large to achievable from Earth. The Hubble Space Telescope, long since rescued from its early problem with a defective primary mirror, is now contributing importantly to Solar System science out to Pluto, at least on those rare occasions when control of HST can be wrested from the deep-space astronomers. New ground-based search programs of very high sensitivity are contributing almost daily discoveries of new bodies in the outer reaches of the Solar System. Beyond Cassini, there are no approved missions to the outer Solar System: our progress in understanding this vast region and its significance for Earth will remain limited without new spacecraft missions to expand our frontiers. As a reminder of this principle, we close our discussion of the outer planets and their satellites and rings with a photograph of the launch of Voyager 1 (Fig. VI.42).

**Figure VI.42** Launch of Voyager 1. The Voyager spacecraft is launched by a Titan 3D Centaur booster from the Eastern Test Range on Cape Canaveral, Florida. Further exploration of the outer Solar System is heavily dependent upon future spacecraft missions—the Voyagers have now experienced their last planetary encounters and are departing at about 2 AU per year into interstellar space.

## Exercises

### *Introduction*

VI.1 Using the elemental abundances tabulated in Table II.4, calculate the density of (a) dry rock and (b) cometary rock-plus-ice solids. Use the equilibrium condensation model as outlined down to step 12 (step 4 for dry rock) in Table IV.7 and mineral densities from the *Handbook of Chemistry and Physics* or similar source. Neglect all elements that are less abundant than iron. (*Hint*: Find the total mass and the total volume using the following format:)

| Element/ compound | Abundance | # moles | Molecular weight | Mass | Density | Volume |
|---|---|---|---|---|---|---|

### *Surfaces of Icy Satellites*

VI.2 Observations of a nearby asteroid at a wavelength of 15 $\mu$m, covering a wide range of phase angles, permit accurate evaluation of the phase integral and measurement of the geometric albedo at that wavelength. From these values of $p$ and $q$, the monochromatic Bond albedo is calculated to be much greater than 1. Describe clearly how this is possible.

VI.3 A number of asteroids of the same spectral class and albedo are observed in roughly circular orbits at many different distances from the Sun. A graduate student draws up a database of all the known properties of these asteroids and runs a statistical analysis program to search for unexpected correlations between the entries in his database. The program tells him, much to his astonishment, that the temperatures (K) of these bodies are linearly related to their circular orbital velocities (km s$^{-1}$) by the equation $T = 10.5V_{\text{circ}}$.

a. Derive an explanation for this peculiar correlation.
b. What would the dependence of $T$ on $V$ be for identical asteroids in orbit about an M5 red dwarf?

VI.4 Suppose that the thermal emission of the Saturnian satellite Pseudoiapetus is found to be strongly correlated with the visual brightness, so that the ratio of the infrared flux to the visible flux is nearly constant over the course of an orbit, while both fluxes vary cyclically over a range of 2.0 magnitudes. What can you deduce quantitatively about this bizarre satellite?

### *Eclipse Radiometry*

VI.5 In recent decades the technique of eclipse radiometry has been applied many times to the satellites of Jupiter and Saturn, but never to the satellites of Uranus. Why?

VI.6 a. Explain why the infrared flux observed during an eclipse of a satellite of Jupiter (see Fig. VI.2a) is not symmetrical.
b. Explain why the visual flux observed during an eclipse of Ganymede is symmetrical.

### *Surface Temperatures*

VI.7 A Saturn-family comet has an orbital semimajor axis of 5.7 AU and an orbital eccentricity of 0.982. Assuming the comet nucleus has a Bond albedo of 0.055 and a thermal emissivity of 0.92, what is its steady-state surface temperature at aphelion? At perihelion?

VI.8 a. The mean or effective temperature of the surface of a sphere is $2^{-1/2}$ times the temperature of a nonrotating body at its subsolar point. Why? Does this rule hold for all possible spin rates and axial orientations? Why or why not?

b. What is the relationship between the (mean) equatorial temperature $T_{e,m}$ of a rapidly rotating body and the temperature $T_{ss}$ of a nonrotating body at its subsolar point? Assume an axial inclination (obliquity) of zero relative to the pole of the orbit. Does this result depend on the orbital inclination relative to the ecliptic?

*Surface Morphology of the Galilean Satellites*

VI.9 The *minimum* energy required to excavate a crater is just the amount of energy that must be provided to lift all the excavated mass to the height of the surrounding surface. Suppose that a crater 20 km in diameter and 3 km deep is excavated by a comet impact on Ganymede. Given an impact velocity of $20\,\mathrm{km\,s^{-1}}$, what is the minimum mass of the comet needed to produce the crater?

*Density and Composition of Icy Satellites*

VI.10 Suppose that Ganymede formed by collapse of a very large sphere of ice particles to the present radius of Ganymede. Using Eq. (II.73), and taking $8 \times 10^6\,\mathrm{erg\,g^{-1}\,K^{-1}}$ as the heat capacity of cold ice, estimate how much energy is released as a result of conversion of gravitational potential energy into heat during accretion.

VI.11 From Fig. VI.11, which is denser, ice VI or ice VIII? II or V?

VI.12 Using the phase diagram of ice in Fig. VI.11, estimate the enthalpy of the reaction ice VI $\rightarrow$ ice VIII at a pressure of 16 kbar. For simplicity, assume that the density difference between these phases is 0.1 and the average density of the two is about $1.2\,\mathrm{g\,cm^{-3}}$.

*Internal Thermal Structure of Galilean Satellites*

VI.13 The short-lived radionuclide $^{26}\mathrm{Al}$($t_{1/2} = 7.2 \times 10^5$ years) could have been an important heat source during the earliest days of the Solar System. The accretion time scale for the planets is estimated by many authors as 100 million years. By what factor would the rate of energy production by $^{26}\mathrm{Al}$ decay have decreased after 100 million years?

VI.14 Consider a small ice-plus-rock body at 160 K. If the ice is pure water ice, then the melting temperature will be that of ice I at low pressures (273 K). But if a few weight percent of $NH_3$ is present, first melting will occur at the $NH_3$–$H_2O$ eutectic temperature (Fig. V.14).

a. Using Eq. (VI.28), estimate the viscosity of ice for these two different melting temperatures.

b. How important is it to know whether ammonia is present in an icy body when calculating its viscous behavior?

VI.15 Calculate the adiabatic gradient for convecting ice in the crust of Ganymede at 150 K. Consult the *Handbook of Chemistry and Physics* or other reference materials for the data you need.

*Dynamical Interactions of the Galilean Satellites*

VI.16 Consider the surface of Io to be rigid.

a. What is the mean acceleration of gravity on Io's surface?

b. Including the gravity of Jupiter as well as Io, calculate the acceleration of gravity at Io's mean surface elevation at the sub-Jupiter point.

c. Calculate the correction to Io's surface gravity due to the centrifugal force associated with its orbital motion.

VI.17 Suppose that Titan, at the time of its formation, had a rotation period of 30 hours. Tidal interactions with Saturn then rapidly despun Titan. Assuming all the rotational kinetic energy of Titan was converted into heat, how much could the spin energy have raised the internal temperature of Titan?

*Thermal and Tectonic Evolution of Icy Satellites*

VI.18 In Fig. VI.16, calculate the heat released by Ganymede during evolution from stage b to d.

a. Assume the outer layer that differentiates during this stage of evolution is 500 km thick.

b. How does the heat released by foundering of a 100-km undifferentiated crust (as between stages c and d) compare to the heat required to melt the foundered ice?

VI.19 An ice-plus-rock body with a diameter of 300 km orbits the Sun at about the same distance as Saturn. It starts out cold, in radiative steady state with the solar radiation field, at about 60 K. Assume a heat capacity of $2 \times 10^7$ erg

$g^{-1}\,K^{-1}$ for ice and $0.8 \times 10^7$ erg $g^{-1}\,K^{-1}$ for the rocky component,

a. How much heat is needed to warm the entire body to the point of incipient melting of ice I?

b. How much heat is needed to warm the body up to the point of incipient melting of ammonia-water eutectic solution?

c. The assumption of a constant heat capacity is not very good over such a wide range of temperatures. Given that the heat capacity of ice I fits the equation $C_p = 1.49 \times 10^6 + 7.17 \times 10^4 T(\mathrm{erg\,g^{-1}\,K^{-1}})$ over the temperature interval from 60 to 273 K, reevaluate the heat required to raise the initially cold ice to the melting point of ice I.

d. Do the same for warming to the ammonia–water eutectic temperature.

VI.20 Compare the amount of heat needed to warm 1 g of ice from 60 to 273 K to the amount of heat needed to melt the ice. Because the heat capacity of ice varies greatly over this temperature range, use the equation for the heat capacity given in Exercise VI.19 rather than assuming $C_p$ is constant.

*Minor Satellites of Jupiter*

VI.21 a. At what distance from Jupiter would a satellite in a prograde equatorial circular orbit have an orbital period of 1 Jupiter day (Jovi-synchronous orbit)?

b. Of 1 Earth year?

c. Of 1 Jupiter year?

d. Assuming the body in c was observed to be directly between the Sun and Jupiter at time $t_0$, sketch its position at $t_0 + 0.25$, at $t_0 + 0.50$, at $t_0 + 0.75$, and at $t_0 + 1.00$ Jovian years.

e. At $t_0$, where would this body be located relative to the point at which the gravitational force of the Sun and of Jupiter on the body are equal?

*Planetary Rings*

VI.22 Jupiter's ring, viewed in normal incidence, has an optical depth (determined by measuring the attenuation of the light from stars occulted by the ring) of $3 \times 10^{-5}$.

a. If the ring is composed entirely of opaque meter-sized bodies, what is the area density $\sigma$ of the ring (in, say, grams per square kilometer)?

b. What is $\sigma$ if the particles are 1 cm in size?

c. 100 $\mu$m?

d. 1 $\mu$m?

e. From these results, suppose that the ring is optically dominated by 1-$\mu$m dust. How much larger might the true mass of the ring be if 10% of its cross-section area is provided by meter-sized objects?

VI.23 Suppose particles in Saturn's rings have a mean out-of-plane speed of about 1 cm $s^{-1}$ relative to local circular orbit velocity. Collisions will "thermalize" these particles so that they will also have a mean in-plane random speed of the same magnitude.

a. What would be the orbital eccentricity of the particle?

b. In the B ring, the mean optical depth in the normal direction is 1.5 and the mean ring thickness is about 100 m. What is the mean distance the particle can travel along such an in-plane eccentric orbit before suffering a collision?

c. Compare this distance to the circumference of ring B.

VI.24 Earth satellites with high densities and smooth, highly laser-reflective surfaces have been proposed for geodetic purposes: their high densities make aerodynamic drag negligible compared to the varying gravitational accelerations produced by irregularities in the mass distribution of Earth. Suppose a research team conceives the idea of placing a meter-sized sphere of liquid mercury (density = 13.55 g $cm^{-3}$) in a low circular orbit about Earth for use as a geodetic probe for Earth-based laser tracking. Are there any practical limits on the altitude of the orbit of this geodetic satellite imposed by the Roche limit for this mercury drop in Earth's gravity field?

*Titan*

VI.25 What are the surface acceleration of gravity and escape velocity of Titan?

VI.26 What are the thermal speeds of $H_2$, methane, and nitrogen at 100 K?

VI.27 What is the mass of the atmosphere of Titan?

VI.28 What would be the scale height of the lower atmosphere of Titan if it were made of hydrogen? Of pure methane? Pure nitrogen? Pure radiogenic argon?

VI.29 Suppose the last 0.01% of the mass of Titan was accreted after the outgassing of the body to produce its atmosphere. Would the accretion of this last trace of mass be enough energy to heat the entire atmosphere up to a high enough temperature for shockwave production of complex organic molecules (roughly a temperature of 2000 K)?

VI.30 In deriving the hydrostatic equation for the structure of a planetary atmosphere, we assumed that the scale height $H$ was much smaller than the radius of the planet; that is, that $g$ can be considered a constant throughout the atmosphere. But early Titan may have had sufficient hydrogen in its atmosphere so that this assumption would no longer be valid. Rederive the hydrostatic equation for this case, taking into account the variation of $g$ with planetocentric distance. [*Hint*: Follow the derivation that starts with Eq. (V.13), but do *not* assume $g =$ constant.]

### *The Intermediate-Sized Saturnian Satellites*

VI.31 The solid, brittle crust of Tethys, a satellite of Saturn, is split by a deep fissure, called Ithaca Chasma, which is tens of kilometers wide and stretches for over 2000 km. One observer suggests that the crack developed as a result of shrinkage of the crust as it cooled through the freezing point. Another speculates that the cooling of Tethys' interior caused the rift. A third attributes the crack to recent warming of the interior due to decay of long-lived radionuclides. A fourth claims that the crack formed after the surface was thoroughly frozen, produced by further cooling. Critically assess these competing theories in light of the phase diagram of water and the crater density data described in the text.

### *Minor Satellites of Saturn*

VI.32 According to Fig. VI.30, small bodies following horseshoe orbits about a primary body P can approach a secondary body S and be "repelled" from collision by the gravity of S. Minor perturbations of small bodies in one of the coorbital Lagrange swarms can result in promotion of the body from a tadpole orbit about that point to a horseshoe orbit.

a. If one of these small bodies should happen to suffer a minor collision while close to S, it may be captured (perhaps temporarily) into an orbit about S. Would such an orbit be prograde or retrograde?
b. Would you expect to find major compositional differences between Jupiter's Trojan and Greek asteroid swarms? Between them and the outermost Jovian satellites? Explain your answers.

### *Satellites of Uranus*

VI.33 A spacecraft flies by Uranus, passing at a minimum distance of $d_0$. The along-track velocity $V$ of the spacecraft after the encounter (relative to Uranus, of course), is the same as it was at the same distance from the planet on the inbound leg. If the deflection is not too large, we can approximate the cross-track velocity change of the spacecraft as $\Delta V = a\Delta t$. The time scale for the encounter is roughly $\Delta t = 2d_0 V$. Calculate the cross-track velocity change and estimate the deflection angle $x\vartheta$ experienced by the spacecraft. Note the functional dependence of $\vartheta$ on $d$ and $V$.

VI.34 A spacecraft (or comet), heading almost radially outward from the Sun as seen from Uranus, passes behind Uranus as Uranus moves along its orbit. The spacecraft is strongly deflected, through an angle (as seen from Uranus!) of about 90°, so that it exits from the Uranus system heading in the same direction as Uranus. In the Uranus-centered coordinate system, the approach velocity at large distance ($V_\infty$) and the departure velocity at large distance are equal. But as seen from the Sun, the spacecraft picked up during the encounter a component of velocity tangential to Uranus' orbit that is equal to the orbital velocity of Uranus.

a. In a Sun-centered system, how much velocity must the spacecraft pick up in order to escape from the Solar System after the Uranus encounter?
b. What is the approach velocity the spacecraft must have in order to use this effect to achieve Solar System escape velocity?
c. What is the actual encounter velocity of the spacecraft with Uranus in the Sun-centered frame of reference?
d. Would the spacecraft with that approach velocity have been able to escape from the Solar System *without* the Uranus encounter?
e. If it were necessary for some reason to increase the spacecraft approach velocity, how would you have to change the aim point ($d_0$) of the spacecraft in order to ensure the same angular deflection?

VI.35 Examine the Voyager image of Umbriel in Fig. VI.34. Umbriel appears to have an ancient, undifferentiated surface and crust. Suggest an explanation for the two craters discussed in the figure caption in terms of impacts on a body with a structure similar to that shown in Fig. VI.16c. Can you think of any other explanation for these two features?

### *Satellites of Neptune*

VI.36 Do you think that Eqs. (VI.43) and (VI.44) are exact and general, or just useful approximations for small values of sin $L$? Consider, for example, how the tidal system would behave for lunitidal intervals of $L = 0$ and $L = 90°$.

VI.37 Using the data in the text on the pole orientation and albedo of Triton, calculate the maximum temperature of Nereid's north (summer) pole.

VI.38 How much does the orbital velocity of Nereid vary from periapsis to apoapsis?

VI.39 When Uranus is nearly pole-on to the Sun, its satellites orbit in a bull's-eye pattern as seen from the Sun. Their orbital velocities sometimes add to, and sometimes subtract from, the orbital motion of Uranus about the Sun.

a. Calculate the maximum and minimum velocities of Miranda about the Sun.

b. Sketch the apparent motion of Miranda *relative to Uranus* and, separately, *relative to the stars*, as seen by an observer close to the Sun (such as on Earth).

*The Pluto–Charon System*

VI.40 Suppose that a spectroscopic study of Pluto in 2022 A.D. shows a methane column abundance over the surface of 8 cm agt. Calculate the surface partial pressure of methane on Pluto at that time.

*The Neptune–Pluto Resonance*

VI.41 a. Calculate the velocity change associated with "flipping" Triton from an equatorial prograde circular orbit to Triton's present orbit, assuming for convenience that the orbital semimajor axis remained roughly constant during the maneuver.

b. How large a velocity change $\Delta V$ would this entail for the Pluto–Charon pair if they were responsible for this deflection?

c. Is this $\Delta V$ compatible with Pluto's present orbit about the Sun?

*Spacecraft Exploration*

VI.42 Calculate the additional velocity (above Earth's orbital velocity) that must be imparted to a spacecraft in Earth's orbit about the Sun (but out of Earth's gravity well) to permit it to reach Neptune. Assume that the orbits are coplanar, and use the minimum energy (Hohmann) ellipse, which is tangent to Earth's orbit at departure (perihelion) and tangent to Neptune's orbit at arrival (aphelion).

VI.43 Calculate the velocity change necessary for a vehicle to depart from the atmosphere of Uranus with enough velocity so that, after escaping from Uranus' gravity, it will have a large enough velocity to follow a heliocentric orbit that drops in to Earth's orbit at perihelion. Use the principle of conservation of energy.

*Chapter V Revisited*

VI.44 A recent cover story in a national popular science magazine featured the proposal that radioactive decay of uranium and thorium is responsible for the thermal luminosity of Jupiter and the other giant planets.

a. Assuming a fivefold enhancement in the abundances of all heavy elements relative to solar proportions, calculate the *present-day* steady-state radiogenic heat flux from Jupiter. Consult Table VI.3 and Fig.VI.12 for crucial data.

b. Assume, as in that story, that a "natural reactor" is responsible for enhancing the fission decay rate and energy production rate to match the internal heat flux of Jupiter (see Fig. V.30). How long could this reactor sustain the observed luminosity of Jupiter before all the *original* U and Th is exhausted?

# VII. Comets and Meteors

## Historical Perspectives

Comets are among the most spectacular and memorable phenomena of nature. Several times per century a Great Comet arrives, appearing almost anywhere in the heavens, brightening to outshine any star or planet, and growing a luminous tail that may stretch more than 90° across the sky. After a few weeks the comet fades and dashes off in another apparently random direction.

It is striking enough that there are comets of such exceptional brilliance that they can be seen in the daytime sky. But comets are even more remarkable for another reason: since prehistoric times, comets have provided the most extraordinary example of irregular and unpredictable phenomena in the heavens. Millennia ago, naked-eye astronomers learned to predict the motions of the Moon and the planets with good precision, and to use their skills to predict eclipses of the Sun and Moon. But until the 17th century comets defied prediction. Because of this lack of orderly, predictable behavior, medieval Europeans viewed bright comets as "unnatural" phenomena, and hence as warnings from God. Invariably, within a few years after the appearance of a bright comet, kingdoms would fall, or great, decisive battles would be fought, famine or pestilence would rage, or an emperor or king would die. Halley's comet was seen in 66 AD, several years before the destruction of Jerusalem by emperor Titus. The apparition of Halley's comet in 1066 closely coincided with the Battle of Hastings (and is even immortalized in the Bayeux Tapestry). But the Halley's apparition of 1456 occurred 3 years after the fall of Constantinople to the Turks, an event with which it was often linked, and seems a somewhat tardy portent! A comet was often described as resembling a gigantic sword reaching across the heavens: such a harbinger surely must portend a great disaster. Indeed, the very word we still use to describe the visit of a comet, "apparition," carries the emotional implication of a messenger from the spirit world.

Any casual modern student of medieval European history might come to a very different conclusion: there was *always* a war, famine, battle, pestilence, or the death of a great prince within a few years of any date. Besides, wars and battles have winners as well as losers: what portends ill for one ought to portend victory for the other. Why do we not hear of this interpretation as well? It is a common truism that history is written by the victors: have we here found a counterexample? But if people believe, following Shakespeare's King Lear, that "'tis the stars, the stars above us, govern our condition," then comets, with all their brilliance and unpredictability, *must* be important. Assigning a particular meaning to each event then becomes a challenge. Who could admit to being not clever enough to interpret so magnificent a sign?

The 19th century was exceptionally rich in spectacular comet sightings. Great comets were recorded in 1811, 1843, 1858, 1861, and 1882. Of these, the most astonishing was the Great Comet of 1882, which seems to correlate with no great tragedy. It is true that some editions of the prophetic verses of the English prophetess Agatha Shipton (the prototype of the caricature used by the British humor magazine *Punch*) predict that

> The world to an end shall come
> In eighteen hundred and eighty one.

However, this prediction is believed from other evidence to have been inaccurate.

All of these comets were more spectacular than any seen in the last century, except perhaps for the Great Daylight Comet of 1910 (a near-contemporary of the 1910 return of Halley's comet) which was, as its name suggests, bright enough to be seen in the daytime sky.

The development of the Theory of Universal Gravitation by Isaac Newton paved the way to a conceptually simple, very accurate understanding of the motions of planets and satellites. Newton's theory captured the popular imagination of his time much as the Apollo expeditions to the Moon did in the 20th century. A young follower of Newton, Edmond Halley, felt that a stringent test of the new theory would be to bring order to the apparently chaotic and unpredictable world of comets. Halley therefore set himself the task of examining all available records on the apparitions of comets, with the ultimate purpose of finding a physical explanation for the motions recorded. Halley hoped that the orbits of certain comets could be calculated with sufficient precision to permit a prediction of their next appearance date. The comet he chose to study was that of 1682, which was then the most recent and best-tracked bright comet.

As Halley searched back through the records he found, as he hoped, tantalizing evidence of earlier comets in similar orbits. However, the times elapsed between consecutive perihelion passages were not exactly the same. A comet in a closely similar orbit had been seen to pass perihelion in 1607; another in 1531; yet others in 1456 and 1378. Thus, if these were all the same comet, the orbital periods were 75, 76, 75, and 78 years. There could be no doubt: if these were indeed all observations of the same comet, then the orbit was not exactly the same from one apparition to the next. But, as Halley attempted to refine the orbit mathematically, it became clear that the comet changed its orbit slightly when (and only when) it chanced to pass close to one of the giant planets. The gravitational effects of these encounters, which had hitherto always been ignored, could be estimated from Newton's laws. Halley calculated the effects of these encounters, and the conclusion was clear: the observations did indeed refer to the same comet! Halley announced his results in 1705, including the prediction that the comet would next return in the winter of 1758–59.

Sadly, Newton did not live to see this spectacular vindication of his theory of gravitation. After Newton's death in 1727, Halley lived on alone in the hope of seeing the comet, but Halley himself died in 1742. The world, however, did not forget the prediction. When the comet returned on schedule, on Christmas day of 1758, the comet was named after the man who had spent much of his life pursuing it through old records.

A contemporary painting shows a graveyard scene at night, with Halley's comet blazing bright across the sky. A beam of light from the comet plays upon Halley's grave, illuminating his name on the tombstone. Next to the grave stands a beautiful angel clothed in white, with one hand raised to indicate the comet, and the other reaching down toward the shade of Edmond Halley, half-emergent from the grave. The painting conveys the clear and popular message that universal truth has been revealed by divine inspiration to men of science. It was perhaps this schmaltzy and inappropriate reverence that led the irreverent William Blake to paint his famous portrait of Newton, all aglow with a regal and almost seductive physical beauty that the real Newton surely did not possess. Sadly, this wholly mistaken view of science (like the equally wrong-headed view that attributes the inspiration for science to what Shaw called "the alternative establishment") survives to our time.

Some readers will no doubt recall the media hype surrounding the discovery of a new bright comet. Comet scares recur almost as frequently as prophecies of the end of the world, and indeed the two are often bundled together. I especially recall the events following the discovery of one fairly middling comet (Kohoutek) in 1973. Experts, questioned about the expected performance of comet Kohoutek as it approached the Sun, responded honestly that the performances of comets show little respect for human predictions: it could be so faint as to be hard to see with the naked eye, or bright enough to outshine Venus. The media faithfully reported that Kohoutek would be brighter than Venus. I recall waiting for a train in the Park Street subway station in Boston on the evening when Kohoutek was at its brightest (but still very much fainter than Venus). A girl about 16 years of age approached me with some flyers from an organization called the Children of God and attempted to convince me that the world was coming to an end. I asked why she thought that Kohoutek, of all comets, was a harbinger of doom. She responded that the comet was unusually spectacular, and besides, it was the official doctrine of her sect. I asked her whether she was aware that there was a comet every few years as bright as Kohoutek. She was inclined to doubt my veracity; she pointed out that Kohoutek *must* be spectacular because the media said so. Abandoning this tack, I asked her whether she had seen the infamous comet. She had not. "Why not?" I asked. She had been too busy in the subway handing out flyers. I suggested that she spare five minutes from her appointed rounds to go outside and look at the sky, where she would assuredly not see the comet. As my train pulled in she was still frozen in indecision. Perhaps she feared that anyone faithless enough to doubt the predictions of chiliastic seers and Boston newspapers might be justly punished by being denied the sight of the putative Great Comet. I do not know whether my cheerful empiricism shortened her proselytizing career, but I can affirm that her sect, like many others before and since, faded quickly from public view when the world failed to come to an end on schedule.

## Nature and Nomenclature of Comets

A comet is a small ice-rich body, rarely larger than a few kilometers in diameter, in an eccentric orbit about the Sun. The phenomena that may make the comet spectacular (and indeed, that make it observable) are associated with the progressively rapid evaporation of ices from its solid *nucleus* during its precipitous fall toward the Sun. These gases, and some entrained dust, expand outward around the nucleus to form a roughly spherical envelope, or *coma*, about the nucleus. The coma and nucleus are collectively called the *head* of the comet. At a distance of thousands to about 100,000 km from the nucleus the expanding coma becomes so tenuous that the gas and dust in it become uncoupled and begin to stream systematically away from the nucleus, albeit in different directions. The dust roughly follows ballistic trajectories, whereas the gas flows in the direction radially outward from the Sun. Such a long, tenuous streamer of gas or dust is called a *tail*. Not surprisingly, many comets have two visible tails, one of gas and one of dust, pointing in different directions (see Fig. VII.1).

Many comets have orbits that lie well within the Solar System. Frequently the aphelia of their orbits are near the orbit of Jupiter or Saturn. Their orbital periods are all below 200 years, which implies two things. First, their orbital periods are short enough so that they can be observed on two or more perihelion passages, and hence their orbital elements can be refined and their future positions predicted with good accuracy. Second, because they spend so much time close to the Sun and pass through perihelion so frequently, they lose ices at a great rate. They therefore cannot survive long in such orbits. For obvious reasons, these are called *short-period comets* or *periodic comets*.

The other comets have orbits that are nearly randomly distributed in space, have enormous orbital eccentricities and semimajor axes, and typically pass perihelion only about once every few million years. These orbits are very vulnerable to perturbations by nearby stars, and of course no predicted return dates for such comets have ever been tested observationally. Ten million years is too long for even an astronomer to wait. With a sigh of resignation, astronomers refer to them as *long-period comets*.

Historically, when a comet is spotted, as usually happens several times a year, it has at once been named after its discoverer. For example, Alain Maury found the eighth comet discovered (or recovered) in 1994. The comet was then called Comet Maury 1994h. This designation persisted until the comet's next perihelion passage. At that time the comet received a new designation which put the comet in the order in which it passed through perihelion. Thus, if 1994h is tracked, its orbit calculated, and its perihelion date found to be 29 January 1995 (the third of the year to pass perihelion), the same comet would also bear the designation 1995III. Thus, under this system, a short-period comet may bear a large number of different designations assigned at times of different recoveries or perihelion passages! Partly to warn about this possibility, short-period comets are commonly assigned the prefix P/ (for periodic), as in P/Halley. Since there are many observers who have found more than one comet, it has become expedient to affix a catalog number to each periodic comet, as in 51 P/Harrington or 19 P/Borelly. In the past few years the naming convention for new comets has also been changed and simplified. Suppose that a new comet is found on an image taken by the Solar and Heliospheric Observatory (SOHO) spacecraft. The comet is given a preliminary designation such as C/2001 Q3 (SOHO). The prefix C/ denotes a long-period comet (P/ for a periodic comet), 2001 is the date of discovery, Q designates the half-month in which the discovery was made (A for the first half of January, B for the second half, and so on, skipping I), and 3 is the serial number of the discovery in that time interval. The name of the discoverer is appended in parentheses.

Occasionally a comet goes dormant or, for lack of sufficiently accurate orbital data, is lost. The comets that have suffered such a fate bear the designation D/, such as

**Figure VII.1** Comet West, showing a broad and striated white dust trail and a narrow blue gas tail in the predawn sky. This photo is a time exposure taken with a fixed camera, causing the images of the stars and comet to be smeared due to the Earth's rotation. Photo courtesy of Dennis and Betty Milon.

11 D/Tempel–Swift. That comet was later reacquired under the "new discovery" alias of P/2001 R3.

Highly sensitive asteroid search programs such as LINEAR, Spacewatch, and LONEOS (see Chapter VIII) have generated numerous comet discoveries. The Infrared Astronomical Satellite (IRAS) was discoverer or co-discoverer of several comets. In addition, SOHO images posted on the Web are routinely analyzed by amateur astronomers from their home computers, resulting in dozens of new discoveries of Sun-grazing and Sun-impacting comets. Thus comets named LINEAR, SOHO, etc., are now common.

It is not unusual for a comet to be discovered by two or more observers on the same night, an event that produces names such as Barnard–Hartwig and P/Churyumov–Gerasimenko, or even Mori–Sato–Fujikawa and IRAS–Araki–Alcock. Finally, certain observers have a clear gift for finding comets. Thus one finds comets with names like P/Brooks 2, Shoemaker 4, and P/Schwassmann–Wachmann 3. Again, the current designations add a catalog number, such as 29 P/Schwassmann–Wachmann 1.

The appellation Great Comet is reserved for visually spectacular sightings. The hazard of being the discoverer of a Great Comet is that it will not bear your name!

## Cometary Orbits

The orbits of the short-period comets are ellipses of moderate eccentricity and inclination (Fig. VII.2). Almost all have inclinations of less than 20° relative to the ecliptic plane. Their orbital eccentricities lie mostly between 0.2 and 0.7. Only four of 113 short-period comets with well-known orbits have periods over 20 years (P/Halley, 76 years; P/Brorsen–Metcalf, 70.6 years; P/Tempel–Tuttle, 33.2 years; P/Crommelin, 27.4 years). These are also the only four with eccentricities greater than 0.9. P/Halley, P/Temple-Tuttle, and P/Tuttle are the only three with inclinations greater than 32°, of which the first two are retrograde. The shortest orbital period is 3.3 years for P/Encke (Fig. VII.3), a body whose orbit falls within the range of near-Earth asteroid orbits (Chapter VIII). Most short-period comets approach to within 1 to 2 AU of the Sun at perihelion, although perihelia as close as a searing 0.33 AU can be found (P/Encke). Two extraordinary objects, P/Kowal 1 and 29 P/Schwassmann–Wachmann 1, have perihelia beyond the asteroid belt (4.67 AU for Kowal; 5.77 AU for S-W 1). The longitudes of the nodes and the longitudes of the perihelia of short-period cometary orbits are roughly evenly distributed around the ecliptic. There is also a striking tendency

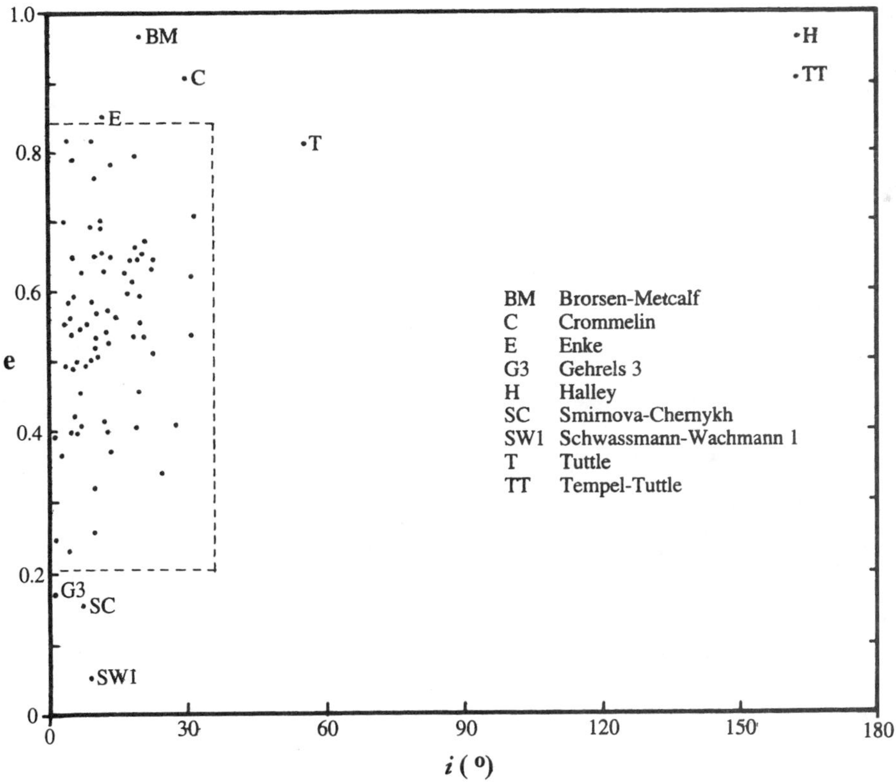


**Figure VII.2** Eccentricity–inclination plot for short-period comets. All but nine of the known periodic comets lie within the indicated box.

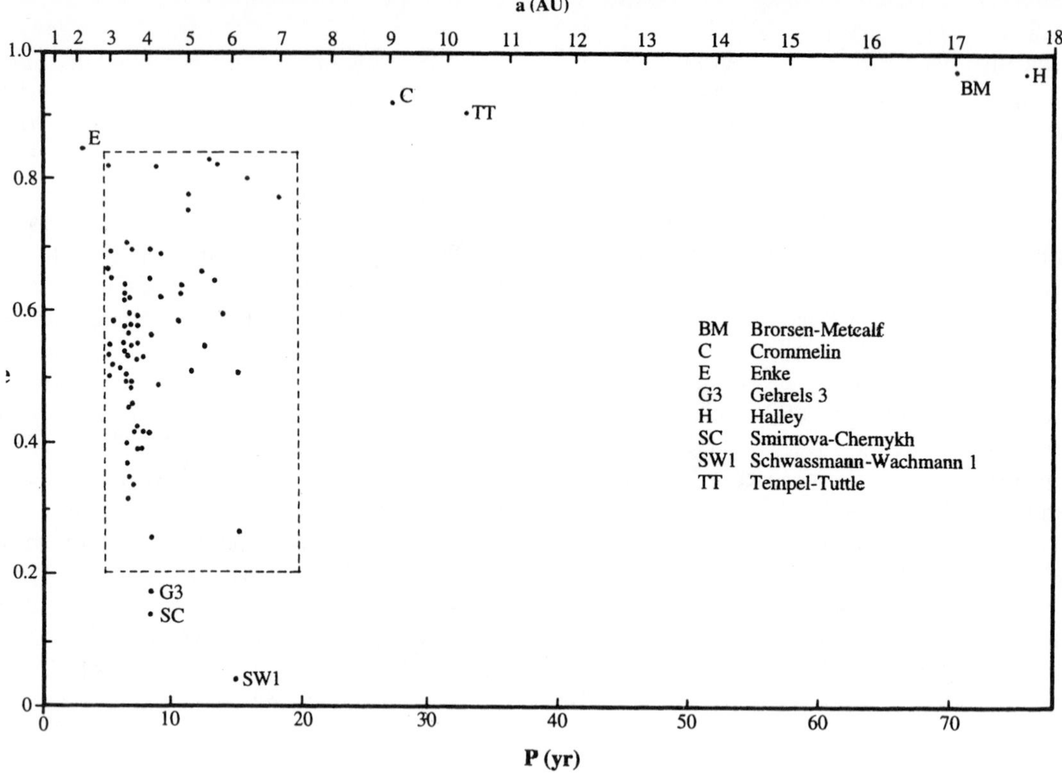


**Figure VII.3** Eccentricity–period plot for short-period comets. The orbital semimajor axis is also given. All but eight comets lie within the indicated box.

for the periodic comets to have aphelia near the orbit of Jupiter and, to a lesser degree, Saturn.

Both families of comets are depleted by two important dynamical factors: first, comets that cross the orbits of several planets must from time to time collide with one, resulting in the destruction of the comet (and possible serious damage to the planet). Second, less obvious but still important, is the possibility that an encounter with a massive planet may eject the comet from the Solar System at a speed above the local escape velocity of the Sun, thus causing the comet to be lost forever from the Sun's family.

A long-period comet will not be recognized unless it approaches closely enough to the Sun for it to begin to evaporate and grow a coma and tail. Let us suppose that the critical perihelion distance is about 2 AU. Thus virtually every known comet will pass through a sphere, centered on the Sun, with a cross-section area of $\pi(2 \times 1.5 \times 10^{13})^2 = 3 \times 10^{27}\,\text{cm}^2$. The planets within that volume (Mercury, Venus, Earth and Mars, with the Moon thrown in for good measure) have a total cross-section area of only $3 \times 10^{18}\,\text{cm}^2$. Thus a comet randomly crossing the inner Solar System once has a probability of about $10^{-9}$ of striking a planet. A long-period comet with a period of $10^6$ years that has been in its present orbit for $10^9$ years will have passed through the inner Solar System $10^3$ times and incurred a probability of only $10^{-6}$ of being destroyed by a collision with a planet. Those that approach only to within 10 AU of the Sun (and which therefore seldom "light up" and become detectable) still have the option of collision with Jupiter or Saturn, with an aggregate cross-section area 100 times that of the terrestrial planets, superimposed on a target area that is also 100 times as large. The probability of collision per perihelion passage is therefore roughly the same as for those long-period comets that penetrate to inside 1 AU.

The biggest target in the Solar System is the Sun itself. Many comets approach the Sun so closely that they do not survive perihelion passage. These bodies, called *Sun-grazing* comets, are often dynamically related to each other, most belonging to a group called the Kreutz family of Sun-grazers. Of the more than 500 faint comets discovered very close to (or on an impact course with) the Sun by SOHO, fully 94% are members of the Kreutz group, which appear dynamically to have originated in the catastrophic breakup of a giant cometary progenitor roughly 12,000 years ago. Having fallen all the way to the surface of the Sun, these bodies have acquired an enormous amount of kinetic energy.

If a comet is gravitationally bound to the Sun, the greatest velocity that it may have upon crossing any heliocentric distance (such as the orbit of a planet or the surface of the Sun) is the escape velocity of the Sun from that point. The escape velocity is

$$v_{esc} = (2GM_{\odot}/R)^{1/2} \qquad \text{(VII.1)}$$

which is about 42 km s$^{-1}$ at Earth's orbit and an astonishing 619 km s$^{-1}$ at the surface of the Sun (a distance equal to the radius of the Sun in 15 minutes). This velocity vector may be oriented in any direction. The approach direction that minimizes the encounter velocity relative to a planet is for the comet to be overtaking the planet from behind, in a prograde orbit with zero inclination. The encounter velocity is then the vector difference of $v_{esc}$ and $v_{orb}$ for the planet. For a planet in circular orbit, $v_{orb} = v_{circ} = v_{esc}/2^{1/2} = 0.7071 v_{esc}$. The closing velocity is therefore 0.2929 $v_{esc}$. The highest possible relative velocity is achieved when the comet is in a retrograde orbit of 180° inclination and strikes the planet head-on. The closing velocity is then 1.7071 $v_{esc}$. Thus, at Earth, the closing velocity for a long-period comet can range from 12.3 to 71.7 km s$^{-1}$.

Once close to Earth, the comet will begin to be accelerated by Earth's gravity. The velocity increase experienced by an infalling comet is not a fixed quantity, since a fast-moving comet (which of course feels the same acceleration at each distance from Earth as a slow-moving comet) spends less time passing through that distance. As a consequence, the total velocity increase is much smaller for faster-moving comets. This can be seen quantitatively from the principle of conservation of energy:

$$v_{imp}^2 = v_{app}^2 + v_{Eesc}^2, \qquad \text{(VII.2)}$$

where the subscript Eesc denotes Earth's surface escape velocity, 11.2 km s$^{-1}$, imp means impact, and app denotes approach. The minimum impact velocity of *any* body from outside Earth's own gravitational well is thus 11.2, the minimum impact velocity for a long-period comet is 16.6, and the maximum $v_{imp}$ is 72.6 km s$^{-1}$. The velocity increment experienced by the comet due to Earth's gravity is thus between 0.9 and 4.3 km s$^{-1}$, compared to an 11.2 km s$^{-1}$ increment for a slow-moving piece of nearby space debris.

The origin of the long-period comets presents an interesting problem. It would be very helpful to explore the distribution of these comets over their entire range of orbital semimajor axes ($a$), from about 100 AU on up. Of course, for very eccentric long-period orbits, even small observational errors introduce significant errors into the determination of $a$. A useful representation of these orbits that does not exaggerate these errors is a plot of number vs orbital energy ($1/a$) (Fig. VII.4). The salient features are that there is a strong peak in the distribution of observed long-period comets at $1/a = 0$ (parabolic heliocentric orbit) and a dearth of comets with energies greater than parabolic; that is, "undomesticated" comets wandering into our Solar System from interstellar space must be rare. This observation is especially striking when we consider that comets may achieve escape velocity due to gravitational perturbations and to rocket-like jetting of gases caused by solar heating near perihelion. We would expect *some* runaway comets from our own system, and there may well be vast numbers of comets in interstellar space that have escaped from other stars.

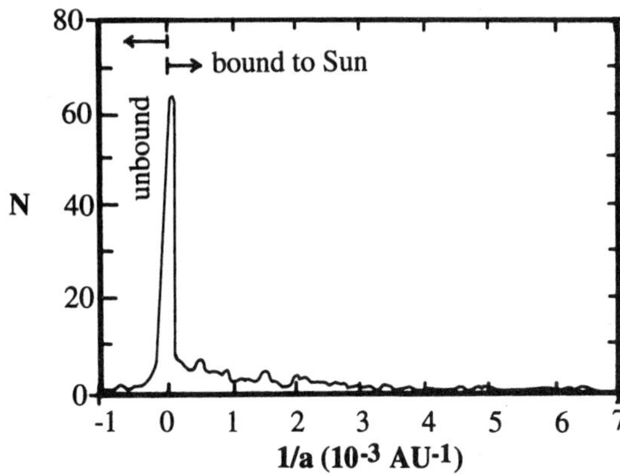


**Figure VII.4** Frequency–energy plot for long-period comets. Most long-period comets have nearly parabolic orbits. Those that depart significantly from parabolic motion are clearly in elliptical orbits, and none are found to have velocities large enough to suggest that they are interstellar "wanderers."

Let us suppose that interstellar space is generously populated by comets that have escaped from other stars. The mean velocity of these comets with respect to the Sun will necessarily be the same as the mean velocity of nearby stars relative to the Sun, about 20 km s$^{-1}$. These comets, if detected by astronomers on Earth, would have very high excess energies above parabolic velocity, and would plot near −0.4 to −0.5 on the $1/a$ scale. Note that the most energetic comet ever observed plots on Fig. VII.4 at −0.0007 on the $1/a$ scale. It is therefore quite unreasonable to attribute any of the observed comets to an extrasolar source.

Long-period comets have very small orbital velocities near aphelion. They spend virtually all of their lives at great distances from the Sun, stored at extremely low temperatures. Figure VII.4 suggests that most long-period comets have $1/a$ values less than $10^{-4}$ (semimajor axes greater than $10^4$ AU). A dark object that would have a radiative steady-state temperature of 300 K at 1 AU would be maintained at only about 3 K by solar illumination at $10^4$ AU. Solar heating is so weak at that distance that the Big Bang cosmic background radiation is an important source of heat!

How common are comets with orbital semimajor axes much larger than $10^4$ AU? First, consider that one parsec is about $2 \times 10^5$ AU. This means that these comets are, at aphelion, about 5% of the way to the nearest star. We could scarcely expect them to go out more than 10 times as far and still return, since they would then be closer to (and more strongly attracted to) another star. But this argument treats only average conditions. In fact the Solar System is moving through a kind of "gas" of stars, and it must from time to time pass much closer to some of them. When this happens, vast numbers of comets can be stripped off and lost into the interstellar medium by the gravitational influence of the passing star.

We saw in Chapter II (Fig. II.12) that there are about 556 MS stars visible within 10 pc of the Sun, 0.14 stars per cubic parsec. The Sun's velocity through this medium is 20 km s$^{-1}$, which fortuitously converts to about 20 pc Ma$^{-1}$. A spherical surface of cross-section area $A$ (pc$^2$) traveling at speed $v$ (pc Ma$^{-1}$) through a medium with a stellar density of $d$ (pc$^{-3}$) will have a mean time between encounters $T$ (Ma) of

$$T = 1/(Avd). \qquad \text{(VII.3)}$$

This tells us that a surface of radius $r = 20,000$ AU (0.1 pc) will have a mean time between encounters of 8 million years. That frequently, a star will approach to 0.1 pc and (if, improbably, it is as massive as the Sun) remove all the comets closer to it than to the Sun (0.05 pc; 10,000 AU). About 12 such collisions (> 100My) are required to sweep away most of the comets on all sides of the Sun. The closest single stellar encounter experienced by the Sun in 4.6 Gy of life would be at a distance of about 1800 AU. The most common stars, however, are not Sunlike stars but are M9 dwarfs of mass 0.07 $M_\odot$: their gravitational effect would be roughly comparable to a single G0 star passing at a distance of 7000 AU. These considerations tell us that most comets reaching out to about 10,000 AU at aphelion can survive ejection by stellar perturbations over periods of billions of years.

The emerging picture, first suggested by Dutch astrophysicist Jan Oort in 1950, is that of a huge spheroidal cloud of long-period comets with a radius of roughly 10,000 AU, centered on the Sun. Stellar perturbation events stir up the orbits of comets in this *Oort cloud*, causing some to escape from the Sun, and others to pass through the inner Solar System, where they are observed to have semimajor axes of about 10,000 AU. The orbital velocity of a comet near aphelion can be calculated from

$$v^2 = GM_\odot(2/r - 1/a), \qquad \text{(VII.4)}$$

whence the orbital velocity of a comet of $a = 10,000$ AU is found to be 0.3 km s$^{-1}$ at $r = a$. For an orbital eccentricity of 0.9999 and a semimajor axis of 10,000 AU the perihelion distance, $q$, and the aphelion distance, $Q$, are

$$q = a(1 - e) = 1\,\text{AU} \qquad \text{(VII.5a)}$$

$$Q = a(1 + e)(= 2a - q) = 19,999\,\text{AU}. \qquad \text{(VII.5b)}$$

The orbital velocity calculated for perihelion from Eq. (VII.4) is 42.4253 km s$^{-1}$, compared to the local escape velocity at 1 AU of 42.4264 km s$^{-1}$, meaning that a velocity increment of only 1.1 m s$^{-1}$ caused by a planetary perturbation or gas jetting near perihelion would push the comet above the escape velocity of the Solar System. The orbital velocity at aphelion is 2.1 m s$^{-1}$. (By comparison, the escape velocity at aphelion is 298 m s$^{-1}$. Thus a tiny perturbation near periapsis has a much more profound effect on the orbit than the same velocity change applied near apoapsis.)

The velocity perturbation produced by a passing star can be calculated from the acceleration $g$ felt by the comet:

$$g = GM_*/D^2 \qquad \text{(VII.6)}$$

where $D$ is the distance of the star and $M*$ is its mass. The integrated velocity change is just the product of the acceleration times the duration of the encounter, which is simply given as $t = 2D/V*$, where $V*$ is the velocity of the star relative to the Sun. Thus the velocity change, $\Delta V$, is

$$\Delta V = 2GM_*/V_*D \qquad \text{(VII.7)}$$

which, for a typical (M class) MS star of mass $M* = 0.1\,M_\odot$ and distance 10,000 AU traveling at 20 km s$^{-1}$, is 1.3 m s$^{-1}$. This is closely comparable to the orbital velocity of the comet and is easily sufficient to cause it to change its perihelion distance, orbital period, or inclination dramatically.

The same principle may be applied to a long-period comet that happens to make a pass within, say, $10^6$ km of Jupiter. The orbital velocity of the comet at 5.2 AU is 18.6 km s$^{-1}$. The effect of the Jupiter swingby will be to produce a velocity change of 2 km s$^{-1}$. Added to the comet's speed, it would readily propel the comet out of the Solar System. Subtracted from the comet's speed, it would leave the comet in an orbit with a semimajor axis of less than 13 AU and an orbital period of 47 years. This is typical of a short-period comet. Subsequent encounters with the Jovian planets are unavoidable.

Our study of the icy satellites and Pluto in Chapter VI suggested the possibility that large populations of ice-rich preplanetary bodies may have been necessary to provide the Pluto–Charon system, Triton's encounter with Neptune, and even the axial tilt of Uranus. It was first suggested by Gerard P. Kuiper over 50 years ago that a residual belt of thousands of hundred-kilometer iceballs might still exist beyond the orbit of Pluto, where sweepup mechanisms are extremely ineffective. These bodies, termed the *Kuiper belt*, would have orbital inclinations of order 10° and eccentricities of order 0.2, similar to the asteroid belt. No direct observational evidence existed to establish the real presence of such a belt, or to attribute any known comets to an origin in it, until 1992, when David Jewitt and Jane Luu of the University of Hawaii discovered an "asteroidal" (no tail or coma) object initially named 1992 $QB_1$, at a heliocentric distance of 42 AU. Analogy with Pluto suggests that its brightness is appropriate for a body with a radius of 120 km. Two families of asteroidal bodies with distant orbits, the Centaurs and the trans-Neptunian objects, will be discussed in detail in Chapter VIII.

Thus perturbations of the orbits of unobservably distant Oort-cloud long-period comets by passing stars can divert them into observable orbits, and perturbations of long-period comets and other small bodies by Jupiter and Saturn can change them into short-period comets. Since the impulse imparted by the giant planets is usually in the direction of their orbital motion (they share some of their angular momentum with the comet), short-period comets initially enter prograde orbits of moderate inclination. The destruction of short-period comets must be statistically balanced by the conversion of long-period comets by Jupiter and Saturn.

Major orbital changes require close encounters with planets. Impacts also require that the aim point of the projectile be very close to the disk of the target planet. Above, we estimated the probability that the aim point of a long-period comet would hit a terrestrial planet is only about $10^{-9}$ per perihelion passage, and did so by comparing the geometrical cross-section area of the terrestrial planets to the cross-section area of the terrestrial planet region. In doing so, we neglected the gravitational attraction of the target planet, tacitly assuming that gravitational focussing by the planets is negligible for such fast-moving projectiles. Figure VII.5 shows the geometry of approach of a fast-moving projectile to a planet. The trajectory shown is the one that separates impacts from flybys: it results in a grazing impact with the planetary surface at a velocity $V_{imp}$. At the point of grazing contact the velocity vector must be perpendicular to the radius $R$ of the planet. At great distance from the planet, before gravitation has had any significant effect, the projectile's aim point is offset from the center of the planet by a distance $D$ and the approach velocity is $V_{app}$. By conservation of angular momentum,

$$DV_{app} = RV_{imp\odot} \qquad \text{(VII.8)}$$

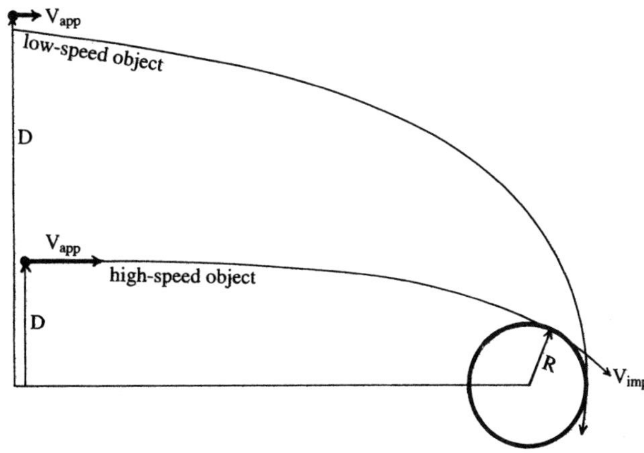


**Figure VII.5** Trajectories of grazing-incidence projectiles vs approach velocity. Note the much more severe bending of the trajectory of slow-moving bodies that spend a long time close to the planet.

Conservation of energy gives us, after Eq. (VII.2),

$$V_{imp}^2 = V_{app}^2 + 2GM/R, \qquad \text{(VII.9)}$$

and combining,

$$D = (R^2 + 2GMR/V_{app}^2)^{1/2}, \qquad \text{(VII.10)}$$

which shows that the critical offset for impact is

$$\begin{aligned} D &= R(1 + 2GM/RV_{app}^2)^{1/2} \\ &= R[1 + (V_{esc}/V_{app})^2]^{1/2}. \qquad \text{(VII.11)} \end{aligned}$$

For a long-period comet encounter with Earth at 55 km s$^{-1}$, $V_{esc}/V_{app}$ is about 0.2, and $D/R = 1.02$. Thus the approximation that the collision cross-section equals the geometrical cross-section is sound. But note that, for a body with an approach velocity of 1.1 km s$^{-1}$, $D/R$ would be 7. The probability of impact of course scales with the effective *gravitational cross-section*, $\pi D^2$, not the geometric cross-section area, $\pi R^2$, so the probability of impact would be 50 times larger for the slower projectile! This same approach can be used when the "projectile" is a fast-moving star and the "targets" are nearly stationary Oort cloud comets.

## Heating by Passing Stars

The prospect that long-period comets may be pristine, unaltered relics that have evaded significant heating since the time of their formation suggests that they may be the best probes of ancient Solar System (and presolar) processes available to us. But the passage of other stars through the outer Oort cloud, described above for its dynamical importance, might give vast numbers of otherwise pristine comets the opportunity to be heated briefly to much higher temperatures than could be provided by the Sun alone. Can passing stars destroy the delicate record borne by long-period comets? At encounter distances close enough for significant heating, would the comet experience gravitational accelerations sufficient to eject it from the Solar System?

Most stellar encounters involve M-class red dwarfs. In order to achieve temperatures high enough for the loss of very volatile species such as $CH_4$ and Ar (roughly 30 K), the encounter distance of the star from the comet must be rather small. The temperature of a black body at distance $r$ from a star of luminosity $L$ (in units of the solar luminosity) is roughly

$$T = 300r^{-1/2}L^{1/4} \qquad \text{(VII.12)}$$

where $r$ is the heliocentric distance in AU. Thus, for an M5 dwarf of $L* = 10^{-4}L_\odot$, a temperature of 30 K is attained at a distance of 1 AU. We may picture a red dwarf passing through the Oort cloud along a chord several thousand kilometers long, heating all the nuclei within 1 AU of its path. Since, as we have seen, long-period comets near aphelion have orbital velocities of several m s$^{-1}$ and the typical star has a velocity of about 20 km s$^{-1}$, the star passes through the comet swarm as if the comets were standing still. The heated volume swept out by the star is about $3 \times 10^4$ AU$^3$, out of a total Oort cloud volume of about $3 \times 10^{12}$ AU$^3$. Thus each encounter with an M star heats about $10^{-8}$ of the Oort cloud population to over 30 K. A typical encounter will involve the star approaching a comet to within about 0.7 AU (a peak temperature of 36 K, similar to the surface temperature of Pluto near aphelion) and traveling 1.4 AU along a chord inside a distance of 1 AU. The total duration of the encounter is then about $1.4\,\text{AU} \times 1.5 \times 10^8\,\text{km AU}^{-1}/20\,\text{km s}^{-1} = 10^7\,\text{s}$. For a comet nucleus with a thermal diffusivity of about 1, this is sufficient time for the heat pulse from the stellar encounter to penetrate to a depth of only about 30 m. Much of this heat may be lost by evaporation of highly volatile species. Thus a nucleus with $r = 1$ km could at most experience heating of a few percent of its mass.

From Eq. (VII.7) we calculate that a star of mass 0.1 $M_\odot$ passing at a distance $D = 1$ AU from a comet nucleus at a speed of 20 km s$^{-1}$ can impart a velocity change of about 10 km s$^{-1}$ (compared to the heliocentric escape velocity of about 300 m s$^{-1}$ for the comet). Thus any comet that gets close enough to an M star to be significantly heated will certainly be ejected from the Solar System.

For a high-luminosity early Main Sequence star the distance out to which comets can be heated to 30 K is about $100L*^{1/2}$ AU, which, for a mass-luminosity relation of $L* = M*^{3.5}$, implies $D_{30K} = 100M*^{1.75}$. However, the distance at which perturbations by the star change the orbital velocity of the comet dramatically (i.e., by more than 2 m s$^{-1}$) is $2 \times 10^{31}\text{GM}_*/1.5 \times 10^{13}V_*$, or $4 \times 10^4 M_*$AU. These distances are equal only for a stellar mass of 3000 $M_\odot$. This is a nonphysical answer, since stars with masses greater than about 60 $M_\odot$ are unstable against their own internal radiation pressure and cannot exist stably on the MS, and hence cannot obey the $M$–$L$ relationship used. We conclude that heating of Oort cloud comets by passing stars almost always occurs under conditions that cause them to be lost from the Solar System.

## Evaporation and Nongravitational Forces

At large distances from the Sun the absorption of sunlight governs the surface temperature of an icy body, and the surface temperature in turn governs the vapor pressure of the ice and its rate of evaporation. The steady-state balance between absorbed and emitted energy is then that given in the previous chapter,

$$(1 - A)F_\odot/R^2 = 4\sigma T_e^4 = 4\varepsilon\sigma T^4, \qquad \text{(VI.3)}$$

where, as before, $F_\odot$ is the solar constant, $A$ is the albedo, and $\varepsilon$ is the emissivity of the ice surface.

Ices, of course, have significant vapor pressures. We saw in Chapter V that the vapor pressure of a substance at low pressures can be compactly represented by

$$\ln(P_{\text{vap}}) = \Delta H^\circ_{\text{vap}}/RT - \Delta S_{\text{vap}}/R, \qquad \text{(V.44)}$$

where the molar enthalpy and entropy of vaporization change only slowly with temperature. Over rather wide ranges of temperature these two variables may be treated as constants, which permits us to express the slope of the vapor pressure curve as

$$\partial \ln(P_{\text{vap}})/\partial T = -\Delta H^\circ_{\text{vap}}/RT^2. \qquad \text{(VII.13)}$$

We can see from Eq. (V.44) why a plot of log vapor pressure vs $1/T$ is extremely close to a straight line.

At higher temperatures, at which the vapor pressure is large, the evaporation of ice can absorb a significant amount of the incident heat flux. Each mole of ice that evaporates (sublimes) absorbs $\Delta H_{\text{vap}}$. We need now only relate the evaporation rate to the temperature in order to quantify this cooling effect.

From the kinetic theory of gases it is possible to show that the rate of evaporation of a substance is related to its vapor pressure by

$$dm/dt\,(\text{g cm}^{-2}\,\text{s}^{-1}) = P_{\text{vap}}[m/2\pi kT]^{1/2}. \qquad \text{(VII.14)}$$

The factor multiplying $P_{\text{vap}}$ is the familiar preexponential factor encountered in the derivation of the Maxwellian velocity distribution. The molecules depart with an average speed that, according to the kinetic theory of gases, is

$$c = [8kT/\pi m]^{1/2}. \qquad \text{(VII.15)}$$

Recalling that $m/k = \mu/R$, the heat flux lost by the vaporization of the ice is

$$\Phi = L\, dm/dt = LP_{\text{vap}}\,[\mu/2\pi RT]^{1/2}. \qquad \text{(VII.16)}$$

Here $L$ is the latent heat of evaporation per gram of ice, $\Delta H_{\text{vap}}/\mu$. The associated momentum flux carried by the gas being lost is

$$\phi = c\, dm/dt = P_{\text{vap}}[4/\pi^2]^{1/2} = 2P_{\text{vap}}/\pi. \qquad \text{(VII.17)}$$

The momentum flux $(1/A)d(mv)/dt$, has units of $\text{g cm}^{-1}\,\text{s}^{-2}$. This is the same as dyn $\text{cm}^{-2}$, the dimensions of pressure. Simply, the escaping gas exerts a back-pressure on the comet surface and imparts to it a potentially very important nongravitational force. This force generally is greatest where the evaporation rate is greatest, in the midafternoon region of the comet nucleus. The force therefore generally has both a radial component (that partially offsets the gravitational attraction of the Sun) and a component normal to the comet–Sun line. Because the spin axis of the comet nucleus may be oriented in any direction, this normal component of the reaction force may accelerate or decelerate the comet in its motion about the Sun. It may also have a powerful influence on the rotation of the nucleus.

The thermal consequences of a high evaporation rate are also very important. Once the comet gets close enough to the Sun so that the heat carried off by evaporating surface ices is a significant part of the total incident solar flux, it becomes necessary to modify the heat balance equation [Eq. (VI.3)] to

$$(1 - A)F_{\odot}/R^2 = 4\varepsilon\sigma T^4 + L\, dm/dt. \qquad \text{(VII.18)}$$

Because evaporation carries off heat at a rate that varies with the vapor pressure and because the vapor pressure is roughly exponentially dependent upon temperature, a comet approaching the Sun experiences a rather abrupt transition from the radiatively controlled regime ($T$ varying with $r^{-1/2}$) to an evaporatively controlled regime, in which the evaporation of ice thermostats the temperature of the comet nucleus. Nucleus-surface temperatures are effectively limited to about 180 K irrespective of the closeness to the Sun, because closer approach in this regime means an enormous increase in evaporation rate and in nongravitational accelerations with only a modest increase of surface temperature.

It is interesting that the eutectic temperature of the ammonia–water system is 173 K. It is therefore marginally possible for water ice I and eutectic liquid aqueous ammonia solution to coexist on the surface of a comet nucleus near perihelion. By the usual criterion ($P_{\text{vap}} > P_{\text{atm}}$) the liquid would be above the boiling point. However, the eutectic solution has properties more similar to those of a glass than to those of a conventional mobile, low-viscosity liquid, and bubble nucleation and boiling would not be expected. The concentration of ammonia in most comets cannot be high, but, if any is present, a liquid phase may exist. In some scenarios for the origin of comets, many new comets are ice balls ejected from the region in which the Jovian planets are forming into the Oort cloud. Those that originate in the Solar Nebula in the Jovian region would be expected to contain little ammonia and methane and to have abundant CO, whereas those that form in gaseous planetary subnebulae and escape from them would be expected to contain much more methane and ammonia.

In the vicinity of the water–ammonia eutectic the vapor pressure is about $1.3 \times 10^{-5}$ bar (13 dyn $\text{cm}^{-2}$) and is dominantly composed of ammonia vapor. Thus, under the most extreme conditions, a 1-km-radius comet nucleus may have pressures of tens of dynes per square centimeter in the subsolar region. The force exerted by the vapor on the dayside of the nucleus must be less than $\pi r^2 P_{\text{vap}}$, and the mass of the nucleus is $4\pi\rho r^3/3$. Thus the jetting of gases from the dayside imparts a maximum nongravitational acceleration of

$$a_{\text{nongrav}} = F/m = 3P_{\text{vap}}/4\rho r, \qquad \text{(VII.19)}$$

which, for a mean nuclear density of $1\,\text{g cm}^{-3}$, is about $10^{-4}\,\text{cm s}^{-2}$. Taking a mean speed of $50\,\text{km s}^{-1}$ and a typical chord length of 2 AU ($3 \times 10^{13}$ cm) for the comet's passage through the inner Solar System, the time ($\tau$) over which this acceleration is acting is $6 \times 10^6$ s, and the total velocity change is $\Delta V = \tau a_{\text{nongrav}} = 6\,\text{m s}^{-1}$. Because the hottest point on the nucleus will be in midafternoon, not at the subsolar point, this acceleration will generally have a large nonradial component that, depending upon the orientation of the spin axis of the nucleus, may increase or decrease the orbital period.

Having portrayed a "clean" scenario for processes on the nucleus of a comet near perihelion, it now behooves us to recognize a host of complicating factors. First, comet nuclei are rotating. Because of their rotation, the rate of heating varies greatly over each cometary "day." Second, the axial tilt of the comet nucleus is random and therefore likely to be large, so that the latitude has a strong effect on the range of daily temperatures. Third, nongravitational forces can of course modify both the spin and the orbit. Preferential evaporation near the equator of a comet nucleus can reduce the principal rotational moment of inertia of the nucleus, making it unstable against tumbling. Fourth, comets are not made of a single piece of pure ice; they contain a substantial component of dust that probably dominates the optical properties of the surface. Fifth, dust may be ejected by the vapors streaming off the nucleus, especially close to the Sun. This dust both screens the surface partially from solar heating and back-warms the surface by its own infrared emission; the dust, especially close to the Sun, can be much warmer than the nucleus. Sixth, prolonged mild heating far from the Sun may release gases so gently that a "lag deposit" of dust is left behind to serve as an insulating layer of the nucleus. Seventh, complex mixtures of ices, including clathrate hydrates, with very different vapor pressures are almost certainly present. Eighth, comet nuclei may contain amorphous ice, which is stable at low temperatures but transforms to crystalline ice I at a temperature of about 155 K. This phase change is exothermic. As a further complication, amorphous ice has an extraordinarily low thermal conductivity. Ninth, temperature gradients at the surface of the nucleus may be very large (near perihelion, temperatures in the dust layer may be above 200 K, whereas those at modest depth are still representative of interstellar space). Thus thermal conduction into the nucleus may be very important. This introduces a lag into the thermal response of the nucleus, causing the temperature maximum to occur after perihelion.

## The Nucleus and Coma of P/Halley

The nucleus of P/Halley was imaged at close range by the European Space Agency's Giotto mission in 1986 (Fig. VII.6). The shape of the nucleus is extremely irregular, due either to gentle accretion of large bodies or to collisional erosion. Strong internal heating would tend to reduce the nucleus to near-spherical shape, but no such effect is evident in the images. Apparent impact craters can be discerned on its surface, which has a very low albedo. Gas release from the nucleus occurs in a very uneven manner, with strong jets of gas and dust. It would be interesting to know whether there is a widespread, thick mantle of dust and whether venting is preferentially associated with impact events that penetrated the dust layer.

Upon expanding into a coma, the gas distribution around the nucleus takes on nearly spherical symmetry. Expansion continues until the dynamic pressure of gas expansion reaches a balance with the dynamic pressure of the solar wind,

$$\rho_g v_g^2 = \rho_{sw} v_{sw}^2. \qquad \text{(VII.20)}$$

With a gas pressure of at most $10^{-5}$ bar near the nucleus, the surface gas density is then less than $3 \times 10^{14}\,\text{cm}^{-3}$. After expansion to a radius of $10^4$ km the number density has dropped to about $3 \times 10^6\,\text{cm}^{-3}$, and, taking $v_g \simeq 1\,\text{km s}^{-1}$ and $v_{sw} \simeq 600\,\text{km s}^{-1}$, this balance is achieved when the solar wind number density is about $10\,\text{cm}^{-3}$, inside the orbit of Venus. The coma expansion time is then about $10^4$ s. Any species in the coma that can photodissociate or photoionize in this time interval may do so. Once ions in the coma reach pressure balance with the solar wind, they become entrapped in the solar wind magnetic field and are driven outward from the Sun to make a plasma tail. Figure VII.7 shows P/Halley in all its glory, with its coma at the lower left completely obscuring the tiny nucleus and the plasma tail streaming outward from the Sun. The short, diffuse dust trail can be seen just below the densest part of the plasma tail. The main molecule responsible for the gaseous display is the most abundant cometary molecule, water.

## Chemistry and Photochemistry of Water

Without doubt the principal volatile constituent of cometary nuclei is water ice. For many comets, it seems likely that water ice makes up more than 50% of the total mass, perhaps even reaching 80% in some rare cases. This underscores the importance of water ice evaporation in regulating the temperature of the nucleus near perihelion and raises the issue of the importance of solid hydrates, especially clathrate hydrates, as carriers of other volatile species. But it also clearly implies that the chemistry of the coma and tail cannot be understood without careful study of the photochemistry of water vapor.

The initial reactions that occur when water vapor is exposed to the solar UV flux are

$$H_2O + h\nu \rightarrow$$

$$OH + H \quad \text{(photodissociation)} \qquad \text{(VII.21a)}$$

$$O + H + H \quad \text{(photodissociation)} \qquad \text{(VII.21b)}$$

$$O + H_2 \quad \text{(photodissociation)} \qquad \text{(VII.21c)}$$

$$O(^1D) + H_2 \quad \text{(photodissociation)} \qquad \text{(VII.21d)}$$

$$H_2O^+ + e^- \quad \text{(photoionization).} \qquad \text{(VII.21e)}$$

The first of these reactions can be driven by photons with wavelengths as long as 1950 Å. Such photons are common in the solar flux (see Fig. IV.36). The second and third reactions are quantitatively much less important because they require (rarer) more energetic photons. Photoionization requires photons with wavelengths less than about 1000 Å, which are several hundred times less abundant than those in the 1800- to 1950-Å region. Almost all solar emission shortward of about 1400 Å is in lines, not continuum.

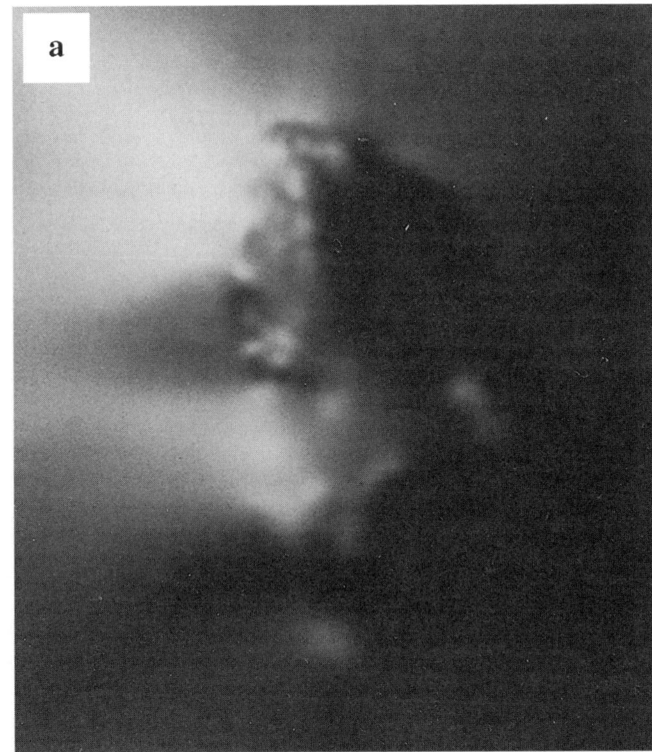


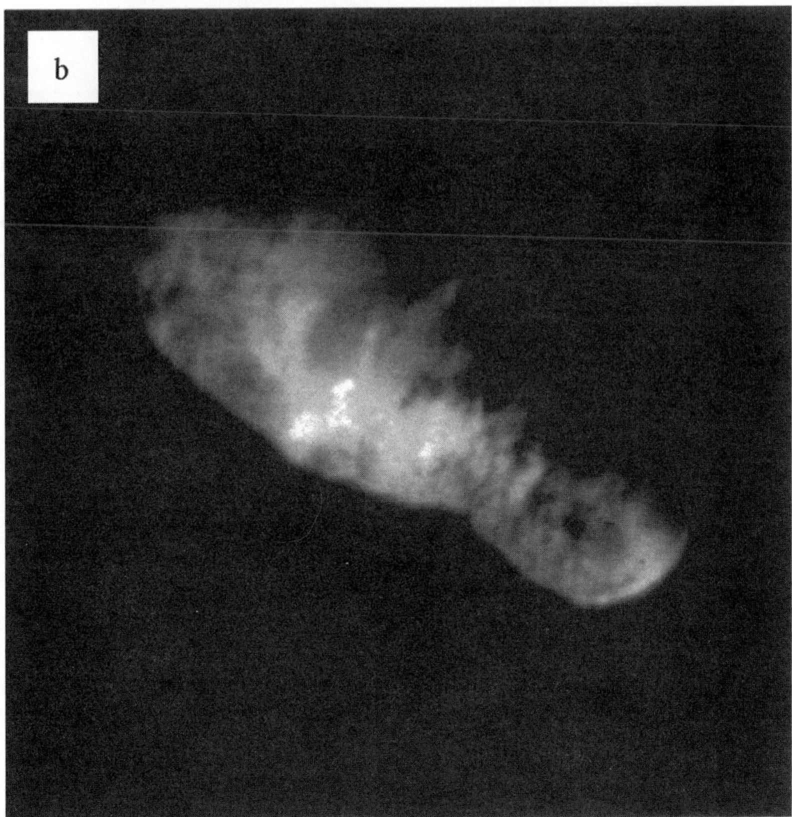


**Figure VII.6** The nuclei of Halley's comet and Comet Borelly. (a) This image of the nucleus of P/Halley is a composite of six frames taken by the Halley Multicolor Camera on the European Space Agency's Giotto spacecraft. The best resolution is about 60 m. Note the irregular shape, the very low albedo, and the pronounced jetting of gases from the sunward side of the nucleus. (b) The nucleus of 19 P/Borelly as seen by NASA's Deep Space 1 spacecraft. Note the extremely irregular "snowman" shape and the hints of impact cratering.

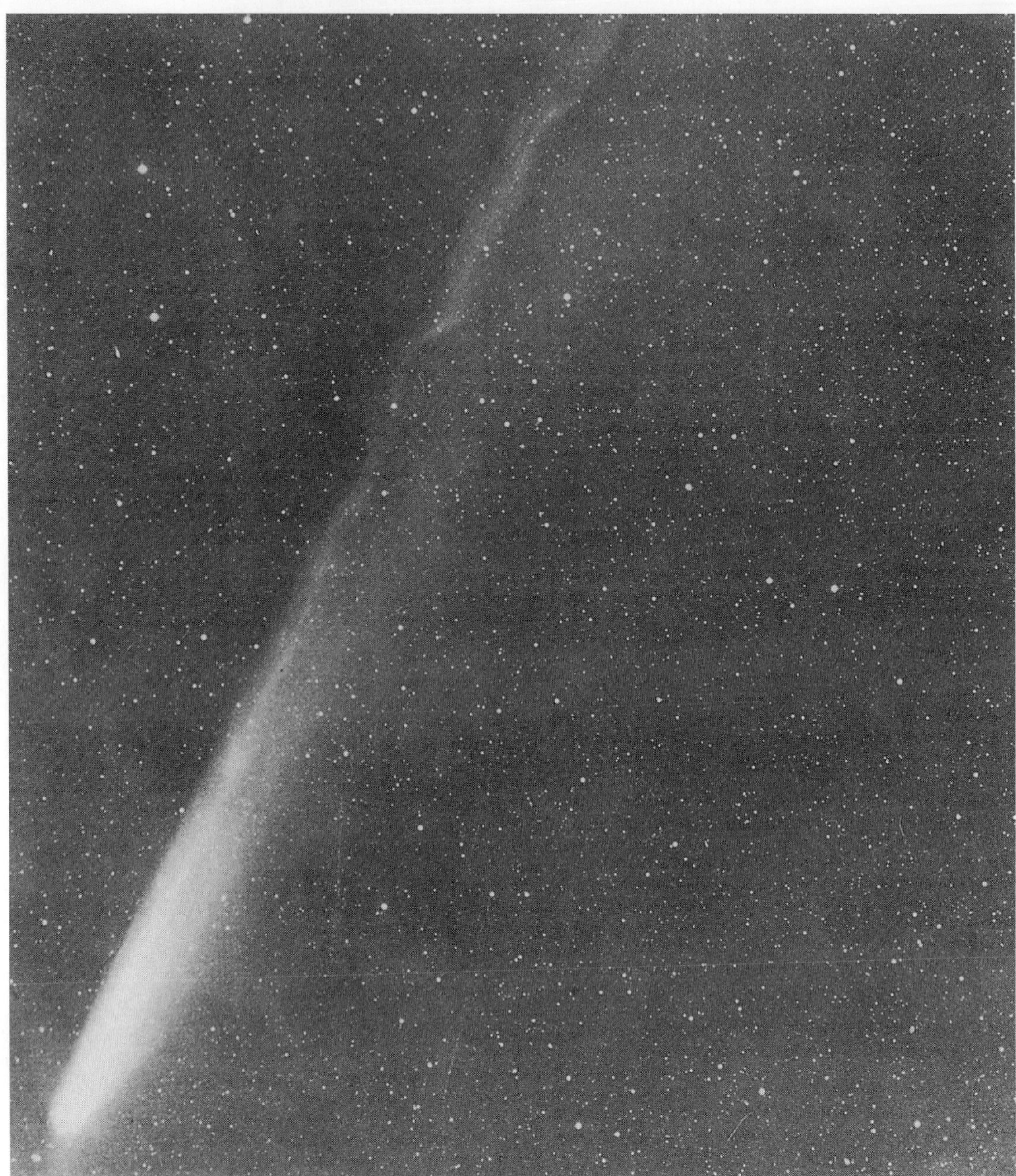

**Figure VII.7** P/Halley in all its splendor. This photograph from the spectacular 1910 apparition gives us a notion of how a bright comet should look. Although there were many bright comets in the 19th century, the 20th century's two best displays were in 1910.

The absorption coefficient for water vapor in this spectral region is wavelength dependent, but for simplicity we shall take the mean absorption coefficient ($\alpha$) as about $10^3\,\text{cm}^{-1}$. This value means that there is about $10^3\,\text{cm}^2$ of effective absorbing cross-section area per cubic centimeter of gas at normal density (standard temperature and pressure, STP: 273.16 K and 1 atmosphere). A gas at STP contains Avogadro's Number of molecules ($N_0 = 6.022 \times 10^{23}$) per molar volume (22.4 liters $= 2.24 \times 10^4\,\text{cm}^3$), or $2.69 \times 10^{19}\,\text{cm}^{-3}$ (Loschmidt's Number). Each molecule then has an effective cross-section area ($\sigma$) of about $10^3/2.69 \times 10^{19} = 4 \times 10^{-17}\,\text{cm}^2$.

From Fig. IV.37 we see that there is a total flux of about 3 to $4 \times 10^9$ photons $\text{cm}^{-2}\,\text{s}^{-1}$ at 5.2 AU, or about $10^{11}\,\text{cm}^{-2}\,\text{s}^{-1}$ at 1 AU for the wavelength region 0 to 1950 Å. The rate of photolysis for a gas of density $n\,(\text{cm}^{-3})$ composed of molecules with a cross-section of

$4 \times 10^{-17}\,cm^2$ exposed to a photon flux of $I$ photons $cm^{-2}\,s^{-1}$ is

$$J(cm^{-3}\,s^{-1}) = \sigma n I = 4 \times 10^{-6} n. \qquad (VII.22)$$

The solar photons can penetrate through the vapor until they have encountered enough of the molecular cross-section to completely intercept the beam. Each cubic centimeter of gas contributes $\sigma n$ $cm^2$ of the cross-section area. The mean solar photon will penetrate the gas to the distance $X$, at which each square centimeter of the beam has encountered $1\,cm^2$ of the total molecular cross-section. In the formalism of Chapter IV,

$$I(X) = I_0\, e^{-\sigma n X}, \qquad (IV.173)$$

and the intensity of the incident light is largely used up when $I/I_0 = 1/e$; that is, $\sigma n X = 1$. We then can say that the average photon in the beam is effectively stopped at a depth of $X = 1/\sigma n$.

The vapor density is of course at its highest immediately adjacent to the nucleus when the comet is at its hottest (at or slightly after perihelion). The highest temperature attainable is probably close to 200 K. The vapor pressure of ice is roughly given by $\log p_v\,(bar) = 5.62 \times 2100/T$, which is about $10^{-5}$ atm at 200 K. The associated vapor density is $10^{-5}\,(273/200) = 1.37 \times 10^{-5}$agt, or $n = 1.37 \times 10^{-5} \times 2.69 \times 10^{19} = 3.7 \times 10^{14}\,cm^{-3}$. For this densest possible vapor, $\sigma n$ is $4 \times 10^{-17} \times 3.7 \times 10^{14} = 0.014$, and $X$ is only $1/\sigma n = 70$ cm. Thus the gas near a comet nucleus may be completely opaque to chemically active ultraviolet radiation.

In the near-nucleus regime, where chemical reactions are unimportant, the cometary vapors expand outward at a speed given by Eq. (VII.15). For plausible constituents of ices at reasonable temperatures, expansion velocities are a few hundred $m\,s^{-1}$. Because a typical cometary nucleus has a radius ($a$) on the order of 1 km and an escape velocity ($v_e = (2GM/a)^{1/2}$) on the order of $1\,m\,s^{-1}$, the coma behaves like a freely expanding gas with a nearly constant velocity of expansion, little retarded by the gravitational attraction of the nucleus. The density of the expanding gas that makes up the coma drops off rapidly according to an inverse-square law. One might expect that the gas, which is expending irreversibly into a vacuum, might cool off rapidly as it converts its internal thermal energy into expansion ($Pv$) work. Indeed, because the expansion is irreversible, the initial thermal energy goes not only into $Pv$ work, but also into accelerating the expanding gas, the expansion should be accompanied by even more cooling than a reversible (adiabatic) expansion through the same pressure interval. In practice, however, cometary envelopes contain significant amounts of dust, and dust can both absorb and emit sunlight over time scales that are short compared with the lifetime of neutral coma gases. The dust, further, may be so abundant near perihelion (when evaporation is most vigorous) that it may severely reduce the intensity of sunlight falling upon the nucleus. It may therefore play an important role in moderating the heating of the nucleus as well as in governing the temperature of the coma.

Because solar photons with sufficient energy to ionize cometary gases are very rare relative to those capable of dissociating these same gases, the most probable process destroying a polyatomic molecule in a cometary coma is photodissociation. Supposing, as a crude approximation, that the molecular cross-section is about $4 \times 10^{-17}\,cm^2$ over the entire wavelength range in which photodissociation and photoionization occurs, we can calculate how far from the nucleus a primary water vapor molecule can travel before being photolyzed or ionized. At a typical point in the inner Solar System (Earth's orbit) the solar constant is $1.25 \times 10^6\,erg\,cm^{-2}\,s^{-1}$, of which about 1% is at wavelengths short enough to break chemical bonds. The typical solar photon that dissociates water vapor is at about 1800 Å; the cross-section of water vapor drops off rapidly toward longer wavelengths and the solar flux drops off rapidly toward shorter wavelengths. A wavelength of 1800 Å ($1.8 \times 10^{-5}$ cm) corresponds to a frequency of $\nu = c/\lambda = 1.6 \times 10^{15}$ Hz and a photon energy of $\varepsilon = h\nu = 6.625 \times 10^{-27} \times 1.6 \times 10^{15} = 10^{-11}$ erg. Thus a typical UV flux in the inner Solar System, about $10^4\,erg\,s^{-1}$, corresponds to a flux of about $10^{15}$ photons $cm^{-2}\,s^{-1}$. The photolysis rate for an optically thin gas of molecules with cross-sections of $\sigma$ $cm^2$, given in Eq. (VII.22), is $\sigma n I$, and the lifetime for an average molecule is $1/\sigma I$. For the conditions we are considering here, that time (the *photodissociation lifetime*) is about 25 s. Thus, once a molecule gets far enough from the nucleus to be exposed to nearly unattenuated sunlight, it is likely to be broken apart very quickly. By comparison, the *photoionization lifetime* is governed by the very much lower flux of photons energetic enough to ionize molecules and is typically on the order of hours near 1 AU.

This chain of reasoning leads us to expect that the relatively dense molecular gas near the coma will be surrounded by a shell of photodissociated molecular fragments, such as OH and H, which becomes progressively more highly ionized into $H^+$, free electrons, etc., as it continues outward from the nucleus. Atomic H can in principle absorb and resonantly scatter solar Ly $\alpha$ radiation (1216 Å), but the high orbital velocities of comets can Doppler shift the cometary absorption line well off the center of the solar emission feature.

Photolysis is accompanied by several important phenomena. First, the photon energy required to break a bond of known strength always greatly surpasses the actual bond energy. This is because the incident energy goes not only into bond breaking but also into rotational, vibrational, electronic, and translational excitation of the molecular fragments produced. This phenomenon is important because, for example, the H atoms produced by water vapor photodissociation will be traveling much faster than their parent water molecules. A second important result is that photolysis causes the deposition of important amounts of energy within the gas, heating it strongly. Third, radicals are often produced in excited, highly reactive states that themselves have short lifetimes against chemical reaction.

Before introducing the many other species that have been observed, deduced, or theoretically expected in comet nuclei, it is wise to consider how complex the chemistry of pure water vapor can be. The major photochemical reactions destroying water were given above [Eqs. (VII.21a)–(VII.21d)]. $H^+$ ions (and free electrons) are produced by photoionization of some of the atomic H made by photodissociation of water and destroyed by recombination with free electrons:

$$H + h\nu \rightarrow H^+ + e^- \qquad \text{(VII.23)}$$

$$H^+ + e^- \rightarrow H + h\nu. \qquad \text{(VII.24)}$$

The latter reaction is very slow because of the necessity of losing a highly energetic photon during the brief duration of the collision between the atomic hydrogen ion and the electron.

Both $H^+$ and $H_2O^+$ are very reactive:

$$H_2O^+ + H_2O \rightarrow H_3O^+ + OH \qquad \text{(VII.25a)}$$

$$H_2O^+ + e^- \rightarrow H_2 + O(^1D) \qquad \text{(VII.25b)}$$

$$OH + H \qquad \text{(VII.25c)}$$

$$H_2 + O \qquad \text{(VII.25d)}$$

$$H^+ + H_2O \rightarrow H_2O^+ + H. \qquad \text{(VII.25e)}$$

The last of these, the charge exchange reaction between the atomic hydrogen ion and water vapor, is rapid in a water-rich gas such as a cometary coma.

Note that electronic recombination of the positive water vapor ion releases so much energy that it causes dissociation of the molecule. Two of these *dissociative recombination* reactions produce atomic oxygen in different electronic states, $^1D$ and $^3P$. Of these the latter (triplet) state is the ground state. The singlet $D$ state lies enough higher in energy so that it is much more reactive.

## Further Chemical Processes in the Coma and Tail

Spectroscopic observations of numerous comets have revealed the presence of nearly three dozen gaseous species. Visible and UV emission features show the presence of H, NH, $NH_2$, O, OH, and S in cometary comas. These may be thought of as the direct photolysis products from the solar UV irradiation of primary gases such as ammonia, water vapor, and hydrogen sulfide. Indeed, microwave observations directly confirm the presence of the $H_2O$ molecule. It is at least qualitatively easy to understand why these species should be present in low-temperature Solar System ices and their irradiation products.

A number of carbon-bearing species (sometimes loosely referred to as "organic") have also been observed in the coma. Visible/UV spectroscopy reveals C, $C_2$, $C_3$, CH, CN, CO, and CS. Several recent bright comets have been studied by microwave and infrared techniques. Methyl alcohol, $CH_3OH$, a prominent constituent of the interstellar medium, has been found in 21 P/Giacobini–Zinner in concentrations of about 1% of the water abundance. The HCN abundance in the coma of C/1995 O1 Hale–Bopp has likewise been measured as $0.40 \pm 0.05\%$. The abundances of $CH_3CN$, $HC_3N$, $H_2CS$, and NS have also been found to be near 0.03%. Farther from the nucleus, where the gas density is much lower and the neutral coma gases have had sufficient time to be photoionized, the ions $C^+$, $CH^+$, $CO^+$, $CO_2^+$, $H_2O^+$, $OH^+$, $N_2^+$, and $CN^+$ are seen via their UV and visible emission features. These ions are entrained by the magnetic field of the solar wind and, instead of forming an outer spherical "ionosphere" about the coma, stream out anti-Sunward to form the glowing plasma tail. It is not unusual to see the Na $D$ lines in the spectra of comets. Indeed, sun-grazing comets experience such high temperatures that a host of other visible and UV lines of the other rock-forming elements, K, Ca, Fe, Co, Ni, V, Mn, and Cu, can be observed in their comas. The $Ca^+$ ion is also found in many comet tails near perihelion. Mass spectra of the atmosphere of Comet Halley made by the European Space Agency's Giotto spacecraft in 1986 provided an exciting new addition to the list: polymers of formaldehyde. The formaldehyde unit ($CH_2O$), at high concentration, forms a polymer called polyoxymethylene $(-CH_2-O-)_n$, which evaporates to form a variety of fragments that may cleave apart at any C–O bond. Thus any fragment containing $n$ $CH_2$ groups (mass $= 14$ AMU) may contain $n - 1, n,$ or $n + 1$ O atoms (mass 16).

It is obvious that comas are dominated by photodissociated fragments of molecules and thermally dissociated vapors of minerals, not by the molecular species that actually compose the solid nucleus. It is even more obvious that the species in the plasma tail are derived from further processing of the coma species (principally by photoionization and subsequent ionic reactions). What is not so obvious is just exactly what primary species are present in the nucleus before evaporation and photolysis. A number of increasingly complex chemical schemes, some involving hundreds of gas-phase reactions, have been proposed in an attempt to find the connection between nucleus composition and coma (and tail) spectral data.

The tentative reconstruction of parent molecules must begin with $H_2O$. Ammonia, though not abundant, seems necessary to explain the $NH_2$ and NH fragments. The observed presence of HCN, CN, and $N_2^+$ suggests, however, that most of the nitrogen is in the form of unsaturated, high-temperature species. On the basis of available evidence, $CS_2$ seems reasonable as the source of both CS and S, but other sulfur-bearing parent species such as COS, $H_2S$, solid sulfur, and organic SH compounds may emerge as reasonable contributors to S. HCN, $CH_3CN$, and $CH_2O$ are all readily produced by disequilibrating processes in moist reducing atmospheres, and are all well-known constituents of the interstellar medium. A wide variety of complex biochemicals, including organic acids, aldehydes, alcohols, ketones, esters, amino acids, peptides, proteins, and sugars, may be made from these starting materials by almost shockingly simple reactions.

The absence of $CH_4$, $CH_3$, $CH_2$, and saturated light hydrocarbons (such as $C_2H_6$ and $C_3H_8$) is reminiscent of the chemistry of giant molecular cloud complexes in the interstellar medium (ISM). The observed presence of multiply bonded, unsaturated carbon species such as $CH_3CN$, HCN, CN, CO, CS, $C_2$, $C_3$, $CN^+$, $CO^+$, and $CO_2^+$ confirms that the parent molecules in the nucleus are chemically far removed from the methane expected for chemical equilibration of solar material at low temperatures. Comet expert Armand H. Delsemme of the University of Toledo suggested in 1976 that comet ices are derived from relatively high-temperature gases rich in CO, not low-temperature equilibrium gases rich in methane and ammonia. A theoretical paper by Ronald Prinn and myself in 1980 showed that, under the low-temperature conditions under which methane and ammonia are thermodynamically stable, the rates of the reactions making them from high-temperature gases (CO and $N_2$) are extremely slow. In a turbulent solar nebula, gas parcels that experienced high temperatures would have readily equilibrated to make abundant CO and $N_2$, but the same parcels, when transported outward to cool regions in which ammonia and methane were thermodynamically stable, would have reaction time scales very much longer than the lifetime of the Solar System. Thus the gases available for condensation in the outer part of the Solar Nebula would have resembled a high-temperature gas mixture from the inner Solar System (or a sample of unequilibrated material from the ISM), not a methane- and ammonia-rich low-temperature equilibrium mixture. Thus, instead of having equilibrium proportions of reduced ices (roughly 20 water molecules and 12 methane molecules with about 2 ammonia molecules) we would expect to find about 12 CO molecules and 8 water molecules with 1 nitrogen molecule. Making some of the CO into carbon dioxide would further deplete the water abundance. Indeed, Delsemme and others find that the abundance of oxygen compounds in cometary nuclei is very close to the amount expected based on solar elemental abundances, and nitrogen and sulfur both appear to be present in near-solar proportions. Carbon, however, is severely depleted in the observed coma gases released by the evaporation of ices. A reasonable explanation for the paucity of carbon is that comets, like the interstellar molecular clouds, have hidden about 75% of their total carbon in the form of involatile organic polymers.

The confirmation of the extremely low albedo of the nucleus of P/Halley by the Soviet Vega and European Giotto spacecraft is compatible with the idea that the dust in Halley is extremely carbonaceous, possibly even darker and more carbon rich than the carbonaceous meteorites (see Chapter VIII). The organic chemistry of this involatile material is complex and interesting. Comet infall may provide an efficient means to "seed" all the planets with copious amounts of prebiological organic matter, but only in an interestingly indirect manner: direct impact of a large comet is so energetic that the fireball thermally destroys all organic matter, but the infall of cometary dust is so gentle that organics can sometimes survive.

The mass spectrometric analysis of gases in the coma of P/Halley that revealed the presence of polymeric formaldehyde, $(CH_2O)_x$, has interesting implications. Formaldehyde, a product of irradiation of mixtures of CO with $H_2O$ or $H_2$, is an excellent starting material for use in the synthesis of complex organic matter, including sugars. We also mentioned in Chapter VI a mechanism for production of polyformaldehyde chains by irradiation of ammonia-formaldehyde mixtures in ice at temperatures as low as 40 K. The principal limitation on this method of making sugars and other biochemical species is the problem of making carbon–carbon bonds.

## Behavior of Small Particles

A small nonvolatile particle resting upon, but not "welded" to, the surface of a comet nucleus experiences two forces strong enough to influence its behavior. The first force is the gravitational attraction of the nucleus, which depends on the density and radius of both the grain and the nucleus. The second is the drag force caused by the outward flow of gases from evaporating ices in the nucleus. This force depends on the cross-section area of the grain and the momentum flux in the gas stream, which in turn depends on the temperature, molecular weight, and vapor pressure of the evaporating ices. The temperature is a complex function of heliocentric distance, albedo, position on the nucleus, the spin vector of the nucleus, and time. Both of these forces may be comparable to the centripetal force required to keep the particle in circular motion around the center of the rotating nucleus.

In general, very small grains are especially vulnerable to expulsion by the gas flux. Since the only attractive force, gravity, is proportional to the cube of the grain radius, and the major repulsive force, gas drag, is proportional to the square of the grain radius, sufficiently large grains will be massive enough for the gravitational attraction to overcome the gas drag and prevent the grain from blowing away. The full force balance on a grain at the surface of the nucleus is given by

$$GMm/R^2 - \phi_g VA - mR(\Omega \cos l)^2 = ma, \qquad \text{(VII.26)}$$

where $M$ and $R$ are the mass and radius of the comet nucleus, $m$ is the mass of the particle, $A$ is the cross-section area of the particle, $\phi_g$ is the gas flow rate ($\text{g}\,\text{cm}^{-2}\text{s}^{-1}$), $V$ is the velocity of the gas stream, $\Omega$ is the rotational angular velocity of the nucleus, and $l$ is the latitude at which the grain rests. The inward acceleration of the grain is $a$. For any given set of environmental constraints ($l, \Omega, M, V$, and $\Phi$) there is a critical size for the grain (given by $m$ and $A$) below which there is a net outward acceleration and the grain blows away and above which the grain experiences a net inward acceleration and hence remains in place. This critical size obviously is that at which $a = 0$.

A long-period comet in a highly eccentric orbit with a perihelion distance of only a few solar radii will experience very diverse conditions as it traverses its orbit. Almost all of the time it will be so far from the Sun that it will be too cold for evaporation of ices. As it passes through the terrestrial planet region, evaporation will be rapid enough to blow away small grains, but the largest grains may resist entrainment and stay behind on the nucleus, beginning the formation of a dust layer. Near perihelion the heating and evaporation rate will be so severe that no dust could be retained. It is not hard to envision the accumulation of thick volatile-depleted dust layers covering the nuclei of comets that have frequent perihelion passages, but do not approach the Sun too closely at perihelion. This is a good description of virtually all of the periodic (short-period) comets. This problem was first studied by David Brin and Asoka Mendis of the University of California at San Diego (UCSD) and has been modeled in greater detail by Fraser Fanale and James Salvail of the University of Hawaii, Harry Houpis of UCSD, Paul Weissman of the Jet Propulsion Laboratory (JPL), Hugh Kieffer of the United States Geological Survey (USGS), Flagstaff, and others. The general conclusion of all these studies has been that dust layers with thicknesses on the order of 10 cm to a few meters may commonly accumulate on the surfaces of short-period comet nuclei. A dust layer both inhibits the heating of the cometary ices by the Sun (the dust is very opaque and a very poor conductor of heat) and slows the loss of gases from below. A comet nucleus in an orbit typical of periodic comets would rather quickly decrease its evaporation rate to the point that neither coma nor tail would be detectable. Such a body, if discovered by terrestrial astronomers, would be described as a low-albedo near-Earth asteroid. It would be able to retain much of its store of ices over its entire dynamical lifetime of roughly 100 Ma.

## Dynamical Behavior of Dust in Space

In addition to the gravitational attraction of the Sun, small particles also experience forces due to their interaction with the Solar radiation field. The gravitational force is proportional to the mass of the particle (that is, proportional to its volume or to $a^3$), whereas interactions with the flux of sunlight increase in proportion to $a^2$. Thus, for smaller particles the *relative* importance of radiation forces is greatest. The first of these forces is radiation pressure. It arises from the momentum flux carried by incident sunlight. Every photon of frequency $\nu$ carries a momentum of

$$p = h\nu/c, \qquad \text{(VII.27)}$$

and the pressure exerted by a plane-parallel flux of $n(\nu)$ photons of frequency $\nu$ per square centimeter per second is

$$P = n(\nu)h\nu. \qquad \text{(VII.28)}$$

For an isotopic radiation field with the same total flux, only $\frac{1}{3}$ of the total flux is traveling normal to any randomly oriented surface, and $P = n(\nu)h\nu/3$. In this case, the pressure on the "back" side of the surface is exactly balanced by the pressure on the "front."

Most natural light sources are spectrally complex, and almost all stars can be approximated rather closely as Planckian sources. The total pressure due to a spectrally complex radiation field is calculated by integrating over wavelength.

Since $h\nu$ is the energy of the individual photon, we can recognize Eq. (VII.28) as equating the energy density to the pressure. The radiation pressure force on a surface is the product of the radiation pressure times the cross-section area of the irradiated surface.

Radiation pressure forces are only rarely of any importance in nature; only two examples are at all familiar. The first of these arose in the early 1960s, when several extremely large, light inflated balloon-like satellites were launched to serve as sensitive monitors of atmospheric density at high altitudes or as passive reflectors for long-distance radio communications experiments. The first of these, Echo 1, was launched from Cape Canaveral on a Thor Delta booster on 12 August 1960 into a nearly circular orbit with a mean altitude of 1600 km. When the orbit plane was oriented so that it was roughly face-on to the Sun, the center of the orbit was displaced outward (away from the Sun) by a small but measurable amount. The weight of Echo 1 was about 80 kg, and its diameter when fully inflated was 30.5 m, for an area loading ($\sigma = M/A$) of about 11 mg cm$^{-2}$. By comparison, a meteoroid with a density of 2 g cm$^{-3}$ and a radius of 0.003 cm (30 $\mu$m) has an area loading ($\sigma = M/A = 4\pi/3$) or 8 mg cm$^{-2}$. The second familiar example, the device called a radiometer, is actually a cheat: it is propelled not by radiation pressure, but by the recoil of gas molecules from the black (solar-heated) sides of its vanes.

A small particle in orbit about the Sun is moving relative to the solar radiation field. We may view this from the perspective of the grain; it sees the incident radiation coming not from the true direction of the Sun, but from an angle, $\sin^{-1}\Psi = v_{orb}/c$, slightly ahead of the Sun. This phenomenon is called *aberration* ($\Psi$ is the *angle of aberration*). The particle emits its own thermal radiation isotropically. Alternatively, we may view the particle as seen from the Sun; the grain absorbs light moving straight out from the Sun, but emits its thermal radiation in a way that is isotropic in the grain's coordinate frame, but anisotropic as viewed from the Sun. Note that in Eq. (VI.21), $p = h\nu/c$, the momentum (which is classically equal to the product of mass times velocity) is $mc$ for a photon; thus the equivalent mass of a photon is $m = h\nu/c^2$. (Alternatively we may say that the photon energy ($\varepsilon$) $= h\nu = mc^2$. This was first pointed out by Einstein.)

From the Sun-fixed perspective, radiation emitted from the grain carries off angular momentum ($mvr$), which is lost from the total angular momentum content ($L$) of the grain. For a grain in a circular orbit at distance $R$ from the Sun, $v_c = (GM_\odot/R)^{1/2}$, and the "mass" emitted with each photon is its equivalent mass. The change in angular momentum of the grain is then

$$\delta L = h\nu v_c R/c^2 = h\nu(GM_\odot R)^{1/2}/c^2, \qquad \text{(VII.29)}$$

compared with the total angular momentum of the grain, $m_g v_c R$. This retarding force is called the *Poynting–Robertson effect.*

For a photon emission rate of $n(\mathrm{s}^{-1})$, the rate of loss of angular momentum is increased by a factor of $n$. The ratio of the total angular momentum content of the grain to its rate of loss of angular momentum gives the time scale for the grain to lose all of its orbital momentum and spiral into the Sun:

$$t = m_g v_c Rc^2/nh\nu(GM_\odot R)^{1/2}. \qquad \text{(VII.30)}$$

The rate of energy emission by the grain, $nh\nu$, is limited by the steady-state balance between absorbed solar energy and emission. The flux of absorbed energy is of course proportional to the luminosity of the Sun and inversely proportional to the square of the distance from the Sun: $nh\nu \propto L(/4\pi R^2$. For cometary dust particles in highly eccentric orbits, this force retards the motion of the grain very strongly near perihelion, but only weakly elsewhere. Thus the Poynting–Robertson force tends to *reduce the eccentricity* of the orbit as it reduces the orbit's total energy.

One final factor influences the acceleration of grains by sunlight: the emission from a grain is isotropic only if the surface temperature of the grain is uniform within a few degrees. Supposing that we have a simple cubical grain of length $l$, with one "hot" side toward the Sun, opposite the "cold" side, the conducted flux (given by $F = K(\partial T/\partial l)$ must be comparable to the absorbed flux on the Sunward side in order to maintain nearly uniform temperatures over the grain surface. The temperature difference between the hot and the cold sides is

$$T = Fl/K. \qquad \text{(VII.31)}$$

Near 1 AU the flux ($F$) is on the order of $10^6$ erg cm$^{-2}$s$^{-1}$, and typical silicates have thermal conductivities ($K$) near $10^5$ erg cm$^{-1}$K$^{-1}$s$^{-1}$. The temperature contrast is then less than 10 K for objects of normal rocky composition with sizes less than 1 cm. For larger objects the temperature may be maintained uniform either by exceptionally high thermal conductivity (as for metals) or by rapid rotation. Most objects with dimensions of a few centimeters up to 1 km are, like meteorites, of rocky or cometary rather than metallic composition. They, like small asteroids, generally have rotation periods of a few hours and hence have maximum temperatures near the equator in midafternoon, with minimum temperatures at the winter pole and (at low latitudes) just before local dawn. Thus the emitted thermal flux will be strongly anisotropic. For prograde rotators with modest axial tilts, more momentum will be radiated in the trailing hemisphere than in the leading hemisphere. These larger bodies may then actually be accelerated in their orbital motion, causing them to retreat slowly from the Sun. Retrograde rotators are probably common in this size range, because most of these bodies are collision fragments. Retrograde rotation places the hot region in the leading hemisphere and thus enhances the Poynting–Robertson deceleration. The phenomenon of differential acceleration by emission of radiation from regions of different temperatures is called the *Yarkovsky effect*. The effect is unimportant for bodies larger than a few kilometers because of their very large mass per unit area ($10^6\,\mathrm{g\,cm^{-2}}$ or more) and is especially unimportant for comets that spend little time under intense illumination close to the Sun. In the cometary case, the forces generated by temperature-dependent differences in evaporation rate are vastly larger than those due to any form of radiation pressure.

The solar wind particle and energy flux also exerts a force upon small grains. The protons, electrons, and embedded magnetic fields of the solar wind flow radially outward from the Sun. Small grains generally carry a small positive electric charge because of the photoejection of electrons from their surfaces by solar UV photons. The grains then experience Lorentz ($q\mathbf{V} \times \mathbf{B}$) forces. In this case $V$ is the velocity of the grain relative to the solar wind, which is the difference between the outflow velocity of the wind (radial at several hundred $\mathrm{km\,s^{-1}}$) and the orbital velocity of the grain (transverse at tens of $\mathrm{km\,s^{-1}}$). The $B$ field is almost perfectly radial, so the force on the grain acts to perturb the grain out of its orbital plane. We recall from the discussion of the solar wind that the magnetic field polarity has a sector structure, so that the direction of the perturbing Lorentz force reverses many times per orbital revolution of the grain. Thus this small force is rather efficiently played off against itself and averages out very close to zero. Dust in a planetary magnetosphere experiences a largely dipolar field, roughly aligned with the spin axis, that maintains the same polarity. The $q\mathbf{V} \times \mathbf{B}$ force on a charged dust grain in a low-inclination orbit about a planet is approximately radial. The Lorentz acceleration ($q\mathbf{V} \times \mathbf{B}/m$) is vectorially added to the gravitational acceleration, so that the motions of particles with different $q/m$ ratios at the same planetocentric distance can imitate Keplerian motion about planets with different effective masses. The effective mass may be less than or greater than the actual mass, depending upon the direction of orientation of the planetary dipole field relative to the direction of orbital motion.

Because of the initial velocity differences caused by ejection and because of the small, progressive changes in the orbital elements of grains emitted by comet nuclei caused by radiation effects, dust particles spread out to fill out an elongated ellipse about the nucleus, stretched along the orbit. After many orbits, the dust cloud stretches all the way around the orbit, although of course the episodic nature of active dust release (only near the time of perihelion passage) causes the dust distribution along the orbit to be very clumpy. Small velocity variations at perihelion make large differences in orbital period, but have only modest effects on orbital inclination.

## Meteors

On almost any clear night an observer can see several *meteors* per hour. They appear, dash often tens of degrees across the sky, and vanish in a fraction of a second. From time to time exceptionally bright meteors are seen, with brightnesses rivaling that of the Moon. A very small percentage of meteors leave luminous trails that may persist for seconds or even a few minutes before fading to invisibility. Triangulation of meteors by two or more observers spaced 10 to 50 km apart reveals a very pronounced parallax. The point at which the meteor first became luminous and the vanishing point can be located quite precisely by careful observations even without instruments. Observers find that meteors usually first become luminous at heights of 60 to 120 km and that they travel tens of kilometers before burning out. Timing the meteors is difficult, because few last longer than a second. But this fact alone tells us that meteors must be traveling at speeds of tens of kilometers per second, typical of the orbital speeds of bodies moving about the Sun independently of Earth.

Accurate timing is usually accomplished by taking long photographic time exposures of a large area of the sky with a fast revolving shutter in front of the lens. The trail of the meteor on the film is "chopped" many times by the shutter, perhaps at 0.01-s intervals. From the locations of the beginning and ending points (triangulation) and the elapsed time along the trail (from the chopped photographic trail) the speed may be calculated with good precision. The speeds thus found range from about 12 to about $70\,\mathrm{km\,s^{-1}}$, which confirms that they are not native to the Earth–Moon system. In medieval times it was widely believed that meteors were stars that had been ejected from Heaven and fell to Earth (although no one ever reported one of the previously mapped stars missing!). This belief gave rise to the name "shooting stars," an anachronistic and misleading name for meteors that is unfortunately sometimes still used by journalists. Further confusion arises from the similarity of the word "meteor" to "meteorite" and "meteoroid." A meteor is an optical display caused by a small solid particle entering the atmosphere at such high speeds that it, and the air that it collides with, are heated to incandescence at high altitudes. A *meteorite* is a solid body from space, usually of fairly strong rocky material, that partially survives entry into an atmosphere and lands on a planetary surface.

A *meteoroid* is a small rocky body traveling through space. Meteorites, which are usually of asteroidal origin, are discussed in detail in Chapter VIII. Meteors are treated here because of their association with comets.

Any naked-eye observer who persists for a few hours in mapping the paths of meteors on the sky will begin to discern certain regularities in their motions. On most nights when more than a dozen meteors are seen, one will find that the majority of the visible meteors seem to emanate from one or two specific points in the sky. The groups of meteors that are associated in this manner are called *meteor showers*. The fact that they stream from a particular point (radiant) on the sky attests to the fact that their orbits about the Sun are extremely similar. The orbitally related swarms or clouds of particles that generate meteor showers upon encounter with a planet are called *meteor streams*. Many such streams may be emitted by a comet over many different perihelion passages, each pursuing a somewhat different orbit due to perturbations by solar radiation forces and by gravitational interactions with the planets.

Meteor showers are usually of short duration, ranging from a few hours to about 10 days, and recur every year on the same dates. Dozens of faint showers exist, but only a few of these are readily recognizable to a casual observer. The most prominent showers are listed in Table VII.1.

The intensity of a shower may vary markedly from year to year. A young swarm of dust and pebbles (plus rare large chunks) that has not had time to become widely dispersed in space by Poynting–Robertson and Yarkovsky forces will be traversed by the Earth in a few hours. The swarm is so small in cross-section area (and even in extent along the swarm's orbit) that slight differences in the Earth–swarm encounter geometry from year to year may cause striking variations in the peak meteor flux experienced by Earth. By contrast, an older

**Table VII.1** Prominent Meteor Showers

| Date/peak | Shower name | Radiant RA (h, min) | Radiant dec (°) | Rate (*n*/h) | Source |
|---|---|---|---|---|---|
| Jan 1–4/3 | Quadrantids | 15 25 | 51 | 40 | |
| Jan 17 | Kappa Cygnids | 19 40 | 52 | Low | |
| Feb 5–10 | Alpha Aurigids | 5 0 | 40 | Low | |
| Mar 10–12 | Zeta Bootids | 14 32 | 12 | Low | |
| Apr 19–23 | Lyrids | 18 10 | 33 | 12 | 1861 I |
| May 1–6/4 | Eta Aquarids | 22 23 | 1 | 20 | Halley |
| May 11–24 | Zeta Herculids | 16 28 | 28 | Low | |
| May 30 | Eta Pegasids | 22 0 | 28 | Low | |
| Jun 2–17 | Alpha Scorpiids | 16 48 | −23 | Low | |
| Jun 27–30 | Pons–Winnecke | 15 12 | 58 | Low | Pons–Winnecke |
| Jul 14 | Alpha Cygnids | 20 56 | 47 | Low | |
| Jul 18–30 | Alpha Capricornids | 15 10 | 58 | Low | |
| Jul 25–Aug 4 | Alpha/Beta Perseids | 3 12 | 43 | Low | |
| Jul 26–31/30 | Delta Aquarids | 22 36 | −15 | 20 | |
| Aug 10–14/12 | Perseids | 3 6 | 58 | 50 | Tuttle 1862 |
| Aug 10–20 | Kappa Cygnids | 19 20 | 52 | Low | |
| Aug 21–23 | Omicron Draconids | 19 24 | 60 | Low | |
| Aug 21–31 | Zeta Draconids | 17 28 | 63 | Low | |
| Sep 7–15 | Epsilon Perseids | 4 4 | 36 | Low | |
| Oct 2 | Quadrantids | 15 20 | 52 | Low | |
| Oct 9 | Giacobinids | 17 40 | 55 | 10 | 1933 III[a] |
| Oct 12–23 | Epsilon Arietids | 2 48 | 21 | Low | |
| Oct 18–23/21 | Orionids | 6 18 | 15 | 20 | |
| Oct 31–Nov 6 | Taurids | 3 32 | 14 | 15 | [b] |
| Nov 14–18/16 | Leonids | 10 6 | 22 | 15[c] | Tempel 1866 |
| Nov 26–Dec 4 | Andromedes/Bielids | 1 40 | 45 | [d] | Biela |
| Dec 11–13/13 | Geminids | 7 32 | 32 | 50 | 3200 Phaeton |
| Dec 22 | Ursids | 14 28 | 76 | 15 | |

[a] 1933 III is also called Comet Giacobini–Zinner.

[b] The Taurid stream is a complex combination of several weak streams that are associated with the orbits of at least four near-Earth asteroids (Oljato, 1982TA, 1984KB, and 5052 P-L).

[c] The Leonids are a narrow, young stream that varies enormously in intensity from year to year. The rate of 15/h is for an average year; immense fluxes, as high as 150,000/h, are seen every 33 years.

[d] The Bielids are a very narrow, young stream that has not been observed for decades, but still must pass very close to Earth's orbit. Because of weak planetary perturbations, it could reappear at any time.

meteor stream may stretch all the way around its orbit, will generally be much more diffuse than a younger stream, may have discernibly different orbital parameters, and may be much larger in cross-section area. Streams originating from the same comet, but having widely different ages, may have orbits that are different enough so that the orbits of the various streams precess at different rates, which changes the longitude of perihelion and the date of crossing of Earth's orbit. Allowing for differential precession not only makes it easier to link meteor swarms to their parent bodies, but also enables detailed predictions of future meteor showers. At the cores of many of the densest (and hence youngest) streams we may find an active or recently deceased comet. The Geminid meteor shower, for example, follows an orbit that is clearly associated with the near-Earth asteroid (3200) Phaethon. Phaethon is therefore a prime candidate for a spacecraft mission to an extinct comet.

Two of the narrowest swarms, the Leonids and Bielids, have shown quite extraordinary behavior over the years. The Bielids are associated with the orbit of Biela's comet, a member of the short-period family with an orbital period of 7 years. Biela was observed to split into two similar-sized comets during its 1845 apparition, and both pieces were seen again near the time of the next perihelion passage in 1852. At the time of the next subsequent perihelion passage, no comet was seen. Instead, a bright, brief meteor shower occurred when Earth came closest to the orbit of Biela. The Bielid shower has, however, not been seen for several decades. Minor planetary perturbations have offset the Bielids' orbit slightly from that of Earth, so that the core of the swarm no longer can bombard our skies at night. The Bielids may eventually return as a result of further perturbations, but radiation forces may by then have dispersed the swarm to the point that it is no longer capable of creating spectacular displays.

The Leonids, leftover dust and debris from Tempel's comet of 1866, also form an extremely narrow swarm that does not stretch far along its orbit. The orbit reliably intersects Earth's every November 16, with the earliest and latest swarm members arriving only 2 days earlier and 2 days later, respectively. Because of the nonuniform distribution of the Leonid meteoroids along their orbit and the very different orbital periods of the swarm and Earth, the Earth's annual crossing of Tempel's orbit usually traverses a rather sparse portion of the stream, yielding only 10 to 20 visible meteors per hour. However, every 33 years the dense part of the swarm crosses Earth's orbit just as Earth arrives, producing a brilliant meteor shower. In 1833 the Leonids performed spectacularly, with a single observatory reporting 250,000 meteors in the space of a few hours. The peak rate was about 1000 per minute, and the display was so spectacular that many observers took it for the end of the world. But Earth survived nicely. The Leonids were uninteresting for the next 32 years, but November of 1866 brought an almost equally brilliant display. The displays in 1900 and 1933 were bright but not overwhelming, and it appeared that the orbit of the Leonid swarm may be shifting slightly away from its point of intersection with Earth's orbit.

On the evening of November 16, 1966, observers all over western North America were frustrated in their attempts to observe the Leonids by heavy, widespread cloud cover. I recall spending several boring hours in the California desert under dense, unbroken clouds before giving up. A team of 13 amateur observers from Tucson had better luck. Dennis Milon, one of the Tucson observers, related to me that, as they drove up Kitt Peak around midnight, the skies began to clear. They counted a respectable 33 Leonids in the hour between 1:30 and 2:30 AM in a clear sky. Between 2:50 and 3:50 one observer recorded 192 Leonids, of which 30 were as bright as the brightest stars (negative magnitudes). By 4:10 AM the count was up to 30 per *minute*. A fireball of approximate magnitude −8 (similar to a first-quarter moon) then exploded overhead, and the skies almost literally began to rain light. Milon reports,

> By 4:30 there were several hundred per minute. At 4:45 the meteors were so intense we guessed how many were seen by a sweep of the head in one second. The fantastic rate of about 40 per *second* was reached at 4:54 AM. ... Some of the brighter ones left trains for several minutes, and were photographed. Fully half left trains. We took many photos ... a 43 second exposure of the Big Dipper ... shows 43 Leonids. ... The view was so spectacular that we didn't know where to look ... looking directly at the radiant gave the effect in depth of the Earth moving through space. ... Sometimes we would spin around, taking in the whole sky. Or we alternated with looking toward the western horizon (where it was very clear) and gazing right at the radiant. Different parts of the sky would light up, and we would glance here and there. Everyone was yelling and laughing at the incredible, dazzling sight, and at our luck in seeing it.

The meteor count rate was over 500 per minute for almost exactly 1 h, peaking at about 2500 per minute (150,000/h). These few observers had witnessed the most spectacular meteor shower in recorded history (Fig. VII.8), surpassing even the 1833 Leonids.

The Leonids, on their most recent return, gave good displays in November 1999 and 2000, and a very good showing in November 2001. Predictions of the display intensity, taking into account many individual meteor

**Figure VII.8** The Leonid meteor shower of November 17, 1966. The original negative shows 70 meteor trails. This photo was taken by Dennis Milon, a member of the observing team atop Kitt Peak. All star images are trailed by the time exposure: the two bright points near the center are "head-on" entering meteors, which serve to define the radiant of the shower quite closely. At the time of the photo, visual observers were counting about 40 meteors per second! Photo courtesy of Dennis Milon.

streams, were generally useful as a guide to observers, but still fell short of predicting the behavior of the shower in detail.

Several general features of meteor showers bear comment. First, the overwhelming majority of all meteor observers live in the Northern Hemisphere. For this reason, radiants at positive declinations are easily observed, whereas those at negative (southern) declinations are dramatically underrepresented. Second, inspection of the right ascensions of observed radiants in Table VII.2 shows that there is a strong correlation of the RAs of low-declination radiants with the time of year. (Radiants with high declinations, especially in the circumpolar region, with dec $>66.5°$, are visible from northern mid-latitudes almost all year long.) The clear, but not especially profound, conclusion is that meteor showers have to occur at night to be observed!

The third generalization is that observers of the predawn sky see more (and more spectacular) meteor displays than observers of the evening sky. The reason for this is sketched out in Fig. VII.9: the predawn observer is on the leading side of Earth in its orbital motion about the Sun and can see meteors of almost any speed and orbit being swept up by Earth. Meteors in retrograde orbit encounter the Earth head-on, at spectacular speeds ranging up to $70\,\mathrm{km\,s^{-1}}$ (for the Leonids). This of course imparts an enormous amount of extra energy to the incoming particles, making them much brighter. Consider Earth's interaction with two low-inclination meteors that are both traveling at $40\,\mathrm{km\,s^{-1}}$ at Earth's orbit, one of which is in a prograde orbit and the other, in a retrograde orbit. Earth's Keplerian velocity is about $30\,\mathrm{km\,s^{-1}}$, so the prograde object will approach Earth from behind at a closing speed of $10\,\mathrm{km\,s^{-1}}$, whereas the retrograde object will approach the dawn side of Earth at a relative speed of $70\,\mathrm{km\,s^{-1}}$. The kinetic energies per gram ($0.5v_{\mathrm{rel}}^2$) are then $5 \times 10^{11}$ and $2.5 \times 10^{13}$ erg, respectively. Adding to each the gravitational energy per gram picked up by falling into Earth's gravitational well ($0.5v_{\mathrm{esc}}^2$), the kinetic energies of the two particles upon striking the upper atmosphere are $1.2 \times 10^{12}$ erg (prograde) and $2.6 \times 10^{13}$ erg (retrograde). Therefore relatively faint displays will be expected for evening meteor showers.

Table VII.2 Right Ascensions of Low-Declination Meteor Showers

| Date | RA (dec < 40) | Date | RA (dec < 40) |
|---|---|---|---|
| 3/10 | 14 32 | 9/7 | 4 4 |
| 4/19 | 18 10 | 10/12 | 2 48 |
| 5/1 | 22 23 | 10/18 | 6 18 |
| 5/11 | 16 28 | 10/31 | 3 32 |
| 5/30 | 22 0 | 11/17 | 10 6 |
| 6/2 | 16 48 | 12/11 | 7 32 |
| 7/26 | 22 36 | | |

Most meteors are weakly consolidated, possibly highly porous chunks of rocky material with masses under 1 g. Milligram meteors are visible under good seeing conditions. Truly bright meteors, with masses of kilograms to many tons, are sometimes observed in conjunction with meteor showers, but not a single meteorite has ever been recovered in connection with a meteor shower. Even the brightest cometary meteors seem to be so fragile that they are utterly consumed by their explosive interaction with the upper atmosphere. Considering that roughly $10^9$ erg will crush a gram of strong rock, $10^{10}$ erg will melt it, and $10^{11}$ erg will vaporize it, the entry of weak rocks with kinetic energy densities of up to about $2.7 \times 10^{13}$ erg per gram should not often permit solid material to survive and land intact on earth as a meteorite.

Ironically, very small and fluffy meteoric particles, such as cometary dust particles, can survive entry intact. Simply because they have so much surface area per gram, they are exceptionally efficient radiators of their entry heat load. They avoid heating to high temperatures and therefore do not evaporate before decelerating to the speed of sound. The particles that survive entry with minimal damage are typically 2 to 50 μm in diameter; smaller particles are extremely hard to identify, and larger particles tend to melt or break up. It is obviously futile to expect to find 10-μm meteorites lying in a field; it would be far easier to find a needle in a haystack! Similarly, airborne dust particles of extraterrestrial origin are vastly outnumbered by native terrestrial dust particles throughout the lower atmosphere. It seems that any attempt to capture these particles intact on an Earth-orbiting spacecraft would be a dismal failure: the instantaneous deceleration of the dust particle from speeds of tens of kilometers per second at the moment of impact with the spacecraft would lead to complete vaporization of the dust particle (and part of the spacecraft as well!). In fact, collectors on the trailing ends of satellites will be overtaken by particles with relative velocities as low as the difference between escape velocity and orbital speed (about $11.2 - 8.0 = 3.2$ km/s), below the speed of sound in many geological materials. Even particles that strike at high enough velocities to vaporize themselves can leave a trace of recondensed material at and near their impact point that is sufficient for chemical analysis. Unfortunately, most of the impact craters seen on recovered spacecraft such as the Long-Duration Exposure Facility (LDEF) and the Space Shuttle orbiters are produced by tiny pieces of man-made space debris in low orbit about the Earth. Craters have even been found in the windows of the Space Shuttles, requiring over 100 STS window replacements to date.

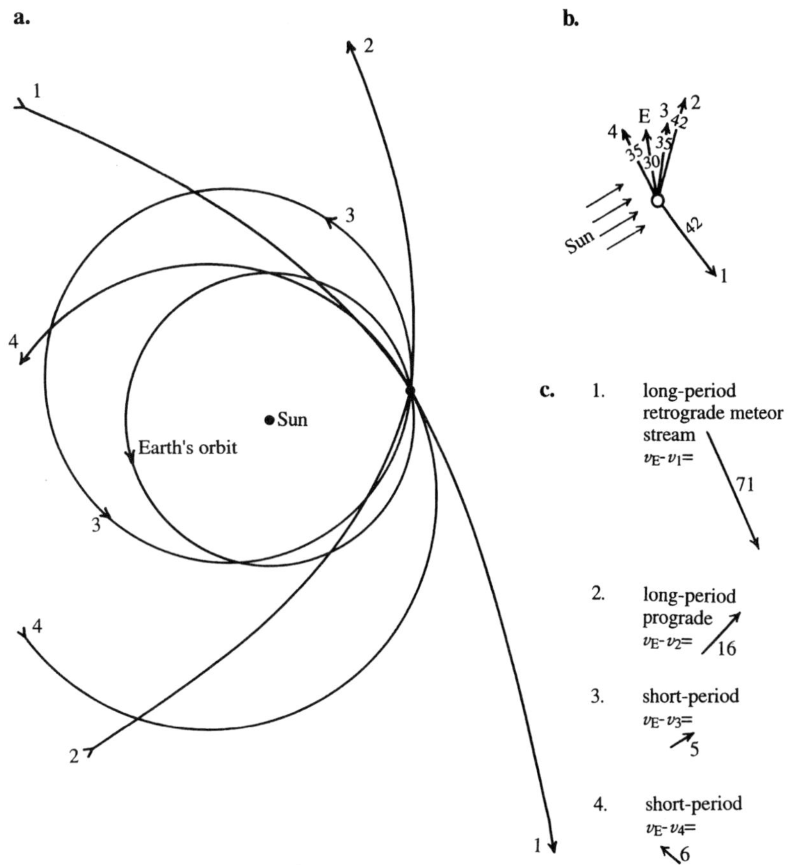


**Figure VII.9** Meteor velocities relative to Earth. a shows the orbit of a retrograde long-period comet (1), a prograde long-period comet (2), and two short-period (and hence prograde) comets (3 and 4) relative to the orbit of Earth (E). b shows the absolute velocities of these bodies at the point where their orbits intersect, and c shows the vector velocity difference between these bodies and Earth before acceleration by Earth's gravity. Short-period bodies mostly overtake Earth from behind at low closing velocities, whereas long-period comets may hit from any direction; those that encounter Earth head-on (on the dawn side of the planet) may carry more than 100 times more kinetic energy per gram than the typical body that overtakes Earth.

Above the convectively mixed, turbulent portion of Earth's atmosphere (the troposphere), in the much less turbulent and almost dust-free stratosphere, the population of terrestrial dust particles is very low. Stratospheric hazes of distinctive composition, especially aerosols of ammonium sulfate or sulfuric acid droplets, could not possibly be confused with silicaceous or carbonaceous extraterrestrial dust. The main terrestrial contaminant in the micrometer size range is spherules of aluminum oxide produced in the burning of the solid-propellant rocket boosters (SRBs) of the Space Shuttle and the Titan 3 and 4, which are readily identifiable and very distinct from nonterrestrial dust.

As a further bonus, dust sediments down through the stratosphere at very low speeds, making it possible to collect the dust grains without destroying them in the attempt. Thus high-flying jet aircraft may collect meteoric dust samples with only modest contamination by terrestrial dust. Many stratospheric flights by U-2 aircraft have successfully collected meteoric dust samples and returned them for analysis. Further, extraterrestrial particles can be identified in deep-sea sediments derived from areas very far from continental sources of dust and river-borne sediments, where the total sedimentation rate is very low and the concentration of extraterrestrial particles is therefore unusually high. These are quenched droplets of melt from the aerodynamic ablation and strong heating of entering objects, with severely depleted abundances of volatile elements such as sodium or sulfur. The identifiable nonterrestrial materials in these sediments have of course also been leached and hydrated by exposure to seawater, and only the largest (0.1 to 3 mm) can be identified readily.

In general, most of the nonterrestrial particles collected in Earth's stratosphere are very fine-grained, black composite particles with very high abundances of volatile elements including C, S, Na, and Zn, similar in major-element abundances to grains from the more volatile-rich varieties of meteorites, moderately heated by the energy of their entry into the upper atmosphere. (We shall discuss such material in more detail in our discussion of meteorites in Chapter VIII.) A large proportion of these fine, unaltered high-altitude dust particles that have been characterized in the laboratory are plausible Solar System materials that are not found in museum collections of macroscopic meteorites. Over 1500 *interplanetary dust particles* (IDPs) have been analyzed in the laboratory, beginning with the pioneering work of Donald R. Brownlee of the University of Washington and P. Fraundorf and Robert M. Walker of Washington University in St. Louis. They find that the nonmeteoritic dust particles are invariably fragile and grossly similar in overall elemental composition to volatile-rich meteorites, but distinct in detail. Many of these particles have a complex, porous structure reminiscent of a raspberry or a bunch of grapes. Both the chemical and the physical evidence show that they have not been strongly heated or melted. Because a large proportion of the dust entering Earth's atmosphere is known to be associated with the orbits of comets, the association of these nonmeteoritic particles with comets is natural.

The principal difference between the suspected cometary particles and laboratory samples of volatile-rich meteorites is that the meteoritic particles are rich in magnetite ($Fe_3O_4$) and calcium and magnesium sulfates, whereas the cometary particles are utterly devoid of these phases. Also, hydroxyl silicates and carbonates, very abundant in certain meteorites, are quite rare in suspected cometary IDPs. Further, the very porous, weak structures of the cometary IDPs are not compatible with the survival of meteorite-size pieces during atmospheric entry. Thus we do not expect macroscopic pieces of IDP material to survive entry as meteorites. The implication of these observations is that the cometary IDPs, unlike volatile-rich (carbonaceous) meteorites, are tiny, weakly aggregated pieces of minerals that did not experience the chemical alteration undergone by the meteorites. Comet nuclei therefore seem to be mixtures of ancient high-temperature and low-temperature condensates that have never been reheated sufficiently to permit extensive chemical reactions between these phases.

Some further information on the physical properties of interplanetary dust particles can be gleaned from studies of the deceleration and breakup of visible meteors. Such meteors are usually a few millimeters to centimeters in diameter, although rare large fireballs of meter size have also been studied. These particles are chemically complex aggregates of micrometer-sized grains; a 1-cm meteor could contain up to $10^9$ 10-$\mu$m grains if it is well compacted. The meteor observations show that there are systematic differences in density and strength from one comet to another. Perseid, $\delta$ Aquarid, Orionid (from P/Halley), Lyrid, and most sporadic meteors have densities in the range from 0.25 to 0.40 g $cm^{-3}$. A few showers, including the $\kappa$ Cygnids, Quadrantids, and $\alpha$ Capricornids, have slightly lower densities of 0.14 to 0.20 g $cm^{-3}$. The Draconids, derived from P/Giacobini–Zinner, have spectacularly low densities of only 0.01 g $cm^{-3}$, suggestive of loosely compacted, magnetically bonded strings of particles, effectively cosmic dust-bunnies. The Geminids, which are debris from the asteroid 3200 Phaethon, have an unusually high density of 1.06 g $cm^{-3}$, not surprising for a rocky source material.

Laboratory studies on the evaporation of comet-like ice–rock mixtures often produce fluffy composite grains similar to meteors. But they also produce a weakly consolidated crust of material with higher density and strength. Meter-sized meteors may be pieces of ice-deficient crust disl- odged by eruptions due to impacts or perihelion passage.

Analysis of data from mass spectrometers carried aboard the Soviet VEGA 1 and VEGA 2 spacecraft during the 1986 flyby of P/Halley show that some of the cometary dust particles that impacted these spacecraft during the high-speed flyby of Halley's nucleus produced vapors rich in Mg and C and probably have compositions close to $MgCO_3$, suggestive of some aqueous alteration in the nucleus. Other impacting particles produced vapors that contain abundant Fe but no Ni, typical of only one known meteoritic mineral, magnetite. Interestingly, some spectral observations of the dust tail of P/Halley show a weak feature at a wavelength of 6.8 $\mu$m, attributable to crystalline carbonates. Thus application of the thermal evolution models used for icy satellites in Chapter VI to comet nuclei would seem fruitful. Unfortunately, we have seen that the critical size for melting of bodies with low boundary temperatures and ice-plus-rock compositions is several hundred kilometers. Barring the unlikely possibility that the nucleus of P/Halley is the evaporation (or collisional fragmentation) remnant of an ancient body of several hundred kilometers diameter, the most appealing route to generate internal temperatures sufficient for melting would be to have the nucleus covered with a meters-thick layer of materials with extremely high porosity and low thermal conductivity for almost all the age of the Solar System. This insulating layer would severely retard conductive or radiative loss of heat from the decay of long-lived radionuclides in the interior of the nucleus, possibly permitting melting to occur. Alternatively, another early short-lived heat source, such as T Tauri phase solar wind inductive heating or decay of the short-lived supernova-produced radionuclide $^{26}$Al, might have been responsible.

## Cometary Fireballs

Fireballs are extremely brilliant meteors with peak magnitudes $< -4$, often comparable in brightness to the full Moon, which has an apparent magnitude of about $-12.5$. They are produced by meteors with masses of at least 100 g and rarely as high as $10^6$ g (1 Mg = 1 tonne). Fireballs exhibit a wide range of behavior during atmospheric entry, reflecting a great variety in their density and mechanical strength. The strongest fireball materials give rise to meteorite falls (see Chapter VIII) and are not associated with cometary orbits. They are presumed to be asteroidal in origin. However, some fireballs are associated with cometary meteor showers and may bear evidence of the conditions on the surfaces of the nuclei of a number of different comets.

The fireballs that are of unambiguously cometary origin are those that are in retrograde heliocentric orbits, have aphelia beyond 5 AU, and are associated with known meteor streams. These fireballs have been studied by two photographic networks, the Prairie Network in the central United States and Canada and the European Network, which began as two stations in Czechoslovakia and now comprises many stations throughout western and central Europe.

The results of such observations include the magnitude, altitude, velocity, and deceleration history of each fireball. Accurate deceleration measurements are extremely difficult for objects that are observed for less than 1 or 2 s, and the data therefore discriminate against meteors of unusually high velocities and low strengths. Despite this bias, it is found that almost all of these "trans-Jovian fireballs" are weaker than any known meteorite. Figure VII.10 shows the behavior of several dozen of these objects, which all disappeared at velocities greater than 8 km s$^{-1}$, suggesting breakup or burnup rather than successful deceleration. These objects therefore would not generally be expected to yield recoverable meteorites. The dynamic pressure at the point of disappearance is approximately equal to the bulk crushing strength of the fireball material. The data in the figure are sorted according to their deceleration behavior into low-density and high-density fireballs. Note that the terminal dynamic pressures for these objects (at disappearance) cover a vast range from $10^4$ to $3 \times 10^7$ dyn cm$^{-2}$ (1 bar is $10^6$ dyn cm$^{-2}$, and the most

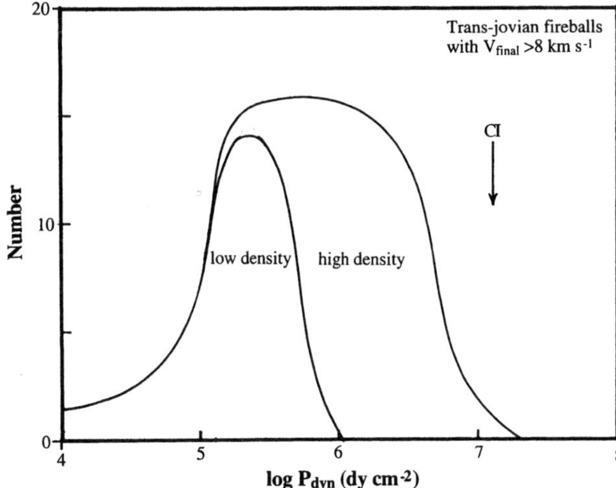


**Figure VII.10** Dynamic pressures for fragmentation of trans-Jovian fireballs. The observed frequency of fragmentation loads for fireballs brighter than $-5$ mag with distant aphelia ($Q > 5$ AU) and terminal velocities greater than 8 km per second are shown. The fireball population is divided according to deceleration behavior into high-density and low-density components. The crushing strength of CI carbonaceous meteorites is shown for comparison.

fragile meteorites that survive to the Earth's surface have crushing strengths of about $10^7$ dyn cm$^{-2}$). The dynamic pressure generated by the meteor's flight through the upper atmosphere is related to its velocity ($V$) by

$$P_{\text{dyn}} = \rho_{\text{a}} C_{\text{D}} V^2 \qquad \text{(VII.32)}$$

where $\rho_{\text{a}}$ is the local atmospheric density and $C_{\text{D}}$ is the drag coefficient. The aerodynamic deceleration of the meteor is then

$$dV/dt = -0.5 C_{\text{D}} V^2 A/m \qquad \text{(VII.33)}$$

where $A$ is the cross-section area of the meteor and $m$ is its mass (a function of time).

In addition, there is an along-track acceleration due to Earth's gravity:

$$dV/dt = (g \sin \Theta)/m \qquad \text{(VII.34)}$$

where $\Theta$ is the entry angle (the angle below the horizontal).

The total rate of dissipation of kinetic energy by the meteor is the differential of its kinetic energy,

$$dE = 0.5 V^2 dm + mV dV. \qquad \text{(VII.35)}$$

The rate of ablation (aerodynamic mass erosion) from the meteor is strongly velocity-dependent:

$$dm = (0.413 A \Gamma \rho_{\text{a}} V^3 / H_{\text{vap}}) dt, \qquad \text{(VII.36)}$$

where $\Gamma$ is the ablation coefficient and $H_{\text{vap}}$ is the heat of vaporization of the meteor material.

Meteors are an optical phenomenon: the luminous flux of the meteor as seen on the surface of the planet directly beneath a fireball at altitude $Z$ is

$$F = -(\tau V^2/8\pi Z^2)dm/dt \qquad \text{(VII.37)}$$

where $\tau$ is the luminous efficiency.

The strongest 25% of the observed trans-Jovian fireballs have strengths greater than 1 bar. Even the low-density objects have typical strengths of $10^5$ dyn cm$^{-2}$, showing that meteors of this size are clearly *not* fairly described as "dust balls," but better resemble loosely consolidated clods of dirt. Meteor experts discriminate fireball observations into three major classes according to the dynamic behavior: Type I fireballs, which are as strong and dense at meteorites, are not found among the trans-Jovian subset. Type II fireballs, which typically have strengths near $10^6$ to $10^7$ dyn cm$^{-2}$, are usually identified with the most fragile types of known meteorites, the CI and CM chondrites, which are discussed in some detail in Section VIII. Only three of the trans-Jovian fireballs fell into this category, and all three decelerated to less than 8 km s$^{-1}$ before disappearance. They may have yielded recoverable meteorite fragments. The weak Type III fireballs are subdivided into Types IIIA (intermediate) and IIIB (very weak), both of which are clearly weaker than any known meteorite. The CI and CM meteorites are known from their cosmic ray exposure histories to survive exposure in space for times of only about $10^6$ years. Materials 100 times weaker would be eroded away by micrometeoroid bombardment in times of about $10^4$ years.

It should not be assumed that the strength of materials is determined by composition alone. Natural materials contain cracks at all size scales. Large pieces of material therefore have more cracks, and lower strength, than would be found in a million fragments of the same material, each with a millionth of the total mass. The simplest and most widely used treatment of this phenomenon is the Weibull strength law,

$$S = S_0(m/m_0)^n, \qquad \text{(VII.38)}$$

where the standard strength measurement is usually made on laboratory samples of about 1 g. The exponent $n$ is about $-1/12$ for most common geological materials. Obviously this is not a major correction for meteors with masses of $10^{-6}$ to $10^6$ g, but becomes very important for comet- or asteroid-sized impactors.

The total mass flux from fireballs is not very accurately known, in part because most of the mass is carried by very rare, massive objects whose arrival rates are poorly constrained by the observations: the longer you wait, the higher the average flux is. For objects with masses less than about 3 tonnes, the total mass flux onto Earth is about $10^9$ g per year. Only a very small fraction, about $10^{-3}$, of the total cometary meteoric mass flux is strong enough to survive the rigors of atmospheric entry.

The trans-Jovian fireballs are the least ambiguous cometary fireballs, but many others with similar properties are found among the populations of prograde meteors and those with aphelia inside 5 AU. Many of these must originate from short-period comets or from extinct comet cores among the near-Earth asteroid population. Chapter VIII will further illustrate how our increasing knowledge of comets and asteroids blurs the traditional boundaries between these groups.

## Cometary Impacts on Jupiter

One of the strangest discoveries in the recent study of the Solar System occurred on March 24, 1993, when Carolyn and Gene Shoemaker and David Levy, who have together discovered numerous comets, found a very peculiar comet-like blur of light on a photographic plate. The image showed what looked like a string of pearls embedded in a bright haze. Their discovery was immediately reported to the International Astronomical Union's Central Bureau for Astronomical Telegrams, located at the Smithsonian Astrophysical Observatory in Cambridge, Massachusetts, where it was named Comet P/Shoemaker–Levy 9 (familiarly, S–L 9) after its discoverers. Alerted by electronic mail messages from Brian Marsden in Cambridge, many astronomers around the world rushed to follow up on this strange body. Images taken by several astronomers at different observatories showed up to 21 "pearls" lying in a nearly straight line, with a highly asymmetric haze of light around them. These fragments were assigned labels in the order in which they progressed along their common orbit, running from A to W (I and O were skipped to avoid confusion with numbers). Some of the bodies disappeared while still in space, and others split into pairs while under surveillance. Spectroscopic searches of the diffuse halo for the emission lines and bands of many common cometary molecules failed: there was apparently no gas present. The bright halo about the discrete bodies must therefore be a cloud of reflecting dust, not glowing gas. The so-called "comet" thus more closely resembled a disrupted asteroid or satellite than a normal comet (Fig. VII.11).

Intensive observations of S–L 9, carried out by many astronomers during the next few weeks, provided enough information to calculate a crude orbit for the "comet" about the Sun. Interestingly, the calculation showed that S–L 9 had an orbital period very similar to Jupiter's. The calculation also revealed that S–L 9 had made an extremely close approach to Jupiter in July

**Figure VII.11** This Spacewatch image captures the bizarre early appearance of Comet Shoemaker–Levy 9. Later images resolved a long "string of pearls" within the densest dust concentrations seen here. This Spacewatch image, taken March 30, 1993, is reproduced here by courtesy of Jim Scotti.

1992, well inside the Roche limit (see discussion in Chapter VI). Once inside that point, in the region where the tidal acceleration of Jupiter at the surface of the body exceeds its own gravitational acceleration, S–L 9 was literally pulled apart by Jupiter's gravity. The dust-embedded "string of pearls" seen several months after passage by Jupiter was then a spray of fragments of the original multikilometer comet or asteroid that broke up at that time, and the dust was fine debris from the breakup process. But by May of 1993, Marsden reported that S–L 9 was actually in a highly eccentric orbit about Jupiter, not the Sun, with an orbital period of about 2 Earth years. It was due to return to perijove in July of 1994. Astonishingly, the calculated perijove distance was 0.7 Jupiter radii from the center of Jupiter: S–L 9 was clearly on a collision course with Jupiter.

Capture of a comet into the Jovian system is not without precedent. P/Oterma was briefly captured by Jupiter in the early 1960s, and a decade later the short-period comet P/Gehrels 3 was temporarily captured by Jupiter, reaching a perijove of only 0.0014 AU ($2.8R_J$), and then lost by Jupiter after about 3 years. Between about 1976 and 1983 P/Helin–Roman–Crodselt executed several orbits about Jupiter, reaching only as close as 0.02 AU. But Shoemaker–Levy 9 was the first known case of capture leading to impact on a planet. Brian Marsden has estimated that S–L 9 may have been in orbit about Jupiter since before the 1920s, but was detected only after an extraordinarily close approach broke it up and generated a bright dust cloud.

Unfortunately, the predicted location of the impact sites of the individual fragments was on the dark side of Jupiter. Although the impact site would not be visible from Earth, instruments aboard both the retreating Voyager and the approaching Galileo Probe could view the crash—but only from a great distance. The Galileo spacecraft, closest to Jupiter, had a clear view of the impact site, which was located over the horizon as seen from Earth.

The energy of the impact events could be predicted only if the masses and velocities of the fragments were known. The orbit determination gave very precise velocities for the individual "pearls," giving an impact velocity of about 60 km s$^{-1}$ for every piece, but the sizes of these bodies proved very difficult to measure. All the bodies were far too small to be resolved in images taken from observatories on Earth or from the Hubble Space Telescope. Their brightness could only be measured approximately because they were all embedded in the extensive, bright cloud of dust. And their diameters could only be estimated crudely from their brightness because their albedos were completely unknown. Estimates of the diameter of the largest pearl ranged from about 0.5 to about 4 km, the latter based on a laborious attempt by Zdenek Sekanina of the Jet Propulsion Laboratory to extract the light contribution of the solid nuclei from the general background of light scattered by the surrounding dense dust cloud. Thus the smallest credible mass (and impact energy) of the largest impactors was nearly 1000 times smaller than the largest credible mass.

Before the impacts it was suggested that the flash of the explosion would be so bright that, if they were to hit on the day side of the planet, observers on Earth would see the flashes with the naked eye. Given the unfavorable geometry of the impact calculated by Marsden, however, the best that could be hoped was to see Jupiter's Galilean satellites brighten briefly as each of the largest fragments crashed. With luck, we might see the fireballs of the largest dark-side impacts rise above the horizon. With an impact speed of 60 km s$^{-1}$, the energy content of the largest fragments would be similar to a long-period comet impacting on Earth. The entire sequence of impact events was predicted to extend over several days.

Guessing that the largest individual fragments were about 3 km diameter, their kinetic energy of impact on Jupiter was estimated at about 10,000,000 megatons. This compares to about 5 megatons for the explosion of Mount St. Helens, or roughly 100 megatons for the 1885 explosion of the Indonesian volcano Krakatoa. But the altitude at which the entering bodies would disrupt in the atmosphere was dependent on their strength, which was very poorly known. The tensile strength suggested by the observed ease of breakup of S–L 9 when it passed inside the Roche limit was close to zero, but it is not obvious that this is the correct strength to use in modeling the comet's entry into Jupiter's atmosphere. Proponents of the most extreme "weak-comet" ideas, based on the observed breakup behavior of cometary "fluff balls," expected complete disintegration at very high altitudes, with negligible effect on the cloud layers and atmospheric circulation. "Hard-rock" proponents argued that the larger pearls, if they were in fact strong asteroidal bodies, might penetrate all the way to depths where the atmospheric pressure is several times as high as on Earth, close to or below the hypothesized water-cloud layer. Such deep explosions could disrupt the cloud layers over a vast area, blast condensable gases such as hydrogen sulfide and water vapor up to altitudes where they could be detected spectroscopically from Earth, and even disturb the large-scale circulation of the atmosphere.

Theoretical estimates of the size of the parent body were made by Jay Melosh and Jim Scotti of the University of Arizona on the basis of the breakup and dispersal behavior of the body at the perijove passage immediately before discovery. They calculated a maximum size for the fragments of about 0.7 km and an original diameter of the parent body of about 1.5 to 2.0 km. The breakup scenario would work only if the body failed at a stress level of $10^3$ dyn cm$^{-2}$ or less, an amount reminiscent of cometary dustballs. Apparently the binding force was self-gravitation, not the tensile strength of the material.

When impact week arrived (July 16 to 22, 1994), the truth was found to lie well in the middle ground between these extremes. Water vapor was readily detected in the fireballs of the largest impacts, but not enough to be certain that the water belonged to Jupiter. Evaporation of even dry asteroidal solids in the hydrogen-rich Jovian atmosphere would produce abundant water by reduction of metal oxides. The Hubble Space Telescope found large amounts of sulfur vapor ($S_2$) in the larger events. Several other sulfur compounds, including COS, CS, and $CS_2$, were seen in quantities that could apparently have been supplied by an impactor with a chondritic or cometary sulfur abundance (about 6% by weight). The only other species seen was CO, which could be made by oxidation of Jovian atmospheric methane by oxygen from the projectile, and methane and ammonia. Theoretical calculations by Kevin Zahnle of NASA Ames Research Center show that, in addition to CO, a very water-rich fireball should also contain detectable traces of $CO_2$ and $SO_2$, and even NO and SO, but none of these were in fact observed. Ammonia is, in the normal course of events, not present in Jupiter's upper atmosphere because it condenses at the altitude of the main cloud layer, and because it is extremely vulnerable to solar UV photolysis, as was discussed in Chapter V. Very hot shock waves at very low ambient pressures are not effective at synthesizing ammonia from nitrogen and hydrogen, and indeed the abundance of nitrogen gas at altitudes above the ammonia cloudtops is quite unknown at present. The hot ammonia gas seen radiating in the post-shock gas has most likely been excavated from below the ammonia cloud base and entrained in the rising plume from the explosion. Somewhat deeper penetration, into the $NH_4SH$ clouds (at about 220 K), would also entrain and lift sulfur gases, and even deeper penetration, to the water-cloud layer (about 300 K), would lift large amounts of water. Since sulfur and water (oxygen) are both also components of the impactor, it is difficult to prove that the projectile penetrated deeply simply from the observed chemical signature of the fireball, especially the absence of massive amounts of water and other oxides. Water in the plume could be supplied from the Jovian water cloud layer or by the impactor, although the latter is far more reasonable. Ann Sprague and her co-workers have measured the water content of the *R* impactor (with a diameter of about 300 m) and find it to be roughly equivalent to a sphere of ice with a diameter of 60 m. This water content could be supplied by carbonaceous asteroid material. Thus the very identity of S–L 9 as a comet can reasonably be questioned.

The impacts were certainly spectacular enough in most respects to satisfy even the most jaded critic. The German–American ROSAT X-ray astronomy satellite saw X-ray bursts corresponding to the impacts, and paired bursts from the point in the opposite hemisphere of Jupiter (same magnetic longitude; opposite magnetic latitude) where the magnetic field lines from the impact site reenter the Jovian upper atmosphere, evidently produced by bursts of energetic electrons ejected by the impact fireball. The impact explosion fireballs from the larger pearls launched immense plumes of hot gases that rose thousands of kilometers above the planet some tens of seconds later, all in plain view of the solid-state imaging camera and photopolarimeter experiment on Galileo. These plumes soon rose into view of telescopes on Earth, then arched thousands of kilometers high above the cloudtops and splashed violently back into the Jovian atmosphere some 10 minutes later. Ultraviolet observations detected atomic silicon, magnesium, and iron vapor in the plume. The near-infrared mapping spectrometer on Galileo observed brilliant infrared emission from the reentering plume of shocked gas. Some of the plume gas, striking at a shallow angle, bounced off the atmosphere in a kind of skip-glide reentry, then entered a second time farther from the blast site.

As the string of impactors fell in, one blast after another could be clearly seen by observers on Earth as the fireballs emerged over the horizon. The dust embedded in the reentering plume formed a huge roughly circular blanket above the bright ammonia clouds. Wave-like rings of dark and light could be followed radiating out from the center of each blast. Several of these dust clouds were larger in area than the surface area of Earth, even though dispersal of these dust clouds on Jupiter occurs hindered by a gravitational acceleration of about $2500\,\mathrm{cm\,s^{-2}}$. The dark dust features remained visible for many months after the impacts, despite the dissipative effects of sedimentation, turbulence, and wind shear.

The latest estimate of the total explosive yield is several times $10^5$ megatons. The largest individual explosions were of order $10^4$ Mt TNT.

How rare an event was the S–L 9 impact? Motivated by the discovery of S–L 9, Shoemaker and others have calculated that cometary breakup events should occur near Jupiter about once per century. If such breakup events are so common, then freshly fragmented comets and asteroids leaving Jupiter's system must occasionally collide with one of the Galilean satellites. Jay Melosh, in a reexamination of the Voyager photographs of the Galilean satellites, has found at least 13 examples of chains of impact craters on Ganymede and Callisto, the two largest and most stable targets. (See Fig. VII.12.) Of course, most such disrupted bodies would not be gravitationally bound to Jupiter, and would remain in correlated heliocentric orbits. Close encounters with Jupiter may therefore be an important source of comet families. There is an extensive family of Sun-grazing comets, at least 29 in number, that appears to have been derived from a single giant precursor by a series of disruption events beginning at least 150 years ago. Breakup of such giant comets may occur mostly as a result of close encounters with Jupiter or the Sun itself.

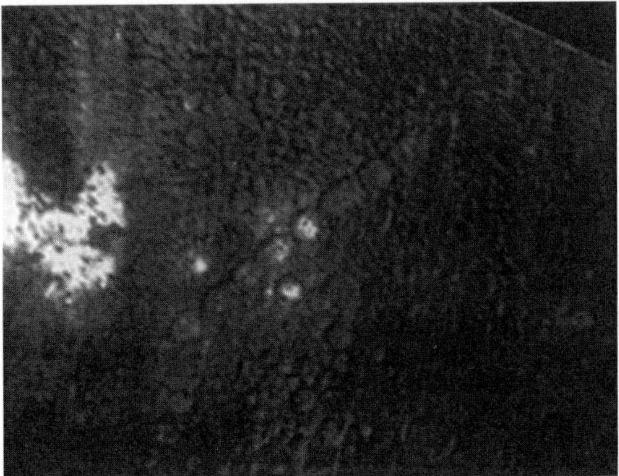

**Figure VII.12** This Voyager image of Callisto shows Gipul Catena, a series of at least 14 precisely aligned craters with similar diameters of about 40 km. This crater chain is one of 13 examples attributed by Jay Melosh to the sequential impact of the fragments of an asteroidal or cometary body that was disrupted by Jupiter's gravity.

## Exercises

***Historical Perspective***

VII.1 Suppose that there are eight nations in medieval Europe, each of which experiences the death of a sovereign every 12 years on average. Each nation engages in one war per century, each of which has a beginning, an end, and several memorable battles. The consort of a reigning sovereign dies with the same frequency as sovereigns themselves. Once every 50 years the sovereign of a given nation marries. Crown princes are born about once every 10 years in each nation. Major natural disasters occur several times per century in each nation. This gives us a total of $64 + 48 + 64 + 16 + 80 + 40 = 312$ significant events per century. About 10 bright comets appear at random times in each century. What is the probability that a bright comet will appear within one month of one of these memorable events sometime during the century?

***Cometary Orbits***

VII.2 Suppose that each of the $2 \times 10^{11}$ stars in our Galaxy loses one comet into interstellar space every year. Further assume that the average age of stars in the Milky Way is $10^{10}$ years. In order to be seen from Earth, an interstellar comet would have to pass within about 3 AU of the Sun to be heated strongly enough to develop a visible coma and tail. Assuming random motions of these comets, how frequently should we see one?

VII.3 Many science fiction stories have featured roving "factory ships" that patrol the Oort cloud, moving from comet to comet and "grazing" on their content of ices. If there are $10^{12}$ long-period comet nuclei within 20,000 AU of the Sun, what is the mean distance of nearest neighbor comets from each other?

VII.4 Suppose that there is one Earth mass of 100-km bodies in the Kuiper Belt, which we shall assume reaches from about 40 to 100 AU from the Sun.

a. Using the data in Exercise VII.3, how many Oort-cloud comets pass through the Kuiper Belt in the average year?

b. How frequently should collisions between long-period comets and Kuiper belt bodies occur?

VII.5 a. How fast would a long-period comet be traveling as it passes the orbit of Jupiter?

b. For a comet with an orbital inclination of 90°, what would its encounter velocity with Jupiter be? (The encounter velocity is the resultant of the orbital velocity vectors of the two bodies, ignoring their mutual gravitational attraction.)

c. For such an encounter, how would the gravitational cross-section of Jupiter compare to its geometrical cross-section?

*Evaporation and Nongravitational Forces*

VII.6 Suppose that an irregularly shaped comet nucleus ($2 \times 1 \times 1$ km), or one that has a very patchy exposure of evaporating ices (see Fig. VII.6), passes through the inner Solar System in a manner similar to the example given in the text after Eq. (VII.19). A significant fraction of the velocity increment provided by evaporation can then act to spin up the nucleus. How fast a spin rate is needed to permit this snowball (with no tensile strength) to break up?

VII.7 Evidence has recently emerged for the presence of a large population of ice-plus-rock (cometary, or Ganymedian) bodies of a wide range of sizes, the Kuiper Belt, orbiting beyond the outermost planets. Suppose that many of these bodies are so far from the Sun that solar heating is negligible compared to their internal source of heat [see Eq. (VI.22)]. Thus the surface temperatures of these bodies may be dominated by the heat released by internal decay of radionuclides. Assume that the present rate of radioactive heat production is the same as for Ganymede ($1.6 \times 10^{-7}$ erg $g^{-1}$ $s^{-1}$).

a. Derive the equation for the dependence of the surface temperatures of these bodies on their mass.

b. Below what mass would the effective temperature of the surface be regulated by the leftover Big Bang radiation field (Fig. II.3)?

*Chemistry and Photochemistry of Water*

VII.8 Atomic hydrogen at a temperature of 8000 K at the base of the solar corona emits Lyman-alpha (Ly-$\alpha$) photons. The width of the Ly-$\alpha$ line is increased by the Doppler effect caused by the random thermal motions of these atoms.

a. Estimate the thermal velocity of the hot H and the half-width of the solar Ly-$\alpha$ line.

b. For a long-period comet with a perihelion at 0.1 AU, calculate its velocity at both 1 AU and at perihelion.

c. How large is the Doppler shift of the solar Ly-$\alpha$ line due to the motion of the comet at these two locations compared to the width of the solar line?

VII.9 The absorption cross-section for ionizing solar radiation is about the same as for photodissociating radiation. Why then is ionization so unimportant close to the comet nucleus?

*Further Chemical Processes in the Coma and Tail*

VII.10 According to Table IV.4, the ionization potential of Ca is only 6.11 eV, less than that of iron (and the other iron-group elements). But the iron-group (transition metal) ions show up in comet tails much farther from the Sun than $Ca^+$ does. Why?

*Behavior of Small Particles*

VII.11 Suppose that a dynamic pressure of 1 dyn $cm^{-2}$ is maintained by gas outflow from a nonrotating evaporating comet nucleus that is 1 km in diameter. What size particles will be blown away from its surface by the gas flow, and what will stay behind?

*Dynamical Behavior of Dust in Space*

VII.12 Suppose that you were to build a *real* radiometer that has four small panels arranged 90° apart about an axis. Each panel is blackened on its "leading" side and is highly reflective on its "trailing" side, and its center lies 3 cm from the axis. The thermal emissivity of both sides of the plate is 1.000. Unlike a commercial "radio meter," air is rigorously excluded from the device by the simple expedient of placing it in a high orbit around Earth without its glass envelope. The device is then exposed to direct, unattenuated sunlight with the axis of the radiometer pointed at right angles to the Sun.

a. Assume that each panel has an area of 1 $cm^2$ and a thermal conductivity of zero. What is the net torque (magnitude and direction) on the device when it is oriented so that the left (black) panel and the right (perfectly reflective) panels both face the Sun?

b. If the thermal conductivity of the panel were infinite rather than zero, how would the torque be changed?

VII.13 What is the cyclotron radius for a 10-$\mu$m dust particle from which photoelectric ejection has removed 10 electrons, traveling at 100 cm $s^{-1}$ and trapped in a uniform magnetic field of 10 G?

*Meteors*

VII.14 a. Devise an experiment in which you and one other observer measure the altitude at which a meteor becomes luminous and the altitude at which it burns out. Be specific about the *minimum* equipment needed, the layout of the equipment, the desired characteristics of the sites, and the best time to carry out the experiment.

b. How would you need to change this experimental design in order to measure the velocity of the entering meteor?

VII.15 Suppose a 10-km-radius comet nucleus has a 10-m surface layer of fluffy dust with a density of $0.01\,g\,cm^{-3}$, a thermal conductivity of $10^2\,erg\,cm^{-1}\,s^{-1}\,K^{-1}$, and a heat capacity of $10^7\,erg\,g^{-1}\,K^{-1}$, overlying a compacted rock-plus-ice interior of Ganymede-like composition.

a. What was the steady-state heat flux through the surface of that body 4.6 Ga ago?

b. What would the temperature difference across the surface dust layer be at steady state? Is this enough to permit internal melting?

*Cometary Fireballs*

VII.16 Why do you think we have never recovered a meteorite from a meteor shower? Make your answer as quantitative as possible.

*Cometary Impacts on Jupiter*

VII.17 The atmosphere of Jupiter near the ammonia cloudtops has a density of about $10^{-4}\,g\,cm^{-3}$ and a scale height of about 8 km. An asteroidal fragment of stainless steel, similar to an iron meteorite, with a crushing strength of $3.6 \times 10^9\,dyn\,cm^{-2}$, is traversing this portion of the atmosphere at a speed of $60\,km\,s^{-1}$. At what altitude (relative to the cloudtops) will this body be crushed by the aerodynamic ram pressure caused by its motion?

# VIII. Meteorites and Asteroids

## Introduction

The recognition of meteorites as genuine extraterrestrial material and the discovery of the asteroid belt both date from the first decade of the 19th century. From then until as recently as the 1960s the only samples of extraterrestrial material available on Earth were meteorites. *Meteoritics*, the laboratory study of meteorites, was regarded as our best window on events at the time of the origin of the planets even before their age could be measured. The connections between meteorites and asteroids have gradually become more clear over time, and it is now recognized that most meteorites originated on or within asteroids and were ejected from their surfaces by violent collisions. Recently, discoveries in both meteoritics and asteroid research have brought forward several interesting connections with comets, and even the Voyager revelation of the nature of ice-rich satellites has influenced our understanding of meteorite origins. At least 18 meteorites have been linked to origins on the Moon, where they sample regions not visited by the Apollo missions. Another 15 meteorites have been traced back to an origin on Mars. They constitute an invaluable record of conditions and processes on Mars, from which no space mission has yet returned any samples. Both lunar and Martian meteorites were ejected into space by comet and asteroid impacts on these planetary bodies. More than ever, meteorite and asteroid studies are an integral part of planetary science, with singular potential for informing us about ancient environments and processes.

## Introduction to Meteorites

*Meteorites* are solid objects of extraterrestrial origin, commonly between a gram and 10 tonnes in mass, that have been found or observed to fall upon the surface of the Earth. They have been subjected to severe selection effects during atmospheric entry on the basis of their mechanical strength. Smaller samples of extraterrestrial material found on Earth, called *micrometeorites* or *cosmic dust*, are discussed in Chapter VII.

Meteorites larger than a few tens of tonnes reach the surface of the Earth at very high speeds because they cannot be effectively decelerated by the atmosphere. They therefore commonly explode violently upon impact and leave a crater and numerous strongly shocked fragments to attest to their arrival. Meteorite-sized objects in near-Earth space that have not yet encountered Earth are called meteoroids. There is no universally accepted way of describing "meteorites" found on other planets: some refer to pieces of the Moon that have been ejected by violent collisions and that have later fallen on Earth

**Table VIII.1** Frequency of Meteorite Finds

| Stones 52.3% | | Stony-irons 5.4% | | Irons 42.1% | | |
|---|---|---|---|---|---|---|
| Chondrites | Achondrites | Mesosiderites | Pallasites | Hexahedrites | Octahedrites | Ataxites |
| 51.3% | 1.0% | 1.6% | 3.8% | 12.0% | 29.0% | 1.1% |

as "lunar meteorites." Others refer to small asteroidal fragments recovered from the lunar regolith as "lunar meteorites." The simplest means of describing these objects without ambiguity would be to call them "meteorites from the Moon" and "meteorites on the Moon," respectively.

Since some 30,000 pieces of meteorites from at least 20,000 separate falls are known, some method of naming, describing, and categorizing them must be adopted for general use. The established method of naming meteorites is to use the name of the town or geographical feature nearest to the point of recovery. Thus the stone meteorite that fell in Wethersfield, Connecticut, in 1803 is called the Wethersfield meteorite. Other names illustrative of this principle include Odessa, Bacubirito, Tucson, Norton County, Revelstoke, Orgueil, Jajh deh Kot Lalu, Black Mountain, and Great Bear Lake. When more then one meteorite is found in a particular locality, a number is usually assigned, either in serial order of discovery, or the year of discovery. Examples are Hale Center No. 1, Akron 1954, and Grady 1933. In a few cases there have been meteorites of very different types recovered from the same locality, or meteorites have fallen near widely separated localities that unfortunately bear the same name. Such cases are handled as they arise by application of common sense. Examples include Gladstone iron, Gladstone stone (not a typographical error!), Edmonton Canada, and Edmonton Kentucky.

By definition, every known terrestrial meteorite has fallen from the sky and been found on the ground. It therefore comes as something of a surprise to the novice to discover that meteorites are divided into two groups called *falls* and *finds*. It would not be unreasonable to infer from this distinction that some of the meteorites found had not fallen, and the rest of the named meteorites had fallen but not been found! Actually, the intended distinction is between those that had been *observed to fall* (falls) and those whose fall was not observed, but were simply *found on the ground* (finds). A breakdown of finds is given in Table VIII.1, and the statistics of total falls plus finds (all known meteorites) are given in Table VIII.2. The high proportion of irons among finds is a result of the distinctive appearance and high density of metallic ("natural stainless steel") meteorites and their good resistance to weathering. Weathered stones, by comparison, rapidly lose their surface texture and disintegrate.

These two tables distinguish several broadly defined classes of meteorites, based on general appearance and density rather than upon careful chemical analysis and petrographic study. The most spectacular and memorable of these from the viewpoint of the museum visitor are the *irons*, which are made of natural metal alloys with densities of about 7.5 $g\,cm^{-3}$ and the toughness and corrosion resistance of stainless steel. The largest iron meteorites on Earth, such as Hoba in Namibia and Bacubirito in Sinaloa, Mexico, still lie where they were first found. They are almost impossible to move and extremely frustrating to cut. The hard, tough metal wears out and gums up saw blades with incredible ease. Hammering on them flattens them out a bit, but does not knock off pieces. These meteorites are dominated by metallic alloys of iron, nickel, and cobalt, with traces of many other elements in solid solution, and small amounts of a wide variety of accessory minerals. Three structural types of irons are recognized.

There are virtually no occurrences of native iron on Earth. If you should stumble across an extremely dense, tough, magnetic rock that shows bare metal when abraded with sandpaper or a hardened steel knife point, it is quite likely that you have found a meteorite. If you have instead found a piece of native terrestrial iron, the sample would probably be even more valuable! The greatest hazard in field identification of iron meteorites is posed by the widespread occurrence of weathered metallic waste of industrial origin. Most of the false meteorites ("meteorwrongs") brought to meteorite experts for authentication are metals or slag from industrial processes.

**Table VIII.2** Frequency of Meteorite Falls plus Finds

| Stones 94.1% | | Stony-irons 0.7% | | Irons 5.2% | | |
|---|---|---|---|---|---|---|
| Chondrites | Achondrites | Mesosiderites | Pallasites | Hexahedrites | Octahedrites | Ataxites |
| 90.7% | 3.4% | 0.4% | 0.3% | 1.6% | 3.4% | 0.2% |

The second general category of meteorites is *stones*. Stony meteorites have 0 to about 30% metal grains and up to several percent of sulfides disseminated in a body that is dominantly composed of silicate minerals. Some of these are dark, crumbly stones full of clay minerals and water-soluble salts; others are very hard igneous rocks (formed by crystallization of a silicate melt) dominated in some cases by a single silicate mineral. The stones are very diverse in composition, appearance, and history. Stones are broadly subdivided into those of igneous origin (called *achondrites*) and assemblages of materials that have never experienced melting (called *chondrites*). Most chondrites contain small glassy or crystalline spheroidal beads, called *chondrules*, with diameters commonly in the range 0.1 mm to a few mm.

The third main category, meteorites with nearly equal amounts of metal and silicates, is the *stony-irons*, which are classed as *mesosiderites* or *pallasites* according to their structure. The pallasites have silicates embedded in a continuous metal matrix, whereas the mesosiderites are made of chunks of metal and silicates.

It is often easy to make a petrological identification of a stony-iron by examination in the field. However, this is not always true of stones. First, there is the problem of their great diversity: only a person with an extensive familiarity with meteorites could expect to identify many of them on sight. Second, some of the classes of stones bear close similarity to terrestrial rock types. This is especially true for meteorites that have been badly weathered. Third, there is the problem of weathering mentioned earlier.

A very fresh fall, regardless of its composition, is usually easy to distinguish from terrestrial rocks: the fall will be completely or almost completely covered by a thin glassy *fusion crust* produced by severe aerodynamic heating during passage through the atmosphere. The fusion crust can usually (especially on stones and stony-irons) be easily chipped off to reveal the thin glassy surface layer coating an unmelted interior. The fusion crust usually shows patterns due to the flow of molten droplets away from the aerodynamic stagnation point, which is the point of highest aerodynamic pressure and (except for small protuberances) the point of most severe heating during entry into the atmosphere.

An easy test to identify and discard most of the forms of terrestrial materials commonly mistaken for metal-bearing meteorites (slag, chunks of manufactured metal, dark or unusual-looking rocks) is to dissolve a small piece of the suspect metal in nitric acid, neutralize the unreacted acid with a slight excess of aqueous ammonia solution, and add to this clear solution a few drops of alcoholic dimethylglyoxime. The appearance of a voluminous red precipitate demonstrates the presence of nickel, a ubiquitous component of meteoritic metal. This is not absolute proof of the meteoritic nature of the metal, since many nickel alloys are made commercially. However, the absence of nickel is proof of the nonmeteoritic nature of the sample and can save the finder further disappointment and wasted effort. Any rock that looks like a metal-bearing meteorite and also contains nickel should be taken to a museum or university geology department for further study. Objects with apparent fusion crusts which do not contain any visible metal should also be brought in for expert assessment if they are distinct in internal appearance from common local rock types. Rocks long exposed to weathering in deserts may develop a black alteration crust, called *desert varnish*, which is not made of glass, but which may present the superficial appearance of a fusion crust. Glass and flow markings are absent from such weathering crusts. Heavily weathered meteorites which no longer have discernible fusion crusts or metal in them are very difficult to recognize. We may take some solace from the fact that virtually all the scientific value of the meteorite has been lost by the time the meteorite becomes unrecognizable.

The study of meteorites as samples of extraterrestrial matter is not an ancient art. The oldest observed fall for which we have both a record of the fall and a surviving meteorite was that of the Ensisheim stone, which fell in Alsace on 16 November 1492. The famous Black Stone in the Kaaba in Mecca, dating to before 600 AD, may also be a meteorite, although eyewitness testimony on its appearance is mixed and removal of samples of the rock for analysis is infeasible for obvious reasons. Upon the capture of Mecca in 630 AD, Mohammed destroyed a pagan shrine built about the Black Stone, and in his later years he reoriented the direction his followers faced in prayer from the Rock of Abraham in Jerusalem to the newly cleansed Kaaba.

Peter Pallas' 1776 study of meteorites from Siberia and Ernst Chladni's early treatise of 1794 brought attention to the possible extraterrestrial origin of meteorites. However, the first widespread scientific acceptance of the fall of meteorites from the heavens arose from the investigation of the L'Aigle meteorite fall of 26 April 1803 by J. B. Biot, who reported his findings to the French Academy of Science. Soon thereafter, the fall of the Weston meteorite in Connecticut on 14 December 1807 was investigated by the distinguished geologist Benjamin Silliman of Yale, who wrote a report on the fall in the journal he founded, the *American Journal of Science*. But acceptance was not universal. Another prominent American naturalist of the period, Thomas Jefferson, is said (according to a delightful, persistent, but undocumented tradition) to have expressed his doubts eloquently: "I would find it easier to believe that two Yankee professors would lie, than that stones should fall from the sky."

Since that time, a number of museums have built distinguished collections of meteorites, and hundreds of scientists throughout the world now devote substantial portions of their time to studies of these and other samples of extraterrestrial material. The reader interested in seeing meteorites or in finding an expert to authenticate a suspected meteorite sample should have little difficulty doing so. At worst, under unfavorable circumstances, it should not be too difficult to find an expert by telephone and mail him a few grams of samples for his inspection.

In the Western Hemisphere, the largest collections of both meteorites and meteoriticists are located at the American Museum of Natural History in New York City, the Smithsonian Institution in Washington, D.C., the Field Museum in Chicago, and the Center for Meteorite Studies at Arizona State University in Tempe. There are about a dozen smaller but good collections ranging from Harvard University's Peabody Museum to the research collection at the University of California at Los Angeles. Reports of meteorite falls are collected and disseminated by the Center for Short-Lived Phenomena at the Smithsonian Astrophysical Observatory in Cambridge, Massachusetts. Certain kinds of research, such as study of the short-lived radioactivity induced in meteorites by their exposure to cosmic rays, can only be done on very fresh samples, and there is consequently some interest in acquiring falls as quickly as possible after their arrival on Earth.

In Europe there are superb collections at the British Museum of Natural History in London, the Musee National d'Histoire Naturelle in Paris, the Naturhistorisches Museum in Vienna, and the Russian (formerly Soviet) Academy of Sciences in Moscow. A worldwide directory of meteorite collections and researchers is available from UNESCO in Paris or New York. Inquiries can also be directed to any mineralogical museum or university geology department. Astronomy departments are likely to be of little help beyond suggesting the names of meteorite experts, since very few astronomers study meteorites.

A number of books have been written about meteorites since Peter Pallas, a German scientist in the employ of the Russian Tsaritsa Katherine the Great, wrote the first treatise on the subject in 1776. Most of these books are either so old or so idiosyncratic that they are of little use to other than historians of science. Brian Mason's excellent introductory book *Meteorites*, published by J. Wiley in 1962, was long the best available general introduction to the subject. It is, of course, dated in its technical and interpretive details, but still repays reading. The more recent *Meteorites: Classification and Properties* by John T. Wasson (published by Springer-Verlag in 1974) is a rich source of raw data and references with special emphasis on irons. V. F. Buchwald's monumental *Handbook of Iron Meteorites* (University of California Press, Berkeley, 1975) is a definitive study of the irons known at that time. The only other general books on meteorites from the last 25 years that we would recommend are Robert T. Dodd's *Meteorites: A Petrologic–Chemical Synthesis* (Cambridge Univ. Press, Cambridge, 1981) and Harry Y. McSween's *Meteorites and their Parent Planets* (Cambridge University Press, New York, 1999). Dodd's book is, as the subtitle suggests, especially strong in dealing with petrological considerations, but omits discussion of the physical properties of meteorites and the phenomena associated with atmospheric entry and impacts, and it touches only lightly on stable-isotope geochemistry. McSween's book is a highly readable and broad survey of modern meteorite lore that provides a compact introduction to virtually every area of current interest in meteorite research. A much more technical and much longer multiauthor book from an earlier perspective is *Meteorites and the Early Solar System*, edited by John F. Kerridge and Mildred S. Matthews (University of Arizona Press, Tucson, 1988). Finally, the current standard catalogue of meteorites is M. M. Grady's *Catalog of Meteorites* (Cambridge University Press, New York, 2000). Her compilation includes the circumstances of fall or discovery, a physical description of the meteorite, present locations of samples in museum collections, references to critical literature, and a statistical overview of meteorite types and places of fall.

There are two principal journals devoted to the subject, the Russian-language *Meteoritika* and the English-language *Meteoritics*. Several other journals publish large amounts of related material, notably *Geochimica et Cosmochimica Acta* (not, despite its name, actually written in Latin, and usually called simply GCA) and *Earth and Planetary Science Letters* (familiarly EPSL). The most useful abstracting services are *Mineralogical Abstracts* and *Chemical Abstracts*. Many excellent but more narrowly focused treatises and reviews are also available.

## Meteorite Orbits

Accurate orbit determinations on meteorites are potentially of great value in establishing where the members of particular meteorite classes come from. Unfortunately, the difficulties surrounding this effort are immense. The principal difficulty is that meteorite falls are not very frequent, and the entire process of entry into the atmosphere and deceleration to subsonic speeds is extremely fast. The entering body cannot be observed until it and its trail are glowing from the frictional dissipation of the body's kinetic energy in the upper atmosphere. Entry speeds of $30\,\mathrm{km\,s^{-1}}$ are not unusual, and the entire useful portion of the interaction of the meteorite with the atmosphere before it has been too severely decelerated may last no more than a few seconds. If the body can be photographed during entry by two or more cameras with synchronized rotating shutters to "chop" the light trail at precisely determined time intervals, then the velocity of the meteorite at the point of first contact with the atmosphere can be determined. It is then not hard to calculate the orbit that the body was following before encounter with the Earth.

The first recovered meteorite to have been photographically tracked during entry into the atmosphere was the Příbram stone from Czechoslovakia. The semimajor axis of its orbit around the Sun was 2.42 AU. The orbital eccentricity was 0.674, and the point of closest approach to the Sun, $q = a(1 - e)$, was 0.79 AU. Thus perihelion lay not far outside the orbit of Venus. The aphelion distance, $Q = a(1 + e)$, was 4.05 AU, very close to the Hilda family of asteroids beyond the outer edge of the main Asteroid Belt. Its orbital inclination, 10.4°, lay well within the range typical of asteroids and short-period comets. It is obvious that the Příbram meteorite may have originated anywhere within the vast range of heliocentric distances spanned by its orbit. It is less obvious, but equally true, that the orbit being pursued by Příbram at the time of its encounter with Earth could not have been occupied for very long: the orbit crossed the orbits of both Mars and Earth, and it may also have interacted strongly with Venus in the not too distant past. It is not possible to narrow down the ultimate point of origin of Příbram in any useful way from its observed orbit at encounter with Earth.

Soon after the fortuitous observations of the entry of the Příbram stone in 1959, and partly inspired by it, both the United States and Czechoslovakia established photographic networks for the purpose of tracking entering meteorites. The All-Sky network in Czechoslovakia did not produce any more recovered meteorites, although it did track one small iron meteorite that must have survived entry and landed in a forested area in the southern part of the German Democratic Republic. That meteorite was identified as an iron by the emission spectrum of its fireball, which was rich in atomic emission lines of iron and nickel. In recent years the Czech tracking network has served as the nucleus for the European Network, with numerous stations in Central and Western Europe.

A similar but even more disappointing fate awaited the Prairie Network in the central United States. Hopes were raised by an early success, the tracking and subsequent discovery of the Lost City stone in 1970. Several hundred bright fireballs were tracked, of which about a dozen should have resulted in the fall of meteorites with masses in excess of a kilogram. Unfortunately, not one of the other fireballs yielded a recovered meteorite. Lost City was found to have a semimajor axis of 1.66 AU, an eccentricity of 0.417 (perihelion of 0.97 AU; aphelion of 2.35 AU), and an inclination of 12.0°. The Prairie Network has since been shut down. A Canadian network began operation, and added a third recovered meteorite, Innisfree, in 1977 (Fig. VIII.1). Two other meteorite falls have since been observed with sufficient precision to permit a determination of their preatmospheric trajectories. Of these, Dhajala has an aphelion in the heart of the Belt and a perihelion between the orbits of Venus and Earth, and Farmington has an aphelion near the inner edge of the Belt and a perihelion inside the orbit of Venus.

Despite the rather different orbital parameters of Lost City, Příbram, Dhajala, Farmington, and Innisfree, these meteorites have certain features in common: their orbits are subject to strong perturbations by Jupiter, such that the eccentricities of their orbits varied rapidly and widely just prior to encounter with Earth. Further, all five had modest orbital inclinations and aphelia located within the asteroid belt.

Photographic tracking is not the only method of estimating the preencounter orbits of meteorites, but it

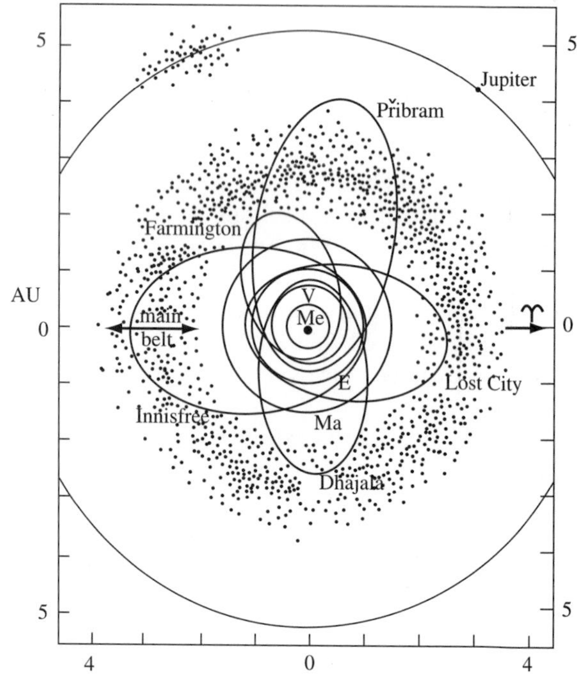


**Figure VIII.1** Orbits of recovered meteorites. The only five recovered meteorites whose entry into Earth's atmosphere was observed with good precision are ordinary chondrites. They entered from very different heliocentric orbits, all of which passed through the Asteroid Belt, and all of which are typical of the orbits of Near-Earth Asteroids.

is surely the most accurate. More than a dozen observed meteorite falls have yielded sufficiently numerous and reliable eyewitness descriptions of the fireball trajectory to place useful constraints on the preatmospheric orbit, but it is difficult with this kind of data to do more than establish the direction of the velocity vector at entry: in the absence of precise timing data, the magnitude of the entry velocity is difficult to constrain. This means that one cannot as a rule calculate real heliocentric orbits for these bodies, but one can, for example, screen the entry trajectories for compatibility with origin from known meteor streams. This has been done, and no such relationship between recovered meteorites and meteor streams has been found.

Other even more limited information on the preatmospheric trajectories of meteorites can be deduced from the statistical data on the dates and times of fall of meteorites. These data suggest that the large majority of the observed meteorite falls, like the fireballs observed by the Prairie Network, come from prograde orbits of modest inclination and eccentricity, similar to the orbits pursued by Earth-crossing asteroids.

## Phenomena of Fall

The most frequently observed phenomenon associated with the entry of meteorites is the formation of a brilliantly luminous trail of shock-heated atmospheric gases and meteorite vapor. The vapor trail is usually white or bluish, but yellow or reddish trails are not uncommon. On rare occasions, green fireballs are reported, possibly due to copper in the body. Only a very small proportion of observed bright fireballs (bolides) is associated with the recovery of meteorites: only about 30% of the fireballs observed by the Prairie Network are made of strong, rocky material comparable to most stony meteorites (type I fireballs), and fully 60% are like the most fragile meteorites (type II).

Occasionally observers report faint hissing, humming, buzzing, or sizzling sounds during the visible display. These sounds may be "electrophonic" effects due to electromagnetic interactions of a poorly understood nature, involving direct excitation of the tympanum by audiofrequency electromagnetic waves generated by turbulence in the plasma wake left by an entering meteorite. More commonly, sound effects are reported to occur 2 or 3 minutes after the end of the visual display. These sounds. which are described by observers as resembling distant gunfire or prolonged rolling thunder, are surely shock waves propagating at the ambient speed of sound.

A typical meteorite may enter at, say, a 45° angle to the horizontal plane. It first begins to encounter significant traces of atmosphere, and may begin to glow visibly, at an altitude of about 150 km. The minimum speed at the top of the atmosphere is set by Earth's escape velocity, $v_e = 11.2\,\mathrm{km\,s^{-1}}$, while the maximum possible speed would correspond to head-on impact with a body in a parabolic orbit with an inclination of 180° (a retrograde orbit in the plane of the ecliptic with an eccentricity of 1.000). The speed of the meteorite relative to Earth just prior to encounter is the sum of Earth's orbital velocity $v_o$ ($30\,\mathrm{km\,s^{-1}}$) and the escape velocity from Earth's orbit, $1.414v_o$. The additional energy added by the gravitational acceleration of the meteorite by Earth up to the moment of impact with the atmosphere is $0.5v_e^2$ per gram. Thus the total energy per gram of material has a maximum value of

$$v_{max}{}^2 = (2.414v_o)^2 + v_e^2, \qquad \text{(VIII.1)}$$

where a factor of $\frac{1}{2}$ has been removed from each term. Solving for $v_{max}$, we find that the highest possible Earth entry speed for a body in a bound heliocentric orbit is $73.28\,\mathrm{km\,s^{-1}}$. In practice, only a small proportion of the observed fireballs have speeds over $40\,\mathrm{km\,s^{-1}}$.

A body entering at the maximum speed carries a kinetic energy density of $2.6 \times 10^{13}\,\mathrm{erg\,g^{-1}}$. By comparison, the energy required to heat a gram of typical rock from room temperature to its melting temperature is about $1.2 \times 10^{10}$ erg. The energy required to crush the same material to a fine powder is less than a tenth of the amount needed for melting. Clearly, bodies entering at the highest speeds have little prospect of survival, irrespective of their strength.

We will take $30\,\mathrm{km\,s^{-1}}$ as a typical "fast" entry speed for survivable meteorites. Taking our example of entry at 45° to the horizontal, the slant range from the point of first heating and deceleration to the projected impact point is about 150 km. An average deceleration of 150 Earth gravities would be required to stop the meteorite before it collides with the ground. Of course, the density of the atmosphere is increasing very rapidly (almost exponentially) with the depth of penetration of the meteorite, so that the peak deceleration occurs rather late in the trajectory and is far greater than this average. Crushing is obviously a serious problem, and of course aerodynamic heating must be extremely severe. The thermal conductivity of rock is low, and the time of exposure of the meteorite to heating is necessarily very short. Conductive heating of the interior of the meteorite is therefore usually limited to a millimeter or so in depth in stones, or about a centimeter in irons. The atmosphere strikes the meteorite like a blowtorch with a temperature of over 20,000 degrees. Extremely thin layers of rock are heated very rapidly to the melting temperature, and tiny droplets of liquid rock mixed with shock-heated atmosphere and meteorite vapors are left as an incandescent trail behind the entering body, later to fall as part of the "cosmic dust" flux. This process of aerodynamic erosion of melt droplets is called *ablation*. Predicting the fate of an entering meteorite requires knowledge of several of the physical properties of the body, including crushing strength, melting point, density, and thermal conductivity.

About 30% of all fireballs survive this process until they penetrate to depths where the dynamic pressure reaches a maximum. The peak dynamic pressure exceeds the crushing strength of the fastest-moving stony bodies, and they quite suddenly disintegrate into a number of tiny fragments which promptly are vaporized. The slower rocky bodies may survive the peak deceleration forces and fall to the ground at subsonic speeds as meteorites. Large, strong, and slow bodies will survive peak deceleration, but may strike the ground at supersonic speeds. At impact speeds larger than the speed of sound in rock, the impactor explodes, showers fragments about, and excavates an impact crater.

On rare occasions the entering body will strike the atmosphere at a shallow angle, within a few degrees of horizontal, and pass through the atmosphere for very long distances. Some have even been observed to "skip out" and exit from the atmosphere at a speed in excess of escape velocity. Perhaps the rarest outcome is to exit from such an "aerobraking" trajectory with a speed above circular orbital velocity but less than escape velocity. Such a body would briefly orbit the Earth, but would almost certainly be destroyed by aerodynamic heating or fall to Earth on its next perigee passage.

In the course of ablation, a very fine dust of glass and oxide grains and droplets is left behind in the upper atmosphere as a dust trail. This trail is often visible for minutes to hours after very large daytime fireballs, and a glowing trail of gases may similarly be seen for minutes after very bright nighttime fireballs. The author has had the privilege of witnessing two fireballs that were brighter than a full moon, nearly $-15$ visual magnitude, but has not seen a persistent trail.

It is common for stony meteorites to contain fractures, and thus it is not at all unusual for a stony body to fragment partially at aerodynamic pressures far below the crushing strength of rock. As a result, stone meteorites often occur as "showers" of dozens, hundreds, or even thousands of individual stones. The Pultusk fall in Poland may have comprised more than 100,000 individual small stones. Such fragmentation showers may cover an area of a few square kilometers up to 10,000 square kilometers. This phenomenon is unrelated to meteor showers, from which meteorites are not recovered, and is entirely distinct from impact explosion events associated with craters.

Because most stone meteorites are aerodynamically decelerated to below the speed of sound at altitudes in excess of 15 km, and also because the time scale for strong heating and subsequent cooling of the thin hot fusion crust is so short, meteorites are generally cold when they land. Meteorites that land in summer sometimes form thin coats of frost on their fusion crusts! Of course, meteorites that reach the ground still traveling at supersonic speeds may have warm surfaces, but it is very difficult to take their temperature before they impact! Meteorites of unusually high thermal conductivity, especially irons, may be quite hot, but reports of incandescent meteorites landing on the ground are rare and poorly documented and are probably exaggerations.

Numerous large impact craters have been identified on Earth. Indeed, several crater-forming impacts have occurred in historical times. The famous Meteor Crater in Arizona, 1.3 km in diameter, is the most accessible and best-preserved large crater. Satellite photography of Earth has revealed nearly 200 very large and very ancient craters. These craters are especially abundant on the most stable and oldest portions of continents (*cratons*), attesting to the severity of the bombardment that Earth has undergone. The rarity of cratering effects that are readily visible to observers on Earth's surface is due to the powerful erosive effects of wind and water: the most recent estimates of the cratering rate on Earth are several times higher than that on the Moon, which however retains easily visible evidence of its bombardment history because the only effective erosive process is bombardment itself! As we shall see in Chapter X, these rare but extremely violent cratering events have played a major role in the geological and biological history of our planet.

Occasionally, large objects disrupt completely in the atmosphere, liberating an explosive power equivalent to millions of tons of high explosive. A major event of this type occurred in Siberia on 30 June 1908. This Tunguska event was an air burst that did not produce a crater or leave any meteorites, but did knock down and ignite 2000 square kilometers of forest. One cannot help but wonder what might have happened if the Tunguska object had arrived somewhere over Central Europe in the 1960s, liberating its enormous explosive power indiscriminately in the midst of the standoff between NATO and the Warsaw Pact, without any clear evidence of its nature and origin.

It now seems probable that the Tunguska body was stony, not cometary in origin. The conditions under which it broke up seem much too severe for icy cometary solids, but appropriate for meteoritic rock.

Often, small bodies will be decelerated to subsonic speed before reaching the ground. About the largest body that could possibly reach the ground in one piece, without exploding upon impact, would be a monolithic chunk of iron, preferably extremely flattened, traveling at about the lowest possible entry speed, and entering the atmosphere at a very shallow grazing angle. The limiting size can be calculated to lie in the range of 30 to 100 tonnes. (One metric tonne is 1000 kg, or $10^6$ g.) The upper limit is more easily attainable for flattened, distinctly nonspherical bodies. In view of this theoretical limit, it is interesting to note that the largest meteorite known on Earth, the Hoba iron from Namibia (Southwest Africa), weighs roughly 60 tonnes and may have weighed about 100 tonnes at the time it landed. The kinetic energy of a 100 tonne object traveling at 70 km $s^{-1}$ is close to 60,000 American tons (0.06 Mt) of TNT.

Deceleration is important for entering objects if their mass per unit area (essentially $\rho d^3/d^2$, or $\rho d$) is less than that of the atmosphere above Earth's surface ($\rho_s H$). For a rock with a density of 3 g $cm^{-3}$ this criterion corresponds to a diameter of about 3 m (a mass of about 40 tonnes). Fragmentation is most important for weak, fast-moving objects, but fragmentation is unimportant for very large impactors, irrespective of size, because the fragments of large bodies do not have time to get away from each other before impact with the surface. A useful criterion for neglecting fragmentation is to do so if $\rho d^2 > \rho_s H^2$. For typical stony material, this gives a critical size of 300 m diameter (40 million tonnes). Traveling at a modest 20 km $s^{-1}$, this is $8 \times 10^{25}$ erg of kinetic energy. Equating the chemical potential energy of 1 Mt of high explosive to $4.185 \times 10^{22}$ erg, this energy content translates into a 2000 Mt (2 Gt) explosion. For comparison, the largest man-made explosion in history, a Soviet hydrogen bomb test on the Arctic island of Novaya Zemlya in 1960, had a yield of 57 Mt.

Large impactors are, fortunately, rare. For meter-sized bodies (1 to 4 Mg), the flux onto Earth is only a few bodies per year. Such a body will dissipate its entire kinetic energy of $4 \times 10^{18}$ erg (0.1 kt TNT) in about $\tau = H/v = 3$ seconds, so that an observer at a typical distance of $D = 100$ km would see a peak energy flux of $4 \times 10^{18}/4\pi D^2 \tau = 1000$ erg $cm^{-2}$ $s^{-1}$. This is about $-18.7$ mag if all the energy appears as visible light. The average annual mass flux onto Earth is of order 1000 Mg, but the rate of fall of meteorites is only one meteorite per $10^6$ $km^{-2}$ per year, or 500 per year landing on Earth (only a small percentage are recovered because most land in the oceans or in remote, uninhabited areas). Allowing 20 kg per average meteorite, that comes to only 10 Mg per year. The rest of the entering material is crushed or vaporized and is seen only as extraterrestrial dust.

The Spacewatch asteroid search program at the University of Arizona has recently become able to find very small, nearby asteroids which were previously too small for astronomical detection. From the fluxes of 10-meter to 1000-meter objects they observe, they estimate the size of the largest annual impact event on Earth to be 10 to 100 kt (200 to 2000 Mg per event). Since most of the mass in a natural mass distribution is in the two or three largest objects, the total mass flux predicted on this basis would be 600 to 6000 Mg per year, which agrees reasonably well with the cosmic dust estimate of roughly 1000 Mg per year. Using 300 Mg as the mean largest annual event, this means that there is an aerial explosion of about the size of the Hiroshima nuclear blast somewhere on Earth every year. About once per century such an air blast should occur over a densely populated area. The largest impactor expected in the average century, according to the Spacewatch data, has about 30 Mt explosive yield. The Tunguska event in Siberia in 1908 was about 15 Mt. Since about 75% of all impacts occur in the oceans, Arctic, or Antarctic, such a large explosion is not an improbable fluke.

Interestingly, a crater chain with meteorite fragments in it was recently discovered in Argentina. The three largest craters are between 3 and 4 km long, and very elongated. The impact appears to have been nearly horizontal. The age is only crudely known at present (1000 to 10,000 years), and the explosive yield required to excavate the craters is about 350 Mt. The Spacewatch statistics say that such an event should occur somewhere on Earth about every 2000 years, and about one such event on land every 8000 years. A yield of 350 Mt (at 20 km $s^{-1}$) suggests an entry mass of $7 \times 10^{12}$ g and a diameter of 200 meters for the impactor, which is far too large for deceleration and marginal for successful aerial disruption.

Many well-documented meteorite falls have caused damage to buildings, and several have fallen within a few meters of people. Over many centuries there have been numerous reports of injuries and fatalities to people and animals, but there are only two recent examples of meteorites striking people that can be vouched for by both medical doctors and meteoriticists. A woman in Sylacauga, Alabama, was struck by a stone that pierced the roof of her house in 1954 and hit her as she was reading in bed, and in 1992 a boy in Uganda was struck by a small stone that passed through a tree above his head. American experts tend to dismiss eyewitness reports that lack authentication by medical authorities and licensed meteoriticists, especially if they were published in a foreign language or if no surviving sample of the offending meteorite is available for study. A compilation of over 150 actual reports of deaths, injuries, property damage, and near misses, drawn from English, Latin, Greek, Italian, French, German, Spanish, Russian, and Chinese sources, appears in my book *Comet and Asteroid Impact Hazards on a Populated Earth* (Academic Press, San Diego, 2000).

If you should be fortunate enough to witness a meteorite fall, you will not know it at once. All the visual phenomena will be over before any meteorite reaches the ground. It is therefore best to record with care any and all fireballs seen. Only a tiny proportion of them may yield meteorites (those that disappear with a brilliant terminal flare are very poor prospects), but the value of a good observation is large enough to warrant the effort. You should note the exact date, time, and location. You should, whenever possible, count out the duration of the fireball. Guesses made after the fact by different observers may be wildly divergent and quite useless. Note the maximum brightness as compared to the brighter stars, planets, or the Moon. Note whether the fireball exhibits bursts of light, and whether more than one fragment can be seen. Describe the length, color, and persistence of any trail that may form. Whenever possible, note the exact points of appearance and disappearance among the stars. In the daytime, note the path relative to any fixed points of reference, such as trees, buildings, and the Sun, and mark precisely where you were standing when you saw the fireball. Time the interval between the visual fireworks and the first arrival of sound, and sketch out how the intensity of the sound varied with time. Note the total duration of the sound. If you should then see the meteorite fall, you should note its exact point of landing. If possible, you should pick it up at once with, and seal it into, a polyethylene self-sealing bag to protect it against further contamination. You can also feel through the bag whether the meteorite is warm or cold without introducing organic contaminants from your hands. If you are lucky enough to find a meteorite too heavy to treat in this manner, you should seek expert assistance at once. The scientific value of a meteorite depends critically upon how fresh and uncontaminated it is when brought into the laboratory, and whether the circumstances of fall were carefully observed and recorded.

## Physical Properties of Meteorites

There is good reason to believe that the frequency of meteorite falls diminishes smoothly with increasing mass. Irons and stones have very different crushing strengths, which biases our meteorite collections in the direction of overrepresenting the stronger irons, and favoring the survival of iron meteorites that are substantially larger than the largest stones. The weakest stony material will also be strongly discriminated against by both breakup during entry and weathering once on the ground. A second factor biasing our collections is the very distinctive density and high corrosion resistance of meteoritic iron. After Earth surface exposure times sufficient to destroy stones completely, or to at least render them unrecognizable as meteorites, irons retain their identity and remain recognizable. Thus irons and stones both ought to increase dramatically in abundance toward smaller sizes, but with irons proportionately overabundant, especially at the largest sizes. This trend is clearly evident in the mass statistics on meteorites in our collections.

The second major factor in determining the mass distribution of recovered meteorites is the difficulty of finding and identifying very small meteorites. At the site of a meteorite shower, with knowledge that many larger stones have fallen in a limited area, it is sometimes possible to find individual stones as small as a milligram, especially when the fall occurs upon clean snow. It should be obvious that a milligram meteorite falling in isolation will not be found.

The ability of irons to survive weathering is documented in several ways. First, many large irons have been examined for their content of cosmic-ray-produced radionuclides. These irons have generally been on Earth so long that these radionuclides have decayed away, leaving only nonradioactive cosmic ray products and the decay products of the radionuclides as evidence of the former cosmic ray exposure of the meteorite while in space. Second, some very large irons with long terrestrial residence times, including the Bacubirito iron in Sinaloa, Mexico, and the Hoba nickel-rich iron in Namibia (southwest Africa), are resting on the surface, exposed by erosion, with no evidence of a crater. Some, such as Hoba, are surrounded by as much as 50 cm of iron-rich oxide and clay weathering products, reflecting thousands or even tens of thousands of years of exposure to the elements. Finally, there is the well-documented rarity of irons among observed falls (Table VIII.2), in contrast with the excellent representation of irons among finds (and, by extension, in museum collections). This is easily understood as a consequence of the relative ease of identification of weathered irons and their superior resistance to weathering, as remarked above. Even irons that have been severely shocked can survive intact, as the picture of one of the Gibeon irons from Namibia (Fig. VIII.2) shows.

The densities of stony meteorites range from a low of about 2.3 for the most highly oxidized, volatile-rich objects to a high of about 4.0 for the most metal-rich stones. Densities of stones reflect not only the overall mineralogy, but also the porosity of the material: the more volatile-rich meteorites are also among the most porous. Shock damage also increases porosity. Stony irons generally have densities of 5 to 6, and irons always have densities close to 7.5, ranging up to a maximum of about 7.9. These densities, which closely reflect the total iron content of the meteorite and the oxidation state of the iron, are relatively insensitive to the silicate mineralogy.

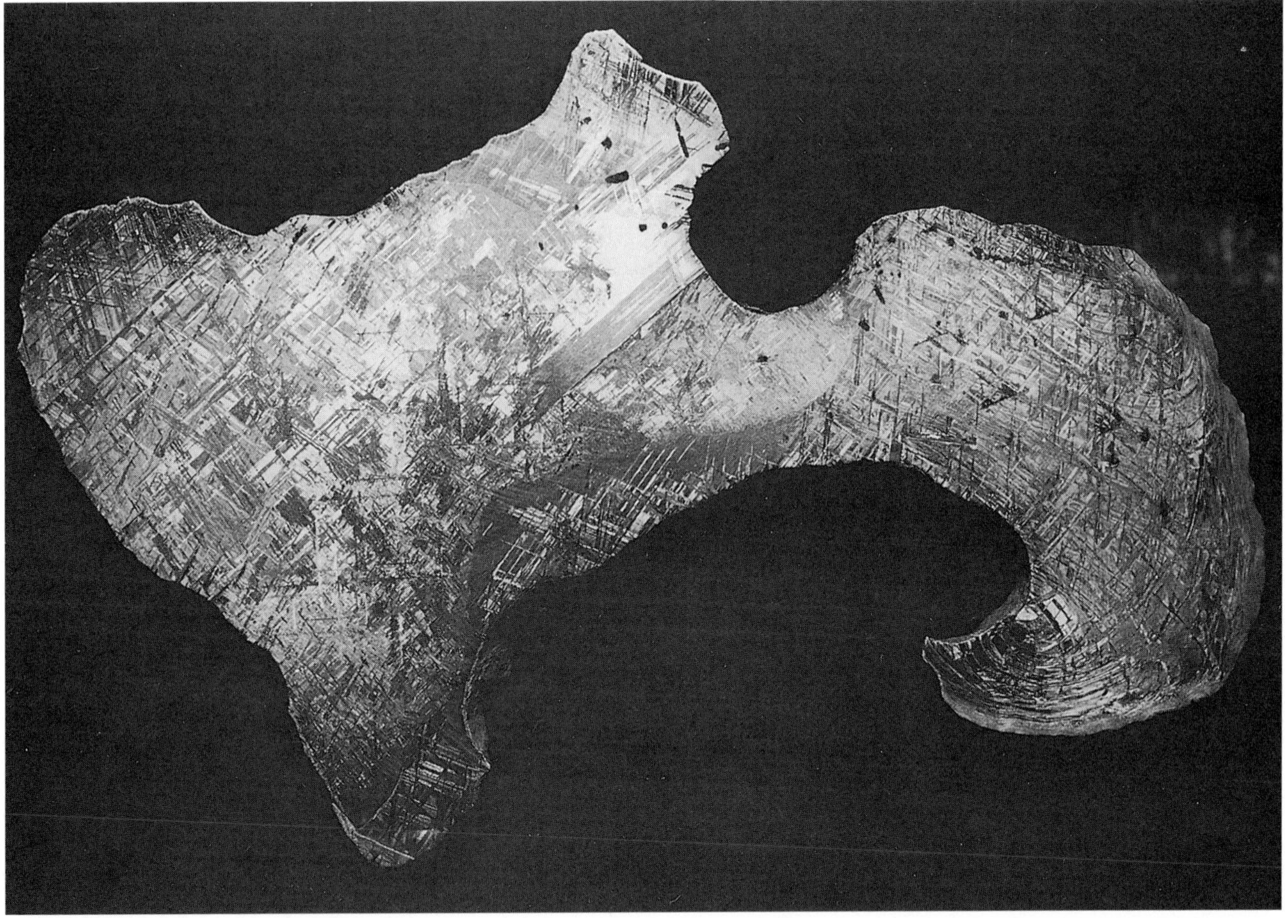

**Figure VIII.2** A cut, polished, and etched section through one of the Gibeon irons. The obvious Widmanstätten pattern is typical of an octahedrite. Severe distortion of the kamacite/taenite intergrowth pattern was caused by a violent impact event. Photo courtesy of the U.S. National Museum.

The sizes and shapes of meteorites provide us little new insight into their intrinsic properties, but provide several useful types of evidence about the physics of meteorite interaction with the atmosphere and the surface of the Earth at the time of fall. The very largest known meteorites are, of course, the most massive irons. These have dimensions ranging up to 4 meters in length. Many of the large irons are very irregular in shape. A number of the largest irons (Willamette, Goose Lake, etc.) have very deep cavities and "wormholes" that riddle their interiors. The most striking and extreme example of these cavities is the famous Tucson iron ring, a meteorite about 140 cm in diameter with a 70-cm hole through it! There are several conceivable explanations for these large cavities: they may be "primordial" features of the breakup of the iron meteorite parent body, they may be formed by ablation during entry into the Earth's atmosphere, or they could be made by corrosion during residence on the Earth's surface. The most likely explanation seems to be that the large cavities were filled with silicates or sulfides at the time of formation of the iron meteorite, and that any process that can remove these materials, including crushing, weathering, etc., can provide the observed textures.

The mechanical strengths of meteorites vary strikingly with composition. The weakest meteorites are the clayey and volatile-rich CI carbonaceous chondrites, which can literally be crushed between the fingers. A reasonable estimate of the range of their crushing strengths is from about 0.1 to 10 bar. The ordinary chondrites have strengths that vary very widely with composition, shock history, oxidation state, volatile content, and degree of terrestrial weathering. Only limited data are available, but strengths ranging from 62 bar to 3.7 kbar have been documented. Curiously the only unweathered chondrite tested, the Holbrook stone, was found to be the weakest of the 10 chondrites so far examined. The achondritic stones have not been tested, but they are hard igneous rocks with bulk compressive strengths of 2.5 kbar to perhaps 4 kbar. Several irons show strengths of 3.2 to 4.4 kbar, but we should keep in mind the possibility, suggested by Aviva Brecher, that hydrogen embrittlement may affect some irons. She suggests that such irons might be very brittle at low temperatures, near 100 K. The tensile strength of one of the Gibeon irons has been measured at 3.2 kbar. No other strength data exist for recovered meteorite samples.

Interestingly, we also have some useful constraints on the strength of pristine meteoritic material derived from its fragmentation behavior during entry into the atmosphere. Both radar and photographic tracking of fireballs provide data on their brightness and altitude during atmospheric entry. Dozens of very bright fireballs have been observed among the vastly larger number of faint events in meteor showers. These meteor-shower fireballs invariably fragment at very high altitudes, where dynamic pressures due to impact with the Earth's upper atmosphere are very small. The dynamic pressure on a spheroidal body traveling through the atmosphere at hypersonic speeds (as are all entering objects) is given by

$$P = 0.6\rho_a v^2 \qquad \text{(VIII.2)}$$

where $\rho_a$ is the density of the atmosphere and $v$ is the speed of the projectile. Thus a body observed entering from a short-period comet orbit with a speed of 30 km/s$^{-1}$ at the top of the atmosphere, and which is observed to break up at the level where the density is $10^{-7}$ g cm$^{-3}$, must have a crushing strength of $6 \times 10^5$ dyn cm$^{-2}$, or 0.6 bar. Actual crushing strengths of meteor-shower fireballs calculated in this manner are generally in the range from 0.1 bar to 1 bar. Sporadic meteors (fireballs not associated with meteor showers) are less frequently observed. One of these fireballs, PN40503, had a calculated crushing strength of 30 bar, and another, EN160166, had a strength of 50 bar. These higher strengths resemble those of actual recovered meteoritic material and remind us that no observed meteorite fall has ever been associated with the orbit of a meteor swarm. The great Tunguska explosion of 1908 occurred at a dynamic pressure of about 200 bar. For comparison, the photographically observed and recovered Lost City and Innisfree chondrites followed trajectories in which the dynamic pressure never passed 200 bar.

This discussion points out an interesting feature of the survival of meteorites during passage through the atmosphere: that meteoroids originating in highly eccentric or highly inclined comet-like orbits, which necessarily encounter Earth at very high speeds, are much less likely to survive entry (and hence give rise to a known class of meteorites) than meteoroids from orbits very similar to that of Earth. Thus our meteorite collections simply cannot be a democratic sample of small bodies in space near Earth. Thus, since cometary material is not only much weaker than asteroidal rock, but also traveling much faster, meteorites that reach the ground must have survived a selection process that discriminates very strongly against cometary solids.

There are also some very limited data on the speed of sound in several ordinary chondrites. Transverse (shear-wave, or *S-wave*) speeds of 600 to 1200 m s$^{-1}$ were measured in six chondritic stones, and longitudinal (pressure-wave, or *P-wave*) speeds ranging from 2000 to 4200 m s$^{-1}$ were found in eight ordinary chondrites. These speeds are generally well below the range found for terrestrial igneous rocks. No clear evidence of any correlation of sound speed with the petrologic type of the meteorite was seen.

There have been a number of studies of the magnetization of meteorites. Since meteorites are a complex mixture of phases with uncertain thermal and shock histories, there is a potential wealth of information stored in the internal magnetization of these bodies. But by the same argument it follows that we must learn a great deal about these meteorites before we can hope to understand anything with confidence.

Most minerals in meteorites are *diamagnetic*; that is, an imposed strong magnetic field has no effect upon the mineral except to induce a very weak opposing field. Most other meteoritic minerals are *paramagnetic*, meaning that atomic magnetic moments are present, but with random orientation. The sample shows no external magnetic field, but is readily attracted to a magnet. In the presence of an orienting force (that is, an externally applied strong magnetic field), some small proportion of these atoms temporarily align so as to reinforce the imposed field. These two types of behavior are uninformative regarding the history of the mineral sample.

A few meteoritic minerals, however, such as metallic iron and nickel and their phosphides and carbides, contain transition metal ions with magnetic moments that are organized into small regions of uniform orientation, called *domains*. They show a large net magnetization and tend to orient themselves in alignment with any externally imposed strong dipole field. These magnetic domains retain their coherent behavior up to some maximum temperature (the *Curie point*), above which the thermal energy becomes large enough to overcome the alignment energy of the individual atoms, and the organized structure breaks down. This causes the domains to lose their permanent magnetism and behave paramagnetically. Such materials when below their Curie temperatures are said to be *ferromagnetic*.

Another highly ordered configuration of atomic moments, in which the individual atoms alternate in the direction of their moments, is called *antiferromagnetic*. Such materials of course have no net external magnetic field and do not respond ferromagnetically to an external magnetic field. Their orderly structure of atomic moments breaks down at high temperatures, at what is called the *Neel point*. Note that, since antiferromagnetic materials have no net magnetization, they do not have a Curie point. The only important meteoritic minerals that display antiferromagnetic behavior are FeS (troilite) and ferrous silicates.

Finally, some transition-metal minerals in meteorites have individual magnetic domains that can exhibit either fully aligned (ferromagnetic) or fully cancelled (antiferromagnetic) atomic moments. These minerals, which include magnetite ($Fe_3O_4$) and the sulfide pyrrhotite, behave like weak ferromagnets. These minerals are said to be *ferrimagnetic*. At elevated temperatures the spin alignments in both of the domain types break down and multidomain grains convert to paramagnetic behavior. The temperature at which this breakdown occurs is called by some authors the Curie point, and by others the Neel point. Clearly both are correct, and we shall not engage in a semantic debate over the relative merits of the two terms.

The relative contributions of the minerals in a given meteorite sample to the overall magnetization of the rock can be determined by heating experiments. The strength of the magnetization carries valuable information on the strength of the external field under which the meteorite was magnetized. However, different mechanisms for magnetization have different efficiencies. Thus, in order to determine that field strength we must know the nature of the mechanism that gave rise to the magnetism. This is unfortunately not known. However, the results from the study of meteorites are so striking as to deserve our attention.

Briefly, there are several plausible mechanisms for imposing a magnetic moment on a meteorite. The most obvious of these, cooling of the meteorite through the Curie temperature of one of the major minerals in the presence of an external field, is called *thermal remanent magnetization* (TRM). Bulk magnetism resulting from a chemical reaction producing a ferromagnetic or ferrimagnetic mineral is called *chemical remanent magnetization* (CRM). Formation of a magnetized rock by the accumulation of partially oriented magnetized grains under the influence of an external orienting field is called *depositional remanent magnetization* (DRM). Magnetization that "diffuses" into a rock as a result of extremely long low-temperature exposure to a strong external field is called *isothermal remanent magnetization* (IRM). Finally, magnetization may be acquired by cooling after shock heating in the presence of an external field. An impact event may in some circumstances drive a highly conducting plasma at very high pressures against a solid body with a preexisting weak external field. This may very briefly produce field strengths orders of magnitude larger than those previously present. This process is termed *shock remanent magnetization* (SRM).

Magnetization under extremely strong fields (thousands of gauss) can orient all of the domains. Clearly we will be unable to distinguish the strengths of different imposed fields if they exceed the level necessary for such saturation magnetization. Such strong fields, however, are seen to occur in the present Solar System only in solar flares and terrestrial laboratories.

If we make the reasonable assumption that magnetized igneous meteorites owe their magnetization to TRM, we can use their observed magnetization to calculate the strengths of the fields under which they were cooled through the Curie temperature. The observed magnetizations are of course highest for those meteorite types that exhibit the highest concentration of ferromagnetic phases. These are the irons and stony irons, which have magnetizations within a factor of 10 of $0.1\,\mathrm{G\,cm^3\,g^{-1}}$. Ordinary chondrites span a wider range, with most falling between $10^{-3}$ and $10^{-2}$ and all falling between $10^{-4}$ and $10^{-1}\,\mathrm{G\,cm^3\,g^{-1}}$. Most achondrites and metal-free chondrites lie below $10^{-3}\,\mathrm{G\,cm^3\,g^{-1}}$. The magnetizing fields estimated for all these classes lie close to 1 G. This is an interesting and impressive result (if we are correct about the magnetization mechanism), since it is by no means easy to make such strong fields. We saw in Chapter IV that the solar wind magnetic field in the inner Solar System is of order $1\gamma$ ($= 10^{-5}$ G). In the entire inner Solar System, only the surface of Earth, with a polar field intensity of about 0.6 G, is presently close to this figure. It has been suggested several times that there were extremely high field strengths and solar wind densities during the earliest evolution of the Sun; indeed, Charles Sonett of the University of Arizona has suggested that such an intense solar wind during the T Tauri phase of the Sun was responsible for the heating and melting of the parent bodies of the igneous meteorites. It is, however, difficult to imagine that these high field strengths persisted long enough to be present when the iron and stony-iron meteorites cooled through the Curie point. This issue is inextricably tied up with the question of the time scale of events during the early history of the Solar System, and we shall return to it later.

The thermal evolution of meteorite parent bodies depends on the nature and abundance of heat sources and the thermal properties of meteoritic solids. The principal heat sources in the Solar System today are long-lived radionuclides, especially isotopes of potassium, uranium, and thorium. Gravitational potential energy is important as a heat source in large bodies such as the terrestrial planets and lunar-sized bodies during and immediately after the accretionary era. Early melting of small parent bodies, however, cannot be attributed to either of these two heat sources. The two best possibilities seem to be the short-lived radionuclide $^{26}$Al and the T Tauri phase solar wind heating mentioned above.

The heat capacities of most chondrites are close to $7 \times 10^6$ erg g$^{-1}$ K$^{-1}$ and their thermal conductivities are near $2 \times 10^5$ erg cm$^{-1}$ s$^{-1}$ K$^{-1}$. Individual stones with very porous structures may well have much lower thermal conductivities, whereas irons and stony irons will be the only classes with significantly higher thermal conductivities. The electrical conductivities of meteorites are very important in solar wind heating scenarios: high metal-like conductivities will prevent penetration of T Tauri solar wind magnetic fields into the meteorite parent body, and very low conductivities (such as that of dry enstatite) would render the parent body wholly transparent to the magnetic fields. In either case, significant interaction and heating would not be possible. It is therefore interesting to note that the electrical conductivities of meteorites range over many orders of magnitude. The chondritic stones alone span more than a factor of $10^5$ in electrical conductivity. The thermal evolution of meteorite parent bodies will be explored later in this chapter.

The issue of porosity has arisen in our discussions of meteorite densities and thermal conductivities. Since 1997 several researchers have carried out porosity measurements on about 100 meteorites. Guy Consolmagno of the Vatican Observatory and Dan Britt of the University of Arizona have summarized, extended, and interpreted these measurements. For CI, CO, and CV carbonaceous chondrites the porosity ranges from low (<3%) to as high as 35%. Ordinary chondrites span the range from 0 to 15% porosity, with no evident correlation with petrographic grade. The very limited data on achondrites are no more informative: the Juvinas eucrite has <1% porosity, whereas the Nakhla SNC, which survived ejection from Mars, has $6 \pm 2.8\%$ porosity. The reasonable supposition that igneous meteorites have very tight structures with little porosity is not substantiated by observation. Some porosity is expected due to the different coefficients of thermal expansion of the component minerals, but the silicate assemblages found in achondrites have little variation in expansion coefficients. Porosities no larger than about 1% would be expected. Thus these observed porosities, if representative, must be due mostly to large-scale cracks rather than to lack of grain compaction.

Stony irons, like chondrites, range from 0 to 13% porosity, and several irons show insignificant porosities (a few percent with an uncertainty comparable to the observed value). Comparison with the densities of asteroids is rendered doubly difficult by the fact that only 15 asteroids have usefully measured densities; however, the meteorites of each composition class appear to be denser than the corresponding type of asteroid. Meteorites are thus sensibly viewed as unusually coherent fragments of bodies that are pervasively fractured.

## Meteorite Minerals

The minerals present in meteorites reflect both the relative abundances of the elements and the conditions under which these minerals formed. Based on our understanding of the cosmic abundances of the elements (Chapter II) and the chemistry of solar material (Chapter IV), it should come as no surprise that the principal minerals in meteorites are compounds of iron, silicon, magnesium, oxygen, sulfur, calcium, aluminum, sodium, and a few other relatively abundant and involatile elements. The important minerals already encountered in our discussion of solar material are included, along with many of the other minerals that have been found in meteorites, in Table VIII.3.

The presence of a mineral in the list does not imply that that mineral is either abundant or widely distributed over many different meteorite classes. Some minerals, such as melilite, spinel, and perovskite, are found only in small refractory inclusions in a few meteorites that belong to rare classes. Some, such as the phyllosilicates serpentine and chlorite, are simply approximations to the composition of poorly characterized phases that are very abundant in fewer than a dozen known meteorites. The phyllosilicates, which are distinguished by their layered (mica-like) structures and high bound water content, are found only in the very volatile-rich *carbonaceous chondrites*. They are found in association with organic matter, sulfur, sulfates, carbonates, magnetite, and other oxidized and volatile-rich materials. Some other minerals, such as osbornite, niningerite, oldhamite, perryite, and djerfisherite, are characteristic of highly reduced mineral assemblages in which no FeO-bearing minerals are present. Meteoritic silicate assemblages in which iron is highly reduced are, for reasons discussed in Chapter IV, dominated by enstatite. Thus these latter minerals are characteristic of the meteorite classes called *enstatite chondrites* and *enstatite achondrites*.

Several solid-solution series are important in meteorites. We have already discussed the iron–nickel, plagioclase, pyroxene, and olivine series in some detail in Chapter IV.

## Taxonomy and Composition of Chondrites

There is a vast literature on chemical analyses of meteorites and their interpretation. Because of the statistical and genetic importance of the unmelted (*chondritic*) meteorites, we shall begin with a general overview of the chemistry of chondrites.

The name "chondrite" originally meant strictly that the meteorite contained small (usually 0.1 mm to several mm)

**Table VIII.3** Selected Meteorite Minerals

| | | Notes | | | Notes |
|---|---|---|---|---|---|
| Metals | | | Oxysalts | | |
| Kamacite | Fe, Ni (<6% Ni) | O | Whewillite | $CaC_2O_4 \cdot H_2O$ | C |
| Taenite | Fe, Ni (>6% Ni) | O | Calcite | $CaCO_3$ | C |
| (Plessite) | Intergrowth of Kam/Tae | I | Aragonite | $CaCO_3$ | C |
| Tetrataenite | Fe, Ni | I | Vaterite | $CaCO_3$ | C |
| Awaruite | $Ni_3Fe$ | C | Dolomite | $CaMg(CO_3)_2$ | C |
| Copper | Cu | | Magnesite | $(Mg, Fe)CO_3$ | C |
| Nonmetals | | | Bassanite | $CaSO_4 \cdot 0.5H_2O$ | C |
| Sulfur | $S_8$ | C | Gypsum | $CaSO_4 \cdot 2H_2O$ | C |
| Graphite | C | I | Epsomite | $MgSO_4 \cdot 7H_2O$ | C |
| Diamond | C | | Bloedite | $Na_2Mg(SO_4)_2 \cdot 4H_2O$ | C |
| Lonsdalite | C | | Stansfieldite | $Ca_4(Mg, Fe)_5(PO_4)_6$ | C |
| Sulfides | | | Farringtonite | $Mg_3(PO_4)_2$ | |
| Troilite | FeS | O | Merrillite | $(Ca, Mg)_3(PO_4)_2$ | C |
| Mackinawite | $FeS_{1-x}$ | C | Sarcopside | $(Fe, Mn)_3(PO_4)_2$ | C |
| Pyrrhotite | $Fe_{1-x}S$ | C | Graftonite | $(Fe, Mn)_3(PO_4)_2$ | C |
| Marcasite | $FeS_2$ | C | Brianite | $CaNa_2Mg(PO_4)_2$ | C |
| Pyrite | $FeS_2$ | C | Panethite | $(Ca, Na)_2(Mg, Fe)_2(PO_4)_2$ | C |
| Pentlandite | $(Fe, Ni)_9S_8$ | C | Buchwaldite | $NaCaPO_4$ | C |
| Smythite | $Fe_9S_{11}$ | C | Apatite | $Ca_5(PO_4)_3(OH, F)$ | C |
| Chalcopyrite | $CuFeS_2$ | C | Chlorapatite | $Ca_5(PO_4)_3Cl$ | |
| Valeriite | $CuFeS_2$ | C | Halides | | |
| Cubanite | $CuFe_2S_3$ | C | Lawrencite | $FeCl_2$ | I |
| Brezinaite | $Cr_3S_4$ | | Silicates | | |
| Daubreelite | $FeCr_2S_4$ | | Olivine S.S. | $(Mg, Fe)_2SiO_4$ | O |
| Caswellsilverite | $NaCrS_2$ | E | Fayalite | $Fe_2SiO_4$ | |
| Alabandite | (Mn, Fe)S | E | Forsterite | $Mg_2SiO_4$ | |
| Gentnerite | $Cu_8Fe_3Cr_{11}S_{18}$ | | Pyroxene S.S. | | O |
| Djerfisherite | $K_3CuFe_{12}S_{14}$ | E | Orthopyroxene | $(Mg, Fe)SiO_3$ | |
| Heazlewoodite | $Ni_3S_2$ | | Enstatite | $MgSiO_3$ | |
| Molybdenite | $MoS_2$ | | Ferrosilite | $FeSiO_3$ | |
| Niningerite | (Mg, Fe)S | E | Clinopyroxene | $(Ca, Mg, Fe)SiO_3$ | |
| Oldhamite | CaS | E | Wollastonite | $CaSiO_3$ | |
| Sphalerite | (Zn, Fe)S | | Augite | $Mg(Fe, Ca)Si_2O_6$ | |
| Carbides, phosphides, etc. | | | Hedenbergite | $CaFeSi_2O_6$ | |
| Cohenite | $(Fe, Ni)_3C$ | I | Pigeonite | $(Fe, Mg, Ca)SiO_3$ | |
| Haxonite | $Fe_{23}C_6$ | I | Feldspar S.S.[a] | | O |
| Perryite | $(Ni, Fe)_5(Si, P)_2$ | E | Plagioclase | | |
| Suessite | $Fe_3Si$ | E | Anorthite | $CaAl_2Si_2O_8$ | |
| Barringerite | $(Fe, Ni)_2P$ | I | Albite | $NaAlSi_3O_8$ | |
| Schreibersite | $(Fe, Ni)_3P$ | I | K-spar | | |
| Carlsbergite | CrN | E | Orthoclase | $KAlSi_3O_8$ | |
| Osbornite | TiN | E | Sanidine | $KAlSi_3O_8$ | |
| Roaldite | $(Fe, Ni)_4N$ | E | Melilite S.S. | | R |
| Sinoite | $Si_2N_2O$ | E | Åkermanite | $Ca_2MgSi_2O_7$ | |
| Oxides | | | Gehlenite | $Ca_2Al_2SiO_7$ | |
| Armalcolite | $FeMgTi_2O_5$ | | Diopside | $CaMgSi_2O_6$ | |
| Corundum | $Al_2O_3$ | R | Monticellite | $Ca(Mg, Fe)SiO_4$ | |
| Perovskite | $CaTiO_3$ | R | Roedderite | $(K, Na)_2Mg_5Si_{12}O_{30}$ | E |
| Scheelite | $CaWO_4$ | | Cordierite | $Mg_2Al_4Si_5O_{18}$ | |
| Hibonite | $CaAl_{12}O_{19}$ | R | Nepheline | $NaAlSiO_4$ | |
| Hematite | $Fe_2O_3$ | C | Ureyite | $NaCrSi_2O_6$ | |
| Maghemite | $Fe_2O_3$ | C | Krinovite | $NaMg_2CrSi_3O_{10}$ | |
| Magnetite | $Fe_3O_4$ | C | Zircon | $ZrSiO_4$ | |
| Spinel S.S. | | R | Hydroxyl silicates | | |
| Spinel | $MgAl_2O_4$ | | Serpentine | $(Mg, Fe)_6Si_4O_{10}(OH)_8$ | C |
| Hercynite | $(Fe, Mg)Al_2O_4$ | | Chamosite | $Fe_6Mg_3[(Si_4O_{10})(OH)_8]_2$ | C |
| Chromite | $FeCr_2O_4$ | | Montmorillonite | $Al_4(Si, Al)_8O_{20}(OH)_4Mg_6$ $(Si, Al)_8O_{20}(OH)_4$ | C |
| Magnesiochromite | $MgCr_2O_4$ | | | | |
| Quartz | $SiO_2$ | E | Kaersutite | $Ca_2(Na, K)(Mg, Fe)_4TiSi_6Al_2O_{22}$ $(F, OH)_2$ | C |
| Tridymite | $SiO_2$ | E | | | |
| Cristobalite | $SiO_2$ | E | Muscovite mica | $(K, Na, Ca)_2Al_4[Si_6Al_2O_{20}]$ $(OH, F)_4$ | C |
| Rutile | $TiO_2$ | | | | |
| Baddeleyite | $ZrO_2$ | | Sodalite | $Na_8Al_6Si_6O_{24}(Cl, OH, F)_2$ | C |
| Ilmenite | $FeTiO_3$ | | Richterite | $Na_2CaMg_5Si_8O_{22}(F, OH)_2$ | C |

[a] Usually only a single feldspar is found in stony meteorites.

*Note.* O, major minerals in ordinary chondrites and achondrites; C, found (almost) exclusively in carbonaceous chondrites; I, found in irons or with the metal phase; E, found in enstatite chondrites and achondrites; R, found in refractory inclusions in some C chondrites; S.S. = solid solution.

spheroidal glass-bearing beads called chondrules. The name now is taken to mean that the meteorite either contains chondrules or is closely similar in structure and composition to other meteorites that do. In practice, all chondrites are primitive, unmelted samples of ancient solar nebula solids that have not undergone density-dependent phase separation (*geochemical differentiation*).

The first general systematization of the compositions of chondrites was attempted by Harold C. Urey and Harmon Craig more than 40 years ago. Their discovery of discrete compositional classes of chondrites set the stage for much of the modern study of meteorites. Urey and Craig used as convenient variables the abundance of reduced iron (metal plus troilite) and the abundance of oxidized iron (Fe oxide in silicates and magnetite). They expected to find that higher oxidized iron concentrations would be correlated with lower reduced iron concentrations in such a way that the total iron content (as measured, for example, by the bulk Fe:Si or Fe:Mg atomic ratio) would be about the same for all chondrites. This relationship, and also the correlation between nickel content of the metal and FeO content of the silicates, was first suggested by G. R. Prior in 1916 on the basis of a small number of relatively low-precision chemical analyses. For that reason they are known as *Prior's Rules*.

The results found by Urey and Craig, now greatly extended and refined by another entire generation of analysts with modern research facilities, have shown that the chondrites can be subdivided into at least five major groups on the basis of their contents of reduced and oxidized iron. Further, they demonstrate clearly that the relationships between these groups cannot be described adequately by Prior's Rules alone: some of these chondrite groups clearly differ substantially in total iron content. A modern version of the *Urey–Craig diagram* is given in Fig. VIII.3. Part a is in the format used by Urey and Craig, with the total Fe plus FeS content graphed against the FeO content. The second is in the format of meteoriticist Robert Dodd of the State University of New York at Stony Brook, with total reduced iron (Fe metal) graphed against the sum of all oxidized iron (FeO + FeS). The left side of the diagram, with its vanishing FeO content, has a ratio of divalent cations ($Mg^{2+}$ and $Ca^{2+}$) to silicon that is only slightly smaller than 1. $MgSiO_3$ (enstatite) is the dominant silicate. Thus the chondrites in this group are called E (enstatite) chondrites. The members of this class span a very wide range of total iron contents, and they may by this and other criteria be subdivided into two subgroups, EH (high iron) and EL (low iron). Grady's *Catalog of Meteorites* lists 125 EH and 38 EL chondrites, with another 38 E chondrites (most of them highly weathered or very small) of uncertain affinity.

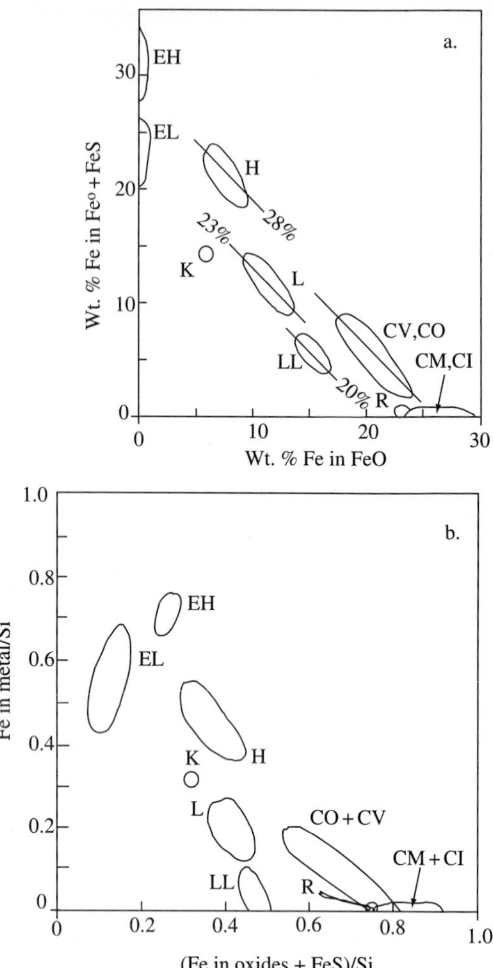


**Figure VIII.3** Two representations of the major-element compositional variations of chondritic meteorites. (a) This graph of weight percentage iron in Fe metal and FeS vs weight percentage iron in Fe oxides, called a Urey–Craig plot, clearly resolves the ordinary chondrites into three classes, H (high iron), L (low iron), and LL (very low iron), based on differences in both oxidation state and total iron content. Diagonal lines of slope −1 on this diagram are contours of constant total iron content. (b) The same analytical data are replotted on a diagram of (Fe metal):(total silicon) vs (Fe in FeO and FeS):(total silicon) as suggested by meteoriticist Robert Dodd of SUNY at Stony Brook.

In the center of the diagram we find the large majority of all known chondrites. These common types, with significant FeO contents coexisting with significant amounts of free metallic iron, are collectively referred to as the *ordinary chondrites*. The ordinary chondrites fall into three distinct classes on the Urey–Craig diagram, differing in total iron content by quite significant amounts. The ordinary chondrite group with the highest total iron content (and also the lowest FeO and highest metal content) is called the H (high-iron) group. The even more common group with somewhat less total iron concentration is termed the L (low-iron) group, while the less populous (and most recently characterized) ordinary chondrite group is called the LL (low-low-iron) group. These have the highest FeO and lowest metallic Fe content of the ordinary chondrite groups. Grady's catalogue lists 14,265 ordinary chondrites, including 6,902 H chondrites, 6,213 L chondrites, and 1,048 LL chondrites.

There is a progression in major-element mineralogy that parallels these chemical trends. First, the abundance of free metal declines markedly from H through L to LL. Second, the ratio of divalent cations ($Mg^{2+}$, $Fe^{2+}$ and $Ca^{2+}$) to silicon increases substantially due to iron oxidation along this series. Thus olivine, which is very minor or absent in the E chondrites, is present in all ordinary chondrites and increases in abundance along the sequence H to L to LL. Pyroxene decreases in importance as olivine increases.

The highest FeO contents are associated with the most volatile-rich meteorites, most of which have easily detectable amounts of organic matter, and many of which are devoid of free metal. These meteorites are collectively referred to as C (carbonaceous) chondrites. All C chondrites have total iron contents at least equal to that of the H chondrites. The C chondrites are subdivided on the basis of a number of criteria into four major groups, each named after a type meteorite. The Vigarano type (CV) and the Ornans type (CO) contain metallic iron, while the Mighei (CM) and Ivuna (CI) types, with progressively higher FeO and volatile-element contents, are devoid of metal. There appears to be slightly more total iron in the CO type than in the CV type, and also slightly more total iron in CI chondrites than in any of the other C or ordinary chondrites. Other carbonaceous chondrites fall into the Karoonda (CK), Renazzo (CR), and provisional CH types. The CH chondrites exhibit extremely high iron abundance, with an Fe:Si elemental abundance ratio of about 1.7:1. Grady lists five CI chondrites, 161 CM chondrites, 78 CRs, 85 COs, 49 CVs, 73 CKs, and 11 CHs.

Note that, although almost all the FeO in an ordinary chondrite is in the pyroxene and olivine phases, the concentration of FeO is not the same in both phases: there are differences in the Gibbs free energy of formation of $Fe_2SiO_4$ and $FeSiO_3$ (the latter is actually unstable as a pure phase), and olivine is unusual in that it has two crystallographically and energetically distinct sites that can be occupied by divalent cations. The compositions of olivine and pyroxene in equilibrium with each other are given in Fig. VIII.4. At low FeO contents (essentially a solid

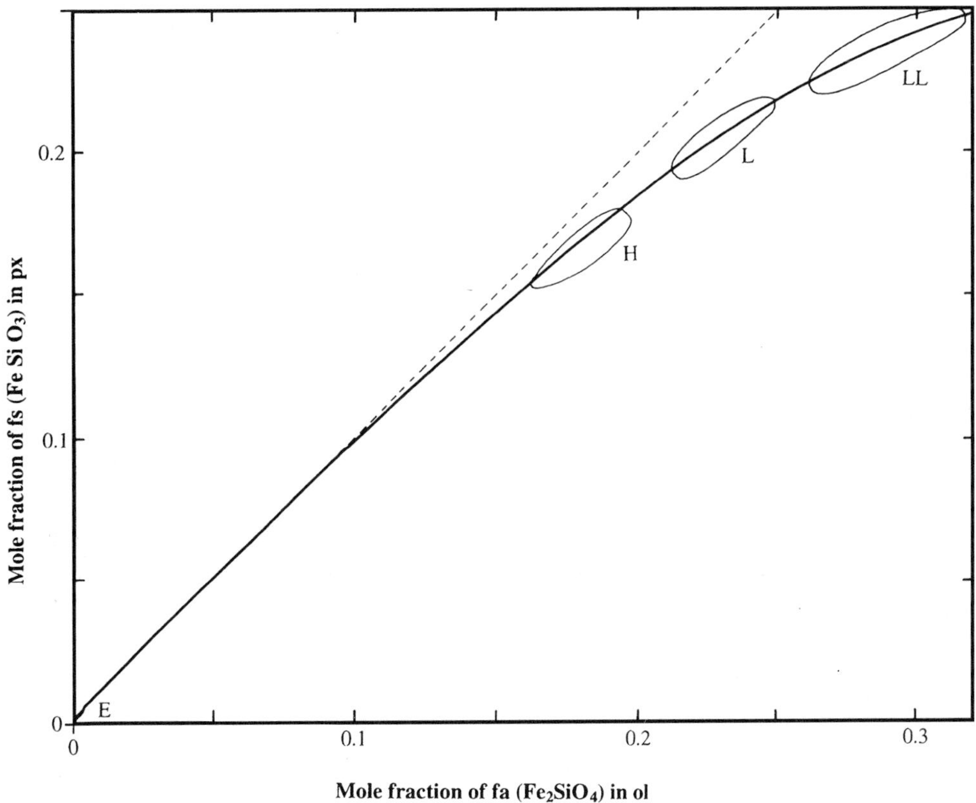


Figure VIII.4 Variation of FeO contents of coexisting pyroxene and olivine in equilibrated chondrites. The diagonal dashed line corresponds to equal ferrous silicate mole fractions in both phases. The equilibrium line is strongly curved toward higher FeO contents.

solution of $FeMgSiO_4$ with $Mg_2SiO_4$) the FeO content increases linearly with the FeO activity (and therefore the fayalite content of olivine increases linearly with the ferrosilite content of coexisting pyroxene). At higher degrees of oxidation the olivine becomes effectively a solid solution of $Fe_2SiO_4$ with $Mg_2SiO_4$ in which the fayalite content increases with the square of the FeO activity.

In addition to the three ordinary chondrite groups recognized in the Urey–Craig system, another group of closely similar chondrites has recently become accepted. These are the three known K chondrites, named after the Kakangari stone, which is the type example of the group. Because the group is so small, meteoriticists are reluctant to call it "ordinary." Its total iron content is slightly less than that of the typical LL stone, but it is less oxidized (has a lower FeO content) than any of the ordinary chondrite groups.

Finally, another group with affinities to both the ordinary and carbonaceous chondrites has been documented. These are the Rumurutiites, named after the type example, Rumuruti. Almost all of the 19 known Rumurutiites are from Antarctica. They contain no more than a bare trace of free metal and have total iron contents at the very low end of the carbonaceous chondrite range, similar to that in the H-group ordinary chondrites. The low metal content, suggestive of a high degree of oxidation, is indeed accompanied by a high FeO content (38% fayalite in olivine). Sulfides are largely oxidized as well, and NiO is present in the olivine. The surviving sulfides are, like those in the C chondrites, copper-rich.

It is interesting to note that, although Prior's Rules are clearly not valid as a means of relating different classes of chondrites, the members within any one group, which presumably arise from the same parent body, do fit Prior's Rules quite well.

There are several other major elements that exhibit clear compositional variations from one Urey–Craig compositional class to another. The data for aluminum, magnesium, and calcium are given in Fig. VIII.5. The magnesium abundances, as represented by the Mg:Si atomic ratios, are lowest in E chondrites (near 0.8), higher in LL and L chondrites (about 0.93), slightly higher yet in H chondrites ($\sim$0.97), and highest in C chondrites ($\sim$1.05). Both aluminum and calcium show rather similar trends, except that the variations are larger than for magnesium: the most magnesium-rich

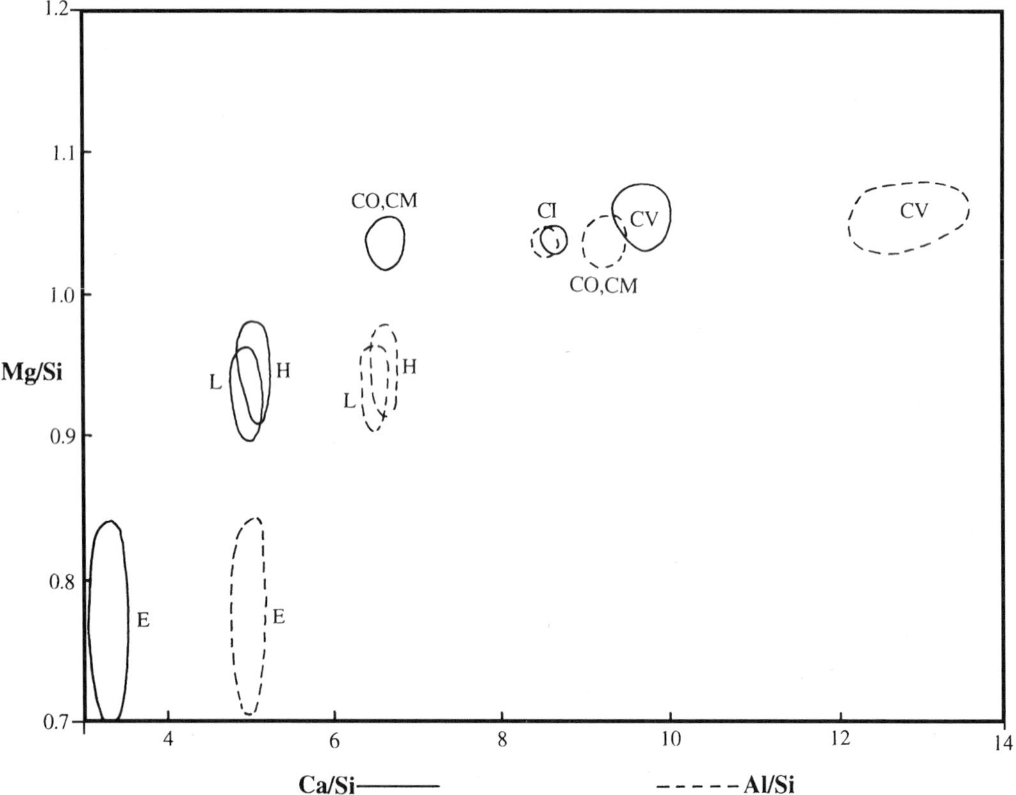


**Figure VIII.5** Magnesium, calcium, and aluminum variations in chondrites. The solid lines plot the Mg vs Ca distribution, and the dashed lines show the Mg vs Al data. Calcium and aluminum are given in percent of the silicon abundance. Note that the Ca:Al ratio is nearly constant in chondrites except for those in the CI group, which have an enhanced Ca:Al ratio.

chondrites contain 1.5 times as much as the most magnesium-poor, while the maximum calcium and aluminum concentrations are about 2.8 times the minimum amounts. In every case the lowest Mg, Ca, and Al concentrations are found in E chondrites, the LL, L, and H chondrites are next (and very close together), and the C chondrites are highest. The CV type is the highest among the C group. The Ca:Al ratio is remarkably constant for almost all the chondrites studied.

The sequence of total iron concentration (Fe:Si atomic ratio) is from the lowest, LL and EL, through the L chondrites, the closely similar H, CV, and CM, to the highest CO, CI, and EH classes. The trend of oxygen content (from E to H to L to LL to C) also appears to be independent of the other fractionation trends. Since the trends of fractionation of these elements are (except for calcium and aluminum) very different, it is clear that no simple or single process, such as silicon fractionation or addition of one element or mineral, can explain these observations. The question immediately arises whether these fractionation processes could be due to variable efficiency of accretion of one or two minerals. The number of different, independent components needed is at least five (to account for the independent behavior of oxygen, iron, magnesium, and the calcium–aluminum pair relative to silicon), and this conclusion rests on an examination of only six elements, one of which was taken as the reference standard! Further data on other elements cannot simplify this conclusion and can add even more complexity.

After these six, the next major element to consider is sulfur. The sulfur mineralogy reflects the overall oxidation state of the meteorite, whereas the sulfur elemental abundance is apparently unrelated to the oxidation state. The most strongly reduced meteorites have, as we saw above, a variety of unusual sulfide minerals containing elements such as manganese, chromium, magnesium, calcium, and even potassium. The ordinary chondrites contain only one important sulfur-bearing mineral, troilite (FeS). Chromium, manganese, magnesium, calcium, and potassium are all found as oxide components of silicates, etc., in the ordinary chondrites. The C chondrites contain minerals with more highly oxidized sulfur, such as elemental S and the sulfates epsomite, gypsum, and bloedite. The sulfates are found in abundance in CI chondrites, and S is found in both CM and CI stones. A significant amount of sulfur is also bound in organic matter in the CM and CI groups. The abundance of sulfur (the atomic S:Si ratio) is highest in the CI chondrites (S:Si = 0.28). The ordinary chondrites decline slightly in sulfur abundance in the sequence from H (S:Si = 0.133) through L and LL to the "most ordinary" of the C chondrites, the CV group (S:Si = 0.117). The E chondrites have sulfur contents near S:Si = 0.168, higher than any of the ordinary chondrites, but far lower than the CI value. Note that the ratio of the highest sulfur content to the lowest, 2.4:1, shows that the moderately volatile element sulfur is slightly less variable in abundance between chondrite classes than are the refractory elements aluminum and calcium.

## Metamorphic Grades of Chondrites

It is now reasonable to ask whether volatile elements might behave similarly to one another, while metallic elements vary coherently as a second group, refractories vary as a third, and so on. Is there any evidence for covariation of such geochemically coherent groups of elements? If so, how many such groups can be identified? Do the elemental variations observed arise from different conditions of temperature and pressure at the time of formation, from physical sorting of minerals or mineral groups during accretion, or from reheating during the thermal evolution of meteorite parent bodies? The last process, called *thermal metamorphism*, can not only recrystallize minerals and crystallize (devitrify) glasses, but also lead to the loss of volatile materials into space. In the extreme case of strong heating, partial or total melting and density-dependent geochemical differentiation may occur, and igneous rocks such as achondrites, stony irons, and irons may be formed.

We shall begin our investigation of these phenomena by examining the structure and textures of chondrites, then reviewing the general abundance trends for a number of elements with markedly different geochemical properties and volatilities.

First, let us recall that the chondrite groups we have so far identified were distinguished only on the basis of bulk chemical composition. We have not considered physical (textural) differences between meteorites that are chemically similar. We must now fill in this gap.

Examination of a number of members of a particular group, such as L chondrites, shows that the members have a number of physical features that serve to distinguish several "petrologic grades," evidently due to different degrees of metamorphism. Most chondrites, as we discussed earlier, contain small glassy particles called chondrules. Indeed, the only chondrite group from which chondrules are altogether absent are the very volatile-rich and highly oxidized CI chondrites. Other chondrites display a range of textures, ranging from those with very sharply defined glassy chondrules through those exhibiting all degrees of devitrification up to and including such thorough crystallization and recrystallization that no glass remains and the boundaries between chondrules and the surrounding matrix materials are rather diffuse. The low-temperature CM chondrites show astonishing contrasts from grain to grain. Neighboring particles may have compositions that are grossly out of equilibrium with each other. Metal grains and magnetite spherules, FeO-rich olivine, pure forsterite, and phyllosilicates containing ferric ion may coexist. These meteorites have clearly never been exposed to temperatures high enough to permit equilibration of the coexisting high-temperature and low-temperature materials.

In order to systematize the sequence of petrologic types with different degrees of thermochemical equilibration, Van Schmus and Wood have proposed a six-level scheme of petrologic types, with type 1 (CI chondrites) as the one least affected by thermal metamorphism. The most strongly affected chondrites, of type 6, have been thoroughly recrystallized and have in some cases even experienced incipient melting (but not differentiation). They thus are close to the borderline between chondrites and achondrites, where the degree of melting becomes sufficient (probably $>1\%$) so that density-dependent geochemical differentiation, and hence large-scale chemical fractionation, begins. The distinguishing traits of these petrologic grades of chondrites may be briefly summarized:

1. No chondrules; very fine-grained and opaque matrix material; 3 to 5% carbon and 18 to 22% bound water (though about half of that water has terrestrial isotopic composition and is due to either contamination by or exchange with terrestrial water)
2. Very sharp chondrules in a matrix rich in opaque minerals; little or no metal, typically devoid of taenite; more than 0.5% Ni in the sulfide minerals; heterogeneous olivine and pyroxene compositions (variations $>5\%$); low-Ca pyroxene mostly monoclinic; no secondary (recrystallized) feldspar; 0.8 to 2.6% C and 2 to 16% water; clear and uncrystallized glass
3. Similar to type 2 except that kamacite and taenite are both present; Ni content of sulfides is $<0.5\%$; C content is 0.2 to 1% and water content is typically 0.3 to 3%
4. Variations in pyroxene and olivine compositions $<5\%$; both monoclinic and orthorhombic low-Ca pyroxene; microcrystalline aggregates of secondary feldspar grains; glass is turbid or extensively devitrified if present; transparent and very finely crystalline matrix; $<0.2\%$ C and no detectable water
5. Olivine and pyroxene uniform; low-Ca pyroxene is orthorhombic; glass absent; chondrule rims somewhat blurred but still readily distinguishable; thoroughly recrystallized matrix
6. Similar to type 5 except that secondary feldspar has formed large, clear grains; chondrules have become poorly defined because of recrystallization; sometimes evidence of incipient melting

These petrological grades are combined with the names of the chemical groups to indicate the precise nature of each chondrite, such as L6, H5, LL3, CM2 (sometimes written C2M), or E4.

We have so far mentioned a few distinctions between CV and CO chondrites, and it is desirable to make the list complete: the CV chondrites have higher Al:Si ($>0.11$) and Ca:Si ($>0.08$) ratios than the CO chondrites, lower Fe:Mg ratios ($<0.76$), and larger chondrules ($>0.5$ mm), and they have Ar:Xe ratios near 100 rather than 200 as found in CO chondrites.

The E chondrites span grades 3, 4, 5, and 6. The EH chondrites are mostly EH3, whereas EL chondrites are dominantly EL6. Most H group members are H5, and most L and LL chondrites are of type 6. Although some representatives of grade 3 are found for all the ordinary chondrite classes, they are not common. The Karoonda group shows the highest grades among the C chondrites, ranging from CK3 to CK6. The CV, CO, and CH groups contains only grade 3 (although one or two meteorites may be better described as CO4). The Mighei group meteorites are all CM2, the Renazzo group are all CR2 or CR3, and the Ivuna group are all CI1. Since there is no real ambiguity, these designations are often written simply as CM and CI (C2 and C1 in the older literature). The grade 1 and 2 chondrites are frequently described as "primitive," meaning that they have experienced little or no thermal alteration since they accreted. The textures and compositions of both grades, however, suggest instead low-temperature aqueous alteration.

What are we to make of this proliferation of types? Are there really different parent bodies for every one of the discernible types? At present, it seems that the chemical evidence favors two parent bodies for the E group (EH and EL), one each for the H, L, LL, K, and R groups, and seven for the very diverse C family, the CI, CM, CR, CO, CK, CH, and CV groups. The different petrologic grades within each group probably reflect both minor differences in accretion temperature and major differences in postaccretion thermal evolution. We can deduce something about these differences by measuring the abundances of many moderately volatile elements in chondrites of different groups and classes.

The most important rock-forming elements are condensed out of a gas of solar composition at temperatures over about 1000 K. These elements include Fe, Mg, Si, the refractory oxide components Ca, Al, Ti, etc., the alkali metals Na and K, Ni, Cr, Mn, and P. At temperatures below about 160 K a variety of ices containing $H_2O$, $CO_2$, $NH_3$, $CH_4$, CO, $N_2$, etc., condense. Only one important element condenses in the wide intermediate temperature range: sulfur reacts with previously condensed Fe metal to form FeS (troilite) near 680 K. The minor and trace elements that also condense between about 1000 K and 160 K include, in rough sequence of condensation, Tl, In, Bi, Cl, Pb, Zn, Cd, Sn, Br, Te, I, and Ge. The rare gases Ar, Kr, and Xe would, at equilibrium, condense to form solid clathrate hydrates in water ice well below 160 K. Mercury also would properly belong on this list, but the prevalence of laboratory contamination of samples by vapor from mercury diffusion pumps and the mobility of Hg under ordinary temperature conditions make it hard to attach any great significance to the scattered and inconsistent data available for Hg in meteorites.

The analytical data on these volatile elements in chondrites show that their abundances, as well as those of the rare gases, are closely correlated. It is quite impossible to accept that all these very diverse elements condense at the same temperature, and very difficult to believe that they all reside in the same mineral. The abundances of these elements are highest in the CI matrix material, and lowest in chondrites of petrologic grade 6. Within each group there is a range of about a factor of 100 in the abundances of these elements from grade 3 to grades 5 and 6. It is entirely plausible to assert that a minor component of very volatile-rich low-temperature dust from the solar nebula was accreted in similar amounts by all the meteorites in a given group, but that the lower petrological grades accreted at lower temperatures. It is also plausible to assert that the meteorites of different petrologic grades accreted at similar temperatures with similar volatile element contents, but that subsequent thermal evolution led to extensive loss of these elements by outgassing to space.

If some version of this thermal metamorphism hypothesis is correct, it still leaves unanswered the question of what heat source was responsible. External heating may lead to a body that is layered with the volatiles in the deep interior, but core heating will lead to transfer of volatiles from the deep interior to the near-surface layers and to space. That thermal metamorphism would deplete, say, xenon and indium, by the same factor despite their enormous difference in volatility seems very unlikely. It seems easier to believe in the existence of a volatile-element carrier of nebular origin.

In response to these data, Edward Anders and his co-workers at the University of Chicago proposed a series of multicomponent models for the bulk composition of chondritic meteorites. These components are high-temperature refractory condensate, iron–nickel alloy, remelted silicate, unremelted silicate, troilite, remelted metal, and low-temperature volatile-rich condensate. The two remelted components mimic the chemistry of chondrules and chondrule fragments, whereas the other five components are condensate fractions with coherent geochemical properties and similar condensation temperatures.

## Taxonomy and Composition of Achondrites

The igneous stony meteorites, called achondrites, are those that have experienced such high temperatures that they have at least partially melted, and as a result have undergone extensive geochemical differentiation. They represent material that has progressed beyond chondritic petrologic grade 6. With few exceptions, they have radiometric solidification ages of about 4.5 Ga.

The compositional distinctions that we encountered for the chondrites are principally due to nonigneous processes that occurred in the solar nebula, and of course igneous meteorites made from these chondrites will reflect many of the compositional trends evidenced by chondrites, such as very large differences in oxidation state. However, igneous meteorites will also exhibit very large fractionations due to segregation of melts of different densities. Strong heating of an ordinary chondrite, sufficient to produce extensive partial melting (but not complete melting) will permit upward segregation of a low-density "crustal" silicate melt rich in alkali metals, calcium, volatiles, and rare heavy metals that readily enter into silicates (called large-ion lithophiles, or LILs). A second melt, formed from metallic iron–nickel and troilite, will also appear at about the same temperature. This metal-sulfide melt will be much denser than the silicates, and hence will sink toward the center of the parent body. This dense Fe–S–Ni–Co liquid does not appear at the melting temperature of any of the pure phases, such as kamacite, taenite, or troilite. Instead, melting begins at a temperature significantly below any of these one-component melting points. The same is true of the formation of the first silicate melt: it will first appear at temperatures lower than any of the melting temperatures of pure albite, anorthite, orthoclase, pyroxenes, or olivine. This behavior, known as eutectic melting, is another example of the phenomenon we first encountered in our study of melting of ice mixtures inside icy satellites in Chapter VI. Figure VIII.6 illustrates melting behavior in the Fe–FeS system. The composition and appearance temperature of the eutectic melt are indicated on the diagram.

The upward-flowing melt will eventually solidify to form a crust rich in feldspars, with a high silica content. The sinking core-forming melt will, upon cooling, produce two layers with an outer core of troilite (or a eutectic mixture of Fe and FeS) resting on top of an inner core of metal. The middle region of the body,

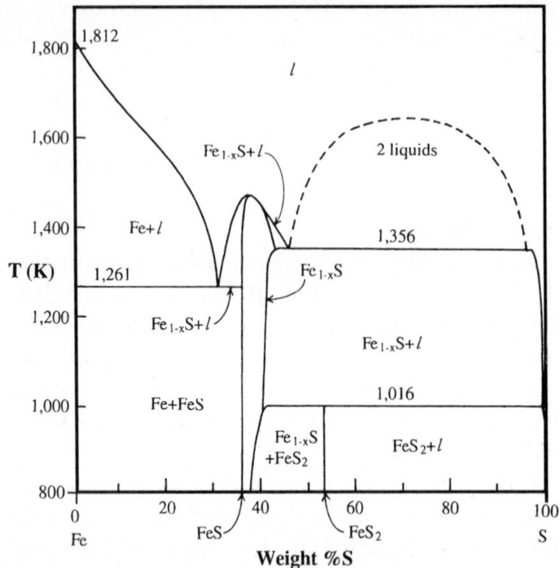


**Figure VIII.6** The iron–sulfur phase diagram. The regions labeled $l$ and $Fe_{1-x}S$ (pyrrhotite) are one-phase regions of freely variable composition. The other regions, including the two-liquid dome at high sulfur abundances, are two-phase regions in which the compositions of coexisting phases are connected by horizontal tie lines (isotherms). Note the incongruent melting of pyrite ($FeS_2$) at 1016 K.

below the crust and outside the core, is dominated by olivine, which has a high melting temperature and a density ($\sim$3.6) intermediate between those of the crust and the core (about 2.7 and 5.7, respectively). The olivine content can range from negligible in highly reduced meteorites to very high in strongly oxidized meteorites. The simple, single-step melting and cooling process will generate rocks ranging from feldspars through pyroxenes and olivine to troilite and metal. Cyclic or episodic remelting of any or all of these layers can generate an even wider diversity of rock types. It is, however, at least clear that achondritic, stony-iron and iron meteorite types can be generated by simple mechanisms. It is now desirable to look in greater detail at the compositions of the known classes of achondrites to see whether the simplest such igneous evolution scenarios are sufficient, and to establish possible genetic relationships between igneous meteorite groups and particular classes of chondritic parent material.

As with chondrites, major-element compositional trends are easy to determine and provide a convenient means of distinguishing several different achondrite groups. The Ca and Fe abundances in a variety of achondrites (and chondrites) are plotted in Fig. VIII.7. Since these are stony differentiated meteorites, they are virtually devoid of metallic iron and troilite. Thus the total iron abundance is essentially identical to the FeO content. There is one class of achondrite with a strikingly low FeO content and a CaO content similar to or slightly lower than that in chondrites. Not surprisingly, the dominant mineral is enstatite, and this group is

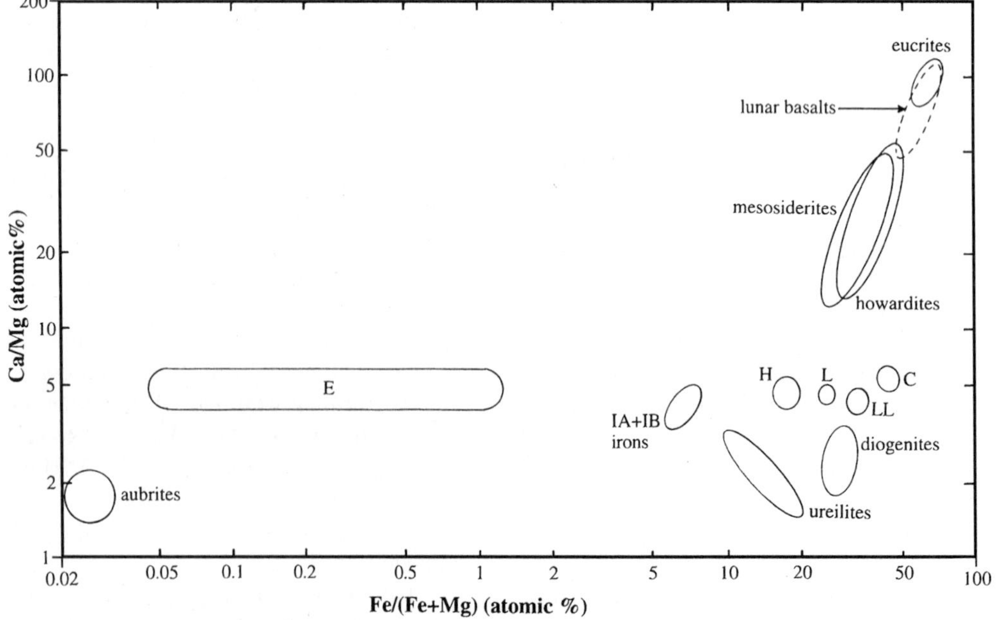


**Figure VIII.7** Major-element compositional variations in meteorites. The Ca:Mg ratio is plotted against the oxidation state (Fe:(Fe+Mg) ratio in the silicates). The chondrites differ in oxidation state, but have closely similar Ca contents. Note, however, the C chondrite data showing modest elevation of Ca. Ca-rich ("crustal") basaltic differentiates cluster in the upper right corner (both basaltic achondrites and lunar basalts), and the nearly pure-$MgSiO_3$ enstatite achondrites (aubrites) cluster in the lower left corner.

termed the enstatite achondrites, of which 46 are known. The type meteorite for this group is Aubres, and the enstatite achondrites are therefore often called aubrites. Note the relationship between aubrites and E chondrites in Fig. VIII.7: it is not hard to imagine a genetic relationship between these classes. One of the novel features of the aubrites is that they are so highly reduced that there is up to ~1.5% of elemental silicon in solid solution in the metal phase.

Another class of achondrites with a low Ca content and an FeO content similar to that in H chondrites is the ureilites, named after the type meteorite Novo Urei, of which 92 are known. These meteorites, which are rather diverse in composition, contain an apparent "seasoning" of volatile-rich primitive material. Some of the ureilites have high carbon contents and have also experienced severe mechanical shock: they contain up to 1% by weight of black, shock-produced diamonds. The ureilites were often referred to as olivine–pigeonite achondrites in the older literature.

There are only three other numerically important classes of achondrites, the howardites (93 in Grady's catalog), eucrites (200 known), and diogenites (94 known). The three classes are collectively referred to as HED achondrites. The eucrites and howardites are dominated by pyroxene and plagioclase, which provides their old name, "pyroxene-plagioclase achondrites." Because the mineralogy of the HED meteorites and their bulk compositions are closely similar to those of terrestrial and lunar basalts, another common designation of the eucrites and howardites is "basaltic achondrites." All three of these classes have major amounts of FeO, with the ferrosilite (fs) content of the low-Ca pyroxene ranging from about 26% in the diogenites to 45–70% in the eucrites. Howardite pxroxene is are even more variable, ranging over fs contents of about 15 to 60%. There is also a striking difference in Ca content between these three classes. Diogenites have a low CaO content similar to that in the aubrites and ureilites, lower than in ordinary chondrites, with an atomic Ca:Mg ratio of about 0.02 to 0.035. The howardites have substantially higher Ca contents (Ca:Mg atomic ratios from about 0.2 to 0.6), and the eucrites have yet more (Ca:Mg ratios from about 0.8 to 1.2).

We will recall that, in the more highly oxidized chondrites, iron is oxidized out of the metal before nickel is. Thus the Ni content of the metal is correlated roughly with the growth of the FeO content of the silicates, and hence of the olivine:pyroxene ratio (Prior's Rules). This should, and indeed may, be generally true of the small amount of residual metal in the achondrites. There are, however, few analyses of their metal phase. Of the known achondrites, only three of the ureilites have metal concentrations greater than 1%. Many achondrites contain 0.1% metal or less, making its separation and analysis difficult.

In addition to the aubrites, ureilites, diogenites, eucrites, and howardites, there are a few smaller but very interesting achondrite classes. These include the acapulcoites (12 known), lodranites (14), winonaites (11), brachinaites (7), and angrites (4). All seem to be plausible products of geochemical differentiation in asteroids.

However, three of these sparse classes, the shergottites (notably Shergotty and Zagami), nakhlites (Nakhla and Lafayette), and the unique chassignite (Chassigny), share a number of features in common and are almost certainly genetically related. Among the more astonishing features of the shergottite/nakhlite/chassignite (SNC) meteorites are their igneous solidification ages of about 1.3 Ga (compared to 4.6 Ga for almost every other achondrite) and their content of traces of noble gases and nitrogen bearing isotopic and elemental abundance signatures strikingly reminiscent of the atmosphere of Mars. At present, 15 SNC meteorites are known. A detailed discussion of these meteorites is reserved for Chapter X, where they are used to provide insight into the composition and history of Mars.

At least 18 other achondrites recovered recently in Antarctica and in the Sahara have chemical and isotopic compositions indistinguishable from returned lunar samples. They are fragments of the Moon that were ejected from the Moon's gravity well by major impact events. Their lunar context is described Chapter IX.

## Taxonomy and Composition of Stony-Irons

The overwhelming majority of the known stony irons belong to two distinct groups with very different internal structures. The larger group, the pallasites, have a continuous matrix of metal surrounding centimeter-size grains of olivine. The fayalite content of the olivine lies between 11 and 20%. The pallasites are dominated by kamacite, taenite, and olivine, with minor amounts of troilite, schreibersite, and farringtonite. Coexistence of the phosphide schreibersite and the phosphate farringtonite defines the oxygen partial pressure in the assemblage.

The 40-plus pallasites are not uniformly distributed over the entire range of olivine compositions: 33 fall between 10.8 and 13.8% fayalite, one lies between 13.8 and 17.8%, and 6 lie between 17.8 and 20.0%. This strongly bimodal distribution suggests that there may be two separate pallasite parent bodies. For quick comparison, the fayalite content of H chondrite olivine lies in the range from 17 to 20%, almost exactly the same as the more FeO-rich pallasites subgroup. The low-FeO pallasite subgroup falls in the hiatus between the E and H chondrites.

The low-FeO group typically has about 55% metal with a nickel content of about 10%. The high-FeO pallasites are near 32% metal, with a Ni content of about 15%.

The only other major group of stony irons, the mesosiderites, contains chunks of metal in a silicate matrix. Our museum collections contain 41 pallasites and only 20 mesosiderites, but six mesosiderites and only two pallasites are observed falls. This suggests to us that pallasites, like irons, can remain intact and recognizable after very long terrestrial residence times, but that mesosiderites are presently more common in near-Earth space. The structural integrity of the iron matrix in pallasites is the likely reason for their persistence.

Mesosiderite silicates are quite different in composition from the almost pure olivine of the pallasites. The silicates are instead plagioclase and pyroxene, often with the visual appearance of crushing and comminution. The plagioclase is highly calcic (anorthite or bytownite) and the pyroxene has a ferrosilite content typically between 20 and 40%.

The Ca content of the silicates in the pallasites is of course very low. The Ca:Mg ratio in the mesosiderites is usually between 0.20 and 0.40, which lies in the same range as the howardite basaltic achondrites. The mesosiderite metal phase contains 6 to 10% Ni and makes up about 30 to 50% of the mass of the meteorite.

After kamacite, taenite, plagioclase, and pyroxene, there are minor amounts of troilite, schreibersite, chromite, apatite, merrillite, and olivine. The olivine occurs as rare large grains that contain only about 10% fayalite, far lower than the ferrosilite content of neighboring pyroxene grains. Aside from the metal and olivine (which may be associated with each other), the remainder of the material of a mesosiderite closely resembles basaltic achondrites. It is a reasonable working hypothesis that these is a close genetic relationship among the eucrites, howardites, and mesosiderites.

There are only a few stony irons that fall outside the bounds of the pallasites and the mesosiderites. One unique object, the Soroti meteorite, is about half metal, but lies outside the usually defined classes because the other "stony" half is made mostly of troilite, not silicates. Soroti is sometimes called an iron, but more often simply ignored. Because Soroti is our only sample of what appears to be a metal-plus-sulfide asteroidal differentiate, it deserves a better fate than it has heretofore received.

## Taxonomy and Composition of Irons

The two principal minerals in the iron meteorites are the iron–nickel alloys kamacite ($\alpha$ iron, with low Ni content) and taenite ($\gamma$ iron, with high Ni content). The three principal structural classes of irons are determined by the relative abundances of these two minerals, which are in turn determined by the bulk Fe:Ni ratio. Members of the first structural class, the *hexahedrites*, contain 4 to 6% Ni, with almost all individuals clustering very close to 5.5% Ni. The alpha iron mineral, kamacite, is able to accommodate only a few percent of Ni in solid solution. The exact solubility of Ni depends on the temperature, as shown in Fig. IV.23. Thus the hexahedrites are distinguished by not having a high enough Ni content to nucleate the formation of taenite: they are made of essentially pure kamacite. The structure of the metal is determined by the body-centered cubic lattice of kamacite. Many of the smaller hexahedrites consist of a single crystal of b.c.c. alpha iron. Upon polishing and etching, systems of fine, parallel lines (called *Neumann lines*) can be seen. The Neumann lines are evidence of twinning of kamacite crystals induced by moderately severe mechanical shock in the 10-kbar range.

The largest class of iron meteorites is the *octahedrites*, which contain between 6 and 14% Ni. They lie in the range of compositions within which kamacite is incapable of accommodating all of the Ni, and thus taenite must be present. However, the iron content also lies above its solubility limit in taenite, and thus both kamacite and taenite must coexist. Commonly, the octahedrites have Ni contents quite close to 8%. Polished and etched sections of octahedrites display a striking multidirectional banded pattern due to the ordered intergrowth of kamacite and taenite crystals. This prominent feature is referred to as the *Widmanstätten structure* (Fig. VIII.2). The texture of octahedrites is closely related to their nickel content. Those with only 6 or 7% Ni show broad bands of kamacite (often displaying their Neumann lines) interrupted by narrow lamellae (layers) of taenite. These give the visual impression of very coarse structure and are called coarse octahedrites. As the nickel content is raised the widths of the kamacite and taenite bands become more closely comparable, and plessite (a very fine-grained intergrowth of taenite and kamacite) becomes progressively more important. The visual impression is one of a less coarse structure. The medium octahedrites, with about 8% Ni, grade into the fine octahedrites, which mostly have 9 to 14% Ni. The fine octahedrites have narrow kamacite bands in a field dominated by plessite.

At Ni concentrations of about 12 to 15% the kamacite bands become very thin and discontinuous, and the structure ceases to show the very striking Widmanstätten pattern. Beyond about 16% nickel taenite is capable of accommodating all the iron at low-temperature equilibrium, but kinetic limitations on the process of diffusive separation of iron and nickel lead to a plessite-like texture rather than a single, clean taenite crystal. These very nickel-rich meteorites have little obvious structure when cut, polished, and etched with acid. For that reason they are called *ataxites*, which inaccurately connotes that they are truly devoid of structure. There are occasional low-nickel irons that show a similar lack of structure, due either to severe shock metamorphism or to anthropogenic tampering (usually by blacksmiths). These meteorites are often referred to as nickel-poor ataxites, but since their texture is not primary, it seems best not to confuse them with the genuine class of ataxites. Instead, they should be assigned to groups according to their chemical composition alone. The highest nickel contents found in irons range through 30% (the Twin City, Lime Creek, and Santa Catharina irons) to 42% (Dermbach) and even 60–62% (the Lafayette iron and Oktibbeha County).

Because of the nature of the melting relations in the Fe–Ni system (see Fig.VIII.8 ) it is appealing to attribute the differences in nickel content between the various irons to partial melting or fractional crystallization of metal. Indeed, nickel contents ranging up to about 68% are possible to derive from a chondritic melt. This is far from a unique explanation, however, since differences in the temperature of equilibration of the solids with the gases in the solar nebula can also lead to large differences in the Ni content of the metal: we know that early metal condensates are somewhat nickel-rich, and that the nickel concentration fall to the cosmic Ni:Fe ratio at temperatures near 1100 K. Low-temperature oxidation of metal extracts iron into FeO-bearing silicates, and troilite formation at and below 680 K also removes iron from the metal. At equilibrium, the residual metal approaches 60% Ni at temperatures near 500 K. Thus radically different scenarios can be invoked to explain Ni variations.

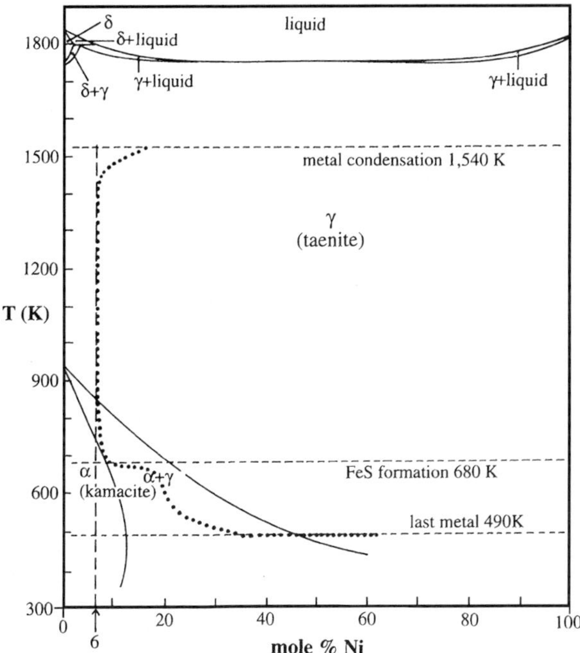


**Figure VIII.8** The iron–nickel phase diagram. The history of Ni content vs temperature for cooling solar-composition material is indicated by the dotted line. Most iron meteorites (octahedrites) are in the two-phase $\alpha + \gamma$ region. Note that low-temperature equilibration can produce very Ni-rich metal (via oxidation of iron to FeO).

This general structural classification of irons into hexahedrites, octahedrites, and ataxites has obvious utility to anyone attempting to identify an iron meteorite from a hand specimen, but we should recall that the only compositional factor we have so far considered is the nickel content. It is in fact not difficult to analyze irons for a number of other elements, as has indeed been done by a number of workers. The most extensive and systematic program of analysis of irons has been carried out by John Wasson and a number of co-workers at UCLA. They have concentrated on a few elements that are relatively easy to measure, for which both sensitivity and precision of analysis are good, and which belong to a variety of different geochemical and volatility classes. They have concentrated their efforts on Ni, Ga, Ge, and Ir. Of these, iridium is the most refractory, and nickel is slightly more refractory than iron. Germanium and gallium vary rather similarly in most compositional groups of irons. The gallium, germanium, and iridium abundances in irons vary over an enormous range, covering a factor of $10^3$ for gallium, a factor of $10^4$ for iridium, and at least a factor of $10^5$ for germanium. Germanium reaches a concentration of 0.2% in the Butler iron! The Ga:Ge ratio is constant within many of the compositional groups, but the different groups cover a significant range of Ga:Ge ratios.

The evidence from the study of these four elements is suggestive of at least 16 different compositional classes of irons. These groups contain about 85% of the ~500 irons studied to data. The remaining irons may belong to distinct groups that for various reasons are poorly represented in our collections; however, for the present, they are referred to as "anomalous." They are very diverse in composition, and cannot be considered a "class."

Figure VIII.9 displays the Ga, Ge, and Ir groups to which the iron meteorites belong. The original distinctions drawn on the basis of the Ga and Ge abundances permitted the identification of only four clusters: group I (>200 ppm Ge), group II (50–200 ppm Ge), group III (0.5–50 ppm Ge), and group IV (0.02–0.2 ppm Ge). The present body of data shows a much more complex behavior, with groups IA-C, IIA-E, IIIA-F, and IVA-B readily resolvable.

The most common compositional groups of irons are IIIAB (230 listed in Grady's catalog), IAB (131 members), IIAB (103 known members), IVA (64 members)

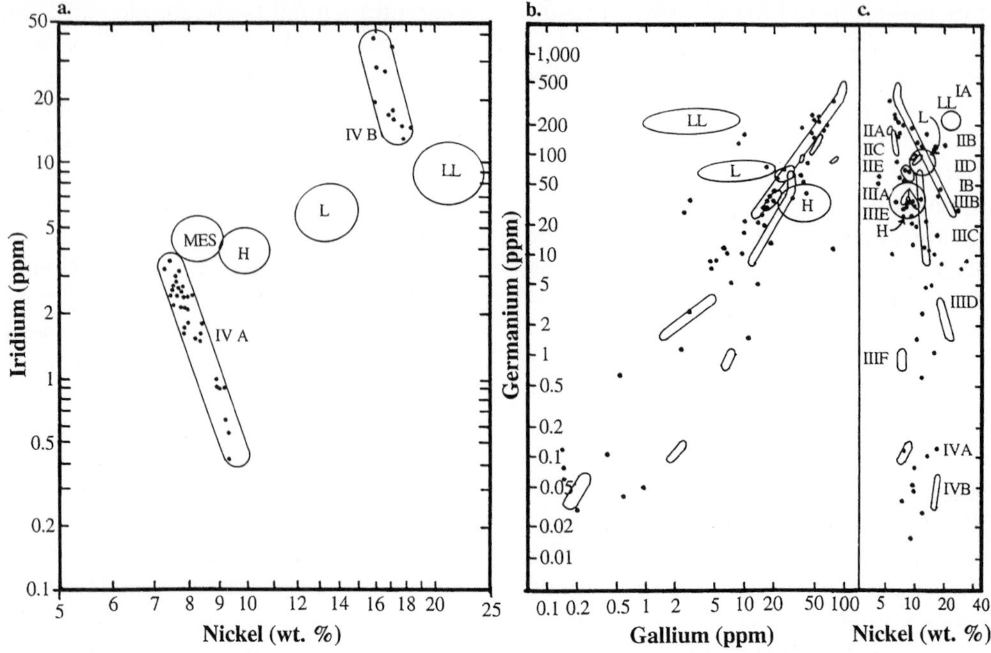


**Figure VIII.9** Gallium, germanium, and iridium classification of meteoritic metals. The H–L–LL sequence in each panel shows the effect of oxidation state. Dispersion roughly perpendicular to the oxidation sequence line is due to fractional crystallization of metal. Well-defined clusters are outlined; unruly individuals are shown as dots. Roughly 15% of all irons studied do not belong to well-defined classes.

and IIICD (41). Far behind in numbers are IIE (18), IID (16), IIIE (13), IVB (13), IC (11), IIC (8), IIIF (6), and IIF (5). But these groups, representing perhaps 13 distinct parent bodies, are far from the whole story. There are in fact 95 analyzed irons that fall into none of these groups, and another 111 irons for which there are insufficient analytical data for classification. Traditionally, a meteorite group or class is designated by name only if it contains five or more members. The 95 ungrouped irons therefore require the existence of at least 24, and probably more than 40, additional parent bodies.

Among the other elements present in the metal phase of iron meteorites, silicon, cobalt, and phosphorus also show interesting variations. Some meteorites, such as the E chondrites and enstatite achondrites, are so highly reduced that appreciable traces of elemental silicon, up to over 1%, are found in the metal. Phosphorus, which is volatile in the solar nebula down to about 1000 K, is difficult to determine as a component of the metal phase because of the prevalence of tiny schreibersite inclusions. Cobalt generally parallels the behavior of nickel both during condensation of the metal phase and during low-temperature oxidation. Both cobalt and nickel enter into sulfides at the endpoint of oxidation of the metal.

The ages of those groups of irons that have silicate inclusions have been determined by rubidium-strontium dating to be $4.45 \pm 0.01$ Ga. Rhenium-osmium dating (see Eqn. II.147 and related discussion) of the IAB, IIAB, IIIAV, IVA and IVB irons lie on the same isochron with a precision of $\pm 0.01$ Ga, but slight uncertainty in the decay constant of $^{187}$Re renders the absolute age estimate of 4.61 Ga uncertain by tens of millions of years.

Another novel isotopic dating technique is provided by the beta decay of hafnium-182

$$^{182}\mathrm{Hf} \longrightarrow {}^{182}\mathrm{W} + e^- + \nu \ (t_{1/2} = 9\,\mathrm{Ma}) \qquad \text{(VIII.3a)}$$

Both hafnium and tungsten are refractory early condensates. Tungsten is strongly siderophile and hafnium is a lithophile. Therefore, if a chondritic body were to melt, differentiate, and generate core and mantle melts many tens of millions of years after accretion, radiogenic tungsten would be efficiently segregated into the core. "Instant" differentiation, however, would segregate the siderophile elements into a core before any substantial fraction of the radioactive hafnium could decay. Both hafnium and its daughter tungsten isotope would then remain in the silicate portion of the parent body, hafnium because it is lithophile, and $^{182}$W because it has been "stranded" without a siderophile melt available to transport it into the core. The metal from such an early melting event would then be depleted in $^{182}$W. Interestingly, studies of the hafnium-tungsten isotope systematics in the IAB and IIAB irons show that their metal phase has so low a $^{182}$W content that core formation must have occurred within 15 Ma of the time of accretion.

A closely similar isotopic system, the beta decay of $^{107}Pd$ to $^{107}Ag$,

$$^{107}\mathrm{Pd} \longrightarrow {}^{107}\mathrm{Ag} + e^- + \nu \qquad \text{(VIII.3b)}$$

has been studied in the IIAB, IIIAB, IVA, and IVB irons and in several ungrouped irons. Here palladium is a lithophile and silver is siderophile. The results show that the differentiation and core-formation process must have been substantially complete within 12 Ma, in accord with the conclusions of the hadnium-tungsten study.

We shall later return to the question of the possible genetic relationships among the irons, stony irons, achondrites, and primitive chondrites.

## Isotopic Composition of Meteorites

There are a number of interesting isotopic anomalies associated with meteorites. The study of elemental and isotopic abundance variations of the noble gases in meteorites has become a major cottage industry, and large isotopic variations in the reactive volatile elements H, C, N, and O are well documented. Although the isotopic compositions of most major and minor elements in bulk meteorite material are quite uniform and uninteresting, there is a class of highly refractory inclusions found in CV and CO chondrites within which many elements exhibit large and unusual isotopic variations. We shall begin with oxygen, the most abundant element that shows significant variations in bulk meteorite material.

Robert Clayton and his co-workers have measured the abundances of the three stable oxygen isotopes ($^{16}O$, $^{17}O$ and $^{18}O$) in a large number of meteorites of many different classes. Their results are summarized in Fig. VIII.10. Isotopic fractionation can occur within any given body because of either equilibrium partitioning or kinetic fractionation during chemical reactions or phase changes. In either equilibrium or kinetic fractionation the 17:16 and 18:16 ratios vary in concert, with the magnitude of the fractionation for each isotope pair being proportional to the mass difference between the isotopes. Thus a plot of the 17:16 fractionation vs the 18:16 fractionation has a slope of almost exactly 0.50. Internal processes within a single isotopically closed system always fractionate along such a line. Two systems that differ in their overall isotopic composition will occupy parallel fractionation lines

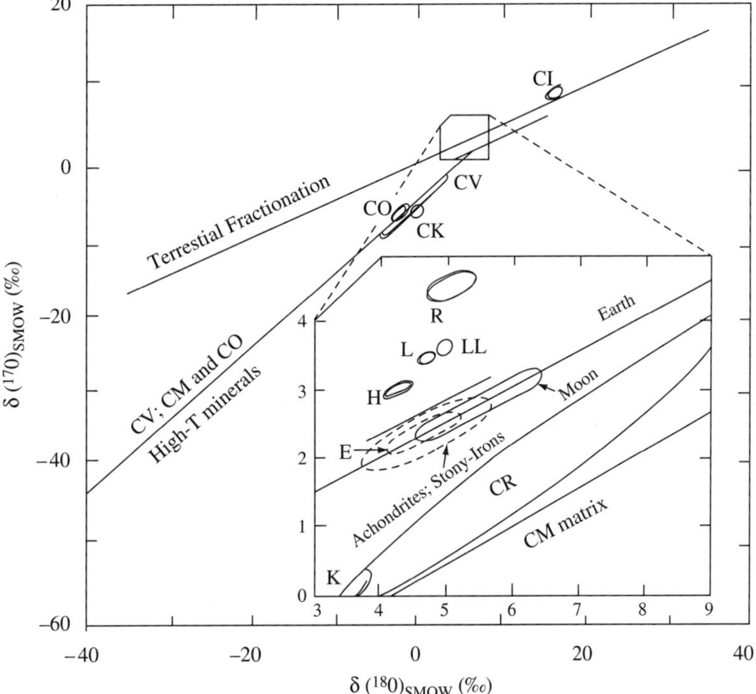


**Figure VIII.10** Oxygen isotope systematics of meteorites. Chemical and physical mass-dependent fractionation effects follow lines with a slope of about 0.5 on this diagram. Note that the high-temperature refractory inclusions in carbonaceous chondrites follow a line of slope 1, indicating addition of variable amounts of pure $^{16}O$. Isotope fractionations are in parts per thousand (permil; % per thousand) relative to the isotopic standard for oxygen, standard mean ocean water (SMOW). E chondrites lie in the cluttered region on the Earth fractionation line. From Lewis and Prinn (1984).

with identical slopes. Figure VIII.10 indeed shows exactly this kind of behavior in almost every respect. The fractionation lines occupied by different minerals in each class of meteorite, for the Moon and for Earth, all have the expected slopes. The oxygen in the R chondrites is heaviest, followed by the L and LL chondrites, which also have a distinctly higher 17:16 ratio than the other classes studied, with the H chondrites next and Earth, Moon, achondrites, and stony irons all displaying indistinguishable overall oxygen isotopic composition. The fine, low-temperature matrix material from the CM2 chondrites lies even lower on the diagram. These differences attest to a lack of any simple genetic relationship between certain of these classes. The converse is not necessarily true: Although enstatite chondrites and achondrites lie on the terrestrial oxygen fractionation line, the elemental abundance ratios and oxidation state of these meteorite classes depart radically from those of Earth. It is therefore clear that materials lying on the same fractionation line need not have any simple genetic relationship to one another.

The oxygen isotope data on pallasites confirm the division into two families suggested by the FeO content data discussed earlier. Two pallasite parent bodies seem to be required by the chemical and isotopic data. The most striking feature of the oxygen isotope data is the line occupied by oxygen from the high-temperature minerals in CM2 and C3 chondrites. That line, with its slope of 1.0, cannot be explained by either equilibrium or kinetic isotope fractionation. Indeed, a line of slope 1.0 on this diagram connects samples with identical 17:18 ratios but variable 16 content. The simplest explanation of this trend is that it represents a mixing line occupied by samples containing variable proportions of two distinct components, one being isotopically normal Solar System material near the upper end of the line, and the other composed of nearly pure oxygen 16. This is presumably the signature of a component derived from the oxygen-rich layer of a supernova explosion.

The most profound implication of the oxygen isotope data is that the material of the inner Solar System was not fully vaporized and homogenized in the solar nebula: if that had happened, then all the solid material in planets, satellites, asteroids, and meteorites would be derived by chemical and physical mass-dependent fractionation from a single isotopically homogeneous pool of oxygen. All these materials would then lie on the same fractionation line. This consideration accords nicely with the "equilibrium condensation" interpretation of Solar System chemistry, which attributes different local condensate compositions to equilibration of previously condensed solids with nebular gases at the highest temperature reached locally in the nebular gas–dust mixture. Equilibration is of course most effective at the highest temperatures close to the Sun: there the composition of the condensate will be well described by the assumption of local chemical equilibrium, and stable-isotope variations will be erased by mixing. Lower-temperature regions of the nebula will experience little volatilization and equilibration and will preserve more of the chemical, mineralogical, and isotopic chaos of the original presolar grains.

The abundance and isotopic composition of carbon have been measured in a number of meteorites. The highest carbon contents are found in the CI and CM chondrites, with lesser amounts in the E, CV, and CO groups and the lowest concentrations in the ordinary chondrites. The isotopic composition of the carbon in the E, CM, and CI groups is rather uniform and only slightly lighter than terrestrial crustal carbon. The C3 chondrites are even lighter, and the ordinary and CO4 chondrites, which have bulk compositions and oxidation states more similar to those of Earth than the other chondrite groups, have the lightest carbon, least like that of Earth's crust. The available data are summarized in Fig. VIII.11.

The nitrogen content of the CM and CI chondrites is quite high, and both have essentially identical overall nitrogen isotopic composition, significantly heavier than terrestrial atmospheric nitrogen (Fig. VIII.12). The nitrogen contained in the organic matter in these meteorites ranges up to even higher 15:14 ratios. The E chondrites have much higher nitrogen contents than the ordinary chondrites because of the presence in them of mineral nitrides, most notably sinoite, a silicon oxynitride ($Si_2N_2O$). Interestingly, the nitrogen in the E chondrites is isotopically much lighter than in the CI and CM chondrites. Even more puzzling is the fact that the C3 chondrites have nitrogen isotopes covering the same range as the E chondrites, averaging 60% lighter than the CI and CM chondrites. The small nitrogen content of the ordinary chondrites is still larger than the visible nitrogen content of the Earth, Mars, or Venus. The

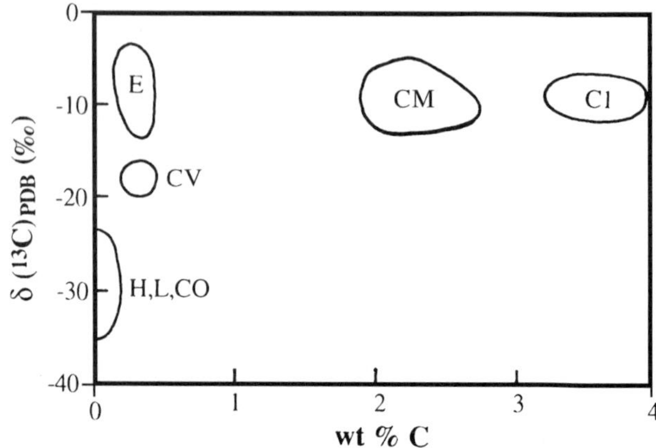


**Figure VIII.11** Carbon abundance and isotopic composition in chondrites. The $^{13}C$ fractionation is in permil relative to the Pee Dee Belemnite (PDB) standard. From Lewis and Prinn (1984).

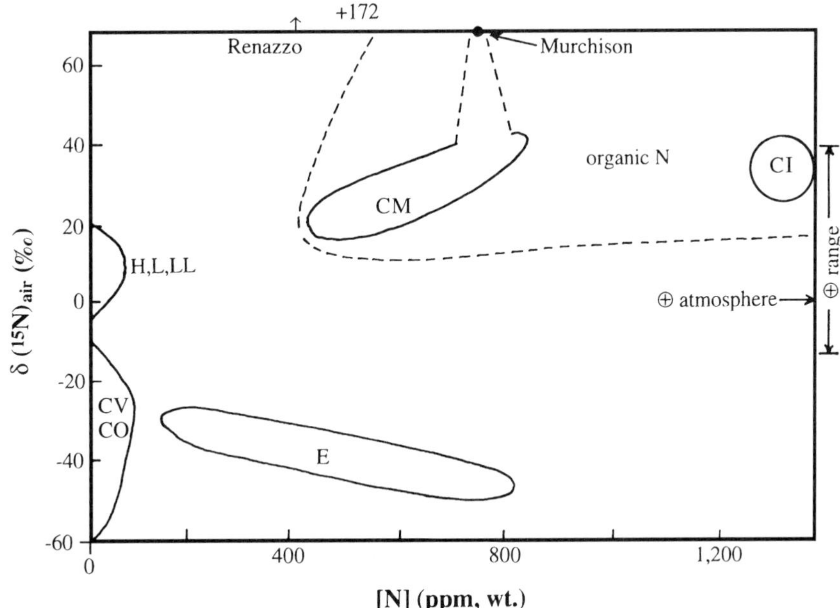


**Figure VIII.12** Nitrogen abundance and isotopic composition in chondrites. The $^{15}N$ fractionation is in permil relative to terrestrial atmospheric nitrogen. Note the extraordinary $^{15}N$ enrichment in the Renazzo carbonaceous chondrite, a unique anomalous metal-bearing meteorite that is otherwise similar to CM chondrites. From Lewis and Prinn (1984).

isotopic composition of ordinary chondrite nitrogen is closely similar to or very slightly heavier than terrestrial atmospheric nitrogen.

Sulfur is found as sulfides in almost all chondrites; indeed, the only abundant sulfur mineral in E, H, L, and LL chondrites is troilite, FeS. The carbonaceous chondrites are a striking exception. Here elemental sulfur coexists with sulfide and a variety of sulfates. The sulfur isotopic composition of these coexisting phases is very far from isotopic equilibrium (see Fig. VIII.13), and an explanation must be sought in nonequilibrium (kinetic) fractionation. The isotopic composition of these phases extracted from the Orgueil CI chondrite shows sulfate to be isotopically lightest, which may be achieved by making the sulfate the last step in a series of kinetically limited processes. Mechanistically, the results can be understood if isotopically light $SO_3^-$ is produced by oxidation of FeS, followed by further incomplete (kinetically limited) oxidation to make $SO_4^-$. The sulfate then ends up very light. Laboratory simulation of the oxidation of FeS, as shown in the figure, produces all three oxidation states with isotopic compositions similar to the corresponding phases in Orgueil if the oxidation takes place at low temperatures, at or below 0 °C. At higher temperatures the isotopic compositions of the phases tend progressively toward equilibrium.

Little information is available on the isotopic composition of hydrogen in meteorites. Water is normally found in abundance only in the carbonaceous chondrites. CI chondrites contain about 20% water by weight, CMs have about half as much, and the CV and CO chondrites have only 0.3 to 3% water. The isotopic composition of the water in the CI and CM chondrites is distinctly heavier than the terrestrial standard (ocean water) by about 15%. Simply detecting native water in other meteorites is extremely difficult, since they are very dry and terrestrial contamination is common.

A number of sources contribute to the rare gas content of chondrites. There is a rare gas component that was captured, with strong elemental fraction, by some poorly understood process from solar nebular gas at the time of formation of the meteorites. This component is called primordial rare gas. Radioactive decay of now-extinct short-lived radioactive nuclides produced certain rare gas isotopes, such as $^{129}Xe$ from decay of $^{129}I$. Other isotopes are produced by beta decay ($^{40}Ar$ from $^{40}K$) or alpha decay ($^{4}He$ from members of the uranium and thorium decay chains). Some chondrites have been exposed to irradiation by the solar wind on the surfaces of parent bodies and have been impregnated with large quantities of solar-wind rare gases. Another rare gas component arises from binary fission of uranium and heavier elements.

The highest concentrations of all volatile elements, the rare gases included, are found in the carbonaceous chondrites. The rare gas abundances in the C chondrites range from 30 to 200 times as high as in the ordinary chondrites. The concentrations of the rare gases in the ordinary chondrites are very similar to those in the Earth.

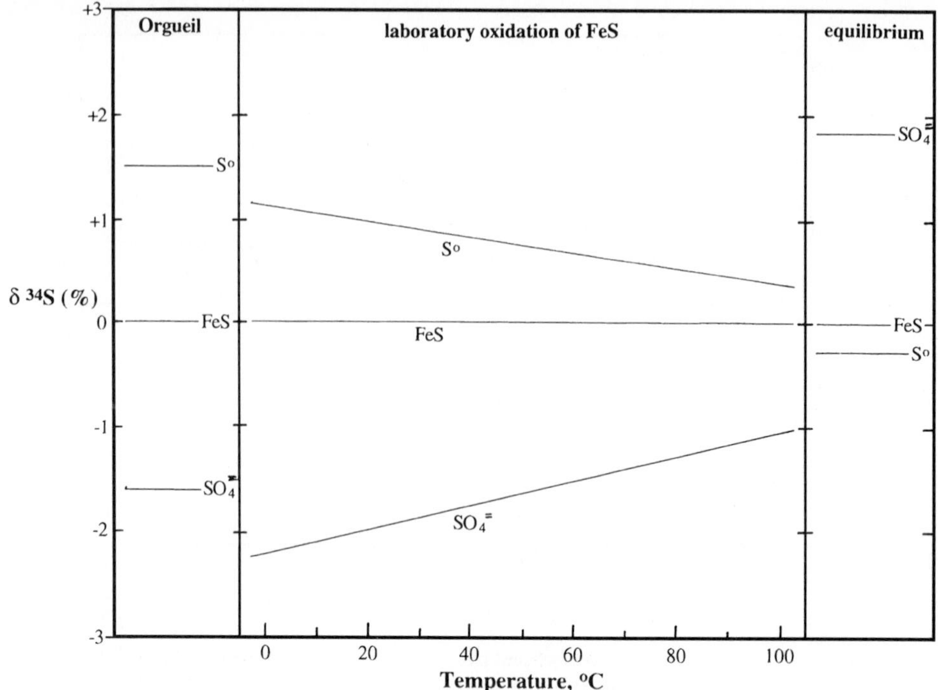


**Figure VIII.13** Sulfur isotopes in the Orgueil CI chondrite and in laboratory simulations. Fractionations for coexisting sulfur, troilite, and sulfate are relative to the Canyon Diablo troilite standard. Low-temperature unidirectional oxidation of FeS provides a reasonable explanation for the observed highly disequilibrium isotopic composition.

A number of differences in elemental and isotopic ratios serve to distinguish the various chondrite groups. Figure VIII.14 compares the elemental abundances of the rare gases in several classes of chondrites. The ordinary chondrites, represented here by the H group, display a rather large Ar:Ne ratio of about 20, and the CO chondrites have an even higher Ar:Ne ratio near 40. The cosmic abundance of neon is of course much greater than that of argon: we saw in Table II.4 that the solar or cosmic Ne:Ar ratio is 30:1. It is evident that whatever process is responsible for implanting the rare gases in these meteorites must fractionate these very gases very strongly in favor of retention of the heavier gas. The cosmic Ar:Kr ratio is about 2500, but the corresponding ratio in chondrites is near 100. Likewise, the Kr:Xe ratio in chondrites, which is in the range between 1 and 2, is far smaller than the cosmic ratio of 9. Thus this strong mass-dependent depletion of the lighter rare gases extends over the entire group of elements. Even the heaviest stable rare gas, xenon, is depleted in ordinary chondrites (and on the Earth) by a factor of about $10^7$ relative to the amount expected from the cosmic Xe:Si ratio. The rare gas retention process must therefore not only lead to strong fractionation, but also be very inefficient. When the rare gases in meteorites were first being systematically studied, some scientists called attention to the resemblance between the chondritic

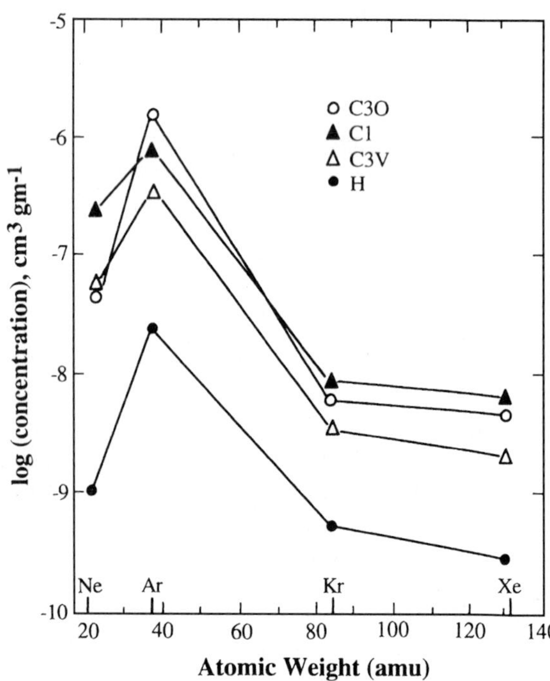


**Figure VIII.14** Abundances of the rare gases in chondrites. Note the very high concentration of rare gases (in $cm^3\ g^{-1}$ at standard temperature and pressure) in the carbonaceous chondrites. From Lewis and Prinn (1984).

and terrestrial rare gases. They chose to refer to this rare gas component in meteorites as the "planetary component." Now that we have rare gas data on the atmospheres of Venus, Earth, and Mars, it is clear that there is a great variety of planetary rare gas compositions, and that the description of that meteorite component as "planetary" is not very useful. The more neutral term "fractionated" has come into use to describe this component.

Another interesting feature of Fig. VIII.14 is the relatively low Ar:Ne ratio in the CI chondrites. This distinctive feature makes it hard to believe that the rare gas content of the ordinary chondrites (or of the Earth) is derived from a small component of CI-type volatile-rich material. As we shall see when we study the Earth, other important differences emerge when the isotopic compositions of terrestrial and meteoritic rare gases are compared.

Any comparison of the isotopic compositions of the rare gases requires that we have some standard for comparison. Unfortunately, the only rare gas samples that are readily obtained are either in meteorites or in the atmospheres of the terrestrial planets. All the samples we can obtain, including modern solar-wind gases implanted in clean surfaces exposed in space to the solar wind, have undergone some type of fractionation. There is also strong evidence that there have been very important changes in the composition of the rare gases in the solar wind since the time of formation of the chondrites. For reasons of convenience, two types of reference rare gas compositions have come into use. One of these is terrestrial atmospheric rare gas; the other is average carbonaceous chondrite (AVCC) rare gas. There are important isotopic differences between these two types, as illustrated by Figures VIII.15 (krypton) and VIII.16 (xenon). There is a grossly linear

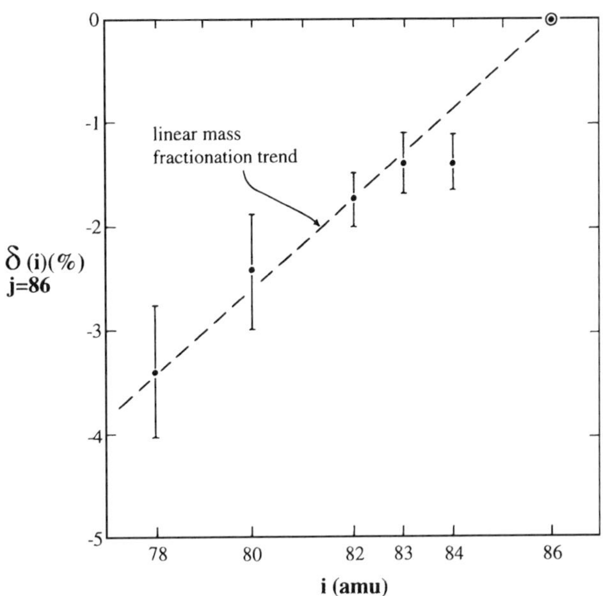


**Figure VIII.15** Isotopic composition of carbonaceous chondrite krypton. The standard is terrestrial atmospheric Kr (see Fig. VIII.16). From Lewis and Prinn (1984).

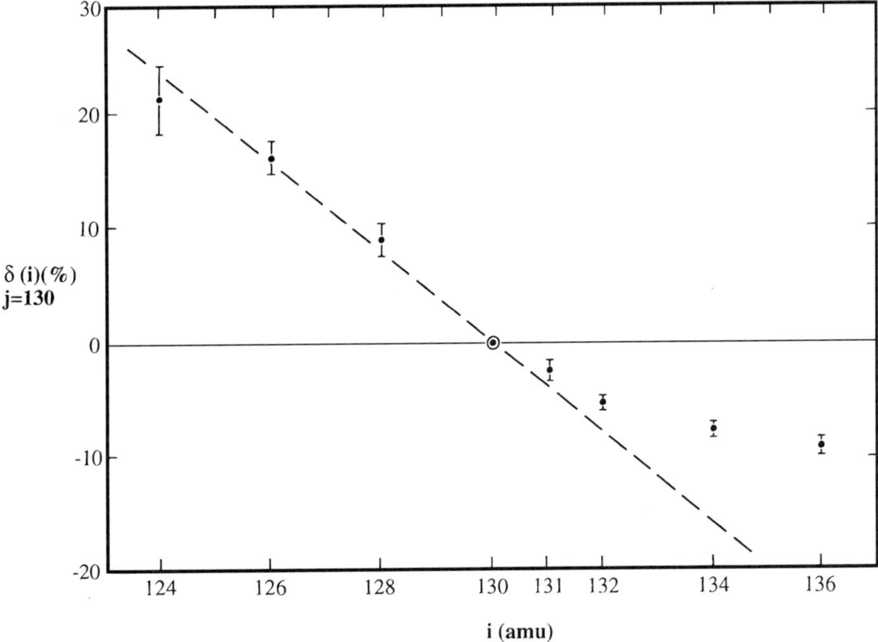


**Figure VIII.16** Isotopic composition of carbonaceous chondrite xenon. The standard is terrestrial atmospheric Xe. The dashed line is a linear fractionation line. The departure at masses 131–136 is due to the presence of a larger component of fission-produced xenon in the meteorite. Note that the xenon fractionation has a slope opposite that of the krypton fractionation! From Lewis and Prinn (1984).

mass fractionation between carbonaceous chondrite krypton and terrestrial atmospheric krypton, with the AVCC krypton being isotopically heavier than its terrestrial counterpart. There is also a linear fractionation trend in xenon, but strangely this trend is in the direction of making the AVCC xenon isotopically lighter than the terrestrial xenon. Relative to this overall trend, the C chondrites have a clear excess of the isotopes 132, 134, and 136. This is yet another component, due to the presence of decay products from spontaneous fission, that enriches the heavier isotopes of xenon.

A few meteorites of a wide variety of classes contain high concentrations of rare gases very similar in composition to the present rare gases in the solar wind. It is almost certain that these gases were implanted directly by solar wind irradiation into material exposed at the surface of meteorite parent bodies. Some meteorites contain both solar wind irradiated (gas rich) and unirradiated materials mixed together in chunks within a single stone. These structures bear evidence of the complex history of collisions, fragmentation, solar wind exposure, refragmentation, compaction, and ejection experienced by the surface material of asteroids.

The radiogenic rare gases in meteorites, those produced by the radioactive decay of uranium and thorium, potassium, and $^{129}$I, carry valuable quantitative information on the histories of the meteorites within which they reside. The decay chains of $^{235}$U, $^{238}$U, and $^{232}$Th involve numerous alpha and beta decay events. The alpha particles are stopped after traveling very short distances, often within the grain in which they were emitted, and then very rapidly pick up electrons to become neutral helium atoms. Thus radiogenic $^{4}$He is frequently trapped so tightly inside mineral grains that the meteorite material will release gaseous helium only if heated to a temperature close to the melting point. This is in contrast to the behavior of solar-wind gases, which are implanted into grain surfaces at such low energies that they are never deeply buried. Mild chemical etching that dissolves a thin rim off of each mineral grain will release all of the solar-wind gases, but only a tiny proportion of the radiogenic helium. These gas components are easily distinguished, since $^{3}$He is utterly absent in the radiogenic component.

Beta decay of the long-lived radioisotope $^{40}$K generates important amounts of $^{40}$Ar, an isotope that is exceedingly rare in solar argon because of its instability during nuclear processing in stellar interiors (it is an odd–odd nuclide). Radiogenic argon atoms have extremely short ranges and are essentially trapped where they are formed in normal igneous minerals. Argon, because of its large atomic radius, diffuses much less readily than helium in mineral grains.

Radioactive decay dating of the (U + Th)/He and K/Ar systems has been carried out on a large number of meteorites. As a rule, almost all chondrites have Ar gas retention ages indistinguishable from $4.6 \times 10^9$ years. The few chondrites with somewhat younger apparent ages all show clear evidence of reheating by severe mechanical shock. It is instructive to compare the K/Ar gas retention ages to those deduced from the (U + Th)/He system (Fig. VIII.17). The helium retention ages are mostly in close accord with the argon retention ages, especially for those with argon retention ages of $4.6 \times 10^9$ years. However, in meteorites that have experienced shock heating and have had their argon accumulation clocks reset, the helium retention ages are often shorter than the argon ages. This is because moderate heating will permit helium to diffuse out of the mineral grains in a time too short for major loss of the more slowly diffusing argon. Severe heating may cause total loss of both gases, which causes the helium and argon retention ages to be concordant, with both clocks reset to zero at the time of the heating event. The rare gas evidence attests to the great age of chondrites, several hundred million years older than the oldest surviving rocks on Earth. Those that have been severely shocked and outgassed are almost all L chondrites: they usually have concordant helium and argon retention ages of

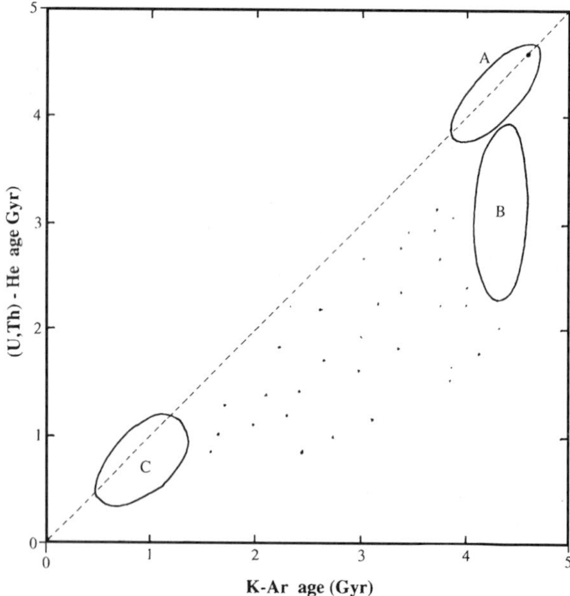


**Figure VIII.17** Gas retention ages of chondrites. The $^{4}$He gas retention age (due to the $\alpha-\beta$ decay series of uranium and thorium) is plotted against the $^{40}$Ar gas retention age (due to decay of $^{40}$K). Most chondrites have concordant ages that approach 4.6 Ga as a limit (A). Most of the points that deviate from the concordance line do so in the direction of lower helium retention ages, an effect easily explained by partial diffusive loss of helium under mild heating (B). A small proportion of the chondrites have concordant short gas retention ages, reflecting a late thermal event (shock heating in a major collision) that reset both the helium and the argon concentrations to zero.

0.4 to about $1.3 \times 10^9$ years, apparently with a strong peak at an age of 520 million years. This is most easily interpreted as a severe asteroidal collision that resulted in major disruption of the L chondrite parent body. There is some evidence that the higher metamorphic grades of the unshocked chondrites are very slightly younger than the lower metamorphic grades of the same classes.

Another extremely useful probe of the history of meteorites is rubidium–strontium dating, based on the beta decay of $^{87}Rb$ to $^{87}Sr$. This dating technique was outlined in Fig. II.26. By reference to this figure, we can see that the minerals in any rock, which are characterized by different elemental Rb:Sr ratios (usually given as the $^{87}Rb$:$^{87}Sr$ ratio), have strontium isotopic compositions that serve to define an isochron. The slope of the isochron defines the evolutionary age of the mineral assemblage, and the intercept of the isochron at zero rubidium content defines the initial strontium isotopic composition. Rocks formed or metamorphosed at a later date will reset their rubidium/strontium clocks to zero age. Such resetting is accomplished by isotopic reequilibration of both elements, and hence begins a new evolutionary history for which all isochrons will originate from an intercept with a high $^{87}Sr$:$^{86}Sr$ ratio. Thus a comparison of the intercepts of the Rb/Sr isochrons for a number of different meteorite types will reveal which are truly the oldest. The lowest initial 87:86 ratios found are 0.6988 in the high-temperature minerals in the Allende CV3 chondrite and in the Angra dos Reis achondrite, 0.6990 in the basaltic achondrites, about 0.6986 in the H chondrites, 0.699 in E chondrites, and 0.700 in the L and LL chondrites. These correspond to ages of 4.46 to 4.60 Ga. The most strongly shocked L chondrites show resetting of the Rb/Sr clock.

Most of the C chondrites seem to follow the pattern of the ordinary chondrites; however, several CI and CM chondrites have apparently experienced alteration as recently as 3.6 Ga ago. Since there is no evidence of significant heating and thermal metamorphism in these meteorites, the most likely explanation seems to be an episode of aqueous alteration at that time.

The decay chains of uranium and thorium isotopes terminate at different lead isotopes: $^{235}U$, after seven alpha emission events, decays to stable $^{207}Pb$; $^{238}U$ decays via eight alpha emission steps to $^{206}Pb$, and $^{232}Th$ decays after eight alpha emissions to $^{208}Pb$. There is also a nonradiogenic isotope of lead, $^{204}Pb$. The decay constants for these three radionuclides are well known, and study of the relative abundances of these four isotopes (called lead–lead dating) is yet another important tool for meteorite dating, giving results independent of those from the other isotopic systems available.

Lead–lead ages of chondrites generally run between 4.50 and 4.58 Ga. Rubidium–strontium ages of chondrites that show no textural evidence of later reheating cluster closely around 4.56 Ga. There is some slight evidence that the higher petrologic grades are discernibly younger than the lower grades. In both the lead–lead and Rb–Sr systems it appears that the ages that have been determined for chondrites refer to the time of metamorphism shortly after accretion. The ages calculated from these two systems seem very closely concordant, so that it is likely that they date the same event or process. The "time of formation" of the chondrites is apparently not a precisely defined point in time, and we can only conclude that accretion took place over a period of a few million years roughly 4.56 Ga ago. The duration of the metamorphic event that set the Rb–Sr and lead–lead clocks seems to have been less than about 100 Ma and cannot be attributed to any long-lived radioactive heat source. Chondrule formation seems by this evidence to have occupied some tens of millions of years.

A short heating time scale for chondritic meteorites suggests that the activity of the early Sun, the release of accretion energy by growing parent bodies, or short-lived radionuclides might be responsible for their metamorphism. There is in fact evidence for the presence of small amounts of such short-lived radionuclides in the early Solar System. This qualitative conclusion can be reached by noting that the gas $^{129}Xe$, is the decay product of the 17 Ma half-life beta emitter $^{129}I$. A quantitative estimate of the importance of this process requires a knowledge of the abundance of iodine in the meteorite. This is most easily accomplished by irradiating the sample with neutrons in a high-flux reactor to convert some of the stable iodine isotope $^{127}I$ into gaseous $^{128}Xe$. The sample is then heated stepwise to release both $^{129}Xe$ and the $^{128}Xe$ iodine tracer. From the 129:128 ratio in the xenon released at high temperatures (where release of both gases is nearly complete) the 129:127 ratio in the initially retained iodine can be determined. Data on the iodine abundance in meteorites suggest that the iodine is condensed into (and along with) the sulfide troilite.

Consider a nucleosynthetic event occurring shortly before the accretion of meteorite parent bodies. A small amount of $^{127}I$ is made, and it becomes isotopically homogenized into the iodine already present (which is essentially pure $^{127}I$), which at this time resides in the gas phase. When temperatures fall low enough for troilite formation and hence for iodine retention in the solids, some substantial fraction of the original radioiodine has already decayed into $^{129}Xe$. Thus only the remnant of undecayed $^{129}I$ is incorporated into the sulfide grains with the rest of the iodine. Xenon, including the $^{129}Xe$ formed recently by iodine decay, is isotopically homogenized in the gas phase and remains almost entirely in the gaseous state down to very low temperatures.

Thus the 129:127 ratio in the sulfide grains is a strong function of the length of the time interval between the cessation of nucleosynthesis and the condensation of the iodine-bearing sulfide grains. Gaseous $^{129}$Xe formed prior to troilite formation becomes a very minor part of background Solar System xenon. The concentration of xenon retained in minerals as they form is extremely small, typically about 1 part in $10^7$ of the cosmic Xe abundance, in ordinary chondrites. Thus the radiogenic xenon liberated in the grain can produce a very substantial isotopic anomaly even though it is not abundant in an absolute sense.

The results of study of this system are interesting and tantalizing. First, the "oldest" materials in the carbonaceous chondrites are the high-temperature refractory inclusions. The magnetite in CI and CM chondrites appears nearly as old. The evidence suggests that magnetite formation occurred about 8 Ma after the refractories condensed. The time intervals (from the end of nucleosynthesis) recorded by other meteorite samples seem to vary without obvious reason over a period of about 25 Ma, with the different groups and grades of chondrites showing no clear trends or regularities. There are even two ordinary chondrites that appear older than the CV refractory inclusions.

The presence of these xenon isotope anomalies clearly demonstrates that short-lived radionuclides were made shortly before the time of formation of meteorite parent bodies, and were still present at that time. Radioiodine was not present in sufficient quantity to provide a significant heat source, but we have strong evidence of one short-lived heat source that was indeed present in some meteorites. This is the isotope $^{26}$Al, a beta emitter with a half-life of 0.72 Ma. It decays to the stable but minor isotope $^{26}$Mg, causing detectable changes in the abundance of this isotope relative to $^{24}$Mg, the alpha-process nuclide that dominates normal magnesium.

The conditions that favor detection of a magnesium isotope anomaly due to aluminum decay are that the aluminum concentration and the isotopic abundance of $^{26}$Al be high in the parent material, and that the concentration of background nonradiogenic magnesium be very low. These conditions are best met in certain high-temperature refractory inclusions in the CV and CO chondrites. For example, corundum (alumina) and gehlenite (the magnesium-free end member of the melilite solid solution series) are excellent minerals in which to search for magnesium isotopic anomalies, and it was in grains rich in these minerals that the presence of the decay product of $^{26}$Al was first detected. In these rare refractory inclusions there is no doubt that the amount of heat liberated by aluminum decay was more than sufficient to cause thorough melting of any kilometer-sized body with their composition. There are some serious questions, however, regarding the significance of this observation. First, there is the problem that we known nothing about the abundance of $^{26}$Al in the most abundant classes of meteorites. Any body that has undergone sufficiently strong metamorphism to homogenize its isotopic composition and bring about chemical equilibration between grains (such as most ordinary chondrites and all the igneous achondrites, stony irons, and irons) will lack identifiable magnesium-poor "carrier" grains that could have preserved the anomaly. Second, we do not know how the radioactive aluminum was implanted in the calcium–aluminum inclusions (CAIs), and hence cannot guess whether it ought also be present in the ordinary chondrites. Third, in order for this heat source to be effective in melting parent bodies, these parent bodies must have accreted to at least kilometer size within two or three half-lives after the cessation of synthesis of the aluminum, which is a mere 2 million years. We shall return to this issue in our discussion of the thermal evolution of asteroids.

The CAIs also exhibit a number of other isotope anomalies that suggest nonuniform mixing of materials with different nucleosynthetic histories. The elements involved, in addition to oxygen and magnesium, include silicon, calcium, strontium, barium, and neodymium. Xenon in C chondrites also contains a component that appears to be due to fission of a transuranium element. Thus the evidence at hand suggests both a complex nucleosynthetic and mixing history for ancient Solar System material and the absence of an extreme and all-pervasive high-temperature phase in the early Solar System: had such a phase occurred, it would have led to isotopic homogenization of the condensible elements. This clearly did not occur.

## Genetic Relationships between Meteorite Classes

We must consider several distinct but related issues. First, what meteorite classes have compositional relationships that suggest that there may be genetic relationships between them? What parent body conditions are required to effect the necessary transformations? What was the agent responsible for these transformations, and how should the importance of this heating agent vary from one body to another? How many parent bodies are required to provide the known classes of meteorites? Finally, on which Solar System bodies did the known meteorite classes originate?

Compositional arguments suggest that the CI, CM, and CO classes represent, to first approximation, a mixing sequence from pure low-temperature, volatile-rich matrix material (CI) to high-temperature minerals not greatly different from the CO chondrites. This sequence of classes may represent three discrete parent bodies sampled from a near-continuum of bodies that are not represented in terrestrial meteorite collections. Alternatively, they may represent layers or regions on a single parent body. The C chondrites have also been suggested by some meteoriticists as the chemical precursors of the ureilite achondrites. Thermal metamorphism and partial melting, perhaps triggered by severe impact shock, might have carried out this transformation. The oxygen isotope data on the ureilites places them in a cluster about the mixing line upon which the refractory CAIs lie, well removed from all other classes of meteorites. Olivine in the ureilites has a high calcium content, suggesting a high formation temperature, and a high chromium content, diagnostic of extremely reducing conditions. The matrix is a quite odd mixture of diamond and other carbon polymorphs, metal, troilite, oldhamite and niningerite, and halite (NaCl) and sylvite (KCl). It seems best, at our present state of knowledge, to relegate all the C chondrite classes and the ureilites to separate parent bodies until more substantial evidence of genetic relationship becomes available.

The only meteorite class that seems derivable directly from the H chondrites is the IIE iron group. This relationship may, however, be circumstantial, and it is difficult to place too much reliance on the skimpy data. The L and LL chondrites may well be the parent materials of the pallasites and IIIAB irons; however, irrespective of their chondritic precursors, the pallasite–IIIAB connection seems well established. The highly reduced IIAB irons may be associated with the E chondrites and achondrites. The latter classes seem to form a well-defined sequence from EH to EL to enstatite achondrite, possibly differing only in the peak temperatures reached during their evolution. The eucrites, diogenites, and howardites form a clear mixing sequence and may be closely linked with the mesosiderites. The howardites are a mixture of eucritic (basaltic) and diogenitic material, formed by brecciation on a parent body surface. Mesosiderites are also eucrite-containing breccias. The shergottites, nakhlites, and chassignites (SNCs) share a very recent origin in a setting strongly suggestive of the crust of Mars. At least 18 other achondrites are clearly of lunar origin. A few other genetic connections have been suggested, but are of lesser likelihood than those discussed above.

Leaving aside those achondrites that appear to be from the Moon or Mars, we find that there is little firm evidence constraining the sizes and locations of their parent bodies or the nature of the heat sources that powered their evolution. The presence of the low-pressure silica polymorph tridymite in the E chondrites constrains their formation pressure to be less than 3 kbar. Minerals indicative of high static pressures are absent. Diamond, common in the ureilites, is black and clearly shock-produced. The heat source responsible for the early metamorphism and melting of many classes, as attested by the radiometric age data detailed above, must have been operative on a time scale of no more than 10 million years. Two such heat sources come to mind, $^{26}$Al decay and T Tauri phase solar wind heating. The CAIs demonstrate that the aluminum activity did exist, but the evidence fails to prove that its decay was important in large parent bodies. If the parent bodies on which the carbonaceous chondrites formed contained abundant $^{26}$Al, then unaltered carbonaceous asteroids larger than about 1 km diameter should not exist.

The very high remanent magnetic fields in many meteorites strongly suggest that gauss field strengths were present at the surfaces of parent bodies, and ancient solar wind gases with unusual isotopic composition are found in gas-rich regolith meteorites. Nonetheless, proof of the importance of this irradiation as a heat source is still lacking. For reasonable-sized parent bodies, accretion energy seems to be of negligible importance.

Few meteorites show evidence of late heating and metamorphism. The lunar and Martian examples are obvious, but other tantalizing hints exist. The Weekeroo Station IIE iron has a Rb/Sr age of 4.38 Ga, compared to about 4.5 to 4.6 Ga for basaltic achondrites. Even more striking, the Kodaikanal IIE iron has a Rb/Sr age of only 3.80 Ga and a roughly concordant K/Ar gas retention age. The IAB irons, which are breccias composed of iron and chondritic material, are very ancient. The formation of most irons took place promptly at the time of origin of the Solar System, but some were formed as long as 800 Ma later. It is very difficult to attribute such late melting (or, more properly, such late cooling) to any short-lived heat source. Accretion of planetary bodies apparently took about 100 Ma, but the heating effect of accretion is most important in large bodies and negligible in small asteroids. The half-lives of $^{40}$K and $^{235}$U are slightly longer than 1 Ga, and we cannot dismiss the possibility that Kodaikanal was heated by these nuclides. However, parent bodies heated by such long-lived radionuclides will not heat up unless they are large enough to insulate their interiors against heat loss by conduction over these very long warmup times. It will be very useful to apply and extend the tools of thermal modeling we introduced in

Chapter VI to the evolution of asteroids and meteorite parent bodies.

## Introduction to Asteroids

Asteroids are small solid bodies, ranging in size from meters to 1000 km, that orbit the Sun in the prograde direction. Larger bodies in heliocentric orbit are called planets; smaller rocky bodies are termed meteoroids. The large majority of all known asteroids orbit in a region extending from approximately 2.2 to 3.3 AU from the Sun. This region is called the *Asteroid Belt*. The distribution of asteroids over orbital semimajor axis is extremely uneven, and many families of asteroid orbits can be discerned.

It is conventional to assume that all asteroids are rocky in composition, but there are good reasons to suspect that ice is an important mineral at least in the outer part of the belt, and that many asteroids have at least experienced the presence of large amounts of water or ice.

Until the 1990s, all knowledge of asteroids was derived from Earth-based astronomical observations or inferred from studies of meteorites. Historically, one of the principal goals of the study of asteroids has been to establish a connection between our vast body of knowledge about meteorites and the equally vast but wholly separate body of astronomical data on asteroids and other Solar System bodies.

The scope of our study of asteroids has in recent years been substantially broadened by the advent of unmanned spacecraft missions. The first spacecraft visit to an asteroid was the high-speed flyby of (951) Gaspra by the Jupiter-bound Galileo mission in 1991. The view of Gaspra seen by Galileo is shown in Fig. VIII.18. In 1993 Galileo also flew by the asteroid (243) Ida, which was found to have a tiny moon, (243) 2 Dactyl. The Near-Earth Asteroid Rendezvous mission (NEAR) flew by the Belt asteroid (253) Mathilde in 1997, en route to its eventual rendezvous with the Near-Earth asteroid (433) Eros. These missions have provided in-depth local coverage of bodies of diverse types, thereby helping in the interpretation of remote-sensing observations of thousands of other asteroids.

Asteroids are named according to two different conventions. The first, suitable for the provisional naming of newly discovered bodies, gives the year, half-month, and discovery order of the asteroid. The name assigned is of the form 1989 FB, which means that this asteroid was the second discovered during the sixth half-month (the last half of March) of the year 1989 (I and Z are omitted). The rate of asteroid discoveries in recent years has accelerated almost exponentially as several successive generations of detection technology have matured in quick succession, beginning with the photographic searches by Eleanor Helin and Eugene and Carolyn Shoemaker in the 1970s, then the 1024 × 1024 CCD detector of the Spacewatch camera with its highly automated computerized search and data analysis system introduced by Tom Gehrels. Now second- and third-generation search systems such as the Lincoln Laboratories' LINEAR, Lowell Observatory's LONEOS, the University of Arizona's Spacewatch II, and several others are in highly productive use. Many smaller observatories, although they lack the equipment for efficient discovery, are well suited for follow-up observations of asteroids discovered by the principal search efforts. Through their involvement, newly discovered asteroids can be tracked long enough, and with high enough precision, that their orbits can be well established. It has become rare for an asteroid to be lost. One consequence of the phenomenal recent discovery rate has been the exhaustion of the alphabet: the 26th asteroid discovered in the second half of October 2002 would be provisionally named 2002 UZ, but what of the 27th? The solution, arrived at under the pressure of necessity, has been to recycle the last letter and add a number to it to tell how many times the alphabet has been cycled; thus we have catalog numbers such as 2001 $SL_9$ (often written 2001 SL9) and 2000 $QT_{297}$.

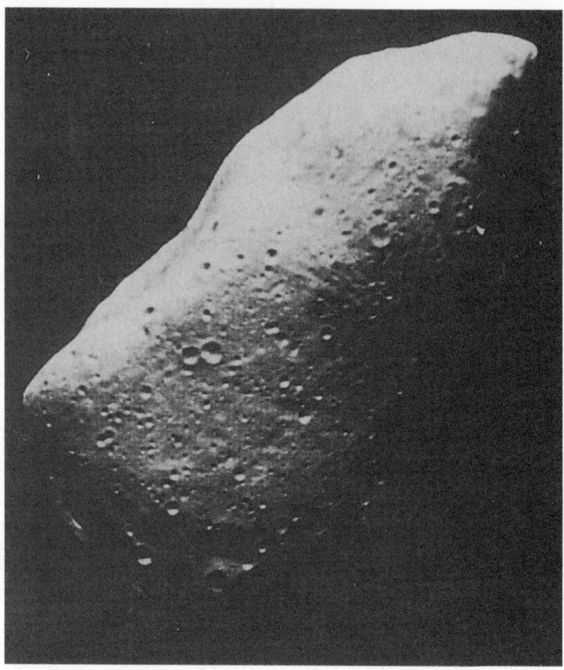

**Figure VIII.18** The asteroid 951 Gaspra as seen by the Galileo spacecraft en route to Jupiter. The surface appears surprisingly smooth, suggesting more regolith than was previously expected on so small a body. Multispectral imaging of Gaspra reveals easily detectable compositional variations over its surface.

Once an asteroid is confirmed and its orbit established to reasonable precision, it is appropriate to honor it with a somewhat more memorable name. The right to assign a name belongs to the discoverer of the asteroid. When the first asteroid was discovered (on the first night of the 19th century!) it was graced with a classical name, Ceres. This tradition was continued with the naming of Juno, Vesta, Pallas (after Athena), and so on, until the successes of asteroid hunters had become so numerous that the lexicon of classical names had been exhausted. Despite the existence of other Solar System bodies with names like Ganymede, Io, Europa, Uranus, Amalthea, Callisto, and the like, asteroids were christened Ganymed, Io, Europa, Urania, Amalthea, and Kallisto. When classicism failed, patriotism emerged: Austria, Polana, Gallia, Russia, Germania, Brasilia, Hungaria, Argentina, Italia, Philippina, Nipponia. Then names became more provincial: Virginia, Bavaria, California, Bohemia, Burgundia, Georgia, Vaticana, Croatia, Transvaalia, Armenia, and Arizona. Then even more local expressions of pride emerged: Bamberga, Heidelberga, Chicago, Padua, Pittsburghia, Tokio, Princetonia, Berkeley, Harvard, Pulkova, Moskva, and Vassar. Virtuous-sounding asteroids appeared: Hygiea, Fortuna, Concordia, Felicitas, Abundantia, sad Lacrimosa, stern Veritas, Philosophia and Justitia, friendly Fraternitas, then practical Geometria, Industria, Photographica, and Stereoskopia, all showing evidence of the semantic Perseverantia of astronomers. Scientists were enshrined as Franklina, Schwassmania, Flammario, Hale, Gaussia, Piazzia; wives and girlfriends became immortalized in the heavens. As the number of asteroids grew and semantic desperation mounted, the very names took on odder and odder overtones: dignified Eunomia and Minerva gave way to Una, Eva, Ino, Ate, Ida, Iva, Ute, Ani, Ara, and Ada; Dejopeja, Anna, Emma, and Olga; Svea, Yrsa, and Ingeborg; Eros; musical Fidelio, Carmen, and Aida; sultry Lola; Preziosa, Phyllis, Agnes, Scheherezade, and far-straying Siwa; Ulula, Oskar, and Cupido; erstwhile Academia and Universitas; Mildred, Sphinx, Susi, Begonia, and Piccolo. More recent discoveries feature names such as John, Paul, George, Ringo, and Jerrygarcia. And then there are the vacation paradises Grubba, Hel, and Dudu.

This Babel of nomenclature, replete with ambiguity and unpronounceability, is put into some vestige of order by requiring that all named asteroids also bear a catalog number prefixed to their names. The first discovered is thus called (1) Ceres, then (2) Pallas, (3) Juno, (4) Vesta, and so on through (1972) Yi Xing, (2001) Einstein, (2100) Ra-Shalom, (3352) McAuliffe, (4179) Toutatis, and (5143) Heracles. Thus we can anticipate that a trip to (439) Ohio might be substantially more expensive than to plain old Ohio, even if we change planes at (334) Chicago, that (291) Alice is not the girl next door, and that (85) Io does not orbit Jupiter.

Another consequence of the extremely high rate of discovery of asteroids has been the development of a substantial gap between the authentication of new asteroids, as symbolized by their receiving a catalog number, and their naming. Thus many asteroids are promoted to adulthood still bearing their "baby names," such as (7735) 2001 OE84. To add insult to injury, many asteroids have been found to have satellites. Newly discovered satellites are given provisional designations of the form S/xxxx Yn, where xxxx is the year, Y is the designator of the body about which the satellite orbits (J for Jupiter, S for Saturn, etc.), and n is the number of the satellite in the list discovered that year. Examples include S/2000 J4, S/1986 U10, and S/2001 (22)1. The latter indicates the first satellite of the asteroid (22) Kalliope discovered in 2001. The designator S/2002 (3749)1 indicates a satellite of the asteroid (3749) Balam, a body whose trivial name seems fated to be Ass (see *Numbers* 22). A satellite with a short orbital period may have a well-determined orbit before its parent asteroid does. When an asteroid that has not yet won a catalog number is found to have a satellite, designations such as S/2000 (1998 $WW_{31}$)1 or S/2001 (1998 $SM_{165}$)1 must be used. Since 1998 $SM_{165}$ has recently received a catalog number (26308), the same satellite must now be referred to as S/2001 (26308)1. No snappy nickname for this body comes to mind.

It appears that several percent of all asteroids have detectable companions, which in most cases means that the companion must be at least a few kilometers in diameter. Satellites have already been found orbiting Belt, NEA, Trojan, plutino, and cubewano bodies.

Occasionally a body discovered as a comet (asteroid) later appears to be an asteroid (comet). Comet Wilson-Harrington 1949 III, after being lost for 30 years, reappeared without evidence of cometary activity and was redesignated as an asteroid. More recently the asteroid 2001 $OG_{108}$, which was later found to exhibit cometary activity, was renamed C/2001 OG 108 (LONEOS).

The number of catalogued asteroids is now nearing 30,000, with several thousand new ones of all sizes (down to about 5 m diameter!) added each year. There are surely many thousands more with sizes greater than 10 km, most of which are so dark and so distant that they have so far evaded detection. The simple question "How many asteroids are there?" cannot be readily answered, since almost all of them are probably too small to have been detected. There is also the semantic question of how small a body must be before it is no longer considered an asteroid, but is instead labeled a meteoroid. There are probably many millions of asteroids with diameters greater than 100 meters. Conventional textbook glosses such as "there are 30,000 asteroids" are useless without a clear statement of the lower size limit that was assumed. A figure of 40,000 is approximately correct for bodies larger than 10 km.

Statistics on the Belt population are now emerging from the ambitious Sloan Digital Sky Survey. This enormous effort has unfortunately become better known for its erroneous use of revised Belt statistics to reassess the asteroid impact rate on Earth, a phenomenon that has nothing to do with Belt statistics.

Despite the great difficulty of observing asteroids, there are still a number of kinds of observations that can and have been made from Earth without the benefit of spacecraft missions. First, the path of the asteroid on the celestial sphere can be determined, and from that positional information the orbital elements may be calculated. An absolute minimum of three well-determined positions, not too close together, is required for an orbit determination. Second, the time variation of the brightness of the asteroid carries information on its rate of rotation. Third, simultaneous observation of the integrated visual and infrared brightness of the asteroid permits determination of its albedo and radius, in the manner discussed for the satellites of the outer planets in Chapter VI. Fourth, fortuitous close passes of two asteroids may permit determination of the size of the gravitational perturbation of the smaller by the larger, and hence may give an estimate of the mass of the larger. Fifth, the spectrum of sunlight reflected from the surface of the asteroid can be used to deduce what minerals are present, and, in conjunction with laboratory reflection spectroscopy of meteorite samples, may permit the identification of the asteroid surface material. Such studies may be of limited resolution (simple UBV photometry) for faint bodies, whereas high-resolution spectra may be obtainable for a few of the brightest asteroids. Sixth, the polarization of reflected sunlight may be measured as a function of wavelength and of the solar phase angle. Again, this topic was discussed in Chapter VI. Seventh, recent advances in radar astronomy have placed a number of asteroids within reach of Earth-based radar studies. Information on the size, shape, roughness, and dielectric constant of the asteroid surface can be deduced from radar observations. Eighth, optical interferometric techniques such as speckle interferometry and single-aperture interferometry hold the promise of providing crude images of a number of asteroids. Ninth, observations of the passive thermal emission of asteroids at radio wavelengths can provide information on the thermal and electrical properties of asteroid surfaces down to depths of a few centimeters.

These Earth-based observational techniques can in some cases be supplemented by *in situ* spacecraft observations of selected asteroids. Numerous studies of multiple-asteroid flyby missions have confirmed that several asteroids can be visited (albeit at rather high speeds) by a single spacecraft. It is not difficult to arrange these missions so that several asteroids of very different size, composition, and heliocentric distance may be visited within a mission lifetime of a few years. As the number of known asteroids grows, the number of opportunities for multiple flybys escalates dramatically. Further, using advanced propulsion techniques such as ion engines or solar sails, it is possible to design missions that can rendezvous with several asteroids over a similar span of time, dropping into orbit around them and even depositing small landing capsules on their surfaces. Missions to explore asteroids and the Martian satellites Phobos and Deimos have been under study for many years, and instruments for *in-situ* study of asteroid surfaces are in an advanced state of development. NASA's Solar System Exploration Committee (SSEC) has endorsed exploratory missions both to the Belt and to near-Earth asteroids. The USSR sent two asteroid-science payloads to Phobos in 1988, and the Galileo and NEAR spacecraft missions were remarked earlier.

Finally, we must mention that a wide range of theoretical studies of asteroid orbital and collisional evolution, physical, optical, and chemical properties are under way and are making valuable contributions to the interpretation and planning of observations. The book *Asteroids*, edited by T. Gehrels (Univ. of Arizona Press, Tucson, 1979) devoted 1181 pages to a synthesis of asteroid observation and theory. This book has been extensively supplemented by the more recent *Meteorites and the Early Solar System*, edited by J. R. Kerridge and M. S. Matthews (Univ. of Arizona Press, Tucson, 1988), and by *Asteroids II*, edited, by R. P. Binzel, T. Gehrels, and M. S. Matthews (Univ. of Arizona Press, Tucson, 1989). By an amusing quirk of fate, Mildred Matthews, perennial editor of books in the U. of A. Space Science Series, is the namesake of the asteroid Mildred mentioned above.

## Asteroid Orbits

The simplest indication of the complexity of the spatial distribution of asteroids comes from examining their distribution over orbital semimajor axis (Fig. VIII.19). The strong concentration of asteroids into the range from 2.06 to 3.65 AU, usually called the Main Belt, and the sharp peaks and valleys in the distribution are both evident. Based mainly on orbital semimajor axis ($a$), several distinct groups of asteroids stand out. Starting nearest the Sun, we encounter the Aten group, which have $a < 1.00$ AU and thus orbit the Sun in less

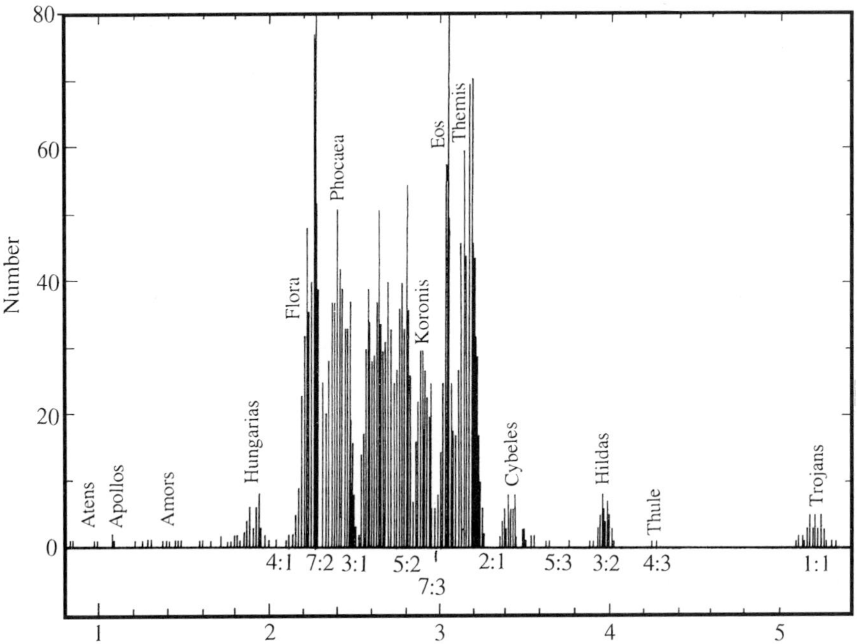


**Figure VIII.19** Distribution of asteroid orbit semimajor axes. The Main Belt, from 2.1 to 3.3 AU, is obvious. The resonance structure due to perturbations by Jupiter is indicated by the ratios of orbital frequency to that of Jupiter (7:2 means an asteroid that orbits the Sun seven times for every two Jupiter orbits).

than 1 Earth year. Next we have those Earth-crossing asteroids with $a > 1.00$, the Apollo group. Next are the Earth-approaching Amors, which do not at present cross Earth's orbit, but do have perihelion distances $q$ inside 1.3 AU (and outside 1.017 AU, the Earth's aphelion distance). As of May 2002, the number of known Atens was 143, compared to 862 Apollos and 837 Amors. Asteroids that always stay inside Earth's orbit are strongly discriminated against by present search techniques, but may exist in modest numbers.

Next in order of distance is the Hungaria family between 1.82 and 2.00 AU, well outside the orbit of Mars (1.5237 AU). Because of the eccentricity of its orbit, Mars presently polices the region from 1.3814 to 1.6660 AU. The Main Belt divides into four different families, I ($2.06 < a < 2.50$), II ($2.50 < a < 2.82$), III ($2.82 < a < 3.27$), and IV ($3.27 < a < 3.65$). The Hilda family lies well detached from the Main Belt near 4 AU, and the Trojans, of which more than 1200 are known, lie in two clusters on Jupiter's orbit, equidistant from the Sun and Jupiter ($a = 5.2$ AU). (944) Hidalgo ($a = 5.86$) has so large an eccentricity that it ranges from $q = 2$ AU to $Q = 9.6$ AU, an orbit reminiscent of short-period comets. (279) Thule lies at $a = 4.26$ AU in splendid isolation.

The existence of Trojan asteroids on Jupiter's orbit has led to several searches for asteroids trapped in similar L4 and L5 locations on the orbits of other planets. There are severe observational difficulties associated with the search for Mercury and Venus Trojans. No true Earth Trojans have yet been found at a detection limit of about 1 km diameter (but also dependent on albedo); however, several asteroids have been found in or very near 1:1 orbital resonance with Earth. The NEA (3753) Cruthina pursues a "horseshoe" orbit that slowly oscillates back and forth between Earth's L4 and L5 points, and the asteroid 2002 $AA_{29}$ has recently been found to follow a similar orbit that loiters within 0.2 AU of Earth for up to 50 years at a time, making it a virtual satellite of Earth. The discovery of a Mars Trojan asteroid, appropriately named (5261) Eureka, several years ago has renewed interest in the search for Earth Trojans. No Saturn Trojans have yet been identified, though a new search using the technologies that have recently been applied to the search for small Saturnian and Uranian satellites is certainly in order.

We should note that Trojan asteroids also may have satellites: the asteroid (617) Patroclus was recently found by William J. Merline to have a similar-sized companion, S/2001 (617)1, only 0.2 mag fainter than its primary. William K. Hartmann suggested many years ago, based on the lightcurve of Hektor, that it may be a contact binary with equal-sized lobes.

When the other orbital parameters are taken into account, more families emerge within these groups. Three of these new Belt families form very tight clusters on a plot of inclination vs semimajor axis (Fig. VIII.20). The Koronis family, near $a = 2.85$ and $i = 2°$, has an

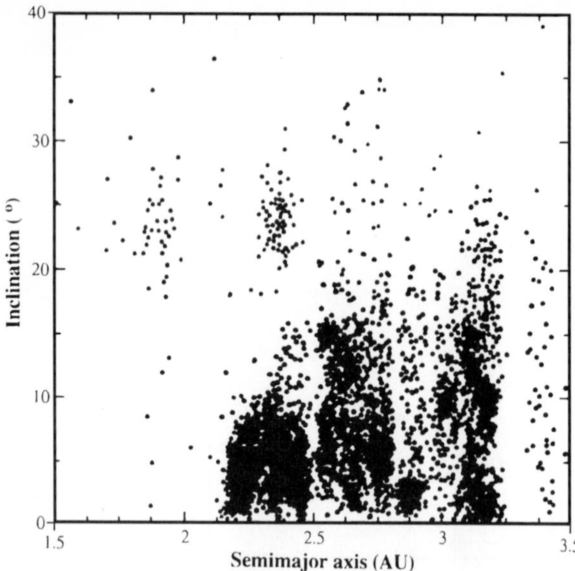


**Figure VIII.20** Inclination and semimajor axis plot of asteroids in and near the Belt. Clustering of orbital properties reflects regions that are relatively stable against perturbations by Jupiter. Some of the groups may be descendants of a single collisionally disrupted large asteroid. The asteroids in a cluster on this diagram are spread all around the Sun and have different eccentricities; no physical clusters of asteroids in space are stable except Jupiter's L4 and L5 Trojan families (see Fig. VIII.21).

eccentricity of about 0.05 and lies within main-belt group III. The Eos family, with $a = 3.01, i = 10°$, and $e = 0.17$, and the Themis family, with $a = 3.13, e = 0.15$, and $i = 1°$, also lie within group III. Just outside the Hungaria group are the Phocaea family ($a = 2.4, i = 23°, e = 0.25$ on average) and the Flora family ($2.06 < a < 2.3, i < 10°$, and $0.08 < e < 0.20$). Another small but sharply defined cluster, the Nysa family, lies near $a = 2.43, e = 0.17$, and $i = 3°$, also within main-belt group I. The study of clustering of asteroids into families has been underway since Hirayama's pioneering work early in the 20$^{th}$ century, and many other less obvious families have been suggested in addition to those described here.

It is easy to pay attention to the orbital parameters of the asteroids, but the subtler tack of asking where the asteroids *aren't* is just as rewarding. It is clear that asteroids avoid certain orbital semimajor axes. There are pronounced gaps in the semimajor axis distribution near 2.5, 2.8, 2.9, 3.3, 3.5 to 3.9, and of course from the Hildas to the Trojans (4.1 to 5.1 AU), a region populated only by 279 Thule. These gaps, discovered in 1867 by D. Kirkwood and explored in more detail a century later by Dirk Brouwer, represent forbidden orbital periods, not forbidden heliocentric distances: most asteroids have sufficient orbital eccentricity so that they cross one or more of these gaps frequently. Indeed, a snapshot of the distribution of asteroid heliocentric distances at any moment would show hardly any evidence of the existence of these gaps! It is very instructive to calculate the exact values of the forbidden periods. These periods are 3.954, 4.745, 5.084, 5.931, and approximately 7.1 years for the best-defined gaps. It takes little effort to show that these periods are resonances of the orbital period of Jupiter, 11.8622 years. Specifically, the major *Kirkwood gaps* correspond to 1/3, 5/2, 3/7, 1/2, and 3/5 of the orbital period of Jupiter. A host of low-order commensurabilities exist beyond the Hilda family (4/5, 5/6, 6/7, 7/8, 8/9 etc.; 6/8, 7/9, 8/10, etc.; 6/9, 7/10, 8/11, etc.), accounting for the dearth of asteroids in such orbits. The phenomenon by which Jupiter orbital resonances deplete asteroids from these gaps is simply this: if there exists an orbital resonance between an asteroid and Jupiter, the geometries of their orbits will follow a repeating pattern during times of closest approach, and as a result the orbital perturbations of the asteroid will accumulate and reinforce each other. On the other hand, if no Jupiter commensurability exists, the perturbations will tend to be random and will average out over long runs of time, with little or no net effect on the orbit of the asteroid.

Asteroidal bodies, solid in appearance but with probably extensive internal reservoirs of ices, are also found in profusion much farther from the Sun. Some of these bodies pursue eccentric paths, highly vulnerable to perturbations, that cross the orbits of one or more of the Jovian planets. These bodies, termed *Centaurs*, including (2060) Chiron, (5145) Pholus, and dozens of others, are certainly transient visitors to that region, derived from other, more stable locations, and fated to depart soon for potentially even more exciting orbital adventures. (2060) Chiron, with $a = 13.69$ AU, crosses the orbit of Saturn and grazes the orbit of Uranus at aphelion, and Pholus (formerly 1991 DA) ranges from Mars to beyond Uranus. Many of these bodies are very large, over 100 km diameter, and some, according to numerical integrations of their orbits, may wander from time to time into Earth-crossing orbits. Even farther from the Sun we run into an immense population of very dark bodies that orbit in asteroidal (low eccentricity, low inclination) orbits beyond Neptune. Pluto is arguably the largest body in this population. These *trans-Neptunian objects* (TNOs), which constitute the *Kuiper Belt*, can be further subdivided into two major categories, those that partake of Pluto's peculiar 3:2 orbital resonance with Neptune (called *plutinos*), and those with nonresonant distant orbits similar to that of 1992 $QB_1$ (whimsically called *cubewanos*). Thus the recently discovered "little Pluto," 2002 LM60, alias Quaoar, with its 288-year period, is a cubewano, not a plutino. These distant bodies, partaking of both asteroidal and cometary characteristics, are treated separately later in this chapter.

The phenomena and data discussed here serve to give a good general guide to the orbital distribution of the asteroids. The problem is, however, massive and complex enough to occupy the attentions of celestial mechanicians for a long time to come. One of the most interesting problems posed by the observed orbits is the stability of the Trojan asteroid families on Jupiter's orbit and the plutinos in their orbit–orbit resonances with Neptune.

## Stability of Trojan and Plutino Orbits

The first treatment of orbits of this type was published by J.-L. Lagrange in 1772, long before the discovery of the first Trojan asteroid, (617) Patroclus, in 1906. Lagrange considered the stability of a three-body system in which $M_1 >> M_2 >> M_3$ and in which Jupiter ($M_2$) follows a circular orbit about the Sun ($M_1$). There are certain configurations of this system in which the relative positions of the three bodies (that is, their positions as seen in a rotating coordinate system in which the Sun–Jupiter line is fixed) do not change with time. For example, a body poised exactly at the equipotential point between $M_1$ and $M_2$ will follow a circular path around $M_1$ with the same orbital period as $M_2$: it will therefore always stay on the Sun–Jupiter line at the same distance from the Sun. The prospect of two bodies with different semimajor axes orbiting the Sun with the same period is rather startling, but the explanation is simply that $M_3$ is so located that the gravitational force of Jupiter partially offsets the Sun's gravitational force, so that the body $M_3$ experiences a net force exactly sufficient to keep it in a circular orbit. Note the delicacy of this arrangement: if the body were displaced very slightly toward the Sun, the net gravitational force on the body would increase and it would speed up relative to the motion of $M_2$. If it were displaced very slightly toward $M_2$, it would feel a significantly diminished Sunward force and would tend to fall into orbit about $M_2$. Thus this "inner Lagrange point," the L1 point (see Fig. VIII.21), is unstable against perturbing forces.

A small body of mass $m$ in circular orbit about the Sun ($M_1$) experiences a gravitational force of $GmM_1/r^2$. The centripetal force necessary to restrain the body in circular motion (the fictitious "centrifugal force") is $mV^2/r$, and the condition of circular motion is simply that the velocity $V$ be such that the gravitational force is just equal to that required for circular motion:

$$GmM_1m/r^2 = mV^2/r \qquad \text{(VIII.4a)}$$

whence

$$V = (GM_1/r)^{0.5}. \qquad \text{(VIII.4b)}$$

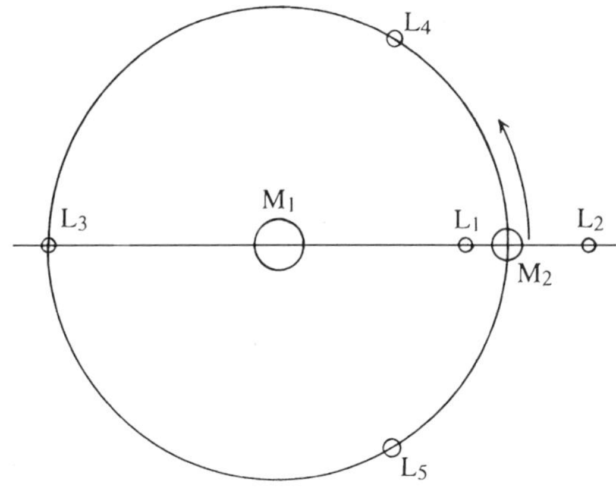


**Figure VIII.21** The Lagrange points. The L4 and L5 points are dynamically stable, and groups of asteroids fill a large volume about these points on Jupiter's orbit ($M_1$ = Sun; $M_2$ = Jupiter). There is only one known Trojan asteroid for another planet, the asteroid 5261 1990 MB, on the orbit of Mars ($M_2$ = Mars).

At the L1 point the net gravitational force is the difference between the forces exerted by the Sun and the planet. The planet, with mass $M_2$, orbits at distance $r$ from the Sun. The distance of $m$ from the planet is $d$. The speed that the body must have to remain in circular motion is $v$, and the condition for stable motion at the L1 point is

$$GmM_1/(r-d)^2 - GmM_2/d^2 = mv^2/(r-d). \qquad \text{(VIII.5)}$$

The speed here is constrained to be such that the periods of the planet ($M_2$) and $m$ are the same.

$$P(M_2) = 2\pi r/V = P(m) = 2\pi(r-d)/v \qquad \text{(VIII.6)}$$

or

$$v = V(r-d)/r. \qquad \text{(VIII.7)}$$

The last four equations permit a solution for $d$ for any given combination of masses and planetocentric distance $r$.

Similarly, on the extension of the Sun–Jupiter line outside of Jupiter's orbit the total Sunward gravitational force is everywhere greater than if Jupiter were absent, and the speed of a body in circular heliocentric orbit is somewhat greater than it would be in a two-body system. It can therefore follow a larger orbit about the Sun than Jupiter, but with the same orbital period, if it is situated at just the right distance. If it is displaced slightly closer to Jupiter it will feel a much larger Sunward force and will tend to drop in closer to Jupiter and orbit it. If displaced slightly outward, it will experience a much smaller Sunward force and will be unable to maintain a circular orbit with Jupiter's period, but will instead fall behind Jupiter's position. This second "outer Lagrange point," called the L2 point, is therefore unstable in the same way that the L1 point is.

The force balance at the L2 point is

$$GM_1/(r+d)^2 + GM_2/d^2 = v^2/(r+d), \qquad \text{(VIII.8)}$$

and the distance $d$ may be calculated as a function of $r$ and the masses of the Sun and the planet in the same manner as above. The point on Jupiter's orbit on the opposite side of the Sun from Jupiter experiences a gravitational force directed toward the center of the Sun, but very slightly stronger than if Jupiter did not exist. Thus a slight outward displacement of a body from this point would put it in an orbit with the same period as Jupiter. Any other heliocentric distance than this would leave the body in an orbit about the Sun with an orbital period different from Jupiter's. This "opposite Lagrange point," L3, is therefore also unstable with respect to infinitesimal perturbations.

The force balance at the L3 point is then

$$GM_1/(r+d)^2 + GM_2/(2r+d)^2 = v^2/(r+d), \qquad \text{(VIII.9)}$$

which may be solved in the same manner as for the L1 and L2 points.

Despite the instability of these locations, spacecraft may be stationed at the L1, L2, or L3 points if they are equipped with propulsion systems capable of delivering very low thrust levels from time to time. Note that the above description of the first three Lagrange points is oversimplified in that it assumes perfectly circular and coplanar orbits for all bodies and neglects the gravitational perturbations exerted by other bodies. Real situations are always more complex and less stable than these idealized cases.

Lagrange also found two regions located near the orbit of $M_2$ at the vertices of the two equilateral triangles that have the $M_2$–$M_1$ (Jupiter–Sun) line as their common base. These locations, 60° ahead of and behind Jupiter, are termed the "leading" (L4) and "trailing" (L5) Lagrange points. A body located precisely at these equidistant points on Jupiter's orbit would feel a gravitational force that is slightly higher than if Jupiter were absent. The force is directed at the center of mass of the Sun–Jupiter system, which is near the surface of the Sun on the side toward Jupiter. At L4 or L5 the force required to hold the body to a circular orbit would be slightly less than the actual total gravitational force, and it would be necessary to displace the body slightly outward away from the center of mass of the system in order to bring these forces into balance.

A body placed near the L5 point with a velocity deficit relative to comovement with Jupiter will fall inward toward the Sun along a slightly eccentric orbit traveling faster than Jupiter itself. As the body approaches inferior conjunction it is accelerated by Jupiter's gravity to a speed greater than Jupiter's orbital speed. It then crosses Jupiter's orbit outbound from the Sun before reaching Jupiter and heads outward from Jupiter following an eccentric orbit with a longer orbital period than Jupiter. It then slowly retreats from Jupiter back to and beyond the L5 point. During this retreat it reaches aphelion and then drops back in to cross Jupiter's orbit behind the L5 point. It continues in this fashion, executing an orbit with the same average period as Jupiter. As seen in a coordinate system rotating with the Sun–Jupiter line, the body exhibits a large retrograde loop about the L5 point. Exactly the same reasoning applies to the motion of bodies slightly displaced from the L4 point. Thus the L4 and L5 points are the "high ground" at the centers of two very large regions of stability, within which bodies will continue to orbit with the same average period as Jupiter despite small perturbations. The two groups of Trojan asteroids orbit within these two stable regions. This behavior should remind us of the behavior of the coorbital satellites in the Saturnian system, as discussed in Chapter VI.

It is an interesting and challenging question whether the Trojan asteroids may escape from these stable regions over long periods of time. In an idealized planetary system, with a single planet (Jupiter) in a perfectly circular orbit, and with no collisions to disturb the motions of the trapped bodies, it is easy to believe that these orbits might persist for very long periods of time. But the presence of other massive planets, finite and variable orbital eccentricity, and chance collisions with asteroidal and cometary debris all serve to make predictions more difficult and less certain. Similarly, we must wonder whether asteroids initially in non-Trojan orbits may wander into the 1:1 resonance and become Trojans. Even if the L4 and L5 points were potential minima (and they are not), it is not obvious how capture could take place without some form of energy change. This could be a collision with an existing Trojan asteroid, a "propulsive maneuver" executed by an expiring comet, or a complex multibody gravitational interaction with massive but remote planets. Each of these capture scenarios suffers from statistical problems, but none of them is physically impossible. Finally, the issue of capture and loss of L4 and L5 bodies is closely linked to the question of capture and loss of planetary satellites. The Lagrange and satellite populations may exchange with one another, so that these bodies collectively constitute a single dynamically related family.

In our discussion of the Neptune–Pluto resonance in Chapter VI, we commented on the apparent improbability that Pluto and Charon may be escaped satellites of Neptune. The duplicity of the system made it even harder to envision such an escape scenario. It has become apparent over the past few years that there is a vast population of plutinos in 3:2 and even 2:1, 4:3, and 5:3

orbit–orbit resonances with Neptune. Some of these, including the TNO body (26308) 1998 $SM_{165}$ mentioned earlier, also have satellites. Another example, 1999 $TC_{36}$, is a plutino that, like all plutinos, orbits at a mean distance of $a = 40$ AU from the Sun, well beyond Neptune. The TNO satellite systems seen to date show a wide variety of separations, periods, and mass ratios. One binary TNO, 2001 $QW_{322}$, has an angular separation of 4″ arc between its two components. At a distance of 40 AU, that is a separation of 116,000 km. The two bodies have virtually indistinguishable brightness, which, assuming equal albedos and densities, implies equal diameters and masses. The orbital period estimated for this system is 4 Earth years. The orbital speed is of order $3\,\mathrm{m\,s^{-1}}$, and the escape velocity from that orbit is of order $1\,\mathrm{m\,s^{-1}}$. Such a fragile system is a real wonder of nature. In any event, the existence of several documented binaries among the plutino and cubewano populations makes untenable any scenarios involving escape from Neptune.

## Sizes, Shapes, and Albedos of Asteroids

Factual knowledge of the sizes, shapes, and absolute reflectivities of asteroids dates back only to the early 1970s. Although reference books published prior to that time often contained tables of asteroid diameters and masses, close examination of the sources of these "data" reveals that some authors simply guessed what seemed to them a reasonable albedo (for example, that of the Moon) and then guessed a density that they believed to be representative of all asteroids (the density of the Moon, Earth, or the author's favorite class of meteorite), then converted visual magnitude to mass and radius. Other unwary authors sometimes took radii or albedos from one such table and masses from another (both based on different false assumptions) and calculated densities that were doubly imaginary. As a result, almost all published "data" on asteroid sizes, masses, and densities prior to 1970 are completely wrong.

These early tables of asteroid sizes and figures came mainly from photometric measurements of the visual brightness. These measurements are often reported as the mean visual magnitude at opposition. This is easily recalculated as a standard magnitude, which is the visual magnitude an asteroid would have if it were one astronomical unit from both the Earth and the Sun with its disk fully illuminated (as it is at opposition). The standard magnitude is then proportional to the negative log of the product of the (unknown) albedo and the (unknown) cross-section area. No means were then available to distinguish large, dark asteroids from small, bright ones. The measured magnitudes, which were often quite reliable and precise, have survived to the present; the interpreted albedos, diameters, densities, and masses have not.

These early magnitude measurements revealed both a surge of brightness of these bodies at very small phase angles (close to opposition) and periodic time-dependent variations in the brightness of many asteroids, evidently due to zonal variations in either albedo or cross-section area of the bodies as they rotated. Again, with these data alone it was not possible to decide whether a given asteroid was irregular in shape or whether it had large-scale albedo variation. Whichever interpretation is correct, the time-dependent brightness variations (the *lightcurves*) of the asteroids tell us their rotation periods. Many asteroids are elongated or irregular in shape, frequently showing two unequal brightness peaks for each rotation. The Centaur 1997 $SE_5$, for example, has the spectrum of a D asteroid, a three-peak lightcurve.

The average rotation speed for an asteroid is about 10 hours, but periods as short as 2.4 and as long as 50 hours are known (see Fig. VIII.22). Rotation periods tend to be slightly shorter for smaller (<30 km) and the largest (>140 km) asteroids (Fig. VIII.23). The size of the brightness variation, which may correlate with changes in the cross-section area, is somewhat larger for smaller (<50 km) asteroids. Typical variations over a single rotation are ~0.2 magnitudes. The amplitude of the brightness fluctuations is greatest for small Earth-crossing asteroids. The paucity of observed 10-km bodies in the Main Belt makes it impossible to be certain whether this effect is also present in the Belt or is peculiar to the NEAs. There is little information on the orientation

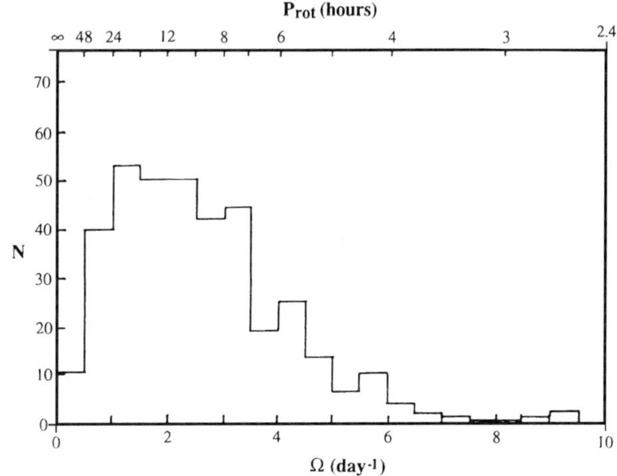


**Figure VIII.22** The rotation periods of asteroids. The spin periods show a broad maximum between 7 and 30 h. At a spin period of 1.5 to 3 h (depending on density) the equatorial centrifugal force equals the gravitational attraction, and the asteroid becomes rotationally unstable and begins to fragment.

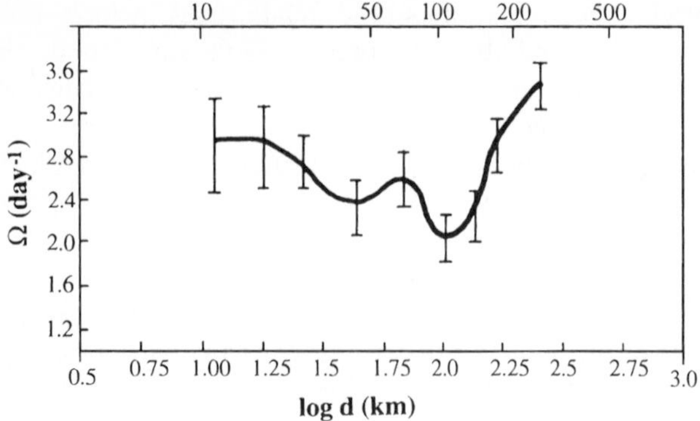


**Figure VIII.23** Rotation periods of asteroids of different sizes. Asteroids over about 125 km diameter form a coherent, relatively rapidly rotating group, which may reflect the properties of an early, small population of large asteroids that has in part survived collisional disruption. The smaller asteroids have more slow rotators and more dispersion of rotation rate, probably reflecting violent disruption events. Faster rotators also tend to have larger-amplitude light curves (they are very irregular in shape).

of the spin axes. Only a few dozen asteroids have been studied, but the data hint at a tendency for the largest asteroids to rotate in the prograde sense.

There are five modern methods available for direct or indirect determination of radii and albedos. First, there are occasional occultations of stars by asteroids. With a sufficiently accurate prediction of the ephemeris of an asteroid, chains of observing stations can be set up across the path of the asteroid's shadow on the Earth. Each station observing an occultation can time the immersion and emersion of the star with good precision, and the lengths of a number of chords across the asteroid can then be calculated. This provides not only the radius of the asteroid, but also some information on its figure. Since occultations by any given asteroid are rare, and since a large proportion of all the occultations that are in principle observable on Earth are actually impractically remote or unobservable because of cloud cover, these data accrue very slowly. Second, there is the polarimetric technique for determining albedos and radii, which we described in some detail in Chapter VI. Third, and also described in Chapter VI, is the infrared technique, in which simultaneous visible and infrared flux measurements are used to determine both the albedo and radius. Fourth, there is optical interferometry, which had to date been used on very few asteroids. Fifth, there is Earth-based radar, which provides estimates of the total effective cross-section area and the time-delay and Doppler widths of the reflected pulse. Because the transmitted pulse drops off in intensity with the asteroid–Earth distance $D$ as $D^{-2}$ and the (imperfectly) reflected signal also drops off as $D^{-2}$, this technique has an overall sensitivity that drops off as $D^{-4}$. It is therefore best applied to close (and therefore small) asteroids. A sixth technique become available for the first time in 1991: direct imaging of an asteroid by a spacecraft camera system.

Once data on either the albedo or the diameter of an asteroid are available, it becomes possible to determine the cause of the lightcurve brightness variations. If, for example, the infrared technique shows that the apparent radius of a body varies by 50% over a revolution, while the apparent albedo remains constant (that is, the visual and infrared fluxes vary in concert with a constant ratio), then we clearly must attribute the brightness variations to a highly nonspherical shape. Conversely, if the sum of the visual (reflected) and infrared (absorbed and re-emitted) fluxes is nearly constant, while the ratio of the two varies widely with time, we must infer dramatic albedo variations from one hemisphere to the next. Such is in fact the case with Saturn's satellite Iapetus, but evidence of large albedo variations with phase is weak for asteroids. On the other hand, there are many asteroids known which clearly have very irregular shapes. Several have cross-section areas that vary by more than a factor of 2 as they rotate.

The albedos of asteroids vary over a wide range, from a low of 0.02 (596 Scheila) to a high of 0.48 (44 Nysa), and the distribution of albedos is clearly bimodal (Fig. VIII.24). There is a peak at an albedo 0.033, a minimum near 0.07, and a second peak near 0.15. The albedo evidence alone clearly demonstrates the existence of at least two optically distinguishable classes of asteroids without recourse to any spectral information. It is tempting even with this modest amount of data to speculate that the extremely dark class of asteroids might be carbonaceous chondrite material, while most other chondrites and achondrites might fall into the high-albedo category.

The diameters of several hundred asteroids can now be estimated by at least one of the techniques listed above. There are approximately 30 known Belt asteroids with diameters greater than 200 km, of which only seven are larger than 300 km. These are (1) Ceres (1020 km), (4) Vesta (550 km), (2) Pallas (540 km), (10) Hygiea (440 km), (704) Interamnia (340 km), (511) Davida (335 km), and (65) Cybele (310 km). The errors in these diameters are commonly about 10%. The size distribution of asteroids shows a rapid dropoff in number above a diameter of about 80 km. However, as Fig. VIII.25 shows, there is good evidence for a surplus of asteroids with diameters in the 60 to 120 km range. The size distribution of asteroids can be only poorly approximated by a single straight line of constant slope. Recent studies of near-Earth asteroids shows even more complexity in the size–frequency relation at sizes under 100 m.

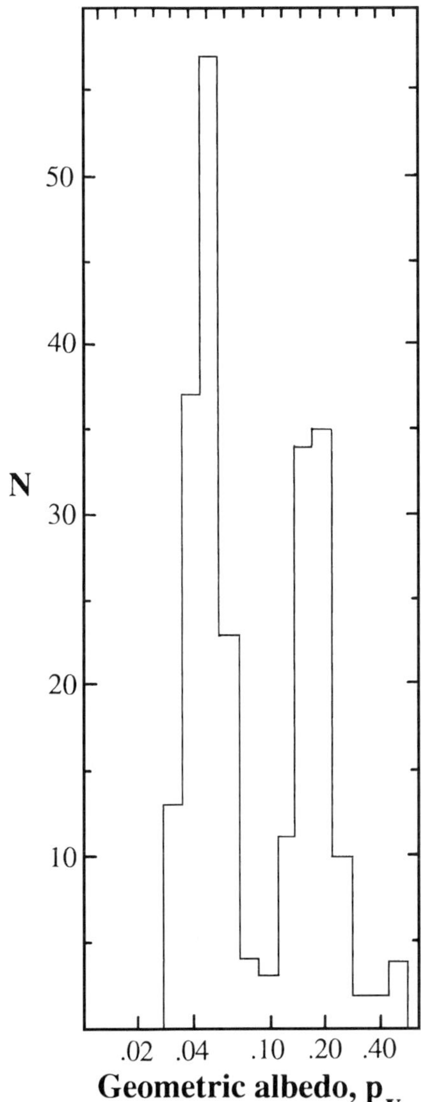


**Figure VIII.24** Geometric albedos of asteroids. The distribution is strongly bimodal, betraying the existence of at least two major classes of asteroid materials, one very dark and one rather bright.

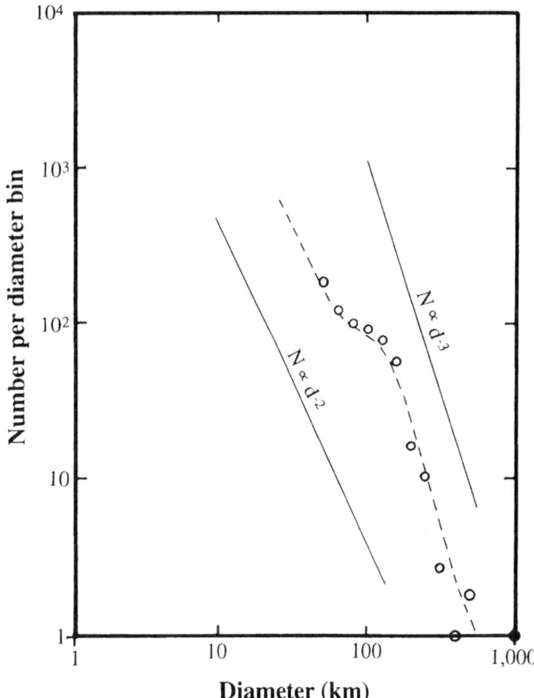


**Figure VIII.25** Sizes of asteroids. The number of asteroids in mass bins that span a factor of 2 is plotted vs mass. There is a striking excess of asteroids at and above approximately 125 km in diameter. Note that Ceres (in the $d = 1000$ km bin) lies well above a line of slope $-3$ (log $N$ vs log $d$), which is the line of constant total mass in each bin, and hence the mass of Ceres dominates the entire Belt. Statistics on Belt asteroids are very difficult to obtain at sizes below 50 km because they are too small to be reliably detected at such a great distance. The Near-Earth Asteroid (NEA) population, which contains only objects under $d = 10$ km, looks very different (Fig. VIII.41).

## Masses and Densities of Asteroids

The conditions necessary for determination of the mass of an asteroid are very stringent. Indeed, the first realization that there were close flybys of a large asteroid by a smaller one came as late as 1949, when G. Fayet found that (197) Arete had been very close to (4) Vesta in 1939. In 1968 it was discovered that Arete and Vesta have an encounter to within 0.04 AU every 18 years. Even later it was found that (1) Ceres and (2) Pallas, which have remarkably similar orbital semimajor axes (2.768 and 2.773 AU, respectively) have an observable effect upon each other's motion because of the long duration of their encounters. A lengthy search has turned up very few prospects for additional mass determinations.

Fairly accurate mass determinations based on mutual perturbations of asteroids are presently available for (1) Ceres, (2) Pallas, and (4) Vesta. But the list of asteroids with experimentally determined masses has grown dramatically over the last few years due in part to the Galileo and NEAR spacecraft missions, and in part to the discovery of natural satellites in orbit around a number of asteroids. The additional asteroids for which mass and density data have been published are (16) Psyche, (45) Eugenia, (90) Antiope, (121) Hermione, (243) Ida, (253) Mathilde, (433) Eros, and (762) Pulcova. Many other asteroids, including those in the Main Belt, NEA, Trojan, plutino, and cubewano populations, have been found to have satellites and should soon provide mass estimates as the orbits are refined by further observations. The densities estimated by Daniel T. Britt and Guy J. Consolmagno are summarized in

Fig. VIII.26. Except for the three most massive asteroids in the set, the deduced bulk densities are lower than expected from laboratory studies of spectrally similar meteorite types. The highest asteroid density found was $3.3 \pm 0.1\,\mathrm{g\,cm^{-3}}$ for the basaltic asteroid (4) Vesta. The other two large asteroids also had densities closely similar to the laboratory data on carbonaceous meteorites, $2.15 \pm 0.15$ for (1) Ceres and $2.6 \pm 0.1$ for (2) Pallas.

The small low-albedo asteroids in the data set, presumably related to the carbonaceous chondrite meteorites (Eugenia, Pulcova, Mathilde, Antiope, and Hermione), have an average bulk density of only $1.3 \pm 0.2$, compared to laboratory densities of 1.58 to 2.2 reported for CI chondrites. The stony asteroids (243) Ida and (433) Eros both have densities close to $2.6\,\mathrm{g\,cm^{-3}}$, compared to laboratory densities of about 3.8 for ordinary chondrites, which have similar mineralogy. The low densities of these smaller asteroids imply substantial porosity, on the order of 30%, presumably due to extensive fracturing and a possible rubble-pile structure.

The density ranges found for all these asteroids lie below the densities of all of the terrestrial planets, even when the latter are corrected for self-compression by their high internal pressures. The density range of (4) Vesta, however, overlaps the density of the Moon.

For purposes of comparison, we should add that the visual Bond albedos of Ceres (0.060) and Pallas (0.066) both lie in the sparsely populated gap between the two prominent albedo peaks. This suggests that there may be something odd about both bodies. Vesta, with the highest density found, has an albedo of 0.216, on the high side of the high-albedo peak. Note that a 5% uncertainty in the radius, which is about the best obtainable for an asteroid, causes a 10% error in the calculated albedo.

## Photometry and Spectroscopy of Asteroids

The simplest, and first-tried, source of spectral information on asteroids was UBV photometry, which has been a standard tool of astronomers for decades. UBV photometric data exist for more than 800 asteroids, and numerous attempts have been made to display and interpret this vast body of low-resolution evidence. The variables at our disposal are the geometric albedo $p_V$, the U, B, and V magnitudes, and the U-B, B-V, and U-V colors. We will here always give the color differences in the sense of short-wave magnitude minus long-wave magnitude. Thus a positive number

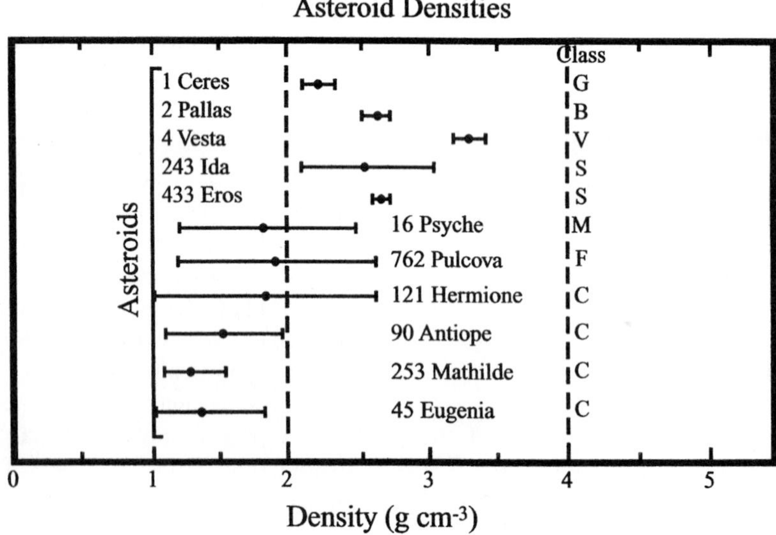


Figure VIII.26 Asteroid densities. Those asteroids for which useful density information is available, from mutual perturbations, the presence of a satellite, or perturbation of the trajectory of a flyby spacecraft, are summarized after the compilation by Britt and Consolmagno. Those with spectral affinities to the carbonaceous chondrites (C, F, and G classes) are systematically less dense than the S-type (ordinary chondrite, achondrite, or stony-iron) and V-type (basaltic achondrite) asteroids. The position of (16) Psyche is ambiguous because of the large error bars, which allow a dry (M) or water-bearing (W) classification.

indicates that the redder magnitude is smaller, and hence that the longer-wave flux is higher: a positive number means that, over the wavelength region considered, the body observed is "red." A simple plot of geometric albedo vs U-V is given in Fig. VIII.27. As shown on this diagram, the asteroids can be resolved into several distinct groups. The low-albedo asteroids, with both albedos and colors similar to laboratory samples of C chondrites, are called C asteroids. The C asteroid (253) Mathilde, found by NEAR to have a density of only $1.3 \pm 0.3$, strangely has no 3-$\mu$m water feature in its spectrum.

Above the "albedo gap" we find two classes with very similar albedo distributions but distinctly different U-V colors. The redder group, with a color characteristic of stony irons or of some as yet unknown very metal-rich stony material, is termed the S type. Their less-red counterparts, which show little evidence of ferrous-iron-bearing silicate minerals, have spectra closely similar to those of iron meteorites or enstatite chondrites. These are called M asteroids because of the apparent dominance of metal in their spectral properties. Their spectra (slightly red, with no distinctive silicate absorption bands) are too simple to allow confident compositional typing, especially for those asteroids for which we have no albedo and size data.

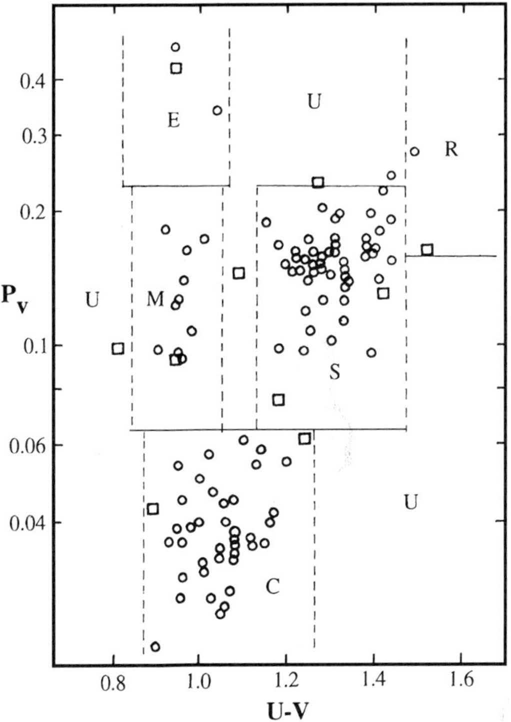


**Figure VIII.27** Albedo–color systematics of asteroids. This two-dimensional classification system resolves several asteroid classes. The regions labeled U contain only unclassifiable outliers.

At least a third of the original set of M asteroids have been found to have an absorption feature near 3 $\mu$m in the near infrared, where water has a strong absorption feature. They have been redesignated as W asteroids, the symbol intended to be mnemonic for water and a reminder that they are "inverted M" asteroids. Since troilite (FeS) is a ubiquitous component of meteoritic iron, and also has an absorption feature in the 3-$\mu$m region, it is quite possible that the full story of the W asteroids and their relationship with the M class remains to be told.

A few extraordinarily bright bodies with less-red U-V color resemble the laboratory reflection spectra of enstatite achondrites. These are called the E group. The R asteroids have moderate to high albedos and very red reflection spectra, with a clear absorption due to $Fe^{2+}$ in silicates. Other groups include the D and P groups, which lie in the outer part of the Belt. At least the D group, which is abundant among the Trojans, appears to be simply related to C asteroids, perhaps with a more volatile-rich ("super-carbonaceous") composition. Asteroids that are spectrally unusual in any way are labeled U (for unusual); for example, an asteroid with unique properties would be called simply U, but one just outside the C class and not near any other family would be labeled CU. Ceres, for example, is G, and Pallas is B, both plausibly described as highly altered carbonaceous bodies.

The V class, originally invented to label the distinctive spectrum of (4) Vesta (see Fig. VIII.32), has been found by Richard Binzel of M.I.T. to include dozens of small bodies in orbits that appear to be dynamically related to Vesta, in the sense that they may have originated as a "spray" of fragments from a major impact on Vesta. The V asteroids provide a very close match to the laboratory reflection spectra of basaltic (HED) achondrites. In 2000, an outer-Belt asteroid, (1459) Magnya, far beyond the family of Vesta-related small bodies, was found to have a V-type spectrum, and in 2002, Akiro Yamaguchi and colleagues of the National Institute of Polar Research in Tokyo, a leading center for the study of Antarctic meteorites, identified the meteorite Northwest Africa 011 as a basaltic achondrite, a eucrite, with many chemical dissimilarities to the HED family usually linked to Vesta, including a distinctly different oxygen isotope signature. Thus both the meteorite and asteroid data suggest at least two independent places of origin for basaltic achondrites.

A number of other asteroid spectral classes with few members exist, as detailed in Table VIII.4.

Since 1970, photometric reflectivity measurements with a spectral resolution of approximately 0.04 $\mu$m have been reported for more than 600 asteroids.

**Table VIII.4** Asteroid Spectral Classes

| Class | Albedo | Minerals | Meteorite analogues |
|---|---|---|---|
| A[a] | High | Olivine; olivine + metal | Achondrite, pallasite |
| B[a] | Low | OH silicates, carbon, organics | Altered CI or CM |
| C | Very low | OH silicates, carbon, organics | CI or CM |
| D | Very low | Carbon, organics | Dry C-like |
| E[a] | Very high | FeO-free enstatite, forsterite | Enstatite achondrites |
| F | Very low | OH silicates, carbon, organics | Altered CI or CM |
| G[a] | Very low | OH silicates, carbon, organics | Altered CI or CM |
| M | Moderate | Metal; metal + enstatite? | Irons |
| P | Very low | Carbon, organics | Dry C-like |
| Q[a] | High | Olivine, pyroxene, metal | Ordinary chondrites |
| R[a] | High | Pyroxene, olivine | px-ol achondrite |
| S | Moderate | Metal, olivine, pyroxene | Pallasites |
| T[a] | Very low | OH silicates, carbon, organics | Very altered CI or CM |
| V[a] | High | Pyroxene, feldspar | Basaltic achondrites |

[a] Rare class; only a few members known.

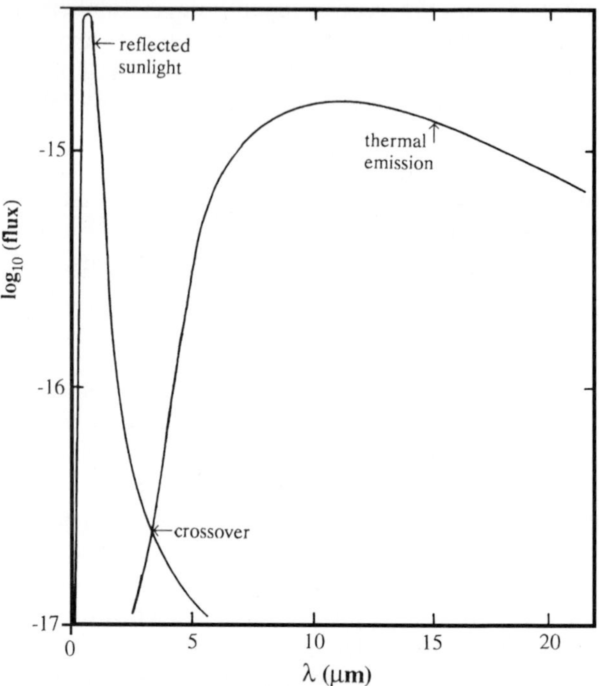


**Figure VIII.28** Reflected sunlight and reemitted thermal radiation. The spectral crossover (and flux minimum) shifts to longer wavelengths farther from the Sun, where the steady-state temperatures are low and the Planck peak shifts to longer wavelengths.

At least 80 distinguishable varieties of spectra have been found, most of which are not identical to those of any known type of meteorite. These observations have sufficient spectral resolution to permit the identification of several major meteoritic minerals. Laboratory spectra of meteorite samples have been studied extensively to assist in learning how to interpret the asteroid observations. The most fruitful spectral region to date has been the visible and near infrared (0.4 to 1.1 $\mu$m): this region is readily observable and contains several strong absorption bands due to major rock-forming minerals. Other bands of interest occur out to about 5 $\mu$m, beyond which the Earth's atmosphere becomes opaque out to the next transmission window, which extends from about 8 to 14 $\mu$m. The flux from an asteroid at wavelengths less than about 3 $\mu$m is dominated by reflected sunlight (note that the exact wavelength of the crossover depends on the albedo and heliocentric distance of the body). The reemitted thermal radiation from a body with a mean surface temperature of 200 K peaks at a wavelength of about 2900/200 or near 15 $\mu$m. Earth-based observers normally see only the dayside of Belt asteroids, and hence see somewhat higher surface temperatures. The observed emission from a typical Belt asteroid peaks between 10 and 15 $\mu$m (see Fig. VIII.28). The region of the flux minimum at the crossover of the reflected and emitted fluxes, usually near 4 $\mu$m, is much harder to observe than the 1-$\mu$m or 10-$\mu$m regions, and is often missed in spectral studies of small or distant asteroids.

The spectral features found in studies of meteorites and asteroids are few in number, but great in importance. It is convenient to distinguish three classes of spectral behavior of materials. First we have opaque minerals, such as magnetite, graphite, amorphous carbon, or tarry organic polymers, which severely darken the entire spectrum but provide little diagnostic information such as discrete absorption bands. Second, there is free metal in the form of kamacite and taenite, which exhibits a smooth and rather distinctive increase of reflectivity toward longer wavelengths (Fig. VIII.29). On a plot of reflectance vs wavenumber these metal alloys exhibit a graph that is concave upward. The higher the nickel content of the metal, the more nearly linear the plot appears. Third, there are the numerous and very diverse silicate minerals. The principal silicates in meteorites and asteroids (olivine, pyroxenes, plagioclase feldspar, and phyllosilicates), especially those containing high concentrations of the transition metals iron and titanium, have strong and distinctive absorption features in the visible and near infrared spectral regions. Long ignored by asteroid spectroscopists has been the ubiquitous sulfide mineral troilite, FeS, which has $Fe^{2+}$ absorption features in the 0.9-$\mu$m region and another absorption near 3 $\mu$m.

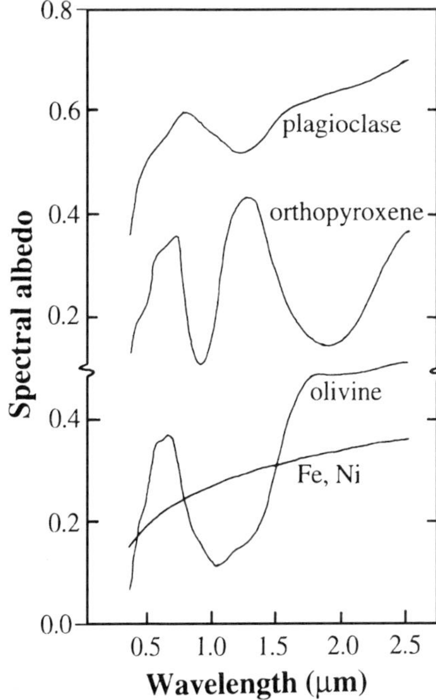


**Figure VIII.29** The wavelength dependence of albedos of selected important asteroidal minerals. Note the change of scale (to avoid the confusion of plotting the four curves on top of each other). Metallic surfaces are red but free of absorption bands. Most normal asteroids contain mixtures of these and other, less abundant, minerals, making spectral interpretation more difficult.

Silicate minerals, which are responsible for essentially all the band structure seen in asteroid and meteorite reflection spectra, have many different modes of absorption of electromagnetic radiation. In the ultraviolet spectral region alone there are several modes of absorption. Electronic excitation to produce an electron–hole pair or to promote a valence electron to a delocalized (conduction band) state are both possible in a wide variety of minerals. Charge transfer between ions, which is especially important for ferrous–ferric ion pairs, also contributes importantly in the ultraviolet. At sufficiently short wavelengths, virtually all minerals are strong UV absorbers. The transition elements from titanium through copper all have incompletely filled *d* orbitals, within which electrons can be distributed over a number of different energy levels. The exact energies and spacings of these levels are determined by the geometry of the compound or crystal within which the metal ion resides. Excitation of the highest-lying electron in the ground electronic state to the lowest-lying unfilled orbital commonly requires an energy corresponding to a photon of visible light. As a result, most of the compounds of these elements are very strong band absorbers in the visible and near-IR regions. As a result, these minerals are visibly colored. The exact band position and shape tell us what ion is responsible for the absorption band and often permit us to identify unambiguously the mineral in which that ion resides. Because these electronic transitions are so strong, minerals containing minor amounts of transition metals, such as plagioclase and magnesian pyroxenes and olivine, can often be identified reliably even though absorption features due to the major constituents of the mineral are not observed.

Ferrous iron (see Fig. VIII.29), a very minor constituent of plagioclase, has a strong absorption feature in the 1.1- to 1.3-$\mu$m wavelength region in the near infrared. The band center lies near 1.1 $\mu$m in the sodic plagioclase endmember, albite, and shifts toward longer wavelengths as the mole fraction of calcic plagioclase, anorthite, increases. Also, of course, the strength of the absorption band increases with the concentration of ferrous ion in the mineral. Ferrous iron in pyroxene produces an analogous narrow and symmetric absorption band centered very close to 0.90 $\mu$m plus another centered at 1.9 $\mu$m. Ferrous ion in olivine produces a very broad absorption feature extending from about 0.7 to 1.6 $\mu$m, with a noticeably asymmetric profile and a center near 1.0 $\mu$m. Olivine exhibits only the barest hint of absorption in the 2 $\mu$m region. Many meteorites contain olivine–pyroxene mixtures: the compositions and proportions of these minerals can be determined by deconvolving the overlapping ferrous iron bands in the 1-$\mu$m region.

Another strongly absorbing and distinctive species in meteoritic and asteroidal mineral assemblages is water. Water vapor has three strong fundamental absorption bands, all located in the easily accessible near infrared. The first of these ($\nu_1$) is the symmetric stretching mode at 2.73 $\mu$m, the second ($\nu_2$) is the H–O–H bending mode at 6.27 $\mu$m, and the third ($\nu_3$) is the asymmetric stretching mode at 2.66 $\mu$m. These bands are shifted and modified in shape in liquid water and solid ice, but are still strong. In ice, these bands lie at 3.11, 6.06, and 2.94 $\mu$m, respectively. In minerals, water may be bound as $H_2O$ molecules (as in hydrated salts or interlayer water in clays), or as -OH groups attached directly to metal ions in the mineral lattice (as in phyllosilicates such as serpentine, chlorite, and amphiboles). In hydrated minerals there is a very strong absorption feature in the region of the $\nu_1$ and $\nu_3$ fundamentals, centered at a wavelength that varies, according to the nature of the mineral, from about 2.9 to 3.3 $\mu$m. There are also combination (overtone) bands due to simultaneous excitation of two or more vibrational transitions by a single photon of higher energy. In solid hydrates, such bands exist near 1.4 $\mu$m (the $2\nu_3$ combination) and 1.9 $\mu$m ($\nu_2 + \nu_3$). Since hydroxyl silicates do not contain H–O–H molecules, the $\nu_2$ bending mode of that species is absent, and thus the 1.9-$\mu$m combination band is also absent. An infrared spectrum of Ceres is shown in Fig. VIII.30. A recent 2.5- to 4-$\mu$m survey of the outer Belt by T. D. Jones and co-workers of the University of Arizona has found water-bearing asteroids belonging to the C, G, T, F, B, and P taxonomic classes. Interestingly, only one of six P asteroids showed evidence of water, and none of the four D asteroids studied showed any sign of the 3-$\mu$m absorption of bound water. They suggest that water is not a primordial constituent of asteroidal silicates, but that hydroxyl silicates are formed as a result of mild thermal processing within asteroids. The energy source for melting appears to drop off rapidly with heliocentric distance, and hence they suggest T Tauri phase solar wind heating as the driving force behind aqueous alteration of asteroidal materials. We recall that laboratory data on the sulfur isotopic composition of minerals in carbonaceous chondrites strongly suggest low-temperature (near 0°C) oxidation and hydration of a high-temperature mineral assemblage as an essential part of generating the observed meteorite materials.

Determining the actual proportions of minerals in an optically characterized asteroid is a complex business. Consider three samples containing magnesian pyroxene (enstatite) with a few percent of ferrosilite. One of these samples consists of centimeter-sized single crystals of pyroxene and millimeter-sized chunks of carbon black piled up to considerable depth. In the second, the pyroxene is crushed to a very fine powder with the large chunks of carbon mixed randomly through it. The third is also crushed into a fine mixture of particle sizes and shapes but with the powdered carbon black homogeneously mixed through it.

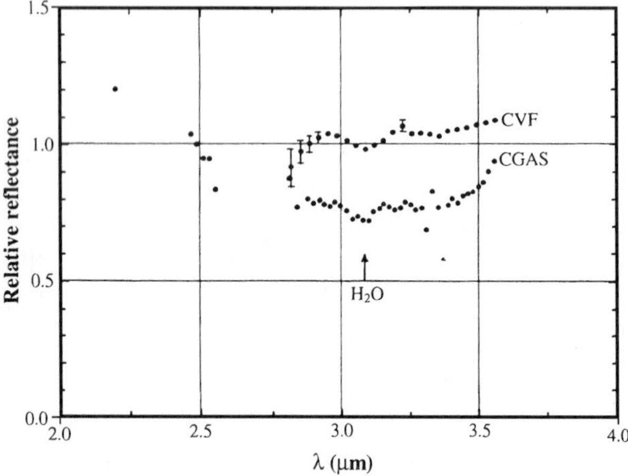


**Figure VIII.30** Spectra of Ceres in the vicinity of the 3-$\mu$m feature of water ice and hydroxyl silicates. Spectra are from two instruments, a circular variable filter (CVF) and the cooled grating array spectrometer (CGAS), at the Infrared Telescope Facility on Mauna Kea, Hawaii. Spectra courtesy of Tom Jones and Larry Lebofsky.

The first sample will permit deep penetration of photons at all wavelengths except where the ferrous iron absorbs strongly. Most of the photons that escape the surface do not encounter carbon grains. The reflection spectrum will show extremely deep absorption features in the 0.9- and 1.9-$\mu$m regions. The overall reflectivity will not be high, because the only important mechanism for sending light back toward the observer is reflection off of the vacuum–pyroxene interfaces, where there are very large differences in refractive index. The second sample will be very bright because photons will constantly be encountering such interfaces. The average path length traversed by a photon within the mineral grains will be very short. As a result, the reflectivity will be high and the absorption bands will be rather weak. Most of the observed photons will be multiply scattered. Hardly any of the incident photons will interact with carbon. The third sample will be severely darkened by the continuum absorber. Average path lengths of photons through pyroxene will be even shorter than they were in the second sample. Thus the reflection spectrum will be dominated by the black absorber, and the absorption feature due to ferrous iron will be very weak.

If iron–nickel alloy is substituted for carbon black in the thought experiment described above, the metal will impart a strong slope to the spectrum in the third case and a weak slope in the first case, but with no observable consequences in the second.

This discussion reveals very clearly that the strength of an absorption feature is influenced not only by the concentration of the absorbing ion in its host mineral and the abundance of that mineral, but also by the grain size distribution of the mineral, the abundance and grain size distribution of opaque continuum absorbers, and the abundance and size distribution of metal grains.

The spectra of several representative meteorites are shown in Fig. VIII.31. The darkening effect of carbon and the reddening of the spectrum by metal can both be seen clearly. Note that all the meteorite samples show precipitous dropoffs in the ultraviolet. The absorption bands in the 2-$\mu$m region (olivine) and the 0.9-$\mu$m region (olivine and pyroxene) are also clearly visible in Fig. VIII.32. The spectral reflectivities of almost all known classes of meteorites have been examined in the laboratory, and almost all of the features seen in the visible and near infrared can be at least qualitatively understood.

The situation with regard to asteroid spectra is less satisfactory. First, most asteroids are so dark, small, and remote that there are both significant statistical errors in their spectra and significant limitations on the spectral range covered. Many spectra have only

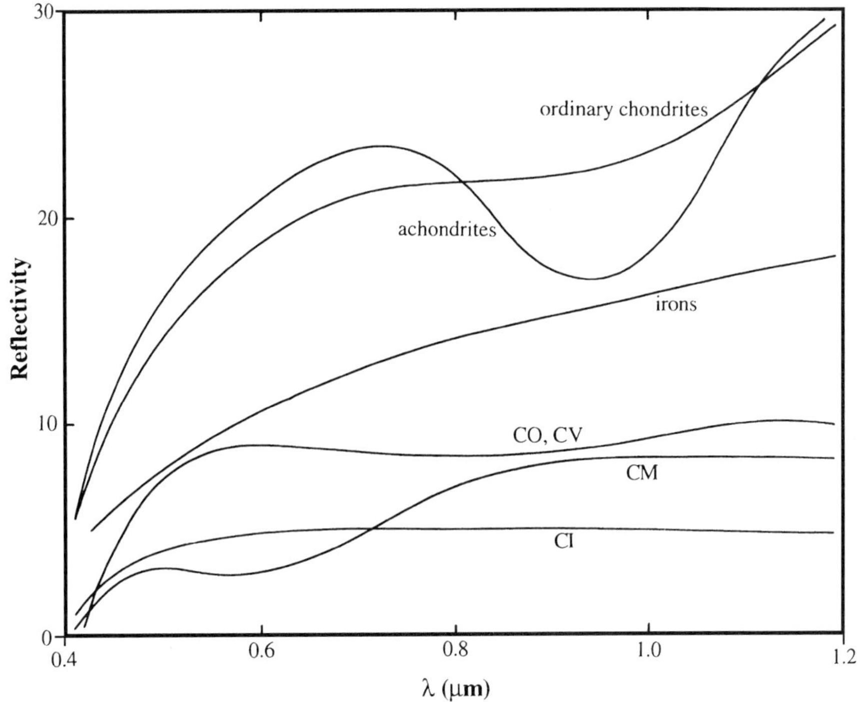


**Figure VIII.31** Spectral reflectivity of meteorite samples. Laboratory studies of meteorite samples and mineral separates provide the basis for identification of asteroid spectra.

very limited spectral resolution. Indeed, hundreds of asteroids have so far been spectrally characterized only by UBV photometry.

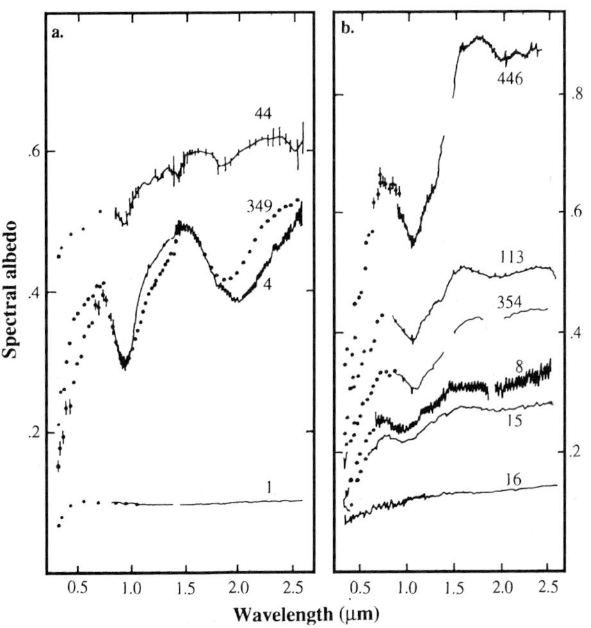


**Figure VIII.32** The wavelength dependence of albedos of several selected asteroids. The asteroid numbers are for (1) Ceres, (4) Vesta, (8) Flora, (15) Eunomia, (16) Psyche, (44) Nysa, (113) Amalthea, (349) Dembowska, (354) Eleonora, and (446) Aeternitas.

Fig. VIII.32 illustrates the reflection spectra of a number of selected asteroids of diverse types, as determined by the Eight-Color Asteroid Survey (ECAS) and its successors. Many of the brightest asteroids clearly contain both ferromagnesian silicates and metallic iron–nickel. An even larger number of asteroids have very flat spectra with little or no suggestion of absorption bands, and hence appear to be carbonaceous in nature. This conclusion is based on very little spectral information (the spectra of carbonaceous chondrites are, after all, virtually featureless), but the albedos of these "C" asteroids are also so low that it is difficult to imagine any naturally occurring material besides carbon black that is so dark. The C asteroids share their very low albedos with only the D and P classes, which are found farther from the Sun than the Cs, and the small B, F, and G classes, which appear to be thermally altered C material (see Table VIII.4).

Asteroids with spectra dominated by metal (irons or enstatite chondrites) are found almost exclusively near the inner edge of the Belt. They are called M asteroids. Those with spectra similar to enstatite achondrites and with very high albedos are called E asteroids. Those with spectra characteristic of stony meteorites (chondrites and achondrites) and stony irons have albedos that are several times higher than those of the C asteroids: these are the S asteroids. The distribution of asteroid spectral classes over geometric albedo, which is a very useful first-order discriminant of composition, is given in Fig. VIII.33. The taxonomy here is due to Edward Tedesco of the Jet Propulsion Laboratory (JPL).

The other taxonomic classes described in Table VIII.4 are generally rare in the Main Belt (but the D and P "super-carbonaceous" classes are dominant beyond the Main Belt). The distribution of the major asteroid spectral classes over heliocentric distance, as determined by Jonathan Gradie (U. of Hawaii) and Edward Tedesco, is shown in Fig. VIII.34.

One of the more interesting results of spectral reflectivity studies of Belt asteroids is the poor correspondence between their spectra and those of the most common types of meteorites arriving on Earth, the ordinary chondrites. For many years, some asteroid astronomers have argued that the spectral properties of the S asteroids may be reconcilable with those of the ordinary chondrites through some as yet unknown process of surface modification. Michael Gaffey of Rensselaer Polytechnic Institute has carefully analyzed the spectra of the S asteroids, making precise measurements of the wavelength of the 1-$\mu$m band produced by FeO in olivine and pyroxene and of the relative strengths of the 1-$\mu$m and 2-$\mu$m bands (the latter, as shown in Fig. VIII.32, due to FeO in orthopyroxene). He finds eight distinct clusters of asteroids in this analysis, of which only one class is compatible with the laboratory spectra of the ordinary chondrites (Fig. VIII.35). Thus the great prominence of S asteroids, second in abundance only to the C asteroids, does not constitute proof of the dominance of ordinary chondrite materials in the Belt. It seems likely that a wide range of parent bodies of achondrites and of chondrite classes that are rare or absent in terrestrial collections must account for the other spectral subtypes of S asteroids.

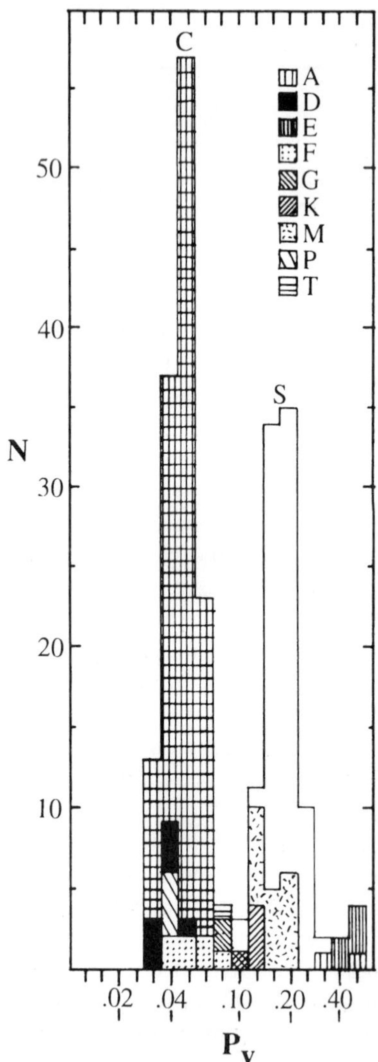


**Figure VIII.33** Relationship of asteroid taxonomic class to albedo. The individual classes are *overlaid*, not stacked, in this diagram.

We know, of course, that fragments knocked off of S asteroids in the Main Belt do not go directly into Earth-crossing orbits, but instead tend to reside in orbits close to the orbits of the asteroids from which they were ejected by impacts. The near-Earth asteroids (NEAs) are a more plausible immediate source of meteorites because ejecta from impacts on them can readily reach Earth. It is thus important to know what spectral classes are present among the near-Earth asteroid population.

To date, fewer than 100 of the near-Earth asteroids have been studied photometrically, of which a substantial fraction remain unpublished. Of these, as Table VIII.5 shows, many have ambiguous spectral classifications, i.e., they have no close equivalent among the classes of meteorites falling on Earth. This effect may be due in part to the extreme faintness of these very small asteroids, which introduces statistical noise into the results. Nonetheless, about 20% of the NEAs appear to be C asteroids, and most of the rest are metal-bearing S asteroids. No P asteroids have yet been identified among the near-Earth population, but P material may be too volatile to survive so close to the Sun. Also, we know that we have so far sampled only a tiny fraction of the near-Earth asteroid population: the search techniques in use up to 1973 discriminated very strongly against small, fast-moving bodies such as the near-Earth asteroids, and both photographic and electronic asteroid searches are still conducted in visible light, which discriminates strongly against discovery of very dark (C, D, P) bodies. Estimates made in the 1980s by Eugene Shoemaker of the U.S. Geological Survey (Flagstaff, Arizona) suggested that the real bias-corrected frequency of dark asteroids in the NEA population is probably 30 to 60%. Shoemaker also estimated that the total NEA population as about 2000 kilometer-sized and 300,000

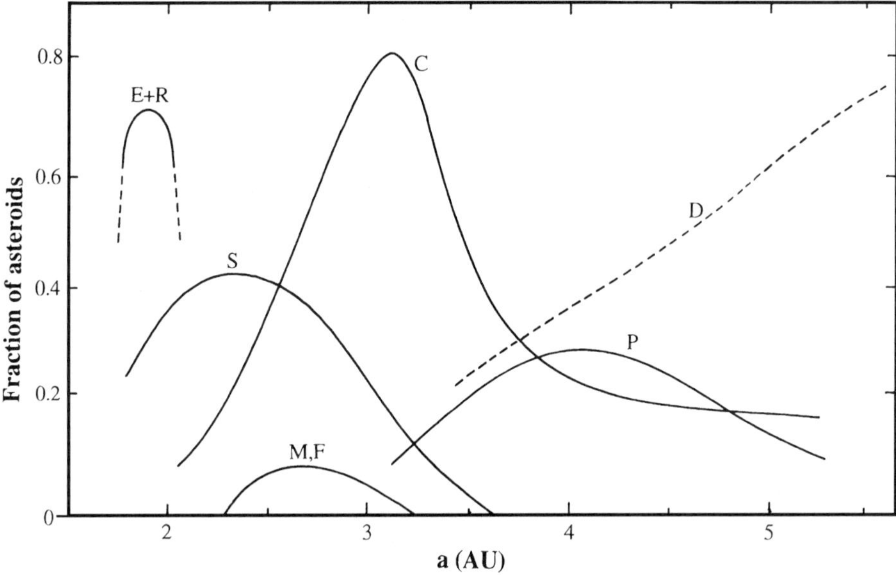


**Figure VIII.34** Spectral classes of asteroids as a function of heliocentric distance. Strong compositional zoning of the Belt is evident. Strong thermal alteration of asteroids near the inner edge of the Belt has produced achondrites, stony-irons, and irons.

100-m-sized objects, with probable errors of a factor of 2. The most recent (2001) data on the statistics of the NEAs suggest a total NEA population of about 1000 ± 400 and a "dark" (C and related types) component of about 30% of that population.

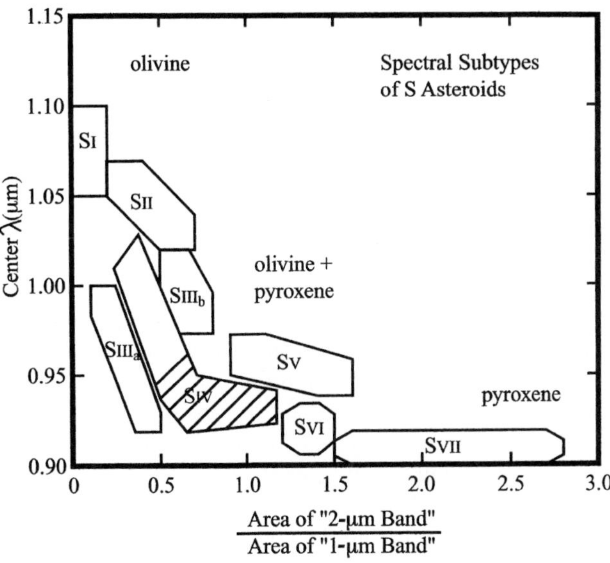


**Figure VIII.35** Spectral discrimination of subcategories within the S asteroid family. This diagram, based on work by Michael Gaffey of R.P.I., separates the S asteroids into eight groups based on the relationship between the wavelength of the center of the "1-μm band" produced by $Fe^{2+}$ in silicates and the relative strength of the "2-μm" and "1-μm" bands. Note that long central wavelengths correspond to olivine-rich rocks. Since olivine shows no 2-μm feature (Fig. VIII.29), a high ratio of band areas implies pyroxene-rich and olivine-poor composition. The ordinary chondrites generally fall in the cross-hatched portion of region $S_{IV}$.

Interestingly, several NEAs (including 1862 Apollo) have spectra closely similar to that of an ordinary chondrite. Thus the proposition that meteorites sample the near-Earth asteroid population is on much stronger ground than the older supposition that they sample the asteroid belt democratically.

## Thermal Evolution of Asteroids

The radiometric age data reviewed earlier show convincingly that meteorites formed by agglomeration of solid grains about 4.56 Ga ago. Most of the known meteorite classes were at least mildly thermally altered in a short period of time after accumulation. Except for those few meteorites that seem to have originated on Mars or on the Moon, the thermal evolution of meteorites appears to have run its course in about 100 Ma. Almost all achondrites and irons were formed in that narrow interval of time.

Decay of the long-lived radionuclides $^{40}K$ and $^{235}U$ releases heat on a time scale of 1 Ga, and the other long-lived nuclides have even longer half-lives. They are thus incapable of providing the observed distribution of ages. Two alternatives for short-lived heating are both supported by circumstantial evidence only. Decay of $^{26}Al$ clearly did take place in certain refractory grains in CV and CO meteorites, but we have no direct evidence that this isotope was widely distributed in the early Solar

Table VIII.5 Frequency of Spectral Classes of Near-Earth Asteroids

| Class | No. known | Score[a] | % of total | Meteorite analogues |
|---|---|---|---|---|
| S | 19 | 23.0 | 48.0 | Pallasites, ureilites, CV, CO |
| SQ | 3 | | | |
| SU | 2 | | | |
| SG | 1 | | | |
| SMU | 1 | | | |
| CSU | 1 | | | |
| QRS | 1 | | | |
| C | 6 | 11.8 | 24.6 | CI, CM, and altered CI, CM |
| CF | 1 | | | |
| CSU | 1 | | | |
| TCG | 1 | | | |
| XC | 1 | | | |
| D | 1 | | | |
| F | 1 | | | |
| CF | 1 | | | |
| G | 0 | | | |
| SG | 1 | | | |
| TCG | 1 | | | |
| T | 0 | | | |
| TCG | 1 | | | |
| Q | 1 | 3.3 | 6.9 | H, L, and LL chondrites |
| QRS | 1 | | | |
| QU | 1 | | | |
| SQ | 3 | | | |
| M | 2 | 2.3 | 4.8 | Irons |
| SMU | 1 | | | |
| V | 3 | 3.0 | 6.3 | Basaltic achondrite |
| A | 1 | 1.0 | 2.1 | Pallasite; ol-rich achondrite |
| E | 1 | 1.0 | 2.1 | Enstatite achondrite |
| R | 0 | 0.3 | 0.6 | Pyroxene-olivine achondrite |
| QRS | 1 | | | |
| U | 0 | 2.2 | 4.6 | Unclear affinity |
| SU | 2 | | | |
| SMU | 1 | | | |
| CSU | 1 | | | |
| QU | 1 | | | |
| Total | | 48.0 | 100.0 | |

[a] 1 for a pure type (such as C), 1/2 for a doubly ambiguous type (such as SU), 1/3 for a triply ambiguous type (such as QRS).

System. Such evidence may be compromised by two factors: first, the isotopic composition of most Solar System materials may have been homogenized by isotopic equilibration after $^{26}$Al decay was essentially complete, and second, the refractory aluminum-rich and magnesium-poor phases (such as corundum, $Al_2O_3$) that distinguish the refractory inclusions are ideal for detection of small $^{26}$Mg anomalies, whereas aluminum in ordinary chondrites and achondrites is adrift in a sea of magnesium in which radiogenic magnesium cannot be distinguished. T Tauri phase solar wind activity apparently is a ubiquitous phenomenon in pre–Main Sequence stars, and strong magnetic fields were present at the time of meteorite formation in the Solar Nebula, but we have no proof that induction heating by the T Tauri phase of the Sun was an important heat source in meteorite parent bodies. Nonetheless, one or both of these two alternatives is almost certainly correct.

The final short-lived heat source, accretion heating, poses several dilemmas that appear to discredit it as an important mechanism for heating asteroids. Consider the accretion of the largest differentiated asteroid, Vesta. During the early stages of accretion the energy released by conversion of gravitational energy into heat is quite negligible. However, during the terminal stages of accretion the infalling bodies are accelerated by virtually the entire present mass of Vesta. Let us consider a small body impacting on the surface of Vesta and depositing 100% of its incident kinetic energy into stored heat. The energy per unit mass of infalling material is

$$E/m = (GM/r^2)dr$$
$$= -GM/r_s = 6.4 \times 10^8 \text{ erg g}^{-1}. \qquad \text{(VIII.10)}$$

This E/m ratio is just $0.5\, v_{esc}^2$. For the present surface radius and mass of Vesta, this gives an escape velocity of about $0.36\,\text{km s}^{-1}$. The maximum temperature rise possible if all the accretion energy were stored internally is

$$T_{max} = (E/m)/(C_P/\mu) = 50\,\text{K}. \qquad \text{(VIII.11)}$$

More realistically, we should allow for the loss of half or more of this accretion energy by radiation into space, and we should allow for impacting bodies to carry kinetic energy in excess of that due simply to gravitational attraction. The mean impact speed cannot be very much larger than the escape velocity, because such violent collisions would have resulted in net erosion of mass (and impact heat!) rather than net accretion. Since these two corrections are of comparable magnitude and work in opposite directions, there is little need to develop them in detail. It is clearly evident that accretion of even a very large asteroid has little effect on the internal temperature. It is also informative to peruse the list of asteroid spectral types: (4) Vesta is *by far the largest* of the differentiated asteroids. Since accretion heating produces a temperature rise that is proportional to the mass of the body, we can see that the problem of explaining how these smaller bodies melted and differentiated is very severe. Richard Binzel of M.I.T. has found a large number of tiny V-type asteroids indistinguishable in spectral properties from Vesta itself. These small bodies pursue orbits suggestive of an origin in a large impact event on Vesta. Therefore some significant fraction of the small differentiated asteroids may simply be collision fragments eroded from much larger differentiated bodies.

Heating by a uniformly distributed short-lived radionuclide is extremely similar to heating by long-lived radionuclides as described in Chapter VI. The main numerical difference is in the decay constant of the radioisotope, which is almost 1000 times larger than for a cosmic mix of uranium, potassium, and thorium. Virtually all the energy from $^{26}Al$ decay is released within 2 million years, and hence the amount of heat lost by conduction to the surface of the asteroid during the decay time can be very small.

The heat flux conducted through the near-surface regions of a spherical body is proportional to the thermal conductivity $k$ and to the vertical temperature gradient:

$$F = -k\partial T/\partial z. \qquad \text{(VIII.12)}$$

For long-lived radionuclide decay within small solid bodies it is often adequate to represent the rate of heat production as being in a steady-state balance with the rate of conductive heat loss. This is equivalent to assuming that the time scale for conduction is short compared to the time scale for heat generation. For short-lived radioactive decay, the opposite will be true for large bodies. Let us define a conductive cooling time scale $t_{\text{cond}}$ as the ratio of the heat content of the solid body to its luminosity:

$$t_{\text{cond}} = C_p M(T - T_s)/L \qquad \text{(VIII.13)}$$

where $T$ is a typical internal temperature, $T_s$ is the mean surface temperature, the thermal luminosity $L$ is

$$L = 4\pi r^2 F = 4\pi r^2 k(T - T_s)/r, \qquad \text{(VIII.14)}$$

and the mass $M$ is $4\pi\rho r^3/3$, whence

$$t_{\text{cond}} = C_p \rho r^2/3k. \qquad \text{(VIII.15)}$$

For heat capacities and thermal conductivities characteristic of chondrites ($k = 2 \times 10^5$ erg cm$^{-1}$ s$^{-1}$ K$^{-1}$; $C_p = 10^7$ erg g$^{-1}$K$^{-1}$), the conductive cooling time scale becomes about $50\,r^2$ seconds. A body of radius 10 km would have a conductive cooling time scale of about $2 \times 10^6$ years. Placing a poorly conducting regolith on the body (lowering $k$ by a factor of, say, 100) could permit much smaller bodies to melt; however, such small bodies probably have great difficulty retaining a thick regolith. In effect, rapid decay is very effective in heating even rather small bodies because the energy is delivered by decay much more rapidly than it can be lost. The body then departs very strongly from steady-state behavior.

The specific issue of the differentiation of Vesta has been studied by Anitabha Ghosh and Harry Y. McSween, Jr., of the University of Tennessee. Following the suggestion by John Wood of the Harvard-Smithsonian Center for Astrophysics, Ghosh and McSween incorporated the temperature dependence of heat capacity into asteroid thermal evolution models. These models provide a natural explanation both for differentiation on a time scale of 4.5 to 6.5 Ma and for the persistence of high internal temperatures in the deep interior of Vesta for about 100 Ma after differentiation, assuming an initial H-chondrite composition and accretion within four half-lives of $^{26}Al$.

Heating by uniformly distributed radionuclides always favors differentiation of the largest bodies, with little dependence on heliocentric distance. This is not true of solar wind heating.

Heating by the T Tauri phase of the Sun is especially complex because there are two different heating modes, and because the interaction of the solar wind with an asteroidal body depends very sensitively on the electrical conductivity, radius, and heliocentric distance of the asteroid. The principle can be illustrated by considering the TM (transverse magnetic) heating mode, in which the passage of the embedded intense magnetic field in the solar wind induces currents in the asteroid. Ohmic losses then generate heat within the interior of the body.

Consider first the case of a sphere of solid metal (or a spherical body completely covered with a layer of metal). The magnetic field lines will scarcely penetrate into the metal, and the lines of force will tend to slide around the exterior of the conducting shell while depositing almost no energy in the body. Any heat that is generated by eddy currents near the surface of the shell will be rapidly conducted to the surface and radiated into space. Since the ambient temperatures in the asteroid belt were probably close to their present value during the T Tauri phase, and since the energy flux carried by the solar wind was surely smaller than the radiative energy flux, this mechanism cannot have much effect on asteroid surface temperatures, and even less effect on internal temperatures.

The second case to consider is that of a sphere of material that is an excellent insulator, such as pure enstatite. In this case, the magnetic lines of force pass through the asteroid almost as if it were not there. No currents are induced because there are no conduction-band electrons available. The surface temperature is again dominated by radiative heating by the Sun.

The electrical conductivities of meteorites actually span a range of many orders of magnitude, from enstatite achondrites on the lower extreme to irons at the upper. High conductivities are associated with high concentrations of metal, of course, but also with high amounts of FeO and volatiles, especially water. High temperatures promote electrons to higher-lying orbitals, and hence an FeO-bearing silicate will conduct much better when warm. The electrical conductivities of the C chondrites range over a factor of $10^5$, with the CI chondrites highest and the 4, 5, and 6 metamorphic grades of CO and CV chondrites lowest.

Another consequence of the immense range of meteorite conductivities is that two bodies of very similar composition may interact in very different ways with the solar wind. Further, two bodies of identical composition and heliocentric distance but different sizes may interact very differently. A poorly conducting material assembled into a 1000-km sphere will be heated in its deep interior, while a 100-km sphere of the same material will undergo almost no heating. If the material is a rather good conductor, the strongest solar wind coupling may occur at a depth of only 50 or 100 km in the 1000-km body, but at the center of a 100-km body. Thus the larger body could develop a layered, differentiated exterior, while the smaller should be thoroughly differentiated except for a thin, unaltered surface layer.

The strength of the heating drops off with distance from the Sun, since the solar wind energy flux drops off inversely with the square of the heliocentric distance. Thus our observation that the S asteroids (achondrites and stony irons) and the M asteroids (irons and stony irons) dominate the inner asteroid belt, while primitive-seeming C, D, and P asteroids dominate at greater distances, makes sense for solar wind heating, but not for radioactive decay heating. Finally, we should recall that the two largest asteroids show no evidence of differentiation, whereas many smaller asteroids from the inner Belt are clearly igneous. This again argues against the importance of radionuclide heating.

These arguments do not address the issue of whether the $^{26}$Al distribution might not have been very erratic. Indeed, there is nothing to prevent one from asserting that the only asteroids that melted were those that received unusually large amounts of $^{26}$Al, and that these happened (for unknown reasons) to be the ones that were closer to the Sun. Although this assertion is purely *ad hoc*, we know so little about the source and implantation mechanism of $^{26}$Al in asteroids that we cannot rule it out.

We are tempted to rule out the possibility that all the aluminum in the asteroid belt started out with the same isotopic composition as in the refractory grains in which daughter magnesium has been found. In these inclusions, $^{26}$Al/$^{27}$Al ratios of about $6 \times 10^{-5}$ are found. This is easily sufficient to melt every asteroid down to about 1–2 km in diameter. This did not happen.

Of course, if the aluminum were allowed to decay in very small (subkilometer) bodies before the asteroids had accreted to their present sizes (that is, if the accretion times of the largest asteroids are longer than the four half-lives used as in the models of Ghosh and McSween), then it is possible to render aluminum decay completely negligible as a heat source in large bodies. One might argue that the accretion of asteroids proceeded most rapidly at the inner edge of the Belt because the spatial density of solids was highest and the synodic periods of nearby bodies were shortest there. The more distant asteroids, accreting more slowly, were thus doomed to geologically uninteresting fates because their main heat source became extinct before they were fully formed. This argument would suggest that the asteroids closest to the Sun would differentiate, and the largest asteroids in that region should differentiate most readily and thoroughly. The same conclusion would apply to T Tauri solar wind heating, in which the presence of larger, warmer bodies closer to the Sun would greatly enhance the rate of heating. Thus accretion rate may play an important and subtle role in asteroid thermal evolution by both solar wind and $^{26}$Al heating, even when accretion heating is itself negligible.

Attempts to deduce the sizes of the asteroidal parent bodies of the common classes of meteorites from petrological considerations have met with little success. No minerals that require high static pressures for their synthesis have been identified despite intensive efforts. There is, however, one interesting and rather indirect method of estimating the sizes of parent bodies by deducing the cooling rate of the meteorite and then comparing that cooling rate to those calculated in mathematical thermal history models of asteroids. The principle is simple. First, consider the iron–nickel phase diagram (Fig. VIII.8) in the region in which the nickel content is similar to that in most meteorites. A piece of metal cooling through the 1200 K temperature region will consist of a single phase, taenite (gamma iron). Kamacite is never stable above 1170 K, the point at which iron and nickel become completely mutually soluble. A metal grain containing 10% Ni will first nucleate kamacite at about 960 K, and the mass fraction of kamacite will rapidly grow with decreasing temperatures. At about 910 K the masses of kamacite and taenite would be equal, and by 800 K taenite will be reduced to a minor phase. This equilibrium behavior is approached during cooling at a finite rate, but the sluggishness of the diffusion of Ni and Fe through the metal prevents perfect equilibrium from being attained. The Ni diffusion coefficient in taenite is approximately 100 times smaller than in kamacite, which leads to rather large gradients in the nickel concentration within taenite bands. Figure VIII.36 shows how the Ni concentration profiles evolve across a kamacite–taenite interface as the temperature falls. Note that taenite evolves toward higher Ni concentrations (i.e., approaches equilibrium with kamacite more closely) only in a thin layer near the interface of the two bands. The composition of the interior of the shrinking taenite band lags far behind and retains a Ni concentration appropriate to a higher temperature. Precise measurement of the Ni concentration profile across these bands combined with laboratory measurement of the diffusion coefficients in the two alloys permits a calculation of the rate of cooling of the metal in the

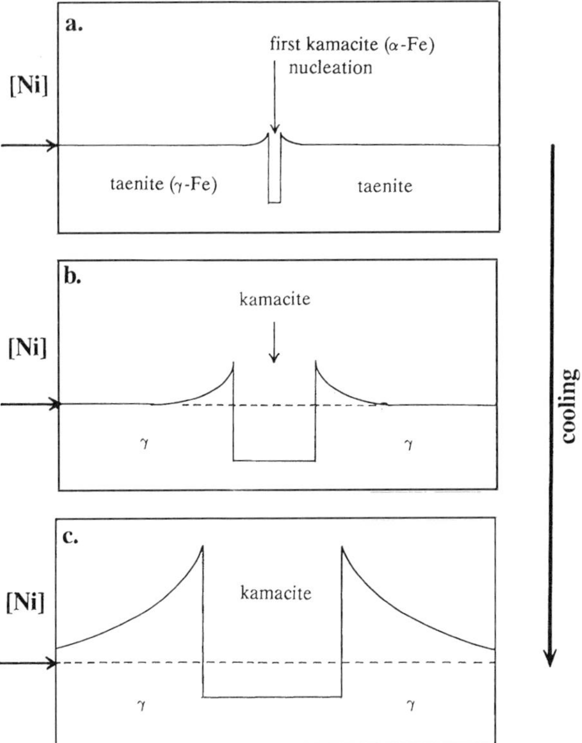


**Figure VIII.36** Nickel concentration profiles across kamacite and taenite bands. The cooling sequence begins with the first nucleation of kamacite (a) during cooling and proceeds to fully developed kamacite at lower temperatures (b and c). Oxidation at even lower temperatures will extract iron from the metal and force the structure back to pure taenite (see Fig. VIII.8).

temperature range in which the kamacite–taenite phase equilibration is occurring.

The calculated cooling rates for most iron meteorites are very low, only a few kelvins per million years. Pallasites seem to have cooled slightly more slowly than most irons. Some of the unequilibrated chondrites and most of the mesosiderites have cooling rates as low as $0.2\,K\,Ma^{-1}$. Many equilibrated chondrites and the IIC, IVA, and IIIAB irons have much higher cooling rates, 10 to over $3000\,K\,Ma^{-1}$.

High cooling rates are expected for small parent bodies and for the outer few kilometers of larger bodies. Thus an iron meteorite that has experienced a cooling rate of $30\,K\,Ma^{-1}$ (at temperatures near 750 to 800 K) may originate in a metal pool formed by near-surface melting on a lunar-sized body, or in the core of an asteroid 100 km in diameter. Many irons indeed cannot originate at great depth in a large parent asteroid, but must have formed and cooled within 10 km of the surface.

Low cooling rates require a considerable depth of burial, and hence are strong evidence for derivation of the meteorite from the deep interior of a large parent body. Consider the case of the mesosiderites: cooling rates of $0.2\,K\,Ma^{-1}$ can be attained at depths of at least 60 km in bodies 800 km in diameter, and at similar depths in larger bodies. Asteroids 400 km or less in diameter are incapable of delivering cooling rates below about $1\,K\,Ma^{-1}$. Figure VIII.37 shows that cooling rates below about $0.1\,K\,Ma^{-1}$ are not attainable in bodies of any size, in agreement with the meteorite data. However, the mesosiderites present us with a dilemma that must force us to wonder whether we really have understood the cooling rate data properly. If we calculate the cooling history of a mesosiderite parent body that starts out at the minimum temperature that is compatible with the petrological evidence that they have experienced at least partial melting (approximately 1250 K), subject to the constraint that the cooling rate near 770 K is $0.2\,K\,Ma^{-1}$, we find that after 0.9 Ga the temperature would still be near 1100 K, well above the appearance temperature of kamacite. This is a problem because the $^{40}Ar$ gas retention age of the Estherville mesosiderite is only 3.9 Ga! The paradox can be resolved if the parent body suffered catastrophic reheating and loss of radiogenic gases at that time. Another related dilemma is that the mesosiderites have textures suggestive of formation in an impact melt deposit in a regolith breccia. Thus they, like the unequilibrated chondrites, seem to combine the textures of near-surface deposits with the thermal histories of very deeply buried locations.

In general, the problems outlined here with the mesosiderites and unequilibrated ordinary chondrites are also echoed by the IAB irons, the pallasites, and the equilibrated ordinary chondrites: all appear to have cooled at so slow a rate that they should not have been able to retain radiogenic argon as early as they did. The cooling rates calculated from the Ni gradients in the metal would have to be increased by a factor of 5 or 6

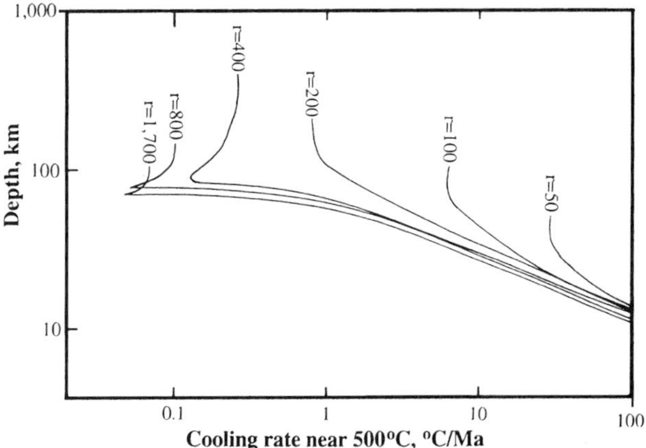


**Figure VIII.37** Cooling rates through 500° C for cooling asteroids as a function of depth and asteroid radius. These models, by John Wood, assume an initial temperature of 600° C and the normal cosmic abundances of long-lived radionuclides.

to be consistent with the observed $^{40}Ar$ ages. John Wood of the Harvard-Smithsonian Center for Astrophysics has suggested that the "heating, then cooling" scenario is too simple to permit a correct interpretation of the metallographic cooling rate data. He suggests that an insulating blanket of dust a few kilometers in thickness covers many asteroids. The radial temperature gradient in this dust mantle is large because of the low thermal conductivity, and the temperature gets high enough for sintering of the dust into a compact solid with a thermal conductivity close to that of normal rocks at the base of the dust layer. Such bodies give very low and stable cooling rates even at shallow depths because of the strong insulating effect of the dust.

There are two ways that such models of asteroid regoliths can be tested with present observational techniques. First, the thermal inertia of the asteroid surface can be estimated from infrared radiometry. Second, the radar reflectivity of a few asteroids can be determined: if the geometric cross-section area of the asteroid is known from radar, infrared, or polarimetric data, then the intrinsic reflectivity of the surface can be determined. A high reflectivity connotes a dense, rocky surface, while a low reflectivity suggests a fluffy, porous regolith of dust. The evidence from both lines of investigation suggests that most larger asteroids do have a dusty regolith, but neither technique can begin to plumb the depths of a 1-km or 3-km dust layer.

Thermal evolution of C asteroids and related classes, according to models calculated by Barbara Cohen and co-workers, has two other features that add interest and diversity to the results: the warming solids absorb a very large amount of heat upon reaching the dehydration temperature of hydrous silicates and hydrated salts, and the initial "melt" is an aqueous solution. Further, upward migration of water can trigger hydration reactions which evolve large amounts of latent heat, which prolongs the lifetime of the liquid phase.

## Dynamical Evolution of the Asteroid Belt

One of the most venerable questions about asteroids is whether they are the results of incomplete accretion or of the disruption of a few larger bodies. There was a close coincidence in time between the first few discoveries of asteroids and the realization that meteorites were of extraterrestrial origin (both occurring within a few years of 1800). After a brief flurry of speculation that all meteorites were of lunar origin, the suggestion was inevitably made that meteorites were asteroidal debris. Because of the novelty of the idea that small (1000 km or less) rocky bodies should be present in space, it was tacitly assumed that all small bodies, including both meteorites and asteroids, were derived from planet-sized bodies. This prejudice was unfortunately strongly reinforced by the prevalent interpretation of Bode's Law, which explicitly predicted that a planet should exist in the region where Ceres was later discovered. The prevailing synthesis of these ideas was to attribute all meteorites and asteroids to the explosive disruption of a "lost planet" that had once orbited where the asteroid belt is now. There was a lengthy search for meteorite minerals that required high formation pressures, such as would be expected in the interior of a planet. The general realization that no high-pressure minerals were present in meteorites did not come until the 1960s. By then it was quite obvious that the total mass in the asteroid belt was less than twice the mass of Ceres: about 5% of the mass of the Moon, or less than 0.0006 Earth masses. This is hardly sufficient to constitute a credible planet.

Nonetheless, it is still unclear whether it was fragmentation or accretion that determined the present properties of the asteroids. Two of the most useful discriminants of the collisional and accretional histories of small bodies are the present size and spin-rate distributions.

We have already seen that there is an excess of asteroids in the 70 to 120 km diameter range relative to a straight line fitted to the larger and smaller asteroids. The products of collisional fragmentation would follow a power law rather closely (a straight line on a log $n$ vs log $M$ plot), with the statistical deviations being most obvious in the poorly populated large-size classes, especially those with diameters over about 200 km. Accretion would presumably also produce a steep, monotonic distribution with the largest asteroids being the rarest.

Another difficulty with a power-law size distribution is that it predicts that the cross-section area of the vast number of smaller bodies will be immense. A crude linear fit to the high end of the asteroid size distribution would give $n$ varying very roughly as $1/d^2$ (Fig. VIII.25). The cross-section area of each body goes as $d^2$, so in each successive uniform interval of log $d$ the total cross-section area is constant. Thus, by extending the size distribution to small enough sizes, one would be led to predict that the asteroid belt should be seen as a diffuse band of light across the sky, following the ecliptic. This is certainly not the case.

The two most immediate lessons to learn from these observations are that power law distributions do not fit well in either the 100-km range or the subkilometer range. This is qualitatively explainable by postulating that the excess moderately large bodies are remnants of a primordial size distribution that has not been wholly erased by subsequent collisional evolution, and that there exists some mechanism for the depletion of the population of very small bodies. We have already encountered such depletion mechanisms in our treatment of dust dynamics in Chapter VII. The most important are the Poynting–Robertson and Yarkovsky effects, both due to the interaction of these small bodies with sunlight.

The spin rates of asteroids show some possible systematics that may be of use to us: C asteroids tend to spin a little more slowly than S asteroids. Since the average C asteroid is both darker and more distant from us than the average S, the known C asteroids tend to be larger than the average known S. There are three possible conclusions that we might be able to extract from these data: first, the C asteroids spin more slowly than S asteroids; second, large asteroids spin more slowly than small asteroids; or third, asteroids far from the Sun tend to spin more slowly than those at smaller heliocentric distance. The picture is somewhat clouded by three factors. First, there is a very wide range of spin rates within each size range. Second, statistics are poor for the very large (very rare) and for the small (very faint) asteroids. Third, what little evidence exists suggests that the largest C asteroids may rotate faster than the average C asteroid. Both M asteroids and the small Earth-crossers have high average spin rates, but the statistics are again rather poor.

In general, bodies that belong to the C and S classes have mean rotation periods near 9.5 and 9.2 hours, respectively. However, asteroids in the inner belt that do *not* belong to the C or S classes have mean rotation periods near 6.3 hours!

Perhaps the most interesting result of recent studies of asteroid lightcurves has been the extension of these observations to the NEA population, with its abundant small bodies. The results to date show that every asteroid studied that has a diameter less than 100 m, starting with 1999 $TY_2$, 1999 $SF_{10}$, and 1998 $WB_2$, is a very rapid rotator, with equatorial rotation speeds so high that the acceleration due to gravity would be insufficient to retain regolith. The Amor asteroid 2001 $OE_{84}$ has recently been found to have a rotation period of $29.190 \pm 0.002$ minutes. The entire surface, except in a tiny patch a few meters in diameter near each pole, should be bare of regolith. Such a body would in general be under tension, which makes it impossible to sustain a rubble-pile structure. The small asteroids are all clearly monolithic collision fragments. In contrast, every asteroid larger than 300 m in diameter is a slow rotator with the ability to bind regolith at its equator.

Further, radar studies of kilometer-sized NEAs show a striking departure from spherical shape. These bodies appear more like stacks of boulders or "snowmen" than spheres. Even rather large asteroids may be highly irregular: the Trojan asteroid (1437) Diomedes, studied during a stellar occultation in 1997, was found to approximate a triaxial ellipsoid with dimensions $284 \pm 61, 126 \pm 35$, and $65 \pm 24$ km, roughly in the proportion 4:2:1. Some, such as the near-Earth asteroid (2063) Bacchus, may be contact binaries. Lacking the gravity and thermal softening of the largest asteroids, they preserve a highly angular, rubble-pile structure. In this perspective, the smallest asteroids are individual pieces of rubble.

The equatorial orbital velocity and escape velocity of a solid spherical body are both proportional to $(GM/r)^{1/2}$, which, for an ensemble of bodies with the same composition and density, is proportional to $(\rho r^3/r)^{1/2}$, and thus directly proportional to $r$. The orbital period of a surface-grazing satellite is the circumference divided by orbital velocity, or $2\pi r/(G^{1/2}\rho^{1/2}r)$, which is independent of $r$. For ordinary chondritic material, the equatorial orbital period is about 2 hours, irrespective of size. Thus the existence of asteroids with rotation periods less than 10 minutes is proof of the unimportance of gravitation in holing these bodies together.

Studies of the spin rates expected for bodies accreting out of vast swarms of smaller bodies in the early Solar System show that the accreted bodies tend to have prograde rotation. There is a tendency for the larger bodies to have somewhat faster rotation rates. The rotation rate is ultimately limited by the fact that the spin rate eventually approaches orbital speed. The body approaches rotational instability as the centrifugal acceleration balances out most of the gravitational force.

Once an asteroid is accreted, it still can suffer collisions with other bodies of comparable size. Although collisions are infrequent (a 100-km asteroid may be hit by a 10-km asteroid about once every 100 million years), they can have substantial effects on the orientation of the spin axis of the asteroid because impact velocities are large. Further, $\sim 50$ such collisions would be enough to affect every 100-km asteroid. The effect of such post-accretion bombardment on the asteroid spin rate is a sensitive function of the size of the asteroid. A 1000-km asteroid has little chance of being strongly perturbed, since 100-km projectiles are about 100 times rarer than 10-km bodies. By the same argument, 10-km asteroids will be struck by very large numbers of 1-km bodies and may have their spin axes thoroughly randomized.

This discussion tacitly assumes that each collision is effective in transferring angular momentum. In fact, however, kilometer-sized loosely aggregated bodies may simply lose some material that is splashed off by an impact. The angular momentum of the impactor is efficiently transmitted to the loose debris that is ejected by glancing collisions, but the remaining material of the asteroid is little affected. An asteroid that is large enough so that the gravitational binding forces are large compared to the strength of its component materials will tend to retain the incident angular momentum because even splashed material will be gravitationally bound in orbits that will mostly reintersect the surface. Well-compacted asteroids of even small size will lose less material upon impact, and hence their angular momenta will be more strongly affected by impacts.

Very large numbers of impacts with very small bodies will also occur. Statistically, the angular momenta imparted by these impacts averages out, but the mass brought in can be quite appreciable. Accretion of these very small bodies has the effect of diluting down the original angular momentum of the asteroid over more mass, and hence lowers the spin rate.

One of the consequences of major collisional perturbation of the spin axis is that asteroids will commonly be set spinning about axes that do not coincide with one of their principal moments of inertia. The stresses set up by this kind of rotation cause the internal dissipation of energy due to the inelastic behavior of asteroidal solids, and the rotation and moment of inertia axes tend to approach alignment in a time of about a million years. For kilometer-sized asteroids, collisions will disturb the spin so frequently that complete relaxation of the axes will not occur in the time between collisions. Indeed, several of the NEAs studied by radar have compound, tumbling rotation about two different axes.

The consequences of large collisions include a kind of random walk of the spin angular momentum, in which the spin may even be reversed, but the mean spin rate will increase with the square root of the number of major collisions. Occasionally an asteroid may wander into the region of rotational instability and fall apart into two or more major fragments. Although these fragments will often depart from each other into separate heliocentric orbits, they will sometimes remain gravitationally bound as a double- or multiple-asteroid system. Some evidence has existed for many years from stellar occultation observations that several of the asteroids studied may be multiple. This issue has in the past few years been unambiguously resolved: many asteroids, perhaps as many as 10% of the entire population, have companions.

Fragmentation of an asteroid into a group of separate bodies with similar orbits about the Sun may lead to the formation of a family with causally related spectral properties. Thus disruption of a homogeneous undifferentiated body into a family of small bodies with similar orbits will provide a family with very coherent spectral properties. Disruption of a differentiated asteroid may provide a family of asteroids with the spectral properties of irons, stony irons, and achondrites with radically different spectra. The relationship between these bodies is that they are complementary geochemical differentiates made from a single type of primitive undifferentiated material.

Studies of the spectral properties of a number of dynamically related families of asteroids show that several of the more prominent families do indeed have correlated compositions. The asteroid (24) Themis and some 16 other members of its family all have UBV colors that loosely correspond with the C class; however, Themis lies at $a = 3.129$ AU in a region of the Belt in which C asteroids are very common. There is a significant dispersion of both UBV colors and albedos, so that it seems unlikely that we are looking at material that is uniform in its physical and chemical properties. The family of (221) Eos also displays a rather wide dispersion of albedos and colors, with most of the 20 bodies lying within the boundaries of the S group. A few of the Eos family transgress the S/C albedo boundary near 0.06. The heliocentric distance of this group (3.012 AU for Eos) places it within the part of the Belt that is dominated by C asteroids, and we cannot help but wonder whether we are here seeing the thermal alteration of C material to produce achondritic S-type material. Disruption of a partially differentiated C asteroid may then produce fragments that are spectrally identifiable as S and C types.

The family of (158) Koronis (2.867 AU) contains 16 spectrally characterized bodies with rather tightly clustered UBV colors and albedos in the S category. The available evidence strongly suggests fragmentation of a very homogeneous parent body. Koronis is further distinguished by its proximity to the 2:5 Kirkwood gap. It is reasonable to expect that small pieces of debris from the Koronis family may find its way into the gap and then be strongly perturbed by Jupiter. Thus Koronis and its family members are reasonable sources of Earth-approaching asteroids and recoverable meteorites.

A number of other families are known for which there exists some spectral and albedo data. The family of (44) Nysa (2.423 AU) appears to be dominated by E-type material, and (12) Victoria (2.334 AU) has a family that is largely of C type. Many other families have only two or three spectrally characterized members. M-type material is seen in the Nysa and (338) Budrosa families and in a few others, always associated with other types.

The asteroid belt is an example of a dynamically stable region in which small bodies, barring collisions, can persist for many billions of years. Theoretical studies have identified several regions in which a lesser but significant stability can be achieved. One example is the region inside Mercury's orbit, sometimes called the Vulcanoid region in reference to the hypothetical and often-sought planet Vulcan, which was long supposed to orbit in this region. Myron Lecar of the Harvard/Smithsonian Center for Astrophysics and his co-workers have shown that there is a relatively stable "refugium" for asteroidal bodies between 0.09 and 0.21 AU from the

Sun. However, a large population of asteroids in this region is infeasible for several reasons: perturbations by the planets should clear out the region with a half-life of about a billion years; the volume of available space is tiny compared to the Belt; high spatial densities combined with short synodic periods and high relative velocities create a very hostile collisional environment; and bodies may easily drift into orbits in which the insolation is so intense that they would evaporate, becoming in effect rocky comets. The dayside temperature of a typical airless rocky body at 0.09 AU from the Sun is roughly 1000 K, almost high enough for evaporation to be the principal limiting factor on its lifetime. The SOHO (Solar and Heliospheric Observatory) spacecraft has searched for Vulcanoids in the region from 0.07 to 0.21 AU from the Sun and found no bodies larger than their detection limit, which is a diameter of 20 to 60 km, depending on albedo. One 60-km asteroid would have a mass of $2 \times 10^{-4}$ the mass of Ceres. The total mass of the Vulcanoid population is unlikely to be as large as $10^{-3}$ of the mass of the asteroid belt, and probably far smaller.

At present, this body of data on the compositions of asteroid family members remains very incomplete and very hard to interpret. The most reasonable hypothesis is surely that the asteroids represent the collisionally evolved fragmentation products of an earlier population of $\sim$100- to 1000-km bodies. Attempts to run the evolution of the Belt backward from the present state to its "pristine" state at the end of accretion can be made plausible, but it seems likely that a wide range of earlier states could be generally compatible with the present state.

The degree of compositional mixing in the Belt can be limited by the observation of the well-defined compositional zoning that persists to the present. Clearly there has not been major long-distance transfer of material from one region of the Belt to another. On the other hand, the substantial departures of the Belt asteroids from coplanar circular orbits attests to the presence of strong disturbing forces at some point in their orbital evolution. For example, (2) Pallas has an orbital inclination of 34.8°, and (33) Polyhymnia has an orbital eccentricity of 0.3408. Circular orbital velocity at 2.8 AU (near both Pallas and Polyhymnia) is about $18.5\,\mathrm{km\,s^{-1}}$, and hence velocity changes on the order of $10\,\mathrm{km\,s^{-1}}$ must have been imparted to them by past encounters. The rms (root mean square) velocity departure of the average asteroid from circular coplanar orbit in the ecliptic plane is 4 or $5\,\mathrm{km\,s^{-1}}$. In order to explain such large perturbations, many authors have invoked gravitational scattering by massive bodies that passed through the Belt in the distant past. However, the mass of the necessary body must have been surprisingly large: numerical calculations show that the perturbational velocities produced by such encounters were on the order of the escape velocity of the perturbing body. This suggests that the body had an escape velocity of roughly $5\,\mathrm{km\,s^{-1}}$, which places it in the size range of the terrestrial planets. Thus a roughly Mars-sized object may once have "stirred" the Belt. Where is it now?

It is not a matter of great surprise that high-speed bodies may have transgressed the Belt in the distant past. In Chapter IV we saw that the radial distribution of mass in the Solar System shows a smooth decline from Venus to Pluto except for the region of Mars and the asteroid belt. That region is severely depleted of mass. Calculations by Stuart Weidenschilling show that the acceleration of planetesimals from the region in which Jupiter accreted must have subjected a wide region of the Solar System, extending in almost as far as the orbit of the Earth, to a heavy bombardment flux of fast-moving bodies. He attributes the mass deficiency of both Mars and the Belt to such a strong early bombardment. The total amount of missing mass is on the order of one or two Earth masses, and the mass of planetesimals that Jupiter accelerated into the affected region must have been at least comparable to an Earth mass.

## Centaurs and Trans-Neptunian Objects

In the past decade the number of known *Centaurs* and *trans-Neptunian objects* has grown from one to roughly 300. These classes are distinguished primarily on the basis of orbital properties. The Centaurs are in modest-inclination orbits with sufficient eccentricity to cross the orbits of two or more of the giant planets. Such orbits are unstable against both impact on one of these planets and perturbation into orbits of other types. There is no clear orbital distinction between Centaurs and periodic comets; indeed, 29 P/Schwassmann–Wachmann and 39 P/Oterma both follow typical Centaur orbits. The prototypical Centaur asteroid, (2060) Chiron, has been observed to generate a cometary coma, for which it has received the designation 95 P/(2060 Chiron). By all standards, Chiron would make a truly spectacular comet if it were to pass through the inner Solar System. Numerical integrations of Chiron's orbit show that it may be an episodic Earth-crosser. Given Chiron's size (a diameter of $148 \pm 8$ km) and orbit, it could deliver to Earth 2 million times the mass of a 1-km diameter asteroid at twice the impact speed, for an impact energy of roughly $10^{12}$ megatons, nearly 10,000 times the energy of the K/T impactor that ended the Cretaceous Era. This is sufficient energy to boil away all of Earth's oceans. From this perspective it is somewhat unsettling to realize that 39 P/Oterma has also partaken of the Centaur Wanderlust:

before about 1950 it was in an orbit with an aphelion distance $Q < 5.2$, with orbital parameters well represented among the bodies we now call near-Earth asteroids.

The orbits of the Centaurs are contrasted to those of the plutinos, cubewanos, and scattered disk objects (SDOs) in Fig. VIII.38. The cloud of low-inclination bodies beyond 40 AU is the cubewano family, whereas those with generally higher eccentricities (0.05 to 0.34) and semimajor axes close to the 3:2 Neptune resonance are plutinos.

The geometric albedos of the Centaurs studied to date are low but not exceptional, $0.17 \pm 0.02$ for (2060) Chiron and $0.12 \pm 0.03$ for (8405) Asbolus. The Bond albedo of (5145) Pholus is low, $0.04 \pm 0.02$, corresponding to a diameter of $185 \pm 16$ km. The Centaurs are spectrally bimodal, consisting of a neutral or slightly reddish group and a highly reddened group, which may reflect either different places of origin or different evolutionary (thermal) histories. By analogy with Pluto, Triton, and other large bodies far from the Sun, many Centaurs may have a distilled frost component on their surfaces.

Like the near-Earth asteroids, the Centaurs are dynamically vulnerable and short-lived. Again like the NEAs, the Centaurs may be fed into their present orbits from a belt farther from the Sun, in this case, the Kuiper Belt and the Neptune-resonant bodies, which together constitute the class of trans-Neptunian objects, or TNOs. All TNOs have $a > 30.2$ AU. Note that the designation TNO does not include bodies that spend part of each orbit outside Neptune, but have a shorter orbital period than Neptune. Indeed, fully half of the known Centaurs have aphelia that lie at $Q > 30.2$ AU. Numerical integrations of the orbits of 32 Centaurs have been carried out for a 100-Ma period. Of these, 14 were found to have orbits that were unstable over that time interval, and 18 had orbits that were stable, just as Pluto's orbit is stable.

The type example of the classical Kuiper Belt object is 1992 $QB_1$. Like the bodies predicted a half-century ago

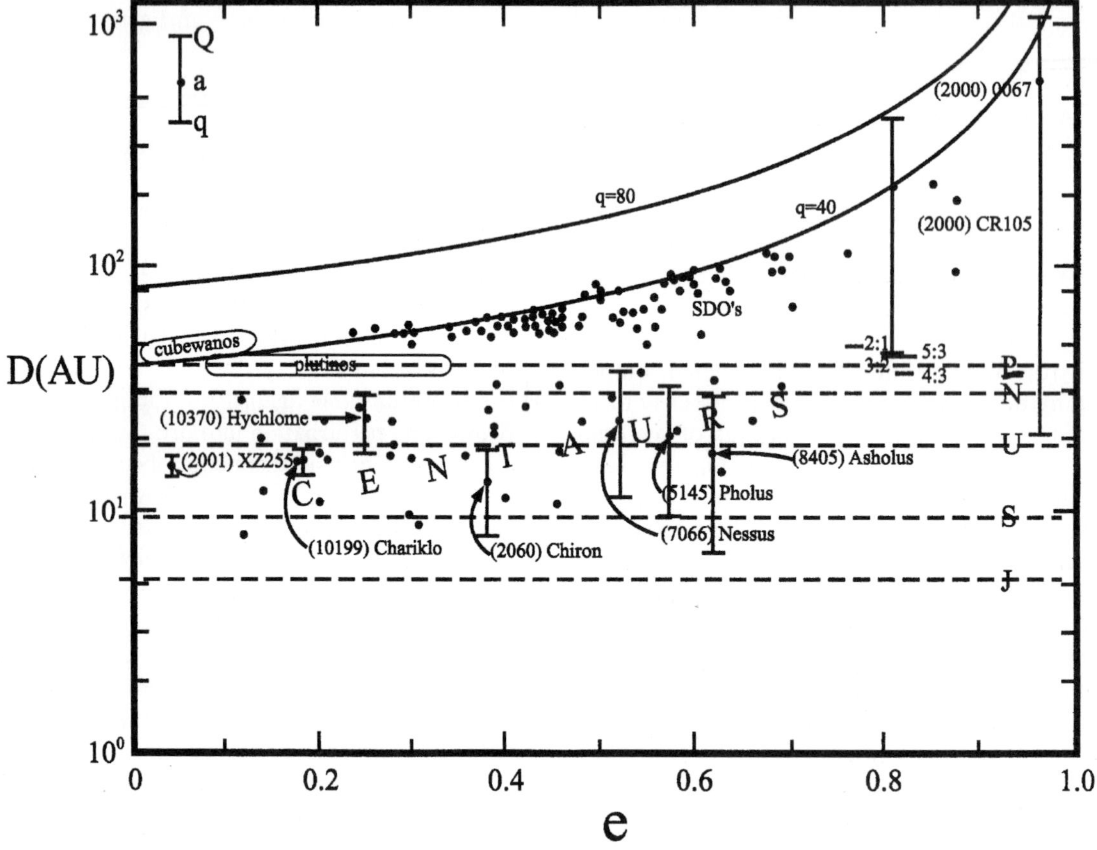


**Figure VIII.38** Distance vs eccentricity plot of outer Solar System bodies. The cubewano and plutino populations occupy the small marked regions (see Fig. VIII.39 for details). Each dot represents the semimajor axis of one body. The "error bars" on selected points actually represent the range of heliocentric distances, from perihelion ($q$) out to aphelion ($Q$). The large region beyond $a = 40$ AU is occupied by scattered disk objects, all of which have undergone severe perturbations by one or more of the Jovian planets. Inside 40 AU is the dynamically unstable Centaur region, within which most bodies are multiple planet-crossers.

by Gerard Kuiper, these objects pursue grossly circular orbits in the "empty" space beyond Neptune and Pluto. These bodies, named after "cue bee one," are now called *cubewanos*. They typically have low orbital eccentricities (of order 0.1) and moderate inclinations (mostly 0° to 30°), so that they are dynamically only weakly influenced by Neptune. They make up half of all the known TNOs. Those bodies that have strong dynamical links to Neptune make up the other half of the TNO population.

The KBOs, like the Centaurs, have a wide range of spectral properties. Some are extremely red. Most KBOs, however, have colors similar to the less dynamically extreme Centaurs. At visible and infrared wavelengths, no systematic relationship between color and heliocentric distance has yet been identified.

The largest known KBO is 1995 $SM_{55}$, at a diameter of 813 km. The largest plutino, 1998 $SN_{165}$, has a diameter of 709 km. The nearly 700 KBOs known as of August 2002 are surely a tiny proportion of the total population. The size of the TNO population has been estimated in several different ways. David Jewitt estimates that there are at least $10^5$ KBOs larger than 100 km in diameter with an aggregate mass similar to the mass of Earth. Since the Pluto-Charon system is dynamically related to the plutino population, some would describe Pluto, with a diameter of 2300 km, as the largest plutino. For a sense of scale, the total water content of Earth's oceans is equivalent to a sphere with a diameter of 860 km.

The population statistics of these distant populations may soon be much better understood as a consequence of the new Deep Ecliptic Survey, led by R. L. Millis and Mark W. Buie, which uses an $8192 \times 8192$ CCD mosaic detector on the 4 m Mayall telescope located at Kitt Peak National Observatory in Arizona. The first results from this team include discovery of 69 Centaurs and KBOs, ranging down to apparent visual magnitudes as low as 24. About ten of these bodies are plutinos, and four have high eccentricities. The orbital inclinations of KBOs are closely similar to those of short-period comets, which reinforces the idea that these comets largely originate in the Kuiper belt. The Kreutz family of comets may originate from a Centaur that developed a dangerously high orbital eccentricity.

The high eccentricity TNOs are not Neptune-resonant, but bear evidence of strong perturbations in the past. These bodies, the Scattered Disk Objects (SDOs) include examples as extreme as 1996 $TL_{66}$, which has an eccentricity of $e = 0.59$ ($a = 85$ AU, $q = 35$, $Q = 135$). They all have $a > 41$ AU and high eccentricities.

The majority of the strongly interacting bodies have orbits that, like Pluto, are at or near the 3:2 orbit-orbit resonance with Neptune. These bodies, the plutinos, make up about 40% of the TNO population. True plutinos have $a = 39.33$ to 40.05 AU, inclinations of about 2 to 20°, and eccentricities of 0.1 to 0.35. In addition, over a dozen of the known TNOs are in or very close to the 2:1 Neptune resonance (corresponding to $P = 330$ years and $a = 48$ AU), and several more are very near the 4:3 and 5:3 resonances. The eccentricities and orbital semimajor axes of more than 125 plutinos and cubewanos are given in Fig. VIII.39. The contours of perihelion distance are given for reference. Note that Neptune's semimajor axis is at 30.2 AU, so all the plutinos above the $q = 30$ contour are Neptune crossers. Note also that there are 30 or more recently discovered cubewanos of low eccentricity, marked "+30", but whose orbits are not yet sufficiently well determined to plot them precisely on this diagram.

Another interesting dynamical feature of these remote bodies was remarked earlier: despite the infancy of the search for satellites of these bodies, a number of cubewanos and plutinos have already been found to have companions within a few thousand kilometers.

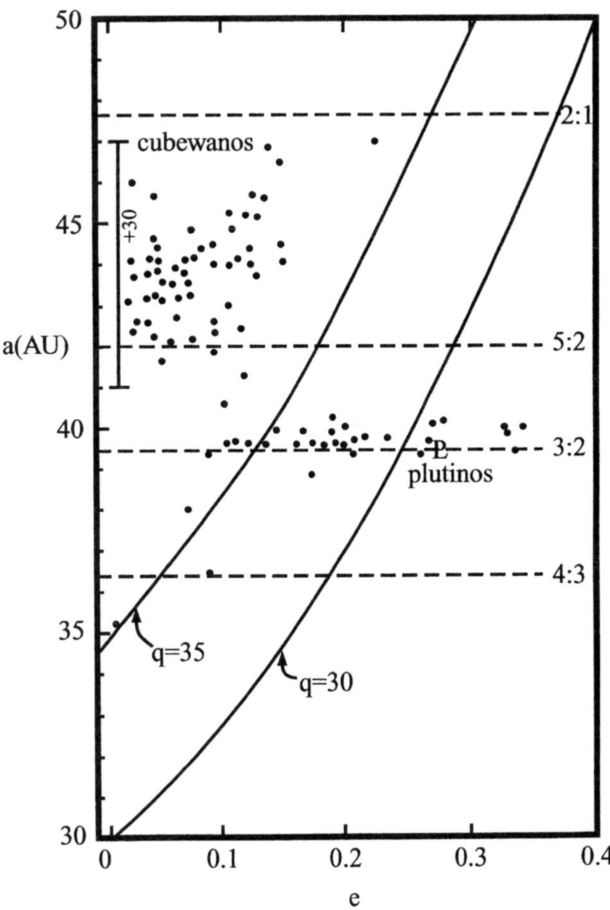


**Figure VIII.39** Semimajor axes and eccentricities of the plutino and cubewano populations. Most plutinos have eccentricities between 0.1 and 0.35, and all cluster tightly around the 3:2 Neptune resonance. All plutinos to the right of the $q = 30$ line are Neptune-crossers. The position of Pluto on the diagram is marked. The 5:2 and 4:3 Neptune resonances are lightly populated.

Recent additions to the list include a satellite of 2000 $CF_{105}$ and another within about 6000 km of (26308) 1998 $SM_{165}$. It will be interesting to see whether the Centaurs have the same proportion of satellites as their parent population in the Kuiper belt.

Capture of small bodies into resonant relationships with Neptune would seem to be a very improbable event. To transfer from a nonresonant orbit to a resonant orbit, some small energy transfer must occur. Frictional dissipation by passage through an extended planetary atmosphere, or a fortuitous collision with a much smaller body at the right place and time, could lead to capture, but both types of capture would generally lead to unstable orbits. When Pluto was the only known body in a plutino (3:2) resonant orbit, one could accept the idea that this single capture event occurred with some philosophical reserve. But now that more than 100 plutinos have been discovered it seems wise to try to replace philosophy with physics. Renu Malhotra has looked in some detail at the early orbital evolution of giant gaseous protoplanets embedded in a (disturbed) solar nebula disk. She finds that such a body, call it proto-Neptune, must slowly spiral inward as long as the nebula is present. The reason for this phenomenon is that the protoplanet, under the influence of the gravitational force of the disk, attempts to orbit at the local Keplerian orbital velocity, whereas the nebular gas experiences a pressure gradient force that partially offsets the gravitational force, causing the gas to travel at slightly less than circular orbit velocity [see discussion in Chapter IV, especially Eq. (IV.194)]. The protoplanet therefore is always opposed in its motion by a "headwind" that constantly saps the protoplanet's orbital momentum, forcing it to spiral slowly inward, while the subsonic passage of the protoplanet drives turbulence in the surrounding gas. Small future plutinos and Pluto itself migrate at different speeds, and they all migrate at a different speed than proto-Neptune. Therefore the 2:3 (and 3:4, 4:5, etc.) Neptune resonance sweeps slowly across the small-body population and captures into resonant orbits bodies that were initially in rather different orbits, all the while pumping up their eccentricities and inclinations. This capture process is so gentle that satellites of the captured bodies are not lost; hence the Pluto–Charon pair, the 1999 $TC_{36}$, system, and many others.

## Relationships among Asteroids, Meteorites, and Comets

The most reasonable correlations to make between asteroid and meteorite compositions have already been suggested: C chondrites are C asteroids; ordinary chondrites are rare in the Belt, constituting only a small subset of the S asteroids; most S asteroids are unfamiliar classes of chondrites, stony irons, or achondrites; M asteroids are irons or enstatite chondrites; E asteroids are enstatite achondrites; D and P asteroids are super-carbonaceous volatile-rich materials; U (unclassifiable) asteroids are a very heterogeneous assortment of odd types. The assignment of these classes is inferential, and we possess ground truth only on those few asteroids that have been visited and analyzed by spacecraft. There are those who believe, for example, that some subtle and unknown influence has systematically altered the spectra of the other S asteroids, and that they are all in fact chondritic, presumably ordinary chondrites. Although this seems improbable, we are certainly not in a position to prove that it is impossible. The evidence for the reddening of asteroidal material after long exposure in space, often called *space weathering*, is strong: indeed, the spectra of the S(IV) asteroids can be fully reconciled with the ordinary chondrites only if this phenomenon is real.

The compositional zoning of the Belt seen today, despite all the disturbances that have wracked the Belt since the time of accretion, bears strong testimony to the variation of primordial formation conditions with distance from the Sun. This zoning strongly reinforces our idea of the systematic decrease of formation temperatures with distance, from rock condensation near Mercury's orbit through the formation of FeS in the terrestrial planet region, serpentine formation in the heart of the Belt, and ice formation near 4 AU at the outer edge of the Belt. But, as we have seen, spectral searches for water of hydration features show that the distribution of hydroxyl silicates is not this simple: in the domain of the C asteroids in the outer Belt, bound water is common—but farther from the Sun, among the D and P populations, it is rare or absent. The apparent presence of substantial water in the W asteroids lends further confusion to the issue. The most straightforward interpretation seems to be that the hydroxyl silicates are alteration products made by reaction of melted ices with anhydrous silicates in asteroid interiors. In this case, the other distinctive and diagnostic materials that accompany phyllosilicates in carbonaceous chondrites, including magnetite, water-soluble salts, and massive and diverse organic matter, may also owe their origins to parent-body processes. Carbonaceous chondrites are therefore not primitive in the normal genetic sense, in that they are not the starting materials or precursors for the synthesis of other meteorite types. Although they have escaped igneous processes, their textures and mineralogy testify that they have been severely altered by what we might well call brine magmatism. They also, like the ordinary chondrites, are not directly diverted to Earth from distant Belt asteroids, but come via the NEA population, which we now know contains great numbers of dark asteroids of C and even D types. If 30% of the NEA population by mass is CI material, then about 3% of the mass of the NEA swarm is water. This has implications for the infall of water on Mercury and the Moon. This estimate, however, ignores the extinct short-period comet component of the NEA population. P/comets may contain up to 60% water, hidden beneath a black carbonaceous lag deposit left by evaporation and loss of near-surface volatiles.

In general, one-step derivation of the meteorites that fall on Earth from asteroids in the Belt seems very difficult. Direct explosive ejection of material from asteroids in the Belt into Earth-crossing orbits would require peak shock pressures so intense that the material would be vaporized rather than accelerated as an intact meteorite-sized body. Two-stage scenarios for transport of Belt material to Earth are quantitatively more attractive. The simplest idea is that debris from asteroid collisions may be ejected at relatively low speeds into nearby Jupiter resonances, whence the fragments (whether of meter or kilometer size) are smoothly accelerated by Jupiter perturbations up to high enough eccentricities so that they cross the orbit of Earth. By this mechanism, any asteroid located near a Jupiter resonance could feed material to Earth. This argument places a high premium upon understanding the compositions and orbits of the Earth-approaching asteroids. A more practical consideration is that this mechanism makes available for our inspection, either on or near the Earth, asteroidal material from formation locations so remote and so diverse that they would otherwise remain inaccessible to us for many decades. The other side of this coin is that the present orbits of such samples of remote asteroidal material will probably not be diagnostic of their place of origin.

Recent studies of the evolution of the orbits of comets show that short-period comets emplaced in near-Earth orbits by Jupiter perturbations have orbits similar to those of many near-Earth asteroids. Parallel studies of the thermal evolution of cometary cores in near-Earth orbits show that the outer layers of these cores may outgas to a depth of 1–10 m to leave a very fluffy, insulating dust layer above a thermally unaltered ice-rich interior that preserves its cometary composition. Thus the present NEA population should include bodies rich in native ice.

Attempts to classify the near-Earth asteroids (and, indeed, to find them in the first place!) are still in their infancy. The Spacewatch program at the University of Arizona, under the direction of Tom Gehrels, began using a 2048 × 2048 CCD detector array in 1989 on a fully computerized 36-inch (92 cm) telescope to detect asteroids of unprecedentedly small size. This program and photographic searches by Eleanor Helin, Eugene Shoemaker, and others, which were of high relevance to the problem of the genesis of meteorites and NEAs, successfully paved the way for the present generation of detection systems such as LINEAR, LONEOS, NEAT, and Spacewatch II.

While perusing these data on NEAs it is good to keep in mind the fact that they are all planet-crossers, and therefore can and often do experience disastrous encounters with planets. A typical NEA has an orbital half-life of $10^7$ to $10^8$ years before colliding with a planet or being ejected from the Solar System. Thus the large present NEA population *requires* the existence of mechanisms for replenishing the supply. Perturbation of asteroids out of the Belt appear dominated by the wandering of asteroids (under Poynting–Robertson forces or collisional deflection) into Jupiter resonances, after which they are quickly pumped up in orbital eccentricity. Many become Earth-crossing Atens or Apollos, and many others become Amors or Mars-grazers. But calculations of the rate of supply of Belt asteroids into the NEA population fall well short of providing the needed number of new NEAs. It is logical to ask whether comets might also contribute to the NEA population.

First, a number of asteroids with orbits similar to comets are now known. The Centaur (5145) Pholus (formerly 1991 DA), with a Chiron-like orbit that reaches aphelion beyond Uranus at $Q = 22.120$ AU, approaches as close as $q = 1.548$ AU, making it a Mars-grazer. The SDO 2000 $OO_{67}$, with an eccentricity of 0.962, reaches out to an astonishing 1063 AU at aphelion. Several NEAs have aphelia beyond 4 AU. The NEA 5370 1986 RA reaches out to $Q = 5.466$ AU, and 1984 QY1 reaches $Q = 6.981$ AU (with a perihelion at 0.219 AU!), making it a Mercury-, Venus- Earth-, Mars-, and Jupiter-crosser. Also, 1982 YA reaches aphelion at $Q = 6.291$ AU, and Tom Gehrels' Palomar–Leiden asteroid survey in 1960 found one body (P-L 5025) that has never been reacquired, with $Q = 7.961$ AU.

The Slovak astronomer Lubos Kresek has compared the orbits of short-period comets and asteroids and has found a near-continuum of orbital properties (Fig. VIII.40). Perturbation of a short-period comet into an NEA-type orbit should not be difficult, and, according to Eugene Shoemaker, a major source of NEAs should be the short-period comet population.

An extraordinary twist to this argument was provided by the reacquisition of the Apollo asteroid 1979 VA in August of 1992. This asteroid had been lost since shortly after its discovery in 1979, and the new observations in 1992 permitted a very precise new determination of its orbital parameters. The asteroid's improved orbit was good enough to justify giving it a catalog

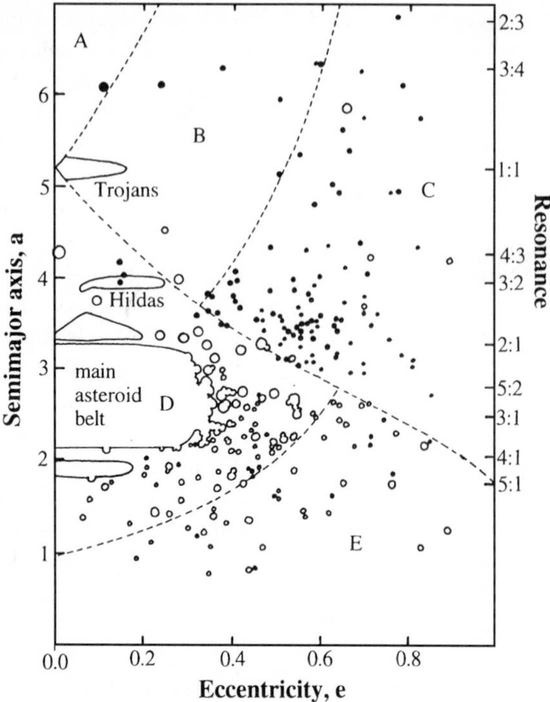


**Figure VIII.40** Semimajor axis vs eccentricity plot of comets and asteroids. Solid dots are comets and open circles are asteroids. Region A contains transjovian comets and asteroids, B is a region of weak Jupiter-related cometary activity, C is the region of strong Jupiter-related comet activity (most of the short-period comets), D is the "normal" asteroid regime, including the Belt, and E contains the Apollo and Aten asteroids. The Amors lie just inside D along its border with E. Note that the boundary between comets and asteroids is not inviolable.

number (it is now 4015 1979 VA). Armed with this improved orbit, a table of positions (an *ephemeris*) was calculated and extended back to prediscovery times in hopes that some earlier single-event observations of this asteroid might be found in the files. What was found was much more spectacular than anyone would have dared to hope: 4015 was linked firmly to a short run of 1949 observations (too short a run to give a good orbit) of Comet Wilson–Harrington 1949 III! The 1949 photographs clearly show this body with a coma and faint tail, but now it looks just like a typical asteroid. We have therefore had the extraordinary luck to observe a comet turning into an asteroid. There is another, less dramatic, indication of a NEA–comet connection: the NEA 3200 Phaethon is the source of the Geminid meteor shower.

One of the central contributions of Spacewatch was its unique ability to discover and track extremely faint asteroids. The Spacewatch discovery 1990 UN was, at a diameter of 100 meters, the smallest asteroid ever discovered. A few months later, Spacewatch found 1991 BA, an incredible 9 meters in diameter. The only reason that so small a body could be found is that it was far closer than any asteroid ever before seen: at its closest point it was only 160,000 km from Earth, less than half the distance of the Moon! Soon thereafter, Spacewatch added 1991 VA and 1992 DU, both 8 to 9 meters in diameter, and dozens more in this size range are now known.

Statistical study of NEA discovery rates of small (< 100 m) bodies by David Rabinowitz and co-workers has shown that bodies smaller than 100 m are progressively more abundant than a simple extrapolation of a Belt asteroid size distribution ($n \propto r^{-2}$ for asteroid diameters less than about 60 km) would predict. This analysis shows that 100-m bodies lie only about a factor of 2 above the Belt asteroid trend line, whereas 30-m bodies are about a factor of 20 above the line, and 10-m bodies must be over 100 times as numerous as the extrapolation suggests. This is most comforting in two ways, because the data on bright fireballs compiled by Czech astronomer Zdenek Ceplecha using data from the European Fireball Network give a flux of 1-m bodies that is roughly a factor of 1000 above the extrapolated Belt numbers, and because Shoemaker's analysis of lunar cratering statistics also suggests a similar upturn for bodies under 100 m diameter. The relevant observational data are compared in Fig. VIII.41. The fireball data clearly show that the large majority of meter-sized fireballs are made of very weak material which is destroyed in the atmosphere during entry, resulting in no meteorite fall.

Another unexpected result of recent asteroid discoveries is that many NEAs have extraordinarily Earth-like orbits. For example, the typical NEA has an orbital eccentricity of 0.4, but the 250-m Spacewatch asteroid 1990 UP has an eccentricity of only 0.169; the 900-m asteroid 1991 CS has $e = 0.165$; the 500-m 1991 JG1 has $e = 0.184$; 520-m 1991 JW has $e = 0.118$; the 20-m diameter 1991 TT has $e = 0.161$; the 1300-m 1991 VH has $e = 0.144$; the 250-m diameter 1992 BA has an eccentricity of an incredible 0.067; little 9-m 1992 DU has $e = 0.175$; and 1992 JD has an even more incredible $e$ of 0.032. The last is a 30-m rock with $q = 1.001$ and $Q = 1.068$. A few tiny rocks in such orbits might be rationalized as possible ejecta from a recent cratering event on the Moon, expelled from Earth's gravity well with little energy to spare. But 250-m to 1300-m asteroids are simply too large for this to be a credible explanation. A source of small asteroids from disruption of weak comet nuclei is very plausible, but why such bodies have such Earth-like orbits is a mystery.

One consequence of the presence of large numbers of NEAs in low-eccentricity, low-inclination orbits is that many of these bodies are energetically accessible for spacecraft launched from Earth. This facilitates

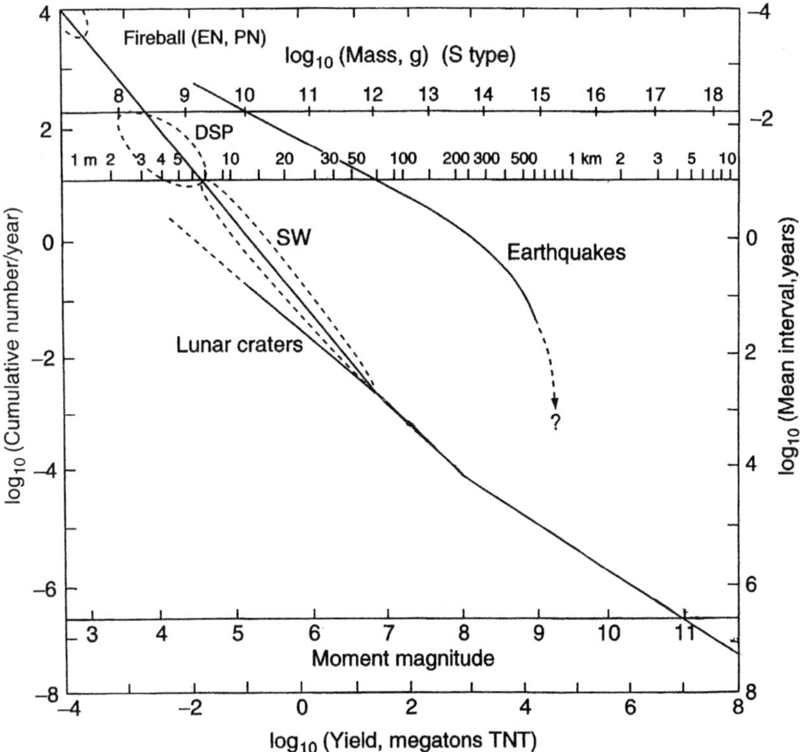


**Figure VIII.41** Near-Earth asteroid frequencies. The long solid line is a fit to the Belt size distribution, supplemented by the Spacewatch LINEAR, and LONEOS data on near-Earth asteroids (SW). The fireball data from Defense Support Program satellite observations (DSP) and Earth-based photographic data from the European Network (EN) and the North American Prairie Network (PN) extend the line to smaller masses. The energy-frequency relation for terrestrial earthquakes is included for comparison. The horizontal scales give log (mass), radius, moment magnitude, and explosive yield as functions of the main interval between events.

scientific exploration, since a rocket capable of landing a tonne of equipment on the Moon (while fighting the Moon's gravity all the way to the surface) could land twice as much mass on one of the better NEAs, for which "landing" is basically a matter of rendezvous and docking, using very little propellant in the process. Even more interesting is the return of material from the surface of the asteroid to Earth or to an Earth-orbiting space station. Return out of the Moon's gravity well requires 6 km s$^{-1}$ of velocity from the rocket engines ($\Delta V = 6.0\,\mathrm{km\,s}^{-1}$). But for some NEAs the required $\Delta V$ is less than a tenth of this number. Indeed, return from the surface of 4660 Nereus can be accomplished with a $\Delta V$ as low as 0.060 km s$^{-1}$, 100 times less than for return from the Moon! Thus it is easier to land on the best 25% of the NEAs than to land on the Moon, and it is vastly easier to get back. The trip time to the Moon is a few days, whereas the trip time to a NEA is several months. This suggests that unmanned missions to retrieve asteroidal material may be remarkably easy to execute.

## Radar Observations of Near-Earth Asteroids

From time to time, NEAs pass close enough to Earth to be tracked by radar. The $1/D^4$ dependence of returned signal strength on distance remarked on earlier places a high premium on proximity. The population of potential targets is large and rapidly growing, affording Steven J. Ostro of the Jet Propulsion Laboratory and his co-workers rapidly growing opportunities for observations with large Earth-based radars. The most widely used are the giant planetary radar at Arecibo, Puerto Rico, and the NASDA Goldstone radar in California's Mojave Desert. Using variants of the technique of delay-Doppler mapping described in Chapter IX, they have found that extreme departures from spherical figure are common among these small bodies. Bodies smaller than about 100 km in diameter not only have small internal pressures and low differential stress levels due to their low gravity, but they also have internal temperatures that are close to the radiative steady-state temperatures of their surfaces, averaged over their often rather eccentric orbits. Penetration of the diurnal and annual thermal waves into these bodies causes extreme temperature only in the outermost centimeters of their surfaces. Radiogenic heat from their interiors is conducted out to the surface and radiated off into space on time scales much shorter than the decay half-life of $^{40}K$ or $^{235}U$. Thus the interior is kept so cold that its viscosity is essentially that of rock at terrestrial room temperature. For these reasons, the usual mechanism of viscous relaxation to near-spherical shape is not effectual.

In addition, these small bodies have typically suffered several collisions with other bodies of at least 1% of their size over the age of the Solar System. We must suppose that the typical NEA, having spent only 1% of its lifetime in an Earth-crossing orbit, would have suffered almost all of these collisions while still within the main Belt. Given average eccentricities of 0.1 and average orbital velocities of $20\,\mathrm{km\,s^{-1}}$ for belt asteroids, mutual collisions of these bodies must occur with typical relative speeds of $2\,\mathrm{km^{s-1}}$. Since a 100-km asteroid with a density of $3.5\,\mathrm{g\,cm^{-3}}$ has an escape velocity of only $70\,\mathrm{m\,s^{-1}}$, mutual acceleration by the gravitational attraction of the colliding bodies during their encounter and gravitational focusing of their trajectories to enhance their collision cross-section are both negligible. In contrast, an NEA with an aphelion in the Belt will pass through the densely populated Belt at a relative speed of about $10\,\mathrm{km\,s^{-1}}$. It will sweep out volume, and incur collisions, at about five times the rate of a body that orbits wholly within the Belt. Further, the energy of each colliding projectile will be 25 times as high for the interloping NEA as for a resident of the Belt. Thus collisional erosion, shattering, and disruption must be much more important for an NEA than for a Belt asteroid—and we know that most Belt asteroids are products of collisional disruption. Therefore, there is an exceptionally high probability that NEAs will be thoroughly shattered bodies, often reaccumulated as random piles of irregular fragments produced by collisions almost energetic enough to disrupt the asteroid completely. Also, because of the violent history of these small bodies, they have had exceptional opportunities to lose their regolith, or at least to have it regionally stripped off or extensively redistributed by small impacts.

The radar data collected by Ostrow and co-workers strongly support this rubble-pile model of NEAs. Figure VIII.42, a set of radar images of the NEA (4179) Toutatis, clearly shows the complex, lumpy structure of this 4.60-km-long body. Note that its longest axis is about 2.5 times as large as its shortest axis. "Necking" of the structure is evident. To compound its oddity, it was found from the radar data that Toutatis does not rotate about its principal axis, that about which it has the greatest moment of inertia. The period of rotation was found to be 5.41 days (relative to its long axis), with a precession period of only 7.35 days for that axis. All these oddities point to the extreme importance of collisions in Toutatis' history.

## Asteroid Resources

The near-Earth asteroids have several properties that make them highly desirable prospects for future exploration and exploitation. First, they are samples of a very wide variety of types of both undifferentiated and differentiated material originating over a very wide range of heliocentric distances. They thus constitute a research collection of incalculable scientific value. Second, they are, both geometrically and energetically, the most accessible bodies in the Solar System. Some of these bodies actually pass through the Earth–Moon system. The most common fate of near-Earth asteroids is eventually to collide with the Earth. Perhaps 20% of all the near-Earth asteroids are so energetically accessible that it actually would require less energy to fly from the Earth to the surface of these asteroids than it did for the Apollo spacecraft to fly to and land on the Moon. Third, the near-Earth population of asteroids contains a wide variety of materials of untold economic value in space. A quarter of all near-Earth asteroids may be partially devolatilized cores of short-period comets, containing substantial amounts of ice beneath a surface lag deposit of fluffy, insulating dust. The principal economically attractive materials are water (found as ice or as water of hydration in C asteroids) and free metals (found as the native Fe–Ni–Co–Pt group alloys kamacite and taenite in almost all classes of non-C material). These two materials can provide the large majority of all the mass needed for large-scale space development in near-Earth space, including propellants, life-support fluids, and structural components such as plates, beams, fixtures, wire, and cables. Fourth, a technique has been developed for the efficient return of nonterrestrial material to Earth or to low Earth orbit (LEO) in which the excess energy of the returning vehicle is dissipated by aerodynamic braking during passage through Earth's upper atmosphere. This technique, when used to effect orbit changes, is called *aerobraking*. When used to capture an incoming spacecraft with hyperbolic geocentric velocity into a bound (elliptical) orbit, it is called *aerocapture*. This technique permits very large orbit changes with the expenditure of extremely small amounts of propellant. Finally, it is clear from statistical arguments that the number of nearby and accessible asteroids is very large: vast numbers are easier to reach than the Moon (Table VIII.6). Of these, probably 30% are extinct

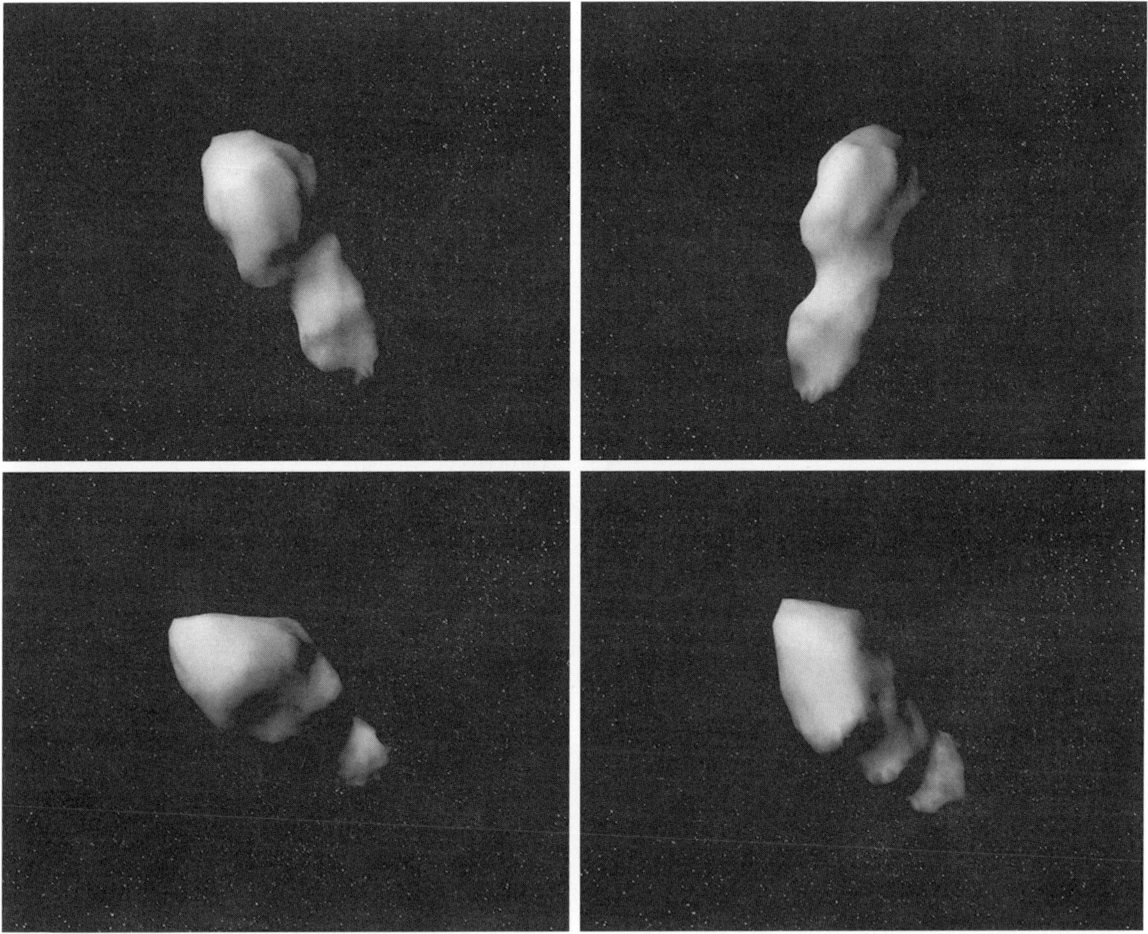

**Figure VIII.42** These radar images of the NEA 4179 Toutatis, showing its great elongation and multimodal character, were created by Drs. Scott Hudson and Steven Ostro from data obtained with the Goldstone (California) and Arecibo (Puerto Rico) radars. The maximum dimension of the asteroid is about 4600 m. Toutatis tumbles like a badly thrown football, with two periods of 5.4 and 7.3 Earth days. This image, which appeared in *Science* on October 6, 1995, is reproduced with the permission of Steve Ostro.

**Table VIII.6** Near-Earth Asteroid Population

| | | | | No. easier to reach than Moon ($\Delta V_{out} < 6.0\,km\,s^{-1}$ from LEO) | | |
|---|---|---|---|---|---|---|
| **Asteroid class** | **No. known (All sizes)** | **Projected ($d > 1$ km)** | **Projected ($d > 0.1$ km)** | **No. known (All sizes)** | **Projected ($d > 1$ km)** | **Projected ($d > 0.1$ km)** |
| Aten ($a < 1.000$ AU) | 143 | 80 | 22,000 | 30 | 15 | 4,000 |
| Apollo ($q < 1.017$ AU) | 862 | 500 | 148,000 | 90 | 100 | 30,000 |
| Amor Earth-crosser ($1.017 < q < 1.300$) | 837 | 370 | 110,000 | 120 | 65 | 18,000 |
| Total | 1842 | 950 | 280,000 | 240 | 180 | 52,000 |

comet cores with icy interiors or volatile-rich carbonaceous asteroids of the C, D, P, or related classes, containing from 5% (CM) to 20% (CI) by mass of water, up to as high as 60% to 80% water for comet core material. Under the conservative assumption that 30% of the NEAs are volatile-rich, the number of kilometer-sized C-like bodies that are more accessible than the Moon is about 120 (Table VIII.7).

**Table VIII.7** Outbound Total $\Delta V$ from LEO to the Surface of NEAs

| $\Delta V$(km s$^{-1}$) | Projected number with $d > 1$ km | Projected number of C types ($d > 1$ km) |
|---|---|---|
| 6.1 | (lunar landing) | |
| 6.0 | | |
| | 122 | 36 |
| 5.5 | | |
| | 40 | 12 |
| 5.0 | | |
| | 12 | 4 |
| 4.5 | | |
| | 5 | 2 |
| 4.0 | | |
| | 1 | 0 |
| 3.5 | | |
| 3.2 | (Earth escape velocity) | |

The optimum means by which these resources may be exploited are not yet known, but we already can demonstrate that there are simple schemes that can greatly multiply the amount of mass available in Earth orbit. One such scheme follows:

> A spacecraft with a hydrogen-oxygen propulsion system is lifted into LEO. The spacecraft carries as its payload a small mining module and a large inflatable solar collector. The spacecraft departs from the Space Station in LEO under H/O propulsion and quickly achieves Earth escape velocity. It then adjusts its speed to intersect the orbit of the target asteroid and, after several months of transit time, it reaches and carries out a rendezvous with the asteroid. The landing speed of several meters per second (the escape velocity of a typical near-Earth asteroid) is easily cushioned by shock absorbers. For several months the miner picks up regolith dust and ices with a scoop (or a magnet, if only metals are desired), heats the dust/ice mixture gently with the solar collector to release water, and loads it into a return container. The return vehicle, carrying perhaps 100 tonnes of water or metals, is lifted free of the asteroid surface by a solar thermal rocket which uses the solar collector to boil water (a "steam rocket"), and given the small velocity necessary to cause its orbital path to intersect the Earth. This velocity change is as low as 60 m s$^{-1}$ for the asteroid 1982 DB, the best candidate presently known.
>
> After a coast time of a few more months the vehicle encounters the Earth and the steam rocket is fired to permit capture into a highly eccentric Earth orbit. The miner device is then left in that orbit, and the main water tank, with only light thermal protection, then makes successive gentle passages through the upper atmosphere until the orbital apogee drops to the altitude of the Space Station. Then a tug from the Space Station makes rendezvous with the tank, and a short engine firing lifts the perigee out of the atmosphere and matches orbits with the Space Station. The cargo is unloaded and part of the water is electrolyzed into hydrogen and oxygen using electricity made by solar cells. The chemical rocket engine is then refueled and dispatched into the transfer orbit where the miner is stored, docks with the miner, and departs on another mission.

This scheme, operated cyclically between eccentric Earth orbit and the surface of a "good" asteroid, is capable of returning more than 100 tons of extraterrestrial material to LEO for every ton that must be launched from Earth. We say that it has a *mass payback ratio* of 100:1. Ultimately, with extensive use of nonterrestrial materials, mass payback ratios over 1000 may be possible. Since all present space endeavors are predicated upon a real Space Shuttle launch cost of about \$6000/lb ($\approx$ \$13,000/kg), this means that the cost of materials in near-Earth space can be lowered at least to \$60 and possibly as low as \$6 per pound. The advent of large, economical launch vehicles capable of delivering payload to low Earth orbit (LEO) at a price as low as \$600 per pound would not diminish the attractiveness of extraterrestrial materials, since their costs are also proportional to Earth launch costs. In this case, a mass payback ratio of 100 would make materials available in LEO for \$6/lb, and a MPBR of 1000 would reduce materials costs to \$0.60/lb. The materials for a 100-ton "house" in space would then cost about \$120,000. (By comparison, America's 82-ton Skylab space station cost about \$2,500,000,000 when it was launched in 1973, and the price tag for the International Space Station, which is two or three times the mass of Skylab, is currently estimated at \$15 billion.) A wide range of space activities that are economically marginal or impossible at present launch costs, such as large space structures, interplanetary expeditions, Mars colonies, and solar power satellites (SPSs), would become highly affordable.

Water-bearing asteroids need only solar or nuclear energy to permit the extraction of the water and the production of hydrogen and oxygen by electrolysis. Most of the energetically attractive NEAs have aphelia in the Belt: indeed, at any moment, most NEAs are actually located in the Belt, as Figure VIII.43 shows. These asteroids commonly cross the orbits of other planets, conveniently traveling back and forth along the main interplanetary thoroughfares. A water-bearing NEA is therefore not only a "flying hotel" that will take a passenger to the heart of the Belt, but also a "flying gas station" that provides the propellants needed to visit and exploit main-Belt bodies that never approach Earth.

Some other "asteroidal" bodies, such as the Martian moons Phobos and Deimos, are ideally situated to serve the needs of the next generation of space activities. They represent enabling resources for the spread of humankind throughout the inner Solar System at least out to the orbit of Jupiter. All we need to do is find and use them before they find and hit Earth (Fig. VIII.44).

## Exercises

### *Introduction to Meteorites*

VIII.1 Use any newspaper index or online clipping service to find an eyewitness account of the fall of a meteorite. Write a brief report on all the *observed phenomena* described in the article. If there are any unjustified *assumptions or conclusions* given in the report, explain why you question them. (Hint: The index of *The New York Times* is perhaps the most widely available in the United States. *The Times* of London, England, is another excellent source, with an index reaching back to the eighteenth century.)

VIII.2 Describe in detail the field marks by which a freshly fallen meteorite may be distinguished.

### *Meteorite Orbits*

VIII.3 A meteorite falls vertically close to Earth's equator just as the Sun is setting. A network of tracking stations set up to photograph meteor trails measures an entry speed (at 100-km altitude) of $15.0\,\text{km s}^{-1}$.

a. Calculate the approach velocity of the meteorite before it was accelerated by Earth's gravity.

b. Calculate the body's orbital properties, including its perihelion and aphelion distances, its orbital semimajor axis, its orbital period, and the eccentricity of its orbit.

### *Phenomena of Fall*

VIII.4 A fireball at an altitude of 80 km above the ground has the apparent brightness of the full Moon (say, $m_V = -15$). The apparent magnitude of the Sun is −26.1. What was the luminosity of the fireball ($\text{erg s}^{-1}$)?

VIII.5 Describe in detail the following:

a. The information that you should collect and record if you should happen to witness a meteorite fall.

b. The actions that you should take if you find a rock that appears to be a meteorite.

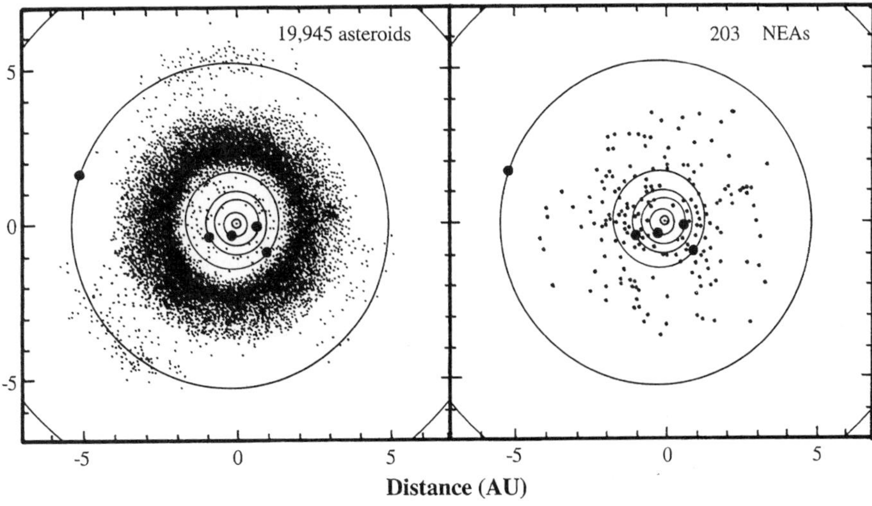


**Figure VIII.43** A "snapshot" of the locations of all asteroids and all NEAs on April 15, 1992. Mark Sykes and Elizabeth Alvarez produced this remarkable display. Note that the resonance gaps in the Belt (which involve semimajor axis, not instantaneous heliocentric distance) are completely undetectable. Note also the Trojan and Greek asteroids on Jupiter's orbit. Most of the NEAs at any moment are in the Belt, but the terrestrial planet region is very busy!

**Figure VIII.44** A typical Spacewatch image showing discovery of two asteroids. The image is a computer superposition of three images taken about an hour apart; the trios of boxes show the three sequential positions of two moving bodies, one a slow-moving Belt asteroid and the other a fast-moving NEA.

VIII.6 A freighter belonging to an interstellar mining consortium, with an empty weight of 90,000 tonnes ($9 \times 10^{10}$ g) and a cargo load of 10,000 tonnes of TNT destined for use blasting asteroids, collides with an uncharted Kuiper Belt body on its way into the Solar System at a speed of 140 km s$^{-1}$.

a. How big an explosion results from releasing the kinetic energy of the impactor? (One megaton, the explosive power of 1 million American tons of TNT, is equivalent to $10^{18}$ calories or $4.2 \times 10^{25}$ ergs.)

b. By what factor is the explosive yield increased by the chemical detonation of the TNT payload on impact?

*Physical Properties of Meteorites*

VIII.7 Laboratory measurements on meteorite rocks and minerals show that many stones have crushing strengths of 100 to 1000 bars ($10^8$ to $10^9$ dyn cm$^{-2}$). The strongest stones are still several times weaker than iron meteorites, which have crushing strengths up to 3.5 kbar. Bodies of uniform composition are known to have cracks on all size scales: larger bodies can accommodate a wider range of crack scales, and hence are slightly weaker than smaller pieces of the same material. This phenomenon is described empirically by the Weibull strength law, which states

$$s = cm^{-1/12}.$$

A 1-cm laboratory sample of a stony meteorite has a measured crushing strength $s = 1$ kbar. What would be the crushing strength of a 30-m asteroidal fragment made of the same material?

*Meteorite Minerals*

VIII.8 Why are highly reduced meteoritic mineral assemblages dominated by enstatite?

*Taxonomy and Composition of Chondrites*

VIII.9 Under what circumstances do Prior's Rules have validity?

VIII.10 Metamorphism in terrestrial rocks is almost inseparably linked with migration and reaction of volatile elements brought about by changes in the temperature and pressure to which the rock is exposed during long periods of deep burial. To what extent do the metamorphic grades of chondrites defined in this section reflect the terrestrial experience? Does the presence of meteoritic rocks of the six different metamorphic grades prove that these rocks were metamorphosed during deep burial by changing environmental conditions?

VIII.11 Although we have meteorites that have been blasted off the surfaces of the Moon and Mars, meteorites that can be geochemically linked to Venus are unknown in our collections. Suppose an enormous impact event on Venus produces great numbers of rock fragments that are traveling at speeds above the escape velocity of Venus. How high would their crushing strength have to be to survive passage through the near-surface atmosphere of Venus, which is made of nearly pure carbon dioxide at a temperature of 750 K and a pressure of 92 bars? Is this calculation an adequate "impossibility proof"?

*Taxonomy and Composition of Achondrites*

VIII.12 Show how the coexistence of schreibersite + farringtonite defines an oxygen buffer. Specify what other *reasonable* minerals are required to give a balanced buffer reaction.

VIII.13 From Fig. VIII.7, assume that the eucrites are derived from a calcium-rich partial melt of C-type chondritic material. What would be the composition of the "complementary differentiate," that is, the material left behind by extraction of the eucritic melt? Can you identify the possible complementary differentiate in Fig. VIII.7?

*Taxonomy and Composition of Stony-Irons*

VIII.14 The aubrites, ureilites, and diogenites, which otherwise have almost nothing in common, are sometimes lumped together as "Ca-poor achondrites." Would the silicate component of either of the classes of stony-irons fit into this category?

*Taxonomy and Composition of Irons*

VIII.15 Follow the dotted line in Fig. VIII.8 from 1540 K down to its end, explaining each aspect of the behavior of the nickel content of meteoritic metal shown in the graph.

VIII.16 Some of the Type IVB irons contain extremely high concentrations of platinum-group metals such as platinum, osmium, and iridium, in some cases reaching 100 ppm by weight. The market value of the mixed platinum metals on Earth is about \$600 per ounce (\$20,000 per kilogram). A nearby metallic asteroid, Amun, with a mass of about $10^{16}$ g, is more accessible to spacecraft missions than the Moon.

a. If Amun is a Type IVB iron, what is the highest plausible market value of the platinum-group metals contained in it?

b. If Amun, an Earth-crossing asteroid, should collide with Earth at 20 km s$^{-1}$, how big an explosion would result (in megatons)? To calibrate your answer, an all-out nuclear exchange between all the nuclear powers on Earth (World War III) would liberate about 15,000 megatons. (One megaton = $4 \times 10^{22}$ erg.

c. The dust from the impact of Amun, if spread uniformly over the entire Earth, would produce an "iridium-rich layer" of sediments to witness that such an impact has occurred. What would be the concentration of platinum metals in this dust layer (g cm$^{-2}$)?

d. Is Amun a threat or an opportunity?

*Isotopic Composition of Meteorites*

VIII.17 Draw a small region of the oxygen 3-isotope plot surrounding the (0,0) point in Fig. VIII.10 and indicate the effects of adding to the (0,0) oxygen (SMOW) a small amount of the following:

a. Oxygen-16.

b. Oxygen-17.

c. Oxygen-18.

VIII.18 A gram of ordinary chondrite rock at 0° C is enclosed in a perfectly insulating container and allowed to heat up due to the decay of the short-lived radionuclide $^{26}$Al. Each decay event releases about 1 MeV. (See conversion factors in Table IV.3.)

a. How much $^{26}$Al must be present initially for its complete decay to heat the rock up to its melting temperature (assume this is 1273 K)?

b. What would the isotopic composition of magnesium be after complete decay of radio-aluminum if it was pure $^{24}$Mg originally?

*Genetic Relationships between Meteorite Classes*

VIII.19 It has been proposed on the basis of major-element abundances that the C3 chondrites are basically a mixture of C1/C2 and ordinary chondrite materials. Assess this proposal in light of the C and N isotopic data in Figs. VIII.11 and VIII.12.

*Introduction to Asteroids*

VIII.20 The newspapers carry word of the antics of the asteroid 1994 FD. What does the name tell you about its discovery circumstances?

*Asteroid Orbits*

VIII.21 Consider the data given in Fig. VIII.19 on orbital semimajor axes of the asteroids.

a. How wide a range of heliocentric distances is actually spanned by the average members of the Eos and Themis families?
b. Why is there a distinct gap visible between them in Figure VIII.19?

*Stability of Trojan and Plutino Orbits*

VIII.22 In Eq. (VIII.9), is $d$ a positive or negative number? Why?

*Sizes, Shapes, and Albedos of Asteroids*

VIII.23 Suppose that a planetary radar is built that has the ability to detect Ceres at opposition. How small an asteroid could it detect at a range of $10^6$ km?

*Masses and Densities of Asteroids*

VIII.24 It was long thought that the dust generated by collisions of asteroids would pervade the Solar System and coat all the solid bodies, including the asteroids, with a layer of uniform dust. How does Fig. VIII.24 show that reaccretion of asteroid dust is very unimportant?

*Photometry and Spectroscopy of Asteroids*

VIII.25 A C asteroid (albedo 0.04) and an S asteroid (albedo 0.20) orbit at the same distance from the Sun. What is the ratio of their effective temperatures?

VIII.26 A C asteroid ($A = 0.04$) and an S asteroid ($A = 0.20$) with equal apparent visual magnitudes orbit the Sun at $a = 3.1623$ AU. What is the ratio of their masses?

*Thermal Evolution of Asteroids*

VIII.27 We have earlier seen that the crustal temperature gradient in a solid body is

$$\partial T/\partial r = -(S\rho r_s/3\kappa) \qquad \text{(VI.25)}$$

Derive an analytical expression for $T(r)$ when $S$, $\rho$, and $\kappa$ are constants throughout the body; that is, for a completely homogeneous, undifferentiated body.

VIII.28 Suppose a T Tauri phase solar wind with a mean field strength of $|B| = 1$ G and a stream velocity of 400 km s$^{-1}$ strikes the surface of Ceres.

a. What is the maximum possible heating rate of Ceres in watts?
b. At this heating rate, and assuming uniform heating, how warm would Ceres be after being heated at this rate for the $10^7$-year duration of the T Tauri phase of the Sun? Ignore radiative cooling.

*Dynamical Evolution of the Asteroid Belt*

VIII.29 A small satellite travels in a low circular orbit 20 km above the surface of a C asteroid that has a radius of 100 km.

a. What is its approximate orbital period?
b. If this asteroid pair is unresolved by telescopes on Earth, so that our only evidence of duplicity is the lightcurve of the system, are we likely to recognize this as a binary system? Consult Fig. VIII.22 in deciding on your answer.
c. An S-band radar operates at a wavelength of 1.8 cm. What is the maximum possible offset in wavelength between the reflected pulses from the main asteroid and its satellite?
d. For a typical rotation period, might the Doppler effect due to the rotation of the main asteroid overwhelm the faint return from the little satellite?

*Relationships among Asteroids, Meteorites, and Comets*

VIII.30 Is it possible for a near-Earth asteroid to turn into a comet? Describe such a scenario using only reasonably probable circumstances and events.

*Asteroid Resources*

VIII.31 It is sometimes argued that the cost of launching payloads into space must always remain high because of the enormous amount of energy needed to get a payload into orbit. Suppose you had a method of using electric power to accelerate payloads and launch them into space:

a. Using a market price of \$0.10 per kilowatt-hour, what would be the cost of launching a 70-kg adult into LEO? Neglect the cost of the extension cord.

b. The present cost of launching astronauts into space is governed by launch costs for the Space Shuttle, which run from \$400 million to \$900 million per launch (depending on how you account for infrastructure costs). STS has the capability to launch eight astronauts per trip. Alternatively, the ticket for a seat on a Russian Soyuz TM vehicle to the Mir space station costs about \$15 million. Which of these three alternatives—the Space Shuttle, the Soyuz, or the electronautics from part (a)—would you prefer to pay for?

VIII.32 A spacecraft returning from a near-Earth asteroid has an approach velocity of 5.00 km $s^{-1}$ before acceleration by Earth's gravity. It encounters the top of the atmosphere at a shallow angle and is slowed down by aerodynamic drag (aerobraking) until it is gravitationally bound to Earth.

a. What is the velocity of the vehicle as it enters the atmosphere at an altitude of 100 km?
b. What is the velocity change that must be brought about by the aerobraking maneuver in order to ensure capture into a closed orbit about Earth?
c. What is the approximate average acceleration required to produce this velocity change while the vehicle is passing through the upper atmosphere?

# IX. The Airless Rocky Bodies: Io, Phobos, Deimos, the Moon, and Mercury

## Introduction

We have followed the evolution of matter from the hydrogen and helium ashes of the Big Bang through galaxy and star formation, heavy-element nucleosynthesis and supernova explosions, the formation of the Solar System, and evolutionary processes in the realm of the outer planets. We then examined the chemical and physical differentiation of nearly homogeneous preplanetary material from the point of view of the Jovian and Uranian planets, ice-rich satellites, and comets. Next we surveyed the Centaur and TNO groups, then moved closer to the Sun to study the Asteroid Belt, Trojan, and NEA populations and relate these asteroidal bodies to the laboratory study of meteorites, for the purpose of using them as guides to the materials and processes that gave rise to the terrestrial planets. Before encountering the terrestrial planets Mars, Venus, and Earth in their full glory and complexity, we have only to study the smaller, airless bodies in the inner Solar System. These bodies bridge the gap in size, degree of evolution, and complexity between meteoritic samples of asteroids and the three large terrestrial planets with atmospheres.

The general trend of planetary compositions and densities clearly favors the idea of formation of the Solar System out of a flattened, dusty gas disk within which the density, pressure, and temperature all dropped off steeply with increasing distance from the center. The bodies discussed in this chapter all permit us to test our ideas of planetary formation and evolution in specific but different ways. Phobos and Deimos appear to be relatively unaltered samples of the small bodies that passed through Mars' accretion zone late in the era of planetary formation. They therefore may contain much useful information on local environmental conditions during the formation of Mars.

Mercury orbits closer to the Sun than any other known Solar System body and serves as a probe of the high-temperature extreme of conditions encountered during planetary formation. As recently as 1966 it was widely believed that the Moon would prove to be a rather primitive object similar to typical preplanetary solids, possibly wholly undifferentiated, like chondritic meteorites. But we now know that Earth's Moon has a composition that deviates markedly from those of the terrestrial planets, in that the Moon seems to be strikingly deficient in native metals. Further, the dynamical history of the Moon may be unique among the known bodies of the Solar System.

Io presents us with our best example of large, rocky bodies in the outer Solar System. Of all the six lunar-sized satellites of the Jovian planets (Io, Europa, Ganymede, and Callisto about Jupiter; Titan at Saturn; Triton at Neptune) and the Pluto–Charon system, it is the only one with rock-like density (3.52) and an ice-free surface. Even Europa, with its similar density of 3.45, has a deep ice layer, perhaps 50 km in thickness, that bars us from studying its rocky component. Io is also "Jupiter's Mercury"; it is the high-temperature end member of planetary formation in the Jovian subnebula, which spawned four bodies of planetary dimensions. The less massive, colder Saturnian subnebula produced no rocky satellites and no obvious systematic density trend with Saturnocentric distance. The Uranian satellites likewise have low density and no discernible trend, and the Neptunian system is unfortunately severely disturbed by the possible capture of Triton. Thus Io plays a unique role in our study of planetary formation: it represents a test of the generality of our ideas of planetary formation from what is effectively another nebular system.

The tale told about these small rocky bodies by even the very simplest indicator of planetary composition, bulk density, is already complex. Figure IX.1 shows the densities of the bodies in the inner Solar System, as corrected for gravitational self-compression. These corrected uncompressed (zero-pressure) densities are to some degree model dependent, but they at least reflect an intrinsic property of the solid raw materials out of which the planets are made. The apparent similarity among the observed densities of Mercury, Venus, and Earth hides a very large compositional difference: at zero pressure, Mercury ($5.3\,g\,cm^{\neq 3}$) is *much* denser than Venus and Earth (both about 4.4, with Earth marginally denser by about 1%). The simplest explanation for the density difference is that Mercury's mass is about 70% metal, compared with about 30% for Venus and Earth. Mars has an observed density of 3.93, which converts to an uncompressed density of 3.74, distinctly less than any of the other terrestrial planets. The Moon's observed density of $3.34\,g\,cm^{\neq 3}$ is essentially unaffected by compression because of the Moon's small mass and low interior pressures. The two small satellites of Mars, Phobos and Deimos, have densities of only about $2\,g\,cm^{\neq 3}$, similar to the most volatile-rich carbonaceous chondrites

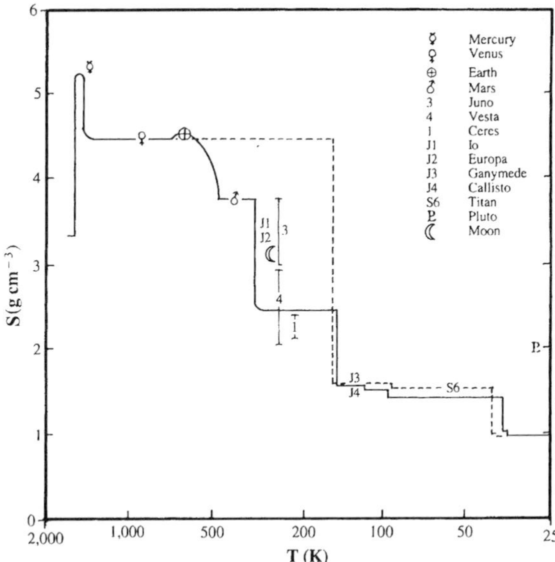


**Figure IX.1** Uncompressed densities of solid Solar System bodies. The solid line is the theoretical dependence of density on temperature for an equilibrium condensation scenario with a slightly elevated Fe:Si ratio. The dashed line is the dependence of density on accretion temperature for a rapid-accretion scenario, in which each condensate accretes into large bodies as it condenses, without further reaction with nebular gas. Note that, even with no allowance for radial mixing of nebular solids with different condensation temperatures, the predicted density of Mercury is too low.

(type CI). Thus the seven bodies that orbit "permanently" in the inner Solar System represent a *minimum* of five distinctly different compositions.

We should recall that many other small bodies are presently in orbits in the same volume of space. The approximately 200 nearest asteroids include one with a strikingly Earth-like orbit (1991 JW) and one that follows a Lagrange point on the orbit of Mars (1990 MB). The asteroids of the Aten family have orbital periods less than 1 year and patrol the space from 0.292 to 1.513 AU from the Sun, and at least 25% of the NEAs have orbital periods less than that of Mars. These bodies, however, are only temporary sojourners in the terrestrial planet region, exiles from the cold reaches of the asteroid belt or retired short-period comets. After a few tens of millions of years they will have suffered collision with one of the terrestrial planets or experienced perturbation into Jupiter-crossing orbits and then ejection from the Solar System. Their presence should remind us that there is a constant supply of new interlopers of extremely diverse composition careering through the space occupied by the terrestrial planets. Indeed, Phobos and Deimos may be refugees from the Belt that have taken up long-term residence in orbits about Mars. Their compositions and histories, if known, might reveal the satellites to have either little or great relevance to the history of Mars.

Because of their widely different locations, the bodies in this chapter experience a wide variety of collisional environments. The location of Phobos and Deimos in orbit about Mars causes their impact evolution to be radically different from that of the other bodies. Mercury experiences impacts that are more energetic than those upon any other body in the inner Solar System, excepting only the Sun itself. The general idea of the dominant role of impacts in governing the surface evolution of small rocky bodies seems challenged by Io, which exhibits not a single identifiable impact feature. Thus a comparison of the cratering histories of these bodies is of great importance.

It is also interesting to note that every one of these bodies has a peculiar spin state: Phobos, Deimos, the Moon, and Io are all locked into a 1:1 spin–orbit resonance with their primaries. Mercury, with the most eccentric orbit of the group, is locked in a 3:2 resonance.

## Orbits and Physical Structure of Phobos and Deimos

Phobos and Deimos are difficult to study from Earth because of their faintness and proximity to Mars. Imaging by Earth-based telescopes cannot resolve their sizes and shapes, and spectra are heavily contaminated by light from Mars that is scattered during passage through Earth's atmosphere.

However, both satellites have been extensively tracked by Earth-based telescopes since the time of their discovery by the American astronomer Asaph Hall in 1877. Their orbits are therefore well known. Phobos, the closer of the two to Mars, has an orbital semimajor axis of 9378 km, compared with 3396 km for the equatorial radius of Mars, whereas Deimos orbits at 23,459 km distance. Both are therefore very deep in the gravity well of Mars. Their orbital periods are 9.5738 h for Phobos and 30.2986 h for Deimos, compared with 24.6229 h for the Martian day, and both are rotationally locked onto Mars; their "day" and "month" are identical, as with so many other satellites in the Solar System. Thus, as seen from the surface of Mars, Phobos moves rapidly across the sky from west to east, and Deimos drifts more slowly from east to west. Synchronous altitude for Mars, the circular orbit location with an orbital period exactly equal to the length of the Martian day, lies between the two satellites but much closer to the orbit of Deimos. Both orbits have small inclinations ($i = 1.0°$ for Phobos; $i$ varies from 0.9° to 2.7° for Deimos as a result of solar gravitational perturbations). Phobos' orbital eccentricity is 0.015, and Deimos' 0.0005.

Because of the close proximity of both Martian satellites to Mars, the influence of tidal forces on the evolution of the orbits of Phobos and Deimos is likely to be great. Any tidal bulge raised on Mars by Phobos will be dragged along by the rotation of the planet, which has a lower angular velocity than Phobos does on its orbit. Thus the lunitidal interval is *negative*, and the tidal torque on Phobos (the gravitational attraction of the Martian bulge on Phobos) will tend to decelerate Phobos, causing its orbit to spiral in toward the Martian equatorial surface. By the same reasoning, Deimos, with its orbital period longer than a Martian day, will be accelerated in its orbital motion by the tidal bulge it raises on Mars and will steadily recede from the planet. Thus the orbits of the two Martian moons are evolving away from each other. Clearly, in the past they were closer together. But how close? Could they have once orbited together? Might they be the largest fragments from a collisionally disrupted ancient satellite that orbited near synchronous-orbit level? And, given the very large density difference between these satellites and Mars itself, might they not have originated elsewhere in the Solar System and experienced capture by Mars? Clearly any additional information we can garner on their chemical and physical properties may have profound genetic significance. But such information is very difficult to get if we stay on Earth.

The American Mariner 9 Mars Orbiter and the two Viking Orbiters flew by these satellites, photographed them, and experienced orbital perturbations by their masses. A Mariner 9 image of Phobos is given in Fig. IX.2. The Soviet Phobos 1 and 2 missions, using a new interplanetary spacecraft design, were dedicated both to remote sensing of Phobos (and possibly Deimos) and to *in situ* study of the surface of Phobos by small automated landers. Unfortunately, Phobos 1 was lost en route to Mars as a result of a controller error, and Phobos 2 mysteriously expired as it was approaching Phobos for the first time, but after returning useful photographic and spectral data.

The Mariner and Viking data reveal Phobos and Deimos as highly irregular, battered, generally asteroidal bodies with thick mantles of regolith material. Very complex figures were deduced from the images of Phobos, which reveal a body that very approximately fits a triaxial ellipsoid of $13.5 \times 10.8 \times 9.4$ km, with the longest axis pointed at Mars. This deduced volume, combined with the perturbational mass determined by tracking the spacecraft, gives an estimated density of $2.0\ \mathrm{g\,cm^{-3}}$, slightly less than CI chondrites. Deimos, with a shape approximating a $7.5 \times 6.5 \times 6.0$ km triaxial ellipsoid, was also found to have a density of about $2.0\ \mathrm{g\,cm^{-3}}$. Wide phase angle coverage was achieved by the spacecraft observations, permitting determination of

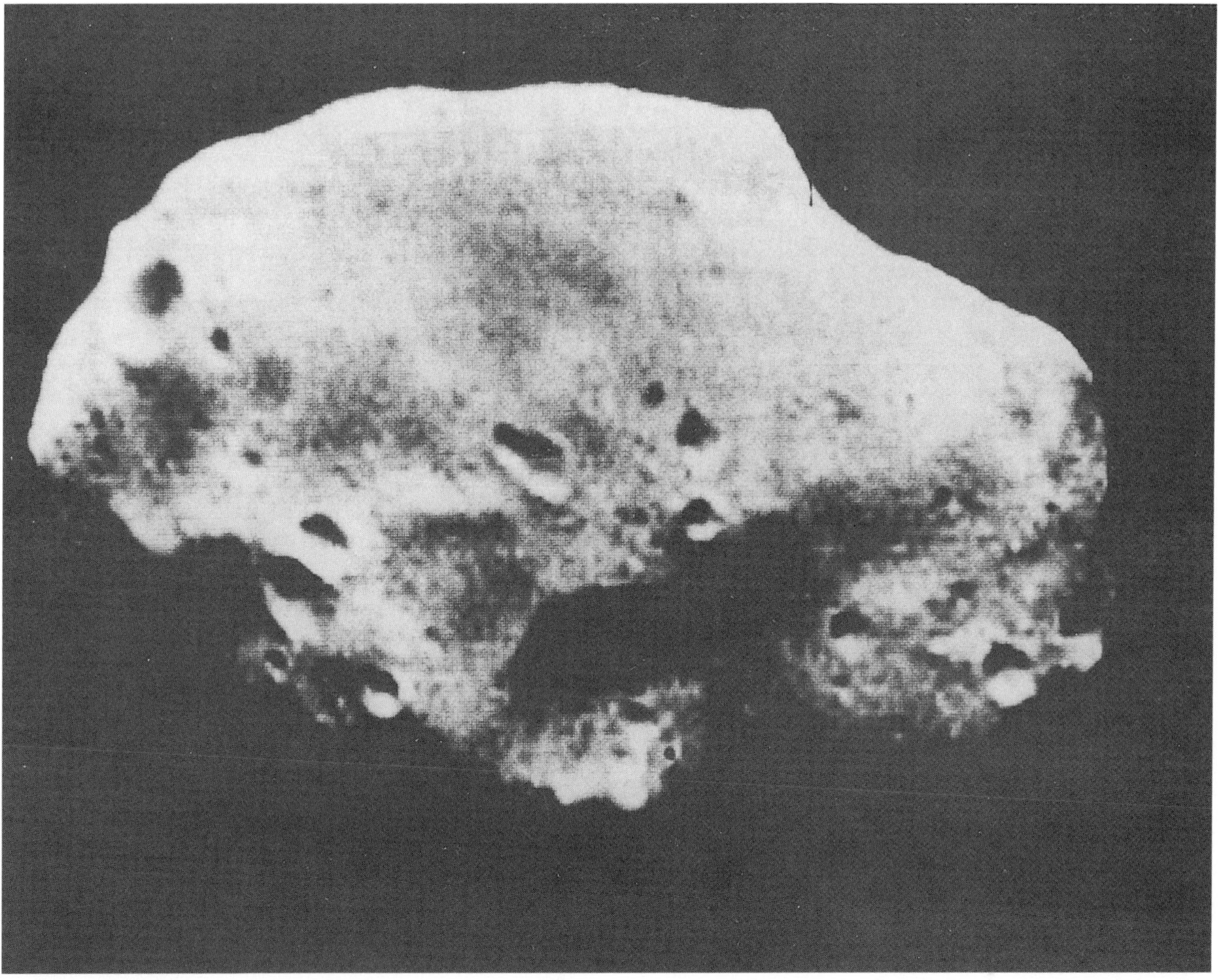

**Figure IX.2** A Mariner 9 image of Phobos. The highly irregular outline is a consequence of impact fragmentation. Note the soft, mantled appearance of the surface.

the phase functions ($p$) and geometric albedos ($q$) of both Phobos ($q = 0.27$) and Deimos ($q = 0.32$). For comparison, the surface of the Moon has a geometric albedo of $q = 0.6$. The Bond albedos of both bodies were found to be very low, with $A_B = pq = 0.018$ for Phobos and 0.022 for Deimos. Thus the density and albedo agree: Phobos and Deimos are most similar to C-type asteroids and hence to carbonaceous chondrites. However, Earth-based spectra of the Martian moons reveal a dark, almost featureless spectrum, with no evidence of an absorption feature due to ice or bound water near 3.0 $\mu$m.

The dynamical environment of these moons differs greatly from that of asteroids in nearby orbits about the Sun. Debris produced by collisional erosion of an asteroid is dispersed into an extremely large volume of space, forming a dust band that initially brackets and follows the position of the asteroid. Because ejection velocities of $200\ \mathrm{m\,s^{-1}}$ are easily achievable in impacts (compared with a typical asteroidal orbital speed of $20\ \mathrm{km\,s^{-1}}$), the dust commonly has an orbital period that differs by about 1% from that of the damaged asteroid from which it originated. Thus over the rather brief time interval of 100 orbital periods (a few centuries) the dust band spreads along the orbit to reach all the way around the Sun. Fragments in slightly different orbits will feel tidal forces slightly differently, and the lines of apsides of the orbits of particles in the debris swarm will then tend to precess at different rates, to produce a shell of dust no longer confined closely to a single orbital plane. On a similar time scale, Poynting–Robertson and Yarkovsky forces caused by solar radiation can cause major changes in orbital geometry, culminating in evaporation during passage close to the Sun, perturbation into very unstable planet-crossing orbits, or even ejection from the Solar

System. The asteroidal debris cloud, when it first reaches all the way around the orbit of an asteroid, may fill a toroidal (doughnut-shaped) volume with a cross-section diameter on the order of $d = 0.01a$, which is typically 0.03 AU, or about $5 \times 10^8$ km, reaching about an orbital circumference of $C = 2\pi a$ and thus occupying a volume of $\pi d^2 C/4 = 2 \times 10^{-5}\pi^2 a^3 = 2 \times 10^{18}\,\text{km}^3$. Supposing 2000 tonnes of ejecta was liberated by the impact, this would amount to a debris density of $10^{-9}\,\text{g}\,\text{km}^{-3}$. The synodic period of a typical piece of debri (the time taken for it to pass completely around the Sun relative to the asteroid and "lap" it) is about 300 years. An asteroid with $1\,\text{km}^2$ of cross-section area traveling at $200\,\text{m}\,\text{s}^{-1}$ relative to the dust swarm would sweep out $2 \times 10^7\,\text{km}^3\text{a}^{-1}$, for a reaccretion rate of $0.02\,\text{g}\,\text{a}^{-1}$. At that rate the time required to sweep up the 2000 Mg of debris would be of order $10^{11}$ a, vastly longer than the orbital lifetime of particles under the influence of solar radiation forces. Reaccretion of this ejected debris is obviously negligible.

But now consider a similar impact on Phobos that dislodges 2000 tonnes of debris with the same velocity dispersion. The orbital velocity of Phobos is $1.71\,\text{km}\,\text{s}^{-1}$, and the debris band will span about 1200 km. The volume over which the band will disperse is then $6 \times 10^{10}\,\text{km}^3$, the spatial density of debris is $3 \times 10^7$ times as high as in the asteroidal case, and the synodic period is about 80 h, not centuries. The sweepup time would then be roughly 3 ka. Under these circumstances, low-velocity reaccretion of ejecta is very probable.

Net erosive loss will still occur due to ejection from Mars or orbital decay into the Martian atmosphere, but this is now an inefficient process. Over long periods of time, impact cratering must eject most of the surface material of Phobos and Deimos over and over again. Shock heating will partially devolatilize carbonaceous dust, dehydrating hydroxyl silicates, pyrolyzing organic matter, and leaving a partly graphitized residue with a very high carbon content but little water. Thus the albedo of the shocked material, and perhaps even its gross color, may continue to resemble carbonaceous chondrites even after loss of most of the water (and loss of the spectral signature of bound water). Strong, prolonged heating of this "dry" material will permit reaction of the partly graphitized polymer with FeO, leading to a formation and release of abundant carbon dioxide and some water vapor. Thus even if "water" (as –OH silicates and bound $H_2O$) is absent, hydrogen and oxygen are not. The state of the pristine deep interior, which is slowly being "unpeeled" by net erosion, may be very much wetter than the shock-heated, reaccreted surface debris.

Viking Orbiter images of Phobos and Deimos do not reveal exposed bedrock and suggest that the deep interior of Phobos at least has been pervaded by impact-induced faulting (Fig. IX.3). The surface of Phobos has numerous pronounced grooves in the form of *crater chains*, formed by the subsidence of regolith into an extensive system of subsurface cracks. Are these moons indeed mere rubble piles, so full of fractures that an explorer could literally walk through them? Such a structure would be quite compatible with the idea that they are derived from a collisional breakup of a larger body. How much water-rich bedrock might be present within these moons? How deeply buried is it? How accessible is it? What is the total water content of Phobos? These questions need to be answered by appropriate spacecraft experiments on missions not yet planned by any nation.

The consequences of the presence of abundant, readily accessible volatiles in the Martian moons are far-reaching. Missions to the Mars system are limited by the necessity of carrying enough rocket propellant to return to Earth. But if the propellants required for excursions within the Mars system and for return to Earth are made out of locally derived resources, such as water and carbonaceous material from Phobos, then the mass

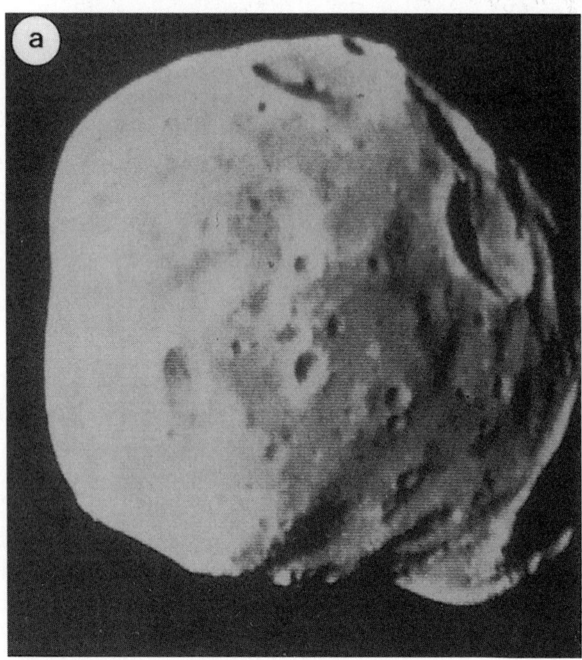


**Figure IX.3** Improved imaging of Phobos. a is an image taken by the Soviet Phobos 2 spacecraft, showing large craters and deep regolith mantling. Image courtesy of the former DDR Institut für Kosmosforschung. b is a Viking high-resolution image showing "grooves" in the regolith of Phobos, consisting of lines of rimless craters formed by subsidence of deep regolith into cracks in bedrock. It is likely that the interiors of both Phobos and Deimos are shattered remanents from near-catastrophic collisions.

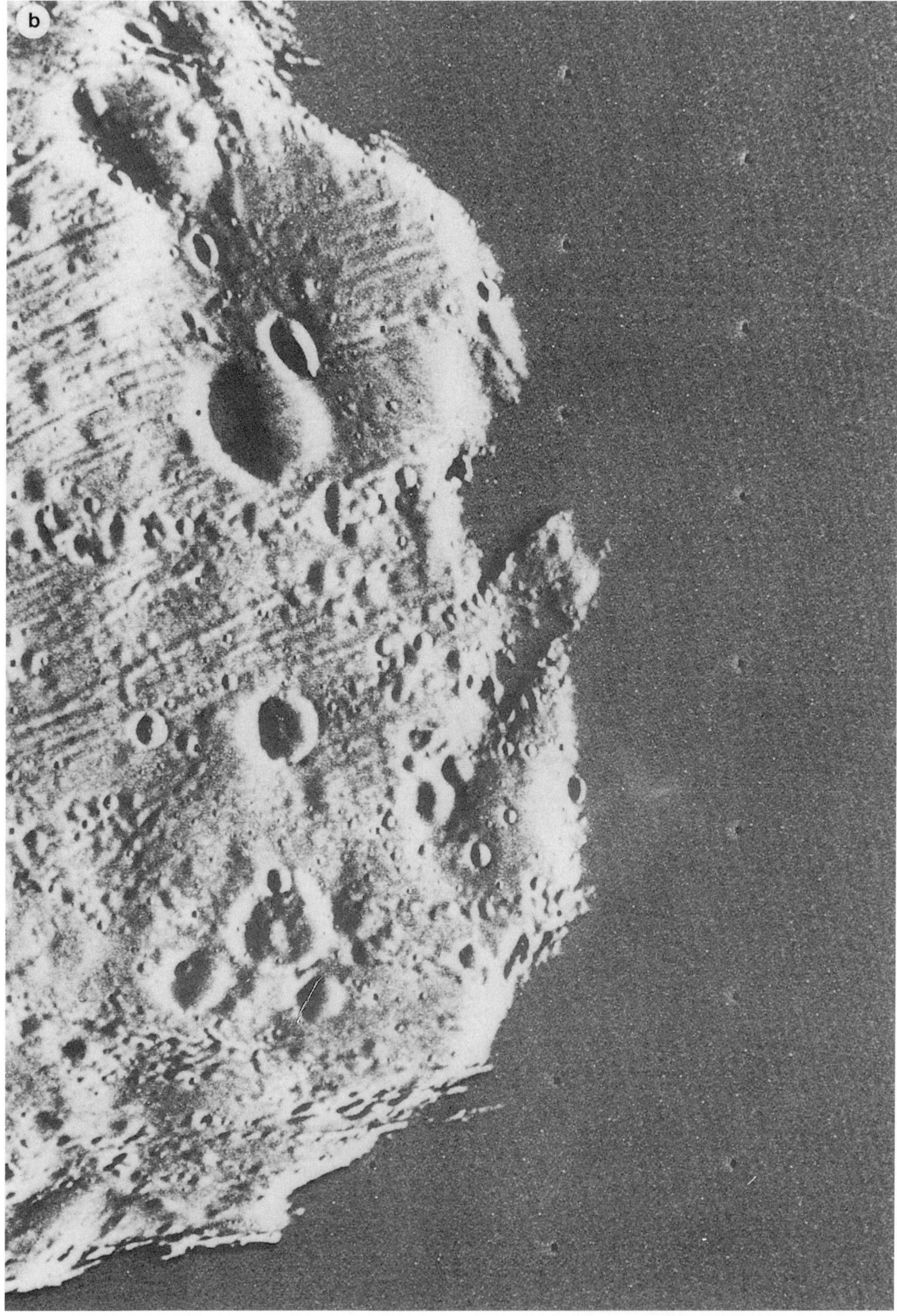


**Figure IX.3** *Continued*

of material lifted from Earth can be reduced by a factor of several, with attendant savings of tens of billions of dollars of launch costs.

## Io: General Properties

The most intensely disputed of the many controversies concerning Jupiter's innermost Galilean satellite, Io, is the proper pronunciation of its name. English-speaking modernists generally prefer EYE-oh, whereas classicists favor EE-oh, apparently under the supposition that the Greeks and Romans had a poor command of English. An occasional reporter contributes to this phonetic confusion by coining another variant; my favorite is "Jupiter's red satellite ten."

Io (yes, EE-oh) is scarcely red, or even orange. It is mostly yellow, locally shading to white, black, and orangish yellow and greenish yellow (Fig. IX.4). The overall albedo is a shockingly high 0.6, a number approached only by the cloud layer of Venus and very fresh snow surfaces, similar to a sheet of glazed white book paper. Although the density of Io is clearly within the range of rocky bodies, the spectrum shows no evidence of exposed silicate materials. Further, Io is unique among the rocky bodies of the Solar System in that its surface shows no evidence of impact craters. There must therefore exist a mechanism for obliterating evidence of Io's impact craters on an even shorter time scale than on Earth, where water and wind erosion and glaciation are very active.

The density of Io, as remarked above, is suggestive of an ice-free body. Both condensation theory and experience with carbonaceous meteorites suggest that material with solar proportions of rock-forming elements may hydrate in two distinct stages (see Chapter IV), the first characterized by formation of amphiboles containing essential calcium, and the second, by aqueous alteration ("serpentinization") of the very abundant ferromagnesian minerals, notably iron-rich olivine and pyroxene, to make not only true serpentine, but also chlorite and other phyllosilicates. The amphibole stage would contain less than 1% water and would have a density similar to that of an ordinary chondrite, about $3.5\,\mathrm{g\,cm^{-3}}$. Full hydration of the silicates adds about 20% to the rock mass while decreasing its density to about 2.4 (Chapter VIII). The observed density of Io, 3.52, suggests an uncompressed density near 3.5 and assures us that it *cannot* contain a large amount of bound water.

Thermal infrared observations of the surface of Io show a puzzling and strong dependence of flux on orbital phase angle (Fig. IX.5). Observations at 8.7 and $10\,\mu\mathrm{m}$ show flux variations with phase of factors of 5 and 3.5, respectively, whereas simultaneous observations at $20\,\mu\mathrm{m}$ show variations of only $\pm15\%$. A Planck peak at $20\,\mu\mathrm{m}$ corresponds to an emitter with an effective temperature near 148 K, and $8.7\,\mu\mathrm{m}$ is the Planck peak for a 340-K emitter. Thus if a small surface feature with a temperature on the order of 300 K is present on a surface with a mean temperature of 135 K (the steady-state temperature for solar insolation of a surface with Io's albedo), a large flux variation will be seen at 8.7 and $10\,\mu\mathrm{m}$, and only a ripple at $20\,\mu\mathrm{m}$.

The dilemma was soon exacerbated by infrared observations showing peak temperatures of 1500 to 1800 K, far too hot for liquid sulfur. Galileo, with the excellent spatial resolution afforded by close flybys of Io, found that the highest temperatures were in the range 1700 to 2000 K, well above the range for basalt magmas on Earth, 1300 to 1450 K. Generating such high temperatures without thinning the crust and collapsing the mountains is very difficult unless heating is a local phenomenon driven by concentrated energy sources such as electrical currents. As an interesting sidelight, the radiation belt environment at Io is so intense that Galileo actually systematically avoided it for most of the orbiter's mission life.

## Io: Surface Processes

Craters a few kilometers in diameter, with depths of a few hundred meters, should be formed at a rate of about one every 10,000 years, but are not seen; this argument crudely suggests that the surface is buried or relaxes viscously at a rate on the order of 100 m per 10,000 years, or about $1\,\mathrm{cm\,a^{-1}}$. But kilometers-high mountainous features are seen, which suggest that rapid viscous relaxation may not be possible. Further, mere viscous relaxation of the surface is not sufficient to remove the albedo features and sprays of small ejecta craters associated with major impacts; either the surface must be buried by debris deposited from the atmosphere or the crust must be subducted on a time scale of 10,000 years or less. The absence of a dense and obvious atmosphere, coupled with the fact that the much larger Earth has a continental drift time scale in excess of $10^8$ years, makes it clear that something extraordinary is happening on Io.

Reflection spectroscopy of Io's surface confirms the absence of ices. The bright surface deposit, which is concentrated in the leading hemisphere of Io as it orbits Jupiter, is found to have the near infrared absorption spectrum of solid sulfur dioxide, $SO_2$ (Fig. IX.6). The dominant, yellow, surface material is elemental sulfur, possibly in a variety of different allotropic forms with different colors.

**Figure IX.4** A Voyager 1 image of Io near zero phase angle. The yellowish plains material is sulfur, and the white mantling deposits are solid sulfur dioxide. Black spots and streaks may be quenched molten polymeric sulfur, and the diffuse round features are the plumes from major volcanic eruptions (See color plate 3).

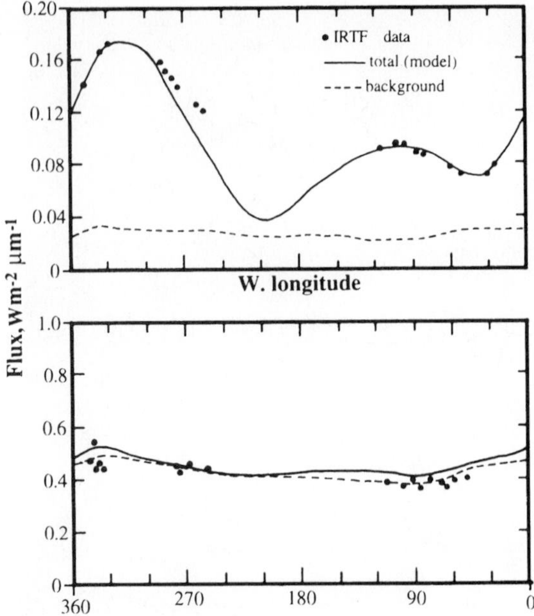


**Figure IX.5** Variation of Io's thermal emission with orbital position. Io's orientation is given as the longitude of the central meridian of the Earth-facing hemisphere. The data taken at 10 $\mu$m or less show the dominant role of local hot spots in providing the emitted flux. The 20-$\mu$m data reflect the Planck peak emission for the average temperature of the surface, with little longitude variation. Data are from the Infrared Telescope Facility (IRTF) in Hawaii.

## Io: Internal Energy Sources

The present rate of radiogenic heat production in Io, estimated by assuming solar proportions of rock-forming elements, is about $6 \times 10^{11}$ W. This is small compared

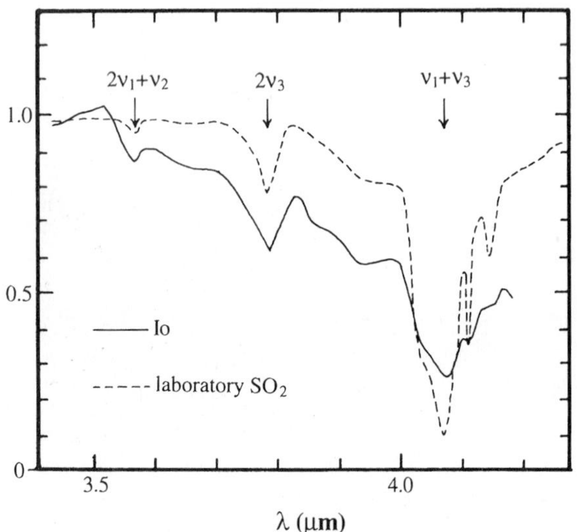


**Figure IX.6** Reflectance of the white frost on Io. The dashed curve is a laboratory reflection spectrum of sulfur dioxide frost, and the solid curve is an Earth-based spectrum of Io. The strong features exhibited by the white frost on Io are clearly due to solid $SO_2$.

with the solar insolation; a radius of 1815 km, a solar flux of $3 \times 10^4$ erg cm$^{-2}$ s$^{-1}$, and a Bond albedo of 0.6 give an absorbed power of $1.2 \times 10^{14}$ W. But the location of the heat source matters greatly. A source on the surface influences the boundary temperature but generates no internal temperature gradients. Because of the unusual geochemistry of Io it is difficult to be certain that we know where the long-lived radionuclides are located. The bulk density suggests a degree of oxidation higher than that of Mars, with a core dominated by FeS rather than Fe and a mantle with a high FeO content. Potassium has demonstrated chalcophilic ("sulfur-loving") tendencies in the extremely dry enstatite meteorites. These same meteorites contain the unusual mineral oldhamite, CaS, which is of interest because uranium and thorium have a well-known geochemical tendency to follow calcium. Thus it is quite conceivable that a significant fraction of the long-lived radioactive energy sources in Io may have been extracted into a sulfide-rich melt and sunk into the core. But K entry into the core would require a low degree of oxidation, contrary to the bulk density evidence.

But bizarre chemistry is not the only distinctive trait of Io. Its location as the innermost of the four Galilean satellites ensures that its rotation is strongly locked onto Jupiter. This location is also close enough to the heart of the Jovian radiation belts (which peak near Amalthea's orbit) to ensure an extremely high flux of energetic protons on its surface. Further, the Jovian magnetic field through which Io orbits is quite intense. Electric fields are induced by the motion of Io relative to the field, assuming Io is a conductor. We are safe in assuming that it interacts rather strongly with the magnetosphere because of the strong modulating effect the position has on the beaming of the decametric radio emission from Jupiter, as we discussed in Chapter V. The $B \times V$ Lorentz force exerted on conduction-band electrons inside Io and on ions and electrons in Io's ionosphere by the rotation of the Jovian magnetosphere sweeping through it induces a $10^6$-A current running through Io and its atmosphere along the Jupiter–Io line (Fig. IX.7). Estimates of the total power dissipated (not all inside Io) by this current run near $10^{12}$ W ($10^{19}$ erg s$^{-1}$). It is most improbable that this energy is dissipated uniformly throughout Io's interior because even a low, uniform electrical conductivity would result in weaker heating in the deep interior (because of shielding by overlying layers). A simple temperature-dependent conductivity profile would be more realistic and would result in the near-surface region being weakly heated because of its low temperature, most of a warm intermediate-depth shell being heated strongly, and the deep interior being nearly perfectly shielded. Of course, compositional layering of Io would further complicate the conductivity profile and energy dissipation rate. A shallow melted

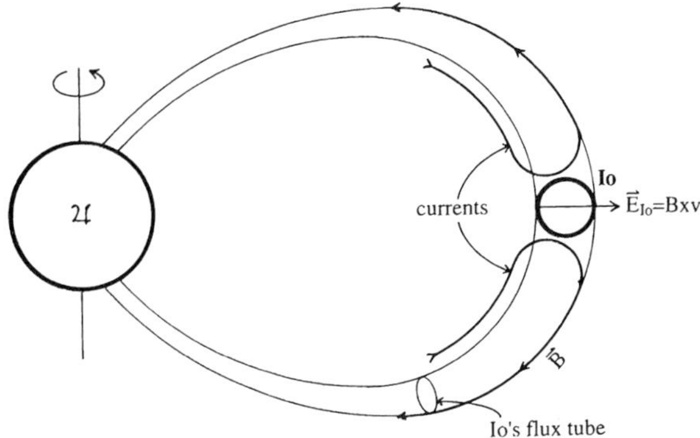


**Figure IX.7** Induction of an electrical current through Io. The rapid rotation of Jupiter drags its strong magnetic field through Io's location at a relative speed of about $50\,km\,s^{-1}$, inducing a very large voltage difference across Io's equator from the sub-Jovian to anti-Jovian points. This potential difference drives an electrical current through Io and its tenuous atmosphere that dissipates roughly $10^{12}$ W of thermal power, much of it probably inside Io.

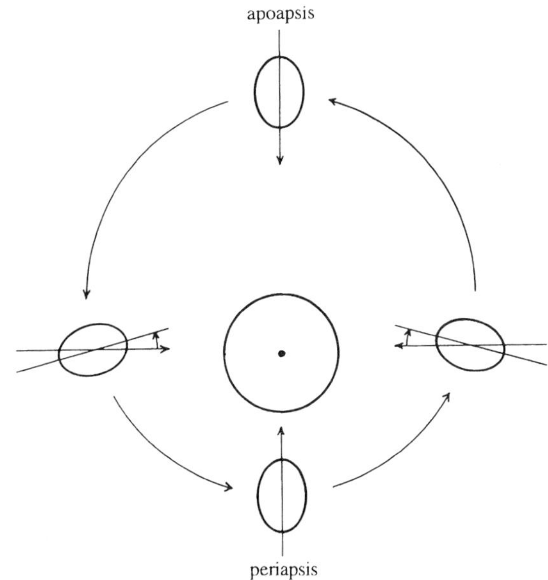


**Figure IX.8** Io's tidal interaction with Jupiter. Io is tidally locked on Jupiter, so that its rotation period equals its orbital period. The eccentricity of Io's orbit is pumped up by the perturbations of the other Galilean satellites, especially Europa and Ganymede, with which Io is resonantly locked. Jupiter's tidal forces accelerate Io on its orbit, and the eccentricity induced in Io's motion by the other satellites causes the orbital angular speed to vary while the rotational angular speed remains constant. This pendulum-like (librating) motion of Io's permanent tidal bulge under Jupiter's tidal force pumps the crust up and down and dissipates large amounts of energy at shallow depths, sufficient to maintain melting and drive extensive volcanism.

layer with good electrical conductiviy could be maintained in a molten state by this energy source.

Another remarkable heating mechanism was suggested on the eve of the Voyager 1 Jupiter encounter by Stanton Peale of the University of California at Santa Barbara and Pat Cassen and Ray Reynolds of NASA Ames Research Center. They argued that the tidal bulge on Jupiter (rotation period, 9.842 h) raised by Io (orbital period, 42.456 h) will always be pulled ahead of the position of Io by Jupiter's rotation, and hence Io will always be subject to acceleration along its orbit. This torque would normally move Io away from Jupiter, except that the situation in the Galilean system is complicated by the commensurability of the orbital angular frequencies of the three inner Galilean satellites, which link their motions together. Further, because the orbit of Io is slightly eccentric ($e = 0.0041$), the angular rate of motion along its orbit will vary sinusoidally over the orbit, whereas the angular rate of rotation is constant. Thus the tidal bulge must oscillate back and forth (librate) relative to the Io–Jupiter line, and the instantaneous tidal forces on Io will not in general correspond to equilibrium with the figure presented by the permanent tidal bulge at any given moment (Fig. IX.8). Jupiter will exert a torque on this oscillating tidal bulge that will effectively pump the surface of Io up and down through one entire cycle per libration period, which is once per orbit. This process dissipates energy inside Io and, to a lesser degree, inside its dynamically linked neighbor Europa.

If Io were alone in orbit about Jupiter, the tidal effects of its orbital eccentricity would evolve in isolation; this energy dissipation would diminish Io's orbital eccentricity until it vanished, which would put an end to the libration and to the energy dissipation caused by it. The internal heat source would die and $e$ would be indistinguishable from zero. But, in the Galilean system, the mutual gravitational interactions keep the eccentricity from dying out, and the energy dissipation persists. Peale and his colleagues concluded that this mechanism could provide a heat source of about $10^{13}$ W inside Io, sufficient to drive strong volcanic activity. Three days after their prediction was published, the first volcanic eruption was observed by Voyager 1!

Almost simultaneously, Fred Witteborn and coworkers at NASA's Ames Research Center happened to observe a dramatic increase in Io's flux between 2 and 5 $\mu$m, which they attributed to a new spectral component approximating a 600 K black body, covering a small fraction of Io's surface area. More recent observations from Earth and Galileo show four distinctive temperatures on Io's surface: 140, 250, 650, and 1800 K.

## Io: Geology

Most of the surface of Io is dominated by obviously volcanic landforms. Volcanic collapse craters (calderas), multicolor flows emanating from surface vents, and extremely bright local frost deposits, presumably of $SO_2$,

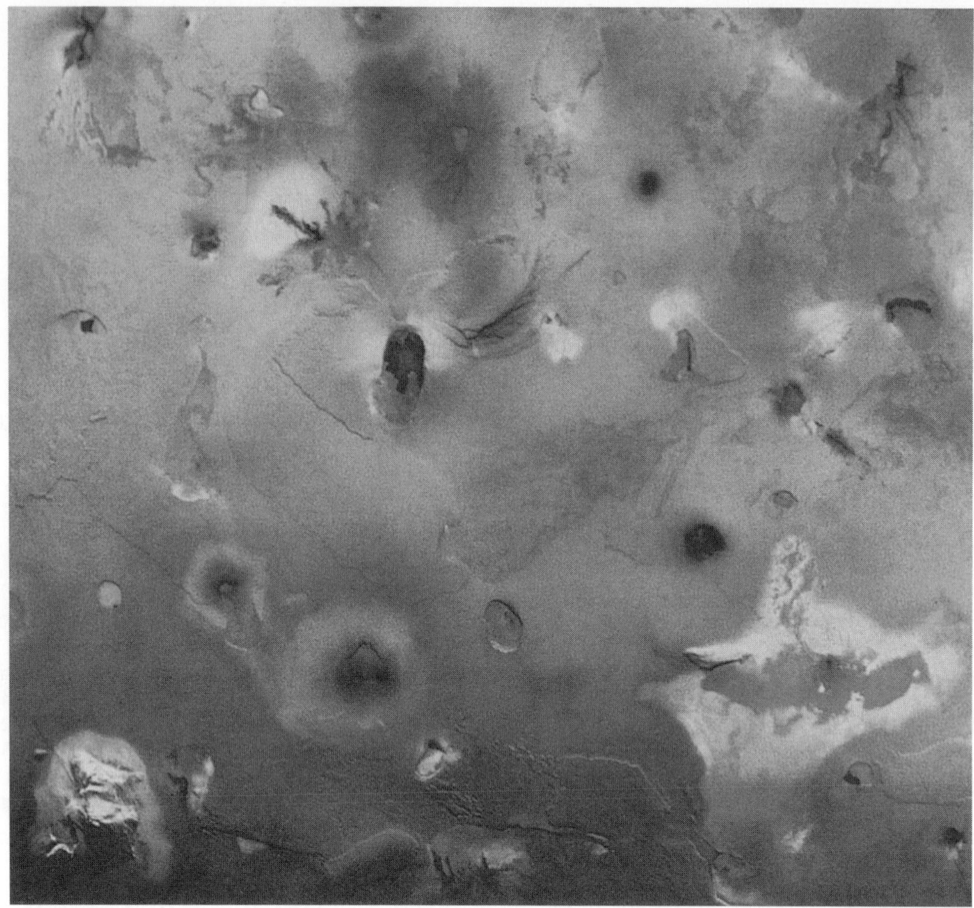

**Figure IX.9** High-resolution color image of Io. The association of sulfur flow, black spots, and white sulfur dioxide deposits with active eruptive centers is evident. Note the high, rugged mountain at the lower left, which must be a silicate structure that emerges through the sulfur layer. It is heavily decorated with sulfur dioxide frost. Numerous shallow, rimless collapse craters are visible in the sulfur plains (See color plate 4).

are widespread (Fig. IX.9). The topography is mostly rather subdued in appearance, but occasional prominent mountain peaks thrust up through the plains. These peaks resemble *inselbergs*, isolated steep features not obviously part of a range. The total topographic relief is about 9 km, which, in light of the furious rate of crustal overturn through volcanism, suggests a surprisingly firm (cold) foundation for these features. Steep scarps about 2 km high confirm this idea. Sulfur, the dominant surface material revealed by spectroscopy, is not a strong material, and the general consensus is that these peaks are silicate features lightly frosted with sulfur compounds, including both volcanic sulfur dust and frosts, principally sulfur dioxide. It is likely that both silicate and sulfur magmatism occur at different depths on Io (Fig. IX.10). Extrusion of silicate magma from the rocky crust (implying intrusion into sulfur lava and sediment layers) may generate vast quantities of gaseous sulfur, molten sulfur, and sulfur dioxide. Because silicate magmas are generally at 1200 K or higher, whereas liquid sulfur can exist above 386 K (and boils at 718 K), the logical candidate for the hotspot material is flows of molten sulfur.

Combining the surface expressions of very active magmatism with the density data, we expect a core of FeS (perhaps with some metallic iron) with a deep mantle of very FeO-rich silicates, overlaid by a strongly heated, low-viscosity layer of molten silicates with a silicate crust floating on it (Fig. IX.11). The physical state and structure of the deep interior are virtually unconstrained by observation, and we should not be surprised to find that this model is misleading or wrong in some of its major features.

The heat flux through the surface of Io has been measured as $4 \times 10^{13}$ to $8 \times 10^{13}$ W, with the flux sometimes varying sharply over a time scale of days. Equating the mean heat flux to the crustal temperature gradient times the thermal conductivity of sulfur (which is very low), steep crustal temperature gradients are calculated, leading to temperatures of over 1000 K at a depth of only 8 km. Such high temperatures would quickly erase even silicate mountain features by viscous creep. Clearly

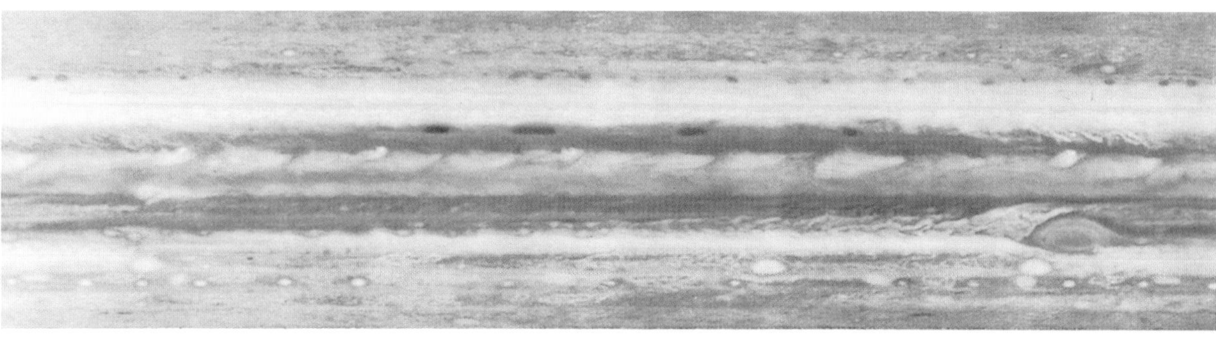

**Plate 1** Voyager 1 and 2 photomosaic maps of the Jovian clouds. The two images are aligned so that the planetocentric longitude (in the radio system) is the same. Note the westward drift of the GRS over the 4-month interval. The white ovals have moved eastward and changed substantially in appearance in the same time. Note also the red "barges" in the NEB that have also drifted eastward and changed dramatically (See Figure V.32).

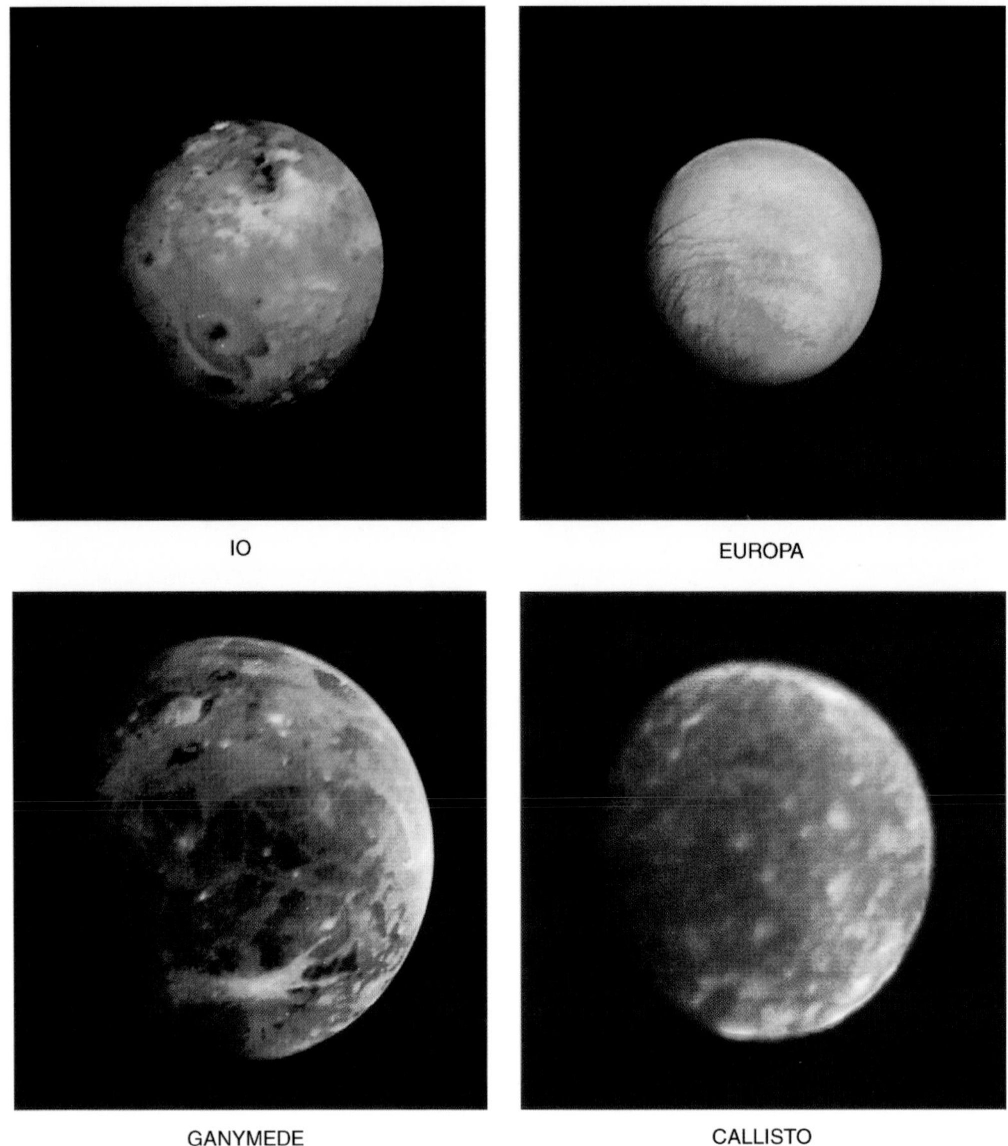


**Plate 2** General comparison of the Galilean satellites (to scale). The intense orange coloration of Io (here exaggerated) is faintly mirrored by Europa. Callisto is most heavily cratered and "moonlike" in appearance (See Figure VI.5).

**Plate 3** A Voyager 1 image of Io near zero phase angle. The yellowish plains material is sulfur, and the white mantling deposits are solid sulfur dioxide. Black spots and streaks may be quenched molten polymeric sulfur, and the diffuse round features are the plumes from major volcanic eruptions (See Figure IX.4).

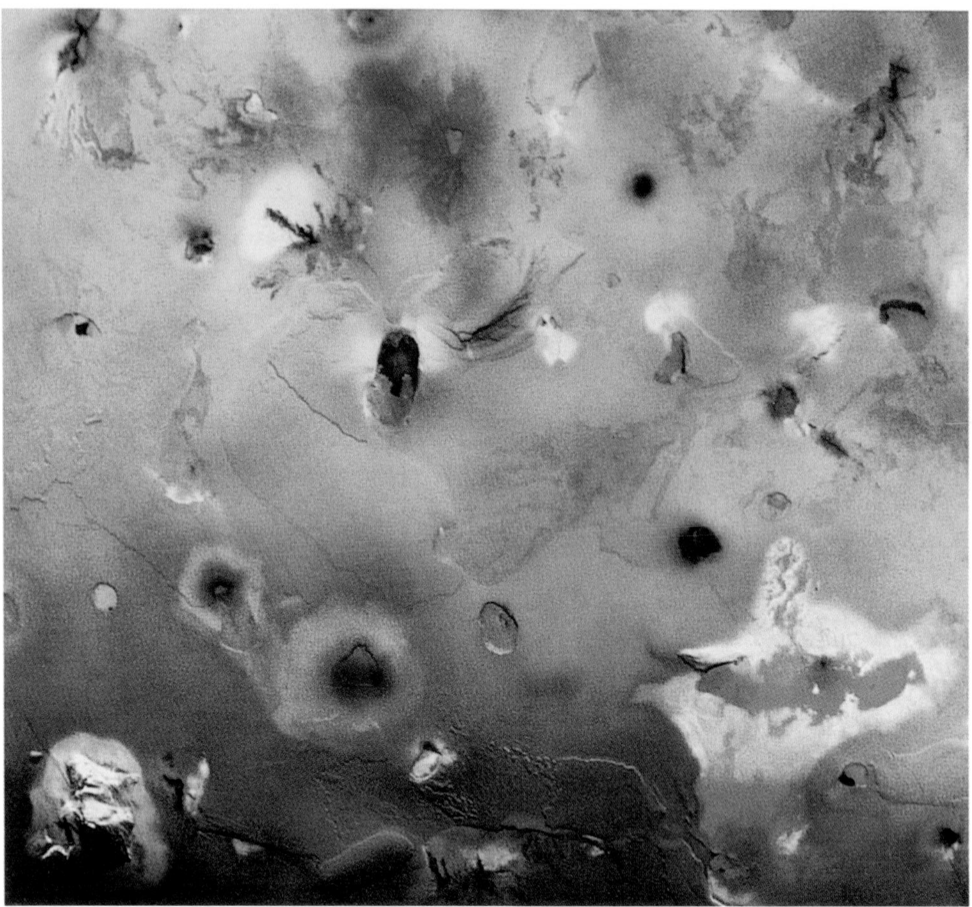

**Plate 4** High-resolution color image of Io. The association of sulfur flow, black spots, and white sulfur dioxide deposits with active eruptive centers is evident. Note the high, rugged mountain at the lower left, which must be a silicate structure that emerges through the sulfur layer. It is heavily decorated with sulfur dioxide frost. Numerous shallow, rimless collapse craters are visible in the sulfur plains (See Figure IX.9).

**Plate 5** View from the surface of Mars. This Viking 1 Lander image shows a pale pink sky and a rocky, desolate surface mantled with red dust. The rocks are probably volcanic in origin, broken up by impacts and shaped by both wind erosion and chemical weathering (See Figure X.18).

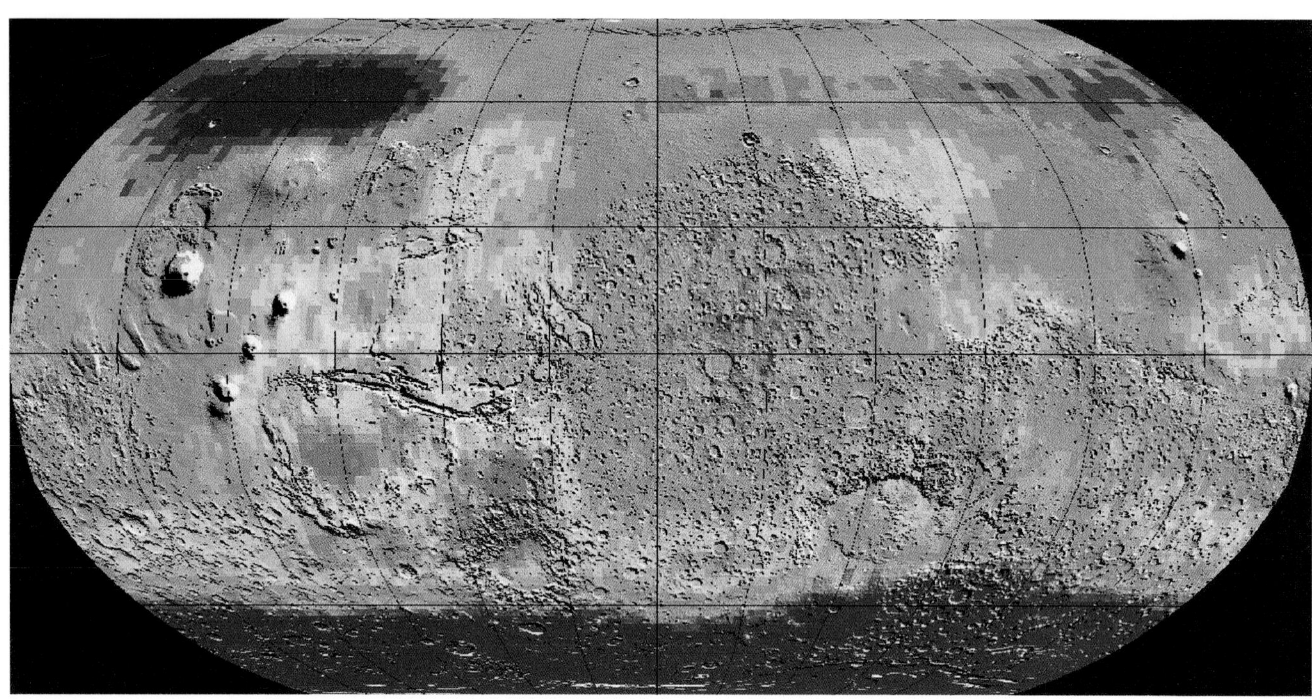

**Plate 6** The distribution of water on the Martian surface as determined by the GRS experiment on the Mars Odyssey spacecraft, superimposed on a MOLA topographic map of Mars. Note the extensive "wet" areas in the south polar region and the northern plains. See Chapter X, on which the water-rich regions are coded blue and the driest regions are coded in red (See Figure X.17).

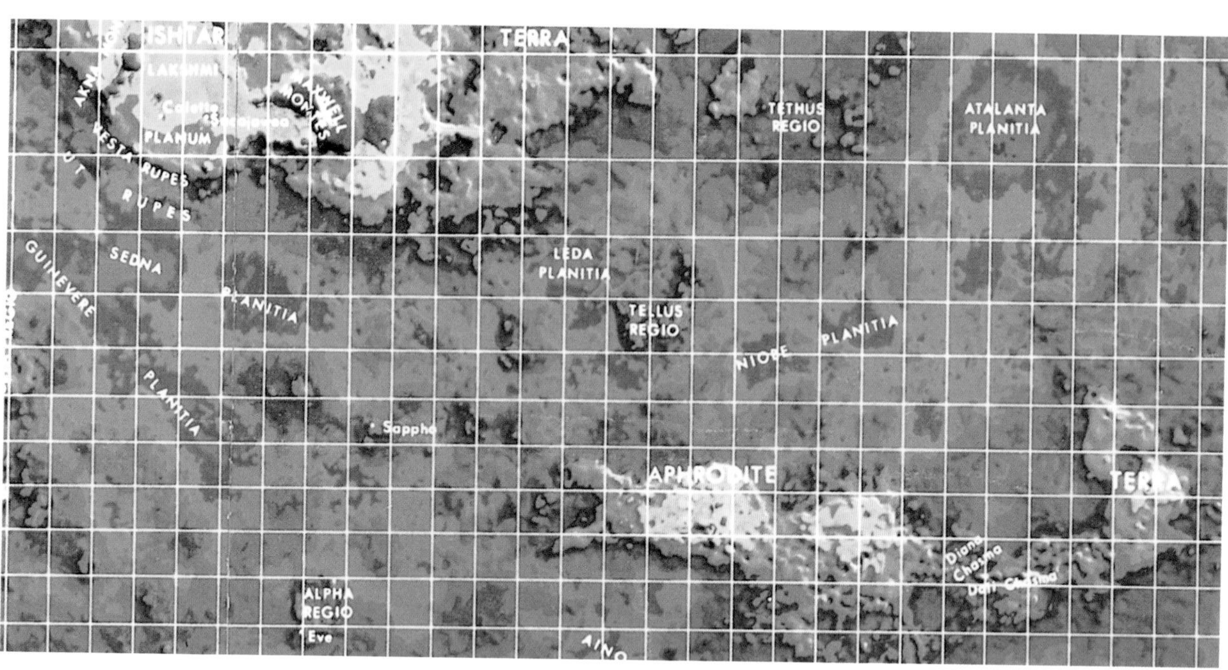


**Plate 7** Topography of Venus. The two major continents, Ishtar and Aphrodite, occupy only a few percent of the total surface area. Most of the planet is covered by low, rolling plains. These data were collected by the Pioneer Venus radar altimeter experiment (See Figure X.25).

**Plate 8** The Hubble space telescope's deep field view (See Figure XII.3).

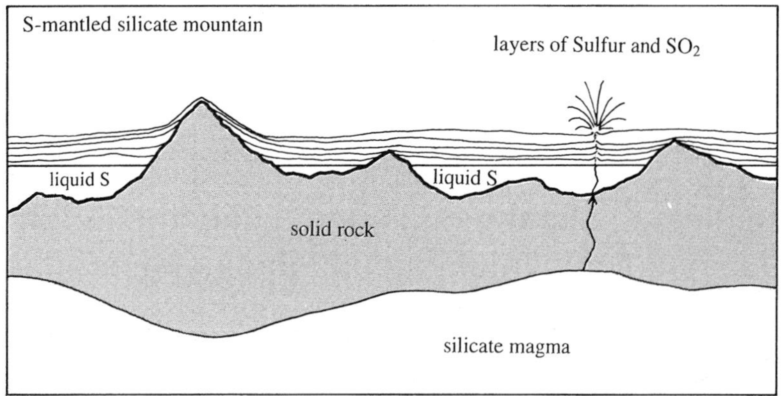


**Figure IX.10** Schematic representation of the near-surface region of Io. Provisions exist for both silicate-driven and sulfur-driven volcanism in which water and other hydrogen compounds are absent and sulfur-bearing gases are dominant. The only solid material strong enough to be capable of supporting large elevation differences is solid silicate.

the large majority of the heat flux must be carried by sulfur flows that are near 650 K, or silicate flows that erupt at 1500–2000 K, and cover only a small part of the surface. The main mechanism for vertical heat transport in Io's crust is eruption, not conduction.

## Io: Atmospheric and Volcanic Gases

The first observational evidence for a possible atmosphere on Io was reported in 1964 by Alan Binder and Dale Cruikshank, who observed a startling excess brightening of Io when it emerged from eclipse by Jupiter. The excess brightness always disappeared within a matter of minutes. Perhaps the most surprising aspect of this discovery was that Io, which has an extremely high albedo, *could* be brightened by anything! The most reasonable explanation seemed to be that traces of atmospheric gases condensed as frost upon the surface when the temperatures plunged during eclipse. The identity of the possible atmospheric gas was completely unknown.

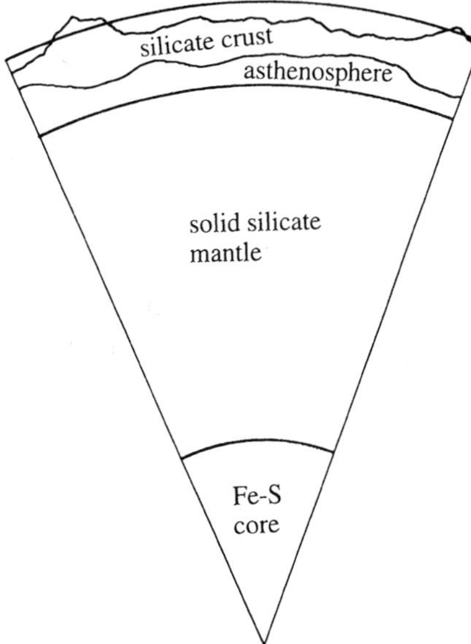


**Figure IX.11** Io's internal structure. This schematic rendering shows a solid silicate mantle surmounting an Fe–S core. Tidal and electrical induction heating at relatively shallow depths maintains a viscous, partially melted region (the *asthenosphere*), above which the silicate crust and sulfur layer float as in Fig. IX.10.

One of the most startling discoveries of the Voyager flybys of Jupiter was the existence of spectacular umbrella-like plumes over nine points on Io's surface. These plumes extend as high as 300 km from Io's surface, far in excess of anything ever seen on Earth. Clearly the gases in these plumes are vented from the crust at a significant fraction of Io's escape velocity. The gas, once expanded and cooled, is principally $SO_2$, with entrained particles of sulfur. We therefore now think of the atmosphere of Io as a tenuous sulfur dioxide gas, probably in vapor pressure equilibrium with the surface except where perturbed by eruptions. There is still debate about whether $SO_2$ frost is the proper phase to regulate the sulfur dioxide vapor pressure, which would be about $10^{-8}$ bar at the prevailing surface temperatures of Io. First, saturation of the major atmospheric constituent would occur first at the poles, but no polar caps are observed. Second, the gaseous $SO_2$ abundance seems inadequate to equal the vapor pressure of solid sulfur dioxide, but entirely adequate to provide adsorbed monolayers of $SO_2$ on surface minerals (about $10^{-11}$ bar). Nonetheless, the spectral signature of the condensate is unmistakable, and a dilemma clearly exists.

It is interesting to consider the consequences of heating grossly chondritic material that is more oxidized than ordinary chondrites. Suppose we start with an assemblage containing abundant FeO and FeS and a

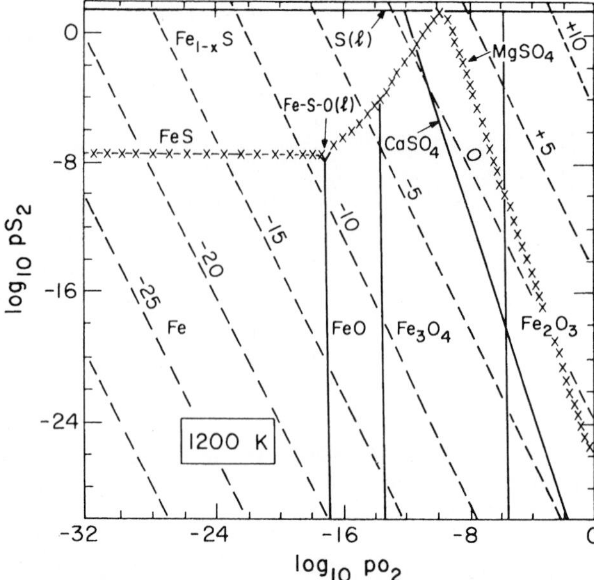


**Figure IX.12** Major phases as a function of oxidation state in Io at 1200 K. The solid lines are phase boundaries for a number of crucial phases in the chemistry of sulfur and oxygen. The × marks show the course of oxidation beginning with a mixture of FeS and Fe. The format of the diagram is a plot of the sulfur and oxygen pressures; note that oxidation of the mineral assemblage leads sequentially to the disappearance of metal, formation of FeO-rich silicates, formation of magnetite, and appearance of $CaSO_4$. This mineral, anhydrite, contains essential calcium, which is much less abundant than sulfur. Thus anhydrite formation cannot accommodate all the available sulfur. Further oxidation leads to the appearance of hot liquid sulfur and $MgSO_4$, with a subsequent rapid decline of the sulfur pressure with further oxidation. The diagonal dashed lines are contours of equilibrium $SO_2$ pressure in bars. At the appearance point of liquid sulfur the sulfur gas pressure is about 30 bar and the sulfur dioxide pressure is about 1 kbar. On Io, with a mean crustal density of about 2.5 and an acceleration of gravity of 150 cm s$^{-2}$, the hydrostatic pressure increases with depth at the rate of about 38 bar km$^{-1}$. Thus this gas mixture could erupt easily from a depth of 25 km.

small amount of metallic iron (Fig. IX.12). We can represent the stability of the relevant minerals conveniently on a diagram of $p(S_2)$ vs $p(O_2)$. Recalling that the Fe–FeO buffer defines a pressure of $O_2$ gas that is a function of temperature alone, and by analogy that the Fe–FeS buffer similarly defines the $S_2$ pressure, we can set up the phase boundaries for these species as buffer lines on an isothermal diagram. The FeS–$FeS_2$ boundary is also a sulfur buffer, and the FeO–$Fe_3O_4$ and $Fe_3O_4$–$Fe_2O_3$ boundaries are oxygen buffers. At elevated temperatures, above its incongruent melting temperature (Fig. VIII.6), $FeS_2$ is not present, and FeS, troilite, grades slowly into $Fe_{1-x}S$, pyrrhotite. Oxidation beyond the point of disappearance of metallic iron causes the phase assemblage to follow the diagonal FeO–FeS border, as indicated by the × marks in Fig. IX.12. FeO converts to magnetite; then $CaSO_4$ (anhydrite) becomes stable. Because the cosmic abundance of Ca is much less than the abundance of sulfur, anhydrite formation cannot use up all the sulfur. Further oxidation progresses to the point of formation of magnesite, $MgSO_4$, and the appearance of impure liquid sulfur. The diagonal dashed lines on the figure are contours of constant $SO_2$ pressure in atmospheres; note that, at the point of appearance of liquid sulfur, the magma at equilibrium can support a sulfur dioxide pressure of about 1 kbar, containing about 30 bars (3 vol%) of gaseous sufur vapor, which at these elevated temperatures is almost all $S_2$. Thus a crustal mineral assemblage (or an ancient mantle mineral assemblage) that contains highly oxidized chondritic material with no metallic iron will generate a high-pressure $SO_2$–$S_2$ vapor phase and liquid sulfur magma and vent it energetically onto the surface. The critical components of the melting rock or magma are pyrrhotite and sulfates, and the equivalent low-temperature assemblage contains pyrite.

Any parent material that contains enough FeO or oxidizing agent to enter the metal-depleted geochemical realm invoked above is likely to be carbonaceous chondrite material. CI or CM parentage of course implies the presence of at least several weight percent water. Yet Io today is distinctive in that it alone of the bodies of the outer Solar System seems to contain no hydrogen compounds. The simplest approach to this problem seems to be to imagine a CV or CO composition, with FeS and FeO-rich silicates coexisting with a trace of metal. In this system, the water abundance would be on the order of 0.1%, and heating (by formation in a hot subnebula or by release of the gravitational energy of accretion) to the temperature of first melting would generate an Fe–FeS–FeO ternary eutectic melt. Equilibration of water vapor with this dense melt would generate a gas in which the $H_2$:$H_2O$ ratio is much greater than unity. A hot volcanic gas of this composition, with a mean molecular weight between 2 and 3, would escape from Io with great ease. Any lightweight gases entrapped in this gas, such as ammonia and methane, would also escape. A negligible portion of the total reactive oxygen (that contained in water vapor and FeO) would be lost, but the global hydrogen inventory could be lost easily. Oxygen from the water will react with and oxidize the crust, leading to the sequence of oxidation reactions discussed above.

Whether or not we accept this explanation of the provenance of sulfur on Io, observations assure us that it is there. Melting of this surface constituent, either by intrusion of a silicate magma at 1200 K or because of a very steep crustal temperature gradient caused by its low thermal conductivity, can easily generate a liquid sulfur magma. Liquid sulfur is a strange magmatic material because of the peculiar structural changes that occur in the liquid as a function of temperature. Just-melted rhombic sulfur is a mobile liquid of puckered $S_8$ rings, with a viscosity that at first decreases in a normal fashion with increasing temperature. But at elevated temperatures the sulfur rings begin to break open, permitting the formation of very long chains of sulfur atoms. Just as solutions of proteins or DNA in water have extraordinary viscosities because of the tangling of their long chains, so liquid sulfur during heating exhibits an abrupt increase of viscosity of about a factor of $10^3$, followed by rapid decrease toward even higher temperatures as the mean sulfur chain length decreases. The chains have unpaired electrons due to bond breakage, and the liquid material exhibits a wide range of colors depending on the temperature and distribution of ring and chain species. Rapidly quenched sulfur preserves these unusual colors, including orange, red, and black.

The composition of Io's atmosphere is sketched out very schematically in Fig. IX.13. The $SO_2$ abundance varies dramatically with time and location. Molecular nitrogen is hard to detect and has never been detected: it would not be surprising if its abundance were far less than the indicated upper limit in the figure. Atomic H is a rare, transient species, also probably far less abundant than the indicated upper limit. The species actually observed on Io, most of them very variable in abundance, are atomic H, O, S, Na, Cl, and molecular $SO_2$ containing about 0.1% NaCl vapor. SO is seen in the plume of the giant Pele eruption. Note the complete absence of hydrogen- and carbon-containing molecules, including water, CO, and $CO_2$.

## Io: Escape and the Plasma Torus

UV spectroscopy of the space about Io reveals the presence of a number of different lower ionization states of oxygen and sulfur. Photographs of the Jovian system at the sodium D line show a remarkable, very large comet-like feature about Io, orbiting tail-first about Jupiter (see Fig. IX.14). The source of sodium in the torus is a matter of interest, because Na is not normally an atmospheric species on a planet with a surface temperature of 140 K! Only two types of origin seem plausible: ejection of a species such as $Na_2S$, $Na_2O$, or NaCl in eruptive plumes and *sputtering* of Na atoms from surface minerals by the impact of energetic magnetospheric protons. The plasma torus contains Na, Cl, $Cl^+$, $Cl^{2+}$, and K. Calculations suggest that NaCl, Na, Cl, KCl, and K are all more important than NaS, NaO, $NaS_2$, $NaO_2$, NaOS, $NaSO_2$, SCl, ClO, $Cl_2$, $S_2Cl$, and $SO_2Cl_2$ in Ionian gases. The prominence of Na atoms in the emission spectrum of the torus need not suggest any preferential injection of elemental sodium into the torus. Sodium atoms have the most intense emission lines of any common atom: the emission probabilities (oscillator strengths) of atomic sodium at the yellow sodium D lines are close to 1. A gas containing, for example, equal amounts of Na and its chemical cousin K would look like pure Na. A massive $SO_2$ atmosphere in vapor pressure equilibrium with frost on the surface ($10^{-8}$ bar) would be a very effective screen against the sputtering of surface atoms by magnetospheric particles. Thus the issue of the source of sodium (and potassium, etc.?) vapor is intertwined with the question of $SO_2$ condensation. It has been suggested several times that sodium and potassium might be erupted or extruded along with sulfur. This is very differen from terrestrial experience, but plausible in light of the absence of gaseous $O_2$ and water on Io.

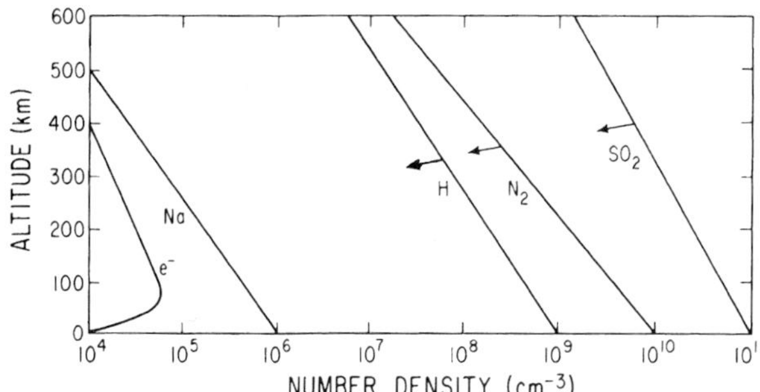


**Figure IX.13** Schematic rendering of the atmosphere of Io. Almost all the information we have about Io's atmosphere is derived from study of the gas torus on its orbit about Jupiter. Sulfur dioxide is the main atmospheric constituent, and hydrogen compounds and nitrogen may be absent. Sputtered atoms of Na, K, etc., are a very minor constituent, but are very readily observed.

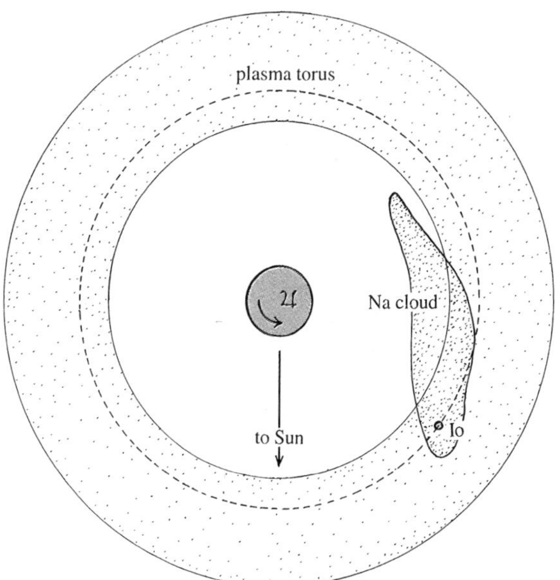


**Figure IX.14** Io's neutral sodium cloud and plasma torus. The torus contains ions entrained in Jupiter's magnetosphere and corotating with Jupiter.

## Io: Genetic Relationships

The presence of a steep temperature gradient in the Jovian subnebula is required by the large density differences of the Galilean satellites. Inward extrapolation of this gradient into the realm of the inner four small satellites suggests Venus- or Mercury-like formation temperatures with a large proportion of reduced iron and a low content of volatiles, especially water. Crude spectra of Amalthea reveal that, not surprisingly, it looks like Io; sulfur from Io's plasma torus has coated Amalthea's surface, and indeed it even tints Europa a pale yellow. Thus attempts to deduce Amalthea's intrinsic composition by remote sensing are doomed to failure.

In drawing an analogy between the Jovian subnebula and the Solar Nebula, there is an implicit suggestion that we may be able to identify closely similar bodies in the two sequences of formation temperature. The closest planetary body in density and oxidation state to Io is Mars, but Mars has a much higher escape velocity and evidence of abundant water in the past. Comparisons with Mars will be made in Chapter X.

## Impact Cratering

In one important respect the study of Io prepares us poorly for the comparative study of the other bodies in this chapter: Io has no impact craters. Cratering is, however, a universal process. The surfaces of all solid Solar System bodies *except* Io, including all the terrestrial planets, bear evidence of large cometary and asteroidal impact events. We have already seen that Phobos and Deimos, although both are heavily cratered and are probably collision fragments, lie in a unique dynamical environment in which reaccretion is very important. For that reason, the landforms created on Phobos and Deimos by cratering deviate in very important ways from those on other airless bodies.

An impact with a body in the inner Solar System occurs with a speed that is at least as great as the escape velocity of the target, but is limited by the maximum possible orbital velocity the impactor can have as a bound member of the System. Typical asteroidal impact velocities on the Moon range from about $5\,\mathrm{km\,s^{-1}}$ to about $35\,\mathrm{km\,s^{-1}}$ (roughly 2 to 14 times the lunar escape velocity). Comets strike the Moon with speeds from about $5\,\mathrm{km\,s^{-1}}$ (for low- inclination, prograde, short-period comets) to as high as $72\,\mathrm{km\,s^{-1}}$ (for retrograde, long-period comets). Impact velocities on Mars begin at a higher threshold velocity because of the higher escape energy of Mars, but do not reach such high maximum velocities because of the lower orbital speeds of both Mars and the projectiles that strike it. Mercury, with its Mars-like escape velocity and its location deep in the Sun's gravity well, can experience impacts up to over $110\,\mathrm{km\,s^{-1}}$. In addition to the fact that Phobos and Deimos efficiently reaccrete their own ejecta, these bodies are unique in that most of the mass that impacts them *is* their own ejecta, which hits at a relative velocity of a few hundred meters per second and less.

Cratering is caused by the partial conversion of the kinetic energy of a fast-moving impactor into the crushing, melting, vaporization, and acceleration of target material. The most useful criterion for "fast-moving" is that the closing speed of the target and projectile is greater than the speed of compressive or pressure sound waves (P waves) in the target material. These are then properly termed supersonic impacts. Impacts at higher speeds set up a strong shock wave that destroys the projectile and target material alike.

Consider a 100-m chunk of asteroidal material encountering the surface of a rocky planet at a speed of $20\,\mathrm{km\,s^{-1}}$. The kinetic energy density of the impactor is $\frac{1}{2}(2 \times 10^6)^2 = 2 \times 10^{12}\,\mathrm{erg\,g^{-1}}$. The energy required to crush a typical rock is a little above $10^8\,\mathrm{erg\,g^{-1}}$. To heat it to its melting point requires about $10^{10}\,\mathrm{erg\,g^{-1}}$ and to vaporize it requires less than $10^{11}\,\mathrm{erg\,g^{-1}}$. Thus the impactor carries enough kinetic energy to not only vaporize itself completely, but also crush up to roughly $10^4$ times its own mass of target rock, melt roughly 100 times its own mass, or vaporize about 10 times its own mass. Alternatively, it carries enough kinetic energy to accelerate 100 times its own mass to a speed of 0.1 times its impact speed ($0.1 \times 20 = 2\,\mathrm{km\,s^{-1}}$). In reality, an impact does all of these things to some degree and divides its energy over all these possible outcomes. Thus an impactor may crush 1000 times its own mass of rock, melt 10 times its mass, vaporize a few times its own mass, and eject 100 times its mass at speeds of tens to hundreds of meters per second and still give off a substantial amount of energy as seismic waves and radiation from the fireball. On small bodies, or in especially energetic impacts, much mass may be accelerated to escape velocity.

Suppose a rocky projectile of density 3.5 impacts a surface that also has a density of 3.5. The momentum content of the impactor will be largely dissipated by the time it has penetrated to a depth of its own diameter into the target surface. At $20\,\mathrm{km\,s^{-1}}$ a 100-m projectile has only $t = D/v = 0.005\,\mathrm{s}$ to decelerate, and the deceleration is

$$a = 2D/t^2, \qquad \text{(IX.1)}$$

or $4 \times 10^6\,\mathrm{cm\,s^{-2}}$, which is 4000 standard Earth gravities. The peak pressure, which is given by the rate of momentum exchange per unit area, is on the order of

$$P_{\max} = \Delta p/At \approx mv/At \approx Dv/t, \qquad \text{(IX.2)}$$

which is about $4 \times 10^{12}\,\mathrm{dyn\,cm^{-2}}$. Thus transient pressures of megabars ($1\,\mathrm{Mbar} = 10^{12}\,\mathrm{dyn\,cm^{-2}}$) can easily be produced in impacts that occur at speeds far above the speed of sound in rock. These are termed *hypersonic* or *hypervelocity* impacts. Fireball temperatures commonly reach tens of thousands of degrees, but of course the fireball expands very rapidly and radiates powerfully at visible and ultraviolet wavelengths, so the duration of the high temperatures is of order $D_c/v_e$, where $D$ is the diameter of the crater and $v_e$ is the expansion velocity. An impactor that crushes $10^2$ to $10^3$ times its volume of target rock will excavate a crater that has a diameter $(10^2)^{1/3}$ to $(10^3)^{1/3}$ (5 to 10 times) the diameter of the impacting body. The expansion time is generally less than 1 s except for impact events involving kilometer-sized projectiles.

The early stages of a cratering event are shown in Fig. IX.15. The impactor is crushed and flattened by the enormous transient pressures caused by its deceleration, pushing a cavity into the target. The region of the target closest to the projectile is heated so strongly that, as the peak shock is unloaded by the expansion of the debris from the explosion, it vaporizes. A much larger volume, usually with several times the radius of the vaporized zone, is thoroughly crushed and partially melted. Since the effective center of the explosion is well below the surface of the target, the explosion blasts out a substantial mass of rim material in a cone above the explosion site. The target-rock ejecta, mingled with a small fraction of impactor material, departs at a velocity that is critically sensitive to the ejection angle. Some mass emerges at a speed comparable to that of the incident projectile and can escape even from a massive planet like Mercury or Mars, but most of the ejecta is traveling at much lower speeds and does not go far. Its impact on the surface of the target body creates low-speed impact craters, called *secondary craters* for obvious reasons. In assessing the cratering history of a body and the mass distribution of the impactors that struck it, it is important to be aware of the presence of these numerous, small secondary craters and omit them from the statistics.

The simplest measures of craters are the diameter, $D_r$ (from rim crest to rim crest), and the depth of the crater floor relative to the mean rim height, $R_i$. The ratio of the two, called the depth–diameter ratio, varies slowly with crater size and the surface gravity, $g$, of the target body according to

$$R_i/D_r = 0.193 + 0.05R_i^{0.12}(g/g_E)^{0.157}, \qquad \text{(IX.3)}$$

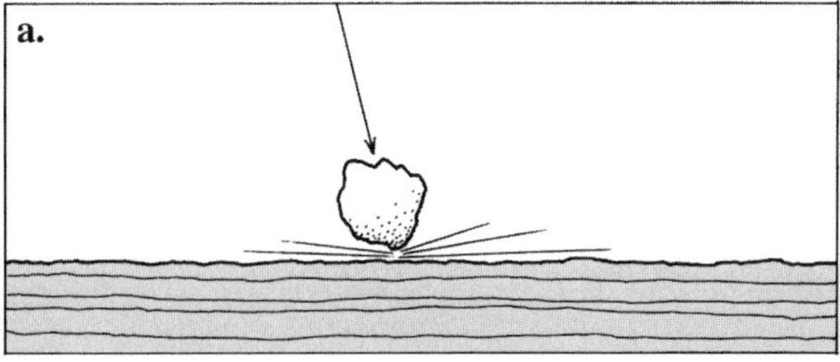


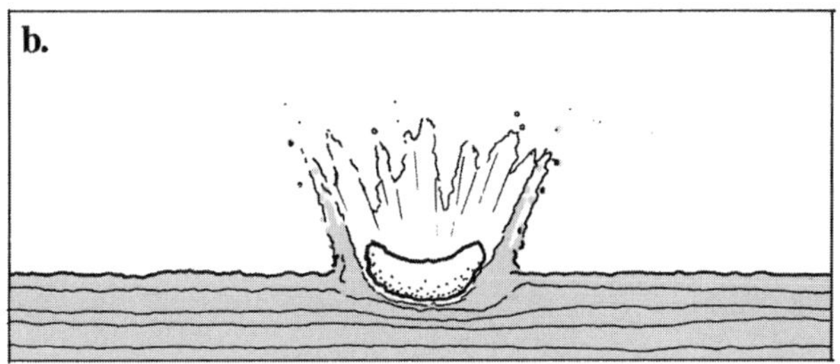


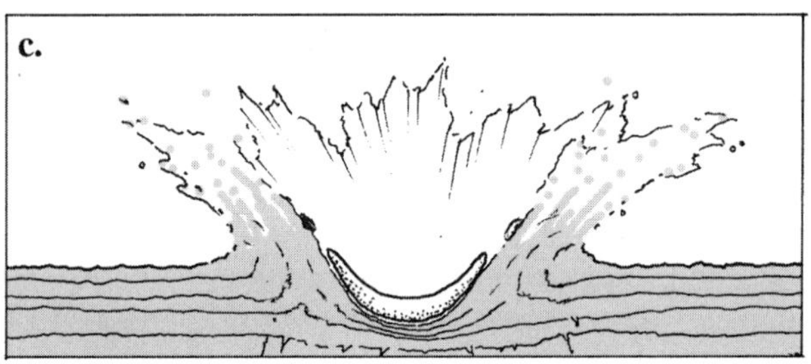


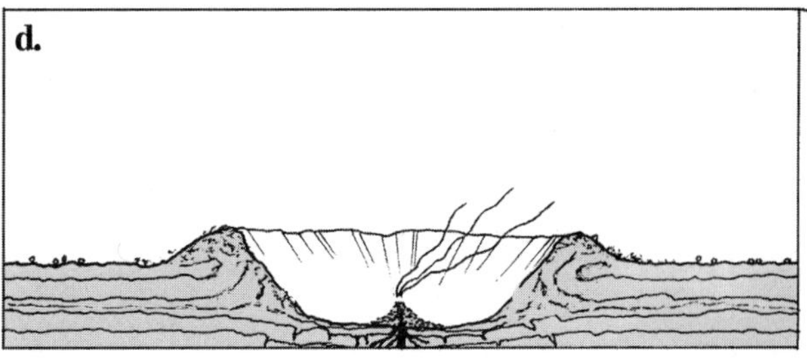


**Figure IX.15** Stages in the crater-forming process. In a the projectile begins to impact the target surface at a speed far above the speed of sound in the solids, causing jetting of a small mass of very-high-velocity material nearly in the horizontal plane. b shows the extreme flattening of the projectile as it is penetrated by the impact shock wave and the beginning of excavation of the crater. c shows the emergence of the compressive shock from the top side of the impactor and the beginning of the destruction of the superheated, high-density projectile by the rarefaction wave. The ejecta from the crater have velocities that vary with elevation angle. Bedrock beneath the crater is shattered and partially melted. The transient crater cavity is here nearly fully excavated, and filling of the crater has not begun. In d the overturned strata near the crater rim and the low-speed ejecta that form nearby secondary craters have come to rest, and the eruption of magmas from the impact melt (and from deep magma reservoirs unleashed by the fracturing of bedrock) produces a central peak as landslides partially fill in the crater from all sides with crushed rim material.

which is generally applicable to airless bodies in the inner Solar System. Crater shape data for the Moon and Mercury are graphed in Fig. IX.16. The results are more draconian for small targets with diameters not much larger than the crater diameter; catastrophic fragmentation is the rule. Other complications arise due to the velocity of the impactor (i.e., its kinetic energy per unit mass) and, especially in the case of icy or metallic impactors and icy targets, the relative densities of the impactor and target. The effects of impact velocity have been incorporated into an expression suggested by William K. Hartmann:

$$R_i/D_r = 0.308 V_i^{-0.279} + 0.129(g/g_E)^{0.208} V_i^{-0.301}. \quad \text{(IX.4)}$$

One of the most interesting features of the data shown in Fig. IX.16 is the sharp knee in the depth–diameter curve near a crater diameter of 10 km. This break is a general feature of cratering wherever it occurs (except, of course, for bodies the size of Phobos and Deimos, which are not large enough to accommodate craters larger than 10 km) and reflects an important break in the morphology of craters that occurs at that size. Basically, the craters smaller than 10 km are simple bowls with rather flat, featureless floors. The larger craters exhibit progressively larger and more complex central peaks, as well as very complex wall terracing due to massive slumping of the rim material into the interior basin. Very large impact

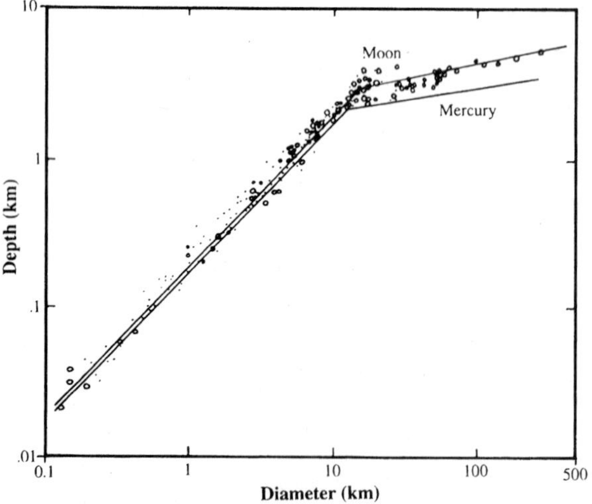


**Figure IX.16** Depth–diameter relations for craters on the Moon and Mercury. Simple bowl-shaped craters scale nicely up to a diameter of about 10 km, beyond which volcanic magma flooding and isostatic adjustment (viscous rebound) of the crater floor cause all larger craters to be relatively shallow in profile. These complex craters have extensive wall collapse, internal lava flows, and central peak clusters. The higher acceleration of gravity on Mercury flattens out the profiles of large craters more effectively than on the Moon. The largest impacts form even more complex multiringed basins.

structures are also characterized by a series of one to five concentric mountain rings, similar in appearance to the transient circular ripples surrounding the point at which a pebble has been dropped into water. Spectacular examples are known on the Moon and Mercury. One of the most striking is the region about Mare Orientale on the western limb of the Moon as seen from Earth. The Orientale basin is shown in Fig. IX.17 from almost directly overhead, as seen by the Lunar Orbiter 4 spacecraft.

Crater sizes are of course generally related to the kinetic energy content of the impactor. For relatively small impacts the critical factor in determining the target's resistance to the explosion is the strength of the material, $S$ (dyn cm$^{-3}$). If $S > \rho g D_a$, where $\rho$ is the target density and $D_a$ is the crater diameter at the level of the original target surface, then the crater excavation process is strength limited. In this case, the diameter scales as

$$D_a(km) \approx W^{1/3}, \quad \text{(IX.6)}$$

where $W$ is the explosion energy in units of millions of tons of TNT equivalent (megatons; Mt). For very large impacts, no material has enough strength to matter, and the cratering process depends only on the gravitational environment in which it occurs:

$$D_a \approx (W/g)^{1/4}. \quad \text{(IX.7)}$$

From this we see that a body with, for example, a surface gravity of $10^{-4}$ g would suffer formation of a crater 10 times wider than an equivalent explosion on Earth would produce. Recall that $g = GM/r^2 = 4\pi\rho Gr/3$, so a body of Earth-like composition will have $g = 10^{-4}$ if $r = 10^{-4} r_E = 0.64$ km, a typical size for a near-Earth asteroid. The more extensive crater would lead to poor retention of regolith on such a small asteroid.

The sizes of craters produced by rocky impactors on rocky targets with $g = 0.1\, g_E$ are illustrated in Fig. IX.18.

Another factor contributing to the morphologies of impact craters is the angle of incidence of the impactor. Intuitively a high-angle impact should produce a near-circular crater, and such is indeed the case. But at extremely shallow angles of incidence the impactor may break up as it grazes the surface and launch a spray of large fragments in the forward direction. Such elongated craters are rare but well documented on the Moon, Mars, and even Earth. They have sometimes been taken as evidence for deorbiting of small planetary satellites, but the inference is without basis. The probability of a given impact angle is easy to calculate from basic principles. Consider a gravitating spherical target that intercepts all the incoming projectiles that are aimed at a

**Figure IX.17** The Orientale basin on the Moon. This is a view from directly overhead by the Lunar Orbiter 4 spacecraft. At least three mountain rings can be seen about the basin. This basin is at the edge of the disk as seen from Earth. Oceanus Procellarum is the dark area in the upper right.

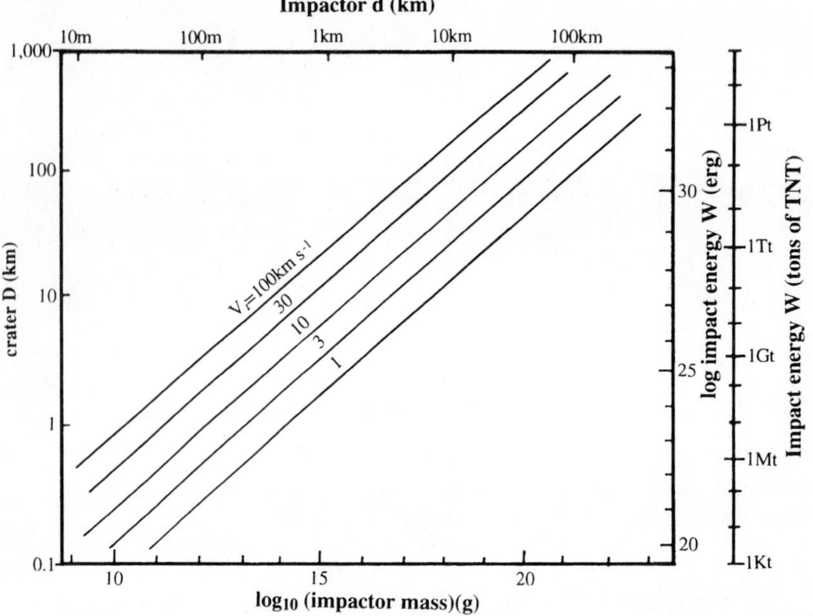


**Figure IX.18** Energetics of impact crater formation. The relations among the size, the mass, the velocity, and the kinetic energy of rocky impactors can be determined from this diagram. The explosive yield in tons of high explosive (TNT) is also given on right: 1 kt is one kiloton ($10^3$ tons) of TNT; 1 Mt is one megaton ($10^6$ tons); 1 Gt is one gigaton ($10^9$ tons); 1 Tt is one teraton ($10^{12}$ tons); and 1 Pt is one petaton ($10^{15}$ tons). The largest man-made explosion ever, a Soviet atmospheric hydrogen bomb test on Novaya Zemlya, had a yield of 60 Mt. World War III, an extensive nuclear exchange of the Great Powers, would release about 10 Gt.

circle of radius $R$ (the capture radius) centered on the planet. The size of the circle depends on the velocity of the incoming projectile and approaches the geometric cross-section area of the target body as the projectile speed approaches infinity. The total probability of impact for a projectile inside $r$ is exactly unity, so

$$P = \pi r^2 = 1. \tag{IX.8}$$

The differential probability that the projectile will pass through an annulus of width $dx$ at distance $x$ from the center of the capture cross-section circle is

$$dP = 2\pi x \, dx. \tag{IX.9}$$

The velocity at impact is related to the approach velocity by conservation of angular momentum,

$$mxV_a = mr_pV_i, \tag{IX.10}$$

where $r_p$ is the radius of the target planet. The velocity at the moment of impact, $V_i$, has a tangential component, $V_t$, and a radial component, $V_r$,

$$V_t/V_i = \sin i, \tag{IX.11}$$

where $i$ is the zenith angle of the trajectory of the projectile at impact ($i = 90°$ is the vertical incidence), as in Fig. IX.19. Then

$$x = r(V_i/V_a)\sin i, \tag{IX.12}$$

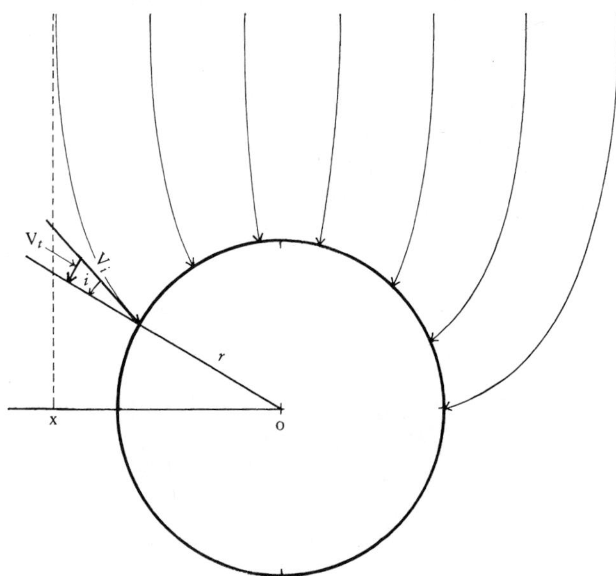


**Figure IX.19** Geometry of surface impact of infalling bodies. The radius of the target planet is $r$, the impactor strikes the surface at speed $V_i$ at an angle of $i$ degrees relative to the zenith, and the transverse component of the impact speed is $V_t$. The impact parameter (the distance the projectile would have passed from the center of the target if it had continued traveling in a straight line) is $x$. A massless target would have $x = r$, and all impacting trajectories would be parallel straight lines. The phenomenon by which massive targets collect impactors from $x > r$ is called *gravitational focusing*.

and thus

$$dx = r(V_i/V_a)\cos\ i\ di. \qquad \text{(IX.13)}$$

Then

$$dP = 2\pi[(rV_i/V_a)\sin i][(rV_i/V_a)\cos\ i\ di], \qquad \text{(IX.14)}$$

$$= \pi r^2(V_i/V_a)^2 \sin\ 2i\ di. \qquad \text{(IX.15)}$$

Because $x = r$ when $i = 90°$, we find that

$$r = r_p(V_i/V_a). \qquad \text{(IX.16)}$$

Substituting into Eq. (IX.8), we conclude that

$$\pi r^2(V_i/V_a)^2 = 1, \qquad \text{(IX.17)}$$

and,

$$dP = \sin\ 2i\ di. \qquad \text{(IX.18)}$$

Thus the most probable angle of incidence must be that for which $\sin 2i = 1$, which is exactly 45°. Note that, for a nongravitating target, $V_i = V_a$, and $\pi r^2 = 1$, so the same conclusion applies.

The probability of near-vertical impact ($i > 80°$) is small because only a very small portion of the celestial sphere lies at so high an elevation; the probability of near-grazing impact ($i < 10°$) is small not because so little of the sky is near the horizon but because the target surface is almost edge-on to the approaching impactor, so that its effective cross-section area is very small.

The relative abundances of craters of different sizes are controlled by the energy distribution of the impactors. The size distribution of NEAs was given in Fig. VIII.41, with the lunar cratering and Spacewatch data compared with the extrapolated Belt asteroid size distribution and the fireball data for meter-sized objects. At sizes over 1 km the role of comets in the cratering flux becomes progressively more important. For 100-km craters, roughly 20% of the impactor flux is due to long-period comets; 20%, to short-period comets; and 60%, to asteroids, including extinct comet nuclei in the near-Earth "asteroid" population. In Chapter VIII we reviewed the arguments suggesting that well over 25%, and possibly nearly 50%, of the kilometer-size NEAs are extinct comet cores or debris from the breakup of comet nuclei. The excess of NEAs smaller than 100 m relative to the Belt size distribution may be principally due to cometary fragmentation, which agrees with the fireball data (for bodies $\leq$ 1 m in diameter) that suggest that most of these small bodies are very fragile and are definitely not strong rock. These complexities of the size–frequency relationship for impactors affect the size distribution and morphology of craters produced by their impacts.

The fluxes and energies of impactors of course vary from one location to another in the Solar System. For example, Jupiter acts as an extremely powerful gravitational "vacuum cleaner" to concentrate the fluxes of Jupiter-crossing objects. In principle, Io must have a higher cratering *mass flux* and *energy flux* than any of the other Galilean satellites, due to the concentration and acceleration of comets and stray asteroids by Jupiter. Thus the absence of impact craters is especially remarkable. Cratering expert William K. Hartmann of the Planetary Science Institute in Tucson, Arizona, has estimated that the relative fluxes of kilometer-sized bodies onto the surfaces of solid inner Solar System bodies vary roughly as in Fig. IX.20.

Using the best estimated cometary and asteroid fluxes, the observed crater densities on Phobos and Deimos, Io, Mercury, and the Moon can be used to estimate the ages of regions of their surfaces. These results are presented in Fig. IX.21. The exceptionally high impact flux recorded by the Martian moons reflects the large-scale reaccretion of collision ejecta, and the upper limit on the crater abundances on Io illustrates the necessity that the surface be very young. The data on Mercury and the Moon are subdivided to reflect the geological realities; both bodies have very ancient, heavily cratered highlands and younger, smoother basins filled by volcanic material. For Mercury, the age of the highlands can be seen to be significantly larger than the age of the huge multiringed Caloris Basin. The ranking of age for the lunar features shown here is that the highlands are oldest, Orientale basin is next, the Imbrium basin is younger, and the fresh ray crater Tycho is even younger. Note that the cratering age of Tycho, one of the youngest features on the Moon, is about $2 \times 10^8$ a, so freshness is clearly relative.

## Motions of the Moon

The Moon orbits at a mean distance of 384,402 km from Earth (0.00256 AU) with a mean orbital eccentricity of $e = 0.0549$ and an inclination (to the ecliptic) of 5.142°. Under the influence of Solar and other gravitational forces, the instantaneous eccentricity can range from about 0.04 to 0.07.

The Moon is a triaxial ellipsoid that is rotationally locked on to Earth, so that the lunar day coincides with the lunar orbital period (the month). But because of the eccentricity of the orbit of the Moon, its distance from Earth ranges from $q = a(1 - e) = 384{,}402 \times 0.9451 = 363{,}299$ km at perigee to $Q = a(1 + e) = 405{,}507$ km at apogee. The acceleration due to Earth's gravity distorts the shape of the Moon so that its figure is elongated along the Earth–Moon line, with a (sub-Earth) tidal bulge directed at Earth and another at the antipodal (anti-Earth) point. The eccentricity of the orbit causes

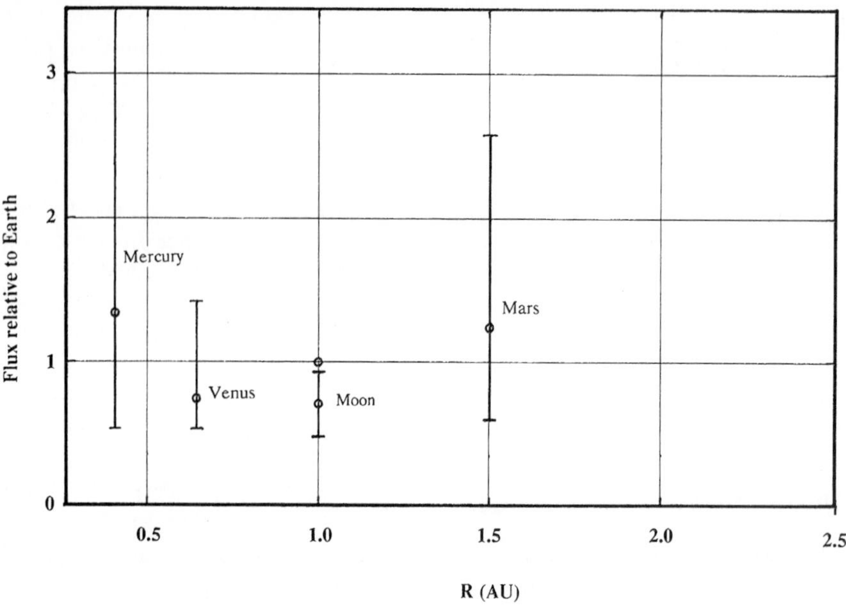


**Figure IX.20** Estimates of the cratering flux of large impactors in the inner Solar System. The Moon's impact flux is less than Earth's because it has too small a mass for effective gravitational focusing.

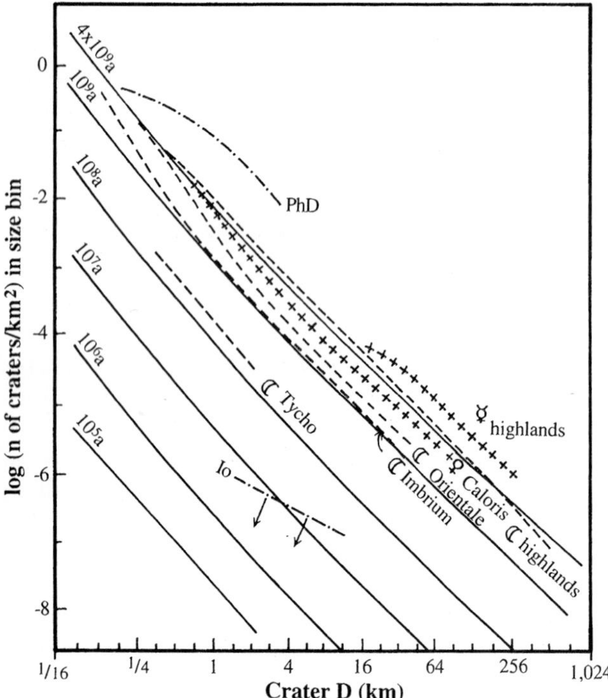


**Figure IX.21** Crater densities on rocky bodies. Crater densities on the Moon (highlands, Orientale basin, Mare Imbrium, and the floor of the very fresh crater Tycho); Mercury (highlands and Caloris basin); Io; and Phobos and Deimos (PhD) are shown as a function of size. Note the great crater density on the Martian moons due to reaccretion of impact ejecta and the upper limit on crater densities on Io.

a variation of the angular rate of motion of the Moon along its orbit, with an orbital velocity a maximum at perigee and a minimum at apogee. Because the Moon rotates at a nearly constant speed, but orbits with a sinusoidally varying speed, the center of the near-side tidal bulge sometimes lags behind the position of Earth in the lunar sky and sometimes runs ahead. This *libration in longitude* causes the Moon to oscillate back and forth by ±7.71° each month as seen from Earth.

The rotational axis of the Moon does not coincide with the axis of its orbit, but is tipped by about 6.5°. This *libration in latitude* permits observers on Earth to see the lunar polar regions alternately each month. Also, recall that Earth's rotational pole is tipped 23.45° with respect to the pole of its orbit (which is also, by definition, the pole of the ecliptic). The Moon can be as far as 28.78° north or south of the equator (a *declination* of ±28.78°). An observer at a latitude greater than 61.2° can sometimes see the Moon circle the sky without setting, making it for a brief time a *circumpolar* object.

Combining both forms of libration and taking into account the finite diameter of Earth, which helps us look around the edges of the lunar disk, some 60% of the surface area of the Moon can at some time be seen. About 40% is always within sight of Earth.

The plane of the Moon's orbit, being tipped relative to the plane of the ecliptic by more than 5°, intersects the plane of the Sun's motion (the plane of the ecliptic) in two regions 180° apart in the sky. The points of crossing, the *nodes* of the orbit, are where eclipses of the Sun and

Moon can occur. Indeed, this is the origin of the word *ecliptic*. The node at which the Moon crosses from the southern to the northern celestial hemisphere (the northbound ecliptic crossing) is called the *ascending node* of the orbit, and the other is of course the *descending node*.

The time required for the Moon to complete one orbit about Earth with respect to the stars averages 27.32166 days, but varies by several hours because of perturbations, mostly by the Sun. This period is called the *sidereal month*. The time required to complete an orbit about Earth with respect to the Sun is of course different (and longer), because the Sun has an apparent westward motion caused by Earth's orbital motion about it. This *synodic month*, effectively the mean time between successive new Moons, is 29.53059 days. This number is not an integral divisor of the year, so phases of the moon do not repeat on the same dates each year. However, intervals of 19 years ($19 \times 365.256 = 6939.864$ days) are almost exactly the same as 235 sidereal months ($235 \times 29.53059 = 6939.688$ days). Accordingly, lunar eclipses repeat in a pattern that cycles almost exactly every 19 years (although offsets of a day or so can occur as a result of the timing of leap years, which do *not* occur regularly every 4 years).

The tidal bulges on the Moon, which on average align along the Earth–Moon line, must oscillate back and forth (librate) as a result of the mismatch between orbital angular velocity and spin angular velocity mentioned above. But the gravitational force of Earth on the tidal bulges exerts a torque that opposes this librational movement. This force would, on a fully fluid Moon, keep the tidal bulge neatly aligned with Earth. However, the Moon is a rather cold silicaceous body, with an extremely high viscosity. The "pumping" of the lunar crust by these Earth tides must cause stress on the lunar rocks, causing occasional weak earthquakes and dissipating energy. This dissipation must continue as long as the Moon's orbit has nonzero eccentricity.

## Physical Properties of the Moon

The mean radius of the Moon is 1738 km. Its mass is almost exactly 1/81.3 of Earth's mass, making the Earth–Moon system in a meaningful sense a double planet, almost as alike in mass as the members of the Pluto–Charon system. The relatively large mass of the Moon contributes to the fact that its orbital angular momentum is larger than Earth's rotational angular momentum. The mean surface gravity of the Moon, $GM/r^2$, is 0.165 Earth gravities, and the escape velocity is $(2GM/r)^{1/2}$, or about 2.38 km s$^{-1}$.

The elongation of the axis of the Moon that points at Earth, which is in gross hydrostatic balance under the tidal effects of Earth's gravity, is about 2.1 km. The other two axes of the Moon differ by only a third as much.

The actual topography of the Moon has a greater amplitude than these variations. Long before the existence of any chemical data on the Moon it was widely recognized that there were two distinct types of lunar terrain and presumably also of lunar surface materials. The two principal terrain types were the elevated, heavily cratered, and relatively bright lunar highlands (originally called *terra*, the Latin word for land) and the low, dark, smooth *mare basins* (after the Latin word for sea). Most of these maria are fairly circular (Mare Serenitatis, Mare Imbrium, Mare Tranquillitatis, Mare Fecunditatis, Mare Crisium, Mare Nubium, etc.), although the largest dark feature, Oceanus Procellarum, is very irregular in outline. Early spacecraft imaging experiments, beginning with the Soviet Luna 3 flyby, showed that the mare basins are abundant on the near side, but very minor on the far side (Mare Moscoviense and the crater Tsiolkovskii only).

Tracking of Lunar Orbiter and Apollo spacecraft in orbit about the Moon permitted the mapping of the lunar gravity field, principally on the near side, with the interesting result that large mass excesses, called mascons, are found to coincide with the circular maria. The size and shape of the mass excess in these mascons are roughly equivalent to $10^5$ g cm$^{-2}$. This mass could, in principle, be a basin fill tens of kilometers thick with a density contrast of 0.3 g cm$^{-3}$ or, at the opposite extreme, a layer of iron–nickel metal about 100 m thick.

## Elemental Composition of the Moon's Surface

Spacecraft exploration of the Moon has established that there is indeed a major chemical difference between the mare basins and the highlands. The first analytical data on the lunar surface were returned by a gamma-ray spectrometer aboard the Soviet Luna 10 Orbiter in early 1966. This instrument measured with very low spectral resolution the natural emission of gamma rays by decay of radioactive uranium, thorium, and potassium on the lunar surface. Since these measurements were made from lunar orbit they had poor spatial resolution (roughly 100 to 200 km), but they generally showed that the abundances of these three elements were low, with proportions similar to those found in terrestrial basalts or basaltic meteorites, not chondritic meteorites or terrestrial granites.

*In situ* analyses of the surface of the Moon were first conducted during the unmanned Surveyor lander missions in 1966–1968. These analyses were conducted by an instrument that irradiated the surface with a source of alpha particles ($^{242}Cm$) and measured the energy spectrum of the alphas that were back-scattered into a detector.

Surveyor 1 landed inside a 100-km crater in Oceanus Procellarum, Surveyor 2 crashed on the Moon, Surveyor 3 landed within a 200-m crater in Oceanus Procellarum, Surveyor 4 landed safely but failed to return scientific data, Surveyor 5 landed in a 10-m crater in Mare Tranquillitatis, Surveyor 6 completed the survey of "safe and sane" Apollo candidate landing sites in the lunar maria by landing in Sinus Medii, and Surveyor 7, the last of the family, landed in the highlands on the rim of the spectacular and relatively young crater Tycho.

The analytical results from these missions confirmed that the mare basins were basically basaltic in character. Only the most abundant elements could be resolved by this technique, but confident detections of oxygen (the most abundant element in silicate rocks), sodium, magnesium, aluminum, silicon, calcium, titanium, and iron were achieved. These are sufficient for very useful chemical "fingerprinting" of the lunar rock types for comparisons with terrestrial and meteoritic samples. The analytical results from the Surveyor alpha-backscatter experiments are summarized in Table IX.1a. Generally, the mare surfaces were found to be compositionally similar to one another. Sodium concentrations are quite low compared to terrestrial basalts, and titanium dioxide is very variable, ranging to well over 7% in Mare Tranquillitatis. The highland sample showed lower iron and higher aluminum and calcium than the mare materials, suggesting high abundances of calcium aluminosilicates in the highland regolith. Indeed, the most distinctive feature of highland material is its high abundance of feldspar. Because of the low abundance of the moderately volatile metals Na and K on the Moon, the feldspar is mainly the calcic end member of plagioclase, $CaAl_2Si_2O_8$, the mineral anorthite. The rock dominated by anorthite and pyroxene is called anorthosite.

Much more detailed chemical and isotopic analyses of lunar surface materials were made possible by the development of techniques to return lunar samples to Earth. The Soviet Union pursued, as a part of the Luna spacecraft series, a vigorous development program on a small automated unmanned lander capable of returning a few hundred grams of material to Earth. At the same time the American Apollo manned lunar landing program was under rapid development, and a Soviet manned landing program was in an earlier stage of experimentation in low Earth orbit.

**Table IX.1a** Surveyor Alpha-Backscatter Analyses

| | Weight % of element oxides | | |
|---|---|---|---|
| Component | Surveyor 5 (Tranquillitatis) | Surveyor 6 (Sinus Medii) | Surveyor 7 (Tycho rim) |
| $SiO_2$ | 46.4 | 49.1 | 46.1 |
| $Al_2O_3$ | 14.4 | 14.7 | 22.3 |
| CaO | 14.5 | 12.9 | 18.3 |
| FeO | 12.1 | 12.4 | 5.5 |
| MgO | 4.4 | 6.6 | 7.0 |
| $TiO_2$ | 7.6 | 3.5 | 0.0 |
| $Na_2O$ | 0.6 | 0.8 | 0.7 |

The first lunar samples were returned on the Apollo 11 mission, which landed not far from the area inspected by Surveyor 5, in a region of deep, mature regolith. During their exploration of the lunar surface near their landing site, astronauts Neil Armstrong and Edwin A. (Buzz) Aldrin collected a wide range of regolith and rock samples. Because the regolith is rich in rock chips excavated by cratering events over a wide area of the lunar surface, many rock fragments were found that did not originate in the local bedrock, including anorthosites and anorthositic pyroxenites of highland origin. About half the rocks found were finely crystalline igneous rocks and half were a welded mass of disparate rock and mineral fragments and fine, glass-rich dust. These rocks, called microbreccias, are essentially shock-lithified samples of the lunar regolith.

The lunar regolith is extremely complex. It is dominated by the products of violent cometary and asteroidal impacts on lunar igneous rocks and on the regolith itself. Rocks of all sizes, from house-sized boulders down to tiny rock chips containing only a few mineral grains, are mixed together with tiny glass droplets, similar to small chondrules, and dust composed of very finely crushed grains. Impacts in the regolith have shock-lithified large chunks of this mixture, and eclectic mixtures of smaller particles are often found welded together by melt glass from more recent impacts. These welded, highly heterogeneous lumps are called *agglutinates*. In some places the lunar regolith has been reprocessed so thoroughly that it has been mostly converted into agglutinates. Such regolith samples are said to be mature.

The second Apollo landing mission, Apollo 12, landed about 180 m from Surveyor 3 in Oceanus Procellarum. During their exploration on the Moon, astronauts Pete Conrad and Alan Bean visited Surveyor 3 and removed parts of it for an investigation of the effects of exposure to the lunar environment on selected materials. The rock types found were quite similar to those seen on the Apollo 11 mission, except that crystalline rocks (usually basalts) were much more common than microbreccias. These igneous rock fragments were excavated by the impact event that formed the crater within which Surveyor 3 landed.

After a near-disaster on the Apollo 13 mission, when an oxygen tank exploded in the service module while en route to the Moon, the landing missions resumed with the successful flights of Apollos 14, 15, 16, and 17, which sampled a variety of landing sites. The last three of these missions carried roving vehicles that greatly extended the area covered by the astronauts during their exploration of the lunar surface.

No samples have ever been returned from the far side of the Moon, and the lunar highlands remain very poorly sampled. Most of our knowledge concerns the mare basins, chosen as landing sites because of the relative smoothness and safety of their surfaces.

Samples returned by the Apollo missions (a total of nearly 400 kg) and by the three successful Luna missions (Lunas 16, 20, and 24) are identified by a simple five-digit code, of which the first two digits describe the mission which returned that sample: "11" for Apollo 11; "12" for Apollo 12; "14" and "15" for Apollos 14 and 15, respectively; and "6" and "7" for Apollo 16 and 17. The Luna samples bear their mission numbers, "16," "20," and "24." These returned samples are supplemented by over a dozen meteorites that have been found on Earth, almost all lying on the ice in Antarctica. Chemical and isotopic analyses and radiometric dating have confirmed them to have lunar origin. Their virtue is that they sample portions of the Moon not visited by landing missions; their greatest drawback is that their exact sites of origin are not known and not documented.

Chemical analyses of lunar samples show a variety of distinct rock types. The materials of the lunar highlands, the relatively bright, heavily cratered terrain that covers most of the Moon, are characteristically rich in aluminum and calcium. The dark, low, relatively sparsely cratered mare basins exhibit high concentrations of titanium and iron.

## Lunar Rock Types

Rocks are assigned to various types according to their elemental composition, mineralogy, and texture. Because of the enormous geological diversity of Earth, it is not surprising that names already existed for most of the important lunar rock types before chemical investigation of lunar materials became possible. In many cases the lunar use of a term such as "basalt" simply means that the lunar rock of this name is more similar in bulk composition, mineralogy, and texture to the terrestrial rock of the same name than to other known terrestrial rock types.

The most general categorization of rocks is into three broad categories, igneous, metamorphic, and sedimentary. Igneous rocks are formed by crystallization from an extensively molten initial state, such as magma extruded from a volcano or intruded between preexisting layers or rock. The most common lunar igneous rock is basalt. Metamorphic rocks have experienced substantial reheating and recrystallization with little mass gain or loss except for fluids. The most common metamorphic process on the Moon is impact shock heating, which can weld regolith materials together into a fairly strong rock. Sedimentary rocks as we understand them on Earth are essentially absent on the Moon, where wind and water are perpetually absent. The regolith is a sediment, composed of layers of comminuted ejecta from impacts of all sizes and distances, but the regolith forms a rock only under the influence of shock metamorphism.

Chemical analyses of representative lunar igneous rocks, reported in Table IX.1b, show that most have silica abundances that are near the low end of the terrestrial range, usually with less than 50% by weight of $SiO_2$, within the range for terrestrial basalts. The lunar basalts differ from the terrestrial basalts in that they have alkali metal abundances that are several times lower and titanium abundances that are several times higher than the usual terrestrial examples. Since the very high titanium abundance is accounted for by large amounts of ilmenite, $FeTiO_3$, the iron abundance is also abnormally elevated compared to terrestrial basalts.

In texture, we can distinguish a range from the most fine-grained lunar basalts, which cooled rapidly upon extrusion as lava flows onto the surface of the Moon, to coarser grained basalts that crystallized more slowly and completely in a more protected (often intrusive) environment beneath the surface, forming volcanic sills or dikes or the bottom of thick lava flows. Extruded basalts often show vesicles, essentially bubbles inflated by magmatic gases such as CO, which unfortunately have long since diffused away. Liquids of lunar basalt composition have very low viscosities and are therefore very poorly suited for building tall volcanic structures about their magma vents, but ideal for making lava

**Table IX.1b** Compositions of Representative Lunar Igneous Rocks

| | (weight %) | | | | | | |
|---|---|---|---|---|---|---|---|
| **Oxide** | **70215** | **15545** | **14053** | **15386** | **62237** | **72415** | **14321c** |
| $SiO_2$ | 37.8 | 45.2 | 46.4 | 50.8 | 41.9 | 40.3 | 74.2 |
| MgO | 6.7 | 10.3 | 8.5 | 9.34 | 5.11 | 44.8 | 0.07 |
| FeO | 19.8 | 22.1 | 16.8 | 10.4 | 5.89 | 11.7 | 5.6 |
| CaO | 11.1 | 9.8 | 11.2 | 9.60 | 16.3 | 1.1 | 8.8 |
| $Al_2O_3$ | 10.4 | 8.6 | 13.6 | 15.0 | 29.6 | 1.3 | 12.5 |
| $TiO_2$ | 10.5 | 2.4 | 2.6 | 2.06 | — | 0.03 | 0.33 |
| $Na_2O$ | 0.40 | 0.31 | — | — | 0.2 | 0.13 | 0.52 |
| $K_2O$ | 0.06 | 0.04 | 0.10 | 0.61 | — | — | 8.1 |
| $Cr_2O_3$ | 0.25 | 0.68 | — | — | — | 0.3 | — |

*Key:* 70215: Apollo 17 high-titanium basalt; 15545: Apollo 15 olivine basalt; 14053: Apollo 14 low-titanium aluminous basalt; 15386: Apollo 15 KREEP basalt; 62237: Apollo 16 ferroan anorthosite; 72415: Apollo 17 Mg-rich dunite; 14321c: Apollo 14 granite

tubes, which may later drain and collapse to make features looking like the lunar rilles. These basaltic liquids also have densities of only about 3.0, compared to about 3.3 for the solidified basalt. Note that the observed density of the entire Moon is only 3.34.

The lunar highlands are dominated by very old ("pristine"), relatively light-colored, less dense rocks in which plagioclase feldspar is the principal mineral. The general depletion of alkali metals and other moderately volatile elements in lunar materials biases the composition of lunar plagioclase strongly in the direction of the calcium aluminosilicate end member, anorthite. Rocks in which anorthite is the dominant mineral are called anorthosites. These aluminum- and calcium-rich rocks have a density of about 2.9, very slightly less than the density of basaltic magma, but much less than that of solid basalt. These rocks are well represented at the Apollo 16 landing site. The second most abundant mineral in these rocks is generally a low-magnesium pyroxene, with olivine also present. These rocks are referred to as ferroan anorthosites. They are among the most ancient rocks on the Moon, dating back at least 4.4 Ga.

The second most abundant highland rock type is the class of magnesium-rich rocks, which are themselves compositionally diverse. These rocks have plagioclase contents that range down to nearly zero in the rock dunite, which is nearly pure olivine. Although we know of surface exposures of olivine on the Moon from remote sensing data, very olivine-rich rock chips make up a minor part of the Apollo and Luna sample collection. Olivine-plagioclase rocks are called norites, and pyroxene and Ca-poor plagioclase rocks are classified as gabbros or norites. Many of the Mg-suite highland rocks also date to before 4.4 Ga.

An unusual lunar highland rock type, containing enhanced abundances of potassium (K), the rare earth elements (REE), and phosphorus (P), found only as small chips in the regolith, has been named KREEP. In major-element chemistry, KREEP is a basalt. As on Earth, the rare earths concentrate strongly into phosphate minerals. Gamma-ray spectroscopy conducted from lunar orbit during the Apollo program tentatively identifies the KREEP source region on the edge of the Mare Imbrium basin, based on the enhanced gamma emission associated with high concentrations of K, U, and Th.

The partitioning of the rare earth elements between different minerals, and hence between different rock types, carries useful information regarding the history of the Moon. The rare earths comprise the inner transition elements with atomic numbers $Z = 57$ (lanthanum) to 71 (lutetium). All these elements are rather efficiently expelled from ferromagnesian minerals (olivine and pyroxene) during crystallization: the rare earths, like the alkali metals and many other elements having ionic charges and radii that are not closely similar to those of iron, magnesium, or silicon, are termed *incompatible elements* because they are both structurally incompatible with these dense mantle minerals that crystallize at high temperatures. All the incompatible elements tend to become concentrated in residual melt. The bulk of the incompatible material consists of relatively abundant, low-density oxides, and hence the rare earths are enriched in low-density, crust-forming liquids. The REEs generally exist in the 3+ valence state; however, europium ($Z = 63$) is extensively reduced to the 2+ state under lunar conditions. $Eu^{2+}$ substitutes freely for $Ca^{2+}$ in minerals, causing a substantial accumulation of europium in anorthite. Therefore, crustal rocks are generally enriched in rare earths compared to the whole Moon, and europium is enriched *relative to the rest of the rare earths* in anorthite-rich rocks. All the rare earths except europium generally fractionate in a fairly coherent manner: the departure of europium from the general trend of the other rare earths is, for historical reasons, termed the *europium anomaly*. It is customary to report the abundances of the rare earths relative to some supposed universal standard, which is taken to be chondritic meteorites. Plots of the rare earth abundances relative to a chondritic standard often show modest differences in slope between the light rare earths (below europium) and the heavy rare earths (above Eu). The rare earths in general are enriched 10- to 100-fold in basaltic lunar rocks, all of which show a modest depletion of europium (i.e., a smaller enrichment) relative to the other rare earths, a phenomenon called a *negative europium anomaly*. (Note that, by elementary considerations of mass conservation, rocks enriched 100-fold in REEs must make up less than 1% of the mass of the Moon.) The very highest REE enrichments, in excess of a factor of 100, are found for KREEP basalts, whose name reflects this distinction. The KREEP basalts also have a larger negative europium anomaly of about a factor of 4. Highland rocks, with their higher plagioclase abundances, have rare earth abundances that are enriched severalfold relative to the chondritic standard and a small positive europium anomaly.

Trace elements carry a wealth of information regarding the history of the rocks in which they reside. Each trace element, because of its ionic charge and radius, and sometimes influenced by the state of oxidation of its environment (which can change the ionic charge of some elements), is partitioned between the coexisting solid and liquid phases. An example is U, which substitutes reasonably well for Ca, and therefore preferentially partitions into many Ca-bearing minerals. The ratio of the equilibrium concentrations of trace element E in two coexisting phases A and B is called the *partition coefficient*, $D_{E(AB)} = [E]_B/[E]_A$. This partition coefficient is simply an equilibrium constant, and as such it may vary substantially with temperature. Partition coefficients may be inferred from studies of natural rocks, or they may be experimentally determined in the laboratory. Partitioning of siderophile metals between liquid and solid metals, between liquid silicates and solid metals, between liquid silicates and liquid metals, between two coexisting metal phases (such as kamacite and taenite), and so on may be determined in the laboratory and used to help reconstruct the conditions of origin of natural samples within which the abundances and distributions of the siderophile trace elements have been measured. Distributions of trace chalcophiles between core-forming metal–metal sulfide melts and silicate phases are also studied in the same manner. Similarly, partition coefficients for rare earth elements between olivine and pyroxene, pyroxene and plagioclase, plagioclase and melt, and so on may be determined. Generally, siderophiles and chalcophiles are strongly concentrated into the core and REEs are strongly enriched in feldspars. All of these partitioning reactions occur subject to the constraint of conservation of mass. When certain knowledge of the original bulk composition or the present bulk composition of a planet is lacking (as is usually the case for large planets, for which we cannot sample the core), simplifying assumptions can and must be made regarding the bulk composition. For the Moon, this may consist of assuming grossly chondritic overall initial composition, or assuming similarity to Earth's mantle. Although such assumptions are not likely to be wildly wrong, they cannot at present be verified experimentally.

The crust of the Moon is significantly enriched in REEs relative to a chondritic standard. The bulk Moon, if assumed to be grossly chondritic in overall composition, must therefore contain a large reservoir of REE-depleted rocks with very low abundances of all incompatible elements. These are ferromagnesian mantle rocks, complementary differentiates of the crust. These rocks, which are too dense to extrude onto the lunar surface, are usually referred to as the *basalt source region.* In some circumstances, especially on Earth and Mars, pressures high enough for the formation of garnet are realized. Garnet is unusual in that it accommodates the lighter rare earths more readily than the heavy ones, creating a REE pattern with a generally negative slope. As we shall see in Chapter X, garnet formation has been invoked on Mars to explain the general REE fractionation trends observed in the Martian (SNC) meteorites but not seen on the Moon.

Silica- and alkali-rich rocks, analogous to terrestrial granites, are quite rare on the Moon.

**Table IX.2** Selected Native Lunar Minerals

| | | Notes |
|---|---|---|
| Metals | | |
| Kamacite | Fe, Ni (<6% Ni) | a |
| Taenite | Fe, Ni (>6% Ni) | a |
| Copper | Cu | a |
| Nonmetals | | |
| Graphite | C | a |
| Sulfides | | |
| Troilite | FeS | a |
| Chalcopyrite | $CuFeS_2$ | a |
| Carbides, phosphides, etc. | | |
| Cohenite | $(Fe, Ni)_3C$ | a |
| Schreibersite | $(Fe, Ni)_3P$ | a |
| Oxides | | |
| Armalcolite | $FeMgTi_2O_5$ | a |
| Perovskite | $CaTiO_3$ | a |
| Spinel S.S. | | a |
| Spinel | $MgAl_2O_4$ | |
| Hercynite | $(Fe, Mg)Al_2O_4$ | |
| Chromite | $FeCr_2O_4$ | |
| Magnesiochromite | $MgCr_2O_4$ | |
| Quartz | $SiO_2$ | a |
| Cristobalite | $SiO_2$ | m |
| Tridymite | $SiO_2$ | m |
| Rutile | $TiO_2$ | a |
| Baddeleyite | $ZrO_2$ | a |
| Ilmenite | $FeTiO_3$ | M |
| Oxysalts | | |
| Apatite | $Ca_5(PO_4)_3(OH, F)$ | a |
| Chlorapatite | $Ca_5(PO_4)_3Cl$ | a |
| Whitlockite | $Ca_3(PO_4)_2$ | a |
| Silicates | | |
| Olivine S.S. | $(Mg, Fe)_2SiO_4$ | m |
| Fayalite | $Fe_2SiO_4$ | |
| Forsterite | $Mg_2SiO_4$ | |
| Pyroxene S.S. | | M |
| Orthopyroxene | $(Mg, Fe)SiO_3$ | |
| Enstatite | $MgSiO_3$ | |
| Ferrosilite | $FeSiO_3$ | |
| Clinopyroxene | $(Ca, Mg, Fe)SiO_3$ | |
| Wollastonite | $CaSiO_3$ | |
| Augite | $Mg(Fe, Ca)Si_2O_6$ | |
| Hedenbergite | $CaFeSi_2O_6$ | |
| Pigeonite | $(Fe, Mg, Ca)SiO_3$ | |
| Feldspar S.S. | | M |
| Plagioclase | | |
| Anorthite | $CaAl_2Si_2O_8$ | |
| Albite | $NaAlSi_3O_8$ | |
| K-spar | | a |
| Orthoclase | $KAlSi_3O_8$ | |
| Sanidine | $KAlSi_3O_8$ | |
| Pyroxferroite | $CaFe_6(SiO_3)_7$ | m |
| Zircon | $ZrSiO_4$ | a |

*Note.* M, major mineral (>10% in some rocks); m, minor mineral (1 to 10% in some rocks); a, accessory (always <1%).

## Lunar Minerals

A list of lunar minerals is given in Table IX.2. Only seven minerals are found in abundances greater than about 1% in lunar rocks. These include pyroxene, $(Ca,Fe,Mg)SiO_3$; calcic plagioclase, $(Ca,Na)(Al,Si)_4O_8$, close to anorthite composition; ilmenite, $FeTiO_3$; olivine, $(Mg,Fe)_2SiO_4$; pyroxferroite, $CaFe_6(SiO_3)_7$; and two polymorphs of silica, $SiO_2$, cristobalite and tridymite.

Pyroxene compositions in lunar rocks trespass into regions of the pyroxene quadrilateral not previously populated by other Solar System materials (Fig. IX.22). Some lunar pyroxenes are fairly standard augites, but others (often the outer layers of normal augite crystals) range far into the ferrosilite corner of the quadrilateral, perilously close to pure $FeSiO_3$, which is thermodynamically unstable with respect to its component oxides. This new material crystallizes in the triclinic system, unlike monoclinic augite, and hence is given a new mineral name, pyroxferroite. Pigeonite, with its low calcium content, is also present in smaller amounts.

Feldspars are present as an unusually calcium-rich plagioclase (from about 60% up to 99% anorthite) and a potassium feldspar. The K-spar, which has sanidine structure, is crystallized in very small amounts out of the residual melt in rapidly crystallized basalts.

Given the low silica content of most lunar rocks, the presence of small amounts of olivine is hardly surprising. Most lunar olivine is fairly iron-rich ($fa_{20}$ to $fa_{50}$), but virtually pure fayalite is sometimes observed. Fayalite, unlike pure ferrosilite, is thermodynamically stable. Many lunar basalts are close to or above silica saturation, often with a few percent cristobalite or tridymite or even a trace of quartz.

Ilmenite occurs in basalts with abundances from a few percent to more than 20%. There is an important difference in oxidation state between terrestrial ilmenite, which is a solid solution of $FeTiO_3$ with $Fe_2O_3$, rich in ferric iron, and lunar ilmenite. Ferric iron is absent from lunar igneous rocks, and in fact small amounts of metallic iron are often found in lunar basalts. The ferric mineral FeOOH, ubiquitous in trace amounts in lunar samples, is an alteration product produced by the attack of terrestrial atmospheric water vapor on lunar lawrencite, $FeCl_2$.

Metallic iron in lunar rocks is usually found in association with troilite and is usually nearly pure iron, with less than 1% nickel. The molar ratio of metal to troilite always lies close to the eutectic composition for an Fe–FeS melt. On rare occasions, tiny traces of metallic copper are also found in association with the metal and troilite in basalts.

By contrast, metallic iron found in the regolith and in the shock-lithified microbreccia contains up to 30% Ni and 1% Co. The metal in the basalt seems to have been made by reduction of FeO during melting, whereas the regolith metal is clearly dominated by asteroidal debris. This view is reinforced by the common occurrences of traces of cohenite $Fe_3C$ and schreibersite $(Fe, Ni)_3P$ in association with Ni-bearing metal in the regolith and breccias.

Several interesting oxide minerals besides ilmenite are also found on the Moon. Rare spinel, nearly stoichiometric $MgAl_2O_4$, has been found in breccias, chromite $FeCr_2O_4$ has been found within regolith nickel–iron particles of apparent asteroidal origin, and ulvospinel, nearly stoichiometric $Fe_2TiO_4$, has been found in trace amounts exsolved from or replacing ilmenite. Other titanates include armalcolite $(Fe, Mg)Ti_2O_5$, usually found within ilmenite grains, and one report of perovskite $CaTiO_3$, containing high concentrations of REEs, in the late-crystallizing component of the coarser basalts. Rutile $TiO_2$ and both baddeleyite $ZrO_2$ and zircon $ZrSiO_4$ are also found in tiny quantities.

Few elements serve as important markers of the oxidation state of lunar material. We have seen lunar basalts lie close to the Fe–FeO buffer. Sulfur is fully reduced to sulfide (indeed, almost exclusively as troilite), and carbon is found as meteoritic carbide. Phosphorus, which is found as coexisting phosphide and phosphate in some meteorites, is found in accessory amounts in lunar basalts as apatite $Ca_5(PO_4)_3X$. Fluorapatite ($X = F$)

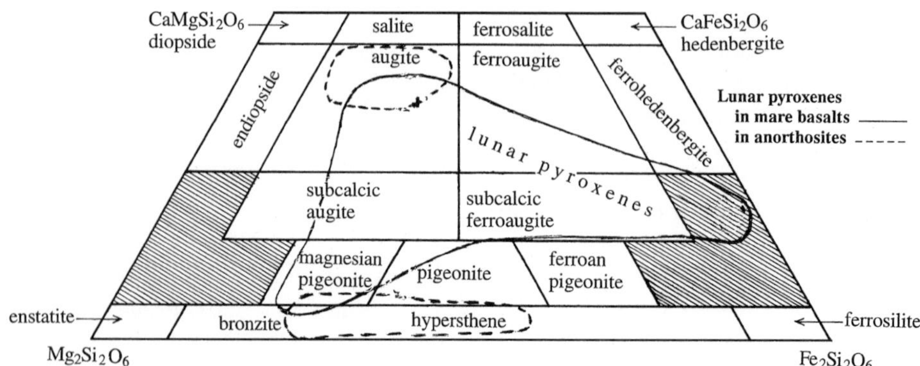


**Figure IX.22** Compositions of lunar pyroxenes. Lunar pyroxenes occupy an FeO-rich portion of the pyroxene quadrilateral (see Fig. IV.20) previously not inhabited by any known natural minerals, attesting to the high degree of oxidation of the lunar rocks relative to terrestrial rocks.

seems to be more common, but chlorapatite (X = Cl) has also been reported. Hydroxylapatite (X = OH) has not been reported. Small amounts of an amphibole with fluoride instead of hydroxyl have also been reported. Whitlockite $Mg_3(PO_4)_2$ with very high concentrations of rare earths (to about 10%) has also been found as a component of KREEP.

## Lunar Elemental Abundance Patterns

The general trends exhibited by elemental abundance patterns in lunar basalts contain important clues to the origin and history of the Moon. The first noteworthy pattern is that many refractories are enriched in the lunar basalts relative to CI chondritic material by a factor of up to 100. These highly enriched elements include the rare earths, the chemically similar actinides U and Th, and Ti, Zr, Hf, Ta, Ba, Sr, and Sc. Calcium, aluminum, and lithium are enriched by about a factor of 10. All of these strongly enriched elements are lithophiles. These elements are poorly accommodated in pyroxenes, plagioclase, and olivine and are expected to be highly enriched in the residual liquid after crystallization of the large majority of the abundant lunar minerals. Their high abundance in lunar rocks may be due entirely to their partitioning behavior during lunar evolution, not to any enrichment of them in the early Moon.

Chalcophiles such as S, Se, Te, Co, Cu, and Zn are generally depleted by about a factor of 10 relative to CI material, consistent with extraction into core-forming metal/sulfide melts during differentiation. Lead, depleted by only a factor of 2, is the exception that proves the rule: the lead found in lunar igneous rocks is highly radiogenic, with an isotopic composition very different from that of primordial lead. The lead observed in lunar basalts was made there by radioactive decay, after differentiation and chalcophile extraction. A number of volatile elements, including H, C, N, Cl, and Br, are depleted by about the same amount as the chalcophiles.

Several siderophiles, including Ge, Co, As, and Sb, are depleted by about a factor of 10. Silver, cadmium, ruthenium, rhodium, palladium, and indium are depleted by about a factor of 100. Gold, iridium, and probably osmium and platinum are depleted by at least a factor of 1000 relative to CI. These depletions were clearly caused by fractionation into core-forming materials.

A number of elements of exogenic origin have left their signature in the regolith and the shock-lithified breccias. Meteorite infall has enriched the abundances of Ni, silver, cadmium, indium, tellurium, Re, Os, Ir, and Pt and, to a lesser degree, germanium, palladium, molybdenum, and cobalt. These enrichments generally suggest about 2% of asteroidal material mixed into the lunar regolith. Several of the most volatile elements are also strongly enriched in the regolith, most notably hydrogen, the rare gases, carbon, and nitrogen. The isotopic composition of lunar regolith hydrogen is incompatible with terrestrial contamination, since the D:H ratio is three to four times smaller than for terrestrial hydrogen. The helium isotopic composition is compatible with an origin from the solar wind.

## Geology of the Moon

On the global scale, the most striking feature of the Moon's geology is the dichotomy between the near side and the far side. The near side contains numerous deep, basalt-flooded basins and a generally lower range of elevations than the far side. The far side, dominated by a thick anorthosite crust, lies generally several kilometers higher than the near side. Post-Apollo theories of the origin and evolution of the Moon have generally assumed that there was a large systematic difference in crustal thickness between the two sides. The topographic variations were attributed to early formation of a hot, relatively thin and readily deformable crust, followed by intense bombardment by large bodies. The craters made by these large impacts were assumed to have attained isostatic compensation (hydrostatic equilibrium) rather closely at the base of the crust. The large impacts resulted in many cases (almost exclusively on the near side) in flooding of the impact basins with basaltic magma melted out of the olivine-rich mantle. The crust then continued to cool in a state of near isostacy, eventually becoming cold and rigid, as we see it today.

These views have required significant reassessment as a result of the Clementine mission in 1994. First, the Clementine laser altimeter experiment found more vertical relief than expected, spanning an altitude range of about 16 km. The far side exhibits far wider variability than the near side; indeed, the hypsograms of the two sides are very different, with that of the near side not only lying lower, but also displaying a far narrower range of altitudes. The Clementine gravity data show that the highlands are generally well compensated, but that large impact basins deviate from isostacy by amounts that are often large, and which show no clear patterns with respect to age or crater size. The huge South Pole–Aitken basin exhibits a topographic low about 8 km below the mean geoid, but shows negligible basalt flooding. In addition, the floor of this enormous depression contains abundant ferromagnesian minerals, suggesting extensive exposure of mantle materials.

The mean crustal thickness is still found to be greater on the far side, but by only about 8 km, less than expected based on Apollo experience. The range of crustal thicknesses found for the far side is an astonishing 10 to 110 km. The absence of basalt-filled major basins on the far side seems to reflect a profound global asymmetry in the lunar mantle, perhaps due to large differences in the temperature profiles in the two hemispheres. Such a true, deep-seated global asymmetry seems to require a special explanation, such as extensive redistribution of the crust by one or more very large early impacts, or by global-scale single-cell convection, which collects anorthositic float rocks above the downwelling region and thins the crust in the upwelling region. The latter hypothesis provides a ready explanation for the prevalence of basaltic eruption on the near side, which would be the region of hot mantle upwelling. It is possible that such single-cell convection may require that the entire early Moon be involved in convection: a cold, primitive interior overlaid by a relatively thin soft, convective layer would probably exhibit numerous uncorrelated regions of upwelling, capable neither of imposing a hemispheric asymmetry on the crust, nor of steering basaltic magmatism into a single hemisphere. In general, correlation of the gravity and topography data shows a puzzling lack of any simple explanation: the lunar lithosphere seems stronger, under more strain, and less homogeneous than previously supposed.

The intermediate-scale structure of the lunar surface is dominated by impact cratering and by basalt-flooded impact basins. Some of the most recent impact craters are marked by bright, extensive systems of *rays*, sprays of secondary craters made by ejecta from the impact event. In the mare basins one frequently sees *rilles*, narrow, often lengthy channels evidently formed by fluid flow. Also in the maria one finds a wide variety of *wrinkle ridges*, irregular and usually sinuous raised features with heights up to about 100 m. Various interpretations of these wrinkle ridges have been proposed and are generally associated with lava flow activity. *Scarps*, vertical discontinuities caused by faults in the lunar crust, are somewhat rarer. Occasional chains of rimless craters are found, as on Phobos, attesting to the presence of open faults beneath the regolith, into which regolith material has drained. Some crater chains, with raised rims, are apparently also produced by sprays of ejects from large impacts.

About 40 of the largest impact basins, such as Mare Orientale and Mare Imbrium, have multiple rings of mountains surrounding the central, basalt-flooded basin. Many of these rings are no longer complete as a result of postimpact processes, including flooding and subsequent impact events. The Caloris Basin on Mercury is another example of such a multiringed basin. Only a few of the largest terrestrial impact craters show evidence of multiringed structure.

The interiors of large craters often contain, in addition to systems of central peaks, networks of narrow rilles. Sometimes these rilles are associated with very dark, roughly circular smudges on the surface, which have been attributed to the venting of volatiles through parts of the rille system.

The production of landforms by large impacts, which generate impact craters, secondary craters, ejecta blankets, central peaks, and fracturing of bedrock, also liberates volcanic processes by opening pathways through the thick, cold upper lunar crust and by depositing large amounts of impact energy in the fractured zone under large craters. The smaller impact events strike the rubble from large impacts and cannot penetrate through that rubble layer. The debris layer is constantly stirred ("gardened") by small impacts, which shock, comminute, melt, and weld together the debris fragments until a near steady state is achieved. This highly fragmented, reworked, partially melted layer is called the *regolith*. Its presence is compounded by two factors. First, regolith materials must be made by the countless small impacts that unavoidably accompany the rare large impacts. Second, the Moon, unlike asteroids, has sufficient gravity to prevent the efficient ejection of regolith material at escape velocity, and the regolith is therefore retained. Much of the geology of the Moon is the study of the formation, distribution, and properties of regolith. A human-scale appreciation of the lunar regolith is offered in Fig. IX.23.

## Geophysics of the Moon

We have few effective tools at our disposal for probing the lunar interior. Seismic studies were conducted by seismometers deployed during the Apollo program (Fig. IX.24). The Moon is extremely inactive compared to Earth, and the few weak quakes that are observed can originate down to depths of 800 to 900 and even 1000 km. If the lunar interior were as hot as Earth's, stresses would be relieved by viscous relaxation, not brittle fracture, at such great depths.

The seismic wave energy from a geological disturbance can be propagated as transverse (shear, or S waves) and longitudinal (pressure, or P waves), as illustrated in Fig. IX.25a. Typically the speed of P waves is considerably larger than that of S waves. Liquids do not transmit shear waves at all. Figure IX.25b shows how the fracture of a fault segment can generate both kinds of waves, with very different intensities when observed from different directions.

**Figure IX.23** Apollo 11 exploring the lunar surface. The rolling regolith surface contains many boulders ejected from craters. Astronauts Edwin A. "Buzz" Aldrin and Neil Armstrong set about deploying instruments on the lunar surface. The finely powdered regolith bears the astronauts' weight and preserves their footprints excellently.

**Figure IX.24** The Apollo 11 ALSEP being deployed on the Moon. The Apollo Lunar Surface Experiments Package was left behind to carry out a variety of experiments, including seismology, after the departure of the astronauts.

a.

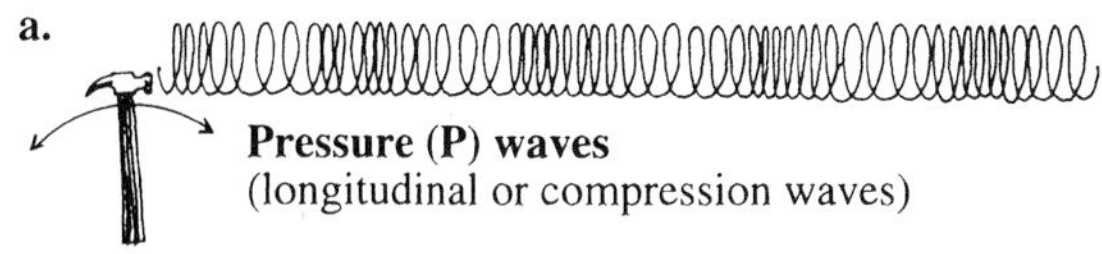


b.

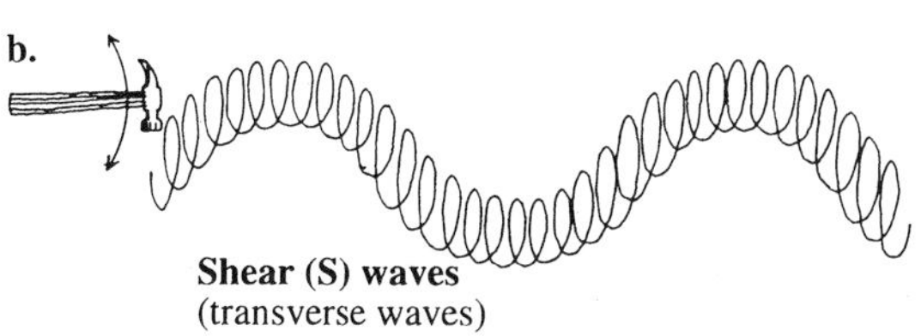


**Figure IX.25** Seismic wave terminology. P waves (a) generally travel much faster than S waves (b). In a hot, "soft" rock, there is also rapid attenuation of S waves. S waves are not transmitted at all by fluids.

Since natural moonquakes are so rare, it is useful to seek other ways of generating seismic waves to probe the lunar interior. Nature provides one excellent but wholly unpredictable source of seismic energy through meteoroid impacts. During the Apollo program, scientists and engineers combined to design an opportunistic lunar seismic experiment: the S-IVb stage of a Saturn V booster rocket that carried a later crew of Apollo astronauts to the Moon was aimed to impact the lunar surface, so that seismometers emplaced on earlier missions could observe the propagation of the seismic waves from a source whose total energy was precisely known.

The results of these seismic probes of the lunar interior suggest that the crust becomes well compacted at a depth of about 20 km, below which the pressure of the overlying rock is sufficient to squeeze out the vast numbers of cracks and voids made by large impact events. The mean thickness of the lunar crust is found to be about 50 km (roughly 8% of the volume of the Moon). Only about 1% of the crust is made of basalt, which is familiar to us mainly because it is widely exposed on the near-side surface, not because it is of great absolute volume. Through the regolith and crust the seismic velocities increase rapidly with depth, reaching $V_S = 4.2\,\mathrm{km\,s^{-1}}$ and $V_P = 7.6\,\mathrm{km\,s^{-1}}$ below 50 km. These velocities remain constant within the precision of the measurement down to a depth of about 1100 km. Natural moonquakes are not unambiguously observed from a greater depth. A single fortuitous impact of a meter-sized meteoroid on the far side of the Moon in 1972 gives us a deeper view: below 1100 km, where the S-wave velocity drops abruptly to about $2.5\,\mathrm{km\,s^{-1}}$, the attenuation of S waves also increases dramatically. This is strongly suggestive of partial melting. One issue of great geochemical importance is whether the Moon has a metallic core. Unfortunately the seismic data do not provide strong evidence for or against the presence of a core. An upper limit on the size of the core from seismic data is 500 to 600 km (2% to 4% of the volume of the Moon).

The magnetic properties of the Moon in principle provide us with another opportunity to detect a lunar core. Orbital magnetometer measurements show that the Moon has no significant external field. It was thus a considerable surprise when laboratory studies of lunar samples returned by the Apollo astronauts showed the presence of a substantial NRM remanent field. The implication is that, at the time of crystallization of the lunar rocks, a strong lunar magnetic field existed. That strong field is, however, clearly no longer present.

The principal magnetic material in lunar samples is metallic iron. Most of this iron is chemically quite pure, with a very low Ni content, and is associated with impact glass. It clearly lies outside the compositional range of meteoritic native metals and is evidently produced by autoreduction of FeO-bearing minerals by embedded hydrogen (which comes ultimately from the solar wind) during the impact shock heating of lunar regolith. Welding of regolith fragments produces breccia, consisting of loosely consolidated angular fragments. Only the impact breccias with the highest degree of thermal processing and recrystallization contain iron particles big enough for multidomain grains.

A sample of a recent impact glass found by the Apollo 17 astronauts (age less than 3 Ma) bearing a field of 0.025 G raises some very interesting complications. A possible mechanism for generating very local high field strengths is a violent impact event in which an extremely hot, dense plasma compresses a weak preexisting ambient field into the rock surface while melting it. Expansion of the fireball then allows the thin melt layer to quench very quickly to a glass film which "remembers" only the enhanced transient field. Other, older (3.9 Ga) lunar samples show the presence of magnetizing fields as high as 1.4 G. These data seem to require the presence of a strong early field, and hence imply the presence of a conductive core. Early T Tauri phase solar wind fields may conceivably have approached $5000\,\gamma$, which is almost certainly too small to provide the observed effect, and which lasts only for the first $10^7$ a of the Sun's lifetime. Thus the T Tauri phase was long over by the time of formation of these rocks at a Solar System age of about $7 \times 10^8$ a.

Magnetic field measurements on the surface of the Moon during the Apollo program demonstrated the presence of fields with strengths ranging from $6\,\gamma$ to more than $300\,\gamma$ ($1\,\gamma = 10^{-5}$ G). Such fields, although feeble by normal planetary surface standards, are nonetheless marginally capable of shielding the surface from the solar wind by generating tiny "magnetospheric cavities" or "bubbles" over some localities on the lunar surface.

The third line of evidence regarding lunar magnetism was provided by small, sensitive magnetometers carried aboard "subsatellites" launched into lunar orbit from Apollo command and service modules as they orbited about the Moon. These subsatellites detected distinct localized ripples in the lunar magnetism as they passed over large craters and over swirl-like surface markings associated with large craters.

The electrical conductivity profile of the Moon can to some degree be reconstructed from the observed response of the Moon to variations in the solar wind magnetic field strength. Interpretation of these measurements of the electrical conductivity of the deep lunar interior place a limit of about 360 km on the radius of a metallic core. This is only about 1% of the volume of the Moon. A "weak consensus" model of the lunar interior is presented in the pie diagram in Fig. IX.26.

Another important geophysical constraint on the lunar interior arises from measurements of the heat flux through the lunar crust. Such measurements were carried out on the Apollo 15 and 17 missions by drilling holes (1.3 and 2.3 m in depth, respectively), removing the core material, and measuring the temperature gradient in the hole. The core sample was returned to Earth and its thermal conductivity was measured. Regolith material in a vacuum is a poor thermal conductor, so the temperature gradient in it is large enough to be measured easily. The heat flux is easily calculated from the temperature gradient and the conductivity:

$$F = -k(\partial T/\partial z). \qquad \text{(VIII.12)}$$

The measured heat fluxes were 2.1 and 1.6 $\mu$W cm$^{-2}$. Calculating the internal thermal structure of the Moon from these fluxes is still a rather uncertain business, because the full thermal conductivity profile is not known. However, constraining the temperatures using the seismic structural data and the observed topography (which would quickly relax through viscous relaxation if the temperatures were too high) leads to models in which the interior temperature asymptotically approaches 1200 to 1500 °C in the deep interior. This cool interior is consistent with the high surface to volume ratio of the Moon. The purely conductive time scale

$$t_{\text{cond}} = C_p \rho r^2/3k \qquad \text{(VIII.15)}$$

is $5 \times 10^{10}$ a, but suppose the interior temperatures were once high enough to permit convective heat transport and cool the Moon on a time scale of about $10^9$ a. Cooling would occur, increasing the internal viscosity, until the cooling time scale was in step with the heat production rate. This feedback mechanism leads to a present-day approach to steady-state conditions.

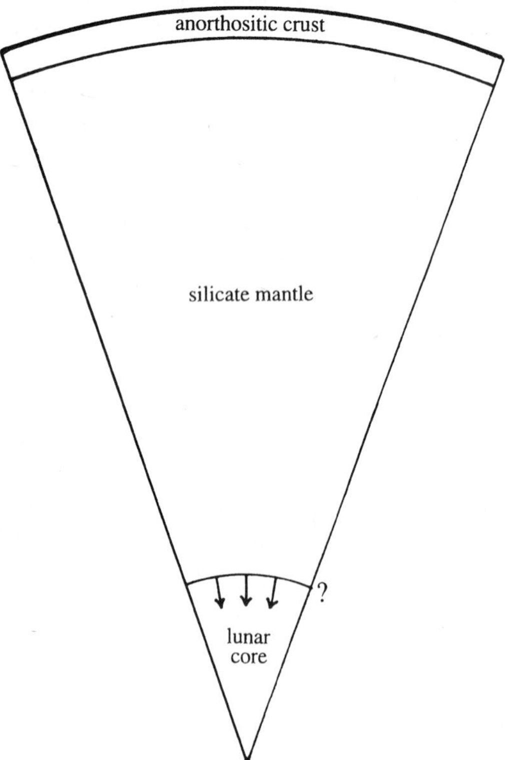


Figure IX.26 Lunar internal structure. The Moon appears to be thoroughly differentiated, but the core material complementary to the crust and mantle may not be present inside the Moon and has never been convincingly detected.

## History of the Earth–Moon System

Observations of the present orbit of the Moon show a low eccentricity and a modest inclination (5°) to the Earth's equator. The rotation of the Earth and the orbital motion of the Moon are, however, coupled weakly by tidal interactions between the two bodies. The familiar oceanic tides cause the figure of Earth's ocean surface to deviate from a sphere by roughly 1 m. The ocean surface directly below the Moon's position experiences a smaller net gravitational acceleration than elsewhere on Earth because the direction of the lunar gravitational force acts to offset a small part of Earth's gravitational attraction. Recall that the Earth and Moon orbit about their common center of mass (barycenter), which, by the lever rule, lies 1/81.3 of the way from the center of the Earth to the center of the Moon. This is roughly 400,000/80 = 5000 km from the center of Earth under the sublunar point, actually within Earth's mantle. The centrifugal force due to Earth's motion about the barycenter thus also helps to raise the water level there. Thus for two separate reasons the ocean surface there rises higher than elsewhere.

But consider the point on Earth's surface opposite the position of the Moon (Fig. IX.27). The water there is farthest from the Moon and experiences less Moonward force than elsewhere, and the distorting force due to Earth's motion about the barycenter is greatest there. Thus the Moon raises two tidal bulges in the oceans, one near the sublunar point, and the other near the antilunar point. These tides drag large masses of water back and forth over shallow ocean floor over the continental shelf and in the Arctic Ocean, and this motion is resisted by friction between the ocean and the solid Earth. The rotation of the Earth drags the tidal bulges slightly ahead of the Moon's position. This causes the rise and fall of the tides to be slightly advanced, so that high tide occurs slightly before the Moon passes a meridian. In other words, the tidal bulges are slightly asymmetric about the Earth–Moon line. The gravitational attraction of the Moon on these tidal bulges exerts a force that has a small component orthogonal to the line of centers of the bodies, and hence exerts a torque on Earth. The friction dissipates energy, which opposes the Earth's rotation and lengthens the day. The dissipative tidal force drops off with the cube of the Earth–Moon distance. But Newton assured us that every force has an equal and opposite force: the reaction force (the gravitational attraction of the Earth's asymmetric tidal bulge on the Moon) works to accelerate the Moon in its orbital motion about Earth. This causes the Moon to retreat slowly from Earth, which gives it an ever longer orbital period, which causes there to be ever fewer months per year.

These oceanic tides have a weaker counterpart within the body of the Earth. Because the Earth is viscous and elastic rather than completely rigid, these tidal forces also pump the crust up and down twice daily with an amplitude of about 1 cm. Thus even in the absence of oceans, evolution of Earth's spin and the Moon's orbit would continue into the indefinite future.

Conversely, if we extend this understanding of the Earth–Moon system back into the past, we see at once that the Moon formerly must have been closer to the Earth, and the month and the day must once have been very much shorter than they are now. Observations of eclipses reveal a small deceleration of the Earth's spin over historical times: if the present rate of energy dissipation is typical, and if the force is simply scaled with $r^{-3}$, the rate of energy dissipation must have been very much greater in the past when the Moon was closer.

The displacement of the tidal bulge by Earth's rotation is, on the average, in the plane of Earth's equator. Thus the effect of the tidal torque is to accelerate the Moon about Earth's rotational axis. Earlier in time, the Moon had less angular momentum about this axis, but the angular momentum associated with its out-of-plane motion (its orbital inclination) must have been essentially constant. Thus, at earlier times, ever larger proportions of the total orbital angular momentum of the Moon must have been out of plane. That is, the orbital inclination of the Moon must have been much greater in the past. Extrapolation back in time takes the Moon back to a close, highly inclined orbit, then to an even closer polar orbit, then back to a retrograde orbit of arbitrary inclination. If run even farther back in time, the Moon escapes from Earth. Reasonable (but untestable) calculations show that the time scale for evolution of the Moon's orbit is 1.5 or 2 billion years. This number is of course highly uncertain.

The obvious literal conclusion is that the Moon was captured by Earth from an independent heliocentric orbit, following a very slow retrograde first approach that dissipated enough energy to leave the Moon with

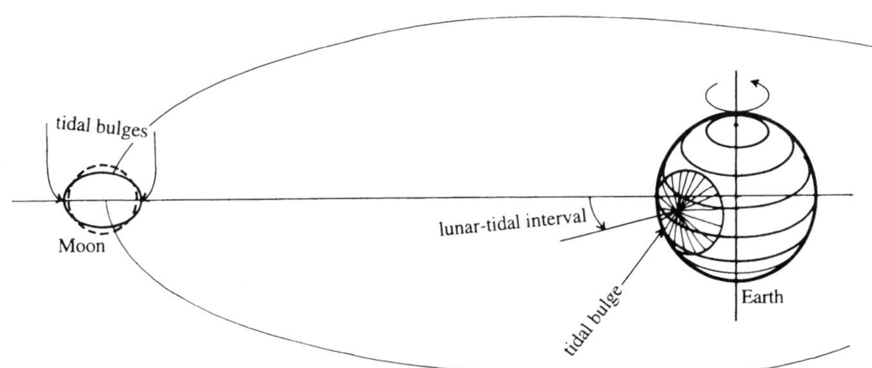


**Figure IX.27** Tidal interaction of Earth and Moon. The tidal bulges raised by Earth's gravity point on average directly at Earth, but the eccentricity of the lunar orbit causes the tidal bulge to librate about the sub-Earth point. The rotation of Earth is much faster than the orbital angular speed of the Moon, so the tidal bulge on Earth's surface always leads the sub-Moon point. Tides in the open ocean are on the order of 1 m in height, and solid-body tides in the crust have an amplitude of a few centimeters. The gravitational attraction of the tidal bulge on Earth always acts to accelerate the Moon in its orbit about Earth and causes it to retreat to more distant orbits with longer periods.

less than escape velocity. Such a perfect encounter appears very improbable, but at least has the virtue that it is not demonstrably impossible. A corollary of this scenario is that there must have been extremely intense internal energy dissipation inside both Earth and Moon within known geologic time. Estimates of the amount of energy dissipated are so large that both bodies should have been profoundly altered by the capture and close-approach events. Further, the oceanic tides raised at the time of closest approach are predicted to be so high (kilometers!) that the ocean would violently override the continents twice daily. The length of the day at that time is estimated to be about 9 hours. But examination of the geological history of the Earth and the ages of lunar rocks show no such event in the last 3.5 or 4 billion years. If this happened at all, it must have happened very early in the history of the Solar System.

## Origin and Internal Evolution of the Moon

Speculations regarding the origin and evolutionary history of the Moon have been rife since well before the Apollo program. Before any chemical data were available on the Moon there was considerable uncertainty about how large a Solar System body could be without melting and differentiating. The reasoning was simple. Consider a grossly chondritic body that formed directly out of the Solar Nebula and that, once formed, evolved as a closed system: the rate of production of heat by radioactive decay is proportional to the mass of the body (i.e., to $r^3$) and declines exponentially with time. The rate of heat loss by conduction through the body's crust is proportional to the surface area (i.e., to $r^2$). The larger the volume-to-surface ratio (i.e., the larger $r$), the higher the peak temperature reached in the interior. A body the size of the Moon, starting cold, with chondritic abundances of the long-lived radioactive elements, could possibly avoid extensive internal melting. It therefore seemed reasonable that the present surface of the Moon would have primitive chondritic composition, that no melting and differentiation had occurred, and hence that the Moon was the largest surviving primitive body in the Solar System, a veritable Rosetta stone of early Solar System processes. The craters on its surface must therefore be due to impacts, not internal volcanic processes. The observation that the rotational moment of inertia of the Moon was indistinguishable from that of a uniform sphere seemed to confirm the reasonableness of this theory.

But what of the calculations of the history of the lunar orbit that suggest that the Moon was very close to the Earth long after the origin of the Solar System? Might not the answer to our problem be that the Moon and the Earth were once a single body that split apart due to rotational instability? The Moon then could be (for example) the missing part of the Earth's crust and upper mantle from the Pacific Basin. But, dynamically, this fission process must always give rise to a Moon in the Earth's equatorial plane. The calculations on the early orbit of the Moon, contrary to this picture, suggest a very highly inclined orbit at the point of the Moon's closest approach to Earth.

The chemical composition of the lunar surface as determined by the Surveyor and Luna spacecraft and the Apollo missions is quite incompatible with this simple view. Indeed, the very observation that many meteorites had experienced melting and differentiation led to deserved skepticism: how could many of the meteorite parent bodies, which were presumably mere asteroids less than 1000 km in diameter, have melted without the much larger Moon (3476 km in diameter; more than 30 times the mass of Ceres) also melting thoroughly? But what heat source could possibly melt asteroids less than 1000 km in diameter? Long-lived radioactive decay of isotopes of potassium, uranium, and thorium simply could not do it. Attention focused on alternative heat sources in the early Solar System, including short-lived radionuclides, especially $^{26}Al$, and induction heating by the early T Tauri phase solar wind. But if these mechanisms could melt many small bodies in the asteroid belt, why would they not melt the Moon, which is both larger (a better conserver of radiogenic heat) and closer to the Sun (subjected to a much more intense T Tauri solar wind)? And if the Moon melted, where did its core go? The moment of inertia and the seismic data both give an upper limit on the mass of the core of about 2% of the Moon's total mass. Is it possible that the early Moon was so oxidized that it not only had the right density, but also had the ferrous metals fully oxidized? Our understanding of the theory of chemical processes in the Solar Nebula (Chapter IV), as well as our practical experience with meteorites (Chapter VIII), lead us to associate complete oxidation of metal with significant retention of water in phyllosilicates. If so, a Moon that has a much higher FeO content than Earth should also contain large amounts of water.

This argument puts the geochemical evidence for severe volatile- and siderophile-element depletion and great enrichment of refractory elements in the forefront. The significant elevation of FeO in the Moon above terrestrial norms also takes on new significance. Briefly, refractory element enrichment is *prima facie* evidence for high-temperature condensation of lunar material. But there is no clear evidence that the entire Moon has an unusual amount of refractories: they may simply be efficiently partitioned into basalts by internal processes.

The elevated FeO abundance suggests a significantly lower formation temperature than Earth's. The severe depletion of volatiles, including the alkali metals, requires high-temperature condensation or later severe baking, and the absence of a detectable core combined with the severe depletion of siderophiles tells us that melting and differentiation occurred: the failure to detect a core would mean that it was present but too small to detect (requiring a very small mass fraction of free metals, and hence low condensation temperatures), or that the core was somehow "stolen."

The simplest way (at least, "simplest" within the elastic rules of this extravagant intellectual game) to differentiate the Moon and steal its core is to make the Moon by rotational fission of Earth, leaving the core material inside Earth; but we have seen that the dynamical objections to this scenario are powerful.

If fission and capture are such unconvincing explanations of the origin and nature of the Moon, what other possibilities are there? Early in the debate it was sometimes suggested that the Earth and Moon simply accreted together as a binary system, a double planet. If this were to happen, we surely would expect that the material falling on both Earth and Moon would be essentially identical. Why then would there be such a large density difference between these bodies? From our post-Apollo perspective, we must also ask why such a Moon should have a higher FeO content than Earth. The answer is far from obvious.

The last of the pre-Apollo theories postulates yet another very specific link between the formation of the Moon and Earth: during the formation of the Earth, very high temperatures were developed by the kinetic energy of the infalling material, sufficient to generate a huge silicate vapor atmosphere. Cooling and shrinkage of the hot atmosphere left behind an equatorial condensate ring, analogous to the ring system of Saturn, from which the Moon eventually accreted. This scenario then explicitly derives the Moon from the most volatile materials of Earth. How we end up with a Moon that is at least as refractory as Earth, and much poorer in volatiles, is far from obvious. Also, of course, this scheme shares with the fission hypothesis the problem of the inclination of the Moon's orbit.

The great intellectual triumph of the post-Apollo years is the invention of a fifth hypothesis for the origin of the Moon. In this scenario, the early Earth orbits alone late in the accretionary era, when it is struck by a planet of approximately the mass of Mars. Most of the mass of the impactor is captured into Earth, which, if not already molten, is thoroughly melted by the impact energy. The core of the impactor sinks into Earth's core, liberating additional gravitational potential energy. Some of the impactor splashes out and reaches escape velocity, and some is left in a variety of eccentric orbits about the Earth. This material is mostly very hot and is derived from the material of Earth's upper mantle as well as the material of the impactor. Thus we have capture, fission, two planets, and a sediment ring all wedded to a giant impact. Everyone ought to be happy. But not everyone is happy. So messy is this scenario, and so poorly constrained by either chemistry or physics, that it is quite impossible to demonstrate that it has unique explanatory virtues. A more eclectic, or less testable, theory is hard to imagine. To postulate two planets, of unknown initial composition, already differentiated to an unknown degree, fragmented and mixed in an unknown manner, and heated to such high temperatures that all the chemistry of the Moon (at least) is completely reworked, scarcely seems a tractable approach—and yet it is the best we have.

Even if we declare the giant-impact theory triumphant, the derivation of lunar rocks from primitive materials still presents problems. Many petrologists, especially in the immediate wake of the Apollo expeditions, have described scenarios in which the outer layers of the Moon to a depth of 300 to 400 km were completely molten, constituting a "magma ocean" with about half the mass of the entire Moon. The most extreme versions of this hypothesis postulate depths of the magma ocean ranging from 100 to 1700 km. Cooling of this putative ocean of magma leads to the early crystallization of dense, iron-rich olivine, which settles to the bottom of the melted region to form olivine-rich "cumulates." Low-density plagioclase, carrying incompatible elements, floats to the top and forms an anorthite-rich first crust, which incorporates magnesium-rich pyroxenes in its deeper regions. Convection currents in the cooling mantle may then deliver floats of plagioclase-rich crust above downwelling regions to form "rockbergs." Later versions of this model, while still preserving the picture of an early global magma ocean, have emphasized the derivation of mare basalts from the sunken olivine-rich cumulates.

The existence of a magma ocean has been proposed to explain why a nearly global anorthosite-rich crust is present, and why the *relative* abundances of the incompatible elements in late-crystallizing rocks are nearly identical regardless of location: these phenomena would come about naturally as a consequence of cooling of a homogeneous global magma reservoir. But D. Walker has pointed out that the essential element of these explanations is global homogeneity, not simultaneous global melting. He has devised a scenario in which numerous sequential episodes of melting generate upwardly mobile plagioclase-rich melts and dense deposits of olivine-rich cumulates that sink to the floors of magma chambers. This "serial magmatism" hypothesis not only generates the gross chemical features of the lunar igneous rocks, but even reproduces the REE abundance patterns, including the Eu anomaly. Thus the very issue of whether a global magma ocean was ever present stands in doubt: in effect, Walker has pointed out a route to reducing the depth of the ocean to zero. Still, the magnitude of the fractionation factors and the mass of the crust require at least a few hundred kilometers of the outer Moon to have been molten at some early time (not necessarily the same time).

The interpretation of the REE pattern is based on the preferential partitioning of Eu (relative to the other REEs) into plagioclase, with its consequent enrichment in the highland crust. Very late melt residues (especially KREEP basalts), those with the highest concentrations of incompatible elements, have enormous overall REE enrichments and enormous negative Eu anomalies, evidently due to efficient extraction of Eu-rich plagioclase in the course of crystallization. Of course the major-element abundances tell the same tale. The cumulate region, whether a single layer of global extent or numerous regional layers, is depleted in plagioclase and in all REEs, with a modest negative Eu anomaly complementary to but smaller than the Eu enhancement in the anorthositic crust. The role of the cumulate layer as the source region of mare basalts is indicated by the basalts' inheritance of the negative Eu anomaly. Thus, in any model, plagioclase flotation must precede basalt production. As we have seen, the earliest melting must date back to at least 4.4 Ga.

## Solar Wind Interaction with the Moon and Mercury

Surface magnetic fields on both the Moon and Mercury are weak. Except for small local "bubbles" in which the field offers local partial protection, it is likely that the solar wind strikes the surface of the Sunward side of the Moon unimpeded. Protons and other ions in the solar wind strike the lunar regolith at several hundred kilometers per second and are implanted in the exposed surfaces of grains. The solar wind magnetic field sweeps through the Moon, but the ions do not. In the lee of the Moon there is a cavity, carrying slightly disturbed solar wind magnetic fields, but no ions. Protons and electrons cycling about magnetic lines of force in the solar wind eventually diffuse back into the swept region, closing the plasma cavity far downwind from the Moon.

A similar phenomenon presumably occurs on Mercury, but with the added complication that a planetary magnetic field does exist and has sufficient strength to stand off the solar wind slightly from the surface. The geometry of the three Mariner 10 flybys of Mercury is such that the spacecraft was targeted to photograph the dayside of the planet, and hence barely crossed the magnetospheric boundary, but did twice pass through the plasma cavity.

The Mariner 10 magnetometer observations show a field extending out to about 1.5 Mercury radii on the dayside. The maximum field strength was reported as about 200 nT ($1\ \text{Tesla} = 10^4\ \text{G}; 1\ \text{nT} = 10^{-9}\ \text{T} = 10^{-5}\ \text{G} = 1\gamma$). The integrated magnetic field strength is variously estimated at 200 to 400 nT $R_M^3$ ($2$ to $3 \times 10^{-3}\ \text{G}\ R_M^3$), depending on the method of analysis of the data. Essentially, there is insufficient three-dimensional coverage of the magnetosphere to permit a firm acceptance or rejection of any of the calculated values. Nonetheless, the presence of a weak but unambiguously detected dipole field requires the existence of some internal mechanism for generation of the field. A core dynamo, the most likely possibility, requires present-day fluid motions in Mercury's core. Since such a small planet has difficulty maintaining high internal temperatures, it is easiest to conclude that some impurity in the core, such as sulfur, is responsible for producing a low-temperature eutectic melt.

The interaction of planetary surfaces with the solar wind entails more than simple implantation of solar ions. For example, $^{40}$K on the Moon decays to form $^{40}$Ca and $^{40}$Ar. The radiogenic argon diffuses slowly out of the regolith in response to solar heating, and neutral argon atoms bounce about on ballistic trajectories until they are photoionized by solar UV radiation. The $Ar^+$ ions are then trapped by the solar wind magnetic field, which flows almost unimpeded through the Moon and is heading toward the lunar surface on the dayside. These ions are then slammed into the surface at solar wind speeds of 300 km $s^{-1}$ or more, to become embedded along with the solar wind gases. The regolith on the lunar surface contains a substantial excess of $^{40}$Ar over the amount attributable to decay of potassium within the regolith grains, and this gas is physically correlated with grain surfaces.

The nature of the embedded gases in the lunar regolith, and their compositional differences from those in the lunar rocks, is made clear in Table IX.3. The rare gas abundances in the regolith reflect the solar abundances quite well. Note that the $^{40}$Ar abundance in the fine regolith particles is a factor of 10 higher than in the regolith's parent rocks. All the other rare gas isotopes are enhanced by at least a factor of 100 by implantation of solar gases (which of course contain very little radiogenic argon). In the most extreme cases, $^3$He concentrations can reach $10^{-8}$ per gram of regolith. Curiously, this is the most concentrated source of $^3$He in the inner Solar System. Few lunar materials are potentially of practical

**Table IX.3** Noble Gases in Lunar Regolith Fines and an Igneous Rock

| | Gas concentration ($cm^3\,g^{-1}$ STP) | | | | | | | | |
|---|---|---|---|---|---|---|---|---|---|
| **Sample** | $^{3}$He | $^{4}$He | $^{20}$Ne | $^{22}$Ne | $^{36}$Ar | $^{38}$Ar | $^{40}$Ar | $^{84}$Kr | $^{132}$Xe |
| Fines | 7.5E-5 | 1.9E-1 | 3.1E-3 | 2.4E-4 | 3.4E-4 | 6.6E-5 | 4.1E-4 | 2.0E-7 | 4.1E-8 |
| 10058[a] | 5.8E-7 | 2.1E-4 | 6.4E-7 | 1.2E-7 | 1.5E-7 | 1.2E-7 | 3.7E-5 | 1.5E-10 | 7.0E-11 |

[a]Apollo 11 rock number 10058.

interest, of which $^{3}$He is probably at the top of the list: the controlled fusion reaction of $^{3}$He with D to make $^{4}$He and H is unusually attractive as a source of power because it produces no neutrons, and hence no radioactive waste. Further, all the energy of reaction is carried off by the energetic charged products, from which the energy can readily be recaptured. Deuterium is an inexpensive commodity on Earth, where it is extracted commercially from seawater, but $^{3}$He is very rare on Earth (indeed, its largest and most concentrated terrestrial source is from the decay of tritium in thermonuclear warheads). A second interesting lunar resource is hydrogen (or water), whose presence on the Moon is due to solar wind bombardment and asteroid and comet infall. It is ironic that the most attractive materials on the metal-poor, volatile-poor Moon are not native to the lunar environment!

## The Planet Mercury

Mercury is the high-temperature end member of the Solar System; at least, it is presently by far the closest planet to the Sun. What do we truly know and understand about its nature?

Because of Mercury's small size and angular proximity to the Sun (always less than about 28°), it is never high in the sky unless the Sun is above the horizon, and never observable in a dark sky. As a result of this limitation imposed by orbital geometry, Earth-based astronomical studies of Mercury are very difficult. Very little was learned about its intrinsic properties before the advent of planetary spacecraft. (The history of spacecraft exploration of Mercury, the Moon, Phobos, and Deimos is summarized in Appendix III.) The overwhelming majority of what we know about Mercury was contributed by the American Mariner 10 mission in 1973–1975. Aside from the three encounters of Mariner 10, no other spacecraft has ever been sent to Mercury. The photographic evidence from Mariner 10 showed a landscape very similar to that of the Moon, heavily cratered, with volcanic and tectonic modification, and wholly devoid of evidence for the former presence of an atmosphere or oceans. This superficial similarity contrasts with major geophysical differences, most notably Mercury's much higher density.

## Motions of Mercury

Mercury orbits the Sun as the innermost planet ($a = 0.387$ AU), with the shortest orbital period of the known members of the Solar System ($P = 86$ days). We have seen that its orbit is quite eccentric ($e = 0.206$), taking it from a perihelion distance of $q = a(1 - e) = 0.387(1 - 0.206) = 0.307$ AU to an aphelion distance of $Q = a(1 + e) = 0.467$ AU. The intensity of solar radiation falling upon Mercury therefore varies by a factor of $(0.467/0.307)^2 = 2.314$. The standard for measuring the intensity of sunlight is the Solar Constant, $F_{\odot}$, %, the mean intensity of sunlight at a distance of 1 AU from the Sun, unattenuated by passage through an atmosphere. The Solar Constant is close to 1370 $\mathrm{W m^{-2}}$. The intensity of sunlight experienced by Mercury ranges between 4.59 and 10.61 times the Solar Constant.

Radar observations, including measurement of the Doppler width of reflected radar pulses, disclosed that Mercury rotates on its axis, which is closely normal to the plane of its orbit, with a period of 57.3 days, $\frac{2}{3}$ of the orbital period. Mercury is locked in a 3:2 spin–orbit resonance, rotating relative to the stars exactly three times during two orbits about the Sun. Thus the same region on the surface of Mercury faces the Sun on alternate perihelion passages. Because solid planets are never perfect spheres, we are justified in concluding that Mercury's figure is slightly elongated along the Sun–planet line at the time of perihelion passage, when the solar tidal forces are at their greatest, and that alternate ends of the long axis of the planet point at the Sun at consecutive perihelion passages. Tidal torques exerted on Mercury's oblate figure by the Sun drop off in intensity with the cube of the distance of Mercury from the Sun and hence vary by a factor of $(0.467/0.307)^3 = 4.52$ between perihelion and aphelion.

The peculiar spin–orbit resonance leads to a most unusual distribution of solar heating on the Mercurian surface. There are two "hot poles" on the equator at opposite ends of the long axis. Other locations on the equator are never exposed to an overhead Sun near the time of perihelion passage. The average angular rate of motion of Mercury due to its orbital motion is $360/86 = 4.18°$ per (standard Earth) day as seen from the Sun. This rate (called the *proper motion*) is the same as the average rate of progress of the Sun through the zodiac as seen from the surface of Mercury. The angular rate of rotation of Mercury is $\frac{3}{2}$ times as large, 6.27° per (Earth) day. But the rotation rate is constant, whereas the rate of orbital motion varies greatly from perihelion to aphelion. The apparent rate of motion of the Sun in the sky as seen from the surface of Mercury is the difference of these two rates. However, near the time of perihelion passage the angular rate of movement of Mercury along its orbit is greater than its average value for two reasons: it is traveling at a higher speed because it is deeper in the Sun's gravitational field, and it is closer to the Sun. In general, the angular rate is proportional to $v/r$, where $v$ varies according to the vis viva equation (conservation of energy),

$$v = \left[\mathrm{GM}\left(\frac{2}{r} - \frac{1}{a}\right)\right]^{1/2}, \qquad \text{(VII.4)}$$

and $r$ varies from $q$ to $Q$. The orbital angular rate (as seen from the Sun) varies from 2.81° per day at aphelion to 6.50° per day at perihelion.

The apparent angular rate of the Sun across the sky as seen from the surface of Mercury, taking into account both the (constant) rotation rate and the (variable) orbital motion is $6.27 - 2.81 = 3.46°$ per day compared with the average value of $6.27 - 4.18 = 2.09°$ per day. But at perihelion the apparent rate of motion of the Sun is $6.27 - 6.50 = -0.23°$ per day. The very small negative value means that the Sun not only stops its normal east-to-west progress across the sky, but also actually temporarily reverses its apparent direction (we would say that the Sun "retrogrades" at this time) and drifts slightly eastward before resuming its westward motion shortly after perihelion. The average angular rate of the Sun over the time around perihelion passage (when the solar tidal torques are the greatest) is zero; the permanent equatorial bulge of Mercury is locked firmly onto the Sun during the time when the tidal forces are greatest. A point on the equator 90° east of the perihelion subsolar point would see the Sun progress across the afternoon sky at a steadily decreasing rate, getting constantly larger and brighter as sunset approaches, and set just before perihelion. The swollen Sun would then briefly rise just above the horizon at the time of perihelion passage and set again immediately thereafter! A point on the equator 90° west of the perihelion subsolar point would see the brilliant Sun rise sluggishly above the horizon at dawn, halt and set again, and then rise a second time to accelerate across the sky, becoming ever smaller and fainter as noon approaches!

In the polar regions of Mercury there may be permanently illuminated mountaintops and permanently shadowed crater basins. The angular diameter of the Sun as seen from Mercury varies from about 64 arc minutes at aphelion to about 98 arc minutes at perihelion, compared with about 30 arc minutes as seen from Earth. At the exact pole (assuming zero axial tilt) and with a flat horizon, the horizon would bisect the solar disk, leaving an effective solar intensity of 2.3 to 5.3 times the solar constant on surfaces normal to the direction of sunlight (i.e., vertical surfaces). The ground would on average be very cold because of the grazing incidence of the sunlight, but steep slopes (and solar collectors on spacecraft!) could easily be maintained near 500 K.

## Composition and Structure of Mercury

No spacecraft has ever orbited about or landed on Mercury, Phobos, or Deimos. All of our meager compositional data on these bodies are derived from a very limited body of remote-sensing observations from Earth or from a few spacecraft flyby missions. In comparison with these bodies, the Moon is extremely well studied. Not only have many spacecraft orbited the Moon, but also a number of others have carried out one-way landing missions, several manned and unmanned roving vehicles have roamed its surface, and a few missions have brought back samples collected on the lunar surface (Appendix III).

The most direct single observation bearing on the surface composition of Mercury is the marginal detection of an absorption feature in the 0.9-$\mu$m region, where ferrous iron in silicate minerals absorbs. Unfortunately, observations by the same observer under essentially identical conditions sometimes show a barely detectable trace (a few percent) of ferrous iron and sometimes do not. The strongest reliable conclusion is that the overall spectrum of Mercury is almost featureless and quite similar in slope to that of the lunar highlands. Information on the spatial variation of Mercury's surface composition is almost impossible to obtain from Earth; Mercury is lost in the Sun's glare almost all of the time. Even at the times of its maximum angular separation from the Sun (eastern and western elongation) it is generally visible only in an incompletely dark sky and through large masses of terrestrial air.

Besides the bulk density and the marginal detection of FeO, almost no information on the composition of Mercury exists. Tiny amounts of atomic sodium and potassium vapor have been seen in the atmosphere of

Mercury, but it is simply impossible to deduce the concentration of these elements in the planetary surface from observations of the trace of atmosphere. Ann Sprague and co-workers have searched spectra of the surface of Mercury in the mid infrared region (8.1 to 12.5 $\mu$m) for the lattice bands of surface minerals with somewhat surprising results: they report evidence for the presence of bronzite (see Fig. IV.20), a pyroxene that contains 10 to 20% of the ferrosilite ($FeSiO_3$) component, along with the sodic feldspar albite, $NaAlSi_3O_8$, and sodalite, $Na_4Al_3Si_3O_{12}Cl$. Bronzite is unexpected because of its significant FeO content on a planet that generally shows no evidence of FeO. Albite is normally found in very silica-rich rocks such as granites, in solid solution with anorthite as plagioclase feldspar. The presence of albite requires a high degree of differentiation. Sodalite, whose formula can be rewritten $3NaAlSiO_4 \cdot NaCl$, is found in silica-poor leucite-bearing rocks saturated with sodium chloride, a mineral that could be made by baking an *evaporite* deposit left behind by a vanished ocean in contact with undifferentiated silica-poor rocks. That Na and Cl minerals should be common enough to detect on the much hotter Mercury than they are on the Moon, where they are severely depleted, is most surprising of all.

In the absence of evidence demonstrating the existence of a differentiated crust on Mercury, it is difficult to be sure whether the planet has in fact undergone extensive internal melting, density-dependent phase separation, and formation of a crust, mantle, and core. Thermal history models show that the planet can readily melt and differentiate if a significant portion of the accretion energy is stored internally. Early extensive melting is also abetted by the presence of even small amounts of FeS, which greatly lowers the temperature of first melting of core-forming material through its eutectic interaction with metallic iron–nickel alloy. Of course, because no spacecraft has ever landed upon or orbited about Mercury, geophysical data on the internal structure are as limited as the geochemical evidence described above. Only one piece of evidence seems to be relevant: the magnetometer experiment on Mariner 10 detected an apparent small planetary dipole field.

## Noncrater Geology of Mercury

Mercury's reputation as a heavily cratered planet will not be seriously endangered if we briefly survey the geological features that are not directly due to normal cratering events. As far back as we can see in Mercury's tectonic history the planet has had a very thick, rigid crust. Global-scale features, and features caused by global-scale processes, are rare. A few of these features are related to "supernormal" cratering events, especially that which excavated the huge Caloris Basin (Fig. IX.28) near one of the hot poles. This multiringed structure was formed by an event so violent that its effects are visible all over the surface of the planet. Inside the basin itself are grabens (tensional block-faulted depressions) due to readjustments of the crustal load following the impact.

One surprising class of features discovered in the Mariner 10 images is the globally distributed system of lobate scarps (see Fig. IX.29). These scarps may in fact be related to the Caloris event, because some evidence suggests that they are preferentially oriented radially to the Caloris Basin. Whatever their cause, these faults are compressional thrust faults that, over the circumference of the planet, suggest about a 4-km shrinkage in the diameter of Mercury. Two causes for this shrinkage suggest themselves: tidal despinning, which causes the equatorial bulge to subside while the polar diameter increases, and planetary cooling, which places the entire surface in nearly uniform compression. The expected global uniformity of the latter is, however, significantly tempered by the steep gradient in mean surface temperature between the equator and poles. Internal isotherms must be much closer to the surface near the hot equator than near the poles, so that a uniformly applied stress might be reflected by very uneven evidence of strain. But in either case, the most heavily stressed region, and the first to fail in response to these stresses, is the equatorial region.

A variety of linear features also form a coherent planet-wide pattern. Numerous linear scarps, ridges, and troughs are found to have a preferred orientation, favoring the NW-to-SE and SW-to-NE directions. These reflect the change in the pattern of stresses attendant upon tidal despinning. Another unusual feature is the "weird terrain," a large area that shows irregular hills and depressions, which is directly antipodal to the Caloris Basin and was probably produced by the transmitted shock as modified and focused by passage through Mercury's core.

## Geophysics of Mercury

Temperatures on the surface of Mercury vary in a complex manner that reflects the peculiar variation of solar illumination imposed by the spin–orbit resonance and the orbital eccentricity. Daytime temperatures near the subsolar point at perihelion reach about 700 K, but near aphelion the lower solar flux maintains a temperature of only 550 K. Nighttime temperatures drop to

**Figure IX.28** The Caloris Basin on Mercury. Rings surrounding the basin and the heavy ejecta blanket from the basin formation event are clearly visible.

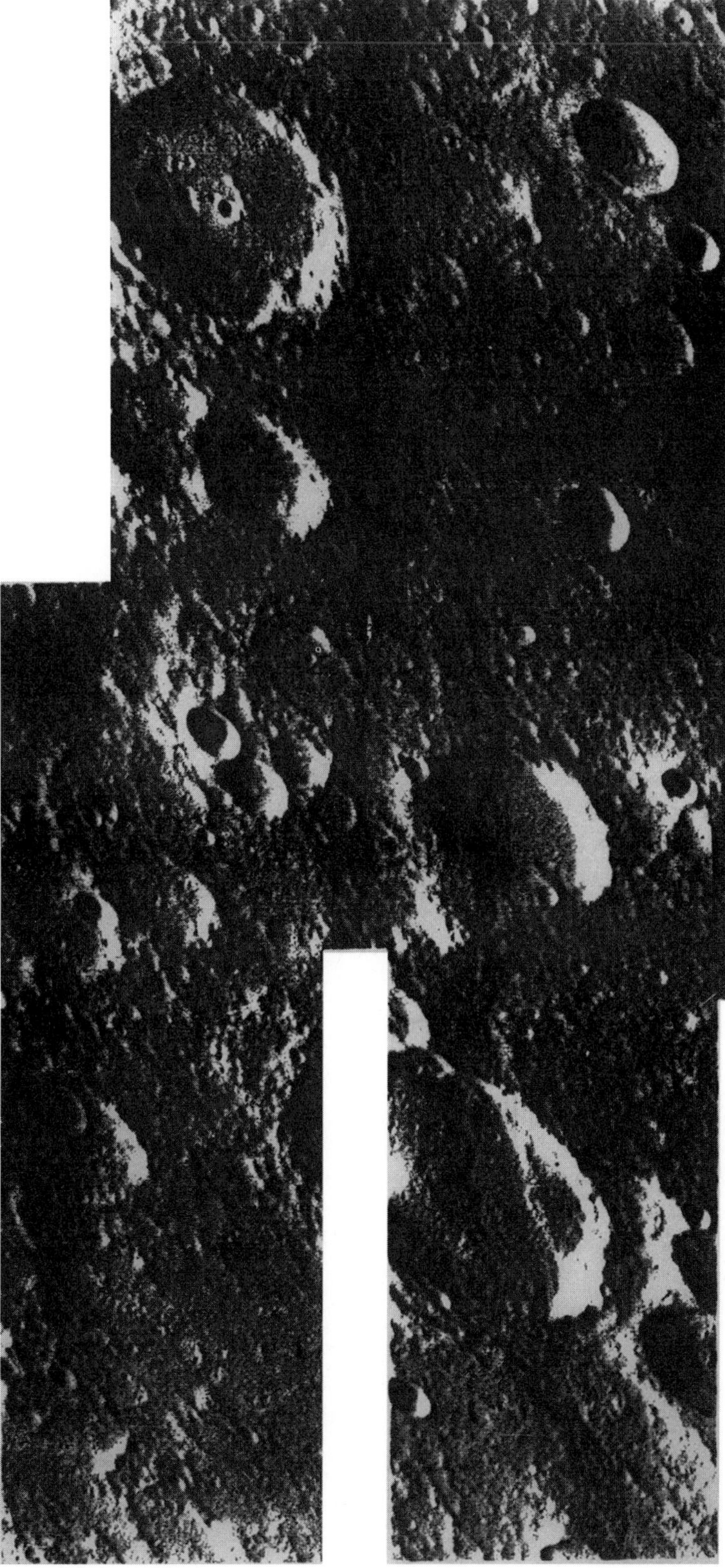

**Figure IX.29** The Discovery Scarp on Mercury. This Mariner 10 image shows one of a global system of unconnected compression features caused by global cooling and, in the equatorial region, by tidal despinning.

about 100 K, low enough to serve as a very effective but transient cold trap.

The internal temperatures of Mercury remain quite unknown; indeed, there is no proof that Mercury has differentiated and formed a core. In addition, thermal models of Mercury cannot be constrained by heat flow data, because the crustal heat flux has never been measured. Nonetheless, a number of plausible conclusions regarding the thermal evolution and present internal structure can be made on the basis of general principles.

First, because of its proximity to the Sun, the material at Mercury's orbit should be at a rather elevated temperature even before accretion. Models of the Solar Nebula suggest temperatures on the order of 1400 K at 0.4 AU during the lifetime of the nebula, a few times $10^5$ years, which is probably too short a time for accretion. Second, during the next few million years the surface temperature will be governed by the (constantly changing) luminosity of the T Tauri phase of the Sun. Third, the internal temperatures will be enhanced by T Tauri phase solar wind heating. This electromagnetic induction heating will probably be deposited at relatively shallow depths in Mercury because its initial composition must have been very rich in metallic iron. Fourth, the energy of accretion of Mercury is available to raise internal temperatures during and after accretion, a time period of about $10^7$ a. Fifth, the gravitational potential energy released by the sinking of dense core material also becomes available once melting begins. The question is whether these heating mechanisms can lead to the initiation of melting of a major component such as iron or ferromagnesian silicates. Of course, our knowledge of the phase composition of the primitive, undifferentiated planet is very uncertain. Models of condensation and accretion suggest a sulfur-poor interior, but even a fraction of a percent of sulfur can greatly assist in core formation via generation of an Fe–FeS eutectic melt.

We cannot hope to quantify all of these processes at our present level of knowledge, but a few simple arguments may relieve us of the necessity of doing an exhaustive treatment. First, the mean luminosity of a T Tauri phase Sun is about $2L_\odot$, and the mean surface temperature of an airless body at Mercury's distance of 0.4 AU must be given by the radiative balance

$$2F_\odot(1 - A)\pi r^2 = 4\pi r^2\sigma T^4 \qquad \text{(IX.19)}$$

or

$$T = [F_\odot(1 - A)/2\sigma]^{1/4}, \qquad \text{(IX.20)}$$

where $2F_\odot$, twice the present solar flux at Mercury, is $1.69 \times 10^7$ erg cm$^{-2}$ s$^{-1}$, $A$ is the Bond albedo, $(1 - A)$ is about 0.92, and $\sigma$ is the Stefan–Boltzmann constant $(5.67 \times 10^{-5})$. The effective temperature of Mercury is then about 740 K. At perihelion ($q = 0.30750$ AU) the mean effective temperature over the entire surface would have been 850 K, but, since the rotation period is comparable to the orbital period, this mean has little significance. The peak equatorial daytime temperature near perihelion, neglecting "solar furnace" effects associated with heat collection in craters, would be higher by a factor of $2^{1/2}$, an astonishing 1200 K. The Fe–FeS eutectic temperature is only 1261 K (1210 K for the Fe–FeS–FeO eutectic), so a temperature rise of about 520° will permit large-scale melting.

The gravitational binding energy of Mercury, which is the minimum amount of energy produced during accretion (it neglects the kinetic energy of materials that enter at speeds above escape velocity), is given by

$$E = -(16/9)\pi^2\rho^2 Gr^5, \qquad \text{(IV.9)}$$

which is $3 \times 10^{37}$ erg. If all this energy is stored internally and spread uniformly over Mercury's mass of $3.30 \times 10^{26}$ g, with an average heat capacity of $7.5 \times 10^6$ (for a mixture of two parts metallic iron with one part magnesium silicates at about 800 K), the temperature rise of the planet would be about 15,000°, certainly high enough to ensure melting! Less than perfect efficiency in retaining the accretion energy would be more realistic, but so would excess kinetic energies of the accreting fragments. Furthermore, if giant collisions have eroded much of the mass of silicate material from Mercury's mantle and crust, the kinetic energy input from their impacts must have been much larger than the accretion energy of the planet as it is today. Another important consideration is that the accretion energy is not provided uniformly throughout the interior, but is concentrated toward the surface: late-accreting fragments experience a much higher acceleration of gravity, and impact at much higher speed, than the bodies that form the deep interior. The kinetic energy density of the impactors varies as $v^2$, and the minimum accretion energy case has $v$ equal to the escape velocity, $(2Gm/r)^{1/2}$. For a growing, homogeneous, incompressible planet, the instantaneous accretion heating then varies as

$$E(r) = 4\pi\rho Gr^2/3. \qquad \text{(IX.21)}$$

Thus, neglecting the excess velocities of the accreting fragments above Mercury's instantaneous escape velocity, the severity of heating should increase with $r^2$. Melting should begin at relatively shallow depths.

Once melting of the homogeneous planet begins, sinking of dense metal to the core ($\rho_c \simeq 7.5$) and rising of mantle silicate rocks ($\rho_m \simeq 3.5$) cause an internal release of gravitational potential energy,

$$\Delta E = -GM_c^2(1 - \rho_c/\rho_m)^2[1/r_c - 1/r_s], \qquad \text{(IX.22)}$$

where $r_s$ is the radius of the planet and $r_c$ is the final radius of the metallic core ($0.75r_s$). The core mass, $M_c$, is taken as 65% of the mass of the planet. The energy released by differentiation is then $1.3 \times 10^{36}$ erg, which is enough to raise the temperature of the entire planet by 500 K. Thus melting, once begun, should accelerate itself to completion. Given all these considerations, it is natural to conclude that Mercury is a thoroughly differentiated planet. The presence of a massive core then would be inevitable, and the generation of a planetary magnetic field would at least be plausible, if not inevitable.

Early formation of a pure-metal (Fe–Ni) core would have led to effective heat transport by both conduction and convection within the core (which fills about 75% of the planetary radius), and the subsequent thermal history of Mercury would have been dominated by rapid cooling, freezing of the core, and loss of any mechanism for generating a planetary dipole field through convective motions in the core. If a small proportion of a fluxing agent such as FeS is present, most of the core could freeze to produce a massive, solid, and dense inner core of Fe–Ni, surrounded by a thin liquid shell of Fe–FeS melt. Such a relatively cool eutectic melt layer would have a lower electrical conductivity than pure metal, but would be much more easily maintained in the molten state by, for example, radioactive decay or tidal heating. In any case, freezing of most of the core appears unavoidable. The $\Delta V$ of the phase change, combined with the general shrinkage of the cooling planet, leads to compressive stresses that afford a natural explanation of the tectonic structure of the crust. A somewhat simplified structural model of Mercury as it is today is given in Fig. IX.30.

The presence and nature of a magnetosphere on Mercury is clearly of great importance in constraining the interior structure of the planet, and the only source of data on Mercury's magnetic field and magnetosphere is the Mariner 10 mission. Figure IX.31 shows the geometry of the three Mariner 10 flybys of Mercury. Ideally, the structure of a magnetosphere is determined by one or more orbiting spacecraft in highly eccentric orbits that pass through the magnetosphere many times and map out the field strengths and the fluxes, energies, and directions of motion of trapped charged particles. Passing very close to the planetary surface on the dayside and examining the structure of the nightside magnetosphere out to great distances are both essential parts of this process. But the needs of the imaging experiment on Mariner 10 were incompatible: with only two or three opportunities to fly by the planet; mapping as much of the surface as possible required that the approaches to Mercury be on the dayside and high enough so that spacecraft motion would not smear the precious mapping imagery. Since the orbital

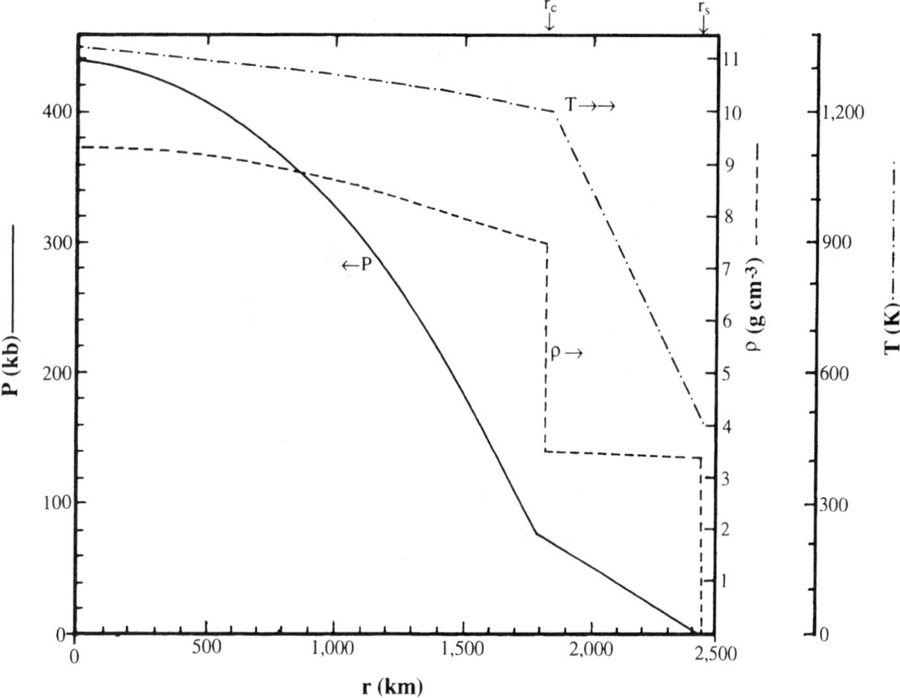


**Figure IX.30** Model of the interior of Mercury. Profiles of pressure, temperature, and density are drawn according to the general wisdom, which reflects much intelligent speculation and little direct data. Note the immense size of Mercury's core.

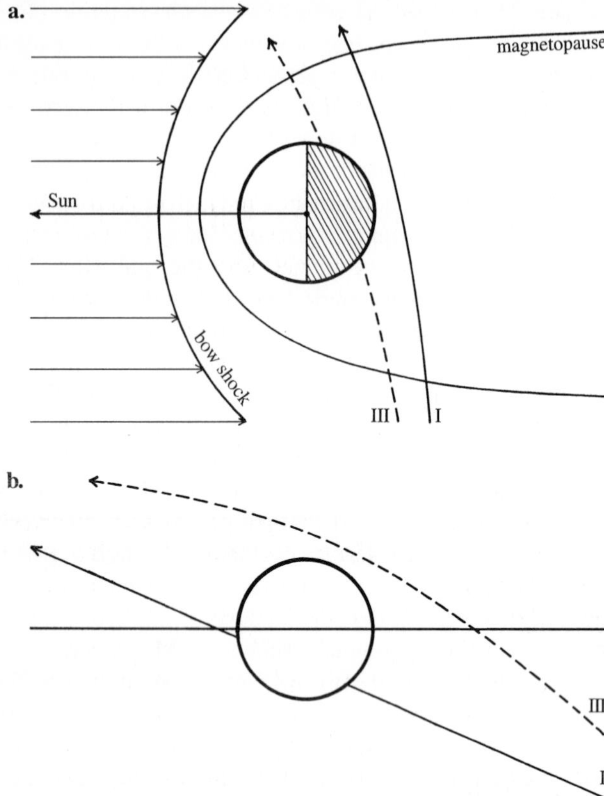


**Figure IX.31** Mariner 10 Mercury flyby geometry. The first and third Mercury flybys by Mariner 10 (29 March 1974 and 16 March 1975) passed close to the planet and penetrated the magnetosphere. a shows the flyby trajectories as projected onto the plane of the ecliptic. b shows the view of these trajectories as seen from the Sun. The solar wind flow is offset about $8^b$ from the radial direction by the high orbital velocity of Mercury.

period of Mariner 10 around the Sun was 176 days, it intercepted Mercury at intervals of 2 Mercury years (3 Mercury rotation periods), at which times Mercury presented exactly the same hemisphere to the Sun, ruling out both photographic mapping of one entire hemisphere and in-depth exploration of the magnetosphere. Nonetheless, data collected by the magnetometer while passing through the bow shock strongly suggest the presence of a weak planetary field. The field may be pressed down to ground level by the high solar wind pressure, creating a "bald spot" in the planetary field on the dayside, but it extends out into space and defines a small magnetosphere on the nightside. Dayside field lines originating at $>50°$ latitude in both hemispheres are swept into the magnetotail and remain open. The origin of the field seems to require a weak core dynamo. That a planet rotating so slowly may possess a magnetic field is moderately surprising; that it might do so without a core would be completely astonishing.

## Atmospheres of Mercury and the Moon

Given that this chapter is devoted to "airless" planets, the inclusion of a discussion of atmospheres may seem gratuitous. But in fact both Mercury and the Moon have extremely tenuous, essentially collisionless traces of atmosphere (Table IX.4). Both planets are bombarded by the solar wind, and some proportion of the hydrogen and helium ions that reach their surfaces will undergo charge-exchange reactions and emerge as neutral atoms. Mercury's magnetosphere is large enough to protect the surface rather well from direct solar wind impact, but the solar wind gases pervade and constantly flush out the magnetosphere and are in good supply at the surface. Helium atoms are in fact observed on both the Moon and Mercury in amounts that suggest a steady-state balance between solar wind capture and escape. Hydrogen is likewise seen on Mercury, and the observational upper limit on the H atom abundance on the Moon is close to what we would expect based on experience with Mercury.

The sources of Na and K in the atmospheres of the Moon and Mercury are presumed to be sputtering (ejection by impact of energetic ions) from surface minerals, but the mechanism remains obscure. The Na and K contents of the surface rocks on the Moon are very low, and, despite the reported detection of albite and sodalite by Sprague et al., nothing is known regarding their elemental abundances on the surface of Mercury. It is likewise unknown whether a meteoritic source of the alkalis might suffice even in the absence of feldspars or other alkali-bearing minerals. One curious feature of the observations is that the Na:K ratio on Mercury is about 60:1, but only about 4:1 on the Moon. Temperatures near the subsolar point on Mercury are high enough (575 to 700 K) so that diffusion and vaporization of alkali metals from plausible silicate sources are conceivable. On the other hand, the nightside temperatures (110 K at midnight) permit efficient trapping of any hot dayside atoms that land there, at least until local sunrise. Permanent cold traps near the pole could serve as one-

**Table IX.4** Atmospheric gases on Mercury, Io, and the Moon

| | | $n_0$(cm$^{-2}$) | | |
|---|---|---|---|---|
| **Gas** | **Wavelength (nm)** | **Io** | **Moon** | **Mercury** |
| H | 121.6 | (Torus) | <10 | 23–230 |
| He | 58.4 | | 2,000 | 6,000 |
| Ar | 86.9 | | 1,600 | $<6 \times 10^6$ |
| Na | 589.0; 589.6 | 30 (Torus) | 60 | 30,000 |
| K | 766.4; 769.9 | 3 (Torus) | 15 | 500 |
| O | 130.4 | 1600 (Torus) | <500 | 44,000 |
| S | Various | 1900 (Torus) | — | — |

way sinks of alkalis. In addition, photoionization of these species permits "pickup" of their ions by the magnetic field of the solar wind. Ionization of both Na and K is a minor process compared with fluorescent (resonant) scattering of sunlight; nonetheless, once an alkali atom becomes ionized it will be accelerated radially outward from the Sun to several hundred kilometers per second and either be embedded in the dayside surface of the planet or be swept away around the terminator.

There are major differences between the circumstances on Io and those on Mercury and the Moon. First, the sputtering agent on Io is ions trapped in and corotating with the Jovian magnetosphere, striking Io at $55\,km\,s^{-1}$, not solar wind protons and helium nuclei traveling 10 times as fast. Second, the Io environment is dominated by volcanic ejection of $SO_2$ and its fellow travelers, which some workers believe may include the alkali metals. Third, Io's magnetospheric environment traps any ions that are formed near Io, accelerates them to corotation speed, and confines them in a torus that intersects Io. Recombination of one of these ions with a passing electron produces a neutral atom that no longer feels any $B \times qv$ force from the magnetosphere and is traveling well above Jupiter's escape velocity at Io's orbit. Such "atomic bullets" can travel in nearly straight lines until they become ionized by solar UV light, after which they are picked up by the solar wind and lost from the Solar System.

## Polar Deposits on Mercury and the Moon

The search for polar volatiles on the Moon and Mercury sounds at first like an exercise in futility. I refer not to the expected very low abundance of volatiles in both bodies, but to the logic of detection: the only places on the Moon and Mercury that are cold enough to permit the trapping and long-term retention of volatiles (roughly 100 K) are permanently shadowed regions, such as crater bottoms, very close to the poles. Because the ice must be permanently shadowed to survive, it is always in the dark and cannot be photographed. Further, the presence of small very cold regions on a generally very hot planet is difficult to detect in the infrared, where the long wavelengths of thermal infrared radiation degrade our spatial resolution and the high fluxes from hot spots completely swamp the tiny fluxes from the cold regions. The intensity of emitted thermal radiation from 600-K areas is greater than that from 100-K areas by a factor of $(T_{hot}/T_{cold})^4$, or about 1300:1. Temperatures below 110 K are necessary to keep the evaporation rate low enough to preserve water ice for billions of years.

In 1991 an ingenious experiment was carried out by Martin Slade of the Jet Propulsion Laboratory and Bryan Butler and Duane Muhleman of the California Institute of Technology. They used the 70-m radar transmitter at Goldstone in the Mojave Desert to illuminate Mercury with monochromatic (single-frequency) radar pulses at a wavelength of 3.5 cm. The reflected signal from Mercury was then observed by the Very Large Array (VLA) of radio telescopes in New Mexico. The returned pulses are spread out in frequency by the Doppler effect due to Mercury's rotation and also spread out in time delay by the finite "depth" of the target. By sorting the returned signal into a large number of bins spanning a range of frequencies and arrival times, it is possible to develop a map of the surface reflectivity of the planet. This technique, illustrated in Fig. IX.32, is called *delay-Doppler mapping*. By further separation of the signal in these time–frequency bins into components with different polarizations, it is also possible to discriminate between surface regions with very different physical properties.

Slade and his co-workers transmitted a right circularly polarized (RCP) radar signal, which would reflect from a flat mirror-like surface with left circular polarization (LCP). Because the center of the disk of Mercury is essentially a flat, though rough, mirror, there is a large intensity spike in the returned signal that has minimum time delay, zero Doppler shift (after allowing for the relative motion of the transmitter and the center of Mercury), and strong left-hand polarization. Extremely complex scattering from a very rough surface tends to depolarize the returned signal, which means a relative increase in the RCP component. When the Goldstone/VLA data were analyzed, a bright RCP component was found at zero Doppler shift and maximum range, which requires that it originate at a pole. Fortunately, Mercury's orbit is significantly tilted with respect to the plane of the ecliptic, affording observers on Earth frequent opportunities to see either pole. Since, at the time of observation, the north pole of Mercury was slightly tipped toward the observers and the south pole was invisible, the feature clearly must be associated with the north polar regions (Fig. IX.33). Further, the radar scattering properties observed for this feature are unique to ice-covered surfaces. The strong implication is that Mercury has polar ice. Since these original observations, studies at different Earth–Mercury geometries have revealed that a similar feature is present near the south geographic pole as well. That area of anomalous reflection appears to be mostly confined to the floor of the large Chao Meng-Fu impact crater (Fig. IX.34), which is centered at 87.5°S latitude.

The ice need not be exposed on the surface. Tens of centimeters of dry, porous dirt could overlay the ice

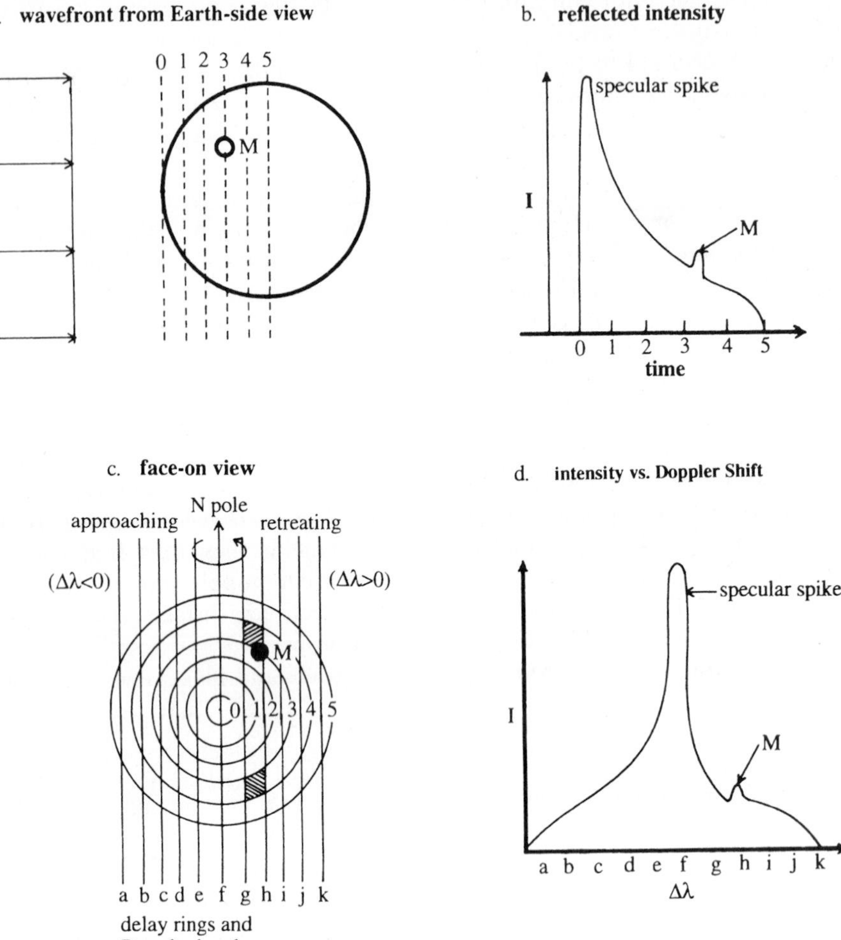


**Figure IX.32** Delay-Doppler mapping. a shows a side view of the passage of a radar wavefront across a spherical planet with a "mountain" area (M) of enhanced radar scattering. b shows the intensity of the returned radar pulse as a function of delay time. The spike at the shortest time delay is due to the mirrorlike (specular) reflection of radar waves from the perpendicular surface at the subradar point. A spike in the returned signal due to the mountain feature is shown in the third time bin. c shows a face-on view of the reflection of a radar pulse from the disk of a rotating planet. The delay rings are marked 0 to 5. Lines of constant Doppler shift are also shown. Note that the delay-Doppler bin 3h contains the mountain. The intensity of the returned pulse as a function of Doppler shift is shown in d. The Doppler shift of the specular point (after correction for motion of the radar telescope relative to the center of Mercury) is zero. Note from comparison of b, c, and d that the location of feature M has a twofold ambiguity (shaded areas in c) that can be removed if the orbit of the target is markedly inclined relative to Earth's orbit or if the obliquity of the planet's pole is large enough.

without degrading the distinctive radar reflection properties of the ice. The thickness of the ice layer is poorly constrained by the observations. A few tens of centimeters or a few meters of ice would suffice to explain the observations, but a few kilometers of ice would look just the same. Further, water ice and solid carbon dioxide (dry ice) would look the same, as may other ices also. Recently the Arecibo radio telescope, transmitting at X band ($\lambda = 12.6\,\mathrm{cm}$), has imaged the radar-bright depolarizing deposits on Mercury with 10 times the spatial resolution of the previous work, showing a complex pattern of bright spots extending, in the most extreme case, down to a latitude as low as 75° N. This discovery presents a challenge to thermal models for the preservation of permanent cold spots.

Thermal modeling of the polar regions by David Paige and co-workers at UCLA shows that flat regions near the poles can be as cold as 167 K, and that the observed cratering of the polar regions favors the existence of small, permanently shadowed enclaves inside

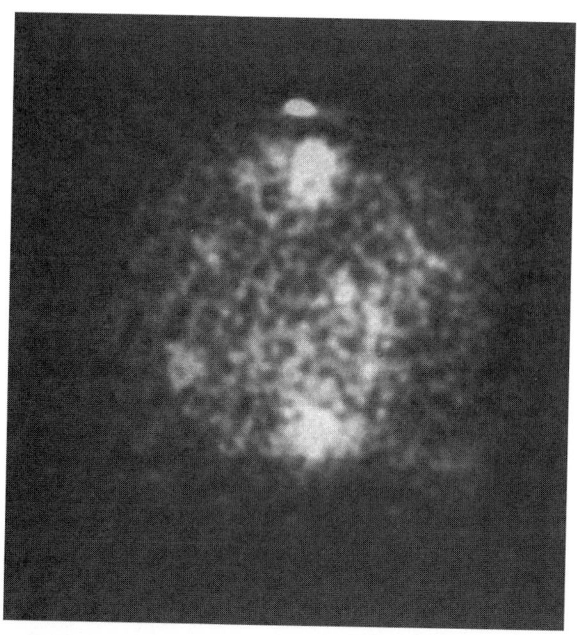

**Figure IX.33** Radar image of Mercury with its north polar ice cap. The bright spot at the top of the image is the polar ice deposit.

craters with large depth:diameter ratios. We have seen in Fig. IX.16 that the craters that exhibit the highest depth–diameter ratios are the simple bowl-shaped craters with rimcrest diameters of 10 km or less. Unfortunately, because of the geometry of the Mariner 10 Mercury encounters, our photographic coverage of the polar regions has only an extremely narrow range of solar longitudes, leaving nearly half of the polar regions unimaged. The available coverage does, however, suggest areas that may be perpetually colder than 100 K and even as low as 60 K. The coldest spot inside Saturn's orbit may be on Mercury!

The knowledge that ices can be stable in Mercury's polar regions is a useful sanity check, but does not enlighten us about the source of the volatiles that are condensed there. Outgassing from Mercury's interior seems an unlikely source if the planet is indeed made of high-temperature condensate; however, the mass of ice required to explain the observations may be as low as a few cubic kilometers, an amount so small as to make it impossible to rule out the presence of that much intrinsic water in the planet. Further, cometary impacts over billions of years provide vastly larger fluxes of water onto Mercury's surface than the amount required by the radar data.

As we discussed in Chapter VII, long-period comet impacts can be enormously energetic events. At Mercury's mean orbital speed of about $V_{orb} = 48\,km\,s^{-1}$, a low-inclination retrograde parabolic comet could encounter Mercury at a speed as high as $(1+\sqrt{2})V_{orb}$, about $116\,km\,s^{-1}$, a kinetic energy density of $6.7\times10^{13}\,erg\,g^{-1}$. Mercury's escape velocity is only $4.3\,km\,s^{-1}$ (an escape energy of $9\times10^{10}\,erg\,g^{-1}$), so an impacting long-period comet carries enough energy to remove its own mass, plus over 740 times its mass of Mercury's regolith, from the planet. This is not a good way of emplacing water on Mercury. Short-period comets, however, have average encounter speeds that are several times lower, and minimum encounter speeds can be on the order of $10\,km\,s^{-1}$. More important, extinct periodic comets (with dust-insulated ice cores) and C-type planet-crossing asteroids may occasionally approach with velocities not much greater than escape velocity. Such impacts not only carry vastly less energy per gram, but also have higher mean atomic weights in their fireballs and hence expand more slowly. Thus they may be the most credible source of water and other volatiles on Mercury.

Unfortunately, our knowledge of the velocity distribution of Mercury-crossing asteroids is certainly very incomplete. Recent NEA discoveries, mentioned in Chapter VIII, show a surprising number of 10- to 500-m bodies in extremely Earth-like orbits of low eccentricity and modest inclination. The origin of this newly discovered class of bodies is not understood, and hence the likelihood of finding such a swarm of low-velocity bodies near Mercury cannot be assessed. Note also that, although there are many known Mercury-crossers among the near-Earth asteroid population (Table IX.5), they represent an extremely biased set: all of them were discovered in the night sky from Earth and therefore must be in high-eccentricity orbits. Conversely, any low-eccentricity bodies orbiting near Mercury (or Venus) could not be discovered by present asteroid search techniques. Further, all Mercury-crossing asteroids so far discovered, because they must also cross Venus and Earth, are strongly depleted in bodies whose inclinations are so low that they make frequent close approaches to these three planets.

One important implication of the discovery of polar ice on Mercury was that it suggested the possibility of a similar deposit on the Moon. The Clementine spacecraft, sent to orbit the Moon in 1994, illuminated the lunar surface with radio waves that were detected by receivers on Earth. This experiment found anomalous depolarization of radio signals apparently reflected from ice deposits of unknown thickness in permanently shadowed crater bottoms at very high latitudes, near where permanently illuminated mountain tops may be found. Given massive deposits of polar volatiles, life-support fluids ($H_2O$ and $O_2$) could be made for use by a polar lunar base. Lunar-derived propellants might also decrease the cost of launch operations from the Moon.

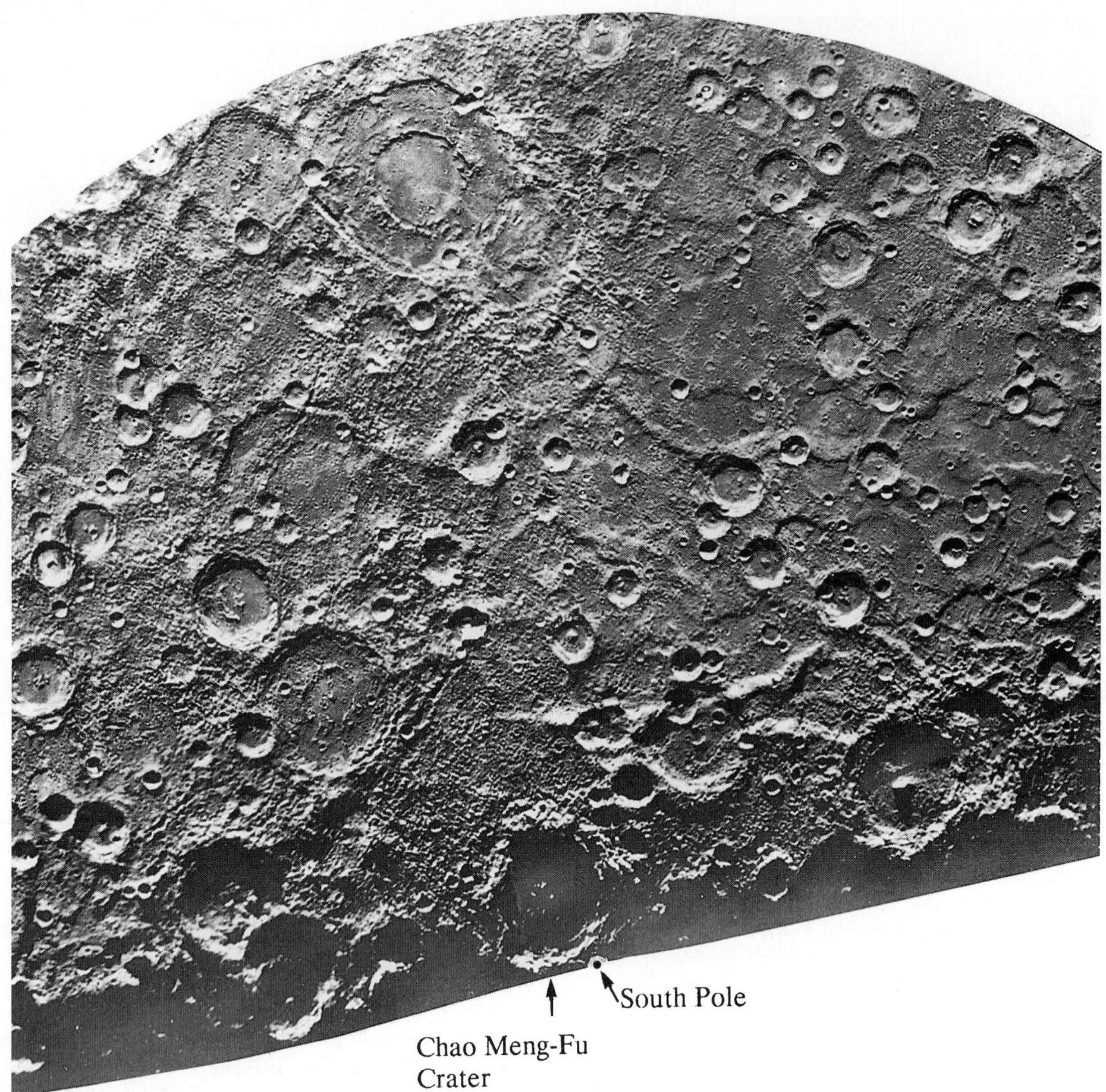


**Figure IX.34** Mariner 10 photomosaic of the south polar region of Mercury. The south geographic pole is marked by the dot at the bottom. The large crater (Chao Meng-Fu) adjacent to the pole is the location of the south polar ice deposit found by radar. The floor of this crater appears to be in permanent darkness. The poles of Mercury can alternately be viewed from Earth (and the delay-Doppler ambiguity can be resolved) because of the high orbital inclination of Mercury.

## Unfinished Business

The only missions that promised a better look at the Martian moons Phobos and Deimos than had been provided by the Viking Orbiter spacecraft were the Russian Phobos 1 and 2 missions of 1988. These were ambitious, well-conceived missions with broad scientific interest and international participation. The in-flight failure of Phobos 1 en route to Mars, due to an unfortunate controller error, and the later loss of Phobos 2 in orbit about Mars while approaching Phobos were therefore extremely disappointing. Later attempts to convince NASA's Mars Observer project to schedule flybys of the Martian moons before dropping into low Mars orbit met with no interest. The very ambitious

**Table IX.5** Mercury-Approaching NEAs

| Asteroid | $q$ | $a$ | $Q$ | $e$ | $i$ |
|---|---|---|---|---|---|
| | | Mercury crossers ($q < 0.467$ AU) | | | |
| 3200 Phaethon | 0.140 | 1.271 | 2.470 | 0.890 | 22.029 |
| 1566 Icarus | 0.187 | 1.078 | 1.969 | 0.827 | 22.912 |
| 1991 RC | 0.188 | 1.081 | 1.973 | 0.826 | 23.148 |
| 1984 QY1 | 0.219 | 3.600 | 6.981 | 0.939 | 17.737 |
| 1989 VA | 0.292 | 0.728 | 1.164 | 0.599 | 28.400 |
| 1991 VE | 0.299 | 0.890 | 1.482 | 0.664 | 7.206 |
| 1990 UO | 0.306 | 1.229 | 2.153 | 0.751 | 28.312 |
| 2212 Hephaistos | 0.357 | 2.165 | 3.972 | 0.835 | 11.878 |
| 1991 LH | 0.364 | 1.352 | 2.340 | 0.731 | 52.071 |
| 1991 TB2 | 0.394 | 2.397 | 4.400 | 0.836 | 8.648 |
| 1991 VL | 0.419 | 1.834 | 3.249 | 0.771 | 9.195 |
| 1974 MA | 0.420 | 1.775 | 3.130 | 0.762 | 37.780 |
| 5025 P–L | 0.441 | 4.201 | 7.961 | 0.895 | 6.200 |
| 2101 Adonis | 0.443 | 1.875 | 3.308 | 0.764 | 1.362 |
| 3838 1986 WA | 0.450 | 1.505 | 2.551 | 0.701 | 29.313 |
| 2340 Hathor | 0.460 | 0.844 | 1.220 | 0.450 | 5.856 |
| | | Mercury grazers ($0.467 < q < 0.550$ AU) | | | |
| 2100 Ra-Shalom | 0.471 | 0.832 | 1.190 | 0.436 | 15.800 |
| 1990 SM | 0.477 | 2.157 | 3.837 | 0.775 | 11.568 |
| 3753 1986 TO | 0.480 | 0.998 | 1.513 | 0.515 | 19.814 |
| 1991 AQ | 0.499 | 2.158 | 3.818 | 0.769 | 3.190 |
| 1954 XA | 0.509 | 0.777 | 1.045 | 0.345 | 3.900 |
| 1991 AM | 0.517 | 1.694 | 2.871 | 0.695 | 30.011 |
| 4197 1982 TA | 0.521 | 2.297 | 4.073 | 0.773 | 12.203 |
| 1984 KB | 0.523 | 2.216 | 3.909 | 0.764 | 4.846 |
| 3362 Khufu | 0.530 | 0.990 | 1.450 | 0.469 | 9.923 |
| 4769 1989 PB | 0.547 | 1.063 | 1.577 | 0.483 | 8.888 |

Mars '94 mission, planned by the USSR with strong French support, soon fell on hard economic times, was dramatically reduced in scope, including the removal of balloons and surface rovers from the payload, was rescheduled for launch in 1996, and then was lost in a launch failure. Though this mission would not have advanced the study of Phobos and Deimos, its loss was the final straw for the Russian planetary program, making a revival of Phobos-like missions impossible.

The Galileo spacecraft entered an orbit about Jupiter that first flew by Io, then repeatedly encountered the other three Galilean satellites, returning to revisit the hazardous radiation environment of Io only after its principal mission objectives had been met. No other mission to Jupiter's system is presently scheduled by any nation. There has been much discussion of landing probes for exploration of the Galilean satellites, including a Europa lander that would melt its way through the ice crust into its global ocean, but no serious mission plan yet exists, in part because NASA has focused strongly on the exploration of Mars and the reconnaissance of asteroids and comets. An Io orbiter or lander, braving the local intense energetic-particle environment, seems too technically challenging and expensive to contemplate.

The emergence of a satisfactory global view of the Moon has also been hindered by limitations on the data available to us. Even after the Galileo flyby of the Moon and the Clementine and Lunar Prospector missions, there are many missing pieces to the lunar puzzle. Seismic and gravity-field evidence bearing on the size of the putative lunar core is extremely sparse. The lunar polar regions, where deposits of asteroidal or cometary ice might reside in permanently shadowed regions, have provided tantalizing radar and radio-science evidence of accumulations of some form of hydrogen, widely presumed to be water ice. Outcrops of ore-grade mineral deposits are unknown, largely because of limitations on the spatial resolution of spectrophotometric surveys. Photographic mapping of the Moon remains incomplete. The local geology, geochemistry, and solar-wind-implanted volatile content of virtually all of the Moon remains unknown because of the complete absence of lander, rover, and sample-return missions since 1976.

In a more practical vein, there are also serious shortcomings in our ability to plan an optimal location for a lunar base and to find locations on the lunar surface where native lunar resources are most accessible. A simple example must suffice: $^3$He implanted in the lunar regolith by the solar wind is the only significant source of this rare isotope between the asteroid belt and the Sun itself. Experiments with fusion reactors suggest that the controlled fusion of $^3$He with D, which produces no neutrons, might be a clean source of power for 21st-century Earth. Understanding the distribution of implanted helium over the lunar surface and its concentration in regolith grains of different size and composition may hold the key to providing all of Earth's electrical power needs for the next several centuries.

Russian plans for a modern lunar orbiting mission have been widely discussed since about 1984, but no approved mission has materialized. In this case the culprit is *glasnost*, which led to the first publication of the budget of the Soviet space program, abetted by *perestroika*, which recast the nature of the Soviet civil space program as a competitor in the world marketplace for launch, communications, micrograv-ity-processing, and Earth-observing services—in short, the quest for the almighty dollar. The European Space Agency has studied a Polar Lunar Orbiter mission named Polo, but the mission has never achieved flight status. In 1986, Japan's NASDA launched a small lunar probe, Sakigake, which injected an even smaller subsatellite into orbit about the Moon. Another much larger and more ambitious Japanese lunar mission is being planned for launch in the near future aboard the large new H2 booster.

Plans for future exploration of Mercury have also languished since the Mariner 10 mission. NASA's scientific advisory panels have several times endorsed a Mercury Orbiter to fill the same general role as the lunar orbiter discussed at more length above. Ingenious multiple flybys of Earth, Venus, and Mercury can be used to pump the orbit of a spacecraft down into a heliocentric orbit close to that of Mercury, after which a single rocket engine burn can insert the spacecraft into orbit about Mercury. Technically, a Mercury Orbiter seems well within reach. The Mercury Messenger mission has been thoroughly studied and is clearly technically feasibly and scientifically justified. Future potential missions are further described in Chapter XII.

## Exercises

*Introduction*

IX.1 Some theories of the accretion of the terrestrial planets predict that each planet will sample preplanetary solids from a very wide range of heliocentric distances and formation temperatures, and that bodies up to the size of Mars may wander widely throughout the inner Solar System under the effect of planetary gravitational interactions during accretion. If such theories are correct in some planetary system, how would you expect Fig. IX.1 to look for that system?

*Orbits and Physical Structure of Phobos and Deimos*

IX.2 From a point on the equator of Mars, how long does it take for Phobos to complete one circuit of the sky?

IX.3 From a point on the equator of Mars, how long does it take for Deimos to complete one circuit of the sky?

IX.4 What is the synodic period of Deimos as seen from Phobos?

*Io: General Properties*

IX.5 Among the few geochemically plausible materials that can provide Io's very high albedo are a number of soluble salts, including halite (NaCl). Discuss the implications of a global surface layer of salt on a body with no present-day water.

*Io: Surface Processes*

IX.6 a. If the entire surface of Io to a depth of several kilometers were subducted every 10,000 years, what is the rate at which new crustal material must be generated to maintain a steady state?

b. How much energy would be released per second by crustal solidification with such a recycling rate?

c. What would the effective temperature of the surface of Io be if resurfacing were the only source of energy at the surface of Io?

*Io: Internal Energy Sources*

IX.7 Estimate the magnitude of Jupiter's magnetic field at Io's orbit [see Eq. (V.250)] and compare it to that calculated for Earth's magnetic field strength at the mean distance of the Moon. What does this comparison show regarding induction heating of the Moon?

*Io: Atmospheric and Volcanic Gases*

IX.8 Prove that the slopes of the contours of constant $SO_2$ pressure in Fig. IX.12 have a slope of exactly −2.

IX.9 Suppose that a mixture of pyrite ($FeS_2$) and magnetite ($Fe_3O_4$) were heated to the point at which a significant gas pressure is developed.

a. Prove that neither oxygen nor sulfur is buffered in this system.

b. Derive an expression for the slope of the pyrite–magnetite boundary on a diagram such as Fig. IX.12.

*Impact Cratering*

IX.10 Graph the depth–diameter relationship given in Eq. (IX.4) versus velocity for several different solid bodies (Phobos, the Moon, Mercury, Mars, and Earth) and comment on the differences in their behavior. A reasonable first approach is to set the impact velocity equal to the escape velocity.

IX.11 The Tunguska explosion of 30 June 1908 was caused by the aerial explosion of a small asteroid over central Siberia. The explosion, which had a yield of about 10 megatons of TNT, crushed and burned some 2000 $km^2$ of boreal forest. If it had survived passage through the atmosphere and delivered its kinetic energy to the ground, how large a crater would it have produced?

*Motions of the Moon*

IX.12 What is the minimum angle that can occur between the pole of the ecliptic and the pole of the Moon's rotation?

IX.13 How large is the gravitational attraction of the Sun on the Moon compared to Earth's attractive force at a time when the Moon is eclipsing the Sun?

IX.14 Explain why the Moon sometimes completely eclipses the disk of the Sun (total eclipse), but at other times may leave a visible ring of the Sun's photosphere all around the periphery of the lunar disk (annular eclipse).

*Composition of the Moon*

IX.15 Use the Surveyor 5 data in Table IX.1a to calculate the atomic abundances of the elements listed in the analysis and plot these data onto Figs. VIII.5 and VIII.7. What conclusions can you reach regarding the composition of the lunar crust in comparison with the compositions of meteorite classes?

IX.16 The abundance of glass in lunar regolith samples correlates with the abundance of hydrogen and helium in them. Explain why.

*Lunar Rock Types*

IX.17 Suppose a layer of solid basalt 3000 m thick on the Moon is resting atop a region in which basaltic liquids are being generated. An impact opens a crack from the surface down to the liquid layer, and it rises until it achieves hydrostatic equilibrium with the overlying rock. To what height could the column of melt in the crack rise?

*Lunar Minerals*

IX.18 a. If the Moon originally had formed from a high-temperature condensate at 1000 K and $10^{-3}$ bar [see Fig. (IV.7)], what changes would occur in the abundances and compositions of the major (M) and minor (m) lunar minerals listed in Table IX.2?

b. How would the plot in Fig. IX.22 be changed?

*Lunar Elemental Abundance Patterns*

IX.19 Suppose some future mission carries out active seismic sounding (using explosives, not waiting for rare moonquakes) of the deep interior of the Moon and finds that there is no lunar core. How would this discovery constrain chemical models for the origin of the Moon?

*Geophysics of the Moon*

IX.20 What is the upper limit on the fraction of the Moon's mass that is contained in a metallic core, based on the limit of a 360-km radius?

IX.21 Let us take the mean heat flux out of the Moon's interior to be $2 \times 10^{-6}$ W cm$^{-2}$ s$^{-1}$. If the thermal conductivity of basalt is $2 \times 10^{5}$ erg cm$^{-1}$ s$^{-1}$ K$^{-1}$, what would you expect the vertical temperature gradient to be in the basalt fill of a mare basin?

IX.22 If most of the radionuclides in the Moon were concentrated in the outermost 30 km, how would you modify Eq. (VIII.15) to estimate the time scale for conductive cooling? Note the similarity to what happens during T Tauri phase solar wind heating of a fairly good conductor.

*History of the Earth–Moon System*

IX.23 The Moon at present raises open-ocean tides with amplitudes of about 1 m. If the Moon was once close to Earth's Roche limit, what would the approximate tidal amplitude have been at that time?

*Origin and Internal Evolution of the Moon*

IX.24 The inclination of the Moon's orbit plays a major role in theories of the history of the Earth–Moon system.

a. Assuming conservation of momentum and a perfectly inelastic collision, how large an impact would be required to change the inclination of the Moon's present orbit by 20°?

b. How large a crater would such an impact produce? Explain your assumptions regarding the velocity of impact.

*Solar Wind Interaction with the Moon and Mercury*

IX.25 Estimate the aberration angle of the solar wind at Mercury (the angular offset in its apparent direction caused by Mercury's high orbital velocity).

*Motions of Mercury*

IX.26 a. Write a program that calculates the elevation of the Sun as viewed from any fixed point on Mercury's equator, assuming zero axial tilt. You should start at local noon and assume any location on Mercury's orbit.

b. Present the results of that program for two Mercury years.

*Composition and Structure of Mercury*

IX.27 Assume the rocky component of a fully differentiated Mercury has a density of 3.2 g cm$^{-3}$ and the dense (metallic) component has a density of 7.5 g cm$^{-3}$.

a. Neglecting self-compression, what fraction of the mass of Mercury is metal if the overall density of the planet is 5.43 g cm$^{-3}$?

b. What is the fractional radius ($r_c/r_s$) of the core in this model?
c. What is the coefficient of rotational moment of inertia of this model?
d. Qualitatively, how would including self-compression effects change the answers to these three questions?

IX.28 Suppose a family of small (uncompressed) planets is made of a mixture of two components: metal with a density of 7.8 and silicates with a density of 3.5.
a. Show how to calculate the volume percent of each component in one of these planets given its bulk density.
b. Show how to calculate the mass percent of each component of such a planet if its bulk density is given.

*Geophysics of Mercury*

IX.29 If the entire silicate mantle of Mercury were stripped off by impacts, how would the escape velocity of the planet be changed? Use the "weak consensus" model of Mercury in Fig. IX.30 as your point of departure.

IX.30 a. Recall our earlier discussions of the relationship between the energy density of the solar wind with velocity $v$ and its dynamical (ram) pressure. Show that the solar wind incident on the equatorial regions of a planet with a symmetrical dipole field of surface strength $B_0$ reaches pressure balance with the planetary magnetosphere at a distance

$$L = r/r_s = (B_0/v)^{1/3}[1/4\pi n m_p]^{1/6}.$$

b. If the early T Tauri phase solar wind was $10^6$ times as intense as the present wind, how much stronger would $B_0$ have to be to stand off the solar wind at the same distance from the planet?
c. If you should calculate that $r/r_s < 1$ for the early solar wind approaching Jupiter, what factor would then be responsible for stopping the solar wind before it hits Jupiter?

*Atmospheres of Mercury and the Moon*

IX.31 Using standard sources of physical data, calculate the temperature at which elemental sodium would have a vapor pressure sufficient to provide the observed gaseous atomic sodium abundance on Mercury.

*Polar Deposits on Mercury and the Moon*

IX.32 A flat crater floor at 85° N on Mercury may contain ice on the permanently shadowed crater floor. How important is it to consider the effects of *indirect* illumination by light scattered from the crater walls? Choose a reasonable depth–diameter ratio for large craters on Mercury and a reasonable albedo and do a simple calculation.

*Unfinished Business*

IX.33 A spacecraft is launched from a circular heliocentric orbit at 1.000 AU from the Sun into an elliptical heliocentric orbit that grazes Earth at aphelion and grazes Mercury's orbit at perihelion. Its mission is to match speeds with Mercury for the purpose of entering orbit about that planet. If the orbit is designed so that it grazes Mercury's orbit at Mercury's perihelion distance, the spacecraft will require a larger injection $\Delta V$, but it will overtake Mercury when Mercury is traveling faster, and therefore need a smaller second $\Delta V$ to match speeds. If it is instead targeted to intercept Mercury at aphelion, the launch from 1 AU circular orbit will be easier, but Mercury will be traveling slower when the spacecraft overtakes it. Which mission has the lower *total* $\Delta V$ requirement?

# X. The Terrestrial Planets: Mars, Venus, and Earth

## Introduction

Mars is, as the traditional description has it, a brother of Earth. For a century writers of fiction and fact alike have tried to draw a moral from the study of Mars. Some saw it as a parable for the old age and decline of Earth: a still-habitable (but ancient) or once-habitable planet exhausted by the decline of its precious atmosphere, irretrievably lost both in the weathering of surface rocks and in escape from the planet, covered with the fossils of once-abundant life, and even the ruins of advanced civilizations. Venus, cast as the twin sister of Earth in such parables, was seen as the image of the early Earth, in the evolutionary sense, at least, a younger, dynamically evolving planet on the road to full-fledged habitability. Some saw Venus as a cloud-shrouded hot global ocean awaiting the origin of life; some as a sweltering carboniferous swamp teeming with alien saurischians and ornithischians.

In large part these fancies arose quite outside the mainstream of modern science, and it should be little more than a matter of amusement to us that our grandparents thought of the planets forming close to the Sun, gradually moving outward and aging as they receded into the cold and dark, to be supplanted by yet more newly formed young planets emerging from the fires of the Sun. But, like most fantasies that lack any quantitative basis, there is nonetheless a valid qualitative concern that underlies its excesses. Thus today, when we look at Mars and Venus, we still think in terms of two overriding questions about them and their relationship to Earth. First, we wonder how the present diversity of these three neighboring planets may have been influenced by their different conditions of origin, and, second, we wonder how much of their differences are due to divergent evolutionary paths.

There may indeed be powerful and meaningful cautionary tales underlying the conditions we observe today. Consider: Mars, a planet of grossly Earth-like composition with a somewhat lower density and 10-fold lower mass than Earth, is located much closer to the realm of volatile-rich comets and asteroids. The lower density strongly implies a greater degree of oxidation and a higher abundance of volatiles than we have on Earth, fully in keeping with a cooler place of origin, farther from the Sun. Mars has an atmosphere, polar caps, and widespread geological evidence of the former presence of rivers, glaciers, and even oceans. Yet the atmosphere today is a frigid veil of carbon dioxide, less than 1% of the atmospheric pressure of Earth, so cold that water is found as vast deposits of ice and wisps of water vapor. Liquid water is certainly very rare, and possibly absent, on the surface of Mars today. Where are the volatiles that Mars should have had at the time of its formation? Are our theories so poor that they have misled us, and Mars formed poor in volatiles despite our expectations? If so, what are we to make of the geological features that attest to the former presence of vast amounts of water? Or did Mars indeed have abundant water and a much more massive atmosphere in its early history, only to lose it? If so, was the atmosphere lost gradually over very long periods of time, giving life an excellent opportunity to originate, evolve, and eventually adapt to the change, or was it lost at such a rate as to prevent the origin and development of native life? How was the atmosphere lost? Where did the water go? How much is still there? Could a stable biosphere be introduced on Mars? Is there primitive life on Mars today? If there is no life, then are there fossils? If either, then how are Martian life-forms related to terrestrial life? *What caused Mars to turn out as it did? Could the same thing happen to us?*

And what of Venus, with its crushing atmospheric pressure, its surface hotter than a self-cleaning oven, and its clouds of sulfuric acid? Venus, located closer to the fires of the Sun than Earth, ought by right to have partaken of less of the volatile elements than did Earth, and yet its atmospheric pressure is nearly 100 times Earth's. Curiously, the atmosphere is nearly pure carbon dioxide. Not only are water and ice impossible on the surface, but the atmosphere is extremely dry. If so much carbon dioxide was available to Venus, why is there not a comparable amount of water? Where is the water that Venus should have accreted with its carbon dioxide at the time of its formation? Are our theories so poor that they have misled us, and Venus formed rich in volatiles despite our expectations? Did Venus indeed have abundant water and a much more massive atmosphere in its early history, only to lose it? If so, was the water lost gradually over very long periods of time, giving life an opportunity to originate, or was it lost so fast as to prevent the origin and development of native life? How was the water lost? Where did the water go? How much is still there? *What caused Venus to turn out as it did? Could the same thing happen to us?*

## Mars

Mars presents a small target to terrestrial telescopes. Observations are very difficult except at the time when Earth passes directly between Mars and the Sun, and Mars is opposite the Sun in the sky (i.e., when Mars is at *opposition*). Even then the planet is most easily observed only at times when Mars is near perihelion at the time of opposition and hence unusually close to Earth. Early visual observations clearly showed a small reddish planet with white, seasonally varying polar caps and extensive dark markings, which were commonly referred to as green in color. Local springtime in each hemisphere was usually marked by a distinct darkening of the surface that progressed outward from the polar caps as they receded. This phenomenon was termed the *wave of darkening*. Oceans and lakes were not seen, and the red surface seemed to signal extensive rusting of iron-bearing surface minerals by atmospheric oxygen. At times when Mars was brightest in the sky and easiest to observe, Mars was frequently covered by vast dust storms, which frustrated observers by obscuring virtually all the surface features they were intent upon studying.

But despite the observational difficulties, long before the beginning of spacecraft exploration, Mars was already a popular topic of scientific discussion and public speculation. It has played a major role in science fiction writings from the time of H. G. Wells and Edgar Rice Burroughs to the present. If one central reason can be found for the great popular interest in Mars, it is because it has many times been suggested as a possible home for life. The astronomical basis for this suggestion extends far back in the literature. Angelo Secchi, an Italian priest, reported as early as 1869 that there were long dark streaks on the Martian surface, which he named *canali*, which is the Italian word for "channels." There was then, of course, no possible way of knowing whether these features were raised, depressed, or flat, but the name stuck. Just a few years later, these markings were widely popularized by Giovanni Schiaparelli, a Milanese astronomer, who drew maps decorated with great numbers of very narrow, straight dark lines in a complex global network. The American astronomer Percival Lowell further incited public discussion by not only mapping over a hundred of these markings, but insisting on calling them not channels, the correct translation of *canali*, but "canals," a word with the strong connotation of artificial origin. Lowell went on to describe these canals as irrigation channels built by intelligent life on Mars for the express purpose of distributing precious springtime meltwater from the polar caps to agricultural areas and cities over the entire planet. The areas referred to as "dark" or "green" were, in Lowell's interpretation, vegetation. Where vegetation flourishes, who could doubt the presence of herbivores? Where herbivores flourish, what could be more natural than the evolution of carnivores? And so, by simple increments of credulity, a planet with faint dark markings became home to a complex biosphere with intelligent, technologically proficient life at the top of the food chain.

Photographs of Mars, however, resolutely failed to show any canals. They appeared only on drawings made by visual observers. The canal-mappers asserted that they drew only what they saw during rare, fleeting moments of perfect seeing, too brief to photograph. But the majority of astronomers claimed never to see such features, and feared that they were illusory.

From the 1920s, astronomers knew that the Martian atmosphere contained abundant carbon dioxide, but estimates of the water content were generally small and discordant. It was common to suggest the possibility that liquid water could run on the Martian surface, at least in the daytime. But astronomers studying high-resolution infrared spectra of Mars in the mid 1960s discovered that the atmospheric pressure on Mars was less than 1% of the atmospheric pressure on Earth, below the vapor pressure of water at the ice I–liquid water–water vapor triple point, so low that the presence of exposed liquid water on the surface was not possible. Thus the era of spacecraft observations of Mars opened in a climate of opinion that rated the biological interest of Mars at an all-time low. With greatly subdued interest, scientists turned to research on the physical and chemical nature of the planet rather than its suitability for life.

## Motions of Mars

Mars follows an elliptical orbit about the Sun. The orbital semimajor axis of 1.5237 AU and eccentricity of 0.0934 give a perihelion distance of $q = a(1 - e) = 1.3814$ AU and an aphelion distance of $Q = a(1 + e) = 1.6660$ AU. The inclination of the orbit of Mars is a modest 1.85°. The intensity of sunlight at perihelion exceeds that at aphelion by a factor of $(1 + e)^2/(1 - e)^2 = 1.4545$, which leads to a strong annual climate cycle that affects the entire planet, with all temperatures elevated near the time of perihelion passage and depressed near aphelion. To distinguish these global synchronized temperature variations from normal seasonal variations (which are due to axial tilt), we may call these *eccentricity seasons*. Over the Mars year of 1.8809 Earth years the planet goes through one complete warming and cooling cycle. Meteorological activity is most vigorous near perihelion, at which time vast dust storms may envelop the planet.

Mars rotates on its axis relative to the stars in 24 h, 37 min, and 22.663 s, surprisingly close to Earth's rotation period. Mars has at present an axial inclination of 25.19°, also surprisingly similar to Earth. The axial inclination, however, oscillates over a range of 25.2° ± 10.3° with a period of about 125,000 years. The rotation axis also precesses about the pole of Mars' orbit in about 175,000 years.

As we discussed at length in Chapter IX, Mars has two natural satellites, Phobos and Deimos, that are rotationally locked onto Mars. Phobos orbits inside synchronous altitude and is spiraling slowly inward due to the tidal force of Mars. Deimos, well beyond synchronous orbit, is slowly receding from Mars. The only other synchronous relationships of Mars with other Solar System bodies are a few asteroids in 1:1 orbit–orbit resonances with Mars. The prototype of this class is the Mars Trojan asteroid (5261) Eureka. This small asteroid, only about 2 km in diameter, was the first Trojan asteroid to be found in association with any planet besides Jupiter.

## Density and Figure of Mars

Some early structural models of Mars had the entire planet composed of rocky (silicaceous) material, with no core present. Other authors, including British geophysicist Sir Harold Jeffreys, the American Nobel Laureate Harold C. Urey, and American geophysicist Gordon MacDonald, contended during the 1960s that Mars could be assembled of a mixture of Earth-like core and Earth-like mantle materials in proportions different from those found on Earth. One such model, using 3306 km (the pre-Mariner 6 and 7 value) for the radius of Mars, found a metal abundance for Mars not very different from that for the Earth. However, JPL's Arvydas Kliore and his co-workers showed that the mean radius of Mars compatible with the Mariner 6 and 7 radio occultation measurements was in fact 3394 km, which led to a substantial revision in the mass of the core of Mars and the density of both its core and mantle. One model, calculated using a density of 8 g cm$^{-3}$ for the zero-pressure density of Martian core material, found a core mass of only 2.7–4.9% of the planetary mass. However, this is surely a substantial overestimate of the core density, and a value near that of FeS is more likely than one suggestive of Ni-rich Fe–Ni alloy.

Perhaps the most startling difference between early models of Mars and contemporary Earth models was the far lower density of Mars, which in every case required that the core mass be proportionately far smaller in Mars than in Earth. Urey attributed this density difference to physical accretionary processes that discriminated between metal and silicate grains prior to the formation of planet-sized bodies. This model was proposed at a time when it was the unanimous opinion of astrophysicists and solar spectroscopists that the abundance of iron-group elements in the Sun (relative to, say, silicon) was about a factor of 10 less than that in chondritic meteorites and the terrestrial planets. Thus, Urey was forced to invent a process for extreme enrichment of the metal phase in accreting solid bodies. Applying this hypothetical enrichment process to different degrees in different planets accounted for the apparent fractionation of metal relative to silicates found by comparing the uncompressed densities of Mercury (then poorly known), Venus, Earth, and Mars.

One worker, geochemist Ted Ringwood of Australian National University, rashly (but correctly!) chose to disbelieve the spectroscopic iron abundance. He proposed that all of the terrestrial planets originally formed with the same bulk composition, which he assumed to be CI carbonaceous chondritic. The present large density differences among these planets were then assumed to be due to extreme heating, devolatilization, and atmospheric escape, which most severely depleted the volatiles from the largest planets. In this view, Earth has the most devolatilized interior and hence the lowest FeO content. Ringwood pictured the bulk composition of Mars as more FeO-rich than Earth's mantle, with a core composed of a high-pressure polymorph of magnetite. Others, such as Don Anderson of CalTech's Seismological Laboratory, favored models containing an Fe–Ni–S core with about 12% of the planetary mass, with a mantle containing a high concentration of FeO or metal. A total Fe content of about 25% was preferred. Anderson, taking the core and mantle zero-pressure densities as constrained to be the same as Earth's, was led by the high value of the moment of inertia to conclude that Mars was an "incompletely differentiated planet." Since core formation is a self-accelerating process, this conclusion is evidently unacceptable.

However, the theory of condensation of preplanetary material discussed in Chapter IV ascribes the density differences among the terrestrial planets mainly to the temperature gradient in the solar nebula at the time of last equilibration of grains and dust. In this equilibrium condensation model, Mercury should have an elevated density because of incomplete silicate condensation at the high temperatures prevalent very close to the Sun. All the other terrestrial planets are pictured as having very nearly the same relative abundances of rock-forming elements, which are fully condensed below about 1000 K. Venus, Earth, and Mars are therefore expected to differ in composition and density only with respect to the volatile components, especially water and sulfur. This model predicts an iron:silicon atomic ratio of about unity for all three of these planets, with the iron becoming progressively more oxidized with increasing heliocentric distance. Mars should then differ from Earth principally in having more FeO and less Fe metal, thus a denser mantle and less-dense core than Earth. Core formation would then involve Fe–FeS eutectic melting.

The presently accepted density of Mars, deduced from the motions of its natural satellites and orbiting spacecraft sent from Earth, is $3.93\,g\,cm^{-3}$. The density corrected for self-compression, which is called the zero-pressure density, is $3.73\,g\,cm^{-3}$, a number similar to the density of ordinary chondrites. If the Fe:Si ratio is the same for Venus, Earth, and Mars, then the much lower zero-pressure density of Mars reflects a much higher degree of oxidation, with all or nearly all metal oxidized to sulfides and FeO-bearing silicates.

The polar diameter of Mars is close to 6754 km and the equatorial diameter is 6792 km. The oblateness of Mars ($\varepsilon = (6792 - 6754)/6792 = 0.0055$) and the principal moment of inertia, $C$ (the *reduced moment of inertia*, $C/MR^2$, is 0.365), deduced from tracking of its satellites, combine to suggest a differentiated planet, with a moment of inertia significantly less than that of a compositionally uniform planet with the same dimensions, for which $C/MR^2$ would be slightly under 0.40. Attempts to model the observed moment of inertia and density by assuming a two-layer model with an Earth-like core composition, Earth-like mantle composition, and unknown core radius produce quite unsatisfactory results, such as Anderson's above-mentioned conclusion that Mars was incompletely differentiated.

Models with a core sized to give the right density have much too low a moment of inertia, and models with a core sized to give the right moment of inertia have much too large a density. The only modeling strategy that resolves this dilemma is to admit a core zero-pressure density that is substantially less than 7.5 (pure iron) and a mantle density that is higher than that of the terrestrial mantle. These are entirely consistent conditions, because both require a higher degree of oxidation than Earth (less Fe metal, therefore less core mass and a higher proportion of light elements such as sulfur in the core; more FeO, therefore a denser mantle). Since neither the core density nor mantle density is known, and the core radius is not known independently (such as through seismic studies), a unique solution for all three variables cannot presently be obtained. A reasonable physical constraint to narrow down the field of competing models would be to recognize that differentiation is a major source of heat, and that differentiation is self-accelerating; therefore, partially differentiated models make no sense and should be avoided. Additional constraints, such as the geochemically plausible assumption that the core is made of Fe–S, provide a set of consistent solutions for the other variables. From this we conclude that an Fe–S core is plausible, but we lack the data to determine whether any particular plausible solution is in fact the correct one.

Geophysical models with a higher mantle density and lower core density than Earth can readily be fitted to Mars. They generally give an FeO/(FeO + MgO) ratio close to 0.25, about twice the terrestrial mantle value, with an allowable range of mantle zero-pressure densities from 3.44 to 3.54 g cm$^{-3}$. Several geochemists have also proposed geochemical models for Mars that have mantle FeO contents similar to those geophysical models which have metal-free FeS cores. Kenneth A. Goettel, then at Washington University, has carried out an error analysis for interior models of Mars. He finds a zero-pressure mantle density of $3.44 \pm 0.06$ g cm$^{-3}$, with a density near 3.41 favored for models with FeS cores. He finds also that the density of Mars is compatible with the assumption that the planet contains the major rock-forming elements in their solar proportions, but differs from Earth in that some 30% of the total Fe content of Mars is tied up as FeO in mantle silicates. These models use a reduced moment of inertia of 0.365, in agreement with the determinations mentioned above, and further assume a bulk density of $3.933 \pm 0.002$ g cm$^{-3}$ for the planet. His conclusions are generally compatible with both the model that attributes the high degree of oxidation to the small mass of the planet and that which ascribes a higher FeO content to Mars because it accreted at greater heliocentric distance, and hence at lower temperatures, than Earth.

Range and Doppler tracking of the Mars Pathfinder spacecraft has had little effect on these conclusions: $C/MR^2$ is now quoted as $0.366 \pm 0.0017$. We shall revisit this issue after our discussion of the compositional constraints provided by analyses of the Martian SNC meteorites.

One further method of investigating the internal structure and mass distribution in a planet is by the mapping of its external gravitational field through precise tracking of the Mariner 9 and Viking 1 and 2 Orbiters. We have already mentioned some of these results, such as the finding that the Tharsis region is characterized by a mass excess, with a thick and partially compensated crust. Tectonic scenarios have been generated to account for the present gravity field and surface morphology, such as the interpretation of the Valles Marineris–Coprates trough system as a tensional feature caused by polar wandering. Perhaps the most enthusiastic assessment of crustal mobility is that of V. E. Courtillot and co-workers of the University of Paris, who claimed that they had identified transform faults and even a possible triple plate junction. In light of their claims to have found a number of areas of crustal extension, the absence of any evidence for convergence is most striking. It is most reasonable to conclude that continental drift ceased at a very early stage on Mars.

However carefully interior structure models are generated and however accurately the equations of state and phase stability relations of geochemically plausible materials are known, it is simply not possible to produce unique and reliable models of the internal composition and structure of a planet without additional, usually chemical, constraints. Further, especially in the case of a planet as small as Mars, extreme assumptions about the accretionary history and radioactive element content can generate very different predictions of the present thermal state and tectonic activity of the planet.

## Geophysical Data on Mars

Two seismometers have landed on the surface of Mars, both during the Viking program in the 1970s. The seismometer on the Viking 1 lander failed to deploy properly because of the failure of the packing restraints, which protected it from vibration during launch, the flight to Mars, and landing, to retract and "uncage" the seismometer mechanism. As a result, no useful data could be obtained from that instrument. The Viking 2 instrument was successfully uncaged and functioned for several years on the surface of Mars. On both Viking landers the seismometer was mounted well above the ground on the lander body. For this reason the instrument was extremely vulnerable to wind buffeting. Seismic data could be collected only at night or during unusually quiet daytime conditions. Several disturbances that were potentially attributable to seismic activity were all found to correlate with wind gusts measured by the meteorological package. Only one other event that could be attributed to Martian seismic activity was observed, and this event happened to occur at a time when the wind speed was not being monitored. The interpretation of the data from this event cannot be used with confidence to deduce anything about the interior of Mars. Perhaps the firmest conclusion is negative: the seismic activity of Mars is less than that of Earth. This is scarcely a surprising conclusion.

As we saw in our discussion of the interior of Mercury in Chapter IX, measurement of the magnetic field of a planet can help place useful constraints on the presence and properties of a core. The early American Mariner 4 mission encountered and crossed a bow shock indicative of a small magnetosphere (Fig. X.1), and the Soviet Mars 2, 3, and 5 missions also detected evidence of a bow shock and magnetopause. A solar wind dynamic pressure balance with the energy density of a planetary magnetic field gives

$$P_{\text{dyn}} = \tfrac{1}{2}\rho v^2 = \mathrm{B}^2/8\pi, \tag{X.1}$$

which, at the stagnation point of the putative Martian magnetosphere (at which there is normal incidence of the

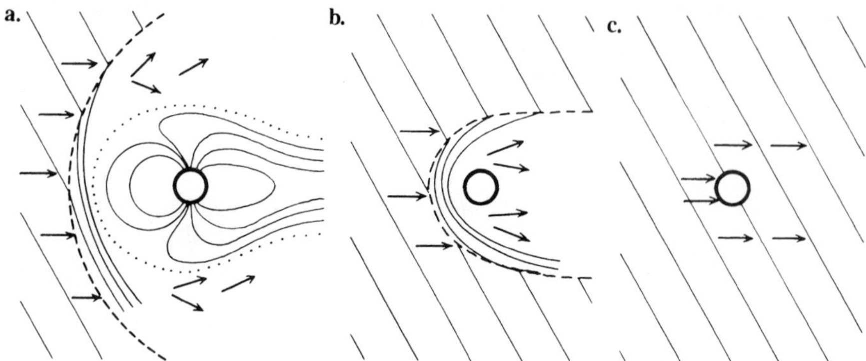


**Figure X.1** Interaction of the solar wind with planets. The solar wind, flowing from left to right, interacts with a planetary magnetosphere in part a. The region of continuous, undisturbed solar wind flow is terminated at the bow shock (dashed line). Inside the dotted line (the magnetopause) the magnetic field is dominated by the planet, and field lines with low magnetic shell numbers are closed. The high-latitude field lines are blown outward from the sun by the dynamic pressure of the solar wind and are open. The two regions of open field lines of opposite polarity are separated by a neutral sheet. Part b illustrates the interaction of the solar wind with an electrically conducting sphere with no intrinsic magnetic field. Solar wind field lines, which cannot penetrate the conducting surface (which may be a hot iron-bearing silicate, a metallic core a conductive ocean, or a dense ionosphere), pile up before the obstacle and slip around its edges. An insulating cold silicate or ice body, shown in part c, permits free passage of the solar wind magnetic field.

solar wind momentum flux), is about $2.5 \times 10^{-9}$ dyn cm$^{-2}$. The magnetic field required to stand off the solar wind is then $2.5 \times 10^{-4}$ G, or 25 $\gamma$. However, the bow shock is rather close to the top of Mars' ionosphere (at a height of 5000 km), implying that the magnetopause lies in a region in which a significant pressure of electrons and positive ions can be present. In addition, the ionosphere is an electrical conductor, so that the embedded magnetic field in the solar wind can induce currents in the ionosphere, which in turn generate magnetic fields. Thus 25 $\gamma$ is in fact an *upper limit* on the magnetic field strength due to Mars' intrinsic field. After correcting for the ionospheric gas pressure, a field strength of 5 to 15 $\gamma$ seems reasonable. The volume integral of the planet's magnetic field would then be in the range $2 \times 10^{21}$ to $6 \times 10^{21}$ G cm$^3$, which is between $2.5 \times 10^{-5}$ and $7.5 \times 10^{-5}$ that of Earth. A reasonable estimate of the height of the magnetopause is then only 400 km. To what degree this field is due to ionospheric electrical currents remains undecided. If the field is indeed planetary in origin, then the surface field is of order 20 $\gamma$.

The magnetometers on the Mars Global Surveyor (MGS) spacecraft found a complex magnetic signature from its vantage point in low Mars orbit. MGS found no evidence for either a present-day dynamo or a distinct dipolar component to the measured field. Instead, it found a complex variation of fields due to remanent magnetization of the Martian crust, with little evidence of contributions from ionospheric currents. Most striking was the discovery of a pattern of magnetic bands of alternating polarity, reminiscent of the magnetic banding of Earth's ocean floor, in the southern hemisphere. The remanent fields are surprisingly strong, about 10 times as high as in terrestrial rocks, requiring the former presence of an episodically reversing strong dipolar field, probably quite early in the history of the planet. We shall see in our discussion of the SNC meteorites that very early melting and differentiation of Mars and emplacement of the core is required by the isotopic data.

If the field is due to a core dynamo, then the core must remain at least partially melted. A pure Fe or Fe–Ni core would have so high a melting temperature that mantle convection should quickly freeze the core. An FeS core would have a much lower melting temperature (Fig. VIII.6) and would make maintenance of core fluid motions much easier. Best, however, is an FeS core with an admixture of metallic iron, which would begin to melt at the Fe–S eutectic temperature. Thus, if we knew with certainty that Mars had a planetary dipole field, such as an unambiguous record of magnetometer data from a lander, uncontaminated by magnetic fields generated by electrical equipment on the lander, we could then draw important conclusions about the planetary core.

A second aspect of the magnetic field strength is also of great importance: if the planetary field is too weak, the solar wind can sweep ions from the upper atmosphere and remove them from the planet. Thus magnetic fields may have an important influence on the evolution of abundances of gases that have too high a molecular weight to be vulnerable to thermal (Jeans) escape.

## Gravity and Tectonics of Mars

Very precise tracking of the Mariner 9, Viking 1 and 2, and MGS Orbiters and the Mars Pathfinder lander have permitted determination of the gravity field of Mars up to spherical harmonics of degree 12 and order 12 (see Appendix V). Generally the gravity field is well compensated on large scales, but some prominent local features, especially the largest volcanoes, remain uncompensated, presumably because they were erupted rather late in Martian geological history onto a thick, cold crust with high strength. Several of the largest and most recent of these great volcanoes lie, in apparent sequence of age, along a line called the Tharsis Ridge.

Another line of photogeological evidence for global-scale tectonic activity, possibly related to core formation or to changes in the orientation of the spin axis, is found in the widespread fracture systems in the crust. Apparent global-scale sets of lineaments were found in Mariner 4, 6, and 7 photography, including two different generations of lineaments in the Valles Marineris–Labyrinthus Noctis region. The older fractures, which trended along WSW–ENE directions, are of tensional origin. The later, less well-defined, lineaments trend WNW–ESE and N–S and appear to be associated with the opening of the Valles Marineris canyon system. Many of the largest channels may be initiated by tensional fracturing, but the origin of such huge extensional displacements is not easy to understand. Sets of concentric and radial fractures have also been found associated with both the Hellas and the south polar basins, presumably due to local rather than global stresses.

The frequency distribution of altitudes (the *hypsometric curve*) for Mars is unimodal, with one enormous topographic high associated with the Tharsis volcanic complex and bulge and with two low areas, the great Coprates canyon system and the Hellas basin (Fig. X.2). The large majority of the surface of the planet has elevations that are distributed fairly symmetrically about the mean. The simplicity and unimodal character of the hypsometric curve do not, however, imply a simple or uninteresting geology at intermediate elevations.

Precision tracking of the MGS spacecraft and altimetry by its Mars Orbiter Laser Altimeter (MOLA) have demonstrated an offset of the center of figure of the planet from its center of mass of 2986 m, of which about half can be attributed to the Tharsis Ridge volcanoes. Most of the elevation variability can be attributed to Tharsis, the Hellas basin, and massive sediment transport into the Chryse region. The north pole is found to lie 6 km lower than the south pole, which helps explain the lack of symmetry in seasonal behavior between the two poles. Slopes found by MOLA show that some 75% of the surface area of the planet drains into the northern lowlands. The southern high, cratered terrain was found to have a thicker crust close to isostacy. The uncompensated gravity anomalies in the northern hemisphere reflect a thinner and stronger crust in the lowlands.

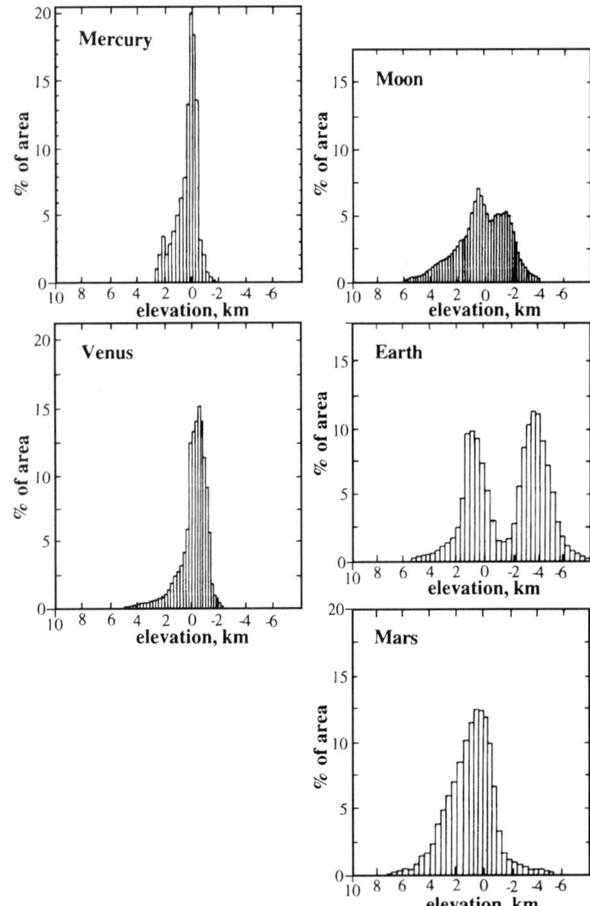


**Figure X.2** Hypsograms of the terrestrial planets. The elevation distributions of Mercury, the Moon, Venus, Earth, and Mars relative to the their mean radii are shown. Note the "tail" of the distribution, reflecting the small amount of elevated terrain on Venus. The weakly bimodal distribution for the Moon reflects the highland–mare dichotomy, and the strongly bimodal distribution for Earth corresponds to the division of Earth's surface into two preeminent classes, abyssal plains (near −4 km) and continental platforms (near +1 km).

## Geology of Mars

Another crucial line of evidence, not available until the spacecraft missions of recent years, concerns the surface morphology of the planet. One of the principal contributions of the Mariner and Viking missions was the compilation of a vast library of photographs and maps, which reveal a great wealth of detail concerning the tectonic activity of Mars. Only a few of the detailed geological studies based on these images and maps can be discussed here. Only those that have some evident relationship to the overall tectonic history and structure of Mars or to the release, storage, or cycling of volatiles will be emphasized.

The first great surprise generated by spacecraft studies of Mars was the heavily cratered appearance of the provinces seen by Mariner 4. Crater counts quickly revealed that the region seen must be extremely ancient (several billion years) and that degradation of these craters by wind and thermal stresses caused them to have more subdued profiles than lunar craters of comparable diameter. Most of the features found on earlier Mars maps, which had been based largely on drawings made by visual observers with moderate-sized telescopes, could not be identified in the enormously superior flyby photographs. However, certain of the *canali* did seem to be associated in a crude way with broad, dark, apparently tectonic features. The sharp, narrow, straight lines of Schiaparelli and Lowell were nowhere to be seen, an evident optical illusion.

The much wider but still very unrepresentative coverage of the Mariner 6 and 7 missions revealed, in addition to the heavily cratered terrain, other regions with "chaotic" or essentially featureless appearance. The huge Hellas basin was found to be almost completely devoid of discernible features, possibly because of erosional obliteration. The *chaotic terrain* (Fig. X.3) seemed to have been produced by slumping and collapse on a vast scale, possibly related to melting or evaporation of ground ice or to geothermal activity. It was generally accepted that the cratered regions were representative of Mars. Several large craters also were identified and studied by Earth-based radar. The general results of the Mariner 4, 6, and 7 imaging was a confirmation and extension of the attitudes generated by the spectroscopic data, which showed only a very thin atmosphere. Both lines of evidence contributed to the depressing prevailing image of Mars as a dead, Moon-like planet.

One interesting issue addressed by the early spacecraft imaging is the relationship between albedo and elevation, which is tied up with the interpretation of the dark alleged *canali*. Carl Sagan and James Pollack, then at Harvard College Observatory, claimed that the dark areas tend to lie higher than the bright areas. Others, measuring the surface relief by means of the horizontal variations in the $CO_2$ column abundance above the surface, found no correlation between albedo and height. Radar observations by MIT's Gordon Pettengill and co-workers tended to confirm this generalization. The radar reflectivity was found to vary between bright and dark regions in a manner suggesting that the bright "desert" regions were rougher. Other authors claimed that the bright desert areas were high, that the dark maria were low or on slopes, that *canali* occurred in broad and deep valleys, and that the dark areas were probably of biological origin. More extensive IR $CO_2$ abundance mapping, both from the Mariner 6 and 7 IR spectrometer and from Earth-based IR studies, showed that the high areas are generally bright, whereas low areas can be dark, light, or mixed. The dark Syrtis Major region was found to have a large mean slope. The Russian astronomer V. I. Moroz found by the same technique that dark areas could be either low or high, whereas several other workers independently concluded that no correlation between albedos and topography could be found!

The first three successful Mars flyby missions, Mariners 4, 6, and 7 (see Appendix III for a complete historical summary) thus returned very limited imaging data that, by an improbable coincidence, provided coverage of heavily cratered, very ancient terrains. The consensus understanding of the nature of Mars was so strongly affected by these flybys that it became widely accepted that Mars was a geologically dead planet, more like the Moon than like Earth.

One of the most tantalizing results of these early flyby missions was the hint by CalTech's Bob Leighton that the Tharsis region, which had not been well observed by the Mariner flybys, might depart dramatically from the character of the regions seen most clearly. Indeed, the thorough mapping of the Martian surface by the very successful Mariner 9 Orbiter mission revealed a rich variety of previously unknown or poorly characterized terrain types and phenomena. Also, because Mariner 9 orbited about Mars rather than flying by at high speeds, it was an excellent platform for study of time-dependent changes on the surface over many months.

One of Mariner 9's most striking discoveries was a chain of massive volcanoes, the Tharsis Ridge, several of which are larger than the biggest volcanic constructs on Earth. Further progress in crater-count dating was made possible by the very high resolution and nearly complete areal coverage of the Mariner 9 photography. This study permitted the determination of the relative ages of different parts of the surface, and, with certain assumptions about the calibration of the flux of asteroidal and cometary bodies at the orbit of Mars, absolute ages could be estimated. A smooth decay of the cratering flux and a constant degradation rate for the impact craters seem to fit the data well. Early analysis of the Mariner 9 crater density data by Bill Hartmann of Tucson's Planetary

**Figure X.3** The rim of Ganges Chasma on Mars. Note the huge landslide features on the canyon walls in this Viking 1 image. The dark area at the lower left is a dune field. Wind erosion has sculptured the rim of the chasm, and numerous impact craters dot the ancient high terrain. The existence of a kilometers-thick layer of weakly consolidated sediments is attested by the depth of the excavation of the chasm. The presence of abundant volatiles, presumably mostly water ice, is revealed by the splash marks around the largest and deepest impact craters.

Science Institute (PSI) suggested that the ancient heavily cratered terrain was about 3–4 billion years old, whereas the large volcanic constructs were only 300 million years old. (Note, however, that the flux of comets and asteroids across the orbit of Mars is not yet very tightly constrained.) The most ancient areas were found to be fully as heavily cratered as the lunar highlands. Degradation processes, initially (as on the Moon) largely due to the bombardment process itself, later became dominated by weathering and wind erosion.

The distribution of heavily cratered terrain on Mars is far from uniform; like the Moon, the Martian surface falls naturally into two rough hemispheres with very different histories. Just as the Moon has a side (the far side) that is heavily cratered and very ancient and a side that bears almost all of the large impact basins that have been thoroughly resurfaced, so also Mars divides into two hemispheres with a similar age dichotomy. The older terrain on Mars corresponds roughly with the southern hemisphere. The northern hemisphere is noteworthy for both extensive plains and gigantic volcanic structures. Mars is also noteworthy for two types of features that appear to have been cut by running water. First, there are dry dendritic valley systems, sometimes approaching in complexity the structure of terrestrial river systems. Second, there are very wide (up to 100 km width), intensely eroded outflow or flood-cut channels, apparently cut during brief and catastrophic episodes, which have very little dendritic structure and run for distances as large as 2000 km. Both of these types of valley preferentially originate in the ancient cratered terrains.

Martian craters exist in a wide range of morphologies. Small craters are bowl-shaped, with a transition to complex morphologies (central peaks, slump terraces on the crater walls, and multiple-ring structures) at a mean diameter of about 7 km (Fig. X.4). Mercury, with very nearly identical surface gravity, shows the same transition at a mean crater diameter of about 14 km. The general trend of the transition is for it to occur at smaller sizes on bodies with larger surface gravities, but this simple correlation masks a variety of more complex effects. The mean density of the impactors may vary systematically from one location to another. Larger planets should generally have a higher threshold velocity for impacts; larger planets should generally have steeper temperature gradients in their crusts and hence be capable of rapid viscous relaxation at shallower depths; larger planets may have higher concentrations of volatiles in their crusts, etc.

Martian craters exhibit a much wider range of morphologies than lunar or Mercurian craters. Many are surrounded by "splash" marks, as if the crater were made by an impact in mud or wet sand. Note in Fig. X.5 the 20-km splashes surrounding the crater Arandas. This crater, with a rim-to-rim diameter of 25 km, has terraced, severely collapsed walls and a large central peak, as shown on this Voyager 1 image. The floor of the largest known crater on Mars, the Hellas basin, which lies fully 6 km below the level of the surrounding rim, appears nearly featureless in many photographs because of atmospheric dust and haze. Pre-Viking studies of the origin of large craters and of their subsequent evolution led to the early suggestion that extensive isostatic adjustment after the Hellas impact had modified the terrain extensively, more than in the comparably sized craters on the Moon. A number of related degradational and igneous processes that are widely associated with large Martian craters have been documented. For example, extensive fracturing of crater floors, often with intrusion of magma along crustal faults (*dike* activity), attests to volcanic activity stimulated by impacts. Many internal modifications appear to be due to a draining (*sapping*) of meltwater from the shock melting of mineral ice in the preimpact rock. Long sinuous channels are often found associated with large impact craters, especially at low latitudes

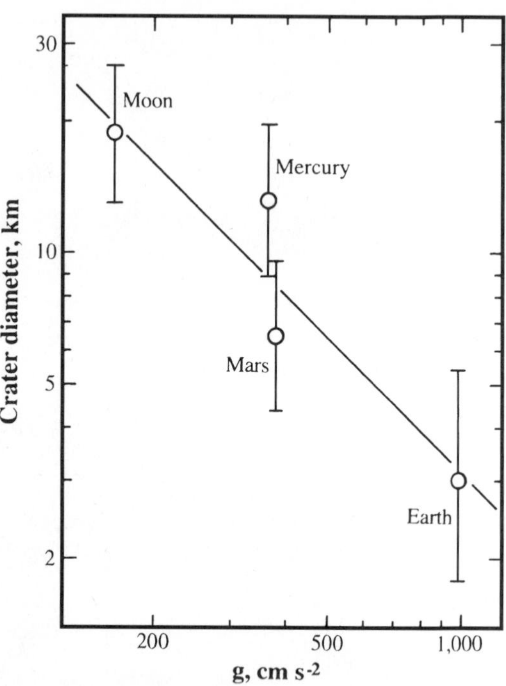


**Figure X.4** The dependence of impact crater morphology on gravity. The critical size for the transition from simple bowl-shaped craters to complex crater morphology (central peaks, terraces, and pronounced wall collapse) is graphed for unambiguous impact craters on the Moon, Mercury, Mars, and Earth. The general effect of increasing gravity in suppressing surface relief is evident, despite the large differences in surface temperature, presence of water, and crustal temperature gradient in these bodies.

**Figure X.5** The Crater Arandas on Mars. This crater, about 24 km in diameter, has a central peak complex and extensive wall collapse, as expected from its size. The most striking feature in this Viking 1 Orbiter image is the radially striated ejecta blanket surrounding the crater. The appearance of this material suggests a surface surge of highly water-rich material rather than ballistic transport of fragmental debris.

and in the vicinity of the huge Coprates canyon complex. Lobate external flow structures (Fig. X.5) are also commonly found around craters in the Coprates province, suggestive of the flow of a fluidized material derived from shock heating of a volatile-rich surface.

An even more striking type of evidence for the existence of massive deposits of volatiles (presumably water ice) in the Martian crust is the prevalence of large areas of tumbled "chaotic" and deeply depressed troughed terrain. The volume of ice that must have been withdrawn from beneath the surface in order to produce the observed depressions is roughly $1.6 \times 10^6$ km$^3$. The greatest puzzle attaches to the greatest valley of all, the Valles Marineris complex, of which Coprates is a major component. There is simply no plausible mechanism for excavating the vast amounts of missing material. The enormous landslides along the walls of the valley appear to have had the character of mud flows, presumably originating in the release of water from subsurface aquifers at the heads of the slide regions. No such features are known on either the Moon or Mercury. A portion of the great canyon complex is shown in Fig. X.6. A detailed view of several giant landslides is also given in Fig. X.7.

Of all the features seen on Mars, by far the most exciting evidence for the presence of water is the channeled terrain, which is laced with large *dendritic* (branched) valley systems. These valleys are generally very much less branched than river systems on Earth that derive their water from widespread precipitation. Some of the larger sinuous, braided channels can be traced for 2000 km and have channel widths up to tens of kilometers. Victor Baker of the University of Arizona argues that the only reasonable interpretation of the morphology of these channels seems to be that they were cut by catastrophic flooding by liquid water (Fig. X.8). Widespread presence of ground ice seems essential to provide the observed flows. But despite the numerous riverine structures, there is little evidence for drainage of surface water or for precipitation on a large scale. Stream valleys are generally very immature and have apparently been formed in one or a few major catastrophic episodes, perhaps lasting only a few days. These rare, brief flooding episodes may have been spaced widely over billions of years. Russian geochemist C. P. Florenskii has claimed that flooding may have persisted down to fairly recent times, but the weight of the evidence from counts of superimposed craters is that most of the channels are very old. Some of the smaller Martian channels may have been formed by lava, but the most impressive channels were not: rather, they were probably formed by melting or evaporation of permafrost or massive deposits of ground ice. The antiquity of the channels is also attested to by the fact that they are found in ancient cratered terrain. The late Harold Masursky dated the channels as between 0.5 and 3.5 billion years old.

Another ice-related feature, an area of numerous small (1 km diameter) domes in the Cydonia region, is apparently due to volcanic eruptions through ice-saturated ground, analogous to similar features seen in Iceland.

These enormous surface expressions of the presence and mobility of water are important because modern Mars surface conditions cannot support the presence of liquid water. The surface atmospheric pressure of Mars is less than the vapor pressure of $H_2O$ at the triple point; the presence of large, open expanses of liquid water would require an atmospheric pressure, surface temperature, and atmospheric IR opacity much higher than those we find at present. It is of great importance to determine whether there has been a secular decline in the atmospheric mass or whether periodic or random fluctuations occur by which the atmospheric mass is changed by a factor of several. In either case, the channels clearly attest to the presence in Mars of a reservoir of volatiles vastly larger than the atmospheric inventory and show that far more volatiles were exposed at the surface at times in the past than can be currently observed.

An ironic aspect of the discovery of the enormous chasms and apparently flood-cut valleys is that they remind us forcibly of the prespacecraft days when Mars was covered with canals. It is, however, clear that there is essentially no correspondence between the features mapped by Lowell and those seen by Mariner 9 and the Viking Orbiters.

From the time of the first tentative identification of volcanic features in 1972, the age and nature of the volcanic activity on Mars have been of great interest. Michael Carr reported finding remnants of a number of very ancient large volcanoes lying along several lineaments. Of the several huge, fresh (nearly uncratered) volcanic piles seen along the Tharsis Ridge, the largest and youngest, Olympus Mons (Fig. X.9), has a volume of at least $8.5 \times 10^5$ km$^3$, comparable to the total volume of all the volcanoes in the Hawaii-Emperor seamount chain. The great height of the rim of its summit crater (*caldera*), about 27 km, presents a severe challenge for sources of magma. Either a very low-density magma was extruded or the base of the magma column must have extended to very great depths within the planet. The weight of a fluid magma column is in balance with the hydrostatic pressure exerted by the nearby solid mass of the volcano:

$$\rho_{crust}\, g H_{crust} > \rho_{magma}\, g H_{magma}. \tag{X.2}$$

Here $H_{magma}$ is the height of the magma column and $H_{crust}$ is the thickness of the solid crust (the *lithosphere*). For hydrostatic extrusion of magma, the pressure exerted by the magma column must be less than that exerted by the adjacent crust and the height of the magma column must equal the thickness of the crust, which leaves simply $\rho_{magma} < \rho_{crust}$ as the criterion for eruption. The buoyancy of the magma, $(\rho_{crust} - \rho_{magma})gH$, must be positive for eruption to occur. The low viscosity required for such long surface flows as those seen on Olympus Mons suggests a basaltic, rather dense magma, which in turn implies an extremely deep magma source and a very thick crust. The mobility of vast amounts of lava is further attested to by the presence of great

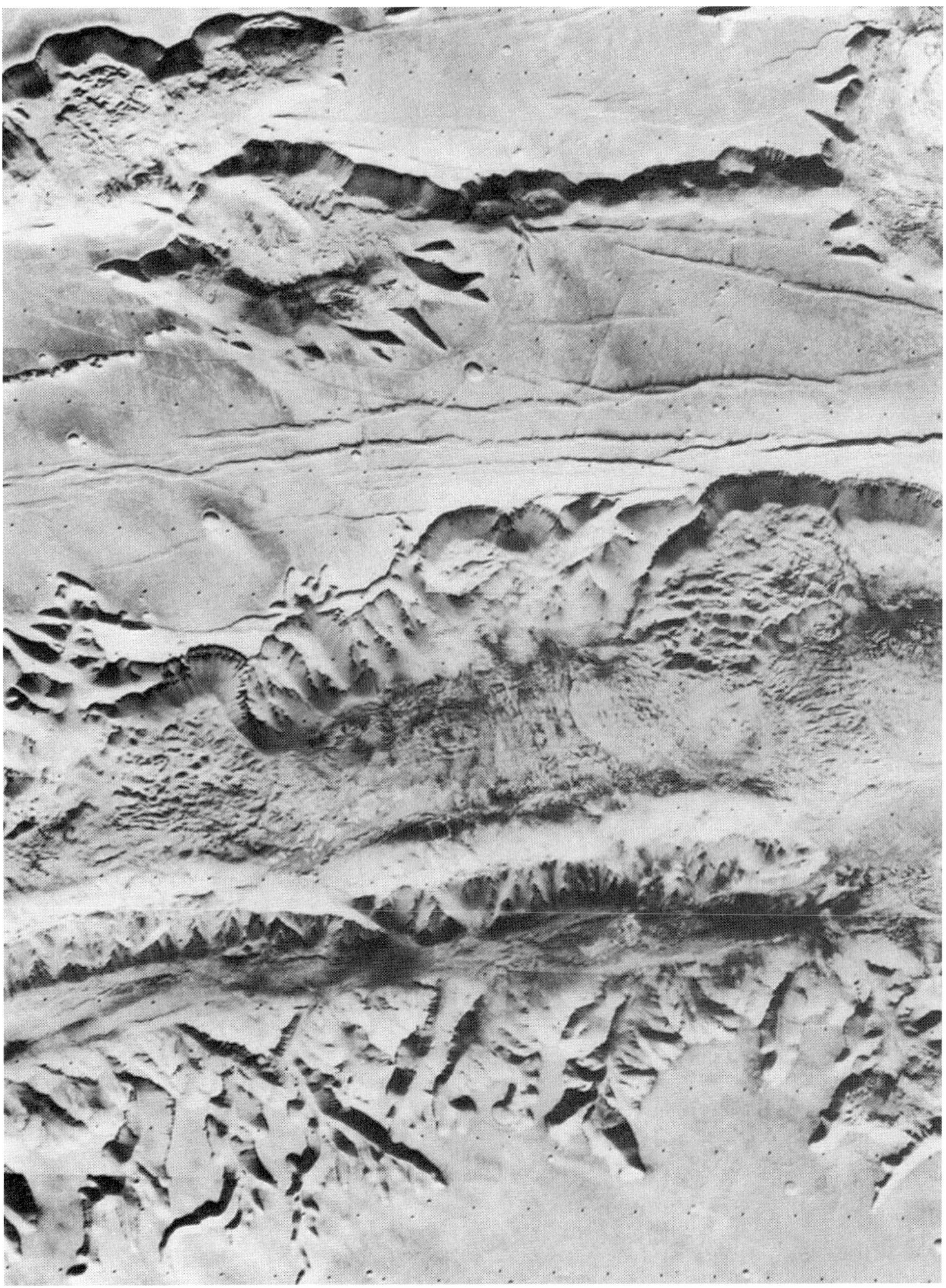

**Figure X.6** A section of Valles Marineris. This enormous canyon complex spans some 30% of the circumference of Mars. The floor of the main canyon contains a long ridge system that parallels the axis of the canyon. Note the subsidiary canyons to the north (the top of this Viking 1 Orbiter image) and the tensional grabens that also parallel the canyon walls. Giant landslide features are evident in the main and secondary canyons. The stubby dendritic erosional systems on the canyon edges are presumably due to wind erosion. Careful examination of the original image suggests the widespread presence of a thin, moderately consolidated surface layer in the highlands overlying deep, weak deposits of sediments.

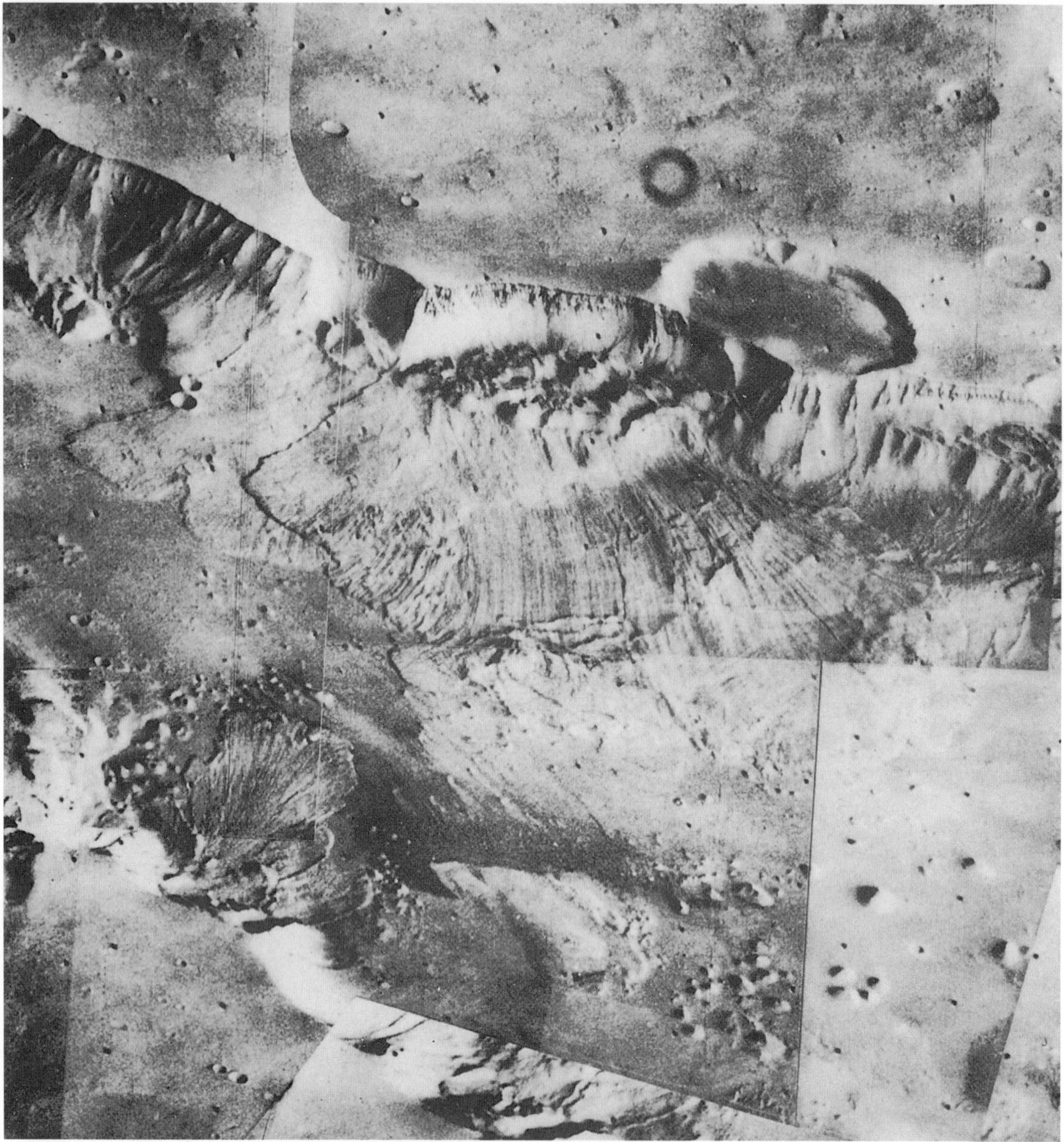

**Figure X.7** Landslides in Valles Marineris. This Viking 1 image reveals several generations of spectacular landslides and a partially destroyed large impact crater. Note the numerous small hillocks on the canyon floor, possible evidence of small-scale volcanism.

complexes of collapse craters (calderas) at the summits of Olympus Mons and the Tharsis ridge volcanoes. The 100-km summit caldera of the great Tharsis volcano Arsia Mons, complete with lava flow features and ring-shaped faults, suggests the withdrawal of tens of thousands of cubic kilometers of magma from beneath the summit (Fig. X.10).

Morphologically, the volcanoes seen by Mariner 9 seem more analogous to intraplate (continental) volcanism on Earth than to activity above crustal subduction zones, and they qualitatively suggest the presence of a very few extremely intense "hot spots" in the Martian mantle. (As important as these hot spots are to the volcanic geology, the actual magma extrusion rates

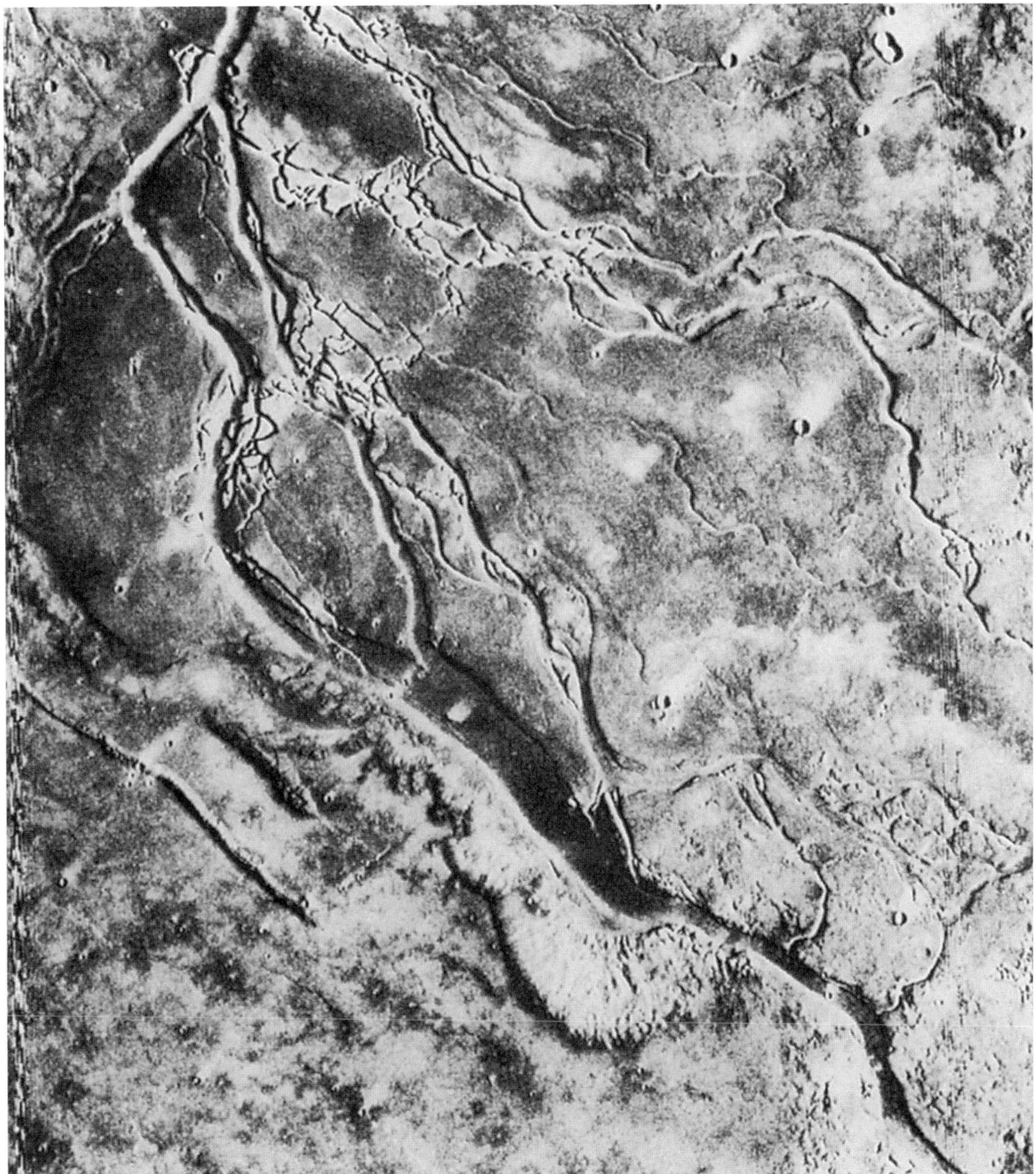

**Figure X.8** Complex drainage systems on Mars. These features seem to reflect two or more generations of erosional activity, with the larger channels cutting the smaller (and hence younger) ones. The system is much more braided than dendritic in character, and well-defined stream heads are rare or absent. The numerous small impact craters (many detectable from their accompanying wind streaks) attest to the great age of the uneroded terrain.

averaged over billions of years are not great; the volcanoes are spectacular, but virtually all of the heat escaping from the interior of Mars is carried by conduction through the lithosphere.) The Tharsis ridge is intimately connected to a vast, domelike expression of the Martian crust (called the Tharsis bulge), covering over 20% of the surface area of the planet and centered somewhat to the southeast of the line of volcanoes, which seem to have formed at a later time than the bulge. The Tharsis bulge is seen as both a topographic and a gravitational excess. The youngest volcanic constructs are only partially compensated at depths on the order of 100 km and may possibly be more fully compensated by low-density roots at greater depth.

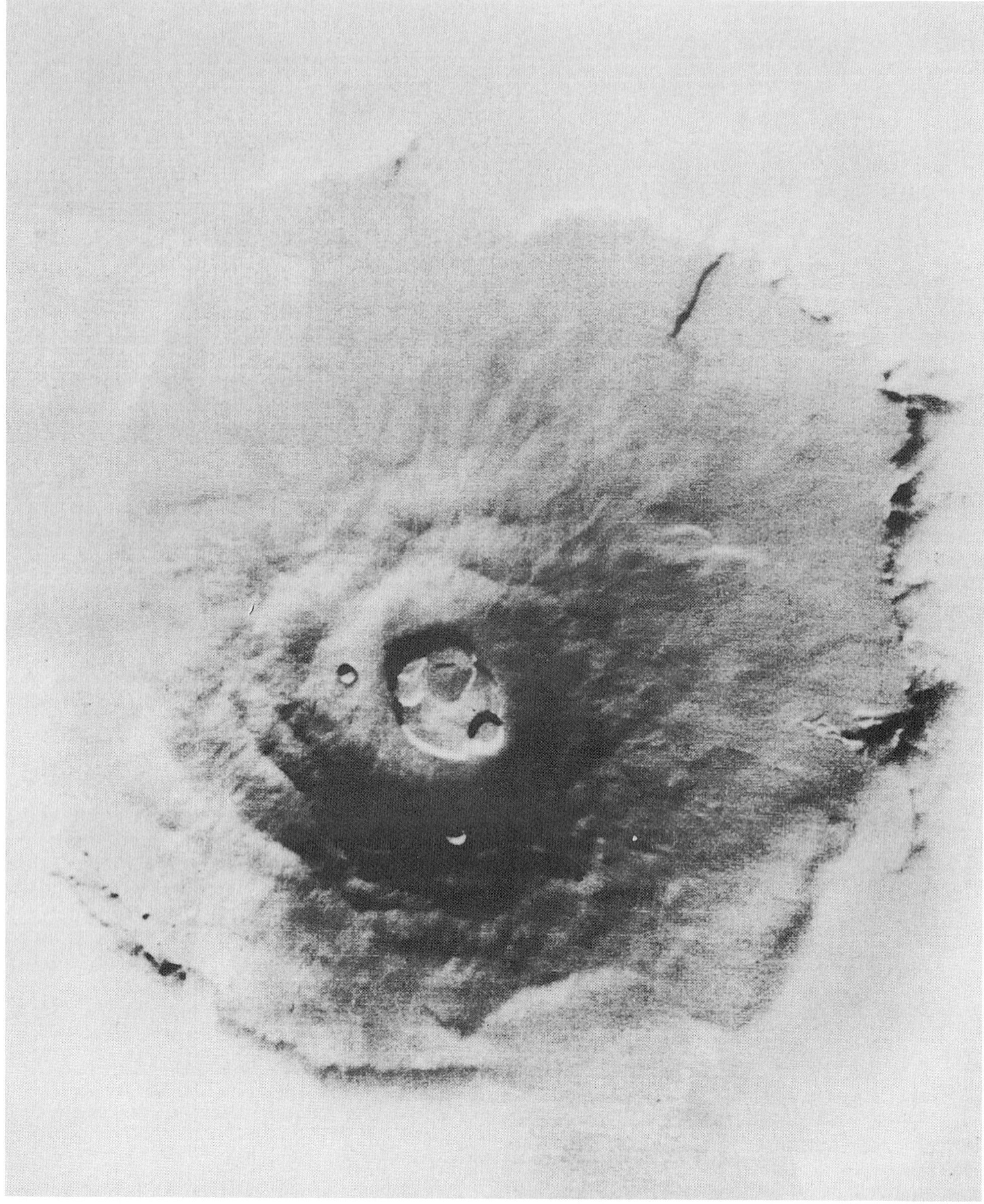

**Figure X.9** Olympus Mons. The largest known volcano in the Solar System, Olympus Mons rises nearly 26 km above the surrounding plains, has an average base diameter of about 500 km, and has a complex caldera system about 65 km wide at its summit. Steep scarps surround most of the periphery of the cone, suggesting erosive removal of kilometers of sediments from the surrounding lands since the time of the main eruption. Two large impact craters can be seen on the flanks of the volcano in this computer-processed Mariner 9 image.

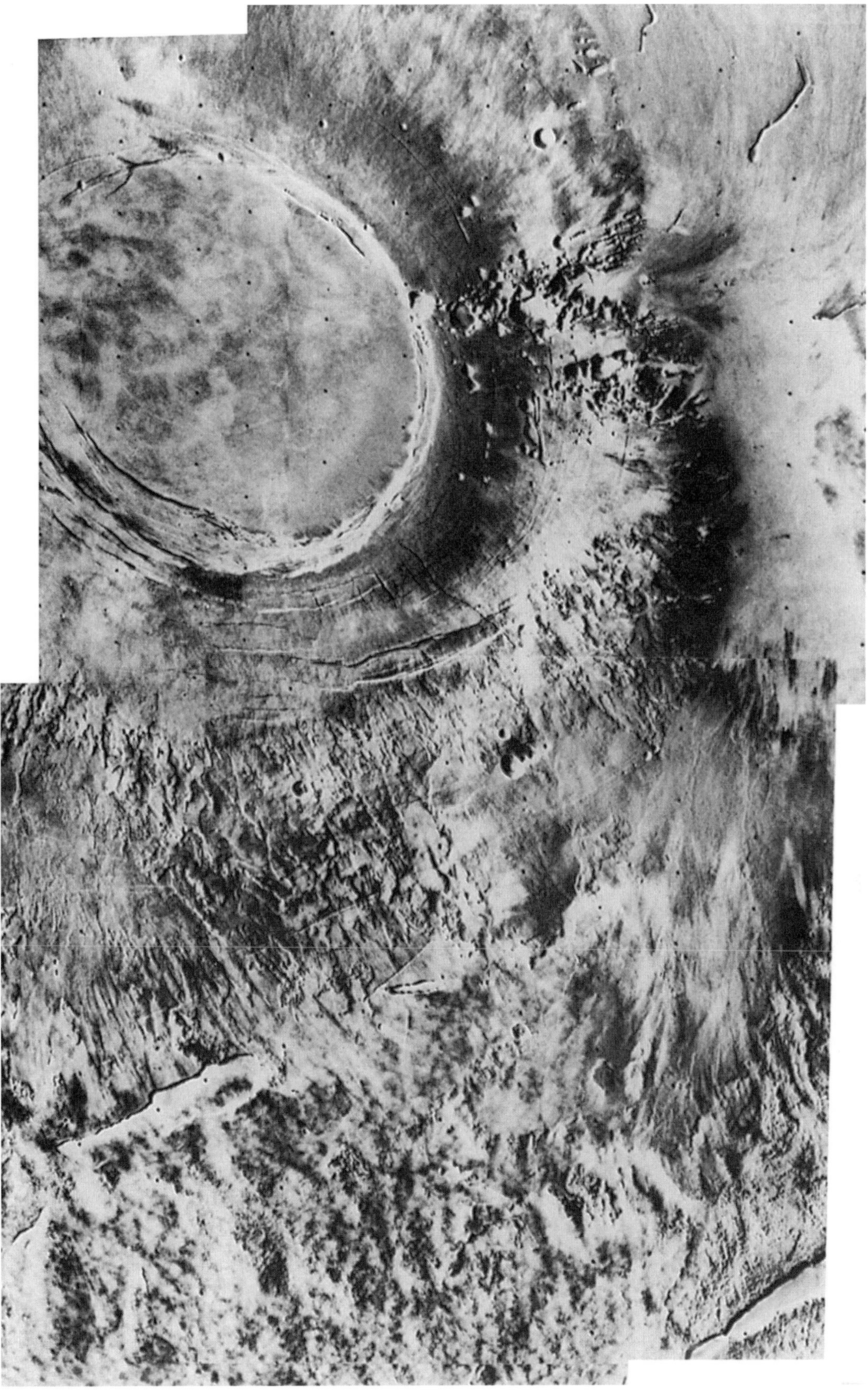

**Figure X.10** Arsia Mons in the Tharsis Ridge. A portion of Arsia Mons, including most of the 100-km summit caldera complex and extending out to the edge of the cone, is shown in this Viking 1 Orbiter image. Roughly three times the height of the largest volcano on Earth (Mauna Loa, which rises 9 km from the Pacific floor), it rivals Olympus Mons in sheer mass. Several scarps and lava channels can be seen on the flanks of the cone. The numerous small impact craters suggest an age of well over a billion years.

One interesting feature of the structure of Olympus Mons is the high scarp that surrounds it (Fig. X.9). The scarp has been interpreted in terms of ash flow emplacement mechanisms that cause the deposited ash to become less dense and physically weaker when deposited (at lower temperatures) far from the vent. Alternatively, erosion of a soft older substrate by winds may be responsible for the scarp. Violent volcanic (*pyroclastic*) activity and formation of ash flows are apparently widespread phenomena. The numerous narrow lava channels seen on the slopes of Olympus Mons attest to the ability of the lava extruded from the summit to flow for distances of hundreds of kilometers. The flow behavior deduced from study of these features departs from that of the oldest terrestrial analogue, Hawaii, in that the Martian lava was apparently more silicic and its eruption rate was much larger than that of any single terrestrial volcano. A cratering age of about 2.5 billion years was originally estimated for Olympus Mons based on the crater density, but several workers have found extensive areas near the summit of Olympus Mons and in the calderas of that volcano and of Arsia Mons with crater counts suggestive of an age of about 100 million years. Geologist Don Wise, commenting on the earlier view that Mars is a dead planet, remarks that Tharsis "must be regarded as at least a warm corpse." It is likely that active volcanism has spanned most of the history of Mars, and rare episodes of present-day volcanism cannot be ruled out.

Evidence of wind erosion is ubiquitous. Wind streaks are seen more often in the lee of craters or other topographic features. The orientation of the wind streaks often changes from season to season, reflecting changes in the direction of the prevailing winds. Some depressions have served as collectors of wind-blown debris, as the 130-km × 60-km dune field in the Mariner 9 image of a crater bottom shows (Fig. X.11). The giant dunes pictured here are roughly 100 m high and have a wavelength of about 1500 m.

Dune fields are widespread on Mars and probably have been for billions of years. Figure X.12 shows two examples of dune fields as seen by the MGS imaging system, as well as the effects of dust devils and evidence that layered deposits extend back for very long periods of time.

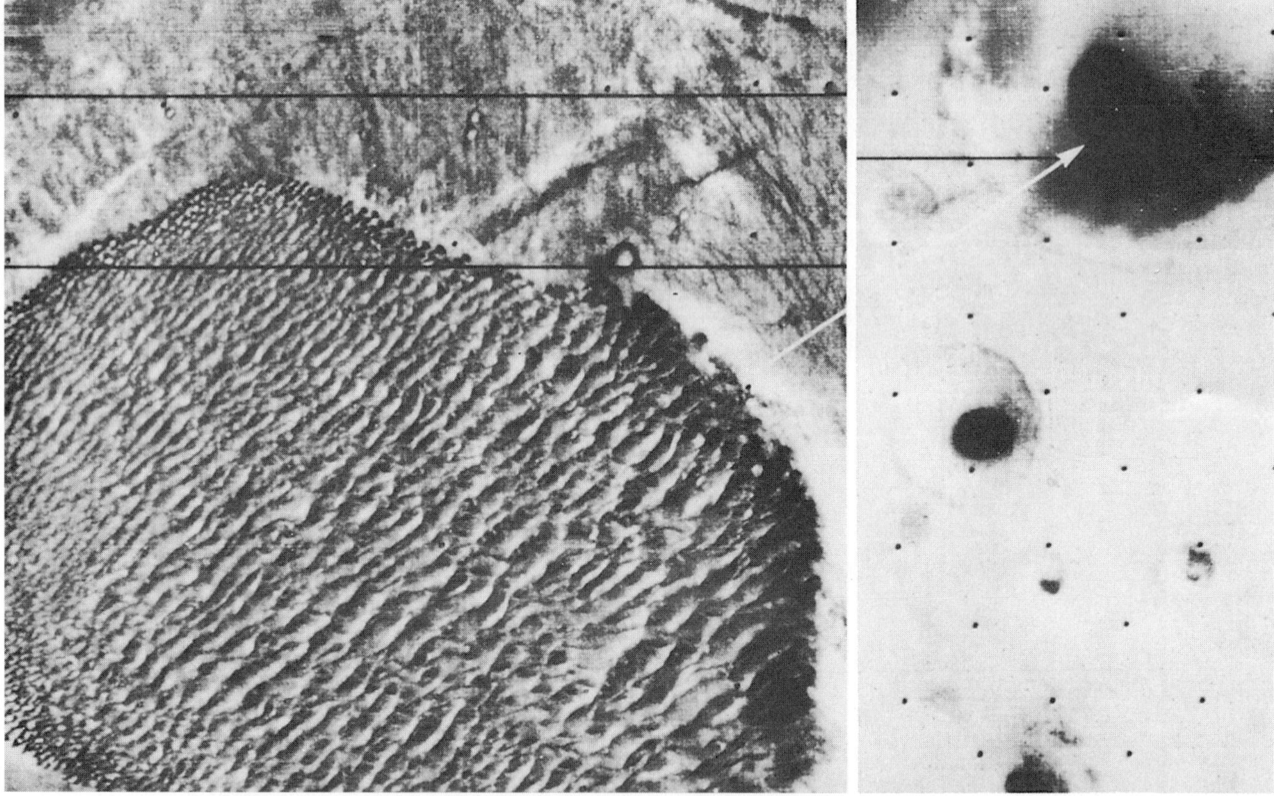

**Figure X.11** A dune field in the Hellespontus region of Mars. The floor of a 150-km crater contains a dune field about 120 × 60 km in extent. The wavelengths of the dunes are typically 1 to 2 km in this Mariner 9 image. The structural and orientational uniformity of the dune field implies long-term consistency in the prevailing wind patterns. As on Earth, dunes near the margin of the field (where the thickness of the layer of windblown sediments is smallest) have smaller heights and wavelengths than those in the interior. Such dune fields are common on Mars.

**Figure X.12** Massive dust transport on the surface of Mars as seen by MGS. Part a shows an enormous arcuate barchan dune many kilometers long and nearly a kilometer wide advancing across a mottled terrain with small patches of bright sediment of unknown origin. Part b shows two families of dunes propagating independently. One set of dunes is made of very dark material, possibly larger grains of basalt, and the other of possibly fine-grained bright dust. Note that the two sets of dunes have different lengths, heights, and wavelengths. Part c shows the tracks left by numerous dust devils as they scour fine bright dust from the underlying darker surface. Part d, a high-resolution view of a canyon wall in the Coprates complex, reveals not only extensive dune fields at the base of the wall, but also, through the layering evident in the wall, strong evidence that the canyon was cut through countless layers of ancient sediments laid down by winds or, possibly, water.

**Figure X.12** *Continued*

The high-resolution view of a portion of the wall of the Coprates valley (part d) shows distinct layering in the sediments through which the canyon was cut. Figure X.13 shows another segment of canyon wall with startling evidence of possible drainage of water or brine from a layer close to the surface. It is also conceivable that these drainage patterns may be produced by small landslides. Note the sediment fans at the end of each channel and the extensive dune fields on the canyon floor.

Finally, the geology of terrains at high Martian latitudes is dominated by recent processes associated with deposition and erosion of volatiles. Both polar caps are surrounded by a broad band of subdued topographic relief and softened detail. The polar caps themselves are thin but widespread (Fig. X.14). During the spring warming, ice disappears first at lower altitudes, giving a striking banded appearance with bright elevations and dark depressions, as shown in the Viking 2 Orbiter mosaic in Fig. X.15. Water ice, with its relatively low volatility, dominates the residual cap. The winter cap, with complete coverage of the ground at all altitudes, is dominated by a surface layer of fresh, bright solid $CO_2$ frost. It is unfortunately impossible to determine how thick the polar ice deposits are.

Deposition of dust after great dust storms must be a planetwide phenomenon. In both polar regions, outside the ice-covered cap, are vast areas of *laminated terrain* that appear to be the result of wind erosion of very thick deposits made up of alternating layers of dark and light material. The individual layers, which are on the order of 10 m thick, are stacked to make deposits that are at least several hundred meters thick. It seems possible that the mass of volatiles buried in the laminated terrains is much larger than the mass of water ice and dry ice in the bright

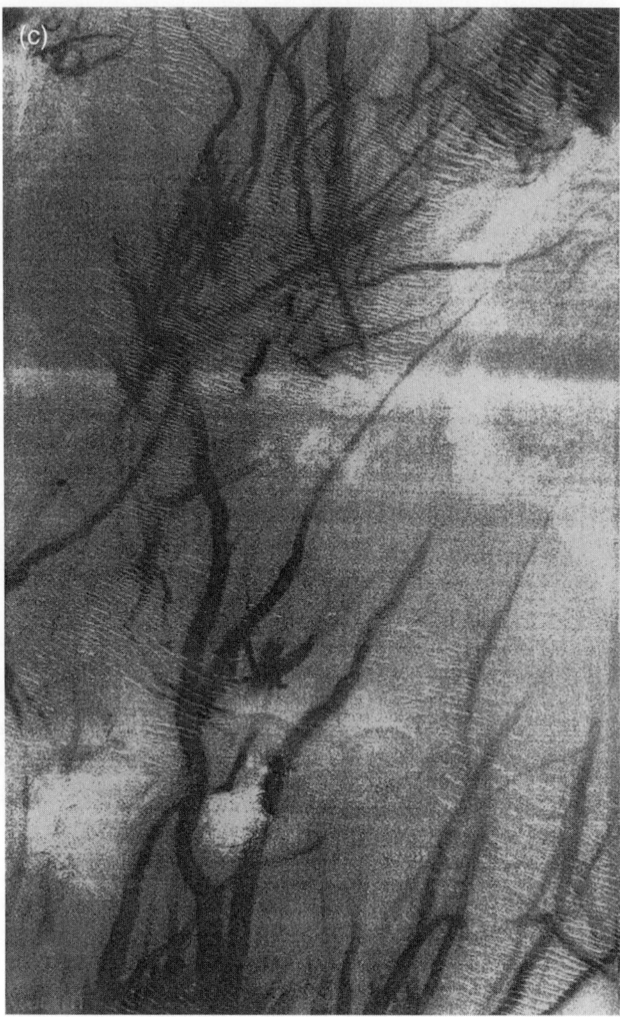


**Figure X.12** *Continued*

**Figure X.12** *Continued*

polar caps. Each layer is much too thick to be produced by annual transport of volatiles, and hence the formation of a layer must reflect a longer-term cycle. One possibility is that the precession cycle of Mars modulates the deposition of volatiles. Figure X.16 shows two scenes of polar laminated terrain, hinting at dozens of cycles of deposition. If the deposition cycle is indeed related to the precession of the spin axis, then the wind-cut valleys seen in Fig. X.16b expose the climate history of Mars for a span of several million years.

## Surface Composition

Prior to the Viking landing missions, our main source of information concerning the chemical composition of the crust of Mars, and hence the best available indicator of

**Figure X.13** Possible water seepage from near-surface sediments in a canyon wall. The interpretation of these features is uncertain, but water seepage is perhaps the easiest explanation of the erosion channels.

the degree of geochemical differentiation of the interior, was Earth-based reflection spectroscopy of the surface.

The first observational fact about the surface of Mars that inspired comment was the pervasive red color of the bright areas. As far back as 1934, the German-born physicist Rupert Wildt published a chemical and physical discussion of Mars in which he attributed the red color to oxidation of iron-bearing igneous rocks by photochemically produced $O_2$ and $O_3$. The reflection spectrum of the surface of Mars, however, was not studied until the 1960s. It appeared from the first spectroscopic data that the water abundance in the atmosphere of Mars was about what would be expected in equilibrium with coexisting goethite and hematite.

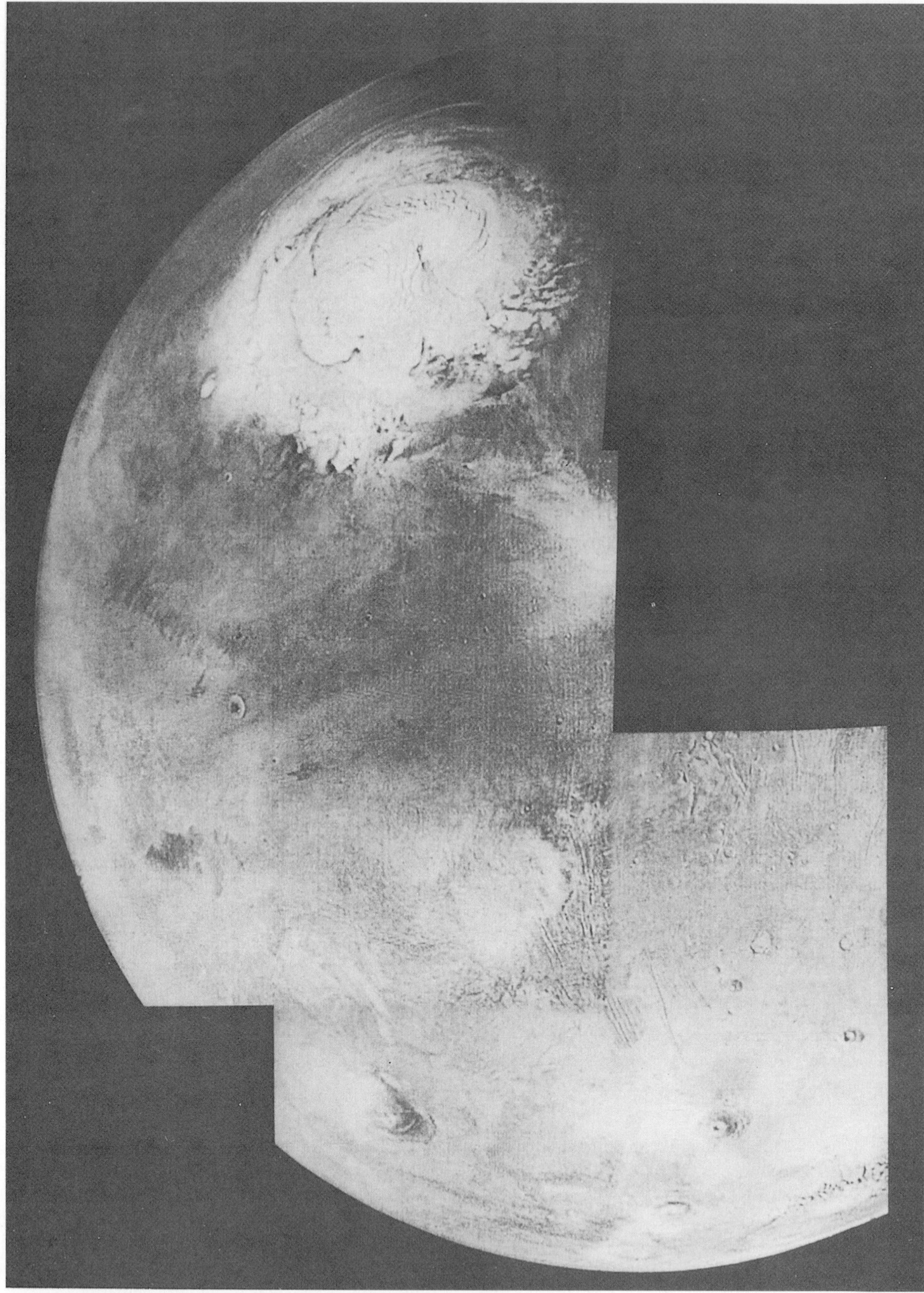

**Figure X.14** Photomosaic of the north polar regions of Mars. The complex structure, wide areal extent, and thinness of the polar seasonal deposits are evident in this Mariner 9 image taken in local springtime. Olympus Mons and the three great Tharsis Ridge volcanoes can be seen near the lower edge of the mosaic. The northern edge of Valles Marineris is barely discernible southeast of the Tharsis complex.

**Figure X.15** The northern polar cap of Mars in summer. This Viking 2 Orbiter mosaic shows wide lanes of dark ground exposed between frost-covered ridges. The residual summer polar cap is made principally of water ice. b. An MGS high-resolution image of the south polar cap in springtime. The winter snow cover (mesas) is rapidly eroding due to evaporation in circular patterns.

$$\underset{\text{goethite}}{2\text{FeOOH}} = \underset{\text{hematite}}{\text{Fe}_2\text{O}_3} + \text{H}_2\text{O} \quad (X.3)$$

$$K_3(T) = p_{\text{H}_2\text{O}}(a_{\text{hem}}/a_{\text{goe}}). \quad (X.4)$$

It now appears from laboratory simulations that goethite would be unstable on the surface in the daytime, but should exist at nighttime temperatures and persist at depths great enough to escape the diurnal thermal wave, which penetrates a few centimeters.

A low-resolution IR spectrum of Mars in the 0.8- to 1.6-$\mu$m region can be very approximately fitted to a laboratory sample of limonite (a natural mixture of goethite and hematite). This was in accord with the idea that the limonite constituted a weathering crust produced by atmospheric oxidation of primary ferrous-iron-bearing minerals. The brightly colored (red) minerals need not be the dominant surface material in order to dominate the color; the large majority of the mass of the surface might be transparent or very-fine-grained silicates, such as feldspars, which would contribute little or nothing to the spectrum. A minuscule mass fraction of limonite stain could provide a rather intense reddish coloration if applied to the surfaces of transparent grains. One possible implication was that the oxidation of the surface was at the expense of vast quantities of water. In that case, the entry of oxygen into the crust was accompanied by the escape of an ocean's worth of hydrogen.

An alternative interpretation was offered by Carl Sagan in response to the failure of the early Mariner flybys to confirm a planetary magnetic field and their photographic evidence for very heavy cratering of the Martian surface. He suggested that Mars had never differentiated (a conclusion later ruled out by the revised moment of inertia) and therefore had never possessed oceans. The red coloring matter he attributed to limonite produced by photochemically oxidized iron-rich minerals.

Early thermodynamic calculations were based on spectroscopic upper limits on the abundance of CO on Mars, considering H, C, O, Ca, Mg, Al, and Fe, and on the very improbable assumption that the equilibrium oxygen fugacity of the atmosphere, $f_{O_2}$, could be calculated from the CO:$CO_2$ ratio at temperatures well below 300 K. Because $f_{O_2}$ is photochemically regulated and independent of equilibrium regulation by CO and $CO_2$ for strong kinetic reasons, conclusions regarding oxidation behavior reached in this manner are not reliable.

**Figure X.15** *Continued*

These calculations used the $CO:CO_2$ ratio to deduce an oxygen fugacity of about $10^{-80}$ bar! But CO is photochemically produced from carbon dioxide, which means that $O_2$ is also photochemically produced in at least roughly comparable amounts via

$$CO_2 \rightarrow CO + O \qquad (X.5)$$

$$O + O + M \rightarrow O_2 + M, \qquad (X.6)$$

where $M$ is any other molecule (such as $CO_2$) capable of carrying off some of the reaction energy of the O atoms. A CO pressure of about $10^{-6}$ bar implies about the same $O_2$ pressure, some 74 orders of magnitude more than the equilibrium assumption predicts! This vastly larger $O_2$ pressure easily stabilizes ferric iron, and goethite becomes stable. However, the portion of these calculations that did not deal with oxidation processes did show that feldspars, corundum ($Al_2O_3$), and boehmite (AlOOH) are plausible minerals and found the calcite + quartz and magnesite + quartz assemblages to be stable.

Meanwhile, an unsuccessful search for the 0.87-$\mu$m absorption band of ferric oxide in the Mars spectrum argued strongly against the presence of large amounts of limonite and indicated that the ferric coloring agent was present as a stain with a rather small mass fraction. The mean particle sizes in the soil were based on interpretation of the photometric and polarimetric properties of the surface in terms of a systematic particle size difference between bright and dark areas. A 3-$\mu$m spectral feature was found that could be interpreted as due to the presence of goethite, but could also be explained by adsorbed surface water or other hydrous minerals. But laboratory reflection spectra of limonite powders do not support the idea that a grain size variation alone suffices to fit the spectra of light and dark areas, which again suggests the presence of hematite-stained transparent mineral grains.

Another interesting possibility compatible with the IR spectrum of Mars is that the clay mineral montmorillonite could be the principal dust component raised during dust storms. The 3-$\mu$m absorption band is probably due to chemically bound water, most likely in phyllosilicates. The strength of the absorption suggests a water content of about 1%. The presence of a significant mass of goethite and clay weathering products is a reasonable expectation based on those physicochemical weathering processes known to exist on the Martian surface. The nature of these weathering processes will be discussed later.

Thomas B. McCord of M.I.T. reported the interesting observation that the spectral change undergone by regions of the surface affected by the "wave of darkening" could be simply described: the albedo changed by the same factor at all wavelengths. Thus, the visual darkening is not accompanied by a discernible color change. Reports by visual observers that the dark areas are "green" are reporting a phenomenon that cannot be verified by instrumentation: the greenness is due to retinal fatigue under intense red illumination. McCord and co-workers also studied the spectral reflectivity of selected small areas of the surface over the 0.3- to 2.5-$\mu$m spectral region and were able to confirm that the most pronounced "green" area on Mars, Syrtis Major, is in fact dark and red rather than green or neutral in color.

**Figure X.16** Eroded layered deposits near the south pole of Mars. (a) The large mesa appears to be made of dozens of discrete, continuous layers of alternating dark and bright materials. These may be wind-blown deposits of alternating ice-rich and dust-rich sediments, reflecting long-term climatological cycles driven by astronomical effects such as cyclic variation of Mars' axial obliquity and orbital eccentricity. The layered terrain, which is common over a vast area near both poles, overlies a basement of very rough terrain. b gives a Viking high-resolution view of a small portion of the layered terrain, showing its great number of alternating layers.

The IR spectroscopic experiment on the Mariner 9 Orbiter found from the thermal IR spectrum of the surface that the best single-component fit to the silicate bands was for a silicious rock of intermediate silica content. This argues against rock types with high magnesium and iron contents (*mafic* rocks, from the words *ma*gnesium and *f*err*ic*) and especially against rocks dominated by olivine (*ultramafic* composition). The production of silica-rich rocks requires extensive geochemical evolution of the Martian crust.

The observed seasonal and regional differences in reflection spectra are probably due to several different factors, including particle size, degree of oxidation, and aeolian transport of dust. In general, the high-albedo areas, although active and variable in albedo, are spectrally very similar to the dust clouds observed during great dust storms. Although some of the lighter areas contain detectable traces of ferrous iron, the general rule is that ferrous iron absorption near 0.95 $\mu$m is strongest, and ferric absorption in the darker surface units is weakest. The position and shape of this band suggest that the $Fe^{2+}$ resides in olivine and pyroxene, probably in the form of a primary basaltic rock. Thus the dark area may be relatively mafic primary rocks.

The silicate reststrahlen band near 9 $\mu$m, originally attributed to montmorillonite, is also consistent with other silica-rich minerals, such as feldspars or micas, present in major amounts in the mobile dust. Overall,

**Figure X.16** *Continued*

the spectrum suggests that the dust could not be pure montmorillonite and may be dominated by mixed intermediate and acidic rock dust, containing no more than a few percent of limonite, carbonates, and nitrates. We conclude that the evidence from spectrophotometric and polarimetric studies strongly indicates the presence of silica-rich dust. The simplest explanation, that these silica-rich materials are produced by extensive geochemical differentiation and require density-dependent fractionation of the core, mantle, and crust, may well be correct. However, it is also important to consider the nature of photochemically driven weathering reactions on the surface to see whether they are capable of generating such dust by direct attack of water and oxidizing agents on primary mafic (or even ultramafic) rocks.

The distribution of water in the present Martian surface has been clarified by the hydrogen maps generated by the Gamma Ray Spectrometer (GRS) experiment on the Mars Odyssey 2001 spacecraft. High hydrogen (and, by inference, water) concentrations are observed in the south polar region, where water surely constitutes the major component of the permanent underlying ice cap. The high elevations near the south pole, found by the Mars Observer Laser Altimeter (MOLA) experiment on MGS to average 6 km above the mean level of the north polar region, are so cold that extensive deposition of carbon dioxide frost occurs during local winter. The underlying ice is seen clearly in the water map shown in Fig. X.17.

Note the second broad band of water-bearing rocks and weathering products in the low northern plains, bordering the north polar cap. This is the same region that has been suggested as the site of a former sea by some geologists, based on geomorphology alone. Note also the very dry highlands and the absence of any clear water signature in the deep Hellas basin (lower right center).

## Viking Lander Investigations

These remote, global-scale investigations of the geology and tectonic evolution of Mars were given a wholly new dimension by the *in situ* surface observations of the Viking landers. Viking 1 landed on volcanic terrain in the Chryse region, where the terrain is broken and blocky, a consequence of both faulting and impacts, and is weathered by the formation of a ferric oxide stain and iron-bearing clays (Fig. X.18). A wide variety of rock types are suggested by the lander photography. The intrusion of dike complexes and the ballistic transport of impact ejecta from nearby craters may contribute to this diversity. Drifts of fine-grained dust are small but numerous.

The Viking 2 landing site lies on the outskirts of an ejecta debris apron surrounding the crater Mie, in Utopia Planitia. The site is quite flat and rich in wind-transported sediments and is covered moderately densely and uniformly with boulders with dimensions of about 1 m. Most of the rocks seen are visibly *vesiculated* (filled with voids), although it is not clear whether these voids are caused by gas bubbles in volcanic extruded lava, by solvent extraction of water-soluble salts from shock-lithified or compacted dirt, or even by differential mechanical and chemical weathering. Loosely cemented clods of dirt and thin sheets of apparent *duricrust* appear to be held together by a chemical sediment such as

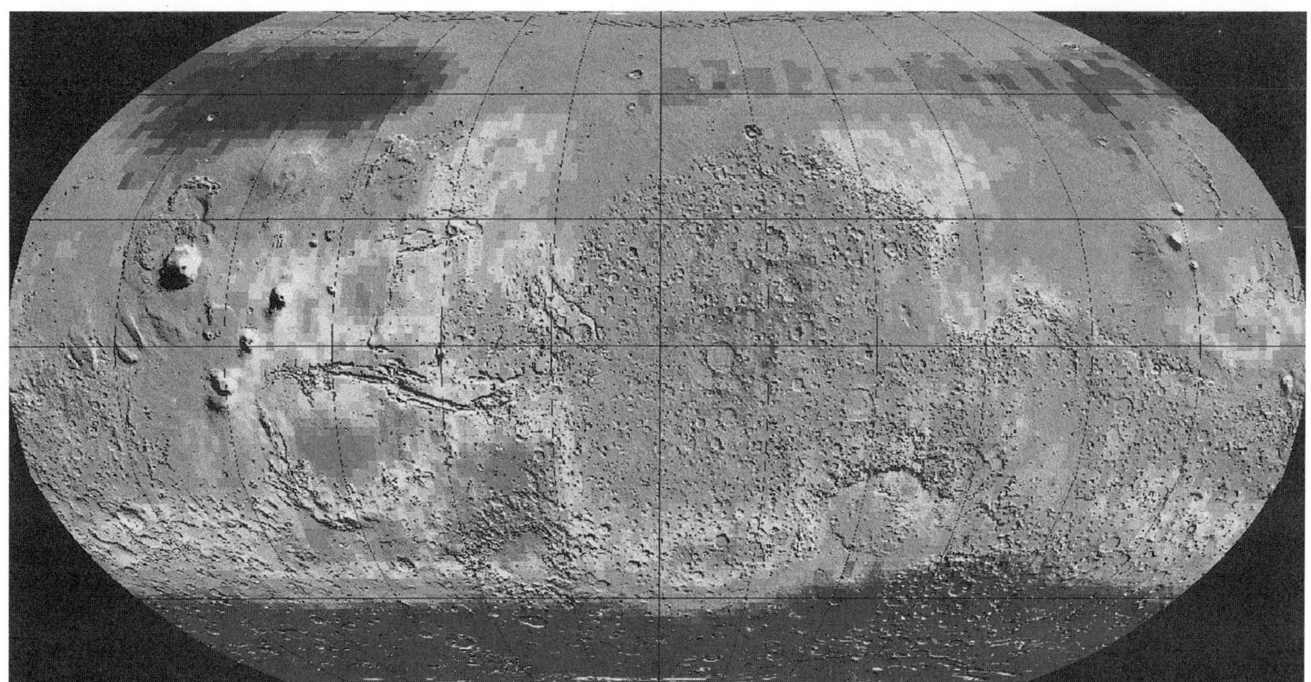

**Figure X.17** The distribution of water on the Martian surface as determined by the GRS experiment on the Mars Odyssey spacecraft, superimposed on a MOLA topographic map of Mars. Note the extensive "wet" areas in the south polar region and the northern plains. See color plate 6, on which the water-rich regions are coded blue and the driest regions are coded in red.

**Figure X.18** View from the surface of Mars. This Viking 1 Lander image shows a pale pink sky and a rocky, desolate surface mantled with red dust. The rocks are probably volcanic in origin, broken up by impacts and shaped by both wind erosion and chemical weathering (See color plate 5).

calcium sulfate, which is the principal cementing agent of *caliche* in terrestrial desert settings.

The geological, geophysical, and geochemical experiments conducted by the Viking landers were the same at both landing sites. The first experiment is implicit in the landing site descriptions given above: imaging. The physical properties of the Martian surface were studied via the dynamics of the landing legs upon impact with the surface, the penetration of objects dropped by the lander, and the mechanical behavior of the surface sampling arm. Magnets attached to the sampling arm head were used to search for and detect magnetic materials.

Certainly one of the most important instruments on the Viking landers was the X-ray fluorescence (XRF) spectrometer for analysis of the surface and atmosphere. Results reported for the Viking 1 landing site reveal that the fine-grained soil contains a high amount of iron and sulfur, with calcium, aluminum, and silicon also easily discerned (Fig. X.19). Only a minor amount of titanium was detected, and a very high Ca:K ratio was indicated. Early data from the Viking 2 lander showed a distinct resemblance. The sulfur concentration at both sites ranged from 10 to 100 times higher than that in terrestrial crustal rocks, and the potassium abundance was clearly

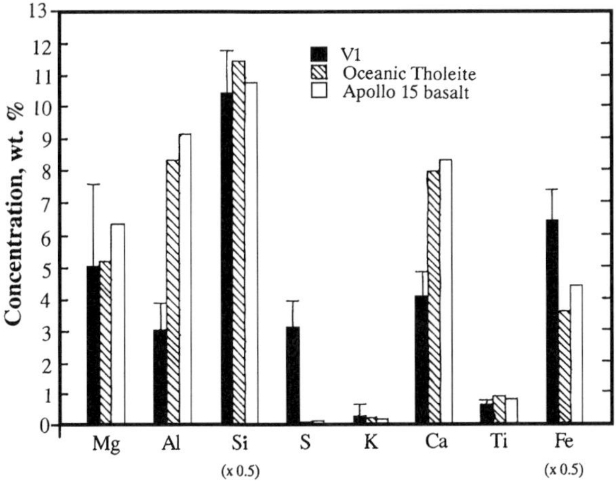


**Figure X.19** Composition of Martian surface materials. The abundances of eight major elements determined by the X-ray fluorescence experiment on the Viking 1 Lander are compared with those of Lunar and terrestrial basalts. The Martian regolith is markedly poorer in calcium and aluminum than either the terrestrial or the lunar material. The Martian surface also contains a distinctly higher abundance of iron and a very much higher abundance of sulfur.

several times lower than that expected in comparison with Earth's crust. Suspected samples of duricrust and cemented clods exhibited an even higher sulfur content, in keeping with the idea of cementing by a sulfate. Other general features of the analysis suggest that the soil was produced by weathering of a mafic igneous rock. An approximate modal interpretation of the analysis gave about 80% iron-rich clays, 10% magnesium sulfate, and about 5% each of calcite and higher oxides of iron, such as hematite, goethite, and maghemite. The Martian surface samples all exhibit low trace element, alkali, and alumina contents, and no evidence for the presence of acidic (high-silica) rock types was found. The observed weathering products can easily be derived from mafic rocks, but not from acidic rocks such as granites.

It is amusing that the observed surface composition of Mars is rather similar to the major-element composition of CI chondrites. The limited scope of the XRF experiment makes it impossible to test this similarity very closely, and important issues such as the water content of the Martian surface remain unresolved.

## The Shergottite, Nakhlite, and Chassignite Meteorites

Three rare classes of meteorites, the shergottites, nakhlites, and chassignites (the SNC meteorites), represented by more than a dozen different individual stones, have been found to have extremely small formation ages of about 1.3 Ga or less, compared to the normal age of asteroidal meteorites, about 4.55 Ga. Further, shock-melted glass in these meteorites contains unmistakable traces of chemically and isotopically distinctive gases from the atmosphere of Mars. It is now widely accepted that these meteorites are Martian igneous rocks that were ejected from Mars by one or more major impact events.

The meteorites that appear to have an origin on Mars include eight shergottites, three nakhlites, a unique chassignite, and another unique object, ALH84001. These stones are listed in Table X.1.

The shergottites in turn are subdivided into a basaltic group and a lherzolitic group. The basaltic shergottites are dominated by plagioclase and the pyroxenes pigeonite and augite (see Fig. IV.20), supplemented by minor ilmenite, rare chromite, and titanomagnetite (a solid solution of ilmenite, $FeTiO_3$, and magnetite, $Fe_2O_3$), and traces of pyrrhotite, the phosphates apatite and whitlockite, quartz, fayalitic (fe-rich) olivine, and baddelyite. These minerals reflect formation under more strongly oxidizing conditions than those experienced by the basaltic achondrites from the asteroid belt. The pyroxene crystals are zoned, with FeO-rich layers surrounding a relatively magnesian core. Trapped melt droplets deep within the cores of the pyroxene crystals contain spinel, sulfides, and the amphibole kaersutite. Kaersutite and apatite both are willing hosts of hydroxyl ions, but also accommodate fluoride and, to a lesser degree, chloride.

The lherzolitic shergottites are dominated by olivine accompanied by minor chromite, supplemented by a plagioclase–pigeonite–augite mixture closely similar to the basaltic shergottites. The olivine and chromite are largely contained inside large, later-forming orthopyroxene crystals. The interstices between the orthopyroxene crystals are filled by a mixture of pigeonite, augite, chromite, ilmenite, ulvospinel, whitlockite, sulfides, and maskelynite (plagioclase glass), plus a small amount of titanium-bearing magnetite. Again, as in the basaltic shergottites, ferric iron is present in the chromite and titanomagnetite, reflecting a higher degree of oxidation than comparable terrestrial lherzolites. Textures suggest a cumulate of crystals gravitationally segregated from a magma or aligned by shear forces associated with flowing.

All the shergottites have rather small grain sizes, characteristic of erupted lava flows or intrusion at quite shallow depths. Also, they all show strong evidence of severe mechanical shock. Thus they are most easily thought of as near-surface igneous rocks that were ejected from Mars by a major impact event.

Whole-rock dating of the shergottites by Rb–Sr or U–Th–Pb systematics give model ages of about 4.5 Ga, suggesting that these techniques measure the time of

**Table X.1** The SNC Meteorites

| Class | Rock Type* | Name | Mass (kg) | Minerals (wt. %)* | | | | |
|---|---|---|---|---|---|---|---|---|
| | | | | **Pig** | **Aug** | **Opx** | **Ol** | **Plag** |
| Shergottites | bas. | Shergotty | 5 | 36.3 | 33.5 | | | 23.3 |
| | bas. | Zagami | 23 | 36.5 | 36.5 | | | 21.7 |
| | lherz. | ALHA77005 | 0.48 | 26. | 11. | | 52. | 10. |
| | bas. | EETA79001 | 7.90 | 59.3 | 6.1 | 5.4 | 5.9 | 17.1 |
| | lherz. | LEW88516 | 0.013 | + | + | | ++ | + |
| | bas. | QUE94201 | 0.012 | ++ | ++ | | | ++ |
| | lherz. | Y793605 | 0.018 | + | + | | ++ | + |
| | bas/lh | Dar al Gani 476 | | | | | | |
| | lherz. | Los Angeles | 0.1 | + | + | | ++ | + |
| Nakhlites | cpxt | Nakhla | 40.0 | | 78.6 | | 15.5 | 3.7 |
| | cpxt | Lafayette | 0.80 | | ++ | | + | + |
| | cpxt | Gov. Valadares | 0.16 | | ++ | | + | + |
| Chassignites | dun. | Chassigny | 4.0 | | 3.8 | 4.0 | 88.5 | 2.6 |
| Unclassified | opxt | ALH84001 | 1.9 | | | 90. | | |

* *Key:* bas. = basalt; lherz. = lherzolite; cpxt = clionopyroxenite; opxt = orthopyroxenite; dun. = dunite; ++ = major mineral; + = minor mineral; Pig = pigeonite; Aug = augite; Opx = orthopyroxene; Ol = olivine, Plag = plagioclase

planetary differentiation, which must have coincided closely with the accretion of Mars. The preservation of such an ancient date points out an interesting difference between Mars and Earth: billions of years of subduction and mantle convection have extensively remixed Earth's mantle and crust, leaving an apparent average age of about 2 Ga. The crust of Mars, however, survives unhomogenized since very early in its history. Such early melting requires a transient heat source more powerful than decay of long-lived radionuclides. The most obvious source is the gravitational potential energy and kinetic energy of accretion. The actual crystallization ages of the shergottites has been a matter of active debate and no little confusion. The Nd–Sm "isochron" suggests a crystallization age of 1.3 Ga. Rb–Sr dating of mineral separates give an apparent age of 180 ± 20 Ma, consistent with the 190 ± 30 Ma given by the U–Th–Pb system. These dates are sometimes

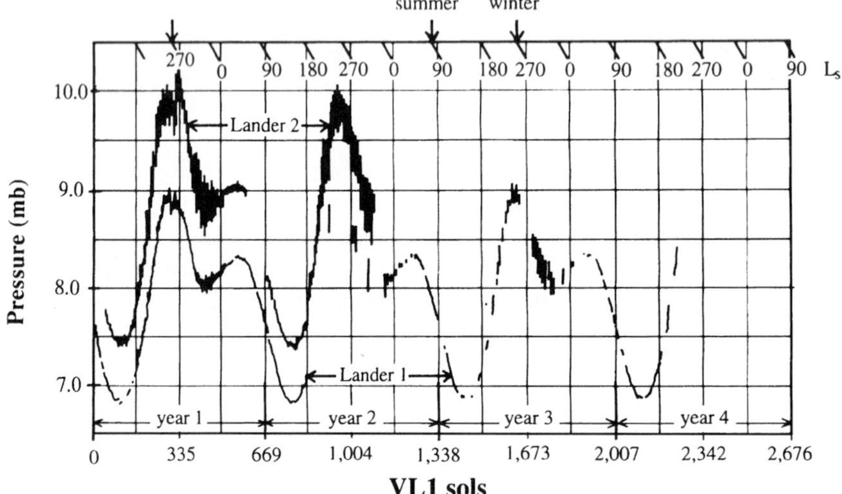


**Figure X.20** Seasonal surface pressure variations on Mars. The total atmospheric pressure varied from 7.4 to 10.2 mbar at the Viking 2 landing site during the first Martian year as a consequence of cyclic condensation of carbon dioxide at the winter pole and evaporation at the summer pole. There is a strong asymmetry between the behavior of the two hemispheres, due in part to the eccentricity of Mars' orbit and in part to systematic elevation differences between the two polar regions. The strong inflection near Martian day (sol) 330 is correlated with a major dust storm in 1977.

interpreted as the time of shock metamorphism that reset these radionuclide clocks, but shock experiments have so far failed to duplicate the effect. The basaltic shergottite QUE94201 has a measured crystallization age of 300 Ma, and the new Los Angeles lherzolitic shergottite has a reported crystallization age of 170 Ma. These ages, if correctly interpreted, suggest a very real possibility that volcanic activity persists on Mars down to the present.

The shock event that ejected these meteorites from Mars is usually dated by a variant on K–Ar dating called $^{40}Ar/^{39}Ar$ dating. A sample containing potassium and radiogenic argon is irradiated by a known neutron fluence (flux integrated over time; particles per square centimeter) in a reactor. A small portion of the potassium is converted to $^{39}Ar$ by neutron capture followed by beta decay. The sample is then melted in a vacuum system attached to a mass spectrometer, and the 40:39 ratio in the released gases gives the ratio of radiogenic argon to potassium, and hence the age of the sample. However, all SNC meteorites have been exposed to the Martian atmosphere, and all shergottites have been severely shocked in that atmosphere. Thus the shock event, whether or not it heats the stone severely enough to release old radiogenic argon, also shocks it severely enough to implant atmospheric gases. Rb–Sr dating of ALHA77005 gives a shock metamorphism age of less than 15 Ma. This age can be compared to the cosmic ray exposure ages of these meteorites during their sojourn in interplanetary space. Cosmic ray irradiation produces distinctive inert gas nuclides (notably $^{3}He$ and $^{21}Ne$) and short-lived radionuclides (especially $^{10}Be$, $^{26}Al$, $^{36}Cl$, and $^{53}Mn$). The production of these isotopes depends on time, depth of burial, and the chemical composition of the target rock: typical results give an irradiation age of 10 to 15 Ma and a depth of burial of some tens of centimeters (under material that was part of the rock while in space, but removed by ablation upon atmospheric entry on Earth). Interestingly, the results available to date show no evidence for significant cosmic ray irradiation on Mars, suggesting a depth of burial of at least 1–2 m while on the Martian surface.

The nakhlites are mainly composed of Mg-rich augite plus minor fayalitic olivine, two feldspars, and titanomagnetite. The fine-grained material between the larger augite crystals contains plagioclase and alkali-rich feldspars, with FeO-rich augite, pigeonite, titanomagnetite/ilmenite intergrowths, the sulfides pyrite, troilite, and chalcopyrite, plus chlorapatite and rare trace silica. Again the mineral content proves formation under oxidizing conditions. The texture suggests that they are cumulates. Cooling times have been estimated as a few years, compatible with intrusive origin in a dike or sill, but not with an eruption. The rare earth element (REE) pattern of the nakhlites is unusual, enriched in the light REEs and without a europium anomaly.

Chassigny is a dunite, nearly 90% fayalitic olivine with minor augite, plagioclase, pigeonite, and chromite. Traces of the sulfides troilite and pentlandite, chlorapatite, and the oxide minerals ilmenite, rutile, and baddeleyite are also found. Ferric iron is present on the ilmenite and chromite, demonstrating oxidizing formation conditions. Rare and very small melt inclusions contain kaersutite and biotite, both hydroxyl-bearing minerals. The REE pattern grossly emulates that of the nakhlites (enriched in light REEs), but absolute abundances are lower and the details of the pattern are sufficiently different to make it dubious whether Chassigny and the nakhlites were derived from the same parent melt. Further, Chassigny and the nakhlites share a crystallization age of 1.3 Ga. In Chapter VIII we saw that the basaltic HED meteorites have crystallization ages of about 4.5 Ga, very soon after the origin of the primitive chondrites themselves. Thus, for the NC portion of the SNCs, we must seek a parent body that is large enough to sustain extensive near-surface melting for more than two-thirds of the age of the Solar System. This time constraint requires a parent body of planetary dimensions, much larger than Earth's Moon, whose igneous rocks are all at least 3 Ga old.

Finally, the unique orthopyroxenite meteorite ALH84001 is dominated by large Mg-rich orthopyroxene crystals making up 90% of the rock, with 4% enclosed minor chromite containing ferric iron. Augite, pyrite, and apatite are interspersed with maskelynite glass in the groundmass. Traces of eroded olivine, the phosphate merrillite, $Ca_3(PO_4)_2$ (which bears most of the REE), and carbonates of Fe, Ca, and Mg are found as alteration products. The composition of the principal minerals resembles both the nakhlites and the lherzolitic shergottites. The light REEs are depleted, with a small whole-rock negative europium anomaly. Maskelynite contains generally depleted REEs, but with a 300-fold enrichment of Eu. This stone has achieved fame, or perhaps notoriety, because of a vigorous debate concerning the possibility that some of its chemical and mineralogical features are of preterrestrial biological origin. The organic matter in this meteorite has been found to have terrestrial, not Martian, carbon isotopic composition. Though the evidence for the presence of fossils of ancient microorganisms is weak, the meteorite is interesting for another reason: several lines of evidence suggest (but do not prove) that ALH84001 is very old, roughly as old as Mars.

Despite their petrological diversity, the SNC meteorites share a common oxygen isotope fractionation line, distinct from the Earth–Moon–enstatite achondrite line. They also contain adsorbed gases that match the distinctive chemical and isotopic signature of the atmosphere of Mars to high precision.

These meteorites fill a very important niche in our understanding of Mars. Based on the small inventories of volatile elements in the Martian atmosphere, the suggestion was made that Mars is strikingly deficient in volatiles relative to Earth. Meteoriticist Edward Anders of the University of Chicago and astronomer Tobias Owen of the State University of New York at Stony Brook used the visible atmospheric abundances to argue that the volatile contents of the terrestrial planets dropped off smoothly from Venus to Earth to Mars. Their meteorite-based composition model then required that Mars also be deficient in a host of other moderately volatile elements. For example, the low atmospheric abundance of $^{40}Ar$ and other rare gases seemed double proof that potassium and the other alkali metals were also depleted on Mars relative to Earth.

This suggestion contradicts the predictions of both Ringwood, who held that the volatile content of the terrestrial planets decreased with increasing mass, and Lewis, who argued that the volatile content increased with increasing heliocentric distance. Orbital gamma-ray spectroscopy of Mars (which samples the soil almost exclusively) found an atomic K:U ratio of about 300:1 for Mars (similar to lunar rocks), compared to about 10,000:1 for Earth's crust and mantle and about 64,000:1 for "chondrites." Such an astonishingly low K abundance would require 99.5% effective removal of potassium from Mars, which would in turn require even more efficient loss of the elements that, like sulfur, are more volatile than the alkali metals. This view is not only contradicted by the extremely high observed sulfur content on the surface of Mars (Fig. X.19), but also by the analyses of SNC meteorites, which actually have a higher K:U ratio than Earth. The depletion of potassium in the soil that has been derived from weathering of Martian igneous rocks is probably due to aqueous leaching of highly soluble salts from the soil. The existence of such an effect requires that liquid water was once widely present on Mars, and that there presently exists a reservoir of potassium salts and geochemically similar materials, presumably as a brine solution, at sufficient depths that cratering does not mix it back into the surface soil. Among the other volatile elements that exhibit higher abundances in Martian (SNC) rocks than on Earth are P, Rb, Cs, F, Cl, Br, I, Zn, and Ga. Almost all of these are strong candidates for leaching into brine.

Reconstruction of the bulk silicate composition of Mars by G. Dreibus and Heinrich Waenke of the University of Mainz confirms the idea that volatile elements are more abundant on Mars than on Earth. Their interpretations, based on analyses of the SNC meteorites, further find that several chalcophile elements, notably Ni, Co, Cu, and In, are less abundant in Martian rocks than in terrestrial rocks, whereas slightly chalcophile siderophilic elements such as Cr, Mn, and W are more abundant in the Martian silicates. This appears to be another strong indication of the formation of the core of Mars by sinking of a sulfide-rich melt closer to Fe–S eutectic composition than Earth's core. Current core models for Mars range from about 12% to 22% S, corresponding to an atomic Fe:S ratio between 3 and 2.

With respect to the major radioactive heat sources, the K:U ratio in the models by both Harry Y. McSween, Jr., and Dreibus and Waenke are consistent with a K/U ratio on Mars that is the same as on Earth. Strangely, the U:Th ratio on Mars appears to be about four times the terrestrial value. The reason for this discrepancy is unknown.

The SNC meteorites are also noteworthy for their very high FeO and $Fe^{3+}$ contents. As we saw in our survey of meteorites in Section VIII, formation of FeO is a result of gas-dust equilibration at low nebular temperatures, and is correlated with high volatile content: the oxygen in FeO is, in effect, one of these volatiles. Thus the FeO content of the silicate portion of Mars is about 18%, compared to about 8% in Earth's silicate portion.

## Atmospheric Structure

Before the spacecraft exploration of Mars, the surface pressure had been deduced to be less than about 10 mbar. The thermal structure of the Martian atmosphere was investigated at four points by the Mariner 6 and 7 radio occultations, and Earth-based high- resolution IR spectroscopy of the dayside provided mean rotational temperatures, averaged in a complex way over altitude, latitude, and time of day, that were known to reach as high as about 0°C. Earth-based radio astronomical brightness temperature measurements and remote sensing and *in situ* observations by the Mars 2, 4, 5, and 6 spacecraft also enriched our understanding.

During the Viking missions, thermal IR mapping of both the daytime and the nighttime surface by the Viking Orbiters was combined with radio occultation data, probe entry deceleration measurements, and *in situ* surface temperature measurements by the lander meteorology experiment package to permit synoptic study of the thermal structure and motions of the atmosphere. The radiometry results show a range of surface brightness temperatures from as low as 130 K to as high as 290 K. It appears that there is no simple argument by which the very low winter polar temperatures can be explained away. No combination of low emissivity, high-altitude $H_2O$ ice clouds, and high altitude can fully resolve the problem posed by such low apparent temperatures. Even at 140 K the vapor pressure of $CO_2$ is below 2 mbar, and it is very likely that the near-surface atmosphere near the winter pole has a composition that reflects the thermodynamic necessities; the $CO_2$ partial pressure must be very severely depressed by condensation, whereas the total atmospheric pressure must be roughly the same. Therefore, the mole fraction of inert, noncondensible gases at the pole may be in excess of 0.5. But the resulting reduction of the mean molecular weight of the atmosphere from 43.5 to less than 36 caused by $CO_2$ precipitation leaves the residual air buoyant. Advection of the "dry" parcel would draw in "wet," dense, $CO_2$-rich air at ground level, the condensation of which provides an important flux of latent heat into the polar regions to mitigate cooling.

At temperatures low enough to condense $CO_2$ (about 148 K at typical polar surface elevations), the vapor pressures of a number of potential atmospheric constituents are high enough so that they would remain wholly in the gas phase. At this temperature, the vapor pressures of some of these species are as follows: Xe, 240 mbar, and $SO_2$, 10 $\mu$bar (1 $\mu$bar = 1 dyn cm$^{-2}$). Ar, CO, $N_2$, and $O_2$ would be completely evaporated. It is thus clear that any retention of "inert" gases such as $N_2$ or Ar in the polar regions would have to be due to adsorption or trapping, not condensation.

The Viking lander and Mars Pathfinder meteorological packages have provided, in addition to the direct measurement of diurnal temperature variations from about 190 to 240 K, a direct measurement of long-term seasonal variations in the total atmospheric pressure. The atmospheric pressure varies by 20% or more over a year (Fig. X.20), and thus the mole fractions of noncondensible gases must vary inversely by the same amount.

The variation of temperature with altitude was deduced from the entry of the Viking landers. Viking 1 entered at 1600 h local time (20 July 1976) and Viking 2 entered at 0900 h (3 September 1976). Both found temperatures at high altitudes to be rather low, typically 120 to 180 K, with evidence of a strong wavelike structure to the temperature at altitudes above 40 km. The surface temperatures at the times of landing were about 230 and 220 K, respectively. The temperature gradient in the troposphere was in both cases close to half the adiabatic gradient, and even in daytime the temperature in the stratosphere approached within 20 K of the saturation temperature of carbon dioxide. These data are summarized in Fig. X.21.

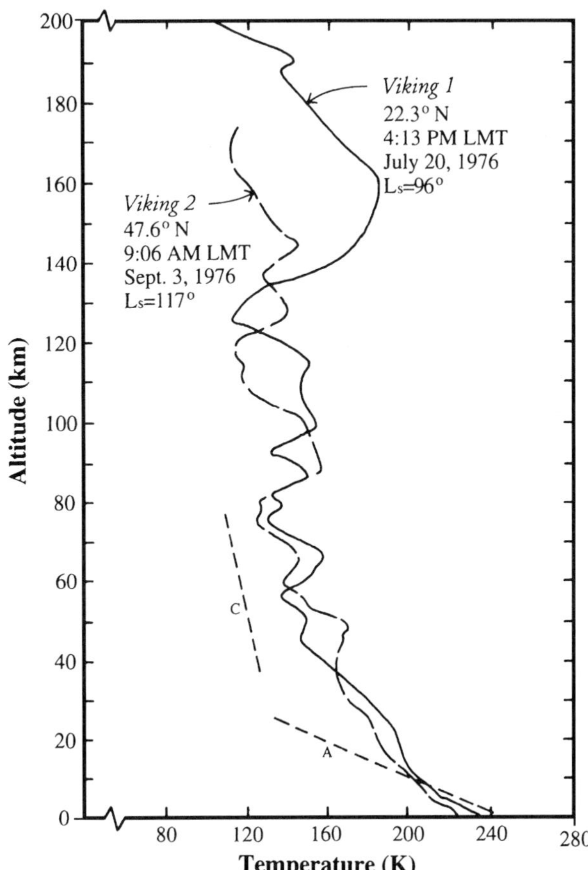


**Figure X.21** Temperature profiles through the Martian atmosphere. The temperature profiles deduced from entry data on the two Viking landers are compared. The dashed line labeled A gives the slope of the adiabatic lapse rate near the surface. Note that, above a few kilometers in altitude, the temperature gradient is clearly subadiabatic at both entry locations. The line marked C is the locus of $CO_2$ condensation, at which the ambient $CO_2$ pressure is equal to the vapor pressure of dry ice. The wavelike structure seen above about 40 km is surely a real, time-dependent wave phenomenon. Note that the "thermospheric" temperature is extremely low.

## Atmospheric Circulation

The similarity of Mars' rotation period, axial tilt, and atmospheric transparency to Earth's suggests a valuable parallel to our experience with terrestrial meteorology, but certain features unique to Mars lead to novel departures from familiar behavior. First, as noted above, the main atmospheric gas on Mars condenses on the surface, leading to large variations in surface pressure. Second, the enormous thermal inertia of Earth's ocean and the massive transport of latent heat by both oceanic circulation and evaporation of ocean water are not present on Mars to moderate day–night temperature variations and latitude-dependant temperature gradients. Also, the eccentricity of Mars' orbit causes "seasonal" temperature variations that heat and cool the entire planet in concert.

Monitoring of wind streaks on the Martian surface by the Viking Orbiters provides us with useful data on prevailing wind directions, and both Viking landers carried small meteorological packages that measured the local wind speeds. Observations of cloud motions provide sporadic data on wind directions and speeds at certain locations where clouds are common. Most useful are "yellow" clouds of wind-blown dust.

Temperature mapping by Mars orbiters, which reveals the driving force behind planetary-scale winds, also shows that the surface has an extremely small thermal inertia, requiring the presence of a fine-grained, poorly conducting dust with a very low pore gas pressure. The surface temperature therefore responds quickly to changes in insolation (irradiation by the Sun).

Circumstances are different when the atmospheric dust content is high. Dust is always a factor on Mars; even in the absence of dust storms, the vertical optical depth of the atmosphere was always at least on the order of 0.5 during the entire time of Viking lander measurements. Absorption and scattering of sunlight is never truly negligible and is sometimes a dominant effect. Absorption of sunlight by dust can lead to formation of a strong *inversion layer* (a warm layer well above the ground that inhibits convection from below), which not only affects the daytime vertical motions and limits the daytime solar heating rate, but also radiates heat to the ground at night, further diminishing the day–night temperature range. Dust storms also dramatically warm the poles by the same mechanism.

The general (planetary-scale) circulation of the atmosphere is characterized by the average zonal (east–west) wind speeds, the average meridional (north–south) motions, near-stationary wave structures, and seasonally varying features due to traveling waves. Small-scale processes are also important, as is the rapid inflation and deflation of the atmosphere during the diurnal cycle. This phenomenon of *atmospheric tides* is exacerbated by the rapid radiative response of the atmosphere. Also, the profound range of topography, from the deep Hellas basin to the Olympus and Tharsis-ridge summits (which penetrate most of the depth of the atmosphere), has a powerful steering effect on surface winds.

The Martian atmosphere exhibits a wide range of dynamical behavior that depends on the dust content of the atmosphere. The dynamics of the rise, maintenance, and decay of giant dust storms are poorly understood, although much interesting work is in progress on at least five major competing theories. One intriguing factor is an apparently chaotic factor governing the onset of severe dust storms: very similar seasonal and meteorological conditions will generate severe dust storms in some years, transient storms in others, and no storms at all in others. It is likely that the process of storm growth is highly nonlinear and sensitive to rapid divergence in response to extremely small differences in initial conditions. This behavior may be deeply chaotic in the mathematical sense, subject only to statistical description over time scales longer than a few days.

Initial dust raising may occur by means of disturbance of dust over large areas by elevated steady wind speeds; the mechanism involves the initiation of *saltation* (wind-driven grain hopping of sand-sized particles) in a very thin layer at the surface, which in turn dislodges fine dust particles that sediment out of the atmosphere very slowly. Alternatively, the dust may be mainly raised by funnel-shaped thermals of the sort known in desert regions on Earth as *dust devils*. About 100 dust devils with funnel diameters up to 1 km and heights up to 6.8 km have been identified on Viking Orbiter high-resolution images (See also Figure X.12c).

The dust-free atmosphere has a circulation pattern with a wavelike structure similar to that known on Earth, with three to six waves circling the planet. As on Earth, these wavelike structures enhance the meridional (poleward) transport of heat. The familiar surface expression of these *baroclinic waves* is the formation of cyclonic and anticyclonic systems in concert with a quasiperiodic succession of low- and high-pressure areas passing over local observers. These structures cause local periodic oscillations in surface pressure and wind speed that can be identified by Fourier analysis of the Viking 2 lander data.

## Atmospheric Composition

Recalling that $CO_2$ features in Mars' IR spectrum have been under study for decades, we should remind the reader that the deduced surface pressures were in the range 90–200 mbar until the early 1960s. The reinterpretation of the $CO_2$ spectrum first suggested low surface pressures, and the radio occultation experiments on Mariners 4, 6, and 7 confirmed that surface pressures were generally in the range of 6–7 mbar at the mean surface level. These conclusions were further borne out by a host of Earth-based spectroscopic and polarimetric studies. These measurements generally found 50–90 m agt (1 m agt = 1 m of path length through a gas at the standard temperature and pressure, 298 K and 1 bar) of $CO_2$ and revealed no evidence for the presence of any abundant spectroscopically inert gas. This implied mean surface pressures of about 6 mbar. Had there been a discrepancy between the pressure that would be provided by the observed abundance of $CO_2$ and the pressure deduced from the $CO_2$ linewidths, it would have been necessary to postulate a large mole fraction of, say, $N_2$ or Ar.

After $CO_2$, the first molecular species to be sought in spectra of Mars were $H_2O$ and $O_2$. Both species were tentatively identified in 1926, with a column abundance of water vapor on Mars equal to 0.03 of that in the terrestrial atmosphere and an $O_2$ abundance equal to 0.16 of that in Earth's atmosphere. Both of these estimates are enormously greater than the amount actually present. Water vapor was repeatedly and unsuccessfully sought in the following decades and was finally marginally identified in 1963.

Because of a nomenclatural oddity, the water vapor abundance is customarily reported in strange units that make direct comparison with observations of other gases unnecessarily difficult; it is usual to report how thick a layer of liquid $H_2O$ would be formed by condensation of the observed amount of vapor. The vapor abundance is then reported in units of micrometers (or centimeters) of precipitable water, as, for example, "15 $\mu$m ppt $H_2O$." The fact that ppt is a common abbreviation for "parts per thousand" should be sufficient grounds to ban this usage. The units can be converted as follows: 1 $\mu$m ppt $H_2O$ is equivalent to $10^{-4}$ g cm$^{-2}$ or 0.124 cm agt. The vague "cm atm" and the correct but needlessly awkward "cm atm NTP" or "cm atm STP" should be abandoned to minimize confusion.

Water vapor was first convincingly demonstrated in the atmosphere of Mars by Lewis D. Kaplan and co-workers in 1964. They estimated a water abundance of $1.7 \pm 0.9$ cm agt. It is well established that the $H_2O$ vapor abundance over the entire disk of Mars is variable and sometimes undetectably small. Seasonal variations caused the abundance to vary from below 1.2 cm agt to about 6.0 cm agt. These results have since been confirmed qualitatively by both Soviet and American spacecraft observations. The Mars 3 Orbiter found peak water vapor abundances near 2.5 cm agt at low latitudes in the northern hemisphere, less than half of the southern hemisphere summer figures cited above. Thus, typical planetary average water abundances are a few hundred parts per million of the $CO_2$ abundance based on the assumption of uniform mixing of these gases. However, because of the temperature gradient in the atmosphere, the water is strongly concentrated into the lowest 6–10 km, with even further enhancement likely in the lowest 1 or 2 km. Observations throughout the year-long Mariner 9 mission disclosed a maximum water vapor content of 2.5–3.7 cm agt during local spring over the north polar cap. The Viking Orbiters in 1976 found an average of only 0–0.4 cm agt. The water vapor abundance was found to be a smooth function of latitude, ranging almost linearly from 0 near 60°S to 1 cm agt near 20° N.

Carbon monoxide, which is an obviously likely component of any $CO_2$ atmosphere exposed to solar UV irradiation, was first detected and measured in 1969. Present estimates of the CO abundance are near 19 cm agt, equivalent to a mole fraction of $2.7 \times 10^{-3}$. Photolysis of $CO_2$ will, of course, generate not only CO but also atomic O, which will rapidly recombine via three-body reactions to form molecular oxygen, $O_2$. The earliest claimed detection of oxygen by St. John and Adams in 1926 was clearly erroneous. Many other measurements have followed, and observers have succeeded in establishing the presence of a small but significant amount of $O_2$, about $9.5 \pm 1.0$ cm agt. The mole fraction of $O_2$ is then about $1.4 \times 10^{-3}$. In the photolysis of $CO_2$ the molar ratio $O_2$:CO in the products is 0.5 ($CO_2 \rightarrow CO + \frac{1}{2}O_2$). Ozone is also present in detectable amounts in the polar regions.

The search for nitrogen compounds has been under way for nearly 20 years. Numerous papers dealing with nitrogen dioxide have been published without a positive detection. The UV spectrometers carried by Mariners 6 and 7 failed to disclose any nitrogen-containing atmospheric species. Based on these data, Alexander Dalgarno and Michael McElroy deduced a firm upper limit of about 5 mol% on $N_2$ and possibly less than 0.5% under certain circumstances. As we shall see, the only nitrogen compound detected to date in the troposphere of Mars is the molecular nitrogen found by the Viking analytical experiments, at a level of about 2.5%.

Another lengthy debate surrounded the "Sinton bands" near 3.46 $\mu$m. American astronomer William Sinton argued that they were produced by the stretching mode of the C–H bond. This proposed identification was later refined to the aldehyde group and then to acetaldehyde. One author reinterpreted the bands as due to enormous concentrations of HDO in the Martian atmosphere, a phenomenon so striking as to virtually unambiguously require the loss of oceanic masses of hydrogen from Mars with attendant deuterium enrichment. However, the Sinton bands are in fact due to terrestrial atmospheric HDO in entirely unsurprising amounts. Preliminary analysis of the data from the Mariner 6 and 7 IR spectrometer indicated significant amounts of ammonia and methane in the reflection spectrum of the polar caps, but the features are now agreed to be due to solid $CO_2$.

Two distinct theories attempted to explain the red coloration of Mars on the basis of photochemical reactions of atmospheric gases. One model attributed the red color to the dimer of $NO_2$, nitrogen tetroxide, whereas the other ascribed the red color to polymers of $C_3O_2$, carbon suboxide. However, spectroscopic data set an extremely low upper limit on the abundance of gaseous $C_3O_2$ monomer, only 0.02 cm agt. This is some $10^7$ times less than the vapor pressure of $C_3O_2$ monomer in the regions of Mars surveyed.

The first attempted *in situ* chemical analysis of the Martian atmosphere was by a mass spectrometer carried by the Mars 6 lander. The mass spectrometer unfortunately failed to operate, but the ion pump current was successfully telemetered back to Earth. The ion current was observed to be large and to decay slowly, suggestive of a large mole fraction of a poorly pumped inert gas, which, by analogy with Earth, was assumed to be radiogenic argon. From that time until the Viking mission, discussions of the origin and evolution of the atmosphere were predicated upon the apparent necessity of explaining the presence of about 30% $^{40}$Ar in the Martian atmosphere. The basic conclusion was that the high amount of outgassed $^{40}$Ar required the release of proportionately large amounts of other volatiles from the interior. So much $^{40}$Ar requires essentially complete outgassing of a chondritic Mars. If one assumes (not because it is obviously true, but because, in the absence of any other evidence, it seems like a plausible first guess) that the relative abundances of volatile elements in Mars are similar to those in the Earth, masses of $H_2O$ and $CO_2$ sufficient to produce kilometer-thick ice layers would have been released. If one allows for the lower formation temperature, higher oxidation state, and higher volatile content of preplanetary Mars material favored by many authors, then even more $H_2O$ and $CO_2$ must have been released relative to $^{40}$Ar. But such enormous masses of volatiles are hard to find on Mars today. Without the most extreme exertions, such a large amount of $H_2O$ cannot be accommodated in permafrost and hydrous minerals. Hiding a comparable amount of $CO_2$ is an even more severe problem, requiring the presence of several kilometers of $CaCO_3$ over the entire surface of the planet. The existing data on the composition of the surface, as reviewed above, show clearly that carbonates cannot be more than a very minor surface constituent. The need thus arises for escape of vast amounts of $CO_2$ (molecular weight, 44) with simultaneous complete retention of Ar (atomic weight, 40). This is a most unpalatable state of affairs.

Fortunately, the Viking landers carried three experiments capable of detecting and measuring atmospheric argon. The X-ray fluorescence (XRF) experiment, although having only sufficient sensitivity to detect argon down to the 2–3% mole fraction level, was able to demonstrate that the argon abundance was less than the detection limit. The much more sensitive entry mass spectrometer (EMS) experiment also quickly established that argon was in fact a minor constituent. The early gas chromatograph–mass spectrometer (GCMS) data gave a total argon abundance of about 2% and showed that the argon was overwhelmingly composed of $^{40}$Ar. The $^{40}$Ar:$^{36}$Ar ratio was found to be $2750 \pm 500$, about nine times higher than the terrestrial ratio of 296. Thus, the absolute abundance of radiogenic argon in the atmosphere of Mars is about 20 times less than the Mars 6 data at first implied. At the same time, the very high $^{40}$Ar:$^{36}$Ar ratio clearly demonstrates that $^{40}$Ar is *relatively* more important on Mars than on Earth!

The Viking GCMS data on the nitrogen abundance further underline the dilemma by bracketing the $N_2$ abundance between 2 and 3 vol%. Thus, the $^{40}$Ar:$N_2$ ratio on Mars is about 0.4, whereas in Earth's atmosphere the ratio is near 0.0125. On the other hand, the $^{36}$Ar:$N_2$ ratios on Earth and Mars differ by less than a factor of 3.

The EMS data also provided a reliable measurement of the isotopic composition of nitrogen. The $^{15}$N:$^{14}$N ratio was found to be about 70% higher than the terrestrial or chondritic meteoritic value. This fractionation effect is a logical consequence of the existence of a mechanism for nonthermal escape of nitrogen atoms, involving dissociation of $N_2$. (We shall return to these processes below.) The large observed enrichment of $^{15}$N on Mars implies the escape of large quantities of N atoms, so that initial $N_2$ partial pressures (with presumed "normal" isotopic composition) of several millibars would be required.

The abundances of the Martian rare gases display some interesting similarities and differences relative to those seen on Earth and Venus and in meteorites. The early Viking data placed restrictive upper limits on the abundances of neon, krypton, and xenon, and further interpretation of these data demonstrated the positive detection of both krypton and xenon. Detection and positive identification of Ne is difficult because of the masking of the $^{20}Ne^{+}$ ions by $^{40}Ar^{2+}$ at high ionizing electron voltages. However, the neon abundance was successfully measured at the level of 2.5 ppm. The absolute primordial rare gas concentrations in the atmospheres of Venus, Earth, and Mars are compared in Fig. X.22. The elemental abundance pattern for Mars is strikingly similar to the terrestrial pattern. The large excess of $^{40}$Ar relative to the primordial rare gases is, of course, not visible in such a display. Likewise, there is a large positive anomaly for $^{129}$Xe in the Martian atmosphere. Thus, the two radiogenic nuclides that are not capable of thermal escape are both enriched severalfold on Mars relative to the nonradiogenic rare gas nuclides. Neither isotope would be depleted by early atmosphere loss. The $^{20}$Ne:$^{22}$Ne ratio on Mars was not measured, and the total Ne abundance was estimated by multiplying the $^{22}$Ne abundance by 11.

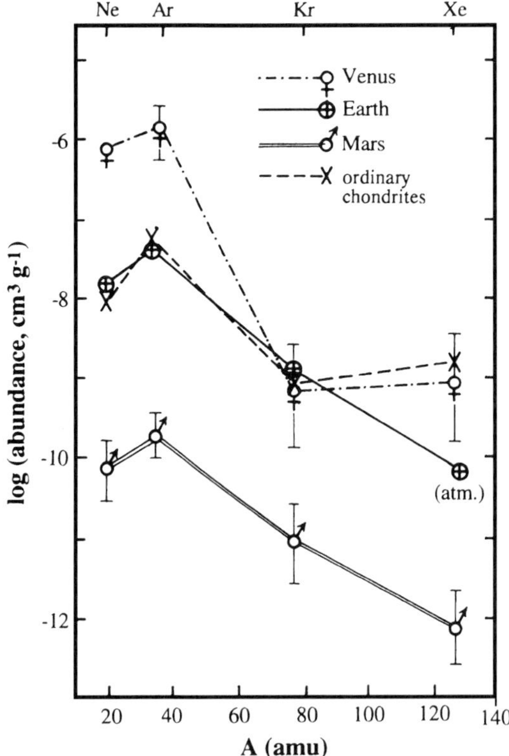


**Figure X.22** Abundances of the rare gases on the terrestrial planets. The general abundance trend for the primordial rare gases on Mars is similar to that on Earth, but lies lower by a factor of 100 to 200. Both Mars and Earth are deficient in xenon relative to ordinary chondrites. After Lewis and Prinn.

The Viking GCMS was also used to analyze gases released by stepwise heating of surface dirt samples. By this means, extremely small concentrations of a very wide variety of organic molecules could have been detected. Detection limits for a large number of species ranged from 0.1 to 1.0 ppb. Surface samples heated to 500°C were observed to release abundant $CO_2$ and $H_2O$, as expected, but no evidence was found for the presence of any organic material on Mars.

We summarize our post-Viking knowledge of the atmospheric composition of Mars in Table X.2. All species listed are known to be present and have been directly measured. Of these 11 species, six (nitrogen, neon, krypton, xenon, and the two argon isotopes) were discovered by the Viking mission.

**Table X.2** Composition of the Atmosphere of Mars

| Species | Abundance (mole fraction) | Species | Abundance (mole fraction) |
|---|---|---|---|
| $CO_2$ | 0.953 | Ne | 2.5 ppm |
| $N_2$ | 0.027 | $^{36}Ar$ | 0.5 ppm |
| $^{40}Ar$ | 0.016 | Kr | 0.3 ppm |
| $O_2$ | 0.15% | Xe | 0.08 ppm |
| CO | 0.18% | $O_3$ | 0.01 ppm (varies) |
| $H_2O$ | $<0.03\%$ (varies) | | |

Upper limits on a number of possible atmospheric constituents besides $C_3O_2$ have been set by spectroscopic observations. These studies also set an upper limit of 3 cm agt for $H_2S$ and considered a number of other species such as formaldehyde ($<5 \times 10^{-3}$ cm agt) and formic acid ($<7 \times 10^{-3}$ cm agt). The best present upper limit on $NO_2$ is 8 $\mu$m agt (0.12 ppm).

## Photochemical Stability and Atmospheric Escape

It seems likely that the Martian atmosphere has always consisted largely of $CO_2$ with minor amounts of $H_2O$. We must then address the question of the stability of such an atmosphere against photochemical decomposition and escape of volatiles.

The abundance and stability of carbon dioxide present similar problems on both Mars and Venus. How do we explain the stability of $CO_2$ with respect to conversion to CO, O, and $O_2$? Because there is so little $CO_2$ on Mars (about $10^4$ times less than on Venus), the time required for the Martian atmosphere to be converted to a predominantly CO–$O_2$ atmosphere in the absence of recombination mechanisms is very short, as little as 3500 years. Even more troublesome is the fact that the observed amounts of CO and $O_2$ on Mars could be supplied by UV photolysis of carbon dioxide in only 3 and 8 years, respectively.

Before discussing modern theories of Martian photochemistry, we will briefly review the principal constraints which must be applied to photochemical models of the Martian atmosphere. The constraints are as follows:

(i) The column abundances of $O_2$ and CO measured by ground-based IR spectroscopy are $(2.4 \text{ to } 3.1)\times 10^{30}$ molecules cm$^{-2}$ and $(1.5 \text{ to } 5.4) \times 10^{30}$ molecules cm$^{-2}$, respectively, corresponding to average number mixing ratios of $x_{O_2} = (15 \pm 2) \times 10^{-4}$ and $x_{CO} = (18 \pm 10) \times 10^{-4}$. These numbers allow $O_2$:CO ratios ranging from 0.5 to 2.0, compared to the ratio of 0.5 expected for $CO_2$ photolysis, suggesting an oxygen production that may significantly exceed that due to $CO_2$ photolysis alone.

(ii) From rocket-borne UV spectrometer observations the column abundance of $O_3$ outside the polar regions is $< 5.4 \times 10^{15}$ molecules cm$^{-2}$, corresponding to an average number mixing ratio $< 2.9 \times 10^{-8}$. In the polar regions the Mariner 7 and 9 UV spectrometers suggest ozone column abundances of about $2.7 \times 10^{16}$ molecules cm$^{-2}$ for an average ozone mixing ratio of $1.5 \times 10^{-7}$.

(iii) From Mariner 6 and 7 UV airglow observations the CO number mixing ratio is $(0.3 - 1) \times 10^{-2}$ and the

O atom number mixing ratio is $(0.5 - 1.0) \times 10^{-2}$ at 135 km altitude. In the absence of *in situ* catalysis of CO and O recombination at high altitudes, this constraint necessitates very strong downward mixing in the upper Martian atmosphere. We shall see that the upper atmospheric situation on Venus is very similar.

The first constraint, the possible significant excess of oxygen over carbon monoxide, suggests the importance of reactions other than

$$CO_2 + h\nu \rightarrow CO + O \tag{X.7}$$

as sources of oxygen. Carbon dioxide photolysis is energetically possible out to a wavelength of 2270 Å, at which the photon energy ($h\nu$) is exactly equal to the CO–O bond strength. But real dissociation processes put some energy into modes other than bond breaking, such as vibrational excitation of the CO fragment or translational kinetic energy of both fragments; thus in reality photons with wavelengths longer than about 1700 Å have no practical opportunity to dissociate $CO_2$.

Dissociation of $CO_2$ is followed by

$$O + O \rightarrow O_2 \tag{X.8}$$

as the main source of oxygen, because this sequence provides a molar $O_2$:CO ratio of 1:2, whereas observations suggest a number as much as four times higher. This two-body reaction is very slow because of the very difficult requirement that the energy of bond formation be dissipated over the duration of a single O–O collision. Such reactions can be greatly accelerated in the presence of a third collision partner (such as a $CO_2$ molecule), which can help dissipate the released energy through translational, vibrational, rotational, and even electronic excitation:

$$O + O + M \rightarrow O_2 + M. \tag{X.9}$$

Reconstitution of carbon dioxide via

$$CO + O \rightarrow CO_2 \tag{X.10}$$

is also terribly slow because the bond energy released during O addition is very large; disposing of this energy during a simple two-body CO–O collision is impractical. Thus here also a three-body process is favored:

$$CO + O + M \rightarrow CO_2 + M. \tag{X.11}$$

Some O must always be rendered unavailable for CO + O reactions by forming oxygen, so some CO (in stoichiometric proportions of one CO molecule per O atom or of two CO molecules per $O_2$ molecule) must be produced. Thus, because of the basic reaction stoichiometry, even if this mechanism were very effective, it would still not explain the observed excess of molecular oxygen over CO.

Dissociation of water is the only other plausible source of oxygen. In the presence of water,

$$H_2O + h\nu \rightarrow OH + H \tag{X.12}$$

$$\rightarrow O + H + H \tag{X.13}$$

$$\rightarrow O + H_2, \tag{X.14}$$

of which the first is by far the most important. This reaction can be driven by solar UV photons with wavelengths out to about 1900 Å. Hydroxyl radical can react with H atoms to reconstitute water via the three-body reaction

$$OH + H + M \rightarrow H_2O + M, \tag{X.15}$$

which is again faster than the corresponding two-body reaction. Oxygen atoms can react with molecular oxygen via

$$O + O_2 \rightarrow O_3 \tag{X.16}$$

to make ozone or with hydroxyl radical via

$$O + OH \rightarrow HO_2 \tag{X.17}$$

to make hydroperoxyl radical, which reacts with H or O:

$$H + HO_2 \rightarrow H_2O_2 \tag{X.18}$$

$$H + HO_2 \rightarrow OH + OH, \tag{X.19}$$

$$O + HO_2 \rightarrow OH + O_2. \tag{X.20}$$

OH can help oxidize CO by

$$OH + CO \rightarrow H + CO_2, \tag{X.21}$$

which is much faster than the reaction between CO and $HO_2$,

$$HO_2 + CO \rightarrow OH + CO_2. \tag{X.22}$$

Both $O_2$ and $O_3$ are vulnerable to UV photolysis,

$$O_2 + h\nu \rightarrow O + O \tag{X.23}$$

$$O_3 + h\nu \rightarrow O_2 + O, \tag{X.24}$$

with $O_2$ photolysis effective out to a wavelength of about 1800 Å. Ozone is energetically very vulnerable, because even photons of 1-$\mu$m wavelength (10,000 Å) have enough energy to remove an O atom; in practice, however, ozone is destroyed by photons in the 2000- to 4000-Å wavelength range, long enough so that shielding by the much more abundant $CO_2$ and $H_2O$ is impossible. The solar flux is large and increasing very rapidly with wavelength in this region.

However, in order to obtain the high O atom concentrations at low altitudes that are necessary for efficient operation of Reaction (X.20), a vertical eddy diffusion coefficient, $K = 1.5 \times 10^8\ cm^2\ s^{-1}$, in the lower atmosphere is required. Such values are about 1000 times higher than those in the terrestrial troposphere.

There have been a number of estimates of vertical mixing rates in the lower atmosphere of Mars. Some workers have inferred the possibility of very strong turbulence near the Martian surface from theoretical considerations. In particular, they compute $K = 3 \times 10^8\ cm^2\ s^{-1}$ in a mixing layer below 12 km for early afternoon in the equatorial equinox. There are, however, very substantial diurnal, seasonal, latitudinal, and height variations in these computed $K$ values. For example, they are orders of magnitude less at night and above the mixing layer, whereas during the midlatitude winter the mixing layer is confined to below 4 km. From these theoretical considerations it is certainly difficult to justify global- and time-averaged $K$ values throughout the lower atmosphere greater than $10^7\ cm^2\ s^{-1}$. The upper limit on $K$ is probably substantially less than $10^7$ above the 10-km level.

However, $K$ cannot be much greater than $10^5\ cm^2\ s^{-1}$ in the 0- to 20-km region if the extinction optical depth due to ice crystals in the troposphere is to be kept low enough to be consistent with observations. Also, the observed dissipation times for dust clouds during the 1971–1972 dust storm seem to require $K < 10^7\ cm^2\ s^{-1}$ in the 0- to 50-km region. Finally, a lower limit to $K$ can be deduced if the driving mechanisms for the Martian circulation are confined at or near the surface; in order for transient, vertically propagating waves to provide the large $K$ values required at high altitudes, we then need $K > 3 \times 10^5\ cm^2\ s^{-1}$ at the surface.

Of course, all the above methods used for deducing $K$ values must be regarded as providing, at best, order-of-magnitude estimates; the diversity of these estimates is therefore not surprising. For example, these estimates of $K$ do not include the planetary-scale circulation or the mechanical effects of the substantial Martian topography. In view of the many uncertainties, we can conclude that $3 \times 10^5 < K < 10^7\ cm^2\ s^{-1}$ provides a reasonable range for the expected vertical mixing rates in the region below 40 km on Mars.

In an alternative model designed to satisfy the first constraint above, T. Parkinson and Donald M. Hunten of the University of Arizona produced the OH necessary for CO oxidation by the reactions

$$HO_2 + HO_2 \rightarrow H_2O_2 + O_2 \qquad (X.25)$$

$$H_2O_2 + h\nu \rightarrow OH + OH \ (\lambda < 3700\ \text{Å}). \qquad (X.26)$$

These reactions do not require large O atom sources at low altitudes, and thus much lower $K$ values are allowed; they used a value of $5 \times 10^6\ cm^2\ s^{-1}$ in the 0- to 40-km region. However, in order to obtain sufficiently high $H_2O_2$ concentrations for efficient operation of Reaction (X.26), the total odd hydrogen (H, OH, and $HO_2$) required was almost four times greater than that found necessary in the earlier scheme. This large amount of odd hydrogen may result in $H_2$ production by the reaction

$$H + HO_2 \rightarrow H_2 + O_2, \qquad (X.27)$$

exceeding the maximum $H_2$ removal rate possible by $O(^1D)$, $CO_2^+$, and escape. The rate constant for this reaction, however, is not sufficiently well known for this conclusion to be definite.

Another interesting model was developed by Robert Huguenin of the University of Massachusetts. In this model, constraint (i) is satisfied by recombination of CO and $O_2$ adsorbed on ferrous minerals on the Martian surface. The chemical reactions involved are the following:

$$CO(g) + \tfrac{1}{2}O_2(g) + O^{2-}\ (\text{in } Fe^{2+} \text{ crystal}) \rightarrow CO_3^{2-}\ (\text{adsorbed on } Fe^{2+} \text{on crystal surface}) \qquad (X.28)$$

$$CO_3^{2-}\ (\text{adsorbed on } Fe^{2+} \text{ on crystal surface}) + h\nu \rightarrow CO_3^{2-}\ (\text{adsorbed on } Fe^{3+} \text{ on crystal surface}) + e^- \qquad (X.29)$$

$$CO_3^{2-}(\text{adsorbed on } Fe^{3+} \text{ on crystal surface}) \rightarrow CO_2 + O^{2-}\ (\text{in } Fe^{3+} \text{ crystal}) \qquad (X.30)$$

$$Fe^{3+} \text{ crystal} + e^- \rightarrow Fe^{2+} \text{ crystal}. \qquad (X.31)$$

The slowest (rate-determining) step is Eq. (X.29); the basic process can therefore be regarded as a photocatalytic surface recombination mechanism. Predicted maximum rates for that reaction imply that surface recombination of CO and $O_2$ can easily compete with CO and $O_2$ production from $CO_2$ photodissociation.

This surface catalysis model is appealing because it requires neither the very large K values necessary in the strong-vertical-mixing model nor the very large odd hydrogen concentrations present in the Parkinson and Hunten model. However, all three models lie within the present realms of possibility. Indeed the actual mechanism for CO and $O_2$ recombination on Mars may involve contributions from all three processes.

Our second constraint on Martian photochemistry involves ozone. Odd oxygen (O and $O_3$) is produced mainly by $CO_2$ photodissociation. In the dry Martian atmosphere these species are removed mainly by

$$O + O_3 \rightarrow O_2 + O_2 \qquad (X.32)$$

or by surface reactions. (Ozone is very easy to detect because of its strong absorption in the near ultraviolet and presents a sensitive test of such models.) This reaction acting alone cannot keep $O_3$ concentrations below observable levels outside of the polar regions. But the presence of water vapor provides an alternative route for removal of odd oxygen:

$$H + O_3 \rightarrow OH + O_2 \qquad (X.33)$$

$$O + OH \rightarrow O_2 + H. \qquad (X.34)$$

The required $x_H$ level is only a few parts in $10^{10}$, which is easily provided by $H_2O$ photodissociation. Note that this abundance of odd hydrogen is not sufficient to catalyze recombination of O or $O_2$ with CO unless $K$ is extremely large.

Because water vapor is the main source of odd hydrogen, we would expect that dry atmospheric regions near the Martian poles would contain much larger quantities of ozone. This is in fact observed; indeed, ozone levels at the poles appear to be close to those expected in an almost completely dry Martian atmosphere (that is, one with no odd hydrogen).

Finally, in order to satisfy constraint (iii), we require only that $K(z)$ values be large in the upper atmosphere, near the turbopause. This rapid vertical mixing is instrumental in transporting O and CO from the high levels down to low altitudes. There the O is partially converted to $O_2$ and then recombined with CO using the mechanisms described above.

The important role of $H_2O$ as a source of odd hydrogen on Mars should now be clear. Water photochemistry also regulates the escape of hydrogen, which will be discussed shortly.

A brief review of the water chemistry is therefore in order. The water vapor concentrations observed in the Martian atmosphere are strongly dependent on the local temperature and, therefore, on altitude, latitude, and time of day. Most atmospheric $H_2O$ is probably confined near the surface by vapor pressure limitations, where it is destroyed by photodissociation [Eq. (X.12)] and by reaction with energetic $O(^1D)$ from ozone photolysis:

$$O(^1D) + H_2O \rightarrow OH + OH. \qquad (X.35)$$

The odd hydrogen produced by these processes is rapidly cycled through H, OH, and $HO_2$ by the reactions summarized in Fig. X.23. The odd hydrogen sinks are

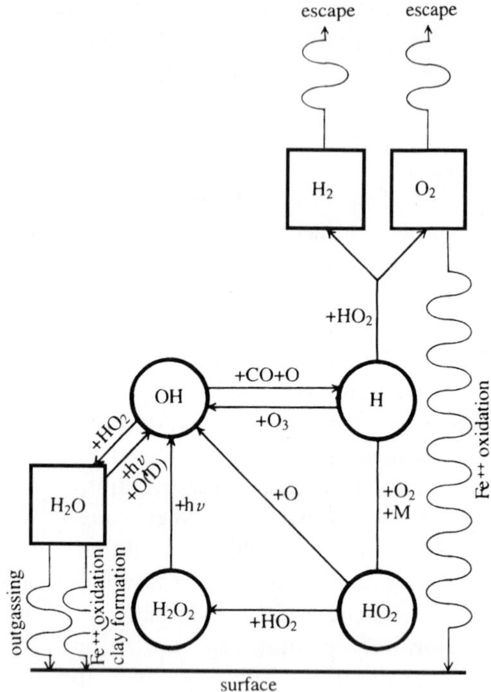


**Figure X.23** Cycles and loss mechanisms for water on Mars. The reactive minor constituents are indicated with circles. Both oxidation and hydration reactions with the surface are important. The rate of oxygen escape is coupled with the rate of hydrogen production and escape.

$$H + HO_2 \rightarrow H_2O + O, \qquad (X.36)$$

$$\rightarrow H_2 + O_2, \qquad (X.37)$$

$$HO_2 + HO_2 \rightarrow H_2O_2, \qquad (X.38)$$

$$HO_2 + OH \rightarrow H_2O + O_2. \qquad (X.39)$$

The hydrogen peroxide produced here may either freeze out in the polar regions (if it can get there) or photodissociate via Reaction (X.26) to reform odd hydrogen. Recall that spectroscopic observations suggest that the $O_2$ abundance exceeds the CO abundance on Mars; the $O_2$ generated by water photochemistry is the only plausible source of this additional oxygen. In a dry $CO_2$ atmosphere we expect the $O_2$ abundance to be one half that of CO.

Reaction cycles that generate hydrogen gas, however, raise the question of atmospheric escape from Mars. Molecular hydrogen is generated by reaction of H with $HO_2$, both of which are products of $H_2O$ photodissociation in the lower atmosphere. Some $H_2$ is removed by

$$O(^1D) + H_2 \rightarrow OH + H \qquad (X.40)$$

near the ground, leading next to reconstitution of water vapor, and also, in the ionosphere, by

$$CO_2^+ + H_2 \rightarrow COOH^+ + H \quad (X.41)$$

$$COOH^- + e^- \rightarrow CO_2 + H \quad (X.42)$$

and

$$H_2 + h\nu \rightarrow H + H. \quad (X.43)$$

Photolysis of molecular hydrogen requires rare photons with $\lambda < 850$ Å, which penetrate poorly through carbon dioxide. Thus almost all $H_2$ photolysis occurs at altitudes above about 80 km, close to the top of the atmosphere.

If we denote the concentration of molecules in the atmosphere as $[M]$ molecules cm$^{-3}$, then the effective escape level lies at an altitude $z_{exo}$ (the *exobase*), which is defined by

$$\sigma \int_{z_{exo}}^{\infty} [M] dz = 1, \quad (X.44)$$

where $\sigma$ is the collisional cross-section of the escaping H atom. Diffusive separation is much faster than turbulent mixing at high altitudes, and each gas behaves independently. The rest of the atmosphere (mostly carbon dioxide) is in hydrostatic balance according to

$$d\ln[M]/dz = -(\mu g/kT) - d\ln T/dz = -1/H_m, \quad (X.45)$$

where $\mu$ is the molecular weight of carbon dioxide [Eq. (X.44)] and $H_M$ is the local scale height of carbon dioxide. Integrating and assuming a constant temperature for the region above the exobase (the *exosphere*), we find

$$[M] \approx [M]_{exo} \exp -(z - z_{exo})/H_M. \quad (X.46)$$

Combining,

$$[M]_{exo} \approx 1/\sigma H_M. \quad (X.47)$$

This relationship defines the atmospheric density at which the exobase occurs.

If escape is not too fast, the velocity distribution of molecules near the exobase will follow a Maxwellian distribution; the number of H atoms with speeds in the interval between $v$ and $v + dv$ is given by

$$f(v) = [H]_{exo} v^2 (c)^{3/2} \exp(-cv^2), \quad (X.48)$$

where $c = m_H/2\pi kT$. The total escape flux is found by integrating the upward-moving component over all velocities greater than $v_{esc}$ and over all directions in the outward-facing hemisphere. This thermal escape flux, first derived by Sir James Jeans in 1928, is called the *Jeans escape flux*, $\Phi_H$ (atoms cm$^{-2}$ s$^{-1}$). This flux is set equal to the product of the H density at the exobase and the net escape velocity: $\Phi = [H]w_H$, which is unrelated to the planetary escape velocity $v_{esc}$.

Mariner 6 and 7 measurements of H Ly $\alpha$ resonance emission from Mars show that the H exobase lies at an altitude of about 250 km, where the H number density is $3 \times 10^4$ atoms cm$^{-3}$. The exospheric temperature was found to be about 350 K, and the H escape flux, using the Jeans formula, is $1.8 \times 10^8$ atoms cm$^{-2}$ s$^{-1}$. At the turbopause, $H_2$ is the dominant form of hydrogen and the escape flux of H is limited by the $H_2$ diffusion flux (the H actually traveling as $H_2$) of about $1.2 \times 10^8$ atoms cm$^{-2}$ s$^{-1}$. The observed escape flux is in fact roughly equal to this value.

In a steady state, this outward flux of hydrogen must be balanced by the production of $H_2$ from $H_2O$. This production rate is hard to estimate because of poorly known rate constants and the unknown catalytic effects of the ground and atmospheric aerosols on several important reactions. The observed escape flux implies that about 30% of the dissociated $H_2O$ produces $H_2$.

The total amount of water vapor that would be destroyed on Mars if the present escape flux of H atoms were sustained over the lifetime of the planet (roughly 5 billion years) is about one thousandth of the water in the oceans on Earth today. This small water loss will certainly not have noticeably depleted Martian water resources. The other product of water photolysis, $O_2$, does present a problem. With the observed escape rate of H, photolysis will produce the observed $O_2$ amount in just $10^5$ years. To resolve this dilemma, we must search for ways to remove oxygen from the atmosphere.

Photoionization of carbon dioxide produces $CO_2^+$ in the ionosphere, and Mariner 6 and 7 measured sufficient O atom concentrations to enable the charge transfer reaction

$$O + CO_2^+ \rightarrow O_2^+ + CO \quad (X.49)$$

to raise $O_2^+$ to the rank of dominant ionospheric positive ion. Michael McElroy of Harvard University pointed out that the two O atoms produced in dissociative recombination of $O_2^+$,

$$O_2^+ + e^- \rightarrow O^* + O, \quad (X.50)$$

have sufficient energy to escape the gravitational field of Mars even if one of these O atoms were produced in the excited $^1D$ state. Thus, the upward-traveling atom of the pair could escape. He also noted a similar situation for recombination of $N_2^+$ and $e^-$, enabling escape of N atoms, and recombination of $CO^+$ and $e^-$, producing C atom escape. His computed escape fluxes for O, N, and

C were $6 \times 10^7$, $3 \times 10^5$, and $4 \times 10^5$ atoms $cm^{-2} s^{-1}$, respectively. The escape of volatile elements from Venus and Earth is not possible by this mechanism. The gravitational fields on these planets are considerably larger and the energies required for O, N, and C atoms to escape are roughly four times those required on Mars. This is considerably more than the amount of energy available from dissociative recombination.

Because momentum is conserved in the dissociative recombination of these molecular ions, the lower-mass atom is ejected with the higher speed. Thus $CO^+$ yields preferential escape of C atoms, $O^{2+}$ yields enhanced escape of $^{16}O$, and $N^{2+}$ favors loss of $^{14}N$ relative to $^{15}N$. The atmospheric reservoirs of C and O are so vast compared to the C and O escape fluxes that no isotopic fractionation is expected (or found) in the remaining atmosphere, but the nitrogen reservoir is much smaller. Strong isotopic fractionation is in fact both expected and observed, with $^{15}N$ enhanced by an impressive 70% relative to normal Solar System nitrogen. McElroy's dissociative recombination mechanism can explain this fractionation if the original isotopic composition were similar to terrestrial nitrogen and if escape has depleted the original nitrogen by at least a factor of 20.

On Earth the $N_2$:$CO_2$ ratio is 0.03:1. It is difficult, however, to accurately assess the planetary $N_2$:$CO_2$ ratio. If we assume this ratio to be the same as that on Earth, then the escape of N atoms by McElroy's mechanism keeps the $N_2$ mixing ratio close to $10^{-2}$ the actual value depending on the outgassing history and the position of the turbopause.

We might also expect $N_2$ to be decomposed in the Martian lower atmosphere by cosmic rays as it is in the Earth's lower stratosphere. Reaction of the resultant N atoms with $O_2$ would lead to small amounts of $NO_2$ and $HNO_3$, which might be irreversibly removed from the atmosphere by formation of surface nitrites and nitrates.

The calculated oxygen atom escape rate is in remarkable agreement with the production rate of O from $H_2O$ photodissociation inferred from the present H escape rate. This balance is no coincidence, but is in fact expected from the buffering effect of molecular hydrogen. If these escape rates have remained constant over geologic time we can compare the total water lost by this mechanism with the total $CO_2$ now present in the atmosphere. The inferred $H_2O$:$CO_2$ ratio in the outgassing of Mars is about 45:1. The ratio of the water content of Earth's oceans to the $CO_2$ content of carbonates in sedimentary rocks yields a ratio of about 5:1. There are, as we shall see, several indications that the $H_2O$:$CO_2$ ratio on Mars should equal or exceed that of Earth. If that is indeed the case, there must have existed water removal mechanisms that were far more potent than present-day H and O escape mechanisms.

The fact that the nonthermal O atom escape rate discussed above is roughly equal to one half the H escape rate does not in fact mean that such escape is the dominant oxygen removal mechanism on Mars. First, the computed escape rates are model dependent. Second, we should also investigate possible surface sinks for oxygen, because most of the total oxygen (O, $O_2$, and $O_3$) is, of course, contained in the lowest few scale heights. We should here recall that the optical properties of the Martian surface are best simulated in the laboratory by olivine- and magnetite-bearing basalts that have been partially oxidized to the extent that they contain between 1% (dark regions of Mars) and 10% (bright regions on Mars) of $Fe_2O_3$. In view of our earlier discussion of the primordial mineralogy of the Martian surface, it is apparent that essentially all of the surface Fe should be in the $Fe^{2+}$ oxidation state (FeO, FeS, etc.). A question then arises regarding the mechanismfor producing the $Fe^{3+}$ that is required to fit the observations. The work by Huguenin described above has demonstrated that the reaction between black magnetite ($Fe_3O_4$) and atmospheric $O_2$ to produce bright red hematite ($Fe_2O_3$) is powerfully stimulated by UV radiation with $\lambda > 2000$ Å. Since the thin Martian atmosphere transmits most of the radiation in this spectral region, this photostimulated oxidation of $Fe^{2+}$ to $Fe^{3+}$ offers both a suitable production method for $Fe_2O_3$ and a suitable sink for atmospheric $O_2$. The possible removal rate for $O_2$ was predicted to lie in the range $5 \times 10^7$ to $5 \times 10^{10}$ molecules $cm^{-2} s^{-1}$. Even the smallest estimate exceeds that required to balance the removal of photochemically produced $H_2$ from the atmosphere by thermal escape.

Oxidation of surface rocks may not have always proceeded by the above mechanism. During the cyclic alternation of the permanent pole caused by the slow precession of the Martian spin axis every 50,000 years, substantial quantities of water vapor would be released into the atmosphere, perhaps enabling the temporary existence of rivers and lakes. Oxidation of $Fe^{2+}$ proceeds readily in this environment and may provide both another sink for oxygen and a production mechanism for $Fe_2O_3$.

The escape of $H_2$, operating together with the above mechanisms for removing $O_2$ from the atmosphere, results in a net loss of $H_2O$. This loss rate is, of course, limited by the average $H_2O$ photolysis rate, $10^9$ molecules $cm^{-2} s^{-1}$.

Hydrogen escape from Mars inevitably favors preferential loss of the lighter isotope, $^1H$. Deuterium in the Martian atmosphere, measured through IR determinations of the HDO:$H_2O$ ratio, is considerably elevated over terrestrial levels, $\delta_D = +4000‰$, relative to terrestrial standard mean ocean water (SMOW). L. L. Watson and co-workers have analyzed water extracted from the hydrous phases (the amphibole kaersutite, the hydroxyl phosphate apatite, and the mica biotite) in the SNC meteorites and found a D enrichment ranging from $\delta_D = 900$ to about 4400‰ (SMOW). These results suggest that late (D-rich) Martian atmospheric water has exchanged partially, but incompletely, with the much older hydrous minerals. The enormous size of the D enrichment (to over a factor of 4.5) implies a large time-integrated escape flux of H atoms. The lowest and therefore most ancient D enrichment, +500‰ (in some kaersutite samples), requires that considerable enrichment occurred early in the history of the planet. The fascinating question of how readily atmospheric water vapor exchanges with mineral ice at high latitudes remains unanswered by experiment.

Solar wind sweeping may also be making a contribution to escape from the Martian atmosphere. Ions are produced by photoionization and by charge exchange with solar wind protons near the planetary limb, specifically at latitudes higher than 70° and altitudes greater than 250 km. These ions, which include $O_2^+$, $CO^{2+}$, $O^+$, $N_2^+$, $NO^+$, and $CO^+$, can be swept up by the electric field of the impinging solar wind. Significant escape rates of $CO_2$ ($5 \times 10^5$ molecules cm$^{-2}$ s$^{-1}$) and $N_2$ ($2 \times 10^5$ molecules cm$^{-2}$ s$^{-1}$) may result.

## Explosive Blowoff

An early study of the effects of large comet and asteroid impacts on the atmospheric evolution of Earth led G. Hampton Watkins and me to suggest that impacts could erode atmosphere from Mars with high efficiency. Indeed, if early melting, differentiation, and outgassing of Mars occurred simultaneously with terminal accretion, it was quite difficult to avoid loss of more than 99% of the primordial atmosphere. Jay Melosh and Ann Vickery of the University of Arizona have explored this idea in detail using hydrodynamic codes appropriate for large explosions and have found that such erosion is quantitatively very important for Mars because the low planetary escape velocity permits easy escape of shocked gases. Such atmosphere-eroding events were also found to be effective at removing strong rocks from the planet, thus offering a reasonable explanation of the ejection of SNC meteorites from Mars.

The most interesting aspect of explosive blowoff as an escape mechanism is that it is entirely mass independent. The blowoff process is hydrodynamic and occurs on a time scale (minutes) too short for mass-dependent fractionation of gases.

## Origin and Evolution of the Atmosphere

The above discussion of the bulk composition of Mars, the photochemical and aeronomical evolution of its atmosphere, volatile-element budgets, and the geological evidence for massive deposits of mineral ice has covered most of the evidence and interpretation appropriate to this subject. We will therefore conclude with only a brief summary and prospectus for future work.

We have surveyed the various mechanisms that have been suggested for depleting volatiles from the atmosphere of Mars. Thermal escape has been found to be satisfactory for helium and hydrogen, whereas nonthermal mechanisms such as dissociative recombination are capable of ejecting atoms of nitrogen, carbon, and oxygen. Solar wind sweeping is effective in removing any charged particles present in the ionosphere. Multiplying the present escape fluxes for these mechanisms times the age of the Solar System, we find that the amount of atmosphere that has been lost by these known processes is probably at least several times larger than the present atmospheric pressure. Explosive blowoff can raise this factor to 100 or more without introducing observable isotopic fractionation effects.

Ringwood's model for the formation of Mars begins with CI chondritic material, from which some gas loss has occurred. Because CI material contains about 20 wt% water, melting and differentiation of Mars would release enough water to cover the entire planet to a depth of 700 km. It was Ringwood's belief that wholesale loss of volatiles occurred on all the terrestrial planets very early in their history, an idea that he supports by reference to McElroy's interpretation of the nitrogen isotope anomaly found by the Viking landers. However, the latter effect is due to a slow, selective escape mechanism operating over the entire lifetime of the planet, not a massive, nonselective mass loss episode. The nitrogen isotope evidence is quite irrelevant to the issue of whether early, massive volatile loss occurred.

Ringwood also accepted the idea that the high oxidation state of Mars requires low formation temperatures, but he also believed that the XRF soil analyses prove that Mars is depleted in potassium relative to chondritic meteorites. He then argued that it is "reasonable to expect a corresponding depletion in sulphur, because the latter element is much more volatile than potassium ... in the solar nebula." Thus, Ringwood, like Anders and Owen, believed that the surface dirt at the Viking landing sites reflects faithfully the overall composition of Mars. In that case, it is important to note that one of the most striking features of the Viking XRF data discussed above is the remarkably high sulfur abundance in every sample. The significance of the high sulfur content is discussed in some detail by the Viking

XRF team, who explore the possibility that the surface dirt is produced by large-scale oxidation and disaggregation of primitive sulfur-rich material. Clearly, we cannot easily accept Ringwood's assertion that sulfur is severely depleted in Mars. Ringwood's conclusions that the core must be $Fe_3O_4$ and not FeS are based on the same logic and thus do not follow from the observational evidence. Further, it is a well-known result of thermodynamic studies of condensation in the nebula that Fe oxides, which are very abundant in Mars, form at temperatures so low that both K and S are fully condensed.

Ringwood then concludes that the low $^{36}Ar$ abundance in the atmosphere (a relative abundance of only 1% of the $^{36}Ar$ abundance in Earth's atmosphere) requires either early massive atmosphere loss or very inefficient outgassing. We accept these conclusions, although we do not accept the arguments for early outgassing discussed above.

Ringwood mentions the fact that the $^{40}Ar$:$^{36}Ar$ and $^{129}Xe$:$^{132}Xe$ ratios are very high on Mars, but does not suggest an interpretation beyond the paradoxical remark that K is depleted in Mars. This, of course, would decrease the 40:36 ratio. This evidence demonstrates that radiogenic rare gases are relatively more abundant compared with primordial rare gases on Mars, which would be expected for volatiles released relatively late in the history of the planet. If Mars were to outgas its volatiles instantly upon accretion, with high efficiency, the initial atmosphere would be very poor in the radiogenic nuclides $^{40}Ar$ and $^{129}Xe$. Further slow outgassing would add small amounts of a rare gas component that becomes progressively richer in radiogenic gases as time passes. Thus, early blowoff of a massive atmosphere formed at the time of planetary formation, followed by slow release of volatiles over a prolonged period of time, would be consistent with the observations. Massive early outgassing is clearly consistent with, but not absolutely proved by, the rare gas data.

The way in which the crustal elemental abundances are interpreted is of major importance in assessing bulk composition models for the planets. The XRF data refer to surface samples found at two widely separated points, both on relatively flat regions with a deep regolith. The dirt analyzed has certainly been subjected to planetary differentiation, possible remelting, possible hydrothermal alteration, extensive oxidation, wind sorting and transport, and probably at least some solvent leaching and deposition of water-soluble salts. The clay, oxide, and sulfate components may have different origins. The very existence of enormous amounts of clays and sulfates implies reactions destroying equally vast amounts of water. Nothing is known of the abundance of carbonates in the surface dirt, although they cannot be a major component.

An ambitious attempt at interpretation of the compositional data on the atmosphere and surface of Mars has been made by Edward Anders and Tobias Owen. They base their interpretation of the composition of Mars on a five-component model developed originally to explain fractionation trends observed in chondritic meteorites. These five components are sensibly defined in terms of their geochemical behavior during condensation from the solar nebula. The five classes are as follows: (1) early condensates, (2) metal, (3) magnesium silicates, (4) moderately volatile elements (condensation temperatures below 1300 K but above 600 K), and (5) volatiles condensing below 600 K. Fractionation among the different classes of carbonaceous chondrites can be rather well described by such a scheme, and the differences between the classes of ordinary chondrites can be rather less satisfactorily explained. The fractionation process is envisioned as the selective accretion of condensate grains by small parent bodies, in which the physical properties of the grains, such as crushing strength, magnetic moment, and electrical conductivity, are more important than gravity in selecting the grains to be retained. We can thus picture small parent bodies accreting with different compositions because of these mechanisms. Such fractionation mechanisms may work for bodies up to 1 km in size, but it is hard to see how ensembles of $10^{12}$ such bodies, when accreted to form an Earth or Venus, could be other than perfectly democratic samples of all the solids available. Gravitation, which dominates accretion of bodies from the size of an average asteroid on up, does not distinguish among grains with different individual physical properties.

Two possible approaches for extending grain-by-grain physical fractionation processes to planetary-scale bodies have been suggested. First, there is the possibility that all solids in the early solar nebula were fully evaporated and then partially condensed during cooling. In this case, radial sorting of grains by volatility can occur, leading to volatile-poor bodies close to the Sun (irrespective of size) and volatile-rich bodies far from the Sun (again, irrespective of size). This is also expected in the chemical equilibrium model of Lewis wherein complete evaporation does not occur anywhere outside about 0.3 AU from the Sun. In either case, a fairly simple monotonic radial trend of Fe oxidation state and volatile content is expected.

The second mechanism for effecting composition differences between planet-sized bodies is aerodynamic sorting based on differences among the mean sizes and densities of grains in the solar nebula, as suggested by Stuart Weidenschilling, which we discussed in connection with the fractionation of metal from silicates in the early Solar System. Both this and the fractional condensation or equilibrium condensation approach predict a smooth and continuous (but not necessarily monotonic) variation of planetary bulk compositions with heliocentric distance.

The idea that the grain-related fractionation processes could shape the compositions of planets was also criticized by Ringwood on five grounds. First, the five-component meteorite model is not obviously so successful on chondrites that it need be accepted. Second, it appears that the ordinary chondrites are derived from very small and quantitatively unimportant asteroid types whose places of origin are not known. Third, accretion of planets from ensembles of $10^{12}$ small bodies that are statistically sampled over a wide range of heliocentric distance would certainly wipe out the compositional variations of single bodies. Fourth and fifth, Ringwood finds that even the later seven-component version of this model does not adequately describe the compositions of the Moon and the Earth.

Anders and Owen applied this model to Mars, and Morgan and Anders have revised this approach for the purpose of comparing the composition of Mars to the other terrestrial planets. They assume strict coherence of the elements in each of the geochemical groups on Mars; an example is their assumption that the atmospheric inventory of $^{36}Ar$ fixes the planetary inventory of thallium and a number of other unobserved elements. The assumption is explicitly made that no large-scale loss of atmosphere has occurred. This assumption is defended by pointing out that the rare gas composition on Mars is very similar to that on Earth, and any thermal loss process would introduce very large (and quite unobserved) mass fractionation effects in, say, depleting $^{36}Ar$ by a factor of 10 while leaving the xenon abundance unchanged. This argument, of course, does not apply to catastrophic loss of an atmosphere by processes other than Jeans escape. Morgan and Anders explicitly claim: "Following Anders and Owen (1977), we are assuming that Mars never lost either $^{36}Ar$ or $^{40}Ar$. First, both isotopes are too heavy to be lost from the present atmosphere. Second, there is no evidence whatsoever, from any planet, for a catastrophic loss of an early atmosphere." They then analyze the GCMS and XRF data on Mars and reach the expected conclusion that Mars is deficient in volatile (condensation temperatures below 600 K) elements by a factor of 40 relative to those on Earth. Moderately volatile (600–1300 K) elements are similarly depleted by a factor of only about 3, but that conclusion depends on guessing the efficiency of release of sulfur from the interior during degassing. They assume the same release efficiency for Mars as for Earth, despite the fact that the total abundance of sulfur-bearing gases in a mixed-volatile system is a sensitive function of both temperature and oxygen fugacity, whereas the sulfur gas composition depends on the water vapor fugacity as well. Such an assumption could easily be wrong by a factor of 100.

Anders and Owen explicitly reject a more volatile-rich Mars by reference to the high $^{40}Ar$:$^{36}Ar$ ratio. Radiogenic argon is derived from moderately volatile (600–1300 K) potassium, whereas primordial argon is highly volatile (a "thallium-group" element condensing below 600 K). If equal release efficiency of the argon isotopes is assumed, then the Tl group could be raised in abundance by a factor of 3, but would still be depleted by a factor of about 12 relative to Earth. Of course, the abundance of the "Tl group" is based on the primordial rare gas abundances, which would have been most severely affected if large-scale blowoff had occurred.

Morgan and Anders argue that the volatile contents of planets increase with increasing planetary mass, whereas the FeO contents, which in meteorites are closely correlated with the abundances of volatiles, increase with heliocentric distance. They then point out that the parent body of the shergottite achondrites is even smaller than the smallest planet, yet has a higher volatile content! Morgan and Anders regard this as "no great calamity, because this trend has no obvious explanation anyway."

The entire argument upon which this analysis is constructed fails if massive, nonselective volatile loss can occur on small planets. If Mars, the Moon, and Mercury have suffered atmosphere blowoff without fractionation, then there is no such thing as a "thallium group" of elements in planets and the logical deductions made by assuming such a creature are invalid. Besides, the shergottites are from Mars!

Although we have been assured by Anders and Owen that there is no evidence whatsoever for catastrophic loss of early atmospheres from planets, however small, it seems clear that a search for evidence of such a mechanism would be very much worthwhile. After all, whatever we believe about the primordial volatile contents of planets, it is obvious that escape is easier from the smallest ones. It is qualitatively appealing to interpret the FeO gradient in the inner Solar System as requiring a smooth increase in volatile content with heliocentric distance in the raw planetary material. By this criterion, Mars should have originally accreted with a higher volatile content than Earth.

The geochemical models of Dreibus and Waenke mentioned in our earlier discussion of the SNC meteorites draw not only upon the limited compositional data gathered by the XRF experiment on the Viking landers, but also on the chemistry of the SNC meteorites. They point out that the Mn:Mg ratio in both ordinary and carbonaceous chondrites is quite constant, apparently a universal condition for primitive (undifferentiated) preplanetary material. Further, they observe that the Fe:Mn ratio in silicates is quite insensitive to igneous fractionation processes: in effect, each planetary body has a "signature" Fe:Mn ratio that pervades its entire suite of igneous rocks. It follows that the Fe:Mn ratio observed in any igneous rock from a planet may be used to constrain the Fe:Mg ratio in its bulk silicates, which in practice means the mantle of the planet. Dreibus and Waenke then construct a model for Mars consistent with the SNC meteorites, containing all the lithophile elements in their chondritic (CI) proportions. The bulk planetary Fe:Si ratio, including the core, is also chondritic. The calculated Fe/(Fe + Mg) ratio for the Martian mantle is then 0.25, equivalent to 18% FeO in the bulk silicate portion of Mars. The bulk sulfur abundance, calculated somewhat less securely from the abundances of other volatile elements in the crust, is sufficient to provide 14.5% S in the Martian core, which they find based on their geochemical constraints to have a radius of 1680 km.

Calculations by petrologists C. M. Bertka and I. Fei, incorporating the Dreibus–Waenke chemical model as well as the value of $C/MR^2 = 0.366 \pm 0.0017$ deduced from the radio science experiment on the Mars Pathfinder lander, find that assuming the chondritic Fe:Si ratio forces a solution with a very thick (up to 320 km) crust with a density in the range 2.7 to 3.0 g cm$^{-3}$. Relaxing the Fe:Si assumption gives models with a much smaller core of radius 1420 km and a crust of density 3.0 that is only 50 km thick. They estimate a pressure of 25 GPa (250 kbar) at the core–mantle boundary, at a temperature of about 1770 K. This is far above the Fe–S eutectic temperature of about 1400 K at this pressure, ensuring extensive core melting for any reasonable sulfur content.

We shall reject thermal escape and dissociative recombination as the agents of such an escape process; they are selective and would produce severe fractionation that is not seen. T Tauri phase solar wind activity and tidal stripping by the early Sun are poorly understood at best, probably occurred too early in the accretion history of the planets to be relevant, and cannot be quantified.

The effect of large impact events is, however, of obvious interest. Mars, the Moon, and Mercury are all heavily cratered by objects that commonly left craters measuring tens of kilometers in diameter. Even small impact events are capable of blowing holes larger than a scale height in size in the lower atmosphere of Mars. Such loss of volatiles by explosive blowoff would be highly nonselective, in that all uncondensable volatiles would be very easily dissipated into space, whereas only the condensed volatiles, $H_2O$ and $CO_2$, would be retained.

As a consequence, the early atmosphere of Mars may not survive the accretionary era. If a primitive atmosphere amounting to 1 ppm of the mass of Mars was present after 99% of the planet had been accreted, it would have taken only 0.01% of the impact energy of the late-accreting material to blow the entire atmosphere off the planet. This "explosive blowoff" or "atmospheric erosion" scenario, suggested by myself and Hampton Watkins in 1982, has been studied by Ann Vickery and Jay Melosh of the University of Arizona. They have found it capable of removing some 99% of the early atmosphere.

In summary, the fundamental issue of the primordial volatile content of Mars thus remains unresolved by the available evidence. No one disputes that the present atmospheric inventory is small; what is at issue is whether Mars has outgassed thoroughly and whether massive loss of volatiles has taken place. The bulk density of the planet and its rotational moment of inertia reveal a high degree of oxidation, a property that, in chondrites, is always correlated with a high volatile content. For this reason we favor compositional models for the early Mars in which the proportions of water, carbon, etc., were higher than in the early Earth.

If Fanale's arguments about early catastrophic outgassing, occurring within about 100 million years of the dissipation of the nebula, are correct, then the accretional energy of the planet was available to dissipate any primary reducing (Stage I) atmosphere. This phenomenon has never been modeled, although studies on blowoff of a carbon dioxide atmosphere described above should provide the necessary modeling techniques. Such dissipation would be of much less importance on planets with more massive atmospheres and deeper gravitational wells, such as Earth and especially Venus.

The suitability of Mars for the origin of life during its earliest history is far from obvious, and the impossibility of producing and accumulating organic matter in the present highly oxidizing photochemical regime is manifest. There remain a number of urgent needs in the study of the evolution of Mars and its atmosphere. First and foremost among these, and at the top of the list in order of expense, is the dating of Martian geological and tectonic processes. Valuable insight into the chemical, petrological, and mineralogical evolution of the planet is being gained by combining elemental composition data from landers, SNC meteorites, and orbital gamma-ray spectrometers (GRS) with mineralogical data from multispectral imaging and the thermal emission spectrometer (TES) on MGS, isotopic data from SNC samples and Earth-based IR spectroscopy, and conventional photogeological mapping.

## Organic Matter and the Origin of Life

At the beginning of the era of spacecraft exploration of the planets it was widely, but by no means universally, accepted that Mars was inhabited by at least primitive life forms. The spirited debate concerning the possible biological significance of the IR "Sinton bands" was finally resolved in the negative by the discovery by Donald Rea that these features were caused by terrestrial HDO. Laboratory experiments were carried out on the reflection spectra of organic and biological materials for comparison with Mars, and the possibility of prebiological synthesis of organic matter on Mars was seriously discussed. The wave of darkening was still under active consideration as a biological phenomenon as late as 1967.

Fraser Fanale's geochemical arguments, based on the presence of ferrous and ferric iron in the crust, imply that the initial products of planetary outgassing may have provided a rather massive reducing atmosphere on Mars. This atmosphere is, of course, unstable against thermal hydrogen escape, and Fanale emphasized in a series of papers that it was plausible that Mars outgassed rapidly and extensively very early in its history, presumably in response to severe accretional heating. Slow outgassing could not keep pace with $H_2$ escape, and the atmosphere would be oxidizing. Rapid outgassing could maintain a substantial reducing atmosphere for only a short time. Fanale also stressed that the present surface conditions on Mars are hostile to the formation and preservation of organic matter. Thus, if Mars ever had the ability to form organic matter and life, that chapter in its history must have been both brief and early in the evolution of the planet. Carl Sagan has also offered the speculation that cyclic climatic excursions on Mars (specifically, those caused by the precessional cycle) would periodically provide benign conditions and that life forms might survive in a freeze-dried state from one climate optimum to the next.

It was in this climate of opinion that the plans for the Viking landers were developed. Although the credibility of Martian life suffered greatly as a result of the Mariner 4, 6, and 7 flybys, which sampled dreary, ancient cratered terrain, the thorough mapping of the surface by the Mariner 9 Orbiter raised new hopes. The photogeological evidence for river beds and flooding caused the freeze-dried hopes of exobiologists to germinate and grow with new vigor. Viking was planned. There was no question about the driving force behind Viking; it was not geophysical interest in whether Mars was an active planet, nor geochemical interest in the composition and history of the crust. There was no interest in whether Mars had a core or a magnetic field. Meteorology was found tolerable in small doses, but the addition of geophysical and geochemical experiments could be carried out only by the most determined interventions. Viking's mandate would be to detect life on Mars.

The biological experiment package was designed to accommodate four experiments. First, there was the carbon assimilation experiment, in which $^{14}C$-labeled CO and $CO_2$ were introduced into a "culture medium" that, by terrestrial standards, would resemble a torture chamber. The reasoning was that, however inhospitable Mars seems to terrestrial organisms, Martian organisms (if any exist) must be adapted to it and can be expected to be very intolerant of the vastly different environments to which terrestrial microorganisms are accustomed. Accordingly, samples of Mars dirt are placed by the Viking lander sampling arm in a small chamber with a window that admits sunlight. Traces of labeled CO and $CO_2$ are admitted and allowed to remain in contact with the dirt for several hours. Unreacted gases are then flushed from the system and the soil sample is heated to temperatures of about 850 K, sufficient to kill any microorganisms and thoroughly pyrolyze their organic remains. Radioactive gases released during pyrolysis are then cleaned of organic molecules, and the CO and $CO_2$ component of the released gas is then counted for its radiocarbon content. Initial experience with this experiment on the Viking 1 lander mission showed that slow assimilation of CO and $CO_2$ took place, and preheating the surface sample or exposing it to moisture inhibited this process. The final report on this experiment shows more clearly how this activity depends on the severity of the preheating: 2 h at 90 °C has no effect, whereas 3 h at 175 °C reduces the rate of assimilation about 90%. Furthermore, the inhibition of assimilation by water vapor turned out not to be verifiable. The experimenters conclude that it is unlikely that the reactions observed, which apparently produced minuscule amounts of organic matter, were of biological origin.

The second major biological experiment, called the labeled release experiment, involved exposing dilute aqueous solutions of $^{14}C$-labeled organic nutrient to a sample of Mars soil. The release of radioactive gases is taken as evidence for metabolic activity. Early Viking results showed a small response, which, like that in the carbon assimilation and pyrolytic release experiment described above, could be prevented by preheating to sterilizing temperatures. Both inorganic and organic processes in the Mars dirt could account for these results. Further experience with this experiment on both Viking landers has led to no further insights into the nature of the process.

The third life-detection experiment involved the use of a gas chromatograph for detection of gas exchange between Martian soils and the gases in contact with it. All experiments performed on the Viking landers showed such composition changes, which apparently are largely due to desorption of atmospheric gases from the soil sample combined with oxidation of organic matter in nutrient media that can be added to the sample. The oxidizing agent could, as suggested by simulations on Earth, be $\gamma$-$Fe_2O_3$. Oxygen is evolved upon the addition of $H_2O$, possibly from the decomposition of photochemically produced superoxides in the surface dirt. $MnO_2$ produces this effect. No evidence requiring the presence of biological activity was found.

The fourth biology experiment originally planned for the Viking missions was a light-scattering experiment that sought to identify turbidity of biological origin in a nutrient solution that had been inoculated with Mars dirt. This experiment encountered severe problems during development and was deleted from the payload.

In a sense, one of the most powerful and general life-detection experiments aboard Viking was the GCMS experiment, which was designed to permit extremely sensitive searches for organic matter in surface samples. The results were, as we have seen, negative at the level of 0.1 – 1.0 ppb for a host of different organic species.

Overviews of the results of the Viking search for life on Mars have reached cautiously dismal conclusions. It is, of course, impossible to rule out the presence of life on Mars on the basis of the Viking results. Those who preferred to spin fantasies in the absence of diagnostic evidence are at liberty to continue to do so. Mars has been left for over 16 years to the attentions of astronomers and theorists. Life on Mars is commonly regarded as a bad joke left over from the 1960s, and it was impossible to generate any new missions to Mars from 1976 to 1992. The political climate was poisoned by the failure of a billion-dollar investment to prove the existence of little green men, and many of our most fundamental questions about Mars had to remain unanswered for many years.

The fundamental questions regarding the organic and biological history of Mars remain unchanged. Was there ever a reducing atmosphere on Mars? If so, how long did it last and what did it produce? Yuk Yung of the California Institute of Technology has shown that an assumed massive, mildly reducing atmosphere can generate vast amounts of hydrocarbons; in fact, he suggests that the fluvial features on Mars were produced by rivers of alkanes, floods of gasoline and lighter fluid that would dwarf the Amazon. Another fascinating question has come to the fore: can organic matter be produced without a reducing atmosphere? The production of organic matter from $CO_2$ in the presence of UV light and ferrous iron has been reported, and there is a thermodynamic rationale for the protection of the organic matter from oxidation by ferrous iron. It has also been shown that UV light can produce formic acid (HCOOH) from a mixture of $CO_2$ and $H_2O$ at temperatures as low as 200 K. But if organic matter is formed, is it stable in the Martian regolith today? The Viking GCMS experiment tells us that organics are absent in the surface dirt, but does not tell us why. The reason is that rapid UV photochemical processes at the Martian surface will produce small amounts of organic matter and larger amounts of very strong oxidizing agents simultaneously. At steady state the concentrations of the interesting organic species are negligible, and buildup of organic products cannot occur because of attack by OH, $O_3$, $O_2$, and atomic O.

However clouded the past history of life on Mars, we may not so readily dismiss the possibility of a glorious future. Joseph Burns and Martin Harwit have proposed a particularly brute-force approach to ameliorating Martian climate by shifting around huge masses of material so as to increase the precession period and thus prolong (perhaps indefinitely) the coming "Martian spring." Elsewhere Sagan has also proposed transporting some $10^{10}$ tons of very dark dust to the polar regions, where it will absorb sunlight, evaporate the polar caps, raise the atmospheric pressure above the triple point of water, and thus permit a greatly enhanced greenhouse effect and a much warmer climatic regime. However, as suggest by British physicist James Lovelace, very much smaller amounts of gaseous halocarbons can provide an enhanced greenhouse effect in a single step, without the problem of condensates coating, or wind-borne dust burying, the dark pigments. Plausible schemes for manufacturing these gases on Mars and releasing them into the atmosphere in part-per-billion quantities (approximately $10^4$ tonnes of halocarbons). could make Mars more Earthlike in climate, a process called *terraforming* the planet. By such means, Mars may be made more hospitable to life than at any time in its past.

## Venus

Our first knowledge of any of the intrinsic properties of Venus dates back to observations made by the Russian astronomer M. V. Lomonosov in 1761. He observed refraction of sunlight around the disk of Venus while observing a transit of Venus across the face of the Sun, thus establishing for the first time the presence of an atmosphere on Venus.

In the 19th and 20th centuries, visual observations and photographs of Venus have revealed a brilliant, pale-yellow, featureless disk, clearly the top of a dense cloud layer through which it is impossible to see. UV photographs show faint V- or Y-shaped markings, symmetric across the presumed equator, that drift around the planet in about 4 Earth days.

A quasimythological understanding of Venus, Earth, and Mars, inherited from the all-too-recent days when we knew virtually nothing about the other planets, assigned the four terrestrial planets to an evolutionary sequence that we now know to be without basis. Mars was the most mature of the four, formed before its siblings and now evolved into an ever more distant orbit from the Sun, its former oceans vanished, and its advanced life forms extinct or degenerate. The attribution of Martian "canals" to extinct or waning Martian intelligent life harmonized well with this naive view, as did the scenario of an invasion of Earth by technically advanced refugees from the dying Mars, as in H. G. Wells' *War of the Worlds*. Earth was of course the type example of a planet in the prime of its life, not yet in danger of losing its oceans and air. Venus was a young planet, not yet home to advanced and intelligent life, but possibly analogous to a Carboniferous Era global swamp: hot, humid, and with dense cloud cover; covered with lakes, oceans, jungle, and swamps; and probably punctuated by violent volcanic eruptions as well. Mercury, just emerging from the fires of the Sun, was an unfinished work, not yet ready for the appearance of atmosphere and oceans. This conceptual model had a strong emotional impact on writers of popular science fiction as well, persisting long after the realization that the planets are all the same age and are not retreating from the Sun.

As early as 1922 it became technically feasible to use infrared spectroscopy to search for water and oxygen on Venus and Mars, but the first searches found neither gas at rather restrictive levels of sensitivity. By 1932 the presence of large amounts of carbon dioxide was well established. This was to remain the only detected gas on Venus until 1965, when high-resolution infrared interferometric spectroscopy was introduced.

In the 1930s and 1940s the leading scholar of Venus was the German-born physicist Rupert Wildt. Wildt moved to Princeton in the postwar years to work with the dean of American astronomers, Henry Norris Russell, and later moved on to an illustrious career at Yale. His first contribution to the study of Venus was to show that a large carbon dioxide pressure caused a strong *greenhouse effect*, in which visible sunlight penetrated the atmosphere with moderate efficiency, was absorbed by the surface, and then was reemitted at thermal wavelengths, which are very strongly absorbed by $CO_2$ and hence trapped near the surface. The surface temperature then rises until the short-wavelength end of the Planck peak emerges sufficiently from the IR region of strong gaseous absorption to dump the heat out at wavelengths at which the gas is rather transparent. He calculated that the surface temperature of Venus must be well above the normal boiling point of water, a conclusion borrowed without attribution by the mid-20th-century pseudoscientist Immanuel Velikovsky. Addition of water vapor (not yet detected at that time) greatly strengthens the greenhouse effect by adding opacity at wavelengths in between the $CO_2$ bands, but water also provides the material for massive, bright clouds that hinder the penetration of sunlight to the surface. Wildt concluded that liquid water was absent on the surface of Venus and therefore that the apparent dryness of the atmosphere meant that Venus must be very deficient in water relative to Earth. He suggested that water-bearing minerals might be the main reservoir of water on the planet. Wildt also noted that the amount of $CO_2$ then believed to be present in carbonate rocks on Earth (about $10\,\mathrm{kg\,cm^{-2}}$) was comparable to the amount he expected to find in the atmosphere of Venus.

In 1952, the American Nobel-Prize-winning chemist Harold C. Urey turned his attention to the Solar System. In his classic book, *The Planets: Their Origin and Development*, he suggested that the pressures of reactive gases in planetary atmospheres might be regulated by chemical reactions with their surfaces. A reaction in which one thermodynamic variable (in this case, the $CO_2$ pressure) is determined by temperature alone is called a *buffer reaction*. On Venus, he proposed that the schematic reaction

$$CO_2 + CaSiO_3 \rightarrow CaCO_3 + SiO_2 \qquad \text{(X.51)}$$

was responsible for regulating the carbon dioxide pressure. Given information about the mineral assemblage on the planetary surface, reactions such as this, involving silicates of several different metals and a number of different reactive volatiles, could regulate the mass and composition of the atmosphere.

Several workers, unprepared to admit very high surface temperatures on the strength of radiative theory alone, offered suggestions about how to keep Venus safe for life. Donald Menzel and Fred Whipple of Harvard College Observatory proposed that the Urey reaction, which associated high $CO_2$ pressures with high surface temperatures, failed on Venus because the surface was covered with deep global oceans. Gases therefore could not react with crustal rocks because there was no weathering on Venus to make ions of calcium, magnesium, etc., available for reaction.

The discovery in 1958 that Venus had a high radio brightness temperature of about 600 K at a wavelength of 3 cm added an important constraint. The simplest interpretation was that the surface temperature of Venus was at least 600 K, and the thermal emission from the surface, weakly attenuated by absorption in the atmosphere, was the radiation we saw at microwave wavelengths. Carl Sagan, in a seminal paper in 1962, revived Wildt's greenhouse theory and showed that the radio wavelength emission from Venus could be explained by a surface temperature of at least 700 K and a surface atmospheric pressure near 100 bar. Further, the absence of any detectable absorption feature near the water vapor rotational absorption line at 1.35-cm wavelength showed that the water content of the Venus atmosphere could not be larger than about 0.1%.

Some resistance to the idea of Venus as a red-hot desert drew upon the idea that the high radio-frequency brightness temperatures could be attributed to free-free emission from electrons in an extremely dense ionosphere or from vast numbers of lightning discharges in an atmosphere filled with torrential rainstorms. The last gasp of those who wanted to preserve Venus as a wet planet was heard in 1968, when Willard Libby argued that Venus had as much water as Earth, but that it was frozen into gigantic 10-km-thick ice caps that covered half the planet. The "failure" of the Urey equilibrium was then attributed to the absence of liquid water, which meant an absence of weathering to release calcium and magnesium ions, etc.

The advent of spacecraft exploration of Venus in the 1960s laid most of these theories to rest. A microwave radiometer on the Mariner 2 Venus flyby in 1963 clearly showed that the microwave emission from Venus was *limb darkened* (dropping off in intensity toward the edges of the visible disk), proving that it was emitted from deep in the atmosphere and attenuated by passage through the atmosphere, not made in the ionosphere or cloudtop regions as some had suggested.

The Soviet Venera 4 entry probe made the first successful entry into the atmosphere of another planet in 1967, reporting crude data on pressure, temperature, and atmospheric composition before being crushed by the atmospheric pressure while still high above the ground. Venera 7 in 1972 was the first spacecraft to reach the surface of Venus in working order, and two landers returned photographs of the surface in 1975, showing a blazing hot, rock-strewn desert. The carboniferous swamps, jungles, herbivores, and carnivores are absent; the lakes, oceans, and ice caps are banished. Venus is the ultimate desert, with a surface temperature near 750 K and a surface pressure near 92 bar. Whether Venus owes the great differences between its present conditions and those of Earth principally to different conditions of origin or to divergent evolutionary paths, the facts are clear: Venus is no twin of Earth, nor is it in an early evolutionary phase that Earth once went through.

## Motions and Dynamics of Venus

Venus orbits in the normal (*prograde*) direction about the Sun in a nearly circular orbit every 224.7 Earth days. It rotates relative to the stars in the retrograde direction every 243.01 Earth days (its *sidereal* rotation period). From these figures it can be seen that the orbital motion of Venus over half a Venus year causes the Sun to move halfway around the sky, whereas the rotation of Venus adds to and roughly doubles that motion. Thus the Sun as seen from the surface of Venus (if it could be seen from the surface!) makes roughly one complete circuit of the sky in half a Venus year. The length of the Venus day is given more precisely by $1/t = (1/224.7 + 1/243.0)$ or $t = 116.75$ Earth days.

The position of Venus in Earth's sky, and hence its observability, goes through a complete cycle in the time taken for Venus to lap Earth in its orbit. This relative or synodic period is given by $1/P_{\text{syn}} = 1/P_{\text{V}} - 1/P_{\text{E}} = 1/224.7 - 1/365.24$, or $P_{\text{syn}} = 583.92$ Earth days.

Three apparent coincidences add interest to these dry statistics. First, the Earth–Venus synodic period is 583.92/116.75, or 5.0014 Venus days. Thus at equivalent relative positions in their orbits (for example, at successive inferior conjunctions of Venus) the same point on the surface of Venus points at Earth, with very high but not perfect precision. As a direct result, radar mapping of Venus, which is much easier when the range is smallest, tends to see the same portion of Venus over and over again with little offset from one inferior conjunction to the next. The second coincidence is that the synodic period is very nearly $\frac{8}{5}$ of the terrestrial year ($1.6000 \times 365.24 = 584.34$ days). Thus every 8 Earth years the positions of Earth and Venus repeat closely with only a slow relative drift. Every 152 synodic periods, 243.01 Earth years elapse, leading to a long cycle of even more precise repetition of the positions of Earth and Venus. Passages of Venus across the disk of the Sun are observed from Earth to follow a pattern in which consecutive transits can be almost exactly 8 Earth years apart. Venus will next transit the disk of the Sun on June 8, 2004, and again on June 6, 2012. Venus will then miss the disk of the Sun until eclipses again occur at the opposite node (point of crossing of the orbital planes of Earth and Venus) a half-cycle later, around the year 2130. The third coincidence, which I have never seen discussed in the literature, is even more strange (and probably less susceptible to physical interpretation): the number of sidereal rotations in a Venus transit cycle is 365.24, the same as the number of days in an Earth year!

The inclination of Venus' rotation pole to the normal to its orbit is only 2.6°. Of the terrestrial planets, only Mercury, which is tidally locked into its 3:2 spin–orbit resonance with the Sun, has a lower axial obliquity.

## Geophysical Data on Venus

The density of Venus has been determined from the gravitational mass, measured by tracking Venus-orbiting spacecraft such as the Pioneer Venus Orbiter and the

Magellan Venus radar mapper, and the radius, determined directly from Earth-based radar studies of the planet. The mass is found to be $4.871 \times 10^{27}$ g. The equatorial radius, 6051.3 km, is taken to be the same as the mean radius because the very slow rotation rate implies an oblateness that is negligibly small. The deduced density of $5.245\,\mathrm{g\,cm^{-3}}$ is about 5% less than Earth's. Correction of the density to zero pressure requires a precise knowledge of the equations of state of the planetary materials, the internal structure (degree of differentiation), and the temperature distribution inside the planet. The only point of agreement between competing models is the assumption that the planet is completely differentiated. Depending upon what set of assumptions is made regarding the other two parameters, zero-pressure densities about 1% less than that of Earth are found, with a range from about 0% to perhaps 3%. Whatever process is responsible for the large enrichment of iron in Mercury relative to Earth, that process cannot have had any comparable effect on Venus; the Earth–Venus and Venus–Mercury density trends have the *opposite sign*.

We are severely limited in our ability to deduce the interior structure of Venus by the absence of seismic data. No spacecraft has ever survived more than a matter of minutes on the Venus surface, and no seismic experiment has ever been attempted. A variety of geochemically plausible models are compatible with a core–mantle boundary near $2800 \pm 100$ km depth, but no means exists to test these predictions.

Data suggestive of a magnetosphere on Venus date back to the Venera 4 and Mariner 5 missions of 1968, but acceptance of the reality of a bow shock did not become universal until the Soviet Venera 9 and 10 orbital missions in 1975 and 1976. Nonetheless, no evidence yet exists for a detectable planetary dipole; the Pioneer Venus Orbiter in 1978 was able to measure the magnetic field strength deep into the ionosphere, through the solar wind bow shock, and found no planetary field. The bow shock exists by virtue of the interaction of the solar wind with the planetary ionosphere, in which currents induced by the solar wind magnetic field flow and in turn generate local magnetic fields. Venus sweeps out a cavity in the solar wind flow that extends outward from its orbit. The interaction of Venus with the solar wind resembles that shown in Fig. X.1b.

The absence of an external dipole field points to several possible defects with a core dynamo source: the core of Venus may be completely solid, in which case core convection cannot occur; the core may be completely molten but devoid of an internal heat source to drive convection; or the very slow rotation of Venus may be insufficient for the Coriolis forces to organize core motions into the large spatially and temporally coherent structures necessary for generation of planetary-scale poloidal or toroidal fields.

A crucial early source of information about the surface conditions of Venus was Earth-based measurements of the thermal emission of the planet in the radio region (wavelengths of 0.2 to 70 cm). The Planck function over wavelength,

$$B_\lambda = (2hc^2/\lambda^5)(\exp[hc/\lambda kT] - 1)^{-1}, \qquad (X.52)$$

can, at long wavelengths ($\lambda >> hc/kT$), be approximated by

$$B_\lambda = (2hc^2/\lambda^5)(\lambda kT/hc) = 2ckT/\lambda^4. \qquad (X.52a)$$

The *brightness temperature*, $T_B$, the effective temperature estimated from a monochromatic long-wavelength flux measurement, is then just

$$T_B = \lambda^4 B_\lambda/2ck. \qquad (X.53)$$

As always, the brightness temperature or effective temperature is related to the true temperature, $T$, of the emitting surface by $\varepsilon T^4 = T_B^4$. Near 6 cm the observed brightness temperature of Venus is near 700 K (Fig. X.24). Reasonable emissivities are not much less

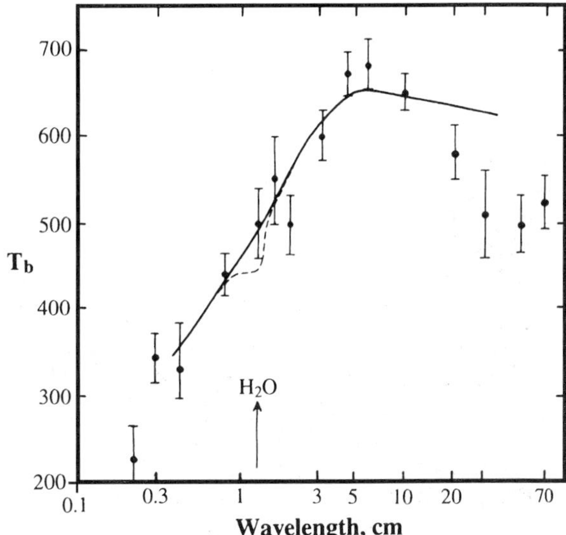


**Figure X.24** Radio-wavelength brightness temperatures of Venus. The atmosphere is most transparent at and beyond several centimeters in wavelength. The decline of $T_b$ toward shorter wavelengths is due to molecular opacity in the atmosphere, and the decline toward longer wavelengths presumably contains information about the particle size distribution and vertical structure of the regolith. Measurement of the radar reflectivity of Venus in the 10- to 100-cm region reveals no significant atmospheric attenuation. The location of the pure rotational band of water, collision broadened to cover the 1- to 2-cm wavelength region, is sketched in to show that the Venus thermal emission spectrum contains no evidence for the presence of water in the lower atmosphere (probably < 0.1% based on this criterion alone).

than 1, and some absorption of radio energy must take place in the atmosphere. The real surface temperature then must be over 700 K.

The precipitous decline of the observed brightness temperature toward shorter wavelengths is due to absorption in the atmosphere. Water vapor has a strong microwave absorption feature due to rotational excitation at 1.35-cm wavelength. Because the spacing of the energy levels is so much less than the translational thermal energy of the gas, collisions can very seriously perturb the rotational energy levels. The width of the 1.35-cm feature is extremely sensitive to the fraction of water vapor molecules that are being perturbed by collision, and hence the feature broadens rapidly toward higher pressures. The complete absence of such a feature in the Venus thermal emission spectrum (Fig. X.24) places an upper limit of about 0.1% (1000 ppm) on the water vapor content of the atmosphere.

The equally steep decline of brightness temperature toward longer wavelengths is not easily explained. These data in isolation might suggest a dominant surface particle radius ($a$) of about 10 cm ($2\pi a \approx \lambda$) combined with a size-dependent emissivity, a strongly wavelength-dependent dielectric constant, or pure-rotational absorption by another species in the atmosphere. The last seems rather interesting in that it implies an unusual rotational moment of inertia for the absorbing species. Recall that

$$\varepsilon_r = r(r+1)h^2/4\pi^2 I, \qquad \text{(IV.161)}$$

so a species with a rotational feature at 50 cm (vs 1.35 cm for water) would have a rotational moment of inertia about 40 times larger than that of water vapor. But radar studies rule out both the strongly wavelength-dependent dielectric constant and the large-molecule absorber hypotheses. Some oddity of the surface structure must be responsible for this effect.

## Geology of Venus

Earth-based delay-Doppler mapping of Venus has revealed two large elevated regions, Ishtar Terra and Aphrodite Terra, and several much smaller elevated areas, most notably Alpha Regio, Beta Regio, and Tellus Regio (Fig. X.25). Accurate altimetry was obtained by a radar altimeter on the Pioneer Venus Orbiter, which provided useful coverage from about 65° to 75 °N. The hypsometric curve of Venus was compared with those of Earth, Mars, Mercury, and the Moon in Fig. X.2. The very highest elevations are the Maxwell Montes in Ishtar, at a latitude of about 65 °N, which rise to an altitude of about 10.5 km above the mean planetary radius. About half of Ishtar is occupied by the Lakshmi Planus, a rather flat, grossly elliptical area at a mean altitude of about 3 km. There are two large, irregular depressions in Ishtar, Sacajawea (about 200 km in diameter) and Colette (about 100 km), both of probable

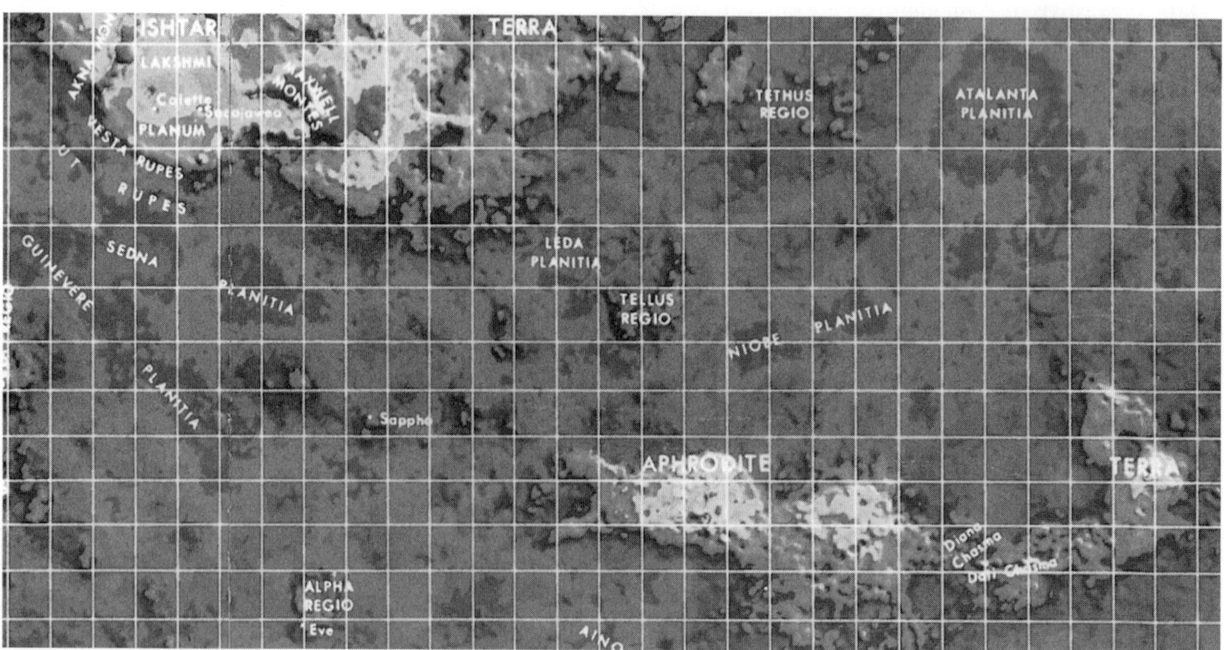


**Figure X.25** Topography of Venus. The two major continents, Ishtar and Aphrodite, occupy only a few percent of the total surface area. Most of the planet is covered by low, rolling plains. These data were collected by the Pioneer Venus radar altimeter experiment (See color plate 7).

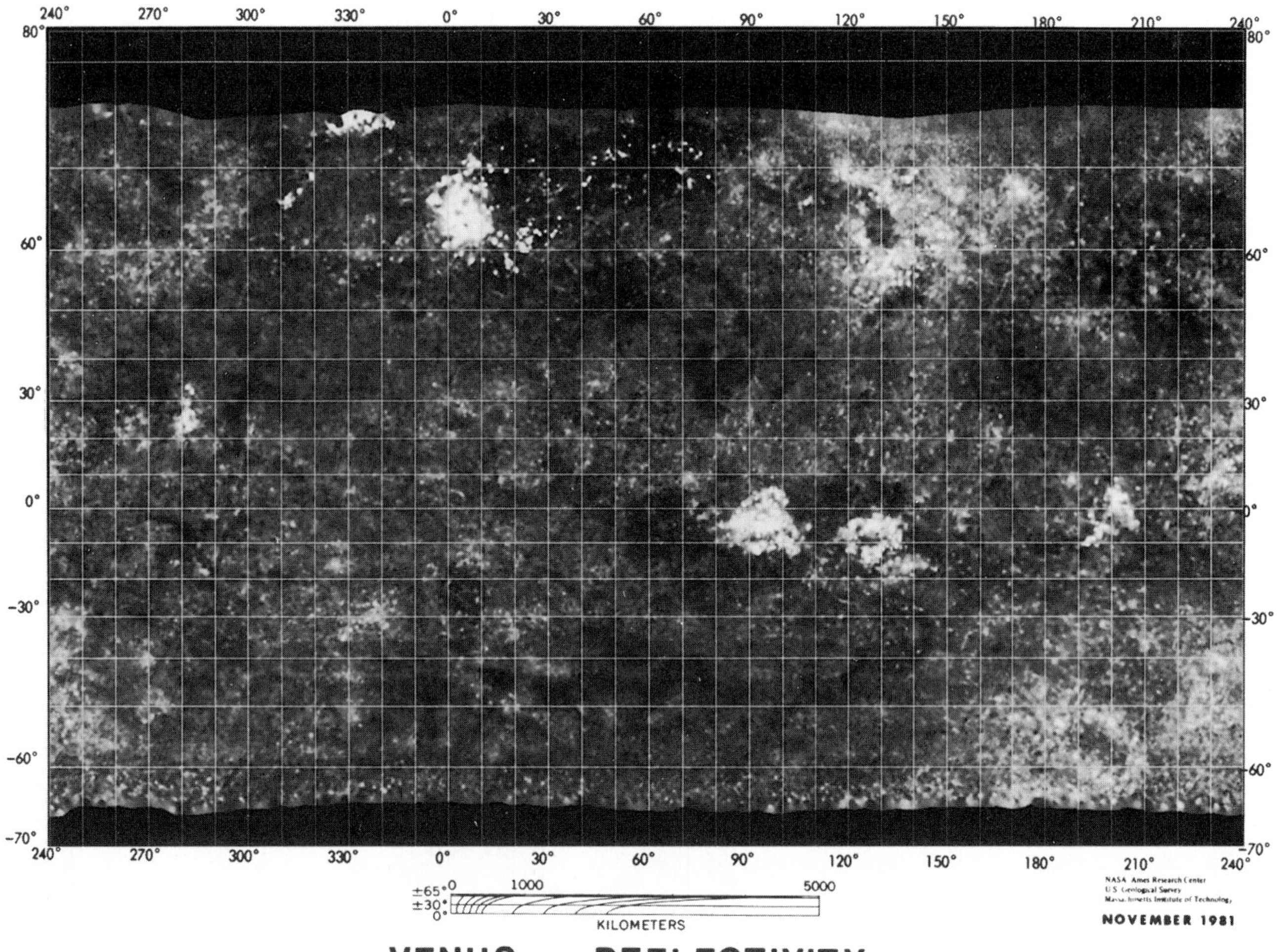


**Figure X.26** Reflectivity of the surface of Venus. The regions of highest reflectivity found by the Pioneer Venus radar altimeter correlate nicely with the highest mountains in the altimetry map. Reflectivity is physically associated both with roughness on the scale of the radar wavelength and with a high surface dielectric constant.

volcanic origin. Ishtar is bounded on the east by a 5-km scarp that rises to the Maxwell Montes and on the west by a 2.5-km scarp that drops down to the plains.

Aphrodite, a long, narrow area at low latitudes bracketing the equator, has a few isolated peaks that reach as high as about 8 km and stretches over 150° of longitude. Aphrodite has about the area of South America, about twice that of Ishtar. These elevated areas together cover only a few percent of the total surface area of the planet; about 5% of the planetary surface is at an altitude greater than 2 km above the planetary mean. Over 90% of Venus consists of low, rolling plains.

In eastern Aphrodite are three deep troughs, Artemis Chasma, Diana Chasma, and Dali Chasma, which plunge as low as 2 km below the planetary mean. They are part of an immense system of parallel ridges and troughs that stretch over 6000 km in the east–west direction. Some individual features are over 2500 km long. The vertical relief in this complex is from −2.5 to +2.5 km altitude.

Areas of enhanced radar reflectivity mapped by both Earth-based radars and the Pioneer Venus radar altimeter include the Maxwell Montes and three large parts of Aphrodite Terra (Fig. X.26). Numerous much smaller radar-bright spots were observed, most associated with major elevations. Radar reflectivity can be enhanced either by great roughness on the same scale as a radar wavelength or by an elevated dielectric constant in the surface material, due to large abundances of dense, moderately conductive minerals such as iron sulfides and iron oxides. These large-scale reflectivity features are generally associated with mountainous areas of expected high roughness. The best overview of Aphrodite comes from the computer mosaic of images from

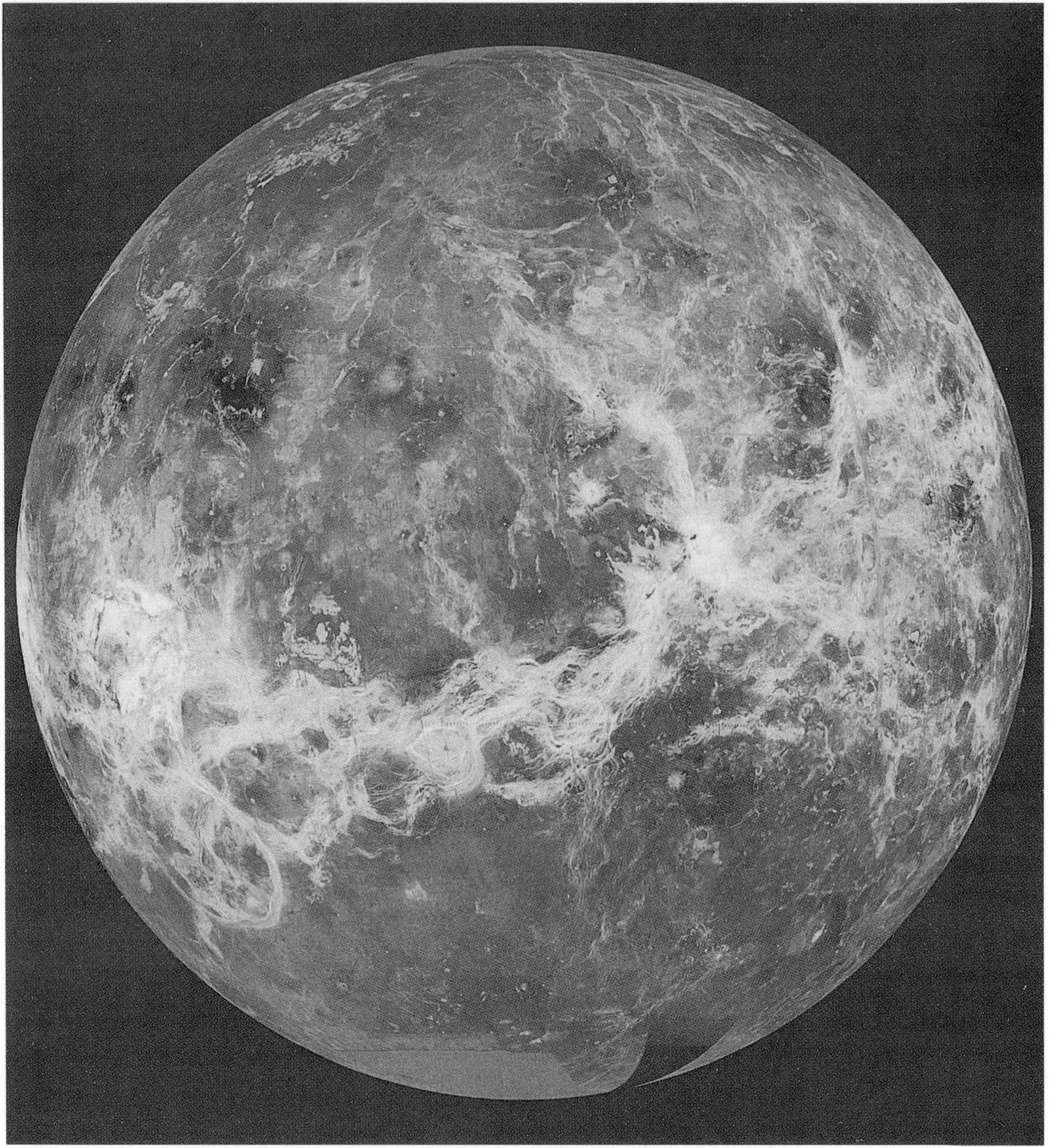

**Figure X.27** Radar image of the Venus surface. The Magellan radar mapper coverage is nearly complete. The bright feature across the center of the disk is Aphrodite Terra. The very large circular feature at lower left is the Artemis corona. Coronae, which seem to appear in developed and unambiguous form only on Venus, are crustal expressions of powerful mantle upwelling.

the Magellan Venus radar mapping mission. Magellan, launched in 1989, was the enormously successful sequel to two pioneering Soviet radar mappers, Venera 15 and 16, which mapped small portions of the Venus surface with very good resolution. The Magellan view of Aphrodite is shown in Fig. X.27. The two bright spots just east of the center of the disk are shield volcanoes. The huge circular feature southwest of Aphrodite is the Artemis *corona*, marked by concentric rings of ridges and troughs. Such features, which are believed to be due to rising currents in the mantle, are unique to Venus.

The surface of Venus is dominated by volcanic processes. The higher elevations are mostly volcanic constructs, and the plains are covered with very extensive lava flows. Magellan reflectivity and altitude data have been combined in a computer model of the topography that permits the computation of views of the surface from angles at which it was not actually observed. One such image, constructed to assist geologists in interpreting the structure of the planetary crust, is shown in Fig. X.28. This image shows the Sapa Mons volcano in the foreground and the 8-km-high Maat Mons volcano in eastern Aphrodite Terra in the background. These are the same two volcanoes seen near the center of Fig. X.27.

The channels visible in Fig. X.28 are but small, local examples of features that are widespread in the volcanic areas of Venus. Some of these narrow channels extend to phenomenal distances from the volcanic vents that feed them, in one case extending roughly 5000 km.

A few large impact craters were recognized on Earth-based radar images as both altitude and roughness features; they have flat, smooth (radar-dark) floors with elevated, radar-bright rims. But the overwhelming majority of the evidence regarding impact cratering of Venus comes from Magellan mapping, which provided data on 842 impact craters. A spectacular trio of impact craters with raised rims, central peak complexes, and flat, radar-dark lava-flooded floors can be seen in Fig. X.29. In general, craters on Venus have ejecta blankets that are limited to the immediate vicinity of the crater because the high density, $\rho = \mu P/RT$, of the atmosphere suppresses ballistic expulsion of ejecta. The surface atmospheric density on Venus is about 54 times that on Earth and 3900 times that on Mars.

The atmospheric density and depth have a further profound effect upon the cratering history of the surface of Venus. The massive atmosphere acts as a protective blanket that exerts powerful aerodynamic forces on high-speed entering objects, exceeding the crushing strength of most natural materials. Small weak bodies as a rule break up completely in the atmosphere and do not produce recognizable surface features. Intermediate-strength bodies explode deep within the atmosphere and scour the surface beneath them with a powerful blast wave, but do not produce impact craters with ballistic ejection of cratering debris. Somewhat larger or stronger bodies will disrupt into multiple fragments while passing through the atmosphere, and the fragments will be sufficiently dispersed by aerodynamic forces before reaching the ground to form a tight cluster of overlapping impact craters. The largest and strongest impactors reach the surface as single bodies or very compact swarms and produce a single large crater. The observed (smoothed) size–frequency distribution of Venus impact craters from the Magellan data is given in Fig. X.30. The $d^{-2.5}$ slope of the size distribution for the larger impactors (dashed line) would predict a density of 300 2-km craters per million square kilometers on an airless Venus. The observed number is near 0.01. Thus destruction of these bodies by atmospheric explosions is about 99.997% effective. Only the very strongest and slowest survive. Craters of this size correspond roughly to 100-m-diameter metallic projectiles. By comparison, the Meteor Crater projectile in Arizona was metallic and 50–60 m in diameter.

The distribution of entering objects over elevation angle was derived in Chapter IX [Eq. (IX.18)], and the dynamic pressure on a spheroidal projectile passing through a gas of density $\rho_g$ at hypersonic speeds is

$$p = 0.6\rho_g v^2. \tag{VIII.2}$$

The projectile decelerates at a rate of

$$a = -F_{\text{drag}}/m = -pA/m \tag{X.54}$$

and ablates at a rate of approximately

$$dm/dt = -(\rho_g v^3/2\Delta H_{\text{vap}})(m/\rho_p)^{2/3}, \tag{X.55}$$

where $\Delta H_{\text{vap}}$ is the heat of vaporization and $\rho_p$ is the density of the projectile material. Deceleration can be neglected when the area loading ($m/A$) of the projectile is much larger than the cumulative mass of atmosphere swept out by the projectile.

The projectile fragments if $p$ is greater than the crushing strength of the projectile, which can be as low as $10^4$ dyn cm$^{-2}$ for fluffy cometary debris to as high as 3 kbar ($3 \times 10^9$ dyn cm$^{-2}$) for an iron meteorite with no cracks. For a typical meter-sized hard rock the ultimate strength of the rock is 1 to 2 kbar, but flaws in the rock structure (cracks produced during its collisional evolution) would commonly lower the strength of larger rocky bodies by a factor of 10. Several bright fireballs that dropped meteorites on Earth experienced peak dynamic pressures near 100 to 200 bar without crushing, and the famous Tunguska bolide in Siberia exploded near a dynamic pressure of 200 bar.

Consider a typical NEA approaching Venus at a relative speed of 16 km s$^{-1}$. The speed at entry, after addition of the escape energy, will be about $(162 + 102)1/2$, or 18.9 km s$^{-1}$. The dynamic pressure at the 1-bar pressure level in the atmosphere of Venus is $0.6\rho_g\ (1.89 \times 10^6)^2$. The local atmospheric temperature at the 1-bar level, 51 km above the surface, is about 330 K, and the molecular weight is 44, so the density is $44(298/330)/2.24 \times 10^4 = 1.77 \times 10^{-3}$ g cm$^{-3}$, giving a dynamic pressure of $3.8 \times 109$ dyn cm$^{-2}$ = 3.8 kbar.

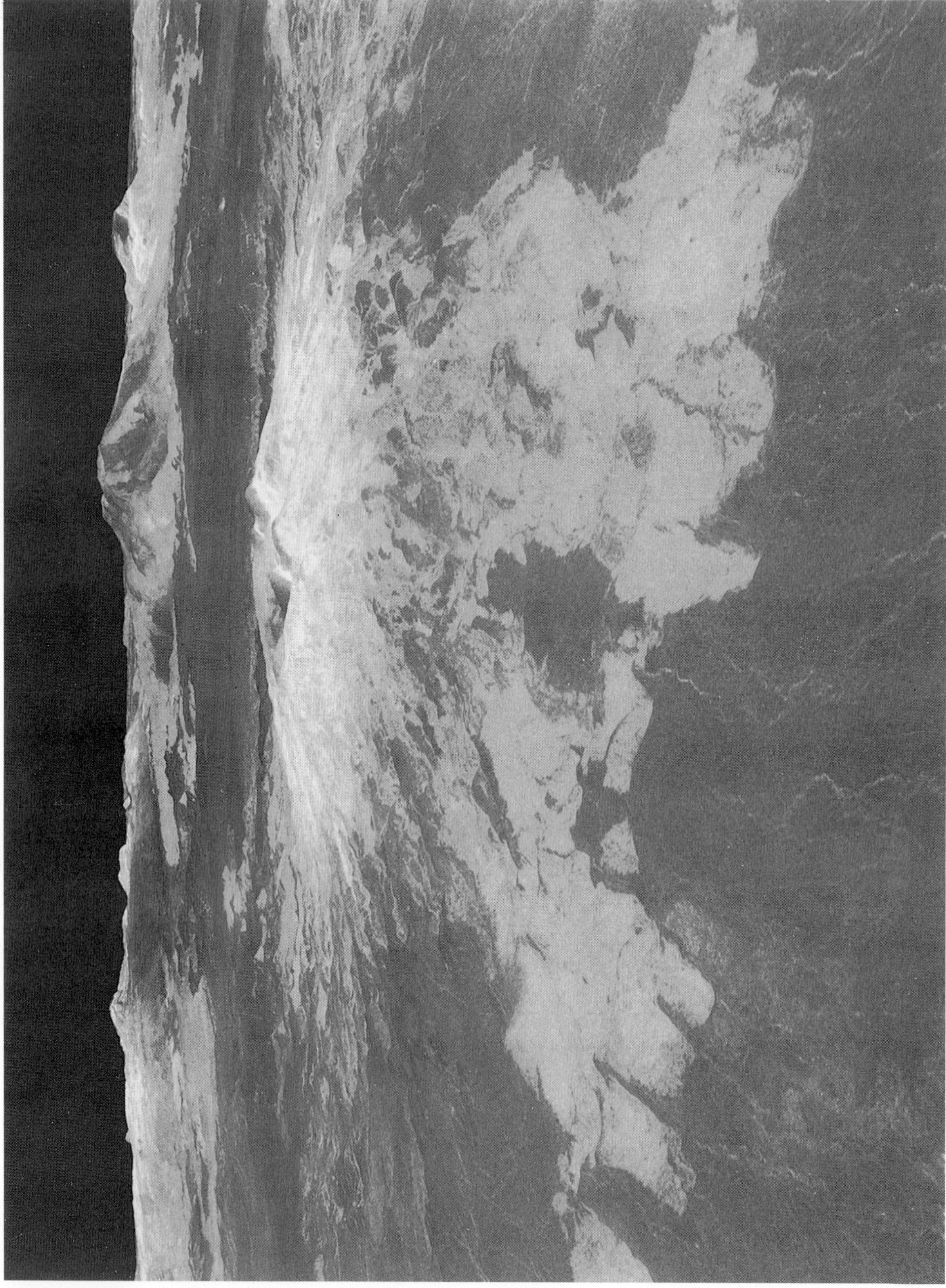

**Figure X.28** Sapas Mons on Venus. This computer-generated three-dimensional oblique view, based on Magellan altimetry and reflectivity data, shows the Sapas Mons volcano in Atla Regio with its associated highly radar-reflective lava flows and complex summit caldera. Maat Mons, the largest shield volcano on Venus, looms on the horizon.

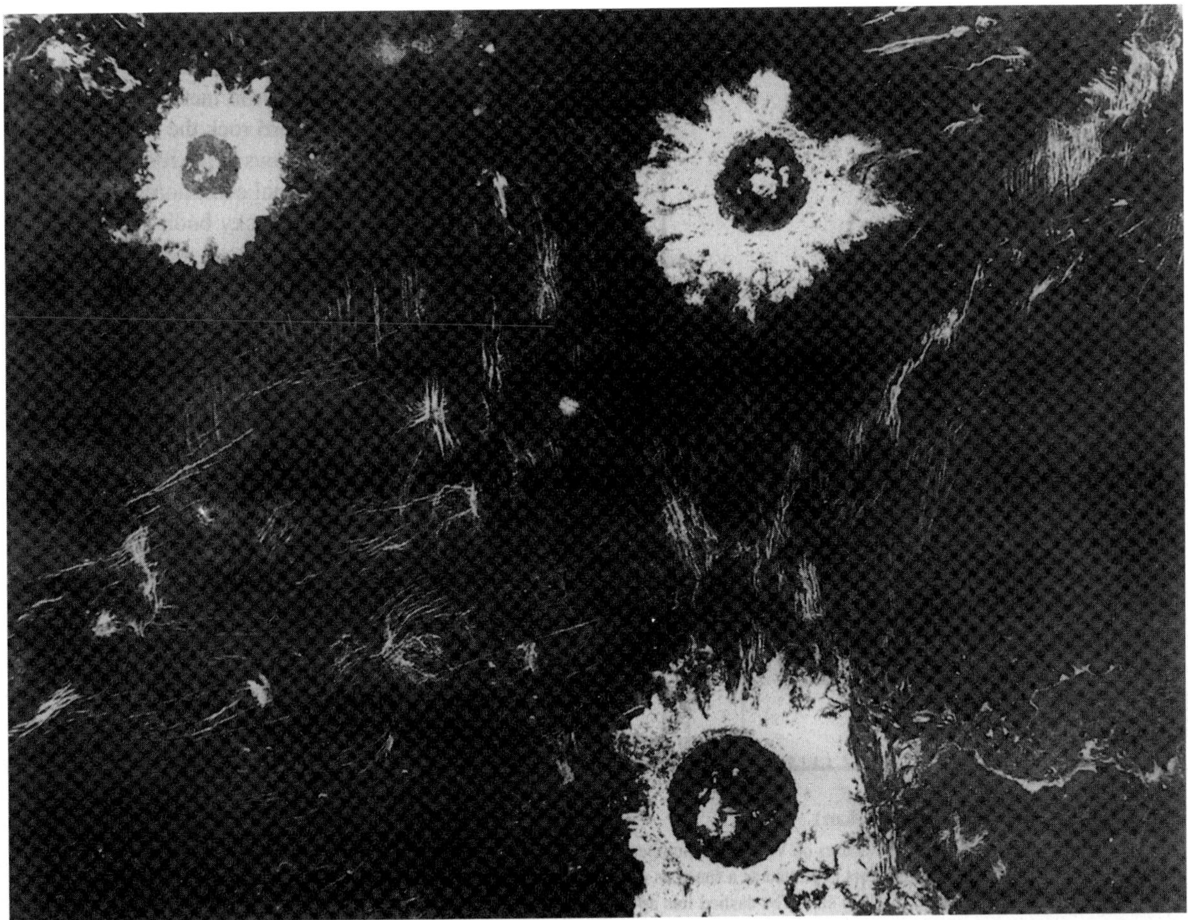

**Figure X.29** A trio of impact craters on Venus. These three large impact craters observed by Magellan seem to be accompanied by an astonishingly small number of smaller craters. It is possible that these craters are the result of a correlated impact of a gravitationally bound cluster of asteroidal or cometary fragments, or that small craters are intrinsically rare on Venus (see Fig. X.30). Note that the rim ejecta blanket is very bright at radar wavelengths because it is rich in angular fragments of centimeter to meter size, whereas the crater floor is flooded with a smooth radar-dark volcanic deposit.

This is sufficient to crush any natural rock, including iron meteorites. By comparison, at the surface of the Earth ($P = 1$ bar, 300 K, zero altitude, and a density of $1.3 \times 10^{-3}$) an average NEA would have a speed of about $(122 + 11.22)1/2 = 16.4\,\mathrm{km\,s^{-1}}$ and a dynamic pressure of 2.1 kbar. A significant fraction of small hard-rock projectiles will therefore be strong enough and slow enough to survive to land on Earth as meteorites.

Kilometer-sized bodies behave differently, in that it is difficult for the fragments of even a thoroughly crushed body to get away from each other in the short time spent transiting the atmosphere. They tend to penetrate to the surface of Earth, or even Venus, as tight swarms with virtually undiminished kinetic energy. The observed crater counts on Venus do show a rapid drop-off of crater densities for impactors below about 1 km diameter and a virtual absence of craters from impactors smaller than 100 m. Several smaller craters on Venus do indeed appear to be the result of impacts of tight swarms of several fragments. Other surface features look as if they had been scoured by the blast wave from a powerful air burst.

The surface of Venus has been photographed by both the Venera 9 and the Venera 10 landers, which were designed and insulated to survive for approximately 1 h under Venus surface conditions without active refrigeration. Both panoramas show a blocky surface largely covered by flattened fragments of bedrock, with soil visible between the blocks. The original raw Venera 9 panorama, complete with both interleaved data (regularly spaced bars) and noise (irregularly spaced "snow"), is shown in Fig. X.31. The horizon appears in the upper left corner, and the base of the spacecraft and a deployed instrument (a gamma-ray densitometer) appear at the bottom. Both Venera 9 and 10 landed on the lower flanks of the volcanic Beta Regio, and both show a terrain suggestive of degraded (weathered and fractured) lava flows. The Venera 13 and 14 Probes, which arrived

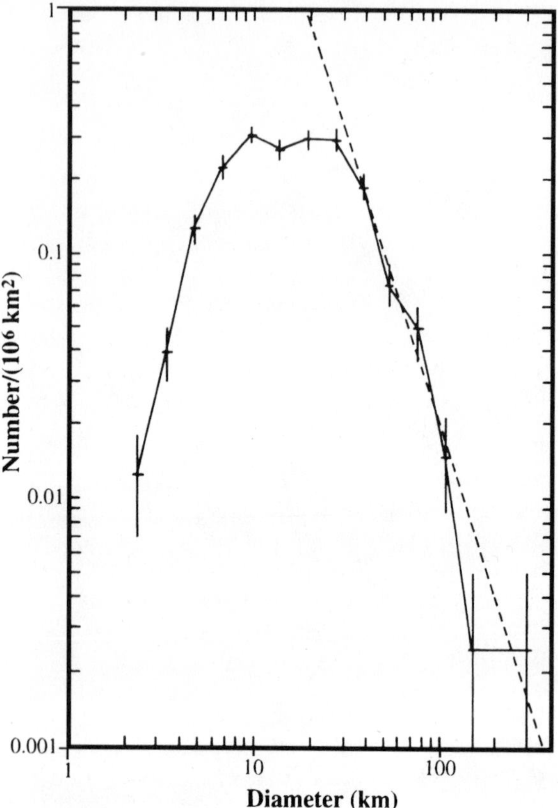


**Figure X.30** Size distribution of impact craters on Venus. The number of mapped craters per million square kilometers is plotted as a function of crater diameter in bins covering a factor of $2\frac{1}{2}$ in size. The dashed line approximates the slope of the size distribution of large impactors. The turnover of the crater density near 20 km diameter is due to severe fragmentation of entering bodies with diameters less than a few kilometers during traversal of the dense atmosphere. This figure is based on the work of Roger Phillips.

at Venus in 1982, also returned surface panoramas showing a striking resemblance to lava flows.

## Venus: Atmospheric Structure and Motions

The early indications of high surface temperatures from the radio-wavelength thermal emission data already implied a massive atmosphere on Venus. The infrared emission from the cloudtops fixes their temperature at about 240 K, and the surface temperature is close to 750 K. Consider an atmosphere with adiabatic structure

$$P/P_0 = (T/T_0)^{C_P/R} \qquad \text{(X.56)}$$

and with the heat capacity of carbon dioxide at 50 K (about $42\,\mathrm{J\,mol^{-1}\,K^{-1}}$). Then $C_P/R$ is about 5. Taking the cloud-layer conditions as 240 K and 100 mbar, the lowest possible surface pressure would be for the atmospheric model in which temperature increases at the adiabatic lapse rate

$$dT/dz = -\mu g/C_P \qquad \text{(X.57)}$$

downward from the clouds. The heat capacity is about $4.1 \times 10^8\,\mathrm{erg\,mol^{-1}\,K^{-1}}$, and $g$ is $887\,\mathrm{cm\,s^{-2}}$ at the surface of Venus, giving a mean temperature gradient of about $9\,\mathrm{K\,km^{-1}}$. The minimum altitude of the clouds is then roughly $(750 - 240)/9 = 57$ km. The minimum surface pressure is $0.1(750/240)^5$, or about 30 bar. It has been well established by entry probes that the temperature gradient is actually subadiabatic immediately below the cloudtops, and of course the heat capacity of the atmosphere varies with temperature; the actual mean surface pressure is about 92 bar. Mean temperatures as a function of altitude are shown in Fig. X.32.

Because of the immense mass, thermal inertia, and infrared opacity of the atmosphere, diurnal temperature changes on the surface of Venus are expected to be very small. The response of the lower atmosphere to day–night insolation changes is further limited by the high visible-wavelength opacity of the cloud layer: the Bond albedo of the planet is about 0.8. The solar constant at Venus, $2620\,\mathrm{W\,m^{-2}}$, is about twice that at Earth, but the absorbed flux of $2620(1 - A_B)$ is only $132\,\mathrm{W\,cm^{-2}}$, compared with $1370(1 - A_B)/4 = 242\,\mathrm{W\,m^{-2}}$ for much less cloudy Earth. Only about 2.5% of the incident sunlight reaches and is absorbed by the planetary surface. Temperatures observed deep in the Venus atmosphere by the Pioneer Venus entry probes show a 5-K range of temperatures at the same time at fixed altitudes, but there is no clear day–night difference. Instead, the temperature contrasts seem linked to the planetary-scale circulation.

Observations of the cloudtops at ultraviolet wavelengths reveal a retrograde rotation with a period of about four Earth days, corresponding to a mean wind speed of $2\pi r_V/P = 100\,\mathrm{m\,s^{-1}}$ relative to the surface. The speed of sound in the lower atmosphere,

$$c_s = (\gamma RT/\mu)^{1/2}, \qquad \text{(V.129)}$$

is about $400\,\mathrm{m\,s^{-1}}$, and hence the cloud-level winds travel at a speed of about Mach 0.25 relative to the surface. However, the great temperature difference between the surface and the atmosphere above the cloudtops is reflected as a decrease of sound speed to only about $210\,\mathrm{m\,s^{-1}}$ above an altitude of 90 km. The zonal wind speed is therefore on the order of Mach 0.5 above the clouds.

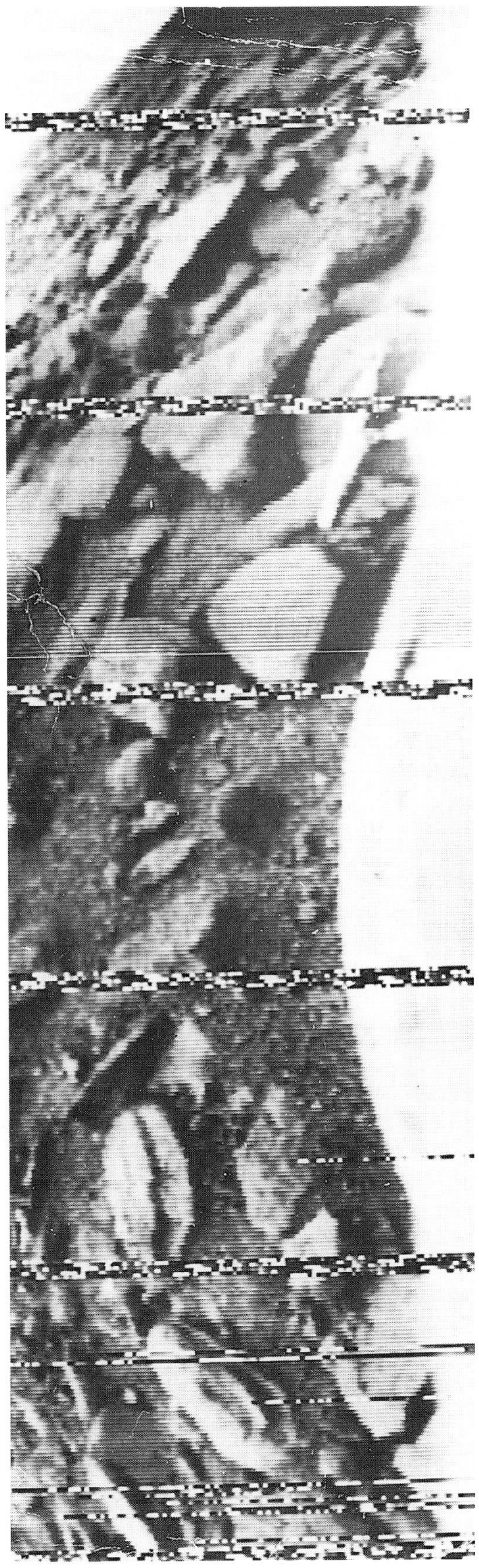

The very slow rotation of the atmosphere suggests that the Rossby number should be quite small. If the Rossby number,

$$\mathrm{Ro} = v_x/fL, \tag{V.164}$$

is << 1, then the Coriolis forces are important. Taking the mean zonal wind speed near the cloudtops as $v_x = 100\,\mathrm{m\,s^{-1}}$, $f$ as $2\Omega \sin l = 3 \times 10^{-7}\,\mathrm{s^{-1}}$ (for a latitude of 30°), and $L = 10^3$ km, we find that Ro is about 300.

The stability of the atmosphere against spontaneous overturn is quantified by the static stability, $S$, which is the difference between the actual local lapse rate and the adiabatic lapse rate,

$$S = \partial\Theta/\partial z = \partial T/\partial z - (\partial T/\partial z)_{\mathrm{ad}}, \tag{V.167}$$

and the static stability is parameterized by the Richardson number,

$$\mathrm{Ri} = (gS/T)/(\partial v_x/\partial z)^2. \tag{V.168}$$

The factor $\partial v_x/\partial z$ is the vertical wind shear, and the factor $gS/T$ has dimensions of $(\mathrm{cm\,s^{-2}})(\mathrm{K\,cm^{-1}})(\mathrm{K^{-1}}) = \mathrm{s^{-2}}$, so the square root of $gS/T$ has dimensions of frequency (Hz). This characteristic frequency is called the *Brunt–Väisälä frequency*,

$$N = (gS/T)^{1/2}; \tag{X.58}$$

large values of $N$ reflect large values of the static stability. If an atmosphere is isothermal and irrotational, $N$ is the maximum allowed frequency for internal gravity waves. In this usage, gravity waves mean free vertical oscillations of the atmosphere and have absolutely *nothing* to do with cosmic gravity waves produced by stellar explosions, etc. On Venus the lowermost troposphere below about 23 km altitude and a narrow interval near 55 km altitude have adiabatic temperature gradients ($S = 0$) and hence have Brunt–Väisälä frequencies of zero.

Line-of-sight wind speeds have been measured by precise Doppler tracking of probes falling through the

**Figure X.31** A rare glimpse of the surface of Venus. This panorama was returned by a camera on the Soviet Venera 9 lander. The swath imaged extends almost from horizon to horizon at about a 45° angle to the horizontal: the base of the spacecraft is visible at the right center, and an instrument deployed from the lander is visible at the upper right. The horizon is clearly seen in the upper left corner. The blocky, layered rocks may be pieces of lava flows. The regular bars of "noise" are interleaved data from other instruments. The ambient conditions on the Venus surface are formidable: the temperature is about 750 K and the pressure is near 92 bars. Wind speeds are near or below 1 m s$^{-1}$ and particulate opacity is virtually absent.

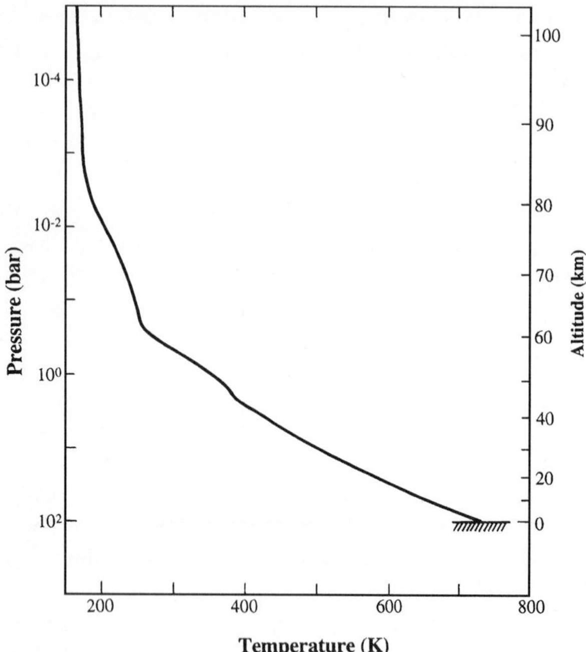


**Figure X.32** Temperature profile of the atmosphere of Venus. The results of a large number of spacecraft measurements are here combined to give a mean temperature structure for the atmosphere. The temperature gradient near the surface is close to adiabatic. Above about 60 km the atmosphere is markedly subadiabatic; indeed, above about 80 km it is nearly isothermal at the very low temperature of about 180 K.

atmosphere of Venus. The Soviet Venera 4 through 6 entry probes, the Venera 7 through 12 landers, and the four Pioneer Venus probes generally agree with the UV observations that zonal wind speeds above about 55 km are on the order of $100\,\mathrm{m\,s^{-1}}$, with the speed dropping off almost linearly with altitude down to about $1\,\mathrm{m\,s^{-1}}$ in the lowermost few kilometers of the troposphere.

Above 9 km altitude, at which the atmosphere is in more rapid motion, the entire atmosphere partakes of nearly the same angular rotation rate; that is, the atmosphere rotates almost as a solid body. Rotation causes an imbalance between the centripetal force necessary to maintain circular motion (which points radially outward from the spin axis) and that exerted by gravity (which points toward the center of mass of the planet). This imbalance can be resolved into two components, one radial from the center of Venus, and one in the horizontal plane. The radial component simply corresponds to a slight reduction in the effective gravitational acceleration of Venus; but the horizontal component, which points toward the equator, must force the lower atmosphere toward the equator until the pressure gradient between pole and equator becomes large enough to balance the equatorward component of the force. The condition in which this balance is achieved is called *cyclostrophic balance*. The equatorward component of the force caused by rotation is $(mv_x^2 \tan l)/r$ and the meridional (poleward) pressure gradient exerts an opposing force of $-m(\partial P/\partial y)/\rho$. Cyclostrophic balance is then achieved when

$$(v_x^2 \tan l)/r = -(\partial P/\partial y)/\rho, \tag{X.59}$$

but

$$\partial P/\partial z = -\rho g, \tag{III.169}$$

and, by the properties of exact differentials (Appendix I),

$$(\partial P/\partial y)_z = -(\partial P/\partial z)_y(\partial z/\partial y)_P, \tag{AI.53}$$

we find

$$(v_x^2 \tan l)/r = -(\partial P/\partial y)/\rho = -g(\partial z/\partial y)_P. \tag{X.60}$$

Introducing the geopotential $\Phi = gz$, we can set

$$(v_x^2 \tan l)/r = -(\partial\Phi/\partial y)_P = -(\partial\Phi/\partial l)/r. \tag{X.61}$$

It is now convenient to introduce a new (altitude-like) pressure coordinate,

$$\zeta = -\ln(P/P_0), \tag{X.62}$$

whence

$$d\zeta = -dP/P. \tag{X.63}$$

Then

$$(\partial P/\partial z)/g = -\rho = \partial P/\partial\Phi, \tag{X.64}$$

and, using the ideal gas law, $P = \rho RT$,

$$P(\partial\Phi/\partial P) = -P/\rho = -RT = -(\partial\Phi/\partial\zeta). \tag{X.65}$$

Then, differentiating Eq. (X.61) with respect to $\zeta$, substituting in from Eq. (X.65), and rearranging, we get

$$2v_x^2(\partial v_x/\partial\zeta) = -(R/\tan l)(\partial T/\partial l). \tag{X.66}$$

This is called the *cyclostrophic thermal wind equation*. It tells us that, if the "horizontal" (i.e., on an isobaric surface) temperature gradient is positive toward the equator, then the zonal wind velocity must increase with altitude. This behavior is indeed seen on Venus at all altitudes up to about 70 km. Above 70 km the sign of the meridional temperature gradient reverses, and the zonal wind speed diminishes with altitude. The winds probably persist to about 100 km in altitude. These winds are so efficient at transporting heat from the dayside to the nightside that the horizontal temperature gradient is kept small at these altitudes. At the cloud layer and near the equator the cyclostrophic balance breaks down, and motions become less dominated by the rotation (and hence less constrained to axial symmetry).

As an interesting aside, Titan, which rotates very slowly, appears to be in the same general cyclostrophically balanced dynamical regime as Venus, with wind speeds on the order of $100\,m\,s^{-1}$.

## Venus: Atmospheric Composition

Infrared spectroscopy of Venus from Earth is limited in its penetration into the atmosphere by the bright, well-defined cloudtops at a temperature of about 240 K and a pressure near 200 mbar. Carbon dioxide was detected in the atmosphere of Venus in 1932 by the American astronomers W. S. Adams and T. Dunham, Jr. Their spectra showed no evidence for water or oxygen, leading Rupert Wildt to argue that the low water abundance and high $CO_2$ abundance combined to suggest a strong greenhouse effect, with a surface temperature above the boiling point of water. In turn, the absence of liquid water on the surface, as implied by such high surface temperatures, suggests that life should not be possible on the surface of Venus. This makes the absence of detectable oxygen easy to understand.

The first credible claim of water in the atmosphere of Venus came as a result of high-altitude balloon-borne spectroscopic observations in 1965, and several ground-based detections followed, observing from high, dry sites at times of maximum orbital Doppler shift of the Venus lines off of their terrestrial counterparts. These observations appear to document real excursions in the cloudtop water abundance from as low as 1 ppm to as high as 100 ppm. The water content of the atmosphere below the clouds has also been investigated by a variety of techniques and appears equally difficult to constrain. The absence of a 1.35-cm absorption feature in the microwave thermal emission places an upper limit on the water vapor content of the lower atmosphere of about 0.1% (1000 ppm). *In situ* measurements by entry probes give water vapor abundances that range from about 10 ppm (near the surface) to 200 ppm near the cloud base, with substantial internal disagreement. For example, an optical spectrometer on the Venera 11 and 12 landers measured the attenuation of sunlight from 62 km altitude down to the surface over the wavelength range from 0.4 to 1.2 $\mu$m. Two strong water bands at 1.13 and 0.94 $\mu$m gave very different results: the 0.94-$\mu$m band showed a peak water abundance of 200 ppm at the cloud base, dropping to 20 $\mu$m at the surface, whereas the 1.13-$\mu$m band gives an estimate of 20 to 30 ppm water throughout the lower atmosphere. The existence of unmodeled sources of opacity would mean these numbers are upper limits to the water abundance; in particular, there is a real possibility that the discrepancy between the two bands is due to a cloud-correlated source of opacity (for example, due to gaseous $H_2SO_4$) contributing to the strength of the 0.94-$\mu$m band. In this case, the bizarre altitude dependence of the water vapor abundance is an experimental artifact, and a nearly constant water abundance in the lower atmosphere is possible. That abundance, about 25 ppm ($2.5 \times 10^{-3}$ bar), is vastly smaller than the terrestrial water inventory of over 250 bar; Venus, Earth's putative twin sister, has $10^{-5}$ times as much water as Earth! The solution to this puzzling discrepancy can be sought in either differing conditions of origin or divergent evolutionary pathways.

The Pioneer Venus large probe carried a mass spectrometer (MS) experiment to analyze the chemical and isotopic composition of the lower atmosphere. Fortuitously, the mass spectrometer ingested a small cloud droplet consisting mainly of sulfuric acid and was able to determine the isotopic composition of water released from the evaporation of the droplet. The results show a deuterium abundance (HDO: $H_2O = 0.032 \pm 0.004$) that is elevated by about a factor of 150 over the terrestrial or meteoritic ratio. This observation is based on the identification of mass 2 in the mass spectrum as being wholly due to the $D^+$ ion, not $H_2^+$, and has not been confirmed independently. Such an enormous enrichment of deuterium would require the loss, by mass-dependent hydrogen-escape mechanisms, of a mass of water at least several hundred times as large as the present atmospheric inventory. Some authors have postulated that Venus formed with a water abundance equivalent to a terrestrial ocean and lost it all through photodissociation and escape of hydrogen. The oxygen released by water photolysis is clearly not present in the modern atmosphere and must be absorbed by the crust. We will return to this issue below.

In attempting to assess the significance of the low water content and unusual hydrogen isotopic composition of the atmosphere of Venus, one very important factor has frequently been overlooked. That is the fact that the total water content of the atmosphere, about $10^{19}$ g, is about the mass of water that would be brought in by the impact of a 20-km-diameter comet nucleus. Such an impact, with an energy of about $10^{32}$ erg, should have a frequency of about $10^{-9}$ per year. Thus the amount of exogenic water brought into Venus by cometary impactors in the last 4.5 Ga must be at least several times the present water content of the atmosphere. Any theory that purports to explain the present water content and isotopic composition without quantitative inclusion of the water impact flux cannot be taken seriously.

It was not until 1967, with the first application of high-resolution infrared interferometric spectroscopy to planetary research, that any additional species were identified in the atmosphere. An international team of spectroscopists, including the French instrumental experts Janine and Pierre Connes and the American

**Table X.3** Composition of the Lower Atmosphere of Venus

| Species | Abundance (mole fraction) | Species | Abundance (mole fraction) |
|---|---|---|---|
| $CO_2$ | 0.965 | He | 12 ppm |
| $N_2$ | 0.035 | Ne | 7 ppm |
| $SO_2$ | 150 ppm | $H_2S$ | 3 ppm |
| Ar | 70 ppm | HCl | 400 ppb |
| CO | 30 ppm | Kr | 30 ppb |
| $H_2O$ | 20 ppm | HF | 5 ppb |

spectroscopists William Benedict and Louis Kaplan, succeeded over a 1-year period in finding not only a dense forest of lines and bands due to many isotopic variations of $CO_2$, but also carbon monoxide, hydrochloric acid, and hydrofluoric acid. The abundances of these and other species are summarized in Table X.3.

The Connes spectra included a number of weak isotopic bands of carbon dioxide that are on the linear portion of the curve of growth [see the discussion in Chapter IV, beginning with Eq. (IV.170)]. They were therefore able to estimate both the column abundance of $CO_2$ above the clouds and the pressure at the cloudtops. Their data implied that the total pressure was very close to the pressure that would be exerted by the observed abundance of carbon dioxide; therefore, they proposed that the mole fraction of $CO_2$ in the atmosphere of Venus must be close to 1.00. The Venera 4, 5, and 6 entry probes carried very simple instruments that captured samples of the atmosphere, extracted carbon dioxide from them using a selective absorber, and measured the residual atmospheric pressure. These measurements bracketed the $CO_2$ mole fraction between 0.93 and 1.00. The Pioneer Venus Probes refined this estimate to $f(CO_2) = 0.965 \pm 0.008$. Oxygen and carbon monoxide are both produced by solar UV photolysis of $CO_2$ and hence should be present above the cloudtops. Further, equilibration of $CO_2$ with the surface can maintain a small CO abundance in the lower atmosphere, deep below levels at which the solar UV flux is absorbed, whereas equilibrium of minerals containing $Fe^{2+}$ with the atmosphere would keep the $O_2$ partial pressure unobservably small. Thus CO unaccompanied by $O_2$ is plausible in the lower atmosphere. Spectroscopic observations of CO above the clouds show an abundance of $51 \pm 1$ ppm, and entry probes find a slightly smaller amount below the clouds, with the CO mole fraction decreasing toward the surface. The Pioneer Venus large probe gas chromatograph (GC) found 10 to 90 ppm CO near 52 km in altitude and $30 \pm 18$ ppm CO at 42 km. The Venera 12 GC found an average of $28 \pm 7$ ppm CO below 42 km, and the PV instrument reported $20 \pm 3$ ppm CO at 22 km in altitude.

Oxygen remains an enigma. Earth-based spectroscopic studies find a firm upper limit of 1 ppm on the oxygen abundance above the clouds. The Pioneer Venus GC experiment, however, claimed $44 \pm 25$ ppm $O_2$ at 52 km in altitude. This is in serious conflict with other, more directly derived composition data on the atmosphere, and the wisest course seems to be to reject this discrepant observation. Ozone, which is produced as a by-product of oxygen photolysis, is an extraordinarily easy molecule to detect in the UV region; spectra of Venus place an upper limit of $f(O_3) < 3 \times 10^{-9}$ above the clouds. Such a very restrictive upper limit cannot be reconciled with the presence of tens of ppm of free oxygen.

After $CO_2$, the principal atmospheric gas is $N_2$. The Pioneer Venus large probe and Venera 11 and 12 mass spectrometer experiments and the PV and Venera 12 GC experiments combine to suggest a nitrogen mole fraction of $0.035 \pm 0.008$. No other nitrogen compound has been detected in the troposphere of Venus.

The rare gases are of extraordinary interest as tracers of atmospheric evolution (Fig. X.33). We shall begin by briefly discussing the relative elemental abundances of the rare gases neglecting radiogenic isotopes. Measurements by the Pioneer Venus large probe mass spectrometer (LMS) and the Venera 11 and 12 mass spectrometers (VMS) agree that the abundances of neon and nonradiogenic argon on Venus are substantially higher than those on Earth. The LMS and VMS data on krypton are in irreconcilable disagreement with each other. Both Venera instruments found Kr abundances of about 0.6 ppm, compared with 0.025 ppm for the LMS experiment. As a result of this discrepancy, the Ar:Kr ratio reported by LMS is about 2700:1, compared with about 180:1 from VMS. This difference leads to dramatically different interpretation of the rare gas data. The rare gas abundance data for Venus are compared with terrestrial and solar rare gas abundances in Fig. X.33. Note that the terrestrial rare gas abundance (omitting radiogenic $^{40}Ar$) is scaled so that the terrestrial $^{36}Ar$ abundance plots on the Venus datum. The VMS elemental abundance data for neon, argon, and krypton then are similar to those in Earth's atmosphere, except that the Ne:$^{36}$Ar ratio is about half as large on Venus as on Earth. Xenon, which would have an abundance of about 0.1 ppm by either the terrestrial or the solar standard, remains poorly characterized; the Venera 13 and 14 MS experiments suggest a Xe:Kr ratio of 0.1:1 or slightly higher. Although a detailed quantitative account of the Venera 13/14 MS data on Kr and Xe is not yet available, a Kr abundance far below the earlier Venera 11 and 12 VMS report is suggested. This change, if it stands the test of detailed scrutiny, would lend further credence to solar-like relative elemental abundances of the heavy rare gases on Venus.

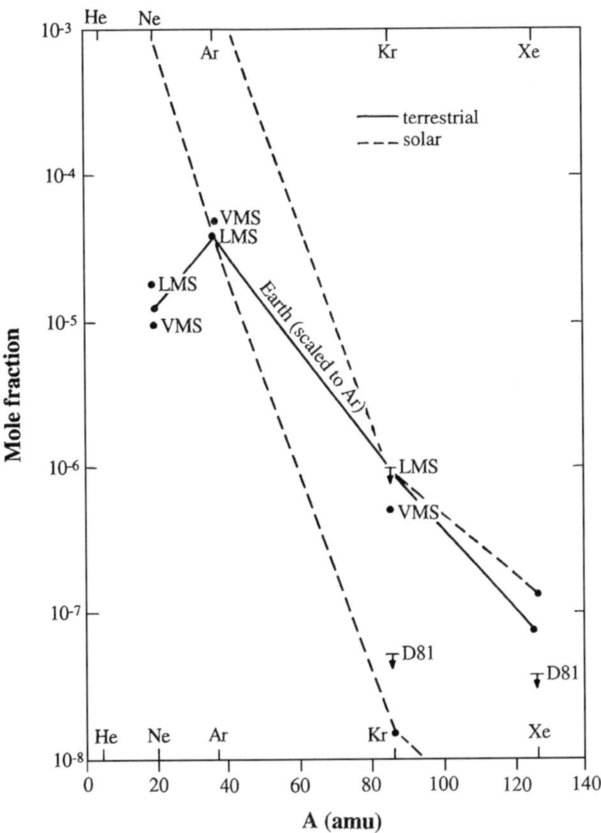


**Figure X.33** Primordial rare gases in the atmosphere of Venus. The solid line gives the terrestrial rare gas abundances scaled to the Venus $^{36}Ar$ abundance. LMS is the Pioneer Venus Large Probe Mass Spectrometer experiment, VMS is the Venera 12 Mass Spectrometer experiment, D81 is a reinterpretation of the LMS data by Thomas Donahue in 1981, and V13 is the preliminary report of data from the Venera 13 MS experiment. The dashed lines show the solar rare gas abundance profiles. Note the very large discrepancies among the early Kr measurements. The abundance of neon on Venus is clearly 100 times lower than the solar Ne:Ar ratio would suggest. The relative abundances of the heavy rare gases now appear to be roughly terrestrial, but uncertainties remain large.

Isotopic abundance data on the rare gases are very limited. The LMS instrument found a $^{22}Ne:^{20}Ne$ ratio of $0.07 \pm 0.02$, slightly less than the terrestrial value of 0.097 (the Martian value is $0.10 \pm 0.03$). The $^{40}Ar:^{36}Ar$ ratio was found to be an astonishing $1.03 \pm 0.04$, compared with about 300:1 for Earth and 3000:1 for Mars, and the $^{38}Ar:^{36}Ar$ ratio was found by both VMS and LMS to be close to 0.19, indistinguishable from the ratios of both Earth and Mars. The radiogenic argon abundance is in reasonable agreement with geochemical expectations, but the nonradiogenic argon is vastly more abundant on Venus than that on the other terrestrial planets. The Kr isotopic composition was crudely constrained by the LMS experiment and looks "normal" within very broad error limits.

The chemistry of the sulfur-bearing gases on Venus is complex, and the reported abundances of the sulfur-bearing gases are full of contradictions that defy explanation. Sulfur dioxide has been detected in the cloud region by Earth-based spectroscopy at abundance levels less than 1 ppm. The abundance rises in a poorly constrained manner at intermediate pressures, reaching $150 \pm 50$ ppm near 22 km in altitude. The very low pressures in the cloud are understood in terms of the photochemical conversion of gaseous $SO_2$ into a cloud layer dominated by sulfuric acid droplets. The photochemical production of the clouds will be discussed later. Carbonyl sulfide, COS, has been reported many times as a minor constituent of the atmosphere of Venus, but no convincing detection remains in the literature. About 3 ppm of hydrogen sulfide, $H_2S$, has been reported by the LMS experiment below the 20-km altitude level. Also, spectrophotometers on the Venera 11 and 12 landers showed very strong atmospheric absorption of sunlight shortward of 600 nm between 10 and 30 km in altitude. Interpretation of these observations implies that the $S_3$ molecule was responsible for the absorption, implying a total column abundance of elemental sulfur gases (mostly $S_2$) of about 0.02 ppm.

## Venus: Atmosphere–Lithosphere Interactions

At the very high temperature and pressure of the Venus surface, chemical reactions between the atmosphere and the surface may be very important. Equilibration times are difficult to calculate because of our ignorance of the surface grain size distribution and the surface abundance of reactive polar molecules (especially water). Although atmosphere–surface equilibration is rapid on geological time scales, it may be slow compared to the rate of photochemical reactions in the atmosphere.

The surface composition of Venus is still poorly constrained by observations. Tables X.4a/b summarizes the available data. The Venera 8, 9, and 10 landers carried gamma ray spectrometers to measure the amounts of K, U, and Th in the surface rocks. The later Venera 13 and 14 landers, and also the Vega 2 lander, carried X-ray fluorescence instruments to determine the abundances of the major elements at their landing sites. In no case have

**Table X.4a** Radionuclide Abundances in the Crust of Venus

| Venera mission | U (ppm) | Th (ppm) | K (wt %) | K/U |
|---|---|---|---|---|
| 8 | $2.2 \pm 0.7$ | $6.5 \pm 0.2$ | $4.0 \pm 1.2$ | $1.8 \times 10^4$ |
| 9 | $0.6 \pm 0.2$ | $3.6 \pm 0.4$ | $0.5 \pm 0.1$ | $0.8 \times 10^4$ |
| 10 | $0.5 \pm 0.3$ | $0.7 \pm 0.3$ | $0.3 \pm 0.2$ | $0.7 \times 10^4$ |

**Table X.4b** Major-Element Composition (wt %) of the Crust of Venus

| Oxide | Venera 13 | Venera 14 | Vega 2 |
|---|---|---|---|
| $SiO_2$ | 45.1 | 48.7 | 45.6 |
| $Al_2O_3$ | 15.8 | 17.9 | 16.0 |
| MgO | 11.4 | 8.1 | 11.5 |
| CaO | 7.1 | 10.3 | 7.5 |
| FeO | 9.3 | 8.8 | 7.7 |
| $K_2O$ | 4.0 | 0.2 | 0.1 |
| $TiO_2$ | 1.6 | 1.2 | 0.2 |
| $SO_3$ | 1.6 | 0.9 | 4.7 |
| MnO | 0.2 | 0.2 | 0.1 |

minerals actually been identified on the surface of Venus. Note that the gamma-ray spectroscopy (GRS) data in Table X.4a show two different types of material. Venera 8, which landed on a cratered, ancient plain southeast of the Beta Regio volcanic complex, reported an analysis that reflects a surface composition similar to an extreme alkali basalt or granite. The Venera 9 and 10 landers analyzed the east slopes of Beta Regio and found a much lower radioactive element content, on the borderline between alkali basalts and terrestrial oceanic basalts. The Venera 13 and 14 landers analyzed the southeast portion of Phoebe Regio (a southward extension of Beta Regio). Venera 11 and 12 also landed in Phoebe Regio very close to the Venera 13 and 14 landing sites, but did not carry out surface analyses. Thus all of our compositional data on the Venus surface except the GRS analysis from Venera 8 refer to a single unambiguously volcanic complex, which, not surprisingly, has a basaltic composition. Interestingly, the Pioneer Venus large probe fell just east of the Venera 10 landing site, also on the edge of Beta Regio.

The oxide abundances reported in Table X.4b are not raw data, but interpretations of *elemental* abundance data. The iron was *assumed* to be found in the form of FeO; S, in the form of $SO_3$ (i.e., sulfates); etc. Reports of the results from this experiment also commonly contain an entry for $Na_2O$; this number, however, is an interpolation, is not based on a measurement (or even a detection) of sodium in the Venus surface, and is deleted here. Thus 100 g of soil analyzed by Vega 2, reported to contain 4.7 g of $SO_3$ and 7.7 g of FeO, actually contains $4.7 \times 32/(32+3 \times 16) = 1.88$ g of S and $7.7 \times 56/(56+16) = 5.99$ g of Fe. If the iron is found in the form of pyrrhotite $Fe_{1-x}S$ (and $x$ is small), then 5.99 g of iron would make $5.99(56+32)/56 = 9.41$ g of pyrrhotite, using up 3.42 g of sulfur. Thus there is insufficient sulfur to tie up all the iron as a sulfide, and at least $100 \times (3.42 - 1.88)/3.42 =$ 45% of the iron must be in the form of oxides or silicates. It could be found as FeO in pyroxenes or olivine, as magnetite $Fe_3O_4$, as hematite $Fe_2O_3$, or even as metallic iron, so far as these analyses are concerned, and the sulfur could be found as sulfides or sulfates. These analyses tell us nothing about the *oxidation state* of the surface.

The composition of the atmosphere, however, can help us constrain the composition of the surface. For example, the reaction

$$\mathrm{Fe} + \mathrm{CO_2} \rightarrow \mathrm{FeO} + \mathrm{CO} \qquad \text{(X.67)}$$

$$K_{67} = (a_{\mathrm{FeO}}/a_{\mathrm{Fe}})(p_{\mathrm{CO}}/p_{\mathrm{CO_2}}) \qquad \text{(X.68)}$$

would permit solid metallic iron and pure FeO (or FeO-rich pyroxene) to be in equilibrium with each other at 750 K if the CO:$CO_2$ ratio (essentially the mole fraction of CO) satisfies the condition

$$f_{\mathrm{CO}} = K_{67}(a_{\mathrm{Fe}}/a_{\mathrm{FeO}}) = K_{67}. \qquad \text{(X.69)}$$

Looking up the thermodynamic data on this reaction, we find that log $K_{67}$ at 750 K is about $10^{0.3} = 2$. Thus a CO:$CO_2$ ratio of $f_{\mathrm{CO}} > 2$ would be required for metallic iron to be stable on the surface of Venus. The actual value of $f_{\mathrm{CO}}$ from Table X.3 is about 20 ppm near the planetary surface, and therefore metallic iron cannot be stable there. If we further allow for the dispersal of FeO in solid silicates, so that the activity of FeO is 0.1 instead of 1.0, this conclusion remains very secure.

Consider the oxidation sequence Fe $\rightarrow$ FeO $\rightarrow$ $Fe_3O_4 \rightarrow Fe_2O_3$. Taking an FeO activity of 0.1, the Fe/sFeO boundary would occur at $f_{\mathrm{CO}} = K_{67}a_{\mathrm{FeO}} = 0.2$. The FeO/magnetite boundary would occur where

$$3\mathrm{FeO} + \mathrm{CO_2} \rightarrow \mathrm{Fe_3O_4} + \mathrm{CO} \qquad \text{(X.70)}$$

$$\mathrm{K}_{70} = f_{\mathrm{CO}}(a_{\mathrm{mag}}/a_{\mathrm{FeO}}) = f_{\mathrm{CO}}(1/0.1) \qquad \text{(X.71)}$$

or $f_{\mathrm{CO}} = 10K_{70}$. From the data tables, $K_{70}$ at 750 K is 0.07, and the CO mole fraction in equilibrium at this boundary would be $0.1K_{70}$, or $7 \times 10^{-3}$. The observed CO mole fraction is $2 \times 10^{-5}$, so the gas at the surface of Venus is too CO-poor (too oxidizing) for this high an activity of FeO. Finally, magnetite oxidizes to hematite via

$$2\mathrm{Fe_3O_4} + \mathrm{CO_2} \rightarrow 3\mathrm{Fe_2O_3} + \mathrm{CO} \qquad \text{(X.72)}$$

$$K_{72} = f_{\mathrm{CO}}(a_{\mathrm{hem}}^3/a_{\mathrm{mag}}^2) = f_{\mathrm{CO}}, \qquad \text{(X.73)}$$

in which both hematite and magnetite are quite pure phases with activities close to 1.00. $K_{72}$ at 750 K is $2.5 \times 10^{-6}$, and therefore the gas is not oxidizing enough to produce hematite. Absorption of oxygen from a putative early ocean would tax the ability of the crust to absorb it; 10% FeO in a crust 30 km thick would be $7.5 \times 10^5$ g of FeO per square centimeter of planetary surface area. An ocean of water would contain about $2.7 \times 10^5$ g cm$^{-2}$ of oxygen. Assuming complete oxidation of FeO all the way to hematite, the stoichiometry of oxidation would be

$$2FeO + \tfrac{1}{2}O_2 \rightarrow Fe_2O_3, \qquad (X.74)$$

using 16 g of oxygen for every 144 g of FeO. Thus the crust could absorb a maximum of $7.5 \times 10^5$ $(16/144) = 8.3 \times 10^4$ g oxygen per square centimeter, less than a third of an ocean equivalent. Beyond the end point of iron oxidation to hematite, oxygen would accumulate in the atmosphere to a pressure of 190 bar. Disposing of this vast amount of unobserved oxygen, and preventing the crust from being fully oxidized to the ferric state, requires somehow dispersing the oxygen throughout the mantle of Venus. Since the crust has a lower density than the mantle and since we see no evidence of crustal subduction on Venus, hiding the oxygen in this manner is at very least a difficult feat.

In this discussion of iron oxidation we have used the relative abundances of CO and $CO_2$ as indicators of the oxidation state of the atmosphere. A more direct indication of oxidation state is the equilibrium partial pressure (fugacity) of oxygen, which is related to the carbon monoxide mole fraction by

$$CO_2 \rightarrow CO + \tfrac{1}{2}O_2, \qquad (X.75)$$

whence

$$K_{75} = p_{O_2}^{1/2}(p_{CO}/p_{CO_2}) = p_{O_2}^{1/2} f_{CO}. \qquad (X.76)$$

$K_{75}$ is about $5.5 \times 10^{-16}$ at 750 K, whence $p_{O_2}$ must be of order $10^{-22}$ atm, an amount far too small to be directly measurable. Use of the oxygen fugacity as the measure of oxidation state permits comparison of the results of calculations on iron-bearing minerals with those on oxidation of sulfur-bearing minerals from sulfides to sulfur and sulfates, etc.

It is attractively simple to suggest that the atmospheric abundances of all the reactive gases on Venus, including oxygen, carbon dioxide, hydrochloric and hydrofluoric acid, and the sulfur compounds, are regulated by buffer reactions with crustal minerals. To order-of-magnitude precision, this is indeed the case, but the details of the chemical composition present a variety of puzzling discrepancies. It is hard to get a high enough total abundance of sulfur gases unless the atmosphere is more reducing than the $f_{CO}$ data suggest. But in any event the atmosphere (and the rocks with which it is in contact) cannot be so highly reduced as to make methane and other hydrocarbons in detectable amounts. Such gases are not observed in the atmosphere of Venus. The implications of a reducing interior and a more oxidizing crust are clear: limited crustal oxidation by loss of a small fraction of an ocean of water is acceptable, with expectably poor chemical communication between the oxidized crust and the reduced mantle.

The hypothesis of regulation of reactive gas pressures is still difficult to test because of the paucity of analyses of the lower atmosphere (about 85% of the atmospheric mass lies below an altitude of 22 km, where our analytical data from Pioneer Venus cease) and the absence of data on surface mineralogy. Still, it is easy to make testable predictions. Several calculations have been done on the equilibration of the atmosphere with basalts, influenced by the above-mentioned tendency of landing probes to set down on the basaltic Beta Regio structure, but the chemistry of a terrestrial planet must be vastly more diverse than this one small area would suggest. Alkali basalts, granites, or both with high silica contents must also be expected, in keeping with the Venera 8 K, U, and Th data on a plains site east of Beta Regio. Enthusiasts of ancient oceans on Venus may wish to consider silicate-poor carbonate sediment beds from ancient times, and all will recognize the importance of altitude-dependent (temperature-dependent) shifts of gas–mineral equilibria, leading to weathering products whose composition depend on altitude. It is a relatively straightforward exercise to calculate equilibria between gases and well-defined rock compositions, but the vast majority of the surface of Venus consists of windblown weathering products of mixtures of rock types. Sediments and sedimentary rocks may have greatly elevated calcium abundances, which enhance the ability of the surface to buffer reactive gases such as $CO_2$, HCl, HF, $SO_2$, and $H_2O$. At the opposite extreme of altitude, high mountains are so much cooler than the lowlands that they can collect a coating of moderately volatile elements that are baked out of the low plains.

The case of carbon dioxide buffering is an interesting one. The Urey-type buffering reaction

$$CaSiO_3 + CO_2 \rightarrow CaCO_3 + SiO_2 \qquad (X.77)$$

is in equilibrium at 90 bar and 750 K, but is not relevant unless all three of the participating minerals are present on the surface. Silica-poor basalts are a poor choice for carbon dioxide buffering, but alkali basalts, granites, and their weathering products are appropriate. Calcium is relatively easily mobilized by weathering because of the stability of its carbonate, sulfate, chloride, and fluoride weathering products, and calcium may therefore be disproportionately abundant in the fine-grained weathered soil (which has a very large surface area per unit mass). Cations other than Ca make poor candidates as gas buffers.

The sulfur gases also present a most interesting situation. Buffers capable of regulating $SO_2$ could include sulfate decomposition reactions such as

$$CaSO_4 + CO \rightarrow CaO + CO_2 + SO_2 \qquad (X.78)$$

or, allowing for the probable presence of silica,

$$CaSO_4 + CO + SiO_2 \rightarrow CaSiO_3 + CO_2 + SO_2. \qquad (X.79)$$

Here it is clear that low values of $f_{CO}$ (high oxygen fugacities) act to suppress the release of sulfur dioxide. Thus a surface oxidation state that favors complete oxidation of sulfides to sulfates leads to an ever-diminishing sulfur gas abundance. Alternatively, two different oxidation states of sulfur may coexist in the surface layer,

$$FeS_2 + 2CaCO_3 + 3\tfrac{1}{2}O_2 \rightarrow 2CaSO_4 + FeO + 2CO_2, \qquad (X.80)$$

where the FeO activity is on the order of 0.1. Note that in this case the sulfur minerals are acting as an oxygen buffer and keeping the oxidation state from drifting into the sulfide-free zone. This sulfide/sulfate oxygen buffer lies near an $O_2$ fugacity of $10^{-22}$ atm at 750 K, very close to the oxidation state indicated by the CO abundance. The COS pressure under these circumstances is about $5 \times 10^{-3}$ atm, or $f_{COS} =$ 50 ppm. (The Venera 11 and 12 analyses claim 30 ppm COS near the surface, and Earth-based infrared spectroscopy claims 0.25 ppm COS near 50 km altitude.) The $H_2S$ pressure can be calculated from the exchange reaction

$$COS + H_2O \rightarrow CO_2 + H_2S \qquad (X.81)$$

$$K_{81} = (p_{H_2S}/p_{COS})(p_{CO_2}/p_{H_2O}) = (p_{H_2S}/p_{COS})/f_{H_2O}, \qquad (X.82)$$

where the water mole fraction is taken to be about 30 ppm. The equilibrium constant at 750 K is found to be $K_{81} = 190$, $p_{H_2S}/p_{COS}$ is then near $6 \times 10^{-3}$, and $f_{H_2S}$ is then 0.3 ppm. The Pioneer Venus LMS experiment reported $3 \pm 2$ ppm, and this result is likely to err on the low side because of the fragility of the $H_2S$ molecule. The $SO_2$ pressure can be calculated from

$$COS + 2CO_2 \rightarrow 3CO + SO_2, \qquad (X.83)$$

which gives an $SO_2$/COS ratio of about 5, for an $SO_2$ mole fraction of about 250 ppm. The total sulfur gas abundance decreases dramatically toward higher oxygen fugacities for reasons outlined above. Both the COS and the $H_2S$ pressures increase toward lower oxygen fugacities until the dominant sulfide becomes pyrrhotite, after which they stay constant at about a factor of 10 higher than the numbers calculated here. The $SO_2$ pressure, however, decreases rapidly toward both the more-oxidizing and the less-oxidizing sides of the sulfide/sulfate buffer. Elemental sulfur vapor, $S_2$, also peaks at the sulfide/sulfate boundary, where it reaches as high as 12 ppm. Unfortunately, the molecular weights of $SO_2$ and $S_2$ are the same (64), and the possibility exists that these species could be confused with each other in a mass spectrometric analysis.

The stability fields of a number of iron and sulfur minerals of potential importance in buffering oxygen are shown in Fig. X.34 on a plot of FeO activity vs oxygen fugacity. The surface of Venus is close to $a_{FeO} = 0.1$ and $f_{O_2} = 10^{-22}$, close, as nature so often seems to arrange it, to the region of maximum complexity in the diagram.

The radar evidence from Magellan that shows enormously long eroded channels emanating from volcanic structures has engendered a lively debate concerning what material could flow such great distances (in some cases, thousands of kilometers) without solidifying. Some of these features seem to be causally associated with impact craters (see, for example, Fig. X.29). Like the rilles on the Moon, there is often no hint of a massive deposit of lava at the termini of the flows. The suggestion that these are simply low-viscosity basaltic magmas

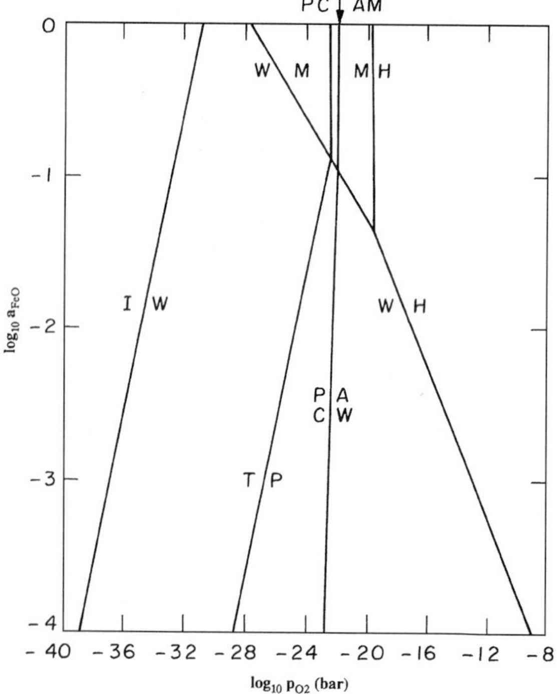


**Figure X.34** Iron-mineral stability on Venus. The stability fields of a number of common iron-bearing minerals are given at 750 K as a function of the FeO activity in the surface rocks and the atmospheric oxygen fugacity. The minerals are I, iron metal; W, wüstite (FeO); M, magnetite; T, troilite (FeS); P, pyrite ($FeS_2$); C, calcite; A, anhydrite; and H, hematite. Each line in the diagram is an oxygen buffer. The surface conditions on Venus appear to lie close to the PC/AM or PC/AW buffer near an FeO activity of 0.1, the region of greatest complexity on the diagram.

is sensible for the smaller features, but strains the credulity when applied to 5000-km flows. A second alternative, that the magmas are carbonatites, with melting points greatly depressed by the presence of large amounts of dissolved water and carbon dioxide in the melt, requires that the source region rocks be rich in carbonates and hydrous silicates. Bruce Fegley of Washington University has argued that the most stable hydrous silicates, such as the amphibole tremolite, may be unstable on the surface of Venus but decompose so slowly that they may persist for geological ages. This argument is consistent with a significant deep-crustal water reservoir, but the generation of carbonatite magmas may require massive deposits of sedimentary carbonates and hydrous silicates. As noted earlier, Magellan provided no evidence for the former presence of oceans on Venus. A third possibility is that the magma is liquid sulfur with dissolved sulfides. This would allow for a liquid that could not freeze on Venus, but which would evaporate to generate sulfur vapor. The residual melt, possibly a pyrite–sulfur melt (see Fig. VIII.6), would precipitate radar-bright pyrite in the channels and flow until evaporative exhaustion of sulfur, leaving no terminal deposit. The pyrite would in turn be subject to oxidation by the atmosphere (to make iron oxides) as it releases vast amounts of sulfur gases into the atmosphere. Episodic injection of sulfur from such flows could explain the evident excess of sulfur gases over the amounts that could be present at equilibrium. Generation of a sulfur magma, however, requires that the crust not be too highly oxidized (i.e., not inside the sulfate stability field).

## Venus: Photochemistry and Aeronomy

At the upper boundary of the troposphere an entirely different suite of reactions, driven by ultraviolet sunlight, comes into play. Sulfur dioxide and other sulfur gases can be chemically processed by reactions with strong oxidizing agents derived from photolysis of water vapor and carbon dioxide. The initial photochemical reactions are as on Mars,

$$CO_2 + h\nu \rightarrow CO + O \qquad (X.7)$$

$$H_2O + h\nu \rightarrow OH + H \qquad (X.12)$$

$$\rightarrow O + H + H, \qquad (X.13)$$

with resultant accumulation of oxygen, CO, and hydrogen. We have seen that oxygen is undetectable in the troposphere, and therefore some process capable of removing oxygen must occur near the site of oxygen production. It would take less than 5 years of exposure to solar UV irradiation to make an observable amount of oxygen, and hence the removal process must be very effective on an even shorter time scale. Also the strong observed gradient in the CO abundance, declining from about 700 ppm at 100 km through 50 ppm at 64 km to about 20 ppm near the surface, suggests a major photochemical contribution at high altitudes and loss in the lower atmosphere.

Two other known constituents of the cloudtop atmosphere are also photochemically active:

$$HCl + h\nu \rightarrow H + Cl \qquad (X.84)$$

$$SO_2 + h\nu \rightarrow SO + O. \qquad (X.85)$$

The first of these is an important source of atomic H, feeding (by diffusion) the upper atmosphere with a steady stream of H. The latter reaction is yet another source of atomic oxygen near the clouds. Sulfur dioxide reacts readily with O via

$$SO_2 + O \rightarrow SO_3 \qquad (X.86)$$

followed by hydration of sulfur trioxide

$$SO_3 + H_2O \rightarrow H_2SO_4 \qquad (X.87)$$

to make sulfuric acid vapor. The sulfuric acid vapor nucleates and condenses readily at the low temperatures of the cloud layer, making droplets of very concentrated sulfuric acid. Thus the main cloud layer is a photochemical haze of sulfuric acid droplets. The reflective index of sulfuric acid is much higher than that of pure water, and estimates of the refractive index made on the basis of measurements of the wavelength and phase angle dependence of polarization of sunlight scattered from the clouds suggested strong acid solutions on Venus well before *in situ* measurements verified the presence of sulfuric acid.

The sulfuric acid clouds are very reflective, giving a bright and very uniform planetary albedo of about 0.84 at visible wavelengths. In the ultraviolet, significant gaseous absorption by sulfur-bearing gases lends a strongly banded appearance to the planet, as the Mariner 10 image in Fig. X.35 attests. It is from tracking these cloud features and from Doppler measurements of narrow spectral lines near the cloud level that the 4-day circulation period of the upper troposphere was first derived.

An instrument for measuring cloud density profiles (a *nephelometer*) was carried by the Pioneer Venus large probe. A dense, sharply defined cloud layer dominated by transparent spherical particles (liquid sulfuric acid droplets) was observed at 50–52 km in altitude. An extended haze with several times lower density extends up to about 68 km in altitude, with a thin detached haze layer near 77 km (Fig. X.36). From about 49 km down to 31 km there is a very tenuous haze of nearly constant

**Figure X.35** Ultraviolet photograph of the clouds of Venus. This Mariner 10 UV image reveals complex structure in the high-altitude clouds of Venus. The patterns are reclining U-, V-, or Y-shapes, oriented so as to lie roughly symmetrically across the equatorial plane.

density. Venera data suggest a thin aerosol of solid particles with a high chlorine content. The identity of this cloud material is not known, but a variety of volatile salts have been suggested, including $AlCl_3$ and $FeCl_2$. None of these seem geochemically plausible, because none of those so far suggested are volatile enough to provide condensation of the deduced cloud mass. It is somewhat annoying that, after all the study of the Venus clouds carried out to date, we cannot prove that sulfuric acid is the major cloud constituent. The PV nephelometer data suggest that the mystery chloride is the more important phase by mass.

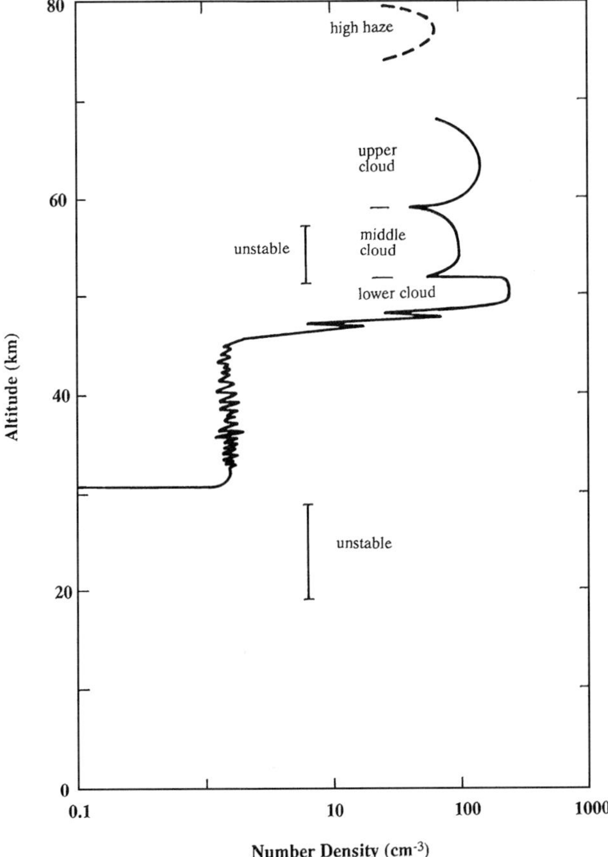


**Figure X.36** Cloud structure of Venus. Most of the cloud mass was found by the Pioneer Venus Large Probe to be provided by micrometer-sized droplets of very concentrated sulfuric acid. Solid particles of other materials, and possibly droplets of other liquids, remain plausible candidates for some of the less-common cloud components. The regions marked as unstable have slightly superadiabatic temperature gradients.

The fate of the sulfuric acid aerosol must be to sediment out of the atmosphere until it encounters regions of high enough temperature to evaporate the droplets. The initial product of evaporation, gaseous $H_2SO_4$, decomposes at slightly higher temperatures to produce water vapor and $SO_3$. Because sulfur trioxide is very unstable relative to sulfur dioxide at high temperatures, it must react with a suitable reducing agent, such as CO, as it mixes down toward the surface. This process will cause a decline in $f_{CO}$ toward the surface, and such a decline is indeed observed. It still seems likely that COS, $H_2S$, and $S_2$ are present in amounts larger than a few ppm near the surface; indeed, both hydrogen sulfide and $S_3$ have been detected. The sulfur cycle on Venus, as suggested by Ronald Prinn of M.I.T., is sketched out in Fig. X.37.

The water vapor profile implied by these considerations should be severely depleted near the main cloud layer by formation of gaseous sulfuric acid and then by cloud droplet condensation. There is a remote possibility that the apparent depletion of water vapor close to the ground has a similar cause, namely the formation of gaseous carbonic acid, $H_2CO_3$, from carbon dioxide and water vapor.

The presence of gaseous sulfur enables an unusual kind of photochemical activity. Thiozone, $S_3$, is readily photodissociated by visible sunlight deep below the clouds, producing free S atoms by

$$S_3 + h\nu \rightarrow S_2 + S, \qquad \text{(X.88)}$$

which releases a supply of very reactive atoms into the atmosphere at levels which otherwise ought to be chemically inactive. Sulfur can react according to

$$S + COS \rightarrow CO + S_2 \qquad \text{(X.89)}$$

$$S + CO + M \rightarrow COS + M \qquad \text{(X.90)}$$

$$S_3 + CO \rightarrow COS + S_2, \qquad \text{(X.91)}$$

which have the net effect of converting CO into COS near the surface. Thus we have two independent methods of depleting CO near the surface, and we find that even atmospheric composition measurements made within 20 km of the surface may sample a gas mixture that is undergoing photochemical processing.

There is an obvious but still unstated principle that unifies this discussion: at all altitudes from the surface to the cloud base the atmosphere is a continuous gaseous mixture with *constant elemental composition but variable molecular composition*. Where condensation occurs, the elemental composition changes, and of course both atmospheric escape and exchange with the surface can also change the composition. There are plausible ways of satisfying the requirement of conservation of carbon (or CO units), but the continuity of hydrogen abundances cannot be demonstrated. No other detected hydrogen compound experiences wide swings complementary to the reported water vapor abundance excursions. If the partitioning of water between vapor and gaseous sulfuric and carbonic acids cannot be verified, hydrogen will remain the most serious problem with our theories of the lower atmosphere.

Atomic chlorine from Reaction (X.84) is very reactive and participates readily in a number of reactions near the cloudtops. The normal reactions of atomic oxygen,

$$O + O_2 + M \rightarrow O_3 + M \qquad \text{(X.92)}$$

$$O + O + M \rightarrow O_2 + M, \qquad \text{(X.93)}$$

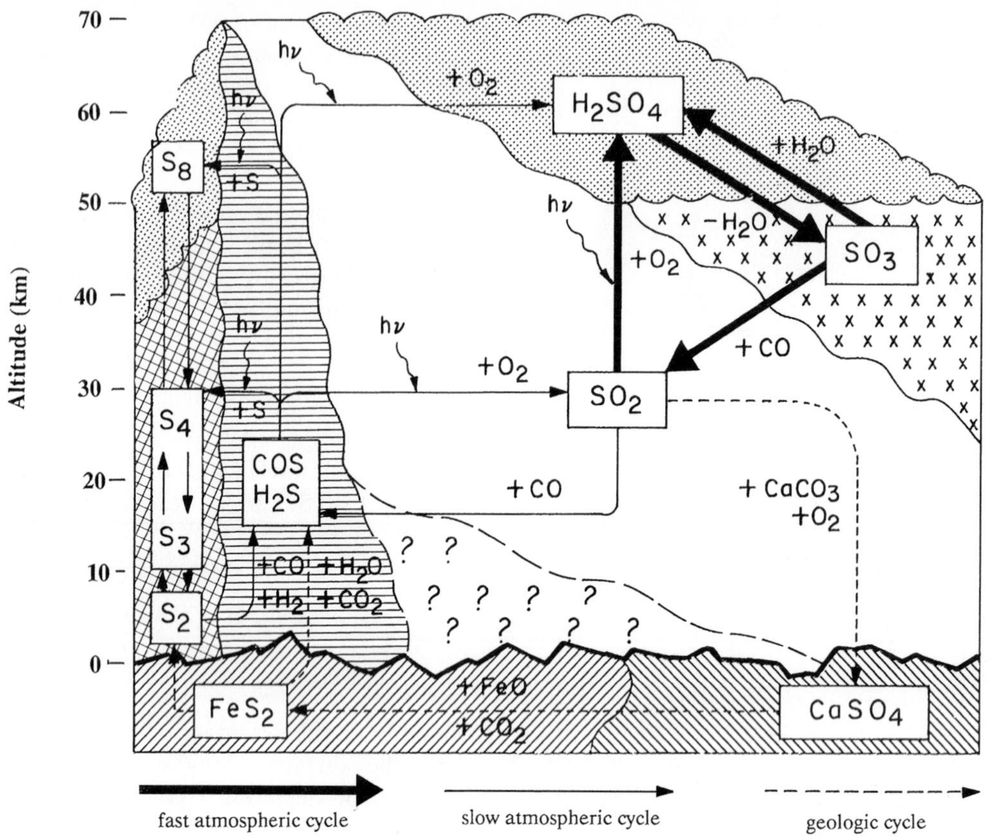


**Figure X.37** The Venusian sulfur cycle. This schematic rendering of the atmospheric and geochemical behavior of sulfur compounds on Venus is from the book by Lewis and Prinn. The roles of sulfur-removal reactions such as pyrite and anhydrite formation are difficult to assess, but must be important if the reported evidence for wild fluctuations in the $SO_2$ abundance is reliable. Note the sulfur gases (left), which are strong UV absorbers.

are supplemented by

$$Cl + O_2 + M \rightarrow ClOO + M \quad (X.94)$$

$$ClOO + CO \rightarrow ClO + CO_2 \quad (X.95)$$

$$ClO + CO \rightarrow Cl + CO_2 \quad (X.96)$$

$$Cl + O_3 \rightarrow ClO + O_2 \quad (X.97)$$

and the like. Thus chlorine from HCl photolysis may play a major role in reforming $CO_2$ near the cloudtops and in keeping the oxygen concentration low. The role of chlorine in destroying ozone, which was originally proposed in the context of Venus photochemistry, has been "brought home" to Earth and is the basis for our present understanding of the effects of Cl (produced by photolysis of anthropogenic chlorofluorocarbons, CFCs) on the ozone layer.

At altitudes above about 100 km on Venus the atomic oxygen abundance climbs rapidly. Diffusive separation becomes competitive with mixing at about 120 km, and molecules and atoms above that level take on individual scale heights, $H_i = RT/\mu_i g$, that reflect collisional equipartition of energy and different molecular weights. O becomes the main constituent of the atmosphere at about 160 km.

Above about 120 km the photoionization of carbon dioxide becomes important. Essentially all photons capable of ionizing relevant molecules are at wavelengths shorter than 1000 Å, at which the Sun is a bright-line emitter. The main emission lines are at 304 Å ($He^+$Ly $\alpha$) and 584 Å (He). Typical molecular cross-sections ($\sigma$) for photoionization are a few times $10^{-17}$ cm$^2$, and hence the incoming photons are mostly absorbed within a scale height of the level at which $H\sigma n = 1$. In round numbers, this occurs at $n = 1/(10^6 \times 10^{-17}) = 10^{11}$ cm$^{-3}$. This occurs at $z = 133$ km on Venus, where the rate of production of $CO_2^+$ is indeed a maximum. The ionosphere is maintained by the flow of energy into a gas with complex elemental and molecular composition, which changes with altitude and with time. Among the most important reactions are

$$CO_2^+ + O \rightarrow O^+ + CO_2 \quad (X.98)$$

$$CO_2^+ + O \rightarrow O_2^+ + CO \quad (X.99)$$

$$O^+ + CO_2 \rightarrow O_2^+ + CO, \quad (X.100)$$

which have the effect of converting carbon dioxide molecular ions into oxygen molecular ions. $O_2^+$ is in fact the dominant ionospheric ion on Venus. Because of the low ionization potential of carbon dioxide, the $CO_2^+$ ion is not effective in ionizing other molecules by charge exchange (except for the relatively minor and very easily ionized NO molecule). Figure X.38 shows the abundance profiles of the most abundant ions in the ionosphere of Venus. Note the much larger scale heights of the atomic ions, especially $He^+$ and $H^+$. Despite the extremely low abundance of water and other hydrogen-bearing molecules above the cold-trap in the clouds and the low abundance of helium in the troposphere (about 12 ppm), these species come to dominate the exosphere because of their very low molecular weights and resulting large scale heights.

Above about 150 km, electrons ejected by photoionization do not immediately thermalize by collisions with molecules. The electron gas typically has a much higher temperature than the ions or neutrals, reaching 3000 to 4000 K above 180 km, compared with about 300 K for neutrals. The ion temperature rises slowly from about 500 K near 200 km to 1500 to 2000 K above 240 km, where helium and hydrogen take over. Escape of light atoms is an important source of secular change and plays a major role in the evolution of the terrestrial planets.

## Venus: Atmospheric Escape

The observation of a high relative abundance of mass 19 (interpreted as $HDO^+$) relative to mass 18 ($H_2O$) by the Pioneer Venus LMS experiment suggests a D:H ratio of $0.016 \pm 0.002$. For comparison, the cosmic D:H ratio is $1.65 \times 10^{-5}$. CI and CM chondrites exhibit D:H ratios from about (1.5 to 2.2) $\times 10^{-4}$. The terrestrial value is D:H $= 1.5 \times 10^{-4}$, about the same as in cometary ices, and D:H on Mars is about $9 \times 10^{-4}$. Understanding the deuterium enrichment of Venus is clearly much more difficult than understanding that of the other terrestrial planets; indeed, if comet and C-type asteroid infall is a major source of water on these planets, then Earth's D:H ratio requires no special explanation. In view of the low escape velocity and low surface water inventory on Mars, enrichment of deuterium by a factor of 6 over cometary source material seems relatively easy. Venus is the problem.

The upper atmosphere of Venus is relatively cool; direct thermal escape of neutral species, even atomic

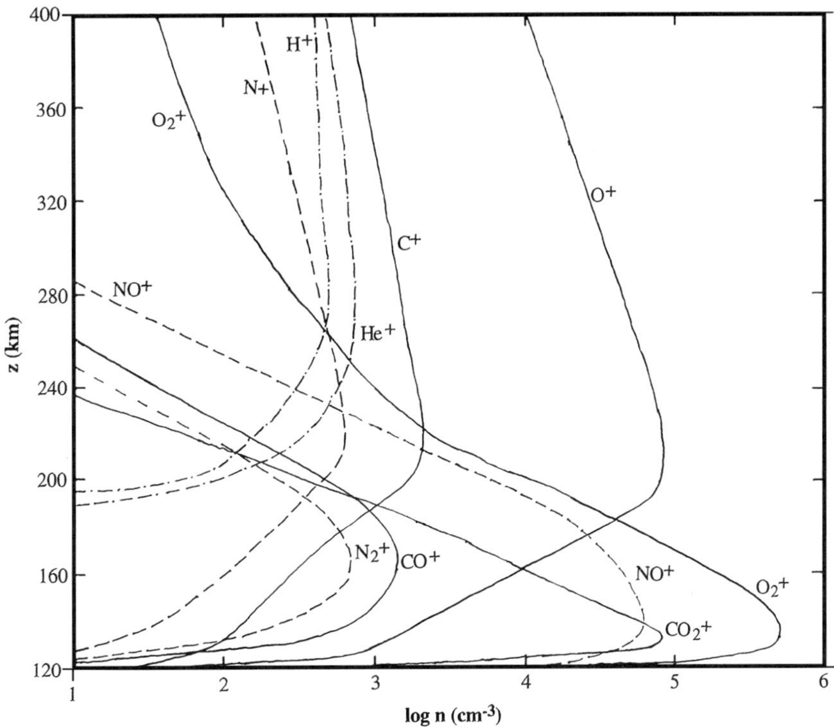


**Figure X.38** Ion abundance profiles on Venus. The chemistry of carbon dioxide and its fragment ions (drawn with solid lines) is of prime importance. Nitrogen compounds (dashed lines), despite the low abundance of molecular nitrogen, are quite significant, especially $NO^+$ because of the low ionization potential of NO. Note the very large scale heights of the light species $H^+$ (— · — ·) and $He^+$ (— — · — — ·) in the ionospheric topside.

hydrogen, is so slow as to be unimportant at the normal dayside temperature of 300 K at the exobase, from which escaping neutral species originate. Still, several energetic processes contribute to nonthermal hydrogen escape. The first of these mechanisms in importance is charge exchange between hot H ions and neutral H atoms,

$$H + H^{+*} \rightarrow H^+ + H^*, \qquad (X.101)$$

where the asterisk denotes translationally excited ("hot") H atoms. This mechanism can provide a present-day escape flux of about $2.8 \times 10^7 \, cm^{-2} s^{-1}$. The second mechanism in importance is collisions of neutral H atoms with hot O atoms produced by photolysis or electron-impact dissociation of $CO_2$:

$$H + O^* \rightarrow H^* + O. \qquad (X.102)$$

This mechanism is effective because the photons actually available in the extreme ultraviolet (EUV) line emission spectrum of the Sun frequently have far more energy than is needed to break the OC–O bond. The excess energy after bond scission shows up as translational energy of the molecular fragments, electronic excitation, or even ionization of one of the fragments. Because of conservation of momentum, the smaller fragment during breakup of $CO_2$ (atomic O) is always ejected at a much greater speed than the heavier fragment (CO). The present hydrogen escape flux from this mechanism is about $0.35 \times 10^7$. The $O^+$ atomic ion is a major constituent of the ionosphere and often is translationally excited, although less so than neutral O. Charge exchange of $O^+$ with H can also produce fast-moving H atoms, but is ineffectual as a source of escaping H atoms. The $H^+$ ion can also be swept away by the solar wind in the magnetotail of Venus. The present escape flux for this process is about $0.5 \times 10^7$, giving a total escape flux of $3.7 \times 10^7$ H atoms $cm^{-2} s^{-1}$.

It is important to realize that the escape rate of hydrogen from Venus is ultimately limited by the supply of hydrogen compounds in the atmosphere and hence that higher escape rates result from higher water abundances. As pointed out by Donald M. Hunten, a cap on the escape rate is supplied by the rate of diffusive transport of water to the escape layer, which depends linearly on the water abundance. The relative importance of different escape mechanisms of course changes depending how much hydrogen is escaping, as shown in Fig. X.39. For a stochastic impact model of the water supply the water content is sometimes, for relatively brief intervals, very much higher than usual, leading to short periods of greatly enhanced escape flux. The *average* water content is larger than the *most probable* water content, and for that reason the time-averaged escape flux is *larger* than the most probable (present-day?) flux by a factor of 1.5 to 2.0 (Fig. X.40).

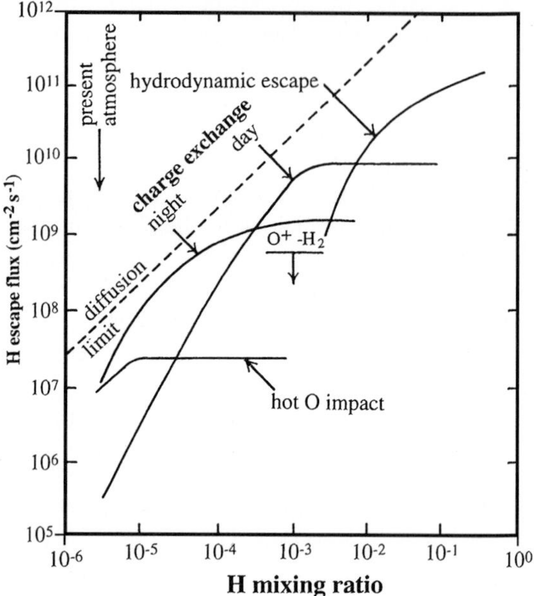


**Figure X.39** Mechanisms for H escape from Venus. High total H concentrations (right) permit hydrodynamic blowoff of hydrogen with little isotopic fractionation. At lower hydrogen concentrations, charge-exchange reactions are the dominant source of fast H atoms. Near present H abundances (arrow), both nightside charge exchange and impact of hot O atoms are important. Some authors, neglecting cometary and asteroidal infall, assume that Venus began with a high total H content (i.e., water rich) and has suffered unidirectional hydrogen loss over its history. They therefore interpret this diagram as a time sequence, starting at the upper right and evolving to the left.

The total inventory of hydrogen in the atmosphere of Venus is about 30 ppm, equivalent to $1 \, g \, cm^{-2}$ of water, or $0.1 \, g \, cm^{-2}$ of hydrogen. This is $6 \times 10^{22}$ H atoms $cm^{-2}$. The time scale for loss of all the water on Venus under present-day conditions is $6 \times 10^{22}/3.7 \times 10^7 = 1.6 \times 10^{15}$ s, or $5 \times 10^7$ years. Thus the atmospheric water inventory on Venus has an expected lifetime of less than 1% of the age of the Solar System. The loss rate of water by the planet is then $9 \times 10^{10} \, g \, a^{-1}$, or $9 \times 10^{19}$ g per billion years. This is essentially identical to the expected water flux onto Venus due to cometary and asteroidal impactors. Thus the rate of hydrogen escape from Venus has profound historical significance, but is probably reflective of recent planetary history, not initial conditions on Venus.

Helium can also escape from Venus. The only helium isotope so far detected on Venus, $^4$He, is radiogenic in origin. Its abundance has been determined only at altitudes above the turbopause, and extrapolation is required to estimate its abundance in the troposphere. The abundances observed at and above 145 km suggest a tropospheric abundance of about 12 ppm with error limits

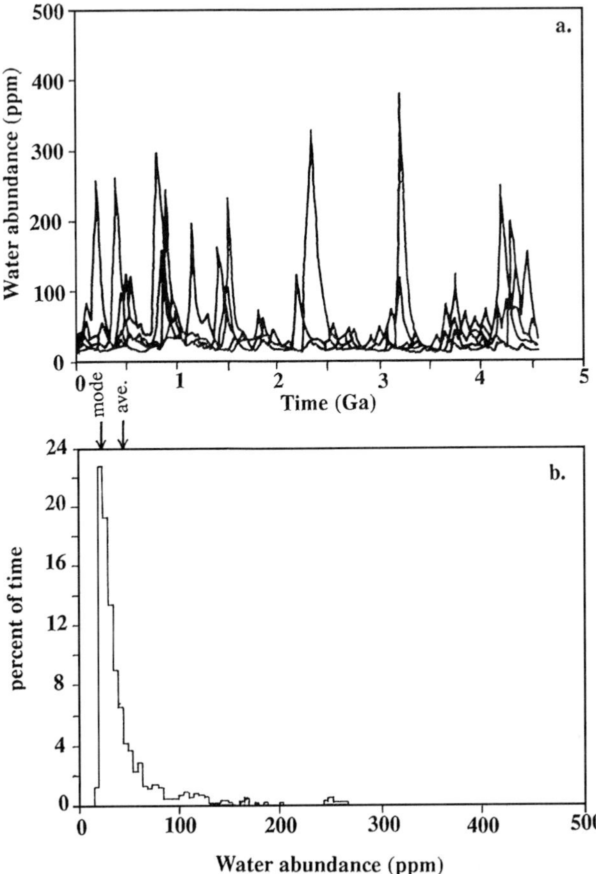


**Figure X.40** Water content of the Venus atmosphere for a cometary infall model. a shows the atmospheric water content during a typical stochastic infall history in which the mean cometary flux has been unchanged at present levels for the entire history of the Solar System. b shows the probability distribution of the atmospheric water content for the infall model in a. The most probable water content is found to be near 20 ppm. Figure courtesy of David Grinspoon (University of Colorado).

running from 4 to 36 ppm. This very uncertain amount of radiogenic helium is several times smaller than the amount that would be produced by the decay chains of uranium and thorium over the age of the Solar System. Charge exchange is ineffectual as a means of producing $He^+$ ions that might be swept away in the magnetotail; the high ionization potential of helium causes helium ions to abstract electrons from other atoms and molecules upon collision. Some neutral helium can be lost by being struck by hot O atoms, analogous to Reaction (X.102).

Escape of other gases from the modern Venus is negligible, but some models for the early evolution of Venus, especially those with high water contents, involve early conditions in which mechanisms other than these might have been effective.

## Venus: Planetary Evolution

Physicist Robert O. Pepin of the University of Minnesota has proposed a complex model to explain the elemental and isotopic fractionations of the rare gases on the terrestrial planets. This model, based upon a theory of fractionation during hydrodynamic escape developed by Donald M. Hunten, envisions much larger abundances of water and primordial rare gases on all the terrestrial planets. The model envisions an early Sun with a greatly elevated ultraviolet flux, as is normal for a star evolving from the T Tauri phase onto the Main Sequence. Water vapor in the atmospheres of newly accreted planets is rapidly photodissociated by the ultraviolet radiation, releasing hydrogen and oxygen. The hydrogen mixes up to the turbopause and diffusively separates to form an extended hydrogen exosphere that is heated strongly by the ultraviolet flux. Meanwhile, oxygen reacts with surface rocks to oxidize iron- and sulfur-bearing minerals. The hot hydrogen exosphere expands outward and flows off the planet hydrodynamically, not by Jeans (thermal molecular) escape. This "planetary wind" is in some ways reminiscent of the Parker model for the solar wind. The hydrogen wind blows along other atoms and molecules in inverse proportion to their mass per unit cross-section area. The atomic cross-section areas of the rare gases are rather similar, but the masses of neon, argon, krypton, and xenon range from 20 to over 130. Xenon is of course least affected, and helium is removed with very high efficiency. This fractionation emulates the observed depletion patterns of the rare gas elements on the terrestrial planets and predicts a more severe gas depletion for Mars than for Venus or Earth.

Since Venus is both less massive and closer to the Sun than Earth, one might naively expect that it would end up with lower abundances of rare gases. The theory is, however, sufficiently elastic to allow any initial endowment of solar-type rare gases, and this problem is solved by simply putting more solar-type rare gases on Venus than Earth to start with. But this requires a much larger mass flux of escaping hydrogen to effect the separation. The estimated mass of water that must be converted into hydrogen and oxygen on Venus is 88 times the present mass of Earth's oceans, $88 \times 280 \times 10^3 = 2.5 \times 10^7\,g\,cm^{-2}$, a layer of water 250 km deep. The mass of "orphaned" oxygen that must be absorbed by the planet is $2.2 \times 10^7\,g\,cm^{-2}$ or $1.0 \times 10^{26}$ g, 2% of the mass of Venus and 3% of the mass of the mantle. This is larger than the mass of the crust, making the idea of absorption of oxygen by the crust entirely ludicrous; instead, the entire mantle must be extensively oxidized. This in turn requires intimate contact between the atmosphere and the deep interior on a truly staggering scale. Even if the crust were completely melted, the fact remains that a crustal melt is stably density stratified and will not readily mix with the mantle. In addition, Magellan data fail to disclose evidence for subduction zones in which crustal material may be forced down into the upper mantle.

Suppose the FeO content of nebular condensate at Venus' distance from the Sun is 2%, and Venus accretes 20% of its mass (15 lunar-sized bodies) from Earth's zone of the nebula, where the FeO content is about 9%. The FeO content of the newly accreted Venus would then be about 4%. The $3.5 \times 10^{27}$-g mantle of Venus would then contain $1.4 \times 10^{26}$ g FeO. Fully oxidizing this FeO to $Fe_2O_3$ will require only another $1.5 \times 10^{25}$ g of oxygen. Raising the FeO content to 18% for the mantle of Venus (a number that exceeds Earth's mantle FeO content by a factor of 2) would accommodate only a third of the orphaned oxygen while oxidizing all of the FeO in the entire planet to the ferric state. This is simply an incredible scenario.

On a more modest level, the D:H ratio on Venus has been interpreted as requiring the loss of an ancient water inventory of at least 0.02 Earth oceans and possibly more than one Earth ocean equivalent. Condensation–accretion models that derive about 20% of the mass of Venus from Earth's vicinity usually end up with about 0.2 Earth oceans of water, although this number is very sensitive to the width of the accretion sampling probability function and the temperature profile in the Solar Nebula. A range from perhaps 0.01 to 0.5 of Earth's water abundance seems possible. The problem with the scenario that sees the present water content and D:H ratio as direct results of billions of years of monotonic escape of hydrogen is that the same results can be achieved by models that assume zero initial water on Venus and consider only the stochastic addition of water by cometary and asteroidal impactors and escape loss of hydrogen. The safest conclusion is not that Venus must have started out wet or that Venus must have started out dry; instead, we conclude that the present water abundance in the atmosphere is *dominated by recent events*. We cannot see back more than a billion years and cannot reach any firm conclusion about the early water abundance. Furthermore, models that neglect water infall cannot be relevant. Early water inventories within a factor of 5 of 0.1 Earth oceans are plausible, but not provable from present data.

It is unfortunate that spacecraft exploration of Venus has languished for over a decade. Many fundamental questions regarding the origin and evolution of the planet require spacecraft data for their resolution.

## Earth

The vast majority of all human knowledge of the planets is knowledge of Earth. To see Earth in a true planetary perspective, in proper relationship to the other planets, requires distilling and abstracting this vast body of terrestrial knowledge. It further requires setting aside the natural presumption that whatever is familiar to us is "normal" and whatever is not native to Earth is "abnormal"; we have the unenviable task of distinguishing between local manifestations of truly universal phenomena and features that are idiosyncratic to Earth. Even worse, to paint so broad a landscape upon so limited a canvas as this book requires omitting detailed discussions of vast amounts of interesting and relevant material, such as the dynamics and thermodynamics of the core, magnetic field generation, mantle convection, the petrology of subduction and crust generation, the history of Earth's climate, the chemistry and physics of the oceans, the circulation of the atmosphere, and the details of the progress and setbacks of life on Earth, all of which would require a hundred volumes to summarize. But it is life that sets Earth apart from the other planets and lends such extraordinary complexity to its behavior: life, especially intelligent life, plays a major role in even the maintenance of the present surface temperature and atmospheric composition.

Earth, as the home of humanity, provides us with our basic definitions of units of measure. The second is the period of oscillation of a pendulum of 1 m length at the surface of the Earth. The day is Earth's rotation period with respect to the Sun, the month was originally the time between successive full moons, and the year is the period of Earth's orbit about the Sun. The kilometer was originally 1/10,000 of the distance from the equator to either pole along a surface at sea level. The mean distance of Earth from the Sun (strictly, the semimajor axis of Earth's orbit) is defined as 1 AU. The unit of density is the density of liquid water, the most distinctive and widespread chemical on Earth's surface. In the same manner, it also provided us with the expectations and vocabulary with which we first went forth to understand the other planets. Earth has been deferred to the end of this book precisely to minimize the tendency to anthropomorphize and "terramorphize" the Solar System, to force us to build a broad perspective that relates a wide range of much less familiar Solar System environments and processes before we delve into the dazzling complexity of Earth.

The most distinctive traits of Earth that are directly observable by remote sensing over interplanetary distances are the presence of vast amounts of liquid water and a high abundance of gaseous oxygen. Closer examination reveals a complex meteorology, massive heat transport by circulation of the oceans, rapidly advancing and retreating ice sheets, and slowly moving continents that open and close entire ocean basins, crumpling the crust, triggering long arcs of volcanoes, and forcing the crust down to destruction in the mantle. Even imaging from Earth orbit reveals the profusion of life both in the oceans and on land. A naked-eye view from space of the nightside of Earth shows vast constellations of lights; of cities, burning natural gas, and brush fires, with aurorae crowning the poles. Radiation belts encircle the equator, and thousands of small artificial satellites populate the space between the Earth and the Moon. Here is a planet truly worthy of study!

It should be obvious, however, that the vast literature on Earth can be only cursorily summarized in a text of this sort. Our purpose must be to sketch out Earth's broadest traits in the context of the other planets. Dozens of recent specialized texts on particular aspects of Earth's nature and history are available to provide depth to this portrait. The best example of an introductory textbook on planet Earth in the modern context of Solar System studies is *Earth: Evolution of a Habitable World*, by Jonathan I. Lunine, Cambridge University Press (1999). The book is written at a low mathematical level, but at a very high intellectual level: crucial concepts are explained and developed clearly, fully, and with logical rigor for an introductory-level audience. It heavily references recent books and reviews on specialized subjects to provide an entry into the quantitatively rigorous specialized literature. The title reveals the author's motivating interest: to understand the fitness of Earth for life and the effects of life on our planet. Chapters XI and XII of this book should whet the reader's appetite for Lunine's approach.

## Earth: Motions

Earth rotates upon its axis relative to the Sun every 24 h, which is the basis of civil or mean solar time. Earth rotates in inertial space (relative to distant stars) in 23 h, 56 min, and 4.091 s of mean solar time. This is equal to the duration of the *sidereal day*. The length of the day is slowly lengthening as a direct result of the dissipation of energy by tides. This deceleration of Earth's rotation is directly related to the acceleration of the Moon's orbital motion discussed in Chapter IX. The change in rotation rate of Earth is clearly discernible through ancient records of eclipses. The apparent length of the day also changes over the year due to the fact that the rotation rate is very nearly constant, but the rate of motion of Earth along its orbit varies cyclically as Earth moves from perihelion to aphelion and back over the course of the year. This phenomenon is complicated by the offset in longitude of the Sun caused by the obliquity of the ecliptic. The combined effect of these two factors causes the time of local meridian crossing of the Sun to vary over a range of about a half an hour from the mean. The offset of true noon (observed meridian crossing of the Sun) from mean solar noon (using a constant length of day averaged over the year) is called the *equation of time*.

The inclination of the rotation axis to the pole of Earth's orbit (the *obliquity of the ecliptic*) is about 23.5°, but varies with time in a complex manner due to tidal forces exerted on the Earth by the Moon and the Sun. Over the short term, the obliquity can be calculated from the linear approximation

$$i = 23^\circ\ 27'\ 8''.26 - 0''.4684(Y - 1900), \qquad \text{(X.103)}$$

where $Y$ is the year. Over the long term, the obliquity varies as a complex sinusoid. The rotation axis of Earth precesses under the long-term influence of lunar and solar tidal torques at a rate of

$$p = 50.2564 + 0.000222(Y - 1900) \qquad \text{(X.104)}$$

seconds of arc per year, so that the precession period is about 26,000 years. The forces exerted by the Moon and Sun that pull Earth's equatorial bulge toward the ecliptic plane are constantly varying, and indeed the net force exerted by the Sun vanishes twice a year when the Sun crosses the celestial equator. The Moon's net force also vanishes twice a month for the same reason. As a result, precessional advance of the direction of the pole is complicated by these high-frequency changes in the tidal torque, and the instantaneous rate of motion of the pole oscillates about the mean motion. This causes a low-amplitude periodic disturbance or *nutation* in the motion of the pole. In addition, seasonal redistribution of air, water, ice, and snow; geological activity; and free oscillations of the body of the planet produce a small-scale "polar wandering" with an amplitude of about 10 m. It is best understood as a slight flexure of the Earth relative to the well-defined geometrical axis; it is the surface of Earth that is wandering, not the pole.

As on Mars, with its similar axial tilt, the obliquity of the pole causes strong seasonal variations in both hemispheres, with opposite phase. Unlike Mars, the orbital eccentricity is so small (0.0167) that the planet-wide in-phase "eccentricity seasons" have negligible effect.

In addition to the sidereal or "true" year, it is convenient to recognize the time between successive passages of the Sun by the vernal equinox. This is called a *tropical year*, with a length of 364.24219879 mean solar days. One sidereal Earth year is 365.25636042 mean solar days. The calendar has long been adjusted for the nonintegral number of days in the year by the insertion of an extra day every four years (*leap year*). This innovation was introduced by Julius Caesar in 45 BC, but the accumulating error in this approximation (0.00636 days per year) had, by the year 1582, resulted in a drift of 0.00636(1582 + 45) or about 10.3 days in the dates recorded by this *Julian calendar*. At the instruction of Pope Gregory, 10 days were dropped at the end of October 4, 1582 (the next day then being October 15). To avoid future drift of the calendar, the only century years henceforth to be leap years were those evenly divisible by 400 (1600, 2000, etc.). This *Gregorian calendar* was immediately adopted by the Catholic nations, but was not accepted in England until 1752 (indeed, England regarded the first day of the year as March 25 up to that time!). The change was greeted by rioting mobs of English subjects who chanted, "Give us back our fortnight!" To this day England still treasures its fortnights but it is only the Americans who hold fervently to other archaic and unwieldy units such as miles, pounds, quarts, and acres. The Orthodox nations were even tardier in their recognition of the superiority of the Gregorian reckoning. Russia accepted the Gregorian calendar in 1918, and Romania followed along in 1919, by which time the accumulated error had reached 13 days. Astronomers measure time in their own manner, counting by "Julian days" (JD) from the arbitrary starting point of January 1, 4713 BC. For example, the eclipse of January 24, 1925, occurred on JD 2,424,175. Time intervals between events are readily computed by simply subtracting their Julian dates. The other peculiarity of Julian days, the fact that they start at noon rather than midnight, is an obvious adaptation for the convenience of astronomers whose "working day" spans midnight.

## Earth: Internal Structure

The density of Earth (about 4 g cm$^{-3}$ uncompressed) is too high to be explained by plausible silicate minerals alone. Like primitive, ancient chondritic meteorites (Chapter VII), Earth must contain a significant percentage of metallic iron–nickel alloy. Indeed, the densest chondrites are about 4% less dense than Earth, suggesting a slightly higher metal content in Earth (Fig. X.41).

Seismic studies of Earth's interior reveal a density that increases steadily inward, with a major density discontinuity at a depth of about 2896 km. The P (pressure)- and S (shear)-wave velocities also show discontinuous breaks at the same depth, attesting to

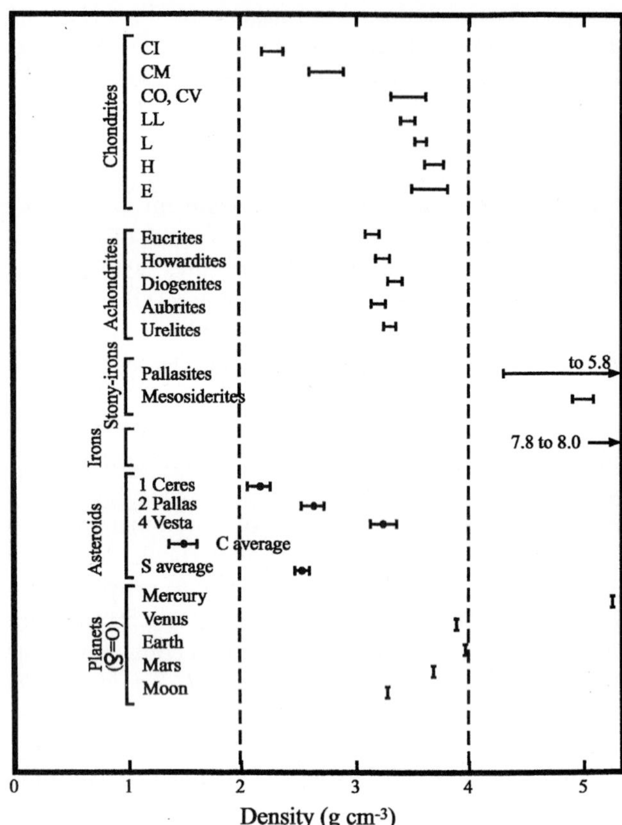


Figure X.41 Densities of meteorites and rocky planets. The ranges of reported densities for members of 16 classes of meteorites are compared with the calculated uncompressed (zero-pressure) densities of the terrestrial planets and the Moon. See the text for further discussion.

the presence of a dense metallic core with a radius of 3473 km (Fig. X.42). The density profile and the P-wave velocity profile both reveal an even more dense inner core with a radius of 1251 km. The core makes up about 32.4% of the total mass of the planet. The low-density crust provides about 0.4% of the mass, and the mantle accounts for the remaining 67.2%. The crust varies markedly in composition and seismic behavior from place to place: the crust under the ocean floors is both thinner and denser than the continental crust.

The disappearance of S-waves in the core shown in Fig. X.42 is a direct consequence of the inability of liquids to transmit shear waves. The density jump at the surface of the inner core suggests a composition or phase change involving the metal-rich core material, from an outer core fluid slightly less dense than pure iron to an inner core slightly denser than pure iron.

The compositions of the interior layers of Earth are of course not determined by seismic techniques; geochemical inference, cosmic elemental abundances, and laboratory studies of the equations of state, melting behavior, phase transitions, and multicomponent phase

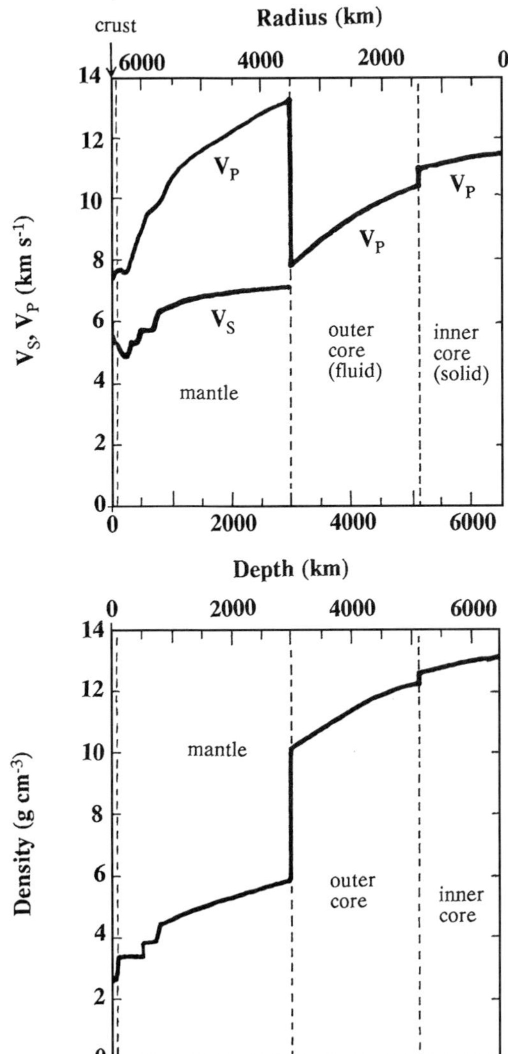


Figure X.42 Seismic wave velocities and density in Earth's interior. The major divisions of the body of the planet are obvious. Shear waves are not propagated through the fluid outer core. Note also the complexity of the outermost few hundred kilometers of the Earth (the crust and upper mantle).

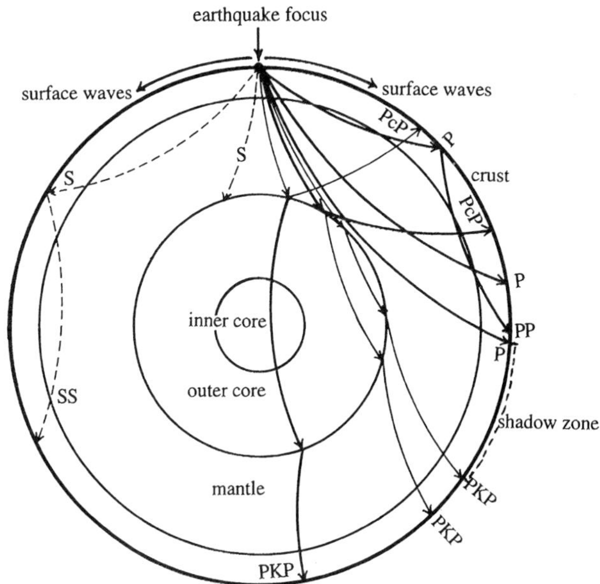


Figure X.43 Ray tracing and nomenclature of seismic wave propagation. Surface waves propagate in the lithosphere and do not penetrate into the body of the planet. Direct pressure and shear waves are marked P and S. PP denotes a P wave reflected from the surface. Reflection from the core is indicated with a c, and transmission through the core is marked with a K. A P wave that grazes the core encounters the surface at a very different location from an adjacent wave that barely enters the core: the PKP wave is strongly refracted in the core and reaches the surface at much greater distance from the source than the adjacent (core-grazing) pure P wave. Between these distances is a "shadow zone" in which P, PcP, and PKP waves cannot arrive, although surface waves, SS waves, and PP waves can be detected.

behavior of silicates lead to more concrete suggestions regarding the elemental and mineralogical composition of these deep regions. The crust and a limited selection of mantle inclusions (erupted onto or extruded into the crust by deep-seated volcanic activity) help us limit the compositions of the deep crust and upper mantle.

Because of the variation of sound speed with depth in the body of Earth and also because of the presence of a cold, hard crust, the ray paths traced out by these seismic waves show strong refraction, reflection, and multipath-propagation effects. Figure X.43 illustrates some of the wave types caused by reflection of S and P from the crust (SS- and PP-waves), by reflection from the core (PcP-waves), and by transmission through the core (PKP-waves). Generally, the signature of a remote earthquake consists of a direct P-wave signal, followed by the PP-wave reflected off the base of the crust and then by the S-wave, with the strong but slow surface waves arriving last. As can be seen from Fig. X.43, direct P-waves are excluded from a "shadow zone" extending from about 102° to 143° from the earthquake source, and direct S-waves are excluded at all points more than 102° from the earthquake focus. The curvature of the internal wave paths clearly reflects the refraction caused by the increase of seismic wave velocity with increasing pressure.

Geochemical models of the core suggest sulfur and oxygen as the two most likely light constituents of the outer core, responsible for both lowering the density and depressing the freezing point so as to permit melting. The high-density inner core is most simply explained as a solid iron–nickel alloy. Growth of the inner core by crystallization of metal out of an iron–nickel–sulfur–oxygen liquid is a reasonable possibility. Such growth releases latent heat, and crystallization can occur only during cooling; thus freezing of the inner core requires net loss of heat from the core into the mantle. The latent heat of fusion of iron is so small that very large cooling (crystallization) rates are required to provide a significant heat flux.

Geochemical models have suggested other possible heat sources in the core, including energetically significant amounts of radioactive elements such as potassium, rubidium, and uranium. All of these elements can enter into sulfide melts in the presence of metallic iron, and hence any or all of them may be present to a small extent in core-forming metal-plus-metal-sulfide melts. Further, the alkali metals K and Rb are both highly compressible, which favors their entry into a metallic melt.

The tidal torques on Earth that cause the spin axis to precess also work upon the core, which would, in isolation, precess at a different rate than the rest of the planet. The differential precession of core and mantle is a source of frictional energy and of shear in the core. The magnitude of this source of heat is surely significant, but it affects only the outermost portion of the outer core.

Finally, growth of the core by melting and differentiation of a silicate–metal mixture is a major heat source during the accretion of the planet. Modern-day heat production by core accretion is possible if disproportionation of $Fe^{2+}$ occurs at high mantle pressures:

$$3Fe^{2+} \rightarrow 2Fe^{3+} + Fe^{0}, \qquad (X.105)$$

but it is not possible for this mechanism to produce more than a few percent of the core mass over the last few billion years.

## Earth: Magnetic Field and Magnetosphere

Direct evidence regarding the rate of heat generation in the core is lacking, but several relevant phenomena help constrain that rate. First, Earth has a rather strong planetary dipole field with a mean surface field strength of about 0.6 G that is roughly aligned with the spin axis (the north magnetic pole is in the Northwest Territories in Canada). A permanent dipole field due, for example, to ferromagnetism in mantle minerals is ruled out by Earth's high internal temperatures, which lie well above the Curie point of any natural geological material. Any imposed field would quickly diffuse out and vanish, and hence the field must be continually produced within Earth. Generation of a magnetic field requires motions in an electrically conducting medium, which is equivalent to saying that the geomagnetic field is produced and maintained in the outer core.

Earth's external magnetic field is easily strong enough to stand off the solar wind from the surface and atmosphere. A crude symmetrical dipole field model (unperturbed by solar wind pressure) would have $B$ dropping off as $R^{-3}$, and the magnetic pressure ($B^2/8\pi$) therefore drops off as $R^{-6}$. Since the solar wind dynamic pressure during quiet-Sun conditions at Earth's orbit is typically $\frac{1}{2}\rho v^2 = \frac{1}{2} \times 7 \times 1.6 \times 10^{-24}(3.6 \times 10^7)^2 = 10^{-8}$ dyn cm$^{-2}$ (Table IV.2), the approximate distance to the point of dominance of the solar wind over Earth's external field (i.e., the distance to the boundary of the magnetosphere) would be roughly at 10 $R_E$, or about 60,000 km. As on the Jovian planets, the solar wind pressure ensures that the magnetosphere is far from spherical. A bow shock shaped roughly like a paraboloid of revolution is filled by a turbulent, postshock plasma, within which is the orderly magnetosphere with its closed but asymmetrical field lines. Trapped radiation belts, called the *Van Allen Belts* after their discoverer, dominated by energetic protons and electrons, lie deep within the magnetosphere (generally within 6000 km of the planetary surface). A comet-like "magnetotail" stretches outward to great distances, carrying open field lines from the immediate vicinity of the north and south geomagnetic poles. These open field lines constitute a path for both injection of solar ions into the inner magnetosphere and escape of terrestrial light ions. A sketch of the structure of Earth's magnetosphere is shown in Fig. X.44.

## Earth: Surface Geology

The visible surface of Earth is divided naturally into oceans and continents. However, a careful study of the distribution of altitudes on Earth show a somewhat more complex and more interesting dichotomy. As we saw in the Earth hypsogram in Fig. X.2, the distribution of surface topography on Earth is in fact bimodal. The most common altitudes are close to the mean depth of abyssal plains on the ocean floor, about 4 km below mean sea level. The second, smaller peak is centered slightly above mean sea level and includes most of the continental platforms and the shallow ocean near continental margins, called the *continental shelf*. High mountains and deep ocean trenches account for the spectacular but rare extremes.

Examination of these altitude data reinforces the significance of the minimum in the frequency distribution of altitudes near a depth of 1 to 2 km below mean sea level. It is at this depth that the true natural division between continents and oceans occurs. Here, just beyond the very gently sloping continental shelf, the ocean floor descends rapidly to the abyssal plains. This steeper, often unstable margin to the continents is called the *continental slope*.

The bimodal altitude distribution of the crust takes on vastly greater significance when we analyze a wide range of continental and oceanic rocks. We find that the continents are a very complex mixture of rocks with high silica, alumina, and alkali metal oxide

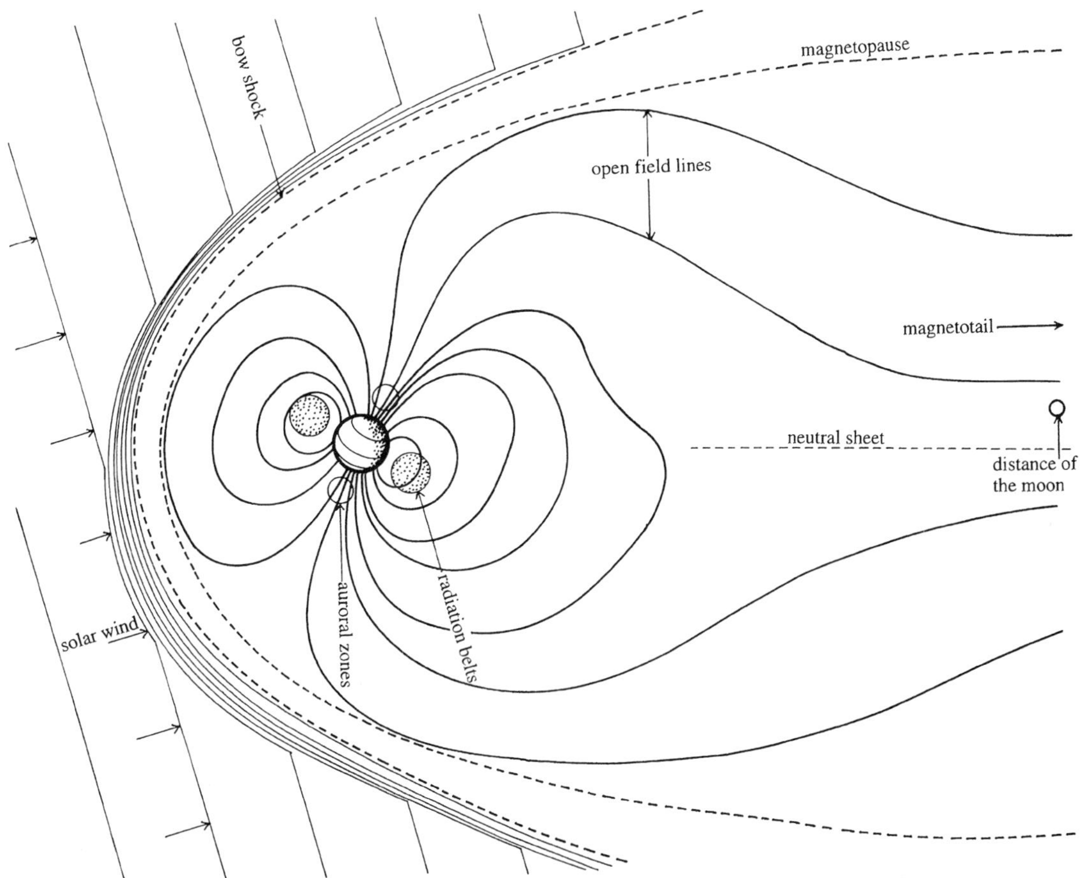


**Figure X.44** Structure of Earth's magnetosphere. The Van Allen radiation belts are deep within the inner magnetosphere and reach down to the top of the ionosphere. Capture of solar ions and loss of light terrestrial ions occur most readily in the vicinity of the magnetic poles. Precipitation of radiation belt particles into the atmosphere is concentrated at the feet of the highest-latitude closed field lines, in two narrow annuli surrounding the magnetic poles (the "auroral zones"). The outer magnetosphere is distorted by the solar wind pressure and blown outward from the Sun. The neutral sheet separates two regions of oppositely directed open (broken) magnetic field lines. Chance reconnection of field lines can release large bursts of energy, causing magnetospheric "storms." The Moon orbits partly within the outer magnetosphere (near full Moon) and mostly outside the bow shock. It traverses the bow shock and magnetopause twice per orbit.

contents in contrast to the ocean basins, which are dominated by dark, dense basaltic material with a high content of ferrous iron and magnesium oxides. These types of crust are often referred to as *sial* (Si–Al rich) and *sima* (Si–Mg rich). It is a tradition in geochemistry to refer to silicon as an "acidic" element (a holdover from an obsolete theory of rock formation in which aqueous solution behavior was believed to be important) because it forms silicic acid, $H_4SiO_4$, in solution. Thus continental rocks, with their very high silica content, are called *acidic rocks*. The ocean-floor rocks, with their low silica content and high magnesium and iron content, are termed *basic rocks*. Rocks dominated by olivine have a very high (Fe + Mg):Si molar ratio of 2:1 and are called *ultrabasic*. It is now common to refer to basic and ultrabasic rocks by the more descriptive terms *mafic* and *ultramafic*, where the name makes direct reference to their magnesium and iron content.

Seismic studies of the continental blocks and ocean floor reveal that the crust is dramatically thicker under the continents, suggesting the view that thick "rafts" of low-density continental sialic material float on the mantle, supporting the elevations of the higher mountains by providing thick buoyant "roots" of low-density continental crust beneath them. By comparison, the ocean floor is thin and made of materials much closer in density to mantle materials.

We are already familiar with *igneous* rocks, which are made by the cooling, and usually the crystallization, of previously molten material. Basalts on asteroids, Mars, Venus, and so on are especially familiar examples of igneous processes elsewhere in the Solar System. But on Earth there are vast amounts of rock that was deposited at low temperatures by chemical and physical weathering processes working on older rocks. The downhill transport of debris from rock weathering by wind and water leads to the formation of vast wedges of sediment on continental platforms and on the ocean floor adjacent to continents. These sediments are often augmented, and sometimes dominated by, carbonates and other materials that have been formed and deposited by biological processes. Under conditions of high pressures and moderate temperatures (not hot enough to melt or extensively recrystallize the sediments) these sediments become compacted into *sedimentary rocks*. Further heating and compression can lead to extensive mineralogical and chemical changes, often accompanied by deformation, to make recrystallized *metamorphic rocks*. In places where volcanic eruptions have penetrated a sedimentary terrain, one may see a complete and continuous gradation from sedimentary to igneous textures. Further heating, leading to extensive melting, density-dependent separation, and recrystallization, produces a new generation of igneous rocks. The cycle igneous–sedimentary–metamorphic–igneous is called the *rock cycle*. Note that it is also possible for igneous rocks to be "cooked" with volatiles under moderate conditions to make metamorphic rocks, and of course sediments may be made by weathering of any rock type. Thus any rock type may be converted into any other, and the rock cycle is not unidirectional.

Clearly, any attempt to describe "static" features of Earth quickly leads us into dynamic geochemical considerations, such as the interconversion of rock types. Similarly, even simple observations of the "static" geology of the continents and abyssal plains lead us into consideration of dynamic geophysical processes. Consider, for example, the observed wedges of sediment on the continental platforms and along continental margins. The removal of this vast mass of rock by weathering at higher elevations must unload the continents. Removal of this weight from the high mountains allows the buoyant continental blocks to rise and expose at the surface materials that have recently been deeply buried and hence subjected to high pressures and temperatures.

The continental blocks are very rich in alkali metals and in heavy atoms of elements that readily form oxides. Thus the continents are very rich in K, U, and Th and are quite powerful sources of radioactive decay energy. The combined effects of weathering and radiogenic heating should rapidly erase continental relief and fill the ocean basins with sediments. But the real world is not like this: a rugged topography persists, and the sediment layer on the ocean floors tapers very smoothly from the continental slopes (where the layer is thickest) down to essentially zero thickness in midocean. Interestingly, the middles of several ocean basins are marked by chains of undersea volcanoes. These volcanic chains mark a line that is equidistant from the bordering continents, where the oceanic basalt crust is thinnest. Thus the Mid-Atlantic Ridge is the site of generation of new oceanic crust at the point of greatest weakness in the crust.

Radiochemical dating of the ocean floor exposes a remarkable pattern: the crust adjacent to the midocean ridges does indeed have nearly zero age. The age of the basaltic basement rock increases steadily toward the continental margins. Further, the age of the sediments atop the basalt forms a sequence from zero age at the top of the sediment layer to an age equal to that of the underlying basalt at the sediment–basalt interface. In the South Atlantic, the maximum age of the sea floor is 190 million years, adjacent to the African and South American continents. The North Atlantic is somewhat younger. In sharp contrast, the ages of continental rocks are found to run from zero (for recent volcanic eruptions) back to over 3 billion years. The oldest known continental rocks, at $3.9 \times 10^9$ a, are in Greenland.

The ages of continental rocks also show an interesting pattern. Continents tend to have a small core of very ancient rock, called a *craton*, with terranes of progressively younger rocks "accreted" onto it. All these observations suggest that the continents can and do move with respect to each other. A newly formed rock unit, such as a lava sheet from a volcanic eruption, of course carries a radiochemical record of its time of origin; for example, the gas-retention clocks based on accumulation of $^4$He and $^{40}$Ar are reset to zero age by melting and gas release. But what can a rock tell us of its original *location*? Not surprisingly, the longitude cannot be determined by any known rock property. Certainly fossil evidence, including marine microfossils and pollen grains, can suggest to us whether the site of deposition of sedimentary rocks was cold or hot, and this information translates in a crude way into an estimate of latitude. The discovery of coal seams full of the fossil remains of tropical plants and animals in Antarctica has led us to conclude that Antarctica was not close to the pole at the time these sediments were laid down. But this evidence is hard to quantify in terms of latitude because the mean temperature of Earth varies with time and the temperature at a given latitude varies with longitude due to the uneven distribution of land masses about the globe. Fortunately, a number of minerals (mostly iron-bearing species such as magnetite) are capable of recording the magnetic field in which they crystallized. The study of the intensity and direction of ancient magnetization (*paleomagnetism*) tells another fascinating story.

In areas where nearly continuous volcanic activity has occurred over the past few million years, such as

Hawaii, Iceland, and various sites in North America and Russia, it is possible to measure the gas-retention ages and paleomagnetic properties of numerous samples closely spaced in time. When this is done, we find a coherent worldwide picture: the locations of the magnetic poles have not varied greatly over the past few million years, but the *polarity* of the field has reversed several times, most recently 850,000 years ago. As a consequence of field reversal, the magnetism of rocks being formed in continuous sequences on both sides of midocean ridges will show reversals of direction that are mirrored by the ridge axis. Rocks of a given age are found at nearly equal distances from and on both sides of the weak *spreading center* at the ridge axis, much as two tapes being fed simultaneously through a tape recorder and pulled out in opposite directions would show symmetrical recordings. A magnetometer dragged from a ship along the sea bottom easily measures the magnetism of the underlying bedrock and does indeed show a remarkable symmetry of the magnetic banding (pole reversals) on both sides of midocean ridges. Further, not only do all active spreading centers show a symmetry of patterns, but in fact they all show the same pattern. The pattern is truly global, and only a global-scale causative agent (i.e., reversals in the planetary dipole field) can be responsible.

The rate of continental drift varies from one spreading center to another. Rates are generally at least 1 cm per year and reach nearly 50 cm $a^{-1}$ in the East Pacific.

The magnetic records of rocks reveal not only the paleofield strength, but also the direction and dip angle of the ancient pole. (The dip angle is zero at the magnetic equator and increases rapidly to about 90° as one approaches the magnetic pole.) Thus a comparison of rocks on different continents, although incapable of rendering true geographic longitudes or latitudes, can nonetheless provide us with *magnetic latitudes* and sometimes, using seafloor magnetic data, with *relative longitudes*. This information, combined with paleontological data that suggest the geographical latitude (temperature) of rocks of various ages, has permitted a rather detailed reconstruction of the history of the positions of the continents over the past few hundred million years. Perhaps the most interesting result is that all the continents were joined together as recently as 225 million years ago. This supercontinent, Pangaea (literally "all Earth"), split eventually into seven major fragments: North and South America, Africa, Australia, Antarctica, India, and Eurasia. Michael Rampino has suggested that the initial breakup of Pangaea was caused by a major impact event near the present location of the Falkland Islands in the South Atlantic.

## Earth: Early Geological History

Reconstructions of the geological and biological history of Earth are closely connected because of the central role of fossils in assigning ages to rock units. Major discontinuities in the rock record and in the corresponding fossil record are used as the dividing lines between geographical ages. Within each age the pace of geological and biological change is modest; at the ends of ages, the changes are often profound. Because of the common use of distinctive biological markers (*index fossils*) to define the ages of ancient rocks, early geologists named the geological ages starting with the first in which fossil evidence of life was clear and unambiguous. The first few geological ages were named after their first-known and best-documented locales in Great Britain; the Cambrian, Ordovician, Silurian, and Devonian periods all were characterized and named at that time. The later Mississippian and Pennsylvanian periods obviously were defined by type examples in the United States. The beginning of the oldest of these periods, the Cambrian, dates back to 570 Ma BP. All rocks earlier than these, which were devoid of fossils that could be identified at that time, were lumped together as the Precambrian. Thus the vast span of time from 4.6 to 0.57 Ga before the present (BP) bore a single name, even though it accounts for some 88% of the age of our planet. For convenience, the Precambrian is now subdivided into the Hadean era, >3.9 Ga BP, from which no rocks survive; the Archaean or Early Precambrian period 3.9 to 2.5 Ga ago, from which rocks but no fossils are known (also called the Azoic, or lifeless, Era); and the Proterozoic (first life) period from 2.5 to 0.57 Ga BP, throughout which simple life forms populated Earth's oceans. The Proterozoic is for convenience further subdivided into the Middle Precambrian, 2.5 to 1.7 Ga BP, and the Late Precambrian, 1.7 to 0. 57Ga BP. In recent years the latest Precambrian, especially in British Columbia and in Australia, has yielded fascinating glimpses of bizarre life forms that preceded the well-documented proliferation of fossil forms in the Cambrian Era.

Geologists in the 18th century assembled enough information about the sequence of fossils in consecutive layers of sedimentary rocks to found both a science of fossils (*paleontology*) and a science of relative geological chronology (*stratigraphy*). Absolute ages were then difficult to estimate, because radioactivity had not yet been discovered. Estimates of sedimentation rates derived from studies of modern processes were used by many geologists to suggest approximate times needed to produce the complete thickness of sediments in the marine geological record; estimates ranged from tens of millions of years upward. Such estimates were severely attacked by the religious establishment, who based their official estimate of the age of Earth on a chronology published by Bishop James Ussher of the Church of Ireland in 1664. Ussher's chronology, based solely upon his personal interpretation of Biblical texts, placed the Creation on 26 October, 4004 BC, at 9:00 AM (is this Eastern Standard Time?). Thus any age for Earth in excess of 6000 years represented an attack on religious orthodoxy (but not on the Scriptures!) and hence was treated as heresy.

As the evidence for vast periods of geological time became more and more persuasive during the 18th century, opinions on this issue became institutionalized under two principal schools of thought. First, those who accepted the Ussher interpretation of biblical chronology attributed the abrupt endings of geological ages to the domination of a handful of catastrophic events (in extreme cases, only the single event of Noah's flood) that did virtually all the work of laying down the kilometers of sediment in the geological record. These *Catastrophists* began with the conclusion that Earth is really only a few thousand years old. Fossils were often dismissed by Catastrophists as the remains of "that accursed race which perished with the Flood" or as artifacts made by Satan to mislead us or by God to "test our faith." At the other extreme were those geologists who accepted that these vast thicknesses of sediments took many millions of years to be produced and many millions of years to be deposited. Without any means of dating geological materials, the best guide to the rate of accumulation of sediments was taken to be the modern rate, derivable by study of everyday processes at work around us. This view gave rise to the *Uniformitarian* school, which attributed all the work of geology to very slow, unspectacular processes working over vast periods of time. The Uniformitarians denied the reality of global catastrophes, which smacked of magic and seemed intended not to explain the observational data, but to provide an excuse for disregarding the data.

The debate was both sharpened and heated in the mid-1860s by Charles Darwin's work on the origin of biological species. Darwin proposed that the proliferation of species occurred as the combined effect of two main factors, chance isolation of populations and natural selection. The Darwinian explanation for the fossil record emphasized migration or division of preexisting species by slow geological change to establish independent gene pools that initially differed only slightly. In these smaller, isolated populations, combinatorial variation of existing genes played out different roles in different environments, where the criteria for survival were different. A trait that is useful to a jungle gene pool might be irrelevant to a grassland gene pool and harmful in a desert setting; therefore, differing environments will favor different traits and lead to genetic divergence even if the original gene pools were identical. Further, mutations occurring randomly and independently within these separate gene pools lead inevitably to genetic drift, because it is biologically improbable that the same mutation would occur in all the isolated gene pools. Most mutations are harmful or nonfunctional, many are lethal, and only a few will confer some functional advantage on their carriers. Individuals with helpful mutations will tend to be more successful and leave more offspring, and hence these new traits will (barring unfortunate accidents) tend to spread throughout the local breeding population. After a series of such mutations, the different populations will cease to be able to interbreed and will become separate species. The natural time scale for such divergence of a cluster of species from a common ancestor by this process of *natural selection* is on the order of $10^5$ to $10^6$ years. The entire biological evolutionary saga from the origin of cells to the present, with its vast numbers of intermediate branches and divisions, must occupy orders of magnitude more time than this: billions of years.

The modern quantitative study of the history of Earth and its life dates back only a few decades. The German physicist C. F. von Weizsäcker proposed the principle underlying potassium–argon dating in 1937 [see Eqs. (II.149)–(II.151)], and dating of geological and meteoritic samples by K–Ar and U/Th–He techniques began in the 1940s, just in time to be seriously interrupted by the Second World War. Carbon-14 dating of recent biological species was begun at nearly the same time. Knowledge of the existence of these dating techniques engendered laboratory experiments directed at measuring and refining the half-lives of geologically interesting radionuclides and also inspired searches for additional radiochemical systems to provide independent measurement techniques. Lead–lead dating, based on the isotopic analysis of radiogenic lead ($^{206}$Pb, $^{207}$Pb, and $^{208}$Pb) from the respective decay chains of $^{238}$U, $^{235}$U, and $^{232}$Th [Eqs. (II.152)–(II.154)], came next, to be followed by the Rb–Sr, Re–Os, and Nd–Sm dating techniques. The results confirm the picture discussed in Chapter II: the chemical elements present in the Solar System were mostly synthesized in supernova explosions between 4.6 and about 12 billion years ago. The formation of the Solar System dates back to about 4.6 billion years before the present (4.6 Ga BP), and the accretion of Earth was about 99% complete by 4.5 Ga BP.

Application of these dating techniques to Earth suggest rapid core formation, with isolation of primordial lead from uranium and thorium (the sources of radiogenic lead) largely complete by the end of accretion. The most ancient crustal material known, from the

Isua formation in Greenland, dates to 3.9 Ga BP, leaving the first half-billion years of Earth's accretion and early evolution completely unsampled. Recall from Chapter IX that most of the samples returned from the Moon by the Apollo and Luna missions date from before 3.9 Ga BP; the geological evolution of the Moon ground to a halt and large-scale endogenic alteration of the lunar surface ceased before the earliest surviving fragment of Earth's crust was formed. The terrestrial and lunar historical records are thus complementary.

Insofar as we can tell from the limited surviving evidence, the crust of our planet dates from very early times. Geochemical cycling of the crust, presumably involving oceans of water, has been the norm for all of recorded geological history.

The early Earth must have been the scene of vigorous geological activity. First, the recent accretion of the planet had released enough heat to melt the interior thoroughly. Indeed, depending upon how and when volatiles were first released into the atmosphere and precisely how and how fast accretion occurred, the entire planetary surface may have been blanketed under a dense steam atmosphere, with mean surface temperatures well above the melting point of rock. Each gram of late-accreting material, impacting at a speed on the order of $15\,km\,s^{-1}$, would carry enough kinetic energy to vaporize itself completely; storage of this heat under an atmosphere with a high infrared opacity could keep the surface molten for many millions of years. In addition, massive impactors would drive strong shock waves through the planetary interior, heating the entire body of the planet. Core formation would then occur naturally during terminal accretion and need not await the slow warming of Earth by decay of long-lived radionuclides.

During the first billion years of postaccretion history, the total rate of energy generation by decay of long-lived radionuclides was nearly an order of magnitude larger than that today (see Fig. VI.12). Both the rate of convection in the mantle and the rate at which the mantle exuded crustal material and volatiles were much greater than those in recent history. Both strong convective overturn and severe impact cratering work strongly against the persistence of a stable crust.

The origin of the Moon dates back to the accretionary era, when the flux of very large bodies through the inner Solar System had not yet been used up by the accretion of the terrestrial planets. An exact age for the Moon is hard to determine, but it is clearly close to 4.5 Ga old. The Earth impact event that presumably ejected the material of the Moon into orbit about Earth was so violent that it must have vaporized a vast mass of material from the impactor and from Earth's crust and upper mantle. Earth must have been left in a molten state, with a massive atmosphere of silicate vapors.

## Earth: Biological History

The earliest evidence of life on Earth is the presence of biogenic organic matter in sedimentary rocks dating back to 3.8 Ga BP. The most distinctive molecules found in these rocks (Fig. X.45) include the normal alkanes, which are linear hydrocarbons of general formula $(H_3C–(CH_2)_x–CH_3$; porphyrins, which are complex polycyclic compounds capable of complexing a metal ion (familiar examples include chlorophyll and the heme group in hemoglobin); and isoprenoids, which are a family of hydrocarbon polymers of the isoprene unit, a common building block of volatile hydrocarbons in conifers and of rubber. The earliest microfossils, the fossilized remains of bacteria and filamentous algae, date back to nearly the same time as these chemical fossils.

The period from the oldest rocks to 2.5 Ga BP is called the Archaean or Early Precambrian (see Fig. X.46). The atmosphere at this early stage was certainly not strongly oxidizing (negligible free oxygen) and may have been a chemically reducing mixture of water vapor, lower hydrocarbons, and ammonia. Such a reducing mixture is termed a Stage I atmosphere. The geochemical plausibility of such an atmosphere is debatable. Early arguments regarding planetary differentiation and outgassing assumed low surface temperatures. Under such circumstances $H_2$-rich magmatic gases, which would be regulated at a very low oxygen pressure by the presence of metallic iron, could react spontaneously as the gases reach the cool surface to make hydrocarbons and ammonia. More geochemically plausible is a phase with a hot (molten) silicate surface with no metallic iron present, giving rise to a Stage II atmosphere dominated by nitrogen, water vapor, and CO or $CO_2$ with only a trace of hydrogen. The thermal instability of hydrocarbons and the thermal and photochemical instability of ammonia militate against their presence in large abundance.

Life may have originated during the sampled portion of the Archaean or possibly earlier in history, when the bombardment rate of very large bodies was so great that the entire surface of the Earth could be sterilized by the largest impacts. It is entirely possible that primitive unicellular life (or virus-like noncellular life) arose and was destroyed repeatedly until the surface environment became stable enough for life to persist. In any event, the first cells were simple lipid membrane bags containing a soup of enzymes, proteins, and DNA. The earliest cells must have lacked nuclei and organelles (*prokaryotic* cells) and must have been *heterotrophic*, meaning that they depended upon fermentation of inorganically produced organic matter for their source of energy. Production of such organic matter is easiest in a Stage I atmosphere, but is also possible in a Stage II atmosphere in the presence of ferrous iron.

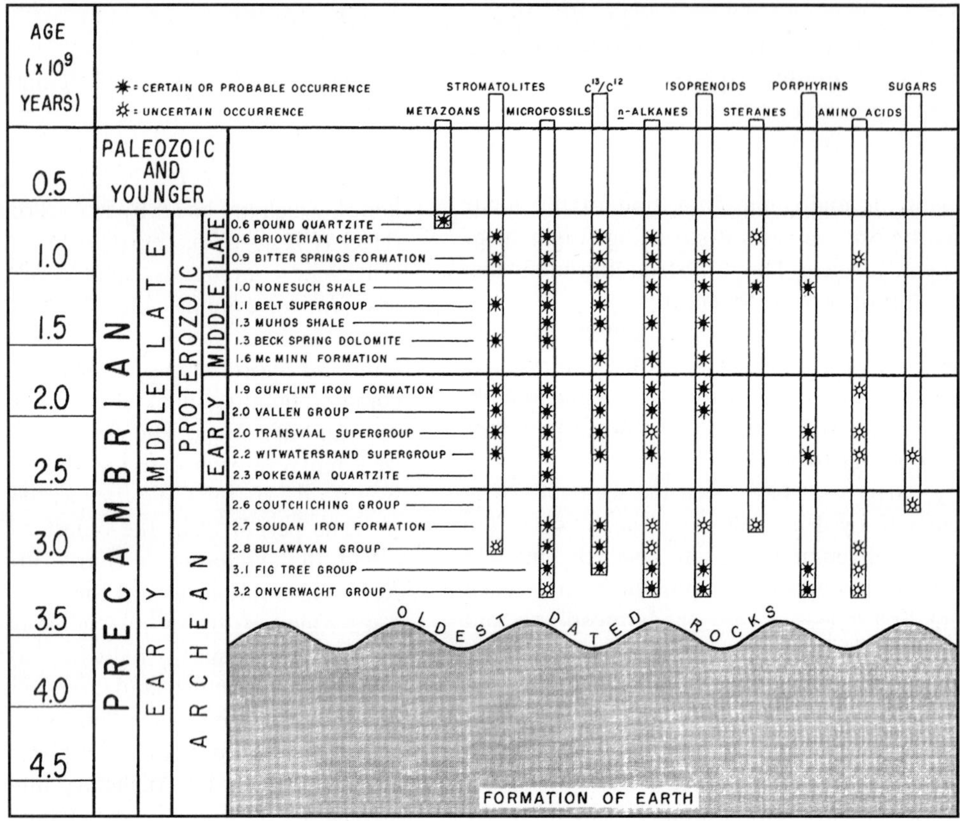


**Figure X.45** Physical and chemical fossil evidence of ancient life. This figure is from the book by Lewis and Prinn. Since that time the age of the oldest known rocks has been pushed back to about 3.9 Ga. (Although it sometimes feels that way, Earth has not actually aged by 0.5 Ga since the publication of that book.)

The next benchmark in the development of life is the rise of *autotrophic* cells capable of synthesizing organic matter directly from inorganic gases and solar energy. It is likely that these earliest photosynthetic organisms were unable to survive in the presence of oxygen (*obligative anaerobes*). Any oxygen liberated by photosynthesis would have been sequestered by reaction with ferrous iron or other inorganic reducing agent.

The earliest organisms to turn to actual oxygen gas production were probably prokaryotic blue–green algae. Their presence has a number of implications for the biological and geochemical evolution of Earth. First, oxygen is a highly toxic substance. The rise of oxygen in the atmosphere would have been in effect a devastating form of chemical warfare; species not equipped with protective enzymes to guard them against oxidation would have had to improvise defenses or die. Second, the anoxic early ocean would have been able to dissolve vast quantities of ferrous iron. Ferrous iron reacts readily with oxygen to form insoluble oxide and hydroxide precipitates. Cells that for any reason maintained a supply of ferrous iron within them would have been able to combat oxygen poisoning. But on a much grander level, the oceans would serve as an oxygen sink on the planetary scale.

The geochemical evidence pertaining to this era in Earth's history is quite remarkable. From the first appearance of blue–green algae around 3.2 Ga BP, precipitation of iron minerals began to contribute to oceanic sediment production. During the Proterozoic Era in the middle and late Precambrian, 2.5 to 0.57 Ga BP, precipitation of "red beds" of oxidized iron, mainly hematite $Fe_2O_3$ and limonite FeOOH, grew in importance. By about 1.9 Ga BP these beds had become very common, forming thick layers of iron oxide sediments that now constitute most of the iron ores mined on Earth. Between about 3.2 and 1.9 Ga BP the biospheric oxidation state oscillated back and forth between oxidizing and reduced states. Thick layers of red ferric oxides alternated with thin layers of green clay containing ferrous iron. These *banded iron formations* (BIFs) are accompanied in the late Precambrian by copious precipitation of calcite $CaCO_3$ and dolomite $CaMg(CO_3)_2$. The increased availability of calcium and magnesium

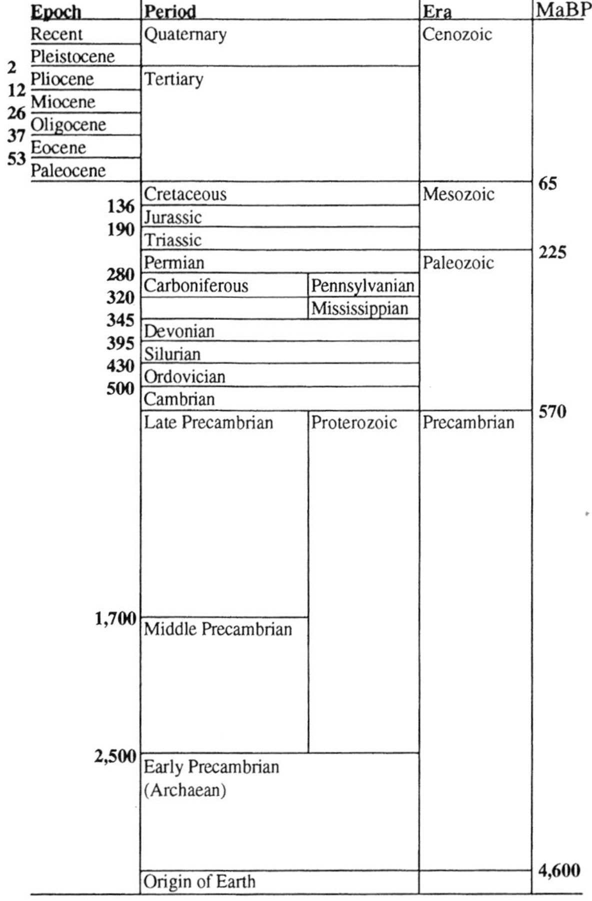


**Figure X.46** The geological ages. Only microfossils, algal colonies (stromatolites), and chemical fossils are known before the very latest Precambrian. Complex life arose and flourished in the sea in the lower Paleozoic, emerged onto the land in the upper Paleozoic, and was decimated by the great extinction event at the end of the Permian. The age of reptiles, the Mesozoic, lasted 160 Ma, and 65 Ma of the age of mammals (Cenozoic) has now elapsed. The Quaternary, dating from the rise of the early hominids and spanning the present Ice Age, is only 2 Ma old. Human-recorded history dates back only some 6000 years. The time elapsed since the blossoming of Egyptian civilization occupies only one millionth of the age of Earth.

ions in the ocean is presumably due to a change in the weathering behavior of crustal rocks caused by increasing oxygen levels.

Hand in hand with the rise of oxygen came the photochemical production of ozone. Until oxygen was well established through photosynthesis, solar UV with wavelengths as short as 200 nm could penetrate to the surface. Such radiation destroys all of the chemicals essential to life, including DNA itself. The emergence of life onto land may have been enabled by the achievement of a critical ozone opacity or by the appearance of protective pigments and enzymes.

Extended colonies of Precambrian lime-secreting (*calcareous*) algae produced large brain-shaped growths called *stromatolites*. In Western Australia, modern stromatolites very similar to the Precambrian fossils can be seen living and growing. There are also Precambrian coal beds that bear evidence of the presence of large masses of rather fragile plant material, poorly suited for the production of distinctive fossils.

In the middle and late Precambrian nucleated *eukaryotic* cells appeared and became very important. According to theoretical biologist Lynn Margulis of Boston University, these cells came about as a result of symbiosis of prokaryotic cells that produced mutually beneficial products. Colonies of such cells (essentially a uniform organic "tissue") eventually gave way to mixed colonies of different types of cells, which came to function as "organs" in larger organisms, both plants and animals. By the late Precambrian, rather complex soft-tissued organisms must have been abundant, but such delicate creatures (imagine a jellyfish or slug) are rarely preserved as fossils. Nonetheless, many late Precambrian jellyfish and segmented worm fossils are now known. Predation must have become a serious problem during the late Precambrian. Two types of hard parts arose and proliferated through natural selection in the early Cambrian: teeth or rasps suitable for attacking other organisms and shells suitable for protecting organisms from attack. The Cambrian period (570 to 500 Ma BP) was marked by the emergence of a wide variety of molluscs with shells of calcium carbonate (the *pelycypods*, or bivalves, and the *gastropods*, or snails) or of calcium phosphate (the *brachiopods*, or lamp shells). Creatures morphologically similar to modern horseshoe crabs, the *trilobites*, abounded in the Cambrian seas, as did sponges, worms, coelenterates similar to corals and anemones, echinoderms rather like modern sea cucumbers, and early scorpion-like arachnids. The middle Cambrian was noteworthy for the extraordinary diversity and richness of life contained in it. The great success of animals at this time was surely related to the rise of oxygen, which made much more energetic modes of metabolism available. In places in western North America, Cambrian sediments attain a thickness of about 4 km.

During the Ordovician period (500 to 430 Ma BP) many species of trilobites, corals, and colony-forming bryozoans flourished. Crinoids (sea lilies; echinoderms related to starfish, sand dollars, sea urchins, and sea cucumbers) were common. The dominant life-forms of the time were large molluscs, including huge nautiloids with straight conical shells extending up to 4 m in length. The nautiloids and other *cephalopods* such as the squid and octopus were to remain very important until as recently as 65 Ma BP.

The Silurian period (430 to 395 Ma BP) saw a continuing proliferation of the families and genera that were established in the Ordovician. Perhaps the most striking innovation of this period was the emergence of plants onto the land beginning in the latest Ordovician, probably as an adaptation fostered by the pressure to escape being eaten by animals. Algal colonies, crinoids, and corals flourished. Huge scorpion-like *eurypterids*, related to modern horseshoe crabs, grew to lengths of about 3 m, and jawless (*agnathan*) fishes gained prominence.

The Devonian period (395 to 345 Ma BP) was distinguished by the great proliferation of fishes, land plants, and amphibians. It is at this time that animals first emerged onto the land. The agnathan fishes (the *ostracoderms*), plated fishes (*placoderms*), bony fishes (*osteichthyes*), and sharks diversified and prospered, culminating in the 10-m placoderm Dinichthys. The lobe-finned fishes (*crossopterygians*) gave rise to the first pioneering amphibians (*ichthyostegids*). True millipedes and spiders appeared among the tree ferns, lycopods (scale trees), and rushes in swamps and on land. The primitive "trees" and "bushes" of this era, lacking true roots and leaves, bore more resemblance to modern tropical tree ferns and gigantic mosses than to modern trees. Molluscs, including brachiopods, cephalopods, and pelycypods, continued to be abundant, and coral reefs were common. Trilobites were driven to the verge of extinction.

During the Mississippian period (345 to 320 Ma BP) the proliferation of corals, crinoids and the closely similar blastoids, brachiopods, and colonies of tiny bryozoans continued. On the land, forests continued to spread throughout all moist areas, and the amphibians followed. The culmination of the spread of land plants in the Pennsylvanian period (320 to 280 Ma BP) led to the formation of vast coal deposits. Forests of giant scale trees and seed ferns (*pteridosperms*) up to 50 m in height were widespread. Insects, including 30-cm cockroaches and protodragonflies with 90-cm wingspreads, lived in these forests and left occasional fossil remains in the coal. Diversification of the amphibians led to the appearance of several types of primitive reptiles.

The Permian period (280 to 225 Ma BP) marked the transition from Pennsylvanian-type wet forests to primitive conifers, with an extension of the range of land plants to ever drier terrains. Insects, fishes, and amphibians did well, and the reptiles proliferated wildly. Reptiles diversified into a variety of forms, most notably the squat *Seymouria*, which looked like a cross between a lizard and a frog, and the sail-backed carnivore *Dimetrodon*. Some reptiles, such as the crocodile-like mosasaurs, found attractive niches in the sea. A minor branch of the reptiles, the *theriodonts*, prospered as small carnivores. Unlike the early amphibians and reptiles, the theriodonts walked erect upon their legs, rather than sprawling on the ground. This mode of locomotion must have enhanced their agility and competitive success. The close of the Permian marks the greatest extinction event in the history of Earth and closes the pages of the Paleozoic ("ancient life") era. This extinction event, which coincided with the beginning of the breakup of the Pangaea supercontinent, saw the disappearance of many families of foraminifera, crinoids, cephalopods, bryozoans, corals, and brachiopods. The trilobites and blastoids utterly vanished.

The Mesozoic ("middle life") Era (225 to 65 Ma BP) was the great age of the reptiles, when dinosaurs ruled the Earth unchallenged for 160 million years. The Triassic period (225 to 190 Ma BP) saw the rise of the reptiles to dominance over the dry land and the emergence of the first dinosaurs. In the wake of massive extinctions in the sea, the coiled cephalopod *ammonites* rose rapidly in prominence and diversified and multiplied to an astonishing extent. *Ichthyosaurs*, looking like toothy dolphins, ruled the seas and ate unwary ammonites. Modern-looking corals, crinoids, and arthropods resembling modern lobsters appeared in the sea. On land, cycads and conifers prospered in dry terrains, as did ferns and rushes in swamps.

During the Jurassic (190 to 136 Ma BP) the three main groups of dinosaurs reached their first culmination. The *sauropods*, including Diplodocus and Apatosaurus (formerly called Brontosaurus), with very long necks and tails, elephantine legs, and negligible intellects, were the largest and tallest herbivores. The *stegosaurs*, with heavily armored and spiked backs and tails and 100-g brains, better adapted to grazing on low vegetation on dry land than to browsing, also failed to develop higher mathematics. The *theropod* carnivores, such as Allosaurus, hung out on street corners and ate the herbivores. The duck-billed herbivorous dinosaurs also appeared at this time, but had not yet reached prominence. The first flying reptiles appeared. In the oceans there was an incredible proliferation of huge marine reptiles. Some ammonites reached nearly 2 m in diameter. *Belemnites*, squidlike cephalopods with internal calcareous shells, appeared in profusion. Rat-sized true mammals put in their first inauspicious appearance, and the first true bird, Archaeopteryx, flew. The forests were dominated by cycads, conifers, and ferns and hosted a vast variety of insects. Many species of ginkgo trees were present, of which only a single species survives today.

The Cretaceous period (136 to 65 Ma BP) was the last great age of the dinosaurs. The varied ranks of dinosaurs were joined by the *ceratopsians* (boneheads) such as Triceratops, tanklike *ankylosaurs*, and large duck-billed *ornithopods*. Huge plesiosaurs up to 13 m in length roamed the seas. The sauropods had by this time passed their prime, but the flying reptiles, such as Pteranodon, prospered; Pteranodon sometimes attained a wingspread of 8 m. The flowering plants (*angiosperms*) appeared in the lower Cretaceous and came to dominate the surface of the planet by the end of the period. Many of today's dominant species of trees, including oaks and maples, willows, birches, poplars, and beeches, originated in the Cretaceous. Animals and insects adapted to the new sources of high-quality food provided by fruits, grains, nuts, berries, and seeds. High-metabolic-rate animals such as small mammals and birds were especially helped by this innovation.

The entire rich tapestry of Cretaceous life was torn asunder by a giant impact event 65 Ga ago. An asteroidal body traveling roughly 20 km $s^{-1}$ struck the continental shelf on the north slope of Mexico's Yucatan peninsula, driving tsunami waves deep into the interior of the continent. The sea surged back into the red-hot 180-km cavity left by the impact and boiled vigorously. Vaporized target rock rich in carbonates and sulfates fed a huge fireball with carbon dioxide and sulfur dioxide. Rock vapor in the fireball began to condense to a fine dust. Haiti, then much closer to the Yucatan than it is today, was battered by towering waves and buried under meters of crater ejecta. Rock ejected at near orbital speed from the impact crater flew all over the Earth, liberating so much energy upon reentry that the sky heated up to over 1000°C. The forests of Earth, grilled by several minutes of intense radiation, burst into flames, raising immense globe-girdling clouds of soot, and burned to the ground. The burning sky went black with dust and soot. Some 90% of the species on Earth were destroyed, including the last dinosaurs. Of all the wealth of animal life of the Cretaceous, only small burrowing animals massing less than about 20 kg survived to face a world permanently and dramatically altered. Among them were our ancestors. Total elapsed time: about 45 min.

Over the following months, acid rain charged with sulfuric, nitric, and nitrous acids deluged the continents and poisoned the seas, washing vast amounts of iridium-bearing asteroidal dust from the skies. Huge amounts of nitrates and organic matter from decaying plants fecundated the oceans, sparking an explosive proliferation and diversification of life. And the lowly mammals, at last given an opportunity, became masters of the world.

The total elapsed time since the Cretaceous, only 65 million years, is less than 1.5% of the lifetime of Earth; indeed, it is shorter than the duration of the Cretaceous. In this *Cenozoic Era* (recent life) the biosphere reestablished its dynamic balance with mammals at the top of the food chain. During the Tertiary period (65 to 2 Ma BP) the whales replaced plesiosaurs; huge grazing mammals such as Megatherium, Titanotherium, and elephants replaced the sauropods; and the great cats such as the saber-toothed tiger, Smilodon, replaced the theropods. The meek inherited Earth—and some of them ceased to be meek.

The last 2 million years constitute the Quaternary period, within which humans became the dominant higher life form on Earth. The anthropoid apes, which had little influence on the world of 5 million years ago, produced one offshoot species called *Homo sapiens sapiens*, the most complimentary name ever given to a species. Typically, they awarded this compliment to themselves.

## Earth: Geochemistry and Petrology

The average elemental composition of the crust is compared with the bulk composition of Earth in Table X.5. We have already remarked on the enhancement of the abundances of Si, Al, Ca, and the alkali metals in the crust relative to the rest of the planet. Consideration of the abundances of all the elements in the crust and in solar material leads to a lengthy list of elements enriched in the crust. Since the crust makes up only about 0.4% of the mass of Earth, elements that are perfectly partitioned into the crust would have crustal concentrations 250 times as high as the planetary average. It would be easier to report compositional differences between the crust and the mantle by emphasizing which elements are enriched in the much more massive mantle; the only major element enriched in the mantle is magnesium. Removal of the core-forming elements from Table X.5 (metallic iron and nickel, P, and S) suggests that the mantle is dominated by magnesium and ferrous iron silicates.

A significant generalization that helps to explain the partitioning of elements between the crust and the mantle is that elements that substitute readily for magnesium in minerals are easily accommodated in the mantle. The species that readily substitute for $Mg^{2+}$, called the *compatible* elements, are usually those that have ionic

**Table X.5** Major-Element Composition (wt %) of Earth's Crust

| Element | Bulk Earth | Earth's crust | Element | Bulk Earth | Earth's crust |
|---|---|---|---|---|---|
| O | 30 | 46.6 | Mg | 14.9 | 2.1 |
| Si | 14.9 | 27.7 | Ti | 0.05 | 0.4 |
| Al | 1.4 | 8.1 | H | 0.003 | 0.14 |
| Fe | 35 | 5.3 | P | 0.35 | 0.12 |
| Ca | 1.6 | 3.6 | Mn | 0.3 | 0.10 |
| Na | 0.2 | 2.8 | S | 1.9 | 0.05 |
| K | 0.02 | 2.6 | Ni | 2.4 | <0.01 |

charges and radii that are closely similar to those of magnesium (2+ and 0.78 Å, respectively). The ferrous ion (2+; 0.82 Å) is a very close match. The only other very abundant divalent cation, $Ca^{2+}$, with an ionic radius of 0.99 Å, is not nearly as good a match. (Note from Table X.5 that Ca evidently is not strongly fractionated between the crust and the mantle.) In principle, any ion closely similar to $Si^{4+}$ (4+; 0.39 Å) should also be accommodated by mantle silicates, but it happens that no other abundant element occurs in this valence state. An interesting aspect of this partitioning behavior is that [$Al^{3+}Al^{3+}$] can substitute for [$Mg^{2+}Si^{4+}$] without causing a charge-balance problem. Thus, if the ionic radii permit, $MgSiO_3$ can serve as a host for dissolved $Al_2O_3$. The aluminum ion (3+; 0.57 Å) is indeed intermediate in size between $Fe^{2+}$ and $Si^{4+}$; in fact, the *average* radius of these two ions is 0.58 Å, virtually identical to that of aluminum. Note from Table X.5 that aluminum is enriched in the crust by only a factor of 5 above the average for the planet, far smaller than that for the truly *incompatible* elements such as the rare earths and alkali metals. Examples of ions strongly incompatible with the mantle include lead (4+; 0.84 Å), barium (2+; 1.35 Å), silver (1+; 1.26 Å), lanthanum (3+; 1.15 Å), mercury (2+; 1.10 Å), lithium (1+; 0.60 Å), sodium (1+; 0.95 Å), cesium (1+; 1.69 Å), and cerium (4+; 1.01 Å). The rare gases are electrically neutral and have large radii (1.12 and 1.90 Å for Ne and Xe, respectively) and hence are also excluded from the mantle. Electronegative elements can be retained in the mantle if they can substitute readily for oxygen (2−; 1.40 Å). The most abundant candidate elements for oxygen substitutes in solar material are sulfur (2−; 1.84 Å in sulfides), chlorine (1−; 1.81 Å in chlorides), and fluorine (1−; 1.36 Å in fluorides). Some fluoride substitution for O is expected, but the other anions are much less compatible.

Because of the abundances of the major elements in the crust and mantle, it is inevitable that most igneous rocks found at Earth's surface should be dominated by a relatively small number of common minerals. These include olivine, pyroxene, free silica, feldspars, and more complex silicates, such as hornblende and biotite mica, which are the principal crustal carriers of igneous water and ferric iron. Igneous rock types are assigned names based on two traits, mineralogical makeup and crystal grain size. The most important variable in the rock composition is the total silica ($SiO_2$) content. Likewise, the most important factor in determining the mineral grain size is the rate of cooling of the parent magma; very slow cooling, as occurs when massive lenses of magma intrude into a deeply buried body of hot rock, leads to large crystal sizes. Conversely, magmas erupted or extruded by volcanic activity have very little time to grow crystals. The rocks formed by extrusion have fine-grained textures, and indeed those formed by violent eruption may solidify with most of their mass remaining as a supercooled liquid (i.e., as glass).

Figure X.47 presents the terminology for the major types of igneous rocks. Mafic rocks, with their low silica content, are already familiar to us in the form of lunar, asteroidal, Martian, and Venusian basalts. Ultramafic rocks such as dunite, which is nearly monomineralic olivine, are rare both among meteorites and on Earth's surface, because these iron-rich rocks are much denser than typical crustal material and cannot rise to the surface unless propelled by violent, deep-seated eruptions. Silica-rich rocks are present on Venus but apparently extremely rare on the Moon; they are probably also present on Mars. But, aside from a single chip of lunar silica-rich rock, all of our practical experience with highly silicic rocks is on Earth. Granites, with very high contents of silica, alkali metals, rare earths, and volatiles, are among the commonest igneous rocks on our planet.

*Petrology*, the study of the composition and origin of rocks, was pioneered by the American geochemist N. L. Bowen early in this century. Bowen was in part motivated by the observation that very similar *magmas* (rock liquids) sometimes gave rise to very different rock types. It seemed clear that temperature, pressure, and cooling rate all had an effect on the mineral content of the forming rock, but the relationships were at first obscure. Bowen chose to investigate the effects of these factors in the laboratory rather than solely by observations in the field. He thus became the founder of experimental petrology.

Bowen knew that rapidly cooled magmas would become very viscous with falling temperature, because the viscosity drops off as $\exp(-\Delta E^*/kT)$, where $\Delta E^*$ is the activation energy for bond breaking in the magma. Crystals grow only slowly from cool magmas because the silicate ions in the melt form ever longer and more complex chains and sheets as the temperature falls, and it becomes ever more difficult to accommodate these complex and disorderly units into orderly crystals. The growth in melt complexity toward lower temperatures is in fact the cause of the increase in viscosity. Thus rapidly cooling melts will have little success in building large crystals, and the small crystals will have little success in separating from the viscous melt in accordance with their densities. The final rock will then have a very fine-grained texture and will be almost identical in composition to the parent magma. A basaltic magma erupted by a volcano might begin to precipitate magnetite and olivine when the temperature drops to about 1400 K and then form small crystals of pyroxene near 1360 K. At about 1250 K hornblende would also crystallize, but hornblende is formed by

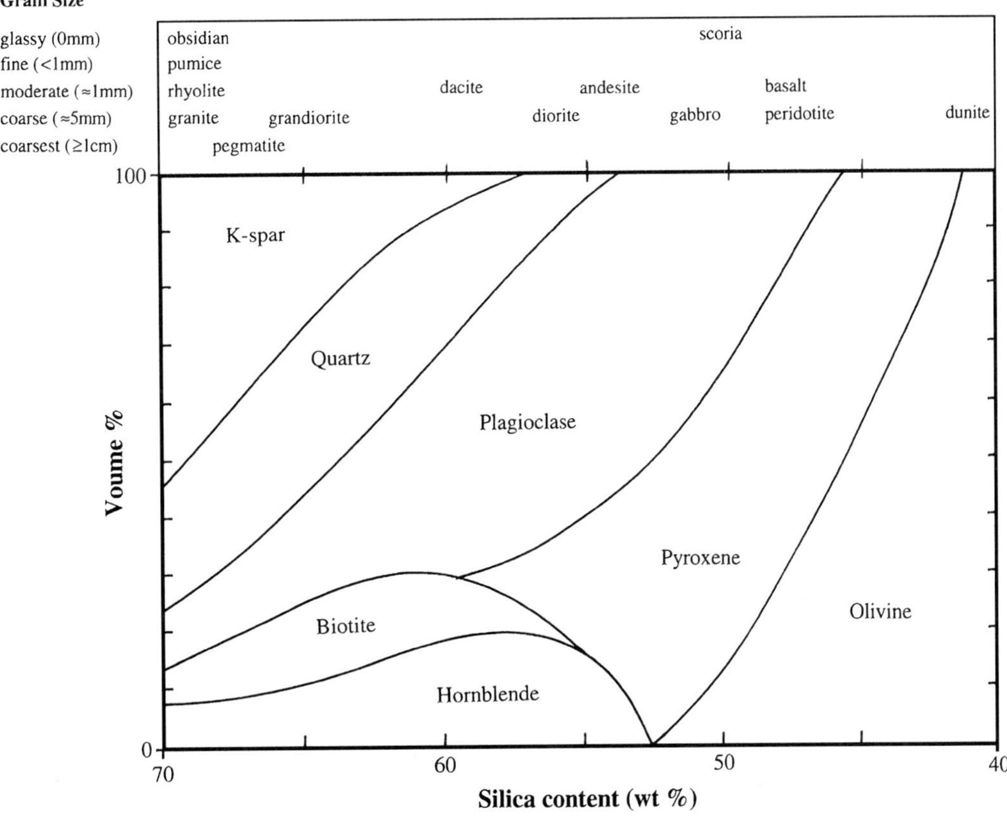


**Figure X.47** Igneous rock types. The mineralogical composition of terrestrial crustal igneous rocks is shown as a function of silica ($SiO_2$) content in the lower panel. The names of the corresponding rock types, reflecting both composition (mineralogy) and grain size (texture), are given at the top. For example, a rock containing 48% silica, consisting of olivine, pyroxene, and plagioclase, with a grain size less than 1 mm, is a basalt. A glass containing 65% silica is called obsidian. The same material, crystallized into centimeter-sized grains of K-spar, plagioclase, quartz, biotite, and hornblende, would be called a pegmatite.

reaction of olivine with the remaining melt; olivine will then disappear as hornblende crystallizes. At about 1200 K the pyroxene disappears through further reaction with the melt, and at about 1150 K the mineral sphene ($CaTiO_3$) appears along with the surviving hornblende and magnetite. Plagioclase forms at 1090 K, and crystallization is complete by about 1050 K. The last-crystallizing material is the low-melting alkali feldspar; during cooling the feldspar composition changes from Ca rich (anorthite) to Na rich (albite + anorthite = plagioclase). The same basic principle applies to crystallization of basaltic magmas on other Solar System bodies, except that elsewhere in the Solar System (the Moon and basaltic asteroids, Mercury, and possibly Venus) ferric iron is normally very rare or absent in igneous magmas. Magnetite and hornblende therefore do not form, and the dominant minerals after cooling are plagioclase and pyroxene, often with some olivine.

Bowen referred to this kind of crystallization sequence, in which the minerals do not separate according to density, but instead stay in intimate contact and continue to react with each other during cooling, as a *continuous reaction series*. This kind of process is appropriate for rapidly cooling magmas extruded onto the surface, but not for magmas that are intruded into hot country rocks at great depth, where cooling is very slow.

Consider a deeply buried, slowly cooling magma chamber of the same initial composition as above. The first minerals to form, magnetite and olivine, are silica poor, very rich in iron, and therefore much denser than the melt. When cooling is very slow, these dense minerals sink to the bottom of the magma chamber and accumulate as an ultramafic layer beneath the cooling body of magma. They are then effectively removed from further chemical interaction with the melt: they can neither be reabsorbed by the melt nor react with it to produce other minerals. A rock formed in this manner is called a *cumulate*.

The next mineral to crystallize during slow cooling is pyroxene, which has a higher silica content than the previous precipitates, but also has a high ferrous iron content. The pyroxene is also denser than the melt and will settle to the bottom of the magma chamber atop the olivine and magnetite. The residual melt, now rich in incompatible elements, also has quite a low density because of its content of silica, sodium, aluminum, and water. Further cooling will precipitate amphiboles, a family of chain silicates with the basic formula $Ca_2Mg_5Si_8O_{22}(OH)_2$, of which the one precipitating in the present example is hornblende. Next comes biotite mica and then K-spar, muscovite mica, and quartz. Thus a granitic rock can be produced by slow cooling of an initially basaltic magma. The high silica content is realized by the simple expedient of precipitating the basic rocks first and removing them from interaction with the cooling, silicic magma. The sequence of rock types laid down by this process, from basalt to andesite and then to rhyolite, constitutes Bowen's *discontinuous reaction series*.

Sedimentary rocks exhibit an extremely wide range of compositions, reflecting the great diversity of processes and source materials contributing to their genesis. The chemical composition of sediments includes chemical sediments formed by evaporation of aqueous solutions (*evaporites*), such as NaCl (the mineral halite) and other salts.

Evaporites are usually formed by evaporation of seawater. In many places on Earth, seawater evaporation is an industrial process for the manufacture of salt, and the process of evaporite formation has been studied extensively both in commercial seawater evaporation basins and in the laboratory. Seawater is chemically very complex, and the detailed sequence of minerals formed during evaporation is somewhat temperature dependent. Nonetheless, the general sequence of mineral products is well established. The first mineral to precipitate during evaporation of normal seawater is calcium carbonate, as the mineral calcite or aragonite. Next to precipitate is calcium sulfate, either as the mineral anhydrite or as the hydrate $CaSO_4 \cdot 2H_2O$, which is the familiar mineral gypsum. When about 90% of the water has been lost, the mineral halite (NaCl; rock salt) precipitates. When the solution has been reduced to only about 1.5% of its original volume, sylvite (KCl), magnesium chloride, and sulfate salts finally precipitate.

The early evaporite minerals, the calcium carbonates and sulfates, are the materials that precipitate from "hard" (mineral-rich) water inside household plumbing. This material is generally referred to as "scale" or, less accurately, "lime." Lime is actually pure CaO, so the use of this word by plumbers must be taken with a grain of salt.

There are also abundant carbonate and phosphate sediments deposited by biological mechanisms such as precipitation by molluscs to form shells and by calcareous algae to form stromatolites. The most common carbonate minerals are $CaCO_3$ (calcite), the dominant mineral in limestone, and $CaMg(CO_3)_2$, dolomite, the principal mineral in the rock of the same name. Phosphates are deposited by both planktonic organisms and brachiopods. The most common phosphate minerals are $Ca_5(PO_4)_3X$ (where $X$ can be OH, Cl, or F), apatite, and various hydrated and anhydrous forms of $Ca_3(PO_4)_2$. Other biogenic sediments include organic matter, often cooked at high pressures into natural gas, crude oil, soft (bituminous) coal, or hard coal (anthracite).

Other substances that were deposited abundantly from the ancient oceans, but which are negligible in modern sediments, include amorphous, partially hydrated silica (opal and chalcedony, found in the rock named chert) and iron oxide minerals, as seen in banded iron formations. The major iron precipitates, formed in BIFs during oxidation of the Precambrian ocean, include hematite ($Fe_2O_3$), limonite (FeOOH), and siderite ($FeCO_3$).

Metamorphic rocks are produced by moderate heating of a previously existing rock at either the local or the regional scale. Local invasion of country rock by an igneous intrusion, called *contact metamorphism*, produces a steep temperature gradient on a small scale. Deep burial of a large body of rock for prolonged periods of time results in *regional metamorphism*. Most products of regional metamorphism are massive bodies of foliated (layered) rocks, most often metamorphosed sediments, in which the recrystallization of relatively low-temperature minerals such as micas marks the rock layers. The most coarse-grained foliated products of regional metamorphism are called *gneiss*, those with intermediate grain size are called *schist*, and those with the finest grain size are called *slate*. The grain size may be determined either by that of the parent rock or by the degree of recrystallization experienced during the time of deep burial. Rocks produced by contact metamorphism are usually not foliated. The finest-grained unfoliated metamorphic rocks are called *hornfels*, and the coarser rocks are called *granulite*. Several other types of metamorphic rocks are found, of which three are made of essentially a single mineral: marble, made of calcite, is simply metamorphosed limestone; quartzite, made from quartz sandstone; and amphibolite, made of amphiboles. Finally, some are strongly influenced by crushing, folding, and faulting, such as the *mylonites*. Stronger heating of metamorphic rocks, sufficient to cause partial melting, transforms them into igneous rocks.

## Weathering in the Rock Cycle

Many minerals, when exposed to water and air, are subject to chemical attack. Iron and aluminum oxides are very insoluble and unreactive in a wet, oxidizing environment in the absence of strong acids or bases. The acidity or alkalinity (basicity) of aqueous solutions is determined by the relative concentrations of hydrogen ions (casually referred to as $H^+$, although this species in water instantly hydrates to form the *hydronium ion* ($H_3O^+$) and hydroxide ions ($OH^-$). Pure liquid water is very slightly ionized according to the equilibrium

$$2H_2O \leftrightarrow H_3O^+ + OH^- \quad (X.106)$$

$$K_{106}(25°C) = [H_3O^+][OH^-] = 10^{-14}. \quad (X.107)$$

Since charge is conserved, the concentrations of hydronium and hydroxyl ions (normally expressed by $M$, moles of solute per liter of solution) must be equal,

$$[H_3O^+] = [OH^-], \quad (X.108)$$

whence

$$[H_3O^+]^2 = 10^{-14}. \quad (X.109)$$

The normal measure of acidity is the pH unit:

$$pH = -\log[H_3O^+]. \quad (X.110)$$

Thus the pH of pure water at room temperature is 7. An aqueous solution containing a strong acid with a hydronium ion concentration of 1 $M$ [and a hydroxyl concentration of $10^{-14}$, according to Eq. (X.107)] would have pH 0, and a strong base with a hydroxyl concentration of 1 $M$ (hydronium ion concentration of $10^{-14}$) would have pH 14.

Dissolved species with higher ionization constants than water's may dominate the pH of their solutions. Normal rainwater is slightly acidic because of the presence of dissolved carbon dioxide:

$$CO_2 + H_2O \leftrightarrow H_2CO_3 \quad (X.111)$$

$$H_2CO_3 + H_2O \leftrightarrow H_3O^+ + HCO_3^- \quad (X.112)$$

$$HCO_3^- + H_2O \leftrightarrow H_3O^+ + CO_3^{2-}. \quad (X.113)$$

In the first reaction, carbon dioxide dissolves in water to make the neutral carbonic acid molecule. This is the reaction by which soft drinks are "carbonated." The equilibrium constant for the second reaction (producing bicarbonate ion) at normal temperatures is about $10^{-3}$, so the spontaneous ionization of a 0.01 $M$ solution of carbonic acid would give

$$K = 0.001 = [H_3O^+][HCO_3^-]/[H_2CO_3] \quad (X.114)$$

$$[H_3O^+][HCO_3^-] = 10^{-5}. \quad (X.115)$$

The equilibrium constant for the third reaction is only $10^{-5}$, so the concentration of carbonate ion is usually low unless the hydronium ion concentration is extremely small (the solution is very alkaline). The acidity of rainwater can be greatly enhanced by natural phenomena such as the emission of $SO_2$ by volcanic activity. Sulfur dioxide is readily oxidized and hydrated in normal air to make $H_2SO_4$, sulfuric acid. Burning fossil fuels releases great quantities of $CO_2$ and smaller but very damaging amounts of sulfuric and nitric acid into the air. Combustion also heats air to a high enough temperature that nitrogen partially oxidizes to make NO and other oxides, which hydrate to make nitric and nitrous acids:

$$N_2 + O_2 \leftrightarrow 2NO \quad (X.116)$$

$$2NO + O_2 \leftrightarrow 2NO_2 \quad (X.117)$$

$$2NO_2 \leftrightarrow N_2O_4 \quad (X.118)$$

$$N_2O_4 + H_2O \leftrightarrow HNO_3 + HNO_2. \quad (X.119)$$

Both nitric and sulfuric acids are *strong acids*, meaning that they are almost completely ionized in aqueous solutions, unlike the *weak acid* $H_2CO_3$. The lowering of the pH of rainwater by such natural and anthropogenic contaminants is called *acid rain*.

Acidic rainwater readily attacks many common minerals. For example, consider calcite in limestone or marble. Calcite spontaneously dissolves to a very limited extent in pure water by forming the calcium and carbonate ions:

$$CaCO_3 \leftrightarrow Ca^{2+} + CO_3^{2-}. \quad (X.120)$$

The equilibrium constant for this type of solution reaction,

$$K_{120} = [Ca^{2+}][CO_3^{2-}], \quad (X.121)$$

is called the *solubility product*. The carbonate ion reacts readily with hydronium to make bicarbonate as above [Eq. (X.111)]. According to Eqs. (X.111) and (X.112), addition of carbon dioxide gas provides more hydronium, which converts more carbonate ion to bicarbonate and hence lets more calcite dissolve.

Feldspars are only slightly less reactive on Earth's surface. Consider pure albite ($M^+ = Na^+$) or orthoclase (K-spar; $M^+ = K^+$) leached by slightly acidic rain to extract alkali metal ions into solution:

$$2H_3O^+ + 2MAlSi_3O_8 \rightarrow 4SiO_2(aq) + 2M^+ + Al_2Si_2O_5(OH)_4 + H_2O. \quad (X.122)$$

Here the clay mineral kaolinite and an aqueous solution containing silica and alkali metal ions are produced.

Because of the limited solubility of silica, it will readily precipitate as gel-like silicic acid $H_4SiO_4$, which in turn dehydrates to make chert or chalcedony and eventually quartz. Note that the acidity of the solution is diminished by the reaction. Both quartz and clay minerals are rather resistant to further weathering reactions.

What igneous minerals would be *least* stable in the moist, oxidizing atmosphere of Earth? Those that contain ferrous iron and no water, including pyroxene and especially olivine. Olivine, because of its density, is hard to transport to the surface. Once on the surface, it rapidly weathers away. One would suspect that olivine would be a relatively rare and valuable substance on Earth's surface, despite its chemical similarity to the mantle. In fact, olivine crystals are sold commercially as the green semiprecious gem peridot.

The products of weathering reactions on rock are called *soils*. Thorough leaching and weathering of surface rock in areas of abundant rainfall remove soluble materials and leave a residue of quartz, clays, and iron oxides. Alkali metals and calcite would be essentially absent. Such soils are known as *pedalfers* as a reminder of their high aluminum and ferric content. Weathering in dry climates fails to remove calcium effectively (the bicarbonate ion in aqueous solution is the agent of weathering of carbonates). Such soils are called *pedocals*. In the tropics the silicates are completely altered, leaving only oxides of aluminum and iron. Even silica has been leached away. This soil type is called *laterite*.

The direct effect of weathering reactions is to consume certain reactive atmospheric and hydrospheric species. Water is consumed by many of these reactions to form hydroxides and clay minerals. Weathering reactions of ferrous minerals also involve consumption of oxygen and oxidation of iron to the ferric state:

$$2FeSiO_3 + \tfrac{1}{2}O_2 + H_2O \rightarrow 2FeOOH + 2SiO_2 \qquad (X.123)$$

$$2Fe_2SiO_4 + O_2 + 2H_2O \rightarrow 4FeOOH + 2SiO_2. \qquad (X.124)$$

Other reactive gases such as $SO_2$ and HCl are removed from the atmosphere by dissolution in rainwater and react with minerals to produce chlorides and sulfates. Oxidation of $SO_2$ to form highly hygroscopic $SO_3$ leads to rapid formation of sulfuric acid, which is the reactive sulfur species in rainwater. Nitrogen oxides, which react with oxygen and water vapor as discussed above to make nitric acid, suffer a very different fate; fixed nitrogen (compounds of nitrogen with other elements) is a limiting nutrient in the biosphere and is scavenged by biological processes to make amino acids and organic bases. The rare occurrence of nitrates as terrestrial minerals tells the story: potassium nitrate (saltpeter) is found as a product of aerobic oxidation of manure in horse stalls, and sodium nitrate (soda nitre) is found in oxidized bat guano on islands off the coast of Chile.

On Earth, the major inorganic sinks of oxygen are ferrous iron and sulfur. Oxidation of organic matter, both living (metabolism) and dead (aerobic decay), is also very important. By far the largest source of oxygen on Earth is photosynthesis. Thus the composition of the atmosphere is strongly influenced, and in many respects dominated, by biological processes.

## Earth: Atmospheric Composition and Cycles

The composition of air in Earth's troposphere is given in Table X.6. Water vapor is highly variable in abundance because of the large temperature variations on Earth's surface. At the extreme low temperatures encountered during winter in Siberia and Antarctica, the water mole fraction may drop as low as a few parts per million. Under tropical monsoon conditions, water vapor mole fractions as high as 4% can be produced locally. Because of this large variability of an important constituent, all the abundances of noncondensible species in Table X.6 are given for *dry air*.

A number of other gases are variable in abundance, especially biogenic and lightning-produced gases such as carbon dioxide, methane, hydrogen, ammonia, hydrogen sulfide, the nitrogen oxides, and nitric acid. The equilibrium composition of heated air containing 1% water vapor is shown in Fig. X.48. Note the predominance of NO over the other nitrogen oxides at high temperatures characteristic of flames or shock waves. Quenching of

**Table X.6** Composition of Earth's Troposphere

| Species | Mole fraction | Species | Mole fraction |
|---|---|---|---|
| $N_2$ | $7.81 \times 10^{-1}$ | $NO_2$ | $<3 \times 10^{-9}$ |
| $O_2$ | $2.09 \times 10^{-1}$ | NO | $<3 \times 10^{-9}$ |
| $^{40}Ar$ | $9.34 \times 10^{-3}$ | HCl | $<1.5 \times 10^{-9}$ |
| $H_2O$ | $0\text{–}4 \times 10^{-2}$ | $HNO_3$ | $<1 \times 10^{-9}$ |
| $CO_2$ | $2\text{–}4 \times 10^{-4}$ | HOOH | $1 \times 10^{-9}$ |
| Ne | $1.82 \times 10^{-5}$ | $CH_3Cl$ | $6 \times 10^{-10}$ |
| $^4He$ | $5.24 \times 10^{-6}$ | COS | $5 \times 10^{-10}$ |
| $CH_4$ | $1\text{–}2 \times 10^{-6}$ | $CF_2Cl_2$ | $2.8 \times 10^{-10}$ |
| Kr | $1.14 \times 10^{-6}$ | $H_2SO_4$ | $1 \times 10^{-10}$ |
| $H_2$ | $4\text{–}10 \times 10^{-7}$ | $CFCl_3$ | $1.7 \times 10^{-10}$ |
| $N_2O$ | $3.0 \times 10^{-7}$ | $CCl_4$ | $1.2 \times 10^{-10}$ |
| CO | $1\text{–}20 \times 10^{-8}$ | $CH_3CCl_3$ | $1.2 \times 10^{-10}$ |
| Xe | $8.7 \times 10^{-8}$ | $CHCl_2F$ | $1.4 \times 10^{-11}$ |
| $O_3$ | $<5 \times 10^{-8}$ | $CH_3I$ | $1 \times 10^{-11}$ |
| $NH_3$ | $<2 \times 10^{-8}$ | $CHCl_3$ | $9 \times 10^{-12}$ |
| $SO_2$ | $<2 \times 10^{-8}$ | $CH_3Br$ | $5 \times 10^{-12}$ |
| $H_2S$ | $2\text{–}20 \times 10^{-9}$ | Rn | $6 \times 10^{-20}$ |
| $CH_2O$ | $<9 \times 10^{-9}$ | | |

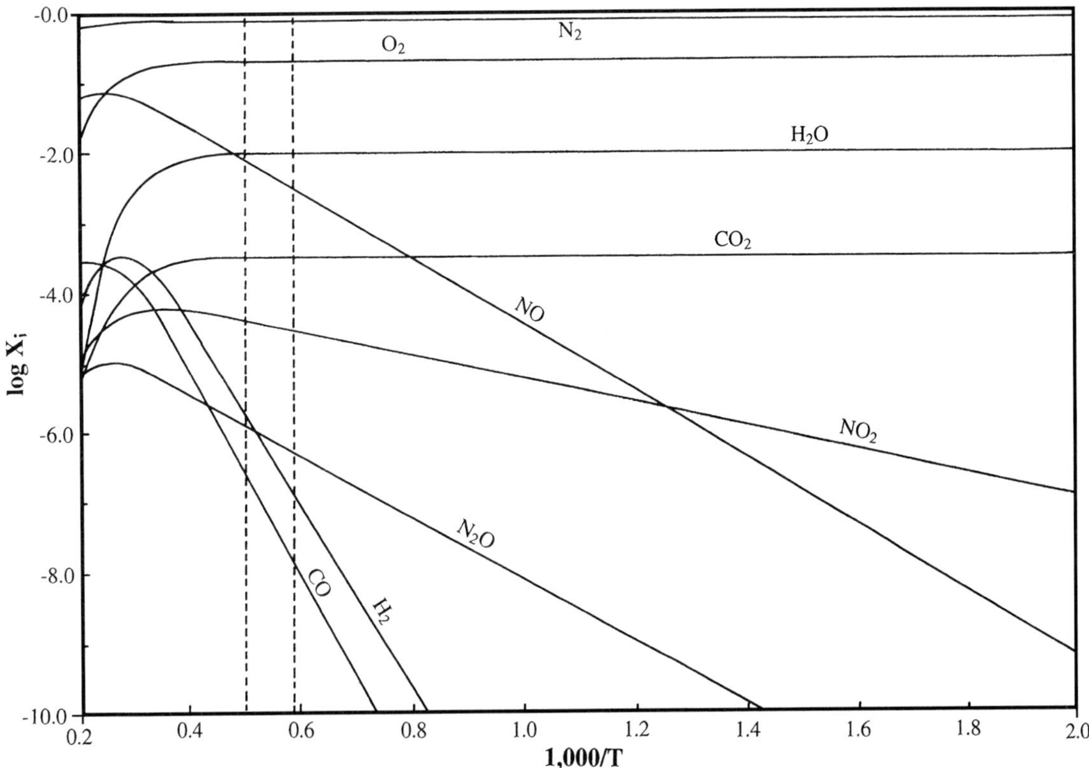


**Figure X.48** Shock-produced gases. The equilibrium mole fractions of gases in strongly heated samples of Earth's atmosphere are graphed as a function of temperature. Rapid cooling will cause the chemical reequilibration of these gases to "freeze in" at a temperature of 1700 to 2000 K, depending on the cooling rate. Thus lightning or shock waves from impact events will produce gases containing 0.3 to 0.9% of NO, a precursor of the other nitrogen oxides and of nitric acid. An impact large enough to shock 1% of the mass of Earth's atmosphere ($5 \times 10^6$ km$^2$ of surface area) would produce a globally averaged $NO_x$ concentration close to 100 ppm.

high-temperature gas composition occurs during rapid cooling at a temperature that depends on the cooling rate, but is commonly between 2000 and 1600 K.

The average abundance of carbon dioxide in the troposphere is about 367 ppm, but the locally measured value near the ground has a pronounced diurnal oscillation due to vigorous removal by daytime photosynthesis and continuous release by plant respiration, and an annual cycle in each hemisphere reflecting the annual cycle of plant growth and decay. The $CO_2$ abundance is also undergoing a secular increase due to the massive burning of fossil fuels.

Anaerobic decay in rice paddies, mud flats, and the guts of termites contributes methane $CH_4$, ammonia $NH_3$, and hydrogen sulfide $H_2S$. The history of the atmospheric abundance of methane can be reconstructed from analyses of bubbles of ancient air trapped in deeply buried ice in the Greenland and West Antarctic ice sheets. Radiocarbon and stratigraphic dating permit direct determination of the age of the ice layer (i.e., the date when it fell as snow) back to about 10,000 years ago. The methane abundance in the atmosphere is found to have averaged about 700 ppb (parts per $10^9$) until about the year 1800. The average slowly grew to about 850 ppb in 1900 and is above 1700 ppb today. The growth in methane reflects a growth in rice paddies caused by the burgeoning population of south and east Asia. Extension of these measurements to longer times is difficult because of the dynamic nature of the large ice sheets: the weight of fresh precipitation constantly depresses the surfaces of the ice sheets, and old ice is extruded from the edges of the sheets into the surrounding oceans, where the ice quickly melts. Ice older than about 10,000 years is very rare and very deeply buried.

A number of anthropogenic gases containing carbon and the halogens fluorine, chlorine, and bromine are listed in Table X.6. These *halocarbons* include methyl chloride $CH_3Cl$, Freon-12 $CF_2Cl_2$, Freon-11 $CFCl_3$, carbon tetrachloride $CCl_4$, methyl chloroform $CH_3CCl_3$, Freon-121 $CHCl_2F$, methyl iodide $CH_3I$, chloroform

$CHCl_3$, and methyl bromide $CH_3Br$. These gases generally have abundances below one part per billion ($1\,ppb = 10^{-9}$), but most of these species are powerful infrared absorbers that contribute significantly to the greenhouse effect. Further, the halogen atoms liberated by UV photolysis of these halocarbons have a powerful effect on the chemistry of the ozone layer. These factors combine to lend these gases an importance far out of proportion to their seemingly minor abundance.

Several other gases in the troposphere are maintained by photochemical processes. Methane oxidation produces carbon monoxide, and hydrogen is produced during UV photolysis of water vapor. Hydrogen peroxide is both produced and destroyed as part of the photochemical processing of water vapor. Ozone $O_3$ is a product of oxygen photolysis and recombination at high altitudes. As on Venus, sulfuric acid is produced by photochemical oxidation and subsequent hydration of sulfur-bearing gases. Finally, volcanic gases contain sulfur dioxide, carbonyl sulfide COS, hydrogen sulfide $H_2S$, and hydrochloric acid and in certain rare locations even ammonia and hydrogen.

Only the rare gases are free of these chemical vagaries; however, radiogenic $^{40}Ar$, $^{4}He$, and fissiogenic krypton and xenon are continuously produced within the Earth and released into the atmosphere, and isotopes of helium can, like hydrogen, escape into space. The time scale for accumulation and escape of radiogenic helium is on the order of $10^6$ years, much longer than the chemical cycling time of a few decades for most of the chemically active species. The active species with the longest overturn time is nitrogen, which cycles in about 17 Ma.

The chemically active atmospheric gases are continually injected into the atmosphere and removed by chemical reactions. For example, we have already introduced several parts of the carbon cycle: dissolution in rainwater, weathering of crustal rocks, inorganic and biogenic precipitation of carbonates, photosynthesis, anaerobic decay, coal and oil formation, fossil fuel burning, and photochemical processing. The flows in this system are summarized in Fig. X.49. Note that the single boxes denote atmospheric species and the double boxes denote surface species.

By far the largest known reservoir of carbon on Earth is the crust, with its enormous supply of carbonate minerals in sedimentary and metamorphic rocks. Approximately 91% of the crustal carbon inventory resides as limestone, dolomite, chalk, and marble, in which the dominant minerals are calcite and dolomite. The other 9% is buried organic matter, mostly coal. Estimates of the total carbonate content of the crust are generally close to 60 bar equivalent pressure of carbon dioxide, compared with 360 mbar of carbon dioxide in the atmosphere. It is very likely that most of the carbon present in Earth is actually in the core; terrestrial smelting operations imply, and metallic asteroids prove, that natural partitioning of the elements during melting and phase separation of metal-bearing rocks leads to efficient extraction of carbon into the metal. Upon cooling, the liquid Fe–Ni–S–O–P–C core-forming material crystallizes to produce major amounts of kamacite, taenite, and troilite (FeS) and minor amounts of schreibersite ($Fe_3P$) and cohenite ($Fe_3C$). Unfortunately the core is not available to us for analysis.

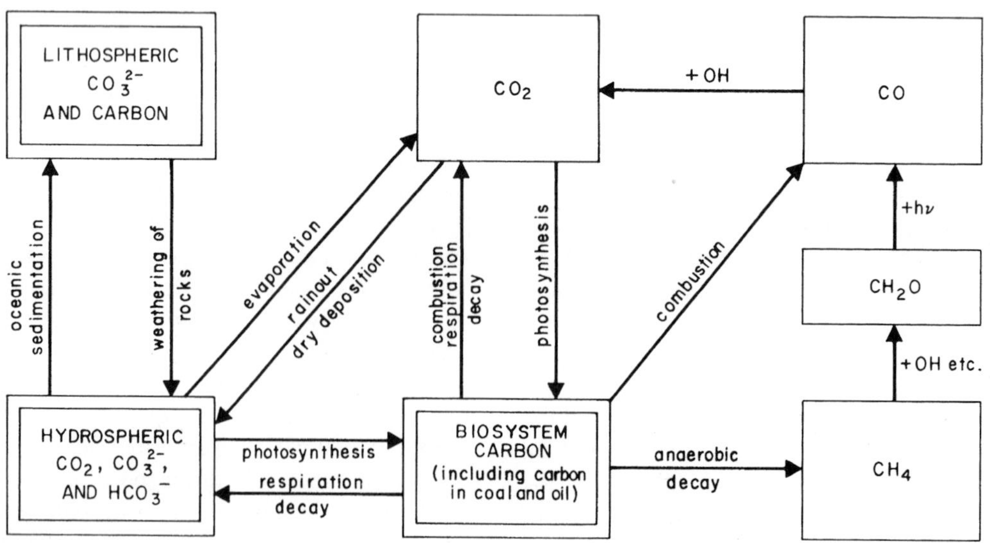


**Figure X.49** The carbon cycle. The single boxes enclose a atmospheric constituents, and the double boxes mark condensed (crustal) species. The production of massive amounts of buried carbon, such as during the Carboniferous, must be accompanied by release of vast amounts of oxygen. From Lewis and Prinn.

Carbonate rocks produced as a result of crustal weathering cannot accumulate indefinitely; all carbon would quickly be sequestered in rocks, and life would cease. Rather, most carbonates deposited in sedimentary settings eventually get buried to great depths by more recent sediment or are forced into subduction zones by plate motions. Regional metamorphism of deeply buried sediments produces marble, and more extreme heating decomposes carbonates to release $CO_2$. Likewise, subducted sediments forced under the margins of continental plates are strongly heated and release volatiles. These volatiles, principally $CO_2$ and water vapor, are released from—and help to drive—volcanic eruptions. The "ring of fire" of volcanoes about the rim of the Pacific basin reflects this activity. In this manner, sedimentary carbon is returned to the biosphere.

Large increases in the abundance of atmospheric $CO_2$ can result either from episodic geological (volcanic) processes or from human industrial activities. The climatological consequences of, for example, doubling the atmospheric carbon dioxide content are poorly understood. Current models suggest that 600 ppm $CO_2$ would enhance the greenhouse effect enough to raise the mean surface temperature of Earth by 2 to 4 K. But these models also suggest that the actual local temperature rise is very dependent upon location and initial modeling assumptions. (James Hansen of the Goddard Institute for Space Studies in New York, a leading climate model-builder, and many colleagues at that and other institutions, recently reported a joint attempt to compare the outputs of competing climate models with the same, or very slightly varied, initial conditions. They found that tiny differences in initial conditions, within the uncertainty of the input data, sometimes led to exponentially divergent predictions.) Further, elevated surface temperatures lead to warming of the oceans and hence to both a lag in the warming of the atmosphere and a secular increase in the rate of evaporation of seawater. This effect in turn will almost certainly lead to enhanced cloud cover and snowfall rates and therefore to a higher albedo for the planet, thus reducing the amount of absorbed solar energy. The net effect of warming of the subarctic and baking of the tropics (if that is what would actually happen) upon the total mass of plants on Earth is not understood. It is simply not clear how effective this "cloud feedback" mechanism is. The enhancement of the greenhouse effect by fossil fuel combustion is real, but the system is so complex that its behavior cannot yet be predicted with confidence.

The geochemical cycle of nitrogen is outlined in Fig. X.50. The cycling of "odd nitrogen," molecular species such as $NO_x$ and ammonia that contain only one N atom, is largely due to microbial processes. Soil bacteria such as *Pseudomonas*, *Bacillus*, *Micrococcus*, and *Thiobacillus* operate under anaerobic conditions to convert rainwater nitrates into $N_2$ (90% yield) and $N_2O$ (10%). This nitrous oxide is destroyed by photolysis at high altitudes in about 50 years, but a small portion is converted into nitric oxide (NO), which in turn oxidizes into nitrogen dioxide ($NO_2$), which ultimately finds its way back down to the troposphere and reacts with water to make nitric acid, which is washed back into the soil by rain. A number of organisms have developed the ability to convert gaseous nitrogen into odd nitrogen. These include *Rhizobium* (which inhabits root nodules on leguminous plants), free-living soil microbes

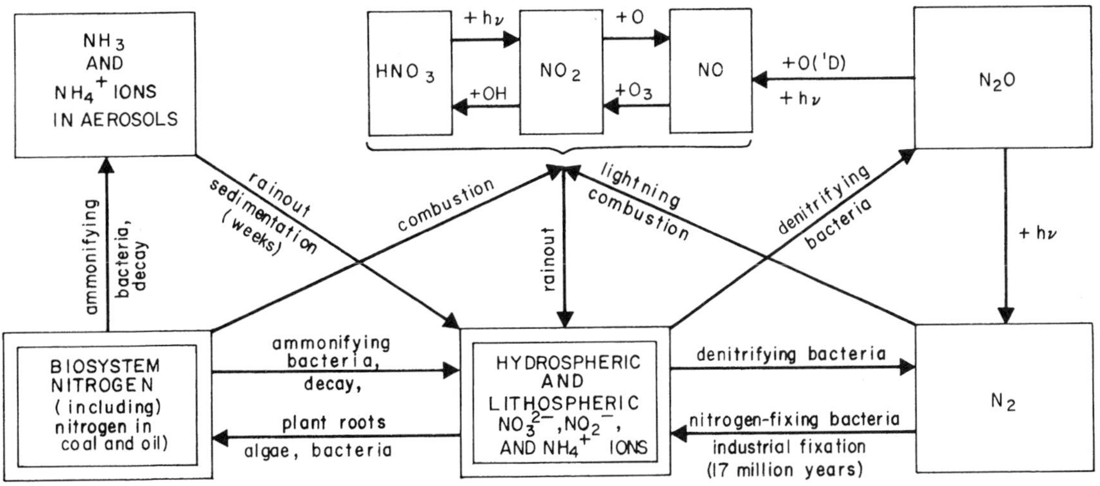


**Figure X.50** The nitrogen cycle. The vast majority of the nitrogen known on Earth is atmospheric $N_2$. The only likely nitrogen reservoir of comparable size is the core, which does not participate in the nitrogen cycle. Note the important role of the rare odd-nitrogen compounds in both inorganic and organic processes. After Lewis and Prinn.

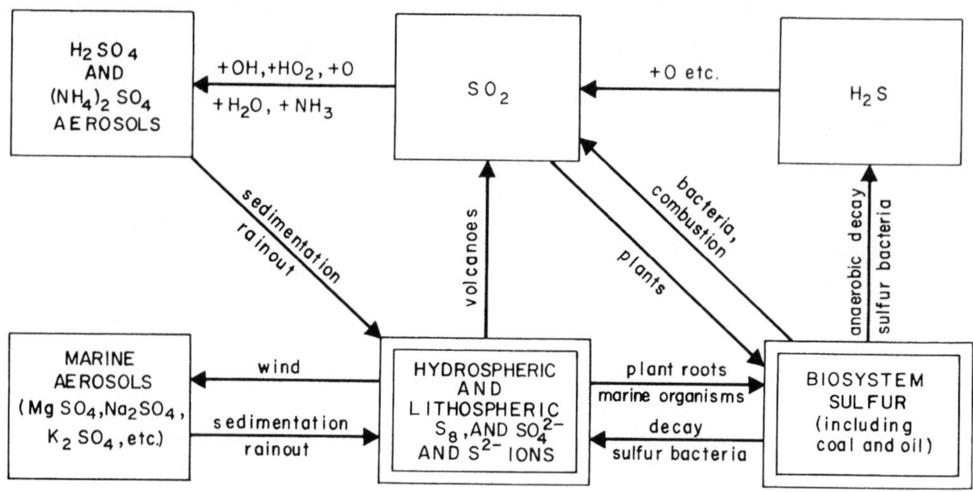


**Figure X.51** The sulfur cycle. The vast majority of sulfur in this cycle is resident in crustal sulfate and sulfide materials. The overwhelming majority of the sulfur in Earth is in the core, where it cannot participate in cycling. As with odd nitrogen, anthropogenic sources are important globally. After Lewis and Prinn.

such as *Clostridium* and *Azobacter*, and the oceanic blue–green algae *Trichodesmium*. At present, such organisms account for about 67% of the global rate of nitrogen fixation. The industrial Haber–Bosch process for manufacture of fertilizers accounts for another 15%, and nitrogen burning in lightning and internal combustion engines, for about 18%.

A similar overview of the geochemical cycle of sulfur is shown in Fig. X.51. The marine aerosols in the figure are sulfate particles derived from drying of droplets of ocean spray.

Biological processing of sulfur is very important because sulfur, like nitrogen, is an essential ingredient of amino acids (cysteine and cystine) and hence of proteins. Anaerobic decay of plant and animal proteins produces abundant $H_2S$, the familiar odor of rotten eggs, and smaller amounts of carbonyl sulfide (COS), carbon disulfide ($CS_2$), methyl mercaptan ($CH_3SH$), dimethyl sulfide ($(CH_3)_2S$), and other highly odorous substances. Conditions suitable for microbial production of ammonia and $N_2O$ also favor the formation of these reduced sulfur species. Some hydrogen sulfide is also produced by anaerobic microbial reduction of sulfate by organisms such as *Desulfovibrio*. Aerobic oxidation of hydrogen sulfide is also used as a source of energy by *Thiobacillus*. Hydrogen sulfide is acutely toxic to animal life because of its ability to bond strongly to iron in hemoglobin and to transition metals in metalloenzymes.

Biologically produced hydrogen sulfide and related organic sulfides and thiols (R–SH compounds) are readily oxidized in the atmosphere by ozone, atomic oxygen, and the hydroxyl radical to provide about 60% of the atmospheric supply of $SO_2$. Nearly 40% of the sulfur dioxide in the present troposphere is anthropogenic, resulting mainly from combustion of fossil fuels. Smelters and volcanic eruptions are also extremely potent local sources of $SO_2$. Both local and regional acid rain is strongly exacerbated by sulfuric acid produced by further oxidation of sulfur dioxide. Under normal atmospheric conditions, $SO_2$ oxidation is slow, and rather wide dispersal of the gas occurs before oxidation and sulfuric acid formation. But in urban photochemical smogs, strong oxidizing agents such as organic peroxides, ozone, and nitrogen oxides dramatically hasten the oxidation process, leading to very severe local acid production with attendant health effects and corrosion. Intense local burning of high-sulfur fossil fuels is the central culprit.

This issue has wide geopolitical significance; the only abundant source of inexpensive energy in Eastern Europe and China is soft (bituminous) coal, which commonly has a very high content of sulfur and a number of other noxious elements, such as selenium, arsenic, and even uranium and thorium. Most of the severe environmental pollution revealed by the collapse of socialism in Eastern Europe is due to use of soft coal for domestic and commercial heating and in the chemical and metallurgical industries. The general linkage of national productivity with energy consumption suggests that improved standards of living for most of the world's population depend upon finding new sources of cheap, clean, and abundant power. Soft coal is cheap and abundant, but its use already has produced a mass health crisis in industrial regions of Poland, the former German Democratic Republic, and the Czech Republic; produced mass arsenic poisoning in China; and threatens third-world food production via acid rain. Economically advanced nations can remove sulfur from crude oil and natural gas at moderate expense; nations that can barely afford power, and nations heavily dependent upon soft coal, have no easy way out. One of the great challenges facing scientists in the coming century is to find sources of energy to raise the standard of living of most of the world while averting an environmental catastrophe.

Coal burning also releases substantial amounts of COS, which is relatively inert in the troposphere. Upon mixing to altitudes above about 25 km, COS is photolyzed by solar UV as on Venus, making CO and atomic S. CO reacts with the OH radical to form $CO_2$, and the atomic sulfur is oxidized stepwise to produce sulfate. At altitudes of about 18 to 25 km in the stratosphere, Earth has a very thin haze layer composed principally of tiny droplets of sulfuric acid. Particles of ammonium sulfate are also present in this *Junge layer*. Coal combustion is already contributing significantly to the optical thickness of this global layer. Thus the environmental effects of coal burning are not purely local.

Figure X.52 shows the chlorine cycle on Earth. Most chlorine in the atmosphere is HCl produced by reaction of sulfuric acid droplets with marine NaCl aerosols. The halocarbons described above are a minor part of the total atmospheric chlorine inventory, but they are insoluble in rainwater and hence can readily mix up to high altitudes. Photolysis of these halocarbons produces chlorine and bromine atoms that, as we shall see, play a major role in the chemistry of the ozone layer. The discovery of the chemistry responsible for this effect was stimulated by spacecraft observations of the atmosphere of Venus in the Mariner program. The main sources of atmospheric chlorofluorocarbons (CFCs) are aerosol propellants and leakage from refrigeration and air conditioning equipment. The usage of CFCs as propellants in aerosol cans has essentially ceased as a result of legislation inspired by this NASA-sponsored research. For a more complete discussion of the geochemical cycles of carbon and the other reactive elements, see the book *Planets and Their Atmospheres* by me and Ronald G. Prinn.

## Radiocarbon Dating

The radioactive isotope $^{14}C$, with a half-life of 5570 years, is produced in the atmosphere by reactions driven by cosmic ray primaries striking nitrogen and oxygen atoms, which together make up 99.5% of the atoms in the atmosphere. The most abundant nuclide in the atmosphere, $^{14}N$, cannot produce $^{14}C$ by spallation, and the heavy natural isotope of nitrogen, $^{15}N$, is rare (0.37% of natural nitrogen). The lightest and most abundant isotope of oxygen, the $\alpha$-process nuclide $^{16}O$, is a possible source of spallation $^{14}C$, but the main source is from secondary reactions of spallation-produced neutrons. The reaction is

$$n + {}^{14}N \rightarrow H + {}^{14}C. \tag{X.125}$$

Radiocarbon decays by $\beta$ emission to the already-abundant nuclide $^{14}N$, which makes the accumulation of the decay product useless as a measure of the age of a sample. Instead, direct measurement of the $\beta$ activity of samples has long been used as a means of dating them. The practical limit to this technique is a sample age of about 30,000 years. A much more sensitive technique using particle accelerators has permitted the extension of radiocarbon dating to samples as old as 50,000 years.

Most living organisms are immersed in the atmosphere. Natural assimilation of atmospheric carbon ($CO_2$) by plants during photosynthesis samples the background carbon isotopic composition of the atmosphere

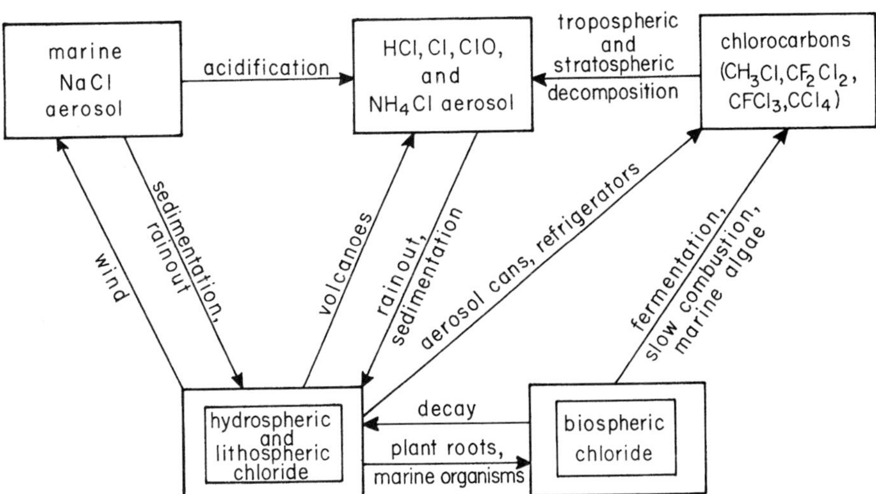


**Figure X.52** The chlorine cycle. Virtually all cyclable chlorine is crustal. Note again the importance of anthropogenic sources. After Lewis and Prinn.

with very little fractionation. Herbivores eat fresh plant matter, usually on a time scale of no more than a few months after the plants sampled the atmosphere, and carnivores eat herbivores, which contain carbon that is rarely more than a few years removed from the atmosphere. Thus the atmosphere and living plant and animal matter have essentially identical $^{14}C$ content. However, the principal human source of carbon dioxide is combustion of fossil fuels. Typically the coal, gas, and crude oil being burned today were removed from isotopic interaction with the atmosphere during the Pennsylvanian, 300 Ma BP. All the radiocarbon in the fossil fuels has long since decayed away, and hence the massive injection of carbon dioxide now under way is accompanied by a steady decrease in the $^{14}C$ content of atmospheric carbon.

The flux of cosmic rays is dominated by relatively low-energy protons, which are especially vulnerable to changes in interplanetary conditions. Changes in the density, speed, and energy density of the solar wind associated with the 22-year sunspot cycle cause a significant periodic fluctuation in the low-energy (1 MeV) proton flux striking the atmosphere and therefore in the $^{14}C$ production rate. Correlation of the $^{14}C$ chronology with tree ring dating (*dendrochronology*) has permitted correction for these effects back to times long before the earliest sunspot observations. Most of the early puzzling discrepancies in radiocarbon dating have been resolved by this technique.

Cosmic ray spallation also produces small amounts of a very distinctive and short-lived light product, T ($^3H$). Tritium decays to $^3He$ with a half-life of about 12.3 years, which renders it useful for measuring time scales of years to about a century, commensurate with the lifetimes of higher organisms. Since the beginning of nuclear weapons testing, tritium from explosive chain reaction of uranium and plutonium has become an important part of the tropospheric inventory. Tritium is used in very small amounts as a biochemical tracer and in larger amounts as the most important of the "herbs and spices" added to the lithium deuteride charge in fusion (hydrogen) bombs to accelerate the fusion process. The tritium content of polar snow records a profile of the history of above-ground nuclear weapons testing. The accumulation of $^3He$, the decay product of tritium, in the atmosphere is severely limited by the ease with which it escapes from the planet.

## Stable Isotope Climate Records

Several of the elements that play a crucial role in the chemical and dynamical behavior of the atmosphere have two or more stable isotopes which undergo physical and chemical fractionation in a manner that depends sensitively on temperature. Examples of physical fractionation mechanisms include distillation, in which the heavier isotopic variants of a particular compound have a slightly depressed vapor pressure, or Jeans escape, in which the lighter isotopic forms travel faster and hence have a higher probability of escape. Chemical fractionation takes two forms: kinetic fractionation, in which the reaction of the lighter isotopic forms of a compound take place more readily than that of the heavier forms (such as the attack of solid sulfides by an oxidizing agent, which preferentially extracts $^{32}S$ while leaving $^{34}S$ and $^{36}S$ behind), and equilibrium fractionation, which partitions isotopes selectively between two different molecules or states of matter (such as water vapor in equilibrium with ice or liquid water, in which the vapor is isotopically lighter than the condensed phase). The difference between the isotopic composition of the condensed phase and the vapor decreases at higher temperatures, where chaos reigns. The thermodynamic argument why there must be isotopic fractionation is simple and elegant: suppose a liquid solution of $H_2$ and HD is in equilibrium with its own vapor. Equilibrium means that the state variables, such as the mole fraction of HD in the liquid and in the gas, are functions of temperature but independent of time. This in turn requires that the rate of collision of HD molecules with the surface of the liquid must equal the rate of escape of HD from the liquid, and the same must be true of $H_2$. But HD molecules have a speed that, from the kinetic theory of gases, is proportional to $(T/\mu)^{1/2}$. Thus, at any temperature, the speed of HD is less than that of $H_2$ by a factor $(\mu_{H2}/\mu_{HD})^{1/2} = 0.8165$, and the rate of evaporation of HD is less by the same factor.

Hydrogen, found principally in water and ice, contains not only normal hydrogen ($^1H$) and heavy hydrogen ($^2D$), but also three stable isotopes of oxygen $^{16}O$, $^{17}O$, and $^{18}O$. Carbon has two stable isotopes, $^{12}C$ and $^{13}C$, which are dramatically fractionated by biochemical processes.

During warm periods, evaporation of water from the warm, low-latitude oceans preferentially leaves deuterium behind in the liquid water and releases moderately deuterium-poor water vapor. On its poleward travels, this vapor partially condenses, precipitating water and snow which is isotopically heavy relative to the vapor and leaving residual vapor that is even more severely depleted in deuterium. In Antarctica or Greenland the last traces of water vapor condense, depositing snow that is isotopically light. During times when Earth is much cooler than average, the isotopic fractionation during evaporation is more severe, and the vapor is isotopically even lighter, resulting in deposition of isotopically lighter snow. Thus both the oxygen and hydrogen isotope ratios in the polar ice packs vary in concert in a way that reflects mean global (oceanic) temperatures, with light isotopic compositions reflecting low temperatures.

The behavior of the isotopes of carbon is greatly complicated by the substantial kinetic isotope effect associated with removal of carbon dioxide from the atmosphere by incorporation into plants. This effect strongly favors uptake of the lighter isotope, $^{12}C$, and enriches the atmosphere in $^{13}C$. Thus the isotopic make-up of carbon reflects complex biological processes as well as climate, rendering it an ambiguous source of climate data: the carbon dioxide sequestered in the shells of ancient organisms yields an oxygen isotope climate record and a much harder to interpret carbon isotope history.

## Photochemistry and Aeronomy

The most important photochemical processes in Earth's atmosphere are those that regulate the abundance of ozone, which is normally concentrated in the *ozone layer*, between 15 and 30 km in altitude (see Fig. X.53). The basic reactions of oxygen and ozone are

$$O_2 + h\nu \rightarrow O + O \quad (\lambda < 2400\ \text{Å}) \qquad (X.126)$$

$$O + O_2 + M \rightarrow O_3 + M \qquad (X.92)$$

$$O + O + M \rightarrow O_2 + M \qquad (X.93)$$

$$O_3 + h\nu \rightarrow O_2 + O \quad (\lambda < 7100\ \text{Å}) \qquad (X.127)$$

$$O + O_3 \rightarrow O_2 + O_2. \qquad (X.128)$$

Of these reactions involving odd oxygen, only the last is misleading: direct reaction of atomic O with ozone accounts for only about 18% of the removal rate of odd oxygen. The main route for removal of odd oxygen is provided by the nitrogen oxides via

$$NO + O_3 \rightarrow NO_2 + O_2 \qquad (X.129)$$

$$NO_2 + O \rightarrow NO + O_2. \qquad (X.130)$$

These reactions proceed efficiently at night; however, in sunlight, nitrogen dioxide is rapidly destroyed by

$$NO_2 + h\nu \rightarrow NO + O \quad (\lambda < 3959\ \text{Å}). \qquad (X.131)$$

This not only hinders destruction of odd oxygen by Reaction (X.129), but also produces more O.

We saw in our study of the photochemistry of the atmosphere of Mars that water vapor can have a closely similar effect on odd oxygen via

$$H + O_3 \rightarrow OH + O_2 \qquad (X.33)$$

$$OH + O \rightarrow H + O_2. \qquad (X.34)$$

In addition, chlorine can destroy ozone, as on Venus, via the additional similar reaction set

$$Cl + O_3 \rightarrow ClO + O_2 \qquad (X.132a)$$

$$ClO + O \rightarrow Cl + O_2. \qquad (X.132b)$$

Thus NO, H, and Cl act as *catalysts* of ozone destruction; the species responsible for ozone removal are regenerated at the end of the destruction cycle. Since ozone is responsible for preventing killing UV radiation ($\lambda < 3000$ Å) from reaching the surface, protection of the ozone layer from these catalysts is an urgent priority. Human production of $NO_x$ has negligible impact on the ozone layer, but high-flying supersonic aircraft inject water vapor into the otherwise very dry upper atmosphere, and CFC emissions inject chlorine into the troposphere in a highly unnatural form very resistant to removal by the normal cleaning process, rainfall. Bromine, which is fortunately a much rarer constituent of the atmosphere, acts even more effectively than chlorine as a catalyst of ozone destruction. Methyl bromide ($CH_3Br$) is in rapidly increasing use as an agricultural fumigant and may eventually present a hazard to the ozone layer.

The ozone layer provides a shield that protects life on the surface, but ozone in the troposphere in excess of about 0.05 ppm is a noxious gas. Global monitoring of the status and behavior of the ozone layer is best done from space, and extremely sensitive satellite instruments have been built to provide us with the data we need. But not all ozone monitoring must be done at high altitudes; photochemical reactions in urban smog produce ozone in exactly the place where this irritant is least welcome: in cities. Electrical discharges, both natural (lightning) and manmade, are effective sources of both ozone and nitrogen oxides.

Generally the volatile reactive elements on Earth participate in closed geochemical cycles. The exceptions are the radiogenic rare gases mentioned above and those species that are capable of escape from the atmosphere, the isotopes of hydrogen and helium. Finally, comet and asteroid impacts may both add and remove planetary atmospheric volatiles. A reminder of the cycles of the major reactive species in the atmosphere is given in Fig. X.54.

## Escape and Infall

Trace light gases in Earth's atmosphere, such as hydrogen and helium, are well-mixed in the atmosphere up to the level (the *homopause*) at which the mean free path of gas molecules has become so large that turbulence can no longer keep the gases homogenized.

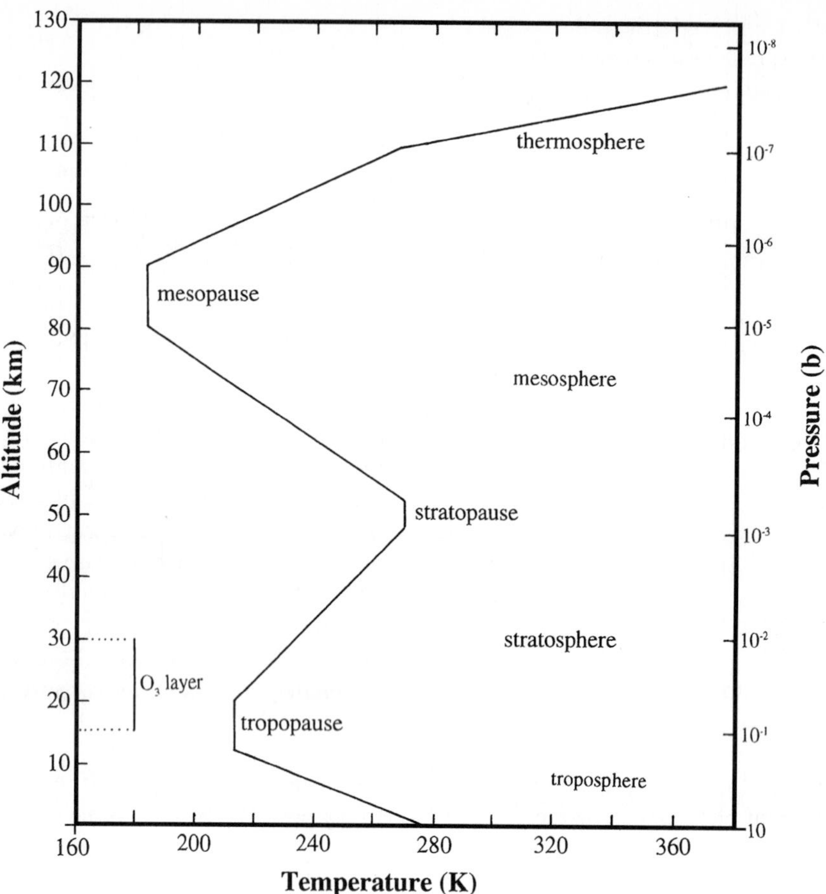


**Figure X.53** Thermal structure of Earth's atmosphere. Heating of the surface of the planet by absorption of visible sunlight drives weather in the troposphere and maintains a steep vertical temperature gradient. The local temperature maximum at the base of the mesosphere is maintained by absorption of solar near IR and UV radiation. The thermosphere is heated in part by solar extreme UV and in part by the breaking of waves propagating up from the lower atmosphere.

Even higher is the *escape layer*, at which an upward-moving particle has a probability of $1/e$ of emerging from the atmosphere without a collision and a probability of $1 - 1/e$ of suffering a collision while heading outward. The altitude of the escape layer is the *exobase*, which marks the dividing line between molecular and viscous flow of gases. The exobase lies nearly at the altitude at which the mean free path of the gas becomes equal to the scale height of the atmosphere. Equations (X.44)–(X.48) describe the basic behavior of thermal (Jeans) escape.

The escape of $^3He$ from Earth has been the subject of a number of studies that suggest a mean rate of escape of about 4 to 6 atoms $cm^{-2}\,s^{-1}$. The average $^3He$ abundance on Earth is $1.3 \times 10^{-6}$ of the $^4He$ abundance ($[^4He] = 5.24$ ppm). With a total atmospheric pressure of $0.98 \times 10^6$ dyn $cm^{-2}$ at the surface, the column abundance of $^3He$ is then about $6.7 \times 10^{-10}\,g\,cm^{-2}$, or $1.3 \times 10^{14}$ atoms $cm^{-2}$. The characteristic escape time for $^3He$ is then about $3 \times 10^{13}$ s, or 1 Ma. The supply of $^3He$ must be similar to the escape flux in order to maintain the observed abundance in the atmosphere.

A small amount of $^3He$ can be made by $(n, \alpha)$ reactions on $^6Li$, for which the main source of neutrons is fission of $^{238}U$. The production rate calculated for this process is only $0.1\,cm^{-2}\,s^{-1}$. Some $^3He$ and T can be made by cosmic ray spallation reactions in the atmosphere, but a calculation suggests a production rate of only about $0.2\,cm^{-2}\,s^{-1}$. A clue to the source of atmospheric $^3He$ is found in the analysis of its abundance in seawater. Harmon Craig of the Scripps Institution of Oceanography and his co-workers have found that, whereas the Atlantic Ocean has a $^3He$ content consistent with simple solubility equilibrium with the atmosphere, some deep Pacific water shows a much higher abundance

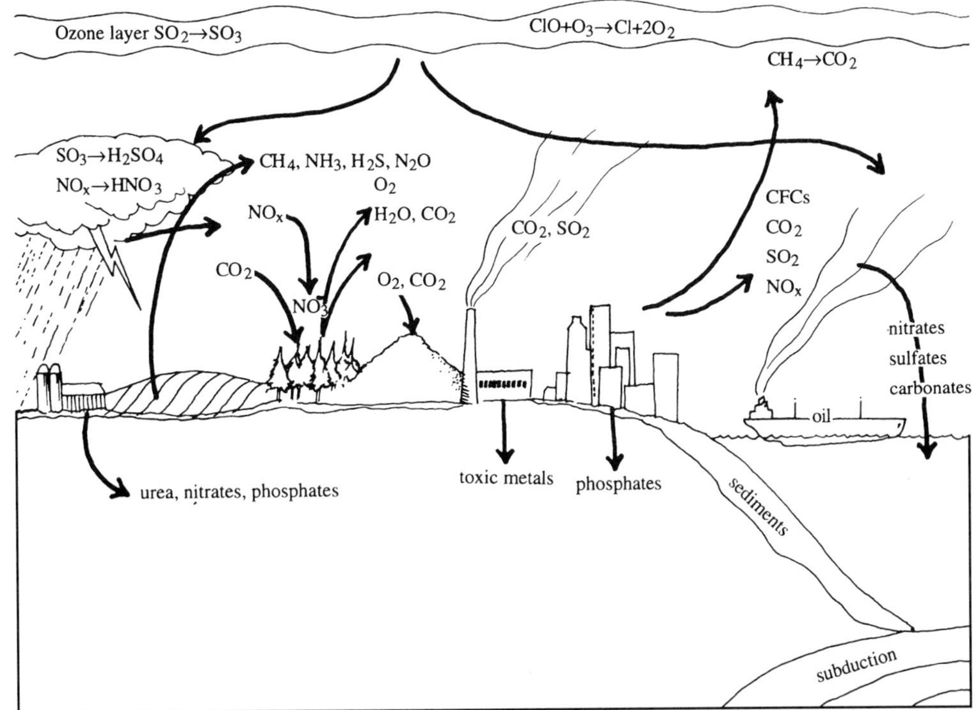


**Figure X.54** The world we live in. This cartoon is intended to remind the reader of the enormously complex interactions that surround, support, and are perturbed by human activities.

than could be provided by interaction with the atmosphere. The gradient of $^3$He concentration with depth shows the existence of an upward flux through the Pacific Ocean, which implies a source with a globally averaged strength of about $2\,\mathrm{cm^{-2}\,s^{-1}}$. Gases of mantle origin, such as volcanic gases from Kilauea in Hawaii, hydrothermal vent fluids from the vicinity of the Galapagos Islands in the eastern Pacific, and hot brines on the floor of the Red Sea all display high $^3$He abundances. With the discovery of the outgassing of $^3$He at rates far in excess of the rate of production by nuclear reactions, the existence of continuing outgassing of primordial rare gases seems securely demonstrated.

The rate of escape of $^4$He is limited by diffusion to about $2 \times 10^6$ atoms $\mathrm{cm^{-2}\,s^{-1}}$. Jeans escape can account for only about 3% of the total escape flux. Chemical reactions such as the collision of an ionized He atom with atomic oxygen release a large amount of energy, which can fire off a neutral helium atom at very high speeds,

$$\mathrm{He^+ + O \rightarrow He^* + O^+}, \qquad \text{(X.133)}$$

and the escape flux is given by the rate of production of upward-moving (the factor of 0.5) hot He atoms by this charge-transfer reaction integrated from the height of the escape layer upward to the top of the atmosphere:

$$\phi_{\mathrm{He^+}} = 0.5 \int_{z_e}^{\infty} k_{133}[\mathrm{He^+}][\mathrm{O}]dz. \qquad \text{(X.134)}$$

We saw in Fig. X.1a that Earth's magnetosphere has open field lines in the vicinities of the magnetic poles. These field lines are swept back into the magnetotail and into interplanetary space by the dynamic pressure of the solar wind. The fraction of Earth's surface over which the magnetic field lines are open is a little over 1% for each polar region. There is always a vertical gradient in the electric field at high altitudes due to diffusive separation of electrons and positive ions in the ionosphere; electrons are so light that their natural scale height is immense, so a negatively charged electron-rich layer tends to lie above the heavier positive ions. The electric fields associated with this partial charge separation accelerate positive ions upwards, and the lightest of these ions may achieve high enough velocities to escape. The observed consequence of this effect is the rapid loss of the lightest ions (hydrogen and helium) from the polar regions. $\mathrm{He^+}$ and $\mathrm{H^+}$, which are the main ionospheric ions at low latitudes, are in fact severely depleted near the magnetic poles. Two mathematical models of this escape mechanism, known as the *polar wind* and *polar breeze* models, have been proposed, in which the protagonists differ in the importance they assign to collisions during the acceleration process. The *polar wind* model bears a close similarity to the classical Parker model of the solar wind described in Chapter IV. In any case, polar escape accounts for some 97% of the total rate of helium escape from Earth.

The rate of hydrogen escape from Earth is limited to about $2.7 \times 10^8\ cm^{-2}\ s^{-1}$ by the ability of water vapor to penetrate the cold-trap of the stratosphere and reach high enough altitudes to be photodissociated and by the ability of hydrogen to diffuse upward to the escape layer. About 60 to 90% of the H escape flux is via nonthermal mechanisms such as charge transfer. All estimates of the present rate of hydrogen escape from Earth are so small that the present water inventory could not be depleted by more than 0.01% over the age of the planet.

Hydrogen escape from the early Earth may have been enhanced by a number of factors. If, contrary to current geochemical opinion, Earth did begin with a strongly reducing Stage I atmosphere with abundant free hydrogen ($XH_2 >> 10^{-2}$), then the early atmosphere could have been strongly heated and distended by the superluminous T Tauri phase of the early Sun, which in other young pre-MS stars is observed to be correlated with a huge excess of UV emission (see Chapter IV). Heating of the upper atmosphere by UV and EUV solar radiation with up to $10^6$ times the present flux would have not only vastly increased the rate of water photolysis, but also driven hydrodynamic escape of the upper atmosphere. In a current theory originated by Donald M. Hunten of the University of Arizona and developed by Robert O. Pepin of the University of Minnesota, escape of hydrogen in amounts equivalent to 30 or 40 times the present mass of the oceans occurred. The upward flow of this (Parker-type) escape flux levitates heavier atoms in accordance with their aerodynamic drag, and these species resist the aerodynamic forces in proportion to their mass. The net effect is that there is a well-defined cutoff mass for any given hydrogen escape flux, below which the other gases are strongly fractionated and extensively lost and above which they are largely unaffected by escape (because the net force on them is downward). An exponentially declining escape flux driven by the declining T Tauri phase of the early Sun could, according to detailed quantitative models by Pepin, fractionate the rare gases sequentially in a manner closely similar to their observed abundances and isotopic compositions on the present Earth.

Such a mechanism releases approximately 10 kbar ($10^7\ g\,cm^{-2}$) of oxygen into the lower atmosphere. The geochemical problem of assimilating this vast amount of oxygen into Earth's interior or losing it to space is extremely daunting. The magnitude of the problem is clear from the fact that the total mass of oxygen that must be hidden is $5 \times 10^{25}$ g, twice the mass of the crust. If all the ferric iron in the lithosphere (6% of the 60-km-deep lithosphere) was made by oxidation of preexisting FeO by atmospheric oxygen, the $1.4 \times 10^{26}$-g lithosphere, containing $8 \times 10^{24}$ g of $Fe_2O_3$, would have absorbed only $8 \times 10^{23}$ g of oxygen, 1/60 of the amount produced by Pepin's mechanism. Loss of oxygen must then be at least 98.4% effective. Why then does any FeO survive?

The problem is exacerbated by the fact that the only known sources of such vast amounts of water, carbonaceous asteroids and comet nuclei, would also have brought in large amounts of ferric iron. Thus a CI meteorite source would bring in some $2.5 \times 10^{25}$ g of $Fe_2O_3$ for every $5 \times 10^{25}$ g of oxygen in water. Where is it? Alternatively, appeal to comet nuclei (60% water instead of 20% water) lowers the relative amount of ferric oxide from about 20% to 8%, an amount closely similar to the observed lithospheric inventory—but this would require that *all* of the vastly larger amount of oxygen liberated by water dissociation must have vanished without a trace! Also, the $2 \times 10^7\ g\,cm^{-2}$ of cometary mass brought in by this postulated "veneering" process ($10^{26}$ g of comets) would nearly equal the entire mass of the lithosphere! Note also that these events must have occurred entirely after the production of the Moon by a giant impact event: the ubiquitous presence of ferric iron in the outer layers of Earth and its equally impressive absence on the Moon clearly imply that lunar formation by a giant impact ($10^{27}$ g) must have antedated the infall of water on Earth via numerous comet impacts involving a total of 10 times less mass.

These impacts would also have brought in at least $10^{24}$ g of oxidized nickel. The 0.007% Ni in the crust corresponds to only $1.7 \times 10^{21}$ g, and the total Ni content of the mantle is about $8 \times 10^{24}$ g, thus requiring homogenization of the mantle. The same impact flux would bring in roughly $5 \times 10^{24}$ g of carbon, compared with $3.5 \times 10^{22}$ g in the crust and $8 \times 10^{22}$ g in the entire silicate portion of Earth. Thus carbon must also be lost from Earth with 98% efficiency. Very strong isotopic fractionation of carbon would be expected, yet the differences between terrestrial and CI carbon are very small, with the CI and CM meteorites depleted in $^{13}C$ by about 8‰ relative to terrestrial carbon. Nitrogen, with an abundance of about 0.13% in CI chondrites and 0.9 ppm in the bulk silicate Earth, should be brought in by CI impacts in amounts in excess of $10^{23}$ g, compared with the $3.5 \times 10^{21}$ g presently found. Thus some 97% of the CI nitrogen must also be lost from Earth and should also be strongly fractionated. The N abundance in comets is less well known, but the corresponding nitrogen loss efficiency for cometary impactors is probably only slightly larger. The isotopic composition of nitrogen in CI and CM chondrites clusters around +30‰ enrichment of $^{15}N$. Thus not only do the observed isotopic fractionations of terrestrial carbon and nitrogen have *opposite signs* relative to CI material, but also the nitrogen is isotopically lighter than the putative unfractionated nitrogen. Clearly mechanisms involving a 98% lower influx of volatiles would be more attractive, but such mechanisms provide only the present oceanic mass of water and are inadequate to provide the massive hydrogen escape flux needed for Pepin's mechanism to work.

Many authors, starting with Fred Whipple of the Harvard–Smithsonian Center for Astrophysics, have speculated that the volatile elements on Earth were supplied by an early era of intense cometary bombardment. During the tail end of the accretion process, volatile-rich objects perturbed by the Jovian planets into highly eccentric orbits must have been very abundant. Although quantitative modeling of the size and time evolution of this flux is very difficult, order-of-magnitude calculations suggest that the present mass of the oceans and atmosphere might plausibly be derived from such a source. (Such a flux still falls far short of providing the 30- to 40-times-larger water inventories necessary for Pepin's hypothesis to work.) Chris Chyba has suggested that fine-grained cometary debris could also provide abundant organic matter to the early Earth in a form that might survive entry and assist in providing raw materials for the origin of life. It is therefore something of a disappointment to find that the isotopic composition of water on Earth appears to be incompatible with a cometary source.

Accumulation of volatiles from cometary and asteroidal impacts during recent Solar System history is not important for Earth because of the large background abundances of most volatiles. We have already seen that certain volatiles, especially water, accumulate on Venus at an important rate from infall events. The main effect of infall on Mars is to erode the atmosphere due to the explosive acceleration of atmospheric gases to speeds in excess of escape velocity by impact fireballs. On Earth, chemical processing of the atmosphere by impact shock waves is the most important single effect of comet and asteroid infall.

## Climate History, Polar Ice, and Ice Ages

At least three times during the history of Earth there have been extended periods of glaciation. The first of these, the Huronian glaciation, occurred about 2.5 Ga BP. The second glacial era, at 0.8 Ga BP, occurred in the late Precambrian. The third great ice age is still in progress. In the present ice age, as many as eight advances of the ice have occurred in the past 2 million years, with interglacial periods that were at least as warm as it is now. The most recent retreat of the great continental ice sheets began only some 11,000 years ago and is still continuing. Mountain glaciers are presently in retreat almost everywhere on Earth and have been retreating almost continually throughout all historic time. Because we live in such an unusual time in Earth's history, it would be well to take a broader look at Earth's climate.

Early in Earth's history, 3.9 Ga ago, the Sun was as much as 25% fainter than today, implying that, all else being equal, the mean temperature of the planet ($T \propto L^{1/4}$) should have been about 6% lower than the present 293.4 K, or 275.8 K: 10 degrees colder than the minimum of 286 K reached in each of the last series of ice ages. The temperature history from earlier (Precambrian) times, virtually devoid of fossil evidence and completely free of data on land environments, is much harder to reconstruct. The oxygen isotopes in chert and amorphous silica deposited in ancient seas seem to tell a story of very hot Precambrian oceans, certainly not a global deep freeze, but the interpretation of these data is controversial. At the same time, some geologists have argued for an early "iceball Earth" phase, in which the oceans were frozen solid to great depths. Sediments from the Precambrian appear to require the presence of a normal rock cycle, with erosion, sedimentation, burial, and metamorphism of oceanic sediments, at least as far back as the oldest known rocks, 3.9 Ga BP. The greenhouse effect could have kept the oceans melted if the carbon dioxide abundance had been $10^3$ to $10^4$ times the present content, but with such greatly elevated carbon dioxide concentrations sedimentary minerals such as siderite, $FeCO_3$, would be stable. Siderite is not found in contemporary Precambrian sediments, leading us to doubt the presence of such large $CO_2$ concentrations. What then kept the early Earth warm?

Moreover, the climate history reconstructed from isotopic data and fossil evidence on the geographical distribution of plant and animal species shows no general warming of Earth. Figure X.55 summarizes climate data from a variety of sources spanning the last 570 Ma. Geological nomenclature and dates may be correlated with Fig. X.46. Figure X.55a spans the Paleozoic (570 Ma to 225 Ma BP) and the Mesozoic (225 to 65 Ma BP). The center of the diagram corresponds to present-day conditions. Note the two ice ages at 280 and 430 Ma BP Since the early Triassic, 225 Ma BP, Earth has been warmer than the present about 90% of the time. Indeed, we see in Fig. X.55b that over the Cenozoic (65 Ma BP to the present) the ice ages of the last 2 Ma are the only times that have been colder than

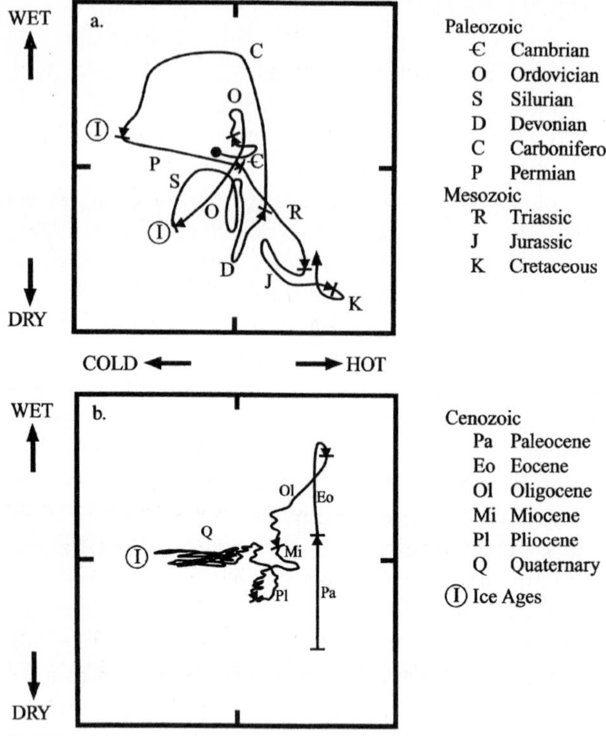


**Figure X.55** The climate history of Earth since the Precambrian. Many sources of data are integrated into a plot of the time evolution of the mean temperature and wetness of the surface of Earth. Both parts of the figure are centered on the present mean conditions on Earth. Part a, detailing the Paleozoic and Mesozoic climate, shows wide excursions in climate, ranging from ice ages (I) to the wet Carboniferous and the hot, dry Cretaceous. Part b, detailing the Cenozoic, shows the Quaternary cycles of ice ages, the only times in the last 225 ma when temperatures were lower than at present.

the present. The last 120 Ma have been a period of general cooling, as the Sun slowly increases in luminosity.

Why then do we think so much of global warming? We often hear it said that "Earth is warmer than any time in the last 1000 (or 800) years." There are many reasonably similar estimates of the temperature history of the past few thousand years, and it seems likely that Earth was warmer than it is now around the time period from about 800 to 1100 AD. It is therefore equally valid to say that "Earth is cooler than it was 1100 years ago." The fact is that we are still emerging from the last ice age. The newspapers tell us that "Antarctica is warming," but in fact only weather stations on the Antarctic Peninsula, which accounts for about 3% of the land area of Antarctica, show a warming trend. The rest of the continent, where the great ice sheets are located, shows no warming and a slow thickening of the ice.

Over the past few hundred ka we have a very detailed oxygen-isotope temperature record from the largest ice caps on Earth, in Greenland and Antarctica. The Antarctic record extends back farther than the Greenland record. Analyses of gases trapped in air bubbles in the ice show a startling and significant correlation between the composition of the trapped air and the temperature recorded by the oxygen isotopes in the ice. Figure X.56 shows the last 400,000 years of data on the $CO_2$ and methane concentrations alongside the temperature data. Abrupt warming episodes end each glacial period, followed by slow and erratic declines in temperature. Temperature maxima in interglacial periods routinely surpass present-day temperatures without the benefit of higher $CO_2$ levels. The $CO_2$ and $CH_4$ abundances in the ice cores, however, correlate closely with temperature, raising an interesting issue. From the perspective of the greenhouse effect, elevated carbon dioxide or methane levels cause an increased infrared opacity in the atmosphere, and hence tend to raise global temperatures. Rising temperatures cause increased evaporation

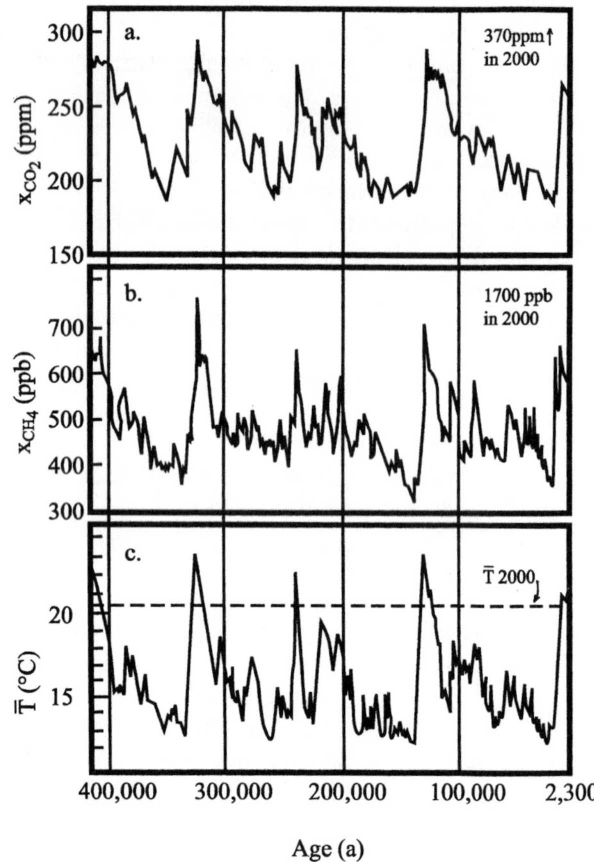


**Figure X.56** Atmospheric carbon dioxide and methane concentrations and temperatures of the past 400 ka. All measurements were carried out on Antarctic ice cores. Part a gives the carbon dioxide abundance in ppm and part b gives the methane abundance in ppb (parts per billion; $10^{-9}$). Part c gives the oxygen-isotope temperature. Note the recurrent pattern of abrupt temperature increases to initiate interglacial periods and slow, erratic temperature declines into ice ages. The $CO_2$ and $CH_4$ abundances and temperature for the year 2000 AD are indicated.

from the oceans, more cloudiness, and more precipitation. Increased cloudiness causes a higher planetary albedo, which reflects more sunlight and moderates warming. Elevated water vapor levels enhance the greenhouse effect, causing further warming. Higher temperatures not only cause higher precipitation, but also higher rates of evaporation from land, and so on. So complex is the network of feedback mechanisms that we cannot confidently decide the sign, let alone the magnitude, of some unquestionably significant effects. Consider, for example, the excellent correlation between the $CO_2$ and $CH_4$ abundances. Why should this be? It would be simpler to attribute the rise of these gases to warming of the oceans, but would this be true? It is reasonable to attribute the climate instability (ice age cycles) of the past 2 million years to the fact that the carbon dioxide level fell too low. But can we be sure that this is correct?

Despite general trends, climatic fluctuations do occur, usually with profound consequences for humankind. The northern hemisphere was unusually warm around 400 to 1000 AD, with a general retreat of oceanic pack ice leaving the sea lanes around Europe, Iceland, and southern Greenland clear. This period was long described as the Little Climatic Optimum. The politically correct name for that period is now the value-neutral term, Medieval Warm Period. The Greenland continental ice sheet retreated, leaving broad green grasslands about its margin (hence the name Greenland). It was during this time that the Vikings extended their range to include Greenland and Newfoundland, founding settlements in both locations.

The general trend of temperatures from about 800 to 1400 AD was one of steady decline. Pack ice again advanced southward to choke the sea routes westward from Iceland. The settlement on the northern tip of Newfoundland was quickly abandoned, and the centuries-long farming of coastal areas of Greenland ceased as the ice sheet again advanced. The Greenland settlements were extinct by about 1400 AD. In the 1600s temperatures in Western Europe dropped so low that normally ice-free rivers such as the Thames froze from bank to bank. This period is known as the Little Ice Age. Warmer conditions prevailed from about 1750 to 1800 AD, then cooler conditions from about 1800 to 1830. The period from 1910 to 1970 was quite warm (the warmest since the Little Climatic Optimum).

In the 20$^{th}$ century, three different source of temperature data have given conflicting messages. Direct surface thermometer data from weather stations (which are concentrated in urban areas and airports, both major "heat islands") show an uneven warming trend. Balloon payloads in the middle atmosphere show very little warming. Weather satellites carrying infrared sensors have provided data since the 1960s that appeared to show no global warming at all. A recent (2003) reanalysis of the satellite data now appears to reconcile the different data sets, but suggests a rate of warming that is several times smaller than that predicted by global climate models. The press gloss, that the data confirm global warming, is misleading, since hardly anyone doubts that some warming occurred in the last century. What the press neglects to mention is that these data show that the magnitude of the warming is so small that it casts serious doubts on the more apocalyptic predictions of most climate models.

It is interesting to compare these temperature trends with the data on the sunspot cycle presented in Fig. IV.5 and the $^{14}C$ history in Fig. IV.6. There is a good correlation between high temperatures, high sunspot maxima, and low $^{14}C$ abundances. As with so many other topics related to Earth's climate history, the broad, wide cycle in radiocarbon production, with a period of about 12,000 years, can be correlated with the strength of Earth's magnetic field, which assists in screening Earth's atmosphere from low-energy charged particle bombardment. The brief, abrupt upward excursions in $^{14}C$ production correlate well with sunspot minima, suggesting that, at times of low solar activity the shielding of Earth from low-energy cosmic rays by the magnetic field in the solar wind is weakened. But what possible mechanistic connection can there be between magnetic screening, cosmic ray effects, and tropospheric temperatures? The solar luminosity changes associated with the sunspot cycle are several times too small to explain the observed temperature variations. Are there as-yet unmodeled effects, such as nucleation of high-altitude clods by ions from cosmic ray bombardment, that may amplify the effect? Any why does the slow variation of Earth's magnetosphere have little or no effect on climate, whereas smaller, briefer excursions in solar activity seem to have much stronger effects?

The study of climate is replete with such examples of nested phenomena and nested cycles. If someone asks whether the climate is getting warmer or colder, one may often be able to respond with something like, "On a 3-year time scale, it's getting cooler; on a 30-year time scale it's getting warmer; on a 300-year time scale it's getting cooler; on a 3000-year time scale it's getting warmer, and on a 30,000 year time scale it's getting cooler, and it's definitely warmer now than it was 300,000 years ago."

The causes of these many fluctuations are complex and inadequately understood. Some of the longer-period fluctuations are cyclic, driven by secular variations in Earth's orbital parameters arising from perturbations by other planets. Some of the briefer excursions are likely produced by dust-induced cooling of Earth resulting from major volcanic eruptions. Major climate excursions in the distant past may have been caused by giant impact events. Solar variability is a tantalizing factor in search of a viable mechanism for driving climate change.

In the present, the scale of human activities is so large that anthropogenic $CO_2$ contributes to the greenhouse effect, CFCs attack the ozone layer, and sulfate aerosols from metals smelting and burning of sulfur-containing fossil fuels screen sunlight from the ground. Many of these disturbances change dramatically on a time scale of a few decades: CFCs and sulfur gases are in decline, whereas carbon dioxide continues to rise. It is no longer clear whether the oscillatory pattern of advances and retreats of the continental ice sheets characteristic of the last 2 Ma will continue in the face of human activities. Neither is it clear that anthropogenic effects are of great significance compared to natural driving forces.

The grandest question of all, however, is why Earth was not locked in a deep ice age for virtually its entire history. The easiest answer is that the greenhouse effect was more efficient in the past. However, there is still a problem accounting for the large size of the effect. Some authors have proposed that strongly reducing gases, such as ammonia and hydrogen sulfide, may have been much more abundant on the early Earth and could have served to enhance the greenhouse effect for billions of years. The problem with this suggestion is that these molecules are extremely vulnerable to solar UV photolysis and would be removed on the time scale of years to centuries. This explanation seems no more plausible than the proposal that vast amounts of carbon dioxide, perhaps 1 to 10 bar, were present in the Precambrian atmosphere.

The answer, although not presently known, may well lie in the interaction of life with the atmosphere. Reactive, polar trace gases are relatively easily controlled by biological processes, and these are precisely the gases that have the strongest effects on the infrared opacity of the atmosphere. The solution to this dilemma may eventually emerge from an improved understanding of the history of life on Earth.

## Life: Origins

Present understanding of the origin of life is based upon certain ideas related to the propagation of order and complexity by chemical mechanisms. Living organisms familiar to us all use the same basic genetic code and the same building blocks for the complex molecules that carry the genetic code and build the structures of living tissue. The basic building blocks are as follows.

**Phosphates** Phosphates are compounds containing the inorganic $(PO_4)^{3-}$ ion. Phosphates participate in forming the bones (endoskeletons) of higher life-forms as minerals in the apatite family, with formulas of the form $Ca_5(PO_4)_3X$. The usual form is hydroxyapatite ($X = OH$), but Cl and F can also substitute for OH. The shells (exoskeletons) of brachiopods are made of phosphates. Fluorapatite ($X = F$) lends strength to tooth enamel. Bones and shells are, however, only the most visible and most easily fossilized phosphate materials. By far the most universal are the phosphate units that link the genetic materials of all forms of life.

**Sugars** Sugars are molecules with the basic formula $(CH_2O)_x$; they may be thought of as polymers containing equal numbers of water molecules and carbon atoms and are generally referred to as *carbohydrates*. A five- or six-carbon backbone with attached hydroxyl groups forms a molecule (a "simple sugar" such as *pentose* or *hexose*) with an aldehyde group (CHO) at one end. These molecules can then be cyclized by splitting out a water molecule to make a five-member ring (containing four C atoms and one O atom). These sugars are both important energy-transporting molecules and building blocks of more complex organic matter. *Disaccharides* consist of two sugars linked by splitting out a water molecule, such as that in cane sugar (sucrose), with the general formula $C_{12}(H_2O)_{11}$. A long polymer of sugar units, called a *polysaccharide*, may have rather different properties; a common example is starch, $\{C_6(H_2O)_5\}_x$, which does not have a sweet taste, but can be *hydrolyzed* in acidic or basic solution to make simple sugars. Some polysaccharides are very toxic. The simple sugar ribose is an essential ingredient of genetic material.

**Organic Bases** A number of compounds of H, C, N, and sometimes O have strongly basic properties (they are excellent proton acceptors, analogous to ammonia, which readily picks up H ions to make the ammonium ion, $NH_4^+$). Several of these organic bases are ubiquitous in genetic material. These important organic bases fall into two groups, the *purines* and the *pyrimidines*, with two and one rings in their structures, respectively. The most important organic bases, adenine (A), guanine (G), thymine (T), and cytosine (C), are shown in Fig. X.57. These bases can bond strongly with the simple sugar 2-deoxyribofuranose to make one-to-one pairs of a base with a sugar. These molecules are called *nucleosides*. The individual nucleosides can then bond through the sugar molecule to phosphate ions to form *nucleotides*. Each nucleoside can bond to exactly two phosphates, and each phosphate can bond to exactly two nucleosides. The ends of the chain are always phosphate (acid) groups. The formation of nucleoside–phosphate bonds therefore leads to formation of extremely long chains, containing millions of nucleoside units, with alternating phosphates and nucleosides. These nucleotide polymers are called *polynucleotides*. The alternating phosphates and sugars serve as the repeating structural framework for a very long sequence of organic bases. Since the molecular "skeleton" is always –ribose–phosphate–ribose–phosphate–, the resulting acidic molecule can be represented in shorthand simply by the sequence of attached organic bases, such as…GCATTAACGTT-

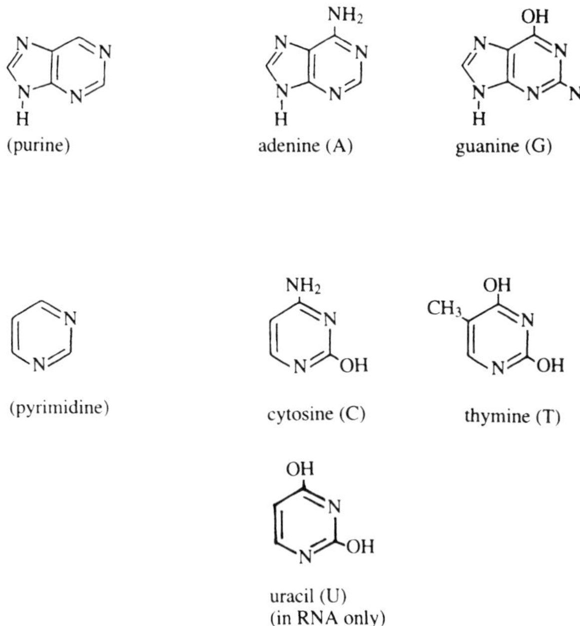


**Figure X.57** The nucleotide bases. The structures of the two fundamental types of nucleotide bases are shown on the left. The four nucleotide bases (A, G, C, and T) based on these purine and pyrimidine structures are used in the genetic code of DNA. The final pyrimidine base illustrated, uracil, is found in RNA only. Note that adenine contains no oxygen: its formula can be written $(HCN)_5$, and it can be made directly by polymerization of HCN.

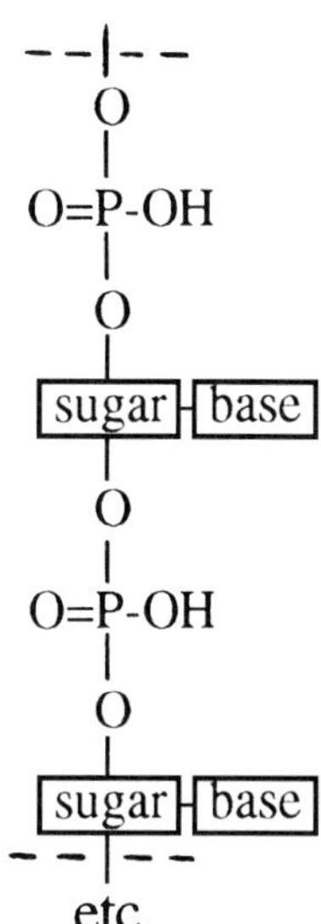


**Figure X.58** Structure of DNA. The basic plan for a single DNA chain is sketched out here to show how the sugar-base (nucleoside) units are linked together by phosphates. The complete DNA molecule has a second chain that is identical in structure except that it has a sequence of complementary bases that are hydrogen bonded to the bases in the first chain (see text for further details). The two chains are wrapped into a rodlike double helix.

CCGCCTTTTAACCAGGGA..., etc. This molecule, schematically shown in Fig. X.58, is *deoxyribose nucleic acid*, or *DNA* for short. Pairs of DNA strands align to for a double helix, with the two strands of DNAbonded lightly together by hydrogen bonds between the base pairs in the different strands. Hydrogen bonds form readily between cytosine in one chain and guanine in the other or between adenine in one chain and thymine in the other. Thus if one strand of DNA carries the sequence given above, its complementary strand must read... CGTAATTGCAAGGCGGAAAATTG-GTCCCT.... The strands may be separated by hydrolysis, and each strand can then serve as a template for the synthesis of its complement. Then two complete two-strand helices of DNA will result.

DNA is the carrier of hereditary information. In it are coded the instructions for manufacture of all the complex chemicals essential to life, especially proteins. The DNA code is read in *triplets* of consecutive bases; GCA codes for one amino acid, TTA codes for another, and so on. So widespread and universal is the chemical apparatus of life that well over 90% of the DNA in every living organism is the same! The difference in genetic code between closely related organisms, such as between sheep and goats and between Norway maples and sugar maples, is found in a tiny percentage of the coded instructions. Humans and chimpanzees have such closely similar genetic code that they have nearly identical chemistry, even down to sharing the same blood types! Comparison of the genetic material of different species is an excellent way of determining their genetic affinities, that is, of reconstructing their phylogenetic trees. The actual production of proteins is relegated to the RNA molecule, which is similar to DNA in structure except that the sugar is D-ribofuranose, and thymine is replaced entirely by uracil (see Fig. X.57).

**Amino Acids** Amino acids all contain a carboxylic acid group, –COOH, and an amine group, $-NH_2$. The essential elements in 21 of the 23 essential amino acids are simply H, C, N, and O. The other two, methionine and cystine, also contain essential S. Tiny quantities of the iodine-substituted amino acids diiodotyrosine and thyroxine are also needed for other purposes. The names, abbreviations, and structures of the 23 essential amino acids (plus thyroxine) are given in Fig. X.59. Small polymers of these amino acids are produced by linking an amine group in one molecule to a carboxylate group in another molecule by splitting out water: $-COOH + H_2N- \rightarrow -CO-HN- + HOH$. The resulting low polymers of amino acids are called *peptides*. Long-chain high polymers of amino acids (polypeptides) are called *proteins*. Proteins are among the most important structural and functional molecules in a living organism. Some proteins "wrap up" metal ions to make extremely

| **Name** | **Abbreviation** | **Formula** |
|---|---|---|
| glycine | Gly | $NH_2CH_2CO_2H$ |
| alanine | Ala | $CH_3CHCO_2H$ / $NH_2$ (on CH) |
| glutamic acid | Glu | $HO_2C(CH_2)_2CHCO_2H$ / $NH_2$ (on CH) |
| glutamine | Glu-$NH_2$ | $NH_2CO(CH_2)_2CHCO_2H$ / $NH_2$ (on CH) |
| $\delta$-hydroxylysine | Lys-OH | $NH_2CH_2CH(CH_2)_2CHCO_2H$ / OH (on first CH), $NH_2$ (on second CH) |
| arginine | Arg | $NH{=}C(NH_2)\text{-}NH(CH_2)_3CHCO_2H$ / $NH_2$ (on CH) |
| aspartic acid | Asp | $HO_2CCH_2CHCO_2H$ / $NH_2$ (on CH) |
| asparagine | Asp-$NH_2$ | $NH_2COCH_2CHCO_2H$ / $NH_2$ (on CH) |
| serine | Ser | $HOCH_2CHCO_2H$ / $NH_2$ (on CH) |
| histidine | His | 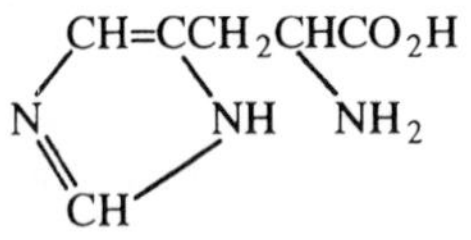 $CH{=}CCH_2CHCO_2H$ with ring N, NH, CH (N=CH–NH); $NH_2$ (on CH) |

Figure X.59 The essential amino acids. An amino acid is an organic molecule that contains an amine group ($-NH_2$) and a carboxylic acid group ($-COOH$). All biologically useful amino acids have the amine group attached to the same carbon atom that the carboxylic acid group is attached to; these are termed alpha amino acids. Many of the amino acids used in making proteins can be synthesized in the human body. Those that cannot be synthesized and must be included in the diet are marked with an asterisk.

important molecules called *enzymes*, which are catalysts of biochemical reactions.

**Mutations** Mutations consist of the alteration of the genetic code by any means, such as cosmic ray impact, attack by a chemical such as a reactive free radical, or even mechanical damage. Errors such as the insertion or deletion of a nucleoside lead to a dramatic change in the reading of the code: ...GCATTAACGTTCCGCCTTT-

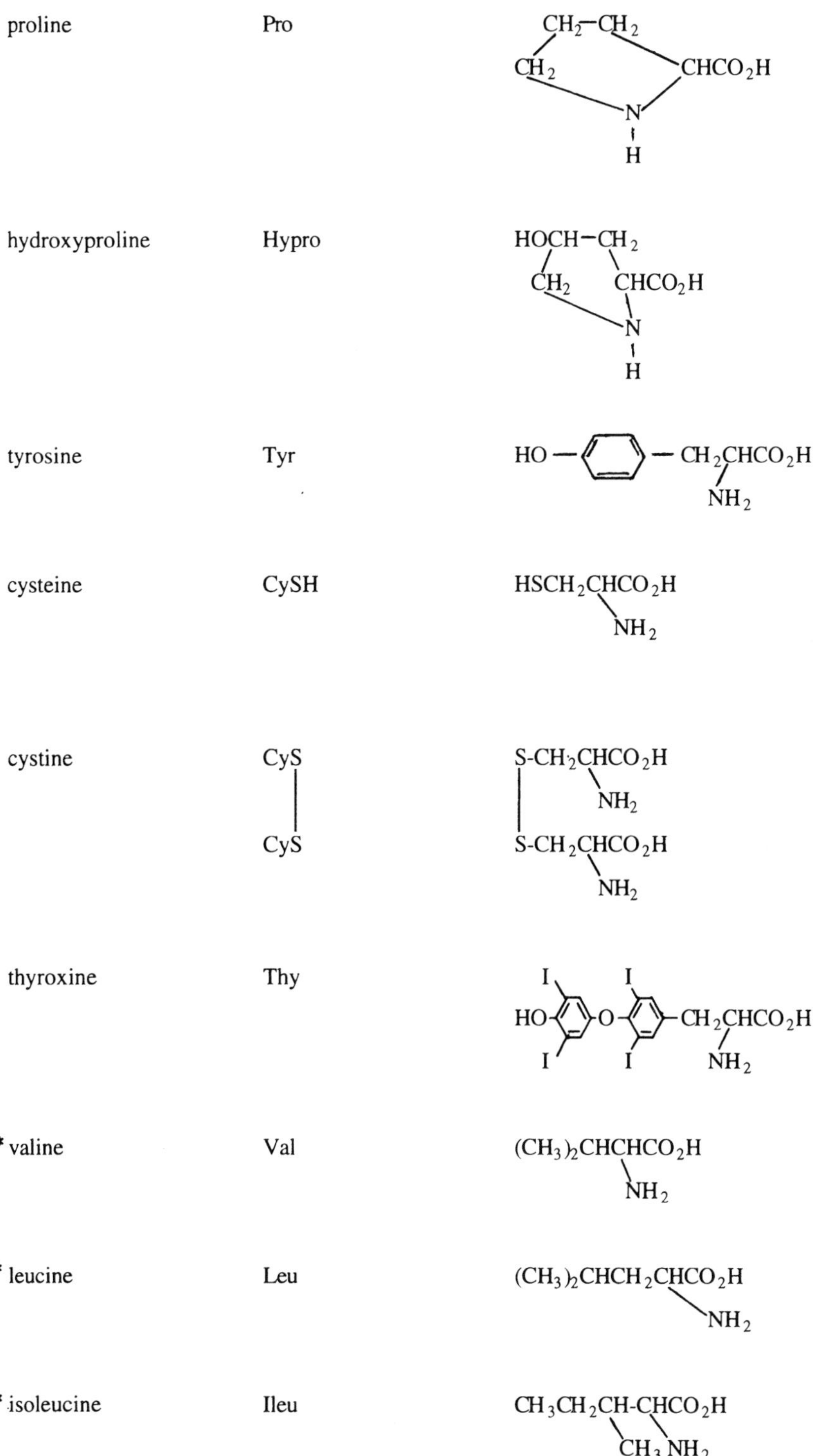


**Figure X.59** *Continued*

TAA ... would normally be read in triplets as GCA TTA ACG TTC CGC CTT TTA ..., etc. Suppose that the third T is accidentally deleted. Then the code would read ... GCATTAACGTCCGCCTTTTAA ... and would be read in triplets as GCA TTA ACG TCC GCC TTT TAA ..., etc. After the first three amino acids, the rebracketed reading of the code is complete nonsense. The resulting error in protein synthesis may be fatal.

| | | |
|---|---|---|
| * phenylalanine | Phe | 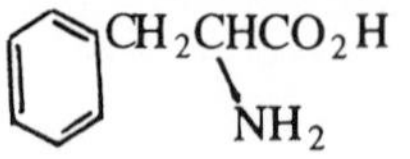 $C_6H_5CH_2CH(NH_2)CO_2H$ |
| * threonine | Thr | $CH_3CH(OH)\text{-}CH(NH_2)CO_2H$ |
| * lysine | Lys | $NH_2(CH_2)_4CH(NH_2)CO_2H$ |
| * methionine | Met | $CH_3S(CH_2)_2CH(NH_2)CO_2H$ |
| * tryptophan | Try | 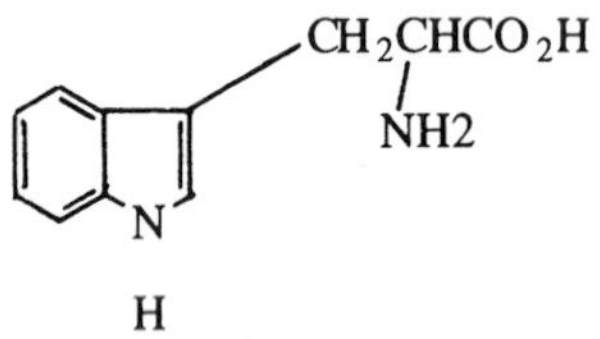 (indole, N–H)$CH_2CH(NH2)CO_2H$ |

**Figure X.59** *Continued*

Even if an organism is born with this defect, it may be unable to synthesize an absolutely essential protein or enzyme. Indeed, many "genetic diseases" are known that are attributable to exactly this kind of defect. These diseases are reflected functionally as the absence of a crucial protein or enzyme, causing serious dysfunction and often leading to early death.

Some mutations lead to differences in the structure or function of the organisms that bear them that are value neutral or even beneficial. The usefulness of such minor changes is often dependent upon details of local conditions. A small desert lizard that, through mutation, emerges from the egg to develop a bright green skin will be easy prey for predators and will be unlikely to survive to pass this mutation down to the next generation. An identical mutation occurring to a small lizard living in grasslands or a jungle may confer a significant degree of camouflage, raising the probability of survival of this mutant and its descendants.

Mutations are not rare. In each generation, approximately 1 billion mutations occur to human genetic material. Of these, most are dramatic and lethal. Most of the others are minor and value-neutral. A few convey significant advantages in certain locales and not in others and are conditionally beneficial. Because species can wander and climates can change, some beneficial mutations may be "sleepers" that are not very common in the gene pool until changing conditions begin to select in favor of them. A frog that accidentally acquires a phenomenal level of resistance to lead poisoning through a mutational change in an enzyme may have no selective advantage until one of his descendants wanders into a polluted drainage basin; there this gene will become universal in the population, because all that do not bear it will die. A few mutations are universally beneficial. The latter are the most likely to expand throughout the gene pool. High-mutation-rate environments (with, for example, high levels of natural or man-made radiation) produce multiple mutations. The probability that a multiply mutated offspring will be viable plummets rapidly with the number of mutations. Any favorable mutation is highly likely to be accompanied by another, lethal mutation. The fertility of the irradiated population declines dramatically, and nonlethal but deleterious mutations become common in their surviving offspring.

Interestingly, the most common mechanism for mutation by radiation or chemicals is through the production of extremely reactive free radical oxidizing agents such as peroxyl, ·OOH. There are enzymes and other substances, such as vitamins A, C, and E, that destroy these dangerous agents that threaten the genetic material. Thus a mutation that produces a more effective antioxidant, or one that cripples production of an antioxidant that is rarely used under normal conditions, may change the adaptive ability of a species in ways that are not at all obvious in the short term. Later, under assault by rising concentrations of oxidizing free radicals (from acid rain, radioactivity, increased exposure to UV radiation due to ozone depletion, etc.), such changes might have strong selective repercussions.

The origin of life requires that conditions conducive to formation of sugars, organic bases, and amino acids were present on Earth. The simple molecules that are most useful as the raw materials for making these substances are hydrogen cyanide (HCN), formaldehyde ($CH_2O$), and water. Mixtures of water with methane or CO and nitrogen or ammonia, devoid of free gaseous oxygen, is capable of making these molecules under the impetus of any of several sources of disequilibrating energy. Transient high temperatures (magma from volcanic eruptions, shock waves from lightning, or entering meteorites); the action of the ultraviolet radiation; alpha, beta, and gamma irradiation from radioactive decay; and ionization by cosmic rays are all potential sources of these products.

The earliest form of life probably used a genetic code much simpler than the present one. Possibly only two organic bases, one purine and one pyrimidine, were used in DNA, and possibly only a few amino acids were coded for by these bases. At the earliest stages, proteins and nucleotides may have been assembled on templates provided by mineral grains. Many minerals, especially phyllosilicates (the common clay minerals), have the ability to catalyze the formation of amino acids and organic bases, to adsorb these species on their surfaces, and to catalyze the polymerization of them into proteins and nucleotides.

At our present level of knowledge, we cannot preclude the origin of life by similar processes elsewhere in the Solar System. The best candidate location for an independent origin of life appears to be Mars, during the era when liquid water was stable on its surface.

## Life: Stability of the Biosphere

The uniformitarian view of Earth, which reflects the "normal" state of affairs, is a good approximation on short time scales. But over extended periods of time, rare events become unavoidable. Because some of these rare events are immensely powerful, they may impose catastrophic changes on environments and populations.

For example, earthquakes display an intensity–frequency relation that has a steep slope. Extremely violent earthquakes (8.0 on the moment magnitude scale) occur only a few times per century. Magnitude 9 earthquakes are rarer still; perhaps only one every few thousand years. But there is an *upper limit* on the size of an earthquake of about magnitude 9.5, corresponding to the release of all the stored energy in the largest fault system in the world. A "billion year event" (the biggest earthquake experienced in a billion-year period) earthquake is little more severe than a "million year event." The same is true of most natural disasters, including floods, tidal waves, landslides, volcanic eruptions, droughts, and hurricanes as well as earthquakes: they have natural limits to their scope and their consequences. Even the most severe disasters are localized, and few have global effects. Thus "natural disasters" as classically defined are not threats to life in general or to mankind's survival in particular.

But no such limitations apply to the impact of large comets and asteroids. A billion-year event is not only more severe than a million-year event, it is roughly 8000 times as severe! (See Fig. IX.1.) The only way to have an earthquake, volcanic eruption, tsunami, or landslide that surpasses the natural geological limits is to provide the energy for it by means of an impact event. Then global consequences will follow.

This means that species and genera with restricted geographical range can indeed be extinguished by disasters native to Earth, but global extinction events are very difficult to arrange without giant impacts. Conversely, once the fall of meteorites on Earth has been accepted (as was generally accomplished by about 1806), it is unavoidable that much larger bodies will impact Earth at rare intervals. Some geologists have argued that the release of vast amounts of carbon dioxide by the eruption of the Deccan basalts in India 70 Ma ago is a more plausible explanation of the Cretaceous extinction event than the putative impact of a comet that released the same amount of carbon dioxide. This conservative, uniformitarian point of view has of course been thrown into disfavor by the discovery of the "smoking gun," the 65-Ma-old Chicxulub crater in the Yucatan, and by the dating discrepancy between the Deccan eruptions and the extinction events 5 Ma later. But a more fundamental problem is that natural rates of native Earth processes, such as volcanism, are not global-scale "events": they are localized or drawn-out processes. A given amount of carbon dioxide has radically different effects when released into the atmosphere in 1 min by an impact than when released over 500,000 years by the Deccan eruption. This is especially true when the carbon dioxide is accompanied by vast amounts of sulfur dioxide, dust, and nitrogen oxides. When a vast mass of these pollutants is injected over hundreds of thousands of years, natural regulatory process with time scales of a few months (dust fallout; acid rain) to a few centuries (overturn of the oceans) remain very effective. But when the same mass of dust and noxious gases is injected and transported worldwide on a time scale of minutes, all natural regulatory mechanisms are overwhelmed. *Native terrestrial events are incapable of "saturating" regulatory processes; impact events have no such difficulty.*

The issue of the stability of the biosphere and the continuity of life boils down to understanding the impact history of Earth. Global extinction events, especially the sudden, simultaneous demise of numerous highly successful, cosmopolitan species, are generally caused by impacts.

The reader interested in exploring the history of Earth and the habitability of its biosphere in greater depth is strongly encouraged to begin with Jonathan I. Lunine's rigorous and readable introduction, *Earth: Evolution of a Habitable World* (Cambridge Univ. Press, 1999).

## Exercises

### *Mars*

X.1 a. What is the acceleration of gravity on the equator of Mars? Remember to allow for the rotation of the planet.
b. What is $g$ at the Martian poles?

### *Motions of Mars*

X.2 By what factor does the effective temperature of Mars change in going from perihelion to aphelion?

### *Geophysical Data on Mars*

X.3 What is the total energy content of the Martian magnetic field?

X.4 An Earth-based spectroscopic measurement of the strength of some weak features in the infrared spectrum of carbon dioxide on Mars gives a column abundance of $CO_2$ of 8000 cm agt. [One amagat is the density of a gas at the standard temperature (273.15 K) and pressure (1 atm)].
a. What is the corresponding partial pressure of carbon dioxide at the Martian surface?
b. The paper reporting this result fails to specify whether this column abundance refers to the sub-Earth region on Mars or to the average over the entire visible disk of the planet. What would the surface pressure be if the reported abundances were actually a whole-disk average?

X.5 The atmosphere of Mars at the Viking 2 landing site has a minimum pressure of 7.4 mbar and a maximum pressure of 10.2 mbar. Assume, to simplify matters, that the atmosphere overlying the landing site is isothermal. The maximum temperature observed by the Viking 2 lander is 250 K and the minimum is about 180 K. The ideal gas law states that $Pv = RT$, so it seems possible to attribute the observed pressure variations to the temperature changes alone. Is this argument correct or fallacious? Explain your answer quantitatively.

X.6 a. An asteroidal body with perihelion at Mars and aphelion at 3.33 AU strikes Mars. What would its impact velocity be?
b. The Martian crater Arandas (Fig. X.5), made by the impact of an asteroidal body traveling at the speed you calculated in part (a), has a diameter of about 24 km. What is the expected depth of this crater? [Recall Eq. (IX.4)!]
c. What was the explosive yield of this impact (in megatons of TNT)?
d. What was the mass of the impactor?
e. How frequent are impacts of this energy on Mars?
f. Assuming the crust of Mars has a density of 2.4 g cm$^{-3}$ and that the brecciated rim material of Arandas has a pore volume of about 16%, what would the hydrostatic pressure be at the base of the rim wall? Compare this to the crushing strengths of the types of natural materials discussed in Chapter VIII.

### *Surface Composition*

X.7 Assuming that the analyses of weathered Martian surface dirt given in Fig. X.19 reflect the presence of anhydrous magnesium sulfate and a magnesium-bearing pyroxene, estimate the relative weight proportions of $MgSO_4$ and $MgSiO_3$ in that material.

X.8 If the K content of Martian igneous rocks is about 1.5% and the K content of weathered Martian soil is about 0.3%, what happened to the "missing" potassium during weathering? In other words, where is it? How would you, as an explorer of Mars, search for it?

### *Atmospheric Composition*

X.9 On a warm summer day in Florida the water vapor mole fraction in the lower troposphere over Orlando is 0.01. Assuming uniform mixing, what is the water content of the atmosphere in the following units of measure?
a. millibars
b. millimeters of mercury
c. weight fraction
d. cm agt
e. micrometers of precipitable water

X.10 Suppose that the rate constant for the reaction

$$A + B \rightarrow AB$$

is $10^{-20}$ cm$^6$ s$^{-1}$ and the rate constant for

$$A + B + M \rightarrow AB + M$$

is $10^{-30}\,cm^9\,s^{-1}$. What concentration of M is required for the two reactions to have equal rates?

X.11 The most volatile-rich meteorites, the CI chondrites, contain $8 \times 10^{-7}$ g of primordial argon per cubic centimeter.

a. If the CI chondrites are 15% void space, what partial pressure of argon would result from releasing all the argon in a sealed sample (closed system)?
b. If Mars were made of CI chondrites and all the primordial argon were outgassed into the atmosphere, what would the argon pressure be on Mars?
c. If Venus were made of CI material and all primordial argon were outgassed, how would that amount of argon compare with the amount actually present?

*Photochemical Stability and Atmospheric Escape*

X.12 A planet with a carbon dioxide atmosphere receives a global average $3 \times 10^5\,erg\,cm^{-2}\,s^{-1}$ of light at wavelengths (below 3700 Å) capable of driving the reaction

$$H_2O_2 + h\nu \rightarrow OH + OH. \qquad (X.26)$$

Hydroxyl is formed two scale heights ($H = 10^6$ cm) above the ground and destroyed by reactions at the planetary surface.

a. What is the maximum average production rate of OH in $cm^{-2}\,s^{-1}$?
b. The eddy diffusion coefficient $K$ in the lower two scale heights of the atmosphere is $10^5\,cm^2\,s^{-1}$. What is the concentration of hydroxide $n_{OH}$ in the lower atmosphere at steady state: that is, when the rate of delivery of OH to the surface to equal to its production rate?

X.13 Dissociative recombination of molecular oxygen [Eq. (X.50)] can produce either $O^*$ ($^1D$) or O ($^3P$). Explain in words why the escape rate of O decreases when the fraction of atoms in the highly energetic $^1D$ state is increased.

X.14 Suppose an Earth-sized planet has a source of H atoms from

$$H_2^+ + e \rightarrow H^* + H,$$

and that the recombination energy of molecular hydrogen (100 kcal per mole of $H_2$) is roughly the same as for $O_2$, $N_2$, or CO (none of which can escape from so large a planet by dissociative recombination). Could this mechanism drive H to escape from the Earth-sized planet?

*Geophysical Data on Venus*

X.15 In the format of Fig. X.24, sketch the $T_b$ distribution for an airless body at 750 K with a surface emissivity of 0.92.

*Atmospheric Structure and Motions*

X.16 Look up tables of the heat capacity of $CO_2$ and use them to calculate the altitude dependence of temperature and pressure for an adiabatic pure carbon dioxide atmosphere above a surface at 9 bars and 750 K. Compare your results to the average $C_p$ model given in the text.

X.17 A mine shaft in South Africa is 3 km deep. The temperature in the rock increases downward due to Earth's internal heat flux. A forced-convection ventilating system with heavily insulated ductwork pumps fresh ambient surface air down to the bottom of the mine at a high flow rate (adiabatically). Will the gas cool or warm the bottom of the mine? Assume the rock and air have a mean surface temperature of 300 K.

X.18 a. Assume that the $CO_2$ adiabat is a smooth extrapolation of the plot in the altitude range 0 to 25 km in Fig. X.32, and use it to estimate the static stability profile of the troposphere of Venus for altitudes from 0 to 60 km.
b. How does the Brunt–Väisälä frequency vary with altitude over this range?

X.19 Laboratory experiments and field tests in terrestrial deserts and streams have shown that wind or water flow can begin to move surface grains at a critical velocity that depends on the local acceleration of gravity $g$, the grain diameter $d$, and the densities of the fluid phase ($f$) and the particle ($p$):

$$\nu_{crit} = [gd(\rho_p - \rho_f)/\rho_f]^{1/2}.$$

a. What minimum speed is needed to move 100 $\mu$m (0.01-cm) particles of density 2.4 at the following places?
   (i) Earth's subaerial surface
   (ii) The surface of Venus
   (iii) The surface of Mars
   (iv) The bottom of a lake or stream on Mars
b. Would particles of this size be mobilized by the measured wind speeds seen near the surface of Venus?
c. Do you see any conditions under which this equation breaks down?

X.20 a. How important is the Coriolis force in controlling motions at the surface of Venus at 90 °N latitude?
b. At 10 °N?

*Venus: Atmospheric Composition*

X.21 a. What is the total mass of HF in the atmosphere of Venus?

b. How thick a layer of sodic plagioclase (albite) would be needed to remove all the fluorine through formation of NaF?

X.22 If the endpoint of FeO oxidation in the crust of Venus is $Fe_3O_4$ instead of hematite [which was used in Eq. (X.74)], how much oxygen ($g\,cm^{-2}$) would be absorbed by the crust of Venus? The reaction is:

$$3FeO + 0.5O_2 = Fe_3O_4.$$

X.23 Explain in simple terms why methane is an extremely implausible component of the atmosphere of Venus.

X.24 a. The reaction $CO_3 + SiO_2 = CaSiO_3 + CO_2$ is studied in the laboratory at 750 K and found to be at equilibrium at a $CO_2$ pressure of 100 bars. What is the Gibbs free energy of this reaction at that temperature?

b. The same reaction is studied at room temperature and found to have an equilibrium $CO_2$ pressure of $3 \times 10^{-4}$ bar. What is the average standard enthalpy change $\Delta H^\circ$ for this reaction over the temperature range studied?

X.25 According to the Clausius–Clapeyron equation, the vapor pressures of moderately volatile substances vary enormously with temperature:

$$d \ln p/dT = \lambda/RT^2. \quad \text{(V.82)}$$

Many volatile minerals, such as sulfides, halides, and selenides, are baked out of the lowland surface of Venus by the 750 K temperatures there. Suppose, for simplicity, that these minerals have typical heats of vaporization of 100 kcal/mol. Mountaintops 10.5 km above the lowlands are colder than the lowlands by an amount you can easily calculate. By how large a factor is the vapor pressure of these minerals reduced at mountaintop elevations? (Interestingly, the lower atmosphere is very clear, quite free of condensate hazes.)

X.26 Derive the slope of the $W/H$ boundary in Fig. X.34.

*Venus: Photochemistry and Aeronomy*

X.27 HBr is an undetected but certain component of the atmosphere of Venus. HBr is photolytically destroyed much more rapidly than HCl because its UV cutoff is at a longer wavelength. Bromine is observed in the laboratory to catalyze the destruction of ozone much more rapidly than even chlorine. Propose a mechanism for the destruction reaction.

X.28 What, in words, does the following "reaction" mean and what evolutionary significance might it have on Venus or Earth?

$$He + O^* \rightarrow O + He^*.$$

*Earth: Internal Structure*

X.29 Earth's solid inner core transmits shear waves readily, but its $V_s$ cannot be determined from seismic data. Explain why.

X.30 Suppose that Earth's mantle contains 1% by weight $Fe_2O_3$. If that ferric iron were formed exclusively by disproportionation of FeO [Eq. (X.105)], how large an increase in Earth's core mass would result from the metallic iron produced by this reaction?

*Earth: Magnetic Fields and Magnetosphere*

X.31 During times of geomagnetic reversals the dipole field strength may drop to zero, allowing the solar wind to impact the low-latitude ionosphere. Suggest several possible consequences of such an event.

*Earth: Biological History*

X.32 a. From Fig. IX.21, estimate the total yield of the K–T impactor, which produced a 200-km crater.

b. Compare this energy to that needed to boil Earth's oceans completely dry.

*Weathering in the Rock Cycle*

X.33 The solubility of anhydrite, $CaSO_4$, is 0.1 g per 50 g of water at 15°C.

a. What is the solubility of $CaSO_4$ in moles per liter?

b. What is the solubility product $K$ for dissolving pure solid $CaSO_4$ to make an ideal solution of $Ca^{2+}$ and $SO_4^{2-}$ ions?

c. A hydrothermally altered lake in West Africa contains 0.1 moles of sodium sulfate, $Na_2SO_4$, per liter. What is the solubility of $CaSO_4$ in that lake? (That is, if you add lots of solid anhydrite to this lake water, how much will dissolve?)

*Escape and Infall*

X.34 a. Estimate the amount of energy released by the charge-exchange reaction

$$He^+ + O \rightarrow He + O^+.$$

b. Assuming conservation of energy and momentum, estimate the speed of the neutral helium atom made by reaction of cold $He^+$ and O and compare that speed to the escape velocities of the terrestrial planets.

### *Life: Origins*

X.35 Many proteins contain more than 100 amino acid units. Suppose we have 23 different kinds of amino acids: how many different proteins with 100 amino acid units each could be made?

# XI. Planets and Life around Other Stars

Are we alone? Is there life elsewhere in our Solar System, or in our galactic neighborhood? Is there intelligent life elsewhere in space? Can we make contact with them or visit them, or they visit us? These are profound questions, ones that we cannot yet answer with confidence. Negative answers would be of as fundamental interest and significance as positive answers.

Tentative answers to these questions, and suggestions of areas for future research on these questions, can conveniently be tendered if we subdivide the question into several parts. First, given what we know of life, where is it most likely to originate? We shall argue that terrestrial-type water- or ice-dominated planetary surfaces are the best known locations. In what astronomical setting, then, do we expect these conditions to be achieved in a reasonably stable way, either in our Solar System or about nearby stars? Are these conditions achievable in the recently discovered, exotic realm of super-Jovian extrasolar planets and brown dwarfs? What might typical extrasolar stellar systems be like? Finally, arguing from this astronomical perspective, how common might life (and intelligence) be?

## Chemical and Physical Prerequisites of Life

An essential prerequisite to life is the availability of a chemical system that can store, read, and write very complex genetic information. Such a high level of complexity requires a chemistry within which stable molecules of intricate structure and high information content can be synthesized, read, and transcribed. Further, the chemical flexibility that is essential to synthesizing and reading information-bearing molecules seems to require that these processes occur in a liquid. Therefore an appropriate solvent for the large information-bearing molecule is also required. It is not sufficient to find an element that can form chains of atoms. Many elements can form chains of atoms that are poorly suited to information storage and retrieval either in solution or in the pure state. Some allow only linear repetitions of a single fundamental unit, as in sulfur chains and polysulfide, polyhalide, polyphosphate, and silicate chain ions:

$$-S-S-S-S-S-S-S-S-S-S-, \text{etc.}(S_x) \quad (XI.1)$$

$$S_8^-, S_9^-, S_{10}^-, S_{11}^-, S_{12}^-(S_x^-\} \quad (XI.2)$$

$$I_3^-, I_5^-, I_7^-, I_9^-, I_{11}^-(I_{2x+1}^-) \quad (XI.3)$$

$$(OPO_3)_x \quad (XI.4)$$

$$O(SiO_3)_y \quad (XI.5)$$

The polysulfide and polyhalide chains rearrange very readily, dropping and adding terminal S atoms or $I_2$ units, or even cleaving spontaneously in mid-chain. Such chains with lengths over 15 or 20 atoms are so unstable that they cannot be isolated. The polyphosphate chains present in concentrated aqueous solutions of phosphates or in molten phosphates are much longer, up to at least 100 units, but they share the problem of an increase in reactivity and instability with increasing chain length. Silicate chains, consisting of $SiO_4$ tetrahedra sharing a single oxygen atom with each of two neighbors, are found in a highly labile state in silicate melts, and in an excessively stable form (not mobile, not readable, and not transcribable) in certain solid silicates (the pyroxenes).

Silicates also form large two-dimensional sheets of repeating, interlinked silicate tetrahedra, both in reactive form in melts, and in ultrastable form in layer-lattice silicates (the phyllosilicates, including micas and clays). Finally, silicates can form extremely unstable three-dimensional grids in melts, and completely information-free three-dimensional lattices in crystals. In aqueous solution these units hydrolyze rapidly to make silicic acid, $H_4SiO_4$. Low-temperature silicate melts can be made by dissolving water in molten silicates at elevated pressure. Such "cool" melts remain fluid at temperatures hundreds of degrees below the normal melting temperature of silicates, but the water hydrolyzes the silicate–silicate bonds, leading to a dramatic decrease in viscosity and a complete destruction of the complex molecular chains. Thus the common science-fiction device of silicate-based life seems most improbable from a chemical perspective.

Clay minerals do have an interesting property that enables them to code information in their structure. The silicate sheets in them are anionic, but the clay particles themselves are electrically neutral. Most clays accommodate a wide range of possible cations in layers between the anionic silicate layers. Growth of the clay particles from an aqueous solution of silicic acid and metal cations builds layers that duplicate the pattern of their template layer. Mechanical cleavage of a clay particle parallel to the silicate sheets liberates two new surfaces which can resume adding surface layers with faithful copying of the cation pattern. Thus clays can replicate. If a chemical event alters the configuration of cations in the active surface layer, further growth of the clay will emulate this altered structure. Thus clays can also mutate. Further, organic monomers, including amino acids, are readily adsorbed on, and even polymerized on, clay surfaces. A. G. Cairns-Smith has argued that the genetic code had an inorganic origin in clays.

Silicon does form another interesting series of long-chain compounds called *silicones*. The silicones are built up of chains of alternating Si and O atoms with two other monovalent substituents on the silicon atoms. Examples include $(H_2SiO)_x$, $(Cl_2SiO)_y$, $[(CH_3)_2SiO]_z$, and $[HCH_3SiO]_x$. Each such polymer can form very long chains with thousands of units, but again the alphabet normally consists of a single letter endlessly repeated, which is not a good way to communicate information. But it is in principle possible to imagine information stored by alternating different units such as those listed above, in the manner of a genetic code. There is, however, the problem of what solvent to use to mobilize these chains. The normal state of silicones is an oily liquid of moderate to high viscosity, or a rubber, depending mainly on the chain length. It is the short chains, with low information content, that are fluid.

Such single-letter alphabets, in which the individual words spontaneously and randomly change their lengths, lack the stability and information-carrying capacity of carbon compounds. Carbon can participate in many bonding schemes that are potentially capable of encoding information. A few examples include:

$$H-CHR-CHR'-CHR''- \quad (XI.6)$$

$$H-C-CR{=}CR'-CR''{=}CR- \quad (XI.7)$$

$$H-[C_2]_x-R \quad (XI.8)$$

$$[(CHR)-O-]_x \quad (XI.9)$$

$$[\text{sugar-phosphate}]_y \quad (XI.10)$$

and so on. The first of these, Eq. (XI.6), is a saturated hydrocarbon with a variety of small substituent groups attached to the carbon-chain backbone. The substituent groups than become the "letters" that encode information, their sequence spelling out "words." Here the various substituent radicals, such as the hydrocarbon radicals –H, $-CH_3$ (methyl), $-CH_2CH_3$ (ethyl), and $-C_6H_5$ (phenyl) and the polar radicals –OH (alcohol), $-NH_2$ (amine), –CN (nitrile), –COOH (carboxylic acid), and –CHO (aldehyde), are symbolized by R, R′, R″, etc. Pure hydrocarbons are soluble principally in nonpolar solvents such as hexane ($C_6H_{14}$), heptane ($C_7H_{16}$), or octane ($C_8H_{18}$), whereas those with substantial numbers of polar substituents may be quite soluble in water. The most abundant natural solvent for hydrocarbons is crude oil, which is vastly less abundant than water in the Universe. An interesting special case is the class of long-chain carboxylic acids, such as stearic acid, $C_{17}H_{35}COOH$, which has a long hydrocarbon chain that is *hydrophobic* (water-fearing) and a carboxylate group on one end that is *hydrophilic* (water-loving). These *fatty acids* can bind oils and greases with their nonpolar hydrocarbon end and dissolve in water on their carboxylate end. They therefore are effective in mobilizing and dispersing oily materials in water. Such a material is called a *soap*. Head-to-tail arrangements of such molecules, with an "oily" or "fatty" layer sandwiched between inner and outer wet surfaces defined by polar functional groups, produce *lipid bilayers*, the prototype material for cell walls.

The second scheme for bonding into chains, Eq. (XI.7), is a conjugated-bond chain of carbon atoms (essentially substituted ethylene units). Third, Eq. (XI.8), is a polyacetylene of the sort found in interstellar molecular cloud complexes. The R terminating the polyacetylene chains seen in the ISM is usually –CN; however, note that, irrespective of R, a molecule with only one substituent "letter" per molecule is poorly equipped to carry information. Polyacetylene hydrocarbons (where R is H or a hydrocarbon radical) are probably even more common than polyacetylene nitriles in nature, but lack a significant dipole moment and are therefore undetectable by rotational spectroscopy. Eq. (XI.9), illustrating a substituted polyformaldehyde chain, is reminiscent of the unsubstituted $-(CH_2O)-$ polyformaldehyde chains found by the Giotto mass spectrometer experiment in the coma of P/Halley. As appealing as this structure seems, with up to two letters of coding as $-(CRR'O)-$ on each carbon atom, these chains suffer the severe disadvantage of being very readily hydrolyzed in water. Their only known biological function on Earth is as a poison for killing slugs and snails. Equation (XI.10), with sugar molecules linking phosphate units into long chains, can encode information either by having different sugar molecules S, S′, S″, etc., in the chain, or by having a single essential sugar that carries a variable substituent that constitutes the encoding letter. The latter is the approach used in DNA and RNA, which encode all genetic information on Earth (see Figs. X.57 and X.58). These molecules are stable enough so that they do not spontaneously hydrolyze in water.

Based on these considerations, and insofar as we understand the nature of life, it appears probable that life will generally be based on the chemistry of carbon, especially polar carbon compounds. Carbon's astonishing ability to produce complex, diverse, and stable long-chain compounds that dissolve in water and in a variety of other polar solvents makes it very attractive. It is premature to rule out other chemistries absolutely, but the fact remains that, after a diligent search, we have not yet found any other molecule/solvent system of sufficient stability, flexibility, and complexity to serve as a plausible basis for advanced life forms.

The necessity for mobilizing polar carbon compounds so as to permit the solution of a wide variety of biochemicals (and to allow physical flexibility of the organism) requires a solvent. By far the most abundant solvent in the Universe is water, a compound of the first and third most abundant elements in the cosmos. We know from terrestrial experience that water works, but we lack direct experience with any other solvent for life. Nonetheless, several other candidate solvents come to mind.

The second most abundant polar molecule in the Universe after water is ammonia. In many respects, liquid ammonia seems to be an attractive substance to serve as the solvent in living tissue. There are, however, three major problems standing in the way of its widespread use in living systems. First, ammonia is highly mutually soluble with the much more abundant water. When temperatures are low enough for ammonia to be condensed in the liquid (or solid) state, water is already condensed, usually in much larger abundances. Thus the fluid phase that results from ammonia condensation is actually a moderately concentrated aqueous solution of ammonia (about 10% ammonia and 90% water, roughly 5 moles of ammonia per liter of solution), not liquid ammonia itself. The eutectic temperature in the ammonia-water system, as we saw in Chapters V and VI, is near 173 K. Second, the liquid range of ammonia at normal pressures is much smaller than that of water, and lies at lower temperatures. Third, our exploration of the Solar System so far has in fact turned up numerous examples of the stable natural occurrence of liquid water, but not a single example of natural liquid ammonia, in agreement with theoretical expectations. A fourth potential objection is sometimes raised: that ammonia in the form of cold aqueous solution is, like liquid ammonia itself, at too low a temperature for normal biological reactions to take place. This latter objection seems to be balanced by two other factors related to aqueous ammonia solutions. First, ammonia assists in broadening the prospects for alien life by extending the liquid range of water by 70%. Second, the biochemical reactions that would evolve in such a solvent would not be identical to those that evolved on Earth in "normal" (i.e., familiar to residents of Earth) water, and hence the awkwardness of cold aqueous ammonia as a solvent for these "normal" reactions is irrelevant. Thus our exploration of liquid ammonia as a medium for life leads us once again to water.

Other polar molecules are known: liquid hydrogen sulfide $H_2S$, phosphine $PH_3$, and hydrogen chloride HCl are all conceivable, albeit far rarer than water and ammonia. Of these, phosphine has only a very small dipole moment. On terrestrial planets Cl, P, and S are strongly bound up in stable chloride, phosphate, phosphide, sulfate, and sulfide minerals, and hence are rapidly removed from the atmosphere by weathering reactions (Chapter X) when injected by volcanism or combustion. Hydrogen sulfide and phosphine both absorb UV very strongly out to such long wavelengths that they cannot be shielded by any plausible molecule except ozone; however, both are extremely vulnerable to oxidation and cannot survive in contact with any oyxgenic atmosphere capable of providing an ozone shield. HCl is so highly soluble in water that it dissolves in any aqueous liquids present. The atomic abundance of Cl is so low relative to oxygen ($Cl:O = 5.2 \times 10^3/23.8 \times 10^6 = 1:4.6 \times 10^3$) that, even if no chloride minerals are formed, HCl would occur as an aqueous solution of approximate concentration 0.012 moles per liter. Thus HCl, like ammonia, would be far more likely to enter into solution in water, or, even better, into aqueous ammonia solution, than to form an independent liquid phase. The presence of ammonium chloride in freshly fallen CI chondrites (Chapter VIII) suggests that this process may indeed have taken place in the early Solar System. Since all volatile molecules with large dipole moments have far lower abundances than water and readily dissolve in aqueous solution, it is clear that water is the preferred solvent for use by life.

## The Planetary Environment

It is now appropriate to ask where liquid water can be found. The most familiar example is of course on or near the surfaces of terrestrial planets, in the region in which the combined effects of insolation and the greenhouse effect provide surface temperatures that are within the liquid range somewhere on the planetary surface. It is not necessary to present a theoretical rationale for why this might be an interesting case, since the readers of this book are likely to be residents of water-bearing planets, and can draw upon their own experience. Note that the range of heliocentric distances within which liquid water might be stable (the liquid water zone) is shifted outward from the central star by greenhouse warming of the planetary surface. The liquid range, usually taken to be 0°C to 100°C, can be significantly extended either by the addition of solutes such as ammonia or of involatile salts that depress the freezing point and elevate the boiling point, or by an elevated total pressure, which likewise depresses the freezing point and raises the boiling point. Highly volatile solutes such as ammonia may, as we have seen, depress the freezing point spectacularly (to –100°C in this case) but have little effect on the boiling point.

A second potential site for water (or for ammonia–water liquids) is within the hydrospheres of large icy satellites. A familiar example is the liquid water layer on Europa, beneath a thick ice crust. There an ocean tens of kilometers in depth can survive for very long periods of time despite the low steady-state temperature of the surface, assisted by tidal heating of Europa's interior. Hydrothermal activity may provide a rich source of disequilibrium organic and inorganic material at the base of the ocean, analogous to the biologically frenetic and bizarre environments seen in the vicinity of hydrothermal vents on Earth's ocean floor. The same argument applies to Earth-sized planets of Ganymede-type composition, even though they are lacking in our own Solar System. Here the internal thermal energy needed to maintain a liquid-water layer beneath a permanent ice crust might come from an appropriate combination of warmer surface temperatures, addition of ammonia antifreeze, and internal decay of long-lived radionuclides. The production rate of organic matter by this mechanism, and the energy flux made possible by it, could be as low as $10^{-4}$ of the solar energy flux at Earth's surface if the only energy source is decay of long-lived radionuclides, or higher if, as on Europa, tidal dissipation is an important source of heat. In general, however, we must expect the biological productivity of such an ecosystem to be hundreds to thousands of times lower than that of Earth's biosphere.

The third natural site of occurrence of liquid water in the Solar System is in the clouds of the Jovian and Uranian planets. Here abundant dissolved ammonia is the norm. Cloud droplets of aqueous ammonia solution do not have very long lifetimes: small droplets are readily lifted by updrafts to altitudes at which the droplets freeze and become exposed to DNA-destroying ultraviolet radiation from the planet's star. Large drops of aqueous salt solution may sediment out of the main cloud layer to fall to levels so warm that the solvent evaporates. Turbulent overturn of the atmosphere will repeatedly mix the desiccated remains of these droplets, both organic and inorganic, down to levels with sterilizing temperatures. Further, the temperatures in the upper tropospheres of giant planets tend to decrease with time as the internal heat of the planet leaks out, driving isotherms deeper into the planet where convective overturn is more important relative to photochemical production rates of organic molecules. Thus chemical evolution in the atmosphere of a gas-giant planet is hindered by the utter absence of stable locations for the accumulation and propagation of prebiological chemicals and simple life forms. Although it is certainly possible to imagine large, complex organisms that ride on hydrogen bladders and maintain fixed altitudes and constant temperatures in the atmosphere (*vide* Carl Sagan's "gas bags"), the evolutionary path to such structures seems fraught with deadly and unavoidable hazards. It would be rash to rule out utterly the origin of complex life under such circumstances, but the difficulties seem immense.

These difficulties may be ameliorated to some degree by postulating super-Jovian planets, which cool more slowly than Jovian planets, preserving liquid water clouds for billions of years. This apparent temporal advantage, however, is largely offset by the very long time required for the most massive super-Jovian planets to cool enough for liquid water to begin to condense.

Whether or not we take seriously the third option above, the existence of life seems to imply the necessary presence of terrestrial-type planets, of icy moons orbiting about giant planets, or of planet-sized, Ganymede-composition ice balls. These bodies must deliver environments in which liquid water or aqueous ammonia solutions are stable. The temperatures in these environments may range throughout the two-component liquid solution stability field in Figure V.14. Since some of these environments can be at pressures of tens of kilobars, an upward extension of the liquid water stability field to the critical temperature might seem warranted. However, because the adiabatic gradient in liquid water is so small, and because the temperature at the top of the liquid water ocean is constrained to lie on the freezing curve of ice (or ammonia–water), it is very difficult to achieve liquid temperatures above 100°C. Indeed, a floating ice crust requires that the crustal phase be ice I (the only form of ice less dense than water), and that the temperature at the crust–mantle boundary lie on the branch of the freezing curve between 0 bar and 2 kbar (see Figure VI.11), with temperatures below 0°C. Thus the basal temperature of the "ocean" is not likely to be much warmer than 10°C. An Earth-sized planet with an ice I crust with a basal pressure of 2 kbar (and temperature of −30°C) will have a crustal thickness of $dz = P/\rho g$, which must be on the order of 20 km. The entire liquid ocean is then at a pressure greater than 2 kbar, and there is the strong likelihood that it will freeze completely to make high-density forms of ice, which would be better described as a dense-ice mantle than as an ocean. The high-density ice will rest on a silicate surface thousands of kilometers deeper than the putative liquid layer, and hence hydrothermal activity will be very thoroughly separated from any liquid layer that may exist at the base of the crust.

Clearly the optimal case is one in which there is just enough ice to generate pressures close to 2 kbar at the base of the ice, with the local temperature there not far below 0°C. The coldest surface temperature that can be achieved without freezing the entire interior is realized if there is no crustal convection and maximal crustal thickness (we will use 20 km). Then the temperature gradient in the crust will be that appropriate for steady-state conductive transport of the planet's long-lived radiogenic heat to the surface. That gradient is dependent upon the size of the planet and the crustal thermal conductivity, but must be less than about 4 K km$^{-1}$. The surface temperature then could lie in the range from 0°C (near zero crustal thickness) to −80°C. For such a range of models, a liquid water ocean could rest on the silicate surface and be directly influenced by hydrothermal action, while the crust would remain frozen. The most promising version of the "Ganymede-like" model therefore leads us to a "Europa-like" composition and structure.

Planets satisfying the terrestrial-planet scenario should have surface temperatures that are somewhere in the range 0°C to perhaps 50°C, well within the range allowed by the vapor pressure equation of water. For a surface above 50°C, the high vapor pressure of water will almost certainly lead to a runaway greenhouse effect and total evaporation of the ocean. For a surface temperature everywhere below 0°C, we have effectively adopted the Europa-type model. It is of course allowable to have a planet with Earth-like surface temperatures and vast amounts of water, so that the oceans are much deeper than on Earth. But if the basal pressure of the ocean is above 2 kbar, there is always the danger that the base of the ocean will freeze to make a layer of high-density ice. Earth itself, where the pressure at the bottoms of the deepest oceanic trenches is about 1 kbar, is not far removed from this fate.

A planet need not deviate strikingly from Earth's composition for it to have a very different hydrosphere. Consider, for example, an "Earthlet" with a bulk composition identical to Earth, but only half the radius. With less self-compression than Earth, this roughly Mars-sized Earthlet has half the radius, a quarter the surface area, an eighth the volume, and a tenth the mass of Earth. Suppose, rather generously, that the outgassing efficiency of water from Earthlet is the same as from Earth. Then Earthlet would have a tenth as much water spread over one quarter as much surface area. The average depth of the water would be only 40% of what it is on Earth, or about 1.2 km instead of the 3 km of water above Earth's abyssal plains. If the outgassing efficiency were less, there would be even less water. The acceleration of gravity on Earthlet, proportional to $M/r^2$, would be $0.4\,g_{\oplus}$. The average pressure at the bottom of the ocean, proportional to the depth of the ocean and to the acceleration of gravity, is only 16% of what it is on Earth, 50 bar instead of 300 bar.

Mountains on Earthlet likewise would weigh only 40% of what they would on Earth. If typical crustal rocks on both planets have a crushing strength of 2000 bar, then mountains could be built 2.5 times taller on Earthlet than on Earth. Mount Everest, which rises over 12 km above the abyssal plains, would pale to insignificance alongside Earthlet's highest mountain, which could rise as high as 30 km! (Note that the highest peak on Mars, which is not an Earthlet in composition although similar in size, is Olympus Mons, 27 km high.) The internal heat flux and the temperature gradient through the crust, according to Eq. (VI.25), are proportional to $\rho V/r^2 \propto \rho r$, or about 50% of Earth's, so that the melting and softening temperatures of rock and the base of the lithosphere (indeed, all isotherms) are about twice as deep on Earthlet. Since the crust is thinner for the same reason the oceans are shallower, the lithospheric base descends to well within the mantle.

The upper limit for the size of a terrestrial planet remains uncertain, but any size from 1.5 to about 2 times Earth's radius seems possible. We shall consider the most extreme plausible case, an "Earthissimo" with twice Earth's radius, four times its surface area, eight times its volume, and (because of greater self-compression) about 12 times Earth's mass. Its acceleration of gravity, proportional to $m/r^2$, would be three times that of Earth. The higher heat flow and crustal temperature gradient raise the lithosphere–asthenosphere boundary up well into the crust. Mountain heights, supported by isostacy, are limited to a third of their height on Earth, about 4 km. But the oceans would have 12 times the mass of Earth's oceans spread over four times the surface area, for a mean depth of about 9 km. Therefore the tallest mountain peaks would be more than 5 km below the surface of the ocean. Earthissimo would be a true ocean world.

Equally interesting is the pressure at the bottom of Earthissimo's oceans. With three times the water depth and three times the acceleration of gravity, the mean pressure at the ocean floor would be 2.7 kbar. According to Fig. VI.11, this ocean could freeze from the bottom up, depositing a layer of dense ice III, if the mean surface temperature of the planet is below about 260 K. The ocean would then be sealed in by a layer of ice I on top, and ice III on the bottom.

Earthlike planets should release other volatiles, such as carbon dioxide, nitrogen, and radiogenic argon, in roughly constant proportion to water. Earthissimos may therefore release the equivalent of 3 to 6 bar pressure of nitrogen and 180 bar of carbon dioxide. The fate of the carbon dioxide is difficult to predict. Normally, $CO_2$ on Earth dissolves in rainwater, forms a dilute solution of carbonic acid, precipitates onto the continents (about 30% of the time), weathers exposed minerals to extract $Na^+, K^+, Ca^{2+}, Mg^{2+}$, and other ions, and washes down rivers into the oceans as a dilute solution of alkali and alkaline carbonates. But on Earthissimo, no precipitation falls on exposed land. Carbon dioxide accumulates in the atmosphere to high pressures and dissolves in the ocean surface to make a much more concentrated solution of carbonic acid. At hydrothermal vents in the ocean floor, carbonate-rich seawater reacts with crustal-composition (alkali-rich) magma to produce abundant carbonate minerals. Since the internal heat source of Earthissimo is 12 times that of Earth, spread over four times the surface area, the density or intensity of hydrothermal vent activity should be several times higher than experienced on Earth's abyssal plains.

Terrestrial-type planets containing sufficient carbon, hydrogen, oxygen, nitrogen, etc., to make abundant prebiological organic matter must be common. The elemental raw materials for life are ubiquitous in star-forming regions, where organic matter is an important constituent of interstellar molecular clouds. Cometary and asteroidal bombardment of planets will provide a rich source of these elements. Although most of the complex organic molecules in comets and asteroids will be destroyed by the explosion of the impacting object, the elements needed for life are nonetheless brought in. A problem arises with Earthlets, planets that are substantially less massive than Earth, which can lose atmospheric gases through impact erosion.

While a young planet is still smaller than Mars, most of the gases released by explosion of volatile-rich impactors will be readily lost from the planet. More massive planets with higher escape velocities ($>5\ \mathrm{km\,s^{-1}}$) will be better suited for retention of an atmosphere. Processing of this $H_2O$–$CO$–$N_2$ atmosphere by disequilibrating processes depends on a number of factors. UV photolysis, which has been very important on Earth throughout the history of the planet, is dependent upon the presence of an adequate UV emission from the planet's star. Lightning discharges are probably much less common when the surface temperature of the planet is close to the freezing point, since little water evaporates and condenses, and latent-heat transport is accordingly very small. Cosmic ray irradiation is also probably very variable, in that stars with intense solar wind emission will be much more effective in screening their inner planetary systems from low-energy cosmic radiation.

Generally, the abundances of the elements and the chemistry of hydrogen-rich preplanetary nebulae will constrain the gross mineralogy and oxidation state of preplanetary solids to be quite similar to those in the Solar System: the oxidation state on the surface of a young planet will generally be dominated by ferrous iron, and any early atmosphere (Chapter X) would be suitable for the origin of organic material.

## The Stellar Environment

The large majority of all stars lie on the Main Sequence. They are the most stable and predictable stars, the ones least likely to destroy life on their planets. T Tauri stars, furious emitters of UV radiation and stellar winds, are rapidly evolving toward the Main Sequence. T Tauri lifetimes are comparable to or shorter than the time scale for planetary accretion. Red giants and supergiants are rapidly evolving post-MS stars, prone to extreme changes in luminosity or to explosive destruction of their planetary systems. White dwarfs and neutron stars are products of dangerous evolutionary processes that are capable of devastating any planets they may once have possessed. Thus the most common stars, those on the Main Sequence, are the most attractive abodes for life.

The most luminous MS stars, of the relatively rare O, B, and A spectral classes, have short MS lifetimes compared to the time it took life on Earth to evolve. It is likely that the last origin of life on Earth was between 4.0 and 4.4 Ga BP. About 0.57 Ga BP, life first became capable of building hard parts, leaving obvious fossils, and colonizing the continental shelves heavily. For roughly 3.6 Ga, therefore, life existed on Earth but remained in a rather primitive state. But the O, B, and A stars have MS lifetimes well under $10^9$ years. Our experience with Earth suggests that life should not be very advanced after $10^8$ years or so of evolution. It would be helpful if we knew what it was that triggered the emergence of terrestrial life from the seas. Was the biochemical evolution of life rather linear, most of it unseen in the fossil record, with the Cambrian appearing to be revolutionary simply because the innovations of that era left readily identifiable fossils? Or was evolution highly nonlinear? Would modestly varied environmental conditions have produced radically different rates of extinction and speciation? How much faster might life have evolved with a different star in place of the Sun? With a different impact history? We simply don't know. Nonetheless, we cannot dismiss our terrestrial history, which implies that the upper MS stars are too short-lived to produce advanced life. Further, such stars are uncommon. They thus appear to have two strikes against them.

The low end of the MS is populated by M-type red dwarfs with luminosities ranging from $10^{-2}$ down to $10^{-4}\,L_\odot$. They have MS lifetimes extending out to about $10^{12}$ years. If time is of the essence in the origin and development of life, then M dwarfs look very attractive. It is sometimes argued that the very low photospheric temperatures of M stars lead to virtually no emission of chemically active UV, and that this lack can impede evolution. However, recent observations of T Tauri stars suggest that young stars of all masses produce abundant UV during the accretionary era. The relative lack of UV from a mature M star is not necessarily a bad thing, since maintaining a planet in habitable condition for $10^{12}$ years or so is made much easier if hydrogen compounds such as water are not photolyzed to release hydrogen. An adequate mutation rate can be maintained in the absence of UV irradiation by decay of $^{40}$K and $^{14}$C and by cosmic rays.

Unfortunately, M dwarfs are active flare stars. From time to time, apparently at irregular intervals, they increase abruptly in brightness to as much as several times their normal luminosity. The mechanism responsible for these flares is not known. Late M stars (M6 to M9) are strong, irregular sources of X-ray bursts originating in these flares.

Situating a planet in orbit around an MS star at a distance that will ensure temperatures in the liquid water range leads to other complications. For example, consider an M0 MS star of luminosity $L_* = 10^{-2}\,L_\odot$. The rate of energy absorption by a planet of radius $r$ at distance $R$ from the star is

$$(dE/dt)_{\rm in} = L_*(1 - A)(r^2/4R^2), \qquad \text{(XI.11)}$$

and the rate of energy emission over the entire surface of the planet is

$$(dE/dt)_{\rm out} = \sigma\varepsilon T^4(4\pi r^2). \qquad \text{(XI.12)}$$

A thermal steady state balance, with $(dE/dt)_{\rm in} = (dE/dt)_{\rm out}$, requires

$$\sigma\varepsilon T^4(4\pi r^2) = L_*(1 - A)(r^2/4R^2) \qquad \text{(XI.13)}$$

or

$$16\pi\sigma\varepsilon T^4 R^2 = L_*(1 - A). \qquad \text{(XI.14)}$$

Earthlike temperatures would be found at 0.1 AU from the star. Leaving aside the question of whether any planet might form so close to a star, there is the problem that solar tidal forces, which are proportional to $M_*/R^3$, are $0.3/(0.1)^3 = 300$ times larger than those exerted by the Sun on Earth, or 16 times as large as the solar tides on Mercury. Thus the planet would certainly be despun, probably locked into a 1:1 spin–orbit resonance if the eccentricity of the orbit is not very large.

The oceans would freeze on the night side, and eventually radiative cooling of the dark hemisphere would cause carbon dioxide, nitrogen, and oxygen to fall as snow. Temperatures on the dark side, maintained only by the internal heat flux, would be on the order of 30 K. The narrow band of hospitable temperatures near the terminator would be immersed in a tenuous atmosphere of neon and helium, robbed of all chemical interest by the nightside cold-trap.

As if this were not bad enough, a planet close enough to a low-luminosity star to have Earthlike temperatures may fall inside that star's Roche limit. Consider the Roche criterion,

$$r_{\rm Roche}/R_* = 2.456(\rho_*/\rho_{\rm p})^{1/3}. \qquad \text{(XI.15)}$$

The density of an M5 star is about 30 g cm$^{-3}$, 6 to 7 times that of a terrestrial planet. With a stellar radius $R_*$ of about $1.5 \times 10^5$ km, the critical Roche distance is about $7 \times 10^5$ km, and, for a luminosity of 0.001 $L_\odot$, Earthlike temperatures would be achieved at a distance of 0.0317 AU, or $48 \times 10^5$ km. Thus the planet, though despun, would be safe from tidal disruption. For an M9 star, teetering at the very threshold of the Main Sequence, with a luminosity of $10^{-4}$ $L_\odot$, Earthlike temperatures would be achieved at a distance of 0.01 AU ($15 \times 10^5$ km), and the Roche limit would be near $4.8 \times 10^5$ km. Slightly less massive bodies, incapable of sustained fusion of hydrogen, would be less luminous and denser. Thus planets with Earthlike temperatures would have to be closer, and the dangers of tidal despinning and disruption are greater, in systems whose central body is a brown dwarf.

The general conclusion, summarized in Fig. XI.1, is that mid-sized MS stars (spectral classes F, G, and K) are apparently the best candidates for maintaining stable planetary conditions for long periods of time. That the Sun lies in the middle of the preferred range may reflect a profound universal principle, or it may merely demonstrate our inability to imagine how life could cope with conditions not already familiar to us.

Finally, one other environmental issue is of great importance: is there a linkage between the locations, sizes, and types of planets formed in a young stellar system and the properties of the central star? Could the mass, spectral type, and angular momentum of the star be closely associated with the general properties of the planetary system that evolves around it? Note that there are strong causal relationships among the mass, spectral type, color, temperature, and luminosity of MS stars. Furthermore, we saw evidence in Chapter II that the rotational velocities of MS stars decline precipitously with spectral type, with the O, B, and A stars rotating up to 100 times as fast as late MS stars. This is suggestive of angular momentum transfer outward from a star with a time scale comparable to the lifetime of a B star (several hundred million years). It may be a consequence of the transfer of momentum to a very dense, T Tauri phase solar wind, and thus be related to dust and gas sweeping from a nebula and angular momentum injection into the nebular disk. For MS stars belonging to the spectral classes of greatest interest to us hereafter, the angular rates are all very small.

Since the majority of all MS stars are members of multiple systems, there is some reason for concern that

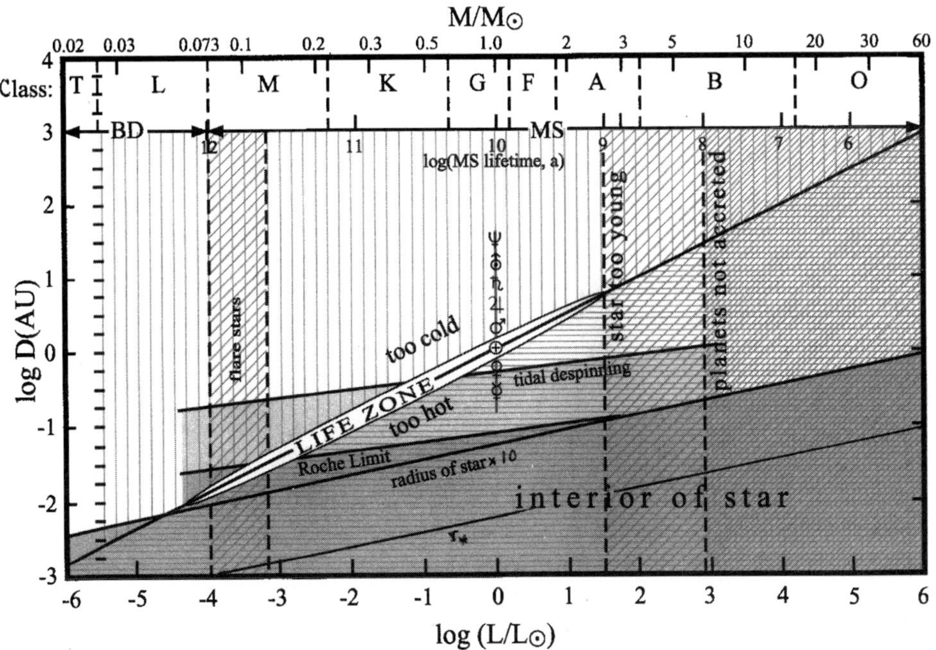


**Figure XI.1** Constraints on the locations of Earth-like planets. The liquid water zone is outlined for a wide variety of MS stars and brown dwarfs with a wide range of masses and luminosities. Constraints limiting the origin and development of advanced life forms include the requirement of sufficient time for evolution (MS star lifetimes of $10^8$ and $10^9$ years are indicated), the onset of tidal despinning for close planets of low-luminosity stars, and even tidal disruption of a planet at the Roche limit. Flare stars on the lower MS, which undergo radical luminosity changes at unpredictable intervals, are also indicated. The MS radius of the star and $10\times$ that radius indicate the volumes excluded by the present and past sizes of the star.

most stellar systems may be unsuitable for sustaining stable planetary orbits. However, a microlensing event, in which the passage of an unseen small companion in front of a star briefly focuses the light from the star, was recently observed in a double-star system, suggesting the presence of a Jovian planet at a distance of about 7 AU from a double star with a separation of about 1.8 AU. Theoretical modeling of planetary orbits in double star systems has also shown that distant planets of close doubles are often stable, as are close planetary orbits in distant doubles. Stellar multiplicity need not be an insuperable barrier to the existence of a stable planetary system.

## Brown Dwarfs

The terminology for describing bodies with masses less that those of the latest M-class red dwarfs is a bit complex. Bodies with masses of 0.001 $M_\odot$ (Jupiter's mass) to 0.013 $M_\odot$ are clearly Jupiter-like in their structure and behavior, devoid of any significant internal fusion energy source. These are best described as super-Jovian planets, an extension of the familiar Jovian planets such as Jupiter and Saturn that lie in the 0.0003 to 0.001 $M_\odot$ range.

Bodies with masses in the range from 0.013 to 0.06 $M_\odot$ are too small, according to present theory, to originate by rotational fission of a protostar, and are therefore expected to originate like planets from circumstellar disks. They are large enough to have experienced a brief episode of deuterium burning (about $10^8$ years at most), and can be regarded as stars that failed. Deuterium is burned principally by

$$^2\mathrm{D} + {}^1\mathrm{H} \rightarrow {}^3\mathrm{He} + \gamma \qquad \text{(II.64)}$$

at core temperatures of at least $4.5 \times 10^5$ K. Upon exhaustion of the deuterium supply, the body ceases to be a star. It then derives all its luminosity from its own slow shrinkage.

We saw in Chapter II that there are only four naturally occurring stable odd–odd nuclides, D, $^6$Li, $^{10}$B, and $^{14}$N, with both unpaired neutrons and protons in their nuclei. Of these, D has the highest abundance (see Table II.5) and the lowest Coulomb barrier, $^6$Li has the second highest abundance and Coulomb barrier, and so on. All isotopes of Li, Be, and B are relatively unstable in terms of nuclear bonding (see Table II.1). Lithium, more reactive than Be or B and more readily detectable than either, is an extremely useful index of how far nuclear processing has progressed in a young or low-mass star. Brown dwarfs between 0.06 and 0.072 $M_\odot$ are massive enough to experience both lithium and deuterium burning. Lithium is destroyed at core temperatures above $2.5 \times 10^6$ K by

$$^7\mathrm{Li} + {}^1\mathrm{H} \rightarrow ({}^8\mathrm{Be}^*) \rightarrow {}^4\mathrm{He} + {}^4\mathrm{He} + \gamma \qquad \text{(XI.16)}$$

It should be noted that burning of $^3$He by reactions such as

$$^3\mathrm{He} + {}^3\mathrm{He} \rightarrow {}^4\mathrm{He} + {}^1\mathrm{H} + {}^1\mathrm{H} + 2\gamma \qquad \text{(II.65)}$$

requires a core $T > 6 \times 10^6$ K for 10 Ga or more ($M > 0.15\, M_\odot$), and hence is ineffectual even in those brown dwarfs that successfully burn D and Li.

Since the time scale for lithium depletion is about $10^8$ a, and star-forming regions and their resulting open clusters have lifetimes of about the same duration, all BDs found in locations such as the Orion Nebula are expected to contain lithium, irrespective of mass. However, "field" BDs, those that are scattered throughout the spiral arms without any association with open clusters, are generally old enough so that all the BDs more massive than the threshold for lithium burning have in fact severely depleted their lithium content. For these older stars, Li content becomes a mass marker.

Finally, bodies with masses above approximately 0.072 $M_\odot$ and core temperatures above $10^6$ K are capable of maintaining nuclear reactions [via the proton–proton chain, Eqs. (II.63)–(II.66)] in their interiors for very long periods of time after passing through deuterium burning. These are simply low-mass MS stars with strong absorption bands due to gases such as TiO and VO. The minimum temperature of an M9 star, about 2200 K, is low enough so that refractory oxides are condensing out of the atmosphere to make fine dust grains.

Core convection is of fundamental importance in all lower MS stars ($M_* < 0.3\, M_\odot$). The convective heat flux

$$F_{\mathrm{conv}} \propto \rho \nu_{\mathrm{conv}} C_{\mathrm{p}} \delta T \qquad \text{(XI.17)}$$

where $\nu_{\mathrm{conv}}$ is the convection velocity (Prandtl mixing length theory) and $\delta T$ is the temperature contrast between an advecting parcel of warm atmosphere and the ambient gas. It can be shown that this expression is equivalent to

$$F_{\mathrm{conv}} \propto (Q/\gamma)^{1/2}(1/H_{\mathrm{P}})^2 \rho C_{\mathrm{V}} c_{\mathrm{s}} T S^{3/2} \qquad \text{(XI.18)}$$

where $Q$ is the isobaric coefficient of thermal expansion, $-(d \ln \rho / d \ln T)_{\mathrm{P}}$, $\gamma$ is the ratio of specific heats, $H_{\mathrm{P}}$ is the pressure scale height, $c_{\mathrm{s}}$ is the speed of sound,

$$c_{\mathrm{s}} = (\gamma RT/\mu)^{1/2}, \qquad \text{(V.129)}$$

and $S$ is the static stability of the atmosphere,

$$S = \partial\Theta/\partial z = \partial T/\partial z - (\partial t/\partial z)_{\mathrm{ad}}, \qquad \text{(V.167)}$$

where $T$ is the potential temperature, and the vertical gradient of potential temperature (called the static stability) is the difference between the ambient lapse rate and the adiabatic lapse rate appropriate to rapidly advecting eddies. The Rayleigh number Ra is the formal criterion of convective instability,

$$\mathrm{Ra} = \alpha g d^3 \delta T / K\nu \qquad \text{(VI.38)}$$

where $a$ is the coefficient of thermal expansion [Eq. (VI.30)], $d$ is a characteristic vertical dimension (in this case, $d = H_P$), $K$ is the thermal diffusivity, and $\nu = \eta/\rho$ is the kinematic viscosity. In lower MS stars and in brown dwarfs, Ra is of order $10^{25}$, assuring free convection and adiabatic structure. The Rossby number Ro, which parameterizes Coriolis forces, is

$$\mathrm{Ro} = v_x / fL \qquad \text{(V.164)}$$

where $v_x$ is the eastward (zonal) component of the convective wind speed (typically of order $10^2\,\mathrm{cm\,s^{-1}}$ for BDs), $f = 2\omega \sin l$ is the Coriolis parameter, and $L$ is the characteristic horizontal length scale, which, for three-dimensional turbulent eddies, is about the same as the vertical length scale, the pressure scale height $H_P$. Rotation speeds of late MS stars and BDs are generally at least $20\,\mathrm{km\,s^{-1}}$ and, in one case as high as $80\,\mathrm{km\,s^{-1}}$, which, for these grossly Jupiter-sized bodies, corresponds to $\omega = 10^{-3}\,\mathrm{rad\,s^{-1}}$ (a rotation period of 1.5 h) The Rossby number is then in the range $10^{-2}$ to $10^{-3}$, small enough to suggest that small-scale turbulence is inhibited. Nonetheless, the observed Li depletion proves that there must be some other mechanism for efficient vertical mixing. Larger-scale mixing cannot be inhibited by viscosity, since the Reynolds number

$$\mathrm{Re} = \rho v_{\mathrm{conv}} D/\eta = v_{\mathrm{conv}} L/\nu \qquad \text{(V.160)}$$

is larger than unity throughout the star. David Stevenson has argued that the magnetic energy density, which must be less than the convective kinetic energy density

$$B^2/4\pi < \rho v_{\mathrm{conv}}^2 \qquad \text{(XI.19)}$$

for convection to occur freely, remains safely small for magnetic field strengths less than $10^4$ G, which is only a little more than the actual fields measured in these bodies and inferred in their interiors. The photospheres of these bodies, however, are dominated by radiative energy fluxes and may be unable to sustain convection. Complete radiative–convective models are required to describe the thermal structure, energy fluxes, and vertical compositional structure accurately.

Maximum-size brown dwarfs (BDs) cool and shrink continually, their luminosities diminishing approximately as $L_* \propto t^{-5/4}$. The luminosity decrease is compounded of a shrinkage in radius and a cooling of its surface ($L \propto T^4 r^2$). Once the surface cools below about 1800 K, the reaction

$$\mathrm{CO} + 3\mathrm{H_2} \rightarrow \mathrm{CH_4} + \mathrm{H_2O} \qquad \text{(IV.104)}$$

generates significant amounts of methane (see Figs. IV.15 and IV.16 for the dependence of the methane abundance on temperature and pressure). By the time the temperature has fallen to about 1300 K, CO is a rare constituent and methane absorption bands, along with those of water vapor, dominate the near-IR spectrum. Those BDs that are rich in CO and cool enough for condensation of refractory oxides of Al, Ca, Ti, etc. (see Fig. IV.21), are called L dwarfs, whereas those with abundant methane are called T dwarfs. As of September 2002, there were 145 known L dwarfs and 31 known T dwarfs.

The one-to-one correspondence between mass and luminosity that we associate with MS stars does not apply to BDs because the absence of a fusion heat source allows the core and surface temperature to fall throughout its lifetime. One consequence of this cooling is that the Hertzsprung–Russell diagrams of BD populations are, unlike the H-R diagram of an MS population, strikingly time-dependent. Another consequence is that an old population of BDs would be very difficult to observe in visible light because these bodies are very faint and have Planck peaks in the infrared:

$$\lambda_{\max} = 2897/T\ \mu\mathrm{m}. \qquad \text{(II.44)}$$

The emission peak is typically longward of 2 $\mu$m for T dwarfs, which in any event are strikingly non-Planckian because of the strong IR absorption bands of water and methane. Dedicated near-IR search programs such as DENIS, SDSS, and 2MASS, which have been designed to allow detection of such cool sources, have succeeded in finding many T and L dwarfs. Any planet close enough to one of these old, cool, dense BDs to have Earthlike temperatures would lie well within the Roche limit.

Temperatures inside low-mass or old BDs may allow condensation of water clouds. The tropopause temperature is determined by a balance between radiation from the lower hemisphere with an intensity proportional to $T_{\mathrm{eff}}^4$ and the "black" (cold) upward hemisphere. The omnidirectional absorbed flux is half that of a gas completely immersed in a sphere of uniform temperature, and hence the temperature at the upper boundary of the troposphere, and the typical stratospheric temperature, is less than $T_{\mathrm{eff}}$ by a factor of $2^{1/4}$ (1.19). Thus even a planet with an effective temperature above the melting point of ice I may have tenuous ice clouds near the tropopause. Figure XI.2 shows the condensation sequence for low-temperature condensates in a system with solar elemental composition. The dashed T-P profiles labeled L and T illustrate the cloud structures at three different stages of cooling

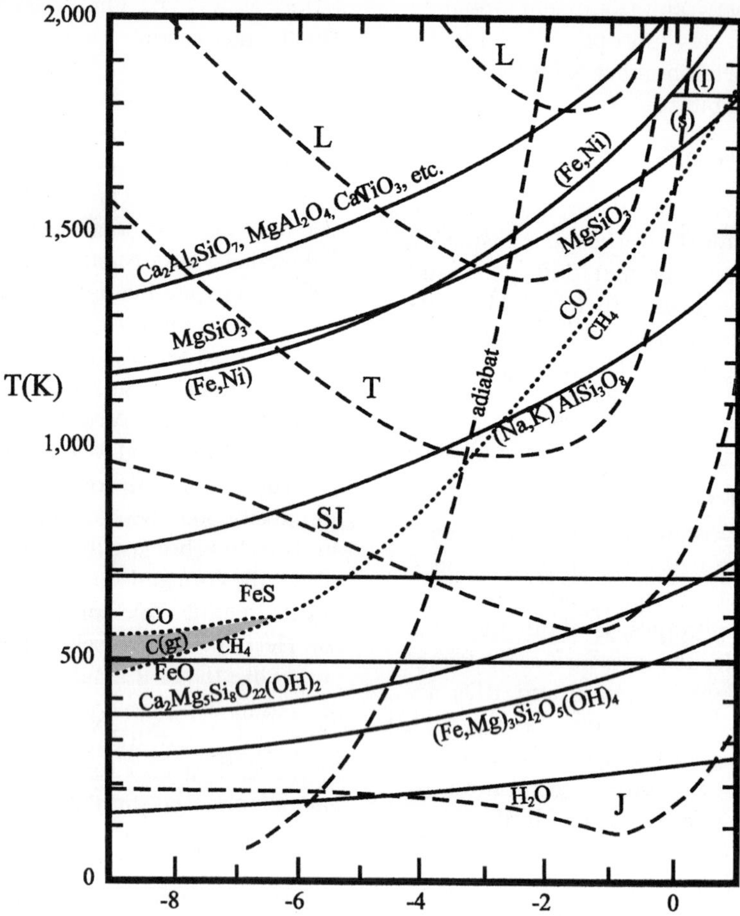


**Figure XI.2** Condensation in the atmospheres of brown dwarfs and Jovian planets. The dashed curves are temperature profiles for a Jupiter-like planet (J), a super-Jovian planet, and brown dwarfs (T, L) of different masses and ages. Young, hot brown dwarfs (L) have cloud layers of refractory oxides, metallic iron–nickel, or magnesian silicates near the photosphere. The dotted line marks equal abundances of CO (dominant at high temperatures) and $CH_4$ (at low temperatures) and the graphite precipitation region near 500 K at low pressures. Note that an extensive portion of the atmosphere of the T dwarf lies within the stability field of methane and below the condensation threshold of alkali metals. .

of brown dwarfs. Note that adiabatic structure ceases to be reasonable at and above the tropopause, where radiative transport dominates and the atmosphere becomes roughly isothermal. The need for a radiative–convective model at high altitudes and low pressures is indicated by truncating the adiabatic profiles at about 300 mbar pressure and sketching in a qualitative transition to the isothermal stratosphere. The lower adiabat in the figure is similar to the present-day structure of Jupiter's atmosphere. Note that it takes even a T dwarf many billions of years to achieve a rich cloud structure with biological interest. Such an atmosphere would require either an external UV source (a star) or abundant lightning discharges in the cloud region to produce prebiological organic matter. Energy fluxes through the atmospheres of field BDs that have reached this stage in their evolution will be similar to those through Jupiter's troposphere [see Fig. (V.30)], of order $10^4\,\mathrm{erg\,cm^{-2}\,s^{-1}}$.

The importance of brown dwarfs in the future development of planetary science cannot be overestimated. The available evidence on the population statistics of brown dwarfs suggests that they may be more abundant than all the visible stars. The large majority of them must be old enough so that they would evade detection by even the most advanced search programs operating in the 2-$\mu$m region, since they may have Planck peaks as far out as 20 $\mu$m. Many nearby stars must have brown dwarf companions, some too small and faint to detect. Also, it is a real possibility that the

Sun's nearest neighbor in the Galaxy is a solitary brown dwarf.

## The Search for Planets of Other Stars

The theoretical complexities surrounding the origin of other planetary systems, the nature and evolution of the bodies in them, and the possible origin and evolution of life on their planets can be minimized by observing other planetary systems directly. It is important to realize that we are just now developing several different techniques for detection of planetary systems about nearby stars. The discovery rate is high and rapidly growing. At the time of this writing (September 2002) more than 100 planets of other stars have been discovered. An "encyclopedia" of extrasolar planet discoveries is maintained by Jean Schneider of the Observatoire de Paris on the observatory Web site, http://obspm.fr/planets.

The first proposed, and first attempted, method of discovery of extrasolar planets was the *astrometric* technique, in which the position of a star is measured with high precision over periods of many years. The periodic motion of the star and its largest planets about the center of gravity of the system causes an observable perturbation in the position of the star if the ratio of the mass of the planets to the star is large and the distance of the system from the Sun is not great. The detectability of a system by this technique, proportional to $m_p/M_*$, is then is maximized when the planet is of at least Jovian mass, the star is an M dwarf, and the star is within a few parsecs of the Sun. At greater distance, this technique fails for two reasons: the likelihood that a star as faint as $10^{-3}$ to $10^{-4}\,L_\odot$ would be catalogued and studied at that distance is quite remote, and the angular amplitude and rate of motion become very small. Clearly observations would have to be extended over excessively long periods of time in order to confirm the presence of a Jovian planet with an orbital period of 100 years!

One crucial factor in assessing the feasibility of such an experiment is the size distribution of planets. As recently as 1995 there was essentially no information on the abundances of bodies intermediate in mass between Jupiter and M8 dwarfs because none are present in the Solar System, and those outside the Solar System are so faint that the likelihood of discovery was nearly nil. The MS luminosity functions derived by Luyten and a generation of successors (see Fig. II.13) could not be extrapolated beyond about spectral class M6 because of the inability to sample fainter stars with photographic techniques (which are poorly suited for detecting cool, faint, very red bodies). Bodies in the mass range below M9 have luminosities of $10^{-10}$ to $10^{-4}\,L_\odot$, with surface areas comparable to Jupiter's and effective temperatures ranging from about 100 to a maximum of about 2100 K. The Planck functions for these temperatures peak in the infrared between 60 and 1.5 $\mu$m wavelength. These bodies therefore cannot be discovered at visible wavelengths. Further, they emit so much less infrared than an M star that they are also hard to detect in the near IR. The best hope for detecting such a "brown dwarf" or Jovian planet was evidently to discover it in orbit around another star, rather than seek it as an independently moving "field" body.

Despite a long history of Earth-based attempts to discover planets using the astrometric method, the first significant fruits were borne by the European Space Agency's Hipparcos mission. Hipparcos has surveyed vast numbers of stars with unprecedented positional accuracy, including a number of systems in which planets have been found using other techniques.

The most successful of these other techniques for detection of planetary systems is to measure the *radial velocity* of the central star as a function of time. By the lever rule, the circumferences of the orbits about the center of mass of the system executed by the star and planet are inversely proportional to the mass of the body. Since the orbital periods are identical, the radial components of the orbital velocities of the two bodies must have opposite signs and magnitudes in the same proportion as the orbital circumferences: $V_p/V_* = M_*/M_p$. For remote detection of the Jupiter-Sun system, $V_* = 13.2/10^3$, or only $13.2\,\mathrm{m\,s^{-1}}$. Such a small signal could barely be extricated from observations of the Sun because the turbulent velocities in the photosphere, which must be averaged out to get the true radial velocity, are much larger than this. A star of 0.1 $M_\odot$ with a planet of 0.01 $M_\odot$ might, depending on the planet–star distance, have a radial velocity amplitude as high as a kilometer per second.

Two additional factors enter into the interpretation of radial velocity measurements. First there is the unknown orientation of the orbital plane of the system. If the planet–star system orbits face-on to the Sun, then no radial velocities will be seen. The measured velocities are in fact proportional to $M_p \sin i$, where $i$ is the inclination of the plane of the planet's orbit (the angle between the pole of the planet's orbit and the direction vector from it to the Sun). Second, there is the interesting question of where giant planets and brown dwarfs might form relative to their primaries. In M-type stellar systems, Jovian planets might form much closer to their star than the Jupiter–Sun distance. Small separations mean high orbital velocities but small amplitudes for the stellar orbital motion. Large separations favor use of the astrometric technique over the radial velocity technique.

In our present ignorance, both approaches must be pursued.

The masses of the planets of MS stars found to date range from about 0.12 $M_J$ to well over 13 $M_J$. The smallest are therefore possibly on the borderline between Jovian and Uranian planets, and the largest are well up into the brown dwarf range, Note, however, that the measured mass is really $m \sin i$, so all of the reported masses are for $\sin i = 1$ and are therefore lower limits. Some of the "super-Jovians" are probably brown dwarfs, and some of the apparent brown dwarfs are probably late M-type MS stars.

Since these detections are based on, at most, 7 years of data, only bodies with orbital periods of a few years or less are easily substantiated from the present data set. Within this bias, the orbital semimajor axes of the planets found by the radial velocity technique range from 0.038 to 5.9 AU and orbital periods range from an incredible 3 days up to 14.7 years. The earliest discoveries necessarily had very short orbital periods, giving the biased impression that the "typical" extrasolar giant planet was very close to its star. The 3-day figure is not a fluke: the orbital period of the planet of HD 83443 is 2.9861 days; the companion of HD 46375 orbits in 3.024 days; HD 187123's planet orbits in 3.097 days, HD179949's planet in 3.093 days, and several others have orbital periods of 3.1 to 5 days. The most recent discoveries, which include up to 7 years of data, tend to have larger orbital periods more similar to those of the giant planets in our Solar System.

Like semimajor axes, orbital eccentricities also range all over the landscape, from values indistinguishable from $e = 0$ up to a startling 0.927. Figure XI.3 displays the eccentricity–semimajor axis data for the first 101 discovered extrasolar planets. The highest eccentricity member of this set, the planet of HD 80606, with $a = 0.439$ AU, actually ranges from $q = a(1 - e) = 0.032$ to $Q = a(1 + e) = 0.846$ AU, so that the planet experiences a variation in the intensity of its insolation by a factor of about $[(1 + e)/(1 - e)]^2 = 720$ over each orbit.

There are several examples of systems with more than one detected planet. Upsilon Andromedae ($\upsilon$ And) and 55 Cancri (55 Cnc) each have three known Jovian planets, and seven other systems have two known planets.

The temperatures of these Jovian and super-Jovian planets can be astonishingly high, especially those that orbit close to F and G stars. The planet of the F6 MS star Tau Bootis ($\tau$ Boo), for example, is a 3.87 $M_J$ Jovian planet orbiting at $a = 0.0462$ AU, with an eccentricity of $e = 0.018$. The intensity of sunlight on the planet, governed by the luminosity of the star (about $9L_\odot$) and its extreme proximity, is about $1.1 \times 10^5$ times the intensity of sunlight on Jupiter. In the absence of its internal heat source, Jupiter would have a radiative steady-state

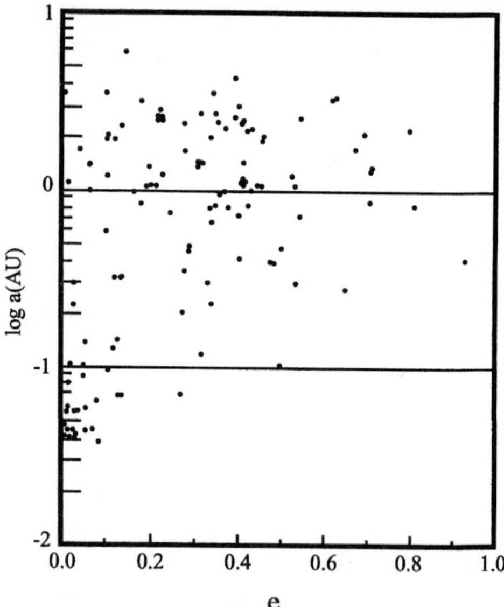


**Figure XI.3** Eccentricity–semimajor axis plot of the orbits of extrasolar giant planets. Note the "wall" near 0.04 AU and the prevalence of high eccentricities. The graph is incomplete on the top because of the finite history of observation. Inclination, either with respect to the line of sight from Earth or relative to the rotational equator of the parent star, is generally unknown.

temperature of only 103 K. Assuming a similar albedo for Jupiter and the planet, the effective temperature of $\tau$ Boo A would be roughly $103(1.1 \times 10^5)^{1/4} = 1875$ K! At these temperatures, clouds of refractory oxide particles and droplets of liquid iron would streak the face of the red-hot planet (see Fig. IV.27). The planet would be distended significantly by its high atmospheric temperature and consequent large scale height, a body nearly as bizarre as an ocean world or a brown dwarf. Thus the radial velocity technique has brought us face-to-face with entire new classes of planets that are without representation in our Solar System.

Clearly, however, these indirect discovery techniques can tell us nothing of the intrinsic properties of a detected planet except its mass (or $m \sin i$), and, by inference, its approximate surface temperature. Correlation of astrometric and radial velocity data on a system, when both are available, permits a good three-dimensional determination of the orbits and real masses of the largest bodies in it. But "seeing" Jovian exoplanets and even finding terrestrial-type planets remain unaccomplished.

The difficulties in the way of direct observation of planets in other stellar systems are immense. For example, consider an attempt by an observer at a distance of 5 parsecs from the Solar System to observe Jupiter directly by reflected sunlight. The greatest possible angular separation of the Sun and Jupiter is only 1 arc second. The intensity ratio Sun:Jupiter is about $10^9$ in visible light, and light diffracted by the telescope mirror edge and scattered by minor imperfections in the figure of the mirror will overlay the image of Jupiter with an approximately $10^4$ times higher intensity of scattered sunlight. The actual apparent magnitude of a Jovian planet orbiting 5.2 AU from a G0 star will be 22.5 magnitudes fainter than its star. Direct visible-light detection of planets and brown dwarfs is therefore not feasible even with a large telescope.

Increasing the aperture $l$ of a telescope causes the resolution to improve by a factor of $l$ and the light-gathering power to increase as $l^2$. Unfortunately, the cost of a telescope goes up as $l^3$ or faster. Actually, not much improvement in light-gathering power is required if the resolution is high enough to isolate the planet from the diffraction rings surrounding the star's image. This increase in resolution can be achieved by combining the light from two large telescopes of aperture $l$ separated by a baseline of length $L$. The resolution of such an instrument (an optical interferometer) is proportional to $L$, whereas the light-gathering power increases as $l^2$. The two 10-m telescopes at the Keck observatory in Hawaii are designed to be used as an optical interferometer. Further, because each telescope has so large an aperture, it is possible to do a very good search by teaming one of the 10-m telescopes with one or more much smaller, cheaper telescopes.

The detection of Earthlike planets at visible wavelengths is also very difficult: a typical terrestrial planet is ten times smaller in radius and in astrocentric distance than a typical Jovian planet. Assuming comparable albedos, the visible-light brightness of the two are about the same, whereas the terrestrial body is at one tenth the angular distance from the star. It is therefore much harder to detect.

Although the planet:star intensity ratio is very unfavorable in visible light, observations near the Planck peak of the planet rather than the star (near 50 $\mu$m rather than 1 $\mu$m for an M star or 0.6 $\mu$m for a G0 star) would find a star:planet intensity ratio of about $10^4$:1. Unfortunately, the resolution of a telescope drops off as $1/\lambda$: a telescope capable of an ultimate resolution of 0.2 arcsec at 0.6 $\mu$m would have a resolution of only 20 arcsec at 60 $\mu$m. Thus the great potential advantage of infrared detection can only be realized by using infrared interferometry.

Infrared emission is more intense from terrestrial planets than from Jovian planets because of their higher temperatures: a 300 K surface radiates 81 times as much heat as a surface with the same emissivity and area at 100 K. Unfortunately, terrestrial planets are about 10 times smaller in radius than Jovian planets, which lowers their thermal luminosity by a factor of 100, neatly compensating for their higher temperature. Although the thermal luminosity of an Earthlike planet and a Jovian planet are comparable, the angular separation of a terrestrial planet from its star is typically 10 times smaller than for a typical Jovian planet. This makes the detection of terrestrial planets difficult even in the infrared.

Nature provides another subtle clue to the presence of planets: occasionally, as seen from a nearby star, a Jovian planet will transit the disk of its star. Jupiter, transiting the disk of the Sun, would eclipse almost exactly 1% of the disk area, leading to a detectable brightness dip. However, the Sun normally has about 1% of its surface area obscured by sunspots, which vary on time scales of days due both to the time development of individual spots and to the rotation of the Sun. The presence and variability of these sunspots greatly complicates the detection of planetary transits. Nonetheless, the optical signatures of apparent planetary transits of the surfaces of other stars by the OGLE team in Poland and at Princeton University suggest roughly 60 more planetary candidates whose orbits are not yet determined. Note that this technique is limited to the small fraction of planetary systems in which the Sun lies close enough to the system's ecliptic that transits are observable from Earth. Ironically, planet–star systems in which the planet is large and the star is small, oriented edgewise to the Sun, at very small orbital separations, are the best targets—essentially the same criteria that favor the radial velocity technique. The strength of the stellar occultation technique is in its extreme photometric sensitivity.

The duration of a partial solar eclipse by Jupiter is always less than the diameter of the Sun divided by Jupiter's orbital velocity; indeed, from almost any external vantage point, Jupiter would *never* be seen to transit the solar disk. The maximum eclipse time is then about $2r_{\odot}/V_J = 1.4 \times 10^{11}\ \text{cm}/1.2 \times 10^6\ \text{cm s}^{-1} = 10^5\ \text{s}$, about 1 Earth day. Shorter chords across the solar disk are much more likely, with corresponding reductions in the duration of the event. Such eclipses by Jupiter and Saturn occur with a maximum frequency of one per 10 years ($3 \times 10^8$ s) for an ideally situated observer. Thus some 6000 observations would have to be made at the rate of one every 12 hours for 10 years in order to observe a single transit event. Of course, two completely unambiguous observations are required to constitute a discovery.

The *a priori* probability that an observer will be able to see *any* transits of Jupiter across the Sun is approximately equal to the ratio of the diameter of the Sun to the diameter of Jupiter's orbit, $1.4 \times 10^{11}/5.2 \times 1.5 \times 10^{13}$, or $1.8 \times 10^{-3}$. Therefore several million high-quality photometric observations (thousands of observations on each of a thousand stars) must be done to have a reasonable chance of detecting Jupiter.

At first sight it would seem advantageous to extend this observing program to M dwarfs, which are common.

In M dwarf systems the stellar diameters are much smaller and the orbital diameters of Jovian-type planets are probably much smaller, making transit events both much deeper and more frequent. However, the abundance and complexity of sunspot phenomena also increase toward the red end of the MS, interfering powerfully with the search for clean transit signals. Close super-Jovian planets greatly enhance the probability of observable transits. Even if such systems are rare, they should be disproportionately detectable.

To detect Earth-like planets around other stars, much more difficult techniques are required. It may come as a surprise to many to learn that three roughly Earth-mass planets have already been detected in orbit around another star—the pulsar PSR 1257 + 12. The first of these, with a mass of 0.015 Earth masses (similar to Earth's Moon), is in a near-circular orbit at 0.19 AU from the pulsar; the second, with $3.4M_\oplus$, has $a = 0.36$ AU and $e = 0.0182$; and the third, with $2.8M_\oplus$, orbits at $a = 0.47$ AU with an eccentricity of $e = 0.0264$. These planets may be either the surviving cores of extinct close Jovian planets, or postcataclysmic stellar debris condensed and accreted in a disk surrounding the dying star.

Detection of Earth by reflected solar radiation is very challenging over interstellar distances because of the nearly $10^9$:1 intensity ratio of the Sun to Earth at visible wavelengths. At a distance of 5 pc, the Earth–Sun system would have a maximum angular separation of 0.2 arcsec, so close that the resolution of the two sources would require a large, ultrastable space-based optical interferometer. (Opinion is divided over whether the best geometry is to attach the interferometer elements by means of a rigid boom, or for them to be free-flyers in an environment with very low gravity gradients.) The flux ratio is more favorable at near-infrared wavelengths close to Earth's Planck peak. With Earth's effective temperature of about 275 K, the Planck peak is close to 10 $\mu$m and the intensity ratio is only about $10^7$:1. Unfortunately, the same telescope that offered a resolution of 0.2″ or better at optical wavelengths can provide only 3″ resolution in the thermal infrared.

Once a terrestrial planet has been detected, spectroscopic analysis holds the promise of providing valuable information on its atmospheric composition. If the planet has an atmosphere within the familiar terrestrial range of compositions, water vapor will render the atmosphere opaque from the tropopause down to the surface at all wavelengths from about 20 $\mu$m to several millimeters. Near 15 $\mu$m is a very strong absorption band of carbon dioxide. Both water vapor and methane have strong absorption bands in the 7.0 to 7.5 $\mu$m region. A truly Earthlike planet will exhibit a strong ozone absorption feature in the 9.0 to 9.5 $\mu$m region. Although the 2 to 4 $\mu$m region is rich in valuable spectral data on trace molecular constituents, this region is so far from the Planck peak of both the MS star and the planet that there is very little light available for observation.

A variety of interferometric instruments and techniques have been suggested to detect and study extrasolar terrestrial planets. Careful study of the light-gathering power, angular resolution, and sensitivity of both Earth-based and space-based systems have already made it clear that the goal of finding terrestrial planets in orbit around stars out to distances of order 10 pc is achievable, albeit technically demanding and expensive. It is interesting that the Next Generation Space Telescope (NGST), the planned replacement for the Hubble Space Telescope (HST), operating at visible and near-IR wavelengths, could with modest enhancements conduct a spectroscopic survey of terrestrial-type planets out to distances of several parsecs, including detection of water vapor and oxygen. More advanced dedicated planet-finder spacecraft operating in the 4 to 20 $\mu$m region could measure carbon dioxide and methane abundances as well. With such information, informed conjectures regarding the evolutionary state of the planet may be made.

Remote detection of Earth from a distance of many parsecs would actually be easier at radio wavelengths than in any other spectral region: the presence of intelligent life on Earth has allowed the development of radio, television, and radar transmitters which inadvertently advertise our existence and location.

## The Search for Extraterrestrial Intelligence

After decades of fiscal starvation and extraordinary technological progress, plans to search for radio-frequency emissions from other intelligent species in our galaxy have finally begun to bear fruit. Large radio telescopes are now being employed in a search for non-random radio emission from many nearby stars under the name "Search for Extraterrestrial Intelligence," or SETI for short. More than a million frequency channels are being monitored by very fast computer systems in an effort to identify intelligent transmissions. Most of the data processing for this effort is done on a volunteer basis on home personal computers under the "SETI at Home" program.

The easiest signal to detect would be one intentionally designed to communicate over interstellar distances. Strong, simple signals, such as chains of pulses counting out the integers from one to 100, then the sequence of prime numbers over the same range, could be used to get attention. Much more detailed information could then be transmitted at a higher bit rate, and the entire sequence repeated indefinitely. The challenge of coding messages in a manner that permits any receiving party to decode the message efficiently has been the subject of many papers and arguments.

Much more likely "first contacts" are accidental transmissions, not intended for alien receivers. Earth is "brightest" (has the highest effective temperature) at the wavelengths at which large radar transmitters operate. The easiest terrestrial signals to detect would be the powerful pulses transmitted by anti-ballistic-missile (ABM) early-warning and space-surveillance radars, which convey no information whatsoever, but have enormous bandwidth power. Such radars serve an inadvertent role as "here I am" signs that, with present-day terrestrial receiver technology, are detectable across the diameter of the Galaxy. A hypothetical observer at the galactic antipodes would receive these content-free signals after some 60,000 years of travel time. The logical response for the alien observer might be, "What? I don't understand you!" One fine Tuesday evening, 120,000 years after the original radar pulses were sent out from Earth, this complaint arrives at Earth. Will anyone be listening? Will anyone remember the original transmission? Will radio technology still be in use? Will anyone still live on Earth? Clearly, this is no way to establish a meaningful dialogue.

Such radar transmissions are Earth-based and Earth-fixed. As Earth rotates and orbits about the Sun, these radars illuminate a haphazard portion of the celestial sphere in nearly inscrutable patterns. A remote observer would see a brief episode of very powerful pulses repeating at intervals of 1 Earth day. Attempts to decipher the presumed "message" would probably be unsuccessful: there *is* no message, but a sufficiently imaginative analyst might invent one! It is very likely that the diurnal variation would quickly be interpreted correctly as due to transmission from a planetary surface with a 24-h rotation period. Prolonged observation would reveal periodic Doppler shifts associated with both rotation and Earth's orbital motion. Integration of the Doppler shift over a day would reveal the radius of Earth (actually, the radius of the circle of latitude on which the transmitter lies), and over a year, the diameter (and period) of Earth's orbit, and hence the mass of the Sun. Radio interferometry would disclose the distance of the Solar System, and encourage discussion of whether to simply observe, send a reply, or dispatch a fumigation team. Most importantly, detection of Earth by this means would encourage examination of Earth's location in space with a number of sensitive instruments operating at many different frequencies. Detection of much fainter, information-rich signals would inevitably follow. These would include transmissions from the most powerful radio and television transmitters, especially national propaganda stations operating in the shortwave region. The expanding wavefronts from our earliest TV transmissions are now over 15 pc from Earth. The amusing picture of alien scientists sitting before their computers, watching in breathless anticipation as single frames of ancient terrestrial television broadcasts emerge, is unfortunately not very realistic. Still, it is amusing to wonder what intelligent aliens would make of Milton Berle, Korean War newscasts, Howdy Doody, Jackie Gleason, the Mickey Mouse Club, and the coronation of Queen Elizabeth—and how their sociologists and theologians would interpret the contents of these very mixed messages. It is also interesting to speculate whether they would find present-day programming to be an improvement. It is sobering to ponder whether our best programming efforts might actually *deserve* a visit from the Pan-Galactic Pest-Control Company.

Interstellar communication is clearly technically feasible. The number of communicating civilizations in the Galaxy is of course wholly unknown, and therefore it is quite impossible to make meaningful statistical estimates of the mean distance between nearest-neighbor intelligent civilizations. How many stellar systems develop planets suitable for the origin of life? On how many of these do life forms arise? How many of these achieve intelligence? Of these, how many develop a technology capable of detection by others, or capable of detecting others? How long does it take for a communicating civilization to arise? How long does it last? Are advanced civilizations likely to destroy themselves shortly after achieving the technical ability to do so? Are they likely to give up relevant technologies voluntarily? Are they likely to achieve nondetectable communications technologies (closed laser circuits, fiber-optic cables, etc.) quickly, and hence abandon radio and television transmission? Will they be sufficiently paranoid—or realistic—to avoid emitting signs of their presence? Are there "berserkers" in space that seek out and destroy upcoming young technological civilizations as a means of self-protection? Are there star-faring civilizations that actually have the means to travel between stars? None of these questions can presently be answered.

More easily answered are questions about which some relevant evidence is available: Can mankind build automated spacecraft capable of interstellar travel? The answer is "Yes." Can mankind build radio equipment capable of transmitting and receiving intentionally broadcast signals from remote locations in our galaxy?

Again, the answer is positive. Can mankind develop "world ships" that can carry a living crew on voyages of hundreds to thousands of years across space to other stellar systems? The answer again is almost certainly "Yes." Can mankind develop the physics and engineering skills to build ships capable of crossing interstellar distances at relativistic speeds within a human lifetime? The answer here is far less certain, but is probably again positive. Can we develop "faster-than-light drives," "hyperspace propulsion," or any of the myriad of other quasi-magical devices employed by science fiction authors to enable their heroes and heroines to flit about the Universe? Considering that much of what we take for granted today would have appeared utterly magical to an observer from only 100 years ago, it would be rash to dismiss this possibility out of hand. As a wise and experienced science fiction writer expressed it, any truly advanced technology looks magical to the uninitiated. Nonetheless, it is perfectly fair to say that the physical and engineering basis of such developments remains completely unknown: there is no rational basis for predicting specific advances of this sort. Neither is there any basis for their categorical rejection.

Something can, however, be said of the near future, and of the possible uses of physics and technology about which we now know something. That will be the theme of our final, brief chapter.

## Exercises

### *Chemical and Physical Prerequisites of Life*

XI.1 How many amino acids could be coded for by triplets of bases, using only ACGT?

XI.2 In a two-letter genetic code (A,B), how many letters per word are needed to permit coding for 30 different products?

### *The Stellar Environment*

XI.3 For planets in circular orbits about four different spectral classes of MS stars, calculate heliocentric distances $a_{min}$ and $a_{max}$ for the presence of liquid water on their surfaces. Assume a planetary albedo of 0.3. This is a good spreadsheet problem.

### *The Search for Planets of Other Stars*

XI.4 Planets are relatively easier to discriminate from their parent stars at infrared wavelengths. At what wavelength does the peak emission of Jupiter occur?

XI.5 Suppose your telescope system has a resolution of $0.2''$ of arc under ideal conditions, observing at 0.6-$\mu$m wavelength.

a. At what maximum distance could a planet and star 5.2 AU apart be resolved?

b. Using the same telescope at a wavelength of 6 $\mu$m, how close would the system have to be to be resolvable?

# XII. Future Prospects

Now that we have surveyed the Solar System and explored a wide variety of theoretical explanations of the phenomena we have encountered, it is useful to look forward to the prospects for answering the great questions we now have about the origin, evolution, composition, and structure of planetary systems. The observational and experimental prospects before us may conveniently be divided into Earth-based and space-based techniques. Of these, the Earth-based techniques are generally more susceptible to prediction. Continuing technological progress in detectors and in computational manipulation of huge data sets assures a constant stream of new data and insights arising from study of that portion of the electromagnetic spectrum that reaches Earth's surface. The Spacewatch, NEAT, LONEOS, and LINEAR asteroid search programs, the 2MASS, DENIS, and SDSS sky surveys, the OGLE stellar occultation detector, the SOHO solar physics and comet mission, and many others generate vast databases appropriate for "mining" by scientifically astute and computer-literate prospectors. Meanwhile, technological advances have increased the sensitivity of radio observations by a factor of 10 every few years since the 1950s, and highly sophisticated million-channel analysis of radio data is now under way (as in SETI). Tracking of distant spacecraft has benefited from these improvements in detector and radio telescope technology to the point that very distant spacecraft can be tracked despite their increasing distance and the slow degradation of their RTG power supplies: Pioneer 10, launched in 1972, is still transmitting at a distance of over 77 AU, and Voyager 1 is still returning data from a distance of 80 AU. Visible and near-infrared spectroscopy, pressing hard against limits imposed by the transmissivity and stability of Earth's atmosphere, continue to make new discoveries in the hands of skilled observers. Two-dimensional detector arrays, especially very large and sensitive mosaic Charge-Coupled Device (CCD) arrays, have become the workhorses of modern astronomy.

But it is still true that the large majority of our recent discoveries in the planetary sciences have arisen not from Earth-based observations, but from spacecraft missions. A wide range of issues of fundamental importance can be addressed only by means of space flight. Space-based remote sensing, as epitomized by the Hubble Space Telescope (HST), the Infrared Astronomical Satellite (IRAS), the Space Infrared Telescope Facility (SIRTF), the Extreme Ultraviolet Explorer (EUVE), the Chandra X-Ray telescope, the Einstein gamma-ray observatory, etc., makes the entire electromagnetic spectrum accessible from vantage points above the top of Earth's atmosphere. Flyby missions to every planet except Pluto have opened our eyes to the astonishing diversity and complexity of the Solar System, and have made it clear that few of our present questions can be answered by further flybys, which by their nature give high-quality data on a planet for only a brief period of time. Such missions give us invaluable insights into the nature of the target planet, sufficient to plan further detailed investigations, but with only a tantalizingly brief glimpse of important dynamical processes. Many urgent issues, such as topographic mapping of the surface and meteorological and aeronomical studies of the atmosphere of Venus, geological and geochemical mapping of the Moon and Mars, studying the magnetospheric dynamics of the Jovian planets, monitoring of volcanism on Io, and meteorological studies of the circulation of the atmospheres of Saturn, Uranus, and Neptune, require remote sensing from planetary orbiters over extended periods of time. Still other issues, such as the deep structure of the atmospheres and heat transport mechanisms within the atmospheres of the Jovian planets, mineralogical and geochronological studies of the Moon, Mercury, and Mars, and seismic studies of the interior structures of rocky and icy planetary bodies, require entry probes, landers, or rovers. In some cases, automated sample returns from a comet or Mars may be the approach of choice. And ultimately, if costs can be contained, human exploration beyond the Moon is a real possibility.

Unfortunately, our growing reliance on planetary spacecraft missions occurs during a prolonged period of high budgetary and political stress. From the 1960s through the 1980s the frequency of spacecraft missions dedicated to the exploration of the Solar System experienced a steady decline, from over a dozen missions per year to an average of less than one. Appendix III documents this history and briefly hints at some of the political context. But what of the future? In the wake of the collapse of international Communism, the fission of the Soviet Union into a number of independent states, and a dramatic decline in the military and space budgets of the former Soviet Republics, much of the traditional political motivation for space exploration has evaporated. It would be pleasant to contemplate a world in which the search for knowledge of the laws of nature and of the history of our planet was conducted on its own merits. Unfortunately, this has generally not been the case. Political leaders, including such otherwise divergent personalities as Adolf Hitler, Joseph Stalin, John Kennedy, and Charles DeGaulle, have seen space missions principally as a means of demonstrating national technical, military, industrial, political, and philosophical superiority.

The fundamental enabling factor for civil deep-space missions was the development of military ballistic missile technology. Most of the world's launch vehicles have been developed directly from military hardware. Until very recently, space activities have been a monopoly of governmental or quasi-governmental agencies with little or no incentive to lower the cost of access to space. When new launch systems enter the world market to compete for business, as Russia and the People's Republic of China found to their chagrin, the economies made possible by very large production runs (Russia) and very low labor costs (China) do not translate into competitive advantage: rather, they stimulate a host of governmental agencies in the Western nations to erect hurdles to suppress competition with their own national launch entities. In several nations, civil and commercial launches serve as a means of keeping missile production lines open, driving technological advances, and training launch personnel. There is more at stake here than mere economics. Thus the U.S. Defense Department hastens to defend the production of launch vehicles by its leading aerospace contractors, and the State Department raises the specter of inadvertent technology transfer in the form of highly sophisticated Western electronics in payloads sent to Kazakhstan or Sinkiang for launch. Politicians, no doubt inspired in part by American commercial interests in their own districts, complain about "dumping" of launch services on the Western market by "politically subsidized" foreign launch services (ESA, Japan, etc.) and worry publicly about how essential aerospace industries would suffer if exposed to the evils of unbridled international competition. Thus protectionist barriers, which should be anathema in a true market economy, obstruct the decline of launch costs by preventing meaningful competition.

In an era of level or declining research budgets (and modest inflation), the intransigent refusal of governments to allow a lowering of launch costs has had the effect of further freezing out exploratory missions. When research launch opportunities became rare, then naturally there was powerful pressure to fly every possible instrument on each future mission. Thus the unit cost of exploratory missions was further inflated, launches became even less frequent, and only a single unique spacecraft was built for each mission opportunity. The consequences of even a single mission failure became so disastrous that there was a strong incentive toward "gold-plating" of the payload and launch vehicle to ensure success. This forced costs even higher. With less frequent launches, the competence of launch teams declined, and the rate of launch failures increased. Insurance costs on commercial payloads soared, and the overhead costs of the manufacturer were spread over fewer launches per year, driving up the cost of boosters for research missions. This spiral continued unbroken until the aftermath of the Challenger disaster.

In the legal and regulatory environment of the past few years, the barriers in the way of competitive pricing of launch services have been reduced. But the psychology of most of the aerospace giants still reflects prior practice. The handful of American corporations that dominate the launch market are reluctant to change their way of doing business until forced to. The financial pressures that have driven several dozen major aerospace companies to acquire and merge with each other until only five survive suggest that such changes are long overdue. But still they are content with the hardware they are accustomed to building: billions of dollars, mostly provided by the government, have gone into the development of their products, and the aerospace companies, realistically expecting declining revenues from military contracts, are not inclined to invest in developing new-technology launch systems. This means that the competitive challenge to existing launch costs must originate with smaller aerospace concerns, especially new startup companies with little or no track record and inadequate capitalization. Insurers, who are naturally highly skeptical of untried boosters and new companies without track records, charge such prohibitive premiums that commercial payload owners rarely can afford to use the new cheaper boosters. Governmental agencies, generally anticipating decreasing budgets, are in no mood to help subsidize the development or flight qualification of new hardware. They feel it necessary to feed all their revenues into maintaining operational programs and retaining essential personnel. But this policy leads to shrinking employment, poor job prospects for new graduates with up-to-date skills, and an unprecedented aging of the labor force.

It is in the context of all of these negative factors that present plans for future exploration of the Solar System have been developed. The disparity between what we want (and know how) to do and what we actually do is wider now than at any time since 1960. Many extremely promising mission studies are carried out, but hardly any are funded. Those that do receive funding may expect serious reductions (euphemistically referred to as "descoping") or outright cancellation before launch. Nonetheless, it is possible to indicate the likely direction of Solar System exploration as presently foreseeable. The reader should keep in mind two overwhelmingly important considerations: first, the exploration of the Solar System may simply coast to a stop if present trends continue. Second, if an American aerospace industry "perestroika" occurs and launch costs are substantially reduced, a modern Age of Exploration could begin at once, building upon new technologies and new low-cost hardware, and featuring large numbers of small, special-purpose spacecraft, run off on production lines, with quick turn-around to capitalize upon new discoveries. To a large extent, this scenario reflects the intentions of Dan Goldin during his tenure as Administrator of NASA.

Let us now tour the Solar System as it might be seen in the coming 10 or 20 years if present levels of effort and present managerial philosophies are maintained, comparing our needs to our expectations for each exploration target in turn.

## Mercury

Mercury, having been visited by only a single flyby mission, poses a vast array of interesting questions. The composition of the surface and the bulk composition of the interior are potentially diagnostic clues to the mode of origin of the planet and the reason for its very high metal content. A highly refractory-rich lithosphere, resulting from massive evaporation of planetary surface material, would differ dramatically from a grossly chondritic crust and mantle, resulting from selective planetary accretion favoring metal. The structures of the gravitational and magnetic fields are poorly known. The abundances of radioactive elements in the crust are by themselves of great interest as clues to the origin of the planet, since potassium is volatile and uranium and thorium are refractory. A Mercury Orbiter could provide us with the geophysical and geochemical mapping data we need to solve most of these problems. A lander or instrumented penetrator would contribute a vastly richer store of compositional data on the surface and atmosphere, and also search for seismic activity driven by the tidal distortion of the planet. Finally, we have little more than a detection of the astounding polar ice deposits. We know almost nothing of their composition, history, and stability.

Studies of a Mercury Orbiter have been carried out sporadically by NASA since the late 1970s, immediately after the very impressive Mariner Venus–Mercury (Mariner 10) triple flyby mission. The principal uses of a Mercury Orbiter include mapping the entire surface of the planet by imaging its surface from a high-inclination orbit over the duration of at least 1 Mercury day, mapping the magnetic and gravitational fields of the planet in three dimensions, studying the interaction of Mercury's magnetosphere with the solar wind, and determining the elemental and mineralogical composition of Mercury's surface for the purpose of unraveling its genetic relationship to the other planets. A Mercury Orbiter requires a very high total propulsion energy in order to cancel out the high relative velocity of approach to Mercury of a spacecraft launched from Earth. Thus payloads are likely to be small even when launched by extremely capable boosters such as the Titan 4 Centaur.

Three independent Mercury missions have received considerable study in the last few years. The MESSENGER mission in NASA's Discovery program, planned for launch in late March of 2004, would be the first of these orbital missions. MESSENGER would arrive in Mercury orbit after a series of gravity-boost encounters with Earth and Venus. The European Space Agency is simultaneously developing plans for the Bepi Colombo mission, which would be launched in 2009. Rendezvous with Mercury and entry into orbit around the planet would be accomplished by means of an ion engine using xenon as the working fluid. The Japanese ISAS agency has also pursued plans for a Mercury Orbiter mission, also proposed for launch in 2009. That mission has been cancelled, but prospects remain open for ISAS collaboration with ESA on the Bepi Colombo mission. As presently conceived, Bepi Colombo would consist of three independent spacecraft elements, which would make collaboration with ISAS relatively easy. The first element, the Mercury Planetary Orbiter, is a low-altitude planet-oriented mapping and analysis mission, stabilized in all three axes, with nadir-pointing remote sensing instrumentation and with ancillary uses in radio science experiments. The second element, the Mercury Magnetospheric Orbiter, would carry an assortment of particles-and-fields instruments in a highly eccentric orbit around Mercury, for the purpose of carrying out a three-dimensional survey of the magnetosphere of Mercury. The third, and most ambitious, element of the mission is the Mercury Surface Element, which could consist either of a single analytical device soft-landed on the surface, or of two instrumented penetrators that would impact at moderate speed and carry out chemical analyses and heat-flow measurements to a depth of several meters.

## Venus

Many features of the atmospheric composition of Venus, such as the detailed elemental and isotopic composition of the noble gases, the reported dramatic variations in the sulfur dioxide abundance, and the hydrogen chemistry of the lower atmosphere, require further study. The implications of some (water-rich) atmospheric evolution models for the oxidation state of the crust and upper mantle could readily be tested by a wider and more detailed survey of the elemental and mineralogical composition of the surface, a difficult and demanding task in light of the severe surface conditions. The details of the mechanisms for hydrogen escape and for isotopic fractionation need further study to see whether Venus might be a former water-rich Earth-like planet that "went bad." The morphologies of impact features below the 20-km diameter crater size peak, the multiplicity of large impact features, and evidence of airburst explosions are all of great interest in assessing the effects of impacts on Earth.

Since the Magellan radar mapper mission in 1989, the only spacecraft to study Venus has been Galileo, as it flew by Venus in 1990 for a gravity assist en route to Jupiter. NASA's Venus Environmental Satellite, which was to be dedicated to the study of the atmosphere and solar wind interactions, has not received approval or funding. Many studies of other future missions have been carried out by NASA, including other meteorologically oriented orbiters and even survivable landers to study the red-hot surface, but no flight program presently exists. The Soviet Union's very long and intensive program of exploration of Venus under the Venera and Vega programs has come to a pause, and no Russian return to Venus is now foreseeable. ESA has tentative plans for the Venus Express mission, to be launched in 2005. The closest thing to a planned launch to Venus is the 2007 Planet-C Venus Orbiter mission under development by Japan's ISAS. It would extend the capabilities demonstrated by Japan in its launches to Halley's Comet and the Moon, giving Japan its first spacecraft mission to another planet. The mission as presently conceived would reach orbit around Venus in 2009 and study the deep atmosphere by means of imaging in narrow, carefully selected transmission bands in the infrared.

## Earth's Moon

So much time has elapsed since the end of the Apollo era that it is hard to believe that we still have many fundamental unanswered questions about the Moon—and that many of these are the same questions we had in the mid-1970s. Indeed, since 1972 we have seen only four spacecraft investigations of the Moon: the launch of the Hiten and Hagormo (MUSES-A) lunar orbiter and subsatellite by Japan in 1990, a flyby of the Moon by the Jupiter-bound Galileo spacecraft in December 1990, the Clementine mission of 1994, and the Lunar Prospector mission in 1998.

After an 18-year gap in lunar missions, we have resumed by sending a few small spacecraft to orbit the Moon. The last landings on the Moon came as part of the Apollo program in 1969 through 1972. As a result, our geochemical and mineral maps of the Moon have been greatly improved, as has our knowledge of the lunar gravity field and of the ice-bearing polar regions. But, in the absence of landers, we still lack useful seismic constraints on the size or even the presence of the lunar core, which geochemical arguments suggest may make up about 2% of the mass of the Moon. Despite our orbital mineral and elemental composition maps, we still have a very limited idea of how variable the local composition may be: whether, for example, there are ore bodies of interesting and useful materials of small horizontal extent but very high concentration. We do not know the composition and origin of the ice in the lunar polar regions.

Several lunar missions have been studied in detail and have approached the status of approved flight missions, but political and fiscal approval has been very difficult to secure. NASA studied the concept of a polar-orbiting spacecraft to study the Moon beginning in the late 1970s, at the end of the Apollo program. Variously named the Lunar Polar Orbiter, Lunar Geosciences Observer, etc., these studies culminated in the Lunar Observer mission. NASA has no pending plans for future lunar missions. One NASA scheme, current in the 1990s under the name of Artemis, was a hotbed of exciting ideas regarding everything from resumption of basic scientific reconnaissance of the Moon to the evaluation, extraction, and use of lunar resources. All it lacked was funding.

A closely comparable mission contemplated by the Soviet Union in the 1980s has also failed to win funding. In the late 1980s a similar ESA mission named Polo was studied extensively, but failed to receive political endorsement. The next lunar mission to fly is likely to be the Lunar-A Japanese orbiter in 2003, to be closely followed by a lunar seismic network deployed by the first Japanese landers on another planetary body, in the Selene mission in about 2005. Alternative concepts for Selene include one in which a large orbiter, with radar and lidar altimetry, X-ray fluorescence and gamma-ray spectroscopy, and multispectral imaging experiments, is accompanied by a small radio relay satellite to maintain uninterrupted communications with Earth and a radar receiver aboard a small satellite dedicated to Very Long Baseline Interferometry (VLBI).

In the same time frame as the Lunar-A and Selene missions is ESA's SMART-1 lunar orbiter. This Small Mission for Advanced Research and Technology is the first of a series initiated by ESA's Horizons 2000 plan. The emphasis of this program, like that of NASA's New Millennium series, is on the development of new propulsion and instrument technologies and their application to science problems.

Various commercial schemes for lunar exploration, including the most recent revival of the Artemis Project and several mission concepts from LunaCorp, are presently seeking funding. None of these missions is sufficiently advanced to have a scheduled launch date. China, meanwhile, plans to send men to the Moon by 2018.

## Mars

The broad success of the Viking program provided us with many useful insights into Mars and has whetted our appetite for information about the early history of the planet. Study of the chemical weathering and volatile-transport processes at work today may help greatly in understanding the evolution of surface conditions. The historical record and significance of the polar layered terrain remains wholly untapped. Atmospheric erosion by relatively small, numerous impactors has emerged as a potentially dominating evolutionary process. Ancient marine and lacustrine sediments may preserve evidence of prebiological chemical evolution. And the Martian environment is especially suitable for the production of rocket propellants and life-support materials to assist unmanned sample-return missions as well as manned expeditions and bases, suggesting a range of engineering experiments on utilization of Martian materials.

After a lengthy romance with Venus, the Soviet Union turned toward exploration of Mars in 1988. An ambitious new spacecraft was developed for the Phobos program, completely supplanting all hardware hitherto used in their ill-fated Mars program. A series of missions to Mars were planned to follow in the next few launch windows after Phobos 1 and 2, with two or more spacecraft to be launched in each launch window. The unfortunate failure of the first two Phobos spacecraft forced a delay in this program. The situation was further exacerbated after 1990 by the declining fiscal fortunes of Russian space activities after the collapse of the Soviet Union.

This new generation of Russian Mars missions features heavy involvement of many other nations, among which Germany has played an especially notable role. In broad outline, the program plans have involved a coordinated use of Russian orbiters, landers, French atmospheric balloons, and surface rovers, including some highly sophisticated tiny American rovers using the latest in artificial intelligence technology. The first launch in this program was the Mars '96 mission, which unfortunately suffered a launch failure. Early French participation now seems unlikely as a result of budgetary strains within ESA. Also, it is certain that, for purely Russian budgetary reasons, subsequent launches will occur no more frequently than alternate launch windows (about every 4 years). Indeed, there is no guarantee that any future Russian Mars missions will fly.

Japan is deeply interested in the deployment of a seismic network on Mars by means of a number of impacting penetrator probes. The lunar seismic network may be regarded as a test of the design and deployment of the Martian system: if this network is not flown in the

Selene mission, plans for a similar network on Mars will be delayed indefinitely.

The billion-dollar American effort to return to Mars, beginning with the Mars Observer (MO) Orbiter, has experienced both brilliant successes and abject failures. MO, launched in 1992, was lost during final approach to the planet. The next American visit to Mars after the MO mission was the Mars Global Surveyor, followed by the Pathfinder mission that landed the Sojourner rover on Mars in 1997.

The first Japanese planetary mission, the Planet-B (Nozomi) Mars Probe, was launched in 1998.

The Mars Climate Orbiter, launched by NASA on 11 December 1998, was lost due to a controller error (a contractor used English system units without converting them to NASA's standard metric units). MCO crashed on Mars on 28 September 1999. Its companion spacecraft, the Mars Polar Lander (MPL), launched on 3 January 1999, flew uneventfully to Mars, but suffered a catastrophic failure during entry into the Martian atmosphere on 3 December 1999. These failures triggered severe criticism of the NASA Mars exploration program. The next Mars mission, however, the Mars Odyssey Orbiter, was launched on 7 April 2001 and achieved Mars orbit successfully. In addition to geochemical mapping of Mars, especially mapping the distribution of water and minerals on the Martian surface, Mars Odyssey is equipped to serve as a communications relay for later Mars landing missions in 2003–2004.

Two rovers are scheduled to be sent to Mars by NASA in 2003 to use surface mobility to extend the reach of landed analytical and imaging experiments. The Mars Airborne Geophysical Explorer (MAGE), carrying the Kitty Hawk Marsplane, was tentatively scheduled for a 2003 launch, but the program is now in doubt. The 2003 launch window will see the launch of ESA's Mars Express orbital mission. Mars Express will deploy a small British-made lander, Beagle II. Further Mars missions in NASA's Mars Surveyor series are tentatively planned for the 2005–2009 time period, and ESA's Netlander seismic array is under consideration for the 2007 launch window. The possibility of collaboration with Japan on the Netlander mission remains open.

## Asteroids

Despite our vast wealth of data on meteorites, our ability to link these familiar objects to particular asteroids is still limited by the small number of spacecraft encounters with asteroids. Asteroids with especially distinctive reflection spectra can be confidently identified with particular meteorite classes, but many asteroids have reflection spectra that suggest familiar minerals in unfamiliar proportions. The lengthy struggle of astronomers to find a "home" for the abundant ordinary chondrites, which now can be identified with a subset of the S asteroids, illustrates these problems. The high-speed flybys of the asteroids Gaspra (in 1991) and Ida (and, fortuitously, Ida's small satellite Dactyl) in 1993 by the Galileo spacecraft on its way to Jupiter provided our first detailed spacecraft data on individual asteroids. The Near Earth Asteroid Rendezvous (NEAR) mission, launched by NASA in 1996, carried out a detailed orbital study of the asteroid Eros, culminating in a landing on the asteroid in 2001.

Asteroid pairs and multiples appear to be common; but evidence regarding the origins of such systems is weak. Some asteroids and asteroid companions may be composites of very different types of material brought together in low-speed collisions. Some must have differentiated (as many meteorites did) and then broken apart in impacts so as to reveal their interior structure. The highly accessible near-Earth asteroids (NEAs) may be an eclectic sampling of bodies from throughout the Belt, including also many extinct comet cores, collected nearby for our easy scrutiny. Drilling into such bodies, or gently blowing away their dust mantle, may permit direct examination of the most ancient and unaltered Solar System materials, complete with unaltered interstellar grains bearing record of the Giant Molecular Cloud complex from which the Solar System formed. And water extracted from carbonaceous or extinct cometary NEAs may be the key to cheap, easy transportation throughout the inner Solar System.

A variety of other asteroid missions have been studied over the past 20 years, but have not succeeded in securing funding. The Soviet Union's plan in the early 1980s was to use the Phobos spacecraft bus as the basis for an ambitious mission to rendezvous with and land upon the main-belt asteroid Vesta. Designing for a Vesta landing ensures that the spacecraft would have the lifetime and propulsion capability to land on any of several thousand asteroids, including a large portion of the NEA population. Early hardware problems with the Phobos spacecraft and more recent fiscal constraints have, however, led to an indefinite postponement of the Vesta mission.

The Piazzi mission plan of ESA has not led to a flight mission. The later versions of the Vesta mission under discussion around 1990 (for 1997 launch) featured a marriage of Russian and ESA hardware derived from both the original Vesta and Piazzi plans. A series of asteroid missions including multiple belt-asteroid flybys and sample return from a NEA have also been studied by NASA. The Deep Space 1 mission, largely oriented toward development and demonstration of new exploration technologies, was launched on 24 October 1998 and successfully flew by the asteroid (9969) Braille.

The most recent entrant in the Solar System exploration business, Japan, first proposed a 1997 mission to visit the near-Earth asteroid Anteros with a small spacecraft related to their early lunar orbiter. That mission evolved into the Muses-C (Nereus) mission, launched 7 January 2002 from Tanegashima. This mission is now slated to visit the asteroid 1998 $SF_{36}$ to pick up a sample for return to Earth in 2007. The return landing site would be the interior of Australia.

ESA's Rosetta mission, scheduled for launch in January 2003, is planned to execute an asteroid flyby, although its central purpose, to study the comet Wild 2, is detailed below. NASA's Dawn mission, slated for launch in 2006, would use an ion engine to rendezvous with and orbit both (1) Ceres and (4) Vesta. In the same time frame, and still lacking funding, would be the privately developed NEAP mission to a selected near-Earth asteroid. This mission, planned and built by the Space Development Corporation, would offer its data sets for sale to government agencies and make mission participation available via the Web and other media.

## Jupiter

The Jovian system is nearly as complex as the Solar System itself. The meteorology of the giant planets, their atmospheric chemistry, the geological evolution of their satellites, the formation and evolution of their ring systems, their immense and dynamically complex magnetospheres, the dynamics of capture and loss of their outer satellites and interaction of their inner satellites, and the events surrounding the formation of their systems are all of great interest. The most urgent issues are to acquire orbital mapping data on the planet and its major satellites and to probe its atmosphere.

A number of ambitious missions to Jupiter's system have been studied. Atmospheric entry probes and orbiters were first planned in the early 1970s. The first implementation of these ideas came with the Galileo spacecraft, which was launched in 1989 and, after a tortuous journey to Jupiter with a disabled high-gain antenna, dropped an entry probe into the atmosphere and entered orbit around Jupiter for a lengthy satellite tour of the Galilean moons and Amalthea. Despite its severe data transmission problem, Galileo achieved virtually all of its planned objectives, including compositional mapping of the atmosphere and clouds of Jupiter and the surfaces of the Galilean satellites, meteorological studies of Jupiter, and mapping of the magnetosphere over many months.

Beyond Galileo, and partially clouded by our uncertainty about its success, the future of Jupiter exploration is obscure. The INSIDE (Interior Structure and Internal Dynamical Evolution of Jupiter) mission received extensive study and was well reviewed, but has so far failed to receive approval from NASA. No specific American missions to Jupiter's system are planned, and no other nation is yet ready to venture into this arena.

A related mission, to place a spacecraft in orbit around Europa to study its surface morphology and seek evidence relevant to Europa's ice-covered ocean, is under study for the 2004–2006 time frame, but has not yet received funding.

## Saturn, Uranus, and Neptune

The only future Saturn mission currently in the plans of the spacefaring nations is the joint ESA–NASA Cassini–Huygens mission. Since launch on 15 October 1997, Cassini has been in cruise mode en route to Saturn. The highlights of this exciting mission are the study of Saturn and its rings and satellites from orbit, combined with a detailed study of the atmosphere and surface of Titan by the Huygens atmospheric entry probe upon arrival in July 2004.

No other missions to the Saturn system are advanced enough to merit serious consideration. No missions to Uranus and Neptune are under serious consideration by any nation.

## Pluto

Pluto, the planet with the greatest mean distance from the Sun, is the only planet in the Solar System not yet visited by a spacecraft from Earth. At the time of the planning of the Grand Tour missions to the outer Solar System in the early 1970s, it was originally proposed that a pair of spacecraft should be sent to fly by Jupiter, Saturn, Uranus, and Neptune, and a second pair of spacecraft should be launched a year later to fly by Jupiter, Saturn, and Pluto. When these plans were firmed up as the Voyager missions, the second pair of spacecraft was deleted. Later negotiation permitted the retargeting of the other two so that one visited Jupiter and Saturn and the other followed the JSUN Grand Tour trajectory. Pluto was deleted from the mission plan.

The wide range of basic questions addressed by the Voyager flybys of the other outer planets remains unanswered for Pluto. Sizes, masses, and densities are now fairly well constrained, but could be greatly improved. Surface morphology is wholly unknown. Evidence for a trace atmosphere is convincing, but hardly anything is known of the compositions and abundances of surface volatiles, and the atmosphere is condensing and precipitating before our eyes as Pluto passes perihelion and again heads outward from the Sun. Possible linkages with the Neptune system remain completely conjectural.

Planning for Pluto missions has passed through several cycles, each with substantial changes in the mission profile and hardware. The current version, called the Pluto–Kuiper Belt Mission, would be launched in 2006 to encounter Pluto and Charon in a high-speed flyby in 2015, and proceed on to encounters with one or two Kuiper Belt objects in about 2026. The selection of the post-Pluto targets will be made easier as the discovery rate of plutinos and cubewanos continues to grow. No firm commitment to a Pluto mission has yet been made by NASA, leaving our survey of the Solar System still incomplete.

## Comets

The United States, conspicuous by its absence among the armada of spacecraft launched from Earth to study Halley's Comet, consoled itself by renaming the already-old International Sun-Earth Explorer (ISEE; changed to International Cometary Explorer, ICE) and diverting it from its station near Earth's inner Lagrange point to fly by the Comet P/Giacobini–Zinner. American hopes for real study of a comet were reserved for a much more ambitious cometary rendezvous mission that could not be made ready in time to visit Halley, and which in fact could not possibly have made rendezvous with such a fast-moving comet. In the past few years we have witnessed the startup, emasculation, and death of this very ambitious Comet Rendezvous/Asteroid Flyby (CRAF) mission. As its name suggests, CRAF was to fly by an asteroid at close range, proceed on to rendezvous with a periodic comet, follow it through perihelion passage, and fire a heavily instrumented penetrator probe into the comet nucleus to determine its composition and thermal behavior. The loss of this program was a severe blow to American interest in cometary exploration.

The Deep Space 1 mission, launched in October 1998 by NASA, was in part a testbed for the New Millennium technology development program. DS1 flew by the comet 19 P/Borelly and returned excellent photographs of its nucleus.

The Stardust mission, launched in February 1999, is due to encounter comet P/Wild 2 in January 2004 and return to Earth in January 2006. The ambitious ESA Rosetta program, which involves active participation by NASA and other parties, would feature a landing on the nucleus of an active comet, extraction of a sample of pristine cometary solids, refrigeration of the sample, and return of the sample to Earth. Rosetta, originally scheduled for launch in January 2003, has been cancelled and the spacecraft has been placed in storage pending discovery of a new cometary target. NASA had made plans for a similar mission, the Champollion/Deep Space 4 comet rendezvous and lander, also intended for launch in 2003. Champollion was to rendezvous with and orbit the nucleus of P/Tempel 1 in April 2006, land on the nucleus in August, drill into the nucleus, and carry out chemical analyses with a gas chromatograph/mass spectrometer, an infrared spectrometer microscope, and a gamma ray spectrometer. The original mission concept was less heavily instrumented, but included a plan to return a refrigerated sample of the nucleus to Earth. The sample return option was abandoned first, and later the entire mission was cancelled.

The CONTOUR (Comet Nucleus Tour) mission, launched in July 2002 as a replacement for the cancelled Champollion, suffered a catastrophic failure during an engine burn on 15 August 2002, after which communications were lost and Earth-based observers saw two independent fragments continuing on its trajectory. Its mission was to fly by both P/Encke and P/Schwassmann-Wachmann 3 at close range.

Finally, the Deep Impact mission, scheduled for launch in January 2004, will rendezvous with P/Tempel 1 and observe the impact of a massive copper sphere that will excavate deeply buried materials from the nucleus and provide some insight into the physical and chemical nature of cometary ices that have never been exposed to solar heating.

## Beyond the Solar System

The search for planets about other stars has already been begun on the ground, as we saw in Chapter XI. Nonetheless, many contributions to this search program can and will be made by space-based interferometry and astrometry. High-precision photometry to search for planets via their partial eclipses of the disks of their stars is also much more sensibly done from space than from the ground. Advisory committees have proposed, among others, an Astrometric Imaging Telescope (AIT), a Precision Optical Interferometer in Space (POINTS), and an Orbiting Stellar Interferometer (OSI). The Terrestrial Planet Finder, sometimes called Kepler, would search in the mid-infrared for terrestrial-type planets in orbit around stars as distant as 15 pc, including collecting IR spectral data on the abundances of oxygen, water, carbon dioxide, and methane in their atmospheres. An alternative mission concept, the Infrared Space

Interferometer (IRSI; formerly Darwin) mission, would be placed in orbit 5 AU from the Sun to minimize interference from the zodiacal light and, like the TPF-Kepler system, search for the same list of constituents in the atmospheres of terrestrial-type planets. The Next Generation Space Telescope (NGST) would have the capability to detect terrestrial-sized planets and search for oxygen and water in their atmospheres out to a distance of 4 to 8 pc.

Figure XII.1 shows the relationships between the astrometric detectability of Jovian-class planetary bodies in other stellar systems and the distance of the system and the stellar mass. For any set of observational constraints (such as periodic amplitude of 0.3 milli arc seconds and 10.0 visual magnitude for the star) the observable range on the diagram is that below the $M_{vis}$ and amplitude curves. The van Rhijn mass function gives the most probable distance of the nearest star in each of 30 mass intervals. The 14 stars that lie below the van Rhijn curve (which include the easiest targets for astrometric planetary searches) have so far yielded no planet detections.

The first crude interstellar probes are already a reality: the Pioneer 10 and 11 and Voyager 1 and 2 spacecraft are already sounding the boundaries of the heliosphere far beyond the orbits of the planets. Dedicated interstellar probes designed for such missions could be built and launched at any time, but the time constraints for interstellar cruise are formidable. At 10 km $s^{-1}$ relative to the Sun, the travel time to a star 1 pc away would be $3 \times 10^{12}$ s, or 100,000 years. For those readers who make long-range plans, Pioneer 11 will pass relatively close to the star λ Aquilae in approximately 4 million AD.

The search for extraterrestrial intelligent life (SETI), presently being undertaken by Earth-based radio telescopes under modest budgets, has attracted little interest in Washington except from one particularly outspoken Senator who feels it is a waste of money, and who almost single-handedly prevented the funding of the SETI program for many years. Presumably the Senator already knows whether or not there are intelligent beings elsewhere in our galaxy, but he has unfortunately not passed his information on to the rest of us. Or perhaps he simply doesn't care and is content to remain in ignorance: indeed, he not only chooses to remain in personal ignorance, which is his prerogative, but he would require that the rest of us also remain ignorant. Since the

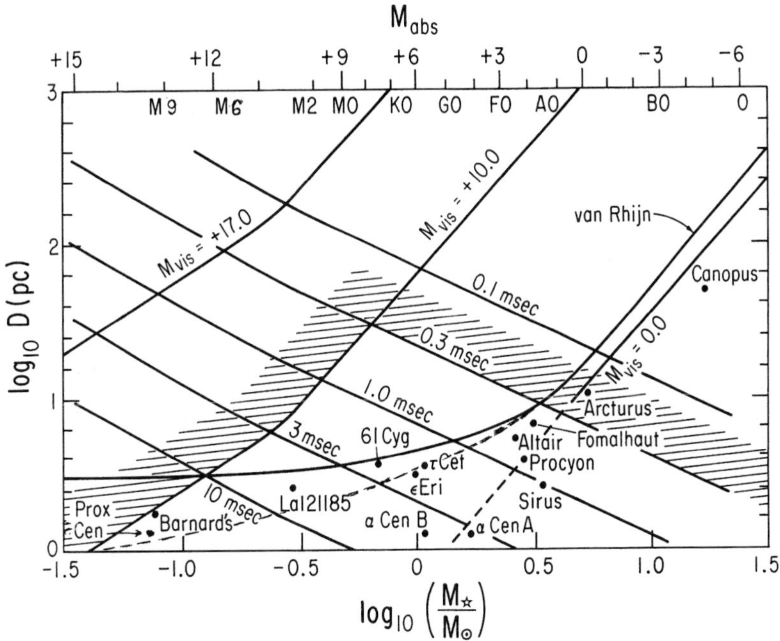


**Figure XII.1** Astrometric searches for planets of nearby stars. The van Rhijn mass distribution curve gives the average nearest-neighbor distance for MS stars of all spectral classes. The only MS stars that lie closer than this curve are the 14 stars indicated in the diagram. The limits imposed by the light sensitivity and the spatial resolution of the observing instrument limit the number and spectral distribution of extrasolar planetary systems that can be detected astrometrically. A hypothetical instrument with an ability to measure positions to 0.3 milli arc second (mas) down to a limiting visual magnitude of $M_{vis} = 10.0$ can detect only systems lying below the shaded boundary. The dashed version of the van Rhijn curve updates the MS population statistics to take into account the current estimates of the abundances of low-luminosity stars.

15th century, scientists have favored observation over introspection as a means of finding out the truth. The scientific attitude is galling to those whose minds are already made up. The Senator's stance brings back memories of the Prince of the Church who refused to look through Galileo's telescope at the wonders it revealed. He argued that the phases of Venus, satellites of Jupiter, and mountains on the Moon alleged by Galileo were inconsistent with the usual interpretations of scripture and the received truths of Aristotle. If he looked and did not see them, it would merely confirm a belief he already wholly accepted, and hence observation would be uselessly redundant. On the other hand, if he did see them, then clearly the telescope was a tool of the devil sent to delude mankind, and the observation would have to be suppressed in the interests of doctrinal purity. Therefore there were no circumstances in which recourse to observation was justified.

In the real world, however, there are many questions that arise from our curiosity and from practical necessity; questions of such interest and importance that nothing short of reliable evidence would satisfy us. Starting close to home, we ask:

How do the natural regulatory processes of Earth work?

What human activities constitute a threat to our planetary environment?

What energy sources for Earth are environmentally benign?

What energy sources for Earth are abundant and affordable?

How is the ozone layer regulated?

How does the greenhouse effect work in detail?

How has Earth's impact history affected geological history?

How have impacts affected the history of life?

When will the next major impact even occur?

How could we avoid an impending major impact?

What are the laws governing the circulation of atmospheres?

When did life originate on Earth?

How did (and does) life originate?

How have planetary surface environments evolved with time?

Did Mars once have lakes, rivers, and oceans?

Did life originate independently on Mars?

Have impacts transported genetic material between planets?

Are there chemical and physical fossils on Mars?

Why did Mars evolve into a parched, frozen desert?

Could the same thing happen to Earth?

What is the composition of Mercury?

How does Mercury relate to the other planets?

Have ancient giant impacts shaped the terrestrial planets?

Did Venus ever have liquid water on its surface?

Why did Venus evolve into a scorched desert?

Could the same thing happen to Earth?

How did the Moon come into existence?

Is there a deep global ocean on Europa?

Did life arise there?

What are the coloring agents in Jupiter's clouds?

How much organic matter is in the atmosphere of Jupiter?

Are there oceans of nitrogen and hydrocarbons on Titan?

How far has chemical evolution progressed on Titan?

Are Pluto and Charon escaped satellites of Neptune?

How are planetary rings made and maintained?

What kinds of planets harbor life?

How similar is the chemistry of life elsewhere?

Is life restricted to planets?

How do planets form?

How is planetary formation linked to star formation?

How common are planetary systems?

How common are planetary systems similar to our own?

What are the laws by which planets evolve?

How important are stochastic events in planetary evolution?

Are there other classes of planets besides those we know?

How common are water-bearing, Earth-like worlds?

How many nearby stars have planets?

How many nearby stars have Earth-like planets?

How common is life on planets around nearby stars?

How common is intelligent life around nearby stars?

Does intelligence necessarily give rise to technology?

How feasible is communication with alien intelligent life?

How feasible is interstellar travel?

How constant is the Sun?

How do changes on the Sun affect Earth?

How does the life history of stars affect their planets?

How do stars and star clusters form?

How were the chemical elements formed?

How has the composition of the Universe changed with time?

How were galaxies formed?

How are galaxies changed by collisions?

How did the Universe begin?

Did the Universe have a beginning?

Will the Universe have an end?

Why?

Such are the questions that motivate our exploration of our Solar System and the astronomical Universe. Many of these are simple questions of fact. Plausible answers to many of these questions are suggested in this book; indeed, we discuss countless observational and theoretical constraints to their answers. But few of these questions can now be answered with a high degree of confidence, and many remain very poorly understood. If we should cease for a time to explore, then we must accept, even endorse, ignorance. But it is the essence of human-ness to wonder and seek knowledge of whence we came, where we are going, and why we are here. Scientific exploration contributes mightily to answering our questions that ask What, When, Where, and How. Questions asking Why are the proper concern of religion and philosophy. But all of these questions are deeply, essentially, human. They are a part of our efforts to understand ourselves, our Universe, and the relationships between them. Understanding How the Universe Works may also have far-reaching practical significance; but we would care even if it did not.

Figure XII.2 portrays the physical limits on direct human exploration of space. On this acceleration-vs-time plot, lines of constant speed have a slope of $-1$ and lines of constant distance have a slope of $-2$. Several realms of spatial exploration are indicated, including low Earth orbit, the Moon, the near-Earth asteroids, the planets, the comets, and the stars. The diagram is truncated at the top by the human tolerance limit for acceleration. The boundary on the right is the normal life span of a human being. At the upper right is the speed of light. A person departing Earth and accelerating at 1 Earth gravity will approach the speed of light in 1 year. At an acceleration of $10^{-2} g_E$, an astronaut would not quite reach relativistic speeds (and experience significant time dilation) within his normal life span. The technological means to achieve such performance remain completely unknown.

Clearly, accessible space is mostly within the Solar System, and perhaps a few of the nearest stars—but note the "exit door" at speeds close to the speed of light that, in principle, allows us to escape the local neighborhood altogether. The immediate task before us, however, is to explore and understand our own local environment, the Solar System, in person, and the few hundred stars within 10 pc of the Sun by remote sensing and perhaps robotic proxy. We can, by choice, seek out thorough knowledge of our own neighborhood. Alternatively, we could turn our backs on the planets and stars and seek solace from introspection and untested speculation.

The penalty for such willful ignorance is the inability to comprehend how planets work or how to correct our misuse of them. The reward of vigorous research is a growing rational and intuitive grasp of how planets

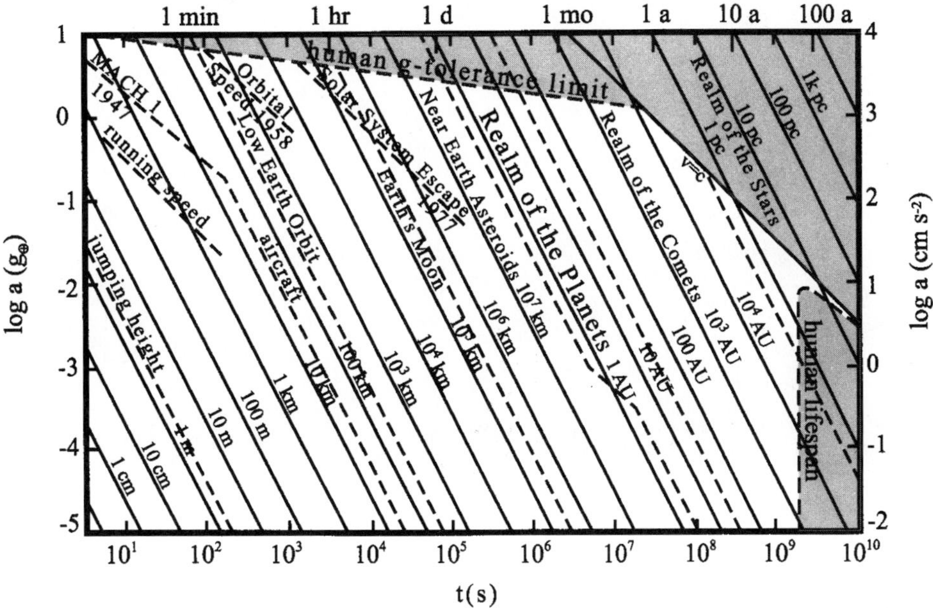


**Figure XII.2** Physical limits to human exploration. Limits imposed by the tolerance of humans to high accelerations, the normal human lifetime, and the finite speed of light are indicated on this plot of acceleration vs time. The diagram allows behavior that is presently unachievable, and perhaps unrealizable. Diagonal dashed lines of slope $-2$ (constant distance) are used to divide accessible space into near-Earth space, the Moon, the near-Earth asteroids, the planets, the Oort cloud of comets, and the nearby stars. Several speed milestones (slope $-1$) are also indicated.

work and an improved ability to harmonize human activities with natural cycles. We seek nothing less than the knowledge to design, build, and repair worlds, beginning with our own. Under the guise of national military preparations for Armageddon, we have paid for the technologies that we will need to explore the planets and learn how they work. The formidable Cold War barriers preventing international collaboration have now, like the infamous Berlin Wall, fallen to dust. It is in the interests of all the residents of Earth to see exploration continue, to see our realm of competence expand to fill the Solar System. Like our ancient ancestors at the time of their emergence from the sea onto the land, we are challenged by events to master this great new environment, to drink of its knowledge and feast on its boundless resources. Let us not squander this golden opportunity.

*There is no way back into the past.*
*The choice is the Universe—or nothing.*

—Herbert George Wells

**Figure XII.3** The Hubble space telescope's deep field view (See color plate 8).

# Appendix I
# Equilibrium Thermodynamics

We present here only a very skeletal outline of the most elementary principles of chemical thermodynamics. More detailed treatments can be found in any physical chemistry or thermodynamics text.

## Heat and Work

Energy, heat, and work are all aspects of energy. Let us begin by defining $E$ as the total internal energy of a system, the sum of the system's kinetic and potential energies. We may choose any of a number of possible reference states for the zero of the energy scale: we will avoid this problem by dealing solely with energy *differences* between the initial (1) and final (2) states of processes, designated $\Delta E$ (where $\Delta E = E_2 - E_1$).

We may increase the total energy of a system by adding heat or by doing work on the system. We may decrease the total energy by removing heat or by using some of the energy to do external work. The work done may be mechanical, electrical, magnetic, chemical, etc. In a general process, the change in internal energy is

$$\Delta E = q - w, \tag{AI.1}$$

where $q$ is the net heat flow into the system and $w$ is the net work done by the system. This equation, which expresses conservation of energy, is one statement of the First Law of Thermodynamics.

The energy of any state is defined totally by the conditions of temperature, pressure, etc., measurable for the system in that state and is independent of the details of how that state was reached. It is thus a *state variable*. $\Delta E$ for any process is a function of only the initial and final states of the system and is thus *path-independent*.

As a simple example of the work done in a process, let us consider an ideal gas being compressed slowly and reversibly by a frictionless piston. The work done on the system (the gas) during an infinitesimal displacement $dl$ of the piston is

$$dw = Fdl, \tag{AI.2}$$

where $F$ is the force exerted by the piston. That force is the product of the pressure $P$ times the face area of the piston, $A$. Since $Adl$ is the volume element $dV$ swept out by the motion of the piston,

$$dw = PdV. \tag{AI.3}$$

The work done in a finite process becomes

$$w = \int_1^2 dw = \int_{V_1}^{V_2} P(V)dV. \quad \text{(AI.4)}$$

If the process is conducted at constant pressure (heat is removed to effect the compression of the gas), then

$$w = P\Delta V \quad \text{(AI.5)}$$

and

$$q = \Delta E + P\Delta V. \quad \text{(AI.6)}$$

Let us now define a new state variable, *enthalpy*, $H$, that is useful for describing isobaric (constant pressure) processes:

$$H = E + PV. \quad \text{(AI.7)}$$

For a finite isobaric process,

$$\begin{aligned}\Delta H &= H_2 - H_1 = E_2 + P_2V_2 - (E_1 + P_1V_1)\\ &= (E_2 - E_1) + (P_2V_2 - P_1V_1)\\ &= \Delta E + P(V_2 - V_1)\end{aligned}$$

or

$$\Delta H = \Delta E + P\Delta V = q. \quad \text{(AI.8)}$$

This $\Delta H$ is equivalent to the heat absorbed during an isobaric process in which the only work done is $P$–$V$ work.

Now let us define a variable called the *heat capacity*, $C$, as $C = dq/dt$; i.e., $C$ is the heat required to raise the temperature of the system by one unit. But, from the First Law, $dq = dE + PdV$, and

$$C = (dE + PdV)/dT. \quad \text{(AI.9)}$$

For heating at a constant volume (denoted by the subscript $v$), $dV = 0$ and

$$C_v = (\partial E/\partial T)_v. \quad \text{(AI.10)}$$

$C_v$ is called the *heat capacity at constant volume*. For heating at constant pressure,

$$C_p = (\partial E/\partial T)_p + P(\partial V/\partial T)_p, \quad \text{(AI.11)}$$

the *heat capacity at constant pressure*. We already know that $H = E + PV$, and hence

$$(\partial H/\partial T)_p = (\partial E/\partial T)_p + P(\partial V/\partial T)_p, \quad \text{(AI.12)}$$

and therefore

$$C_p = (\partial H/\partial T)_p. \quad \text{(AI.13)}$$

For an ideal gas, the difference between $C_p$ and $C_v$ is

$$C_p - C_v = (\partial H/\partial T)_p = (\partial E/\partial T)_v, \quad \text{(AI.14)}$$

and, combining with Eq. (AI.12),

$$C_p - C_v = (\partial E/\partial T)_p + P(\partial V/\partial T)_p - (\partial E/\partial T)_v. \quad \text{(AI.15)}$$

Quite generally, we can write an equation of state for the gas in which the energy $E$ is defined in terms of one intensive state variable (whose value is independent of the size of the system) and one extensive state variable (whose value is proportional to the size of the system). As an example, an equation of state could be of the form $E = E(T, V)$. Then

$$dE = (\partial E/\partial T)_v dT + (\partial E/\partial V)_T dV. \quad \text{(AI.16)}$$

This expression for $dE$ is called a *complete differential*, since it includes changes in $E$ caused by variations in both of the independent variables that define $E$. Dividing by $dT$ and requiring constant pressure, we obtain

$$(\partial E/\partial T)_p = (\partial E/\partial T)_v + (\partial E/\partial V)_T(\partial V/\partial T)_p. \quad \text{(AI.17)}$$

Combining Eqs. (AI.15) and (AI.17),

$$C_p - C_v = (\partial E/\partial V)_T(\partial V/\partial T)_p + P(\partial V/\partial T)_p \quad \text{(AI.18)}$$

which is a quite general expression. If we apply this to one mole of an ideal gas, for which $Pv = RT$, then $P(\partial v/\partial T)_p = R$, and

$$C_p - C_v = R + (\partial E/\partial V)_T(\partial V/\partial T)_p. \quad \text{(AI.19)}$$

The new variable $v$, the *molar volume*, is just $V/n$, where $n$ is the number of moles. We will save this result pending introduction of a few new concepts.

## Adiabatic Processes and Entropy

We define an adiabatic process as one in which no heat enters or leaves the system: $q = 0$. Therefore $w = -\Delta E$ and, if $P - v$ work is done at the expense of the internal energy,

$$Pdv = -dE. \quad \text{(AI.20)}$$

But we know that $(\partial E/\partial T)_v = C_v$, and thus $dE = C_v dT = -Pdv$. By the ideal gas law, $Pv = RT$, and thus

$$C_v dT/RT = -dv/v. \quad \text{(AI.21)}$$

Integrating from $(T_1, v_1)$ to $(T_2, v_2)$,

$$(C_v/R)\ln(T_2/T_1) = -\ln(v_2/v_1),$$

or

$$\ln(T_2/T_1) = (R/C_v)\ln(T_1P_2/T_2P_1). \quad \text{(AI.22)}$$

Now, we already know that $dE = C_v dT$; dividing this expression by $dv$ and holding $T$ constant, we get

$$(\partial E/\partial v)_T = C_v(\partial T/\partial v)_T = 0. \quad \text{(AI.23)}$$

Hence Eq. (AI.19) reduces to very simple form for an ideal gas:

$$C_p - C_v = R. \quad \text{(AI.24)}$$

If we define the ratio $C_p/C_v$ as $\gamma$, then $R/C_v$ in Eq. (AI.22) can be written as $(C_p - C_v)/C_v$, or $\gamma - 1$. Then

$$\ln(T_2/T_1) = (\gamma - 1)[\ln(T_1/T_2) + \ln(P_2/P_1)] \quad \text{(AI.25)}$$

or

$$\ln\gamma\ (T_2/T_1) = (\gamma - 1)\ln(P_2/P_1),$$

which yields

$$P_2/P_1 = (T_2/T_1)^{\gamma/(\gamma-1)}. \quad \text{(AI.26)}$$

This equation relates the temperature and pressure during an adiabatic process on an ideal gas.

Now let us define another new thermodynamic variable, the entropy $S$, by the equation

$$dS = dq_{rev}/T. \quad \text{(AI.27)}$$

Here $dq_{rev}$ is defined as the quantity of heat absorbed during a differential *perfectly reversible* process, i.e., one in which an infinitesimal force can stop and reverse the process. Thus the steady compression of a gas by the addition of infinitesimal weights to a piston may be called perfectly reversible, while the free expansion of a gas from a pressure vessel into a vacuum is *not* reversible.

As a consequence of our definition of $dS$, note that, in a finite reversible adiabatic process, $\Delta S = 0$. In any irreversible process, $\Delta S > 0$. For a finite general reversible process,

$$\Delta S = \int_1^2 (dq_{rev})/T. \quad \text{(AI.28)}$$

## Useful Work and the Gibbs Free Energy

Let us now ask, "What is the amount of energy actually available for performing work in a process?" It is equal to the maximum work, $w_{max}$, less the work expended in internal expansion, $P\Delta v$:

$$w_{useful} = w_{max} - P\Delta v. \quad \text{(AI.29)}$$

We may now define a variable, the Gibbs Free Energy $G$, as

$$G = H - TS. \quad \text{(AI.30)}$$

Then $\Delta G = \Delta H - T\Delta S$ for processes at constant temperature, and we can see from Eq. (AI.28) that, for a finite isothermal reversible process,

$$T\Delta S = q_{rev}. \quad \text{(AI.31)}$$

Also, from (AI.8), $\Delta H = \Delta E + P\Delta v$. Combining,

$$\Delta G = \Delta E + P\Delta v - q_{rev}. \quad \text{(AI.32)}$$

But, by the First Law,

$$w_{max} = q_{rev} - \Delta E, \quad \text{(AI.33)}$$

and

$$-\Delta G = w_{max} - P\Delta v. \quad \text{(AI.34)}$$

We thus can identify $-\Delta G$ as the amount of useful work. In a sense, the Gibbs free energy change is the driving force behind a process. If $\Delta G$ is negative, then the process will occur spontaneously. If $\Delta G = 0$, the process is in a stable equilibrium. Thus we can characterize the equilibrium state as that which minimizes the value of $G$.

Certain differential expressions for the basic heat variables are useful. For example, the differential of Eq. (AI.1) is

$$dE = dq - dw. \quad \text{(AI.35)}$$

For any reversible process, $dq = TdS$. Further, for any process in which only $P$–$V$ work is done, $dw = PdV$. Then

$$dE = TdS - PdV. \quad \text{(AI.36)}$$

By the definition of the enthalpy, $H = E + PV$, and $dH = dE + PdV + VdP$ or, with Eq. (AI.36),

$$dH = TdS + VdP. \quad \text{(AI.37)}$$

Since $G = H - TS$, we have $dG = dH - TdS - SdT$, or, with Eq. (AI.37),

$$dG = VdP - SdT. \quad \text{(AI.38)}$$

For any isothermal process,

$$(\partial G/\partial P)T = V, \quad \text{(AI.39)}$$

or for a finite process,

$$\Delta G = \Delta H - T\Delta S. \quad \text{(AI.40)}$$

For a process that is both isobaric and isothermal, $dG = 0$.

Now let us consider in a little more detail an isothermal process in which pressure is varied from $P_1$ to $P_2$: from (AI.38), $dG = VdP$, and

$$\Delta G = \int_1^2 dG = \int_1^2 V(P)dP. \quad \text{(AI.41)}$$

In 1 mol of ideal gas, $V(P)$ is simply $RT/P$, and

$$\Delta G = \int_1^2 RT(dP/P) = RT\ln(P_2/P_1). \quad \text{(AI.42)}$$

(For "incompressible" materials, $\Delta G$ is approximately $V\Delta P$.)

## Chemical Equilibrium

The expression for $\Delta G$ could be more simply written if we define an appropriate zero for this energy scale. Let us first define the *standard state* of a gas to be $P = 1$ atm. This removes the necessity of carrying a "$P_1$" in all our future equations.

Second, let us recognize that gaseous chemical compounds generally have an energy content relative to the free elements that make up that compound. Thus a mole of $H_2O$ at $P = 1$ atm is very stable relative to its component elements. Rather than define separate and unrelated energy scales for every chemical compound known, we choose instead to relate the Gibbs free energies of all the millions of chemical compounds to some universal reference state. For any compound at $P_1$ and $T_1$ we will define its Gibbs free energy relative to its *component pure elements* in their *stable forms* at $P = 1$ atm and $T_1$ as the reference state. Thus at all temperatures above the normal boiling point of nitrogen, the reference state for all nitrogen compounds will be gaseous nitrogen at 1 atm pressure. At temperatures between the normal (1 atm) melting point and the normal boiling point, liquid nitrogen is the reference state. Below the freezing point, pure solid nitrogen is the reference state. For a chemical compound at $T_1$, the reference energy is the *standard Gibbs free energy of formation* of that compound from the reference state of the component element at $T_1$. This is written $\Delta G^\circ$. Thus, with our conventions for a standard state of 1 atm pressure and a reference state for the elements, we can write

$$\Delta G = \Delta G^\circ + RT \ln P. \tag{AI.43}$$

For any pure gaseous element in the reference state, $\Delta G = RT \ln P$, or for any pure gas at 1 atm pressure, $\Delta G = \Delta G^\circ$. We can now write any general gas-phase reaction as $aA + bB + cC + \cdots = mM + nN + \cdots$, where $a$ is the number of moles of substance $A$ in the balanced reaction, and so on for the other participating compounds.

The free energy of the compounds on the left side of the reaction (process) is $G_{\text{left}} = a(\Delta G_A^\circ + RT \ln P_A) + (\Delta G_B^\circ + RT \ln P_B) + \cdots$, and likewise $G_{\text{right}} = m(\Delta G_M^\circ + RT \ln P_M) + n(\Delta G_N^\circ + RT \ln P_N) + \cdots$.

For the overall reaction (process),

$$\begin{aligned}\Delta G_{rx} &= G_{\text{right}} - G_{\text{left}} \\ &= (m\Delta G_M^\circ + n\Delta G_N^\circ + \cdots - a\Delta G_A^\circ - b\Delta G_B^\circ \\ &\quad + \cdots) + (mRT \ln P_M + nRT \ln P_N + \cdots \\ &\quad - aRT \ln P_A - bRT \ln P_B - \cdots).\end{aligned} \tag{AI.44}$$

We will denote the sum containing all the standard Gibbs free energies of formation (the first parenthesized expression in the expanded form of Eq. (AI.44) above) as $\Delta G_{rx}^\circ$, which is a function of temperature alone. We can also combine all the pressure terms to cast (AI.44) in the form

$$\Delta G_{rx} = \Delta G_{rx}^\circ + RT \ln[(P_M^m P_N^n \cdots)/(P_A^a P_B^b \cdots)]. \tag{AI.45}$$

The quantity in the brackets, containing only pressure variables, is called the pressure *equilibrium constant*, $K_p$, for the temperature $T$. But recall that, *at equilibrium*, the driving force of a chemical reaction is zero and $\Delta G_{rx} = 0$. Thus a very simple expression governs chemical equilibria between ideal gases,

$$\ln K_p = -\Delta G_{rx}^\circ / RT. \tag{AI.46}$$

If pure solids and liquids are involved in the equilibrium, then the pressure of each such species, $P_i$, must be replaced by its thermodynamic activity, $a_i$, which is identically 1 at low pressures. The activity of a pure solid or liquid departs from unity at high pressures because of the contribution of compression ($P$–$V$ work) to the total energy of the phase. Ideal solutions containing a mole fraction $i$ of substance will behave as if they have an activity of $a_i = x_i$. Many solutions are not ideal and require treatment beyond the scope of this brief introduction.

Gases at high pressures begin to deviate from ideal gas behavior because of the finite volume of molecules and complex intermolecular attractive and repulsive forces. The variable used to describe their behavior is an idealized or effective pressure, called the *fugacity*, *f*. The fugacity is defined by

$$(\partial \ln f / \partial G) = 1/RT. \tag{AI.47}$$

## Exact and Complete Differentials

Let us suppose that a state function of form $z = z(x, y)$ describes the behavior of a system. If the equations

$$dz = Mdx + Ndy \tag{AI.48}$$

and

$$(\partial M/\partial y)_x = (\partial N/\partial x)_y \tag{AI.49}$$

are obeyed, where $M = M(x, y)$ and $N = N(x, y)$, then the expression (AI.48) is called an *exact* differential.

A *complete differential* is defined by

$$dz = (\partial z/\partial x)_y dx + (\partial z/\partial y)_x dy. \tag{AI.50}$$

Complete differential forms may be manipulated into useful form by, for example, dividing both sides by $dz$ and holding $y$ constant, or by dividing by $dx$ and holding $z$ constant:

$$1 = (\partial z/\partial x)_y (\partial x/\partial z) \tag{AI.51}$$

or

$$(\partial z/\partial x)_z = 0 = (\partial z/\partial x)_y + (\partial z/\partial y)_x (\partial y/\partial x)_z, \tag{AI.52}$$

whence

$$(\partial z/\partial x)_y = -(\partial z/\partial y)_x (\partial y/\partial x)_z \tag{AI.53}$$

or

$$-1 = (\partial x/\partial z)_y (\partial y/\partial x)_z (\partial z/\partial y)_x. \tag{AI.54}$$

## The Maxwell Relations

However, the energy $E$, the enthalpy $H$, and the Gibbs and Helmholtz free energies ($G$ and $A$, respectively) are state functions and can be expressed completely in terms of complementary intensive and extensive variables such as $P$, $T$, $V$, and $S$. The differentials of all four of these energy functions are exact differentials; hence,

$$dE = TdS - PdV \tag{AI.55}$$

$$(\partial T/\partial V)_S = -(\partial P/\partial S)_V \tag{AI.55a}$$

$$dH = TdS + VdP \tag{AI.56}$$

$$(\partial T/\partial P)_S = -(\partial V/\partial S)_P \tag{AI.56a}$$

$$dG = VdP - SdT \tag{AI.57}$$

$$(\partial V/\partial T)_P = -(\partial S/\partial P)_T \tag{AI.57a}$$

$$dA = -PdV - SdT \tag{AI.58}$$

$$(\partial P/\partial T)_V = -(\partial S/\partial V)_T. \tag{AI.58a}$$

The four differential equations (AI.55a)–(AI.58a) are called the *Maxwell Relations.*

# Appendix II Absorption and Emission of Radiation by Quantum Oscillators

Consider an oscillator with quantized energy levels $E_n = nh\nu$, where $n = 0, 1, 2, \ldots, \infty$. The probability that the oscillator has an energy $E_n$ is proportional to $\exp(-E_n/kT)$, the so-called Boltzmann factor. The *average* energy of the oscillator is the weighted sum of all the possible contributions to the energy, $\Sigma_n E_n \exp(-E_n/kT)$, normalized by the sum of all of the Boltzmann weighting factors, $\Sigma_n \exp(-E_n/kT)$.

Thus,

$$\overline{E} = \frac{\sum_{n=0}^{\infty} nh\nu \exp(-nh\nu/kT)}{\sum_{n=0}^{\infty} \exp(-nh\nu/kT)} \tag{AII.1}$$

$$= \frac{h\nu[\exp(-h\nu/kT) + 2\exp(-2h\nu/kT) + \cdots]}{[1 + \exp(-h\nu/kT) + \exp(-2h\nu/kT) + \cdots]} \tag{AII.2}$$

$$= h\nu \frac{d \ln[1 + \exp(-h\nu/kT) + \exp(-2h\nu/kT) \cdots]}{d(-h\nu/kT)}. \tag{AII.3}$$

where the series in $\exp(-h\nu/kT)$ is simply $1/[1 - \exp(-h\nu/kT)]$. Therefore,

$$\overline{E} = h\nu \frac{d \ln\left[\frac{1}{1-\exp(-h\nu/kT)}\right]}{d(-h\nu/kT)}, \tag{AII.4}$$

and

$$\overline{E} = \frac{h\nu}{\exp(h\nu/kT) - 1}. \tag{AII.5}$$

From this equation the Planck function can be derived. Let us consider a photon gas in equilibrium with the interior of a cubical cavity of side L. For constructive interference, only light standing waves whose wavelengths are integral multiples of $2L$ (the round-trip distance in the cavity) are allowed in each direction,

$$K_x = 2L/\lambda;\ K_y = 2L/\lambda;\ K_z = 2L/\lambda. \tag{AII.6}$$

where each $K_i$ can independently take on any integral value, beginning with 0.

Any general standing wave in three dimensions must satisfy

$$K_x^2 + K_y^2 + K_z^2 = (2L/\nu)^2. \tag{AII.7}$$

The *number* of standing waves in ($K_x$, $K_y$, $K_z$) space over the wavelength interval $d\lambda$ is just the number of different permutations of $K_x$, $K_y$, and $K_z$ that yields wavelengths between $\lambda$ and $\lambda + d\lambda$.

In $K$-space, the vector $K$ ($=2L/\lambda$) is just the distance from the origin to the spherical shell that corresponds to a wavelength $\lambda$. Equation (AII.7) is clearly the formula for a sphere of radius $2L/\lambda$.

The spherical shell from $K$ to $K + dK$, which corresponds to wavelengths between $\lambda$ and $\lambda + d\lambda$, has a volume $4\pi K^2 dK$, of which only the $(+, +, +)$ octant has physical meaning. Thus the number of permutations counted is

$$N(K)dK = \tfrac{1}{2}\pi K^2 dK, \tag{AII.8}$$

but $K = 2L/\lambda$ and $dK = -2L d\lambda/\lambda^2$, and thus

$$\begin{aligned} N(\lambda)d\lambda &= \tfrac{1}{2}\pi(2L/\lambda)^2(2L/\lambda^2)d\lambda & \text{(AII.9)} \\ &= (4\pi L^3/\lambda^4)d\lambda. & \text{(AII.10)} \end{aligned}$$

Since the volume of the cavity is $L^3$, the number of allowed standing waves between $\lambda$ and $\lambda + d\lambda$ *per unit volume*, $n(\lambda)d\lambda$, is

$$n(\lambda)d\lambda = (4\pi/\lambda^4)d\lambda. \tag{AII.11}$$

For each of these waves there are two linearly independent modes which may be present (with orthogonal states of polarization), and therefore the total number of standing waves in Eq. (AII.11) must be doubled.

We now know both the *number* of standing waves allowed per unit volume of the cavity and the average *energy* per mode for light viewed as a quantum oscillator. (It is interesting to note that the classical Rayleigh–Jeans approach simply took $E = kT$ at this point, and ended up in disastrous disagreement with experiment.) All we need to do is convert Eq. (AII.11) into frequency units, double it, and combine it with Eq. (AII.5). We then get for the energy density of the photon gas

$$u(\nu)d\nu = (8\pi h/c^3)\nu^3 d\nu/[\exp(h\nu/kT) - 1]. \tag{AII.12}$$

This is one version of the Planck radiation law.

Now let us imagine any two energy levels in an atom; call them $i$ and $j$. *Absorption* of radiation may take place with excitation of the atom from to $j$. *Spontaneous emission* may take place as the unperturbed excited atom emits a photon and drops from $j$ to $i$. We shall also consider, following Einstein, a possible transition from $j$ to $i$ that is caused by resonant interaction of the excited atom with a photon of frequency $\nu = (E_j - E_i)/h$. This letter process we call *stimulated emission*. (Consider *random* application of resonant-frequency impulses to a large ensemble of children on swings: without care to match phase, fully half of these little oscillators will become *less* excited, leaving their would-be accelerators in a state of shock.) The probability of an induced excitation of a ground-state atom by absorption of a photon is proportional to the energy density available at the resonant frequency $\nu$,

$$P_{i\to j} = B_{ij}u(\nu), \tag{AII.13}$$

where the proportionality constant is $B_{ij}$. The probability of deexcitation of an atom in state $j$ is the sum of two components,

$$P_{j\to i} = A_{ji} + B_{ji}u(\nu), \tag{AII.14}$$

where the first term gives the rate of spontaneous emission and the second the rate of stimulated emission. If we simply denote the number of ground-state atoms as $N_i$ and the number in the excited state as $N_j$, then, at thermal equilibrium, the total rate of excitation $N_i P_{i\to j}$ must be exactly equal to the total rate of deexcitation, $N_j P_{j\to i}$:

$$N_i B_{ij} u(\nu) = N_j[A_{ji} + B_{ji}u(\nu)]. \tag{AII.15}$$

Solving for $u(\nu)$, we find

$$u(\nu) = \frac{A_{ji}/B_{ji}}{(N_i/N_j)(B_{ij}/B_{ji}) - 1}. \tag{AII.16}$$

At equilibrium the relative populations of states $i$ and $j$ are given by the Boltzmann factor,

$$(N_i/N_j) = \exp(-h\nu/kT), \tag{AII.17}$$

and, from Eq. (AII.12), we can see that

$$\begin{aligned} u(\nu) &= \frac{A_{ji}/B_{ji}}{(B_{ij}/B_{ji})\exp(h\nu/kT) - 1} \\ &= \frac{8\pi h\nu^3 d\nu}{c^3[\exp(h\nu/kT) - 1]}. \end{aligned} \tag{AII.18}$$

Therefore it must follow that the Einstein $B$ coefficients are the same,

$$B_{ij} = B_{ji} \tag{AII.19}$$

(we were right about the children on the swings), and

$$A_{ji}/B_{ji} = 8\pi h\nu^3/c^3, \tag{AII.20}$$

a remarkably simple connection between spontaneous and stimulated emission rates.

It can be shown in classical electrodynamics that the lifetime of an excited state of a free atom against spontaneous decay is

$$\tau_A = 3m_e c^3/8\pi^2 e^2 \nu^2, \tag{AII.21}$$

where $m_e$ and $e$ are the electron mass and charge, respectively. The population of state $j$ would then follow a decay law of form

$$N_j = N_j^\circ \exp(-t/\tau). \tag{AII.22}$$

The uncertainty in the lifetime of the state is of the same magnitude as the lifetime itself, and we can approximate the uncertainty in the energy of the state by

$$\Delta E \Delta t \geq h/2\pi \qquad \text{(AII.23)}$$

$$\Delta E \geq (4\pi h e^2 \nu^2)/(3m_e c^3). \qquad \text{(AII.24)}$$

Thus we have derived a measure of the intrinsic width of spectral lines emitted by unperturbed atoms or molecules. If molecules are collisionally deexcited (i.e., $\tau_{\text{coll}} < \tau_A$), then the linewidths must further increase. For a typical visible transition such as a hydrogen Balmer line, $\nu \approx 10^{15}$ Hz. In cgs units, $e = 4.8028 \times 10^{-5}\ \text{g}^{1/2}\,\text{cm}^{3/2}\,\text{s}^{-1}$, and $\tau \approx 4 \times 10^{-9}$ s. (A list of the fundamental physical constants is provided in Appendix IV.) This is about 10 times longer than the mean time between collisions for hydrogen molecules at normal temperature and pressure.

# Appendix III
# Exploration of the Solar System

The first attempts to launch lunar and planetary probes date back to 1958, only a year after the launching of the first artificial Earth satellite, Sputnik 1, by the Soviet Union. At that time, the Soviet Union had a very large booster, known to them as the R-7 "Semyorka," derived from their first-generation intercontinental ballistic missile (ICBM). This rocket was huge by Western standards because the Soviets had not yet developed precise guidance systems. It was therefore necessary to build a booster large enough to lift an atomic bomb warhead with enough explosive power to compensate for the imprecision of its targeting. The development of the R-7 (known as the A1 in the West) was several months ahead of American development of the Atlas ICBM, a vehicle with only about 40% of the weight and thrust of the A1. The first American satellite launches were carried out with launchers based on the smaller Thor and Jupiter intermediate range ballistic missiles (IRBMs) and the even smaller Redstone short range ballistic missile (SRBM) and Vanguard. After numerous early failures and a few successes by both nations, deep-space missions to the Moon and planets moved up to ICBM-class boosters, initially equipped with upper stages built for the smaller Vanguard, Jupiter, and Thor rockets. From late 1959 until May of 1966, all lunar and planetary launches were carried out by ICBM-derived vehicles using low-energy liquid and solid propellants. High-energy hydrogen–oxygen propulsion was introduced by the Atlas Centaur vehicle during the Surveyor missions to the Moon in 1966–1968, providing the Atlas with a payload capacity superior to that of the much larger Soviet A2e booster (an A-class ICBM with enhanced upper stages).

The decision by President Kennedy to assert American excellence in space led to a race between the Soviets and Americans to send astronauts to the Moon. Numerous unmanned precursors were launched in the American Ranger, Lunar Orbiter, and Surveyor programs and in the Soviet Luna program to orbit, map, and land instruments on the Moon in preparation for manned missions.

In 1968, the Soviets introduced a huge new booster, the Proton or D1e, to launch lunar and planetary missions and to flight-test equipment designed for use in manned lunar missions. In the same time frame, the United States introduced the Saturn 1 booster for Earth-orbiting tests of hardware intended for use in a manned lunar landing program. Both nations also had much larger superboosters in development for use in the actual manned lunar missions—the American Saturn 5 and the Soviet M-1 boosters. The Saturn 5, performing flawlessly on every manned flight, ensured the success of the American Apollo program, achieving the first manned mission to the Moon in July of 1969, and culminating with a sixth and final lunar landing mission in 1972. Neither the Saturn 1 nor the Saturn 5 was ever used for any other lunar and planetary program. Shortly after the successful conclusion of the Apollo program both boosters were cancelled and their production facilities sold as scrap metal. The Soviets cancelled their manned lunar program after four disastrous flight explosions during unmanned tests of the giant M-1.

The Soviet D1e booster and the much smaller American Atlas Centaur, with its smaller weight-lifting capability, were the sole vehicles for deep-space missions until the first use of the Titan 3 family on the Viking missions in 1975. The Titan 3E Centaur, with a modified Titan ICBM core stage, two large solid-propellant strap-on boosters, and the Centaur hydrogen–oxygen upper stage, brought American payload capacity up to the level of the D1e.

The debuts of the Japanese and European deep-space programs in 1985 were closely linked with the opportunity to send spacecraft for high-speed flybys of Halley's Comet. Significantly, the waning American space program failed to launch a single spacecraft to Halley to accompany the two Japanese, two Soviet, and one European spacecraft.

The first use of the Space Shuttle to deliver planetary payloads to orbit was in 1989. For safety reasons, a version of the high-energy Centaur stage developed to permit launching of heavy lunar and planetary payloads from the shuttle cargo bay was canceled in the wake of the 1986 Challenger disaster. Thus the United States now finds itself using the outrageously expensive shuttle and completely unnecessarily risking the lives of several astronauts on each planetary mission. This development, based on an attempt to make the shuttle appear indispensable, is one facet of an operating policy that has made American access to space so expensive that few missions ever fly.

The history of all known attempts to launch lunar and planetary payloads is given in Table AIII.1.

**Table AIII.1** Exploration of the Solar System: Spacecraft Launchings

| Spacecraft[1] | Source | Launch[2] | Target | Mission | Vehicle[3] | Remarks |
|---|---|---|---|---|---|---|
| Pioneer 1 | USA | 11 Oct 58 | Moon | flyby | TAb | Reached 70,700 km altitude—missed Moon |
| Pioneer 2 | USA | 8 Nov 58 | Moon | flyby | TAb | Third stage ignition failure |
| Pioneer 3 | USA | 6 Dec 58 | Moon | flyby | J2 | Reached 63,600 km altitude—missed Moon |
| Luna 1 | USSR | 2 Jan 59 | Moon | flyby? | A1 | Flew by Moon at 6000 km distance |
| Pioneer 4 | USA | 3 Mar 59 | Moon | flyby | J2 | Missed Moon by 60,000 km |
| Luna 2 | USSR | 9 Sep 59 | Moon | impact | A1 | Impacted on Moon |
| Luna 3 | USSR | 4 Oct 59 | Moon | flyby | A1 | Photographed far side of Moon |
| (Pioneer) | USA | 26 Nov 59 | Moon | flyby | AAb | Payload shroud failed during launch |
| (Pioneer) | USA | 25 Sep 60 | Moon | flyby | AAb | Second stage malfunction |
| (unnamed) | USSR | 10 Oct 60 | Mars | flyby | A2e | Failed to achieve Earth orbit |
| (unnamed) | USSR | 14 Oct 60 | Mars | flyby | A2e | Failed to achieve Earth orbit |
| (Pioneer) | USA | 15 Dec 60 | Moon | flyby | AAb | First stage exploded |
| Sputnik 7 | USSR | 4 Feb 61 | Venus | flyby | A2e | Failed to depart from low Earth orbit |
| Venera 1 | USSR | 12 Feb 61 | Venus | flyby | A2e | Communications failure in transit |
| Ranger 1 | USA | 23 Aug 61 | Moon | test | AAB | Failed to depart from low Earth orbit |
| Ranger 2 | USA | 18 Nov 61 | Moon | test | AAB | Failed to depart from low Earth orbit |
| Ranger 3 | USA | 26 Jan 62 | Moon | impact | AAB | Missed Moon by 40,000 km |
| Ranger 4 | USA | 23 Apr 62 | Moon | impact | AAB | Impacted Moon with experiments dead |
| Mariner 1 | USA | 22 Jul 62 | Venus | flyby | AAB | Launch phase failure |
| 1962 APi1 | USSR | 25 Aug 62 | Venus | flyby | A2e | Failed to depart from low Earth orbit |
| Mariner 2 | USA | 27 Aug 62 | Venus | flyby | AAB | Flew by Venus at 36,000 km range |
| 1962 AT1 | USSR | 1 Sep 62 | Venus | flyby | A2e | Failed to depart from low Earth orbit |
| 1962 APhi | USSR | 12 Sep 62 | Venus | flyby | A2e | Failed to depart from low Earth orbit |
| Ranger 5 | USA | 18 Oct 62 | Moon | impact | AAB | Missed Moon by 750 km |
| 1962 BI1 | USSR | 24 Oct 62 | Mars | flyby | A2e | Failed to depart from low Earth orbit |
| Mars 1 | USSR | 1 Nov 62 | Mars | flyby | A2e | Communications failure in transit |
| 1962 BX1 | USSR | 4 Nov 62 | Mars | flyby | A2e | Failed to depart from low Earth orbit |
| (Luna) | USSR | 4 Jan 63 | Moon | lander? | A2e | Failed to depart from low Earth orbit |

**Table AIII.1** *Continued*

| Spacecraft[1] | Source | Launch[2] | Target | Mission | Vehicle[3] | Remarks |
|---|---|---|---|---|---|---|
| Luna 4 | USSR | 2 Apr 63 | Moon | lander? | A2e | Missed Moon by 8500 km |
| Kosmos 21 | USSR | 11 Nov 63 | Venus | test | A2e | Failed to depart from low Earth orbit |
| Ranger 6 | USA | 30 Jan 64 | Moon | impact | AAB | Impacted Moon with TV inoperative |
| Kosmos 27 | USSR | 27 Mar 64 | Venus | flyby? | A2e | Failed to depart from low Earth orbit |
| Zond 1 | USSR | 2 Apr 64 | Venus | flyby? | A2e | Communications failure in transit |
| Ranger 7 | USA | 28 Jul 64 | Moon | impact | AAB | Success—returned 4308 photos |
| Mariner C | USA | 5 Nov 64 | Mars | flyby | AAD | Shroud failed to separate after launch |
| Mariner 4 | USA | 28 Nov 64 | Mars | flyby | AAD | First flyby of Mars: 15 July 65 |
| Zond 2 | USSR | 30 Nov 64 | Mars | flyby | A2e | Communications failure in transit |
| Ranger 8 | USA | 17 Feb 65 | Moon | impact | AAB | Success—returned 7137 photos |
| Kosmos 60 | USSR | 12 Mar 65 | Moon | lander | A2e | Failed to depart from low Earth orbit |
| Ranger 9 | USA | 21 Mar 65 | Moon | impact | AAB | Success—returned 5814 photos |
| Luna 5 | USSR | 9 May 65 | Moon | lander | A2e | Landing attempt failed—crashed on Moon |
| Luna 6 | USSR | 8 Jun 65 | Moon | lander | A2e | Missed Moon by 160,000 km |
| Zond 3 | USSR | 18 Jul 65 | Mars | test | A2e | Flew by Moon as test of Mars spacecraft |
| Zond 3 | USSR | 18 Jul 65 | Moon | test | A2e | System test—photographed Moon |
| Luna 7 | USSR | 4 Oct 65 | Moon | lander | A2e | Retros fired early—crashed on Moon |
| Venera 2 | USSR | 12 Nov 65 | Venus | probe | A2e | Communication failure just before Venus arrival |
| Venera 3 | USSR | 16 Nov 65 | Venus | probe | A2e | Communication failure before Venus entry |
| Kosmos 96 | USSR | 23 Nov 65 | Venus | probe | A2e | Failed to depart from low Earth orbit |
| Luna 8 | USSR | 3 Dec 65 | Moon | lander | A2e | Retros fired late—crashed on Moon |
| Luna 9 | USSR | 31 Jan 66 | Moon | lander | A2e | First successful lunar soft landing |
| Kosmos 111 | USSR | 1 Mar 66 | Moon | lander | A2e | Failed to depart from low Earth orbit |
| Luna 10 | USSR | 31 Mar 66 | Moon | orbiter | A2e | First successful lunar orbiter |
| Surveyor 1 | USA | 30 May 66 | Moon | lander | AC | Successful soft landing—sent 11,150 pictures |
| Lunar Orbiter 1 | USA | 10 Aug 66 | Moon | orbiter | AAD | Lunar photographic mapping |
| Luna 11 | USSR | 24 Aug 66 | Moon | orbiter | A2e | Lunar orbit science mission |
| Surveyor 2 | USA | 20 Sep 66 | Moon | lander | AC | Crashed on Moon attempting landing |
| Luna 12 | USSR | 22 Oct 66 | Moon | orbiter | A2e | Lunar photographic mapping |
| Lunar Orbiter 2 | USA | 6 Nov 66 | Moon | orbiter | AAD | Lunar photographic mapping |
| Luna 13 | USSR | 21 Dec 66 | Moon | lander | A2e | Soft lander science mission |
| Lunar Orbiter 3 | USA | 4 Feb 67 | Moon | orbiter | AAD | Lunar photographic mapping |
| Surveyor 3 | USA | 17 Apr 67 | Moon | lander | AC | Surface science mission |
| Lunar Orbiter 4 | USA | 4 May 67 | Moon | orbiter | AAD | Photographic mapping |
| Kosmos 159 | USSR | 17 May 67 | Venus | probe? | A2e | Possible Venera or Molniya failure: inert in high-eccentricity Earth orbit |
| Venera 4 | USSR | 12 Jun 67 | Venus | probe | A2e | Successful atmospheric probe 18 Oct 67 |
| Mariner 5 | USA | 14 Jun 67 | Venus | flyby | AAD | Flew by Venus at 4000 km range |
| Kosmos 167 | USSR | 17 Jun 67 | Venus | probe | A2e | Failed to depart from low Earth orbit |
| Surveyor 4 | USA | 14 Jul 67 | Moon | lander | AC | Communications ceased before landing |
| Lunar Orbiter 5 | USA | 1 Aug 67 | Moon | orbiter | AAD | Photographic mapping |
| Surveyor 5 | USA | 8 Sep 67 | Moon | lander | AC | Lunar surface science |
| Surveyor 6 | USA | 7 Nov 67 | Moon | lander | AC | Lunar surface science |
| Surveyor 7 | USA | 7 Jan 68 | Moon | lander | AC | Lunar surface science |
| Zond 4 | USSR | 2 Mar 68 | Moon | test | D1e | Unmanned test of Soyuz lunar craft |
| Luna 14 | USSR | 7 Apr 68 | Moon | orbiter | A2e | Mapped lunar gravity field |
| Zond 5 | USSR | 15 Sep 68 | Moon | test | D1e | Circumlunar flyby—spacecraft recovered |
| Zond 6 | USSR | 10 Nov 68 | Moon | test | D1e | Lunar flyby—precursor of manned flight |
| Apollo 8 | USA | 21 Dec 68 | Moon | manned | S5 | Manned lunar orbital mission |
| Venera 5 | USSR | 5 Jan 69 | Venus | probe | A2e | Successful atmospheric probe 16 May 69 |
| Venera 6 | USSR | 10 Jan 69 | Venus | probe | A2e | Successful atmospheric probe 17 May 69 |
| Mariner 6 | USA | 24 Feb 69 | Mars | flyby | AC | Flew by 31 Jul 69—returned 75 pictures |
| Mariner 7 | USA | 27 Mar 69 | Mars | flyby | AC | Flew by 5 Aug 69—returned 126 pictures |
| Apollo 10 | USA | 18 May 69 | Moon | manned | S5 | Lunar orbit test for landing mission |
| Luna 15 | USSR | 13 Jul 69 | Moon | return? | D1e | Impacted on Moon after Apollo 11 landing |
| Apollo 11 | USA | 16 Jul 69 | Moon | manned | S5 | First manned lunar landing and return |
| Zond 7 | USSR | 8 Aug 69 | Moon | test | D1e | Unmanned circumlunar flight and return |
| Kosmos 300 | USSR | 23 Sep 69 | Moon | test | D1e | Earth-orbit test of lunar equipment? |
| Kosmos 305 | USSR | 22 Oct 69 | Moon | test | D1e | Earth-orbit test of lunar modules? |
| Apollo 12 | USA | 14 Nov 69 | Moon | manned | S5 | Manned lunar landing |

*Continues*

**Table AIII.1** *Continued*

| Spacecraft[1] | Source | Launch[2] | Target | Mission | Vehicle[3] | Remarks |
|---|---|---|---|---|---|---|
| Apollo 13 | USA | 11 Apr 70 | Moon | manned | S5 | Aborted lunar landing mission |
| Venera 7 | USSR | 17 Aug 70 | Venus | lander | A2e | First successful lander 15 Dec 70 |
| Kosmos 359 | USSR | 22 Aug 70 | Venus | lander | A2e | Failed to depart from low Earth orbit |
| Luna 16 | USSR | 12 Sep 70 | Moon | return | D1e | Returned lunar surface samples to Earth |
| Zond 8 | USSR | 20 Oct 70 | Moon | test | D1e | Unmanned circumlunar flight and return |
| Luna 17 | USSR | 10 Nov 70 | Moon | rover | D1e | Unmanned lunar rover: Lunokhod |
| Apollo 14 | USA | 31 Jan 71 | Moon | lander | S5 | Manned lunar landing |
| Mariner H | USA | 8 May 71 | Mars | orbiter | AC | Second stage failure at launch |
| Kosmos 419 | USSR | 10 May 71 | Mars | lander? | D1e | Failed to depart from low Earth orbit |
| Mars 2 | USSR | 19 May 71 | Mars | orbiter | D1e | First Mars orbiter on 27 Nov 71 |
| Mars 2 lander | USSR | 19 May 71 | Mars | lander | D1e | Crashed in landing attempt |
| Mars 3 | USSR | 28 May 71 | Mars | orbiter | D1e | Orbited Mars on 2 Dec 71 |
| Mars 3 lander | USSR | 28 May 71 | Mars | lander | D1e | Failed immediately after landing |
| Mariner 9 | USA | 30 May 71 | Mars | orbiter | AC | Orbital photographic mapper |
| Apollo 15 | USA | 26 Jul 71 | Moon | lander | S5 | Manned lunar landing |
| P&F satellite | USA | 4 Aug 71 | Moon | orbiter | S5 | Subsatellite deployed by Apollo 15 |
| Luna 18 | USSR | 2 Sep 71 | Moon | return | D1e | Crashed on Moon |
| Luna 19 | USSR | 28 Sep 71 | Moon | orbiter | D1e | Photographic mapping mission |
| Luna 20 | USSR | 14 Feb 72 | Moon | return | D1e | Lunar surface sample return |
| Pioneer 10 | USA | 2 Mar 72 | Jupiter | flyby | AC | Jupiter flyby 3 Dec 73 |
| Venera 8 | USSR | 27 Mar 72 | Venus | lander | A2e | Landed 22 Jul 72 |
| Kosmos 482 | USSR | 31 Mar 72 | Venus | lander | A2e | Failed to depart from low Earth orbit |
| Apollo 16 | USA | 16 Apr 72 | Moon | manned | S5 | Manned lunar landing |
| P&F satellite | USA | 19 Apr 72 | Moon | orbiter | S5 | Subsatellite deployed by Apollo 15 |
| Apollo 17 | USA | 7 Dec 72 | Moon | manned | S5 | Manned lunar landing |
| Luna 21 | USSR | 8 Jan 73 | Moon | rover | D1e | Unmanned lunar rover: Lunokhod 2 |
| Pioneer 11 | USA | 5 Apr 73 | Jupiter | flyby | AC | Jupiter flyby on 4 Dec 74 |
| | | | Saturn | flyby | AC | Saturn flyby on 1 Sep 79 |
| Explorer 49 | USA | 10 Jun 73 | Moon | orbiter | TD | Galactic radio science experiment |
| Mars 4 | USSR | 21 Jul 73 | Mars | orbiter | D1e | Failed to achieve Mars orbit |
| Mars 5 | USSR | 25 Jul 73 | Mars | orbiter | D1e | Orbited Mars on 12 Feb 74 |
| Mars 6 | USSR | 5 Aug 73 | Mars | lander | D1e | Communications lost just before landing |
| Mars 7 | USSR | 9 Aug 73 | Mars | lander | D1e | Engine failure—missed Mars |
| Mariner 10 | USA | 3 Nov 73 | Venus | flyby | AC | Flew by 5 Feb 74 en route to Mercury |
| | | | Mercury | flyby | | Flew by Mercury three times in 1974 |
| Luna 22 | USSR | 29 May 74 | Moon | orbiter | D1e | Photographic mapper |
| Luna 23 | USSR | 28 Oct 74 | Moon | return | D1e | Drill arm damaged—no return attempt |
| Venera 9 | USSR | 8 Jun 75 | Venus | lander | D1e | Landed 22 Oct 75 |
| V 9 orbiter | USSR | 8 Jun 75 | Venus | orbiter | D1e | Orbited about Venus 22 Oct 75 |
| Venera 10 | USSR | 14 Jun 75 | Venus | lander | D1e | Landed 25 Oct 75 |
| V 10 orbiter | USSR | 14 Jun 75 | Venus | orbiter | D1e | Orbited Venus 25 Oct 75 |
| Viking 1 | USA | 20 Aug 75 | Mars | orbiter | T3EC | Orbited Mars 19 Jun 76 |
| Viking 1 lander | USA | 20 Aug 75 | Mars | lander | | Landed on Mars 20 Jul 76—surface science |
| Viking 2 | USA | 9 Sep 75 | Mars | orbiter | T3EC | Orbited Mars 7 Aug 76 |
| Viking 2 lander | USA | 9 Sep 75 | Mars | lander | | Landed on Mars 3 Sep 76 |
| Luna 24 | USSR | 9 Aug 76 | Moon | return | D1e | Lunar surface sample return |
| Voyager 2 | USA | 20 Aug 77 | Jupiter | flyby | T3EC | Flew by Jupiter 9 Jul 79 |
| | | | Saturn | flyby | | Flew by Saturn 26 Aug 81 |
| | | | Uranus | flyby | | Flew by Uranus 10 Jan 86 |
| | | | Neptune | flyby | | Flew by Neptune 25 Aug 89 |
| Voyager 1 | USA | 5 Sep 77 | Jupiter | flyby | T3EC | Flew by Jupiter 5 Mar 79 |
| | | | Saturn | flyby | | Flew by Saturn 12 Nov 80 |
| Pioneer 12 | USA | 20 May 78 | Venus | orbiter | AC | Orbited Venus 8 Dec 78 |
| PV Large Probe | USA | 20 May 78 | Venus | lander | AC | Landed 9 Dec 78 |
| Pioneer 13 | USA | 8 Aug 78 | Venus | landers | AC | Three small probes landed 9 Dec 78 |
| PV bus | USA | 8 Aug 78 | Venus | probe | AC | Probed upper atmosphere 9 Dec 78 |
| Venera 11 | USSR | 9 Sep 78 | Venus | lander | D1e | Landed 25 Dec 78 |
| Venera 12 | USSR | 14 Sep 78 | Venus | lander | D1e | Landed 21 Dec 78 |

**Table AIII.1** *Continued*

| Spacecraft[1] | Source | Launch[2] | Target | Mission | Vehicle[3] | Remarks |
|---|---|---|---|---|---|---|
| Venera 13 | USSR | 30 Oct 81 | Venus | lander | D1e | Landed 1 Mar 82 |
| Venera 14 | USSR | 4 Nov 81 | Venus | lander | D1e | Landed 5 Mar 82 |
| Venera 15 | USSR | 2 Jun 83 | Venus | orbiter | D1e | Radar mapper |
| Venera 16 | USSR | 7 Jun 83 | Venus | orbiter | D1e | Radar mapper |
| Vega 1 | USSR | 15 Dec 84 | Venus | lander | D1e | Landed 10 Jun 85 |
| | USSR | 15 Dec 84 | Venus | balloon | D1e | Deployed in Venus atmosphere |
| | USSR | 15 Dec 84 | Venus | flyby | D1e | Encounter with Comet Halley Mar 86 |
| Vega 2 | USSR | 21 Dec 84 | Venus | lander | D1e | Landed 15 Jun 85 |
| | USSR | 21 Dec 84 | Venus | balloon | D1e | Deployed in Venus atmosphere |
| | USSR | 21 Dec 84 | Venus | flyby | D1e | Encounter with Comet Halley Mar 86 |
| Sakigake | Japan | 7 Jan 85 | Halley | flyby | Mu3S | Distant flyby of nucleus of Comet Halley |
| Giotto | ESA | 2 Jul 85 | Halley | flyby | Ar3 | Halley nucleus flyby, Jan 86 |
| Magellan | USA | 4 May 89 | Venus | orbiter | STS | Venus radar mapping |
| Galileo Orbiter | USA | 17 Oct 89 | Jupiter | orbiter | STS | Orbited Jupiter Dec 95 |
| Galileo Probe | USA | | Jupiter | probe | | Entered Jupiter's atmosphere 7 Dec 95 |
| Phobos 1 | USSR | 7 Jul 88 | Phobos | orbiter | D1e | Killed by controller error on way to Mars |
| Phobos 2 | USSR | 12 Jul 88 | Phobos | orbiter | D1e | Electronics failure in Mars orbit |
| Hiten | Japan | 24 Jan 90 | Moon | flyby | M3S2 | Flew by Moon |
| Hagormo | Japan | | | orbiter | | Deployed into lunar orbit by Hiten |
| Ulysses | ESA/USA | 6 Oct 90 | Sun | orbiter | STS | Flew by Jupiter 8 Feb 92 |
| Mars Observer | USA | 22 Sep 92 | Mars | orbiter | STS | Communications lost en route to Mars |
| Clementine | USA | 25 Jan 94 | Moon | orbiter | T2 | Orbited Moon Feb 95 |
| NEAR | USA | 17 Feb 96 | Eros | orbiter | D2 | Flew by Mathilde; orbited Eros |
| Mars Global Surveyor | USA | 7 Nov 96 | Mars | orbiter | D2 | Orbited and mapped Mars |
| Mars 96 | CIS | 16 Nov 96 | Mars | multiple | D1e | Orbiter, lander, 2 penetrators: launch failure |
| Pathfinder | USA | 2 Dec 96 | Mars | orbiter | D2 | Lander with deployed rover |
| Sojourner | USA | 2 Dec 96 | Mars | rover | D2 | Rover deployed from Pathfinder |
| Cassini | USA/ESA | 15 Oct 97 | Saturn | orbiter | T4BC | En route to Saturn orbit |
| Huygens | USA/ESA | 15 Oct 97 | Titan | probe | T4BC | En route to Titan |
| Lunar Prospector | USA | 7 Jan 98 | Moon | orbiter | D2 | Lunar geochemical mapping |
| Planet-B (Nozomi) | Japan | 3 Jul 98 | Mars | orbiter | MV-3 | First Japanese Mars mission |
| Deep Space 1 | USA | 24 Oct 98 | Braille | cruise | D2 | Flew by asteroid (9969) Braille |
| Mars Climate Orbiter | USA | 11 Dec 98 | Mars | orbiter | D2 | Crashed on Mars 28 Sep 99 |
| Mars Polar Lander | USA | 3 Jan 99 | Mars | lander | D2 | Crashed on Mars 3 Dec 99 |
| Mars Odyssey | USA | 7 Apr 01 | Mars | orbiter | D2 | Mars geochemical mapping |
| CONTOUR | USA | 3 Jul 02 | Comets | flyby | D2 | Flyby to Encke and S-W 3; broke up 15 Aug 02 during rocket firing |

[1] Spacecraft: NEAR, Near–Earth Asteroid Rendezvous; P&F, Particles and Fields; PV, Pioneer Venus.

[2] Launch Sites: All USA lunar and planetary probes are launched from Kennedy Space Center, Cape Canaveral, Florida. All Soviet lunar and planetary probes are launched from "Baikonur Cosmodrome," which is near Tyuratam, Kazakhstan, not near Baikonur. Japanese deep space missions are launched from Kagoshima. The European Space Agency's Giotto mission was launched from Kourou, French Guiana.

[3] Vehicle: AAb, Atlas Able (Atlas ICBM first stage with upper stage stack from Vanguard booster); D2, Delta 2; J2, Juno 2 (Army Jupiter IRBM with three solid-propellant upper stages from the Jupiter C satellite launch vehicle); STS, Space Transportation System, the Space Shuttle; S5, Saturn 5 superbooster (7.5 million pound thrust liquid propellant S1c first stage, high-energy hydrogen–oxygen second and third stages, and lunar descent and ascent stages); TAB, Thor Agena B (Air Force Thor IRBM with Agena B upper stage derived from the bomb pod engine on the B-58 Hustler); TD, Thor Delta (Thor first stage and Delta second stage); T2, modified Titan 2 ICBM; T3EC, Titan 3E Centaur (strengthened two-stage Titan 2 ICBM core with two large solid-propellant strap-on boosters and high-energy Centaur upper stage); T4BC, Titan 4 B Centaur; A2e, A-class booster with added escape stage; D1e, D-class (Proton) booster with K escape stage; M3S2, Mu3S2 Japanese booster based on Delta.

# Appendix IV
# Basic Physical Constants*

| | | |
|---|---|---|
| Speed of light | $c$ | $2.997925 \times 10^{10}$ cm s$^{-1}$ |
| Gravitational constant | $G$ | $6.673 \times 10^{-8}$ cm$^3$ s$^{-2}$ |
| Avogadro's number | $N_0$ | $6.0222 \times 10^{23}$ mole$^{-1}$ |
| Gas constant | $R$ | $8.3144 \times 10^7$ erg K$^{-1}$ mole$^{-1}$ |
| Boltzmann factor | $k$ | $1.38062 \times 10^{-16}$ erg K$^{-1}$ |
| Atomic mass unit | $AMU$ | $1.66053 \times 10^{-24}$ g |
| Planck's constant | $h$ | $6.62620 \times 10^{-27}$ erg s |
| Electron charge | $e$ | $4.80280 \times 10^{-10}$ esu |
| Proton mass | $m_p$ | $1.67261 \times 10^{-24}$ g |
| Neutron mass | $m_n$ | $1.67492 \times 10^{-24}$ g |
| Stefan–Boltzmann constant | $\sigma$ | $5.6696 \times 10^{-5}$ erg cm$^{-2}$ s$^{-1}$ K$^{-4}$ |
| Wien displacement constant | $\lambda_{max} T$ | 0.28978 cm K |

* The accuracy of the values of these constants is indicated by the number of significant figures quoted: all errors are in the last significant figure only. Note that the universal gravitational constant and the Stefan–Boltzmann constant are the least reliably known, and the speed of light in a vacuum is the best known.

# Appendix V
# Gravity Fields

For a central inverse-square gravity field, the gravitational force is

$$F = GMm/r^2 \quad \text{(AV.1)}$$

and the gravitational potential energy is

$$U = -GMm/r. \quad \text{(AV.2)}$$

The potential energy per unit mass, $V$, also called the *potential*, is merely

$$V = -GM/r. \quad \text{(AV.3)}$$

The surface of any reasonably fluid planet is isostatically adjusted, lying on an equipotential surface $V_s = -GM/r_s$. As a result of geochemical differentiation, rotation, mantle convection, and resulting crustal tectonic activity, real planets often differ significantly from a state of spherical symmetry and homogeneity, usually by enough to matter, but not by enough to destroy the usefulness of referring these effects to the standard of an unperturbed sphere. In these cases it is generally quite useful to model $V$ as a Fourier (sine and cosine) function of longitude and latitude. Oblateness caused by rotation perturbs the sphere in a latitude-dependent, longitude-independent manner. Such terms in the equation for $V$ are called *zonal harmonics*. Often, especially for planets with few satellites, natural or artificial, we may have little gravity information beyond the lowest zonal harmonics. Of these, by far the largest is that due to oblateness.

Elongation of one of the equatorial axes of a body, for example, due to the gravitational attraction of a nearby planet (the Moon in Earth's gravity field, for example) introduces powerful longitude-dependent (sectoral) harmonic terms. The distribution of mass (continental blocks, ocean basins, mantle density variations, etc.) over the surface of a planet generally involves complex variations with both longitude and latitude. These features must be analyzed in terms of functions of both latitude and longitude, which effectively subdivide the surface into a large number of longitude–latitude "cells" that, on a map, look like tiles. With sufficiently detailed data on the structure of a planet's gravity field, large numbers of terms in the Fourier series may be necessary to provide a good fit to the data. The similarity of such a map of resolution cells to tiles has given rise to the name *tesseral harmonics* for such functions, from the Latin word *tessera*, meaning "tile."

The most general expression for the potential is similar to that given for magnetic fields by Eq. (V.224). Terms in $1/r$ are unipolar, associated with a $1/r^2$ force field. Higher order terms in the potential, such as $1/r^2$ and so on, are dipolar, quadrupolar, octupolar, and so on.

For the spherical and oblateness terms only, the potential at the planetary surface has the following form:

$$V(r) = -(GM/r_s)[1 - (a/r)^2 J_2 P_2(\cos\Theta)] - \tfrac{1}{2} r_s^2 \omega^2 \sin^2\Theta. \quad \text{(AV.4)}$$

The last term is the contribution of the rotation of the planet (the so-called centrifugal force) to the potential, which vanishes at the poles. Here $P_2(\cos\Theta)$ is a Legendre polynomial of order 2. Note that it is independent of longitude. Tables of all the associated Legendre polynomials can be found in many advanced calculus texts. Here $\Theta$ is the colatitude or polar angle, $90° - l$. Looking up this particular Legendre polynomial, we find that

$$P_2(\cos\Theta) = \tfrac{1}{2}(3\cos^2\Theta - 1). \quad \text{(AV.5)}$$

The surface radius of the planet is related to the equatorial radius, $a$, the oblateness, $f$, and the colatitude by

$$r_s = a(1 - f\cos^2\Theta). \quad \text{(AV.6)}$$

We will use these constraints to solve Eq. (AV.4), which already contains $P_2$; rather than carry terms containing both $P_2$ and $\cos^2\Theta$, we can eliminate the latter in favor of the Legendre polynomial by combining Eqs. (AV.5) and (AV.6):

$$\cos^2\Theta = \tfrac{1}{3} + \tfrac{2}{3} P_2(\cos\Theta). \quad \text{(AV.7)}$$

Hence,

$$r_s = a[1 - f/3 - 2f/3\ P_2(\cos\Theta)]. \quad \text{(AV.8)}$$

Recalling from elementary trigonometry that $\sin^2\Theta + \cos^2\Theta = 1$, we can also replace the $\sin^2\Theta$ term in Eq. (AV.4) with

$$\begin{aligned}\sin^2\Theta &= 1 - \tfrac{1}{3} - \tfrac{2}{3} P_2(\cos\Theta)\\ &= \left(\tfrac{2}{3}\right)[1 - P_2(\cos\Theta)]. \quad \text{(AV.9)}\end{aligned}$$

Combining these equations generates a number of terms, some of which can be disregarded by realizing that both $f$ and $J_2$ are small, and hence terms in $f^2$, $fJ_2$, and $J_2^2$ can be neglected. We also neglect terms in $f\omega^2$ with respect to those in $\omega^2$. Collecting similar terms, we get

$$\begin{aligned}V(r_s) = {} & [-(GM/a)(1 - f/3) - a^2\omega^2/3] + P_2(\cos\Theta)\\ & \times [(GM/a)(J_2 - 2f/3) + a^2\omega^2/3]. \quad \text{(AV.10)}\end{aligned}$$

Since the surface is an equipotential, obviously the potential cannot vary with colatitude. Therefore the coefficient of $P_2$ must be zero. This gives

$$f = \left(\tfrac{3}{2}\right) J_2 + a^3\omega^2/2GM. \quad \text{(AV.11)}$$

This link between $f$ and $J_2$ allows us to use observational constraints on either to estimate the other. In cases in which both can be determined independently, this equation can be used to assess how closely the planet approaches hydrostatic equilibrium.

# Suggested Readings

## Introduction

The emphasis in this list is principally upon collections of papers and overview volumes. Popular or nontechnical sources are cited only in support of background materials that are not covered quantitatively in this book. Individual journal articles are usually not directly cited. To find what you want among the 10,000-plus such articles in print, it is most efficient to select them from the references given in specialized reviews, collections of reports from spacecraft missions, and overview volumes such as those in the University of Arizona Space Science Series. These books and reviews not only are extremely rich sources of primary references, but also place the individual contributions in those primary references in their context relative to other such papers. In addition to books and reviews, many special issues of journals have been devoted to spacecraft encounters with individual planets. These special issues, which are concentrated sources of useful data, are cited collectively (not by individual articles) where appropriate.

## Chapter I—Introduction

As we discussed in Chapter I, the research literature on planetary sciences is vast in volume and scope, spread over many dozens of different journals, and published in many languages. If you are reading this book, then you can read the most widely used language in modern science, English. In my readings I have found important sources in French, German, Russian, Italian, Dutch, Chinese, Czech, Latin, and several other languages, but English is by a wide margin the most universally used. The periodical literature for planetary science was described in some detail in Chapter I, and that discussion will not be repeated here.

## Chapter II—Astronomical Perspective

Dozens of descriptive introductory astronomy texts with little mathematical content are available to provide a general, qualitative overview of the astronomical context of the Solar System. Such books meet the needs of readers craving a rich descriptive overview and willing to get their mathematics elsewhere. Because these books are generally available in fresh editions about every two years, I will not give any particular press date for them: by all means, seek the newest edition. I have found several of these books especially useful: *Astronomy: The Cosmic Journey*, by William K. Hartmann and Chris Impey, from Wadsworth; *Universe*, by William J. Kaufmann, III, published by Freeman; *The Dynamic*

*Universe*, by Theodore P. Snow, and *In Quest of the Universe*, by Karl F. Kuhn, both from West; and *Astronomy: The Evolving Universe*, by Michael Zeilik, from Wiley. At least a dozen other books of comparable scope and level are available but are less familiar to me.

For a mathematical treatment of general principles of astrophysics that assumes a good descriptive background and extends the material given in Chapter II, I would suggest *Astrophysical Concepts*, by Martin Harwit. The book is, as the name suggests, concept-oriented and mathematically sophisticated.

Big Bang nucleosynthesis is discussed in a review by A. M. Boesgaard and G. Steigman in the *Annual Review of Astronomy and Astrophysics*, Vol. 23 (1985). The borderland between large Jovian planets and minimal-mass MS stars is surveyed ably by David J. Stevenson in the *Annual Review of Astronomy and Astrophysics*, Vol. 29 (1991). A more recent survey of low-mass stars and Jovian and super-Jovian planets by G. Chamrier and I. Baraffe can be found in Vol. 38 of the same series (2000).

## Chapter III—General Description of the Solar System

For a broad qualitative introduction to the Solar System, the general astronomy books listed above are all adequate, although I confess a preference for the book by Hartmann and Impey. There are several texts dealing exclusively with the Solar System, of which the most familiar are *Exploration of the Solar System*, by William K. Kaufmann, III (Macmillan, New York, 1978 and later), a nonmathematical survey of the history of planetary exploration; *Moons and Planets*, by William K. Hartmann (Wadsworth, Belmont, CA), a scientific tour of the Solar System with high-school-level mathematical content; and *Meteorites and the Origin of Planets*, by John A. Wood (McGraw–Hill, New York, 1968), a fine but dated qualitative introduction that is similarly sparing of mathematics and physics. Several other nonmathematical texts are available, including *Introduction to the Solar System*, by Jeffrey K. Wagner (Saunders, Philadelphia, 1991); *Exploring the Planets*, by W. Kenneth Hamblin and Eric H. Christiansen (Macmillan, New York, 1990); *The Space-Age Solar System*, by Joseph F. Baugher (Wiley, New York, 1988); and *The Planetary System*, by planetary scientists David Morrison and Tobias Owen (Addison–Wesley, Reading, MA, 1988). An especially interesting and readable but nonquantitative survey of the Solar System is found in *The New Solar System*, 4th ed., edited by J. K. Beatty, C. C. Peterson, and A. Chaikin (Sky Publishing Co., Cambridge, MA, 1999).

## Chapter IV—The Sun and the Solar Nebula

The nature of the Sun is thoroughly explored in *Solar Interior and Atmosphere*, edited by A. N. Cox, W. C. Livingston, and M. S. Matthews (University of Arizona, 1991). The origin, evolution, and fate of the solar nebula is discussed extensively in *The Origin of the Solar System*, edited by S. F. Dermott (Wiley, 1978). For the latest perspective on Solar System origins, and for some taste of the rapidity of change in cosmogonic theory, consult the three University of Arizona Press volumes *Protostars and Planets*, edited by T. Gehrels (1978); *Protostars and Planets, II*, edited by David C. Black and M. S. Matthews (1985); *Protostars and Planets, III*, edited by E. H. Levy and J. I. Lunine (1993), and *Protostars and Planets IV*, edited by U. Manning, A. P. Boss, and S. S. Russell (2000). A wide range of observational and theoretical perspectives is presented. Accretion of solid planets is explored in considerable detail in the review by George W. Wetherill, "Formation of the Earth," in the *Annual Review of Earth and Planetary Science*, Vol. 18, p. 205 (1990), and that by Jack Lissauer in the *Annual Review of Astronomy and Astrophysics*, Vol. 31 (1993). A detailed and up-to-date survey of the solar neutrino problem by W. C. Haxton is in *Annual Reviews of Astronomy and Astrophysics*, Vol. 33 (1995).

A broad introduction to space plasma physics can be found in *Introduction to Space Physics*, by Margaret G. Kivelson and Christopher T. Russell (Kluwer Academic, 1992).

## Chapter V—The Major Planets

The large University of Arizona Space Science Series volumes *Jupiter* (edited by Tom Gehrels, 1976), *Saturn* (edited by Tom Gehrels and Mildred S. Matthews, 1984), *Uranus* (edited by Jay T. Bergstralh, Ellis D. Miner, and Mildred S. Matthews, 1991) and *Neptune and Tristan* (edited by D. P. Quickshank, 1996) are the best collections of papers on the giant planets, providing broad access to the periodical research literature. These thousand-page tomes consist of dozens of detailed reviews of all major aspects of the study of these planets. Each volume is equipped with an extensive bibliography to lead the reader to the primary sources.

The atmospheres of Uranus and Neptune are reviewed by Jonathan I. Lunine in the *Annual Review of Astronomy and Astrophysics*, Vol. 31 (1993). The same volume also contains a review on the Great Red Spot and other large, stable vortical structures in the atmospheres of the Jovian planets by P. S. Marcus. James B. Pollack reviews modern theories for the origin of the outer plants in the *Annual Review of Astronomy and Astrophysics*, Vol. 22 (1984).

The principal sources for Voyager data on the outer planets are special editions of journals dedicated to these missions. The Voyager 1 Jupiter encounter is presented in *Science*, Vol. 204, p. 945 (1979); in *Nature*, Vol. 280, p. 725 (1979); and in *Geophysical Research Letters*, Vol. 7, p. 1 (1980). The early Voyager 2 Jupiter data are found in *Science*, Vol. 206, p. 925 (1979). Mature results of the Voyager Jupiter encounters are given in the *Journal of Geophysical Research*, Vol. 85, p. 8123 (1981).

The Voyager 1 Saturn encounter results are first reported in *Science*, Vol. 212, p. 159 (1981), and *Nature*, Vol. 292, p. 675 (1981). The Voyager 2 Saturn encounter is described in *Science*, Vol. 215, p. 499 (1982). The principal summary papers on Saturn are found in the *Journal of Geophysical Research*, Vol. 88, p. 8625 (1983), and in two special 1983 issues of *Icarus* (Vol. 53, p. 163, and Vol. 54, p. 159).

The Voyager 1 spacecraft did not visit Uranus or Neptune. The Voyager 2 Uranus encounter was first reported in *Science*, Vol. 233, p. 39 (1986), and summarized in the *Journal of Geophysical Research*, Vol. 92, p. 14 (1987). The Voyager 2 Neptune encounter was reported in *Science*, Vol. 246, No. 4936 (1989).

An overview of the goals of the Galileo mission at Jupiter is given by C. T. Russell in *The Galileo Mission* (Kluwer, Dordrecht, 1992). Detailed scientific results from the Galileo remote sensing experiments appeared in September 1998 in Vol. 135 of *Icarus*, and the entry probe data were presented in *Science*, Vol. 272, pp. 837–860.

## Chapter VI—Pluto and the Icy Satellites of the Outer Planets

Several University of Arizona Space Science Series volumes deal exclusively with planetary satellites and ring systems. In chronological order, these include *Planetary Satellites* (edited by Joseph A. Burns, 1977), *Satellites of Jupiter* (edited by David Morrison, 1982), *Planetary Rings* (edited by Richard Greenberg and Andre Brahic, 1984), and *Satellites* (edited by Joseph A. Burns and Mildred S. Matthews, 1986). The repair of the Hubble Space Telescope should permit great advances in the study of the Pluto–Charon system in the near future. A Pluto volume is being planned by the Space Science Series. The best review on Pluto and Charon currently available is by S. A. Stern in the *Annual Review of Astronomy and Astrophysics*, Vol. 30 (1992). The spacecraft data on the satellites of the Jovian planets are found in the special Voyager literature detailed earlier for Chapter V.

A review on the origin and evolution of natural satellites by Stanton J. Peale appears in *Annual Reviews of Astronomy and Astrophysics*, Vol. 37 (1999). A special issue of Icarus (Vol. 146, No. 2, December 2000) treats Pluto and Triton, and the Pluto–Charon system is reviewed by S. A. Stern and D. J. Tholen (eds.), in *Pluto and Charon* (University of Arizona Press, Tucson, 1998), and by M. E. Brown in *Annual Reviews of Earth and Planetary Science*, Vol. 30 (2002).

## Chapter VII—Comets and Meteors

The most important general background source is *Comets*, edited by Laurel L. Wilkening (University of Arizona, 1982). A special issue in *Icarus*, Vol. 47, No. 3 (1981), also provides a broad prespacecraft view of comets. The physics of evaporation, coma formation, photodissociation and ionization, and tail formation is surveyed by D. A. Mendis, H. L. F. Houpis, and M. L. Marconi in *Fundamentals of Cosmic Physics*, Vol. 10 (1985). The investigations of Halley by the Soviet Vega 1 and Vega 2, European Space Agency Giotto, and Japanese Planet A spacecraft in 1986 are summarized in three reviews: the post-Halley literature on cometary nuclei is surveyed by Michael F. Ahearn in the *Annual Review of Earth and Planetary Science*, Vol. 16 (1988), and broad reviews of modern comet science are given by Hyron Spinrad in the *Annual Review of Astronomy and Astrophysics*, Vol. 25 (1987), and by D. A. Mendis in the *Annual Review of Astronomy and Astrophysics*, Vol. 26 (1988).

An interesting overview of comets is given in the book *The Origin of Comets*, by M. E. Bailey, S. V. M. Clube, and W. M. Napier (Pergamon, 1990). The most extensive survey of comets is *Comets in the Post-Halley Era*, edited by R. L. Newburn, M. Neugebauer, and J. Rahe (Kluwer, 1991). A broad, very readable, and less technical review of comets with strong historical interest is *Comets: A Chronological History of Observation, Science, Myth, and Folklore*, by Donald K. Yeomans (Wiley, 1991).

## Chapter VIII—Meteorites and Asteroids

The principal treatises on meteorites are *Meteorites*, by Brian Mason (Wiley, 1962); *Meteorites*, by John T. Wasson (Springer-Verlag, 1974); and *Meteorites: A Petrologic–Chemical Synthesis*, by Robert T. Dodd (Cambridge, 1981). Mason's treatment is very broad and balanced but seriously out of date. Wasson's book is especially strong on iron meteorites. Dodd's treatment emphasizes, as the name reveals, the analytical and petrological aspects of meteoritic material. The links between Solar System origins and meteorite studies are the theme of *Meteorites and the Early Solar System*, edited by John F. Kerridge and Mildred S. Matthews (University of Arizona, 1988).

The most important sources of information on the minor planets are the Arizona volumes *Asteroids*, edited by Tom Gehrels (1979), *Asteroids II*, edited by Richard P. Binzel, Tom Gehrels, and Mildred S. Matthews (1989), and *Asteroids III*, edited by W.F. Bottke, A. Cellino, P. Paolicci, and R.P. Binzer (2002).

The significance of asteroids (and other bodies) as resources for use in exploring the Solar System has been the subject of the nontechnical book *Space Resources: Breaking the Bonds of Earth*, by John S. Lewis and Ruth A. Lewis (Columbia University Press, 1987), and the technical volumes *Space Resources* (NASA SP-509, 1992), edited by Mary Fae McKay, David S. McKay, and Michael B. Duke, and *Resources of Near-Earth Space*, edited by John S. Lewis, Mildred S. Matthews, and Mary L. Guerrieri (University of Arizona, 1993). I have also written two recent popular accounts of impacts on Earth, *Rain of Iron and Ice* (Addison–Wesley, 1996), and the energy and material resources of space, *Mining the Sky* (Addison–Wesley, 1996).

The relationships between nucleosynthesis and the isotopic composition of meteorites is explored in the *Annual Reviews of Earth and Planetary Sciences* (Vol. 26, 1998). The most recent catalog of meteorite classes and descriptions is Marilyn M. Grady's *Catalog of Meteorites* (Cambridge University Press, 2000). A thorough survey of meteorite studies by H. Y. McSween is *Meteorites and Their Parent Planets* (Cambridge, 1999), the most recent and systematic overview presently available.

The results of the encounter of the NEAR spacecraft with the asteroid Mathilde are presented in the July 1999 issue of *Icarus*, and the investigations of Ida and its companion Dactyl by the Galileo spacecraft are detailed in *Icarus* in March 1996.

The rapidly evolving knowledge of the Kuiper Belt is the subject of reviews by Paul R. Weissman in Vol. 33 of the *Annual Reviews of Astronomy and Astrophysics* (1995) and by David Jewitt in Vol. 27 of the *Annual Reviews of Earth and Planetary Sciences* (1998).

## Chapter IX—The Airless Rocky Bodies: Io, Phobos, Deimos, the Moon, and Mercury

A general overview of the geology of rocky planets is found in *Earthlike Planets: Surfaces of Mercury, Venus, Earth, Moon, and Mars*, by Bruce Murray, Michael C. Malin, and Ronald Greeley (Freeman, 1981). With respect to the geology of Mercury and Venus, this book is superseded by *Mercury*, edited by Faith Vilas, Clark R. Chapman, and Mildred S. Matthews (University of Arizona, 1988), and by the radar mapping of Venus by the Magellan mission. The results of the Mariner 10 mission to Mercury are described in the *Journal of Geophysical Research*, Vol. 80 (1975), and in *Icarus*, Vol. 28, No. 4 (August 1976), and summarized in a review by D. E. Gault and coauthors in the *Annual Review of Astronomy and Astrophysics*, Vol. 15, p. 97 (1977).

For an excellent overview of the Moon and its relevance to the rest of the Solar System, I recommend *Planetary Science: A Lunar Perspective*, by S. Ross Taylor (Lunar and Planetary Institute, 1982). Taylor's older but very thorough review of Apollo-era lunar science, *Lunar Science: A Post-Apollo View* (Pergamon, 1975), is also strongly recommended. A thorough coverage of theories of the origin of the Moon is found in the book *Origin of the Moon*, edited by W. K. Hartmann, R. J. Phillips, and G. J. Taylor (Lunar and Planetary Institute, 1986). For the best collections of papers on Io, see the references related to the Voyager missions to Jupiter cited earlier for Chapter V.

A valuable source of critically evaluated data on the Moon is the *Lunar Sourcebook*, edited by G. Heiken, D. Vaniman, and B. M. French (Cambridge University Press, 1991).

## Chapter X—The Terrestrial Planets: Mars, Venus, and Earth

The most important sources dealing with the terrestrial planets are *Venus*, edited by Donald M. Hunten, Larry Colin, Thomas M. Donahue, and V. I. Moroz (University of Arizona, 1983), *Venus II*, edited by S. W. Bougher, D. M. Hunten, and R. J. Phillipps (University of Arizona, 1998); and *Mars*, edited by Hugh H. Kieffer, Bruce M. Jakosky, Conway W. Snyder, and Mildred S. Matthews (University of Arizona, 1992). Journal special issues dealing with Venus include *Science*, Vol. 183, No. 4131 (March 29, 1974), on the Mariner 10 Venus encounter; the *Journal of the Atmospheric Sciences*, Vol. 32, No. 6 (June 1975); *Space Science Reviews*, Vol. 20, Nos. 3 and 4 (May and June 1977); and preliminary reports from the Pioneer Venus missions in *Science*, Vol. 203, No. 4382 (February 23, 1979), and *Science*, Vol. 205, No. 4401 (July 6, 1979). The final results of the Pioneer Venus mission are detailed in the *Journal of Geophysical Research*, Vol. 85, No. A13 (December 30, 1980). *Icarus* published two special issues on Venus in Vol. 51, No. 2 (August 1982), and

Vol. 52, No. 2 (November 1982). Magellan radar mapping of Venus is the subject of two special issues of the *Journal of Geophysical Research*, Vol. 97, Nos. E8 and E10 (1992). The geology of Venus as studied by both Soviet and American spacecraft was surveyed by Basilevskii and J. W. Head, III, in the *Annual Review of Earth and Planetary Science*, Vol. 16 (1988).

The Mariner 6 and 7 missions to Mars are covered in the *Journal of Geophysical Research* in January 1971. The results from the Mariner 9 orbital mapping mission appeared in *Icarus* in October 1972 and in the *Journal of Geophysical Research*, Vol. 78, in July 1973. The ambitious but ill-fated Soviet Mars 4, 5, 6, and 7 missions are reviewed in *Cosmic Research*, Vol. 13, in 1975. The landmark Viking missions to Mars produced three special issues of *Science* [Vol. 193, No. 4255 (August 27, 1976); Vol. 194, No. 4260 (October 1, 1976); and Vol. 194, No. 4271 (December 17, 1976)], the September 30, 1977 issue of the *Journal of Geophysical Research* (Vol. 82, No. 28), and the June 1978 issue of *Icarus*. A major volume on Mars science appeared after the Viking mission in the *Journal of Geophysical Research*, Vol. 84, No. B14 (December 30, 1979). Results from the Mars Data Analysis Program appeared in *Icarus*, Vol. 45, in January and February of 1981, and the climatology of Mars was the subject of papers in *Icarus*, Vol. 50, in 1982.

An excellent source of information on our own planet is the text *Earth*, by Frank Press and Raymond Siever (Freeman, 1974 and later). The literature on Earth alone dwarfs the entire planetary sciences literature. Numerous texts on physical geology, geophysics, geochemistry, petrology, and other branches of Earth science are available in any university library. I would especially recommend *An Introduction to the Rock-Forming Minerals*, by W. A. Deer, R. A. Howie, and J. Zussman (Longman, 1966), and the volume *Plate Tectonics*, edited by J. M. Bird (American Geophysical Union, 1980). For an old but broad survey, see also the book *The Nature of the Solid Earth*, edited by E. L. Robertson (McGraw-Hill, 1972).

Discoveries regarding the importance of impacts in Earth's history are discussed in "Geological Implications of Impacts of Large Asteroids and Comets on the Earth," edited by L. T. Silver and P. H. Schultz, in *Geologic Society of America Special Paper No. 190* (1982).

The composition and evolution of the atmospheres of the terrestrial planets are surveyed by James B. Pollack and Yuk L. Yung in "Origin and Evolution of Planetary Atmospheres," in the *Annual Review of Earth and Planetary Science*, Vol. 8, p. 425 (1980), and in the book *Planets and Their Atmospheres: Origin and Evolution*, by John S. Lewis and Ronald G. Prinn (Academic Press, 1984). The latter is especially heavily referenced. A major compendium of related papers and reviews is found in *Origin and Evolution of Planetary and Satellite Atmospheres*, edited by Sushil K. Atreya, James B. Pollack, and Mildred S. Matthews (University of Arizona, 1989). Two articles of far-reaching importance by R. O. Pepin are "On the Origin and Early Evolution of Terrestrial Planet Atmospheres and Meteoritic Volatiles," in *Icarus*, vol. 92, p. 2 (1991), and a review in the *Annual Review of Earth and Planetary Science*, Vol. 21 (1993). Impact removal of atmospheres is also reviewed by T. J. Ahrens in the same volume.

Climate effects of volcanic eruptions are discussed by M. R. Rampino, S. Self, and R. B. Stothers in the *Annual Review of Earth and Planetary Science*, Vol. 16 (1988), and nuclear winter is reviewed by R. P. Turco *et al.* in the *Annual Review of Earth and Planetary Science*, Vol. 19 (1991). The history of life on Earth, as portrayed in reprints of historical documents from the geological literature, is surveyed in *Adventures in Earth History*, edited by Preston Cloud (Freeman, 1970). A progressive paleontological perspective on the biological effects of large impacts is given by D. J. McLaren and W. D. Goodfellow in the *Annual Review of Earth and Planetary Science*, Vol. 18 (1990).

The physics and aeronomy of terrestrial planet atmospheres are treated in *Theory of Planetary Atmospheres*, by Joseph W. Chamberlain and Donald M. Hunten (Academic Press, 1987). The chemistry and evolution of Earth's oceans are the theme of *The Chemistry of the Atmosphere and Oceans*, by H. D. Holland (Wiley–Interscience, 1978).

Recent concentrated sources for the terrestrial planets include a special issue of *Icarus* in September 1998 (Vol. 139, No. 1), and a special issue on Mars polar science in the same journal in April 2000 (Vol. 144, No. 2). Arden L. Albee has also written a review of the status of Mars investigations in the midst of the current flurry in Mars-related mission activity (*Annual Reviews of Earth and Planetary Science*, Vol. 28, 2000).

## Chapter XI—Planets and Life around Other Stars

The current status of the search for planetary systems about nearby stars is reviewed in *TOPS: Toward Other Planetary Systems*, published by the NASA Solar System Exploration Division (1992). The search for life on other planets, principally by radio techniques, is the subject of *Strategies for Search for Life in the Universe*, edited by M. D. Papagiannis (Reidel, 1980).

Radio techniques for detecting and communicating with other civilizations are discussed in *Interstellar Communication*, edited by A. G. W. Cameron (Benjamin, 1963); *Communication with Extraterrestrial Intelligence*, edited by Carl Sagan (M.I.T. Press, 1973); and *The Search for Life in the Universe*, edited by D. Goldsmith and T. Owen (Benjamin/Cummings, 1980).

Recent observational constraints on brown dwarfs are reviewed by G. Basri in *Annual Reviews of Astronomy and Astrophysics* (2000), and the theory of brown dwarfs and super-Jovian planets is reviewed by G. Chamrier and I. Baraffe in the same volume.

G. W. Marcy and R. P. Butler reviewed the techniques for detection of extrasolar super-Jovian planets in *Annual Reviews of Astronomy and Astrophysics* (1998), and the importance of recently discovered extrasolar planetary systems as constraints on theories for planetary formation is discussed by Peter Bodenheimer and D. N. C. Lin in the *Annual Reviews of Earth and Planetary Science* (2002). A preview of methods for detection of terrestrial-size extrasolar planets can be found in N. Woolf and Roger Angel's chapter in *Annual Reviews of Astronomy and Astrophysics* in 1998.

## Chapter XII—Future Prospects

Plans for future planetary missions are especially unclear in the wake of the collapse of the Soviet Union and the loss of the American Mars Observer spacecraft en route to Mars. There are no definitive planning documents available for the Russian, American, or ESA planetary programs. Revisions of these plans are generally published promptly in *Aviation Week and Space Technology* or in industry publications such as *Space News* and *Space Fax Daily*. Such plans are, however, subject to instant obsolescence. It is far easier to discuss what *could* be done than what *shall* be done.

The search for evidence of extraterrestrial intelligence at radio frequencies is reviewed by Jill Tartar in the 2001 edition of *Annual Reviews of Astronomy and Astrophysics*, and the article by Nick Woolf and Roger Angel cited under Chapter XI also touches on astronomical techniques for detection of life on terrestrial planets by means of the chemical effects of life on the atmosphere of its home planet. The issues involving stable astronomical and planetary settings for life are discussed in a popular volume, *Worlds without End*, by John S. Lewis (Perseus Books, 1998), within the perspective that bodies with disequilibrating energy sources and persistent bodies of liquid water are plausible sites for the origin and persistence of grossly Earthlike, but locally adapted, life. A more or less opposite point of view is presented by Don Brownlee and Peter Ward in the book *Rare Earth*, in which they require truly Edenic conditions for life's milieu. They conclude that life exactly as it evolved on Earth is very unlikely to arise and persist elsewhere.

# Index


A
Absorption and emission of radiation by quantum oscillators, 626–628
Abundances of elements in solar system, 37–39
Accretion models. *See* Rapid accretion, chemistry of
Accretion of rocks, planetesimals, and planets, 136–138
Aerobraking, 416
Aerocapture, 416
Aluminum and calcium, 108–109
Andromeda Nebula, 16
Asteroid Belt, 147, 384, 406–409
Asteroids, 69–71
  albedo–color systematics diagram, 395
  Amor, 69, 387, 407, 413, 414, 417
  Apollo, 69, 70, 387, 413, 414, 417
  Atens, 69, 386, 387, 413, 414, 417, 425
  Belt. *See* Belt asteroids
  books on, 386
  Centaurs and trans-Neptunian object, 409–412
  Ceres. *See* Ceres
  cooling, 405
  debris cloud, 427, 428
  densities diagram, 394
  distribution of orbit semimajor axes, 387
  dynamic evolution of asteroid belt, 406–409
  Eight-Color Asteroid Survey, 399
  frequency of spectral classes of near-Earth asteroids, 402
  future prospects, 614–615
  Gaspra, 179, 384, 614
  geometric albedos of, 393
  Greeks, 69
  Hilda, 69, 70, 354, 387, 388
  introduction, 350, 384–386
  Jovian planets and, 259
  Kuiper belt. *See* Kuiper belt
  Lagrange points, 389, 390
  lightcurves, 391
  Main Belt, 386, 387, 391, 393, 400
  Mars-grazer, 413
  mass of primitive solar nebula, minimum, 119
  masses and densities of, 393–394
  naming, 385
  near-Earth. *See* Near-Earth asteroids (NEAs)
  Near-Earth Asteroid Rendezvous (NEAR) spacecraft, 71, 384, 386, 393, 395, 614
  orbital elements of selected asteroids, 70
  orbital semimajor axes of, 70
  orbits, 386–389
  photometry and spectroscopy of, 394, 395–401
  population of near-Earth asteroids, 417
  radar observations of near-Earth asteroids, 415–416, 417
  relationship among meteorites, comets, and, 412–415
  resources, 416, 417–419
  rotation periods, 391, 392
  satellites of, physical properties of, 67, 68
  sizes, shapes, and albedos of, 391–393
  sizes diagram, 393
  space weathering, 412
  Spacewatch search program, 357, 413, 414, 415, 420, 443
  spectral classes, 396
  spin rates, 407
  stability of Trojan and plutino orbits, 389–391
  thermal evolution of, 401–406
  Trojans. *See* Trojans asteroids
  water-bearing, 418–419
Astronomical perspective, 7–49
Astronomical unit (AU), 7
Atmophile, 112
Atomic ionization potentials, 98
Aufbau principle, 38
Avogadro's number, 4, 5, 79, 182, 330, 634

B
Balmer jump, 81, 87
Balmer line, 628
Barnard's star, 29
Beam splitter, 157
Beattie–Bridgeman equation, 152
Belt asteroids
  compositional zoning, 412
  diameters, 392
  emission from, 396
  orbit, 387
  photometry and spectroscopy, 396
  population, 386
  satellites of, 70
  satellites orbiting, 385
  size distributions, 71, 414

Berthelot equation, 152
Big Bang, 10–15, 270
carbon and, 105
cosmology, 153
energy density of Universe during early stages, 10
evolution of fireball of, 12
long-period comets heating and, 323
masses of light nuclides, 13
nuclear abundances in fireball of, 14
nucleosynthesis, 35
nucleosynthesis limitations, 14–15
Planck function, 11
rest masses of elementary particles, 10–11
Black body emitter, 149, 150, 186, 219, 232
Black hole, 10, 15
Bode's Law, 56, 406
Bohr radius, 153, 181
Bok globules, 31
Boltzmann factor, 11, 80, 626, 627, 634
Boussinesq approximation, 193
Brown and solar system bodies, Harrison, 113
Brown dwarfs, 24, 30, 592, 599, 600–603, 604
Brunt–Väisälä frequency, 535

C
Calcium and aluminum, 108–109
Callisto satellite, 69
color of, 254, 255
craters on, 258, 259, 276, 347
densities of solid bodies versus heliocentric distance, uncompressed, 133
density and composition of, 265
evolution of, 278
internal thermal structure, 272
orbit of, 65, 279, 280
photographs of, 260, 263, 347
physical properties of, 67
radii, masses, and densities of, 265
rotation of, 256
surface morphology, 262–263
surface of, 255, 256
thermal and tectonic evolution of, 276, 278
Carbon, oxygen, and nitrogen, 102–105
Carbon burning, 40, 41
Carbon formation, 35–36
Castor as sextuple star, 29
Catalytic carbon cycle (CC), 35, 36
Centaurs and trans-Neptunian objects, 409–412
Cepheid variables, 8–9, 24
Ceres, asteroid, 147, 409, 458
albedo variation over surface of, 71
catalog numbering and, 385
densities, 394
densities of solid bodies versus heliocentric distance, uncompressed, 133
dynamical evolution, 406
masses and densities, 393, 394
naming of, 385
orbital elements, 70
photometry and spectroscopy of, 395
size, 392, 393
spectra of (diagram), 398
wavelength dependence of albedos (diagram), 399
Chalcophile, 111, 112, 116
Charon satellite, 69, 311
composition and surface color, 310
discovery of, 308
double planet of Pluto–Charon system, 308–310
orbit of, 66
physical properties of, 68
tidal torques, 273
Chemical and physical prerequisites of life, 592–595
Chemical remanent magnetization (CRM), 361
Chromosphere, 84, 87–88
Chromosphores, 213–215
Clathrates, 102–103, 235, 328
Clausius–Clapeyron Equation, 167, 168, 176, 267
Cloud collapse, gas, 15–16
Cloud condensation in $NH_3$–$H_2O$–$H_2S$ system, 165, 166–174
Cloud physics on Jovian planets, 174–179
Coalsack, 30
Cobalt and nickel, 110–111
Collision broadening, 127
Coma Berenices, 16, 17
Comet Maury, 319
Comet West photograph, 320
Cometary fireballs, 343–344
Cometary impacts on Jupiter, 344–347
Comets, 71
chemical processes in coma and tail, 332–333
chemistry and photochemistry of water, 328, 330–332
coma, 319, 328
eccentricity–inclination plot for short-period comets, 321
eccentricity–period plot for short-period comets, 322
evaporation and nongravitational forces, 326–328
frequency–energy plot for long-period comets, 323
future prospects, 616
Great Comet of 1882, 318, 321
Great Daylight Comet of 1910, 318
Halley. *See* Halley's comet
head of, 319
heating by passing stars, 325–326
historical perspectives, 317–319
Ikeya-Saki, 73
Jupiter and impacts of, 344–347
Kohoutek, 318–319
long-period comets, 71, 319, 322, 323, 324, 325, 326
naming of, 319, 320
nature and nomenclature of, 319–321
orbital classifications, 71
orbital elements, selected, 72
orbits, 319, 321–325
periodic comets, 71, 319, 321
relationships among asteroids, meteorites, and, 412–415
short-period comets, 71, 319, 321, 322, 324, 325, 413, 443
small particles behavior, 333–334
Sun-grazing comets, 322
tails, 71, 72, 88, 90, 319, 328
visual appearance of, 73
Compton scattering, 14, 81
Copernicus (Kopernik), Nikolas, 52
Corona, 51, 84, 88–90
Coronal holes, 89, 93
Cosmic dust, 73, 356
Cosmic events and disciplines evolving from, 1–2
Cosmic rays, 73, 74, 85, 86
Crab Nebula, 218
Crater-forming process stages, 439
Crux constellation, 30
Cygnus constellation, 30

D
Deimos, 68, 386
density of, 425
impact cratering and, 438, 440, 443, 444
Mars formation and, 424
orbit of, 65, 479
orbits and physical structure of, 426–430
physical properties of, 67
Delsemme, Armand H., 333
Depositional remanent magnetization (DRM), 361
Diatomic molecules, 122
energy levels, 125
formation, 100
hydrogen as, 149
internal degrees of freedom, 124
Dieterici equation of state, 152
Dimensions and units, 5–6
Disciplines evolving from cosmic events, 1–2
Distance scales in Universe, 7–10
Dreibus–Waenke chemical model, 522
Dust in space, dynamical behavior of, 334–336
Dust opacity, 129–131
Dust sedimentation and turbulence, 134–136

E
Earth, 550–551
appearance, general, 59, 62
atmospheric composition and cycles, 568–573
biological history, 559–563
biosphere stability, 587–588
Cambrian period, 561
carbon cycle diagram, 570
chlorine cycle diagram, 573
climate history, polar age, and ice ages, 579–582
continental shelf, 554

continental slope, 554
Cretaceous period, 562–563
densities of meteorites graph, 552
densities of solid bodies versus heliocentric distance, uncompressed, 133
Devonian period, 562
dimensions, masses, and densities of, 59
effect of gravity of Sun and Moon on equatorial bulge, 58
escape and infall, 575, 576, 577–579
geochemistry and petrology, 563–566
geological ages chart, 561
geological history, early, 557–559
global warming, 580, 581
history of Earth–Moon system, 456–458
hypsogram of, 483
igneous rock types, 565
internal structure, 552–554
isotope climate records, stable, 574–575
Jurassic period, 562
life: stability of biosphere, 587–588
life origins, 582–587
lightening, 216
magnetic field and magnetosphere, 554, 555
magnetosphere structure, 555
major-element composition of crust of, 563
mass of primitive solar nebula, minimum, 119
Mesozoic Era, 562, 579
Mississippian period, 562
Moon. *See* Moon
Moon and orbital motion changes, 57–58
motions, 551–552
nitrogen cycle diagram, 571
orbit of, 56
Ordovician period, 561
Pennsylvanian period, 562
Permian period, 562
photochemistry and aeronomy, 575, 576, 577
physical and chemical fossil evidence of ancient life chart, 560
Quaternary period, 563
radiocarbon dating, 573–574
ray tracing and nomenclature of seismic wave propagation, 553
rotation of, 58
satellites of, orbits of, 65
satellites of, physical properties of, 67
seismic wave velocities and density in interior of, 553
shock-produced gases, 569
Silurian period, 562
solar wind, 554
sulfur cycle diagram, 572
surface geology, 554, 555–557
Tertiary period, 563
thermal structure of atmosphere graph, 576
tidal interaction of Moon and, 457
Triassic period, 562, 579
troposphere composition chart, 568
weathering in rock cycle, 566–568
Eddy diffusion coefficient, 136, 192, 205, 208, 243
Einstein's principle of mass–energy equivalence, 10, 11
Elements, geochemical classification of, 111, 112–115
Elements in solar system, abundance of, 37–39
Energy conversion factors and definitions, 98
Equilibrium thermodynamics, 621–625
adiabatic processes and entropy, 622–623
chemical equilibrium, 623–624
exact and complete differentials, 624
heat and work, 621–622
Maxwell relations, 625
useful work and Gibbs free energy, 623
Euler's equation, 94, 95
Evolution, 2–3
Exploration of solar system, 629–633
Extraterrestrial intelligence, search for, 606–608

F
Feldspars, 109, 110
Fischer–Tropsch process, 130
Fourier transform spectroscopy, 157, 158
Frank–Condon principle, 124
Fraunhofer lines, 51, 84, 88, 89, 231
Free-free transitions, 81, 82, 90
Freezing behavior in $NH_3$–$H_2O$ system, 168
Future prospects, 609–620
asteroids, 614–615
astrometric searches for planets of nearby stars diagram, 617
beyond solar system, 616–620
comets, 616
Earth's Moon, 612–613
Jupiter, 615
Mars, 613–614
Mercury, 611–612
Neptune, 615
physical limits to human exploration, 619
Pluto, 615–616
questions that motivate exploration, 618
Saturn, 615
Uranus, 615
Venus, 612

G
Galaxies, structure and classification of, 16–18
barred spiral, 16
chains, 16
clustering, 16, 17, 18
diagram of classifications, 16
elliptical, 16, 17
spiral, 16, 17
use for classification scheme, 17
Galaxy and star formation, 15–16
Galilean satellites of Jupiter, 227, 252
cometary impacts, 347
dynamical interactions of, 272–275
eclipse radiometry, 256, 257
Galileo and, 615
geology of, 276
impact craters, 347
internal thermal structure of, 267, 268–272
minor satellites and, 278, 279
naming, 69
radii, masses, and densities of, 265
surface, 254, 255, 256
surface morphology of, 258, 259–264
Galileo perspectives on Jovian clouds, 179–180
Ganymede satellite, 69
craters, 263–264, 347
densities of solid bodies versus heliocentric distance, uncompressed, 133
density and composition of, 265, 295
dynamical interactions of, 274, 275
geology of, 276
impact craters on, 347
internal thermal structure of, 270, 272
orbit of, 65
photographs of, 260, 261
physical properties of, 67
planetary environment and, 596
radii, masses, and densities of, 265
radio noise, 278
surface morphology, 259, 260, 262
surface of, 254, 256
thermal and tectonic evolution of, 276, 278
Gas capture from solar nebula, 138–141
Gas cloud collapse, 15–16
Geochemical classification of elements, 111, 112–115
equilibrium condensation, 112–113, 114, 115
state function, 114
types, 111, 112
Giant Molecular Cloud (GMC), 32, 33, 43
Gibbs, J. Willard, 2
Gibbs free energy
change due to pressure and volume changes, 167
of formation of $Fe_2SiO_4$ and $FeSiO_3$, 365
of formation of solid, 267
of liquid, 166
of solid and liquid, 166
of two ensembles of droplets, 178–179
useful work and, 623
of vaporization, 106
Gibbs Phase Rule, 167, 170, 267
Gravity fields, 635–636

H
Hadley regime, 197
Halley, Edmond, 318
Halley's comet
chemical processes in coma and tail, 332, 333
coma of, 594
densities, 342

Halley's comet (*continued*)
eccentricity–period plot, 322
first sighting, 317
future prospects for exploration, 612, 616
naming of, 318
nucleus and coma of, 328
orbital elements, 72
orbits, 321, 322
photographs of, 73, 329, 330
Halogens and phosphorus, 111, 112
Heliosphere, 96
Helium flash, 26, 27, 35, 41, 141
Herbig–Haro (HH) objects, 31
Hertzsprung–Russell (H–R) diagram, 20, 26
brown dwarfs populations, 601
definition, 19
for distant globular cluster, 24
Main Sequence population, 601
with mass and radius data, 22
for Orion Nebula, 29, 30
for stars within 4 pc of Sun, 23
Herzberg, Gerhard, 148
HI regions, 32, 33
HII regions, 32, 33, 34
Hodograph for elliptical motion, 54
Horizontal structure and atmospheric circulation, 187–200, 201, 202, 203
Hubble constant, 9, 27, 46
Hubble Space Telescope (HST), 10, 311, 609
Ceres and, 71
cometary impacts on Jupiter, 346
deep field view from, 620
First Object Spectrometer on, 208
Pluto's two hemispheres images, 309
replacement for, 606
Titan satellite and, 291
Hydrogen
from Big Bang to stars and planets, 144
burning, 34–35, 42, 43
flash, 141
mass–radius diagram for cold hydrogen, 154
rare gases and, 101–102
supercritical, 152

I

Iapetus
hemispheric asymmetry of, 69, 295
orbit, 66, 298
physical properties of, 67
radius of, 295
size, 293
spectrum, 295
surface, 254, 256, 298
Ice minerals, 112, 117
Icy satellites, surfaces of, 253–256
Impact cratering, 438–443, 444, 469
Impulsive perturbations, 57
Infrared Astronomy Satellite (IRAS), 33, 309, 321, 609
Intercloud medium (ICM), 32, 33
Interferogram, 157–158
International Solar Polar Mission (ISPM), 89
Interplanetary current sheet, 93
Interplanetary dust particles (IDPs), 342
Interstellar medium (ISM), 31, 32–33, 34
Interstellar molecules, 31, 32
Introduction, 1–6
Io
atmosphere of, 469
atmosphere schematic, 437
atmospheric and volcanic gases, 435–537
atmospheric gases on, 468
coloring, 430, 431
crater densities and, 443
craters and, 259
density, 430
escape and plasma torus, 437
genetic relationships, 438
geology, 433–435
gravitational interactions, 274, 275
induction of electrical current through, 433
internal energy sources, 432–433
internal structure schematic, 435
introduction to, 425
near-surface region schematic, 435
orbit of, 65
photograph of, 260, 431, 434
properties, general, 430, 431, 432
radii, masses, and densities of, 265
reflectance of white front on, 432
rock-like density, 425
sputtering agent on, 469
surface, 254
surface processes, 430, 431, 432
thermal and tectonic evolution, 278
tidal dissipation in, 276
tidal interaction with Jupiter, 433
Ion production in Jovian atmosphere, 180–183
Ionization, 97–100
Iron, 106–107
effect of high pressure on condensation of, 164
Isothermal remanent magnetization (IRM), 361
Isotropic single-scattering albedo, 184

J

Jean's criterion definition, 15
Jeans escape, 482, 521, 549, 574, 576, 577
Jeans escape flux, 517
Jeans instability, 120
Jovian planets, 147; *see also* Jupiter; Saturn
asteroids and, 259
atmosphere in terms of quantitative physics, 192, 193
cloud condensation in $NH_3$–$H_2O$–$H_2S$ system, 165, 166–174
cloud physics on, 174–179
decimetric radiation, polarization and intensity of, 220
Galileo Probe perspectives on clouds of, 179–180
formation, 34
horizontal structure and atmospheric circulation, 187–200, 201, 202, 203
hydrogen and helium photochemistry, 207
ion production in atmosphere of, 180–183
lifeforms, 217
lightening, 216–217
list of, 59
magnetic field explanations, 228
perspective, 247
photochemistry and aeronomy, 200, 201, 202, 203–217
rotation periods, 58
satellites, 279, 280, 425
search for other planets and, 605, 606
thermosphere, 217–218
trans-Jovian fireballs, 343, 344
troposheric composition and structure theory, 159–165, 166
vertical structure of clouds, 178
water and ice condensation, 166
Jupiter. *See also* Jovian planets
abundances in atmosphere, 159
adiabat for atmosphere of, 159, 163
ammonia photochemistry on, 210
angular momentum, 62
appearance, general, 59, 64
atmospheres of Saturn and, 156, 157–159
belts, 187, 188, 213
Callisto. *See* Callisto satellite
chemistry of atmosphere of, 162, 163
chromophores on, 213–215
cloud bands nomenclature for, 188
cloud structure deep in atmosphere of, 166
cloud structures in upper troposphere of, 172
cometary impacts on, 344–347
composition observed, 156, 157–159
Coriolis force, 196, 278
current speeds, 189
dimensions, masses, and densities of, 59
energy output, 151
Equatorial Zone (EZ), 187, 188
future prospects, 615
Galilean satellites. *See* Galilean satellites of Jupiter
Galileo probe, 159, 178, 179–180, 200
Ganymede. *See* Ganymede satellite
gas abundances in upper troposphere of, 173
Great Red Spit (GRS), 187, 188, 191, 195, 197, 198–199, 211
helium abundance, 149
hydrogen sulfide photochemistry diagram, 212
interior schematic, 157
interiors of Saturn and, 148–151
internal heat sources, 243
Io. *See* Io
isothermal interior models of Saturn and, 151–154
life on, 217
magnetosphere of, 215, 226
magnetospheres and radiophysics, 218–229, 230, 231

mass of primitive solar nebula, minimum, 119
methane photochemistry diagram, 209
modulation of radio emission by Io (diagram), 228
North Equatorial Belt (NEB), 174, 177, 187, 188, 192, 195, 200, 211
orbit of, 56, 57
phosphine photochemistry on, 211
photolysis products accumulation diagram, 214
photolysis rates of major gas species on, 215
planetary rings, 280, 281
radiation belts, 181
radiative and convective fluxes in atmosphere, 192
radio-wavelength emission from (diagram), 220
radiophysics and magnetospheres of Saturn and, 218–229, 230, 231
reflection and emission spectrum, 150
relative cloud masses and moist adiabetic lapse rate in atmosphere of, 177
ring systems comparison, 287
rotation of, 58, 187, 188
satellites of, 65, 69
satellites of, minor, 278–280
satellites of, orbits of, 65
satellites of, physical properties of, 67
solar ultraviolet flux at, 204
South Tropical Disturbance (STD), 191
structure of deep atmosphere of, 161
thermal models of Saturn and, 154–156, 157
thermal structure of upper troposphere and stratosphere–mesosphere of, 186
ultraviolet light depth of penetration diagram, 216
upper atmosphere structure diagram, 207
visual appearance, changing of, 187
zones, 187, 188

K
Kepler, Johannes, 52, 54
Keplerian orbit, 135, 139, 336, 412
Kepler's laws of motion, 71, 311
Kinetic inhibition, 117, 118
Kirkwood gaps, 388, 408
Kronecker delta, 195
Kreutz group, 322
Kuiper, Gerard, 148, 325, 411
Kuiper belt
Centaurs and, 411, 412
components of, 70
example of, classical, 410, 411
future explorations, 616
as iceballs, 325
Pluto and, 148
Quaoar, 311
satellites of, 70
size, 71
trans-Neptunian objects and, 388

L
Lagrangian satellite orbits, 298
Lambert scattering, 184
Lambert surface, 253
Laplace, Pierre-Simon, 274
Laplacian operator, 195
Laws of planetary motion, 52
Le Chatelier's principle, 100
Legendre polynomials, 57, 223
Life, chemical and physical prerequisites of, 592–595
Life origins, 582–587
amino acids, 583–584, 585, 586
DNA structure diagram, 583
mutations, 584, 585, 586, 587
nucleotide bases, 583
organic bases, 582–583
phosphates, 582
sugars, 582
Light absorption and transmission terminology, 131
Limb brightening, 51
Limb darkening, 51
Literature guide, 3–4
Lithophile, 111, 112
Lorentz force, 220, 286, 336, 432
Lorentz line profile, 129
Loschmidt's number, 5, 204, 238, 330
Luminosities, 8, 9
of galaxies, 17
of star, 18, 20
of sun, 18
Luyten's luminosity function, 23–24
Lyman series, 80, 81

M
Magellanic Clouds, 16
Magnesium and silicon, 105–106, 116
Magnitude scale for stars, 18–19
Mars, 478–479
appearance, general, 59, 63
Arandas crater photograph, 487
Arsia Mons photograph, 493
atmospheric circulation, 509–510
atmospheric composition, 510–513
atmospheric structure, 508–509
Deimos. *See* Deimos
densities of solid bodies versus heliocentric distance, uncompressed, 133
density and figure of, 479–481
dimensions, masses, and densities of, 59
drainage systems photograph, 491
dune field in Hellespontus region photograph, 494
dust transfer on (photographs), 495, 496
eccentricity seasons, 479
elliptical orbit, 52
eroded layered deposits near south pole (photographs), 501, 502
explosive blowoff, 519, 522
future prospects, 613–614
Ganges Chasma photograph, 485
geology, 483–496, 497, 498, 499, 500, 501, 502
geophysical data, 481–482
gravity and tectonics, 483
Hellas basin, 483, 484, 486, 503
highest peak, 597
hypsogram of, 483
landslides in Valles Marineris photographs, 490
mass of primitive solar nebula, minimum, 119
motions of, 479
north polar regions photomosaic, 498
Olympus Mons photograph, 492
orbit of, 56
organic matter and origin of life, 522–524
origin and evolution of atmosphere, 519–522
Phobos. *See* Phobos
photochemical stability and atmospheric escape, 513–519
rotation of, 58
satellites of, 279
satellites of, orbits of, 65
satellites of, physical properties of, 67
seasonal surface pressure variations on, 506
shergottite, nakhlite, and chassignite (SNC) meteorites, 505–508, 519, 521, 522
solar wind sweeping, 519
surface composition, 496, 497, 499, 500, 501, 502–503
surface materials composition graph, 505
surface photograph, 504
temperature profiles, 509
terraforming, 524
Tharsis Ridge, 483, 484, 488, 490, 491
Valles Marineris photograph, section of, 489
Viking lander investigations, 503, 504–505
water on, cycles and loss mechanisms for, 516
Mass and angular momentum distribution, 59, 61, 62, 63, 64
Mass and density of solar nebula, 118–121
Mass defect, 36
Maunder sunspot minimum, 85, 86
Maxwellian velocity distribution, 326
Maxwell's equations, 92, 93, 222, 223, 271
Megaparsec (Mpc), 9
Mercury, 461
appearance, general, 59, 60
atmosphere of, 468–469
atmospheric gases on, 468
composition and structure of, 462–463
crater densities and, 443
densities of solid bodies versus heliocentric distance, uncompressed, 133
dimensions, masses, and densities of, 59
flyby geometry, 468
future prospects, 611–612
geophysics of, 463, 466–468
hypsogram of, 483

Mercury, (*continued*)
interior model of, 467
mass of primitive solar nebula, minimum, 119
motions of, 461–462
near-Earth asteroids, approaching, 473
noncrater geology of, 463, 464–465
orbit of, 56
polar deposits on, 469–472, 473
polar ice cap radar image, 471
rotation of, 58
solar winds interactions with Moon and, 460–461
south polar region photomosaic, 472
temperature on, 463, 466
Meteoritics, 350
Meteorites, 72, 73
books on, 353
Caloris Basin photograph, 464
carbonaceous, 118
chondrules, 73, 118
classes, genetic relationships, 382
classes of, 351, 352
collections of, 353
composition of, 73
craters, 73, 75
definitions, 72, 336, 350
density and rocky planets, 552
desert varnish, 352
Discovery Scarp photograph, 465
element compositional variations, major, 370
falls, frequency of, 351
finds, frequency of, 351
fusion crust, 352
gas-retention ages, 45–46
genetic relationships between classes of, 382–384
introduction, 350–353
isotopic composition of, 375–382
journals on, 353
lunar, 351
metamorphic grades of chondrites, 367–369
minerals, 362, 363
naming, 351
Neumann lines, 372
orbits, 353–355
orbits of recovered meteorites, 354
oxygen isotope systematics of, 375
phenomena of fall, 355–358
physical properties, 358–362
properties, 72, 73
relationship among asteroids, comets, and, 412–415
rhenuim–osmium dating, 43
shergottite, nakhlite, and chassignite (SNC), 371, 505–508, 519, 521, 522
solar wind and, 380
sounds associated with, 355
spectral reflectivity of samples, 399
taxonomy and composition of achondrites, 369–371
taxonomy and composition of chondrites, 362, 364–367
taxonomy and composition of irons, 372–375
taxonomy and composition of stony irons, 371–372
thermal evolution, 401, 402
vapor trail colors, 355
Widmanstätten structure, 372
Meteoroids, 72, 337, 338, 350, 384
Meteors, 72, 336–343
cause of, 72
cometary fireballs, 343–344
crater, 75
definition, 336
Geminid shower, 414
Leonid meteor shower photo, 339
photograph of, 74
showers, 72, 337, 340, 342
showers, prominent, 75, 337
storm, 72
streams, 337, 338
timing, 336
velocities relative to Earth (diagram), 341
Mie scattering, 183–184, 185
Milky Way
density of matter within, 16
globular clusters associated with, 17
luminosity of, 17
as spiral galaxy, 17
Mizar–Alcor, 29
Molecules
diatomic. *See* Diatomic molecules
energy diagram, 126
energy levels, solar radiation, and thermal energy diagram, 126
formation and dissociation, 100–101
internal degrees of freedom, 124
interstellar, 32
ionization potentials, first, 183
polyatomic formation, 101
vibrational excitation of, 123
Moment of inertia factor, 58
Moon, 68
atmosphere of, 468–469
atmospheric gases on, 468
crater densities and, 443, 444
elemental abundance patterns, 451
elemental composition of surface of, 445–447
future prospects, 612–613
geology of, 451–452, 453
geophysics, 452, 454–456
history of Earth–Moon system, 456–458
hypsogram of, 483
internal structure, 456
libration in latitude, 444
libration in longitude, 444
Mare Orientale on, 440, 441, 443, 452
minerals, 449–451
moonquakes, 455
motions of, 443, 444–445
native minerals, selected, 449
orbit of, 65, 443
origin and internal evolution of, 458–460
origin of, 559
physical properties, 445
polar deposits on, 469–471
pyroxenes composition, 450
rock types, 447–449
solar wind interaction, 460–461
surface exploration photograph, 453–454
survey or alpha-backscatter analyses, 446
terrain types, 445
tidal interaction of Earth and, 457

N
National Aeronautics and Space Administration (NASA)
Ames Research Center, 346, 433
asteroid exploration, 386, 614, 615
Cassini joint mission, 292, 293, 311
Challenger explosion and, 179
Galileo probe and, 179
ISPM and, 90
Jupiter exploration and, 615
Mars exploration and, 614
Mercury exploration and, 611, 612
missions to Pluto, 311
Moon and, 613
origin of life mandate, 1
Pluto exploration and, 616
Saturn exploration and, 615
solar wind observations, 96
Venus and, 612
Near-Earth asteroids (NEAs), 418
comets and, 413, 414
dynamical evolution of, 407, 408
encounters with planets, 413
exploratory missions to study, 386
frequencies, 415
mapping of, 71
Mercury-approaching, 473
meteorites and, 400
orbits, 69, 354, 387, 413, 414
photometry and spectroscopy, 401
population, 417
radar observations of, 415–416
rotation of, 391, 408
satellites of, 70, 385, 393
shape, 71
size, 393
vulnerability and short life of, 410, 412–413
Neel point, 361
Nephelometer, 543
Neptune
appearance, general, 59, 64
atmospheres of Uranus and, 238–246
cloud condensation diagram, 239
dimensions, masses, and densities of, 59
discovery of, 56
future prospects, 615

Great Dark Spot (GDS), 244–245
infrared and radio brightness temperatures (diagram), 241
interiors of Uranus and, 229, 230, 231–238
internal heat sources, 243
internal structure schematic, 237
mass of primitive solar nebula, minimum, 119
methane abundance, 238, 239
night side photograph, 246
orbit of, 56, 57
planetary rings, 287, 289, 290
ring systems comparison, 287
rotation of, 58, 231, 232
satellites of, 69, 303, 304–308
satellites of, orbits of, 66
satellites of, physical properties of, 68
satellites temperature, 258
temperature–pressure profiles for atmosphere, 242
Triton. *See* Triton
Voyager image of, 244
Neptune–Pluto resonance, 311
Neumann lines, 372
Neutral sheet, 93
Neutrinos, 78–79
Newton, Sir Isaac, 2, 52, 53, 318, 457
Nickel and cobalt, 110–111
Nitrogen, oxygen, and carbon, 102–105
North Equatorial Belt (NEB) of Jupiter, 174, 177, 187, 188
Nova explosions, 34
Nuclear binding energy, 36
Nuclear cosmochronology, 43, 44–47
Nucleosynthetic sources of selected nuclides, 43
Nuclides, stability of, 36
Numbers in science, 4–5

O
Oort cloud, 324, 325, 326, 327
Orbital motion changes, 57–58
Orbits of planets, 52–57
circular, 55
elliptical, 52, 53, 54–55
laws of planetary motion, 52
line of nodes, 55
orientation, 55
shape, 55
table depicting, 56
universal law of gravitation, 52, 53
Orion Nebula, 28, 29, 30
Oxygen, carbon, and nitrogen, 102–105

P
Parallax (p), 7, 9
Parsec (pc), 7, 8
Pele eruption, 437
Phobos, 68, 386, 425
crater chains, 428
density of, 425
future prospects and, 614
impact cratering and, 438, 440, 443, 444
Mars formation and, 424
orbit of, 65, 479
orbits and physical structure of, 426–430
photographs of, 427, 428–429
physical properties of, 67
space exploration and, 472, 473
Phosphorus and the halogens, 111
Photochemistry and aeronomy, 200, 201, 202, 203–217
Physical constants, basic, 634
Planck function and distribution of photon energies in fireball, 11, 19
Planetary environment, 595–597
Planetary magnetosphere schematic, 223
Planetary motion laws, 52
Planetary nebulae, 31, 33, 34
Planetary rings, 280–289, 290
Planetary science(s)
as interdisciplinary science, 2
nature and scope of, 1–3
purpose of, 3
in space age, 74, 75
Planets. *See also* individual planets
angular momentum density of spin of, 64
classes of, 147
dimensions, masses, and densities, 59
Jovian. *See* Jovian planets
major, 147–251
orbital angular movement, 64
orbits. *See* Orbits of planets
properties, 58–59, 60, 61, 62, 63, 64
rotation, 58
terrestrial, 59
Pluto
appearances, general, 59
Bode's Law and, 56
Charon. *See* Charon satellite
densities of solid bodies versus heliocentric distance, uncompressed, 133
dimensions, masses, and densities of, 59
future prospects, 615–616
hemispheres images, 309
mass of primitive solar nebula, minimum, 119
orbit of, 56, 57
rotation of, 58
satellites of, orbits of, 66
satellites of, physical properties of, 68
Pluto–Kuiper Belt Mission, 616
Pluto–Neptune resonance, 311
Polaris as quintuple star, 29
Polyatomic molecules formation, 101
Potassium and sodium, 109–110
Poynting–Robertson effect
dynamical behavior of dust in space and, 335, 336
dynamical evolution of asteroid belt, 407
meteor showers and, 337
orbital geometry and, 427
relationship among asteroids, meteorites, and comets and, 413
Prandtl mixing length theory, 191, 243
Pressure broadening, 81, 127
Prior's Rules, 364, 366, 371

R
R Moncerotis (RMon), 31
Radiation diagram, cyclotron and synchrotron, 221
Radiative transfer, visible and infrared, 183–187
Radio Astronomy Satellite (RAS), 219
Radio wave propagation in space plasmas, 91–92
Radiocarbon dating, 573, 574
Radiolysis, 276
Radiometry, eclipse, 256–257
Rapid accretion, chemistry of, 116–117, 118
heterogeneous accretion model, 116
nonhomogeneous accretion model, 116, 117, 118
Rayleigh scattering
atmosphere of Uranus and, 238, 240
brown dwarfs and, 601
cloud condensation in $NH_3$–$H_2O$–$H_2S$ system and, 174
photochemistry and aeronomy and, 202, 207, 211
scattering versus size of particles, 185
visible and infrared radiative transfer and, 183, 185
Redshift, 9
Refractory oxides, 112, 116
Refractory siderophiles, 112, 116
Reynolds number, 6, 195, 601
Rhenium–Osmium dating, 43, 44
Richardson number, 6, 197, 535
Roche limit, 283, 304, 345, 598, 599, 601
Rossby number, 6, 196, 535, 601
Rubidium–Strontium dating, 45
Rydberg constant, 80

S
Sagan, Carol, 499, 523, 525
Satellites, 63, 65–69
icy, density and composition of, 265–267, 268
icy, surfaces of, 253–256
icy, thermal and tectonic evolution of, 275–278
orbits of, 65–66
physical properties of, 67–68
Saturn. *See also* Jovian planets
abundances in atmospheres, 159
adiabat for atmosphere of, 160
ammonia vapor, 149
appearance, general, 59, 64
atmospheres of Jupiter and, 156, 157–159
composition, observed, 156, 157–159
chromosphores on, 213–215
cloud structure, 178
current speeds, 189
dimensions, masses, and densities, 59

Saturn. *See also* Jovian planets (*continued*)
future prospects, 615
Iapetus. *See* Iapetus
internal heat sources, 243
interior schematic, 157
interiors of Jupiter and, 148–151
isothermal interior models of Jupiter and, 151–154
magnetic field of, 228, 230
magnetosphere of, 228, 229
magnetospheres and radiophysics, 218–229, 230, 231
mass of primitive solar nebula, minimum, 119
orbit of, 56, 57
planetary rings, 280, 281–286
radiation belts, 228
radiophysics and magnetospheres of Jupiter and, 218–229, 230, 231
red spot photograph, 201
Rhea and Dione, 254, 256, 293
ring systems comparison, 287
rings and satellites, 254, 283
rotation of, 58
satellite, intermediate-sized, 293–296, 297
satellites craters, 259
satellites of, minor, 296, 297–299
satellites of, orbits of, 65–66, 257
satellites of, physical properties of, 67
satellites temperature, 258
thermal emission, 151
thermal models of Jupiter and, 154–156, 157
Titan. *See* Titan satellite
wave disturbances in atmosphere photograph, 203
Scattered disk objects (SDOs), 410, 411
Scientific theories flexibility, 3
Scorpius–Centaurus cluster, 28
Search for planets of other stars, 603–606
Shock remanent magnetization (SRM), 361
Siderophile, 111, 112
Silicon and magnesium, 105–106
Silicon burning, 42, 43
Sirius, 19
Skylab solar corono observations, 89
Sloan Digital Sky Survey, 386
Sodium and potassium, 109–110
Solar flares, 85
Solar material, chemistry of, 96–97, 98
Solar Maximum Mission (SMM), 87
Solar nebula
accretion of rocks, planetesimals, and planets, 136–138
adiabatic gas capture from, 140
dust opacity, 129–131
forces on a volume element in, 120
formation of, 120
gas capture from, 138–141
luminosity, 131
mass and density of, 118–121
mass distribution in, 120
opacity of solar material, 139
T Tauri phase, 141–143
thermal opacity in, 121–129
thermal structure, 131–134
turbulence and dust sedimentation, 134–136
turbulence driving forces, 134
Solar power satellites (SPSs), 418
Solar system
description of, general, 50–76
exploration of, 629–633
material chronology, 46
thermal history of early, 143–144
Solar wind, 92–96
acceleration of, 95, 96
definition, 51
discovery of, 90–91
dynamic behavior of, 95
Earth and, 554
heating, 404
inducing heating, 31
interaction with Moon and Mercury, 460–461
interactions with planets, 482
magnetic sector structure, 89, 93
on Mars, 482, 519
Mercury and pressure of, 468
meteorites and, 380
monster streams, 89, 90
observations by Ulysses spacecraft, 96
Parker model for, 549, 578
particles, 93
photosphere and, 92
solar activity and, 85
study of, 89, 90
Sun and, 77
T Tauri phase, 33, 121, 141, 143, 272, 343, 361, 398, 402, 522
Uranus, 245
Space age, planetary science in, 74, 75
Space plasmas, radio wave propagation in, 91–92
Spacecraft explorations, 311–312
Spacecraft launchings table, 630–633
Spectroscopic determination of rotation rates (diagram), 233
Spörer sunspot minimum, 86
Star and galaxy formation, 15–16
Star clusters, 27–29
angular momentum, 28–29
Cepheid variables and, 24
distance measurement, 24
double and multiple stars, 29
globular, 24, 25, 27
gravitational potential energy of, 28
lifetime of young cluster, 30
O associations, 27, 29
Orion Nebula, 28
rotation, 28
Scorpius–Centaurus, 28
T associations, 27
ursa Major, 28, 29
UV brightness, 24
Star formation
energy sources, 25
origin of, 25
outline of, 33–34
Stars, classification of, 18–25
absolute magnitude, 19
Barnard's star, 29
brightness, 18
brown dwarfs. *See* Brown dwarfs
color, 18, 19
colors and temperature, 19
correlation between luminosity and mass, 21, 22
filters to determine color, 19
Hertzsprung–Russell diagram, 19, 20, 22, 23, 24
Luyten's luminosity function, 23–24
magnitude scale, 18–19
mass–luminosity (M–L) diagram, 21, 22
M-class dwarf, 24, 598, 600, 603, 605–606
orbital velocities determination, 21
peculiar A stars, 28
Population I (Pop I), 25
Population II (Pop II), 25
red dwarfs, 21, 22, 23, 27, 30, 326
T Tauri stars. *See* T Tauri stars
UBV filter system, 19
white dwarfs, 21, 22, 23, 27, 598
Wolf–Rayet stars. *See* Wolf–Rayet stars
Stars, random velocities of, 17
Stefan–Boltzmann constant, 20, 254, 634
Stellar environment, 597–600
Stellar evolution, 25–27, 34
Stellar explosion winds, 32
Stellar explosions and nucleosynthesis, 34–43, 44
Stellar origins, 29–33
Stellar winds, 32
Stoke's Law, 135, 179
Stoke's Theorem, 194
Sulfur, 107–108
Sun, 50–52
chromosphere, 51, 84, 87–88
corona of, 51, 84, 88–90
elements in, abundance of, 37
emission spectrum, 51
energy production in, 77–79
energy transport in, 79–83
internal structure, 83–84
IR emission from, 31
luminosity of, 18, 51, 78, 84, 85
magnetic field two-component model, 90
mass and angular momentum, 61, 62
nuclear reactions in, 77–78
photosphere, 51, 79, 81, 82, 83, 87
quiet conditions in solar equatorial plane, average, 94
radical temperature profiles in nebula during T Tauri phase of, 142
rotation of, 50
rotational speed, 17, 18
solar cosmic rays, 51

solar flares, 85
solar wind. *See* Solar wind
sunspots, 50, 84–85, 86
surface, 84–87
surface gravity, 51
temperatures near surface, 87
total eclipse, 87
Supernova
explosions, 33, 41, 42, 45
explosions role in recycling, 46, 47
light curve of, 44
Surface morphology of Galilean satellites, 258, 259–264
Surface temperatures, 257–258, 259

T
T Tauri stars
clusters of, 27
color of most, 19, 30
Herbig–Haro objects and, 31
IR emission from, 31
lifetime of, 598
location, 27
mass loss rates from spectra of, 31
Orion Nebula and, 28
phase, 27, 31, 141–143, 272, 402, 403, 549, 578
phase solar wind, 33, 121, 272, 343, 398, 402, 466, 522
radical velocities, 141
solar wind and, 96, 141, 143, 404, 458
Technetium (Tc), 39
Temperatures, surface, 257–258, 259
Terrestrial planets, 59
Thermal history of early solar system, 143
Thermal inversion, 186
Thermal metamorphism, 367
Thermal opacity in solar nebula, 121–129
Thermal remanent magnetization (TRM), 361
Thermal structure of nebula, 131–134
Thermodynamic equilibrium, 2
Thomson scattering, 81, 89, 90, 183, 185
Tidal torques, 273, 274, 275, 304, 554
Titan satellite, 69, 289, 291–293
atmosphere composition of, 292
atmosphere density of, 259
densities of solid bodies versus heliocentric distance, uncompressed, 133
density, 295
methane in atmosphere of, 236, 306
orbits of, 65, 298
photograph of, 282
physical properties of, 67
planetary rings and, 281
polarization behavior of, 256
rotation, 536
size, 265
surface image of, 291
surface of, 254
Voyager imaging of, 291
wind speeds, 537
Titius–Bode Law, 55, 56
Trans-Neptunian objects (TNOs), 388, 391, 409–412
Triton, 252, 304–306, 410, 425
methane and, 305–306
orbit, 66, 69, 303, 304, 305, 311
photograph of, 307
physical properties, 68
radius of, 304
size, 265
surface, 254, 306, 310
surface temperature, 258
tidal torques on, 304
visible and near-infrared spectrum, 305
Tombaugh, Clyde, 308
Trojans asteroids
dynamical evolution, 407
Jupiter's orbit and, 387
Mars and, 479
orbit, 69, 388
orbit stability, 389–391
photometry and spectroscopy, 395
satellites of, 385, 393
size, 71
Tropospheric composition and structure theory (Jovian planets), 159–165, 166
Turbulence and dust sedimentation, 134–136

U
UBV photometric classification of stars, 19
Units and dimensions, 5–6
Uranian planets, 147, 247; *see also* Neptune; Uranus
Uranus, 56
appearance, general, 59, 64
atmospheres of Neptune and, 238–246
cloud condensation diagram, 59
dimensions, masses, and densities, 59
future prospects, 615
geometry of Voyager 2 flyby, 234
infrared and radio brightness temperatures (diagram), 241
interiors of Neptune and, 229, 230, 231–238
internal heat sources, 243
internal structure schematic, 237
magnetosphere, 237–238, 245
mass of primitive solar nebula, minimum, 119
methane abundance, 238, 239
orbit of, 56, 57
orbits of rings of, 288
photochemistry, 245
planetary rings, 286–287, 288–289
reflection spectrum of, 234
ring systems comparison, 287
rings of, orbits of, 288
rotation of, 58, 230, 231, 232
satellites of, 69, 254, 256, 299–303, 304
satellites of, orbits of, 66
satellites of, physical properties of, 68
satellites temperature, 258
solar wind, 245
temperature–pressure profiles for atmosphere, 242
Voyager 2 views of polar region of, 232
Voyager image of, 244
Urey, Harold, 214, 479, 526
Urey–Craig diagram, 364, 366
Ursa Major, 28, 29

V
Van Allen, James A., 91
van der Waals attraction, 137, 152
Vapor pressure equation, 167
Vapor pressure in $NH_3$–$H_2O$ system, 171
Venus, 477, 478, 524–526
appearance, general, 59, 61
atmosphere–lithosphere interactions, 539–543
atmospheric composition, 537–539
atmospheric escape, 547–549
atmospheric structure and motions, 534, 535, 536–537
cloud structure diagram, 545
clouds ultraviolet photograph, 544
cometary debris, 531
composition of lower atmosphere, 538
densities of solid bodies versus heliocentric distance, uncompressed, 133
dimension, masses, and densities of, 59
future prospects, 612
geology of, 528–534, 535
geophysical data on, 526–528
hypsogram of, 483
impact craters photograph, 533
impact craters size distribution diagram, 534
ion abundance profiles on, 547
iron-mineral stability on (diagram), 542
Magellan Venus radar mapper, 527, 530, 531, 542, 550, 612
major-element composition of crust of, 540
mass of primitive solar nebula, minimum, 119
mechanisms for hydrogen escape from, 548
motions and dynamics of, 526
orbit of, 56
photochemistry and aeronomy, 543–547
Pioneer Venus Orbiter, 526, 527, 528
planetary evolution, 549–550
primordial rare gases in atmosphere, 539
radio-wavelength brightness temperatures of, 527
radionuclide abundances in crust of, 539
reflectivity of surface photograph, 529
rotation of, 58
Sapas Mons photograph, 532
sulfur cycle schematic, 546
surface photograph, 535
surface radar image, 530
temperature profile of atmosphere, 536

Venus, (*continued*)
topography photograph, 528
water content of atmosphere for cometary infall model, 549
Vibration excitation of molecules, 123
Viral theorem, 83
Vis viva equation, 55
Visible and infrared radiative transfer, 183–187

W
Water ice/water/vapor equilibria, 168
Widmanstätten structure, 359, 372
Wildt, Rupert, 148, 156, 497, 525, 537
Wolf–Rayet stars
evolutionary track of star with one solar mass and, 27
interstellar medium and, 31
planetary nebulae and, 33, 34
spectra, 26

Y
Yarkovsky effect, 336, 337, 407, 427

Z
Zurich sunspot numbers, 86

# International Geophysics Series

EDITED BY

**RENATA DMOWSKA**
*Division of Applied Science, Harvard University, Cambridge, Massachusetts*

**JAMES R. HOLTON**
*Department of Atmospheric Sciences, University of Washington, Seattle, Washington*

**H. THOMAS ROSSBY**
*Graduate School of Oceanography, University of Rhode Island Narragansett, Rhode Island*

*Volume 1* Beno Gutenberg. Physics of the Earth's Interior. 1959*
*Volume 2* Joseph W. Chamberlain. Physics of the Aurora and Airglow. 1961*
*Volume 3* S. K. Runcorn (ed.). Continental Drift. 1962*
*Volume 4* C. E. Junge. Air Chemistry and Radioactivity. 1963*
*Volume 5* Robert G. Fleagle and Joost A. Businger. An Introduction to Atmospheric Physics. 1963*
*Volume 6* L. Defour and R. Defay. Thermodynamics of Clouds. 1963*
*Volume 7* H. U. Roll. Physics of the Marine Atmosphere. 1965*
*Volume 8* Richard A. Craig. The Upper Atmosphere: Meteorology and Physics. 1965*
*Volume 9* Willis L. Webb. Structure of the Stratosphere and Mesosphere. 1966*
*Volume 10* Michele Caputo. The Gravity Field of the Earth from Classical and Modern Methods. 1967*
*Volume 11* S. Matsushita and Wallace H. Campbell (eds.). Physics of Geomagnetic Phenomena (In two volumes). 1967*
*Volume 12* K. Ya Kondratyev. Radiation in the Atmosphere. 1969*
*Volume 13* E. Palmån and C. W. Newton. Atmospheric Circulation Systems: Their Structure and Physical Interpretation. 1969*
*Volume 14* Henry Rishbeth and Owen K. Garriott. Introduction to Ionospheric Physics. 1969*
*Volume 15* C. S. Ramage. Monsoon Meteorology. 1971*

*Out of print

Volume 16 JAMES R. HOLTON. An Introduction to Dynamic Meteorology. 1972*
Volume 17 K. C. YEH AND C. H. LIU. Theory of Ionospheric Waves. 1972*
Volume 18 M. I. BUDYKO. Climate and Life. 1974*
Volume 19 MELVIN E. STERN. Ocean Circulation Physics. 1975
Volume 20 J. A. JACOBS. The Earth's Core. 1975*
Volume 21 DAVID H. MILLER. Water at the Surface of the Earth: An Introduction to Ecosystem Hydrodynamics. 1977
Volume 22 JOSEPH W. CHAMBERLAIN. Theory of Planetary Atmospheres: An Introduction to Their Physics and Chemistry. 1978*
Volume 23 JAMES R. HOLTON. An Introduction to Dynamic Meteorology, Second Edition. 1979*
Volume 24 ARNETT S. DENNIS. Weather Modification by Cloud Seeding. 1980*
Volume 25 ROBERT G. FLEAGLE and JOOST A. BUSINGER. An Introduction to Atmospheric Physics, 2nd Edition. 1980*
Volume 26 KUO-NAN LIOU. An Introduction to Atmospheric Radiation. 1980*
Volume 27 DAVID H. MILLER. Energy at the Surface of the Earth: An Introduction to the Energetics of Ecosystems. 1981
Volume 28 HELMUT G. LANDSBERG. The Urban Climate. 1991
Volume 29 M. I. BUDKYO. The Earth's Climate: Past and Future. 1982*
Volume 30 ADRIAN E. GILL. Atmosphere-Ocean Dynamics. 1982
Volume 31 PAOLO LANZANO. Deformations of an Elastic Earth. 1982*
Volume 32 RONALD T. MERRILL AND MICHAEL W. MCELHINNY. The Earth's Magnetic Field: Its History, Origin, and Planetary Perspective. 1983*
Volume 33 JOHN S. LEWIS AND RONALD G. PRINN. Planets and Their Atmospheres: Origin and Evolution. 1983
Volume 34 ROLF MEISSNER. The Continental Crust: A Geophysical Approach. 1986
Volume 35 M. U. SAGITOV, B. BODKI, V. S. NAZARENKO, AND KH. G. TADZHIDINOV. Lunar Gravimetry. 1986
Volume 36 JOSEPH W. CHAMBERLAIN AND DONALD M. HUNTEN. Theory of Planetary Atmospheres, 2nd Edition. 1987
Volume 37 J. A. JACOBS. The Earth's Core, 2nd Edition. 1987*
Volume 38 J. R. APEL. Principles of Ocean Physics. 1987
Volume 39 MARTIN A. UMAN. The Lightning Discharge. 1987*
Volume 40 DAVID G. ANDREWS, JAMES R. HOLTON, AND CONWAY B. LEOVY. Middle Atmosphere Dynamics. 1987
Volume 41 PETER WARNECK. Chemistry of the Natural Atmosphere. 1988*
Volume 42 S. PAL ARYA. Introduction to Micrometeorology. 1988*
Volume 43 MICHAEL C. KELLEY. The Earth's Ionosphere. 1989*
Volume 44 WILLIAM R. COTTON AND RICHARD A. ANTHES. Storm and Cloud Dynamics. 1989
Volume 45 WILLIAM MENKE. Geophysical Data Analysis: Discrete Inverse Theory, Revised Edition. 1989
Volume 46 S. GEORGE PHILANDER. El Niño, La Niña, and the Southern Oscillation. 1990
Volume 47 ROBERT A. BROWN. Fluid Mechanics of the Atmosphere. 1991
Volume 48 JAMES R. HOLTON. An Introduction to Dynamic Meteorology, 3rd Edition. 1992
Volume 49 ALEXANDER A. KAUFMAN. Geophysical Field Theory and Method.
Part A: Gravitational, Electric, and Magnetic Fields. 1992*
Part B: Electromagnetic Fields I. 1994*
Part C: Electromagnetic Fields II. 1994*
Volume 50 SAMUEL S. BUTCHER, GORDON H. ORIANS, Robert J. CHARLSON, AND GORDON V. WOLFE. Global Biogeochemical Cycles. 1992
Volume 51 BRIAN EVANS AND TENG-FONG WONG. Fault Mechanics and Transport Properties of Rocks. 1992
Volume 52 ROBERT E. HUFFMAN. Atmospheric Ultraviolet Remote Sensing. 1992
Volume 53 ROBERT A. HOUZE, JR. Cloud Dynamics. 1993
Volume 54 PETER V. HOBBS. Aerosol-Cloud-Climate Interactions.1993
Volume 55 S. J. GIBOWICZ AND A. KIJKO. An Introduction to Mining Seismology. 1993
Volume 56 DENNIS L. HARTMANN. Global Physical Climatology. 1994
Volume 57 MICHAEL P. RYAN. Magmatic Systems. 1994
Volume 58 THORNE LAY AND TERRY C. WALLACE. Modern Global Seismology. 1995

*Out of print

*Volume 59* Daniel S. Wilks. Statistical Methods in the Atmospheric Sciences. 1995
*Volume 60* Frederik Nebeker. Calculating the Weather. 1995
*Volume 61* Murry L. Salby. Fundamentals of Atmospheric Physics. 1996
*Volume 62* James P. McCalpin. Paleoseismology. 1996
*Volume 63* Ronald T. Merrill, Michael W. McElhinny, and Phillip L. McFadden. The Magnetic Field of the Earth: Paleomagnetism, the Core, and the Deep Mantle. 1996
*Volume 64* Neil D. Opdyke and James E. T. Channell. Magnetic Stratigraphy. 1996
*Volume 65* Judith A. Curry and Peter J. Webster. Thermodynamics of Atmospheres and Oceans. 1998
*Volume 66* Lakshmi H. Kantha and Carol Anne Clayson. Numerical Models of Oceans and Oceanic Processes. 2000
*Volume 67* Lakshmi H. Kantha and Carol Anne Clayson. Small Scale Processes in Geophysical Fluid Flows. 2000
*Volume 68* Raymond S. Bradley. Paleoclimatology, Second Edition. 1999
*Volume 69* Lee-Lueng Fu and Anny Cazanave. Satellite Altimetry and Earth Sciences: A Handbook of Techniques and Applications. 2000
*Volume 70* David A. Randall. General Circulation Model Development: Past, Present, and Future. 2000
*Volume 71* Peter Warneck. Chemistry of the Natural Atmosphere, Second Edition. 2000
*Volume 72* Michael C. Jacobson, Robert J. Charlson, Henning Rodhe, and Gordon H. Orians. Earth System Science: From Biogeochemical Cycles to Global Change. 2000
*Volume 73* Michael W. McElhinny and Phillip L. McFadden. Paleomagnetism: Continents and Oceans. 2000
*Volume 74* Andrew E. Dessler. The Chemistry and Physics of Stratospheric Ozone. 2000
*Volume 75* Bruce Douglas, Michael Kearney, and Stephen Leatherman. Sea Level Rise: History and Consequences. 2000
*Volume 76* Roman Teisseyre and Eugeniusz Majewski. Earthquake Thermodynamics and Phase Transformations in the Interior. 2001
*Volume 77* Gerold Siedler, John Church, and John Gould. Ocean Circulation and Climate: Observing and Modelling The Global Ocean. 2001
*Volume 78* Roger A. Pielke Sr. Mesoscale Meteorological Modeling, 2nd Edition. 2001
*Volume 79* S. Pal Arya. Introduction to Micrometeorology. 2001
*Volume 80* Barry Saltzman. Dynamical Paleoclimatology: Generalized Theory of Global Climate Change. 2002
*Volume 81A* William H. K. Lee, Hiroo Kanamori, Paul Jennings, and Carl Kisslinger. International Handbook of Earthquake and Engineering Seismology, Part A. 2002
*Volume 81B* William H. K. Lee, Hiroo Kanamori, Paul Jennings, and Carl Kisslinger. International Handbook of Earthquake and Engineering Seismology, Part B. 2003
*Volume 82* Gordon G. Shepherd. Spectral Imaging of the Atmosphere. 2002
*Volume 83* Robert P. Pearce. Meteorology at the Millennium. 2001
*Volume 84* Kuo-Nan Liou. An Introduction to Atmospheric Radiation, 2nd Edition. 2002
*Volume 85* Carmen J. Nappo. An Introduction to Atmospheric Gravity Waves. 2002
*Volume 86* Michael E. Evans and Friedrich Heller. Environmental Magnetism: Principles and Applications of Enviromagnetics. 2003
*Volume 87* John S. Lewis. Physics and Chemistry of the Solar System, 2nd Edition. 2004